Quantum Optics

Quantum Optics

J. C. Garrison
Department of Physics
University of California at Berkeley

and

R. Y. Chiao
School of Natural Sciences and School of Engineering
University of California at Merced

OXFORD
UNIVERSITY PRESS

OXFORD
UNIVERSITY PRESS

Great Clarendon Street, Oxford, OX2 6DP,
United Kingdom

Oxford University Press is a department of the University of Oxford.
It furthers the University's objective of excellence in research, scholarship,
and education by publishing worldwide. Oxford is a registered trade mark of
Oxford University Press in the UK and in certain other countries

First published 2008
First published in paperback 2014

British Library Cataloguing in Publication Data
Data available

Library of Congress Cataloging in Publication Data
Data available

ISBN 978-0-19-850886-1 (Hbk.)
ISBN 978-0-19-968999-6 (Pbk.)

This book is dedicated to our wives: Florence Chiao and Hillegonda Garrison. Without their unfailing support and almost infinite patience, the task would have been much harder.

Preface

The idea that light is composed of discrete particles can be traced to Newton's *Opticks* (Newton, 1952), in which he introduced the term 'corpuscles' to describe what we now call 'particles'. However, the overwhelming evidence in favor of the wave nature of light led to the abandonment of the corpuscular theory for almost two centuries. It was resurrected—in a new form—by Einstein's 1905 explanation of the photoelectric effect, which reconciled the two views by the assumption that the continuous electromagnetic fields of Maxwell's theory describe the average behavior of individual particles of light. At the same time, the early quantum theory and the principle of wave–particle duality were introduced into optics by the Einstein equation, $E = h\nu$, which relates the energy E of the light corpuscle, the frequency ν of the associated electromagnetic wave, and Planck's constant h.

This combination of ideas marks the birth of the field now called quantum optics. This subject could be defined as the study of all phenomena involving the particulate nature of light in an essential way, but a book covering the entire field in this general sense would be too heavy to carry and certainly beyond our competence. Our more modest aim is to explore the current understanding of the interaction of individual quanta of light—in the range from infrared to ultraviolet wavelengths—with ordinary matter, e.g. atoms, molecules, conduction electrons, etc. Even in this restricted domain, it is not practical to cover everything; therefore, we have concentrated on a set of topics that we believe are likely to provide the basis for future research and applications.

One of the attractive aspects of this field is that it addresses both fundamental issues of quantum physics and some very promising applications. The most striking example is entanglement, which embodies the central mystery of quantum theory and also serves as a resource for communication and computation. This dual character makes the subject potentially interesting to a diverse set of readers, with backgrounds ranging from pure physics to engineering. In our attempt to deal with this situation, we have followed a maxim frequently attributed to Einstein: 'Everything should be made as simple as possible, but not simpler' (Calaprice, 2000, p. 314). This injunction, which we will call *Einstein's rule*, is a variant of *Occam's razor*: 'it is vain to do with more what can be done with fewer' (Russell, 1945, p. 472).

Our own grasp of this subject is largely the result of fruitful interactions with many colleagues over the years, in particular with our students. While these individuals are responsible for a great deal of our understanding, they are in no way to blame for the inevitable shortcomings in our presentation.

With regard to the book itself, we are particularly indebted to Dr Achilles Speliotopoulos, who took on the onerous task of reading a large part of the manuscript, and made many useful suggestions for improvements. We would also like to express our thanks to Sonke Adlung, and the other members of the editorial staff at Oxford

University Press, for their support and patience during the rather protracted time spent in writing the book.

J. C. Garrison and R. Y. Chiao
July 2007

Contents

Introduction

For the purposes of this book, **quantum optics** is the study of the interaction of individual photons, in the wavelength range from the infrared to the ultraviolet, with ordinary matter—e.g. atoms, molecules, conduction electrons, etc.—described by nonrelativistic quantum mechanics. Our objective is to provide an introduction to this branch of physics—covering both theoretical and experimental aspects—that will equip the reader with the tools for working in the field of quantum optics itself, as well as its applications. In order to keep the text to a manageable length, we have not attempted to provide a detailed treatment of the various applications considered. Instead, we try to connect each application to the underlying physics as clearly as possible; and, in addition, supply the reader with a guide to the current literature. In a field evolving as rapidly as this one, the guide to the literature will soon become obsolete, but the physical principles and techniques underlying the applications will remain relevant for the foreseeable future.

Whenever possible, we first present a simplified model explaining the basic physical ideas in a way that does not require a strong background in theoretical physics. This step also serves to prepare the ground for a more sophisticated theoretical treatment, which is presented in a later section. On the experimental side, we have made a serious effort to provide an introduction to the techniques used in the experiments that we discuss.

The book begins with a survey of the basic experimental observations that have led to the conclusion that light is composed of indivisible quanta—called photons—that obey the laws of quantum theory. The next six chapters are concerned with building up the basic theory required for the subsequent developments. In Chapters 8 and 9, we emphasize the theoretical and experimental techniques that are needed for the discussion of a collection of important experiments in linear quantum optics, presented in Chapter 10.

Chapters 11 through 18 contain a mixture of more advanced topics, including cavity quantum electrodynamics, nonlinear optics, nonclassical states of light, linear optical amplifiers, and quantum tomography.

In Chapter 19, we discuss Bell's theorem and the optical experiments performed to test its consequences. The ideas associated with Bell's theorem play an important role in applications now under development, as well as in the foundations of quantum theory. Finally, in Chapter 20 many of these threads are drawn together to treat topics in quantum information theory, ranging from noise suppression in optical transmission lines to quantum computing.

We have written this book for readers who are already familiar with elementary quantum mechanics; in particular, with the quantum theory of the simple harmonic oscillator. A corresponding level of familiarity with Maxwell's equations for the classical electromagnetic field and with elementary optics is also a prerequisite. On the mathematical side, some proficiency in classical analysis, including the use of partial differential equations and Fourier transforms, will be a great help.

Since the number of applications of quantum optics is growing at a rapid pace, this subject is potentially interesting to people from a wide range of scientific and engineering backgrounds. We have, therefore, organized the material in the book into two tracks. Sections marked by an asterisk are intended for graduate-level students who already have a firm understanding of quantum theory and Maxwell's equations. The unmarked sections will, we hope, be useful for senior level undergraduates who have had good introductory courses in quantum mechanics and electrodynamics. The exercises—which form an integral part of the text—are marked in the same way.

The terminology and notation used in the book are—for the most part—standard. We employ SI units for electromagnetic quantities, and impose the Einstein summation convention for three-dimensional vector indices. Landau's 'hat' notation is used for quantum operators associated with material particles, e.g. $\widehat{q}$, and $\widehat{p}$, but not for similar operators associated with the electromagnetic field. The expression 'c-number'—also due to Landau— is employed to distinguish ordinary numbers, either real or complex, from operators. The abbreviations CC and HC respectively stand for complex conjugate and hermitian conjugate. Throughout the book, we use Dirac's *bra* and *ket* notation for quantum states. Our somewhat unconventional notation for Fourier transforms is explained in Appendix A.4.

1

The quantum nature of light

Classical physics began with Newton's laws of mechanics in the seventeenth century, and it was completed by Maxwell's synthesis of electricity, magnetism, and optics in the nineteenth century. During these two centuries, Newtonian mechanics was extremely successful in explaining a wide range of terrestrial experiments and astronomical observations. Key predictions of Maxwell's electrodynamics were also confirmed by the experiments of Hertz and others, and novel applications have continued to emerge up to the present. When combined with the general statistical principles codified in the laws of thermodynamics, classical physics seemed to provide a permanent foundation for all future understanding of the physical world.

At the turn of the twentieth century, this optimistic view was shattered by new experimental discoveries, and the ensuing crisis for classical physics was only resolved by the creation of the quantum theory. The necessity of explaining the stability of atoms, the existence of discrete lines in atomic spectra, the diffraction of electrons, and many other experimental observations, decisively favored the new quantum mechanics over Newtonian mechanics for material particles (Bransden and Joachain, 1989, Chap. 4). Thermodynamics provided a very useful bridge between the old and the new theories. In the words of Einstein (Schilpp, 1949, Autobiographical Notes, p. 33),

> A theory is the more impressive the greater the simplicity of its premises is, the more different kinds of things it relates, and the more extended is its area of applicability. Therefore the deep impression which classical thermodynamics made upon me. It is the only physical theory of universal content concerning which I am convinced that, within the framework of the applicability of its basic concepts, it will never be overthrown (for the special attention of those who are skeptics on principle).

Unexpected features of the behavior of light formed an equally important part of the crisis for classical physics. The blackbody spectrum, the photoelectric effect, and atomic spectra proved to be inconsistent with classical electrodynamics. In his characteristically bold fashion, Einstein (1987*a*) proposed a solution to these difficulties by offering a radically new model in which light of frequency ν is supposed to consist of a gas of discrete light quanta with energy $\epsilon = h\nu$, where h is Planck's constant. The connection to classical electromagnetic theory is provided by the assumption that the number density of light quanta is proportional to the intensity of the light. We will follow the current usage in which light quanta are called **photons**, but this terminology must be used with some care.[1] Conceptual difficulties can arise because

[1] According to Willis Lamb, no amount of care is sufficient; and the term 'photon' should be banned from physics (Lamb, 1995).

this name suggests that photons are particles in the same sense as electrons, protons, neutrons, etc. In the following chapters, we will see that the physical meaning of the word 'photon' evolves along with our understanding of experiment and theory.

Einstein's introduction of photons was the first step toward a true quantum theory of light—just as the Bohr model of the atom was the first step toward quantum mechanics—but there is an important difference between these parallel developments. The transition from classical electromagnetic theory to the photon model is even more radical than the corresponding transition from classical mechanics to quantum mechanics. If one thinks of classical mechanics as a game like chess, the pieces are point particles and the rules are Newton's equations of motion. The solution of Newton's equations determines a unique trajectory $(q(t), p(t))$ for given initial values of the position $q(0)$ and the momentum $p(0)$ of a point particle. The game of quantum mechanics has the same pieces, but different rules. The initial situation is given by a wave function $\psi(q)$, and the trajectory is replaced by a time-dependent wave function $\psi(q, t)$ that satisfies the Schrödinger equation. The situation for classical electrodynamics is very different. The pieces for this game are the continuous electric and magnetic fields $\boldsymbol{\mathcal{E}}(\mathbf{r}, t)$ and $\boldsymbol{\mathcal{B}}(\mathbf{r}, t)$, and the rules are provided by Maxwell's equations. Einstein's photons are nowhere to be found; consequently, the quantum version of the game requires new pieces, as well as new rules. A conceptual change of this magnitude should be approached with caution.

In order to exercise the caution recommended above, we will discuss the experimental basis for the quantum theory of light in several stages. Section 1.1 contains brief descriptions of the experiments usually considered in this connection, together with a demonstration of the complete failure of classical physics to explain any of them. In Section 1.2 we will introduce Einstein's photon model and show that it succeeds brilliantly in explaining the same experimental results.

In other words, the photon model is *sufficient* for the explanation of the experiments in Section 1.1, but the question is whether the introduction of the photon is *necessary* for this purpose. The only way to address this question is to construct an alternative model, and the only candidate presently available is **semiclassical electrodynamics**. In this approach, the charged particles making up atoms are described by quantum mechanics, but the electromagnetic field is still treated classically.

In Section 1.3 we will attempt to explain each experiment in semiclassical terms. In this connection, it is essential to keep in mind that corrections to the lowest-order approximation—of the semiclassical theory or the photon model—would not have been detectable in the early experiments. As we will see, these attempts have varying degrees of success; so one might ask: Why consider the semiclassical approach at all? The answer is that the existence of a semiclassical explanation for a given experimental result implies that the experiment is not sensitive to the **indivisibility** of photons, which is a fundamental assumption of Einstein's model (Einstein, 1987*a*). In Einstein's own words:

> According to the assumption to be contemplated here, when a light ray is spreading from a point, the energy is not distributed continuously over ever-increasing spaces, but consists of a finite number of energy quanta that are localized in points in space, move without dividing, and can be absorbed or generated only as a whole.

As an operational test of photon indivisibility, imagine that light containing exactly one photon falls on a transparent dielectric slab (a beam splitter) at a 45° angle of incidence. According to classical optics, the light is partly reflected and partly transmitted, but in the photon model these two outcomes are mutually exclusive. The photon must go one way or the other. In Section 1.4 we will describe an experiment that very convincingly demonstrates this all-or-nothing behavior. This single experiment excludes all variants of semiclassical electrodynamics. Experiments of this kind had to wait for technologies, such as atomic beams and coincidence counting, which were not fully developed until the second half of the twentieth century.

1.1 The early experiments

1.1.1 The Planck spectrum

In the last half of the nineteenth century, a considerable experimental effort was made to obtain precise measurements of the spectrum of radiation emitted by a so-called *blackbody*, an idealized object which absorbs all radiation falling on it. In practice, this idealized body is replaced by a **blackbody cavity**, i.e. a void surrounded by a wall, pierced by a small aperture that allows radiation to enter and exit. The interior area of the cavity is much larger than the area of the hole, so a ray of light entering the cavity would bounce from the interior walls many times before it could escape through the entry point. Thus the radiation would almost certainly be absorbed before it could exit. In this way the cavity closely approximates the perfect absorptivity of an ideal blackbody. Even when no light is incident from the outside, light is seen to escape through the small aperture. This shows that the interior of a cavity with heated walls is filled with radiation. The blackbody cavity, which is a simplification of the furnaces used in the ancient art of ceramics, is not only an accurate representation of the experimental setup used to observe the spectrum of blackbody radiation; it also captures the essential features of the blackbody problem in a way that allows for simple theoretical analysis.

Determining the spectral composition (that is, the distribution of radiant energy into different wavelengths) of the light emitted by a cavity with walls at temperature T is an important experimental goal. The wavelength, λ, is related to the circular frequency ω by $\lambda = c/\nu = 2\pi c/\omega$, so this information is contained in the **spectral function** $\rho(\omega, T)$, where $\rho(\omega, T)\,\Delta\omega$ is the radiant energy per unit volume in the frequency interval ω to $\omega + \Delta\omega$. The power per unit frequency interval emitted from the aperture area σ is $c\rho(\omega, T)\,\sigma/4$ (see Exercise 1.1). In order to measure this quantity, the various frequency components must be spectrally separated before detection, for example, by refracting the light through a prism. If the prism is strongly *dispersive* (that is, the index of refraction of the prism material is a strong function of the wavelength) distinct wavelength components will be refracted at different angles.

For moderate temperatures, a significant part of the blackbody radiation lies in the infrared, so it was necessary to develop new techniques of infrared spectroscopy in order to achieve the required spectral separation. This effort was aided by the discovery that prisms cut from single crystals of salt are strongly dispersive in the infrared part of the spectrum. The concurrent development of infrared detectors in

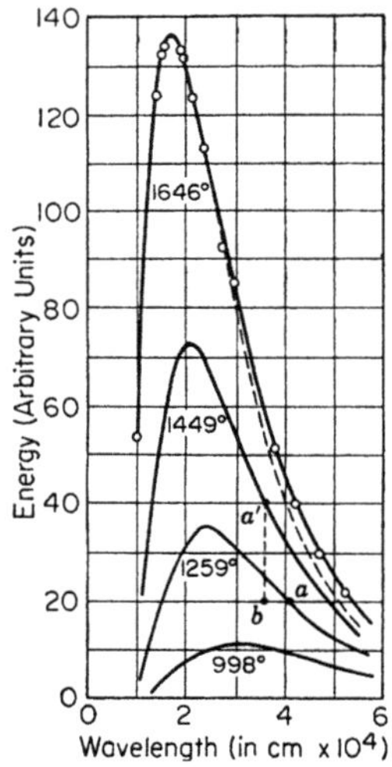

Fig. 1.1 Distribution of energy in the spectrum of a blackbody at various temperatures. (Reproduced from Richtmyer *et al.* (1955, Chap. 4, Sec. 64).)

the form of sensitive bolometers[2] allowed an accurate measurement of the blackbody spectrum. The experimental effort to measure this spectrum was initiated in Berlin around 1875 by Kirchhoff, and culminated in the painstaking work of Lummer and Pringsheim in 1899, in which the blackbody spectrum was carefully measured in the temperature range 998 K to 1646 K. Typical results are shown in Fig. 1.1.

The theoretical interpretation of the experimental measurements also required a considerable effort. The first step is a thermodynamic argument which shows that the blackbody spectrum must be a *universal* function of temperature; in other words, the spectrum is entirely independent of the size and shape of the cavity, and of the material composition of its walls. Consider two separate cavities having small apertures of identical size and shape, which are butted against each other so that the two apertures coincide exactly, as indicated in Fig. 1.2. In this way, all the radiation escaping from each cavity enters the other. The two cavities can have interiors of different volumes and arbitrarily irregular shapes (provided that their interior areas are sufficiently large compared to the aperture area), and their walls can be composed of entirely different materials. We will assume that the two cavities are in thermodynamic equilibrium at the common temperature T.

Now suppose that the blackbody spectrum were not universal, but depended, for example, on the material of the walls. If the left cavity were to emit a greater amount of radiation than the right cavity, then there would be a net flow of energy from left to right. The right cavity would then heat up, while the left cavity would cool down. The flow of heat between the cavities could be used to extract useful work from two bodies at the same temperature. This would constitute a perpetual motion machine of the second kind, which is forbidden by the second law of thermodynamics (Zemansky, 1951, Chap. 7.5). The total flow of energy out of each cavity is given by the integral of

[2]These devices exploit the temperature dependence of the resistivity of certain metals to measure the deposited energy by the change in an electrical signal.

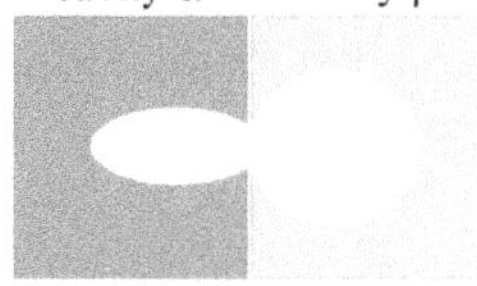

Fig. 1.2 Cavities α and β coupled through a common aperture.

its spectral function over all frequencies, so this argument shows that the integrated spectral functions of the two cavities must be exactly the same.

This still leaves open the possibility that the spectral functions could differ in certain frequency intervals, provided that their integrals are the same. Thus we must also prove that net flows of energy cannot occur in any frequency interval of the blackbody spectrum. This can be seen from the following argument based on the principle of detailed balance. Suppose that the spectral functions of the two cavities, ρ_α and ρ_β, are different in the small interval ω to $\omega + \Delta\omega$; for example, suppose that $\rho_\alpha(\omega, T) > \rho_\beta(\omega, T)$. Then the net power flowing from α to β, in this frequency interval, is

$$\frac{1}{4}c\left[\rho_\alpha(\omega, T) - \rho_\beta(\omega, T)\right]\sigma\Delta\omega > 0\,, \tag{1.1}$$

where σ is the common area of the apertures. If we position absorbers in both α and β that only absorb at frequency ω, then the absorber in β will heat up compared to that in α. The two absorbers then provide the high- and low-temperature reservoirs of a heat engine (Halliday *et al.*, 1993, Chap. 22–6) that could deliver continuous external work, with no other change in the system. Again, this would constitute a perpetual motion machine of the second kind. Therefore the equality

$$\rho_\alpha(\omega, T) = \rho_\beta(\omega, T) \tag{1.2}$$

must be exact, for all values of the frequency ω and for all values of the temperature T. We conclude that the blackbody spectral function is universal; it does not depend on the material composition, size, shape, etc., of the two cavities. This strongly suggests that the universal spectral function should be regarded as a property of the radiation field itself, rather than a joint property of the radiation field and of the matter with which it is in equilibrium.

The thermodynamic argument given above shows that the spectral function is universal, but it gives no clues about its form. In classical physics this can be determined by using the principle of equipartition of energy. For an ideal gas, this states that the average energy associated with each degree of freedom is $k_B T/2$, where T is the temperature and k_B is Boltzmann's constant. For a collection of harmonic oscillators, the kinetic and potential energy each contribute $k_B T/2$, so the thermal energy for each degree of freedom is $k_B T$.

In order to apply these rules to blackbody radiation, we first need to identify and count the number of degrees of freedom in the electromagnetic field. The thermal radiation in the cavity can be analyzed in terms of plane waves $\mathbf{e}_{\mathbf{k}s}\exp(i\mathbf{k}\cdot\mathbf{r})$, where

$\mathbf{e}_{\mathbf{k}s}$ is the unit polarization vector and the propagation vector $\mathbf{k}$ satisfies $|\mathbf{k}| = \omega/c$ and $\mathbf{k}\cdot\mathbf{e}_{\mathbf{k}s} = 0$. There are two linearly independent polarization states for each $\mathbf{k}$, so s takes on two values. The boundary conditions at the walls only allow certain discrete values for $\mathbf{k}$. In particular, for a cubical cavity with sides L subject to periodic boundary conditions the spacing of allowed $\mathbf{k}$ values in the x-direction is $\Delta k_x = 2\pi/L$, etc. Another way of saying this is that each mode occupies a volume $(2\pi/L)^3$ in $\mathbf{k}$-space, so that the number of modes in the volume element d^3k is $2\,(2\pi/L)^{-3}\,d^3k$, where the factor 2 accounts for the two polarizations. The field is completely determined by the amplitudes of the independent modes, so it is natural to identity the modes as the degrees of freedom of the field. Furthermore, we will see in Section 2.1.1-D that the contribution of each mode to the total energy is mathematically identical to the energy of a harmonic oscillator. The identification of modes with degrees of freedom shows that the number of degrees of freedom dn_ω in the frequency interval ω to $\omega + d\omega$ is

$$dn_\omega = 2\int d\theta\,\sin\theta\int d\phi\frac{k^2dk}{(2\pi/L)^3} = \frac{L^3k^2}{\pi^2c}\,d\omega\,, \tag{1.3}$$

where θ and ϕ specify the direction of $\mathbf{k}$. The equipartition theorem for harmonic oscillators shows that the thermal energy per mode is k_BT. The spectral function is the product of dn_ω and the thermal energy density k_BT/L^3, so we find the classical **Rayleigh–Jeans** law:

$$\rho\left(\omega, T\right)\,d\omega = k_BT\,\frac{\omega^2}{\pi^2c^3}\,d\omega\,. \tag{1.4}$$

This fits the low-frequency data quite well, but it is disastrously wrong at high frequencies. The ω-integral of this spectral function diverges; consequently, the total energy density is infinite for any temperature T. Since the divergence of the integral occurs at high frequencies, this is called the *ultraviolet catastrophe*.

In an effort to find a replacement for the Rayleigh–Jeans law, Planck (1959) concentrated on the atoms in the walls, which he modeled as a family of harmonic oscillators in equilibrium with the radiation field. In classical mechanics, each oscillator is described by a pair of numbers (Q, P), where Q is the coordinate and P is the momentum. These pairs define the points of the classical oscillator **phase space** (Chandler, 1987, Chap. 3.1). The average energy per oscillator is given by an integral over the oscillator phase space, which Planck approximated by a sum over phase space elements of area h. Usually, the value of the integral would be found by taking the limit $h \to 0$, but Planck discovered that he could fit the data over the whole frequency range by instead assigning the particular nonzero value $h \approx 6.6\times10^{-34}$ J s. He attempted to explain this amazing fact by assuming that the atoms could only transfer energy to the field in units of $h\nu = \hbar\omega$, where $\hbar \equiv h/2\pi$. This is completely contrary to a classical description of the atoms, which would allow continuous energy transfers of any amount.

This achievement marks the birth of quantum theory, and Planck's constant h became a new universal constant. In Planck's model, the quantization of energy is a property of the atoms—or, more precisely, of the interaction between the atoms and the field—and the electromagnetic field is still treated classically. The derivation of the spectral function from this model is quite involved, and the fact that the result is independent of the material properties only appears late in the calculation. Fortunately, Einstein later showed that the functional form of $\rho(\omega, T)$ can be derived very simply from his quantum model of radiation, in which the electromagnetic field itself consists of discrete quanta. Therefore we will first consider the other early experiments before calculating $\rho(\omega, T)$.

1.1.2 The photoelectric effect

The infrared part of atomic spectra, contributing to the blackbody radiation discussed in the last section, does not typically display sharp spectral lines. In this and the following two sections we will consider effects caused by radiation with a sharply defined frequency. One of the most celebrated of these is the **photoelectric effect**: ultraviolet light falling on a properly cleaned metallic surface causes the emission of electrons. In the early days of spectroscopy, the source of this ultraviolet light was typically a sharp mercury line—at 253.6 nm—excited in a mercury arc.

In order to simplify the classical analysis of this effect, we will replace the complexities of actual metals by a model in which the electron is trapped in a potential well. According to Maxwell's theory, the incident light is an electromagnetic plane wave with $|\boldsymbol{\mathcal{E}}| = c\,|\boldsymbol{\mathcal{B}}|$, and the electron is exposed to the Lorentz force $\mathbf{F} = -e\,(\boldsymbol{\mathcal{E}} + \mathbf{v} \times \boldsymbol{\mathcal{B}})$. Work is done only by the electric field on the electron. Hence it will take time for the electron to absorb sufficient energy from the field to overcome the binding energy to the metal, and thus escape from the surface. The time required would necessarily increase as the field strength decreases. Since the kinetic energy of the emitted electron is the difference between the work done and the binding energy, it would also depend on the intensity of the light. This leads to the following two predictions. (P1) There will be an intensity-dependent time interval between the onset of the radiation and the first emission of an electron. (P2) The energy of the emitted electrons will depend on the intensity.

Let us now consider an experimental arrangement that can measure the kinetic energy of the ejected *photoelectrons* and the time delay between the arrival of the light and the first emission of electrons. Both objectives can be realized by positioning a collector plate at a short distance from the surface. The plate is maintained at a negative potential $-V_{\text{stop}}$, with respect to the surface, and the potential is adjusted to a value just sufficient to stop the emitted electrons. The photoelectron's kinetic energy can then be determined through the energy-conservation equation

$$\frac{1}{2}mv^2 = (-e)\,(-V_{\text{stop}})\,. \tag{1.5}$$

The onset of the current induced by the capture of the photoelectrons determines the time delay between the arrival of the radiation pulse and the start of photoelectron emission. The amplitude of the current is proportional to the rate at which electrons are ejected. The experimental results are as follows. (E1) There is no measurable time delay before the emission of the first electron. (E2) The ejected photoelectron's kinetic energy is independent of the intensity of the light. Instead, the observed values of the energy depend on the frequency. They are very accurately fitted by the empirical relation

$$\epsilon_e = eV_{\text{stop}} = \frac{1}{2}mv^2 = \hbar\omega - W\,, \tag{1.6}$$

where ω is the frequency of the light. The constant W is called the **work function**; it is the energy required to free an electron from the metal. The value of W depends on the metal, but the constant $\hbar$ is universal. (E3) The rate at which electrons are emitted—but not their energies—is proportional to the field intensity. The stark contrast between the theoretical predictions (P1) and (P2) and the experimental results (E1)–(E3) posed another serious challenge to classical physics. The relation (1.6) is called Einstein's photoelectric equation, for reasons which will become clear in Section 1.2.

In the early experiments on the photoelectric effect it was difficult to determine whether the photoelectron energy was better fit by a linear or a quadratic dependence on the frequency of the light. This difficulty was resolved by Millikan's beautiful experiment (Millikan, 1916), in which he verified eqn (1.6) by using alkali metals, which were prepared with clean surfaces inside a vacuum system by means of an *in vacuo* metal-shaving technique. These clean alkali metal surfaces had a sufficiently small work function W, so that even light towards the red part of the visible spectrum was able to eject photoelectrons. In this way, he was able to measure the photoelectric effect from the red to the ultraviolet part of the spectrum—nearly a threefold increase over the previously observed frequency range. This made it possible to verify the linear dependence of the increment in the photoelectron's ejection energy as a function of the frequency of the incident light. Furthermore, Millikan had already measured very accurately the value of the electron charge e in his oil drop experiment. Combining this with the slope h/e of V_{stop} versus ν from eqn (1.6) he was able to deduce a value of Planck's constant h which is within 1% of the best modern measurements.

1.1.3 Compton scattering

As the study of the interaction of light and matter was extended to shorter wavelengths, another puzzling result occurred in an experiment on the scattering of monochromatic X-rays (the K_α line from a molybdenum X-ray tube) by a graphite target (Compton, 1923). A schematic of the experimental setup is shown in Fig. 1.3 for the special

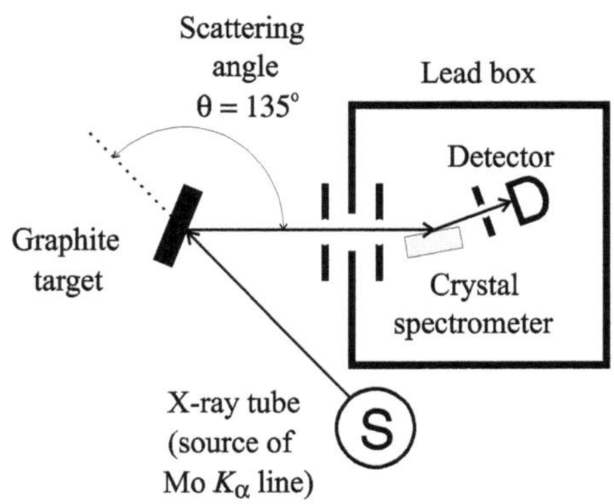

Fig. 1.3 Schematic of the setup used to observe Compton scattering.

case when the scattering angle θ is 135°. The wavelength of the scattered radiation is measured by means of a Bragg crystal spectrometer using the relation $2d\sin\phi = m\lambda$, where ϕ is the Bragg scattering angle, d is the lattice spacing of the crystal, and m is an integer corresponding to the diffraction order (Tipler, 1978, Chap. 3-6). Compton's experiment was arranged so that $m = 1$. The Bragg spectrometer which Compton constructed for his experiment consisted of a tiltable calcite crystal (oriented at a Bragg angle ϕ) placed inside a lead box, which was used as a shield against unwanted background X-rays. The detector, also placed inside this box, was an ionization chamber placed behind a series of collimating slits to define the angles θ and ϕ.

A simple classical model of the experiment consists of an electromagnetic field of frequency ω falling on an atomic electron. According to classical theory, the incident field will cause the electron to oscillate with frequency ω, and this will in turn generate radiation at the same frequency. This process is called **Thompson scattering** (Jackson, 1999, Sec. 14.8). In reality the incident radiation is not perfectly monochromatic, but the spectrum does have a single well-defined peak. The classical prediction is that the spectrum of the scattered radiation should also have a single peak at the same frequency.

The experimental results—shown in Fig. 1.4 for the scattering angles of $\theta = 45°$, 90°, and 135°—do exhibit a peak at the incident wavelength, but at each scattering

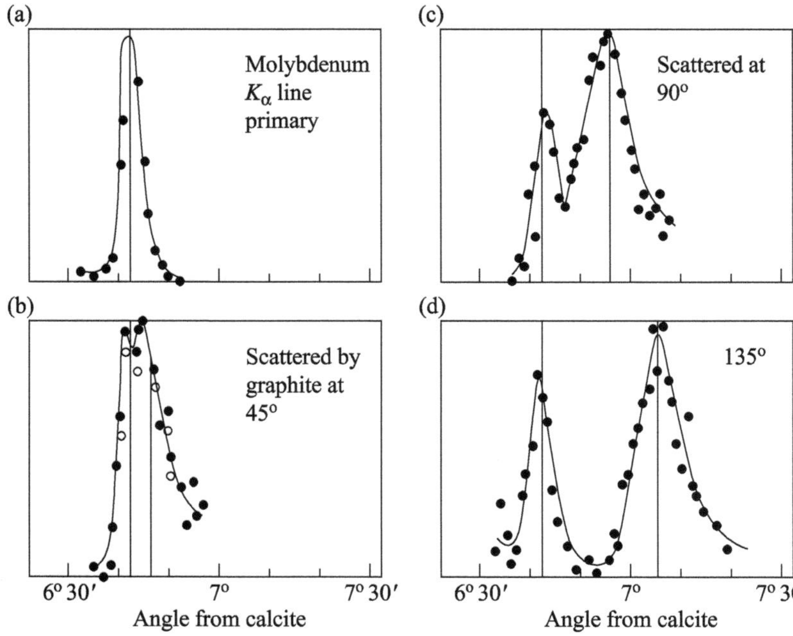

Fig. 1.4 Data from the Compton scattering experiment sketched in Fig. 1.3. A calcite crystal was used as the dispersive element in the Bragg spectrometer. (Adapted from Compton (1923).)

angle there is an additional peak at longer wavelengths which cannot be explained by the classical theory.

1.1.4 Bothe's coincidence-counting experiment

During the early development of the quantum theory, Bohr, Kramers, and Slater raised the possibility that energy and momentum are not conserved in each elementary quantum event—such as Compton scattering—but only on the average over many such events (Bohr *et al.*, 1924). However, by introducing the extremely important method of **coincidence detection**—in this case of the scattered X-ray photon and of the recoiling electron in each scattering event—Bothe performed a decisive experiment showing that the Bohr–Kramers–Slater hypothesis is incorrect in the case of Compton scattering; in fact, energy and momentum are both conserved in every single quantum event (Bothe, 1926). In the experiment sketched in Fig. 1.5, X-rays are Compton-scattered from a thin, metallic foil, and registered in the upper Geiger counter. The thin foil allows the recoiling electron to escape, so that it registers in the lower Geiger counter.

When viewed in the wave picture, the scattered X-rays are emitted in a spherically expanding wavefront, but a single detection at the upper Geiger counter registers the absorption of the full energy $\hbar\omega$ of the X-ray photon, and the displacement vector linking the scattering point to the Geiger counter defines a unique direction for the momentum $\hbar\mathbf{k}'$ of the scattered X-ray. This is an example of the famous collapse of the wave packet.

When viewed in the particle picture, both the photon and the electron are treated like colliding billiard balls, and the principles of the conservation of energy and momentum fix the momentum $\mathbf{p}$ of the recoiling electron. The detection of the scattered X-ray is therefore always accompanied by the detection of the recoiling electron at the lower Geiger counter, provided that the second counter is carefully aligned along the uniquely defined direction of the electron momentum $\mathbf{p}$. Coincidence detection became possible with the advent, in the 1920s, of fast electronics using vacuum tubes (triodes), which open a narrow time window defining the approximately simultaneous detection of a pair of pulses from the upper and lower Geiger counters.

Later we will see the central importance in quantum theory of the concept of an entangled state, for example, a superposition of products of the plane-wave states of two free particles. In the case of Compton scattering, the scattered X-ray photon and the recoiling electron are produced in just such a state. The entanglement

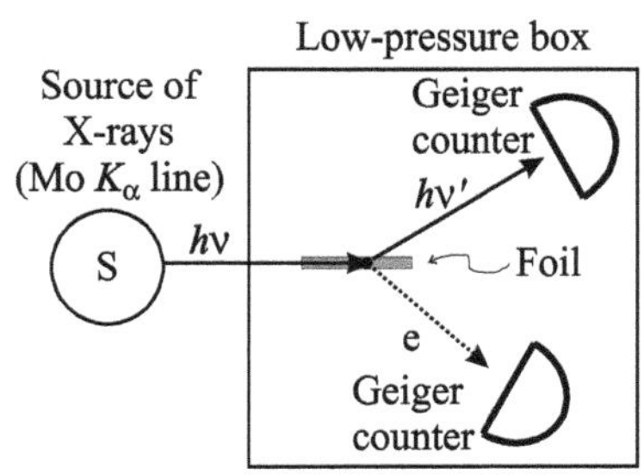

Fig. 1.5 Schematic of Bothe's coincidence detection of a Compton-scattered X-ray from a thin, metallic foil, and of the recoil electron from the same scattering event.

between the electron and the photon produced by their interaction enforces a tight correlation—determined by conservation of energy and momentum—upon detection of each quantum scattering event. It was just such correlations which were first observed in the coincidence-counting experiment of Bothe.

1.2 Photons

In one of his three celebrated 1905 papers Einstein (1987*a*) proposed a new model of light which explains all of the experimental results discussed in the previous sections. In this model, light of frequency ω is supposed to consist of a gas of discrete photons with energy $\epsilon = \hbar\omega$. In common with material particles, photons carry momentum as well as energy. In the first paper on relativity, Einstein had already pointed out that the relativistic transformation laws governing energy and momentum are identical to those governing the frequency and wavevector of a plane wave (Jackson, 1999, Sec. 11.3D). In other words, the four-component vector $(\omega, c\mathbf{k})$ transforms in the same way as $(E, c\mathbf{p})$ for a material particle. Thus the assumption that the energy of a light quantum is $\hbar\omega$ implies that its momentum must be $\hbar\mathbf{k}$, where $|\mathbf{k}| = (\omega/c) = (2\pi/\lambda)$. The connection to classical electromagnetic theory is provided by the assumption that the number density of photons is proportional to the intensity of the light.

This is a far reaching extension of Planck's idea that energy could only be transferred between radiation and matter in units of $\hbar\omega$. The new proposal ascribes the quantization entirely to the electromagnetic field itself, rather than to the mechanism of energy exchange between light and matter. It is useful to arrange the results of the model into two groups. The first group includes the **kinematical** features of the model, i.e. those that depend only on the conservation laws for energy and momentum and other symmetry properties. The second group comprises the **dynamical** features, i.e. those that involve explicit assumptions about the fundamental interactions. In the final section we will show that even this simple model has interesting practical applications.

1.2.1 Kinematics

A The photoelectric effect

The first success of the photon model was its explanation of the puzzling features of the photoelectric effect. Since absorption of light occurs by transferring discrete bundles of energy of just the right size, there is no time delay before emission of the first electron. Absorption of a single photon transfers its entire energy $\hbar\omega$ to the bound electron, thereby ejecting it from the metal with energy ϵ_e given by eqn (1.6), which now represents the overall conservation of energy. The energy of the ejected electron therefore depends on the frequency rather than the intensity of the light. Since each photoelectron emission event is caused by the absorption of a single photon, the number of electrons emitted per unit time is proportional to the flux of photons and thereby to the intensity of light. The photoelectric equation implied by the photon model is kinematical in nature, since it only depends on conservation of energy and does not assume any model for the dynamical interaction between photons and the electrons in the metal.

B Compton scattering

The existence of the second peak in Compton scattering is also predicted by a kinematical argument based on conservation of momentum and energy. Consider an X-ray photon scattering from a weakly bound electron. In this case it is sufficient to consider a free electron at rest and impose conservation of energy and momentum to determine the possible final states as shown in Fig. 1.6.

For energetic X-rays the electron may recoil at velocities comparable to the velocity of light, so it is necessary to use relativistic kinematics for this calculation (Jackson, 1999, Sec. 11.5). The relativistic conservation laws for energy and momentum are

$$mc^2 + \hbar\omega = E + \hbar\omega' \, , \; \hbar\mathbf{k} = \hbar\mathbf{k}' + \mathbf{p} \, , \tag{1.7}$$

where $\mathbf{p}$ and $E = \sqrt{m^2c^4 + c^2p^2}$ are respectively the final electron momentum and energy, $|\mathbf{k}| = \omega/c$, and $|\mathbf{k}'| = \omega'/c$. Since the recoil kinetic energy of the scattered electron ($K = E - mc^2$) is positive, eqn (1.7) already explains why the scattered quantum must have a lower frequency (longer wavelength) than the incident quantum. Combining the two conservation laws yields the **Compton shift**

$$\Delta\lambda \equiv \lambda' - \lambda = \lambda_C \left(1 - \cos\theta\right) , \tag{1.8}$$

in wavelength as a function of the scattering angle θ (the angle between $\mathbf{k}$ and $\mathbf{k}'$), where the electron **Compton wavelength** is

$$\lambda_C = \frac{h}{mc} = 0.0048 \, \text{nm} \, . \tag{1.9}$$

This simple argument agrees quite accurately with the data in Fig. 1.4, and with other experiments using a variety of incident wavelengths. The fractional wavelength shift for Compton scattering is bounded by $\Delta\lambda/\lambda < 2\lambda_C/\lambda$. This shows that $\Delta\lambda/\lambda$ is negligible for optical wavelengths, $\lambda \sim 10^3$ nm; which explains why X-rays were needed to observe the Compton shift.

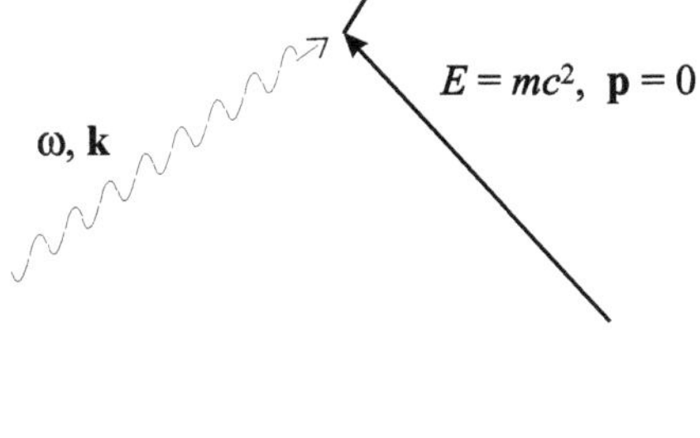

Fig. 1.6 Scattering of an incident X-ray quantum from an electron at rest.

The argument leading to eqn (1.8) seems to prove too much, since it leaves no room for the peak at the incident wavelength, which is also evident in the data. This is a consequence of the assumption that the electron is weakly bound. In carrying out the same kinematic analysis for a strongly bound electron, the electron mass m in eqn (1.9) must be replaced by the mass M of the atom. Since $M \gg m$, the resulting shift is negligible even at X-ray wavelengths, and the peak at the incident wavelength is recovered.

1.2.2 Dynamics

A Emission and absorption of light

The dynamical features of the photon model were added later, in conjunction with the Bohr model of the atom (Einstein, 1987*b*, 1987*c*). The level structure of a real atom is quite complicated, but for a fixed frequency of light only the two levels involved in a quantum jump describing emission or absorption of light at that frequency are relevant. This allows us to replace real atoms by idealized two-level atoms which have a lower state with energy ϵ_1, and a single upper (excited) state with energy ϵ_2. The combination of conservation of energy with the photoelectric effect makes it reasonable (following Bohr) to assume that the atoms can absorb and emit radiation of frequency $\omega = (\epsilon_2 - \epsilon_1)/\hbar$. In this spirit, Einstein assumed the existence of three dynamical processes, absorption, spontaneous emission, and stimulated emission. The simplest cases of absorption and emission of a single photon are shown in Fig. 1.7.

Einstein originally introduced the notion of spontaneous emission by analogy with radioactive decay, but the existence of spontaneous emission is implied by the principle of time-reversal invariance: i.e. the time-reversed final state evolves into the time-reversed initial state. We will encounter this principle later on in connection with Maxwell's equations and quantum theory. In fact, time-reversal invariance holds for all microscopic physical phenomena, with the exception of the weak interactions. These

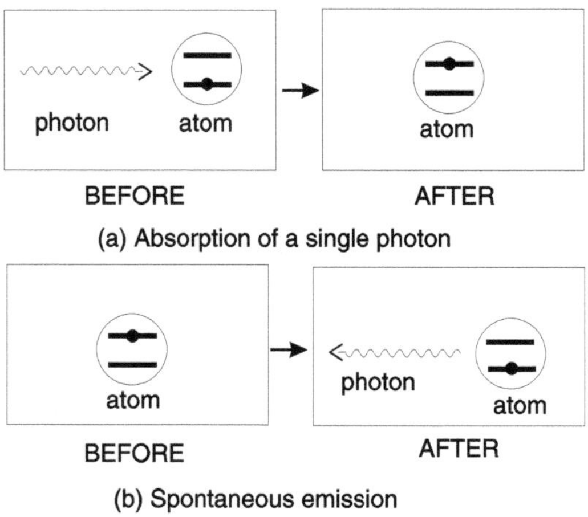

Fig. 1.7 (a) An atom in the ground state jumps to the excited state after absorbing a single photon. (b) An atom in the excited state jumps to the ground state and emits a single photon.

very small effects will be ignored for the purposes of this book. For the present, we will simply illustrate the idea of time reversal by considering the motion of classical particles (such as perfectly elastic billiard balls). Since Newton's equations are second order in time, the evolution of the mechanical system is determined by the initial positions and velocities of the particles, $(\mathbf{r}(0), \mathbf{v}(0))$. Suppose that at time $t = \tau$, each velocity is somehow reversed[3] while the positions are unchanged so that $(\mathbf{r}(\tau), \mathbf{v}(\tau)) \to (\mathbf{r}(\tau), -\mathbf{v}(\tau))$. More details on this operation—which is called time reversal—are found in Appendix B.3.3. With this new initial state, the particles will exactly reverse their motions during the interval $(\tau, 2\tau)$ to arrive at $(\mathbf{r}(2\tau), \mathbf{v}(2\tau)) = (\mathbf{r}(0), -\mathbf{v}(0))$, which is the time-reversed form of the initial state. A mathematical proof of this statement, which also depends on the fact that the Newtonian equations are second order in time, can be found in standard texts; see, for example, Bransden and Joachain (1989, Sec. 5.9).

In the photon model, the reversal of velocities is replaced by the reversal of the propagation directions of the photons. With this in mind, it is clear that Fig. 1.7(b) is the time-reversed form of Fig. 1.7(a). Absorption of light is a well understood process in classical electromagnetic theory, and in principle the intensity of the field can be made arbitrarily small. This is not the case in Einstein's model, since the discreteness of photons means that the weakest nonzero field is one describing exactly one photon, as in Fig. 1.7(a). If we extrapolate the classical result to the absorption of a single incident photon, then time-reversal invariance requires the existence of the process of **spontaneous emission**, pictured in Fig. 1.7(b).

This argument can also be applied to the situation illustrated in Fig. 1.8, in which many photons in the same mode are incident on an atom in the ground state. The absorption event shown in Fig. 1.8(a) is evidently the time-reversed version of the process shown in Fig. 1.8(b). Consequently, the principle of time-reversal invariance implies the necessity of the second process, which is called **stimulated emission**. Since the N photons in Fig. 1.8(a) are all in the same mode, this argument also shows that the stimulated photon must be emitted into the same mode as the $N-1$ incident photons. Thus the stimulated photon must have the same wavevector $\mathbf{k}$, frequency ω, and polarization s as the incident photons. The identical values of these parameters—which completely specify the state of the photon—for the stimulated and stimulating photons implies a perfect amplification of the incident light beam by the process of stimulated emission (ignoring, for the moment, the process of spontaneous emission). This is the microscopic origin of the nearly perfect directionality, monochromaticity, and polarization of a laser beam.

B The Planck distribution

We now consider the rates of these processes. Absorption and stimulated emission both vanish in the absence of atoms and of light, so for low densities of atoms and low intensities of radiation it is natural to assume that the absorption rate $W_{1\to2}$ from the lower level 1 to the upper level 2, and the stimulated emission rate $W_{2\to1}$—from the upper level 2 to the lower level 1—are both jointly proportional to the density of

[3] This is hard to do in reality, but easy to simulate. A movie of the particle motions in the interval $(0, \tau)$ will display the time-reversed behavior in the interval $(\tau, 2\tau)$ when run backwards.

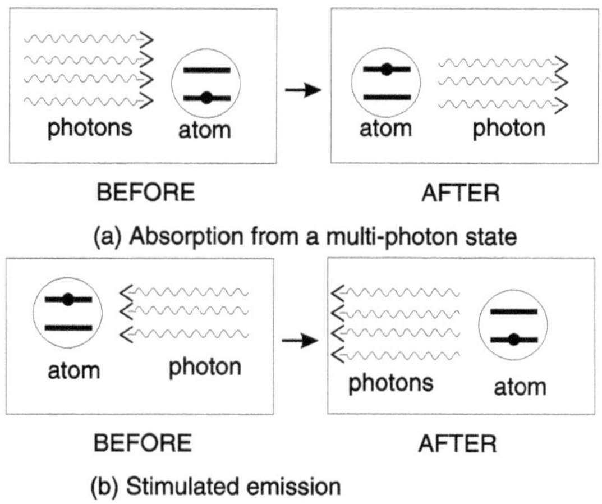

Fig. 1.8 (a) An atom in the ground state jumps to the excited state after absorbing one of the N incident photons. (b) An atom in the excited state illuminated by $N-1$ incident photons jumps to the ground state and leaves N photons in the final state.

atoms and the intensity of the light. We further assume that the two-level atoms are placed inside a cavity at temperature T, so that the light intensity is proportional to the spectral function $\rho(\omega, T)$. Therefore we expect that

$$W_{1\to 2} = B_{1\to 2} N_1 \rho(\omega, T) , \tag{1.10}$$

$$W_{2\to 1} = B_{2\to 1} N_2 \rho(\omega, T) , \tag{1.11}$$

where N_1 and N_2 are respectively the number of atoms in the lower level 1 and the upper level 2. The rate $S_{2\to 1}$ of spontaneous emission can only depend on N_2:

$$S_{2\to 1} = A_{2\to 1} N_2 , \tag{1.12}$$

since spontaneous emission occurs in the absence of any incident photons. The phenomenological Einstein A and B coefficients, $A_{2\to 1}$, $B_{2\to 1}$, and $B_{1\to 2}$, are assumed to be properties of the individual atoms which are independent of N_1, N_2, and $\rho(\omega, T)$.

By studying the situation in which the atoms and the radiation field are in thermal equilibrium, it is possible to derive other useful relations between the rate coefficients, and thus to determine the form of $\rho(\omega, T)$. The total rate $T_{2\to 1}$ for transitions from the upper state to the lower state is the sum of the spontaneous and stimulated rates,

$$T_{2\to 1} = A_{2\to 1} N_2 + B_{2\to 1} N_2 \rho(\omega, T) , \tag{1.13}$$

and the condition for steady state—which includes thermal equilibrium as an important special case—is $T_{2\to 1} = W_{1\to 2}$, so that

$$\left[A_{2\to 1} + B_{2\to 1} \rho(\omega, T)\right] N_2 = B_{1\to 2} \rho(\omega, T) N_1 . \tag{1.14}$$

Since the atoms and the radiation field are both in thermal equilibrium with the walls of the cavity at temperature T, the atomic populations satisfy Boltzmann's principle,

$$\frac{N_1}{N_2} = \frac{e^{-\beta\epsilon_1}}{e^{-\beta\epsilon_2}} = e^{\beta\hbar\omega}, \tag{1.15}$$

where $\beta = 1/k_B T$. Using this relation in eqn (1.14) leads to

$$\rho(\omega, T) = \frac{A_{2\to1}}{B_{1\to2}\exp(\beta\hbar\omega) - B_{2\to1}}. \tag{1.16}$$

This solution has very striking consequences. In the limit of infinite temperature ($\beta \to 0$), the spectral function approaches a constant value:

$$\rho(\omega, T) \to \frac{A_{2\to1}}{B_{1\to2} - B_{2\to1}}. \tag{1.17}$$

On the other hand, it seems natural to expect that the energy density in any finite frequency interval should increase without bound in the limit of high temperatures. The only way to avoid this contradiction is to impose

$$B_{1\to2} = B_{2\to1} = B, \tag{1.18}$$

i.e. the rate of stimulated emission must exactly equal the rate of absorption for a physically acceptable spectral function. This is an example of the principle of detailed balance (Chandler, 1987, Sec. 8.3), which also follows from time-reversal symmetry. Substituting eqn (1.18) into eqn (1.16) yields the new form

$$\rho(\omega, T) = \frac{A}{B}\frac{1}{\exp(\beta\hbar\omega) - 1}, \tag{1.19}$$

where we have further simplified the notation by setting $A_{2\to1} = A$. In the low temperature—or high energy—limit, $\hbar\omega \gg k_B T$ ($\beta\hbar\omega \gg 1$), the energy density is

$$\rho(\omega, T) = \left(\frac{A}{B}\right)\exp(-\beta\hbar\omega). \tag{1.20}$$

This is **Wien's law**, and it indeed agrees with experiment in the high energy limit.

By contrast, in the low energy limit, $\hbar\omega \ll k_B T$ —i.e. the photon energy is small compared to the average thermal energy—the classical Rayleigh–Jeans law is known to be correct. This allows us to determine the ratio A/B by comparing eqn (1.19) to eqn (1.4), with the result

$$\frac{A}{B} = \left(\frac{\hbar\omega^3}{\pi^2 c^3}\right). \tag{1.21}$$

Thus the standard form for the **Planck distribution**,

$$\rho(\omega, T) = \left(\frac{\hbar\omega^3}{\pi^2 c^3}\right)\frac{1}{\exp(\beta\hbar\omega) - 1}, \tag{1.22}$$

is completely fixed by applying the powerful principles of thermodynamics to two-level atoms in thermal equilibrium with the radiation field inside a cavity.

Einstein's argument for the A and B coefficients correctly correlates an impressive range of experimental results. On the other hand, it does not provide an explanation for the quantum jumps involved in spontaneous emission, stimulated emission, and absorption, nor does it give any way to relate the A and B coefficients to the microscopic properties of atoms. These features will be explained in the full quantum theory of light which is presented in the following chapters.

1.2.3 Applications

In addition to providing a framework for understanding the experiments discussed in Section 1.1, the photon model can also be used for more practical applications. For example, let us model an absorbing medium as a slab of thickness Δz and area $\mathcal{S}$ containing $N = n\Delta z\mathcal{S}$ two-level atoms, where n is the density of atoms. The energy density of light in the frequency interval $(\omega, \omega + \Delta\omega)$ at the entrance face is $u(\omega, z) = \rho(z, \omega)\Delta\omega$, where $\rho(z, \omega)$ is the spectral function of the incident light. The incident flux is then $cu(\omega, z)$, so energy enters and leaves the slab at the rates $cu(\omega, z)\mathcal{S}$ and $cu(\omega, z + \Delta z)\mathcal{S}$, respectively, as pictured in Fig. 1.9.

By energy conservation, the difference between these rates is the rate at which energy is absorbed in the slab. In order to calculate this correctly, we must provide a slightly more detailed model of the absorption process. So far, we have used an all-or-nothing picture in which absorption occurs at the sharply defined frequency $(\epsilon_2 - \epsilon_1)/\hbar$. In reality, the atoms respond in a continuous way to light at frequency ω. This is described by a **line shape** function $L(\omega)$, where $L(\omega)\Delta\omega$ is the fraction of atoms for which $(\epsilon_2 - \epsilon_1)/\hbar$ lies in the interval $(\omega, \omega + \Delta\omega)$. In succeeding chapters we will encounter many mechanisms that contribute to the line shape, but in the spirit of the photon model we simply assume that $L(\omega)$ is positive and normalized by

$$\int_0^\infty d\omega L(\omega) = 1\,. \tag{1.23}$$

We first consider the case that all of the atoms are in the ground state, then eqn (1.10) yields

$$[cu(z + \Delta z) - cu(z)]\mathcal{S} = -(\hbar\omega)(B\rho(z, \omega))(L(\omega)\Delta\omega n\Delta z\mathcal{S})\,. \tag{1.24}$$

In the limit $\Delta z \to 0$ this becomes a differential equation:

$$c\frac{du(z, \omega)}{dz} = -\hbar\omega nBL(\omega)u(z, \omega)\,, \tag{1.25}$$

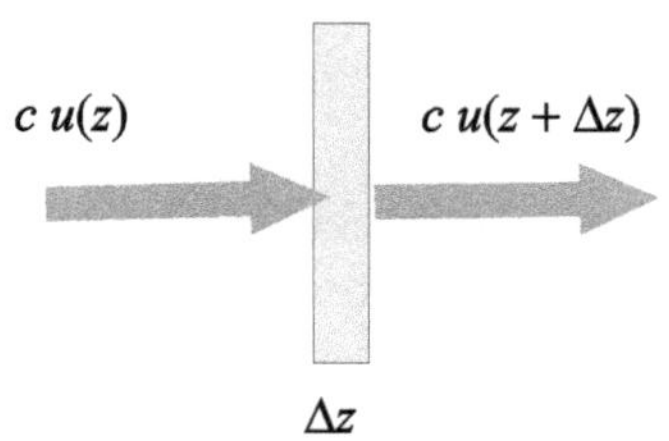

Fig. 1.9 Light in the frequency interval $(\omega, \omega + \Delta\omega)$ falls on a slab of thickness Δz and area $\mathcal{S}$. The incident flux is $cu(z, \omega) = c\rho(z, \omega)\Delta\omega$, where $\rho(z, \omega)$ is the spectral function.

with the solution

$$u(z,\omega) = u(0,\omega)\, e^{-\alpha(\omega)z}, \quad \text{where} \quad \alpha(\omega) = \frac{nL(\omega)\, B\hbar\omega}{c}. \tag{1.26}$$

This is **Beer's law of absorption**, and $\alpha(\omega)$ is the **absorption coefficient.**

In the opposite situation that all atoms are in the upper state, stimulated emission replaces absorption, and the same kind of calculation leads to

$$c\frac{du(z,\omega)}{dz} = \hbar\omega n B L(\omega)\, u(z,\omega), \tag{1.27}$$

with the solution

$$u(z,\omega) = u(0,\omega)\, e^{\alpha'(\omega)z}, \quad \alpha'(\omega) = \frac{nL(\omega)\, B\hbar\omega}{c}. \tag{1.28}$$

In this case we get **negative absorption**, that is, the amplification of light.

If both levels are nondegenerate, the general case is described by densities n_1 and n_2 for atoms in the lower and upper states respectively, with $n_1 + n_2 = n$. In the previous results this means replacing n by n_1 in the first case and n by n_2 in the second. In this situation,

$$\frac{du(z,\omega)}{dz} = g(\omega)u(z,\omega), \quad \text{where} \quad g(\omega) = \frac{(n_2 - n_1)\, L(\omega)\, B\hbar\omega}{c}. \tag{1.29}$$

For thermal equilibrium $n_1 > n_2$, so we get an absorbing medium, but with a **population inversion**, $n_2 > n_1$, we find instead a **gain medium** with gain $g(\omega) > 0$. This is the principle behind the laser (Schawlow and Townes, 1958).

1.3 Are photons necessary?

Now that we have established that the photon model is sufficient for the interpretation of the experiments described in Section 1.1, we ask if it is necessary. We investigate this question by attempting to describe each of the principal experiments using a semiclassical model.

1.3.1 The Planck distribution

This seems to be the simplest of the experiments under consideration, but finding a semiclassical explanation turns out to involve some subtle issues. Suppose we make the following assumptions.

(a) The electromagnetic field is described by the classical form of Maxwell's equations.

(b) The electromagnetic field is an independent physical system subject to the standard laws of statistical mechanics.

With both assumptions in force the equipartition argument in Section 1.1.1 inevitably leads to the Rayleigh–Jeans distribution and the ultraviolet catastrophe. This is physically unacceptable, so at least one of the assumptions (a) or (b) must be abandoned. At this point, Planck chose the rather risky alternative of abandoning (b), and Einstein took the even more radical step of abandoning (a).

Our task is to find some way of retaining (a) while replacing Planck's *ad hoc* procedure by an argument based on a quantum mechanical description of the atoms in the cavity wall. There does not seem to be a completely satisfactory way to do this, so a rough plausibility argument will have to suffice. We begin by observing that the derivation of the Planck distribution in Section 1.2.2-B does not explicitly involve the assumption that light is composed of discrete quanta. This suggests that we first seek a semiclassical origin for the A and B coefficients, and then simply repeat the same argument.

The Einstein coefficients $B_{1\to2}$ (for absorption) and $B_{2\to1}$ (for stimulated emission) can both be evaluated by applying first-order, time-dependent perturbation theory—which is reviewed in Section 4.8.2—to the coupling between the atom and the classical electromagnetic field. In both processes the electron remains bound in the atom, which is small compared to typical optical wavelengths. Thus the interaction of the atom with the classical field can be treated in the dipole approximation, and the interaction Hamiltonian is

$$H_{\text{int}} = -\widehat{\mathbf{d}} \cdot \boldsymbol{\mathcal{E}} \,, \tag{1.30}$$

where $\widehat{\mathbf{d}}$ is the electric dipole operator, and the field is evaluated at the center of mass of the atom. Applying the Fermi-golden-rule result (4.113) to the absorption process leads to

$$B_{1\to2} = \frac{\pi}{3\epsilon_0} \frac{|d_{12}|^2}{\hbar^2} \,, \tag{1.31}$$

where d_{12} is the matrix element of the dipole operator. A similar calculation for stimulated emission yields the same value for $B_{2\to1}$, so the equality of the two B coefficients is independently verified.

The strictly semiclassical theory used above does not explain spontaneous emission; instead, it predicts $A = 0$. The reason is that the interaction Hamiltonian (1.30) vanishes in the absence of an external field. If no external field is present, an atom in any stationary state—including all excited states—will stay there permanently. On the other hand, spontaneous emission is not explained in Einstein's photon model either; it is built in by assumption at the beginning. Since the present competition is with the photon model, we are at liberty to augment the strict semiclassical theory by simply assuming the existence of spontaneous emission. With this assumption in force, Einstein's rate arguments (eqns (1.10)–(1.21)) can be used to derive the ratio A/B. Note that these equations refer to transition rates within the two-level atom; they do not require the concept of the photon. Combining this with the independently calculated value of $B_{1\to2}$ given in eqn (1.31) yields the correct value for the A coefficient. This line of argument is frequently used to derive the A coefficient without bringing in the full blown quantum theory of light (Loudon, 2000, Sec. 1.5).

The extra assumptions required to carry out this semiclassical derivation of the Planck spectrum may make it appear almost as *ad hoc* as Planck's argument, but it does show that the photon model is not strictly necessary for this purpose.

1.3.2 The photoelectric effect

By contrast to the derivation of the Planck spectrum, Einstein's explanation of the photoelectric effect depends in a very direct way on the photon concept. In this case, however, the alternative description using the semiclassical theory turns out to be much more straightforward. For this calculation, the electrons in the metal are described by quantum mechanics, and the light is described as an external classical field. The total electron Hamiltonian is therefore $H = H_0 + H_{\text{int}}$, where H_0 is the Hamiltonian for an electron in the absence of any external electromagnetic field and H_{int} is the interaction term. For a single electron in a weak external field, the standard quantum mechanical result—reviewed in Appendix C.6—is

$$H_{\text{int}} = -\frac{e}{m}\boldsymbol{\mathcal{A}}(\widehat{\mathbf{r}}, t) \cdot \widehat{\mathbf{p}}\,, \tag{1.32}$$

where $\widehat{\mathbf{r}}$ and $\widehat{\mathbf{p}}$ are respectively the quantum operators for the position and momentum. In the usual position-space representation the action of the operators is $\widehat{\mathbf{r}}\psi(\mathbf{r}) = \mathbf{r}\psi(\mathbf{r})$ and $\widehat{\mathbf{p}}\psi(\mathbf{r}) = -i\hbar\boldsymbol{\nabla}\psi(\mathbf{r})$. The c-number function $\boldsymbol{\mathcal{A}}(\mathbf{r}, t)$ is the classical vector potential—which can be chosen to satisfy the radiation-gauge condition $\boldsymbol{\nabla} \cdot \boldsymbol{\mathcal{A}} = 0$—and it determines the radiation field by

$$\boldsymbol{\mathcal{E}} = -\frac{\partial \boldsymbol{\mathcal{A}}}{\partial t}\,, \quad \boldsymbol{\mathcal{B}} = \boldsymbol{\nabla} \times \boldsymbol{\mathcal{A}}\,. \tag{1.33}$$

For a monochromatic field with frequency ω, the vector potential is

$$\boldsymbol{\mathcal{A}}(\mathbf{r}, t) = \frac{1}{\omega}\mathcal{E}_0 \mathbf{e} \exp(i\mathbf{k} \cdot \mathbf{r} - \omega t) + \text{CC}\,, \tag{1.34}$$

where $\mathbf{e}$ is the unit polarization vector, $\mathcal{E}_0$ is the electric field amplitude, $|\mathbf{k}| = \omega/c$, and $\mathbf{e} \cdot \mathbf{k} = 0$. Another application of Fermi's golden rule (4.113) yields the rate

$$W_{fi} = \frac{2\pi}{\hbar} \left|\langle f \left| H_{\text{int}} \right| i \rangle\right|^2 \delta(\epsilon_f - \epsilon_i - \hbar\omega) \tag{1.35}$$

for the transition from the initial bound energy level ϵ_i into a free level ϵ_f. This result is valid for observation times $t \gg 1/\omega$. For optical fields $\omega \sim 10^{15}\,\text{s}^{-1}$, so eqn (1.35) predicts the emission of electrons with no appreciable delay. Furthermore, the delta function guarantees that the energy of the ejected electron satisfies the photoelectric equation. Finally the matrix element $\langle f \left| H_{\text{int}} \right| i \rangle$ is proportional to $\mathcal{E}_0$, so the rate of electron emission is proportional to the field intensity. Therefore, this simple semiclassical theory explains all of the puzzling aspects of the photoelectric effect, without ever introducing the concept of the photon. This point is already implicit in the very early papers of Wentzel (1926) and Beck (1927), and it has also been noted in much more recent work (Mandel *et al.*, 1964; Lamb and Scully, 1969). The energy conserving delta function in eqn (1.35) reproduces the kinematical relation (1.6), but it only appears at the end of a detailed dynamical calculation.

Most techniques for detecting photons employ the photoelectric effect, so an explanation of the photoelectric effect that does not require the existence of photons is a bit upsetting. Furthermore, the response of other kinds of detectors (such as photographic emulsions, solid-state photomultipliers, etc.) is ultimately also based on the photoelectric effect. Therefore, they can also be entirely described by the semiclassical theory. This raises serious questions about the interpretation of some experiments claiming to demonstrate the existence of photons. An early example is a repetition of Young's two slit experiment (Taylor, 1909), which used light of such low intensity that the average energy present in the apparatus at any given time was at most $\hbar\omega$. The result was a slow accumulation of spots on a photographic plate. After a sufficiently long exposure time, the spots displayed the expected two slit interference pattern. This was taken as evidence for the existence of photons, and apparently was the basis for Dirac's (1958) assertion that each photon interferes only with itself. This interpretation clearly depends on the assumption that each individual spot on the plate represents absorption of a single photon. The semiclassical explanation of the photoelectric effect shows that the results could equally well be interpreted as the interference of classical electromagnetic waves from the two slits, combined with the semiclassical quantum theory for excitation of electrons in the photographic plate. In this view, there is no necessity for the concept of the photon, and thus for the quantization of the electromagnetic field.

1.3.3 Compton scattering

The kinematical explanation for the Compton shift given in Section 1.1.3 is often offered as conclusive evidence for the existence of photons, but the very first derivation (Klein and Nishina, 1929) of the celebrated Klein–Nishina formula (Bjorken and Drell, 1964, Sec. 7.7) for the differential cross-section of Compton scattering was carried out in a slightly extended form of the semiclassical approximation. The analysis is more complicated than the semiclassical treatment of the photoelectric effect for two reasons. The first is that the electron motion may become relativistic, so that the nonrelativistic Schrödinger equation must be replaced by the relativistic Dirac equation (Bjorken and Drell, 1964, Chap. 1). The second complication is that the radiation emitted by the excited electron cannot be ignored, since observing this radiation is the point of the experiment. Thus Compton scattering is a two step process in which the electron is first excited by the incident radiation, and the resulting current subsequently generates the scattered radiation. In the original paper of Klein and Nishina, the Dirac equation for an electron exposed to an incident plane wave is solved by using first-order time-dependent perturbation theory. The expectation value of the current-density operator in the perturbed state is then used as the source term in the classical Maxwell equations. The radiation field generated in this way automatically satisfies the kinematical relations (1.7), so it again yields the Compton shift given in eqn (1.8). Furthermore, the Compton cross-section calculated by using the semiclassical Klein–Nishina model precisely agrees with the result obtained in quantum electrodynamics, in which the electromagnetic field is treated by quantum theory. Once again we see that Einstein's quantum model provides a beautifully simple explanation of the kinematical aspects of the experiment, but that the more complicated semiclassical treatment achieves the same end, while also providing a correct dynamical calculation of the cross-section. There is again no necessity to introduce the concept of the photon anywhere in this calculation.

1.3.4 Conclusions

The experiments discussed in Section 1.1 are usually presented as evidence for the existence of photons. The reasoning behind this claim is that classical physics is inconsistent with the experimental results, while Einstein's photon model describes all the experimental results in a very simple way. What we have just seen, however, is that an augmented version of semiclassical electrodynamics can explain the same set of experiments without recourse to the idea of photons. Where, then, is the empirical evidence for the existence of photons? In the next section we will describe experiments that bear on this question.

1.4 Indivisibility of photons

The semiclassical explanations of the experimental results in Section 1.1 imply that these experiments are not sensitive to the indivisibility of photons. Classical electromagnetic theory describes light in terms of electric and magnetic fields with continuously variable field amplitudes, but the photon model of light asserts that electromagnetic energy is concentrated into discrete quanta which cannot be further subdivided. In particular, a classical electromagnetic wave must be continuously divisible at a beam splitter, whereas an indivisible photon must be either entirely transmitted, or entirely reflected, as a whole unit. The continuous division of the classical waves and the discontinuous reflection-or-transmission choice of the photon are mutually exclusive; therefore, the quantum and classical theories of light give entirely different predictions for experiments involving individual quanta of light incident on a beam splitter. The indivisibility of the photon is a postulate of Einstein's original model, and it is a consequence of the fully developed quantum theory of the electromagnetic field. Since even the most sophisticated versions of the semiclassical theory describe light in terms of continuously variable classical fields, the decisive experiments must depend on the indivisibility of individual photons.

Two important advances in this direction were made by Clauser in the context of a discussion of the experimental limits of validity of semiclassical theories, in particular the neoclassical theory of Jaynes (Crisp and Jaynes, 1969). For this purpose, the two-level atom used in previous discussions is inadequate; we now need atoms with at least three active levels. The first advance was Clauser's reanalysis (Clauser, 1972) of the data from an experiment by Kocher and Commins (1967), which used a three-level cascade emission in a calcium atom, as shown in Fig. 1.10. A beam of calcium atoms is crossed by a light beam which excites the atoms to the highest energy level. This

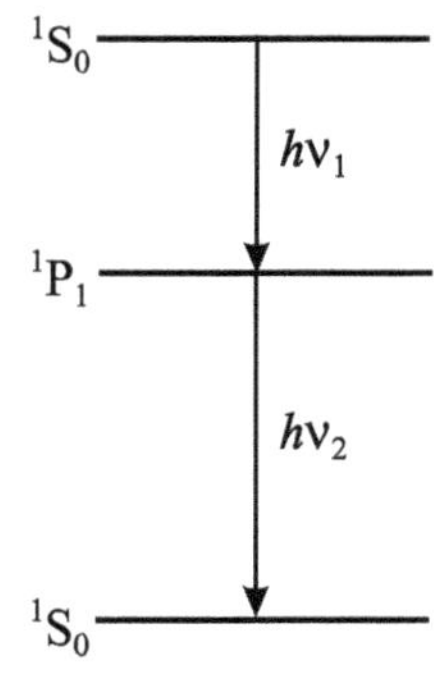

Fig. 1.10 The lowest three energy levels of the calcium atom allow the cascade of two successive transitions, in which two photons $h\nu_1$ and $h\nu_2$ are emitted in rapid succession. The intermediate level has a short lifetime of 4.7 ns.

excitation is followed by a rapid cascade decay, with the correlated emission of two photons. The first $(h\nu_1)$ is emitted in a transition from the highest energy level to the short-lived intermediate level, and the second $(h\nu_2)$ is emitted in a transition from the intermediate level to the ground level. These two photons, which are emitted almost back-to-back with respect to each other, are then detected using fast coincidence electronics. In this way, a beam of calcium atoms provides a source of strongly correlated photon pairs.

The light emitted in each transition is randomly polarized—i.e. all polarizations are detected with equal probability—but the experiment shows that the probabilities of observing given polarizations at the two detectors are correlated. The correlation coefficient obtained from a semiclassical calculation has a lower bound which is violated by the experimental data, while the correlation predicted by the quantum theory of radiation agrees with the data. The second advance was an experiment performed by Clauser himself (Clauser, 1974), in which the two bursts of light from a three-level cascade emission in the mercury atom are each passed through beam splitters to four photodetectors. The object in this case is to observe the **coincidence rate** between various pairs of detectors, in other words, the rates at which a pair of detectors both fire during the same small time interval. The semiclassical rates are again inconsistent with experiment, whereas the quantum theory prediction agrees with the data. The first experiment provides convincing evidence which supports the quantum theory and rejects the semiclassical theory, but the role of the indivisibility of photons is not easily seen. The second experiment does depend directly on this property, but the analysis is rather involved. We therefore refer the reader to the original papers for descriptions of this seminal work, and briefly describe instead a third experiment that yields the clearest and most direct evidence for the indivisibility of single photons, and thus for the existence of individual quanta of the electromagnetic field.

The experiment in question—which we will call the photon-indivisibility experiment—was performed by Grangier *et al.* (1986). The experimental arrangement (shown in Fig. 1.11) employs a three-level cascade (see Fig. 1.10) in a calcium atom located at S. Two successive, correlated bursts of light—centered at frequencies ν_1 and ν_2—are emitted in opposite directions from the source. At this point in the argument, we leave open the possibility that the light is described by classical electromagnetic waves as opposed to photons, and assume that detection events are perfectly describable by the semiclassical theory of the photoelectric effect.

The atoms, which are delivered by an atomic beam, are excited to the highest energy level shortly before reaching the source region S. The photomultiplier PM_{gate} is equipped with a filter that screens out radiation at the frequency ν_2 of the second transition, while passing radiation at ν_1, the frequency of the first transition. The output from PM_{gate}, which monitors bursts of radiation at frequency ν_1, is registered by the counter N_{gate}, and is also used to activate (trigger) a device called a **gate generator** which produces a standardized, rectangularly-shaped gate pulse for a specified time interval, $T_{\text{gate}} = w$, called the **gate width**. The outputs of the photomultipliers PM_{refl} and PM_{trans}, which monitor bursts of radiation at frequency ν_2, are registered by the gated counters N_{refl} and N_{trans} only during the time interval specified by the gate width w.

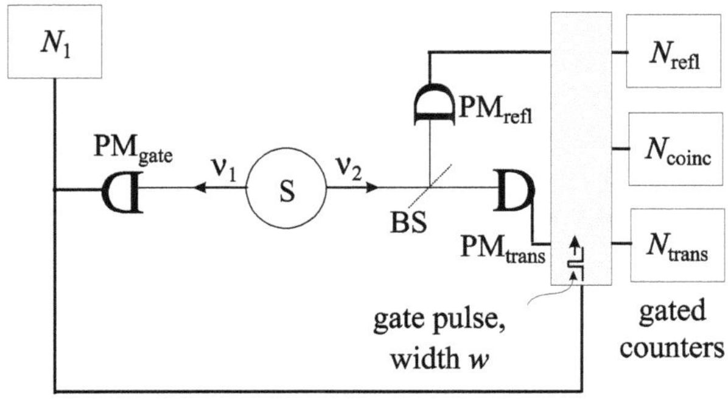

Fig. 1.11 The photon-indivisibility experiment of Grangier, Roger, and Aspect. The detection of the first burst of light, of frequency ν_1, of a calcium-atom cascade produces a gate pulse of width w during which the outputs of the photomultipliers PM_{trans} and PM_{refl} detecting the second burst of light, of frequency ν_2, are recorded by the gated counters. The rate of gate openings is $\dot{N}_{gate} = \dot{N}_1$. The probabilities of detection during the gate openings are $p_{trans} = \dot{N}_{trans}/\dot{N}_1$, $p_{refl} = \dot{N}_{refl}/\dot{N}_1$ for singles, and $p_{coinc} = \dot{N}_{coinc}/\dot{N}_1$ for coincidences. (Adapted from Grangier *et al.* (1986).)

If a burst of radiation at ν_1 has been detected, the burst of radiation of frequency ν_2 from the second transition is necessarily directed toward the beam splitter BS, which partially reflects and partially transmits the light falling on it. The two beams produced in this way are directed toward the two photomultipliers PM_{refl} and PM_{trans}. The outputs of PM_{refl} and PM_{trans} are used to drive the gated counters N_{refl} and N_{trans}, which record every pulse from the two photomultipliers, and also to drive a coincidence counter N_{coinc}, which responds only when both of these two photomultipliers produce current pulses simultaneously within the specified open-gate time interval w. Therefore, the probabilities for the individual counters to fire (*singles probabilities*) are given by $p_{refl} = \dot{N}_{refl}/\dot{N}_{gate}$ and $p_{trans} = \dot{N}_{trans}/\dot{N}_{gate}$, where $\dot{N}_{gate} \equiv \dot{N}_1$ is the rate of gate openings—the count rate of photomultiplier PM_{gate}—and $\dot{N}_{refl}$ and $\dot{N}_{trans}$ are the count rates of PM_{refl} and PM_{trans}, respectively. The coincidence rate $\dot{N}_{coinc}$ is the rate of simultaneous firings of both detectors PM_{refl} and PM_{trans} during the open-gate interval w; consequently, the coincidence probability is $p_{coinc} = \dot{N}_{coinc}/\dot{N}_{gate}$. The experiment consists of measuring the singles counting rates $\dot{N}_{gate}$, $\dot{N}_{refl}$, $\dot{N}_{trans}$, and the coincidence rate $\dot{N}_{coinc}$.

According to Einstein's photon model of light, each atomic transition produces a single quantum of light which cannot be subdivided. An indivisible quantum with energy $h\nu_2$ which has scattered from the beam splitter can only be detected once. Therefore it must go either to PM_{refl} or to PM_{trans}; it cannot go to both. In the absence of complicating factors, the photon model would predict that the coincidence probability p_{coinc} is exactly zero. Since this is a real experiment, complicating factors are not absent. It is possible for two different atoms inside the source region S to emit two quanta $h\nu_2$ during the open-gate interval, and thereby produce a false coincidence count. This difficulty can be minimized by choosing the gate interval $w \ll \tau'$, where τ' is the lifetime of the intermediate level in the cascade, but it cannot be completely removed from this experimental arrangement.

Only three general features of semiclassical theories are needed for the analysis of this experiment: (1) the atom is described by quantum mechanics; (2) each atomic transition produces a burst of radiation described by classical fields; (3) the photomultiplier current is proportional to the intensity of the incident radiation. The first two features are part of the definition of a semiclassical theory, and the third is implied by the semiclassical analysis of the photoelectric effect. The beam splitter will convert the classical radiation from the atom into two beams, one directed toward $\mathrm{PM_{refl}}$ and the other directed toward $\mathrm{PM_{trans}}$. Therefore, according to the semiclassical theory, the coincidence probability cannot be zero—even in the absence of the false counts discussed above—since the classical electromagnetic wave must smoothly divide at the beam splitter. The semiclassical theory predicts a minimum coincidence rate, which is proportional to the product of the reflected and transmitted intensities. The instantaneous intensities falling on $\mathrm{PM_{refl}}$ and $\mathrm{PM_{trans}}$ are proportional to the original intensity falling on the beam splitter, and the gated measurement effectively averages over the open-gate interval. Thus the photocurrents produced in the nth gate interval are proportional to the time averaged intensity at the beam splitter:

$$I_n = \frac{1}{w} \int_{t_n}^{t_n + w} dt I(t) \,, \tag{1.36}$$

where the gate is open in the interval $(t_n, t_n + w)$. The atomic transitions are described by quantum mechanics, so they occur at random times within the gate interval. This means that the intensities I_n exhibit random variations from one gate interval to another. In order to minimize the effect of these fluctuations, the counting data from a sequence of gate openings are averaged. Thus the singles probabilities are determined from the average intensity

$$\langle I \rangle = \frac{1}{M_{\text{gate}}} \sum_{n=1}^{M_{\text{gate}}} I_n \,, \tag{1.37}$$

where M_{gate} is the total number of gate openings. The singles probabilities are given by

$$p_{\text{refl}} = \eta_{\text{refl}} \, w \langle I \rangle \,, \quad p_{\text{trans}} = \eta_{\text{trans}} \, w \langle I \rangle \,, \tag{1.38}$$

where η_{refl} is the product of the detector efficiency and the fraction of the original intensity directed to $\mathrm{PM_{refl}}$ and η_{trans} is the same quantity for $\mathrm{PM_{trans}}$. Since the coincidence rate in a single gate is proportional to the product of the instantaneous photocurrents from $\mathrm{PM_{refl}}$ and $\mathrm{PM_{trans}}$, the coincidence probability is proportional to the average of the square of the intensity:

$$p_{\text{coinc}} = \eta_{\text{refl}} \, \eta_{\text{trans}} \, w^2 \left\langle I^2 \right\rangle , \tag{1.39}$$

with

$$\left\langle I^2 \right\rangle = \frac{1}{M_{\text{gate}}} \sum_{n=1}^{M_{\text{gate}}} I_n^2 \,. \tag{1.40}$$

By using the identity $\left\langle (I - \langle I \rangle)^2 \right\rangle \geqslant 0$ it is easy to show that

$$\langle I^2 \rangle \geqslant \langle I \rangle^2 \,, \tag{1.41}$$

which combines with eqns (1.38) and (1.39) to yield

$$p_{\text{coinc}} \geqslant p_{\text{refl}} \, p_{\text{trans}} \,. \tag{1.42}$$

This semiclassical prediction is conveniently expressed by defining the parameter

$$\alpha \equiv \frac{p_{\text{coinc}}}{p_{\text{refl}} p_{\text{trans}}} = \frac{\dot{N}_{\text{coinc}} \dot{N}_{\text{gate}}}{\dot{N}_{\text{refl}} \dot{N}_{\text{trans}}} \geqslant 1 \,, \tag{1.43}$$

where the latter inequality follows from eqn (1.42). With the gate interval set at $w = 9\,\text{ns}$, and the atomic beam current adjusted to yield a gate rate $\dot{N}_{\text{gate}} = 8800$ counts per second, the measured value of α was found to be $\alpha = 0.18 \pm 0.06$. This violates the semiclassical inequality (1.43) by 13 standard deviations; therefore, the experiment decisively rejects any theory based on the semiclassical treatment of emission. These data show that there are strong anti-correlations between the firings of photomultipliers PM_{refl} and PM_{trans}, when gated by the firings of the trigger photomultiplier PM_{gate}. An individual photon $h\nu_2$, upon leaving the beam splitter, can cause either of the photomultipliers PM_{refl} or PM_{trans} to fire, but these two possible outcomes are mutually exclusive. This experiment convincingly demonstrates the indivisibility of Einstein's photons.

1.5 Spontaneous down-conversion light source

In more recent times, the cascade emission of correlated pairs of photons used in the photon indivisibility experiment has been replaced by spontaneous down-conversion. In this much more convenient and compact light source, atomic beams—which require the extensive use of inconvenient vacuum technology—are replaced by a single nonlinear crystal. An ultraviolet laser beam enters the crystal, and excites its atoms coherently to a virtual excited state. This is followed by a rapid decay into pairs of photons γ_1 and γ_2, as shown in Fig. 1.12 and discussed in detail in Section 13.3.2. This process may seem to violate the indivisibility of photons, so we emphasize that an incident UV photon is absorbed as a whole unit, and two other photons are emitted, also as whole units. Each of these photons would pass the indivisibility test of the experiment discussed in Section 1.4.

Just as in the similar process of radioactive decay of an excited parent nucleus into two daughter nuclei, energy and momentum are conserved in spontaneous down-conversion. Due to a combination of dispersion and birefringence of the nonlinear

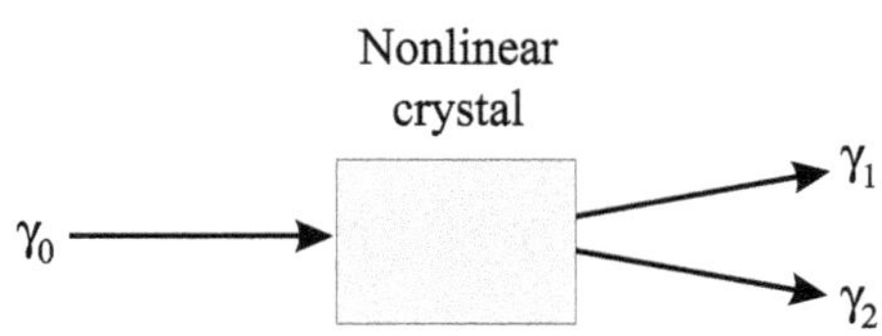

Fig. 1.12 The process of spontaneous down-conversion, $\gamma_0 \rightarrow \gamma_1 + \gamma_2$ by means of a nonlinear crystal.

crystal, the result is a highly directional emission of light in the form of a rainbow of many colors, as seen in the jacket illustration.

The uniquely quantum feature of this rainbow is the fact that pairs of photons emitted on opposite sides of the ultraviolet laser beam, are strongly correlated with each other. For example, the detection of a photon γ_1 by a Geiger counter placed behind pinhole 1 in Fig. 1.12 is always accompanied by the detection of a photon γ_2 by a Geiger counter placed behind pinhole 2. The high directionality of this kind of light source makes the collection of correlated photon pairs and the measurement of their properties much simpler than in the case of atomic-beam light sources.

1.6 Silicon avalanche-photodiode photon counters

In addition to the improved light source discussed in the previous section, solid-state technology has also led to improved detectors of photons. Photon detectors utilizing photomultipliers based on vacuum-tube technology have now been replaced by much simpler solid-state detectors based on the photovoltaic effect in semiconductor crystals. A photon entering into the crystal produces an electron–hole pair, which is then pulled apart in the presence of a strong internal electric field. This field is sufficiently large so that the acceleration of the initial pair of charged particles produced by the photon leads to an avalanche breakdown inside the crystal, which can be thought of as a chain reaction consisting of multiple branches of impact ionization events initiated by the first pair of charged particles. This mode of operation of a semiconductor photodiode is called the Geiger mode, because of the close analogy to the avalanche ionization breakdown of a gas due to an initial ionizing particle passing through a Geiger counter.

Each avalanche breakdown event produces a large, standardized electrical pulse (which we will henceforth call a **click** of the photon counter), corresponding to the detection of a single photon. For example, many contemporary quantum optics experiments use silicon avalanche photodiodes, which are single photon counters with quantum efficiencies around 70% in the near infrared. This is much higher than the quantum efficiencies for photomultipliers in the same wavelength region. The solid-state detectors also have shorter response times—in the nanosecond range—so that fast coincidence detection of the standardized pulses can be straightforwardly implemented by conventional electronics. Another important practical advantage of solid-state single-photon detectors is that they require much lower voltage power supplies than photomultipliers. These devices will be discussed in more detail in Sections 9.1.1 and 9.2.1.

1.7 The quantum theory of light

In this chapter we have seen that the blackbody spectrum, the photoelectric effect, Compton scattering and spontaneous emission are correctly described by Einstein's photon model of light, but we have also seen that plausible explanations of these phenomena can be constructed using an extended form of semiclassical electrodynamics. However, no semiclassical explanation can account for the indivisibility of photons demonstrated in Section 1.4; therefore, a theory that incorporates indivisibility must be based on new physical principles not found in classical electromagnetism. In other words, the quantum theory of light cannot be derived from the classical theory; instead, it must be based on new conjectures.[4] Fortunately, the quantum theory must also satisfy the **correspondence principle**; that is, it must agree with the classical theory for the large class of phenomena that are correctly described by classical electrodynamics. This is an invaluable aid in the construction of the quantum theory. In the end, the validity of the new principles can only be judged by comparing predictions of the quantum theory with the results of experiments.

We will approach the quantum theory in stages, beginning with the electromagnetic field in an ideal cavity. This choice reflects the historical importance of cavities and blackbody radiation, and it is also the simplest problem exhibiting all of the important physical principles involved. An apparent difficulty with this approach is that it depends on the classical cavity mode functions, which are defined by boundary conditions at the cavity walls. Even in the classical theory, these boundary conditions are a macroscopic idealization of the properties of physical walls composed of atoms; consequently, the corresponding quantum theory does not appear to be truly microscopic. We will see, however, that the cavity model yields commutation relations between field operators at different spatial points which suggest a truly microscopic quantization conjecture that does not depend on macroscopic boundary conditions.

1.8 Exercises

1.1 Power emitted through an aperture of a cavity

Show that the radiative power per unit frequency interval at frequency ω emitted from the aperture area σ of a cavity at temperature T is given by

$$\mathcal{P}(\omega, T) = \frac{1}{4} c\rho(\omega, T)\,\sigma\,.$$

1.2 Spectrum of a one-dimensional blackbody

Consider a coaxial cable of length L terminated at either end with resistors of the same small value R. The entire system comes into thermal equilibrium at a temperature T. The dielectric constant inside the cable is unity. All you need to know about this terminated coaxial cable is that the wavelength λ_m of the mth mode of the classical electromagnetic modes of this cable is determined by the condition $L = m\lambda_m/2$, where $m = 1, 2, 3, \ldots$, and therefore that the frequency ν_m of the mth mode of the cable is given by $\nu_m = m\,(c/2L)$.

(1) In the large L limit, derive the classical Rayleigh–Jeans law for this system. Is there an ultraviolet catastrophe?

(2) Argue that the analysis in Section 1.2.2-B applies to this one-dimensional system, so that eqn (1.19) is still valid. Combine this with the result from part (1) to obtain the Planck distribution.

(3) Sketch the frequency dependence of the power spectrum, up to a proportionality constant, for the radiation emitted by one of the resistors.

[4]We prefer 'conjecture' to 'axiom', since an axiom cannot be questioned. In physics there are no unquestionable statements.

(4) For a given temperature, find the frequency at which the power spectrum is a maximum. Compare this to the corresponding result for the three-dimensional blackbody spectrum.

1.3 Slightly anharmonic oscillator

Given the following Hamiltonian for a slightly anharmonic oscillator in 1D:

$$\widehat{H} = \frac{\widehat{p}^2}{2m} + \frac{1}{2}m\omega^2\widehat{x}^2 + \frac{1}{4}\lambda m^2\widehat{x}^4 ,$$

where the perturbation parameter λ is very small.

(1) Find all the perturbed energy levels of this oscillator up to terms linear in λ.

(2) Find the lowest-order correction to its ground-state wave function. (**Hint**: Use raising and lowering operators in your calculation.)

1.4 Photoionization

A simple model for photoionization is defined by the vector potential $\boldsymbol{\mathcal{A}}$ and the interaction Hamiltonian H_{int} given respectively by eqns (1.34) and (1.32).

Assume that the initial electron is in a bound state with a spherically symmetric wave function $\langle \mathbf{r} \,|i\rangle = \phi_i(\mathbf{r})$ and energy $\epsilon_i = -\epsilon_b$ (where $\epsilon_b > 0$ is the binding energy) and that the final electron state is the plane wave $\langle \mathbf{r}\,|f\rangle = L^{-3/2}e^{i\mathbf{k}_f\cdot\mathbf{r}}$ (this is the **Born approximation**).

(1) Evaluate the matrix element $\langle f\,|H_{\text{int}}|\,i\rangle$ in terms of the initial wave function $\phi_i(\mathbf{r})$.

(2) Carry out the integration over the final electron state, and impose the dipole approximation—$k_f \gg |\boldsymbol{k}|$—in eqn (1.35) to get the total transition rate in the limit $\omega \gg \epsilon_b$.

(3) Divide the transition rate by the flux of photons ($F = I_0/\hbar\omega$, where I_0 is the intensity of the incident field) to obtain the cross-section for photoemission.

1.5 Time-reversal symmetry applied to the time-dependent Schrödinger equation

(1) Show that the time-reversal operation $t \to -t$, when applied to the time-dependent Schrödinger equation for a spinless particle, results in the rule

$$\psi \to \psi^*$$

for the wave function.

(2) Rewrite the wave function in Dirac bra-ket notation explained in Appendix C.1, and restate the above rule using this notation.

(3) In general, how does the scalar product for the transition probability amplitude between an initial and a final state $\langle\text{final}|\ \text{initial}\rangle$ behave under time reversal?

2

Quantization of cavity modes

In Section 1.3 we remarked that both classical mechanics and quantum mechanics deal with discrete sets of mechanical degrees of freedom, while classical electromagnetic theory is based on continuous functions of space and time. This conceptual gap can be partially bridged by studying situations in which the electromagnetic field is confined by material walls, such as those of a hollow metallic cavity. In such cases the classical field is described by a discrete set of mode functions. The formal resemblance between the discrete cavity modes and the discrete mechanical degrees of freedom facilitates the use of the correspondence-principle arguments that provide the surest route to the quantum theory.

In order to introduce the basic ideas in the simplest possible way, we will begin by quantizing the modes of a three-dimensional cavity. We will then combine the 3D cavity model with general features of quantum theory to explain the Planck distribution and the Casimir effect.

2.1 Quantization of cavity modes

We begin with a review of the classical electromagnetic field $(\boldsymbol{\mathcal{E}}, \boldsymbol{\mathcal{B}})$ confined to an **ideal cavity**, i.e. a void completely enclosed by perfectly conducting walls.

2.1.1 Cavity modes

In the interior of a cavity, the electromagnetic field obeys the vacuum form of Maxwell's equations:

$$\nabla \cdot \boldsymbol{\mathcal{E}} = 0 \,, \tag{2.1}$$

$$\nabla \cdot \boldsymbol{\mathcal{B}} = 0 \,, \tag{2.2}$$

$$\nabla \times \boldsymbol{\mathcal{B}} = \mu_0 \epsilon_0 \frac{\partial \boldsymbol{\mathcal{E}}}{\partial t} \quad \text{(Ampère's law)} \,, \tag{2.3}$$

$$\nabla \times \boldsymbol{\mathcal{E}} = -\frac{\partial \boldsymbol{\mathcal{B}}}{\partial t} \quad \text{(Faraday's law)} \,. \tag{2.4}$$

The divergence equations (2.1) and (2.2) respectively represent the absence of free charges and magnetic monopoles inside the cavity.[1] The tangential component of the

[1] As of this writing, no magnetic monopoles have been found anywhere, but if they are discovered in the future, eqn (2.2) will remain an excellent approximation.

electric field and the normal component of the magnetic induction must vanish on the interior wall, S, of a perfectly conducting cavity:

$$\mathbf{n}(\mathbf{r}) \times \boldsymbol{\mathcal{E}}(\mathbf{r}) = 0 \quad \text{for each } \mathbf{r} \text{ on } S\,, \tag{2.5}$$

$$\mathbf{n}(\mathbf{r}) \cdot \boldsymbol{\mathcal{B}}(\mathbf{r}) = 0 \quad \text{for each } \mathbf{r} \text{ on } S\,, \tag{2.6}$$

where $\mathbf{n}(\mathbf{r})$ is the normal vector to S at $\mathbf{r}$.

Since the boundary conditions are independent of time, it is possible to force a separation of variables between $\mathbf{r}$ and t by setting $\boldsymbol{\mathcal{E}}(\mathbf{r},t) = \boldsymbol{\mathcal{E}}(\mathbf{r})\mathcal{F}(t)$ and $\boldsymbol{\mathcal{B}}(\mathbf{r},t) = \boldsymbol{\mathcal{B}}(\mathbf{r})\mathcal{G}(t)$, where $\mathcal{F}(t)$ and $\mathcal{G}(t)$ are chosen to be dimensionless. Substituting these forms into Faraday's law and Ampère's law shows that $\mathcal{F}(t)$ and $\mathcal{G}(t)$ must obey

$$\frac{d\mathcal{G}(t)}{dt} = \omega_1 \mathcal{F}(t)\,, \quad \frac{d\mathcal{F}(t)}{dt} = \omega_2 \mathcal{G}(t)\,, \tag{2.7}$$

where ω_1 and ω_2 are separation constants with dimensions of frequency. Eliminating $\mathcal{G}(t)$ between the two first-order equations yields the second-order equation

$$\frac{d\mathcal{F}(t)}{dt} = \omega_1\omega_2 \mathcal{F}(t)\,, \tag{2.8}$$

which has exponentially growing solutions for $\omega_1\omega_2 > 0$ and oscillatory solutions for $\omega_1\omega_2 < 0$. The exponentially growing solutions are not physically acceptable; therefore, we set $\omega_1\omega_2 = -\omega^2 < 0$. With the choice $\omega_1 = -\omega$ and $\omega_2 = \omega$ for the separation constants, the general solutions for $\mathcal{F}$ and $\mathcal{G}$ can written as $\mathcal{F}(t) = \cos(\omega t + \phi)$ and $\mathcal{G}(t) = \sin(\omega t + \phi)$.

One can then show that the rescaled fields[2] $\boldsymbol{\mathcal{E}}_\omega(\mathbf{r}) = \sqrt{\epsilon_0/\hbar\omega}\,\boldsymbol{\mathcal{E}}(\mathbf{r})$ and $\boldsymbol{\mathcal{B}}_\omega(\mathbf{r}) = \boldsymbol{\mathcal{B}}(\mathbf{r})/\sqrt{\mu_0\hbar\omega}$ satisfy

$$\boldsymbol{\nabla} \times \boldsymbol{\mathcal{E}}_\omega(\mathbf{r}) = k\boldsymbol{\mathcal{B}}_\omega(\mathbf{r})\,, \tag{2.9}$$

$$\boldsymbol{\nabla} \times \boldsymbol{\mathcal{B}}_\omega(\mathbf{r}) = k\boldsymbol{\mathcal{E}}_\omega(\mathbf{r})\,, \tag{2.10}$$

where $k = \omega/c$. Alternately eliminating $\boldsymbol{\mathcal{E}}_\omega(\mathbf{r})$ and $\boldsymbol{\mathcal{B}}_\omega(\mathbf{r})$ between these equations produces the Helmholtz equations for $\boldsymbol{\mathcal{E}}_\omega(\mathbf{r})$ and $\boldsymbol{\mathcal{B}}_\omega(\mathbf{r})$:

$$\left(\nabla^2 + k^2\right)\boldsymbol{\mathcal{E}}_\omega(\mathbf{r}) = 0\,, \tag{2.11}$$

$$\left(\nabla^2 + k^2\right)\boldsymbol{\mathcal{B}}_\omega(\mathbf{r}) = 0\,. \tag{2.12}$$

A The rectangular cavity

The equations given above are valid for any cavity shape, but explicit mode functions can only be obtained when the shape is specified. We therefore consider a cavity in the form of a rectangular parallelepiped with sides l_x, l_y, and l_z. The bounding surfaces

[2]Dimensional convenience is the official explanation for the appearance of $\hbar$ in these classical normalization factors.

are planes parallel to the Cartesian coordinate planes, and the boundary conditions are

$$\left.\begin{aligned}\mathbf{n}\times\boldsymbol{\mathcal{E}}_{\omega}&=0\\ \mathbf{n}\cdot\boldsymbol{\mathcal{B}}_{\omega}&=0\end{aligned}\right\}\text{ on each face of the parallelepiped;}\tag{2.13}$$

therefore, the method of separation of variables can be used again to solve the eigenvalue problem (2.11). The calculations are straightforward but lengthy, so we leave the details to Exercise 2.2, and merely quote the results. The boundary conditions can only be satisfied for a discrete set of k-values labeled by the multi-index

$$\kappa\equiv(\mathbf{k},s)=(k_x,k_y,k_z,s)=\left(\frac{\pi n_x}{l_x},\frac{\pi n_y}{l_y},\frac{\pi n_z}{l_z},s\right),\tag{2.14}$$

where n_x, n_y, and n_z are non-negative integers and s labels the polarization. The allowed frequencies

$$\omega_{\mathbf{k}s}=c\left|\mathbf{k}\right|=c\left[\left(\frac{\pi n_x}{l_x}\right)^2+\left(\frac{\pi n_y}{l_y}\right)^2+\left(\frac{\pi n_z}{l_z}\right)^2\right]^{1/2}\tag{2.15}$$

are independent of s. The explicit expressions for the electric mode functions are

$$\boldsymbol{\mathcal{E}}_{\mathbf{k}s}(\mathbf{r})=\mathcal{E}_{\mathbf{k}x}(\mathbf{r})\,e_{sx}(\mathbf{k})\,\mathbf{u}_x+\mathcal{E}_{\mathbf{k}y}(\mathbf{r})\,e_{sy}(\mathbf{k})\,\mathbf{u}_y+\mathcal{E}_{\mathbf{k}z}(\mathbf{r})\,e_{sz}(\mathbf{k})\,\mathbf{u}_z\,,\tag{2.16}$$

$$\begin{aligned}\mathcal{E}_{\mathbf{k}x}(\mathbf{r})&=N_{\mathbf{k}}\cos(k_x x)\sin(k_y y)\sin(k_z z)\,,\\ \mathcal{E}_{\mathbf{k}y}(\mathbf{r})&=N_{\mathbf{k}}\sin(k_x x)\cos(k_y y)\sin(k_z z)\,,\\ \mathcal{E}_{\mathbf{k}z}(\mathbf{r})&=N_{\mathbf{k}}\sin(k_x x)\sin(k_y y)\cos(k_z z)\,,\end{aligned}\tag{2.17}$$

where the $N_{\mathbf{k}}$s are normalization factors. The polarization unit vector,

$$\mathbf{e}_s(\mathbf{k})=e_{sx}(\mathbf{k})\,\mathbf{u}_x+e_{sy}(\mathbf{k})\,\mathbf{u}_y+e_{sz}(\mathbf{k})\,\mathbf{u}_z\,,\tag{2.18}$$

must be transverse (i.e. $\mathbf{k}\cdot\mathbf{e}_s(\mathbf{k})=0$) in order to guarantee that eqn (2.1) is satisfied. The magnetic mode functions are readily obtained by using eqn (2.9).

Every plane wave in free space has two possible polarizations, but the number of independent polarizations for a cavity mode depends on $\mathbf{k}$. Inspection of eqn (2.17) shows that a mode with exactly one vanishing $\mathbf{k}$-component has only one polarization. For example, if $\mathbf{k}=(0,k_y,k_z)$, then $\boldsymbol{\mathcal{E}}_{\mathbf{k}s}(\mathbf{r})=\mathcal{E}_{\mathbf{k}x}(\mathbf{r})\,e_{sx}(\mathbf{k})\,\mathbf{u}_x$. There are no modes with two vanishing $\mathbf{k}$-components, since the corresponding function would vanish identically. If no components of $\mathbf{k}$ are zero, then $\mathbf{e}_s$ can be any vector in the plane perpendicular to $\mathbf{k}$. Just as for plane waves in free space, there is then a polarization basis set with two real, mutually orthogonal unit vectors $\mathbf{e}_1$ and $\mathbf{e}_2$ $(s=1,2)$. If no components vanish, $N_{\mathbf{k}}=\sqrt{8/V}$, but when exactly one $\mathbf{k}$-component vanishes, $N_{\mathbf{k}}=\sqrt{4/V}$, where $V=l_xl_yl_z$ is the volume of the cavity. The spacing between the discrete $\mathbf{k}$-values is $\Delta k_j=\pi/l_j$ $(j=x,y,z)$; therefore, in the limit of large cavities $(l_j\to\infty)$, the $\mathbf{k}$-values become essentially continuous. Thus the interior of a sufficiently large rectangular parallelepiped cavity is effectively indistinguishable from free space.

The mode functions are eigenfunctions of the hermitian operator $-\nabla^2$, so they are guaranteed to form a complete, orthonormal set. The orthonormality conditions

$$\int_V d^3r \boldsymbol{\mathcal{E}}_{\mathbf{k}s}(\mathbf{r}) \cdot \boldsymbol{\mathcal{E}}_{\mathbf{k}'s'}(\mathbf{r}) = \delta_{\mathbf{k}\mathbf{k}'}\delta_{ss'}, \tag{2.19}$$

$$\int_V d^3r \boldsymbol{\mathcal{B}}_{\mathbf{k}s}(\mathbf{r}) \cdot \boldsymbol{\mathcal{B}}_{\mathbf{k}'s'}(\mathbf{r}) = \delta_{\mathbf{k}\mathbf{k}'}\delta_{ss'} \tag{2.20}$$

can be readily verified by a direct calculation, but the completeness conditions are complicated by the fact that the eigenfunctions are vectors fields satisfying the divergence equations (2.1) or (2.2). We therefore consider the completeness issue in the following section.

B The transverse delta function

In order to deal with the completeness identities for vector modes of the cavity, it is useful to study general vector fields in a little more detail. This is most easily done by expressing a vector field $\boldsymbol{\mathcal{F}}(\mathbf{r})$ by a spatial Fourier transform:

$$\boldsymbol{\mathcal{F}}(\mathbf{r}) = \int \frac{d^3k}{(2\pi)^3} \boldsymbol{\mathcal{F}}(\mathbf{k})\, e^{i\mathbf{k}\cdot\mathbf{r}}, \tag{2.21}$$

so that the divergence and curl are given by

$$\nabla \cdot \boldsymbol{\mathcal{F}}(\mathbf{r}) = i \int \frac{d^3k}{(2\pi)^3} \mathbf{k} \cdot \boldsymbol{\mathcal{F}}(\mathbf{k})\, e^{i\mathbf{k}\cdot\mathbf{r}} \tag{2.22}$$

and

$$\nabla \times \boldsymbol{\mathcal{F}}(\mathbf{r}) = i \int \frac{d^3k}{(2\pi)^3} \mathbf{k} \times \boldsymbol{\mathcal{F}}(\mathbf{k})\, e^{i\mathbf{k}\cdot\mathbf{r}}. \tag{2.23}$$

In $\mathbf{k}$-space, the field $\boldsymbol{\mathcal{F}}(\mathbf{k})$ is **transverse** if $\mathbf{k} \cdot \boldsymbol{\mathcal{F}}(\mathbf{k}) = 0$ and **longitudinal** if $\mathbf{k} \times \boldsymbol{\mathcal{F}}(\mathbf{k}) = 0$; consequently, in $\mathbf{r}$-space the field $\boldsymbol{\mathcal{F}}(\mathbf{r})$ is said to be transverse if $\nabla \cdot \boldsymbol{\mathcal{F}}(\mathbf{r}) = 0$ and longitudinal if $\nabla \times \boldsymbol{\mathcal{F}}(\mathbf{r}) = 0$. In this language the $\boldsymbol{\mathcal{E}}$- and $\boldsymbol{\mathcal{B}}$-fields in the cavity are both transverse vector fields.

Now suppose that $\boldsymbol{\mathcal{F}}(\mathbf{r})$ is transverse and $\boldsymbol{\mathcal{G}}(\mathbf{r})$ is longitudinal, then an application of Parseval's theorem (A.54) for Fourier transforms yields

$$\int d^3r \boldsymbol{\mathcal{F}}^*(\mathbf{r}) \cdot \boldsymbol{\mathcal{G}}(\mathbf{r}) = \int \frac{d^3k}{(2\pi)^3} \boldsymbol{\mathcal{F}}^*(\mathbf{k}) \cdot \boldsymbol{\mathcal{G}}(\mathbf{k}) = 0. \tag{2.24}$$

In other words, the transverse and longitudinal fields in $\mathbf{r}$-space are orthogonal in the sense of wave functions. Furthermore, a general vector field $\boldsymbol{\mathcal{F}}(\mathbf{k})$ can be decomposed as $\boldsymbol{\mathcal{F}}(\mathbf{k}) = \boldsymbol{\mathcal{F}}^{\parallel}(\mathbf{k}) + \boldsymbol{\mathcal{F}}^{\perp}(\mathbf{k})$, where the longitudinal and transverse parts are respectively given by

$$\boldsymbol{\mathcal{F}}^{\parallel}(\mathbf{k}) = \frac{\mathbf{k} \cdot \boldsymbol{\mathcal{F}}(\mathbf{k})}{k^2}\mathbf{k} \tag{2.25}$$

and

$$\mathcal{F}^{\perp}(\mathbf{k}) = \mathcal{F}(\mathbf{k}) - \mathcal{F}^{\parallel}(\mathbf{k}) . \tag{2.26}$$

For later use it is convenient to write out the transverse part in Cartesian components:

$$\mathcal{F}_i^{\perp}(\mathbf{k}) = \Delta_{ij}^{\perp}(\mathbf{k}) \mathcal{F}_j(\mathbf{k}) , \tag{2.27}$$

where

$$\Delta_{ij}^{\perp}(\mathbf{k}) \equiv \delta_{ij} - \frac{k_i k_j}{k^2} , \tag{2.28}$$

and the Einstein summation convention over repeated vector indices is understood. The 3×3-matrix $\Delta^{\perp}(\mathbf{k})$ is symmetric and $\mathbf{k}$ is an eigenvector corresponding to the eigenvalue zero. This matrix also satisfies the defining condition for a projection operator: $\left(\Delta^{\perp}(\mathbf{k})\right)^2 = \Delta^{\perp}(\mathbf{k})$. Thus $\Delta^{\perp}(\mathbf{k})$ is a projection operator onto the space of transverse vector fields.

The inverse Fourier transform of eqn (2.27) gives the $\mathbf{r}$-space form

$$\mathcal{F}_i^{\perp}(\mathbf{r}) = \int_V d^3 r \Delta_{ij}^{\perp}(\mathbf{r} - \mathbf{r}') \mathcal{F}_j(\mathbf{r}') , \tag{2.29}$$

where

$$\Delta_{ij}^{\perp}(\mathbf{r} - \mathbf{r}') \equiv \int \frac{d^3 k}{(2\pi)^3} \Delta_{ij}^{\perp}(\mathbf{k}) e^{i\mathbf{k} \cdot (\mathbf{r} - \mathbf{r}')} . \tag{2.30}$$

The integral operator $\Delta_{ij}^{\perp}(\mathbf{r} - \mathbf{r}')$ reproduces any transverse vector field and annihilates any longitudinal vector field, so it is called the **transverse delta function.**

We are now ready to consider the completeness of the mode functions. For any transverse vector field $\mathcal{F}$, satisfying the first boundary condition in eqn (2.13), the combination of the completeness of the electric mode functions and the orthonormality conditions (2.19) results in the identity

$$\mathcal{F}_i(\mathbf{r}) = \int_V d^3 r' \left\{ \sum_{\mathbf{k}s} \left(\mathcal{E}_{\mathbf{k}s}(\mathbf{r})\right)_i \left(\mathcal{E}_{\mathbf{k}s}(\mathbf{r}')\right)_j \right\} \mathcal{F}_j(\mathbf{r}') . \tag{2.31}$$

On the other hand, eqn (2.24) leads to

$$\int_V d^3 r' \left\{ \sum_{\mathbf{k}s} \left(\mathcal{E}_{\mathbf{k}s}(\mathbf{r})\right)_i \left(\mathcal{E}_{\mathbf{k}s}(\mathbf{r}')\right)_j \right\} \mathcal{G}_j(\mathbf{r}') = 0 \tag{2.32}$$

for any longitudinal field $\mathcal{G}(\mathbf{r})$. Thus the integral operator defined by the expression in curly brackets annihilates longitudinal fields and reproduces transverse fields. Two operators that have the same action on the entire space of vector fields are identical; therefore,

$$\sum_{\mathbf{k}s} \left(\mathcal{E}_{\mathbf{k}s}(\mathbf{r})\right)_i \left(\mathcal{E}_{\mathbf{k}s}(\mathbf{r}')\right)_j = \Delta_{ij}^{\perp}(\mathbf{r} - \mathbf{r}') . \tag{2.33}$$

A similar argument applied to the magnetic mode functions leads to the corresponding result:

$$\sum_{\mathbf{k}s} \left(\mathcal{B}_{\mathbf{k}s}(\mathbf{r})\right)_i \left(\mathcal{B}_{\mathbf{k}s}(\mathbf{r}')\right)_j = \Delta_{ij}^{\perp}(\mathbf{r} - \mathbf{r}') . \tag{2.34}$$

C The general cavity

Now that we have mastered the simple rectangular cavity, we proceed to a general metallic cavity with a bounding surface S of arbitrary shape.[3] As we have already remarked, the difference between this general cavity and the rectangular cavity lies entirely in the boundary conditions. The solution of the Helmholtz equations (2.11) and (2.12), together with the general boundary conditions (2.5) and (2.6), has been extensively studied in connection with the theory of microwave cavities (Slater, 1950).

Separation of variables is not possible for general boundary shapes, so there is no way to obtain the explicit solutions shown in Section 2.1.1-A. Fortunately, we only need certain properties of the solutions, which can be obtained without knowing the explicit forms. General results from the theory of partial differential equations (Zauderer, 1983, Sec. 8.1) guarantee that the Helmholtz equation in any finite cavity has a complete, orthonormal set of eigenfunctions labeled by a discrete multi-index $\kappa = (\kappa_1, \kappa_2, \kappa_3, \kappa_4)$ that replaces the combination $(\mathbf{k}, s)$ used for the rectangular cavity. These **normal mode functions** $\boldsymbol{\mathcal{E}}_\kappa(\mathbf{r})$ and $\boldsymbol{\mathcal{B}}_\kappa(\mathbf{r})$ are real, transverse vector fields satisfying the boundary conditions (2.5) and (2.6) respectively, together with the Helmholtz equation:

$$\left(\nabla^2 + k_\kappa^2\right) \boldsymbol{\mathcal{E}}_\kappa = 0\,, \tag{2.35}$$

$$\left(\nabla^2 + k_\kappa^2\right) \boldsymbol{\mathcal{B}}_\kappa = 0\,, \tag{2.36}$$

where $k_\kappa = \omega_\kappa / c$ and ω_κ is the cavity resonance frequency of mode κ. The allowed values of the discrete indices $\kappa_1, \ldots, \kappa_4$ and the resonance frequencies ω_κ are determined by the geometrical properties of the cavity.

By combining the orthonormality conditions

$$\begin{aligned} \int_V d^3r\, \boldsymbol{\mathcal{E}}_\kappa \cdot \boldsymbol{\mathcal{E}}_\lambda &= \delta_{\kappa\lambda}\,, \\ \int_V d^3r\, \boldsymbol{\mathcal{B}}_\kappa \cdot \boldsymbol{\mathcal{B}}_\lambda &= \delta_{\kappa\lambda} \end{aligned} \tag{2.37}$$

with the completeness of the modes, we can repeat the argument in Section 2.1.1-B to obtain the general completeness identities

$$\sum_\kappa \mathcal{E}_{\kappa i}(\mathbf{r})\, \mathcal{E}_{\kappa j}(\mathbf{r}') = \Delta_{ij}^{\perp}(\mathbf{r} - \mathbf{r}')\,, \tag{2.38}$$

$$\sum_\kappa \mathcal{B}_{\kappa i}(\mathbf{r})\, \mathcal{B}_{\kappa j}(\mathbf{r}') = \Delta_{ij}^{\perp}(\mathbf{r} - \mathbf{r}')\,. \tag{2.39}$$

D The classical electromagnetic energy

Since the cavity mode functions are a complete orthonormal set, general electric and magnetic fields—and the associated vector potential—can be written as

[3] The term 'arbitrary' should be understood to exclude topologically foolish choices, such as replacing the rectangular cavity by a Klein bottle.

$$\boldsymbol{\mathcal{E}}(\mathbf{r},t) = -\frac{1}{\sqrt{\epsilon_0}}\sum_\kappa P_\kappa(t)\,\boldsymbol{\mathcal{E}}_\kappa(\mathbf{r}), \tag{2.40}$$

$$\boldsymbol{\mathcal{B}}(\mathbf{r},t) = \sqrt{\mu_0}\sum_\kappa \omega_\kappa Q_\kappa(t)\,\boldsymbol{\mathcal{B}}_\kappa(\mathbf{r}), \tag{2.41}$$

$$\boldsymbol{\mathcal{A}}(\mathbf{r},t) = \frac{1}{\sqrt{\epsilon_0}}\sum_\kappa Q_\kappa(t)\,\boldsymbol{\mathcal{E}}_\kappa(\mathbf{r}). \tag{2.42}$$

Substituting the expansions (2.40) and (2.41) into the vacuum Maxwell equations (2.1)–(2.4) leads to the infinite set of ordinary differential equations

$$\dot{Q}_\kappa = P_\kappa \quad \text{and} \quad \dot{P}_\kappa = -\omega_\kappa^2 Q_\kappa. \tag{2.43}$$

For each mode, this pair of equations is mathematically identical to the equations of motion of a simple harmonic oscillator, where the expansion coefficients Q_κ and P_κ respectively play the roles of the oscillator coordinate and momentum. On the basis of this mechanical analogy, the mode κ is called a **radiation oscillator**, and the set of points

$$\{(Q_\kappa, P_\kappa) \ \text{for} \ -\infty < Q_\kappa < \infty \ \text{and} \ -\infty < P_\kappa < \infty\} \tag{2.44}$$

is said to be the classical oscillator phase space for the κth mode.

For the transition to quantum theory, it is useful to introduce the dimensionless complex amplitudes

$$\alpha_\kappa(t) = \frac{\omega_\kappa Q_\kappa(t) + iP_\kappa(t)}{\sqrt{2\hbar\omega_\kappa}}, \tag{2.45}$$

which allow the pair of real equations (2.43) to be rewritten as a single complex equation,

$$\dot{\alpha}_\kappa(t) = -i\omega_\kappa\alpha_\kappa(t), \tag{2.46}$$

with the general solution $\alpha_\kappa(t) = \alpha_\kappa e^{-i\omega_\kappa t}$, $\alpha_\kappa = \alpha_\kappa(0)$. The expansions for the fields can all be written in terms of α_κ and α_κ^*; for example eqn (2.40) becomes

$$\boldsymbol{\mathcal{E}}(\mathbf{r},t) = i\sum_\kappa \sqrt{\frac{\hbar\omega_\kappa}{2\epsilon_0}}\,\alpha_\kappa e^{-i\omega_\kappa t}\boldsymbol{\mathcal{E}}_\kappa(\mathbf{r}) + \text{CC}. \tag{2.47}$$

One of the chief virtues of the expansions (2.40) and (2.41) is that the orthogonality relations (2.37) allow the classical electromagnetic energy in the cavity,

$$\mathcal{U}_{\text{em}} = \frac{1}{2}\int_V d^3r\left(\epsilon_0\boldsymbol{\mathcal{E}}^2 + \mu_0^{-1}\boldsymbol{\mathcal{B}}^2\right), \tag{2.48}$$

to be expressed as a sum of independent terms: one for each normal mode,

$$\mathcal{U}_{\text{em}} = \sum_\kappa \frac{1}{2}\left(P_\kappa^2 + \omega_\kappa^2 Q_\kappa^2\right). \tag{2.49}$$

Each term in the sum is mathematically identical to the energy of a simple harmonic oscillator with unit mass, oscillator frequency ω_κ, coordinate Q_κ, and momentum P_κ. For each κ, eqn (2.43) is obtained from

$$\dot{Q}_\kappa = \frac{\partial \mathcal{U}_{\rm em}}{\partial P_\kappa} \quad \text{and} \quad \dot{P}_\kappa = -\frac{\partial \mathcal{U}_{\rm em}}{\partial Q_\kappa}\,; \tag{2.50}$$

consequently, $\mathcal{U}_{\rm em}$ serves as the classical Hamiltonian for the radiation oscillators, and Q_κ and P_κ are said to be **canonically conjugate classical variables** (Marion and Thornton, 1995). An even more suggestive form comes from using the complex amplitudes α_κ to write the energy as

$$\mathcal{U}_{\rm em} = \sum_\kappa \hbar\omega_\kappa \alpha_\kappa^* \alpha_\kappa\,. \tag{2.51}$$

Interpreting $\alpha_\kappa^* \alpha_\kappa$ as the number of light-quanta with energy $\hbar\omega_\kappa$ makes this a realization of Einstein's original model.

2.1.2 The quantization conjecture

The simple harmonic oscillator is one of the very few examples of a mechanical system for which the Schrödinger equation can be solved exactly. For a classical mechanical oscillator, $Q(t)$ represents the instantaneous displacement of the oscillating mass from its equilibrium position, and $P(t)$ represents its instantaneous momentum. The **trajectory** $\{(Q(t), P(t)) \text{ for } t \geqslant 0\}$ is uniquely determined by the initial values $(Q, P) = (Q(0), P(0))$.

The quantum theory of the mechanical oscillator is usually presented in the coordinate representation, i.e. the state of the oscillator is described by a wave function $\psi(Q,t)$, where the argument Q ranges over the values allowed for the classical coordinate. Thus the wave functions belong to the Hilbert space of square-integrable functions on the interval $(-\infty, \infty)$. In the Born interpretation, $|\psi(Q,t)|^2$ represents the probability density for finding the oscillator with a displacement Q from equilibrium at time t; consequently, the wave function satisfies the normalization condition

$$\int_{-\infty}^{\infty} dQ\, |\psi(Q,t)|^2 = 1\,. \tag{2.52}$$

In this representation the classical oscillator variables (Q, P)—representing the possible initial values of classical trajectories—are replaced by the quantum operators $\widehat{q}$ and $\widehat{p}$ defined by

$$\widehat{q}\psi(Q,t) = Q\psi(Q,t) \quad \text{and} \quad \widehat{p}\psi(Q,t) = \frac{\hbar}{i}\frac{\partial}{\partial Q}\psi(Q,t)\,. \tag{2.53}$$

By using the explicit definitions of $\widehat{q}$ and $\widehat{p}$ it is easy to show that the operators satisfy the canonical commutation relation

$$[\widehat{q}, \widehat{p}] = i\hbar\,. \tag{2.54}$$

For a system consisting of N noninteracting mechanical oscillators—with coordinates $Q_1, Q_2, \ldots, Q_N$—the coordinate representation is defined by the N-body wave function

$$\psi(Q_1, Q_2, \ldots, Q_N, t)\,, \tag{2.55}$$

and the action of the operators is

$$\begin{aligned} \widehat{q}_m \psi \left(Q_1, Q_2, \ldots, Q_N, t\right) &= Q_m \psi \left(Q_1, Q_2, \ldots, Q_N, t\right), \\ \widehat{p}_m \psi \left(Q_1, Q_2, \ldots, Q_N, t\right) &= \frac{\hbar}{i} \frac{\partial}{\partial Q_m} \psi \left(Q_1, Q_2, \ldots, Q_N, t\right), \end{aligned} \tag{2.56}$$

where $m = 1, \ldots, N$. This explicit definition, together with the fact that the Q_ms are independent variables, leads to the general form of the canonical commutation relations,

$$\left[\widehat{q}_m, \widehat{p}_{m'}\right] = i\hbar\delta_{mm'}\,, \tag{2.57}$$

$$\left[\widehat{q}_m, \widehat{q}_{m'}\right] = \left[\widehat{p}_m, \widehat{p}_{m'}\right] = 0\,, \tag{2.58}$$

for $m, m' = 1, \ldots, N$. This mechanical system is said to have N degrees of freedom.

The results of the previous section show that the pairs of coefficients (Q_κ, P_κ) in the expansions (2.40) and (2.41) are canonically conjugate and that they satisfy the same equations of motion as a mechanical harmonic oscillator. Since the classical descriptions of the radiation and mechanical oscillators have the same mathematical form, it seems reasonable to conjecture that their quantum theories will also have the same form. For the κth cavity mode this simply means that the state of the radiation oscillator is described by a wave function $\psi(Q_\kappa, t)$. In order to distinguish between the radiation and mechanical oscillators, we will call the quantum operators for the radiation oscillator q_κ and p_κ. The mathematical definitions of these operators are still given by eqn (2.56), with $\widehat{q}_\kappa$ and $\widehat{p}_\kappa$ replaced by q_κ and p_κ.

Extending this procedure to describe the general state of the cavity field introduces a new complication. The classical state of the electromagnetic field is represented by functions $\boldsymbol{\mathcal{E}}(\mathbf{r}, t)$ and $\boldsymbol{\mathcal{B}}(\mathbf{r}, t)$ that, in general, cannot be described by a finite number of modes. This means that the classical description of the cavity field requires infinitely many degrees of freedom. A naive interpretation of the quantization conjecture would therefore lead to wave functions $\psi(Q_1, Q_2, \ldots)$ that depend on infinitely many variables. Mathematical techniques to deal with such awkward objects do exist, but it is much better to start with abstract algebraic operator relations like eqns (2.57) and (2.58), and then to choose an explicit representation that is well suited to the problem at hand.

The formulation of quantum mechanics used above is called the **Schrödinger picture**; it is characterized by time-dependent wave functions and time independent operators. The Schrödinger-picture formulation of the quantization conjecture for the electromagnetic field therefore consists of the following two parts.

(1) The time-dependent states of the electromagnetic field satisfy the superposition principle: if $|\Psi(t)\rangle$ and $|\Phi(t)\rangle$ are two physically possible states, then the superposition

$$\alpha\,|\Psi(t)\rangle + \beta\,|\Phi(t)\rangle \tag{2.59}$$

is also a physically possible state. (See Appendix C.1 for the *bra* and *ket* notation.)

(2) The classical variables $Q_\kappa = Q_\kappa(t=0)$ and $P_\kappa = P_\kappa(t=0)$ are replaced by time-independent hermitian operators q_κ and p_κ:

$$Q_\kappa \to q_\kappa \quad \text{and} \quad P_\kappa \to p_\kappa\,, \tag{2.60}$$

that satisfy the canonical commutation relations

$$[q_\kappa, p_{\kappa'}] = i\hbar\delta_{\kappa\kappa'}\,, \quad [q_\kappa, q_{\kappa'}] = 0\,, \quad \text{and} \quad [p_\kappa, p_{\kappa'}] = 0\,, \tag{2.61}$$

where κ, κ' range over all cavity modes.

The statements (1) and (2) are equally important parts of this conjecture.

Another useful form of the commutation relations (2.61) is provided by defining the dimensionless, non-hermitian operators

$$a_\kappa = \frac{\omega_\kappa q_\kappa + i p_\kappa}{\sqrt{2\hbar\omega_\kappa}} \quad \text{and} \quad a_\kappa^\dagger = \frac{\omega_\kappa q_\kappa - i p_\kappa}{\sqrt{2\hbar\omega_\kappa}} \tag{2.62}$$

for the κth mode of the radiation field. A simple calculation using eqn (2.61) yields the equivalent commutation relations

$$\left[a_\kappa, a_{\kappa'}^\dagger\right] = \delta_{\kappa\kappa'}\,, \; [a_\kappa, a_{\kappa'}] = 0\,. \tag{2.63}$$

To sum up: by examining the problem of the ideal resonant cavity, we have been led to the conjecture that the radiation field can be viewed as a collection of quantized simple harmonic oscillators. The quantization conjecture embodied in eqns (2.59)–(2.61) may appear to be rather formal and abstract, but it is actually the fundamental *physical* assumption required for constructing the quantum theory of the electromagnetic field. New principles of this kind cannot be deduced from the pre-existing theory; instead, they represent a genuine leap of scientific induction that must be judged by its success in explaining experimental results.

In the following section, we will combine the canonical commutation relations with some basic physical principles to construct the Hilbert space of state vectors $|\Psi\rangle$, and thus obtain a concrete representation of the operators q_κ and p_κ or a_κ and $a_\kappa^\dagger$ for a single cavity mode. In Section 2.1.2-C this representation will be generalized to include the infinite set of normal cavity modes.

A *The single-mode Fock space*

In this section we will deal with a single mode, so the mode index can be omitted. Instead of starting with the coordinate representation of the wave function, as in eqn (2.53), we will deduce the structure of the Hilbert space of states by the following argument. According to eqn (2.49) the classical energy for a single mode is

$$\mathcal{U}_{\text{em}} = \frac{1}{2}\left(P^2 + \omega^2 Q^2\right), \tag{2.64}$$

where the arbitrary zero of energy has been chosen to correspond to the classical solution $Q = P = 0$, representing the oscillator at rest at the minimum of the potential.

In quantum mechanics the standard procedure is to apply eqn (2.60) to this expression and to interpret the resulting operator as the (single-mode) Hamiltonian

$$H_{\text{em}} = \frac{1}{2}\left(p^2 + \omega^2 q^2\right). \tag{2.65}$$

It is instructive to rewrite this in terms of the operators a and $a^\dagger$ by solving eqn (2.62) to get

$$q = \sqrt{\frac{\hbar}{2\omega}}\left(a + a^\dagger\right) \quad \text{and} \quad p = -i\sqrt{\frac{\hbar\omega}{2}}\left(a - a^\dagger\right). \tag{2.66}$$

Substituting these expressions into eqn (2.65)—while remembering that the operators a and $a^\dagger$ do not commute—leads to

$$\begin{aligned} H_{\text{em}} &= \frac{1}{2}\left\{-\left(\frac{\hbar\omega}{2}\right)\left(a - a^\dagger\right)^2 + \left(\frac{\hbar\omega}{2}\right)\left(a + a^\dagger\right)^2\right\} \\ &= \left(\frac{\hbar\omega}{2}\right)\left\{aa^\dagger + a^\dagger a\right\}. \end{aligned} \tag{2.67}$$

By using the commutation relation (2.63), this can be written in the equivalent form

$$H_{\text{em}} = \hbar\omega\left(a^\dagger a + \frac{1}{2}\right). \tag{2.68}$$

The superposition principle (2.59) is enforced by the assumption that the states of the radiation operator belong to a Hilbert space. The structure of this Hilbert space is essentially determined by the fact that H_{em} is a positive operator, i.e. $\langle \Psi | H_{\text{em}} | \Psi \rangle \geqslant 0$ for any $|\Psi\rangle$. To see this, set $|\Phi\rangle = a|\Psi\rangle$ and use the general rule $\langle \Phi | \Phi \rangle \geqslant 0$ to conclude that

$$\begin{aligned} \langle \Psi | H_{\text{em}} | \Psi \rangle &= \hbar\omega \langle \Psi | a^\dagger a | \Psi \rangle + \frac{\hbar\omega}{2} \\ &= \hbar\omega \langle \Phi | \Phi \rangle + \frac{\hbar\omega}{2} \geqslant 0. \end{aligned} \tag{2.69}$$

In particular, this means that all eigenvalues of H_{em} are nonnegative. Let $|\phi\rangle$ be an eigenstate of H_{em} with eigenvalue W; then $a|\phi\rangle$ satisfies

$$\begin{aligned} H_{\text{em}} a |\phi\rangle &= \left\{[H_{\text{em}}, a] + aH_{\text{em}}\right\}|\phi\rangle \\ &= Wa|\phi\rangle + [H_{\text{em}}, a]|\phi\rangle. \end{aligned} \tag{2.70}$$

The commutator is given by

$$\begin{aligned} [H_{\text{em}}, a] &= \hbar\omega\left[a^\dagger a, a\right] \\ &= \hbar\omega\left\{a^\dagger [a, a] + \left[a^\dagger, a\right] a\right\} \\ &= -\hbar\omega a, \end{aligned} \tag{2.71}$$

so that

$$H_{\text{em}} a |\phi\rangle = (W - \hbar\omega)\, a|\phi\rangle. \tag{2.72}$$

Thus $a|\phi\rangle$ is also an eigenstate of H_{em}, but with the reduced eigenvalue $(W - \hbar\omega)$. Since a lowers the energy by $\hbar\omega$, repeating this process would eventually generate states of negative energy. This is inconsistent with the inequality (2.69); therefore, the Hilbert space of a consistent quantum theory for an oscillator must include a lowest energy eigenstate $|0\rangle$ satisfying

$$a|0\rangle = 0\,, \quad \langle 0| a^\dagger = 0\,, \tag{2.73}$$

and

$$H_{\text{em}}|0\rangle = \frac{\hbar\omega}{2}|0\rangle\,. \tag{2.74}$$

In the case of a mechanical oscillator $|0\rangle$ is the ground state, and a is a **lowering operator**. A calculation similar to eqns (2.70) and (2.71) leads to

$$H_{\text{em}} a^\dagger |\phi\rangle = (W + \hbar\omega)\, a^\dagger |\phi\rangle\,, \tag{2.75}$$

which shows that $a^\dagger$ raises the energy by $\hbar\omega$, so $a^\dagger$ is a **raising operator**. The idea behind this language is that the mechanical oscillator itself is the object of interest. The energy levels are merely properties of the oscillator, like the energy levels of an atom.

The equations describing the radiation and mechanical oscillators have the same form, but there is an important difference in physical interpretation. For the electromagnetic field, it is the quanta of excitation—rather than the radiation oscillators themselves—that are the main objects of interest. This shift in emphasis incorporates Einstein's original proposal that the electromagnetic field is composed of discrete quanta. In keeping with this view, it is customary to replace the cumbersome phrase 'quantum of excitation of the electromagnetic field' by the term **photon**. The intended implication is that photons are physical objects on the same footing as massive particles. The subtleties associated with treating photons as particles are addressed in Section 3.6. Since a removes one photon, it is natural to call it the **annihilation** operator, and $a^\dagger$, which adds a photon, is naturally called a **creation operator**. In this language the ground state of the radiation oscillator is referred to as the **vacuum state**, since it contains no photons.

The **number operator** $N = a^\dagger a$ satisfies the commutation relations

$$[N, a] = -a\,, \quad \left[N, a^\dagger\right] = a^\dagger\,, \tag{2.76}$$

so that the a and $a^\dagger$ respectively decrease and increase the eigenvalues of N by one. Since $N|0\rangle = 0$, this implies that the eigenvalues of N are the the integers $0, 1, 2, \ldots$. The eigenvectors of N are called **number states**, and it is easy to see that $N|n\rangle = n|n\rangle$ implies

$$|n\rangle = Z_n \left(a^\dagger\right)^n |0\rangle\,, \tag{2.77}$$

where Z_n is a normalization constant. The Hamiltonian can be written as $H_{\text{em}} = (N + 1/2)\,\hbar\omega$, so the number states are also energy eigenstates: $H_{\text{em}}|n\rangle = (n + 1/2)\,\hbar\omega\,|n\rangle$. The commutation relations (2.76) can be used to derive the results

$$Z_n = \frac{1}{\sqrt{n!}}\,, \quad \langle n | n'\rangle = \delta_{nn'}\,, \quad a|n\rangle = \sqrt{n}\,|n-1\rangle\,, \quad \text{and} \quad a^\dagger|n\rangle = \sqrt{n+1}\,|n+1\rangle\,. \tag{2.78}$$

The Hilbert space $\mathfrak{H}_F^{(1)}$ for a single mode consists of all linear combinations of number states, i.e. a typical vector is given by

$$|\Psi\rangle = \sum_{n=0}^{\infty} C_n |n\rangle . \tag{2.79}$$

The space $\mathfrak{H}_F^{(1)}$ is called the (single-mode) **Fock space**. In mathematical jargon—see Appendix A.2—$\mathfrak{H}_F^{(1)}$ is said to be *spanned* by the number states, or $\mathfrak{H}_F^{(1)}$ is said to be the *span* of the number states. Since the number states are orthonormal, the expansion (2.79) can be expressed as

$$|\Psi\rangle = \sum_{n=0}^{\infty} |n\rangle \langle n |\Psi\rangle . \tag{2.80}$$

For any state $|\phi\rangle$ the expression $|\phi\rangle\langle\phi|$ stands for an operator—see Appendix C.1.2—that is defined by its action on an arbitrary state $|\chi\rangle$:

$$(|\phi\rangle\langle\phi|)\,|\chi\rangle \equiv |\phi\rangle\langle\phi|\chi\rangle . \tag{2.81}$$

This shows that $|\phi\rangle\langle\phi|$ is the projection operator onto $|\phi\rangle$, and it allows the expansion (2.80) to be expressed as

$$|\psi\rangle = \left(\sum_{n=0}^{\infty} |n\rangle\langle n|\right)|\psi\rangle . \tag{2.82}$$

The general definition (2.81) leads to

$$(|n\rangle\langle n|)\,(|n'\rangle\langle n'|) = |n\rangle\langle n|n'\rangle\langle n'| = \delta_{nn'}\,(|n\rangle\langle n|)\,; \tag{2.83}$$

therefore, the $(|n\rangle\langle n|)$s are a family of orthogonal projection operators. According to eqn (2.82) the projection operators onto the number states satisfy the completeness relation

$$\sum_{n=0}^{\infty} |n\rangle\langle n| = 1 . \tag{2.84}$$

B *Vacuum fluctuations of a single radiation oscillator*

A standard argument from quantum mechanics (Bransden and Joachain, 1989, Sec. 5.4) shows that the canonical commutation relations (2.61) for the operators q and p lead to the uncertainty relation

$$\Delta q \Delta p \geqslant \frac{\hbar}{2} , \tag{2.85}$$

where the rms deviations Δq and Δp are defined by

$$\Delta q = \sqrt{\langle\Psi|q^2|\Psi\rangle - \langle\Psi|q|\Psi\rangle^2}\,, \quad \Delta p = \sqrt{\langle\Psi|p^2|\Psi\rangle - \langle\Psi|p|\Psi\rangle^2}\,, \tag{2.86}$$

and $|\Psi\rangle$ is any normalized vector in $\mathfrak{H}_F^{(1)}$. For the vacuum state the relations (2.66) and (2.73) yield $\langle 0|q|0\rangle = 0$ and $\langle 0|p|0\rangle = 0$, so the uncertainty relation implies that neither $\langle 0 |q^2| 0\rangle$ nor $\langle 0 |p^2| 0\rangle$ can vanish. For mechanical oscillators this is attributed to zero-point motion; that is, even in the ground state, random excursions around the classical equilibrium at $Q = P = 0$ are required by the uncertainty principle. The ground state for light is the vacuum state, so the random excursions of the radiation oscillators are called **vacuum fluctuations**. Combining eqn (2.66) with eqn (2.73) yields the explicit values

$$\langle 0 |q^2| 0\rangle = \frac{\hbar}{2\omega}\,, \quad \langle 0 |p^2| 0\rangle = \frac{\hbar\omega}{2}\,. \tag{2.87}$$

We note for future reference that the vacuum deviations are $\Delta q_0 = \sqrt{\hbar/2\omega}$ and $\Delta p_0 = \sqrt{\hbar\omega/2}$, and that these values saturate the inequality (2.85), i.e. $\Delta q_0 \Delta p_0 = \hbar/2$. States with this property are called minimum-uncertainty states, or sometimes minimum-uncertainty-product states.

The vacuum fluctuations of the radiation oscillator also explain the fact that the energy eigenvalue for the vacuum is $\hbar\omega/2$ while the classical energy minimum is $\mathcal{U}_{\text{em}} = 0$. Inserting eqn (2.87) into the original expression eqn (2.65) for the Hamiltonian yields $\langle 0 |H_{\text{em}}| 0\rangle = \hbar\omega/2$. The discrepancy between the quantum and classical minimum energies is called the **zero-point energy**; it is required by the uncertainty principle for the radiation oscillator. Since energy is only defined up to an additive constant, it would be permissible—although apparently unnatural—to replace the classical expression (2.64) by

$$\mathcal{U} = \frac{1}{2}\left(p^2 + \omega^2 q^2\right) - \frac{\hbar\omega}{2}\,. \tag{2.88}$$

Carrying out the substitution (2.66) on this expression yields the Hamiltonian

$$H_{\text{em}} = \hbar\omega a^\dagger a\,. \tag{2.89}$$

With this convention the vacuum energy vanishes for the quantum theory, but the discrepancy between the quantum and classical minimum energies is unchanged. The same thing can be accomplished directly in the quantum theory by simply subtracting the zero-point energy from eqn (2.68). Changes of this kind are always permitted, since only differences of energy eigenvalues are physically meaningful.

C The multi-mode Fock space

Since the classical radiation oscillators in the cavity are mutually independent, the quantization rule is given by eqns (2.60)–(2.63), and the only real difficulties stem from the fact that there are infinitely many modes. For each mode, the number operator $N_\kappa = a_\kappa^\dagger a_\kappa$ is evidently positive and satisfies

$$[N_\kappa, a_\lambda] = -\delta_{\kappa\lambda} a_\kappa\,, \tag{2.90}$$

$$\left[N_\kappa, a_\lambda^\dagger\right] = \delta_{\kappa\lambda} a_\kappa^\dagger\,. \tag{2.91}$$

Combining eqn (2.90) with the positivity of N_κ and applying the argument used for the single-mode Hamiltonian in Section 2.1.2-A leads to the conclusion that there must be a (multimode) vacuum state $|0\rangle$ satisfying

$$a_\kappa |0\rangle = 0 \quad \text{for every mode-index } \kappa\,. \tag{2.92}$$

Since number operators for different modes commute, it is possible to find vectors $|\underline{n}\rangle$ that are simultaneous eigenstates of all the mode number operators:

$$\begin{aligned} N_\kappa |\underline{n}\rangle &= n_\kappa |\underline{n}\rangle \quad \text{for all } \kappa\,, \\ \underline{n} &= \{n_\kappa \quad \text{for all } \kappa\}\,. \end{aligned} \tag{2.93}$$

According to the single-mode results (2.77) and (2.78) the many-mode number states are given by

$$|\underline{n}\rangle = \prod_\kappa \frac{\left(a_\kappa^\dagger\right)^{n_\kappa}}{\sqrt{n_\kappa!}} |0\rangle\,. \tag{2.94}$$

The **total number operator** is

$$N = \sum_\kappa a_\kappa^\dagger a_\kappa\,, \tag{2.95}$$

and

$$N |\underline{n}\rangle = \left(\sum_\kappa n_\kappa\right) |\underline{n}\rangle\,. \tag{2.96}$$

The Hilbert space $\mathfrak{H}_F$ spanned by the number states $|\underline{n}\rangle$ is called the (multimode) Fock space.

It is instructive to consider the simplest number states, i.e. those containing exactly one photon. If κ and λ are the labels for two distinct modes, then eqn (2.96) tells us that $|1_\kappa\rangle = a_\kappa^\dagger |0\rangle$ and $|1_\lambda\rangle = a_\lambda^\dagger |0\rangle$ are both one-photon states. The same equation tells us that the superposition

$$|\psi\rangle = \frac{1}{\sqrt{2}} |1_\kappa\rangle + \frac{1}{\sqrt{2}} |1_\lambda\rangle = \frac{1}{\sqrt{2}} \left(a_\kappa^\dagger + a_\lambda^\dagger\right) |0\rangle \tag{2.97}$$

is also a one-photon state; in fact, every state of the form

$$|\xi\rangle = \sum_\kappa \xi_\kappa a_\kappa^\dagger |0\rangle \tag{2.98}$$

is a one-photon state. There is a physical lesson to be drawn from this algebraic exercise: it is a mistake to assume that photons are necessarily associated with a single classical mode. Generalizing this to a superposition of modes which form a classical **wave packet**, we see that a single-photon wave packet state (that is, a wave packet that contains exactly one photon) is perfectly permissible.

According to eqn (2.94) any number of photons can occupy a single mode. Furthermore the commutation relations (2.63) guarantee that the generic state $a_{\kappa_1}^\dagger \cdots a_{\kappa_n}^\dagger |0\rangle$ is symmetric under any permutation of the mode labels $\kappa_1, \ldots, \kappa_n$. These are defining properties of objects satisfying Bose statistics (Bransden and Joachain, 1989, Sec. 10.2), so eqns (2.63) are called **Bose** commutation relations and photons are said to be **bosons**.

D Field operators

In the Schrödinger picture, the operators for the electric and magnetic fields are—by definition—time-independent. They can be expressed in terms of the time-independent operators p_κ and q_κ by first using the classical expansions (2.40) and (2.41) to write the initial classical fields $\boldsymbol{\mathcal{E}}(\mathbf{r},0)$ and $\boldsymbol{\mathcal{B}}(\mathbf{r},0)$ in terms of the initial displacements $Q_\kappa(0)$ and momenta $P_\kappa(0)$ of the radiation oscillators. Setting $(Q_\kappa, P_\kappa) = (Q_\kappa(0), P_\kappa(0))$, and applying the quantization conjecture (2.60) to these results leads to

$$\mathbf{E}(\mathbf{r}) = -\frac{1}{\sqrt{\epsilon_0}}\sum_\kappa p_\kappa \boldsymbol{\mathcal{E}}_\kappa(\mathbf{r}), \tag{2.99}$$

$$\mathbf{B}(\mathbf{r}) = \frac{1}{\sqrt{\epsilon_0}}\sum_\kappa k_\kappa q_\kappa \boldsymbol{\mathcal{B}}_\kappa(\mathbf{r}). \tag{2.100}$$

For most applications it is better to express the fields in terms of the operators a_κ and $a_\kappa^\dagger$ by using eqn (2.66) for each mode:

$$\mathbf{E}(\mathbf{r}) = i\sum_\kappa \sqrt{\frac{\hbar\omega_\kappa}{2\epsilon_0}}\left(a_\kappa - a_\kappa^\dagger\right)\boldsymbol{\mathcal{E}}_\kappa(\mathbf{r}), \tag{2.101}$$

$$\mathbf{B}(\mathbf{r}) = \sum_\kappa \sqrt{\frac{\mu_0\hbar\omega_\kappa}{2}}\left(a_\kappa + a_\kappa^\dagger\right)\boldsymbol{\mathcal{B}}_\kappa(\mathbf{r}). \tag{2.102}$$

The corresponding expansions for the vector potential in the radiation gauge are

$$\begin{aligned}\mathbf{A}(\mathbf{r}) &= \sum_\kappa \sqrt{\frac{1}{\epsilon_0}} q_\kappa \boldsymbol{\mathcal{E}}_\kappa(\mathbf{r}) \\ &= \sum_\kappa \sqrt{\frac{\hbar}{2\epsilon_0\omega_\kappa}}\left(a_\kappa + a_\kappa^\dagger\right)\boldsymbol{\mathcal{E}}_\kappa(\mathbf{r}).\end{aligned} \tag{2.103}$$

2.2 Normal ordering and zero-point energy

In the absence of interactions between the independent modes, the energy is additive; therefore, the Hamiltonian is the sum of the Hamiltonians for the individual modes. If we use eqn (2.68) for the single-mode Hamiltonians, the result is

$$H_{\text{em}} = \sum_\kappa \left[\hbar\omega_\kappa a_\kappa^\dagger a_\kappa + \frac{\hbar\omega_\kappa}{2}\right]. \tag{2.104}$$

The previously innocuous zero-point energies for each mode have now become a serious annoyance, since the sum over all modes is infinite. Fortunately there is an easy way out of this difficulty. We can simply use the alternate form (2.89) which gives

$$H_{\text{em}} = \sum_\kappa \hbar\omega_\kappa a_\kappa^\dagger a_\kappa. \tag{2.105}$$

With this choice for the single-mode Hamiltonians the vacuum energy is reduced from infinity to zero.

It is instructive to look at this problem in a different way by using the equivalent form eqn (2.67), instead of eqn (2.68), to get

$$H_{\text{em}} = \sum_{\kappa} \frac{\hbar\omega_\kappa}{2} \left(a_\kappa^\dagger a_\kappa + a_\kappa a_\kappa^\dagger \right) . \tag{2.106}$$

Now the zero-point energy can be eliminated by the simple expedient of reversing the order of the operators in the second term. This replaces eqn (2.106) by eqn (2.105). In other words, subtracting the vacuum expectation value of the energy is equivalent to reordering the operator products so that in each term the annihilation operator is to the right of the creation operator. This is called **normal ordering**, while the original order in eqn (2.106) is called **symmetrical ordering**.

We are allowed to consider such a step because there is a fundamental ambiguity involved in replacing products of commuting classical variables by products of non-commuting operators. This problem does not appear in quantizing the classical energy expression in eqn (2.64), since products of q_κ with p_κ do not occur. This happy circumstance is a fortuitous result of the choice of classical variables. If we had instead chosen to use the variables α_κ defined by eqn (2.45), the quantization conjecture would be $\alpha_\kappa \to a_\kappa$ and $\alpha_\kappa^* \to a_\kappa^\dagger$. This does produce an ordering ambiguity in quantizing eqn (2.51), since $\alpha_\kappa \alpha_\kappa^*$, $\left(\alpha_\kappa^* \alpha_\kappa + \alpha_\kappa \alpha_\kappa^*\right)/2$, and $\alpha_\kappa^* \alpha_\kappa$ are identical in the classical theory, but different after quantization. The last two forms lead respectively to the expressions (2.106) and (2.105) for the Hamiltonian. Thus the presence or absence of the zero-point energy is determined by the choice of ordering of the noncommuting operators.

It is useful to extend the idea of normal ordering to any product of operators $X_1 \cdots X_n$, where each X_i is either a creation or an annihilation operator. The **normal-ordered product** is defined by

$$: X_1 \cdots X_n : \; = X_{1'} \cdots X_{n'} \, , \tag{2.107}$$

where $(1', \ldots, n')$ is any ordering (permutation) of $(1, \ldots, n)$ that arranges all of the annihilation operators to the right of all the creation operators. The commutation relations are ignored when carrying out the reordering. More generally, let Z be a linear combination of distinct products $X_1 \cdots X_n$; then $: Z :$ is the same linear combination of the normal-ordered products $: X_1 \cdots X_n :$. The vacuum expectation value of a normal-ordered product evidently vanishes, but it is not generally true that $Z = \; : Z : + \langle 0 | Z | 0 \rangle$.

2.3 States in quantum theory

In classical mechanics, the coordinate $\mathbf{q}$ and momentum $\mathbf{p}$ of a particle can be precisely specified. Therefore, in classical physics the state of maximum information for a system of N particles is a point $(\underline{q}, \underline{p}) = (\mathbf{q}_1, \mathbf{p}_1, \ldots, \mathbf{q}_N, \mathbf{p}_N)$ in the mechanical phase space. For large values of N, specifying a point in the phase space is a practical impossibility, so it is necessary to use classical statistical mechanics—which describes the N-body system by a probability distribution $f\left(\underline{q}, \underline{p}\right)$—instead. The point to bear in mind here is that this probability distribution is an admission of ignorance. No experimentalist can possibly acquire enough information to determine a particular value of $\left(\underline{q}, \underline{p}\right)$.

In quantum theory, the uncertainty principle prohibits simultaneous determination of the coordinates and momenta of a particle, but the notions of states of maximum and less-than-maximum information can still be defined.

2.3.1 Pure states

In the standard interpretation of quantum theory, the vectors in the Hilbert space describing a physical system—e.g. general linear combinations of number states in Fock space—provide the most detailed description of the state of the system that is consistent with the uncertainty principle. These quantum states of maximum information are called **pure states** (Bransden and Joachain, 1989, Chap. 14). From this point of view the random fluctuations imposed by the uncertainty principle are *intrinsic*; they are not the result of ignorance of the values of some underlying variables.

For any quantum system the average of many measurements of an observable X on a collection of identical physical systems, all described by the same vector $|\Psi\rangle$, is given by the expectation value $\langle \Psi | X | \Psi \rangle$. The evolution of a pure state is described by the Schrödinger equation

$$i\hbar \frac{\partial}{\partial t} |\Psi(t)\rangle = H |\Psi(t)\rangle , \tag{2.108}$$

where H is the Hamiltonian.

2.3.2 Mixed states

In the absence of maximum information, the system is said to be in a **mixed state**. In this situation there is insufficient information to decide which pure state describes the system. Just as for classical statistical mechanics, it is then necessary to assign a probability to each possible pure state. These probabilities, which represent ignorance of which pure state should be used, are consequently classical in character.

As a simple example, suppose that there is only sufficient information to say that each member of a collection of identically prepared systems is described by one or the other of two pure states, $|\Psi_1\rangle$ or $|\Psi_2\rangle$. For a system described by $|\Psi_e\rangle$ $(e = 1, 2)$, the average value for measurements of X is the quantum expectation value $\langle \Psi_e | X | \Psi_e \rangle$. The overall average of measurements of X is therefore

$$\langle X \rangle = \mathcal{P}_1 \langle \Psi_1 | X | \Psi_1 \rangle + \mathcal{P}_2 \langle \Psi_2 | X | \Psi_2 \rangle , \tag{2.109}$$

where $\mathcal{P}_e$ is the fraction of the systems described by $|\Psi_e\rangle$, and $\mathcal{P}_1 + \mathcal{P}_2 = 1$.

The average in eqn (2.109) is quite different from the average of many measurements on systems all described by the superposition state $|\Psi\rangle = C_1 |\Psi_1\rangle + C_2 |\Psi_2\rangle$. In that case the average is

$$\langle \Psi | X | \Psi \rangle = |C_1|^2 \langle \Psi_1 | X | \Psi_1 \rangle + |C_2|^2 \langle \Psi_2 | X | \Psi_2 \rangle + 2 \operatorname{Re} \left[C_1^* C_2 \langle \Psi_1 | X | \Psi_2 \rangle \right] , \tag{2.110}$$

which contains an interference term missing from eqn (2.109). The two results (2.109) and (2.110) only agree when $|C_e|^2 = \mathcal{P}_e$ and $\operatorname{Re}\left[C_1^* C_2 \langle \Psi_1 | X | \Psi_2 \rangle \right] = 0$. The latter condition can be satisfied if $C_1^* C_2 \langle \Psi_1 | X | \Psi_2 \rangle$ is pure imaginary or if $\langle \Psi_1 | X | \Psi_2 \rangle = 0$. Since it is always possible to choose another observable X' for which neither of these conditions is satisfied, it is clear that the mixed state and the superposition state describe very different physical situations.

A The density operator

In general, a mixed state is defined by a collection, usually called an **ensemble**, of normalized pure states $\{|\Psi_e\rangle\}$, where the label e may be discrete or continuous. For simplicity we only consider the discrete case here: the continuum case merely involves replacing sums by integrals with suitable weighting functions. For the discrete case, a **probability distribution** on the ensemble is a set of real numbers $\{\mathcal{P}_e\}$ that satisfy the conditions

$$0 \leqslant \mathcal{P}_e \leqslant 1 \,, \tag{2.111}$$

$$\sum_e \mathcal{P}_e = 1 \,. \tag{2.112}$$

The ensemble may be finite or infinite, and the vectors need not be mutually orthogonal.

The average of repeated measurements of an observable X is represented by the ensemble average of the quantum expectation values,

$$\langle X \rangle (t) = \sum_e \mathcal{P}_e \langle \Psi_e (t) | X | \Psi_e (t) \rangle \,, \tag{2.113}$$

where $|\Psi_e (t)\rangle$ is the solution of the Schrödinger equation with initial value $|\Psi_e (0)\rangle = |\Psi_e\rangle$. It is instructive to rewrite this result by using the number-state basis $\{|\underline{n}\rangle\}$ for Fock space to get

$$\langle \Psi_e (t) | X | \Psi_e (t) \rangle = \sum_{\underline{n}} \sum_{\underline{m}} \langle \Psi_e (t) | \underline{n} \rangle \langle \underline{n} | X | \underline{m} \rangle \langle \underline{m} | \Psi_e (t) \rangle \,, \tag{2.114}$$

and

$$\langle X \rangle (t) = \sum_{\underline{n}} \sum_{\underline{m}} \langle \underline{n} | X | \underline{m} \rangle \left[\sum_e \mathcal{P}_e \langle \underline{m} | \Psi_e (t) \rangle \langle \Psi_e (t) | \underline{n} \rangle \right] . \tag{2.115}$$

By applying the general definition (2.81) to the operator $|\Psi_e (t)\rangle \langle \Psi_e (t)|$, it is easy to see that the quantity in square brackets is the matrix element $\langle \underline{m} | \rho (t) | \underline{n} \rangle$ of the **density operator**:

$$\rho (t) = \sum_e \mathcal{P}_e |\Psi_e (t)\rangle \langle \Psi_e (t)| \,. \tag{2.116}$$

With this result in hand, eqn (2.115) becomes

$$\begin{aligned} \langle X \rangle (t) &= \sum_{\underline{m}} \sum_{\underline{n}} \langle \underline{m} | \rho (t) | \underline{n} \rangle \langle \underline{n} | X | \underline{m} \rangle \\ &= \sum_{\underline{m}} \langle \underline{m} | \rho (t) X | \underline{m} \rangle \\ &= \mathrm{Tr} \left[\rho (t) X \right] , \end{aligned} \tag{2.117}$$

where the trace operation is defined by eqn (C.22). Each of the ket vectors $|\Psi_e\rangle$ in the ensemble evolves according to the Schrödinger equation, and the bra vectors $\langle\Psi_e|$ obey the conjugate equation

$$-i\hbar\frac{\partial}{\partial t}\langle\Psi_e(t)| = \langle\Psi_e(t)|\,H\,, \tag{2.118}$$

so the evolution equation for the density operator is

$$i\hbar\frac{\partial}{\partial t}\rho(t) = [H, \rho(t)]\,. \tag{2.119}$$

By analogy with the Liouville equation for the classical distribution function (Huang, 1963, Sec. 4.3), this is called the **quantum Liouville equation**. The condition (2.112), together with the normalization of the ensemble state vectors, means that the density operator has unit trace,

$$\mathrm{Tr}\,(\rho(t)) = 1\,, \tag{2.120}$$

and eqn (2.119) guarantees that this condition is valid at all times.

A pure state is described by an ensemble consisting of exactly one vector, so that eqn (2.116) reduces to

$$\rho(t) = |\Psi(t)\rangle\,\langle\Psi(t)|\,. \tag{2.121}$$

This explicit statement can be replaced by the condition that $\rho(t)$ is a projection operator, i.e.

$$\rho^2(t) = \rho(t)\,. \tag{2.122}$$

Thus for pure states

$$\mathrm{Tr}\left(\rho^2(t)\right) = \mathrm{Tr}\,(\rho(t)) = 1\,, \tag{2.123}$$

while for mixed states

$$\mathrm{Tr}\left(\rho^2(t)\right) < 1\,. \tag{2.124}$$

For any observable X and any state ρ, either pure or mixed, an important statistical property is given by the **variance**

$$V(X) = \left\langle X^2\right\rangle - \langle X\rangle^2\,, \tag{2.125}$$

where $\langle X\rangle = \mathrm{Tr}\,(\rho X)$. The easily verified identity $V(X) = \left\langle (X - \langle X\rangle)^2\right\rangle$ shows that $V(X) \geqslant 0$, and it also follows that $V(X) = 0$ when ρ is an eigenstate of X, i.e. $X\rho = \rho X = \lambda\rho$. Conversely, every eigenstate of X satisfies $V(X) = 0$. Since $V(X)$ is non-negative, the variance is often described in terms of the **root mean square (rms) deviation**

$$\Delta X = \sqrt{V(X)} = \sqrt{\langle X^2\rangle - \langle X\rangle^2}\,. \tag{2.126}$$

B Mixed states arising from measurements

In quantum theory the act of measurement can produce a mixed state, even if the state before the measurement is pure. For simplicity, we consider an observable X with a discrete, nondegenerate spectrum. This means that the eigenvectors $|x_n\rangle$, satisfying $X|x_n\rangle = x_n|x_n\rangle$, are unique (up to a phase). Suppose that we have complete information about the initial state of the system, so that we can describe it by a pure state $|\psi\rangle$. When a measurement of X is carried out, the Born interpretation tells us that the eigenvalue x_n will be found with probability $p_n = |\langle x_n|\psi\rangle|^2$. The von Neumann projection postulate further tells us that the system will be described by the pure state $|x_n\rangle$, if the measurement yields x_n. This is the reduction of the wave packet. Now consider the following situation. We know that a measurement of X has been performed, but we do not know which eigenvalue of X was actually observed. In this case there is no way to pick out one eigenstate from the rest. Thus we have an ensemble consisting of all the eigenstates of X, and the density operator for this ensemble is

$$\rho_{\text{meas}} = \sum_n p_n |x_n\rangle \langle x_n| . \tag{2.127}$$

Thus a measurement will change the original pure state into a mixed state, if the knowledge of which eigenvalue was obtained is not available.

2.3.3 General properties of the density operator

So far we have only considered observables with nondegenerate eigenvalues, but in general some of the eigenvalues x_ξ of X are degenerate, i.e. there are several linearly independent solutions of the eigenvalue problem $X|\Psi\rangle = x_\xi|\Psi\rangle$. The number of solutions is the degree of degeneracy, denoted by $d_\xi(X)$. A familiar example is $X = J^2$, where $\mathbf{J}$ is the angular momentum operator. The eigenvalue $j(j+1)\hbar^2$ of J^2 has the degeneracy $2j+1$ and the degenerate eigenstates can be labeled by the eigenvalues $m\hbar$ of J_z, with $-j \leqslant m \leqslant j$. An example appropriate to the present context is the operator

$$N_{\mathbf{k}} = \sum_s a^\dagger_{\mathbf{k}s} a_{\mathbf{k}s} , \tag{2.128}$$

that counts the number of photons with wavevector $\mathbf{k}$. If $\mathbf{k}$ has no vanishing components, the eigenvalue problem $N_{\mathbf{k}}|\Psi\rangle = |\Psi\rangle$ has two independent solutions corresponding to the two possible polarizations, so $d_1(N_{\mathbf{k}}) = 2$. In general, the common eigenvectors for a given eigenvalue span a $d_\xi(X)$-dimensional subspace, called the eigenspace $\mathfrak{H}_\xi(X)$. Let

$$|\Psi_{\xi 1}\rangle , \ldots , |\Psi_{\xi d_\xi(X)}\rangle \tag{2.129}$$

be a basis for $\mathfrak{H}_\xi(X)$, then

$$P_\xi = \sum_m |\Psi_{\xi m}\rangle \langle \Psi_{\xi m}| \tag{2.130}$$

is the projection operator onto $\mathfrak{H}_\xi(X)$.

According to the standard rules of quantum theory (see eqns (C.26)–(C.28)) the conditional probability that x_ξ is the result of a measurement of X, given that the system is described by the pure state $|\Psi_e\rangle$, is

$$p(x_\xi|\Psi_e) = \sum_m |\langle\Psi_{\xi m}|\Psi_e\rangle|^2 = \langle\Psi_e|P_\xi|\Psi_e\rangle\,. \tag{2.131}$$

For the mixed state the overall probability of the result x_ξ is, therefore,

$$p(x_\xi) = \sum_e \mathcal{P}_e \sum_m |\langle\Psi_{\xi m}|\Psi_e\rangle|^2 = \mathrm{Tr}\,(\rho P_\xi)\,. \tag{2.132}$$

Thus the general rule is that the probability for finding a given value x_ξ is given by the expectation value of the projection operator P_ξ onto the corresponding eigenspace.

Other important mathematical properties of the density operator follow directly from the definition (2.116). For any state $|\Psi\rangle$, the expectation value of ρ is positive,

$$\langle\Psi|\rho|\Psi\rangle = \sum_e \mathcal{P}_e |\langle\Psi_e|\Psi\rangle|^2 \geqslant 0\,, \tag{2.133}$$

so ρ is a positive-definite operator. Combining this with the normalization condition (2.120) implies $0 \leqslant \langle\Psi|\rho|\Psi\rangle \leqslant 1$ for any normalized state $|\Psi\rangle$. The Born interpretation tells us that $|\langle\Psi_e|\Psi\rangle|^2$ is the probability that a measurement—say of the projection operator $|\Psi\rangle\langle\Psi|$—will leave the system in the state $|\Psi\rangle$, given that the system is prepared in the pure state $|\Psi_e\rangle$; therefore, eqn (2.133) tells us that $\langle\Psi|\rho|\Psi\rangle$ is the probability that a measurement will lead to $|\Psi\rangle$, if the system is described by the mixed state with density operator ρ.

In view of the importance of the superposition principle for pure states, it is natural to ask if any similar principle applies to mixed states. The first thing to note is that linear combinations of density operators are not generally physically acceptable density operators. Thus if ρ_1 and ρ_2 are density operators, the combination $\rho = C\rho_1 + D\rho_2$ will be hermitian only if C and D are both real. The condition $\mathrm{Tr}\,\rho = 1$ further requires $C + D = 1$. Finally, the positivity condition (2.133) must hold for all choices of $|\Psi\rangle$, and this can only be guaranteed by imposing $C \geqslant 0$ and $D \geqslant 0$. Therefore, only the **convex linear combinations**

$$\rho = C\rho_1 + (1 - C)\,\rho_2\,, \quad 0 \leqslant C \leqslant 1 \tag{2.134}$$

are guaranteed to be density matrices. This terminology is derived from the mathematical notion of a **convex set** in the plane, i.e. a set that contains every straight line joining any two of its points. The general form of eqn (2.134) is

$$\rho = \sum_n C_n \rho_n\,, \tag{2.135}$$

where each ρ_n is a density operator, and the coefficients satisfy the convexity condition

$$0 \leqslant C_n \leqslant 1 \quad \text{for all } n \quad \text{and} \quad \sum_n C_n = 1\,. \tag{2.136}$$

The off-diagonal matrix elements of the density operator are also constrained by the definition (2.116). The normalization of the ensemble states $|\Psi_e\rangle$ implies $|\langle\Psi_e|\Psi\rangle| \leqslant 1$, so

$$\begin{aligned} |\langle\Psi|\rho|\Phi\rangle| &= \left|\sum_e \mathcal{P}_e \langle\Psi|\Psi_e\rangle \langle\Psi_e|\Phi\rangle\right| \\ &\leqslant \sum_e \mathcal{P}_e |\langle\Psi|\Psi_e\rangle| \, |\langle\Psi_e|\Phi\rangle| \leqslant 1\,, \end{aligned} \tag{2.137}$$

i.e. ρ is a bounded operator.

The arguments leading from the ensemble definition of the density operator to its properties can be reversed to yield the following statement. An operator ρ that is (a) hermitian, (b) bounded, (c) positive, and (d) has unit trace is a possible density operator. The associated ensemble can be defined as the set of normalized eigenstates of ρ corresponding to nonzero eigenvalues. Since every density operator has a complete orthonormal set of eigenvectors, this last remark implies that it is always possible to choose the ensemble to consist of mutually orthogonal states.

2.3.4 Degrees of mixing

So far the distinction between pure and mixed states is absolute, but finer distinctions are also useful. In other words, some states are more mixed than others. The distinctions we will discuss arise most frequently for physical systems described by a finite-dimensional Hilbert space, or equivalently, ensembles containing a finite number of pure states. This allows us to simplify the analysis by assuming that the Hilbert space has dimension $d < \infty$. The inequality (2.124) suggests that the **purity**

$$\mathfrak{P}(\rho) = \mathrm{Tr}\left(\rho^2\right) \leqslant 1 \tag{2.138}$$

may be a useful measure of the degree of mixing associated with a density operator ρ. By virtue of eqn (2.122), $\mathfrak{P}(\rho) = 1$ for a pure state; therefore, it is natural to say that the state ρ_2 is less pure (more mixed) than the state ρ_1 if $\mathfrak{P}(\rho_2) < \mathfrak{P}(\rho_1)$. Thus the minimally pure (maximally mixed) state for an ensemble will be the one that achieves the lower bound of $\mathfrak{P}(\rho)$. In general the density operator can have the eigenvalue 0 with degeneracy (multiplicity) $d_0 < d$, so the number of orthogonal states in the ensemble is $\mathfrak{N} = d - d_0$. Using the eigenstates of ρ to evaluate the trace yields

$$\mathfrak{P}(\rho) = \sum_{n=1}^{\mathfrak{N}} p_n^2\,, \tag{2.139}$$

where p_n is the nth eigenvalue of ρ. In this notation, the trace condition (2.120) is just

$$\sum_{n=1}^{\mathfrak{N}} p_n = 1\,, \tag{2.140}$$

and the lower bound is found by minimizing $\mathfrak{P}(\rho)$ subject to the constraint (2.140). This can be done in several ways, e.g. by the method of Lagrange multipliers, with the result that the **maximally mixed state** is defined by

$$p_n = \begin{cases} \frac{1}{\mathfrak{N}}, & n = 1, \ldots, \mathfrak{N}, \\ 0, & n = \mathfrak{N}+1, \ldots, d. \end{cases} \tag{2.141}$$

In other words, the pure states in the ensemble defining the maximally mixed state occur with equal probability, and the purity is

$$\mathfrak{P}(\rho) = \frac{1}{\mathfrak{N}}. \tag{2.142}$$

Another useful measure of the degree of mixing is provided by the **von Neumann entropy**, which is defined in general by

$$\mathfrak{S}(\rho) = -\operatorname{Tr}(\rho \ln \rho). \tag{2.143}$$

In the special case considered above, the von Neumann entropy is given by

$$\mathfrak{S}(\rho) = -\sum_{n=1}^{\mathfrak{N}} p_n \ln p_n, \tag{2.144}$$

and maximizing this—subject to the constraint (2.140)—leads to the same definition of the maximally mixed state, with the value

$$\mathfrak{S}(\rho) = \ln \mathfrak{N} \tag{2.145}$$

of $\mathfrak{S}(\rho)$. The von Neumann entropy plays an important role in the study of entangled states in Chapter 6.

2.4 Mixed states of the electromagnetic field

2.4.1 Polarized light

As a concrete example of a mixed state, consider an experiment in which light from a single atom is sent through a series of collimating pinholes. In each atomic transition, exactly one photon with frequency $\omega = \Delta E/\hbar$ is emitted, where ΔE is the energy difference between the atomic states. The alignment of the pinholes determines the unit vector $\widetilde{\mathbf{k}}$ along the direction of propagation, so the experimental arrangement determines the wavevector $\mathbf{k} = (\omega/c)\,\widetilde{\mathbf{k}}$. If the pinholes are perfectly circular, the experimental preparation gives no information on the polarization of the transmitted light. This means that the light observed on the far side of the collimator could be described by either of the states

$$|\Psi_s\rangle = |1_{\mathbf{k}s}\rangle = a^\dagger_{\mathbf{k}s}|0\rangle, \tag{2.146}$$

where $s = \pm 1$ labels right- and left-circularly-polarized light. Thus the relevant ensemble is composed of the states $|1_{\mathbf{k}+}\rangle$ and $|1_{\mathbf{k}-}\rangle$, with probabilities $\mathcal{P}_+$ and $\mathcal{P}_-$, and the density operator is

$$\rho = \sum_s \mathcal{P}_s |1_{\mathbf{k}s}\rangle\langle 1_{\mathbf{k}s}| = \begin{bmatrix} \mathcal{P}_+ & 0 \\ 0 & \mathcal{P}_- \end{bmatrix}. \tag{2.147}$$

In the absence of any additional information equal probabilities are assigned to the two polarizations, i.e. $\mathcal{P}_+ = \mathcal{P}_- = 1/2$, and the light is said to be unpolarized. The opposite extreme occurs when the polarization is known with certainty, for example $\mathcal{P}_+ = 1$, $\mathcal{P}_- = 0$. This can be accomplished by inserting a polarization filter after the collimator. In this case, the light is said to be polarized, and the density operator represents the pure state $|1_{\mathbf{k}+}\rangle$. For the intermediate cases, a measure of the degree of polarization is given by

$$P = |\mathcal{P}_+ - \mathcal{P}_-| \,, \tag{2.148}$$

which satisfies $0 \leqslant P \leqslant 1$, and has the values $P = 0$ for unpolarized light and $P = 1$ for polarized light.

A The second-order coherence matrix

The conclusions reached for the special case discussed above are also valid in a more general setting (Mandel and Wolf, 1995, Sec. 6.2). We present here a simplified version of the general discussion by defining the **second-order coherence matrix**

$$J_{ss'} = \mathrm{Tr}\left(\rho a_{\mathbf{k}s}^{\dagger} a_{\mathbf{k}s'}\right), \tag{2.149}$$

where the density operator ρ describes a **monochromatic state**, i.e. each state vector $|\Psi_e\rangle$ in the ensemble defining ρ satisfies $a_{\mathbf{k}'s}|\Psi_e\rangle = 0$ for $\mathbf{k}' \neq \mathbf{k}$. In this case we may as well choose the z-axis along $\mathbf{k}$, and set $s = x, y$, corresponding to linear polarization vectors along the x- and y-axes respectively. The 2×2 matrix J is hermitian and positive definite—see Appendix A.3.4—so the eigenvectors $c_p = (c_{px}, c_{py})$ and eigenvalues n_p $(p = 1, 2)$ defined by

$$Jc_p = n_p c_p \tag{2.150}$$

satisfy

$$c_p^{\dagger} c_{p'} = \delta_{pp'} \quad \text{and} \quad n_p \geqslant 0\,. \tag{2.151}$$

The eigenvectors of J define eigenpolarization vectors,

$$\mathbf{e}_p = c_{px}^{*}\mathbf{e}_x + c_{py}^{*}\mathbf{e}_y\,, \tag{2.152}$$

and corresponding creation and annihilation operators

$$\mathfrak{a}_p^{\dagger} = c_{px}^{*} a_{\mathbf{k}x}^{\dagger} + c_{py}^{*} a_{\mathbf{k}y}^{\dagger}\,, \quad \mathfrak{a}_p = c_{px} a_{\mathbf{k}x} + c_{py} a_{\mathbf{k}y}\,. \tag{2.153}$$

It is not difficult to show that

$$n_p = \mathrm{Tr}\left(\rho \mathfrak{a}_p^{\dagger} \mathfrak{a}_p\right), \tag{2.154}$$

i.e. the eigenvalue n_p is the average number of photons with eigenpolarization $\mathbf{e}_p$. If ρ describes an unpolarized state, then different polarizations must be uncorrelated and the number of photons in either polarization must be equal, i.e.

$$J = \frac{n}{2}\begin{bmatrix} 1 & 0 \\ 0 & 1 \end{bmatrix}, \tag{2.155}$$

where

$$n = \mathrm{Tr}\left[\rho\left(a_{\mathbf{k}x}^{\dagger} a_{\mathbf{k}x} + a_{\mathbf{k}y}^{\dagger} a_{\mathbf{k}y}\right)\right] \tag{2.156}$$

is the average total number of photons. If ρ describes complete polarization, then the occupation number for one of the eigenpolarizations must vanish, e.g. $n_2 = 0$.

Since $\det J = n_1 n_2$, this means that completely polarized states are characterized by $\det J = 0$. In this general setting, the **degree of polarization** is defined by

$$P = \frac{|n_1 - n_2|}{n_1 + n_2}, \tag{2.157}$$

where $P = 0$ and $P = 1$ respectively correspond to unpolarized and completely polarized light.

B The Stokes parameters

Since J is a 2×2 matrix, we can exploit the well known fact—see Appendix C.3.1—that any 2×2 matrix can be expressed as a linear combination of the Pauli matrices. For this application, we write the expansion as

$$J = \frac{1}{2} S_0 \sigma_0 + \frac{1}{2} S_1 \sigma_3 + \frac{1}{2} S_2 \sigma_1 - \frac{1}{2} S_3 \sigma_2 , \tag{2.158}$$

where σ_0 is the 2×2 identity matrix and σ_1, σ_2, and σ_3 are the Pauli matrices given by the standard representation (C.30). This awkward formulation guarantees that the c-number coefficients S_μ are the traditional **Stokes parameters**. According to eqn (C.40) they are given by

$$S_0 = \mathrm{Tr}\,(J \sigma_0)\,, \quad S_1 = \mathrm{Tr}\,(J \sigma_3)\,, \quad S_2 = \mathrm{Tr}\,(J \sigma_1)\,, \quad S_3 = -\,\mathrm{Tr}\,(J \sigma_2)\,. \tag{2.159}$$

The Stokes parameters yield a useful geometrical picture of the coherence matrix, since the necessary condition $\det (J) \geqslant 0$ translates to

$$S_1^2 + S_2^2 + S_3^2 \leqslant S_0^2 \,. \tag{2.160}$$

If we interpret (S_1, S_2, S_3) as a point in a three-dimensional space, then for a fixed value of S_0 the states of the field occupy a sphere—called the **Poincaré sphere**—of radius S_0. The origin, $S_1 = S_2 = S_3 = 0$, corresponds to unpolarized light, since this is the only case for which J is proportional to the identity. The condition $\det J = 0$ for completely polarized light is

$$S_1^2 + S_2^2 + S_3^2 = S_0^2 \,, \tag{2.161}$$

which describes points on the surface of the sphere. Intermediate states of polarization correspond to points in the interior of the sphere.

The Poincaré sphere is often used to describe the pure states of a single photon, e.g.

$$|\psi\rangle = \sum_s C_s a_{\mathbf{k}s}^\dagger |0\rangle \,. \tag{2.162}$$

In this case $S_0 = 1$, and the points on the surface of the Poincaré sphere can be labeled by the standard spherical coordinates (θ, ϕ). The north pole, $\theta = 0$, and the south pole, $\theta = \pi$, respectively describe right- and left-circular polarizations. Linear polarizations are represented by points on the equator, and elliptical polarizations by points in the northern and southern hemispheres.

2.4.2 Thermal light

A very important example of a mixed state arises when the field is treated as a thermodynamic system in contact with a thermal reservoir at temperature T, e.g. the walls of the cavity. Under these circumstances, any complete set of states can be chosen for the ensemble, since we have no information that allows the exclusion of any pure state of the field. Exchange of energy with the walls is the mechanism for attaining thermal equilibrium, so it is natural to use the energy eigenstates—i.e. the number states $|\underline{n}\rangle$—for this purpose.

The general rules of statistical mechanics (Chandler, 1987, Sec. 3.7) tell us that the probability for a given energy E is proportional to $\exp(-\beta E)$, where $\beta = 1/k_B T$ and k_B is Boltzmann's constant. Thus the probability distribution is $\mathcal{P}_{\underline{n}} = Z^{-1}\exp\left(-\beta E_{\underline{n}}\right)$, where Z^{-1} is the normalization constant required to satisfy eqn (2.112), and

$$E_{\underline{n}} = \sum_{\kappa} \hbar\omega_{\kappa} n_{\kappa} \,. \tag{2.163}$$

Substituting this probability distribution into eqn (2.116) gives the density operator

$$\rho = \frac{1}{Z}\sum_{\underline{n}} e^{-\beta E_{\underline{n}}} |\underline{n}\rangle \langle \underline{n}| = \frac{1}{Z}\exp\left(-\beta H_{\text{em}}\right). \tag{2.164}$$

The normalization constant Z, which is called the **partition function**, is determined by imposing $\text{Tr}\,(\rho) = 1$ to get

$$Z = \text{Tr}\left[\exp\left(-\beta H_{\text{em}}\right)\right]. \tag{2.165}$$

Evaluating the trace in the number-state basis yields

$$Z = \sum_{\underline{n}} \exp\left[-\beta \sum_{\kappa} n_{\kappa}\hbar\omega_{\kappa}\right] = \prod_{\kappa} Z_{\kappa} \,, \tag{2.166}$$

where

$$Z_{\kappa} = \sum_{n_{\kappa}=0}^{\infty} e^{-\beta n_{\kappa}\hbar\omega_{\kappa}} = \sum_{n_{\kappa}=0}^{\infty} \left(e^{-\beta\hbar\omega_{\kappa}}\right)^{n_{\kappa}} = \frac{1}{1 - e^{-\beta\hbar\omega_{\kappa}}} \tag{2.167}$$

is the partition function for mode κ (Chandler, 1987, Chap. 4).

A *The Planck distribution*

The average energy in the electromagnetic field is related to the partition function by

$$U = \left(-\frac{\partial}{\partial\beta}\right)\ln Z = \sum_{\kappa} \frac{\hbar\omega_{\kappa}}{e^{\beta\hbar\omega_{\kappa}} - 1} \,. \tag{2.168}$$

We will say that the cavity is large if the energy spacing $\hbar c\Delta k_{\kappa}$ between adjacent discrete modes is small compared to any physically relevant energy. In this limit the shape of the cavity is not important, so we may suppose that it is cubical, with

$\kappa \to (\mathbf{k}, s)$, where $s = 1, 2$ and $\omega_\kappa \to ck$. In the limit of infinite volume, applying the rule

$$\frac{1}{V}\sum_{\mathbf{k}} \to \int \frac{d^3k}{(2\pi)^3} \tag{2.169}$$

replaces eqn (2.168) by

$$\frac{U}{V} = \frac{2}{(2\pi)^3}\int d^3k \frac{\hbar ck}{e^{\beta\hbar ck} - 1}\,. \tag{2.170}$$

After carrying out the angular integrations and changing the remaining integration variable to $\omega = ck$, this becomes

$$\frac{U}{V} = \int_0^\infty d\omega\, \rho\left(\omega, T\right), \tag{2.171}$$

where the energy density $\rho\left(\omega, T\right) d\omega$ in the frequency interval ω to $\omega + d\omega$ is given by the Planck function

$$\rho\left(\omega, T\right) = \frac{1}{\pi^2 c^3}\frac{\hbar\omega^3}{e^{\beta\hbar\omega} - 1}\,. \tag{2.172}$$

B Distribution in photon number

In addition to the distribution in energy, it is also useful to know the distribution in photon number, n_κ, for a given mode. This calculation is simplified by the fact that the thermal density operator is the product of independent operators for each mode,

$$\rho = \prod_\kappa \rho_\kappa\,, \tag{2.173}$$

where

$$\rho_\kappa = \frac{1}{Z_\kappa}\exp\left(-\beta N_\kappa \hbar\omega_\kappa\right). \tag{2.174}$$

Thus we can drop the mode index and set

$$\rho = \left(1 - e^{-\beta\hbar\omega}\right)\exp\left[-\beta\hbar\omega a^\dagger a\right]. \tag{2.175}$$

The eigenstates of the single-mode number operator are nondegenerate, so the general rule (2.132) reduces to

$$p\left(n\right) = \mathrm{Tr}\left(\rho \left|n\right\rangle\left\langle n\right|\right) = \left\langle n\left|\rho\right|n\right\rangle = \left(1 - e^{-\beta\hbar\omega}\right)e^{-n\beta\hbar\omega}\,, \tag{2.176}$$

where $p\left(n\right)$ is the probability of finding n photons. This can be expressed more conveniently by first calculating the average number of photons:

$$\left\langle n\right\rangle = \mathrm{Tr}\left(\rho a^\dagger a\right) = \frac{e^{-\beta\hbar\omega}}{1 - e^{-\beta\hbar\omega}}\,. \tag{2.177}$$

Using this to eliminate $e^{-\beta\hbar\omega}$ leads to the final form

$$p\left(n\right) = \frac{\left\langle n\right\rangle^n}{\left(1 + \left\langle n\right\rangle\right)^{n+1}}\,. \tag{2.178}$$

Finally, it is important to realize that eqn (2.177) is not restricted to the electromagnetic field. Any physical system with a Hamiltonian of the form (2.89), where the operators a and $a^\dagger$ satisfy the canonical commutation relations (2.63) for a harmonic oscillator, will be described by the Planck distribution.

2.5 Vacuum fluctuations

Our first response to the infinite zero-point energy associated with vacuum fluctuations was to hide it away as quickly as possible, but we now have the tools to investigate the divergence in more detail. According to eqns (2.99) and (2.100) the electric and magnetic field operators are respectively determined by p_κ and q_κ so there are inescapable vacuum fluctuations of the fields. The $\mathbf{E}$ and $\mathbf{B}$ fields are linear in a_κ and $a_\kappa^\dagger$ so their vacuum expectation values vanish, but $\mathbf{E}^2$ and $\mathbf{B}^2$ will have nonzero vacuum expectation values representing the rms deviation of the fields. Let us consider the rms deviation of the electric field. The operators $E_i(\mathbf{r})$ $(i = 1, 2, 3)$ are hermitian and mutually commutative, so we are allowed to consider simultaneous measurements of all components of $\mathbf{E}(\mathbf{r})$. In this case the ambiguity in going from a classical quantity to the corresponding quantum operator is not an issue.

Since trouble is to be expected, we approach $\langle 0 |\mathbf{E}^2(\mathbf{r})| 0\rangle$ with caution by first evaluating $\langle 0 |E_i(\mathbf{r}) E_j(\mathbf{r}')| 0\rangle$ for $\mathbf{r}' \neq \mathbf{r}$. The expansion (2.101) yields

$$\begin{aligned}\langle 0 |E_i(\mathbf{r}) E_j(\mathbf{r}')| 0\rangle &= -\frac{\hbar}{2\epsilon_0}\sum_\kappa\sum_\lambda \sqrt{\omega_\kappa\omega_\lambda}\mathcal{E}_{\kappa i}(\mathbf{r})\mathcal{E}_{\lambda j}(\mathbf{r}') \\ &\quad\times\left\langle 0\right|\left(a_\kappa - a_\kappa^\dagger\right)\left(a_\lambda - a_\lambda^\dagger\right)\left|0\right\rangle,\end{aligned} \tag{2.179}$$

and evaluating the vacuum expectation value leads to

$$\langle 0 |E_i(\mathbf{r}) E_j(\mathbf{r}')| 0\rangle = \frac{\hbar}{2\epsilon_0}\sum_\kappa \omega_\kappa\mathcal{E}_{\kappa i}(\mathbf{r})\mathcal{E}_{\kappa j}(\mathbf{r}')\,. \tag{2.180}$$

Direct evaluation of the sum over modes requires detailed knowledge of both the mode spectrum and the mode functions, but this can be avoided by borrowing a trick from quantum mechanics (Cohen-Tannoudji *et al.*, 1977*a*, Chap. II, Complement B). According to eqn (2.35) each mode function $\boldsymbol{\mathcal{E}}_\kappa$ is an eigenfunction of the operator $-\nabla^2$ with eigenvalue k_κ^2. The operator and eigenvalue are respectively mathematical analogues of the kinetic energy operator and the energy eigenvalue in quantum mechanics (in units such that $2m = \hbar = 1$). Since $-\nabla^2$ is hermitian and $\boldsymbol{\mathcal{E}}_\kappa$ is an eigenfunction, the general argument given in Appendix C.3.6 shows that

$$\left(-\nabla^2\right)^{1/2}\boldsymbol{\mathcal{E}}_\kappa = \sqrt{k_\kappa^2}\boldsymbol{\mathcal{E}}_\kappa = k_\kappa\boldsymbol{\mathcal{E}}_\kappa\,. \tag{2.181}$$

Using this relation, together with $\omega_\kappa = ck_\kappa$, in eqn (2.35) yields

$$\omega_\kappa\mathcal{E}_{\kappa i}(\mathbf{r}) = ck_\kappa\mathcal{E}_{\kappa i}(\mathbf{r}) = c\sqrt{k_\kappa^2}\mathcal{E}_{\kappa i}(\mathbf{r}) = c\left(-\nabla^2\right)^{1/2}\mathcal{E}_{\kappa i}(\mathbf{r})\,. \tag{2.182}$$

Thus eqn (2.180) can be replaced by

$$\langle 0 |E_i(\mathbf{r}) E_j(\mathbf{r}')| 0\rangle = \frac{\hbar c}{2\epsilon_0}\left(-\nabla^2\right)^{1/2}\sum_\kappa \mathcal{E}_{\kappa i}(\mathbf{r})\mathcal{E}_{\kappa j}(\mathbf{r}')\,, \tag{2.183}$$

which combines with the completeness relation (2.38) to yield

$$\begin{aligned}\langle 0\left|E_i(\mathbf{r})\,E_j(\mathbf{r}')\right|0\rangle &= \frac{\hbar c}{2\epsilon_0}\left(-\nabla^2\right)^{1/2}\Delta^{\perp}_{ij}(\mathbf{r}-\mathbf{r}')\\ &= \frac{\hbar c}{2\epsilon_0}\int\frac{d^3k}{(2\pi)^3}k\left(\delta_{ij}-\frac{k_ik_j}{k^2}\right)e^{i\mathbf{k}\cdot(\mathbf{r}-\mathbf{r}')}\,,\end{aligned} \tag{2.184}$$

where the last line follows from the fact that $e^{i\mathbf{k}\cdot\mathbf{r}}$ is an eigenfunction of $-\nabla^2$ with eigenvalue k^2. Setting $\mathbf{r}'=\mathbf{r}$ and summing over $i=j$ yields the divergent integral

$$\left\langle 0\left|\mathbf{E}^2(\mathbf{r})\right|0\right\rangle = \frac{\hbar}{\epsilon_0}\int\frac{d^3k}{(2\pi)^3}k\,. \tag{2.185}$$

Thus the rms field deviation is infinite at every point $\mathbf{r}$. In the case of the energy this disaster could be avoided by redefining the zero of energy for each cavity mode, but no such escape is possible for measurements of the electric field itself.

This looks a little neater—although no less divergent—if we define the (volume averaged) rms deviation by

$$(\Delta E)^2 = \left\langle 0\left|\frac{1}{V}\int_V d^3r\mathbf{E}^2(\mathbf{r})\right|0\right\rangle. \tag{2.186}$$

This is best calculated by returning to eqn (2.180), setting $\mathbf{r}=\mathbf{r}'$ and integrating to get

$$(\Delta E)^2 = \sum_{\kappa}\mathfrak{e}_{\kappa}^2\,, \tag{2.187}$$

where the **vacuum fluctuation field strength**, $\mathfrak{e}_\kappa$, for mode $\boldsymbol{\mathcal{E}}_\kappa$ is

$$\mathfrak{e}_\kappa = \sqrt{\frac{\hbar\omega_\kappa}{2\epsilon_0 V}}\,. \tag{2.188}$$

The sum over all modes diverges, but the fluctuation strength for a single mode is finite and will play an important role in many of the arguments to follow. A similar calculation for the magnetic field yields

$$(\Delta B)^2 = \sum_{\kappa}\mathfrak{b}_\kappa^2\,,\quad \mathfrak{b}_\kappa = \sqrt{\frac{\mu_0\hbar\omega_\kappa}{2V}}\,. \tag{2.189}$$

The source of the divergence in $(\Delta E)^2$ and $(\Delta B)^2$ is the singular character of the the vacuum fluctuations at a point. This is a mathematical artifact, since any measuring device necessarily occupies a nonzero volume. This suggests considering an operator of the form

$$W \equiv -\int_V d^3r\,\boldsymbol{\mathcal{P}}(\mathbf{r})\cdot\mathbf{E}(\mathbf{r})\,, \tag{2.190}$$

where $\boldsymbol{\mathcal{P}}(\mathbf{r})$ is a smooth (infinitely differentiable) c-number function that vanishes outside some volume $V_0 \ll V$. In this way, the singular behavior of $\mathbf{E}(\mathbf{r})$ is reduced by averaging the point $\mathbf{r}$ over distances of the order $d_0 = V_0^{1/3}$. According to the uncertainty principle, this is equivalent to an upper bound $k_0 \sim 1/d_0$ in the wavenumber, so the divergent integral in eqn (2.185) is replaced by

$$\frac{\hbar}{\epsilon_0}\int_{k<k_0}\frac{d^3k}{(2\pi)^3}k = \frac{\hbar}{\epsilon_0}\frac{k_0^4}{8\pi^2} < \infty \,. \tag{2.191}$$

If the volume V_0 is filled with an *electret*, i.e. a material with permanent electric polarization, then $\boldsymbol{\mathcal{P}}(\mathbf{r})$ can be interpreted as the density of classical dipole moment, and W is the interaction energy between the classical dipoles and the quantized field. In this idealized model W is a well-defined physical quantity which is measurable, at least in principle. Suppose the measurement is carried out repeatedly in the vacuum state. According to the standard rules of quantum theory, the average of these measurements is given by the vacuum expectation value of W, which is zero. Of course, the fact that the average vanishes does not imply that every measured value does. Let us next determine the variance of the measurements by evaluating

$$\left\langle 0\right| W^2 \left|0\right\rangle = \int_V d^3r \int_V d^3r' \mathcal{P}_i(\mathbf{r})\,\mathcal{P}_j(\mathbf{r}')\,\langle 0\,|E_i(\mathbf{r})\,E_j(\mathbf{r}')|\,0\rangle \,. \tag{2.192}$$

Substituting eqn (2.180) into this expression yields

$$\left\langle 0\right| W^2 \left|0\right\rangle = \frac{\hbar}{2\epsilon_0}\sum_{\kappa}\omega_\kappa \mathcal{P}_\kappa^2 \,, \tag{2.193}$$

where

$$\mathcal{P}_\kappa = \int_V d^3r\, \boldsymbol{\mathcal{P}}(\mathbf{r})\cdot \boldsymbol{\mathcal{E}}_\kappa(\mathbf{r}) \tag{2.194}$$

represents the classical interaction energy for a single mode. In this case the sum converges, since the coefficients $\mathcal{P}_\kappa$ will decay rapidly for higher-order modes. Thus W exhibits vacuum fluctuation effects that are both finite and observable. It is important to realize that this result is independent of the choice of operator ordering, e.g. eqn (2.105) or eqn (2.106), for the Hamiltonian. It is also important to assume that the permanent dipole moment of the electret is so small that the radiation it emits by virtue of the acceleration imparted by the vacuum fluctuations can be neglected. In other words, this is a *test* electret analogous to the test charges assumed in the standard formulation of classical electrodynamics (Jackson, 1999, Sec. 1.2).

2.6 The Casimir effect

In Section 2.2 we discarded the zero-point energy due to vacuum fluctuations on the grounds that it could be eliminated by adding a constant to the Hamiltonian in eqn (2.104). This is correct for a single cavity, but the situation changes if two different cavities are compared. In this case, a single shift in the energy spectrum can eliminate one or the other, but not both, of the zero-point energies; therefore, the difference between the zero-point energies of the two cavities can be the basis for observable phenomena. An argument of this kind provides the simplest explanation of the Casimir effect.

We follow the approach of Milonni and Shih (1992) which begins by considering the **planar cavity**—i.e. two plane parallel plates separated by a distance small compared to their lateral dimensions—described in Appendix B.4. In this situation edge effects are small, so the plates can be represented by an ideal cavity in the shape of a rectangular box with dimensions $L \times L \times \Delta z$. This configuration will be compared to a cubical cavity with sides L. The eigenfrequencies for a planar cavity are

$$\omega_{lmn} = c\left[\left(\frac{l\pi}{L}\right)^2 + \left(\frac{m\pi}{L}\right)^2 + \left(\frac{n\pi}{\Delta z}\right)^2\right]^{1/2}, \tag{2.195}$$

where the range of the indices is $l, m, n = 0, 1, 2, \ldots$, except that there are no modes with two zero indices. If one index vanishes, there is only one polarization, but for three nonzero indices there are two. We want to compare the zero-point energies of configurations with different values of Δz, so the interesting quantity is

$$E_0\left(\Delta z\right) = \sum_{l,m,n} C_{lmn} \frac{\hbar\omega_{lmn}}{2}, \tag{2.196}$$

where C_{lmn} is the number of polarizations. Thus $C_{lmn} = 2$ when all indices are nonzero, $C_{lmn} = 1$ when exactly one index vanishes, and $C_{lmn} = 0$ when at least two indices vanish. Since this sum diverges, it is necessary to *regularize* it, i.e. to replace it with a mathematically meaningful expression which has eqn (2.196) as a limiting value. All intermediate calculations are done using the regularized form, and the limit is taken at the end of the calculation. The physical justification for this apparently reckless procedure rests on the fact that real conductors become transparent to radiation at sufficiently high frequencies (Jackson, 1999, Sec. 7.5D). In this range the contribution to the zero-point energy is unchanged by the presence of the conducting plates, so it will cancel out in taking the difference between different configurations. Thus the high-frequency part of the sum in eqn (2.196) is not physically relevant, and a regularization scheme that suppresses the contributions of high frequencies can give a physically meaningful result (Belinfante, 1987).

One regularization scheme is to replace eqn (2.196) by

$$E_0\left(\Delta z\right) = \sum_{l,m,n} \exp\left(-\alpha\omega_{lmn}^2\right) C_{lmn} \frac{\hbar\omega_{lmn}}{2}. \tag{2.197}$$

This sum is well behaved for any $\alpha > 0$, and approaches the original divergent expression as $\alpha \to 0$. The energy in a cubical box with sides L is $E_0\left(L\right)$ and the ratio of the volumes is $L^2\Delta z/L^3 = \Delta z/L$, so the difference between the zero-point energy contained in the planar box and the zero-point energy contained in the same volume in the larger box is

$$U\left(\Delta z\right) = E_0\left(\Delta z\right) - \frac{\Delta z}{L} E_0\left(L\right). \tag{2.198}$$

This is just the work done in bringing one of the faces of the cube from the original distance L to the final distance Δz.

The regularized sum could be evaluated numerically, but it is more instructive to exploit the large size of L. In the limit of large L, the sums over l and m in $E_0(\Delta z)$ and over all three indices in $E_0(L)$ can be replaced by integrals over k-space according to the rule (2.169). After a rather lengthy calculation (Milonni and Shih, 1992) one finds

$$U(\Delta z) = -\frac{\pi^2 \hbar c}{720} \frac{L^2}{\Delta z^3}; \tag{2.199}$$

consequently, the force attracting the two plates is

$$F = -\frac{dU}{d(\Delta z)} = -\frac{\pi^2 \hbar c}{240} \frac{L^2}{\Delta z^4}. \tag{2.200}$$

For numerical estimates it is useful to restate this as

$$F[\mu\text{N}] = -\frac{0.13\, L\,[\text{cm}]^2}{\Delta z\,[\mu\text{m}]^4}. \tag{2.201}$$

For plates with area $1\,\text{cm}^2$ separated by $1\,\mu\text{m}$, the magnitude of the force is $0.13\,\mu\text{N}$. This is a very small force; indeed, it is approximately equal to the force exerted by the proton on the electron in the first Bohr orbit of a hydrogen atom.

The Casimir force between parallel plates would be extremely hard to measure, due to the difficulty of aligning parallel plates separated by $1\,\mu\text{m}$. Recent experiments have used a different configuration consisting of a conducting sphere of radius R at a distance d from a conducting plate (Lamoreaux, 1997; Mohideen and Roy, 1998). For perfect conductors, a similar calculation yields the force

$$F^{(0)}(d) = -\frac{\pi^3 \hbar c}{360} \frac{R}{d^3} \tag{2.202}$$

in the limit $R \gg d$. When corrections for finite conductivity, surface roughness, and nonzero temperature are included there is good agreement between theory and experiment.

The calculation of the Casimir force sketched above is based on the difference between the zero-point energies of two cavities, and it provides good agreement between theory and experiment. This might be interpreted as providing evidence for the reality of zero-point energy, except for two difficulties. The first is the general argument in Section 2.2 showing that it is always permissible to use the normal-ordered form (2.89) for the Hamiltonian. With this choice, there is no zero-point energy for either cavity; and our successful explanation evaporates. The second, and more important, difficulty is that the forces predicted by eqns (2.200) and (2.202) are independent of the electronic charge. There is clearly something wrong with this, since all dynamical effects depend on the interaction of charged particles with the electromagnetic field. It has been shown that the second feature is an artifact of the assumption that the plates are perfect conductors (Jaffe, 2005). A less idealized calculation yields a Casimir force that properly vanishes in the limit of zero electronic charge. Thus the agreement between the theoretical prediction (2.202) and experiment cannot be interpreted as evidence for the physical reality of zero-point energy. We emphasize that this does not mean that vacuum fluctuations are not real, since other experiments—such as the partition noise at beam splitters discussed in Section 8.4.2—do provide evidence for their effects.

Our freedom to use the normal-ordered form of the Hamiltonian implies that it must be possible to derive the Casimir force without appealing to the zero point energy. An approach that does this is based on the van der Waals coupling between atoms in different walls. The van der Waals potential can be derived by considering the coupling between the fluctuating dipoles of two atoms. This produces a time-averaged perturbation proportional to $(\mathbf{p}_1 \cdot \mathbf{p}_2)/r^3$, where r is the distance between the atoms, and $\mathbf{p}_1$ and $\mathbf{p}_2$ are the electric dipole operators. This potential comes from the static Coulomb interactions between the charged particles comprising the atoms; it does not involve the radiative modes that contribute to the zero point energy in symmetrical ordering. The random fluctuations in the dipole moments $\mathbf{p}_1$ and $\mathbf{p}_2$ produce no first-order correction to the energy, but in second order the dipole–dipole coupling produces the van der Waals potential $V_W(r)$ with its characteristic $1/r^7$ dependence. The $1/r^7$ dependence is valid for $r \gg \lambda_{\text{at}}$, where λ_{at} is a characteristic wavelength of an atomic transition. For $r \ll \lambda_{\text{at}}$ the potential varies as $1/r^6$.

For many atoms, the simplest assumption is that these potentials are pair-wise additive, i.e. the total potential energy is

$$V_{\text{tot}} = \sum_{m \neq n} V_W\left(\left|\mathbf{r}_n - \mathbf{r}'_m\right|\right), \tag{2.203}$$

where the sum runs over all pairs with one atom in each wall. With this approximation, it is possible to explain about 80% of the Casimir force in eqn (2.200). In fact the assumption of pair-wise additivity is not justified, since the presence of a third atom changes the interaction between the first two. When this is properly taken into account, the entire Casimir force is obtained.

Thus there are two different explanations for the Casimir force, corresponding to the two choices $a^\dagger a$ or $\left(a^\dagger a + a a^\dagger\right)/2$ made in defining the electromagnetic Hamiltonian. The important point to keep in mind is that the relevant physical prediction—the Casimir force between the plates—is the same for both explanations. The difference between the two lies entirely in the language used to describe the situation. This kind of ambiguity in description is often found in quantum physics. Another example is the van der Waals potential itself. The explanation given above corresponds to the normal ordering of the electromagnetic Hamiltonian. If the symmetric ordering is used instead, the presence of the two atoms induces a change in the zero-point energy of the field which becomes increasingly negative as the distance between the atoms decreases. The result is the same attractive potential between the atoms (Milonni and Shih, 1992).

2.7 Exercises

2.1 Cavity equations

(1) Give the separation of variables argument leading to eqn (2.7).

(2) Derive the equations satisfied by $\boldsymbol{\mathcal{E}}(\mathbf{r})$ and $\boldsymbol{\mathcal{B}}(\mathbf{r})$ and verify eqns (2.9) and (2.10).

2.2 Rectangular cavity modes

(1) Use the method of separation of variables to solve eqns (2.11) and (2.1) for a rectangular cavity, subject to the boundary condition (2.13), and thus verify eqns (2.14)–(2.17).

(2) Show explicitly that the modes satisfy the orthogonality conditions

$$\int d^3r \boldsymbol{\mathcal{E}}_{\mathbf{k}s}(\mathbf{r}) \cdot \boldsymbol{\mathcal{E}}_{\mathbf{k}'s'}(\mathbf{r}) = 0 \quad \text{for} \quad (\mathbf{k}, s) \neq (\mathbf{k}', s') .$$

(3) Use the normalization condition

$$\int d^3r \boldsymbol{\mathcal{E}}_{\mathbf{k}s}(\mathbf{r}) \cdot \boldsymbol{\mathcal{E}}_{\mathbf{k}s}(\mathbf{r}) = 1$$

to derive the normalization constants $N_{\mathbf{k}}$.

2.3 Equations of motion for classical radiation oscillators

In the interior of an empty cavity the fields satisfy Maxwell's equations (2.1)–(2.4). Use the expansions (2.40) and (2.41) and the properties of the mode functions to derive eqn (2.43).

2.4 Complex mode amplitudes

(1) Use the expression (2.48) for the classical energy and the expansions (2.40) and (2.41) to derive eqn (2.49).

(2) Derive eqns (2.46) and (2.51).

2.5 Number states

Use the commutation relations (2.76) and the definition (2.73) of the vacuum state to verify eqn (2.78).

2.6 The second-order coherence matrix

(1) For the operators $\mathfrak{a}_p^\dagger$ and $\mathfrak{a}_p$ $(p = 1, 2)$ defined by eqn (2.153) show that the number operators $\mathfrak{N}_p = \mathfrak{a}_p^\dagger \mathfrak{a}_p$ are simultaneously measurable.

(2) Consider the operator

$$\rho = \frac{1}{2} |1_x\rangle \langle 1_x| + \frac{1}{2} |1_y\rangle \langle 1_y| - \frac{1}{4} \left(|1_y\rangle \langle 1_x| + |1_x\rangle \langle 1_y| \right),$$

where $|1_s\rangle = a_{\mathbf{k}s}^\dagger |0\rangle$.

(a) Show that ρ is a genuine density operator, i.e. it is positive and has unit trace.

(b) Calculate the coherence matrix J, its eigenvalues and eigenvectors, and the degree of polarization.

2.7 The Stokes parameters

(1) What is the physical significance of S_0?

(2) Use the explicit forms of the Pauli matrices and the expansion (2.158) to show that

$$\det J = \frac{1}{4}\left[S_0^2 - S_1^2 - S_2^2 - S_3^2\right],$$

and thereby establish the condition (2.160).

(3) With $S_0 = 1$, introduce polar coordinates by $S_3 = \cos\theta$, $S_2 = \sin\theta\sin\phi$, and $S_1 = \sin\theta\cos\phi$. Find the locations on the Poincaré sphere corresponding to right circular polarization, left circular polarization, and linear polarization.

2.8 A one-photon mixed state

Consider a monochromatic state for wavevector $\mathbf{k}$ (see Section 2.4.1-A) containing exactly one photon.

(1) Explain why the density operator for this state is completely represented by the 2×2 matrix $\rho_{ss'} = \langle 1_{\mathbf{k}s} |\rho| 1_{\mathbf{k}s'} \rangle$.

(2) Show that the density matrix $\rho_{ss'}$ is related to the coherence matrix J by $\rho_{ss'} = J_{s's}$.

2.9 The Casimir force

Show that the large L limit of eqn (2.198) is

$$\begin{aligned} U(\Delta z) = {} & \frac{\hbar c}{2}\left(\frac{L}{\pi}\right)^2 \int dk_x dk_y e^{-\alpha k_\perp^2} k_\perp \\ & + \hbar c \left(\frac{L}{\pi}\right)^2 \sum_{n=1}^{\infty} \int dk_x dk_y e^{-\alpha\left(k_\perp^2 + k_{zn}^2\right)} \left[k_\perp^2 + k_{zn}^2\right]^{1/2} \\ & - \frac{\Delta z}{L} \hbar c \left(\frac{L}{\pi}\right)^3 \int dk_x dk_y dk_z e^{-\alpha k^2} k\,, \end{aligned}$$

where $k_\perp = \sqrt{k_x^2 + k_y^2}$, $k = \sqrt{k_\perp^2 + k_z^2}$, and $k_{zn} = n\pi/\Delta z$.

2.10 Model for the experiments on the Casimir force

Consider the simple-harmonic-oscillator model of the Lamoreaux and Mohideen-Roy experiments on the Casimir force shown in Fig. 2.1.

All elements of the apparatus, which are assumed to be perfect conductors, are rendered electrically neutral by grounding them to the Earth. Assume that the spring constant for the metallic spring is k. (You may ignore Earth's gravity in this problem.)

(1) Calculate the displacement Δx of the spring from its relaxed length as a function of the spacing d between the surface of the sphere and the flat plate on the right, after the system has come into mechanical equilibrium.

(2) Calculate the natural oscillation frequency of this system for small disturbances around this equilibrium as a function of d. Neglect all dissipative losses.

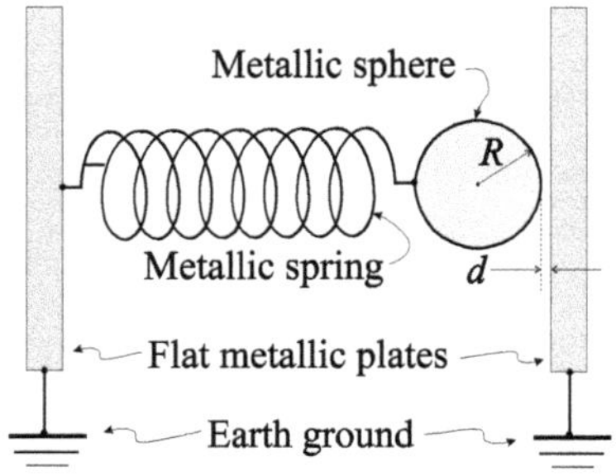

Fig. 2.1 The Casimir force between a grounded metallic sphere of radius R and the grounded flat metallic plate on the right, which is separated by a distance d from the sphere, can be measured by measuring the displacement of the metallic spring. (Ignore gravity.)

(3) Plot your answers for parts 1 and 2 for the following numerical parameters:

$$R = 200\,\mu\text{m}\,,$$
$$0.1\,\mu\text{m} \leqslant d \leqslant 1.0\,\mu\text{m}\,,$$
$$k = 0.02\,\text{N/m}\,.$$

3
Field quantization

Quantizing the radiation oscillators associated with the classical modes of the electromagnetic field in a cavity provides a satisfactory theory of the Planck distribution and the Casimir effect, but this is only the beginning of the story. There are, after all, quite a few experiments that involve photons propagating freely through space, not just bouncing back and forth between cavity walls. In addition to this objection, there is a serious flaw in the cavity-based model. The quantized radiation oscillators are defined in terms of a set of classical mode functions satisfying the idealized boundary conditions for perfectly conducting walls. This difficulty cannot be overcome by simply allowing for finite conductivity, since conductivity is itself a macroscopic property that does not account for the atomistic structure of physical walls. Thus the quantization conjecture (2.61) builds the idealized macroscopic boundary conditions into the foundations of the microscopic quantum theory of light. A fundamental microscopic theory should not depend on macroscopic idealizations, so there is more work to be done. We should emphasize, however, that this objection to the cavity model does not disqualify it as a guide toward an improved theory. The cavity model itself was constructed by applying the ideas of nonrelativistic quantum mechanics to the classical radiation oscillators. In a similar fashion, we will use the cavity model to suggest a true microscopic conjecture for the quantization of the electromagnetic field.

In the following sections we will show how the quantization scheme of the cavity model can be used to suggest local commutation relations for quantized fields in free space. The experimentally essential description of photons in passive optical devices will be addressed by formulating a simple model for the quantization of the field in a dielectric medium. In the final four sections we will discuss some more advanced topics: the angular momentum of light, a description of quantum field theory in terms of wave packets, and the question of the spatial localizability of photons.

3.1 Field quantization in the vacuum

The quantization of the electromagnetic field in free space is most commonly carried out in the language of **canonical quantization** (Cohen-Tannoudji *et al.*, 1989, Sec. II.A.2), which is based on the Lagrangian formulation of classical electrodynamics. This is a very elegant way of packaging the necessary physical conjectures, but it requires extra mathematical machinery that is not needed for most applications. We will pursue a more pedestrian route which builds on the quantization rules for the ideal physical cavity. To this end, we initially return to the cavity problem.

3.1.1 Local commutation relations

In Chapter 2 we concentrated on the operators (q_κ, p_κ) for a single mode. Since the modes are determined by the boundary conditions at the cavity walls, they describe global properties of the cavity. We now want turn attention away from the overall properties of the cavity, in order to concentrate on the local properties of the field operators. We will do this by combining the expansions (2.99) and (2.103) for the time-independent, Schrödinger-picture operators $\mathbf{E}(\mathbf{r})$ and $\mathbf{A}(\mathbf{r})$ with the commutation relations (2.61) for the mode operators to calculate the commutators between field components evaluated at different points in space. The expansions show that $\mathbf{E}(\mathbf{r})$ only depends on the p_κs while $\mathbf{A}(\mathbf{r})$ and $\mathbf{B}(\mathbf{r})$ depend only on the q_κs; therefore, the commutation relations, $[p_\kappa, p_\lambda] = [q_\kappa, q_\lambda] = 0$, produce

$$[E_j(\mathbf{r}), E_k(\mathbf{r}')] = 0, \quad [A_j(\mathbf{r}), A_k(\mathbf{r}')] = 0, \quad [B_j(\mathbf{r}), B_k(\mathbf{r}')] = 0. \tag{3.1}$$

On the other hand, $[q_\kappa, p_\lambda] = i\hbar\delta_{\kappa\lambda}$, so the commutator between the electric field and the vector potential is

$$\begin{aligned}[A_i(\mathbf{r}), -E_j(\mathbf{r}')] &= \frac{1}{\epsilon_0}\sum_\kappa\sum_\lambda [q_\kappa, p_\lambda]\,\mathcal{E}_{\kappa i}(\mathbf{r})\,\mathcal{E}_{\lambda j}(\mathbf{r}') \\ &= \frac{i\hbar}{\epsilon_0}\sum_\kappa \mathcal{E}_{\kappa i}(\mathbf{r})\,\mathcal{E}_{\kappa j}(\mathbf{r}').\end{aligned} \tag{3.2}$$

For any cavity, the mode functions satisfy the completeness condition (2.38), so we see that

$$[A_i(\mathbf{r}), -E_j(\mathbf{r}')] = \frac{i\hbar}{\epsilon_0}\Delta_{ij}^{\perp}(\mathbf{r}-\mathbf{r}'). \tag{3.3}$$

The resemblance between this result and the canonical commutation relation, $[q_\kappa, p_\lambda] = i\hbar\delta_{\kappa\lambda}$, for the mode operators suggests the identification of $\mathbf{A}(\mathbf{r})$ and $-\mathbf{E}(\mathbf{r})$ as the canonical variables for the field in position space. A similar calculation for the commutator between the **E**- and **B**-fields can be carried out using eqn (2.100), or by applying the *curl* operation to eqn (3.3), with the result

$$[B_i(\mathbf{r}), E_j(\mathbf{r}')] = i\frac{\hbar}{\epsilon_0}\epsilon_{ijl}\nabla_l\delta(\mathbf{r}-\mathbf{r}'), \tag{3.4}$$

where ϵ_{ijl} is the alternating tensor defined by eqn (A.3). The uncertainty relations implied by the nonvanishing commutators between electric and magnetic field components were extensively studied in the classic work of Bohr and Rosenfeld (1950), and a simple example can be found in Exercise 3.2.

The derivation of the local commutation relations (3.1) and (3.3) for the field operators in the physical cavity employs the complete set of cavity modes, which depend on the geometry of the cavity. This can be seen from the explicit appearance of the mode functions in the second line of eqn (3.2). However, the final result (3.3) follows from the completeness relation (2.38), which has the same form for every cavity. This feature only depends on the fact that the boundary conditions guarantee the Hermiticity of the operator $-\nabla^2$. We have, therefore, established the quite remarkable result that the local position-space commutation relations are independent of the shape and size of the cavity. In particular, eqns (3.1) and (3.3) will hold in the limit of an infinitely large physical cavity; that is, when the distance to the cavity walls from either of the points $\mathbf{r}$ and $\mathbf{r}'$ is much greater than any physically relevant length scale. In this limit, it is plausible to assume that the boundary conditions at the walls are irrelevant. This suggests abandoning the original quantization conjecture (2.61), and replacing it by eqns (3.1) and (3.3). In this way we obtain a microscopic theory which does not involve the macroscopic idealizations associated with the classical boundary conditions. We emphasize that this is not a *derivation* of the local commutation relations from the physical cavity relations (2.61). The sole function of the cavity-based calculation is to suggest the form of eqns (3.1) and (3.3), which constitute an independent quantization conjecture. As always, the validity of the this conjecture has to be tested by means of experiment. In this new approach, the theory based on the ideal physical cavity—with its dependence on macroscopic boundary conditions—is demoted to a phenomenological model.

Since the new quantization rules hold everywhere in space, they can be expressed in terms of Fourier transform pairs defined by

$$\mathbf{F}(\mathbf{r}) = \int \frac{d^3k}{(2\pi)^3} e^{i\mathbf{k}\cdot\mathbf{r}} \mathbf{F}(\mathbf{k}), \quad \mathbf{F}(\mathbf{k}) = \int d^3r e^{-i\mathbf{k}\cdot\mathbf{r}} \mathbf{F}(\mathbf{r}), \tag{3.5}$$

where $\mathbf{F} = \mathbf{A}$, $\mathbf{E}$, or $\mathbf{B}$. The position-space field operators are hermitian, so their Fourier transforms satisfy $\mathbf{F}^{\dagger}(\mathbf{k}) = \mathbf{F}(-\mathbf{k})$. It should be clearly understood that eqn (3.5) is simply an application of the Fourier transform; no additional physical assumptions are required. By contrast, the expansions (2.99) and (2.103) in cavity modes involve the idealized boundary conditions at the cavity walls.

Transforming eqns (3.1) and (3.3) with respect to $\mathbf{r}$ and $\mathbf{r}'$ independently yields the equivalent relations

$$[E_j(\mathbf{k}), E_k(\mathbf{k}')] = [A_j(\mathbf{k}), A_k(\mathbf{k}')] = 0, \tag{3.6}$$

and

$$[A_i(\mathbf{k}), -E_j(\mathbf{k}')] = \frac{i\hbar}{\epsilon_0} \Delta_{ij}^{\perp}(\mathbf{k}) (2\pi)^3 \delta(\mathbf{k} + \mathbf{k}'), \tag{3.7}$$

where the delta function comes from using the identity (A.96).

3.1.2 Creation and annihilation operators

A Position space

The commutation relations (3.1)–(3.4) are not the only general consequences that are implied by the cavity model. For example, the expansions (2.101) and (2.103) can be rewritten as

$$\mathbf{E}(\mathbf{r}) = \mathbf{E}^{(+)}(\mathbf{r}) + \mathbf{E}^{(-)}(\mathbf{r}), \quad \mathbf{A}(\mathbf{r}) = \mathbf{A}^{(+)}(\mathbf{r}) + \mathbf{A}^{(-)}(\mathbf{r}), \tag{3.8}$$

where

$$\mathbf{A}^{(+)}(\mathbf{r}) = \sum_{\kappa} \sqrt{\frac{\hbar}{2\epsilon_0 \omega_\kappa}} a_\kappa \boldsymbol{\mathcal{E}}_\kappa(\mathbf{r}) = \mathbf{A}^{(-)\dagger}(\mathbf{r}) \tag{3.9}$$

and

$$\mathbf{E}^{(+)}(\mathbf{r}) = i\sum_{\kappa}\sqrt{\frac{\hbar\omega_{\kappa}}{2\epsilon_0}}a_{\kappa}\boldsymbol{\mathcal{E}}_{\kappa}(\mathbf{r}) = \mathbf{E}^{(-)\dagger}(\mathbf{r}). \tag{3.10}$$

Let F be one of the field operators, A_i or E_i, then $F^{(+)}$ is called the positive-frequency part and $F^{(-)}$ is called the negative-frequency part. The origin of these mysterious names will become clear in Section 3.2.3, but for the moment we only need to keep in mind that $F^{(+)}$ is a sum of annihilation operators and $F^{(-)}$ is a sum of creation operators. These properties are expressed by

$$F^{(+)}(\mathbf{r})|0\rangle = 0\,,\quad \langle 0|F^{(-)}(\mathbf{r}) = 0\,. \tag{3.11}$$

In view of the definition (3.9) there is a natural inclination to think of $\mathbf{A}^{(+)}(\mathbf{r})$ as an operator that annihilates a photon at the point $\mathbf{r}$, but this temptation must be resisted. The difficulty is that the photon—i.e. 'a quantum of excitation of the electromagnetic field'—cannot be sharply localized in space. A precise interpretation for $\mathbf{A}^{(+)}(\mathbf{r})$ is presented in Section 3.5.2, and the question of photon localization is studied in Section 3.6.

An immediate consequence of eqns (3.9) and (3.10) is that

$$\left[F^{(\pm)}(\mathbf{r}), G^{(\pm)}(\mathbf{r}')\right] = 0\,, \tag{3.12}$$

where F and G are any pair of field operators. It is clear, however, that $\left[F^{(+)}, G^{(-)}\right]$ will not always vanish. In particular, a calculation similar to the one leading to eqn (3.3) yields

$$\left[A_i^{(+)}(\mathbf{r}), -E_j^{(-)}(\mathbf{r}')\right] = \frac{i\hbar}{2\epsilon_0}\Delta_{ij}^{\perp}(\mathbf{r}-\mathbf{r}'). \tag{3.13}$$

The decomposition (3.8) also allows us to express all field operators in terms of $\mathbf{A}^{(\pm)}$. For this purpose, we rewrite eqn (3.10) as

$$\mathbf{E}^{(+)}(\mathbf{r}) = ic\sum_{\kappa}\sqrt{\frac{\hbar}{2\epsilon_0\omega_{\kappa}}}a_{\kappa}k_{\kappa}\boldsymbol{\mathcal{E}}_{\kappa}(\mathbf{r})\,, \tag{3.14}$$

and use eqn (2.181) to get the final form

$$\mathbf{E}^{(+)}(\mathbf{r}) = ic\left(-\nabla^2\right)^{1/2}\mathbf{A}^{(+)}(\mathbf{r}). \tag{3.15}$$

Substituting this into eqn (3.13) yields the equivalent commutation relations

$$\left[A_i^{(+)}(\mathbf{r}), A_j^{(-)}(\mathbf{r}')\right] = \frac{\hbar}{2\epsilon_0 c}\left(-\nabla^2\right)^{-1/2}\Delta_{ij}^{\perp}(\mathbf{r}-\mathbf{r}')\,, \tag{3.16}$$

$$\left[E_i^{(+)}(\mathbf{r}), E_j^{(-)}(\mathbf{r}')\right] = \frac{\hbar c}{2\epsilon_0}\left(-\nabla^2\right)^{1/2}\Delta_{ij}^{\perp}(\mathbf{r}-\mathbf{r}'). \tag{3.17}$$

In the context of free space, the unfamiliar operators $\left(-\nabla^2\right)^{1/2}$ and $\left(-\nabla^2\right)^{-1/2}$ are best defined by means of Fourier transforms. For any real function $f(u)$ the identity

$-\nabla^2 \exp(i\mathbf{k}\cdot\mathbf{r}) = k^2 \exp(i\mathbf{k}\cdot\mathbf{r})$ allows us to define the action of $f\left(-\nabla^2\right)$ on a plane wave by $f\left(-\nabla^2\right) e^{i\mathbf{k}\cdot\mathbf{r}} \equiv f\left(k^2\right) e^{i\mathbf{k}\cdot\mathbf{r}}$. This result in turn implies that $f\left(-\nabla^2\right)$ acts on a general function $\varphi(\mathbf{r})$ according to the rule

$$f\left(-\nabla^2\right)\varphi(\mathbf{r}) \equiv \int \frac{d^3k}{(2\pi)^3}\varphi(\mathbf{k})\, f\left(-\nabla^2\right) e^{i\mathbf{k}\cdot\mathbf{r}} = \int \frac{d^3k}{(2\pi)^3}\varphi(\mathbf{k})\, f\left(k^2\right) e^{i\mathbf{k}\cdot\mathbf{r}}. \tag{3.18}$$

After using the inverse Fourier transform on $\varphi(\mathbf{k})$ this becomes

$$f\left(-\nabla^2\right)\varphi(\mathbf{r}) = \int d^3r' \left(\mathbf{r}\left|f\left(-\nabla^2\right)\right|\mathbf{r}'\right)\varphi(\mathbf{r}'), \tag{3.19}$$

where

$$\left(\mathbf{r}\left|f\left(-\nabla^2\right)\right|\mathbf{r}'\right) = \int \frac{d^3k}{(2\pi)^3} f\left(k^2\right) e^{i\mathbf{k}\cdot(\mathbf{r}-\mathbf{r}')} \tag{3.20}$$

is the integral kernel defining $f\left(-\nabla^2\right)$ as an operator in $\mathbf{r}$-space. Despite its abstract appearance, this definition is really just a labor saving device; it avoids transforming back and forth from position space to reciprocal space. For example, real functions of the hermitian operator $-\nabla^2$ are also hermitian; so one gets a useful *integration-by-parts* identity

$$\int d^3r\psi^*(\mathbf{r})\, f\left(-\nabla^2\right)\varphi(\mathbf{r}) = \int d^3r \left\{f\left(-\nabla^2\right)\psi^*(\mathbf{r})\right\}\varphi(\mathbf{r}), \tag{3.21}$$

without any intermediate steps involving Fourier transforms.

The equations (3.8), (3.11)–(3.13), (3.15), and (3.16) were all derived by using the expansions of the field operators in cavity modes, but once again the final forms are independent of the size and shape of the cavity. Consequently, these results are valid in free space.

B *Reciprocal space*

The rather strange looking result (3.16) becomes more understandable if we note that the decomposition (3.8) into positive- and negative-frequency parts applies equally well in reciprocal space, so that $\mathbf{A}(\mathbf{k}) = \mathbf{A}^{(+)}(\mathbf{k}) + \mathbf{A}^{(-)}(\mathbf{k})$. The Fourier transforms of eqns (3.12) and (3.16) with respect to $\mathbf{r}$ and $\mathbf{r}'$ yield respectively

$$\left[A_i^{(\pm)}(\mathbf{k}), A_j^{(\pm)}(\mathbf{k}')\right] = 0 \tag{3.22}$$

and

$$\left[A_i^{(+)}(\mathbf{k}), A_j^{(-)}(-\mathbf{k}')\right] = \frac{\hbar}{2\epsilon_0 c}\frac{\Delta_{ij}^{\perp}(\mathbf{k})}{k}(2\pi)^3\delta\left(\mathbf{k}-\mathbf{k}'\right). \tag{3.23}$$

This reciprocal-space commutation relation does not involve any strange operators, like $\left(-\nabla^2\right)^{-1/2}$, but it is still rather complicated. Simplification can be achieved by noting that the circular polarization unit vectors $\mathbf{e}_{\mathbf{k}s}$—see Appendix B.3.2—are eigenvectors of $\Delta_{ij}^{\perp}(\mathbf{k})$ with eigenvalue unity:

$$\Delta_{ij}^{\perp}(\mathbf{k})(\mathbf{e}_{\mathbf{k}s})_j = (\mathbf{e}_{\mathbf{k}s})_i \,. \tag{3.24}$$

By forming the inner product of both sides of eqns (3.22) and (3.23) with $\mathbf{e}_{\mathbf{k}s}^*$ and $\mathbf{e}_{\mathbf{k}'s'}$ and remembering $\mathbf{F}^\dagger(\mathbf{k}) = \mathbf{F}(-\mathbf{k})$, one finds

$$[a_s(\mathbf{k}), a_{s'}(\mathbf{k}')] = \left[a_s^\dagger(\mathbf{k}), a_{s'}^\dagger(\mathbf{k}')\right] = 0 \tag{3.25}$$

and

$$\left[a_s(\mathbf{k}), a_{s'}^\dagger(\mathbf{k}')\right] = \delta_{ss'}(2\pi)^3\delta\left(\mathbf{k}-\mathbf{k}'\right), \tag{3.26}$$

where

$$a_s(\mathbf{k}) = \sqrt{\frac{2\epsilon_0\omega_k}{\hbar}}\mathbf{e}_{\mathbf{k}s}^* \cdot \mathbf{A}^{(+)}(\mathbf{k}) \tag{3.27}$$

and $\omega_k = ck$. The operators $a_s(\mathbf{k})$, combined with the Fourier transform relation (3.5), provide a replacement for the cavity-mode expansions (3.9) and (3.10):

$$\mathbf{A}^{(+)}(\mathbf{r}) = \int \frac{d^3k}{(2\pi)^3}\sqrt{\frac{\hbar}{2\epsilon_0\omega_k}}\sum_s a_s(\mathbf{k})\,\mathbf{e}_{\mathbf{k}s}e^{i\mathbf{k}\cdot\mathbf{r}}, \tag{3.28}$$

$$\mathbf{E}^{(+)}(\mathbf{r}) = i\int \frac{d^3k}{(2\pi)^3}\sqrt{\frac{\hbar\omega_k}{2\epsilon_0}}\sum_s a_s(\mathbf{k})\,\mathbf{e}_{\mathbf{k}s}e^{i\mathbf{k}\cdot\mathbf{r}}. \tag{3.29}$$

The number operator

$$N = \int \frac{d^3k}{(2\pi)^3}\sum_s a_s^\dagger(\mathbf{k})\,a_s(\mathbf{k}) \tag{3.30}$$

satisfies

$$\left[N, a_s^\dagger(\mathbf{k})\right] = a_s^\dagger(\mathbf{k}), \quad [N, a_s(\mathbf{k})] = -a_s(\mathbf{k}), \tag{3.31}$$

and the vacuum state is defined by $a_s(\mathbf{k})|0\rangle = 0$, so it seems that $a_s^\dagger(\mathbf{k})$ and $a_s(\mathbf{k})$ can be regarded as creation and annihilation operators that replace the cavity operators a_κ and $a_\kappa^\dagger$. However, the singular commutation relation (3.26) exacts a price. For example, the one-photon state $|1_{\mathbf{k}s}\rangle = a_s^\dagger(\mathbf{k})|0\rangle$ is an improper state vector satisfying the continuum normalization conditions

$$\langle 1_{\mathbf{k}'s'}|1_{\mathbf{k}s}\rangle = \delta_{ss'}(2\pi)^3\delta\left(\mathbf{k}-\mathbf{k}'\right). \tag{3.32}$$

Thus a properly normalized one-photon state is a wave packet state

$$|\Phi\rangle = \int \frac{d^3k}{(2\pi)^3}\sum_s \Phi_s(\mathbf{k})\,a_s^\dagger(\mathbf{k})|0\rangle, \tag{3.33}$$

where the c-number function $\Phi_s(\mathbf{k})$ is normalized by

$$\int \frac{d^3k}{(2\pi)^3} \sum_s |\Phi_s(\mathbf{k})|^2 = 1 . \tag{3.34}$$

The Fock space $\mathfrak{H}_F$ consists of all linear combinations of number states,

$$|\Phi\rangle = \int \frac{d^3k_1}{(2\pi)^3} \cdots \int \frac{d^3k_n}{(2\pi)^3} \sum_{s_1} \cdots \sum_{s_n} \Phi_{s_1 \cdots s_n}(\mathbf{k}_1, \ldots, \mathbf{k}_n) a^\dagger_{s_1}(\mathbf{k}_1) \cdots a^\dagger_{s_n}(\mathbf{k}_n) |0\rangle , \tag{3.35}$$

where

$$\int \frac{d^3k_1}{(2\pi)^3} \cdots \int \frac{d^3k_n}{(2\pi)^3} \sum_{s_1} \cdots \sum_{s_n} |\Phi_{s_1 \cdots s_n}(\mathbf{k}_1, \ldots, \mathbf{k}_n)|^2 < \infty \tag{3.36}$$

and $n = 0, 1, \ldots$.

3.1.3 Energy, momentum, and angular momentum

A The Hamiltonian

The expression (2.105) for the field energy in a cavity can be converted to a form suitable for generalization to free space by first inverting eqn (3.10) to get

$$a_\kappa = -i\sqrt{\frac{2\epsilon_0}{\hbar\omega_\kappa}} \int_V d^3r \boldsymbol{\mathcal{E}}_\kappa(\mathbf{r}) \cdot \mathbf{E}^{(+)}(\mathbf{r}) . \tag{3.37}$$

The next step is to substitute this expression for a_κ into eqn (2.105) and carry out the sum over κ by means of the completeness relation (2.38); this calculation leads to

$$\begin{aligned} H_{\text{em}} &= \sum_\kappa \hbar\omega_\kappa a^\dagger_\kappa a_\kappa \\ &= 2\epsilon_0 \int_V d^3r \int_V d^3r' E_i^{(-)}(\mathbf{r}) \Delta^\perp_{ij}(\mathbf{r} - \mathbf{r}') E_j^{(+)}(\mathbf{r}') . \end{aligned} \tag{3.38}$$

Since the free-field operator $\mathbf{E}^{(+)}(\mathbf{r}')$ is transverse, the infinite volume limit is

$$H_{\text{em}} = 2\epsilon_0 \int d^3r \mathbf{E}^{(-)}(\mathbf{r}) \cdot \mathbf{E}^{(+)}(\mathbf{r}) . \tag{3.39}$$

This can also be expressed as

$$H_{\text{em}} = 2\epsilon_0 c^2 \int d^3r \mathbf{A}^{(-)}(\mathbf{r}) \cdot \left(-\nabla^2\right) \mathbf{A}^{(+)}(\mathbf{r}) , \tag{3.40}$$

by using eqn (3.15). A more intuitively appealing form is obtained by using the plane-wave expansion (3.29) for $\mathbf{E}^{(\pm)}$ to get

$$H_{\text{em}} = \int \frac{d^3k}{(2\pi)^3} \hbar\omega_k \sum_s a^\dagger_s(\mathbf{k}) a_s(\mathbf{k}) . \tag{3.41}$$

B The linear momentum

The cavity model does not provide any expressions for the linear momentum and the angular momentum, so we need independent arguments for them. The reason for the absence of these operators is the presence of the cavity walls. From a mechanical point of view, the linear momentum and the angular momentum of the field are not conserved because of the immovable cavity. Alternatively, we note that one of the fundamental features of quantum theory is the identification of the linear momentum and the angular momentum operators with the generators for spatial translations and rotations respectively (Bransden and Joachain, 1989, Secs 5.9 and 6.2). This means that the mechanical conservation laws for linear and angular momentum are equivalent to invariance under spatial translations and rotations respectively. The location and orientation of the cavity in space spoils both invariances.

Since the cavity model fails to provide any guidance, we once again call on the correspondence principle by quoting the classical expression for the linear momentum (Jackson, 1999, Sec. 6.7):

$$\begin{aligned}\boldsymbol{\mathcal{P}} &= \int d^3r\; \epsilon_0 \boldsymbol{\mathcal{E}}_\perp \times \boldsymbol{\mathcal{B}} \\ &= \int d^3r\; \epsilon_0 \boldsymbol{\mathcal{E}} \times (\boldsymbol{\nabla} \times \boldsymbol{\mathcal{A}})\,. \end{aligned} \tag{3.42}$$

The vector identity $\mathbf{F}\times(\boldsymbol{\nabla}\times\mathbf{G}) = F_j \boldsymbol{\nabla} G_j - (\mathbf{F}\cdot\boldsymbol{\nabla})\,\mathbf{G}$ combined with an integration by parts and the transverse nature of $\boldsymbol{\mathcal{E}}(\mathbf{r})$ provides the more useful expression

$$\boldsymbol{\mathcal{P}} = \epsilon_0 \int d^3r \mathcal{E}_j(\mathbf{r})\, \nabla \mathcal{A}_j(\mathbf{r})\,. \tag{3.43}$$

The initial step in constructing the corresponding Schrödinger-picture operator is to replace the classical fields according to

$$\boldsymbol{\mathcal{A}}(\mathbf{r}) \to \mathbf{A}(\mathbf{r}) = \mathbf{A}^{(+)}(\mathbf{r}) + \mathbf{A}^{(-)}(\mathbf{r})\,, \tag{3.44}$$

$$\boldsymbol{\mathcal{E}}(\mathbf{r}) \to \mathbf{E}(\mathbf{r}) = \mathbf{E}^{(+)}(\mathbf{r}) + \mathbf{E}^{(-)}(\mathbf{r})\,. \tag{3.45}$$

The momentum operator $\mathbf{P}$ is then the sum of four terms, $\mathbf{P} = \mathbf{P}^{(+,+)} + \mathbf{P}^{(-,-)} + \mathbf{P}^{(-,+)} + \mathbf{P}^{(+,-)}$, where

$$\mathbf{P}^{(\sigma,\tau)} = \epsilon_0 \int d^3r E_j^{(\sigma)} \boldsymbol{\nabla} A_j^{(\tau)} \quad \text{for } \sigma,\tau = \pm\,. \tag{3.46}$$

Each of these terms is evaluated by using the plane-wave expansions (3.28) and (3.29), together with the orthogonality relation, $\mathbf{e}^*_{\mathbf{k}s}\cdot\mathbf{e}_{\mathbf{k}s'} = \delta_{ss'}$, and the reflection property—see eqn (B.73)—$\mathbf{e}_{-\mathbf{k},s} = \mathbf{e}^*_{\mathbf{k}s}$, $(s = \pm)$ for the circular polarization basis. The first result is $\mathbf{P}^{(+,+)} = \mathbf{P}^{(-,-)\dagger} = 0$; consequently, only the cross terms survive to give

$$\mathbf{P} = \int \frac{d^3k}{(2\pi)^3} \frac{\hbar\mathbf{k}}{2} \sum_s \left\{ a_s^\dagger(\mathbf{k})\, a_s(\mathbf{k}) + a_s(\mathbf{k})\, a_s^\dagger(\mathbf{k}) \right\}. \tag{3.47}$$

This is analogous to the symmetrical ordering (2.106) for the Hamiltonian in the cavity problem, so our previous experience suggests replacing the symmetrical ordering by normal ordering, i.e.

$$\mathbf{P} = \int \frac{d^3k}{(2\pi)^3} \hbar \mathbf{k} \sum_s a_s^\dagger(\mathbf{k})\, a_s(\mathbf{k}) \,. \tag{3.48}$$

From this expression and eqn (3.41), it is easy to see that $[\mathbf{P}, H_{\text{em}}] = 0$ and $[P_i, P_j] = 0$. Any observable commuting with the Hamiltonian is called a **constant of the motion**, so the total momentum is a constant of the motion and the individual components P_i are simultaneously measurable.

By using the inverse Fourier transform,

$$a_s(\mathbf{k}) = \sqrt{\frac{2\epsilon_0\omega_k}{\hbar}} \mathbf{e}_{\mathbf{k}s}^* \cdot \int d^3r e^{-i\mathbf{k}\cdot\mathbf{r}} \mathbf{A}^{(+)}(\mathbf{r}) \,, \tag{3.49}$$

which is the free-space replacement for eqn (3.37), or proceeding directly from eqn (3.46), one finds the equivalent position-space representation

$$\mathbf{P} = 2\epsilon_0 \int d^3r E_j^{(-)}(\mathbf{r})\, \nabla A_j^{(+)}(\mathbf{r}) \,. \tag{3.50}$$

C The angular momentum

Finally we turn to the classical expression for the angular momentum (Jackson, 1999, Sec. 12.10):

$$\boldsymbol{\mathcal{J}} = \int d^3r\, \mathbf{r} \times [\epsilon_0 \boldsymbol{\mathcal{E}}(\mathbf{r},t) \times \boldsymbol{\mathcal{B}}(\mathbf{r},t)] \,. \tag{3.51}$$

Combining $\boldsymbol{\mathcal{B}} = \nabla \times \boldsymbol{\mathcal{A}}$ with the identity $\mathbf{F} \times (\nabla \times \mathbf{G}) = F_j \nabla G_j - (\mathbf{F} \cdot \nabla)\,\mathbf{G}$ allows this to be written in the form $\boldsymbol{\mathcal{J}} = \boldsymbol{\mathcal{L}} + \boldsymbol{\mathcal{S}}$, where

$$\boldsymbol{\mathcal{L}} = \epsilon_0 \int d^3r\, \mathcal{E}_j\, (\mathbf{r} \times \nabla)\, \mathcal{A}_j \tag{3.52}$$

and

$$\boldsymbol{\mathcal{S}} = \epsilon_0 \int d^3r\, \boldsymbol{\mathcal{E}} \times \boldsymbol{\mathcal{A}} \,. \tag{3.53}$$

Once again, the initial guess for the corresponding quantum operators is given by applying the rules (3.44) and (3.45), so the total angular momentum operator is

$$\mathbf{J} = \mathbf{L} + \mathbf{S} \,, \tag{3.54}$$

where the operators $\mathbf{L}$ and $\mathbf{S}$ are defined by quantizing the classical expressions $\boldsymbol{\mathcal{L}}$ and $\boldsymbol{\mathcal{S}}$ respectively.

The application of the method used for the linear momentum to eqn (3.52) is complicated by the explicit **r**-term, but after some effort one finds the rather cumbersome expression (Simmons and Guttmann, 1970)

$$\begin{aligned}\mathbf{L} &= -\frac{i\hbar}{2}\int\frac{d^3k}{(2\pi)^3}\left[M_i^\dagger(\mathbf{k})\left(\mathbf{k}\times\frac{\partial}{\partial\mathbf{k}}\right)M_i(\mathbf{k}) - \mathrm{HC}\right] \\ &= -i\hbar\int\frac{d^3k}{(2\pi)^3}M_i^\dagger(\mathbf{k})\left(\mathbf{k}\times\frac{\partial}{\partial\mathbf{k}}\right)M_i(\mathbf{k}),\end{aligned} \tag{3.55}$$

where

$$\mathbf{M}(\mathbf{k}) = \sum_s a_s(\mathbf{k})\,\mathbf{e}_{\mathbf{k}s}\,. \tag{3.56}$$

In this case a substantial simplification results from translating the reciprocal-space representation back into position space to get

$$\mathbf{L} = \frac{2i\epsilon_0}{\hbar}\int d^3r E_j^{(-)}(\mathbf{r})\left(\mathbf{r}\times\frac{\hbar}{i}\nabla\right)A_j^{(+)}(\mathbf{r})\,. \tag{3.57}$$

A straightforward calculation using eqn (3.39) shows that **L** is also a constant of the motion, i.e. $[\mathbf{L}, H_{\mathrm{em}}] = 0$. However, the components of **L** are not mutually commutative, so they cannot be measured simultaneously.

The quantization of eqn (3.53) goes much more smoothly, and leads to the normal-ordered expression

$$\begin{aligned}\mathbf{S} &= \hbar\int\frac{d^3k}{(2\pi)^3}\widetilde{\mathbf{k}}\sum_s s a_s^\dagger(\mathbf{k})\,a_s(\mathbf{k}) \\ &= \hbar\int\frac{d^3k}{(2\pi)^3}\widetilde{\mathbf{k}}\left[a_+^\dagger(\mathbf{k})\,a_+(\mathbf{k}) - a_-^\dagger(\mathbf{k})\,a_-(\mathbf{k})\right],\end{aligned} \tag{3.58}$$

where $\widetilde{\mathbf{k}} = \mathbf{k}/k$ is the unit vector along **k**. Another use of eqn (3.49) yields the equivalent position-space form

$$\mathbf{S} = 2\epsilon_0\int d^3r\mathbf{E}^{(-)}\times\mathbf{A}^{(+)}\,. \tag{3.59}$$

The expression (3.54) for the total angular momentum operator looks like the decomposition into orbital and spin parts familiar from quantum mechanics, but this resemblance is misleading. For the electromagnetic field, the interpretation of eqn (3.54) poses a subtle problem which we will take up in Section 3.4.

D The helicity operator

It is easy to show that **S** commutes with **P** and with H_{em}, and further that

$$[S_i, S_j] = 0\,. \tag{3.60}$$

Thus **S**, **P**, and H_{em} are simultaneously measurable, and there are simultaneous eigenvectors for them. In the simplest case of the improper one-photon state $|1_{\mathbf{k}s}\rangle =$

$a_s^\dagger(\mathbf{k})|0\rangle$, one finds: $H_{\text{em}}|1_{\mathbf{k}s}\rangle = \hbar\omega_k|1_{\mathbf{k}s}\rangle$, $\mathbf{P}|1_{\mathbf{k}s}\rangle = \hbar\mathbf{k}|1_{\mathbf{k}s}\rangle$, $\widetilde{\mathbf{k}}\times\mathbf{S}|1_{\mathbf{k}s}\rangle = 0$, and $\widetilde{\mathbf{k}}\cdot\mathbf{S}|1_{\mathbf{k}s}\rangle = s\hbar|1_{\mathbf{k}s}\rangle$. Thus $|1_{\mathbf{k}s}\rangle$ is an eigenvector of the longitudinal component $\widetilde{\mathbf{k}}\cdot\mathbf{S}$ with eigenvalue $s\hbar$ and an eigenvector of the transverse components $\widetilde{\mathbf{k}}\times\mathbf{S}$ with eigenvalue 0. For the circular polarization basis, the index s represents the helicity, so $\mathbf{S}$ is called the **helicity operator**.

E Evidence for helicity and orbital angular momentum

Despite the conceptual difficulties mentioned in Section 3.1.3-C, it is possible to devise experiments in which certain components of the helicity $\mathbf{S}$ and the orbital angular momentum $\mathbf{L}$ are separately observed. The first measurement of this kind (Beth, 1936) was carried out using an experimental arrangement consisting of a horizontal wave plate suspended at its center by a torsion fiber, so that the plate is free to undergo twisting motions around the vertical axis. In a simplified version of this experiment, a vertically-directed, linearly-polarized beam of light is allowed to pass through a quarter-wave plate, which transforms it into a circularly-polarized beam of light (Born and Wolf, 1980, Sec. 14.4.2). Since the experimental setup is symmetrical under rotations around the vertical axis (the z-axis), the z-component of the total angular momentum will be conserved.

We will use a one-photon state

$$|\psi\rangle = \sum_s \xi_s |1_{\mathbf{k}s}\rangle = \sum_s \xi_s a_s^\dagger(\mathbf{k})|0\rangle\,, \tag{3.61}$$

with $\mathbf{k} = k\mathbf{u}_3$ directed along the z-axis, as a simple model of an incident light beam of arbitrary polarization. A straightforward calculation using eqn (3.55) for L_z shows that $L_z|1_{\mathbf{k}s}\rangle = 0$; consequently, $L_z|\psi\rangle = 0$ for any choice of the coefficients ξ_s. In other words, states of this kind have no z-component of orbital angular momentum. The particular choice

$$|\psi\rangle_{\text{lin}} = \frac{1}{\sqrt{2}}\left[|1_{\mathbf{k}+}\rangle + |1_{\mathbf{k}-}\rangle\right] = \frac{1}{\sqrt{2}}\left[a_+^\dagger(\mathbf{k}) + a_-^\dagger(\mathbf{k})\right]|0\rangle \tag{3.62}$$

defines a linearly-polarized state which possesses zero helicity, i.e. $S_z|\psi\rangle_{\text{lin}} = 0$. Due to the action of the quarter-wave plate, the incident linearly-polarized light is converted into circularly-polarized light. Thus the input state $|\psi\rangle_{\text{lin}}$ changes into the output state $|\psi\rangle_{\text{cir}} = |1_{\mathbf{k},s=+}\rangle$. The output state $|\psi\rangle_{\text{cir}}$ has helicity $S_z = +\hbar$, but it still satisfies $L_z|\psi\rangle_{\text{cir}} = 0$. Since the transmitted photon carries away one unit $(+\hbar)$ of angular momentum, conservation of angular momentum requires the plate to acquire one unit $(-\hbar)$ of angular momentum in the opposite direction. In the classical limit of a steady stream of linearly-polarized photons, this process is described by saying that the light beam exerts a torque on the plate: $\tau_z = dS_z/dt = \dot{N}(-\hbar)$, where $\dot{N}$ is the rate of flow of photons through the plate. The resulting twist of the torsion fiber can be sensitively measured by means of a small mirror attached to the fiber.

The original experiment actually used a steady stream of light composed of very many photons, so a classical description would be entirely adequate. However, if the sensitivity of the experiment were to be improved to a point where fluctuations in the angular position of the wave plate could be measured, then the discrete nature of the angular momentum transfer of $\hbar$ per photon to the wave plate would show up. The transfer of angular momentum from an individual photon to the wave plate must in principle be discontinuous in nature, and the twisting of the wave plate should manifest a fine, ratchet-like Brownian motion. The experiment to see such fluctuations—which would be very difficult—has not been performed.

A more modern experiment to demonstrate the spin angular momentum of light was performed by trapping a small, absorbing bead within the beam waist of a tightly focused Gaussian laser beam (Friese *et al.*, 1998). The procedure for trapping a small particle inside the beam waist of a laser beam has been called an *optical tweezer*, since one can then move the particle around at will by displacing the axis of the light beam. The accompanying procedure for producing arbitrary angular displacements of a trapped particle by transferring controllable amounts of angular momentum from the light to the particle has been called an *optical torque wrench* (Ashkin, 1980). For linearly-polarized light, no effect is observed, but switching the incident laser beam to circular polarization causes the trapped bead to begin spinning around the axis defined by the direction of propagation of the light beam. In classical terms, this behavior is a result of the torque exerted on the particle by the absorbed light. From the quantum point of view absorption of each photon deposits $\hbar$ of angular momentum in the bead; therefore, the bead has to spin up in order to conserve angular momentum.

Observations of the orbital angular momentum, L_z, of light have also been made using a similar technique (He *et al.*, 1995). The experiment begins with a linearly-polarized laser beam in a Gaussian TEM_{00} mode. This beam—which has zero helicity and zero orbital angular momentum—then passes through a computer-generated holographic mask with a spiral pattern imprinted onto it. The linearly-polarized, paraxial, Gaussian beam is thereby transformed into a linearly-polarized, paraxial Laguerre–Gaussian beam of light (Siegman, 1986, Sec. 16.4). The output beam possesses orbital, but no spin, angular momentum. A simple Laguerre–Gaussian mode is one in which the light effectively orbits around the axis of propagation as if in an optical vortex with a given sense of circulation. The transverse intensity profile is doughnut-shaped, with a null at its center marking a phase singularity in the beam. In principle, the spiral holographic mask would experience a torque resulting from the transfer of orbital angular momentum—one unit ($+\hbar$) per photon—to the light beam from the mask. However, this experiment has not been performed.

What has been observed is that a small, absorbing bead trapped at the beam waist of a Laguerre–Gaussian mode—with nonzero orbital angular momentum—begins to spin. This spinning motion is due to the steady transfer of orbital angular momentum from the light beam into the bead by absorption. The resultant torque is given by $\tau_z = dL_z/dt = \dot{N}\,(-\hbar)$, where $\dot{N}$ is the rate of photon flow through the bead. Again, there is a completely classical description of this experiment, so the photon nature of light need not be invoked.

Just as for the spin-transfer experiments, a sufficiently sensitive version of this experiment, using a small enough bead, would display the discontinuous transfer of orbital angular momentum in the form of a fine, ratchet-like Brownian motion in the angular displacement of the bead. This would be analogous to the discontinuous transfer of linear momentum due to impact of atoms on a pollen particle that results in the random linear displacements of the particle seen in Brownian motion. This experiment has also not been performed.

3.1.4 Box quantization

The local, position-space commutation relations (3.1) and (3.3)—or the equivalent reciprocal-space versions (3.25) and (3.26)—do not require any idealized boundary conditions, but the right sides of eqns (3.3) and (3.26) contain singular functions that cause mathematical problems, e.g. the improper one-photon state $|1_{\mathbf{k}s}\rangle$. On the other hand, the cavity mode operators a_κ and $a_\kappa^\dagger$—which do depend on idealized boundary conditions—have discrete labels and the one-photon states $|1_\kappa\rangle = a_\kappa^\dagger|0\rangle$ are normalizable. As usual, we would prefer to have the best of both worlds; and this can be accomplished—at least formally—by replacing the Fourier integral in (3.5) with a Fourier series. This is done by pretending that all fields are contained in a finite volume V, usually a cube of side L, and imposing periodic boundary conditions at the walls, as explained in Appendix A.4.2. This is called **box quantization**. Since this imaginary cavity is not defined by material walls, the periodic boundary conditions have no physical significance. Consequently, meaningful results are only obtained in the limit of infinite volume. Thus box quantization is a mathematical trick; it is not a physical idealization, as in the physical cavity problem.

The mathematical situation resulting from this trick is almost identical to that of the ideal physical cavity. For this case, the traveling waves, $\mathbf{f}_{\mathbf{k}s}(\mathbf{r}) = \mathbf{e}_{\mathbf{k}s}\exp(i\mathbf{k}\cdot\mathbf{r})/\sqrt{V}$, play the role of the cavity modes. The periodic boundary conditions impose $\mathbf{k} = 2\pi\mathbf{n}/L$, where $\mathbf{n}$ is a vector with integer components. The $f_{\mathbf{k}s}$s are an orthonormal set of modes, i.e.

$$(\mathbf{f}_{\mathbf{k}s}, \mathbf{f}_{\mathbf{k}'s'}) = \int_V d^3r\, \mathbf{f}^*_{\mathbf{k}s}(\mathbf{r}) \cdot \mathbf{f}_{\mathbf{k}'s'}(\mathbf{r}) = \delta_{\mathbf{k}\mathbf{k}'}\delta_{ss'}\,. \tag{3.63}$$

The various expressions for the commutation relations, the field operators, and the observables can be derived either by replacing the real cavity mode functions in Chapter 2 by the complex modes $f_{\mathbf{k}s}(\mathbf{r})$, or by applying the rules relating Fourier integrals to Fourier series, i.e.

$$\int \frac{d^3k}{(2\pi)^3} \leftrightarrow \frac{1}{V}\sum_{\mathbf{k}} \quad \text{and} \quad a_s(\mathbf{k}) \leftrightarrow \sqrt{V} a_{\mathbf{k}s}\,, \tag{3.64}$$

to the expressions obtained in Sections 3.1.1–3.1.3. In either way, the commutation relations and the number operator are given by

$$\left[a_{\mathbf{k}s}, a^\dagger_{\mathbf{k}'s'}\right] = \delta_{\mathbf{k}\mathbf{k}'}\delta_{ss'}\,, \quad [a_{\mathbf{k}s}, a_{\mathbf{k}'s'}] = 0\,, \quad N = \sum_{\mathbf{k}s} a^\dagger_{\mathbf{k}s} a_{\mathbf{k}s}\,. \tag{3.65}$$

The number states are defined just as for the physical cavity,

$$|\underline{n}\rangle = \prod_{\mathbf{k}s} \frac{\left(a^\dagger_{\mathbf{k}s}\right)^{n_{\mathbf{k}s}}}{\sqrt{n_{\mathbf{k}s}!}}|0\rangle\,, \tag{3.66}$$

where $\underline{n} = \{n_{\mathbf{k}s}\}$ is the set of occupation numbers, and the completeness relation is

$$\sum_{\underline{n}} |\underline{n}\rangle \langle \underline{n}| = 1\,. \tag{3.67}$$

Thus the box-quantization scheme replaces the delta function in eqn (3.26) by the ordinary Kronecker symbol in the discrete indices $\mathbf{k}$ and s. Consequently, the box-quantized operators $a_{\mathbf{k}s}$ are as well behaved mathematically as the physical cavity operators a_κ. This allows the construction of the Fock space to be carried out in parallel to Chapter 2.1.2-C.

The expansions for the field operators are

$$\mathbf{A}^{(+)}(\mathbf{r}) = \sum_{\mathbf{k}s} \sqrt{\frac{\hbar}{2\epsilon_0 \omega_k V}} a_{\mathbf{k}s} \mathbf{e}_{\mathbf{k}s} e^{i\mathbf{k}\cdot\mathbf{r}}\,, \tag{3.68}$$

$$\mathbf{E}^{(+)}(\mathbf{r}) = \sum_{\mathbf{k}s} i \sqrt{\frac{\hbar\omega_k}{2\epsilon_0 V}} a_{\mathbf{k}s} \mathbf{e}_{\mathbf{k}s} e^{i\mathbf{k}\cdot\mathbf{r}}\,, \tag{3.69}$$

and

$$\mathbf{B}^{(+)}(\mathbf{r}) = \sum_{\mathbf{k}s} \sqrt{\frac{\hbar k}{2\epsilon_0 c V}} s a_{\mathbf{k}s} \mathbf{e}_{\mathbf{k}s} e^{i\mathbf{k}\cdot\mathbf{r}}\,, \tag{3.70}$$

where the expansion for $\mathbf{B}^{(+)}$ was obtained by using $\mathbf{B} = \boldsymbol{\nabla} \times \mathbf{A}$ and the special property (B.52) of the circular polarization basis.

The Hamiltonian, the momentum, and the helicity are respectively given by

$$H_{\text{em}} = \sum_{\mathbf{k}s} \hbar\omega_k a_{\mathbf{k}s}^\dagger a_{\mathbf{k}s}\,, \tag{3.71}$$

$$\mathbf{P} = \sum_{\mathbf{k}s} \hbar\mathbf{k} a_{\mathbf{k}s}^\dagger a_{\mathbf{k}s}\,, \tag{3.72}$$

and

$$\mathbf{S} = \hbar \sum_{\mathbf{k}s} \widetilde{\mathbf{k}} s a_{\mathbf{k}s}^\dagger a_{\mathbf{k}s}\,. \tag{3.73}$$

As always, these achievements have a price. One part of this price is that physically meaningful results are only obtained in the limit $V \to \infty$. This is not a particularly onerous requirement, since getting the correct limit is simply a matter of careful algebra combined with the rules in eqn (3.64). A more serious issue is the absence of the total angular momentum from the list of observables in eqns (3.71)–(3.73). One way of understanding the problem here is that the expression (3.55) for $\mathbf{L}$ contains the differential operator $\partial/\partial\mathbf{k}$ which creates difficulties in converting the continuous integral over $\mathbf{k}$ into a discrete sum. The alternative expression (3.57) for $\mathbf{L}$ does not involve $\mathbf{k}$, so it might seem to offer a solution. This hope also fails, since the $\mathbf{r}$-integral in this representation must now be carried out over the imaginary cube V. The edges of the cube define preferred directions in space, so there is no satisfactory way to define the orbital angular momentum $\mathbf{L}$.

3.2 The Heisenberg picture

The quantization rules in Chapter 2 and Section 3.1.1 are both expressed in the Schrödinger picture: observables are represented by time-independent hermitian operators $X^{(S)}$, and the state of the radiation field is described by a ket vector $\left|\Psi^{(S)}(t)\right\rangle$, obeying the Schrödinger equation

$$i\hbar\frac{\partial}{\partial t}\left|\Psi^{(S)}(t)\right\rangle = H^{(S)}\left|\Psi^{(S)}(t)\right\rangle , \tag{3.74}$$

or by a density operator $\rho^{(S)}(t)$, obeying the quantum Liouville equation (2.119)

$$i\hbar\frac{\partial}{\partial t}\rho^{(S)}(t) = \left[H^{(S)}, \rho^{(S)}(t)\right] . \tag{3.75}$$

The superscript (S) has been added in order to distinguish the Schrödinger picture from two other descriptions that are frequently used. Note that the density operator is an exception to the rule that Schrödinger-picture observables are independent of time.

There is an alternative description of quantum mechanics which actually preceded the familiar Schrödinger picture. In Heisenberg's original formulation—which appeared one year before Schrödinger's—there is no mention of a wave function or a wave equation; instead, the observables are represented by infinite matrices that evolve in time according to a quantum version of Hamilton's equations of classical mechanics. This form of quantum theory is called the **Heisenberg picture**; the physical equivalence of the two pictures was subsequently established by Schrödinger. The Heisenberg picture is particularly useful in quantum optics, especially for the calculation of correlations between measurements at different times. A third representation—called the interaction picture—will be presented in Section 4.8. It will prove useful for the formulation of time-dependent perturbation theory in Section 4.8.1. The interaction picture also provides the foundation for the resonant wave approximation, which is introduced in Section 11.1.

In the following sections we will study the properties of the Schrödinger and Heisenberg pictures and the relations between them. In order to distinguish between the same quantities viewed in different pictures, the states and operators will be decorated with superscripts (S) or (H) for the Schrödinger or Heisenberg pictures respectively. In applications of these ideas the superscripts are usually dropped, and the distinctions are—one hopes—made clear from context.

The Heisenberg picture is characterized by two features: (1) the states are independent of time; (2) the observables depend on time. Imposing the superposition principle on the Heisenberg picture implies that the relation between the time-dependent, Schrödinger-picture state vector $\left|\Psi^{(S)}(t)\right\rangle$ and the corresponding time-independent, Heisenberg-picture state $\left|\Psi^{(H)}\right\rangle$ must be linear. If we impose the convention that the two pictures coincide at some time $t = t_0$, then there is a linear operator $U(t - t_0)$ such that

$$\left|\Psi^{(S)}(t)\right\rangle = U(t - t_0)\left|\Psi^{(H)}\right\rangle . \tag{3.76}$$

The identity of the pictures at $t = t_0$, $\left|\Psi^{(H)}\right\rangle = \left|\Psi^{(S)}(t_0)\right\rangle$, is enforced by the initial condition $U(0) = 1$. Substituting eqn (3.76) into the Schrödinger equation (3.74) yields the differential equation

$$i\hbar\frac{\partial}{\partial t}U(t-t_0)=H^{(S)}U(t-t_0)\,,\quad U(0)=1 \tag{3.77}$$

for the operator $U(t-t_0)$. This has the solution (Bransden and Joachain, 1989, Sec. 5.7)

$$U(t-t_0)=\exp\left[-\frac{i}{\hbar}(t-t_0)\,H^{(S)}\right], \tag{3.78}$$

where the **evolution operator** on the right side is defined by the power series for the exponential, or by the general rules outlined in Appendix C.3.6. The Hermiticity of $H^{(S)}$ guarantees that $U(t-t_0)$ is unitary, i.e.

$$U(t-t_0)\,U^{\dagger}(t-t_0)=U^{\dagger}(t-t_0)\,U(t-t_0)=1\,. \tag{3.79}$$

The choice of t_0 is dictated by convenience for the problem at hand. In most cases it is conventional to set $t_0=0$, but in scattering problems it is sometimes more useful to consider the limit $t_0\to-\infty$. The evolution operator satisfies the group property,

$$U(t_1-t_2)\,U(t_2-t_3)=U(t_1-t_3)\,, \tag{3.80}$$

which simply states that evolution from t_3 to t_2 followed by evolution from t_2 to t_1 is the same as evolving directly from t_3 to t_1. For the special choice $t_0=0$, this simplifies to $U(t_1)\,U(t_2)=U(t_1+t_2)$. The definition (3.78) also shows that $U(-t)=U^{\dagger}(t)$. In what follows, we will generally use the convention $t_0=0$; any other choice of initial time will be introduced explicitly.

The physical equivalence of the two pictures is enforced by requiring that each Schrödinger-picture operator $X^{(S)}$ and the corresponding Heisenberg-picture operator $X^{(H)}(t)$ have the same expectation values in corresponding states:

$$\left\langle\Psi^{(H)}\left|X^{(H)}(t)\right|\Psi^{(H)}\right\rangle=\left\langle\Psi^{(S)}(t)\left|X^{(S)}\right|\Psi^{(S)}(t)\right\rangle, \tag{3.81}$$

for all vectors $\left|\Psi^{(S)}(t)\right\rangle$ and observables $X^{(S)}$. Using eqn (3.76) allows this relation to be written as

$$\left\langle\Psi^{(H)}\left|X^{(H)}(t)\right|\Psi^{(H)}\right\rangle=\left\langle\Psi^{(H)}\left|U^{\dagger}(t)\,X^{(S)}U(t)\right|\Psi^{(H)}\right\rangle. \tag{3.82}$$

Since this equation holds for all states, the general result (C.15) shows that the operators in the two pictures are related by

$$X^{(H)}(t)=U^{\dagger}(t)\,X^{(S)}U(t)\,. \tag{3.83}$$

Note that the Heisenberg-picture operators agree with the (time-independent) Schrödinger-picture operators at $t=0$. This definition, together with the group property $U(t_1)\,U(t_2)=U(t_1+t_2)$, provides a useful relation between the Heisenberg operators at different times:

$$\begin{aligned}X^{(H)}(t+\tau)&=U^{\dagger}(t+\tau)\,X^{(S)}U(t+\tau)\\&=U^{\dagger}(\tau)\,U^{\dagger}(t)\,X^{(S)}U(t)\,U(\tau)\\&=U^{\dagger}(\tau)\,X^{(H)}(t)\,U(\tau)\,.\end{aligned} \tag{3.84}$$

Also note that $H^{(S)}$ commutes with $\exp\left[\pm itH^{(S)}/\hbar\right]$, so eqn (3.83) implies that the Hamiltonian is the same in both pictures: $H^{(H)}(t)=H^{(S)}=H$.

In the Heisenberg picture, the operators evolve in time while the state vectors are fixed. The density operator is again an exception. Applying the transformation (3.83) to the definition of the Schrödinger-picture density operator,

$$\rho^{(S)}(t) = \sum_u \mathcal{P}_u \left|\Theta_u^{(S)}(t)\right\rangle \left\langle \Theta_u^{(S)}(t)\right|, \tag{3.85}$$

yields the time-independent operator

$$\rho^{(H)} = \sum_u \mathcal{P}_u \left|\Theta_u^{(H)}\right\rangle \left\langle \Theta_u^{(H)}\right| = \rho^{(S)}(0), \tag{3.86}$$

which is the initial value for the quantum Liouville equation (3.75).

A differential equation describing the time evolution of operators in the Heisenberg picture is obtained by combining eqn (3.77) with the common form of the Hamiltonian to get

$$\begin{aligned}\frac{\partial X^{(H)}(t)}{\partial t} &= \frac{i}{\hbar} U^{\dagger}(t)\left[H, X^{(S)}\right] U(t) \\ &= \frac{i}{\hbar}\left[H, X^{(H)}(t)\right],\end{aligned} \tag{3.87}$$

where the last line follows from the identity

$$\begin{aligned}U^{\dagger}(t) X^{(S)} Y^{(S)} U(t) &= U^{\dagger}(t) X^{(S)} U(t) U^{\dagger}(t) Y^{(S)} U(t) \\ &= X^{(H)}(t) Y^{(H)}(t).\end{aligned} \tag{3.88}$$

Multiplying eqn (3.87) by $i\hbar$ yields the **Heisenberg equation of motion** for the observable $X^{(H)}$:

$$i\hbar \frac{\partial X^{(H)}(t)}{\partial t} = \left[X^{(H)}(t), H\right]. \tag{3.89}$$

The definition (3.83) provides a solution for this equation. The name 'constant of the motion' for operators $X^{(S)}$ that commute with the Hamiltonian is justified by the observation that the Heisenberg equation for $X^{(H)}(t)$ is $(\partial/\partial t) X^{(H)}(t) = 0$.

In most applications we will suppress the identifying superscripts (H) and (S). The distinctions between the Heisenberg and Schrödinger pictures will be maintained by the convention that an operator with a time argument, e.g. $X(t)$, is the Heisenberg-picture form, while X—with no time argument—signifies the Schrödinger-picture form. The only real danger of this convention is that density operators behave in the opposite way; $\rho(t)$ denotes a Schrödinger-picture operator, while ρ is taken in the Heisenberg picture. This is not a serious problem if the accompanying text provides the appropriate clues.

3.2.1 Equal-time commutators

A pair of Schrödinger-picture operators X and Y is said to be canonically conjugate if $[X, Y] = \beta$, where β is a c-number. Canonically conjugate pairs, e.g. position and momentum, play an important role in quantum theory, so it is useful to consider the commutator in the Heisenberg picture. Evaluating $[X(t), Y(t')]$ for $t \neq t'$ requires a complete solution of the Heisenberg equations for $X(t)$ and $Y(t')$, but the **equal-time commutator** for such a canonically conjugate pair is given by

$$\begin{aligned}[X(t), Y(t)] &= \left[U^\dagger(t) X U(t), U^\dagger(t) Y U(t)\right] \\ &= U^\dagger(t)[X, Y] U(t) \\ &= \beta\,. \end{aligned} \tag{3.90}$$

Thus the equal-time commutator of the Heisenberg-picture operators is identical to the commutator of the Schrödinger-picture operators. Applying this to the position-space commutation relation (3.3) and to the canonical commutator (3.65) yields

$$[A_i(\mathbf{r}, t), -E_j(\mathbf{r}', t)] = \frac{i\hbar}{\epsilon_0}\Delta_{ij}^{\perp}(\mathbf{r} - \mathbf{r}') \tag{3.91}$$

and

$$\left[a_{\mathbf{k}s}(t), a_{\mathbf{k}'s'}^\dagger(t)\right] = \delta_{ss'}\delta_{\mathbf{k}\mathbf{k}'}\,, \tag{3.92}$$

respectively.

3.2.2 Heisenberg equations for the free field

The preceding arguments are valid for any form of the Hamiltonian, but the results are particularly useful for free fields. The Heisenberg-picture form of the box-quantized Hamiltonian is

$$H_{\text{em}} = \sum_{\mathbf{k}s} \hbar\omega_k a_{\mathbf{k}s}^\dagger(t)\, a_{\mathbf{k}s}(t)\,, \tag{3.93}$$

and eqn (3.89), together with the equal-time versions of eqn (3.65), yields the equation of motion for the annihilation operators

$$\left[i\frac{d}{dt} - \omega_k\right] a_{\mathbf{k}s}(t) = 0\,. \tag{3.94}$$

The solution is

$$a_{\mathbf{k}s}(t) = a_{\mathbf{k}s} e^{-i\omega_k t} = e^{iH_{\text{em}}t/\hbar} a_{\mathbf{k}s} e^{-iH_{\text{em}}t/\hbar}\,, \tag{3.95}$$

where we have used the identification of $a_{\mathbf{k}s}(0)$ with the Schrödinger-picture operator $a_{\mathbf{k}s}$. Combining this solution with the expansion (3.68) gives

$$\mathbf{A}^{(+)}(\mathbf{r}, t) = \sum_{\mathbf{k}s} \sqrt{\frac{\hbar}{2\epsilon_0\omega_k V}}\, a_{\mathbf{k}s}\mathbf{e}_{\mathbf{k}s} e^{i(\mathbf{k}\cdot\mathbf{r} - \omega_k t)}\,. \tag{3.96}$$

The expansions (3.69) and (3.70) allow the operators $\mathbf{E}^{(+)}(\mathbf{r}, t)$ and $\mathbf{B}^{(+)}(\mathbf{r}, t)$ to be expressed in the same way.

3.2.3 Positive- and negative-frequency parts

We are now in a position to explain the terms positive-frequency part and negative-frequency part introduced in Section 3.1.2. For this purpose it is useful to review some features of Fourier transforms. For any real function $F(t)$, the Fourier transform satisfies $F^*(\omega) = F(-\omega)$. Thus $F(\omega)$ for negative frequencies is completely determined by $F(\omega)$ for positive frequencies. Let us use this fact to rewrite the inverse transform as

$$F(t) = \int_{-\infty}^{\infty} \frac{d\omega}{2\pi} F(\omega) e^{-i\omega t} = F^{(+)}(t) + F^{(-)}(t), \tag{3.97}$$

where the **positive-frequency part**,

$$F^{(+)}(t) = \int_{0}^{\infty} \frac{d\omega}{2\pi} F(\omega) e^{-i\omega t}, \tag{3.98}$$

and the **negative-frequency part**,

$$F^{(-)}(t) = \int_{-\infty}^{0} \frac{d\omega}{2\pi} F(\omega) e^{-i\omega t}, \tag{3.99}$$

are related by

$$F^{(-)}(t) = F^{(+)*}(t). \tag{3.100}$$

The definitions of $F^{(\pm)}(t)$ guarantee that $F^{(+)}(\omega)$ vanishes for $\omega < 0$ and $F^{(-)}(\omega)$ vanishes for $\omega > 0$.

The division into positive- and negative-frequency parts works equally well for any time-dependent hermitian operator, $X(t)$. One simply replaces complex conjugation by the adjoint operation; i.e. eqn (3.100) becomes $X^{(-)}(t) = X^{(+)\dagger}(t)$. In particular, the temporal Fourier transform of the operator $\mathbf{A}^{(+)}(\mathbf{r}, t)$, defined by eqn (3.96), is

$$\mathbf{A}^{(+)}(\mathbf{r}, \omega) = \int dt\, e^{i\omega t} \mathbf{A}^{(+)}(\mathbf{r}, t) = \sum_{\mathbf{k}s} \sqrt{\frac{\hbar}{2\epsilon_0 \omega_k V}} a_{\mathbf{k}s} \mathbf{e}_{\mathbf{k}s} e^{i\mathbf{k}\cdot\mathbf{r}} 2\pi\delta(\omega - \omega_k). \tag{3.101}$$

Since $\omega_k = c|\mathbf{k}| > 0$, $\mathbf{A}^{(+)}(\mathbf{r}, \omega)$ vanishes for $\omega < 0$, and $\mathbf{A}^{(-)}(\mathbf{r}, \omega) = \mathbf{A}^{(+)\dagger}(\mathbf{r}, -\omega)$ vanishes for $\omega > 0$. Thus the Schrödinger-picture definition (3.68) of the positive-frequency part agrees with the Heisenberg-picture definition at $t = 0$.

The commutation rules derived in Section 3.1.2 are valid here for equal-time commutators, but for free fields we also have the unequal-times commutators:

$$\left[F^{(\pm)}(\mathbf{r}, t), G^{(\pm)}(\mathbf{r}', t')\right] = 0, \tag{3.102}$$

provided only that $F^{(\pm)}(\mathbf{r}, 0)$ and $G^{(\pm)}(\mathbf{r}', 0)$ are sums over annihilation (creation) operators.

3.3 Field quantization in passive linear media

Optical devices such as lenses, mirrors, prisms, beam splitters, etc. are the main tools of experimental optics. In classical optics these devices are characterized by their bulk optical properties, such as the index of refraction. In order to apply the same simple descriptions to quantum optics, we need to extend the theory of photon propagation in vacuum to propagation in dielectrics. We begin by considering classical fields in passive, linear dielectrics—which we will always assume are nonmagnetic—and then present a phenomenological model for quantization.

3.3.1 Classical fields in linear dielectrics

A review of the electromagnetic properties of linear media can be found in Appendix B.5.1, but for the present discussion we only need to recall that the constitutive relations for a nonmagnetic, dielectric medium are $\boldsymbol{\mathcal{H}}(\mathbf{r},t) = \boldsymbol{\mathcal{B}}(\mathbf{r},t)/\mu_0$ and

$$\boldsymbol{\mathcal{D}}(\mathbf{r},t) = \epsilon_0 \boldsymbol{\mathcal{E}}(\mathbf{r},t) + \boldsymbol{\mathcal{P}}(\mathbf{r},t) \,. \tag{3.103}$$

For an isotropic, homogeneous medium that does not exhibit spatial dispersion (see Appendix B.5.1) the polarization $\boldsymbol{\mathcal{P}}(\mathbf{r},t)$ is related to the field by

$$\boldsymbol{\mathcal{P}}(\mathbf{r},t) = \epsilon_0 \int dt' \chi^{(1)}(t-t') \boldsymbol{\mathcal{E}}(\mathbf{r},t') \,, \tag{3.104}$$

where the **linear susceptibility** $\chi^{(1)}(t-t')$ describes the delayed response of the medium to an applied electric field. Fourier transforming eqn (3.104) with respect to time produces the equivalent frequency-domain relation

$$\boldsymbol{\mathcal{P}}(\mathbf{r},\omega) = \epsilon_0 \chi^{(1)}(\omega) \boldsymbol{\mathcal{E}}(\mathbf{r},\omega) \,. \tag{3.105}$$

Applying the definition of positive- and negative-frequency parts, given by eqns (3.97)–(3.99), to the real classical field $\boldsymbol{\mathcal{E}}(\mathbf{r},t)$ leads to

$$\boldsymbol{\mathcal{E}}(\mathbf{r},t) = \boldsymbol{\mathcal{E}}^{(+)}(\mathbf{r},t) + \boldsymbol{\mathcal{E}}^{(-)}(\mathbf{r},t) \,. \tag{3.106}$$

In position space, the strength of the electric field at frequency ω is represented by the power spectrum $\left|\boldsymbol{\mathcal{E}}^{(+)}(\mathbf{r},\omega)\right|^2$ (see Appendix B.2). In reciprocal space, the power spectrum is $\left|\boldsymbol{\mathcal{E}}^{(+)}(\mathbf{k},\omega)\right|^2$. We will often be concerned with fields for which the power spectrum has a single well-defined peak at a **carrier frequency** $\omega = \omega_0$. The value of ω_0 is set by the experimental situation, e.g. ω_0 is often the frequency of an injected signal. The reality condition (3.100) for $\boldsymbol{\mathcal{E}}^{(\pm)}(\mathbf{r},\omega)$ tells us that $\left|\boldsymbol{\mathcal{E}}^{(-)}(\mathbf{r},\omega)\right|^2$ has a peak at $\omega = -\omega_0$; consequently, the complete transform $\boldsymbol{\mathcal{E}}(\mathbf{r},\omega)$ has two peaks: one at $\omega = \omega_0$ and the other at $\omega = -\omega_0$.

We will say that the field is **monochromatic** if the spectral width, $\Delta\omega_0$, of the peak at $\omega = \pm\omega_0$ satisfies

$$\Delta\omega_0 \ll \omega_0 \,. \tag{3.107}$$

We should point out that this usage is unconventional. Fields satisfying eqn (3.107) are often called quasimonochromatic in order to distinguish them from the ideal case in which the spectral width is exactly zero: $\Delta\omega_0 = 0$. Since the fields generated in real experiments are always described by wave packets with nonzero spectral widths, we prefer the definition associated with eqn (3.107). The ideal fields with $\Delta\omega_0 = 0$ will be called **strictly monochromatic**.

The concentration of the Fourier transform in the vicinity of $\omega = \pm\omega_0$ allows us to define the **slowly-varying envelope fields** $\overline{\boldsymbol{\mathcal{E}}}^{(\pm)}(\mathbf{r},t)$ by setting

$$\overline{\boldsymbol{\mathcal{E}}}^{(\pm)}(\mathbf{r},t) = \boldsymbol{\mathcal{E}}^{(\pm)}(\mathbf{r},t)\, e^{\pm i\omega_0 t}, \tag{3.108}$$

so that

$$\boldsymbol{\mathcal{E}}(\mathbf{r},t) = \overline{\boldsymbol{\mathcal{E}}}^{(+)}(\mathbf{r},t)\, e^{-i\omega_0 t} + \overline{\boldsymbol{\mathcal{E}}}^{(-)}(\mathbf{r},t)\, e^{i\omega_0 t}. \tag{3.109}$$

The slowly-varying envelopes satisfy $\overline{\boldsymbol{\mathcal{E}}}^{(-)}(\mathbf{r},t) = \overline{\boldsymbol{\mathcal{E}}}^{(+)*}(\mathbf{r},t)$, and the time-domain version of eqn (3.107) is

$$\left|\frac{\partial^2\overline{\boldsymbol{\mathcal{E}}}^{(\pm)}(\mathbf{r},t)}{\partial t^2}\right| \ll \omega_0 \left|\frac{\partial\overline{\boldsymbol{\mathcal{E}}}^{(\pm)}(\mathbf{r},t)}{\partial t}\right| \ll \omega_0^2 \left|\overline{\boldsymbol{\mathcal{E}}}^{(\pm)}(\mathbf{r},t)\right|. \tag{3.110}$$

The frequency-domain versions of eqns (3.108) and (3.109) are

$$\overline{\boldsymbol{\mathcal{E}}}^{(\pm)}(\mathbf{r},\omega) = \boldsymbol{\mathcal{E}}^{(\pm)}(\mathbf{r},\omega \pm \omega_0) \tag{3.111}$$

and

$$\boldsymbol{\mathcal{E}}(\mathbf{r},\omega) = \overline{\boldsymbol{\mathcal{E}}}^{(+)}(\mathbf{r},\omega - \omega_0) + \overline{\boldsymbol{\mathcal{E}}}^{(-)}(\mathbf{r},\omega + \omega_0), \tag{3.112}$$

respectively. The condition (3.107) implies that $\overline{\boldsymbol{\mathcal{E}}}^{(\pm)}(\mathbf{r},\omega)$ is sharply peaked at $\omega = 0$.

The Fourier transform of the vector potential is also concentrated in the vicinity of $\omega = \pm\omega_0$, so the slowly-varying envelope,

$$\overline{\boldsymbol{\mathcal{A}}}^{(+)}(\mathbf{r},t) = \boldsymbol{\mathcal{A}}^{(+)}(\mathbf{r},t)\, e^{i\omega_0 t}, \tag{3.113}$$

satisfies the same conditions. Since $\boldsymbol{\mathcal{E}}(\mathbf{r},t) = -\partial\boldsymbol{\mathcal{A}}(\mathbf{r},t)/\partial t$, the two envelope functions are related by

$$\overline{\boldsymbol{\mathcal{E}}}^{(+)}(\mathbf{r},t) = i\omega_0\overline{\boldsymbol{\mathcal{A}}}^{(+)} - \frac{\partial}{\partial t}\overline{\boldsymbol{\mathcal{A}}}^{(+)}. \tag{3.114}$$

Applying eqn (3.110) to the vector potential shows that the second term on the right side is small compared to the first, so that

$$\overline{\boldsymbol{\mathcal{E}}}^{(+)}(\mathbf{r},t) \approx i\omega_0\overline{\boldsymbol{\mathcal{A}}}^{(+)}. \tag{3.115}$$

This is an example of the **slowly-varying envelope approximation.**

More generally, it is necessary to consider **polychromatic** fields, i.e. superpositions of monochromatic fields with carrier frequencies ω_β $(\beta = 0, 1, 2, \ldots)$. The carrier frequencies are required to be distinct; that is, the power spectrum for a polychromatic field exhibits a set of clearly resolved peaks at the carrier frequencies ω_β. The explicit condition is that the minimum spacing between peaks, $\delta\omega_{\min} = \min\left[|\omega_\alpha - \omega_\beta|,\ \alpha \neq \beta\right]$, is large compared to the maximum spectral width, $\Delta\omega_{\max} = \max\left[\Delta\omega_\beta\right]$. The values of the carrier frequencies are set by the experimental situation under study. The collection $\{\omega_\beta\}$ will generally contain the frequencies of any injected fields together with the frequencies of radiation emitted by the medium in response to the injected signals. For a polychromatic field, eqns (3.108), (3.113), and (3.115) are replaced by

$$\boldsymbol{\mathcal{E}}^{(+)}(\mathbf{r},t)=\sum_{\beta}\overline{\boldsymbol{\mathcal{E}}}_{\beta}^{(+)}(\mathbf{r},t)\,e^{-i\omega_{\beta}t}\,, \tag{3.116}$$

$$\boldsymbol{\mathcal{A}}^{(+)}(\mathbf{r},t)=\sum_{\beta}\overline{\boldsymbol{\mathcal{A}}}_{\beta}^{(+)}(\mathbf{r},t)\,e^{-i\omega_{\beta}t}\,, \tag{3.117}$$

and

$$\overline{\boldsymbol{\mathcal{E}}}_{\beta}^{(+)}(\mathbf{r},t)=i\omega_{\beta}\overline{\boldsymbol{\mathcal{A}}}_{\beta}^{(+)}(\mathbf{r},t)\,. \tag{3.118}$$

In the frequency domain, the total polychromatic field is given by

$$\boldsymbol{\mathcal{E}}(\mathbf{r},\omega)=\sum_{\beta}\sum_{\sigma=\pm}\overline{\boldsymbol{\mathcal{E}}}_{\beta}^{(\sigma)}(\mathbf{r},\omega-\sigma\omega_{\beta})\,, \tag{3.119}$$

where each of the functions $\overline{\boldsymbol{\mathcal{E}}}_{\beta}^{(\pm)}(\mathbf{r},\omega)$ is sharply peaked at $\omega=0$.

A *Passive, linear dielectric*

An optical medium is said to be passive and linear if the following conditions are satisfied.

(a) **Off resonance.** The classical power spectrum is negligible at frequencies that are resonant with any transition of the constituent atoms. This justifies the assumption that there is no absorption.

(b) **Coarse graining.** There are many atoms in the volume λ_0^3, where λ_0 is the mean wavelength for the incident field.

(c) **Weak field.** The field is not strong enough to induce significant changes in the material medium.

(d) **Weak dispersion.** The frequency-dependent susceptibility $\chi^{(1)}(\omega)$ is essentially constant across any frequency interval $\Delta\omega \ll \omega$.

(e) **Stationary medium.** The medium is stationary, i.e. the optical properties do not change in time.

The passive property is incorporated in the off-resonance assumption (a) which allows us to neglect absorption, stimulated emission, and spontaneous emission. The description of the medium by the usual macroscopic coefficients such as the susceptibility, the refractive index, and the conductivity is justified by the coarse-graining assumption (b). The weak-field assumption (c) guarantees that the macroscopic version of Maxwell's equations is linear in the fields. The weak dispersion condition (d) assures us that an input wave packet with a sharply defined carrier frequency will retain the same frequency after propagation through the medium. The assumption (e) implies that the susceptibility $\chi^{(1)}(t-t')$ only depends on the time difference $t-t'$.

For later use it is helpful to explain these conditions in more detail. The medium is said to be **weakly dispersive** (in the vicinity of the carrier frequency $\omega = \omega_0$) if

$$\Delta\omega_0 \left| \left(\frac{\partial \chi^{(1)}(\omega)}{\partial \omega} \right)_{\omega=\omega_0} \right| \ll \left| \chi^{(1)}(\omega_0) \right| \tag{3.120}$$

for any frequency interval $\Delta\omega_0 \ll \omega_0$. We next recall that in a linear, isotropic dielectric the vacuum dispersion relation $\omega = ck$ is replaced by

$$\omega n(\omega) = ck\,, \tag{3.121}$$

where the index of refraction is related to the dielectric permittivity, $\epsilon(\omega)$, by $n^2(\omega) = \epsilon(\omega)$. Since $\epsilon(\omega)$ can be complex—the imaginary part describes absorption or gain (Jackson, 1999, Chap. 7)—the dispersion relation does not always have a real solution. However, for transparent dielectrics there is a range of frequencies in which the imaginary part of the index is negligible.

For a given wavenumber k, let ω_k be the mode frequency obtained by solving the nonlinear dispersion relation (3.121), then the medium is transparent at ω_k if $n_k = n(\omega_k)$ is real. In the frequency–wavenumber domain the electric field satisfies

$$\left[\frac{\omega^2}{c^2} n^2(\omega) - k^2 \right] \boldsymbol{\mathcal{E}}_{\mathbf{k}}(\omega) = 0 \tag{3.122}$$

(see Appendix B.5.2, eqn (B.123)), so one finds the general space–time solution $\boldsymbol{\mathcal{E}}(\mathbf{r},t) = \boldsymbol{\mathcal{E}}^{(+)}(\mathbf{r},t) + \boldsymbol{\mathcal{E}}^{(-)}(\mathbf{r},t)$, with

$$\boldsymbol{\mathcal{E}}^{(+)}(\mathbf{r},t) = \frac{1}{\sqrt{V}} \sum_{\mathbf{k}s} \mathcal{E}_{\mathbf{k}s} \mathbf{e}_{\mathbf{k}s} e^{i(\mathbf{k}\cdot\mathbf{r} - \omega_k t)}\,. \tag{3.123}$$

For a monochromatic field, the slowly-varying envelope is

$$\overline{\boldsymbol{\mathcal{E}}}^{(+)}(\mathbf{r},t) = \frac{1}{\sqrt{V}} {\sum_{\mathbf{k}s}}' \mathcal{E}_{\mathbf{k}s} \mathbf{e}_{\mathbf{k}s} e^{i(\mathbf{k}\cdot\mathbf{r} - \Delta_k t)}\,, \tag{3.124}$$

where the prime on the **k**-sum indicates that it is restricted to **k**-values such that the *detuning*, $\Delta_k = \omega_k - \omega_0$, satisfies $|\Delta_k| \ll \omega_0$. The wavelength mentioned in the coarse-graining assumption (b) is then $\lambda_0 = 2\pi c / (n(\omega_0)\,\omega_0)$.

For a polychromatic field, eqn (3.108) is replaced by

$$\boldsymbol{\mathcal{E}}^{(+)}(\mathbf{r},t) = \sum_{\beta} \overline{\boldsymbol{\mathcal{E}}}_{\beta}^{(+)}(\mathbf{r},t)\, e^{-i\omega_\beta t}\,, \tag{3.125}$$

where

$$\overline{\boldsymbol{\mathcal{E}}}_{\beta}^{(+)}(\mathbf{r},t) = \frac{1}{\sqrt{V}} {\sum_{\mathbf{k}s}}' \mathcal{E}_{\beta\mathbf{k}s} \mathbf{e}_{\mathbf{k}s} e^{i(\mathbf{k}\cdot\mathbf{r} - \Delta_{\beta k} t)}\,, \quad \Delta_{\beta k} = \omega_k - \omega_\beta \tag{3.126}$$

is a slowly-varying envelope field. The spectral width of the βth monochromatic field is defined by the power spectrum $\left|\overline{\boldsymbol{\mathcal{E}}}_{\beta}^{(+)}(\mathbf{r},\omega)\right|^2$ or $\left|\overline{\boldsymbol{\mathcal{E}}}_{\beta}^{(+)}(\mathbf{k},\omega)\right|^2$. The weak dispersion condition (d) is extended to this case by imposing eqn (3.120) on each of the monochromatic fields.

The condition (3.107) for a monochromatic field guarantees the existence of an intermediate time scale T satisfying

$$\frac{1}{\omega_0} \ll T \ll \frac{1}{\Delta\omega_0}\,, \tag{3.127}$$

i.e. T is long compared to the carrier period but short compared to the characteristic time scale on which the envelope field changes. Averaging over the interval T will wash out all the fast variations—on the optical frequency scale—but leave the slowly-varying envelope unchanged. In the polychromatic case, applying eqn (3.107) to each monochromatic component picks out an overall time scale T satisfying $1/\omega_{\text{min}} \ll T \ll 1/\Delta\omega_{\text{max}}$, where $\omega_{\text{min}} = \min(\omega_\beta)$.

B Electromagnetic energy in a dispersive dielectric

For an isotropic, nondispersive dielectric—e.g. the vacuum—Poynting's theorem (see Appendix B.5) takes the form

$$\frac{\partial u_{\text{em}}(\mathbf{r},t)}{\partial t} + \nabla\cdot\mathbf{S}(\mathbf{r},t) = 0\,, \tag{3.128}$$

where

$$u_{\text{em}}(\mathbf{r},t) = \frac{1}{2}\left\{\epsilon\boldsymbol{\mathcal{E}}^2(\mathbf{r},t) + \frac{1}{\mu_0}\boldsymbol{\mathcal{B}}^2(\mathbf{r},t)\right\} \tag{3.129}$$

is the electromagnetic energy density and

$$\mathbf{S}(\mathbf{r},t) = \boldsymbol{\mathcal{E}}(\mathbf{r},t)\times\boldsymbol{\mathcal{H}}(\mathbf{r},t) = \frac{1}{\mu_0}\boldsymbol{\mathcal{E}}(\mathbf{r},t)\times\boldsymbol{\mathcal{B}}(\mathbf{r},t) \tag{3.130}$$

is the Poynting vector. The existence of an electromagnetic energy density is an essential feature of the quantization schemes presented in Chapter 2 and in the present chapter, so the existence of a similar object for weakly dispersive media is an important question.

For a dispersive dielectric eqn (3.128) is replaced by

$$\mathfrak{p}_{\text{el}}(\mathbf{r},t) + \frac{\partial u_{\text{mag}}}{\partial t} + \boldsymbol{\nabla}\cdot\mathbf{S} = 0\,, \tag{3.131}$$

where the **electric power density**,

$$\mathfrak{p}_{\text{el}}(\mathbf{r},t) = \boldsymbol{\mathcal{E}}(\mathbf{r},t)\cdot\frac{\partial\boldsymbol{\mathcal{D}}(\mathbf{r},t)}{\partial t}\,, \tag{3.132}$$

is the power per unit volume flowing into the dielectric medium due to the action of the slowly-varying electric field $\boldsymbol{\mathcal{E}}$, and

$$u_{\text{mag}}(\mathbf{r},t) = \frac{1}{2\mu_0}\boldsymbol{\mathcal{B}}^2(\mathbf{r},t) \tag{3.133}$$

is the magnetic energy density; see Jackson (1999, Sec. 6.8). The existence of the magnetic energy density $u_{\text{mag}}(\mathbf{r},t)$ is guaranteed by the assumption that the material is not magnetically dispersive. The question is whether $\mathfrak{p}_{\mathrm{el}}(\mathbf{r},t)$ can also be expressed as the time derivative of an instantaneous energy density. The electric displacement $\boldsymbol{\mathcal{D}}(\mathbf{r},t)$ and the polarization $\boldsymbol{\mathcal{P}}(\mathbf{r},t)$ are given by eqns (3.103) and (3.104), respectively, so in general $\boldsymbol{\mathcal{P}}(\mathbf{r},t)$ and $\boldsymbol{\mathcal{D}}(\mathbf{r},t)$ depend on the electric field at times $t' \neq t$. The principle of causality restricts this dependence to earlier times, $t' < t$, so that

$$\chi^{(1)}(t-t') = 0 \quad \text{for } t' > t\,. \tag{3.134}$$

For a nondispersive medium $\chi^{(1)}(\omega)$ has the constant value $\chi_0^{(1)}$, so in this approximation one finds that

$$\chi^{(1)}(t-t') = \chi_0^{(1)}\delta(t-t')\,. \tag{3.135}$$

In this case, the polarization at a given time only depends on the field at the same time. In the dispersive case, $\chi^{(1)}(t-t')$ decays to zero over a nonzero interval, $0 < t-t' < T_{\mathrm{mem}}$; in other words, the polarization at t depends on the *history* of the electric field up to time t. Consequently, the power density $\mathfrak{p}_{\mathrm{el}}(\mathbf{r},t)$ cannot be expressed as $\mathfrak{p}_{\mathrm{el}}(\mathbf{r},t) = \partial u_{\mathrm{el}}(\mathbf{r},t)/\partial t$, where $u_{\mathrm{el}}(\mathbf{r},t)$ is an instantaneous energy density.

In the general case this obstacle is insurmountable, but for a monochromatic (or polychromatic) field in a weakly dispersive dielectric it can be avoided by the use of an appropriate approximation scheme (Jackson, 1999, Sec. 6.8). The fundamental idea in this argument is to exploit the characteristic time T introduced in eqn (3.127) to define the (running) time-average

$$\langle \mathfrak{p}_{\mathrm{el}} \rangle (\mathbf{r},t) = \frac{1}{T}\int_{-T/2}^{T/2} \mathfrak{p}_{\mathrm{el}}(\mathbf{r},t+t')\,dt'\,. \tag{3.136}$$

This procedure eliminates all rapidly varying terms, and one can show that

$$\langle \mathfrak{p}_{\mathrm{el}} \rangle (\mathbf{r},t) = \frac{\partial u_{\mathrm{el}}(\mathbf{r},t)}{\partial t}\,, \tag{3.137}$$

where the effective electric energy density is

$$\begin{aligned} u_{\mathrm{el}}(\mathbf{r},t) &= \frac{d\left[\omega_0 \epsilon(\omega_0)\right]}{d\omega_0}\frac{1}{2}\left\langle \boldsymbol{\mathcal{E}}(\mathbf{r},t)\cdot\boldsymbol{\mathcal{E}}(\mathbf{r},t)\right\rangle \\ &= \frac{d\left[\omega_0 \epsilon(\omega_0)\right]}{d\omega_0}\overline{\boldsymbol{\mathcal{E}}}^{(-)}(\mathbf{r},t)\cdot\overline{\boldsymbol{\mathcal{E}}}^{(+)}(\mathbf{r},t)\,, \end{aligned} \tag{3.138}$$

and $\overline{\boldsymbol{\mathcal{E}}}^{(+)}(\mathbf{r},t)$ is the slowly-varying envelope for the electric field. The effective electric energy density for a polychromatic field is a sum of terms like $u_{\mathrm{el}}(\mathbf{r},t)$ evaluated for each monochromatic component. We will use this expression in the quantization technique described in Section 3.3.5.

3.3.2 Quantization in a dielectric

The behavior of the quantized electromagnetic field in a passive linear dielectric is an important practical problem for quantum optics. In principle, this problem could be approached through a microscopic theory of the quantized field interacting with the point charges in the atoms constituting the medium. The same could be said for the classical theory of fields in a dielectric, but it is traditional—and a great deal easier—to employ instead a phenomenological macroscopic approach which describes the response of the medium by the linear susceptibility. The long history and great utility of this phenomenological method have inspired a substantial body of work aimed at devising a similar description for the quantized electromagnetic field in a dielectric medium.[1] This has proven to be a difficult and subtle task. The phenomenological quantum theory for the cavity and the exact vacuum theory both depend on an expression for the classical energy as the sum of energies for independent radiation oscillators, but—as we have seen in the previous section—there is no exact instantaneous energy for a dispersive medium. Fortunately, an exact quantization method is not needed for the analysis of the large class of experiments that involve a monochromatic or polychromatic field propagating in a weakly dispersive dielectric. For these experimentally significant applications, we will make use of a physically appealing *ad hoc* quantization scheme due to Milonni (1995). In the following section, we begin with a simple model that incorporates the essential elements of this scheme, and then outline the more rigorous version in Section 3.3.5.

3.3.3 The dressed photon model

We begin with a modified version of the vacuum field expansion (3.69)

$$\mathbf{E}^{(+)}(\mathbf{r}) = i \sum_{\mathbf{k}s} \mathcal{E}_k a_{\mathbf{k}s} \mathbf{e}_{\mathbf{k}s} e^{i\mathbf{k}\cdot\mathbf{r}}, \tag{3.139}$$

where $a_{\mathbf{k}s}$ and $a_{\mathbf{k}s}^{\dagger}$ satisfy the canonical commutation relations (3.65) and the c-number coefficient $\mathcal{E}_k$ is a characteristic field strength which will be chosen to fit the problem at hand. In this section we will choose $\mathcal{E}_k$ by analyzing a simple physical model, and then point out some of the consequences of this choice.

The mathematical convenience of the box-quantization scheme is purchased at the cost of imposing periodic boundary conditions along the three coordinate axes. The shape of the quantization box is irrelevant in the infinite volume limit, so we are at liberty to replace the imaginary cubical box by an equally imaginary cavity in the shape of a torus filled with dielectric material, as shown in Fig. 3.1(a).

In this geometry one of the coordinate directions has been wrapped into a circle, so that the periodic boundary conditions in that direction are physically realized by the natural periodicity in a coordinate measuring distance along the axis of the torus. The fields must still satisfy periodic boundary conditions at the walls of the torus, but this will not be a problem, since all dimensions of the torus will become infinitely large. In this limit, the exact shape of the transverse sections is also not important. Let L be the circumference and σ the cross sectional area for the torus, then in the limit of large L a small segment will appear straight, as in Fig. 3.1(b), and the axis of the torus can be chosen as the local z-axis. Since the transverse dimensions are

[1] For a sampling of the relevant references see Drummond (1990), Huttner and Barnett (1992), Matloob *et al.* (1995), and Gruner and Welsch (1996).

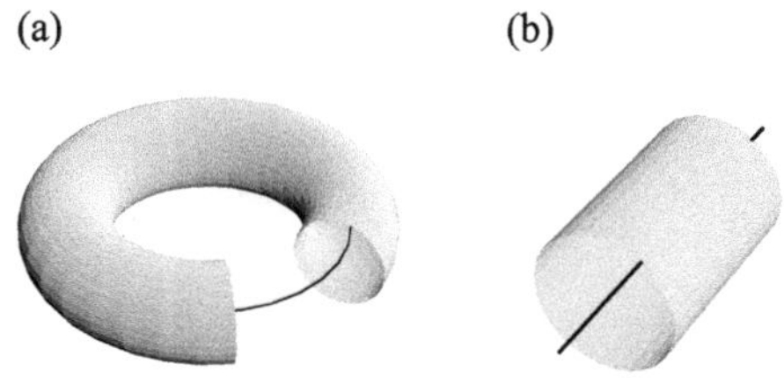

Fig. 3.1 (a) A toroidal cavity filled with a weakly dispersive dielectric. A segment has been removed to show the central axis. The field satisfies periodic boundary conditions along the axis. (b) A small segment of the torus is approximated by a cylinder, and the central axis is taken as the z-axis.

also large, a classical field propagating in the z-direction can be approximated by a monochromatic **planar** wave packet,

$$\boldsymbol{\mathcal{E}}(z,t) = \boldsymbol{\mathcal{E}}_k(z,t)\, e^{i(kz-\omega_k t)} + \mathrm{CC}\,, \tag{3.140}$$

where ω_k is a solution of the dispersion relation (3.121) and $\boldsymbol{\mathcal{E}}_k(z,t)$ is a slowly-varying envelope function. If we neglect the time derivative of the slowly-varying envelope, then Faraday's law (eqn (B.94)) yields

$$\boldsymbol{\mathcal{B}}(z,t) = \frac{1}{\omega_k}\mathbf{k} \times \boldsymbol{\mathcal{E}}_k(z,t)\, e^{i(kz-\omega_k t)} + \mathrm{CC}\,. \tag{3.141}$$

As we have seen in Section 3.3.1, the fields actually generated in experiments are naturally described by wave packets. It is therefore important to remember that wave packets do not propagate at the **phase velocity** $v_{\text{ph}}(\omega_k) = c/n_k$, but rather at the **group velocity**

$$v_g(\omega_k) = \frac{d\omega}{dk} = \frac{c}{n_k + \omega_k (dn/d\omega)_k}\,. \tag{3.142}$$

This fact will play an important role in the following argument, so we consider very long planar wave packets instead of idealized plane waves.

We will determine the characteristic field $\mathcal{E}_k$ by equating the energy in the wave packet to $\hbar\omega_k$. The energy can be found by integrating the rate of energy transport across a transverse section of the torus over the time required for one round trip around the circumference. For this purpose we need the energy flux, $\mathbf{S} = c^2\epsilon_0 \boldsymbol{\mathcal{E}} \times \boldsymbol{\mathcal{B}}$, or rather its average over one cycle of the carrier wave. In the almost-plane-wave approximation, this is the familiar result $\langle \mathbf{S} \rangle = 2c^2\epsilon_0 \operatorname{Re}\{\boldsymbol{\mathcal{E}}_k \times \boldsymbol{\mathcal{B}}_k^*\}$. Setting $\boldsymbol{\mathcal{E}}_k = \mathcal{E}_k\, \mathbf{u}_x$, i.e. choosing the x-direction along the polarization vector, leads to

$$\langle \mathbf{S} \rangle = \frac{2c^2\epsilon_0 k \left|\mathcal{E}_k\right|^2}{\omega_k}\mathbf{u}_z = \frac{2c\epsilon_0 n_k \left|\mathcal{E}_k\right|^2}{\mu_0}\mathbf{u}_z\,, \tag{3.143}$$

where the last form comes from using the dispersion relation. The energy passing through a given transverse section during a time τ is $\langle S_z \rangle \sigma\tau$. The wave packet completes one trip around the torus in the time $\tau_g = L/v_g(\omega_k)$; consequently, by virtue of the periodic nature of the motion, $\langle S_z \rangle \sigma\tau_g$ is the entire energy in the wave packet. In the spirit of Einstein's original model we set this equal to the energy, $\hbar\omega_k$, of a single photon:

$$\frac{2c\epsilon_0 n_k \left|\mathcal{E}_k\right|^2 \sigma L}{v_g\left(\omega_k\right) V} = \hbar\omega_k \,. \tag{3.144}$$

The total volume of the torus is $V = \sigma L$, so

$$\left|\mathcal{E}_k\right| = \sqrt{\frac{\hbar\omega_k v_g\left(\omega_k\right)}{2\epsilon_0 c n_k V}}\,, \tag{3.145}$$

which gives the box-quantized expansions

$$\mathbf{A}^{(+)}\left(\mathbf{r}\right) = \sum_{\mathbf{k}s}\sqrt{\frac{\hbar v_g\left(\omega_k\right)}{2\epsilon_0 n_k \omega_k c V}} a_{\mathbf{k}s}\mathbf{e}_{\mathbf{k}s} e^{i\mathbf{k}\cdot\mathbf{r}} \tag{3.146}$$

and

$$\mathbf{E}^{(+)}\left(\mathbf{r}\right) = i\sum_{\mathbf{k}s}\sqrt{\frac{\hbar\omega_k v_g\left(\omega_k\right)}{2\epsilon_0 n_k c V}} a_{\mathbf{k}s}\mathbf{e}_{\mathbf{k}s} e^{i\mathbf{k}\cdot\mathbf{r}} \tag{3.147}$$

for the vector potential and the electric field. The continuum versions are

$$\mathbf{E}^{(+)}\left(\mathbf{r}\right) = i\int\frac{d^3k}{\left(2\pi\right)^3}\sum_s\sqrt{\frac{\hbar\omega_k v_g\left(\omega_k\right)}{2\epsilon_0 n_k c}} a_s\left(\mathbf{k}\right)\mathbf{e}_{\mathbf{k}s} e^{i\mathbf{k}\cdot\mathbf{r}} \tag{3.148}$$

and

$$\mathbf{A}^{(+)}\left(\mathbf{r}\right) = \int\frac{d^3k}{\left(2\pi\right)^3}\sum_s\sqrt{\frac{\hbar v_g\left(\omega_k\right)}{2\epsilon_0 n_k \omega_k c}} a_s\left(\mathbf{k}\right)\mathbf{e}_{\mathbf{k}s} e^{i\mathbf{k}\cdot\mathbf{r}}\,. \tag{3.149}$$

This procedure incorporates properties of the medium into the description of the field, so the excitation created by $a^\dagger_{\mathbf{k}s}$ or $a^\dagger_s\left(\mathbf{k}\right)$ will be called a **dressed photon.**

A Energy and momentum

Since $\hbar\omega_k$ is the energy assigned to a single dressed photon, the Hamiltonian can be expressed in the box-normalized form

$$H_{\text{em}} = \sum_{\mathbf{k}s}\hbar\omega_k a^\dagger_{\mathbf{k}s} a_{\mathbf{k}s}\,, \tag{3.150}$$

or in the equivalent continuum form

$$H_{\text{em}} = \int\frac{d^3k}{\left(2\pi\right)^3}\sum_s\hbar\omega_k a^\dagger_s\left(\mathbf{k}\right) a_s\left(\mathbf{k}\right). \tag{3.151}$$

We will see in Section 3.3.5 that this Hamiltonian also results from an application of the quantization procedure described there to the standard expression for the electromagnetic energy in a dispersive medium.

The condition (3.144) was obtained by treating the dressed photon as a particle with energy $\hbar\omega_k$. This suggests identifying the momentum of the photon with an eigenvalue of the standard canonical momentum operator $\widehat{\mathbf{p}}_{\text{can}} = -i\hbar\boldsymbol{\nabla}$ of quantum mechanics. Since the basis functions for box quantization are the plane waves, $\exp(i\mathbf{k}\cdot\mathbf{r})$, this is equivalent to assigning the momentum

$$\mathbf{p} = \hbar\mathbf{k} \tag{3.152}$$

to a dressed photon with energy $\hbar\omega_k$. The operator

$$\mathbf{P}_{\text{em}} = \sum_{\mathbf{k}s} \hbar\mathbf{k}a^{\dagger}_{\mathbf{k}s}a_{\mathbf{k}s} \tag{3.153}$$

would then represent the total momentum of the electromagnetic field. In Section 3.3.5 we will see that this operator is the generator of spatial translations for the quantized electromagnetic field.

There are two empirical lines of evidence supporting the physical significance of the canonical momentum for photons. The first is that the conservation law for $\mathbf{P}_{\text{em}}$ is identical to the empirically well established principle of phase matching in nonlinear optics. The second is that the canonical momentum provides a simple and accurate model (Garrison and Chiao, 2004) for the radiation pressure experiment of Jones and Leslie (1978). We should point out that the theoretical argument for choosing an expression for the momentum associated with the dressed photon is not quite as straightforward as the previous discussion suggests. The difficulty is that there is no universally accepted definition of the *classical* electromagnetic momentum in a dispersive medium. This lack of agreement reflects a long standing controversy in classical electrodynamics regarding the correct definition of the electromagnetic momentum density in a weakly dispersive medium (Landau *et al.*, 1984; Ginzburg, 1989). The implications of this controversy for the quantum theory are also discussed in Garrison and Chiao (2004).

3.3.4 The Hilbert space of dressed-photon states

The vacuum quantization rules—e.g. eqns (3.25) and (3.26)—are supposed to be exact, but this is not possible for the phenomenological quantization scheme given by eqn (3.146). The discussion in Section 3.3.1-B shows that we cannot expect to get a sensible theory of quantization in a dielectric without imposing some constraints, e.g. the monochromatic condition (3.107), on the fields. Since operators do not have numerical values, these constraints cannot be applied directly to the quantized fields. Instead, the constraints must be imposed on the states of the field. For conditions (a) and (b) the classical power spectrum is replaced by

$$p_{\mathbf{k}} = \left\langle a^{\dagger}_{\mathbf{k}s}a_{\mathbf{k}s}\right\rangle = \sum_{s} \operatorname{Tr}\left[\rho_{\text{in}}a^{\dagger}_{\mathbf{k}s}a_{\mathbf{k}s}\right], \tag{3.154}$$

where ρ_{in} is the density operator describing the state of the incident field. Similarly (c) means that the average intensity $\langle \mathbf{E}^{(-)}(\mathbf{r})\,\mathbf{E}^{(+)}(\mathbf{r})\rangle$ is small compared to the characteristic intensity needed to produce significant changes in the material properties. For condition (d) the spectral width $\Delta\omega_0$ is given by

$$\Delta\omega_0^2 = \sum_{\mathbf{k}} (\omega_k - \omega_0)^2 p_{\mathbf{k}} \,. \tag{3.155}$$

For an experimental situation corresponding to a monochromatic classical field with carrier frequency ω_0, the appropriate Hilbert space of states consists of the state vectors that satisfy the quantum version of conditions (a)–(d). All such states can be expressed as superpositions of the special number states

$$|\underline{m}\rangle = \prod_{\mathbf{k}s} \frac{\left(a_{\mathbf{k}s}^{\dagger}\right)^{m_{\mathbf{k}s}}}{\sqrt{m_{\mathbf{k}s}!}} |0\rangle \,, \tag{3.156}$$

with occupation numbers $m_{\mathbf{k}s}$ restricted by

$$m_{\mathbf{k}s} = 0 \,, \quad \text{unless } |\omega_k - \omega_0| < \Delta\omega_0 \,. \tag{3.157}$$

The set of all linear combinations of number states satisfying eqn (3.157) is a subspace of Fock space, which we will call a **monochromatic space**, $\mathfrak{H}(\omega_0)$. For a polychromatic field, eqn (3.157) is replaced by the set of conditions

$$m_{\mathbf{k}s} = 0 \,, \quad \text{unless } |\omega_k - \omega_\beta| < \Delta\omega_\beta \,, \quad \beta = 0, 1, 2, \ldots \,. \tag{3.158}$$

The space $\mathfrak{H}(\{\omega_\beta\})$ spanned by the number states satisfying these conditions is called a **polychromatic space**. The representations (3.146)–(3.151) are only valid when applied to vectors in $\mathfrak{H}(\{\omega_\beta\})$. The initial field state ρ_{in} must therefore be defined by an ensemble of pure states chosen from $\mathfrak{H}(\{\omega_\beta\})$.

3.3.5 Milonni's quantization method*

The derivation of the characteristic field strength $\mathcal{E}_k$ in the previous section is dangerously close to a violation of Einstein's rule, so it is useful to give an independent argument. According to eqn (3.138) the total effective electromagnetic energy is

$$\mathcal{U}_{\text{em}} = \frac{d\left[\omega_0 \epsilon(\omega_0)\right]}{d\omega_0} \frac{1}{2} \int d^3r \left\langle \boldsymbol{\mathcal{E}}^2(\mathbf{r}, t) \right\rangle + \frac{1}{2\mu_0} \int d^3r \left\langle \boldsymbol{\mathcal{B}}^2(\mathbf{r}, t) \right\rangle . \tag{3.159}$$

The time averaging eliminates the rapidly oscillating terms proportional to $\boldsymbol{\mathcal{E}}^{(\pm)}(\mathbf{r}, t) \cdot \boldsymbol{\mathcal{E}}^{(\pm)}(\mathbf{r}, t)$ or $\boldsymbol{\mathcal{B}}^{(\pm)}(\mathbf{r}, t) \cdot \boldsymbol{\mathcal{B}}^{(\pm)}(\mathbf{r}, t)$, so that

$$\mathcal{U}_{\text{em}} = \frac{d\left[\omega_0 \epsilon(\omega_0)\right]}{d\omega_0} \int d^3r \boldsymbol{\mathcal{E}}^{(-)}(\mathbf{r}, t) \cdot \boldsymbol{\mathcal{E}}^{(+)}(\mathbf{r}, t) + \frac{1}{\mu_0} \int d^3r \boldsymbol{\mathcal{B}}^{(-)}(\mathbf{r}, t) \cdot \boldsymbol{\mathcal{B}}^{(+)}(\mathbf{r}, t) \,. \tag{3.160}$$

For classical fields given by eqn (3.123) the volume integral can be carried out to find

$$\mathcal{U}_{\text{em}} = \sum_{\mathbf{k}s} \left\{ \omega_k^2 \frac{d\left[\omega_0 \epsilon(\omega_0)\right]}{d\omega_0} + \frac{k^2}{\mu_0} \right\} |\mathcal{A}_{\mathbf{k}s}|^2 \,, \tag{3.161}$$

where $\mathcal{A}_{\mathbf{k}s} = \mathcal{E}_{\mathbf{k}s}/i\omega_0$ is the expansion amplitude for the vector potential. Since the power spectrum $|\mathcal{A}_{\mathbf{k}s}|^2$ is strongly peaked at $\omega_k = \omega_0$, it is equally accurate to write this result in the more suggestive form

$$\mathcal{U}_{\text{em}} = \sum_{\mathbf{k}s} \left\{ \omega_k^2 \frac{d\left[\omega_k \epsilon\left(\omega_k\right)\right]}{d\omega_k} + \frac{k^2}{\mu_0} \right\} \left|\mathcal{A}_{\mathbf{k}s}\right|^2 . \tag{3.162}$$

This expression presents a danger and an opportunity. The danger comes from its apparent generality, which might lead one to forget that it is only valid for a monochromatic field. The opportunity comes from its apparent generality, which makes it clear that eqn (3.162) is also correct for polychromatic fields. It is more convenient to use the dispersion relation (3.121) and the definition $\epsilon\left(\omega\right) = \epsilon_0 n^2\left(\omega\right)$ of the index of refraction to rewrite the curly bracket in eqn (3.162) as

$$\begin{aligned} \left\{ \omega_k^2 \frac{d\left[\omega_k \epsilon\left(\omega_k\right)\right]}{d\omega_k} + \frac{k^2}{\mu_0} \right\} &= 2\epsilon_0 \omega_k^2 n_k \frac{d\left[\omega_k n_k\right]}{d\omega_k} \\ &= 2\epsilon_0 \omega_k^2 n_k \frac{c}{v_g\left(\omega_k\right)} , \end{aligned} \tag{3.163}$$

where the last form comes from the definition (3.142) of the group velocity. The total energy is then

$$\mathcal{U}_{\text{em}} = \sum_{\mathbf{k}s} 2\epsilon_0 \omega_k^2 n_k \frac{c}{v_g\left(\omega_k\right)} \left|\mathcal{A}_{\mathbf{k}s}\right|^2 . \tag{3.164}$$

Setting

$$\mathcal{A}_{\mathbf{k}s} = \sqrt{\frac{\hbar v_g\left(\omega_k\right)}{2\epsilon_0 n_k \omega_k c}} w_{\mathbf{k}s} , \tag{3.165}$$

where $w_{\mathbf{k}s}$ is a dimensionless amplitude, allows $\mathcal{U}_{\text{em}}$ and $\boldsymbol{\mathcal{A}}^{(+)}\left(\mathbf{r}, t\right)$ to be written as

$$\mathcal{U}_{\text{em}} = \sum_{\mathbf{k}s} \hbar\omega_k \left|w_s\left(\mathbf{k}\right)\right|^2 \tag{3.166}$$

and

$$\boldsymbol{\mathcal{A}}^{(+)}\left(\mathbf{r}, t\right) = \sum_{\mathbf{k}s} \sqrt{\frac{\hbar v_g\left(\omega_k\right)}{2\epsilon_0 n_k \omega_k c V}} w_{\mathbf{k}s} \mathbf{e}_{\mathbf{k}s} e^{i\left(\mathbf{k}\cdot\mathbf{r} - \omega_k t\right)} , \tag{3.167}$$

respectively.

In eqn (3.166) the classical electromagnetic energy is expressed as the sum of energies, $\hbar\omega_k$, of radiation oscillators, so the stage is set for a quantization method like that used in Section 2.1.2. Thus we replace the classical amplitudes $w_{\mathbf{k}s}$ and $w_{\mathbf{k}s}^*$, in eqn (3.167) and its conjugate, by operators $a_{\mathbf{k}s}$ and $a_{\mathbf{k}s}^\dagger$ that satisfy the canonical commutation relations (3.65). In other words the quantization rule is

$$\mathcal{A}_{\mathbf{k}s} \to \sqrt{\frac{\hbar v_g\left(\omega_k\right)}{2\epsilon_0 n_k \omega_k c}} a_{\mathbf{k}s} . \tag{3.168}$$

In the Schrödinger picture this leads to

$$\mathbf{A}^{(+)}\left(\mathbf{r}\right) = \sum_{\mathbf{k}s} \sqrt{\frac{\hbar v_g\left(\omega_k\right)}{2\epsilon_0 n_k \omega_k c V}} a_{\mathbf{k}s} \mathbf{e}_{\mathbf{k}s} e^{i\mathbf{k}\cdot\mathbf{r}} , \tag{3.169}$$

which agrees with eqn (3.146). The Hamiltonian and the electric field are consequently given by eqns (3.150) and (3.147), respectively, in agreement with the results of the dressed photon model in Section 3.3.3. Once again, the general appearance of these results must not tempt us into forgetting that they are at best valid for polychromatic field states. This means that the operators defined here are only meaningful when applied to states in the space $\mathfrak{H}(\{\omega_\beta\})$ appropriate to the experimental situation under study.

A Electromagnetic momentum in a dielectric*

The definition (3.153) for the electromagnetic momentum is related to the fundamental symmetry principle of translation invariance. The defining properties of passive linear dielectrics in Section 3.3.1-A implicitly include the assumption that the positional and inertial degrees of freedom of the constituent atoms are irrelevant. As a consequence the generator $\mathbf{G}$ of spatial translations is completely defined by its action on the field operators, e.g.

$$\left[A_j^{(+)}(\mathbf{r}), \mathbf{G}\right] = \frac{\hbar}{i}\boldsymbol{\nabla} A_j^{(+)}(\mathbf{r}) . \tag{3.170}$$

Using the expansion (3.169) to evaluate both sides leads to $[a_{\mathbf{k}s}, \mathbf{G}] = \hbar\mathbf{k}a_{\mathbf{k}s}$, which is satisfied by the choice $\mathbf{G} = \mathbf{P}_{\text{em}}$. Any alternative form, $\mathbf{G}'$, would have to satisfy $[a_{\mathbf{k}s}, \mathbf{G}' - \mathbf{P}_{\text{em}}] = 0$ for all modes $\mathbf{k}s$, and this is only possible if the operator $\mathbf{Z} \equiv \mathbf{G}' - \mathbf{P}_{\text{em}}$ is actually a c-number. In this case $\mathbf{Z}$ can be set to zero by imposing the convention that the vacuum state is an eigenstate of $\mathbf{P}_{\text{em}}$ with eigenvalue zero. The expression (3.153) for $\mathbf{P}_{\text{em}}$ is therefore uniquely specified by the rules of quantum field theory.

3.4 Electromagnetic angular momentum*

The properties and physical significance of H_{em} and $\mathbf{P}$ are immediately evident from the plane-wave expansions (3.41) and (3.48), but the angular momentum presents a subtler problem. Since the physical interpretation of $\mathbf{J}$ is not immediately evident from eqns (3.54)–(3.59), our first task is to show that $\mathbf{J}$ does in fact represent the angular momentum. It is possible to do this directly by verifying that $\mathbf{J}$ satisfies the angular momentum commutation relations; but it is more instructive—and in fact simpler—to use an indirect argument. It is a general principle of quantum theory, reviewed in Appendix C.5, that the angular momentum operator is the generator of rotations. In particular, for any vector operator $V_j(\mathbf{r})$ constructed from the fields we should find

$$[J_i, V_j(\mathbf{r})] = i\hbar\left\{(\mathbf{r}\times\boldsymbol{\nabla})_i V_j(\mathbf{r}) + \epsilon_{ijk}V_k(\mathbf{r})\right\} . \tag{3.171}$$

Since all such operators can be built up from $\mathbf{A}^{(+)}(\mathbf{r})$, it is sufficient to verify this result for $\mathbf{V}(\mathbf{r}) = \mathbf{A}^{(+)}(\mathbf{r})$. The expressions (3.57) and (3.59) together with the commutation relation (3.3) lead to

$$\left[L_i, A_j^{(+)}(\mathbf{r})\right] = i\hbar\int d^3r'\Delta_{kj}^{\perp}(\mathbf{r}-\mathbf{r}')(\mathbf{r}'\times\boldsymbol{\nabla}')_i A_k^{(+)}(\mathbf{r}') \tag{3.172}$$

and

$$\left[S_i, A_j^{(+)}(\mathbf{r})\right] = i\hbar\epsilon_{ikl}\int d^3r'\Delta_{kj}^{\perp}(\mathbf{r}-\mathbf{r}')A_l^{(+)}(\mathbf{r}') , \tag{3.173}$$

so that

$$\left[J_i, A_j^{(+)}(\mathbf{r})\right] = i\hbar \int d^3r' \Delta_{kj}^{\perp}(\mathbf{r}-\mathbf{r}') \left\{(\mathbf{r}' \times \boldsymbol{\nabla}')_i A_k^{(+)}(\mathbf{r}') + \epsilon_{ikl} A_l^{(+)}(\mathbf{r}')\right\}. \quad (3.174)$$

The definition (2.30) of the transverse delta function can be written as

$$\Delta_{lj}^{\perp}(\mathbf{r}-\mathbf{r}') = \delta_{lj}\delta(\mathbf{r}-\mathbf{r}') - \int \frac{d^3k}{(2\pi)^3} \frac{k_l k_j}{k^2} e^{i\mathbf{k}\cdot(\mathbf{r}-\mathbf{r}')}, \quad (3.175)$$

and the first term on the right produces eqn (3.171) with $\mathbf{V} = \mathbf{A}^{(+)}$. A straightforward calculation using the identity

$$k_l k_j e^{i\mathbf{k}\cdot(\mathbf{r}-\mathbf{r}')} = -\nabla_l' \nabla_j' e^{i\mathbf{k}\cdot(\mathbf{r}-\mathbf{r}')} \quad (3.176)$$

and judicious integrations by parts shows that the contribution of the second term in eqn (3.175) vanishes; therefore, eqn (3.171) is established in general.

For a global vector operator $\mathbf{G}$, defined by

$$\mathbf{G} = \int d^3r \mathbf{g}(\mathbf{r}), \quad (3.177)$$

integration of eqn (3.171) yields

$$[J_k, G_i] = i\hbar\epsilon_{kij} G_j \,. \quad (3.178)$$

In particular the last equation applies to $\mathbf{G} = \mathbf{J}$; therefore, $\mathbf{J}$ satisfies the standard angular momentum commutation relations,

$$[J_i, J_j] = i\hbar\epsilon_{ijk} J_k \,. \quad (3.179)$$

The combination of eqns (3.171) and (3.179) establish the interpretation of $\mathbf{J}$ as the total angular momentum operator for the electromagnetic field.

In quantum mechanics the total angular momentum $\mathbf{J}$ of a particle can always be expressed as $\mathbf{J} = \mathbf{L} + \mathbf{S}$, where $\mathbf{L}$ is the orbital angular momentum (relative to a chosen origin) and the spin angular momentum $\mathbf{S}$ is the total angular momentum in the rest frame of the particle (Bransden and Joachain, 1989, Sec. 6.9). Since the photon travels at the speed of light, it has no rest frame; therefore, we should expect to meet with difficulties in any attempt to find a similar decomposition, $\mathbf{J} = \mathbf{L} + \mathbf{S}$, for the electromagnetic field. As explained in Appendix C.5, the usual decomposition of the angular momentum also depends crucially on the assumption that the spin and spatial degrees of freedom are kinematically independent, so that the operators $\mathbf{L}$ and $\mathbf{S}$ commute. For a vector field, this would be the case if there were three independent components of the field defined at each point in space. In the theory of the radiation field, however, the vectors fields $\mathbf{E}$ and $\mathbf{B}$ are required to be transverse, so there are only two independent components at each point. The constraint on the components of the fields is purely kinematical, i.e. it holds for both free and interacting fields, so the spin and spatial degrees of freedom are not independent. The restriction to transverse fields is related to the fact that the rest mass of the photon is zero, and therefore to the absence of any rest frame.

How then are we to understand eqn (3.54) which seems to be exactly what one would expect? After all we have established that $\mathbf{L}$ and $\mathbf{S}$ are physical observables, and the integrand in eqn (3.57) contains the operator $-i\mathbf{r} \times \boldsymbol{\nabla}$, which represents orbital angular momentum in quantum mechanics. Furthermore, the expression (3.59) is independent of the chosen reference point $\mathbf{r} = 0$. It is therefore tempting to interpret $\mathbf{L}$ as the orbital angular momentum (relative to the origin), and $\mathbf{S}$ as the intrinsic or spin angular momentum of the electromagnetic field, but the arguments in the previous paragraph show that this would be wrong.

To begin with, eqn (3.60) tells us that $\mathbf{S}$ does not satisfy the angular momentum commutation relations (3.179); so we are forced to conclude that $\mathbf{S}$ is not any kind of angular momentum. The representation (3.57) can be used to evaluate the commutation relations for $\mathbf{L}$, but once again there is a simpler indirect argument. The 'spin' operator $\mathbf{S}$ is a global vector operator, so applying eqn (3.178) gives

$$[J_k, S_i] = i\hbar\epsilon_{kij}S_j \,. \tag{3.180}$$

Combining the decomposition (3.54) with eqn (3.60) produces

$$[L_k, S_i] = i\hbar\epsilon_{kij}S_j \,, \tag{3.181}$$

so $\mathbf{L}$ acts as the generator of rotations for $\mathbf{S}$. Using this, together with eqn (3.54) and eqn (3.179), provides the commutators between the components of $\mathbf{L}$,

$$[L_k, L_i] = i\hbar\epsilon_{kij}\,(L_j - S_j)\,. \tag{3.182}$$

Thus the sum $\mathbf{J} = \mathbf{L} + \mathbf{S}$ is a genuine angular momentum operator, but the separate 'orbital' and 'spin' parts do not commute and are not themselves true angular momenta.

If the observables $\mathbf{L}$ and $\mathbf{S}$ are not angular momenta, then what are they? The physical significance of the helicity operator $\mathbf{S}$ is reasonably clear from $\widetilde{\mathbf{k}} \cdot \mathbf{S}\,|1_{\mathbf{k}s}\rangle = s\hbar\,|1_{\mathbf{k}s}\rangle$, but the meaning of the orbital angular momentum $\mathbf{L}$ is not so obvious. In common with true angular momenta, the different components of $\mathbf{L}$ do not commute. Thus it is necessary to pick out a single component, say L_z, which is to be diagonalized. The second step is to find other observables which do commute with L_z, in order to construct a complete set of commuting observables. Since we already know that $\mathbf{L}$ is not a true angular momentum, it should not be too surprising to learn that L_z and L^2 do not commute. The commutator between $\mathbf{L}$ and the total momentum $\mathbf{P}$ follows from the fact that $\mathbf{P}$ is a global vector operator that satisfies eqn (3.178) and also commutes with $\mathbf{S}$. This shows that

$$[L_k, P_i] = i\hbar\epsilon_{kij}P_j \,, \tag{3.183}$$

so $\mathbf{L}$ does serve as the generator of rotations for the electromagnetic momentum. By combining the commutation relations given above, it is straightforward to show that L_z, S_z, S^2, P_z, and P^2 all commute. With this information it is possible to replace the plane-wave modes with a new set of modes (closely related to vector spherical harmonics (Jackson, 1999, Sec. 9.7)) that provide a representation in which both L_z and S_z are diagonal in the helicity. The details of these interesting formal developments can be found in the original literature, e.g. van Enk and Nienhuis (1994), but this approach has not proved to be particularly useful for the analysis of existing experiments.

The experiments reviewed in Section 3.1.3-E all involve paraxial waves, i.e. the field in each case is a superposition of plane waves with propagation vectors nearly parallel to the main propagation direction. In this situation, the z-axis can be taken along the propagation direction, and we will see in Chapter 7 that the operators S_z and L_z are, at least approximately, the generators of spin and orbital rotations respectively.

3.5 Wave packet quantization*

While the method of box-quantization is very useful in many applications, it has both conceptual and practical shortcomings. In Section 3.1.1 we replaced the quantum rules (2.61) for the physical cavity by the position-space commutation relations (3.1) and (3.3) on the grounds that the macroscopic boundary conditions at the cavity walls do not belong in a microscopic theory. The imaginary cavity with periodic boundary conditions is equally out of place, so it would clearly be more satisfactory to deal directly with the position-space commutation relations. A practical shortcoming of the box-quantization method is that it does not readily lend itself to the description of incident fields that are not simple plane waves. In real experiments the incident fields are more accurately described by Gaussian beams (Yariv, 1989, Sec. 6.6); consequently, it would be better to have a more flexible method that can accommodate incident fields of various types.

In this section we will develop a representation of the field operators that deals directly with the singular commutation relations in a mathematically and physically sensible way. This new representation depends on the definition of the electromagnetic phase space in terms of normalizable classical wave packets. Creation and annihilation operators defined in terms of these wave packets will replace the box-quantized operators.

3.5.1 Electromagnetic phase space

In classical mechanics, the state of a single particle is described by the ordered pair $(\mathbf{q}, \mathbf{p})$, where $\mathbf{q}$ and $\mathbf{p}$ are respectively the canonical coordinate and momentum of the particle. The pairs, $(\mathbf{q}, \mathbf{p})$, of vectors label the points of the *mechanical phase space* Γ_{mech}, and a unique trajectory $(\mathbf{q}(t), \mathbf{p}(t))$ is defined by the initial conditions $(\mathbf{q}(0), \mathbf{p}(0)) = (\mathbf{q}_0, \mathbf{p}_0)$. A unique solution of Maxwell's equations is determined by the initial conditions

$$\begin{aligned} \boldsymbol{\mathcal{A}}(\mathbf{r}, 0) &= \boldsymbol{\mathcal{A}}_0(\mathbf{r}), \\ \boldsymbol{\mathcal{E}}(\mathbf{r}, 0) &= \boldsymbol{\mathcal{E}}_0(\mathbf{r}), \end{aligned} \tag{3.184}$$

where $\boldsymbol{\mathcal{A}}_0(\mathbf{r})$ and $\boldsymbol{\mathcal{E}}_0(\mathbf{r})$ are given functions of $\mathbf{r}$. By analogy to the mechanical case, the points of **electromagnetic phase space** Γ_{em} are labeled by pairs of real transverse vector fields, $(\boldsymbol{\mathcal{A}}(\mathbf{r}), -\boldsymbol{\mathcal{E}}(\mathbf{r}))$. The use of $-\boldsymbol{\mathcal{E}}(\mathbf{r})$ rather than $\boldsymbol{\mathcal{E}}(\mathbf{r})$ is suggested by the commutation relations (3.3), and it also follows from the classical Lagrangian formulation (Cohen-Tannoudji *et al.*, 1989, Sec. II.A.2).

A more useful representation of Γ_{em} can be obtained from the classical part of the analysis, in Section 3.3.5, of quantization in a weakly dispersive dielectric. Since the vacuum is the ultimate nondispersive dielectric, we can directly apply eqn (3.167) to see that the general solution of the vacuum Maxwell equations is determined by

$$\boldsymbol{\mathcal{A}}^{(+)}(\mathbf{r},t)=\int\frac{d^3k}{(2\pi)^3}\sum_s\sqrt{\frac{\hbar}{2\epsilon_0\omega_k}}w_s(\mathbf{k})\,\mathbf{e}_{\mathbf{k}s}e^{i(\mathbf{k}\cdot\mathbf{r}-\omega_k t)}\,, \tag{3.185}$$

where we have applied the rules (3.64) to get the free-space form. The complex functions $w_s(\mathbf{k})$ and the two-component functions $w(\mathbf{k})=(w_+(\mathbf{k}),w_-(\mathbf{k}))$ are respectively called **polarization amplitudes** and **wave packets**. The classical energy for this solution is

$$\mathcal{U}=\int\frac{d^3k}{(2\pi)^3}\hbar\omega_k\sum_s|w_s(\mathbf{k})|^2\,. \tag{3.186}$$

Physically realizable classical fields must have finite total energy, i.e. $\mathcal{U}<\infty$, but Einstein's quantum model suggests an additional and independent condition. This comes from the interpretation of $|w_s(\mathbf{k})|^2\,d^3k/(2\pi)^3$ as the number of quanta with polarization $\mathbf{e}_s(\mathbf{k})$ in the reciprocal-space volume element d^3k centered on $\mathbf{k}$. With this it is natural to restrict the polarization amplitudes by the normalizability condition,

$$\int\frac{d^3k}{(2\pi)^3}\sum_s|w_s(\mathbf{k})|^2<\infty\,, \tag{3.187}$$

which guarantees that the total number of quanta is finite. For normalizable wave packets w and v the Cauchy–Schwarz inequality (A.9) guarantees the existence of the inner product

$$(v,w)=\int\frac{d^3k}{(2\pi)^3}\sum_s v_s^*(\mathbf{k})\,w_s(\mathbf{k})\,; \tag{3.188}$$

therefore, the normalizable wave packets form a Hilbert space. We emphasize that this is a Hilbert space of classical fields, not a Hilbert space of quantum states. We will therefore identify the electromagnetic phase space Γ_{em} with the Hilbert space of normalizable wave packets,

$$\Gamma_{\text{em}}=\{w(\mathbf{k})\ \text{with}\ (w,w)<\infty\}\,. \tag{3.189}$$

3.5.2 Wave packet operators

The right side of eqn (3.16) is a generalized function (see Appendix A.6.2) which means that it is only defined by its action on well behaved ordinary functions. Another way of putting this is that $\Delta_{ij}^{\perp}(\mathbf{r})$ does not have a specific numerical value at the point $\mathbf{r}$; instead, only averages over suitable weighting functions are well defined, e.g.

$$\int d^3r'\,\Delta_{ij}^{\perp}(\mathbf{r}-\mathbf{r}')\,\mathcal{Y}_j(\mathbf{r}')\,, \tag{3.190}$$

where $\boldsymbol{\mathcal{Y}}(\mathbf{r})$ is a smooth classical field that vanishes rapidly as $|\mathbf{r}|\to\infty$. The appearance of the generalized function $\Delta_{ij}^{\perp}(\mathbf{r}-\mathbf{r}')$ in the commutation relations implies that $\mathbf{A}^{(+)}(\mathbf{r})$ and $\mathbf{A}^{(-)}(\mathbf{r})$ must be operator-valued generalized functions. In other words only suitable spatial averages of $\mathbf{A}^{(\pm)}(\mathbf{r})$ are well-defined operators. This conclusion is consistent with eqn (2.185), which demonstrates that vacuum fluctuations in $\mathbf{E}$ are divergent at every point $\mathbf{r}$. As far as mathematics is concerned, any sufficiently well behaved averaging function will do, but on physical grounds the classical wave packets defined in Section 3.5.1 hold a privileged position. Thus the singular object $A_i^{(+)}(\mathbf{r}) = \mathbf{u}_i \cdot \mathbf{A}^{(+)}(\mathbf{r})$ should be replaced by the projection of $\mathbf{A}^{(+)}$ on a wave packet. This can be expressed directly in position space but it is simpler to go over to reciprocal space and define the **wave packet annihilation operators**

$$a[w] = \int \frac{d^3k}{(2\pi)^3} \sum_s w_s^*(\mathbf{k})\, a_s(\mathbf{k})\,. \tag{3.191}$$

Combining the singular commutation relation (3.26) with the definition (3.188) yields the mathematically respectable relations

$$\left[a[w], a^\dagger[v]\right] = (w, v)\,. \tag{3.192}$$

The number operator N defined by eqn (3.30) satisfies

$$[N, a[w]] = -a[w]\,, \quad \left[N, a^\dagger[w]\right] = a^\dagger[w]\,, \tag{3.193}$$

so the Fock space $\mathfrak{H}_F$ can be constructed as the Hilbert space spanned by all vectors of the form

$$\left|w^{(1)}, \ldots, w^{(n)}\right\rangle = a^\dagger\left[w^{(1)}\right] \cdots a^\dagger\left[w^{(n)}\right] |0\rangle\,, \tag{3.194}$$

where $n = 0, 1, \ldots$ and the $w^{(j)}$s range over the classical phase space Γ_{em}. For example, the one-photon state $|1_w\rangle = a^\dagger[w]|0\rangle$ is normalizable, since

$$\langle 1_w | 1_w \rangle = (w, w) = \int \frac{d^3k}{(2\pi)^3} \sum_s |w_s(\mathbf{k})|^2 < \infty\,. \tag{3.195}$$

Thus eqn (3.192) provides an interpretation of the singular commutation relations that is both physically and mathematically acceptable (Deutsch, 1991).

Experiments in quantum optics are often described in a rather schematic way by treating the incident and scattered fields as plane waves. The physical fields generated by real sources and manipulated by optical devices are never this simple. A more accurate, although still idealized, treatment represents the incident fields as normalized wave packets, e.g. the Gaussian pulses that will be described in Section 7.4. In a typical experimental situation the initial state would be

$$|\text{in}\rangle = a^\dagger\left[w^{(1)}\right] \cdots a^\dagger\left[w^{(n)}\right] |0\rangle\,. \tag{3.196}$$

This technique will work even if the different wave packets are not orthogonal. The subsequent evolution can be calculated in the Schrödinger picture, by solving the Schrödinger equation with the initial state vector $|\Psi(0)\rangle = |\text{in}\rangle$, or in the Heisenberg picture, by following the evolution of the field operators. In practice an incident field is usually described by the initial electric field $\boldsymbol{\mathcal{E}}_{\text{in}}(\mathbf{r},0)$. According to eqn (3.185),

$$\boldsymbol{\mathcal{E}}_{\text{in}}^{(+)}(\mathbf{r},0)=i\int\frac{d^3k}{(2\pi)^3}\sum_s\sqrt{\frac{\hbar\omega_k}{2\epsilon_0}}w_s(\mathbf{k})\,\mathbf{e}_{\mathbf{k}s}e^{i\mathbf{k}\cdot\mathbf{r}}\,,\tag{3.197}$$

so the wave packets are given by

$$w_s(\mathbf{k})=-i\sqrt{\frac{2\epsilon_0}{\hbar\omega_k}}\mathbf{e}_{\mathbf{k}s}^*\cdot\int d^3re^{-i\mathbf{k}\cdot\mathbf{r}}\boldsymbol{\mathcal{E}}_{\text{in}}^{(+)}(\mathbf{r},0)\,.\tag{3.198}$$

3.6 Photon localizability*

3.6.1 Is there a photon position operator?

The use of the term photon to mean 'quantum of excitation of the electromagnetic field' is a harmless piece of jargon, but the extended sense in which photons are thought to be localizable particles raises subtle and fundamental issues. In order to concentrate on the essential features of this problem, we will restrict the discussion to photons propagating in vacuum. The particle concept originated in classical mechanics, where it is understood to mean a physical system of negligible extent that occupies a definite position in space. The complete description of the state of a classical particle is given by its instantaneous position and momentum. In nonrelativistic quantum mechanics, the uncertainty principle forbids the simultaneous specification of position and momentum, so the state of a particle is instead described by a wave function $\psi(\mathbf{r})$. More precisely, $\psi(\mathbf{r})=\langle\mathbf{r}|\psi\rangle$ is the probability amplitude that a measurement of the position operator $\widehat{\mathbf{r}}$ will yield the value $\mathbf{r}$, and leave the particle in the corresponding eigenvector $|\mathbf{r}\rangle$ defined by $\widehat{\mathbf{r}}|\mathbf{r}\rangle=\mathbf{r}|\mathbf{r}\rangle$. The improper eigenvector $|\mathbf{r}\rangle$ is discussed in Appendix C.1.1-B. The identity

$$|\psi\rangle=\int d^3r\;|\mathbf{r}\rangle\langle\mathbf{r}|\psi\rangle\tag{3.199}$$

shows that the wave function $\psi(\mathbf{r})$ is simply the projection of the state vector on the basis vector $|\mathbf{r}\rangle$. The action of the position operator $\widehat{\mathbf{r}}$ is given by $\langle\mathbf{r}|\widehat{\mathbf{r}}|\psi\rangle=\mathbf{r}\langle\mathbf{r}|\psi\rangle$, which is usually written as $\widehat{\mathbf{r}}\psi(\mathbf{r})=\mathbf{r}\psi(\mathbf{r})$.

Thus the notion of a particle in nonrelativistic quantum mechanics depends on the existence of a physically sensible position operator. Position operators exist in nonrelativistic quantum theory for particles with any spin, and even for the relativistic theory of massive, spin-1/2 particles described by the Dirac equation; but, there is no position operator for the massless, spin-1 objects described by Maxwell's equations (Newton and Wigner, 1949).

A more general approach would be to ask if there is any operator that would serve to describe the photon as a localizable object. In nonrelativistic quantum mechanics the position operator $\widehat{\mathbf{r}}$ has two essential properties.

(a) The components commute with one another: $[\widehat{r}_i,\widehat{r}_j]=0$.

(b) The operator $\widehat{\mathbf{r}}$ transforms as a vector under rotations of the coordinate system.

Property (a) is necessary if the components of the position are to be simultaneously measurable, and property (b) would seem to be required for the physical interpretation of $\widehat{\mathbf{r}}$ as representing a location in space. Over the years many proposals for a photon position operator have been made, with one of two outcomes: (1) when (a) is satisfied, then (b) is not (Hawton and Baylis, 2001); (2) when (b) is satisfied, then (a) is not (Pryce, 1948). Thus there does not appear to be a physically acceptable photon position operator; consequently, there is no position-space wave function for the photon. This apparent difficulty has a long history in the literature, but there are at least two reasons for not taking it very seriously. The first is that the relevant classical theory—Maxwell's equations—has no particle concept. The second is that photons are inherently relativistic, by virtue of their vanishing rest mass. Consequently, ordinary notions connected to the Schrödinger equation need not apply.

3.6.2 Are there local number operators?

The nonexistence of a photon position operator still leaves open the possibility that there is some other sense in which the photon may be considered as a localizable or particle-like object. From an operational point of view, a minimum requirement for localizability would seem to be that the number of photons in a finite volume V is an observable, represented by a local number operator $N(V)$. Since simultaneous measurements in nonoverlapping volumes of space cannot interfere, this family of observables should satisfy

$$[N(V), N(V')] = 0 \tag{3.200}$$

whenever V and V' do not overlap. The standard expression (3.30) for the total number operator as an integral over plane waves is clearly not a useful starting point for the construction of a local number operator, so we will instead use eqns (3.49) and (3.15) to get

$$N = \frac{2\epsilon_0}{\hbar c} \int d^3r \mathbf{E}^{(-)}(\mathbf{r}) \cdot \left(-\nabla^2\right)^{-1/2} \mathbf{E}^{(+)}(\mathbf{r}) . \tag{3.201}$$

In the classical limit, the field operators are replaced by classical fields, and the $\hbar$ in the denominator goes to zero. Thus the number operator diverges in the classical limit, in agreement with the intuitive idea that there are effectively an infinite number of photons in a classical field.

The first suggestion for $N(V)$ is simply to restrict the integral to the volume V (Henley and Thirring, 1964, p. 43); but this is problematical, since the integrand in eqn (3.201) is not a positive-definite operator. This poses no problem for the total number operator, since the equivalent reciprocal-space representation (3.30) is nonnegative, but this version of a local number operator might have negative expectation values in some states. This objection can be met by using $\left(-\nabla^2\right)^{1/2} = \left(-\nabla^2\right)^{1/4}\left(-\nabla^2\right)^{1/4}$ and the general rule (3.21) to replace the position-space integral (3.201) by the equivalent form

$$N = \int d^3r \mathbf{M}^{\dagger}(\mathbf{r}) \cdot \mathbf{M}(\mathbf{r}) , \tag{3.202}$$

where

$$\mathbf{M}(\mathbf{r}) = -i\sqrt{\frac{2\epsilon_0}{\hbar c}} \left(-\nabla^2\right)^{-1/4} \mathbf{E}^{(+)}(\mathbf{r}) . \tag{3.203}$$

The integrand in eqn (3.202) is a positive-definite operator, so the local number operator defined by

$$N(V) = \int_V d^3r \mathbf{M}^\dagger(\mathbf{r}) \cdot \mathbf{M}(\mathbf{r}) \tag{3.204}$$

is guaranteed to have a nonnegative expectation value for any state. According to the standard plane-wave representation (3.29), the operator $\mathbf{M}(\mathbf{r})$ is

$$\mathbf{M}(\mathbf{r}) = \int \frac{d^3k}{(2\pi)^3} \sum_s \mathbf{e}_s(\mathbf{k}) a_s(\mathbf{k}) e^{i\mathbf{k}\cdot\mathbf{r}}, \tag{3.205}$$

i.e. it is the Fourier transform of the operator $\mathbf{M}(\mathbf{k})$ introduced in eqn (3.56). The position-space form $\mathbf{M}(\mathbf{r})$ is the **detection operator** introduced by Mandel in his study of photon detection (Mandel, 1966), and $N(V)$ is Mandel's **local number operator**. The commutation relations (3.25) and (3.26) can be used to show that the detection operator satisfies

$$\left[M_i(\mathbf{r}), M_j^\dagger(\mathbf{r}')\right] = \Delta_{ij}^\perp(\mathbf{r}-\mathbf{r}'), \quad [M_i(\mathbf{r}), M_j(\mathbf{r}')] = 0. \tag{3.206}$$

Now consider disjoint volumes V and V' with centers separated by a distance R which is large compared to the diameters of the volumes. Substituting eqn (3.204) into $[N(V), N(V')]$ and using eqn (3.206) yields

$$[N(V), N(V')] = \int_V d^3r \int_{V'} d^3r' \mathcal{S}_{ij}(\mathbf{r}, \mathbf{r}') \Delta_{ij}^\perp(\mathbf{r}-\mathbf{r}'), \tag{3.207}$$

where $\mathcal{S}_{ij}(\mathbf{r}, \mathbf{r}') = M_i^\dagger(\mathbf{r}) M_j(\mathbf{r}') - M_j^\dagger(\mathbf{r}') M_i(\mathbf{r})$. The definition of the transverse delta function given by eqns (2.30) and (2.28) can be combined with the general relation (3.18) to get the equivalent expression,

$$\Delta_{ij}^\perp(\mathbf{r}-\mathbf{r}') = \delta_{ij}\delta(\mathbf{r}-\mathbf{r}') + \frac{1}{4\pi}\nabla_i\nabla_j\frac{1}{|\mathbf{r}-\mathbf{r}'|}. \tag{3.208}$$

Since V and V' are disjoint, the delta function term cannot contribute to eqn (3.207), so

$$[N(V), N(V')] = \int_V d^3r \int_{V'} d^3r' \mathcal{S}_{ij}(\mathbf{r}, \mathbf{r}') \frac{1}{4\pi}\nabla_i\nabla_j\frac{1}{|\mathbf{r}-\mathbf{r}'|}. \tag{3.209}$$

A straightforward estimate shows that $[N(V), N(V')] \sim R^{-3}$. Thus the commutator between these proposed local number operators does not vanish for nonoverlapping volumes; indeed, it does not even decay very rapidly as the separation between the volumes increases. This counterintuitive behavior is caused by the nonlocal field commutator (3.16) which is a consequence of the transverse nature of the electromagnetic field.

The alternative definition (Deutsch and Garrison, 1991*a*),

$$G(V) = \frac{2\epsilon_0}{\hbar\omega_0} \int_V d^3r \mathbf{E}^{(-)}(\mathbf{r}) \cdot \mathbf{E}^{(+)}(\mathbf{r}), \tag{3.210}$$

of a local number operator is suggested by the Glauber theory of photon detection, which is discussed in Section 9.1.2. Rather than anticipating later results we will obtain eqn (3.210) by a simple plausibility argument. The representation (3.39) for the field Hamiltonian suggests interpreting $2\epsilon_0 \mathbf{E}^{(-)} \cdot \mathbf{E}^{(+)}$ as the energy density operator. For a monochromatic field state this in turn suggests that $2\epsilon_0 \mathbf{E}^{(-)} \cdot \mathbf{E}^{(+)}/\hbar\omega_0$ be interpreted as the photon density operator. The expression (3.210) is an immediate consequence of these assumptions. The integrand in this equation is clearly positive definite, but nonlocal effects show up here as well.

The failure of several plausible candidates for a local number operator strongly suggests that there is no such object. If this conclusion is supported by future research, it would mean that photons are nonlocalizable in a very fundamental way.

3.7 Exercises

3.1 The field commutator

Verify the expansions (2.101) and (2.103), and use them to derive eqns (3.1) and (3.3).

3.2 Uncertainty relations for E and B

(1) Derive eqn (3.4) from eqn (3.3).

(2) Consider smooth distributions of classical polarization $\boldsymbol{\mathcal{P}}(\mathbf{r})$ and magnetization $\boldsymbol{\mathcal{M}}(\mathbf{r})$ which vanish outside finite volumes V_P and V_M respectively, as in Section 2.5. The interaction energies are

$$W_E = -\int d^3r \boldsymbol{\mathcal{P}}(\mathbf{r}) \cdot \mathbf{E}(\mathbf{r}), \quad W_B = -\int d^3r \boldsymbol{\mathcal{M}}(\mathbf{r}) \cdot \mathbf{B}(\mathbf{r}).$$

Show that

$$[W_B, W_E] = -\frac{i\hbar}{\epsilon_0} \int d^3r \boldsymbol{\mathcal{P}}(\mathbf{r}) \cdot \nabla \times \boldsymbol{\mathcal{M}}(\mathbf{r}).$$

(3) What assumption about the volumes V_P and V_M will guarantee that W_B and W_E are simultaneously measurable?

(4) Use the standard argument from quantum mechanics (Bransden and Joachain, 1989, Sec. 5.4) to show that W_B and W_E satisfy an uncertainty relation

$$\Delta W_B \Delta W_E \geqslant \hbar K,$$

and evaluate the constant K.

3.3 Electromagnetic Hamiltonian

Carry out the derivation of eqns (3.37)–(3.41).

3.4 Electromagnetic momentum

Fill in the steps leading from the classical expression (3.42) to the quantum form (3.48) for the electromagnetic momentum operator.

3.5 Milonni's quantization scheme*

Fill in the details required to go from eqn (3.159) to eqn (3.164).

3.6 Electromagnetic angular momentum*

Carry out the calculations needed to derive eqns (3.172)–(3.178).

3.7 Wave packet quantization*

(1) Derive eqns (3.192), (3.193), and (3.195).

(2) Derive the expression for $\langle 1_w | 1_v \rangle$, where w and v are wave packets in Γ_{em}.

4

Interaction of light with matter

In the previous chapters we have dealt with the free electromagnetic field, undisturbed by the presence of charges. This is an important part of the story, but all experiments involve the interaction of light with matter containing finite amounts of quantized charge, e.g. electrons in atoms or conduction electrons in semiconductors. It is therefore time to construct a unified picture in which both light and matter are treated by quantum theory. We begin in Section 4.1 with a brief review of semiclassical electrodynamics, the standard quantum theory of nonrelativistic charged particles interacting with a classical electromagnetic field. The next step is to treat both charges and fields by quantum theory. For this purpose, we need a Hilbert space describing both the charged particles and the quantized electromagnetic field. The necessary machinery is constructed in Section 4.2. We present the Heisenberg-picture description of the full theory in Sections 4.3–4.7. In Sections 4.8 and 4.9, the interaction picture is introduced and applied to atom–photon coupling.

4.1 Semiclassical electrodynamics

In order to have something reasonably concrete to discuss, we will consider a system of N point charges. The pure states are customarily described by N-body wave functions, $\psi(\mathbf{r}_1, \ldots, \mathbf{r}_N)$, in configuration space. The position and momentum operators $\widehat{\mathbf{r}}_n$ and $\widehat{\mathbf{p}}_n$ for the nth particle are respectively defined by

$$\begin{aligned} \widehat{\mathbf{r}}_n \psi(\mathbf{r}_1, \ldots, \mathbf{r}_N) &= \mathbf{r}_n \psi(\mathbf{r}_1, \ldots, \mathbf{r}_N), \\ \widehat{\mathbf{p}}_n \psi(\mathbf{r}_1, \ldots, \mathbf{r}_N) &= -i\hbar \frac{\partial}{\partial \mathbf{r}_n} \psi(\mathbf{r}_1, \ldots, \mathbf{r}_N). \end{aligned} \tag{4.1}$$

The Hilbert space, $\mathfrak{H}_{\text{chg}}$, for the charges consists of the normalizable N-body wave functions, i.e.

$$\int d^3r_1 \cdots \int d^3r_N \; |\psi(\mathbf{r}_1, \ldots, \mathbf{r}_N)|^2 < \infty. \tag{4.2}$$

In all applications some of the particles will be fermions, e.g. electrons, and others will be bosons, so the wave functions must be antisymmetrized or symmetrized accordingly, as explained in Section 6.5.1.

In the semiclassical approximation the Hamiltonian for a system of charged particles coupled to a classical field is constructed by combining the correspondence principle with the idea of minimal coupling explained in Appendix C.6. The result is

$$H_{\text{sc}} = \sum_{n=1}^{N} \frac{\left(\widehat{\mathbf{p}}_n - q_n \boldsymbol{\mathcal{A}}(\widehat{\mathbf{r}}_n, t)\right)^2}{2M_n} + \sum_{n=1}^{N} q_n \varphi(\widehat{\mathbf{r}}_n, t), \tag{4.3}$$

where $\mathcal{A}$ and φ are respectively the (c-number) vector and scalar potentials, and q_n and M_n are respectively the charge and mass of the nth particle. In this formulation there are two forms of momentum: the **canonical** momentum,

$$\widehat{\mathbf{p}}_{n,\text{can}} = \widehat{\mathbf{p}}_n = -i\hbar\frac{\partial}{\partial \mathbf{r}_n}, \tag{4.4}$$

and the **kinetic** momentum,

$$\widehat{\mathbf{p}}_{n,\text{kin}} = \widehat{\mathbf{p}}_n - q_n\mathcal{A}\left(\widehat{\mathbf{r}}_n, t\right). \tag{4.5}$$

The canonical momentum is the generator of **spatial translations**, while the classical momentum $M\mathbf{v}$ is the correspondence-principle limit of the kinetic momentum.

It is worthwhile to pause for a moment to consider where this argument has led us. The classical fields $\mathcal{A}(\mathbf{r}, t)$ and $\varphi(\mathbf{r}, t)$ are by definition c-number functions of position $\mathbf{r}$ in space, but (4.3) requires that they be evaluated at the position of a charged particle, which is described by the operator $\widehat{\mathbf{r}}_n$. What, then, is the meaning of $\mathcal{A}\left(\widehat{\mathbf{r}}_n, t\right)$? To get a concrete feeling for this question, let us recall that the classical field can be expanded in plane waves $\exp\left(i\mathbf{k}\cdot\mathbf{r} - i\omega_k t\right)$. The operator $\exp\left(i\mathbf{k}\cdot\widehat{\mathbf{r}}_n\right)$ arising from the replacement of $\mathbf{r}_n$ by $\widehat{\mathbf{r}}_n$ is defined by the rule

$$e^{i\mathbf{k}\cdot\widehat{\mathbf{r}}_n}\psi\left(\mathbf{r}_1, \ldots, \mathbf{r}_N\right) = e^{i\mathbf{k}\cdot\mathbf{r}_n}\psi\left(\mathbf{r}_1, \ldots, \mathbf{r}_N\right), \tag{4.6}$$

where $\psi\left(\mathbf{r}_1, \ldots, \mathbf{r}_N\right)$ is any position-space wave function for the charged particles. In this way $\mathcal{A}\left(\widehat{\mathbf{r}}_n, t\right)$ becomes an operator acting on the state vector of the charged particles. This implies, for example, that $\mathcal{A}\left(\widehat{\mathbf{r}}_n, t\right)$ does not commute with $\widehat{\mathbf{p}}_n$, but instead satisfies

$$\left[\mathcal{A}_i\left(\widehat{\mathbf{r}}_n, t\right), \widehat{p}_{nj}\right] = i\hbar\frac{\partial \mathcal{A}_i}{\partial r_j}\left(\widehat{\mathbf{r}}_n, t\right). \tag{4.7}$$

The scalar potential $\varphi\left(\widehat{\mathbf{r}}_n, t\right)$ is interpreted in the same way.

The standard wave function description of the charged particles is useful for deriving the semiclassical Hamiltonian, but it is not particularly convenient for the applications to follow. In general it is better to use Dirac's presentation of quantum theory, in which the state is represented by a ket vector $|\psi\rangle$. For the system of charged particles the two versions are related by

$$\psi\left(\mathbf{r}_1, \ldots, \mathbf{r}_N\right) = \left\langle \mathbf{r}_1, \ldots, \mathbf{r}_N \middle| \psi \right\rangle, \tag{4.8}$$

where $\left|\mathbf{r}_1, \ldots, \mathbf{r}_N\right\rangle$ is a simultaneous eigenket of the position operators $\widehat{\mathbf{r}}_n$, i.e.

$$\widehat{\mathbf{r}}_n\left|\mathbf{r}_1, \ldots, \mathbf{r}_N\right\rangle = \mathbf{r}_n\left|\mathbf{r}_1, \ldots, \mathbf{r}_N\right\rangle, \; n = 1, \ldots, N. \tag{4.9}$$

In this formulation the wave function $\psi\left(\mathbf{r}_1, \ldots, \mathbf{r}_N\right)$ simply gives the components of the vector $|\psi\rangle$ with respect to the basis provided by the eigenvectors $\left|\mathbf{r}_1, \ldots, \mathbf{r}_N\right\rangle$. Any other set of basis vectors for $\mathfrak{H}_{\text{chg}}$ would do equally well.

4.2 Quantum electrodynamics

In semiclassical electrodynamics the state of the physical system is completely described by a many-body wave function belonging to the Hilbert space $\mathfrak{H}_{\text{chg}}$ defined by eqn (4.2), but this description is not adequate when the electromagnetic field is also treated by quantum theory. In Section 4.2.1 we show how to combine the charged-particle space $\mathfrak{H}_{\text{chg}}$ with the Fock space $\mathfrak{H}_F$, defined by eqn (3.35), to get the state space, $\mathfrak{H}_{\text{QED}}$, for the composite system of the charges and the quantized electromagnetic field. In Section 4.2.2 we construct the Hamiltonian for the composite charge-field system by appealing to the correspondence principle for the quantized electromagnetic field.

4.2.1 The Hilbert space

In quantum mechanics, many-body wave functions are constructed from single-particle wave functions by forming linear combinations of product wave functions. For example, the two-particle wave functions for distinguishable particles A and B have the general form

$$\psi(\mathbf{r}_A, \mathbf{r}_B) = C_1\psi_1(\mathbf{r}_A)\chi_1(\mathbf{r}_B) + C_2\psi_2(\mathbf{r}_A)\chi_2(\mathbf{r}_B) + \cdots . \tag{4.10}$$

Since wave functions are meaningless for photons, it is not immediately clear how this procedure can be applied to the radiation field. The way around this apparent difficulty begins with the reminder that the wave function for a particle, e.g. $\psi_1(\mathbf{r}_A)$, is a probability amplitude for the outcomes of measurements of position. In the standard approach to the quantum measurement problem—reviewed in Appendix C.2—a measurement of the position operator $\widehat{\mathbf{r}}_A$ always results in one of the eigenvalues $\mathbf{r}_A$, and the particle is left in the corresponding eigenstate $|\mathbf{r}_A\rangle$. If the particle is initially prepared in the state $|\psi_1\rangle_A$, then the wave function is simply the probability amplitude for this outcome: $\psi_1(\mathbf{r}_A) = \langle \mathbf{r}_A | \psi_1 \rangle$. The next step is to realize that the position operators $\widehat{\mathbf{r}}_A$ do not play a privileged role, even for particles. The components $\widehat{x}_A$, $\widehat{y}_A$, and $\widehat{z}_A$ of $\widehat{\mathbf{r}}_A$ can be replaced by any set of commuting observables $\widehat{O}_{A1}, \widehat{O}_{A2,}, \widehat{O}_{A3}$ with the property that the common eigenvector, defined by

$$\widehat{O}_{An}|O_{A1}, O_{A2}, O_{A3}\rangle = O_{An}|O_{A1}, O_{A2}, O_{A3}\rangle \quad (n = 1, 2, 3), \tag{4.11}$$

is uniquely defined (up to an overall phase). In other words, the observables $\widehat{O}_{A1}$, $\widehat{O}_{A2}$, $\widehat{O}_{A3}$ can be measured simultaneously, and the system is left in a unique state after the measurement.

With these ideas in mind, we can describe the composite system of N charges and the electromagnetic field by relying directly on the Born interpretation and the superposition principle. For the system of N charged particles described by $\mathfrak{H}_{\text{chg}}$, we choose an observable $\widehat{O}$—more precisely, a set of commuting observables—with the property that the eigenvalues O_q are nondegenerate and labeled by a discrete index q. The result of a measurement of $\widehat{O}$ is one of the eigenvalues O_q, and the system is left in the corresponding eigenstate $|O_q\rangle \in \mathfrak{H}_{\text{chg}}$ after the measurement. If the charges are prepared in the state $|\psi\rangle \in \mathfrak{H}_{\text{chg}}$, then the probability amplitude that a measurement of $\widehat{O}$ results in the particular eigenvalue O_q is $\langle O_q | \psi \rangle$. Furthermore, the eigenvectors $|O_q\rangle$ provide a basis for $\mathfrak{H}_{\text{chg}}$; consequently, $|\psi\rangle$ can be expressed as

$$|\psi\rangle = \sum_q |O_q\rangle \langle O_q | \psi \rangle . \tag{4.12}$$

In other words, the state $|\psi\rangle$ is completely determined by the set of probability amplitudes $\{\langle O_q | \psi \rangle\}$ for all possible outcomes of a measurement of $\widehat{O}$.

The same kind of argument works for the electromagnetic field. We use box quantization to get a set of discrete mode labels $\mathbf{k},s$ and consider the set of number operators $\{N_{\mathbf{k}s}\}$. A simultaneous measurement of all the number operators yields a set of occupation numbers $\underline{n} = \{n_{\mathbf{k}s}\}$ and leaves the field in the number state $|\underline{n}\rangle$. If the field is prepared in the state $|\Phi\rangle \in \mathfrak{H}_F$, then the probability amplitude for this outcome is $\langle \underline{n} | \Phi \rangle$. Since the number states form a basis for $\mathfrak{H}_F$, the state vector $|\Phi\rangle$ can be expressed as

$$|\Phi\rangle = \sum_{\underline{n}} |\underline{n}\rangle \langle \underline{n} | \Phi \rangle ; \tag{4.13}$$

consequently, $|\Phi\rangle$ is completely specified by the set of probability amplitudes $\{\langle \underline{n} | \Phi \rangle\}$ for all outcomes of the measurements of the mode number operators. We have used the number operators for convenience in this discussion, but it should be understood that these observables also do not hold a privileged position. Any family of compatible observables such that their simultaneous measurement leaves the field in a unique state would do equally well.

The charged particles and the field are kinematically independent, so the operators $\widehat{O}$ and $N_{\mathbf{k}s}$ commute. In experimental terms, this means that simultaneous measurements of the observables $\widehat{O}$ and $N_{\mathbf{k}s}$ are possible. If the charges and the field are prepared in the states $|\psi\rangle$ and $|\Phi\rangle$ respectively, then the probability for the joint outcome $(O_q, \underline{n})$ is the product of the individual probabilities. Since overall phase factors are irrelevant in quantum theory, we may assume that the probability amplitude for the joint outcome—which we denote by $\langle O_q, \underline{n} | \psi, \Phi \rangle$—is given by the product of the individual amplitudes:

$$\langle O_q, \underline{n} | \psi, \Phi \rangle = \langle O_q | \psi \rangle \langle \underline{n} | \Phi \rangle . \tag{4.14}$$

According to the Born interpretation, the set of probability amplitudes defined by letting O_q and $\underline{n}$ range over all possible values defines a *state* of the composite system, denoted by $|\psi, \Phi\rangle$. The vector corresponding to this state is called a **product vector**, and it is usually written as

$$|\psi, \Phi\rangle = |\psi\rangle |\Phi\rangle , \tag{4.15}$$

where the notation is intended to remind us of the familiar product wave functions in eqn (4.10).

The product vectors do not provide a complete description of the composite system, since the full set of states must satisfy the superposition principle. This means that we are required to give a physical interpretation for superpositions,

$$|\Psi\rangle = C_1 |\psi_1, \Phi_1\rangle + C_2 |\psi_2, \Phi_2\rangle , \tag{4.16}$$

of distinct product vectors. Once again the Born interpretation guides us to the following statement: the superposition $|\Psi\rangle$ is the state defined by the probability amplitudes

$$\begin{aligned}\langle O_q, \underline{n} | \Psi \rangle &= C_1 \langle O_q, \underline{n} | \psi_1, \Phi_1 \rangle + C_2 \langle O_q, \underline{n} | \psi_2, \Phi_2 \rangle \\ &= C_1 \langle O_q | \psi_1 \rangle \langle \underline{n} | \Phi_1 \rangle + C_2 \langle O_q | \psi_2 \rangle \langle \underline{n} | \Phi_2 \rangle . \end{aligned} \tag{4.17}$$

It is important to note that for product vectors like $|\psi\rangle |\Phi\rangle$ the subsystems are each described by a unique state in the respective Hilbert space. The situation is quite different for superpositions like $|\Psi\rangle$; it is impossible to associate a given state with either of the subsystems. In particular, it is not possible to say whether the field is described by $|\Phi_1\rangle$ or $|\Phi_2\rangle$. This feature—which is imposed by the superposition principle—is called entanglement, and its consequences will be extensively studied in Chapter 6.

Combining this understanding of superposition with the completeness of the states $|O_q\rangle$ and $|\underline{n}\rangle_F$ in their respective Hilbert spaces leads to the following definition: the state space, $\mathfrak{H}_{\text{QED}}$, of the charge-field system consists of all superpositions

$$|\Psi\rangle = \sum_q \sum_{\underline{n}} \Psi_{q\underline{n}} |O_q\rangle |\underline{n}\rangle . \tag{4.18}$$

This definition guarantees the satisfaction of the superposition principle, but the Born interpretation also requires a definition of the inner product for states in $\mathfrak{H}_{\text{QED}}$. To this end, we first take eqn (4.14) as the definition of the inner product of the vectors $|O_q, \underline{n}\rangle$ and $|\psi, \Phi\rangle$. Applying this definition to the special choice $|\psi, \Phi\rangle = |O_{q'}, \underline{n}'\rangle$ yields

$$\langle O_q, \underline{n} | O_{q'}, \underline{n}' \rangle = \langle O_q | O_{q'} \rangle \langle \underline{n} | \underline{n}' \rangle = \delta_{q',q} \delta_{\underline{n}',\underline{n}} , \tag{4.19}$$

and the bilinear nature of the inner product finally produces the general definition:

$$\langle \Phi | \Psi \rangle = \sum_q \sum_{\underline{n}} \Phi^*_{q\underline{n}} \Psi_{q\underline{n}} . \tag{4.20}$$

The description of $\mathfrak{H}_{\text{QED}}$ in terms of superpositions of product vectors imposes a similar structure for operators acting on $\mathfrak{H}_{\text{QED}}$. An operator C that acts only on the particle degrees of freedom, i.e. on $\mathfrak{H}_{\text{chg}}$, is defined as an operator on $\mathfrak{H}_{\text{QED}}$ by

$$C |\Psi\rangle = \sum_q \sum_{\underline{n}} \Psi_{q\underline{n}} \{C |O_q\rangle\} |\underline{n}\rangle , \tag{4.21}$$

and an operator acting only on the field degrees of freedom, e.g. $a_{\mathbf{k}s}$, is extended to $\mathfrak{H}_{\text{QED}}$ by

$$a_{\mathbf{k}s} |\Psi\rangle = \sum_q \sum_{\underline{n}} \Psi_{q\underline{n}} |O_q\rangle \{a_{\mathbf{k}s} |\underline{n}\rangle\} . \tag{4.22}$$

Combining these definitions gives the rule

$$C a_{\mathbf{k}s} |\Psi\rangle = \sum_q \sum_{\underline{n}} \Psi_{q\underline{n}} \{C |O_q\rangle\} \{a_{\mathbf{k}s} |\underline{n}\rangle\} . \tag{4.23}$$

A general operator Z acting on $\mathfrak{H}_{\text{QED}}$ can always be expressed as

$$Z = \sum_n C_n F_n \,, \tag{4.24}$$

where C_n acts on $\mathfrak{H}_{\text{chg}}$ and F_n acts on $\mathfrak{H}_F$.

The officially approved mathematical language for this construction is that $\mathfrak{H}_{\text{QED}}$ is the **tensor product** of $\mathfrak{H}_{\text{chg}}$ and $\mathfrak{H}_F$. The standard notation for this is

$$\mathfrak{H}_{\text{QED}} = \mathfrak{H}_{\text{chg}} \otimes \mathfrak{H}_F \,, \tag{4.25}$$

and the corresponding notation $|\psi\rangle \otimes |\Phi\rangle$ is often used for the product vectors. Similarly the operator product $Ca_{\mathbf{k}s}$ is often written as $C \otimes a_{\mathbf{k}s}$.

4.2.2 The Hamiltonian

For the final step to the full quantum theory, we once more call on the correspondence principle to justify replacing the classical field $\boldsymbol{\mathcal{A}}(\mathbf{r}, t)$ in eqn (4.3) by the time-independent, Schrödinger-picture quantum field $\mathbf{A}(\mathbf{r})$. The evaluation of $\mathbf{A}(\mathbf{r})$ at $\widehat{\mathbf{r}}_n$ is understood in the same way as for the classical field $\boldsymbol{\mathcal{A}}(\widehat{\mathbf{r}}_n, t)$, e.g. by using the plane-wave expansion (3.68) to get

$$\mathbf{A}(\widehat{\mathbf{r}}_n) = \sum_{\mathbf{k}s} \sqrt{\frac{\hbar}{2\epsilon_0 \omega_k V}} \mathbf{e}_{\mathbf{k}s}\, a_{\mathbf{k}s} e^{i\mathbf{k}\cdot\widehat{\mathbf{r}}_n} + \text{HC} \,. \tag{4.26}$$

Thus $\mathbf{A}(\widehat{\mathbf{r}}_n)$ is a hybrid operator that acts on the electromagnetic degrees of freedom ($\mathfrak{H}_F$) through the creation and annihilation operators $a^\dagger_{\mathbf{k}s}$ and $a_{\mathbf{k}s}$ and on the particle degrees of freedom ($\mathfrak{H}_{\text{chg}}$) through the operators $\exp(\pm i\mathbf{k}\cdot\widehat{\mathbf{r}}_n)$.

With this understanding we first use the identity

$$\begin{aligned} \mathbf{A}(\widehat{\mathbf{r}}_n)\cdot\widehat{\mathbf{p}}_n + \widehat{\mathbf{p}}_n\cdot\mathbf{A}(\widehat{\mathbf{r}}_n) &= 2\mathbf{A}(\widehat{\mathbf{r}}_n)\cdot\widehat{\mathbf{p}}_n - [A_j(\widehat{\mathbf{r}}_n), p_{nj}] \\ &= 2\mathbf{A}(\widehat{\mathbf{r}}_n)\cdot\widehat{\mathbf{p}}_n + i\hbar\nabla\cdot\mathbf{A}(\widehat{\mathbf{r}}_n) \,, \end{aligned} \tag{4.27}$$

together with $\nabla \cdot \mathbf{A} = 0$ and the identification of φ as the instantaneous Coulomb potential Φ, to evaluate the interaction terms in the radiation gauge. The total Hamiltonian is obtained by adding the zeroth-order Hamiltonian $H_{\text{em}} + H_{\text{chg}}$ to get

$$H = H_{\text{em}} + H_{\text{chg}} + \sum_{n=1}^{N} \frac{(\widehat{\mathbf{p}}_n - q_n\mathbf{A}(\widehat{\mathbf{r}}_n))^2}{2M_n} + \sum_{n=1}^{N} q_n \Phi(\widehat{\mathbf{r}}_n) \,. \tag{4.28}$$

Writing out the various terms leads to the expression

$$H = H_{\text{em}} + H_{\text{chg}} + H_{\text{int}} \,, \tag{4.29}$$

$$H_{\text{em}} = \frac{1}{2}\int d^3r \left(\epsilon_0 : \mathbf{E}^2 : + \mu_0^{-1} : \mathbf{B}^2 :\right) , \tag{4.30}$$

$$H_{\text{chg}} = \sum_{n=1}^{N} \frac{\widehat{\mathbf{p}}_n^2}{2M_n} + \frac{1}{4\pi\epsilon_0} \sum_{n \neq l} \frac{q_n q_l}{|\widehat{\mathbf{r}}_n - \widehat{\mathbf{r}}_l|}, \tag{4.31}$$

$$H_{\text{int}} = -\sum_{n=1}^{N} \frac{q_n}{M_n} \mathbf{A}(\widehat{\mathbf{r}}_n) \cdot \widehat{\mathbf{p}}_n + \sum_{n=1}^{N} \frac{q_n^2 : \mathbf{A}(\widehat{\mathbf{r}}_n)^2 :}{2M_n}. \tag{4.32}$$

In this formulation, H_{em} is the Hamiltonian for the free (transverse) electromagnetic field, and H_{chg} is the Hamiltonian for the charged particles, including their mutual Coulomb interactions. The remaining term, H_{int}, describes the interaction between the transverse (radiative) field and the charges. As in Section 2.2, we have replaced the operators $\mathbf{E}^2$, $\mathbf{B}^2$, and $\mathbf{A}^2$ in H_{em} and H_{int} by their normal-ordered forms, in order to eliminate divergent vacuum fluctuation terms. The Coulomb interactions between the charges—say in an atom—are typically much stronger than the interaction with the transverse field modes, so H_{int} can often be treated as a weak perturbation.

4.3 Quantum Maxwell's equations

In Section 4.2 the interaction between the radiation field and charged particles was described in the Schrödinger picture, but some features are more easily understood in the Heisenberg picture. Since the Hamiltonian has the same form in both pictures, the Heisenberg equations of motion (3.89) can be worked out by using the equal-time commutation relations (3.91) for the fields and the equal-time, canonical commutators, $[\widehat{r}_{ni}(t), \widehat{p}_{lj}(t)] = i\hbar\delta_{nl}\delta_{ij}$, for the charged particles. After a bit of algebra, the Heisenberg equations are found to be

$$\mathbf{E}(\mathbf{r}, t) = -\frac{\partial \mathbf{A}(\mathbf{r}, t)}{\partial t}, \tag{4.33}$$

$$\nabla \times \mathbf{E}(\mathbf{r}, t) = -\frac{\partial \mathbf{B}(\mathbf{r}, t)}{\partial t}, \tag{4.34}$$

$$\nabla \times \mathbf{B}(\mathbf{r}, t) - \frac{1}{c^2}\frac{\partial \mathbf{E}(\mathbf{r}, t)}{\partial t} = \mu_0 \widehat{\mathbf{j}}^{\perp}(\mathbf{r}, t), \tag{4.35}$$

$$\widehat{\mathbf{v}}_n(t) \equiv \frac{d\widehat{\mathbf{r}}_n(t)}{dt} = \frac{\widehat{\mathbf{p}}_n(t) - q_n \mathbf{A}(\widehat{\mathbf{r}}_n(t), t)}{M_n}, \tag{4.36}$$

$$\frac{d\widehat{\mathbf{p}}_n(t)}{dt} = q_n \mathbf{E}(\widehat{\mathbf{r}}_n(t), t) + q_n \widehat{\mathbf{v}}_n(t) \times \mathbf{B}(\widehat{\mathbf{r}}_n(t), t) - q_n \nabla \Phi(\widehat{\mathbf{r}}_n(t)), \tag{4.37}$$

where $\widehat{\mathbf{v}}_n(t)$ is the **velocity operator** for the nth particle, $\widehat{\mathbf{j}}^{\perp}(\mathbf{r}, t)$ is the transverse part of the **current density operator**

$$\widehat{\mathbf{j}}(\mathbf{r}, t) = \sum_n \delta(\mathbf{r} - \widehat{\mathbf{r}}_n(t))\, q_n \widehat{\mathbf{v}}_n(t), \tag{4.38}$$

and the Coulomb potential operator is

$$\Phi(\widehat{\mathbf{r}}_n(t)) = \frac{1}{4\pi\epsilon_0} \sum_{l \neq n} \frac{q_l}{|\widehat{\mathbf{r}}_n(t) - \widehat{\mathbf{r}}_l(t)|}. \tag{4.39}$$

This potential is obtained from a solution of Poisson's equation $\nabla^2\Phi = -\rho/\epsilon_0$, where the charge density operator is

$$\rho(\mathbf{r},t) = \sum_n \delta(\mathbf{r}-\widehat{\mathbf{r}}_n(t))\, q_n\,, \tag{4.40}$$

by omitting the self-interaction terms encountered when $\mathbf{r} \to \widehat{\mathbf{r}}_n(t)$. Functions $f(\widehat{\mathbf{r}}_n)$ of the position operators $\widehat{\mathbf{r}}_n$, such as those in eqns (4.35)–(4.40), are defined by

$$f(\widehat{\mathbf{r}}_n)\,\psi(\mathbf{r}_1,\ldots,\mathbf{r}_N) = f(\mathbf{r}_n)\,\psi(\mathbf{r}_1,\ldots,\mathbf{r}_N)\,, \tag{4.41}$$

where $\psi(\mathbf{r}_1,\ldots,\mathbf{r}_N)$ is any N-body wave function for the charged particles.

The first equation, eqn (4.33), is simply the relation between the transverse part of the electric field operator and the vector potential. Faraday's law, eqn (4.34), is then redundant, since it is the curl of eqn (4.33). The matter equations (4.36) and (4.37) are the quantum versions of the classical force laws of Coulomb and Lorentz.

The only one of the Heisenberg equations that requires further explanation is eqn (4.35) (Ampère's law). The Heisenberg equation of motion for $\mathbf{E}$ can be put into the form

$$(\boldsymbol{\nabla}\times\mathbf{B}(\mathbf{r},t))_j - \frac{1}{c^2}\frac{\partial E_j(\mathbf{r},t)}{\partial t} = \mu_0 \sum_n \Delta^{\perp}_{ji}(\mathbf{r}-\widehat{\mathbf{r}}_n(t))\, q_n \frac{\widehat{p}_{ni} - q_n A_i(\widehat{\mathbf{r}}_n(t),t)}{M_n}\,, \tag{4.42}$$

but the significance of the right-hand side is not immediately obvious. Further insight can be achieved by using the definition (4.36) of the velocity operator to get

$$\frac{\widehat{\mathbf{p}}_n(t) - q_n\mathbf{A}(\widehat{\mathbf{r}}_n(t),t)}{M_n} = \widehat{\mathbf{v}}_n(t)\,. \tag{4.43}$$

Substituting this into eqn (4.42) yields

$$\begin{aligned}(\boldsymbol{\nabla}\times\mathbf{B}(\mathbf{r},t))_j - \frac{1}{c^2}\frac{\partial E_j(\mathbf{r},t)}{\partial t} &= \mu_0 \sum_n \Delta^{\perp}_{ji}(\mathbf{r}-\widehat{\mathbf{r}}_n)\, q_n v_{ni}(t) \\ &= \mu_0 \int d^3r'\,\Delta^{\perp}_{ji}(\mathbf{r}-\mathbf{r}')\,\widehat{j}_i(\mathbf{r}',t)\,, \end{aligned} \tag{4.44}$$

where $\widehat{j}_i(\mathbf{r}',t)$, defined by eqn (4.38), can be interpreted as the current density operator. The transverse delta function $\Delta^{\perp}_{ji}$ projects out the transverse part of any vector field, so the Heisenberg equation for $\mathbf{E}(\mathbf{r},t)$ is given by eqn (4.35).

4.4 Parity and time reversal*

The quantum Maxwell equations, (4.34) and (4.35), and the classical Maxwell equations, (B.2) and (B.3), have the same form; consequently, the field operators and the classical fields behave in the same way under the discrete transformations:

$$\begin{aligned}&\mathbf{r}\to-\mathbf{r}\ \ (\textbf{spatial inversion}\text{ or }\textbf{parity transformation})\,,\\ &t\to-t\ \ (\textbf{time reversal})\,.\end{aligned} \tag{4.45}$$

Thus the transformation laws for the classical fields—see Appendix B.3.3—also apply to the field operators; in particular,

$$\mathbf{E}(\mathbf{r},t) \to \mathbf{E}^{P}(\mathbf{r},t) = -\mathbf{E}(-\mathbf{r},t) \quad \text{under } \mathbf{r} \to -\mathbf{r}\,, \tag{4.46}$$

$$\mathbf{E}(\mathbf{r},t) \to \mathbf{E}^{T}(\mathbf{r},t) = \mathbf{E}(\mathbf{r},-t) \quad \text{under } t \to -\,t\,. \tag{4.47}$$

In classical electrodynamics this is the end of the story, since the entire physical content of the theory is contained in the values of the fields. The situation for quantum electrodynamics is more complicated, because the physical content is shared between the operators and the state vectors. We must therefore find the transformation rules for the states that correspond to the transformations (4.46) and (4.47) for the operators. This effort requires a more careful look at the idea of symmetries in quantum theory.

According to the general rules of quantum theory, all physical predictions can be expressed in terms of probabilities given by $|\langle \Phi | \Psi \rangle|^2$, where $|\Psi\rangle$ and $|\Phi\rangle$ are normalized state vectors. For this reason, a mapping of state vectors to state vectors,

$$|\Theta\rangle \to |\Theta'\rangle\,, \tag{4.48}$$

is called a **symmetry transformation** if

$$|\langle \Phi' | \Psi' \rangle|^2 = |\langle \Phi | \Psi \rangle|^2\,, \tag{4.49}$$

for any pair of vectors $|\Psi\rangle$ and $|\Phi\rangle$. In other words, symmetry transformations leave all physical predictions unchanged. The consequences of this definition are contained in a fundamental theorem due to Wigner.

Theorem 4.1 (Wigner) *Every symmetry transformation can be expressed in one of two forms:*

(a) $|\Psi\rangle \to |\Psi'\rangle = U|\Psi\rangle$, *where* U *is a unitary operator;*

(b) $|\Psi\rangle \to |\Psi'\rangle = \Lambda|\Psi\rangle$, *where* Λ *is an antilinear and antiunitary operator.*

The unfamiliar terms in alternative (b) are defined as follows. A transformation Λ is **antilinear** if

$$\Lambda\{\alpha|\Psi\rangle + \beta|\Phi\rangle\} = \alpha^*\Lambda|\Psi\rangle + \beta^*\Lambda|\Phi\rangle\,, \tag{4.50}$$

and **antiunitary** if

$$\langle \Phi' | \Psi' \rangle = \langle \Psi | \Phi \rangle = \langle \Phi | \Psi \rangle^*\,, \quad \text{where } |\Psi'\rangle = \Lambda|\Psi\rangle \text{ and } |\Phi'\rangle = \Lambda|\Phi\rangle\,. \tag{4.51}$$

Rather than present the proof of Wigner's theorem—which can be found in Wigner (1959, cf. Appendices in Chaps 20 and 26) or Bargmann (1964)—we will attempt to gain some understanding of its meaning. To this end consider another transformation given by

$$|\Psi\rangle \to |\Psi''\rangle = \exp(i\theta_\Psi)\,|\Psi'\rangle\,, \tag{4.52}$$

where θ_Ψ is a real phase that can be chosen independently for each $|\Psi\rangle$. For any value of θ_Ψ it is clear that $|\Psi\rangle \to |\Psi''\rangle$ is also a symmetry transformation. Furthermore, $|\Psi''\rangle$ and $|\Psi'\rangle$ differ only by an overall phase, so they represent the same physical state.

Thus the symmetry transformations defined by eqns (4.48) and (4.52) are physically equivalent, and the meaning of Wigner's theorem is that every symmetry transformation is physically equivalent to one or the other of the two alternatives (a) and (b). This very strong result allows us to find the correct transformation for each case by a simple process of trial and error. If the wrong alternative is chosen, something will go seriously wrong.

Since unitary transformations are a familiar tool, we begin the trial and error process by assuming that the parity transformation (4.46) is realized by a unitary operator U_P:

$$\mathbf{E}^P(\mathbf{r},t) = U_P \mathbf{E}(\mathbf{r},t) U_P^\dagger = -\mathbf{E}(-\mathbf{r},t) . \tag{4.53}$$

In the interaction picture, $\mathbf{E}(\mathbf{r},t)$ has the plane-wave expansion

$$\mathbf{E}(\mathbf{r},t) = \sum_{\mathbf{k}s} i\sqrt{\frac{\hbar\omega_k}{2\epsilon_0 V}} a_{\mathbf{k}s} \mathbf{e}_{\mathbf{k}s} e^{i(\mathbf{k}\cdot\mathbf{r}-\omega_k t)} + \text{HC} , \tag{4.54}$$

and the corresponding classical field has an expansion of the same form, with $a_{\mathbf{k}s}$ replaced by the classical amplitude $\alpha_{\mathbf{k}s}$. In Appendix B.3.3, it is shown that the parity transformation law for the classical amplitude is $\alpha^P_{\mathbf{k}s} = -\alpha_{-\mathbf{k},-s}$. Since U_P is linear, $U_P \mathbf{E}(\mathbf{r},t) U_P^\dagger$ can be expressed in terms of $a^P_{\mathbf{k}s} = U_P a_{\mathbf{k}s} U_P^\dagger$. Comparing the quantum and classical expressions then implies that the unitary transformation of the annihilation operator must have the same form as the classical transformation:

$$a_{\mathbf{k}s} \to a^P_{\mathbf{k}s} = U_P a_{\mathbf{k}s} U_P^\dagger = -a_{-\mathbf{k},-s} . \tag{4.55}$$

The existence of an operator U_P satisfying eqn (4.55) is guaranteed by another well known result of quantum theory discussed in Appendix C.4: two sets of canonically conjugate operators acting in the same Hilbert space are necessarily related by a unitary transformation. Direct calculation from eqn (4.55) yields

$$\begin{aligned} \left[a^P_{\mathbf{k}s}, a^{P\dagger}_{\mathbf{k}'s'}\right] &= \left[-a_{-\mathbf{k},-s}, -a^\dagger_{-\mathbf{k}',-s'}\right] = \delta_{\mathbf{k}\mathbf{k}'}\delta_{ss'} , \\ \left[a^P_{\mathbf{k}s}, a^P_{\mathbf{k}'s'}\right] &= \left[-a_{-\mathbf{k},-ss}, -a_{-\mathbf{k}',-s'}\right] = 0 . \end{aligned} \tag{4.56}$$

Since the operators $a^P_{\mathbf{k}s}$ satisfy the canonical commutation relations, U_P exists. For more explicit properties of U_P, see Exercise 4.4.

The assumption that spatial inversion is accomplished by a unitary transformation worked out very nicely, so we will try the same approach for time reversal, i.e. we assume that there is a unitary operator U_T such that

$$\mathbf{E}^T(\mathbf{r},t) = U_T \mathbf{E}(\mathbf{r},t) U_T^\dagger = \mathbf{E}(\mathbf{r},-t) . \tag{4.57}$$

The classical transformation rule for the plane-wave amplitudes is $\alpha^T_{\mathbf{k}s} = -\alpha^*_{-\mathbf{k},s}$, so the argument used for the parity transformation implies that the annihilation operators satisfy

$$a_{\mathbf{k}s} \to a^T_{\mathbf{k}s} = a^T_{\mathbf{k}s} = U_T a_{\mathbf{k}s} U_T^\dagger = -a^\dagger_{-\mathbf{k},s} . \tag{4.58}$$

All that remains is to check the internal consistency of this rule by using it to evaluate the canonical commutators. The result

$$\left[a_{\mathbf{k}s}^{T}, a_{\mathbf{k}'s'}^{T\dagger}\right] = \left[-a_{-\mathbf{k},s}^{\dagger}, -a_{-\mathbf{k}',s'}\right] = -\delta_{\mathbf{k}\mathbf{k}'}\delta_{ss'} \tag{4.59}$$

is a nasty surprise. The extra minus sign on the right side shows that the transformed operators are not canonically conjugate. Thus the time-reversed operators $a_{\mathbf{k}s}^{T}$ and $a_{\mathbf{k}s}^{T\dagger}$ cannot be related to the original operators $a_{\mathbf{k}s}$ and $a_{\mathbf{k}s}^{\dagger}$ by a unitary transformation, and U_T does not exist.

According to Wigner's theorem, the only possibility left is that the time-reversed operators are defined by an antiunitary transformation,

$$\mathbf{E}^{T}(\mathbf{r},t) = \Lambda_T \mathbf{E}(\mathbf{r},t)\Lambda_T^{-1} = \mathbf{E}(\mathbf{r},-t)\,. \tag{4.60}$$

Here some caution is required because of the unfamiliar properties of antilinear transformations. The definition (4.50) implies that $\Lambda_T \alpha |\Psi\rangle = \alpha^* \Lambda_T |\Psi\rangle$ for any $|\Psi\rangle$, so applying Λ_T to the expansion (4.54) for $\mathbf{E}(\mathbf{r},t)$ gives us

$$\Lambda_T \mathbf{E}(\mathbf{r},t)\Lambda_T^{-1} = \sum_{\mathbf{k}s} i\sqrt{\frac{\hbar\omega_k}{2\epsilon_0 V}}\left\{-a_{\mathbf{k}s}^{T}\mathbf{e}_{\mathbf{k}s}^{*}e^{-i(\mathbf{k}\cdot\mathbf{r}-\omega_k t)} + \left(a_{\mathbf{k}s}^{\dagger}\right)^{T}\mathbf{e}_{\mathbf{k}s}e^{i(\mathbf{k}\cdot\mathbf{r}-\omega_k t)}\right\}, \tag{4.61}$$

where

$$a_{\mathbf{k}s}^{T} = \Lambda_T a_{\mathbf{k}s}\Lambda_T^{-1}\,, \quad \left(a_{\mathbf{k}s}^{\dagger}\right)^{T} = \Lambda_T a_{\mathbf{k}s}^{\dagger}\Lambda_T^{-1}\,. \tag{4.62}$$

Setting $t \to -t$ in eqn (4.54) and changing the summation variable by $\mathbf{k} \to -\mathbf{k}$ yields

$$\mathbf{E}(\mathbf{r},-t) = \sum_{\mathbf{k}s} i\sqrt{\frac{\hbar\omega_k}{2\epsilon_0 V}}\left\{a_{-\mathbf{k}s}\mathbf{e}_{-\mathbf{k}s}e^{-i(\mathbf{k}\cdot\mathbf{r}-\omega_k t)} - a_{-\mathbf{k}s}^{\dagger}\mathbf{e}_{-\mathbf{k}s}^{*}e^{i(\mathbf{k}\cdot\mathbf{r}-\omega_k t)}\right\}. \tag{4.63}$$

After substituting these expansions into eqn (4.60) and using the properties $\mathbf{e}_{-\mathbf{k},-s} = \mathbf{e}_{\mathbf{k}s}$ and $\mathbf{e}_{\mathbf{k},s}^{*} = \mathbf{e}_{\mathbf{k},-s}$ derived in Appendix B.3.3, one finds

$$a_{\mathbf{k}s}^{T} = -a_{-\mathbf{k},s}\,, \quad \left(a_{\mathbf{k}s}^{\dagger}\right)^{T} = -a_{-\mathbf{k},s}^{\dagger}\,. \tag{4.64}$$

This transformation rule gives us $\left(a_{\mathbf{k}s}^{\dagger}\right)^{T} = a_{\mathbf{k}s}^{T\dagger}$ and

$$\left[a_{\mathbf{k}s}^{T}, a_{\mathbf{k}'s'}^{T\dagger}\right] = \left[-a_{-\mathbf{k},s}, -a_{-\mathbf{k}',s'}^{\dagger}\right] = \delta_{\mathbf{k}\mathbf{k}'}\delta_{ss'}\,; \tag{4.65}$$

consequently, the antiunitary transformation yields creation and annihilation operators that satisfy the canonical commutation relations. The magic ingredient in this approach is the extra complex conjugation operation applied by the antilinear transformation Λ_T to the c-number coefficients in eqn (4.61). This is just what is needed to ensure that $a_{\mathbf{k}s}^{T}$ is proportional to $a_{-\mathbf{k},s}$ rather than to $a_{-\mathbf{k},s}^{\dagger}$, as in eqn (4.58).

4.5 Stationary density operators

The expectation value of a single observable is given by

$$\langle X(t)\rangle = \mathrm{Tr}\left[\rho X(t)\right] = \mathrm{Tr}\left[\rho(t)X\right], \tag{4.66}$$

which explicitly shows that the time dependence comes entirely from the observable in the Heisenberg picture and entirely from the density operator in the Schrödinger picture. The time dependence simplifies for the important class of **stationary** density operators, which are defined by requiring the Schrödinger-picture $\rho(t)$ to be a constant of the motion. According to eqn (3.75) this means that $\rho(t)$ is independent of time, so the Schrödinger- and Heisenberg-picture density operators are identical. Stationary density operators have the useful property

$$\left[\rho, U^{\dagger}(t)\right] = 0 = \left[\rho, U(t)\right], \tag{4.67}$$

which is equivalent to

$$[\rho, H] = 0\,. \tag{4.68}$$

Using these properties in conjunction with the cyclic invariance of the trace shows that the expectation value of a single observable is independent of time, i.e.

$$\langle X(t)\rangle = \mathrm{Tr}\left[\rho X(t)\right] = \mathrm{Tr}\left(\rho X\right) = \langle X\rangle\,. \tag{4.69}$$

Correlations between observables at different times are described by averages of the form

$$\langle X(t+\tau)\,Y(t)\rangle = \mathrm{Tr}\left[\rho X(t+\tau)\,Y(t)\right]. \tag{4.70}$$

For a stationary density operator, the correlation only depends on the difference in the time arguments. This is established by combining $U(-t) = U^{\dagger}(t)$ with eqns (3.83), (4.67), and cyclic invariance to get

$$\langle X(t+\tau)\,Y(t)\rangle = \langle X(\tau)\,Y(0)\rangle\,. \tag{4.71}$$

4.6 Positive- and negative-frequency parts for interacting fields

When charged particles are present, the Hamiltonian is given by eqn (4.28), so the free-field solution (3.95) is no longer valid. The operator $a_{\mathbf{k}s}(t)$—evolving from the annihilation operator, $a_{\mathbf{k}s}(0) = a_{\mathbf{k}s}$—will in general depend on the (Schrödinger-picture) creation operators $a^{\dagger}_{\mathbf{k}'s'}$ as well as the annihilation operators $a_{\mathbf{k}'s'}$. The unitary evolution of the operators in the Heisenberg picture does ensure that the general decomposition

$$F(\mathbf{r},t) = F^{(+)}(\mathbf{r},t) + F^{(-)}(\mathbf{r},t) \tag{4.72}$$

will remain valid provided that the initial operator $F^{(+)}(\mathbf{r},0)$ ($F^{(-)}(\mathbf{r},0)$) is a sum over annihilation (creation) operators, but the commutation relations (3.102) are only valid for equal times. Furthermore, $F^{(+)}(\mathbf{r},\omega)$ will not generally vanish for all negative values of ω. Despite this failing, an operator $F^{(+)}(\mathbf{r},t)$ that evolves from an initial operator of the form

$$F^{(+)}(\mathbf{r},0) = \sum_{\mathbf{k}s} F_{\mathbf{k}s} a_{\mathbf{k}s} e^{i\mathbf{k}\cdot\mathbf{r}} \tag{4.73}$$

is still called the positive-frequency part of $F(\mathbf{r},t)$.

4.7 Multi-time correlation functions

One of the advantages of the Heisenberg picture is that it provides a convenient way to study the correlation between quantum fields at different times. This comes about because the state is represented by a time-independent density operator ρ, while the field operators evolve in time according to the Heisenberg equations.

Since the electric field is a vector, it is natural to define the **first-order field correlation function** by the tensor

$$G_{ij}^{(1)}(x_1; x_2) = \left\langle E_i^{(-)}(x_1) E_j^{(+)}(x_2) \right\rangle, \tag{4.74}$$

where $\langle X \rangle = \text{Tr}[\rho X]$ and $x_1 = (\mathbf{r}_1, t_1)$, etc. The first-order correlation functions are directly related to interference and photon-counting experiments. In Section 9.1.2-B we will see that the counting rate for a broadband detector located at $\mathbf{r}$ is proportional to $G_{ij}^{(1)}(\mathbf{r}, t; \mathbf{r}, t)$. For unequal times, $t_1 \neq t_2$, the correlation function $G_{ij}^{(1)}(x_1; x_2)$ represents measurements by a detector placed at the output of a Michelson interferometer with delay time $\tau = |t_1 - t_2|$ between its two arms. In Section 9.1.2-C we will show that the spectral density for the field state ρ is determined by the Fourier transform of $G_{ij}^{(1)}(\mathbf{r}, t; \mathbf{r}, 0)$. The two-slit interference pattern discussed in Section 10.1 is directly given by $G_{ij}^{(1)}(\mathbf{r}, t; \mathbf{r}, 0)$.

We will see in Section 9.2.4 that the **second-order** correlation function, defined by

$$G_{ijkl}^{(2)}(x_1, x_2; x_3, x_4) = \left\langle E_i^{(-)}(x_1) E_j^{(-)}(x_2) E_k^{(+)}(x_3) E_l^{(+)}(x_4) \right\rangle, \tag{4.75}$$

is associated with coincidence counting. Higher-order correlation functions are defined similarly. Other possible expectation values, e.g. $\left\langle E_i^{(-)}(x_1) E_j^{(-)}(x_2) \right\rangle$, are not related to photon detection, so they are normally not considered.

In many applications, the physical situation defines some preferred polarization directions—represented by unit vectors $\mathbf{v}_1, \mathbf{v}_2, \ldots$—and the tensor correlation functions are replaced by scalar functions

$$G^{(1)}(x_1; x_2) = \left\langle E_1^{(-)}(x_1) E_2^{(+)}(x_2) \right\rangle, \tag{4.76}$$

$$G^{(2)}(x_1, x_2; x_3, x_4) = \left\langle E_1^{(-)}(x_1) E_2^{(-)}(x_2) E_3^{(+)}(x_3) E_4^{(+)}(x_4) \right\rangle, \tag{4.77}$$

where $E_p^{(+)} = \mathbf{v}_p^* \cdot \mathbf{E}^{(+)}$ is the projection of the vector operator onto the direction $\mathbf{v}_p$. For example, observing a first-order interference pattern through a polarization filter is described by

$$G^{(1)}(x; x) = \left\langle \mathbf{e} \cdot \mathbf{E}^{(-)}(x) \, \mathbf{e}^* \cdot \mathbf{E}^{(+)}(x) \right\rangle, \tag{4.78}$$

where $\mathbf{e}$ is the polarization transmitted by the filter.

If the density operator is stationary, then an extension of the argument leading to eqn (4.71) shows that the correlation function is unchanged by a uniform translation, $t_p \to t_p + \tau$, $t'_p \to t'_p + \tau$, of all the time arguments. In particular $G_{ij}^{(1)}(\mathbf{r}, t; \mathbf{r}', t') =$

$G_{ij}^{(1)}(\mathbf{r},t-t';\mathbf{r}',0)$, so the first-order function only depends on the difference, $t-t'$, of the time arguments.

The correlation functions satisfy useful inequalities that are based on the fact that

$$\operatorname{Tr}\rho F^{\dagger}F\geqslant 0\,,\tag{4.79}$$

where F is an arbitrary observable and ρ is a density operator. This is readily proved by evaluating the trace in the basis in which ρ is diagonal and using $\left\langle\Psi\left|F^{\dagger}F\right|\Psi\right\rangle\geqslant 0$. Choosing $F=E^{(+)}(x)$ in eqn (4.79) gives

$$G^{(1)}(x;x)\geqslant 0\,,\tag{4.80}$$

and the operator $F=E_1^{(+)}(x_1)\cdots E_n^{(+)}(x_n)$ gives the general positivity condition

$$G^{(n)}(x_1,\ldots,x_n;x_1,\ldots,x_n)\geqslant 0\,.\tag{4.81}$$

A different sort of inequality follows from the choice

$$F=\sum_{a=1}^{n}\xi_a E_a^{(+)}(x_a)\,,\tag{4.82}$$

where the ξ_as are complex numbers. Substituting F into eqn (4.79) yields

$$\sum_{a=1}^{n}\sum_{b=1}^{n}\xi_a^{*}\xi_b\mathcal{F}_{ab}\geqslant 0\,,\tag{4.83}$$

where $\mathcal{F}$ is the $n\times n$ hermitian matrix

$$\mathcal{F}_{ab}=G^{(1)}(x_a;x_b)\,.\tag{4.84}$$

Since the inequality (4.83) holds for all complex ξ_as, the matrix $\mathcal{F}$ is positive definite. A necessary condition for this is that the determinant of $\mathcal{F}$ must be positive. For the case $n=2$ this yields the inequality

$$\left|G^{(1)}(x_1;x_2)\right|^2\leqslant G^{(1)}(x_1;x_1)\,G^{(1)}(x_2;x_2)\,.\tag{4.85}$$

For first-order interference experiments, this inequality translates directly into a bound on the visibility of the fringes; this feature will be exploited in Section 10.1.

4.8 The interaction picture

In typical applications, the interaction energy between the charged particles and the radiation field is much smaller than the energies of individual photons. It is therefore useful to rewrite the Schrödinger-picture Hamiltonian, eqn (4.29), as

$$H^{(S)}=H_0^{(S)}+H_{\text{int}}^{(S)}\,,\tag{4.86}$$

where

$$H_0^{(S)}=H_{\text{em}}^{(S)}+H_{\text{chg}}^{(S)}\tag{4.87}$$

is the **unperturbed** Hamiltonian and $H_{\text{int}}^{(S)}$ is the **perturbation** or **interaction** Hamiltonian. In most cases the Schrödinger equation with the full Hamiltonian $H^{(S)}$

cannot be solved exactly, so the weak (perturbative) nature of $H_{\text{int}}^{(S)}$ must be used to get an approximate solution.

For this purpose, it is useful to separate the fast (high energy) evolution due to $H_0^{(S)}$ from the slow (low energy) evolution due to $H_{\text{int}}^{(S)}$. To this end, the **interaction-picture** state vector is defined by the unitary transformation

$$\left|\Psi^{(I)}(t)\right\rangle = U_0^{\dagger}(t)\left|\Psi^{(S)}(t)\right\rangle, \tag{4.88}$$

where the unitary operator,

$$U_0(t) = \exp\left[-\frac{i(t-t_0)}{\hbar}H_0^{(S)}\right], \tag{4.89}$$

satisfies

$$i\hbar\frac{\partial}{\partial t}U_0(t) = H_0^{(S)}U_0(t), \quad U_0(t_0) = 1. \tag{4.90}$$

Thus the Schrödinger and interaction pictures coincide at $t = t_0$. It is also clear that $\left[H_0^{(S)}, U_0(t)\right] = 0$. A glance at the solution (3.76) for the Schrödinger equation reveals that this transformation effectively undoes the fast evolution due to $H_0^{(S)}$. By contrast to the Heisenberg picture defined in Section 3.2, the transformed ket vector still depends on time due to the action of $H_{\text{int}}^{(S)}$. The consistency condition,

$$\left\langle\Psi^{(I)}(t)\left|X^{(I)}(t)\right|\Phi^{(I)}(t)\right\rangle = \left\langle\Psi^{(S)}(t)\left|X^{(S)}\right|\Phi^{(S)}(t)\right\rangle, \tag{4.91}$$

requires the interaction-picture operators to be defined by

$$X^{(I)}(t) = U_0^{\dagger}(t)X^{(S)}U_0(t). \tag{4.92}$$

For $H_0^{(S)}$ this yields the simple result

$$H_0^{(I)}(t) = U_0^{\dagger}(t)H_0^{(S)}U_0(t) = H_0^{(S)}, \tag{4.93}$$

which shows that $H_0^{(I)}(t) = H_0^{(S)} = H_0$ is independent of time.

The transformed state vector $\left|\Psi^{(I)}(t)\right\rangle$ obeys the interaction-picture Schrödinger equation

$$\begin{aligned} i\hbar\frac{\partial}{\partial t}\left|\Psi^{(I)}(t)\right\rangle &= -H_0^{(S)}\left|\Psi^{(I)}(t)\right\rangle + U_0^{\dagger}(t)\left(H_0^{(S)} + H_{\text{int}}^{(S)}\right)\left|\Psi^{(S)}(t)\right\rangle \\ &= -H_0^{(S)}\left|\Psi^{(I)}(t)\right\rangle + U_0^{\dagger}(t)\left(H_0^{(S)} + H_{\text{int}}^{(S)}\right)U_0(t)\left|\Psi^{(I)}(t)\right\rangle \\ &= H_{\text{int}}^{(I)}(t)\left|\Psi^{(I)}(t)\right\rangle, \end{aligned} \tag{4.94}$$

which follows from operating on both sides of eqn (4.88) with $i\hbar\partial/\partial t$ and using eqns (4.90)–(4.93). The formal solution is

$$\left|\Psi^{(I)}(t)\right\rangle = V(t)\left|\Psi^{(I)}(t_0)\right\rangle, \tag{4.95}$$

where the unitary operator $V(t)$ satisfies

$$i\hbar\frac{\partial}{\partial t}V(t) = H_{\text{int}}^{(I)}(t)\,V(t)\,, \quad \text{with } V(t_0) = 1\,. \tag{4.96}$$

The initial condition $V(t_0) = 1$ really should be $V(t_0) = I_{\text{QED}}$, where I_{QED} is the identity operator for $\mathfrak{H}_{\text{QED}}$, but alert readers will suffer no harm from this slight abuse of notation.

By comparing eqn (4.92) to eqn (3.83), one sees immediately that the interaction-picture operators obey

$$i\hbar\frac{\partial}{\partial t}X^{(I)}(t) = \left[X^{(I)}(t)\,, H_0\right]. \tag{4.97}$$

These are the Heisenberg equation for free fields, so we can use eqns (3.95) and (3.96) to get

$$a_{\mathbf{k}s}^{(I)}(t) = a_{\mathbf{k}s}^{(S)}e^{-i\omega_k(t-t_0)}, \tag{4.98}$$

and

$$\mathbf{A}^{(I)(+)}(\mathbf{r},t) = \sum_{\mathbf{k}s}\sqrt{\frac{\hbar}{2\epsilon_0\omega_k V}}a_{\mathbf{k}s}^{(S)}\mathbf{e}_{\mathbf{k}s}e^{i[\mathbf{k}\cdot\mathbf{r}-\omega_k(t-t_0)]}\,. \tag{4.99}$$

In the same way eqn (3.102) implies

$$\left[F^{(I)(\pm)}(\mathbf{r},t)\,, G^{(I)(\pm)}(\mathbf{r}',t')\right] = 0\,, \tag{4.100}$$

where F and G are any of the field components and $(\mathbf{r},t)$, $(\mathbf{r}',t')$ are any pair of space–time points.

In the interaction picture, the burden of time evolution is shared between the operators and the states. The operators evolve according to the unperturbed Hamiltonian, and the states evolve according to the interaction Hamiltonian. Once again, the density operator is an exception. Applying the transformation in eqn (4.88) to the definition (3.85) of the Schrödinger-picture density operator leads to

$$i\hbar\frac{\partial}{\partial t}\rho^{(I)}(t) = \left[H_{\text{int}}^{(I)}(t),\,\rho^{(I)}(t)\right], \tag{4.101}$$

so the density operator evolves according to the interaction Hamiltonian.

In applications of the interaction picture, we will simplify the notation by the following conventions: $X(t)$ means $X^{(I)}(t)$, X means $X^{(S)}$, $|\Psi(t)\rangle$ means $\left|\Psi^{(I)}(t)\right\rangle$, and $\rho(t)$ means $\rho^{(I)}(t)$. If all three pictures are under consideration, it may be necessary to reinstate the superscripts (S), (H), and (I).

4.8.1 Time-dependent perturbation theory

In order to make use of the weakness of the perturbation, we first turn eqn (4.96) into an integral equation by integrating over the interval (t_0, t) to get

$$V(t) = 1 - \frac{i}{\hbar}\int_{t_0}^{t}dt_1 H_{\text{int}}(t_1)\,V(t_1)\,. \tag{4.102}$$

The formal perturbation series is obtained by repeated iterations of the integral equation,

$$\begin{aligned} V(t) &= 1 - \frac{i}{\hbar}\int_{t_0}^{t} dt_1 H_{\text{int}}(t_1) + \left(-\frac{i}{\hbar}\right)^2 \int_{t_0}^{t} dt_1 \int_{t_0}^{t_1} dt_2 H_{\text{int}}(t_1) H_{\text{int}}(t_2) + \cdots \\ &= \sum_{n=0}^{\infty} V^{(n)}(t), \end{aligned} \tag{4.103}$$

where $V^{(0)} = 1$, and

$$V^{(n)}(t) = \left(-\frac{i}{\hbar}\right)^n \int_{t_0}^{t} dt_1 \cdots \int_{t_0}^{t_{n-1}} dt_n H_{\text{int}}(t_1) \cdots H_{\text{int}}(t_n), \tag{4.104}$$

for $n \geqslant 1$.

If the system (charges plus radiation) is initially in the state $|\Theta_i\rangle$ then the probability amplitude that a measurement at time t leaves the system in the final state $|\Theta_f\rangle$ is

$$V_{fi}(t) = \langle\Theta_f|\Psi(t)\rangle = \langle\Theta_f|V(t)|\Theta_i\rangle; \tag{4.105}$$

consequently, the **transition probability** is

$$P_{fi}(t) = |V_{fi}(t)|^2. \tag{4.106}$$

4.8.2 First-order perturbation theory

For this application, we choose $t_0 = 0$, and then let the interaction act for a finite time t. The initial state $|\Theta_i\rangle$ evolves into $V(t)|\Theta_i\rangle$, and its projection on the final state $|\Theta_f\rangle$ is $\langle\Theta_f|V(t)|\Theta_i\rangle$. Let the initial and final states be eigenstates of the unperturbed Hamiltonian H_0, with energies E_i and E_f respectively. According to eqn (4.104) the first-order contribution to $\langle\Theta_f|V(t)|\Theta_i\rangle$ is

$$\begin{aligned} V_{fi}^{(1)}(t) &= -\frac{i}{\hbar}\int_0^t dt_1 \langle\Theta_f|H_{\text{int}}(t_1)|\Theta_i\rangle \\ &= -\frac{i}{\hbar}\int_0^t dt_1 \langle\Theta_f|H_{\text{int}}|\Theta_i\rangle \exp(i\nu_{fi}t_1), \end{aligned} \tag{4.107}$$

where we have used eqn (4.92) and introduced the notation $\nu_{fi} = (E_f - E_i)/\hbar$. Evaluating the integral in eqn (4.107) yields the amplitude

$$V_{fi}^{(1)}(t) = -\frac{2i}{\hbar}\exp(i\nu_{fi}t/2)\frac{\sin(\nu_{fi}t/2)}{\nu_{fi}}\langle\Theta_f|H_{\text{int}}|\Theta_i\rangle, \tag{4.108}$$

so the transition probability is

$$P_{fi}(t) = \left|V_{fi}^{(1)}(t)\right|^2 = \frac{4}{\hbar^2}|\langle\Theta_f|H_{\text{int}}|\Theta_i\rangle|^2 \Delta(\nu_{fi}, t), \tag{4.109}$$

where $\Delta(\nu, t) \equiv \sin^2(\nu t/2)/\nu^2$.

For fixed t, the maximum value of $|\Delta(\nu, t)|^2$ is $t^2/4$, and it occurs at $\nu = 0$. The width of the central peak is approximately $2\pi/t$, so as t becomes large the function is strongly peaked at $\nu = 0$. In order to specify a well-defined final energy, the width must be small compared to $|E_f - E_i|/\hbar$; therefore,

$$t \gg \frac{2\pi\hbar}{|E_f - E_i|} \tag{4.110}$$

defines the limit of large times. This is a realization of the energy–time uncertainty relation, $t\Delta E \sim \hbar$ (Bransden and Joachain, 1989, Sec. 2.5). With this understanding of infinity, we can use the easily established mathematical result,

$$\lim_{t\to\infty} \frac{\Delta(\nu, t)}{t} = \lim_{t\to\infty} \frac{\sin^2(\nu t/2)}{t\nu^2} = \frac{\pi}{2}\delta(\nu), \tag{4.111}$$

to write the asymptotic ($t \to \infty$) form of eqn (4.109) as

$$\begin{aligned} P_{fi}(t) &= \frac{2\pi}{\hbar^2} t \left|\langle \Theta_f | H_{\text{int}} | \Theta_i \rangle\right|^2 \delta(\nu_{fi}) \\ &= \frac{2\pi}{\hbar} t \left|\langle \Theta_f | H_{\text{int}} | \Theta_i \rangle\right|^2 \delta(E_f - E_i). \end{aligned} \tag{4.112}$$

The **transition rate**, $W_{fi} = dP_{fi}(t)/dt$, is then

$$W_{fi} = \frac{2\pi}{\hbar} \left|\langle \Theta_f | H_{\text{int}} | \Theta_i \rangle\right|^2 \delta(E_f - E_i). \tag{4.113}$$

This is **Fermi's golden rule** of perturbation theory (Bransden and Joachain, 1989, Sec. 9.3). This limiting form only makes sense when at least one of the energies E_i and E_f varies continuously. In the following applications this happens automatically because of the continuous variation of the photon energies.

In addition to the lower bound on t in eqn (4.110) there is an upper bound on the time interval for which the perturbative result is valid. This is estimated by summing eqn (4.112) over all final states to get the total transition probability $P_{i,\text{tot}}(t) = tW_{i,\text{tot}}$, where the total transition rate is

$$W_{i,\text{tot}} = \sum_f W_{fi} = \sum_f \frac{2\pi}{\hbar} \left|\langle \Theta_f | H_{\text{int}} | \Theta_i \rangle\right|^2 \delta(E_f - E_i). \tag{4.114}$$

According to this result, the necessary condition $P_{i,\text{tot}}(t) < 1$ will be violated if $t > 1/W_{i,\text{tot}}$. In fact, the validity of the perturbation series demands the more stringent condition $P_{i,\text{tot}}(t) \ll 1$, so the perturbative results can only be trusted for $t \ll 1/W_{i,\text{tot}}$. This upper bound on t means that the $t \to \infty$ limit in eqn (4.111) is simply the physical condition (4.110). For the same reason, the energy conserving delta function in eqn (4.112) is really just a sharply-peaked function that imposes the restriction $|E_f - E_i| \ll E_f$.

With this understanding in mind, a simplified version of the previous calculation is possible. For this purpose, we choose $t_0 = -T/2$ and allow the state vector to evolve until the time $t = T/2$. Then eqn (4.107) is replaced by

$$V_{fi}^{(1)}\left(T/2\right) = -\frac{i}{\hbar}\left\langle\Theta_f\left|H_{\text{int}}\right|\Theta_i\right\rangle \exp\left(i\nu_{fi}T/2\right)\int_{-T/2}^{T/2} dt_1 \exp\left(i\nu_{fi}t_1\right). \qquad (4.115)$$

The standard result

$$\lim_{T\to\infty}\int_{-T/2}^{T/2} dt_1 e^{i\nu t_1} = 2\pi\delta\left(\nu\right) \qquad (4.116)$$

allows this to be recast as

$$V_{fi}^{(1)} = V_{fi}^{(1)}\left(\infty\right) = -\frac{2\pi i}{\hbar}\left\langle\Theta_f\left|H_{\text{int}}\right|\Theta_i\right\rangle\delta\left(\nu_{fi}\right), \qquad (4.117)$$

so the transition probability is

$$P_{fi} = \left|V_{fi}^{(1)}\right|^2 = \left(\frac{2\pi}{\hbar}\right)^2\left|\left\langle\Theta_f\left|H_{\text{int}}\right|\Theta_i\right\rangle\right|^2\left[\delta\left(\nu_{fi}\right)\right]^2. \qquad (4.118)$$

This is rather embarrassing, since the square of a delta function is not a respectable mathematical object. Fortunately this is a physicist's delta function, so we can use eqn (4.116) once more to set

$$\left[\delta\left(\nu_{fi}\right)\right]^2 = \delta\left(\nu_{fi}\right)\int_{-T/2}^{T/2}\frac{dt_1}{2\pi}\exp\left(i\nu_{fi}t_1\right) = \frac{T}{2\pi}\delta\left(\nu_{fi}\right). \qquad (4.119)$$

After putting this into eqn (4.118), we recover eqn (4.113).

4.8.3 Second-order perturbation theory

Using the simplified scheme, presented in eqns (4.115)–(4.119), yields the second-order contribution to $\left\langle\Theta_f\left|V\left(T/2\right)\right|\Theta_i\right\rangle$:

$$\begin{aligned} V_{fi}^{(2)} &= \left(-\frac{i}{\hbar}\right)^2\int_{-T/2}^{T/2} dt_1\int_{-T/2}^{t_1} dt_2\left\langle\Theta_f\left|H_{\text{int}}\left(t_1\right)H_{\text{int}}\left(t_2\right)\right|\Theta_i\right\rangle \\ &= \left(-\frac{i}{\hbar}\right)^2\int_{-T/2}^{T/2} dt_1\int_{-T/2}^{T/2} dt_2\theta\left(t_1-t_2\right)\left\langle\Theta_f\left|H_{\text{int}}\left(t_1\right)H_{\text{int}}\left(t_2\right)\right|\Theta_i\right\rangle, \end{aligned} \qquad (4.120)$$

where $\theta\left(t_1 - t_2\right)$ is the step function discussed in Appendix A.7.1 . By introducing a basis set $\{|\Lambda_u\rangle\}$ of eigenstates of H_0, the matrix element can be written as

$$\begin{aligned} \left\langle\Theta_f\left|H_{\text{int}}\left(t_1\right)H_{\text{int}}\left(t_2\right)\right|\Theta_i\right\rangle &= \exp\left[\left(i\nu_{fi}\right)T/2\right]\sum_u\left\langle\Theta_f\left|H_{\text{int}}\right|\Lambda_u\right\rangle\left\langle\Lambda_u\left|H_{\text{int}}\right|\Theta_i\right\rangle \\ &\quad\times\exp\left(i\nu_{fu}t_1\right)\exp\left(i\nu_{ui}t_2\right), \end{aligned} \qquad (4.121)$$

where we have used eqn (4.92) and the identity $\nu_{fu} + \nu_{ui} = \nu_{fi}$. The final step is to use the representation (A.88) for the step function and eqn (4.116) to find

$$V_{fi}^{(2)} = -\frac{i}{2\pi\hbar^2} e^{i\nu_{fi}T/2} \int_{-\infty}^{\infty} d\nu \sum_u \frac{\langle \Theta_f | H_{\text{int}} | \Lambda_u \rangle \langle \Lambda_u | H_{\text{int}} | \Theta_i \rangle}{\nu + i\epsilon} \times 2\pi\delta(\nu_{fu} - \nu) 2\pi\delta(\nu_{ui} + \nu) . \tag{4.122}$$

Carrying out the integration over ν with the aid of the delta functions leads to

$$\begin{aligned} V_{fi}^{(2)} &= -\frac{2\pi i}{\hbar^2} \sum_u \frac{\langle \Theta_f | H_{\text{int}} | \Lambda_u \rangle \langle \Lambda_u | H_{\text{int}} | \Theta_i \rangle}{\nu_{fu} + i\epsilon} \delta(\nu_{fi}) \\ &= -2\pi i \sum_u \frac{\langle \Theta_f | H_{\text{int}} | \Lambda_u \rangle \langle \Lambda_u | H_{\text{int}} | \Theta_i \rangle}{E_f - E_u + i\epsilon} \delta(E_f - E_i) . \end{aligned} \tag{4.123}$$

Finally, another use of the rule (4.119) yields the transition rate

$$W_{fi} = \frac{2\pi}{\hbar} \left| \sum_u \frac{\langle \Theta_f | H_{\text{int}} | \Lambda_u \rangle \langle \Lambda_u | H_{\text{int}} | \Theta_i \rangle}{E_f - E_u + i\epsilon} \right|^2 \delta(E_f - E_i) . \tag{4.124}$$

4.9 Interaction of light with atoms

4.9.1 The dipole approximation

The shortest wavelengths of interest for quantum optics are in the extreme ultraviolet, so we can assume that $\lambda > 100\,\text{nm}$, whereas typical atoms have diameters $a \approx 0.1\,\text{nm}$. The large disparity between atomic diameters and optical wavelengths ($a/\lambda < 0.001$) permits the use of the dipole approximation, and this in turn brings about important simplifications in the general Hamiltonian defined by eqns (4.28)–(4.32).

The simplified Hamiltonian can be derived directly from the general form given in Section 4.2.2 (Cohen-Tannoudji *et al.*, 1989, Sec. IV.C), but it is simpler to obtain the dipole-approximation Hamiltonian for a single atom by a separate appeal to the correspondence principle. This single-atom construction is directly relevant for sufficiently dilute systems of atoms—e.g. tenuous atomic vapors—since the interaction between atoms is weak. Experiments with vapors were the rule in the early days of quantum optics, but in many modern applications—such as solid-state detectors and solid-state lasers—the atoms are situated on a crystal lattice. This is a high density situation with substantial interactions between atoms. Furthermore, the electronic wave functions can be delocalized—e.g. in the conduction band of a semiconductor—so that the validity of the dipole approximation is in doubt. These considerations—while very important in practice—do not in fact require significant changes in the following discussion.

The interactions between atoms on a crystal lattice can be described in terms of coupling to lattice vibrations (phonons), and the effects of the periodic crystal potential are represented by the use of Bloch or Wannier wave functions for the electrons (Kittel, 1985, Chap. 9). The wave functions for electrons in the valence band are localized to crystal sites, so for transitions between the valence and conduction bands even the dipole approximation can be retained. We will exploit this situation by explaining the basic techniques of quantum optics in the simpler context of tenuous vapors. Once these notions are mastered, their application to condensed matter physics can be found elsewhere (Haug and Koch, 1990).

Even with the dipole approximation in force, the direct use of the atomic wave function is completely impractical for a many-electron atom—this means any atom with atomic number $Z > 1$. Fortunately, the complete description provided by the many-electron wave function $\psi(\mathbf{r}_1, \ldots, \mathbf{r}_Z)$ is not needed. For the most part, only selected properties—such as the discrete electronic energies and the matrix elements of the dipole operator—are required. Furthermore these properties need not be calculated *ab initio*; instead, they can be inferred from the measured wavelength and strength of spectral lines. In this semi-empirical approach, the problem of atomic structure is separated from the problem of the response of the atom to the electromagnetic field.

For a single atom interacting with the electromagnetic field, the discussion in Section 4.2.1 shows that the state space is the tensor product $\mathfrak{H} = \mathfrak{H}_A \otimes \mathfrak{H}_F$ of the Hilbert space $\mathfrak{H}_A$ for the atom and the Fock space $\mathfrak{H}_F$ for the field. A typical basis state for $\mathfrak{H}$ is $|\psi, \Phi\rangle = |\psi\rangle |\Phi\rangle$, where $|\psi\rangle$ and $|\Phi\rangle$ are respectively state vectors for the atom and the field. Let us consider a typical matrix element $\langle \psi, \Phi | \mathbf{E}(\mathbf{r}) | \psi', \Phi' \rangle$ of the electric field operator, where at least one of the vectors $|\psi\rangle$ and $|\psi'\rangle$ describes a bound state with characteristic spatial extent a, and $|\Phi\rangle$ and $|\Phi'\rangle$ both describe states of the field containing only photons with wavelengths $\lambda \gg a$. On the scale of the optical wavelengths, the atomic electrons can then be regarded as occupying a small region surrounding the center-of-mass position,

$$\widehat{\mathbf{r}}_{\mathrm{cm}} = \frac{M_{\mathrm{nuc}}}{M} \widehat{\mathbf{r}}_{\mathrm{nuc}} + \sum_{n=1}^{Z} \frac{M_e}{M} \widehat{\mathbf{r}}_n \,, \tag{4.125}$$

where $\widehat{\mathbf{r}}_{\mathrm{cm}}$ is the operator for the center of mass, M_e is the electron mass, $\widehat{\mathbf{r}}_n$ is the coordinate operator of the nth electron, M_{nuc} is the nuclear mass, $\widehat{\mathbf{r}}_{\mathrm{nuc}}$ is the coordinate operator of the nucleus, Z is the atomic number, and $M = M_{\mathrm{nuc}} + ZM_e$ is the total mass.

For all practical purposes, the center of mass can be identified with the location of the nucleus, since $M_{\mathrm{nuc}} \gg ZM_e$. The plane-wave expansion (3.69) for the electric field then implies that the matrix element is slowly varying across the atom, so that it can be expanded in a Taylor series around $\widehat{\mathbf{r}}_{\mathrm{cm}}$,

$$\begin{aligned} &\langle \psi, \Phi | \mathbf{E}(\mathbf{r}) | \psi', \Phi' \rangle \\ &\quad = \langle \psi, \Phi | \mathbf{E}(\widehat{\mathbf{r}}_{\mathrm{cm}}) | \psi', \Phi' \rangle + \langle \psi, \Phi | [(\mathbf{r} - \widehat{\mathbf{r}}_{\mathrm{cm}}) \cdot \nabla] \mathbf{E}(\widehat{\mathbf{r}}_{\mathrm{cm}}) | \psi', \Phi' \rangle + \cdots . \end{aligned} \tag{4.126}$$

With the understanding that only matrix elements of this kind will occur, the expansion can be applied to the field operator itself:

$$\mathbf{E}(\mathbf{r}) = \mathbf{E}(\widehat{\mathbf{r}}_{\mathrm{cm}}) + [(\mathbf{r} - \widehat{\mathbf{r}}_{\mathrm{cm}}) \cdot \nabla] \mathbf{E}(\widehat{\mathbf{r}}_{\mathrm{cm}}) + \cdots . \tag{4.127}$$

The **electric dipole approximation** retains only the leading term in this expansion, with errors of $O(a/\lambda)$. Keeping higher-order terms in the Taylor series incorporates successive terms in the general multipole expansion, e.g. magnetic dipole, electric quadrupole, etc. In classical electrodynamics (Jackson, 1999, Sec. 4.2), the leading term in the interaction energy of a neutral collection of charges with an external electric field $\mathbf{E}$ is $-\mathbf{d} \cdot \mathbf{E}$, where $\mathbf{d}$ is the electric dipole moment. For an atom the dipole operator is

$$\widehat{\mathbf{d}} = \sum_{n=1}^{Z} (-e) (\widehat{\mathbf{r}}_n - \widehat{\mathbf{r}}_{\text{nuc}}) . \tag{4.128}$$

Once again we rely on the correspondence principle to suggest that the interaction Hamiltonian in the quantum theory should be

$$H_{\text{int}} = -\widehat{\mathbf{d}} \cdot \mathbf{E} (\widehat{\mathbf{r}}_{\text{cm}}) . \tag{4.129}$$

The atomic Hamiltonian can be expressed as

$$H_{\text{atom}} = \frac{\widehat{\mathbf{P}}^2}{2M} + \sum_{n=1}^{Z} \frac{(\widehat{\mathbf{p}}_n)^2}{2M_e} + V_C , \tag{4.130}$$

$$V_C = \frac{e^2}{4\pi\epsilon_0} \sum_{n \neq l=1}^{Z} \frac{1}{|\widehat{\mathbf{r}}_n - \widehat{\mathbf{r}}_l|} - \frac{Ze^2}{4\pi\epsilon_0} \sum_{n=1}^{Z} \frac{1}{|\widehat{\mathbf{r}}_n - \widehat{\mathbf{r}}_{\text{nuc}}|} , \tag{4.131}$$

where V_C is the Coulomb potential, $\widehat{\mathbf{P}}$ is the total momentum, and the $\widehat{\mathbf{p}}_n$s are a set of relative momentum operators. Thus the Schrödinger-picture Hamiltonian in the dipole approximation is $H = H_{\text{em}} + H_{\text{atom}} + H_{\text{int}}$.

The argument given in Section 4.2.2 shows that $\mathbf{E} (\widehat{\mathbf{r}}_{\text{cm}})$ is a hybrid operator acting on both the atomic and field degrees of freedom. For most applications of quantum optics, we can ignore this complication, since the De Broglie wavelength of the atom is small compared to the interatomic spacing. In this limit, the center-of-mass position, $\widehat{\mathbf{r}}_{\text{cm}}$, and the total kinetic energy $\widehat{\mathbf{P}}^2/2M$ can be treated classically, so that

$$H_{\text{atom}} = \frac{\mathbf{P}^2}{2M} + H_{\text{at}} , \tag{4.132}$$

where

$$H_{\text{at}} = \sum_{n=1}^{Z} \frac{(\widehat{\mathbf{p}}_n)^2}{2M_e} + V_C \tag{4.133}$$

is the Hamiltonian for the internal degrees of freedom of the atom. In the same approximation, the interaction Hamiltonian reduces to

$$H_{\text{int}} = -\widehat{\mathbf{d}} \cdot \mathbf{E} (\mathbf{r}_{\text{cm}}) , \tag{4.134}$$

which acts jointly on the field states and the internal states of the atom.

In the rest frame of the atom, defined by $\mathbf{P} = 0$, the energy eigenstates

$$H_{\text{at}} |\varepsilon_q\rangle = \varepsilon_q |\varepsilon_q\rangle \tag{4.135}$$

provide a basis for the Hilbert space, $\mathfrak{H}_A$, describing the internal degrees of freedom of the atom. The label q stands for a set of quantum numbers sufficient to specify the internal atomic state uniquely. The qs are discrete; therefore, they can be ordered so that $\varepsilon_q \leqslant \varepsilon_{q'}$ for $q < q'$.

In practice, the many-electron wave function $\psi_q(\mathbf{r}_1, \ldots, \mathbf{r}_Z) = \langle \mathbf{r}_1, \ldots, \mathbf{r}_Z | \varepsilon_q \rangle$ cannot be determined exactly, so the eigenstates are approximated, e.g. by using the atomic shell model (Cohen-Tannoudji *et al.*, 1977*b*, Chap. XIV, Complement A). In this case the label $q = (n, l, m)$ consists of the principal quantum number, the angular momentum, and the azimuthal quantum number for the valence electrons in a shell model description. The **dipole selection rules** are

$$\left\langle \varepsilon_q \left| \widehat{\mathbf{d}} \right| \varepsilon_{q'} \right\rangle = 0 \quad \text{unless } l - l' = \pm 1 \text{ and } m - m' = \pm 1, 0\,. \tag{4.136}$$

The z-axis is conventionally chosen as the quantization axis, and this implies

$$\begin{aligned} &\left\langle \varepsilon_q \left| \widehat{d}_z \right| \varepsilon_{q'} \right\rangle = 0 \quad \text{unless } m - m' = 0\,, \\ &\left\langle \varepsilon_q \left| \widehat{d}_x \right| \varepsilon_{q'} \right\rangle = \left\langle \varepsilon_q \left| \widehat{d}_y \right| \varepsilon_{q'} \right\rangle = 0 \quad \text{unless } m - m' = \pm 1\,. \end{aligned} \tag{4.137}$$

A basis for the Hilbert space $\mathfrak{H} = \mathfrak{H}_A \otimes \mathfrak{H}_F$ describing the composite system of the atom and the radiation field is given by the product vectors

$$|\varepsilon_q, \underline{n}\rangle = |\varepsilon_q\rangle\, |\underline{n}\rangle\,, \tag{4.138}$$

where $|\underline{n}\rangle$ runs over the photon number states.

For a single atom the (c-number) kinetic energy $\mathbf{P}^2/2M$ can always be set to zero by transforming to the rest frame of the atom, but when many atoms are present there is no single frame of reference in which all atoms are at rest. Nevertheless, it is possible to achieve a similar effect by accounting for the recoil of the atom. Let us consider an elementary process, e.g. absorption of a photon with energy $\hbar\omega_k$ and momentum $\hbar\mathbf{k}$ by an atom with energy $\varepsilon_1 + \mathbf{P}^2/2M$ and momentum $\mathbf{P}$. The final energy, $\varepsilon_2 + \mathbf{P}'^2/2M$, and momentum, $\mathbf{P}'$, are constrained by the conservation of energy,

$$\hbar\omega_k + \varepsilon_1 + \mathbf{P}^2/2M = \varepsilon_2 + \mathbf{P}'^2/2M\,, \tag{4.139}$$

and conservation of momentum,

$$\hbar\mathbf{k} + \mathbf{P} = \mathbf{P}'\,. \tag{4.140}$$

The initial and final velocities of the atom are respectively $\mathbf{v} = \mathbf{P}/M$ and $\mathbf{v}' = \mathbf{P}'/M$, so eqn (4.140) tells us that the atomic recoil velocity is $\mathbf{v}_{\text{rec}} = \mathbf{v}' - \mathbf{v} = \hbar\mathbf{k}/M$. Substituting $\mathbf{P}'$ from eqn (4.140) into eqn (4.139) and expressing the result in terms of $\mathbf{v}_{\text{rec}}$ yields

$$\hbar\omega_k = \hbar\omega_{21} + M\mathbf{v}_{\text{rec}} \cdot \left(\mathbf{v} + \frac{1}{2}\mathbf{v}_{\text{rec}} \right), \tag{4.141}$$

where

$$\omega_{21} = \frac{\varepsilon_2 - \varepsilon_1}{\hbar} \tag{4.142}$$

is the Bohr frequency for this transition. For typical experimental conditions—e.g. optical frequency radiation interacting with a tenuous atomic vapor—the thermal velocities of the atoms are large compared to their recoil velocities, so that eqn (4.141) can be approximated by

$$\omega_k = \omega_{21} + \frac{\omega_{21}}{c}\widetilde{\mathbf{k}} \cdot \mathbf{v}\,, \tag{4.143}$$

where $\widetilde{\mathbf{k}} = \mathbf{k}/k$. Since v/c is small, this result can also be expressed as

$$\omega_{21} = \omega_k - \mathbf{k} \cdot \mathbf{v}\,. \tag{4.144}$$

In other words, conservation of energy is equivalent to resonance between the atomic transition and the **Doppler shifted frequency** of the radiation. With this thought in mind, we can ignore the kinetic energy term in the atomic Hamiltonian and simply tag each atom with its velocity and the associated resonance condition.

The next step is to generalize the single-atom results to a many-atom system. The state space is now $\mathfrak{H} = \mathfrak{H}_A \otimes \mathfrak{H}_F$, where the many-atom state space consists of product wave functions, i.e. $\mathfrak{H}_A = \otimes_n \mathfrak{H}_A^{(n)}$ where $\mathfrak{H}_A^{(n)}$ is the (internal) state space for the nth atom. Since H_{int} is linear in the atomic dipole moment, the part of the Hamiltonian describing the interaction of the many-atom system with the radiation field is obtained by summing eqn (4.129) over the atoms.

The Coulomb part is more complicated, since the general expression (4.131) contains Coulomb interactions between charges belonging to different atoms. These interatomic Coulomb potentials can also be described in terms of multipole expansions for the atomic charge distributions. The interatomic potential will then be dominated by dipole–dipole interactions. For tenuous vapors these effects can be neglected, and the many-atom Hamiltonian is approximated by $H = H_{\text{em}} + H_{\text{at}} + H_{\text{int}}$, where

$$H_{\text{at}} = \sum_n H_{\text{at}}^{(n)}\,, \tag{4.145}$$

$$H_{\text{int}} = -\sum_n \widehat{\mathbf{d}}^{(n)} \cdot \mathbf{E}\left(\mathbf{r}_{\text{cm}}^{(n)}\right), \tag{4.146}$$

and $H_{\text{at}}^{(n)}$, $\widehat{\mathbf{d}}^{(n)}$, and $\mathbf{r}_{\text{cm}}^{(n)}$ are respectively the internal Hamiltonian, the electric dipole operator, and the (classical) center-of-mass position for the nth atom.

4.9.2 The weak-field limit

A second simplification comes into play for electromagnetic fields that are weak, in the sense that the dipole interaction energy is small compared to atomic energy differences. In other words $|\mathbf{d} \cdot \boldsymbol{\mathcal{E}}| \ll \hbar\omega_T$, where $\mathbf{d}$ is a typical electric dipole matrix element, $\boldsymbol{\mathcal{E}}$ is a representative matrix element of the electric field operator, and ω_T is a typical Bohr frequency associated with an atomic transition. In terms of the characteristic **Rabi frequency**

$$\Omega = \frac{|\mathbf{d} \cdot \boldsymbol{\mathcal{E}}|}{\hbar}\,, \tag{4.147}$$

which represents the typical oscillation rate of the atom induced by the electric field, the weak-field condition is

$$\Omega \ll \omega_T\,. \tag{4.148}$$

The Rabi frequency is given by $\Omega = 1.39 \times 10^7 d\sqrt{I}$, where Ω is expressed in Hz, the field intensity I in W/cm^2, and the dipole moment d in **debyes** ($1\,\text{D} = 10^{-18}\,\text{esu cm} =$

0.33×10^{-29} C m). Typical values for the dipole matrix elements are $d \sim 1$ D, and the interesting Bohr frequencies are in the range 3×10^{10} Hz $< \omega_T < 3 \times 10^{15}$ Hz, corresponding to wavelengths in the range 1 cm to 100 nm. For each value of ω_T, eqn (4.148) imposes an upper bound on the strength of the electric fields associated with the matrix elements of H_{int}. For a typical optical frequency, e.g. $\omega_T \approx 3 \times 10^{14}$ Hz, the upper bound is $I \sim 5 \times 10^{14}$ W/cm^2, which could not be violated without vaporizing the sample. At the long wavelength limit, $\lambda \sim 1$ cm ($\omega_T \sim 3 \times 10^{10}$ Hz), the upper bound is only $I \sim 5 \times 10^6$ W/cm^2, which could be readily violated without catastrophe. However this combination of wavelength and intensity is not of interest for quantum optics, since the corresponding photon flux, 10^{29} photons/cm^2 s, is so large that quantum fluctuations would be completely negligible. Thus in all relevant situations, we may assume that the fields are weak.

The weak-field condition justifies the use of time-dependent perturbation theory for the calculation of transition rates for spontaneous emission or absorption from an incoherent radiation field. As we will see below, perturbation theory is not able to describe other interesting phenomena, such as natural line widths and the resonant coupling of an atom to a coherent field, e.g. a laser. Despite the failure of perturbation theory for such cases, the weak-field condition can still be used to derive a nonperturbative scheme which we will call the resonant wave approximation. Just as with perturbation theory, the interaction picture is the key to understanding the resonant wave approximation.

4.9.3 The Einstein A and B coefficients

As the first application of perturbation theory we calculate the Einstein A coefficient, i.e. the total spontaneous emission rate for an atom in free space. For this and subsequent calculations, it will be convenient to write the interaction Hamiltonian as

$$H_{\text{int}} = -\hbar \left[\Omega^{(+)}(\mathbf{r}) + \Omega^{(-)}(\mathbf{r}) \right], \tag{4.149}$$

where the positive-frequency **Rabi operator** $\Omega^{(+)}(\mathbf{r})$ is

$$\Omega^{(+)}(\mathbf{r}) = \frac{\mathbf{E}^{(+)}(\mathbf{r}) \cdot \widehat{\mathbf{d}}}{\hbar}, \tag{4.150}$$

and $\mathbf{r}$ is the location of the atom. In the absence of boundaries, we can choose the location of the atom as the origin of coordinates. Setting $\mathbf{r} = 0$ in eqn (3.69) for $\mathbf{E}^{(+)}(\mathbf{r})$ and substituting into eqn (4.150) yields

$$\Omega^{(+)} = \sum_{\mathbf{k}s} i \sqrt{\frac{\hbar\omega_k}{2\epsilon_0 V}} \frac{\mathbf{e}_{\mathbf{k}s} \cdot \widehat{\mathbf{d}}}{\hbar} a_{\mathbf{k}s} \,. \tag{4.151}$$

The initial state for the transition is $|\Theta_i\rangle = |\varepsilon_2, 0\rangle = |\varepsilon_2\rangle |0\rangle$, where $|\varepsilon_2\rangle$ is an excited state of the atom and $|0\rangle$ is the vacuum state, so the initial energy is $E_i = \varepsilon_2$. The final state is $|\Theta_f\rangle = |\varepsilon_1, 1_{\mathbf{k}s}\rangle = |\varepsilon_1\rangle |1_{\mathbf{k}s}\rangle$, where $|1_{\mathbf{k}s}\rangle = a^\dagger_{\mathbf{k}s} |0\rangle$ is the state describing exactly one photon with wavevector $\mathbf{k}$ and polarization $\mathbf{e}_{\mathbf{k}s}$ and $|\varepsilon_1\rangle$ is an atomic state with $\varepsilon_1 < \varepsilon_2$. The final state energy is therefore $E_f = \varepsilon_1 + \hbar\omega_k$. The

Feynman diagrams for emission and absorption are shown in Fig. 4.1. It is clear from eqn (4.151) that only $\Omega^{(-)}$ can contribute to emission, so the relevant matrix element is

$$\left\langle \varepsilon_1, 1_{\mathbf{k}s} \left| \Omega^{(-)} \right| \varepsilon_2, 0 \right\rangle = -i\Omega^*_{21,s}(\mathbf{k}), \tag{4.152}$$

where

$$\Omega_{21,s}(\mathbf{k}) = \sqrt{\frac{\hbar\omega_k}{2\epsilon_0 V}} \frac{\mathbf{d}_{21} \cdot \mathbf{e}_{\mathbf{k}s}}{\hbar} \tag{4.153}$$

is the **single-photon Rabi frequency** for the $1 \leftrightarrow 2$ transition, and $\mathbf{d}_{21} = \left\langle \varepsilon_2 \middle| \widehat{\mathbf{d}} \middle| \varepsilon_1 \right\rangle$ is the **dipole matrix element**. In the physical limit $V \to \infty$, the photon energies $\hbar\omega_k$ become continuous, and the golden rule (4.113) can be applied to get the transition rate

$$W_{1\mathbf{k}s,2} = 2\pi \left|\Omega_{21,s}(\mathbf{k})\right|^2 \delta(\omega_k - \omega_{21}). \tag{4.154}$$

The irreversibility of the transition described by this rate is a mathematical consequence of the continuous variation of the final photon energy that allows the use of Fermi's golden rule. A more intuitive explanation of the irreversible decay of an excited atom is that radiation emitted into the cold and darkness of infinite space will never return.

Since the spacing between discrete wavevectors goes to zero in the infinite volume limit, the physically meaningful quantity is the emission rate into an infinitesimal $\mathbf{k}$-space volume d^3k centered on $\mathbf{k}$. For each polarization, the number of $\mathbf{k}$-modes in d^3k is $Vd^3k/(2\pi)^3$; consequently, the differential emission rate is

$$\begin{aligned} dW_{1\mathbf{k}s,2} &= W_{1\mathbf{k}s,2} \frac{Vd^3k}{(2\pi)^3} \\ &= 2\pi \left|M_{21,s}(\mathbf{k})\right|^2 \delta(\omega_k - \omega_{21}) \frac{d^3k}{(2\pi)^3}, \end{aligned} \tag{4.155}$$

where

$$M_{21,s}(\mathbf{k}) = \sqrt{V}\Omega_{21,s}(\mathbf{k}) = \sqrt{\frac{\hbar\omega_k}{2\epsilon_0}} \frac{\mathbf{d}_{21} \cdot \mathbf{e}_{\mathbf{k}s}}{\hbar}. \tag{4.156}$$

The **Einstein A coefficient** is the total transition rate into all $\mathbf{k}s$-modes:

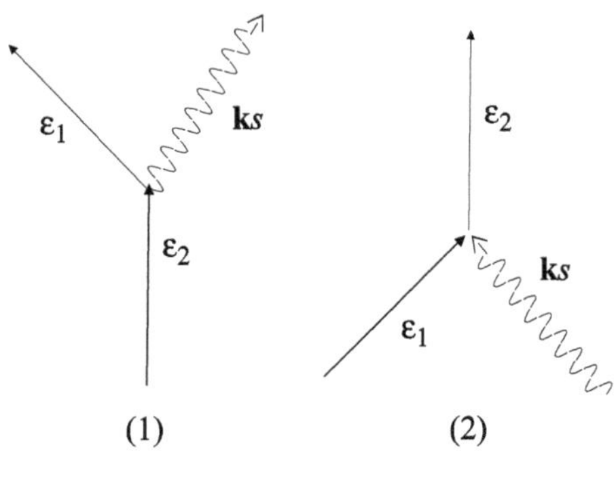

Fig. 4.1 First-order Feynman diagrams for emission (1) and absorption (2). Straight lines correspond to atomic states and wiggly lines to photon states.

$$A_{2\to 1} = \int \frac{d^3k}{(2\pi)^3} \sum_s 2\pi \left|M_{21,s}(\mathbf{k})\right|^2 \delta(\omega_k - \omega_{21}) . \tag{4.157}$$

The integral over the magnitude of $\mathbf{k}$ can be carried out by the change of variables $k \to \omega/c$. It is customary to write this result in terms of the **density of states**, $\mathcal{D}(\omega_{21})$, which is the number of resonant modes per unit volume per unit frequency. The number of modes in d^3k is $2Vd^3k/(2\pi)^3$, where the factor 2 counts the polarizations for each $\mathbf{k}$, so the density of states is

$$\mathcal{D}(\omega_{21}) = 2\int \frac{d^3k}{(2\pi)^3} \delta(\omega_k - \omega_{21}) = \frac{\omega_{21}^2}{\pi^2 c^3} . \tag{4.158}$$

This result includes the two polarizations and the total 4π sr of solid angle, so calculating the contribution from a single plane wave requires division by 8π. In this way $A_{2\to 1}$ is expressed as an average over emission directions and polarizations,

$$A_{2\to 1} = \int \frac{d\Omega_k}{4\pi} \frac{1}{2} \sum_s 2\pi \left|M_{21,s}(\mathbf{k})\right|^2 \mathcal{D}(\omega_{21}) , \tag{4.159}$$

where $d\Omega_k = \sin(\theta_k)\, d\theta_k d\phi_k$. The average over polarizations is done by using eqn (4.153) and the completeness relation (B.49) to get

$$\begin{aligned} \frac{1}{2}\sum_s \left|\mathbf{d}_{21}\cdot\mathbf{e}_{\mathbf{k}s}\right|^2 &= \frac{1}{2}(d_i)_{21}(d_j)_{21}^* \sum_s e_{\mathbf{k}si} e_{\mathbf{k}sj}^* \\ &= \frac{1}{2}\left[\mathbf{d}_{21}^*\cdot\mathbf{d}_{21} - \left(\widetilde{\mathbf{k}}\cdot\mathbf{d}_{21}\right)^*\left(\widetilde{\mathbf{k}}\cdot\mathbf{d}_{21}\right)\right] . \end{aligned} \tag{4.160}$$

In some cases the vector $\mathbf{d}_{21}$ is real, but this cannot be guaranteed in general (Mandel and Wolf, 1995, Sec. 15.1.1). When $\mathbf{d}_{21}$ is complex it can be expressed as $\mathbf{d}_{21} = \mathbf{d}_{21}' + i\mathbf{d}_{21}''$, where $\mathbf{d}_{21}'$ and $\mathbf{d}_{21}''$ are both real vectors. Inserting this into the previous equation gives

$$\sum_s \left|\mathbf{d}_{21}\cdot\mathbf{e}_{\mathbf{k}s}\right|^2 = \left[(\mathbf{d}_{21}')^2 - \left(\widetilde{\mathbf{k}}\cdot\mathbf{d}_{21}'\right)^2\right] + \left[(\mathbf{d}_{21}'')^2 - \left(\widetilde{\mathbf{k}}\cdot\mathbf{d}_{21}''\right)^2\right] , \tag{4.161}$$

and the remaining integral over the angles of $\mathbf{k}$ can be carried out for each term by choosing the z-axis along $\mathbf{d}_{21}'$ or $\mathbf{d}_{21}''$. The result is

$$A_{2\to 1} = \left[\frac{1}{4\pi\epsilon_0}\right] \frac{4\left|\mathbf{d}_{21}\right|^2 k_0^3}{3\hbar} , \tag{4.162}$$

where $k_0 = \omega_{21}/c = 2\pi/\lambda_0$ and $\left|\mathbf{d}_{21}\right|^2 = \mathbf{d}_{21}^*\cdot\mathbf{d}_{21}$. This agrees with the value obtained earlier by Einstein's thermodynamic argument. Dropping the coefficient in square brackets gives the result in Gaussian units.

Einstein's quantum model for radiation involves two other coefficients, $B_{1\to 2}$ for absorption and $B_{2\to 1}$ for stimulated emission. The stimulated emission rate is the rate for the transition $|\varepsilon_2, n_{\mathbf{k}s}\rangle \to |\varepsilon_1, n_{\mathbf{k}s}+1\rangle$, i.e. the initial state has $n_{\mathbf{k}s}$ photons in the mode $\mathbf{k}s$. In this case eqn (4.152) is replaced by

$$\left\langle \varepsilon_1, n_{\mathbf{k}s}+1 \left| \Omega^{(-)} \right| \varepsilon_2, n_{\mathbf{k}s} \right\rangle = -i\Omega^*_{21,s}(\mathbf{k})\sqrt{n_{\mathbf{k}s}+1}\,, \tag{4.163}$$

where the factor $\sqrt{n_{\mathbf{k}s}+1}$ comes from the rule $a^\dagger|n\rangle = \sqrt{n+1}\,|n+1\rangle$. For $n_{\mathbf{k}s}=0$, this reduces to the spontaneous emission result, so the only difference between the two processes is the enhancement factor $\sqrt{n_{\mathbf{k}s}+1}$. In order to simplify the argument we will assume that $n_{\mathbf{k}s} = n(\omega)$, i.e. the photon population is independent of polarization and propagation direction. Then the average over polarizations and emission directions produces

$$\Gamma = [n(\omega_{21})+1]\,A_{2\to1} = A_{2\to1} + n(\omega_{21})\,A_{2\to1}\,, \tag{4.164}$$

where the two terms are the spontaneous and stimulated rates respectively. By comparing this to eqn (1.13), we see that $B_{2\to1}\rho(\omega_{21}) = n(\omega_{21})\,A_{2\to1}$, where $\rho(\omega_{21})$ is the energy density per unit frequency. In the present case this is

$$\rho(\omega_{21}) = (\hbar\omega_{21})\,n(\omega_{21})\,\mathcal{D}(\omega_{21}) = \frac{\hbar\omega_{21}^3}{\pi^2c^3}n(\omega_{21})\,, \tag{4.165}$$

so the relation between the A and B coefficients is

$$\frac{A_{2\to1}}{B_{2\to1}} = \frac{\hbar\omega_{21}^3}{\pi^2c^3}\,, \tag{4.166}$$

in agreement with eqn (1.21). The absorption coefficient $B_{1\to2}$ is deduced by calculating the transition rate for $|\varepsilon_1, n_{\mathbf{k}s}+1\rangle \to |\varepsilon_2, n_{\mathbf{k}s}\rangle$. The relevant matrix element,

$$\left\langle \varepsilon_2, n_{\mathbf{k}s} \left| \Omega^{(+)} \right| \varepsilon_1, n_{\mathbf{k}s}+1 \right\rangle = i\Omega_{21,s}(\mathbf{k})\sqrt{n_{\mathbf{k}s}+1}\,, \tag{4.167}$$

corresponds to part (2) of Fig. 4.1. Since $|\Omega_{21,s}(\mathbf{k})| = \left|\Omega^*_{21,s}(\mathbf{k})\right|$, using this matrix element in eqn (4.113) will give the same result as the calculation of the stimulated emission coefficient, therefore the absorption rate is identical to the stimulated emission rate, i.e. $B_{1\to2} = B_{2\to1}$, in agreement with the detailed-balance argument eqn (1.18). Thus the quantum theory correctly predicts the relations between the Einstein A and B coefficients, and it provides an *a priori* derivation for the spontaneous emission rate.

4.9.4 Spontaneous emission in a planar cavity*

One of the assumptions in Einstein's quantum model for radiation is that the A and B coefficients are solely properties of the atom, but further thought shows that this cannot be true in general. Consider, for example, an atom in the interior of an ideal cubical cavity with sides L. According to eqn (2.15) the eigenfrequencies satisfy $\omega_{\mathbf{n}} \geqslant \sqrt{2}\pi c/L$; therefore, resonance is impossible if the atomic transition frequency is too small, i.e. $\omega_{21} < \sqrt{2}\pi c/L$, or equivalently $L < \lambda_0/\sqrt{2}$, where $\lambda_0 = 2\pi c/\omega_{21}$ is the wavelength of the emitted light. In addition to this failure of the resonance condition, the golden rule (4.113) is not applicable, since the mode spacing is not small compared to the transition frequency.

What this means physically is that photons emitted by the atom are reflected from the cavity walls and quickly reabsorbed by the atom. This behavior will occur for any finite value of L, but clearly the minimum time required for the radiation to be reabsorbed will grow with L. In the limit $L \to \infty$ the time becomes infinite and the result for an atom in free space is recovered. Therefore the standard result (4.162) for $A_{2\to 1}$ is only valid for an atom in unbounded space.

The fact that the spontaneous emission rate for atoms is sensitive to the boundary conditions satisfied by the electromagnetic field was recognized long ago (Purcell, 1946). More recently this problem has been studied in conjunction with laser etalons (Stehle, 1970) and materials exhibiting an optical bandgap (Yablonovitch, 1987). We will illustrate the modification of spontaneous emission in a simple case by describing the theory and experimental results for an atom in a planar cavity of the kind considered in connection with the Casimir effect.

A Theory

For this application, we will assume that the transverse dimensions are large, $L \gg \lambda_0$, while the longitudinal dimension Δz (along the z-axis) is comparable to the transition wavelength, $\Delta z \sim \lambda_0$. The mode wavenumbers are then $\mathbf{k} = \mathbf{q} + (n\pi/\Delta z)\,\mathbf{u}_z$, where $\mathbf{q} = k_x \mathbf{u}_x + k_y \mathbf{u}_y$, and the cavity frequencies are

$$\omega_{qn} = c\left[\mathbf{q}^2 + \left(\frac{n\pi}{\Delta z}\right)^2\right]^{1/2} . \tag{4.168}$$

Both n and $\mathbf{q}$ are discrete, but the transverse mode numbers $\mathbf{q}$ will become densely spaced in the limit $L \to \infty$. The Schrödinger-picture field operator is given by the analogue of eqn (3.69),

$$\mathbf{E}^{(+)}(\mathbf{r}) = i \sum_{\mathbf{q}} \sum_{n} \sum_{s=1}^{C_n} \sqrt{\frac{\hbar\omega_{qn}}{2\epsilon_0}} a_{\mathbf{q}ns} \boldsymbol{\mathcal{E}}_{\mathbf{q}ns}(\mathbf{r}) , \tag{4.169}$$

where the mode functions are described in Appendix B.4 and C_n is the number of independent polarization states for the mode $(n, \mathbf{q})$: $C_0 = 1$ and $C_n = 2$ for $n \geqslant 1$.

Since the separation, Δz, between the plates is comparable to the wavelength, the transition rate will depend on the distance from the atom to each plate. Consequently, we are not at liberty to assume that the atom is located at any particular z-value. On the other hand, the dimensions along the x- and y-axes are effectively infinite, so we can choose the origin in the (x, y)-plane at the location of the atom, i.e. $\mathbf{r} = (\mathbf{0}, z)$. The interaction Hamiltonian is given by eqns (4.149) and (4.150), but the Rabi operator in this case is a function of z, with the positive-frequency part

$$\Omega^{(+)}(z) = i \sum_{\mathbf{q}} \sum_{n} \sum_{s=1}^{C_n} \sqrt{\frac{\hbar\omega_{qn}}{2\epsilon_0}} a_{\mathbf{q}ns} \widehat{\mathbf{d}} \cdot \boldsymbol{\mathcal{E}}_{\mathbf{q}ns}(\mathbf{0}, z) . \tag{4.170}$$

The transition of interest is $|\varepsilon_2, 0\rangle \to |\varepsilon_1, 1_{\mathbf{q}ns}\rangle$, so only $\Omega^{(-)}(z)$ can contribute. For each value of n and z the remaining calculation is a two-dimensional version of the free-space case. Substituting the relevant matrix elements into eqn (4.113) and multiplying by $L^2 d^2q/(2\pi)^2$—the number of modes in the wavevector element d^2q—yields the differential transition rate

$$dW_{2\to 1,\mathbf{q}ns}(z) = 2\pi \left|M_{21,ns}(\mathbf{q}, z)\right|^2 \delta(\omega_{21} - \omega_{qn}) \frac{d^2q}{(2\pi)^2}, \tag{4.171}$$

where

$$M_{21,ns}(\mathbf{q}, z) = \sqrt{\frac{\hbar\omega_{qn}}{2\epsilon_0}} L\mathbf{d}_{21} \cdot \boldsymbol{\mathcal{E}}_{\mathbf{q}ns}(\mathbf{0}, z). \tag{4.172}$$

For a given n, the transition rate into all transverse wavevectors $\mathbf{q}$ and polarizations s is

$$A_{2\to 1,n}(z) = \sum_{s=1}^{C_n} \int \frac{d^2q}{(2\pi)^2} 2\pi \left|M_{21,ns}(\mathbf{q}, z)\right|^2 \delta(\omega_{21} - \omega_{qn}), \tag{4.173}$$

and the total transition rate is the sum of the partial rates for each n,

$$A_{2\to 1}(z) = \sum_{n=0}^{\infty} A_{2\to 1,n}(z). \tag{4.174}$$

The delta function in eqn (4.173) is eliminated by using polar coordinates, $d^2q = qdqd\phi$, and then making the change of variables $q \to \omega/c = \omega_{qn}/c$. The result is customarily expressed in terms of a density of states factor $\mathcal{D}_n(\omega_{21})$, defined as the number of resonant modes per unit frequency per unit of transverse area. For a given n there are C_n polarizations, so

$$\begin{aligned}\mathcal{D}_n(\omega_{21}) &= C_n \int \frac{d^2q}{(2\pi)^2} \delta(\omega_{21} - \omega_{qn}) \\ &= \frac{C_n}{2\pi} \int_{\omega_{0n}} \frac{d\omega\, \omega}{c^2} \delta(\omega_{21} - \omega) \\ &= \frac{C_n \omega_{21}}{2\pi c^2} \theta\left(\Delta z - \frac{n\lambda_0}{2}\right),\end{aligned} \tag{4.175}$$

where $\theta(\nu)$ is the standard step function, $\lambda_0 = 2\pi c/\omega_{21}$ is the wavelength for the transition, and $\omega_{0n} = n\pi c/\Delta z$. This density of states counts all polarizations and the full azimuthal angle, so in evaluating eqn (4.173) the extra $2\pi C_n$ must be divided out. The transition rate then appears as an average over azimuthal angles and polarizations:

$$A_{2\to 1,n}(z) = \mathcal{D}_n(\omega_{21}) \frac{1}{C_n} \sum_{s=1}^{C_n} \int \frac{d\phi}{2\pi} 2\pi \left|M_{21,\mathbf{q}ns}(z)\right|^2. \tag{4.176}$$

According to eqn (4.175) the density of states vanishes for $\Delta z/\lambda_0 < n/2$; therefore, emission into modes with $n > 2\Delta z/\lambda_0$ is forbidden. This reflects the fact that the high-n modes are not in resonance with the atomic transition. On the other hand, the density of states for the $(n = 0)$-mode is nonzero for any value of $\Delta z/\lambda_0$, so this transition is only forbidden if it violates atomic selection rules. In fact, this is the only possible decay channel for $\Delta z < \lambda_0/2$. In this case the total decay rate is

$$A_{2\to1,0} = \left[\frac{1}{4\pi\epsilon_0}\right]\frac{2\pi k_0^2\left|(d_z)_{21}\right|^2}{\hbar\Delta z} = \left[\frac{3\left|(d_z)_{21}\right|^2}{\left|\mathbf{d}_{21}\right|^2}\right]\frac{\lambda_0}{4\Delta z}A_{\text{vac}}\,, \tag{4.177}$$

where A_{vac} is the vacuum value given by eqn (4.162). The factor in square brackets is typically of order unity, so the decay rate is enhanced over the vacuum value when $\Delta z < \lambda_0/4$, and suppressed below the vacuum value for $\lambda_0/4 < \Delta z < \lambda_0/2$.

If the dipole selection rules (4.137) impose $(d_z)_{21} = 0$, then decay into the $(n = 0)$-mode is forbidden, and it is necessary to consider somewhat larger separations, e.g. $\lambda_0/2 < \Delta z < \lambda_0$. In this case, the decay to the $(n = 1)$-mode is the only one allowed. There are now two polarizations to consider, the P-polarization in the $(\widetilde{\mathbf{q}}, \mathbf{u}_z)$-plane and the orthogonal S-polarization along $\mathbf{u}_z \times \widetilde{\mathbf{q}}$. We will simplify the calculation by assuming that the matrix element $\mathbf{d}_{21}$ is real. In the general case of complex $\mathbf{d}_{21}$ a separate calculation for the real and imaginary parts must be done, as in eqn (4.161).

For real $\mathbf{d}_{21}$ the polar angle ϕ can be taken as the angle between $\mathbf{d}_{21}$ and $\mathbf{q}$. The assumption that $(d_z)_{21} = 0$ combines with the expressions (B.82) and (B.83), for the P- and S-polarizations, respectively, to yield

$$A_{2\to1,1} = \frac{3}{2}\left(\frac{\lambda_0}{2\Delta z}\right)\left[1 + \left(\frac{\lambda_0}{2\Delta z}\right)^2\right]\sin^2\left(\frac{\pi z}{\Delta z}\right)\theta\left(\Delta z - \frac{\lambda_0}{2}\right)A_{\text{vac}}\,, \tag{4.178}$$

where we have used the selection rule to impose $\left(d^{\perp}\right)^2 = d^2$. The decay rate depends on the location of the atom between the plates, and achieves its maximum value at the midplane $z = \Delta z/2$. In a real experiment, there are many atoms with unknown locations, so the observable result is the average over z:

$$A_{2\to1,1} = \frac{3}{4}\left(\frac{\lambda_0}{2\Delta z}\right)\left[1 + \left(\frac{\lambda_0}{2\Delta z}\right)^2\right]\theta\left(\Delta z - \frac{\lambda_0}{2}\right)A_{\text{vac}}\,. \tag{4.179}$$

This rate vanishes for $\lambda_0 > 2\Delta z$, and for $\lambda_0/2\Delta z$ slightly less than unity it is enhanced over the vacuum value:

$$A_{2\to1,1} \simeq \frac{3}{2}A_{\text{vac}} \quad \text{for } \lambda_0/2\Delta z \lesssim 1\,. \tag{4.180}$$

The decay rate is suppressed below the vacuum value for $\lambda_0/2\Delta z \lesssim 0.8$.

B Experiment

The clear-cut and striking results predicted by the theoretical model are only possible if the separation between the plates is comparable to the wavelength of the emitted radiation. This means that experiments in the optical domain would be extremely difficult. The way around this difficulty is to use a **Rydberg atom**, i.e. an atom which has been excited to a state—called a **Rydberg level**—with a large principal quantum number n. The Bohr frequencies for dipole allowed transitions between neighboring high-n states are of $O\left(1/n^3\right)$, so the wavelengths are very large compared to optical wavelengths.

In the experiment we will discuss here (Hulet *et al.*, 1985), cesium atoms were excited by two dye laser pulses to the $|n = 22, m = 2\rangle$ state. The small value of the magnetic quantum number is explained by the dipole selection rules, $\Delta l = \pm 1$, $\Delta m = 0, \pm 1$. These restrictions limit the m-values achievable in the two-step excitation process to a maximum of $m = 2$. This is a serious problem, since the state $|n = 22, m = 2\rangle$ can undergo dipole allowed transitions to any of the states $|n', m'\rangle$ for $2 \leqslant n' \leqslant 21$ and $m' = 1, 3$. A large number of decay channels would greatly complicate both the experiment and the theoretical analysis. This complication is avoided by exposing the atom to a combination of rapidly varying electric fields and microwave radiation which leave the value of n unchanged, but increase m to the maximum possible value, $m = n - 1$, a so-called circular state that corresponds to a circular Bohr orbit. The overall process leaves the atom in the state $|n = 22, m = 21\rangle$ which can only decay to $|n = 21, m = 20\rangle$. This simplifies both the experimental situation and the theoretical model. The wavelength for this transition is $\lambda_0 = 0.45\,\text{mm}$, so the mechanical problem of aligning the parallel plates is much simpler than for the Casimir force experiment. The gold-plated aluminum plates are held apart by quartz spacers at a separation of $\Delta z = 230.1\,\mu\text{m}$ so that $\lambda_0/2\Delta z = 0.98$.

The atom has now been prepared so that there is only one allowed atomic transition, but there are still two modes of the radiation field, $\boldsymbol{\mathcal{E}}_{\mathbf{q}0}$ and $\boldsymbol{\mathcal{E}}_{\mathbf{q}1s}$, into which the atom can decay. There is also the difficult question of how to produce controlled small changes in the plate spacing in order to see the effects on the spontaneous emission rate. Both of these problems are solved by the expedient of establishing a voltage drop between the plates. The resulting static electric field polarizes the atom so that the natural quantization axis lies in the direction of the field. The matrix elements of the z-component of the dipole operator, $\langle m|\widehat{d}_z|m'\rangle$, vanish unless $m' = m$, but transitions of this kind are not allowed by the dipole selection rules, $m' = m \pm 1$, for the circular Rydberg atom. This amounts to setting $(d_z)_{21} = 0$. Emission of $\boldsymbol{\mathcal{E}}_{\mathbf{q}0}$-photons is therefore forbidden, and the atom can only emit $\boldsymbol{\mathcal{E}}_{\mathbf{q}1s}$-photons. The field also causes second-order Stark shifts (Cohen-Tannoudji *et al.*, 1977*b*, Complement E-XII) which decrease the difference in the atomic energy levels and thus increase the wavelength λ_0. This means that the wavelength can be modified by changing the voltage, while the plate spacing is left fixed. The onset of field ionization limits the field strength that can be employed, so the wavelength can only be tuned by $\Delta\lambda = 0.04\,\lambda_0$. Fortunately, this is sufficient to increase the ratio $\lambda_0/2\Delta z$ through the critical value of unity, at which the spontaneous emission should be quenched. At room temperature the blackbody spectrum contains enough photons at the transition frequency to produce stimulated emission. The observed emission rate would then be the sum of the stimulated and spontaneous decay rates. In the model this would mean that we could not assume that the initial state is $|\varepsilon_1, 0\rangle$. This additional complication is avoided by maintaining the apparatus at $6.5\,\text{K}$. At this low temperature, stimulated emission due to blackbody radiation at λ_0 is strongly suppressed.

A thermal atomic beam of cesium first passes through a production region, where the atoms are transferred to the circular state, then through a drift region—of length

$L = 12.7\,\text{cm}$—between the parallel plates. The length L is chosen so that the mean transit time is approximately the same as the vacuum lifetime. After passing through the drift region the atoms are detected by field ionization in a region where the field increases with length of travel. The ionization rates for $n = 22$ and $n = 21$ atoms differ substantially, so the location of the ionization event allows the two sets of atoms to be resolved.

In this way, the time-of-flight distribution of the $n = 22$ atoms was measured. In the absence of decay, the distribution would be determined by the original Boltzmann distribution of velocities, but when decay due to spontaneous emission is present, only the faster atoms will make it through the drift region. Thus the distribution will shift toward shorter transit times. In the forbidden region, $\lambda_0/2\Delta z > 1$, the data were consistent with $A_{2\to 1,1} = 0$, with estimated errors $\pm 0.05 A_{\text{vac}}$. In other words, the lifetime of an atom between the plates is at least twenty times longer than the lifetime of the same atom in free space.

4.9.5 Raman scattering*

In Raman scattering, a photon at one frequency is absorbed by an atom or molecule, and a photon at a different frequency is emitted. The simplest energy-level diagram permitting this process is shown in Fig. 4.2. This is a second-order process, so it requires the calculation of the second-order amplitude $V_{fi}^{(2)}$, where the initial and final states are respectively $|\Theta_i\rangle = |\varepsilon_1, 1_{\mathbf{k}s}\rangle$ and $|\Theta_f\rangle = |\varepsilon_2, 1_{\mathbf{k}'s'}\rangle$. The representation (4.149) allows the operator product on the right side of eqn (4.120) to be written as

$$\begin{aligned} H_{\text{int}}(t_1)\, H_{\text{int}}(t_2) = \hbar^2 \left[\Omega^{(-)}(t_1)\,\Omega^{(-)}(t_2) + \Omega^{(+)}(t_1)\,\Omega^{(+)}(t_2)\right] \\ + \hbar^2 \left[\Omega^{(-)}(t_1)\,\Omega^{(+)}(t_2) + \Omega^{(+)}(t_1)\,\Omega^{(-)}(t_2)\right], \end{aligned} \tag{4.181}$$

where the first two terms change photon number by two and the remaining terms leave photon number unchanged. Since the initial and final states have equal photon number, only the last two terms can contribute in eqn (4.120); consequently, the matrix element of interest is

$$\hbar^2 \left\langle \Theta_f \left| \Omega^{(-)}(t_1)\,\Omega^{(+)}(t_2) + \Omega^{(+)}(t_1)\,\Omega^{(-)}(t_2) \right| \Theta_i \right\rangle. \tag{4.182}$$

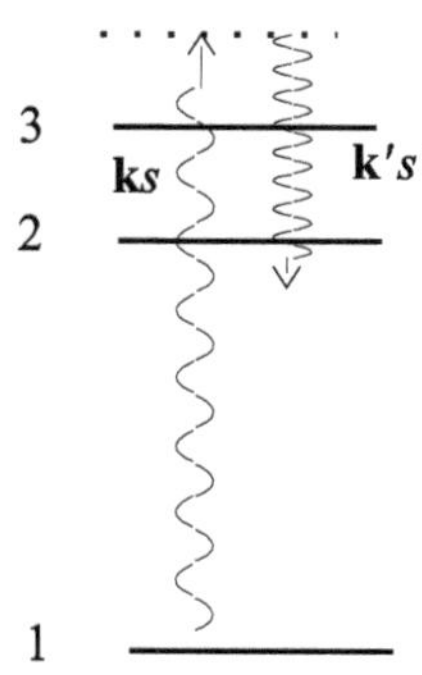

Fig. 4.2 Raman scattering from a three-level atom. The transitions $1 \leftrightarrow 3$ and $2 \leftrightarrow 3$ are dipole allowed. A photon in mode $\mathbf{k}s$ scatters into the mode $\mathbf{k}'s'$.

Since $t_2 < t_1$ the first term describes absorption of the initial photon followed by emission of the final photon, as one would intuitively expect. The second term is rather counterintuitive, since the emission of the final photon precedes the absorption of the initial photon. These alternatives are shown respectively by the Feynman diagrams (1) and (2) in Fig. 4.3, which we will call the *intuitive* and *counterintuitive* diagrams respectively.

The calculation of the transition amplitude by eqn (4.123) yields

$$\begin{aligned} V_{fi}^{(2)} &= -i \sum_u \frac{\left\langle \varepsilon_2, 1_{\mathbf{k}'s'} \left| \Omega^{(-)} \right| \Lambda_u \right\rangle \left\langle \Lambda_u \left| \Omega^{(+)} \right| \varepsilon_1, 1_{\mathbf{k}s} \right\rangle}{\omega_{k'} + \left(\frac{\varepsilon_2 - E_u}{\hbar}\right) + i\epsilon} 2\pi\delta\left(\omega_k - \omega_{k'} - \omega_{21}\right) \\ &\quad - i \sum_u \frac{\left\langle \varepsilon_2, 1_{\mathbf{k}'s'} \left| \Omega^{(+)} \right| \Lambda_u \right\rangle \left\langle \Lambda_u \left| \Omega^{(-)} \right| \varepsilon_1, 1_{\mathbf{k}s} \right\rangle}{\omega_{k'} + \left(\frac{\varepsilon_2 - E_u}{\hbar}\right) + i\epsilon} 2\pi\delta\left(\omega_k - \omega_{k'} - \omega_{21}\right), \end{aligned} \tag{4.183}$$

where the two sums over intermediate states correspond respectively to the intuitive and counterintuitive diagrams. Since $\Omega^{(+)}$ decreases the photon number by one, the intermediate states in the first sum have the form $|\Lambda_u\rangle = |\varepsilon_q, 0\rangle$. In this simple model the only available state is $|\Lambda_u\rangle = |\varepsilon_3, 0\rangle$. Thus the energy is $E_u = \varepsilon_3$ and the denominator is $\omega_{k'} - \omega_{32} + i\epsilon$. In fact, the intermediate state can be inferred from the Feynman diagram by passing a horizontal line between the two vertices. For the intuitive diagram, the only intersection is with the internal atom line, but in the counterintuitive diagram the line passes through both photon lines as well as the atom line. In this case, the intermediate state must have the form $|\Lambda_u\rangle = |\varepsilon_1, 1_{\mathbf{k}s}, 1_{\mathbf{k}'s'}\rangle$, with energy $E_u = \varepsilon_3 + \hbar\omega_{k'} + \hbar\omega_k$ and denominator $-\omega_k - \omega_{32} + i\epsilon$. These claims can be verified by a direct calculation of the matrix elements in the second sum.

This calculation yields the explicit expression

$$V_{fi}^{(2)} = -i \left\{ \frac{M_{32,s'}^{*}(\mathbf{k}')\, M_{31,s}(\mathbf{k})}{\omega_{k'} - \omega_{32} + i\epsilon} + \frac{M_{23,s}(\mathbf{k})\, M_{13,s'}^{*}(\mathbf{k}')}{-\omega_k - \omega_{32} + i\epsilon} \right\} \frac{2\pi}{V} \delta\left(\omega_k - \omega_{k'} - \omega_{21}\right), \tag{4.184}$$

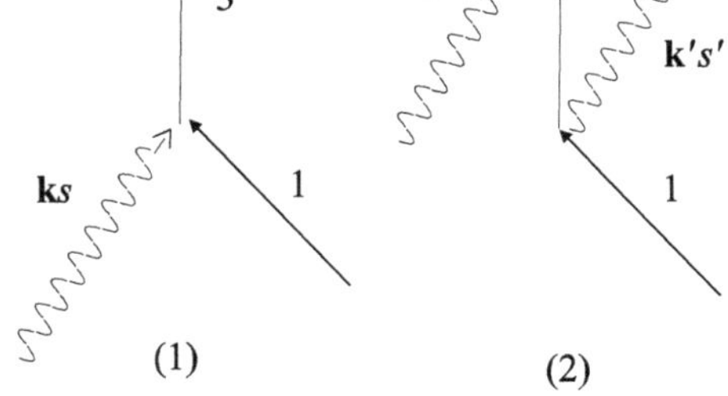

Fig. 4.3 Feynman diagrams for Raman scattering. Diagram (1) shows the intuitive ordering in which the initial photon is absorbed prior to the emission of the final photon. Diagram (2) shows the counterintuitive case in which the order is reversed.

where the matrix elements are defined in eqn (4.156). Multiplying $\left|V_{fi}^{(2)}\right|^2$ by the number of modes $\left[Vd^3k/(2\pi)^3\right]\left[Vd^3k'/(2\pi)^3\right]$ and using the rule (4.119) gives the differential transition rate

$$dW_{3\mathbf{k}s\to 2\mathbf{k}'s'} = 2\pi\left|\frac{M_{32,s'}^{*}(\mathbf{k}')\,M_{31,s}(\mathbf{k})}{\omega_{k'}-\omega_{32}+i\epsilon}+\frac{M_{23,s}(\mathbf{k})\,M_{13,s'}^{*}(\mathbf{k}')}{-\omega_{k}-\omega_{32}+i\epsilon}\right|^2 \times\delta\left(\omega_k-\omega_{k'}-\omega_{21}\right)\frac{d^3k}{(2\pi)^3}\frac{d^3k'}{(2\pi)^3}\,. \tag{4.185}$$

4.10 Exercises

4.1 Semiclassical electrodynamics

(1) Derive eqn (4.7) and use the result to get eqn (4.27).

(2) For the classical field described in the radiation gauge, do the following.

(a) Derive the equation satisfied by the scalar potential $\varphi(\mathbf{r})$.

(b) Show that

$$\nabla^2\frac{1}{|\mathbf{r}-\mathbf{r}_0|} = -4\pi\delta\left(\mathbf{r}-\mathbf{r}_0\right).$$

(c) Combine the last two results to derive the Coulomb potential term in eqn (4.31).

4.2 Maxwell's equations from the Heisenberg equations of motion

Derive Maxwell's equations and Lorentz equations of motion as given by eqns (4.33)–(4.37), and eqn (4.42), using Heisenberg's equations of motions and the relevant equal-time commutators.

4.3 Spatial inversion and time reversal*

(1) Use eqn (4.55) to evaluate $U_P\left|\underline{n}\right\rangle$ for a general number state, and explain how to extend this to all states of the field.

(2) Verify eqn (4.61) and fill in the details needed to get eqn (4.64).

(3) Evaluate $\Lambda_T\left|\underline{n}\right\rangle$ for a general number state, and explain how to extend this to all states of the field. Watch out for antilinearity.

4.4 Stationary density operators

Use eqns (3.83), (4.67), and $U(-t) = U^{\dagger}(t)$, together with cyclic invariance of the trace, to derive eqns (4.69) and (4.71).

4.5 Spin-flip transitions

The neutron is a spin-1/2 particle with zero charge, but it has a nonvanishing magnetic moment $\mathbf{M}_N = -\left|g_N\right|\mu_N\boldsymbol{\sigma}$, where g_N is the neutron gyromagnetic ratio, μ_N is the nuclear magneton, and $\boldsymbol{\sigma} = (\sigma_x,\sigma_y,\sigma_z)$ is the vector of Pauli matrices. Since the neutron is a massive particle, it is a good approximation to treat its center-of-mass motion classically. All of the following calculations can, therefore, be done assuming that the neutron is at rest at the origin.

(1) In the presence of a static, uniform, classical magnetic field $\mathcal{B}_0$ the Schrödinger-picture Hamiltonian—neglecting the radiation field—is $H_0 = -\mathbf{M}_N \cdot \mathcal{B}_0$. Take the z-axis along $\mathcal{B}_0$, and solve the time-independent Schrödinger equation, $H_0 |\psi\rangle = \varepsilon |\psi\rangle$, for the ground state $|\varepsilon_1\rangle$, the excited state $|\varepsilon_2\rangle$, and the corresponding energies ε_1 and ε_2.

(2) Include the effects of the radiation field by using the Hamiltonian $H = H_0 + H_{\text{int}}$, where $H_{\text{int}} = -\mathbf{M}_N \cdot \mathbf{B}$ and $\mathbf{B}$ is given by eqn (3.70), evaluated at $\mathbf{r} = 0$.
   (a) Evaluate the interaction-picture operators $a_{\mathbf{k}s}(t)$ and $\sigma_\pm(t)$ in terms of the Schrödinger-picture operators $a_{\mathbf{k}s}$ and $\sigma_\pm = (\sigma_x \pm i\sigma_y)/2$ (see Appendix C.3.1). Use the results to find the time dependence of the Cartesian components $\sigma_x(t)$, $\sigma_y(t)$, $\sigma_z(t)$.
   (b) Find the condition on the field strength $|\mathcal{B}_0|$ that guarantees that the zero-order energy splitting is large compared to the strength of H_{int}, i.e.

$$\varepsilon_2 - \varepsilon_1 \gg |\langle \varepsilon_1, 1_{\mathbf{k}s} |H_{\text{int}}| \varepsilon_2, 0\rangle| ,$$

   where $|\varepsilon_1, 1_{\mathbf{k}s}\rangle = |\varepsilon_1\rangle |1_{\mathbf{k}s}\rangle$, $|\varepsilon_2, 0\rangle = |\varepsilon_2\rangle |0\rangle$, and $|1_{\mathbf{k}s}\rangle = a^\dagger_{\mathbf{k}s} |0\rangle$. Explain the physical significance of this condition.
   (c) Using Section 4.9.3 as a guide, calculate the spontaneous emission rate (Einstein A coefficient) for a spin-flip transition. Look up the numerical values of $|g_N|$ and μ_N and use them to estimate the transition rate for magnetic field strengths comparable to those at the surface of a neutron star, i.e. $|\mathcal{B}_0| \sim 10^{12}\,\text{G}$.

4.6 The quantum top

Replace the unperturbed Hamiltonian in Exercise 4.5 by $H_0 = -\mathbf{M}_N \cdot \mathcal{B}_0(t)$, where $\mathcal{B}_0(t)$ changes direction as a function of time. Use this Hamiltonian to derive the Heisenberg equations of motion for $\boldsymbol{\sigma}(t)$ and show that they can be written in the same form as the equations for a precessing classical top.

4.7 Transition probabilities for a neutron in combined static and radio-frequency fields*

Solve the Schrödinger equation for a neutron in a combined static and radio-frequency magnetic field. A static field of strength $\mathcal{B}_0$ is applied along the z-axis, and a circularly-polarized, radio-frequency field of classical amplitude $\mathcal{B}_1$ and frequency ω is applied in the (x, y)-plane, so that the total Hamiltonian is $H = H_0 + H_{\text{int}}$, where

$$\begin{aligned} H_0 &= -M_z \mathcal{B}_0 , \\ H_{\text{int}} &= -M_x \mathcal{B}_1 \cos\omega t + M_y \mathcal{B}_1 \sin\omega t , \end{aligned}$$

$M_x = \frac{1}{2}\mu\sigma_x$, $M_y = \frac{1}{2}\mu\sigma_y$, $M_z = \frac{1}{2}\mu\sigma_z$, μ is the magnetic moment of the neutron, and the σs are Pauli matrices. Show that the probability for a spin flip of the neutron initially prepared (at $t = 0$) in the $m_s = +\frac{1}{2}$ state to the $m_s = -\frac{1}{2}$ state is given by

$$P_{\frac{1}{2}\to-\frac{1}{2}}(t) = \sin^2\Theta \sin^2\left(\frac{1}{2}at\right),$$

where

$$\sin^2\Theta = \frac{\omega_1^2}{(\omega_0-\omega)^2+\omega_1^2},$$
$$a = \sqrt{(\omega_0-\omega)^2+\omega_1^2},$$

$\omega_0 = \mu\mathcal{B}_0/\hbar$, and $\omega_1 = \mu\mathcal{B}_1/\hbar$. Interpret this result geometrically (Rabi *et al.*, 1954).

5

Coherent states

In the preceding chapters, we have frequently called upon the correspondence principle to justify various conjectures, but we have not carefully investigated the behavior of quantum states in the correspondence-principle limit. The difficulties arising in this investigation appear in the simplest case of the excitation of a single cavity mode $\boldsymbol{\mathcal{E}}_{\kappa}(\mathbf{r})$. In classical electromagnetic theory—as described in Section 2.1—the state of a single mode is completely described by the two real numbers $(Q_{\kappa 0}, P_{\kappa 0})$ specifying the initial displacement and momentum of the corresponding radiation oscillator. The subsequent motion of the oscillator is determined by Hamilton's equations of motion. The set of classical fields representing excitation of the mode κ is therefore represented by the two-dimensional phase space $\{(Q_{\kappa}, P_{\kappa})\}$.

In striking contrast, the quantum states for a single mode belong to the infinite-dimensional Hilbert space spanned by the family of number states, $\{|n\rangle\, , \; n = 0, 1, \ldots\}$. In order for a state $|\Psi\rangle$ to possess a meaningful correspondence-principle limit, each member of the infinite set, $\{c_n = \langle n\,|\Psi\,\rangle\, , \; n = 0, 1, \ldots\}$, of expansion coefficients must be expressible as a function of the two classical degrees of freedom $(Q_{\kappa 0}, P_{\kappa 0})$. This observation makes it clear that the number-state basis is not well suited to demonstrating the correspondence-principle limit. In addition to this fundamental issue, there are many applications for which a description resembling the classical phase space would be an advantage.

These considerations suggest that we should search for quantum states of light that are **quasiclassical**; that is, they approach the classical description as closely as possible. To this end, we first review the solution of the corresponding problem in ordinary quantum mechanics, and then apply the lessons learnt there to the electromagnetic field. After establishing the basic form of the quasiclassical states, we will investigate possible physical sources for them and the experimental evidence for their existence. The final sections contain a review of the mathematical properties of quasiclassical states, and their use as a basis for representations of general quantum states.

5.1 Quasiclassical states for radiation oscillators

In order to simplify the following discussion, we will at first only consider situations in which a single mode of the electromagnetic field is excited. For example, excitation of the mode $\boldsymbol{\mathcal{E}}_{\kappa}(\mathbf{r})$ in an ideal cavity corresponds to the classical fields

$$\boldsymbol{\mathcal{A}}(\mathbf{r},t) = \frac{1}{\sqrt{\epsilon_0}} Q_\kappa(t)\,\boldsymbol{\mathcal{E}}_\kappa(\mathbf{r}),$$
$$\boldsymbol{\mathcal{E}}(\mathbf{r},t) = -\frac{1}{\sqrt{\epsilon_0}} P_\kappa(t)\,\boldsymbol{\mathcal{E}}_\kappa(\mathbf{r}). \tag{5.1}$$

5.1.1 The mechanical oscillator

In Section 2.1 we guessed the form of the quantum theory of radiation by using the mathematical identity between a radiation oscillator and a mechanical oscillator of unit mass. The real Q and P variables of the classical oscillator can be simultaneously specified; therefore, the trajectory $(Q(t), P(t))$ of the oscillator is completely described by the time-dependent, complex amplitude

$$\mathcal{A}(t) = \frac{\omega Q(t) + iP(t)}{\sqrt{2\hbar\omega}}, \tag{5.2}$$

where the $\hbar$ is introduced for dimensional convenience only. Hamilton's equations of motion for the real variables Q and P are equivalent to the complex equation of motion

$$\dot{\mathcal{A}} = -i\omega\mathcal{A}, \tag{5.3}$$

with the general solution given by the **phasor** (a complex number of fixed modulus)

$$\mathcal{A}(t) = \alpha\exp(-i\omega t). \tag{5.4}$$

The initial complex amplitude of the oscillator is related to α by

$$\mathcal{A}(t=0) = \frac{\omega Q_0 + iP_0}{\sqrt{2\hbar\omega}} = \alpha, \tag{5.5}$$

and the conserved classical energy is

$$E_{\text{cl}} = \frac{1}{2}\left(\omega^2 Q_0^2 + P_0\right) = \hbar\omega\alpha^*\alpha. \tag{5.6}$$

Taking the real and imaginary parts of $\mathcal{A}(t)$, as given in eqn (5.4), shows that the solution traces out an ellipse in the (Q, P) phase space. An equivalent representation is the circle traced out by the tip of the phasor $\mathcal{A}(t)$ in the complex $(\operatorname{Re}\mathcal{A}, \operatorname{Im}\mathcal{A})$ space.

For the quantum oscillator, the classical amplitude $\mathcal{A}(0)$ and the energy $\hbar\omega|\alpha|^2$ are respectively replaced by the lowering operator

$$\widehat{a} = \frac{\omega\widehat{q} + i\widehat{p}}{\sqrt{2\hbar\omega}} \tag{5.7}$$

and the Hamiltonian operator $\widehat{H}_{\text{osc}} = \hbar\omega\widehat{a}^\dagger\widehat{a}$. The Heisenberg equation of motion for $\widehat{a}(t)$,

$$\frac{d\widehat{a}}{dt} = -\frac{i}{\hbar}[\widehat{a}, \widehat{H}_{\text{osc}}] = -i\omega\widehat{a}, \tag{5.8}$$

has the same form as the classical equation of motion (5.3).

We can now use an argument from quantum mechanics (Cohen-Tannoudji *et al.*, 1977*a*, Chap. V, Complement G) to construct the quasiclassical state. According to the correspondence principle, the classical quantities α and E_{cl} must be identified with the expectation values of the corresponding operators, so the quasiclassical state $|\phi\rangle$ corresponding to the classical value α should satisfy $\langle\phi|\widehat{a}|\phi\rangle = \alpha$ and $\langle\phi|\widehat{H}_{\text{osc}}|\phi\rangle = E_{\text{cl}} = \hbar\omega|\alpha|^2$. Inserting $\widehat{H}_{\text{osc}} = \hbar\omega\widehat{a}^\dagger\widehat{a}$ into the latter condition and using the former condition to evaluate $|\alpha|^2$ produces

$$\langle\phi|\widehat{a}^\dagger\widehat{a}|\phi\rangle = \langle\phi|\widehat{a}^\dagger|\phi\rangle\langle\phi|\widehat{a}|\phi\rangle\,. \tag{5.9}$$

The **joint variance** of two operators X and Y, defined by

$$V(X,Y) = \langle(X-\langle X\rangle)(Y-\langle Y\rangle)\rangle = \langle XY\rangle - \langle X\rangle\langle Y\rangle\,, \tag{5.10}$$

reduces to the ordinary variance $V(X)$ for $X = Y$. In this language, the meaning of eqn (5.9) is that the joint variance of $\widehat{a}$ and $\widehat{a}^\dagger$ vanishes,

$$V\left(\widehat{a}^\dagger,\widehat{a}\right) = 0\,, \tag{5.11}$$

i.e. the operators $\widehat{a}$ and $\widehat{a}^\dagger$ are statistically independent for a quasiclassical state. In its present form it is not obvious that $V\left(\widehat{a}^\dagger,\widehat{a}\right)$ refers to measurable quantities, but this concern can be addressed by using eqn (5.7) to get the equivalent form

$$V\left(\widehat{a}^\dagger,\widehat{a}\right) = \frac{\omega}{2\hbar}\left\langle(\widehat{q}-\langle\widehat{q}\rangle)^2\right\rangle + \frac{1}{2\hbar\omega}\left\langle(\widehat{p}-\langle\widehat{p}\rangle)^2\right\rangle - \frac{1}{2}\,. \tag{5.12}$$

The condition (5.11) is the fundamental property defining quasiclassical states, and it determines $|\phi\rangle$ up to a phase factor. To see this, we define a new operator $\widehat{b} = \widehat{a} - \alpha$ and a new state $|\chi\rangle = \widehat{b}|\phi\rangle$, to get

$$\langle\chi|\chi\rangle = \left\langle\phi\left|\widehat{b}^\dagger\widehat{b}\right|\phi\right\rangle = V\left(\widehat{a}^\dagger,\widehat{a}\right) = 0\,. \tag{5.13}$$

The squared norm $\langle\chi|\chi\rangle$ only vanishes if $|\chi\rangle = 0$; consequently, $\widehat{a}|\phi\rangle = \alpha|\phi\rangle$. Thus the quasiclassical state $|\phi\rangle$ is an eigenstate of the lowering operator $\widehat{a}$ with eigenvalue α. For this reason it is customary to rename $|\phi\rangle$ as $|\alpha\rangle$, so that

$$\widehat{a}|\alpha\rangle = \alpha|\alpha\rangle\,. \tag{5.14}$$

For non-hermitian operators, there is no general theorem guaranteeing the existence of eigenstates, so we need to find an explicit solution of eqn (5.14). In this section, we will do this in the usual coordinate representation, in order to gain an intuitive understanding of the physical significance of $|\alpha\rangle$. In the following section, we will find an equivalent form by using the number-state basis. This is useful for understanding the statistical properties of $|\alpha\rangle$.

The coordinate-space wave function for $|\alpha\rangle$ is $\phi_\alpha(Q) = \langle Q|\alpha\rangle$, where $\widehat{q}|Q\rangle = Q|Q\rangle$. In this representation, the action of $\widehat{q}$ is $\widehat{q}\phi_\alpha(Q) = Q\phi_\alpha(Q)$, and the action of the momentum operator is $\widehat{p}\phi_\alpha(Q) = -i\hbar(d/dQ)\phi_\alpha(Q)$. After inserting this into eqn (5.7), the eigenvalue problem (5.14) is represented by the differential equation

$$\frac{1}{\sqrt{2\hbar\omega}}\left(\omega Q + \hbar\frac{d}{dQ}\right)\phi_\alpha(Q) = \alpha\phi_\alpha(Q), \tag{5.15}$$

which has the normalizable solution

$$\phi_\alpha(Q) = \left(\frac{\omega}{\pi\hbar}\right)^{1/4}\exp\left[-\frac{(Q-Q_0)^2}{4\Delta q_0^2}\right]\exp\left[i\frac{P_0 Q}{\hbar}\right] \tag{5.16}$$

for any value of the complex parameter α. The parameters Q_0 and P_0 are given by $Q_0 = \sqrt{2\hbar/\omega}\,\mathrm{Re}\,\alpha$, $P_0 = \sqrt{2\hbar\omega}\,\mathrm{Im}\,\alpha$, and the width of the Gaussian is $\Delta q_0 = \sqrt{\hbar/2\omega}$. We have chosen the prefactor so that $\phi_\alpha(Q)$ is normalized to unity. For $Q_0 = P_0 = 0$, $\phi_0(Q)$ is the ground-state wave function of the oscillator; therefore, the general quasiclassical state, $\phi_\alpha(Q)$, represents the ground state of an oscillator which has been displaced from the origin of phase space to the point (Q_0, P_0). For the Q dependence this is shown explicitly by the probability density $|\phi_\alpha(Q)|^2$, which is a Gaussian in Q centered on Q_0. An alternative representation using the momentum-space wave function, $\phi_\alpha(P) = \langle P|\alpha\rangle$, can be derived in the same way—or obtained from $\phi_\alpha(Q)$ by Fourier transform—with the result

$$\phi_\alpha(P) = (\pi\hbar\omega)^{-1/4}\exp\left[-\frac{(P-P_0)^2}{4\Delta p_0^2}\right]\exp\left[-i\frac{Q_0 P}{\hbar}\right], \tag{5.17}$$

where $\Delta p_0 = \sqrt{\hbar\omega/2}$. The product $\Delta p_0 \Delta q_0 = \hbar/2$, so $|\alpha\rangle$ is a minimum-uncertainty state; it is the closest we can come to the classical description. The special values $\Delta q_0 = \sqrt{\hbar/2\omega}$ and $\Delta p_0 = \sqrt{\hbar\omega/2}$ define the **standard quantum limit** for the harmonic oscillator.

5.1.2 The radiation oscillator

Applying these results to the radiation oscillator for a particular mode $\boldsymbol{\mathcal{E}}_\kappa$ involves a change of terminology and, more importantly, a change in physical interpretation. For the radiation oscillator corresponding to the mode $\boldsymbol{\mathcal{E}}_\kappa$, the defining equation (5.14) for a quasiclassical state is replaced by

$$a_{\kappa'}|\alpha_\kappa\rangle = \delta_{\kappa'\kappa}\alpha_\kappa|\alpha_\kappa\rangle; \tag{5.18}$$

in other words, the quasiclassical state for this mode is the vacuum state for all other modes. This is possible because the annihilation operators for different modes commute with each other. A simple argument using eqn (5.18) shows that the averages of all normal-ordered products completely factorize:

$$\begin{aligned}\left\langle\alpha_\kappa\left|\left(a_\kappa^\dagger\right)^m(a_\kappa)^n\right|\alpha_\kappa\right\rangle &= (\alpha_\kappa^*)^m(\alpha_\kappa)^n \\ &= \left(\langle\alpha_\kappa|a_\kappa^\dagger|\alpha_\kappa\rangle\right)^m\left(\langle\alpha_\kappa|a_\kappa|\alpha_\kappa\rangle\right)^n;\end{aligned} \tag{5.19}$$

consequently, $|\alpha_\kappa\rangle$ is called a **coherent state**. The definition (5.18) shows that $|\alpha_\kappa\rangle$ belongs to the single-mode subspace $\mathfrak{H}_\kappa \subset \mathfrak{H}_F$ that is spanned by the number states for the mode $\boldsymbol{\mathcal{E}}_\kappa$.

The new physical interpretation is clearest for the radiation modes of a physical cavity. In the momentum-space representation, the operator p_κ is just multiplication by the eigenvalue P_κ, and the expansion (2.99) shows that the electric field operator is a function of the p_κs, so that $\mathbf{E}(\mathbf{r})\,\phi_\alpha(P_\kappa) = \mathfrak{E}_\kappa\sqrt{V}\boldsymbol{\mathcal{E}}_\kappa(\mathbf{r})\,\phi_\alpha(P_\kappa)$, where $\mathfrak{E}_\kappa = P_\kappa/\sqrt{\epsilon_0 V}$ is the electric field strength associated with P_κ. The dimensionless function $\sqrt{V}\boldsymbol{\mathcal{E}}_\kappa(\mathbf{r})$ is of order unity and describes the shape of the mode function. The corresponding result in the coordinate representation is $\mathbf{B}(\mathbf{r})\,\phi_\alpha(Q_\kappa) = \mathfrak{B}_\kappa\sqrt{V}\boldsymbol{\mathcal{B}}_\kappa(\mathbf{r})\,\phi_\alpha(Q_\kappa)$ with $\mathfrak{B}_\kappa = k_\kappa Q_\kappa/\sqrt{\epsilon_0 V} = \sqrt{\mu_0/V}\omega_\kappa Q_\kappa$. Eliminating P_κ in favor of $\mathfrak{E}_\kappa$ allows the Gaussian factor in $\phi_\alpha(P)$ to be expressed as

$$\exp\left[-\frac{\epsilon_0 V}{2\hbar\omega_\kappa}\left(\mathfrak{E}_\kappa - \overline{\varepsilon}_\kappa\right)^2\right] = \exp\left[-\frac{\left(\mathfrak{E}_\kappa - \overline{\varepsilon}_\kappa\right)^2}{4\mathfrak{e}_\kappa^2}\right], \tag{5.20}$$

where $\mathfrak{e}_\kappa$ is the vacuum fluctuation strength defined by eqn (2.188). Thus a coherent state displays a Gaussian probability density in the electric field amplitude $\mathfrak{E}_\kappa$ with average $\overline{\varepsilon}_\kappa$, and variance $V(\mathfrak{E}_\kappa) = 2\mathfrak{e}_\kappa^2$. Similarly the coordinate-space wave function is a Gaussian in $\mathfrak{B}_\kappa$ with average β_κ and variance $2\mathfrak{b}_\kappa^2$. The classical limit corresponds to $|\mathfrak{E}_\kappa| \gg \mathfrak{e}_\kappa$ and $|\mathfrak{B}_\kappa| \gg \mathfrak{b}_\kappa$, which are both guaranteed by $|\alpha_\kappa| \gg 1$. As an example, consider $\omega_\kappa = 10^{15}\,\mathrm{s}^{-1}$ ($\lambda_\kappa \approx 2\,\mu\mathrm{m}$) and $V = 1\,\mathrm{cm}^3$, then the vacuum fluctuation strength for the electric field is $\mathfrak{e}_\kappa \simeq 0.08\,\mathrm{V/m}$.

The fact that α_κ is a phasor provides the useful pictorial representation shown in Fig. 5.1. This is equivalent to a plot in the phase plane (Q_κ, P_κ). The result (5.17) for the wave function and the phase plot Fig. 5.1 are expressed in terms of the excitation of a single radiation oscillator in a physical cavity, but the idea of coherent states is not restricted to this case. The annihilation operator a can refer to a cavity mode (a_κ), a (box-quantized) plane wave ($a_{\mathbf{k}s}$), or a general wave packet operator ($a[\mathbf{w}]$), as defined in Section 3.5.2, depending on the physical situation under study. In the interests of simplicity, we will initially consider situations in which only one annihilation operator a (one electromagnetic degree of freedom) is involved. This is sufficient for a large variety

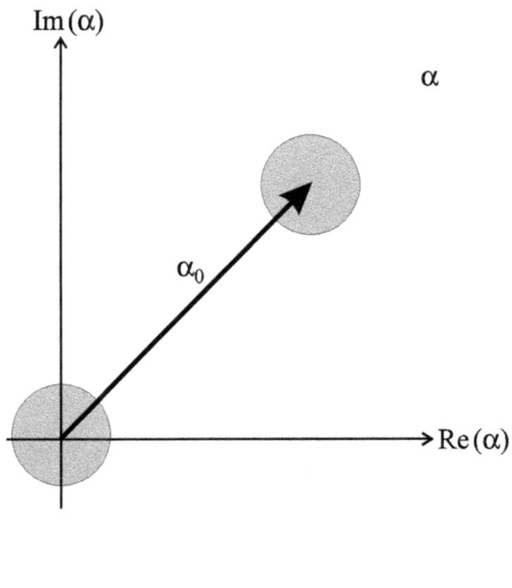

Fig. 5.1 The coherent state (displaced ground state) $|\alpha_0\rangle$ is pictured as an arrow joining the origin to the point α_0 in the complex plane. The quantum uncertainties of the ground state (at the origin) and the displaced ground state are each represented by an error circle (*quantum fuzzball*).

of applications, but the physical justification for isolating the single-mode subspace associated with a is that coupling between modes is weak. This fact should always be kept in mind, since a more complete calculation may involve taking the weak coupling into account, e.g. when considering dissipative or nonlinear effects.

5.1.3 Coherent states in the number-state basis

We now consider a single mode and represent $|\alpha\rangle$ by the number-state expansion

$$|\alpha\rangle = \sum_{n=0}^{\infty} b_n |n\rangle . \tag{5.21}$$

According to eqn (2.78) the eigenvalue equation (5.14) can then be written as

$$\sum_{n=0}^{\infty} \sqrt{n} b_n |n-1\rangle = \alpha \sum_{n=0}^{\infty} b_n |n\rangle . \tag{5.22}$$

Equating the coefficients of the number states yields the recursion relation, $b_{n+1} = \left(\alpha/\sqrt{n+1}\right) b_n$, which has the solution $b_n = b_0 \alpha^n/\sqrt{n!}$. Thus each coefficient b_n is a function of the complex parameter α, in agreement with the discussion at the beginning of the chapter. The vacuum coefficient b_0 is chosen to get a normalized state, with the result

$$|\alpha\rangle = e^{-|\alpha|^2/2} \sum_{n=0}^{\infty} \frac{\alpha^n}{\sqrt{n!}} |n\rangle . \tag{5.23}$$

This construction works for any complex number α, so the spectrum of the operator a is the entire complex plane. A similar calculation for $a^\dagger$ fails to find any normalizable solutions; consequently, $a^\dagger$ has neither eigenvalues nor eigenvectors.

The average number of photons for the state $|\alpha\rangle$ is $\overline{n} = \left\langle \alpha \left| a^\dagger a \right| \alpha \right\rangle = |\alpha|^2$, and the probability that n is the outcome of a measurement of the photon number is

$$P_n = e^{-\overline{n}} \frac{\overline{n}^n}{n!} , \tag{5.24}$$

which is a **Poisson distribution**. The variance in photon number is

$$V(N) = \left\langle \alpha \left| N^2 \right| \alpha \right\rangle - \left\langle \alpha \left| N \right| \alpha \right\rangle^2 = |\alpha|^2 = \overline{n} . \tag{5.25}$$

5.2 Sources of coherent states

Coherent states are defined by minimizing quantum fluctuations in the electromagnetic field, but the light emitted by a real source will display fluctuations for two reasons. The first is that vacuum fluctuations of the field are inescapable, even in the absence of charged particles. The second is that quantum fluctuations of the charged particles in the source will imprint themselves on the emitted light. This suggests that a source for coherent states should have minimal quantum fluctuations, and further that the forces exerted on the source by the emitted radiation—the **quantum back action**—should be negligible. The ideal limiting case is a purely classical current, which is so strong that the quantum back action can be ignored. In this situation the material source is described by classical physics, while the light is described by quantum theory. We will call this the **hemiclassical** approximation, to distinguish it from the familiar semiclassical approximation. The linear dipole antenna shown in Fig. 5.2 provides a concrete example of a classical source.

In free space, the classical far-field solution for the dipole antenna is an expanding spherical wave with amplitude depending on the angle between the dipole $\mathbf{p}$ and the radius vector $\mathbf{r}$ extending from the antenna to the observation point. A receiver placed at this point would detect a field that is locally approximated by a plane wave with propagation vector $\mathbf{k} = (\omega/c)\,\mathbf{r}/r$ and polarization in the plane defined by $\mathbf{p}$ and $\mathbf{r}$. Another interesting arrangement would be to place the antenna in a microwave cavity. In this case, d and ω could be chosen so that only one of the cavity modes is excited. In either case, what we want now is the answer to the following question: What is the quantum nature of the radiation field produced by the antenna?

We will begin with a quantum treatment of the charges and introduce the classical limit later. For weak fields, the A^2-term in eqn (4.32) for the Hamiltonian and the $\mathbf{A}$-term in eqn (4.43) for the velocity operator can both be neglected. In this approximation the current operator and the interaction Hamiltonian are respectively given by

$$\widehat{\mathbf{j}}(\mathbf{r}) = \sum_{\nu} \delta(\mathbf{r}-\widehat{\mathbf{r}}_{\nu})\, q_{\nu} \frac{\widehat{\mathbf{p}}_{\nu}}{m_{\nu}} \tag{5.26}$$

and

$$H_{\text{int}} = -\int d^3r\, \widehat{\mathbf{j}}(\mathbf{r}) \cdot \mathbf{A}(\mathbf{r})\,. \tag{5.27}$$

This approximation is convenient and adequate for our purposes, but it is not strictly necessary. A more exact treatment is given in (Cohen-Tannoudji *et al.*, 1989, Chap. III).

For an antenna inside a cavity, the positive-frequency part of the $\mathbf{A}$-field is

$$\mathbf{A}^{(+)}(\mathbf{r}) = \sum_{\kappa} \sqrt{\frac{\hbar}{2\epsilon_0 \omega_{\kappa}}}\, a_{\kappa} \boldsymbol{\mathcal{E}}_{\kappa}(\mathbf{r})\,, \tag{5.28}$$

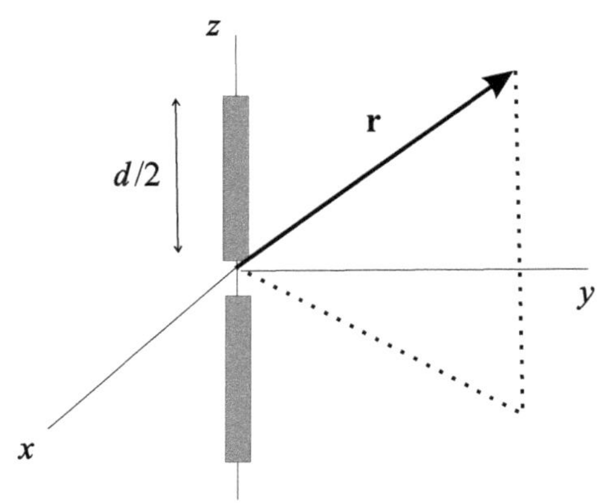

Fig. 5.2 A center fed linear dipole antenna excited at frequency ω. The antenna is short, i.e. $d \ll \lambda = 2\pi c/\omega$.

and the box-normalized expansion for an antenna in free space is obtained by $\boldsymbol{\mathcal{E}}_{\kappa}(\mathbf{r}) \to \mathbf{e}_{\mathbf{k}s} e^{i\mathbf{k}\cdot\mathbf{r}}/\sqrt{V}$. Using eqn (5.28) in the expressions for H_{em} and H_{int} produces

$$H_{\text{em}} = \sum_{\kappa} \hbar\omega_{\kappa} a_{\kappa}^{\dagger} a_{\kappa}\,, \tag{5.29}$$

and

$$H_{\text{int}} = -\sum_{\kappa} \sqrt{\frac{\hbar}{2\epsilon_0 \omega_{\kappa}}} a_{\kappa}^{\dagger} \int d^3r\, \widehat{\mathbf{j}}(\mathbf{r}) \cdot \boldsymbol{\mathcal{E}}_{\kappa}^{*}(\mathbf{r}) + \text{HC}\,. \tag{5.30}$$

In the Heisenberg picture, with $a_{\kappa} \to a_{\kappa}(t)$ and $\widehat{\mathbf{j}}(\mathbf{r}) \to \widehat{\mathbf{j}}(\mathbf{r},t)$, the equation of motion for $a_{\kappa}(t)$ is

$$\begin{aligned} i\hbar\frac{\partial}{\partial t} a_{\kappa}(t) &= \left[a_{\kappa}(t), H\right] \\ &= \hbar\omega_{\kappa} a_{\kappa}(t) - \sum_{\kappa} \sqrt{\frac{\hbar}{2\epsilon_0 \omega_{\kappa}}} \int d^3r\, \widehat{\mathbf{j}}(\mathbf{r},t) \cdot \boldsymbol{\mathcal{E}}_{\kappa}^{*}(\mathbf{r})\,. \end{aligned} \tag{5.31}$$

In an exact treatment these equations would have to be solved together with the Heisenberg equations for the charges, but we will avoid this complication by assuming that the antenna current is essentially classical. The quantum fluctuations in the current are represented by the operator

$$\delta\widehat{\mathbf{j}}(\mathbf{r},t) = \widehat{\mathbf{j}}(\mathbf{r},t) - \boldsymbol{\mathcal{J}}(\mathbf{r},t)\,, \tag{5.32}$$

where the average current is

$$\boldsymbol{\mathcal{J}}(\mathbf{r},t) = \text{Tr}\left[\rho_{\text{chg}} \widehat{\mathbf{j}}(\mathbf{r},t)\right], \tag{5.33}$$

and ρ_{chg} is the density operator for the charges in the absence of any photons. The expectation value $\boldsymbol{\mathcal{J}}(\mathbf{r},t)$ represents an external classical current, which is analogous to the external, classical electromagnetic field in the semiclassical approximation. With this notation, eqn (5.31) becomes

$$\begin{aligned} i\hbar\frac{\partial}{\partial t} a_{\kappa}(t) = \hbar\omega_{\kappa} a_{\kappa}(t) &- \sum_{\kappa} \sqrt{\frac{\hbar}{2\epsilon_0 \omega_{\kappa}}} \int d^3r\, \boldsymbol{\mathcal{J}}(\mathbf{r},t) \cdot \boldsymbol{\mathcal{E}}_{\kappa}^{*}(\mathbf{r}) \\ &- \sum_{\kappa} \sqrt{\frac{\hbar}{2\epsilon_0 \omega_{\kappa}}} \int d^3r\, \delta\widehat{\mathbf{j}}(\mathbf{r},t) \cdot \boldsymbol{\mathcal{E}}_{\kappa}^{*}(\mathbf{r})\,. \end{aligned} \tag{5.34}$$

In the hemiclassical approximation the quantum fluctuation operator $\delta\widehat{\mathbf{j}}(\mathbf{r},t)$ is neglected compared to $\boldsymbol{\mathcal{J}}(\mathbf{r},t)$, so the approximate Heisenberg equation is

$$i\hbar\frac{\partial}{\partial t} a_{\kappa}(t) = \hbar\omega_{\kappa} a_{\kappa}(t) - \sum_{\kappa} \sqrt{\frac{\hbar}{2\epsilon_0 \omega_{\kappa}}} \int d^3r\, \boldsymbol{\mathcal{J}}(\mathbf{r},t) \cdot \boldsymbol{\mathcal{E}}_{\kappa}^{*}(\mathbf{r})\,. \tag{5.35}$$

This is equivalent to approximating the Schrödinger-picture interaction Hamiltonian by

$$H_{\mathcal{J}}(t) = -\int d^3r\, \boldsymbol{\mathcal{J}}(\mathbf{r},t) \cdot \mathbf{A}(\mathbf{r})\,, \tag{5.36}$$

which represents the quantized field interacting with the classical current $\boldsymbol{\mathcal{J}}(\mathbf{r},t)$.

The Heisenberg equation (5.35) is linear in the operators $a_{\kappa}(t)$, so the individual modes are not coupled. We therefore restrict attention to a single mode and simplify the notation by $\{a_{\kappa}, \omega_{\kappa}, \boldsymbol{\mathcal{E}}_{\kappa}\} \to \{a, \omega, \boldsymbol{\mathcal{E}}\}$. The linearity of eqn (5.35) also allows us to simplify the problem further by considering a purely sinusoidal current with frequency Ω,

$$\boldsymbol{\mathcal{J}}(\mathbf{r}, t) = \boldsymbol{\mathcal{J}}(\mathbf{r})\, e^{-i\Omega t} + \boldsymbol{\mathcal{J}}^*(\mathbf{r})\, e^{i\Omega t}\,. \tag{5.37}$$

With these simplifications in force, the equation for $a(t)$ becomes

$$i\frac{\partial}{\partial t} a(t) = \omega a(t) - W e^{-i\Omega t} - W' e^{i\Omega t}\,, \tag{5.38}$$

where

$$\begin{aligned} W &= \sqrt{\frac{1}{2\epsilon_0 \hbar \omega_{\kappa}}} \int d^3r\, \boldsymbol{\mathcal{J}}(\mathbf{r}) \cdot \boldsymbol{\mathcal{E}}^*(\mathbf{r})\,, \\ W' &= \sqrt{\frac{1}{2\epsilon_0 \hbar \omega_{\kappa}}} \int d^3r\, \boldsymbol{\mathcal{J}}^*(\mathbf{r}) \cdot \boldsymbol{\mathcal{E}}^*(\mathbf{r})\,. \end{aligned} \tag{5.39}$$

For this linear differential equation the operator character of $a(t)$ is irrelevant, and the solution is found by elementary methods to be

$$a(t) = a e^{-i\omega t} + \alpha(t)\,, \tag{5.40}$$

where the c-number function $\alpha(t)$ is

$$\alpha(t) = iW e^{-i(\omega+\Omega)t/2} \frac{\sin\left(\frac{\Delta}{2} t\right)}{\frac{\Delta}{2}} + iW' e^{-i\Delta t/2} \frac{\sin\left[\frac{(\omega+\Omega)}{2} t\right]}{\frac{\omega+\Omega}{2}}\,, \tag{5.41}$$

and $\Delta = \omega - \Omega$ is the detuning of the radiation mode from the oscillation frequency of the antenna current. The first term has a typical resonance structure which shows—as one would expect—that radiation modes with frequencies close to the antenna frequency are strongly excited. The frequencies ω and Ω are positive by convention, so the second term is always off resonance, and can be neglected in practice.

The use of the Heisenberg picture has greatly simplified the solution of this problem, but the meaning of the solution is perhaps more evident in the Schrödinger picture. The question we set out to answer is the nature of the quantized field generated by a classical current. Before the current is turned on there is no radiation, so in the Schrödinger picture the initial state is the vacuum: $|\Psi(0)\rangle = |0\rangle$. In the Heisenberg picture this state is time independent, and eqn (5.40) implies that $a(t)|0\rangle = \alpha(t)|0\rangle$. Transforming back to the Schrödinger picture, by using eqn (3.83) and the identification of the Heisenberg-picture state vector with the initial Schrödinger-picture state vector, leads to

$$a|\Psi(t)\rangle = \alpha(t)|\Psi(t)\rangle\,, \tag{5.42}$$

where $|\Psi(t)\rangle = U(t)|\Psi(0)\rangle$ is the Schrödinger-picture state that evolves from the vacuum under the influence of the classical current. Thus the radiation field from a classical current is described by a coherent state $|\alpha(t)\rangle$, with the time-dependent amplitude given by eqn (5.41). According to Section 5.2, the field generated by the classical current is the ground state of an oscillator displaced by $Q(t) \propto \operatorname{Re}\alpha(t)$ and $P(t) \propto \operatorname{Im}\alpha(t)$.

5.3 Experimental evidence for Poissonian statistics

Experimental verification of the predicted properties of coherent states, e.g. the Poissonian statistics of photon number, evidently depends on finding a source that produces coherent states. The ideal classical currents introduced for this purpose in Section 5.2 provide a very accurate description of sources operating in the radio and microwave frequency bands, but—with the possible exception of free-electron lasers—devices of this kind are not found in the laboratory as sources for light at optical wavelengths. Despite this, the folklore of laser physics includes the firmly held belief that the output of a laser operated far above threshold is well approximated by a coherent state. This claim has been criticized on theoretical grounds (Mølmer, 1997), but recent experiments using the method of quantum tomography, explained in Chapter 17, have provided strong empirical support for the physical reality of coherent states. This subtle question is beyond the scope of our book, so we will content ourselves with a simple plausibility argument supporting a coherent state model for the output of a laser. This will be followed by a discussion of an experiment performed by Arecchi (1965) to demonstrate the existence of Poissonian photon-counting statistics—which are consistent with a coherent state—in the output of a laser operated well above threshold.

5.3.1 Laser operation above threshold

What is the basis for the folk-belief that lasers produce coherent states, at least when operated far above threshold? A plausible answer is that the assumption of essentially classical laser light is consistent with the mechanism that produces this light. The argument begins with the assumption that, in the correspondence-principle limit of high laser power, the laser field has a well-defined phase. The phases of the individual atomic dipole moments driven by this field will then be locked to the laser phase, so that they all emit coherently into the laser field. The resulting reinforcement between the atoms and the field produces a mutually coherent phase. Moreover, the reflection of the generated light from the mirrors defining the resonant cavity induces a positive feedback effect which greatly sharpens the phase of the laser field. In this situation vacuum fluctuations in the light—the quantum back action mentioned above—have a negligible effect on the atoms, and the polarization current density operator $\partial\widehat{\mathbf{P}}/\partial t$ behaves like a classical macroscopic quantity $\partial\mathbf{P}/\partial t$. Since $\partial\mathbf{P}/\partial t$ oscillates at the resonance frequency of the lasing transition, it plays the role of the classical current in Section 5.2, and will therefore produce a coherent state.

The plausibility of this picture is enhanced by considering the operating conditions in a real, continuous-wave (cw) laser. The net gain is the difference between the gain due to stimulated emission from the population of inverted atoms and the linear losses in the laser (usually dominated by losses at the output mirrors). The increase of the stimulated emission rate as the laser intensity grows causes depletion of the atomic inversion; consequently, the gain decreases with increasing intensity. This phenomenon is called **saturation**, and in combination with the linear losses it reduces the gain until it exactly equals the linear loss in the cavity. This steady-state balance between the saturated gain and the linear loss is called **gain-clamping**. Therefore, in the steady state the intensity-dependent gain is clamped at a value exactly equal to the distributed loss. The intensity of the light and the atomic polarization that produced it are in turn clamped at fixed c-number values. In this way, the macroscopic atomic system becomes insensitive to the quantum back-action of the radiation field, and acts like a classical current source.

5.3.2 Arecchi's experiment

In Fig. 5.3 we show a simplified description of Arecchi's experiment, which measures the statistics of photoelectrons generated by laser light transmitted through a ground-glass disc. As a consequence of the transverse spatial coherence of the laser beam, light transmitted through the randomly distributed irregularities in the disc will interfere to produce the speckle pattern observed when an object is illuminated by laser light (Milonni and Eberly, 1988, Sec. 15.8). In the far field of the disc, the transmitted light passes through a pinhole—which is smaller than the characteristic spot size of the speckle pattern—and is detected by a photomultiplier tube, whose output pulses enter a pulse-height analyzer.

When the ground-glass disc is at rest, the light passing through the pinhole represents a single element of the speckle pattern.[1] In this situation the temporal coherence of the transmitted light is the same as that of original laser light, so the expectation is that the detected light will be represented by a coherent state. Thus the photon statistics should be Poissonian.

If the disc rotates so rapidly that the speckle features cross the pinhole in a time short compared to the integration time of the detector, the transmitted light becomes effectively incoherent. As a simple classical model of this effect, consider the vectorial addition of phasors with random lengths (intensities) and directions (phases). The resultant phasor is the solution to the 2D random-walk problem on the phasor plane. In the limit of a large number of scatterers the distribution function for the resultant phasor is a Gaussian centered at the origin. The incoherent light produced in this way is indistinguishable from thermal light that has passed through a narrow spectral filter. Therefore, one expects the resulting photon statistics to be described by the Bose–Einstein distribution given by eqn (2.178).

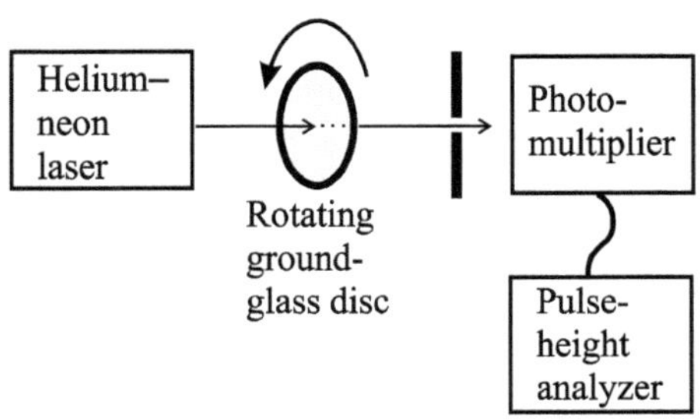

Fig. 5.3 Schematic of Arecchi's photon-counting experiment. Light generated by a cw, helium–neon laser is transmitted through a ground-glass disc to a small pinhole located in the far field of the disc and placed in front of a photomultiplier tube. The resulting photoelectron current is analyzed by means of a pulse-height analyzer. Results for coherent (incoherent) light are obtained when the disc is stationary (rotating).

[1] Murphy's law dictates that the pattern element covering the pinhole will sometimes be a null in the interference pattern. In practice the disc should be rotated until the signal is a maximum.

Photomultiplier tubes are fast detectors, with nanosecond-scale resolution times, so the pulse height (i.e. the peak voltage) of each output pulse is directly proportional to the number of photons in the beam during a resolution time. This follows from the fact that the fundamental detection process is the photoelectric effect, in which (ideally) a single photon would be converted to a single photoelectron. Thus two arriving photons would be converted at the photocathode into two photoelectrons, and so on. In practice, due to the finite thickness of the photocathode film, not all photons are converted into photoelectrons. The fraction of photons converted to photoelectrons, which is called the quantum efficiency, is studied in Section 9.1.3. Under the assumption that the quantum efficiency is independent of the intensity of the light, and that the postdetection amplification system is linear, it is possible to convert the photoelectron-count distribution, i.e. the pulse-height distribution, into the photon-count distribution function, $p(n)$. In the ideal case when the quantum efficiency is 100%, each photon would be converted into a photoelectron, and the photoelectron count distribution function would be a faithful representation of $p(n)$. However, it turns out that even if the quantum efficiency is less than 100%, the photoelectron count distribution function will, under these experimental conditions, still be a faithful representation of $p(n)$.

In Fig. 5.4 the channel numbers on the horizontal axis label increasing pulse heights, and the vertical coordinate of a point on the curve represents the number of pulses counted within a small range (a *bin*) around the corresponding pulse height. One can therefore view this plot as a histogram of the number of photoelectrons released in a given primary event. The data points were obtained by passing the output pulse of the photomultiplier directly into the pulse-height analyzer. This is raw data, in the sense that the photomultiplier pulses have not been reshaped to produce standardized digital pulses before they are counted. This avoids the dead-time problem, in which the electronics cannot respond to a second pulse which follows too quickly after the first one.

Assuming that the photomultiplier (including its electron-multiplication structures) is a linear electronic system with a fixed integration time—given by an RC time constant on the order of nanoseconds—the resulting pulse-height analysis yields a faithful representation of the initial photoelectron distribution at the photocathode, and hence of the photon distribution $p(n)$ arriving at the photomultiplier. Therefore, the channel number (the horizontal axis) is directly proportional to the photon number n, while the number of counts (the vertical axis) is linearly related to the probability $p(n)$. For the case denoted by L (for laser light), the observed photoelectron distribution function fits a Poissonian distribution, $p(n) = \exp(-\overline{n})\,\overline{n}^n/n!$, to within a few per cent. It is, therefore, an empirical fact that a helium–neon laser operating far above threshold produces Poissonian photon statistics, which is what is expected from a coherent state. For the case denoted by G (for Gaussian light), the observed distribution closely fits the Bose–Einstein distribution $p(n) = \overline{n}^n/(\overline{n}+1)^{n+1}$, which is expected for filtered thermal light. The striking difference between the nearly Poissonian curve L and the nearly Gaussian curve G is the main result of Arecchi's experiment.

Some remarks concerning this experiment are in order.

(1) As a function of time, the laser (with photon statistics described by the L-curve) emits an ensemble of coherent states $|\alpha(t)\rangle$, where $\alpha(t) = |\alpha|e^{i\phi(t)}$. The amplitude

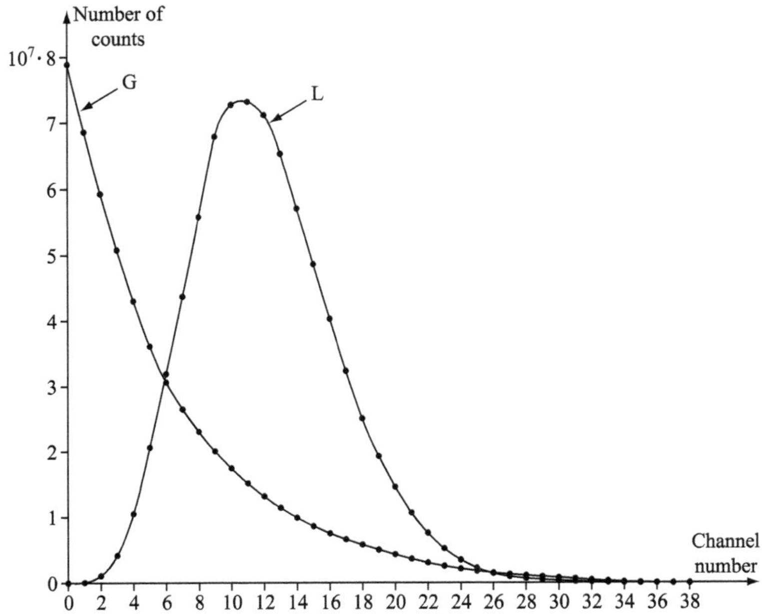

Fig. 5.4 Data from Arecchi's experiment measuring photoelectron statistics of a cw, helium–neon laser. The number of counts of output pulses from a photomultiplier tube, binned within a narrow window of pulse heights, is plotted against the voltage pulse height for two kinds of light fields: 'L' for 'laser light', which closely fits a Poissonian, and 'G' for 'Gaussian light', which closely fits a Bose–Einstein distribution function. (Reproduced from Arecchi (1965).)

$|\alpha(t)| = |\alpha|$ is fixed by gain clamping, but the phase $\phi(t)$ is not locked to any external source. Consequently, the phase wanders (or diffuses) on a very long coherence time scale $\tau_{\text{coh}} \simeq 0.1\,\text{s}$ (the inverse of the laser line width). The phase-wander time scale is much longer than the integration time, $RC \simeq 1\,\text{ns}$, of the very fast photon detection system. Furthermore, the Poissonian distribution $p(n)$ only depends on the fixed amplitude $|\alpha| = \sqrt{\overline{n}}$, so the phase wander of the laser output beam does not appreciably affect the Poissonian photocount distribution function.

(2) For the G-case, the coherence time τ_{coh} is determined by the time required for a speckle feature to cross the pinhole. For a rapidly rotating disc this is shorter than the integration time of the photon detection system. As explained above, this results in incoherent light described by a Bose–Einstein distribution peaked at $\overline{n} = 0$.

(3) The measurement process occurs at the photocathode surface of the photomultiplier tube, which, for unit quantum efficiency, emits n photoelectrons if n photons impinge on it. However, unity quantum efficiency is not an essential requirement for this experiment, since an analysis for arbitrary quantum efficiencies, when folded

in with a Bernoulli distribution function, shows that the Poissonian photoelectron distribution still always results from an initial Poissonian photon distribution (Loudon, 2000, Sec. 6.10). Similarly, a Bose–Einstein photoelectron distribution function always results from an initial Bose–Einstein photon distribution.

(4) The condition that the laser be far above threshold is often not satisfied by real continuous-wave lasers. The Scully–Lamb quantum theory of the laser predicts that there can be appreciable deviations from the exact Poissonian distribution when the small-signal gain of the laser is comparable to the loss of output mirrors. Nevertheless, a skewed bell-shape curve that roughly resembles the Poissonian distribution function is still predicted by the Scully–Lamb theory.

In sum, Arecchi's experiment gave the first partial evidence that lasers emit a coherent state, in that the observed photon count distribution is nearly Poissonian. However, this photon-counting experiment only gives information concerning the diagonal elements $\langle n |\rho| n\rangle = p(n)$ of the density matrix. It gives no information about the off-diagonal elements $\langle n |\rho| m\rangle$ when $n \neq m$. For example, this experiment cannot distinguish between a pure coherent state $|\alpha\rangle$, with $|\alpha| = \overline{n}$, and a mixed state for which $\langle n |\rho| n\rangle$ happens to be a Poissonian distribution and $\langle n |\rho| m\rangle = 0$ for $n \neq m$. We shall see later that quantum state tomography experiments using optical homodyne detection are sensitive to the off-diagonal elements of the density operator. These experiments provide evidence that the state of a laser operating far above threshold is closely approximated by an ideal coherent state.

In an extension of Arecchi's experiment, Meltzer and Mandel (1971) measured the photocount distribution function as a laser passes from below its threshold, through its threshold, and ends up far above threshold. The change from a monotonically decreasing photocount distribution below threshold—associated with the thermal state of light—to a peaked one above threshold—associated with the coherent state—was observed to agree with the Scully–Lamb theory.

5.4 Properties of coherent states

One of the objectives in studying coherent states is to use them as an alternate set of basis functions for Fock space, but we must first learn to deal with the peculiar mathematical features arising from the fact that the coherent states are eigenfunctions of the non-hermitian annihilation operator a.

5.4.1 The displacement operator

The relation (3.83) linking the Heisenberg and Schrödinger pictures combines with the explicit solution (5.40) of the Heisenberg equation to yield $U^{\dagger}(t)\, aU(t) = ae^{-i\omega t} + \alpha(t)$. For $N = a^{\dagger}a$, the identity $\exp(i\theta N)\, a \exp(-i\theta N) = \exp(-i\theta)\, a$ (see Appendix C.3, eqn (C.65)) allows this to be rewritten as

$$U^{\dagger}(t)\, aU(t) = e^{i\omega Nt} a e^{-i\omega Nt} + \alpha(t)\,, \tag{5.43}$$

which in turn implies

$$\left[U(t)\, e^{i\omega Nt}\right]^{\dagger} a \left[U(t)\, e^{i\omega Nt}\right] = a + \alpha(t)\,. \tag{5.44}$$

Thus the physical model for generation of a coherent state in Section 5.2 implies that there is a unitary operator which acts to displace the annihilation operator by $\alpha(t)$.

The form of this operator can be derived from the explicit solution of the model problem, but it is more useful to seek a unitary **displacement operator** $D(\alpha)$ that satisfies

$$D^{\dagger}(\alpha)\, aD(\alpha) = a + \alpha \tag{5.45}$$

for all complex α. Since $D(\alpha)$ is unitary, it can be written as $D(\alpha) = \exp[-iK(\alpha)]$, where the hermitian operator $K(\alpha)$ is the **generator** of displacements. A similar situation arises in elementary quantum mechanics, where the representation $\widehat{p} = -i\hbar d/dq$ for the momentum operator implies that the transformation

$$T_{\Delta q}\psi(q) = \psi(q - \Delta q) \tag{5.46}$$

of spatial translation is represented by the unitary operator $\exp(-i\Delta q\widehat{p}/\hbar)$ (Bransden and Joachain, 1989, Sec. 5.9). This transformation rule for the wave function is equivalent to the operator relation

$$e^{-i\Delta q\widehat{p}/\hbar}\widehat{q}e^{i\Delta q\widehat{p}/\hbar} = \widehat{q} - \Delta q\,. \tag{5.47}$$

The similarity between eqns (5.45) and (5.47) and the associated fact that $[a, a^{\dagger}]$ (like $[\widehat{q}, \widehat{p}]$) is a c-number together suggest assuming that $K(\alpha)$ is a linear combination of a and $a^{\dagger}$:

$$K(\alpha) = g(\alpha)\, a^{\dagger} + g^{*}(\alpha)\, a\,, \tag{5.48}$$

where $g(\alpha)$ is a c-number yet to be determined.

One way to work out the consequences of this assumption is to define the **interpolating operator** $a(\tau)$ by

$$a(\tau) = e^{i\tau K(\alpha)}ae^{-i\tau K(\alpha)}\,. \tag{5.49}$$

This new operator is constructed so that it has the initial value $a(0) = a$ and the final value $a(1) = D^{\dagger}(\alpha)\, aD(\alpha)$. In the τ-interval $(0, 1)$, $a(\tau)$ satisfies the Heisenberg-like equation of motion

$$i\frac{da(\tau)}{d\tau} = [a(\tau), K(\alpha)] = e^{i\tau K(\alpha)}[a, K(\alpha)]\, e^{-i\tau K(\alpha)}\,. \tag{5.50}$$

In the present case, the *ansatz* (5.48) shows that $[a, K(\alpha)] = g(\alpha)$, so the equation of motion simplifies to

$$i\frac{da(\tau)}{d\tau} = g(\alpha)\,, \tag{5.51}$$

with the solution $a(\tau) = a - ig(\alpha)\tau$. Thus eqn (5.45) is satisfied by the choice $g(\alpha) = i\alpha$, and the displacement operator is

$$D(\alpha) = e^{(\alpha a^{\dagger} - \alpha^{*} a)}\,. \tag{5.52}$$

The displacement operator generates the coherent state from the vacuum by

$$|\alpha\rangle = D(\alpha)|0\rangle = e^{(\alpha a^\dagger - \alpha^* a)}|0\rangle . \tag{5.53}$$

The simplest way to prove that $D(\alpha)|0\rangle$ is a coherent state is to rewrite eqn (5.45) as

$$aD(\alpha) = D(\alpha)[a+\alpha], \tag{5.54}$$

and apply both sides to the vacuum state.

The displacement operators represent the translation group in the α-plane, so they must satisfy certain group properties. For example, a direct application of the definition (5.45) yields the inverse transformation as

$$D^{-1}(\alpha) = D^\dagger(\alpha) = D(-\alpha). \tag{5.55}$$

From eqn (5.45) one can see that applying $D(\beta)$ followed by $D(\alpha)$ has the same effect as applying $D(\alpha+\beta)$; therefore, the product $D(\alpha)D(\beta)$ must be proportional to $D(\alpha+\beta)$:

$$D(\alpha)D(\beta) = D(\alpha+\beta)e^{i\Phi(\alpha,\beta)}, \tag{5.56}$$

where $\Phi(\alpha,\beta)$ is a real function of α and β. The phase $\Phi(\alpha,\beta)$ can be determined by using the Campbell–Baker–Hausdorff formula, eqn (C.66), or—as in Exercise 5.6—by another application of the interpolating operator method. By either method, the result is

$$D(\alpha)D(\beta) = D(\alpha+\beta)e^{i\,\mathrm{Im}(\alpha\beta^*)}. \tag{5.57}$$

5.4.2 Overcompleteness

Distinct eigenstates of hermitian operators, e.g. number states, are exactly orthogonal; therefore, distinct outcomes of measurements of the number operator—or any other observable—are mutually exclusive events. This is the basis for interpreting $|c_n|^2 = |\langle n|\psi\rangle|^2$ as the probability that the value n will be found in a measurement of the number operator. By contrast, no two coherent states are ever orthogonal. This is shown by using eqn (5.23) to calculate the value

$$\langle\alpha|\beta\rangle = \exp\left(-\frac{1}{2}|\alpha-\beta|^2\right)\exp\left(i\,\mathrm{Im}\left[\alpha^*\beta\right]\right) \tag{5.58}$$

of the inner product. On the other hand, states with large values of $|\alpha-\beta|$ are approximately orthogonal, i.e. $|\langle\alpha|\beta\rangle| \ll 1$, for quite moderate values of $|\alpha-\beta|$. The lack of orthogonality between distinct coherent states means that $|\langle\alpha|\psi\rangle|^2$ cannot be interpreted as the probability for finding the field in the state $|\alpha\rangle$, given that it is prepared in the state $|\psi\rangle$.

Although they are not mutually orthogonal, the coherent states are complete. A necessary and sufficient condition for completeness of the family $\{|\alpha\rangle\}$ is that a vector $|\psi\rangle$ satisfying

$$\langle\psi|\alpha\rangle = 0 \text{ for all } \alpha \tag{5.59}$$

is necessarily the null vector, i.e. $|\psi\rangle = 0$. A second use of eqn (5.23) allows this equation to be expressed as

$$F(\alpha) = \sum_{n=0}^{\infty} \frac{\alpha^n}{\sqrt{n!}} c_n^* = 0\,, \tag{5.60}$$

where $c_n^* = \langle \psi | n \rangle$. This relation is an identity in α, so all derivatives of $F(\alpha)$ must also vanish. In particular,

$$\left(\frac{\partial}{\partial \alpha} \right)^n F(\alpha) \bigg|_{\alpha=0} = \sqrt{n!} c_n^* = 0\,, \tag{5.61}$$

so that $c_n = 0$ for all $n \geqslant 0$. The completeness of the number states then requires $|\psi\rangle = 0$, and this establishes the completeness of the coherent states.

The coherent states form a complete set, but they are not linearly independent vectors. This peculiar state of affairs is called **overcompleteness**. It is straightforward to show that any finite collection of distinct coherent states is linearly independent, so to prove overcompleteness we must show that the null vector can be expressed as a continuous superposition of coherent states. Let $u_1 = \operatorname{Re}\alpha$ and $u_2 = \operatorname{Im}\alpha$, then any linear combination of the coherent states can be written as

$$\int_{-\infty}^{\infty} du_1 \int_{-\infty}^{\infty} du_2 \; z(u_1, u_2) \, |u_1 + iu_2\rangle\,, \tag{5.62}$$

where $z(u_1, u_2)$ is a complex function of the two real variables $u_1$and u_2. It is customary to regard $z(u_1, u_2)$ as a function of α^* and α, which are treated as independent variables, and in the same spirit to write

$$du_1 du_2 = d^2\alpha\,. \tag{5.63}$$

For brevity we will sometimes write $z(\alpha)$ instead of $z(\alpha^*, \alpha)$ or $z(u_1, u_2)$, and the same convention will be used for other functions as they arise. Any confusion caused by these various usages can always be resolved by returning to the real variables u_1 and u_2.

In this new notation the condition that a continuous superposition of coherent states gives the null vector is

$$|z\rangle = \int d^2\alpha z(\alpha^*, \alpha) \, |\alpha\rangle = 0\,, \tag{5.64}$$

where the integral is over the entire complex α-plane and $z(\alpha^*, \alpha)$ is nonzero on some open subset of the α-plane. The number states are both complete and linearly independent, so this condition can be expressed in a more concrete way as

$$\langle n | z \rangle = \frac{1}{\sqrt{n!}} \int d^2\alpha e^{-|\alpha|^2/2} z(\alpha^*, \alpha) \, \alpha^n = 0 \;\; \text{for all } n \geqslant 0\,. \tag{5.65}$$

By using polar coordinates ($\alpha = \rho \exp i\phi$) for the integration these conditions become

$$\frac{1}{\sqrt{n!}} \int_0^{\infty} d\rho \rho^{n+1} e^{-\rho^2/2} \int_0^{2\pi} d\phi z(\rho, \phi) \, e^{in\phi} = 0 \;\; \text{for all } n \geqslant 0\,. \tag{5.66}$$

In this form, one can see that the desired outcome is guaranteed if the ϕ-dependence of $z(\rho, \phi)$ causes the ϕ-integral to vanish for all $n \geqslant 0$. This is easily done by choosing $z(\rho, \phi) = g(\rho)\, \rho^m \exp(im\phi)$ for some $m > 0$; that is,

$$z(\alpha^*, \alpha) = g(|\alpha|)\, \alpha^m , \quad \text{with } m > 0 . \tag{5.67}$$

The linear dependence of the coherent states means that the coefficients in the generic expansion

$$|\psi\rangle = \int d^2\alpha F(\alpha^*, \alpha)\, |\alpha\rangle \tag{5.68}$$

are not unique, since replacing $F(\alpha^*, \alpha)$ by $F(\alpha^*, \alpha) + z(\alpha^*, \alpha)$ yields the same vector $|\psi\rangle$.

In spite of these unfamiliar properties, the coherent states satisfy a completeness relation, or resolution of the identity,

$$\int \frac{d^2\alpha}{\pi} |\alpha\rangle \langle\alpha| = I , \tag{5.69}$$

analogous to eqn (2.84) for the number states. To prove this, we denote the left side of eqn (5.69) by $\mathfrak{I}$ and evaluate the matrix elements

$$\begin{aligned} \langle n |\mathfrak{I}| m\rangle &= \int \frac{d^2\alpha}{\pi} \langle n |\alpha\rangle \langle\alpha |m\rangle \\ &= \int_0^\infty d\rho\, \rho \frac{\rho^{n+m}}{\sqrt{n!m!}} e^{-\rho^2} \int_0^{2\pi} \frac{d\phi}{\pi} e^{i(n-m)\phi} \\ &= \delta_{nm} . \end{aligned} \tag{5.70}$$

Thus $\mathfrak{I}$ has the same matrix elements as the identity operator, and eqn (5.69) is established.

Applying this representation of the identity to a state $|\psi\rangle$ gives the natural—but not unique—expansion

$$|\psi\rangle = \int \frac{d^2\alpha}{\pi} |\alpha\rangle \langle\alpha |\psi\rangle . \tag{5.71}$$

The completeness relation also gives a useful formula for the trace of any operator:

$$\begin{aligned} \operatorname{Tr} X = \operatorname{Tr}\left\{\left(\int \frac{d^2\alpha}{\pi} |\alpha\rangle \langle\alpha|\right) X\right\} &= \int \frac{d^2\alpha}{\pi} \sum_{n=0}^{\infty} \langle n |\alpha\rangle \langle\alpha |X| n\rangle \\ = \int \frac{d^2\alpha}{\pi} \langle\alpha |X| \alpha\rangle . \end{aligned} \tag{5.72}$$

5.4.3 Coherent state representations of operators

The completeness relation (5.69) is the basis for deriving useful representations of operators in terms of coherent states. For any Fock space operator X, we easily find the general result

$$X = \left(\int \frac{d^2\alpha}{\pi} |\alpha\rangle \langle\alpha| \right) X \left(\int \frac{d^2\beta}{\pi} |\beta\rangle \langle\beta| \right)$$
$$= \int \frac{d^2\alpha}{\pi} \int \frac{d^2\beta}{\pi} |\alpha\rangle \langle\alpha |X| \beta\rangle \langle\beta| . \tag{5.73}$$

Since the coherent states are complete, this result guarantees that X is uniquely defined by the matrix elements $\langle\alpha |X| \beta\rangle$. On the other hand, the overcompleteness of the coherent states suggests that the same information may be carried by a smaller set of matrix elements.

A An operator X is uniquely determined by $\langle\alpha |X| \alpha\rangle$

The diagonal matrix elements $\langle n |X| n\rangle$ in the number-state basis—or in any other orthonormal basis—do not uniquely specify the operator X, but the overcompleteness of the coherent states guarantees that the diagonal elements $\langle\alpha |X| \alpha\rangle$ do determine X uniquely. The first step in the proof is to use eqn (5.23) one more time to write $\langle\alpha |X| \alpha\rangle$ in terms of the matrix elements in the number-state basis,

$$\langle\alpha |X| \alpha\rangle = e^{-|\alpha|^2} \sum_{m=0}^{\infty} \sum_{n=0}^{\infty} \frac{\langle m |X| n\rangle}{\sqrt{m!n!}} \alpha^{*m} \alpha^n . \tag{5.74}$$

Now suppose that two operators Y and Z have the same diagonal elements, i.e. $\langle\alpha |Y| \alpha\rangle = \langle\alpha |Z| \alpha\rangle$; then $X = Y - Z$ must satisfy

$$\sum_{m=0}^{\infty} \sum_{n=0}^{\infty} \frac{\langle m |X| n\rangle}{\sqrt{m!n!}} \alpha^{*m} \alpha^n = 0 . \tag{5.75}$$

This is an identity in the independent variables α and α^*, so the argument leading to eqn (5.61) can be applied again to conclude that $\langle m |X| n\rangle = 0$ for all m and n. The completeness of the number states then implies that $X = 0$, and we have proved that

$$\text{if } \langle\alpha |Y| \alpha\rangle = \langle\alpha |Z| \alpha\rangle \text{ for all } \alpha , \quad \text{then } Y = Z . \tag{5.76}$$

B Coherent state diagonal representation

The result (5.76) will turn out to be very useful, but it does not immediately supply us with a representation for the operator. On the other hand, the general representation (5.73) involves the off-diagonal matrix elements $\langle\alpha |X| \beta\rangle$ which we now see are apparently superfluous. This suggests that it may be possible to get a representation that only involves the projection operators $|\alpha\rangle \langle\alpha|$, rather than the off-diagonal operators $|\alpha\rangle \langle\beta|$ appearing in eqn (5.73). The key to this construction is the identity

$$a^n |\alpha\rangle \langle\alpha| a^{\dagger m} = \alpha^n \alpha^{*m} |\alpha\rangle \langle\alpha| , \tag{5.77}$$

which holds for any non-negative integers n and m. Let us now suppose that X has a power series expansion in the operators a and $a^\dagger$, then by using the commutation relation $\left[a, a^{\dagger}\right] = 1$ each term in the series can be rearranged into a sum of terms in which the creation operators stand to the right of the annihilation operators, i.e.

$$X = \sum_{m=0}^{\infty} \sum_{n=0}^{\infty} \mathcal{X}_{nm}^{A} a^{n} a^{\dagger m} , \tag{5.78}$$

where $\mathcal{X}_{nm}^{A}$ is a c-number coefficient. Since this exactly reverses the rule for normal ordering, it is called **antinormal ordering**, and the superscript A serves as a reminder of this ordering rule. By combining the identities (5.69) and (5.77) one finds

$$\begin{aligned} X &= \sum_{m=0}^{\infty} \sum_{n=0}^{\infty} \mathcal{X}_{nm}^{A} a^{n} \left(\int \frac{d^{2}\alpha}{\pi} \left|\alpha\right\rangle \left\langle\alpha\right| \right) a^{\dagger m} \\ &= \int d^{2}\alpha \mathcal{X}^{A}(\alpha) \left|\alpha\right\rangle \left\langle\alpha\right| , \end{aligned} \tag{5.79}$$

where

$$\mathcal{X}^{A}(\alpha) = \frac{1}{\pi} \sum_{m=0}^{\infty} \sum_{n=0}^{\infty} \mathcal{X}_{nm}^{A} \alpha^{n} \alpha^{*m} \tag{5.80}$$

is a c-number function of the two real variables $\operatorname{Re}\alpha$ and $\operatorname{Im}\alpha$. This construction gives us the promised representation in terms of the projection operators $\left|\alpha\right\rangle \left\langle\alpha\right|$.

5.5 Multimode coherent states

Up to this point we have only considered coherent states of a single radiation oscillator. In the following sections we will consider several generalizations that allow the description of multimode squeezed states.

5.5.1 An elementary approach to multimode coherent states

A straightforward generalization is to replace the definition (5.18) of the one-mode coherent state by the family of eigenvalue problems

$$a_{\kappa} \left|\underline{\alpha}\right\rangle = \alpha_{\kappa} \left|\underline{\alpha}\right\rangle \quad \text{for all } \kappa , \tag{5.81}$$

where $\underline{\alpha} = (\alpha_1, \alpha_2, \ldots, \alpha_{\kappa}, \ldots)$ is the set of eigenvalues for the annihilation operators a_{κ}. The single-mode case is recovered by setting $\alpha_{\kappa'} = 0$ for $\kappa' \neq \kappa$. The **multimode coherent state** $\left|\underline{\alpha}\right\rangle$—defined as the solution of the family of equations (5.81)—can be constructed from the vacuum state by using eqn (5.53) for each mode to get

$$\left|\underline{\alpha}\right\rangle = \prod_{\kappa} D(\alpha_{\kappa}) \left|0\right\rangle , \tag{5.82}$$

where

$$D(\alpha_{\kappa}) = \exp\left(\alpha_{\kappa} a_{\kappa}^{\dagger} - \alpha_{\kappa}^{*} a_{\kappa}\right) \tag{5.83}$$

is the displacement operator for the κth mode. Since there are an infinite number of modes, the definition (5.82) raises various mathematical issues, such as the convergence of the infinite product. In the following sections, we show how these issues can be dealt with, but for most applications it is safe to proceed by using the formal infinite product.

For later use, it is convenient to specialize the general definition (5.82) of the multimode state to the case of box-quantized plane waves, i.e.

$$|\underline{\alpha}\rangle = D(\underline{\alpha})|0\rangle\,, \tag{5.84}$$

$$D(\underline{\alpha}) = \prod_{\mathbf{k}s} D(\alpha_{\mathbf{k}s}) = \prod_{\mathbf{k}s} \exp\left(\alpha_{\mathbf{k}s} a_{\mathbf{k}s}^{\dagger} - \alpha_{\mathbf{k}s}^{*} a_{\mathbf{k}s}\right). \tag{5.85}$$

By combining the eigenvalue condition $a_{\mathbf{k}s}|\underline{\alpha}\rangle = \alpha_{\mathbf{k}s}|\underline{\alpha}\rangle$ with the expression (3.69) for $\mathbf{E}^{(+)}$, one can see that

$$\mathbf{E}^{(+)}(\mathbf{r})|\underline{\alpha}\rangle = \boldsymbol{\mathcal{E}}(\mathbf{r})|\underline{\alpha}\rangle\,, \tag{5.86}$$

where

$$\boldsymbol{\mathcal{E}}(\mathbf{r}) = i\sum_{\mathbf{k}s}\sqrt{\frac{\hbar\omega_k}{2\epsilon_0 V}}\alpha_{\mathbf{k}s}\mathbf{e}_{\mathbf{k}s}e^{i\mathbf{k}\cdot\mathbf{r}} \tag{5.87}$$

is the classical electric field defined by $|\underline{\alpha}\rangle$.

5.5.2 Coherent states for wave packets*

The incident field in a typical experiment is a traveling-wave packet, i.e. a superposition of plane-wave modes. A coherent state describing this situation is therefore an example of a multimode coherent state. From this point of view, the multimode coherent state $|\underline{\alpha}\rangle$ is actually no more complicated than a single-mode coherent state (Deutsch, 1991). This is a linguistic paradox caused by the various meanings assigned to the word 'mode'. This term normally describes a solution of Maxwell's equations with some additional properties associated with the boundary conditions imposed by the problem at hand. Examples are the modes of a rectangular cavity or a single plane wave. General classical fields are linear combinations of the mode functions, and they are called wave packets rather than modes. Let us now return to eqn (5.82) which gives a constructive definition of the multimode state $|\underline{\alpha}\rangle$. Since the operators $\alpha_\kappa a_\kappa^\dagger - \alpha_\kappa^* a_\kappa$ and $\alpha_{\kappa'} a_{\kappa'}^\dagger - \alpha_{\kappa'}^* a_{\kappa'}$ commute for $\kappa \neq \kappa'$, the product of unitary operators in eqn (5.82) can be rewritten as a single unitary operator,

$$\begin{aligned}|\underline{\alpha}\rangle &= \exp\left\{\sum_{\kappa}\left(\alpha_\kappa a_\kappa^\dagger - \alpha_\kappa^* a_\kappa\right)\right\}|0\rangle \\ &= \exp\left\{a^\dagger[\underline{\alpha}] - a[\underline{\alpha}]\right\}|0\rangle\,,\end{aligned} \tag{5.88}$$

where

$$a[\underline{\alpha}] = \sum_{\kappa}\alpha_\kappa^* a_\kappa \tag{5.89}$$

is an example of the general definition (3.191). In other words the multimode coherent state $|\underline{\alpha}\rangle$ is a coherent state for the *wave packet*

$$\mathbf{w}(\mathbf{r}) = \sum_{\kappa}\alpha_\kappa \mathbf{w}_\kappa(\mathbf{r})\,, \tag{5.90}$$

where the $\mathbf{w}_\kappa(\mathbf{r})$s are mode functions. The wave packet $\mathbf{w}(\mathbf{r})$ defines a point in the classical phase space, so it represents one degree of freedom of the field. This suggests changing the notation by

$$|\underline{\alpha}\rangle \to |\mathbf{w}\rangle = D[\mathbf{w}]|0\rangle , \tag{5.91}$$

where

$$D[\mathbf{w}] = \exp\left\{a^\dagger[\mathbf{w}] - a[\mathbf{w}]\right\} \tag{5.92}$$

is the wave packet displacement operator, and $a[\mathbf{w}]$ is simply another notation for $a[\underline{\alpha}]$.

The displacement rule,

$$D^\dagger[\mathbf{w}]\, a[\mathbf{v}]\, D[\mathbf{w}] = a[\mathbf{v}] + (\mathbf{v}, \mathbf{w}) , \tag{5.93}$$

and the product rule,

$$D[\mathbf{v}]\, D[\mathbf{w}] = D[\mathbf{v} + \mathbf{w}] \exp\{i\,\mathrm{Im}(\mathbf{w}, \mathbf{v})\} , \tag{5.94}$$

are readily established by using the commutation relations (3.192), the interpolating operator method outlined in Section 5.4.1, and the Campbell–Baker–Hausdorff formula (C.66). The displacement rule (5.93) immediately yields the eigenvalue equation

$$a[\mathbf{v}]|\mathbf{w}\rangle = (\mathbf{v}, \mathbf{w})|\mathbf{w}\rangle . \tag{5.95}$$

This says that the coherent state for the wave packet $\mathbf{w}$ is also an eigenstate—with the eigenvalue $(\mathbf{v}, \mathbf{w})$—of the annihilation operator for any other wave packet $\mathbf{v}$. To recover the familiar single-mode form, $a|\alpha\rangle = \alpha|\alpha\rangle$, simply set $\mathbf{w} = \alpha\mathbf{w}_0$, where $\mathbf{w}_0$ is normalized to unity, and $\mathbf{v} = \mathbf{w}_0$; then eqn (5.95) becomes $a[\mathbf{w}_0]|\alpha\mathbf{w}_0\rangle = \alpha|\alpha\mathbf{w}_0\rangle$. The inner product of two multimode (wave packet) coherent states is obtained from (5.91) by calculating

$$\begin{aligned}\langle \mathbf{v}|\mathbf{w}\rangle &= \langle 0|D^\dagger[\mathbf{v}]\, D[\mathbf{w}]|\,0\rangle \\ &= \exp\{i\,\mathrm{Im}(\mathbf{v}, \mathbf{w})\}\langle 0|D[\mathbf{w} - \mathbf{v}]|\,0\rangle \\ &= \exp\{i\,\mathrm{Im}(\mathbf{v}, \mathbf{w})\}\exp\left\{-\frac{1}{2}\|\mathbf{w} - \mathbf{v}\|^2\right\} ,\end{aligned} \tag{5.96}$$

where $\|\mathbf{u}\| = \sqrt{(\mathbf{u}, \mathbf{u})}$ is the norm of the wave packet u.

5.5.3 Sources of multimode coherent states*

In Section 5.2 we saw that a monochromatic classical current serves as the source for a single-mode coherent state. This demonstration is readily generalized as follows. The total Hamiltonian in the hemiclassical approximation is the sum of eqns (3.40) and (5.36),

$$H = 2\epsilon_0 c^2 \int d^3r\, \mathbf{A}^{(-)}(\mathbf{r}, t) \cdot \left(-\nabla^2\right) \mathbf{A}^{(+)}(\mathbf{r}, t) - \int d^3r\, \boldsymbol{\mathcal{J}}(\mathbf{r}, t) \cdot \mathbf{A}(\mathbf{r}, t) . \tag{5.97}$$

The corresponding Heisenberg equation for $\mathbf{A}^{(+)}$,

$$i\frac{\partial \mathbf{A}^{(+)}(\mathbf{r},t)}{\partial t} = c\left(-\nabla^2\right)^{1/2}\mathbf{A}^{(+)}(\mathbf{r},t) - \frac{1}{2\epsilon_0 c}\left(-\nabla^2\right)^{-1/2}\boldsymbol{\mathcal{J}}(\mathbf{r},t), \tag{5.98}$$

has the formal solution

$$\mathbf{A}^{(+)}(\mathbf{r},t) = \exp\left[-i(t-t_0)c\left(-\nabla^2\right)^{1/2}\right]\mathbf{A}^{(+)}(\mathbf{r},t_0) + \mathbf{w}(\mathbf{r},t), \tag{5.99}$$

where

$$\mathbf{w}(\mathbf{r},t) = \frac{i}{2\epsilon_0 c}\int_{t_0}^{t} dt' \exp\left[-i(t-t')c\left(-\nabla^2\right)^{1/2}\right]\left(-\nabla^2\right)^{-1/2}\boldsymbol{\mathcal{J}}(\mathbf{r},t'), \tag{5.100}$$

and the Schrödinger and Heisenberg pictures coincide at the time, t_0, when the current is turned on. The classical field $\mathbf{w}(\mathbf{r},t)$ satisfies the c-number version of eqn (5.98),

$$i\frac{\partial \mathbf{w}(\mathbf{r},t)}{\partial t} = c\left(-\nabla^2\right)^{1/2}\mathbf{w}(\mathbf{r},t) - \frac{1}{2\epsilon_0 c}\left(-\nabla^2\right)^{-1/2}\boldsymbol{\mathcal{J}}(\mathbf{r},t). \tag{5.101}$$

Applying this solution to the vacuum gives $\mathbf{A}^{(+)}(\mathbf{r},t)|0\rangle = \mathbf{w}(\mathbf{r},t)|0\rangle$ in the Heisenberg picture, and $\mathbf{A}^{(+)}(\mathbf{r})|\mathbf{w},t\rangle = \mathbf{w}(\mathbf{r},t)|\mathbf{w},t\rangle$ in the Schrödinger picture. The time-dependent coherent state $|\mathbf{w},t\rangle$ evolves from the vacuum state ($|\mathbf{w},t_0\rangle = |0\rangle$) under the action of the Hamiltonian given by eqn (5.97).

5.5.4 Completeness and representation of operators*

The issue of completeness for the multimode coherent states is (infinitely) more complicated than in the single-mode case. Since we are considering all modes on an equal footing, the identity (5.69) for a single mode must be replaced by

$$\int \frac{d^2\alpha_\kappa}{\pi}|\alpha_\kappa\rangle\langle\alpha_\kappa| = I_\kappa, \tag{5.102}$$

where I_κ is the identity operator for the single-mode subspace $\mathfrak{H}_\kappa$. The resolution of the identity on the entire space $\mathfrak{H}_F$ is given by

$$\prod_\kappa \int \frac{d^2\alpha_\kappa}{\pi}|\underline{\alpha}\rangle\langle\underline{\alpha}| = I_F. \tag{5.103}$$

The mathematical respectability of this infinite-dimensional integral has been established for basis sets labeled by a discrete index (Klauder and Sudarshan, 1968, Sec. 7-4). Fortunately, the Hilbert spaces of interest for quantum theory are separable, i.e. they can always be represented by discrete basis sets. In most applications only a few modes are relevant, so the necessary integrals are approximately finite dimensional.

Combining the multimode completeness relation (5.103) with the fact that operators for orthogonal modes commute justifies the application of the arguments in Sections 5.4.3 and 5.6.3 to obtain the multimode version of the diagonal expansion for the density operator:

$$\rho = \int d^2\underline{\alpha}\,|\underline{\alpha}\rangle P(\underline{\alpha})\langle\underline{\alpha}|, \tag{5.104}$$

where

$$d^2\underline{\alpha} = \prod_\kappa \frac{d^2\alpha_\kappa}{\pi}. \tag{5.105}$$

5.5.5 Applications of multimode states*

Substituting the relation

$$a\left[\mathbf{w}\right]=\sqrt{\frac{2\epsilon_0 c}{\hbar}}A^{(+)}\left[\mathbf{w}\right]=\sqrt{\frac{2\epsilon_0 c}{\hbar}}\int d^3r\mathbf{w}^*\left(\mathbf{r}\right)\cdot\left(-\nabla^2\right)^{1/2}\mathbf{A}^{(+)}\left(\mathbf{r}\right)\qquad(5.106)$$

into eqn (5.95) provides the **r**-space version of the eigenvalue equation:

$$\mathbf{A}^{(+)}\left(\mathbf{r}\right)\left|\mathbf{w}\right\rangle=\sqrt{\frac{\hbar}{2\epsilon_0 c}}\mathbf{w}\left(\mathbf{r}\right)\left|\mathbf{w}\right\rangle.\qquad(5.107)$$

For many applications it is more useful to use eqn (3.15) to express this in terms of the electric field,

$$\mathbf{E}^{(+)}\left(\mathbf{r}\right)\left|\mathbf{w}\right\rangle=\boldsymbol{\mathcal{E}}\left(\mathbf{r}\right)\left|\mathbf{w}\right\rangle,\qquad(5.108)$$

where

$$\boldsymbol{\mathcal{E}}\left(\mathbf{r}\right)=i\sqrt{\frac{\hbar c}{2\epsilon_0}}\left(-\nabla^2\right)^{1/2}\mathbf{w}\left(\mathbf{r}\right)\qquad(5.109)$$

is the positive-frequency part of the classical electric field corresponding to the wave packet **w**. The result (5.108) can be usefully applied to the calculation of the field correlation functions for the coherent state described by the density operator $\rho=\left|\mathbf{w}\right\rangle\left\langle\mathbf{w}\right|$. For example, the equal-time version of $G^{(2)}$, defined by setting all times to zero in eqn (4.77), factorizes into

$$G^{(2)}\left(x_1,x_2;x_3,x_4\right)=\mathcal{E}_1^*\left(\mathbf{r}_1\right)\mathcal{E}_2^*\left(\mathbf{r}_2\right)\mathcal{E}_3\left(\mathbf{r}_3\right)\mathcal{E}_4\left(\mathbf{r}_4\right),\qquad(5.110)$$

where $\mathcal{E}_p\left(\mathbf{r}\right)=s_p^*\cdot\boldsymbol{\mathcal{E}}\left(\mathbf{r}\right)$. In fact, correlation functions of all orders factorize in the same way.

Now let us consider an experimental situation in which the classical current is turned on at some time $t_0<0$ and turned off at $t=0$, leaving the field prepared in a coherent state $\left|\mathbf{w}\right\rangle$. The time at which the Schrödinger and Heisenberg pictures agree is now shifted to $t=0$, and we assume that the fields propagate freely for $t>0$. The Schrödinger-picture state vector $\left|\mathbf{w},t\right\rangle$ evolves from its initial value $\left|\mathbf{w},0\right\rangle$ according to the free-field Hamiltonian, while the operators remain unchanged.

In the Heisenberg picture the state vector is always $\left|\mathbf{w}\right\rangle$ and the operators evolve freely according to eqn (3.94). This guarantees that

$$\mathbf{E}^{(+)}\left(\mathbf{r},t\right)\left|\mathbf{w}\right\rangle=\boldsymbol{\mathcal{E}}\left(\mathbf{r},t\right)\left|\mathbf{w}\right\rangle,\qquad(5.111)$$

where $\boldsymbol{\mathcal{E}}\left(\mathbf{r},t\right)$ is the freely propagating positive-frequency part that evolves from the initial ($t=0$) function given by eqn (5.109). According to eqn (5.110) the correlation function factorizes at $t=0$, and by the last equation each factor evolves independently; therefore, the multi-time correlation function for the wave packet coherent state $\left|\mathbf{w}\right\rangle$ factorizes according to

$$G^{(2)}\left(x_1,x_2;x_3,x_4\right)=\mathcal{E}_1^*\left(\mathbf{r}_1,t_1\right)\mathcal{E}_2^*\left(\mathbf{r}_2,t_2\right)\mathcal{E}_3\left(\mathbf{r}_3,t_3\right)\mathcal{E}_4\left(\mathbf{r}_4,t_4\right).\qquad(5.112)$$

5.6 Phase space description of quantum optics

The set of all classical fields obtained by exciting a single mode is described by a two-dimensional phase space, as shown in eqn (5.1). The set of all quasiclassical states for the same mode is described by the coherent states $\{|\alpha\rangle\}$, that are also labeled by a two-dimensional space. This correspondence is the basis for a phase-space-like description of quantum optics. This representation of states and operators has several useful applications. The first is a precise description of the correspondence-principle limit. The relation between coherent states and classical fields also provides a quantitative description of the departure from classical behavior. Finally, as we will see in Section 18.5, the phase space representation of the density operator ρ gives a way to convert the quantum Liouville equation for the operator ρ into a c-number equation that can be used in numerical simulations.

In Section 9.1 we will see that the results of photon detection experiments are expressed in terms of expectation values of normal-ordered products of field operators. In this way, counting experiments yield information about the state of the electromagnetic field. In order to extract this information, we need a general scheme for representing the density operators describing the field states. The original construction of the electromagnetic Fock space in Chapter 3 emphasized the role of the number states. Every density operator can indeed be represented in the basis of number states, but there are many situations for which the coherent states provide a more useful representation. For the sake of simplicity, we will continue to emphasize a single classical field mode for which the phase space Γ_{em} can be identified with the complex plane.

5.6.1 The Wigner distribution

The earliest—and still one of the most useful—representations of the density operator was introduced by Wigner (1932) in the context of elementary quantum mechanics. In classical mechanics the most general state of a single particle moving in one dimension is described by a normalized probability density $f(Q,P)$ defined on the classical phase space $\Gamma_{\text{mech}} = \{(Q,P)\}$, i.e. $f(Q,P)\,dQdP$ is the probability that the particle has position and momentum in the infinitesimal rectangle with area $dQdP$ centered at the point (Q,P) and

$$\int dQ \int dP f(Q,P) = 1\,. \tag{5.113}$$

In classical probability theory it is often useful to represent a distribution in terms of its Fourier transform,

$$\chi(u,v) = \int dQ \int dP f(Q,P)\, e^{-i(uP+vQ)}\,, \tag{5.114}$$

which is called the **characteristic function** (Feller, 1957*b*, Chap. XV). In some applications it is easier to evaluate the characteristic function, and then construct the probability distribution itself from the inverse transformation:

$$f(Q,P) = \int \frac{du}{2\pi} \int \frac{dv}{2\pi} \chi(u,v)\, e^{i(uP+vQ)}\,. \tag{5.115}$$

An example of the utility of the characteristic function is the calculation of the moments of the distribution, e.g.

$$\begin{aligned}\langle Q^2 \rangle &= (i)^2 \left(\frac{\partial^n \chi}{\partial v^2}\right)_{(u,v)=(0,0)} ,\\ \langle QP \rangle &= (i)^2 \left(\frac{\partial^n \chi}{\partial v \partial u}\right)_{(u,v)=(0,0)} ,\\ &\vdots\end{aligned} \tag{5.116}$$

A The Wigner distribution in quantum mechanics

In quantum mechanics, a phase space description like $f(Q,P)$ is forbidden by the uncertainty principle. Wigner's insight can be interpreted as an attempt to find a quantum replacement for the phase space integral in eqn (5.114). Since the integral is a sum over all classical states, it is natural to replace it by the sum over all quantum states, i.e. by the quantum mechanical trace operation. The role of the classical distribution is naturally played by the density operator ρ, and the classical exponential $\exp[-i(uP+vQ)]$ can be replaced by the unitary operator $\exp[-i(u\widehat{p}+v\widehat{q})]$. In this way one is led to the definition of the **Wigner characteristic function**

$$\chi_W(u,v) = \mathrm{Tr}\left[\rho e^{-i(u\widehat{p}+v\widehat{q})}\right] , \tag{5.117}$$

which is a c-number function of the real variables u and v. The classical definition (5.114) of the characteristic function by a phase space integral is meaningless for quantum theory, but the inverse transformation (5.115) still makes sense when applied to χ_W. This suggests the definition of the **Wigner distribution**,

$$W(Q,P) = \hbar \int \frac{du}{2\pi} \int \frac{dv}{2\pi} \chi_W(u,v)\, e^{i(uP+vQ)} , \tag{5.118}$$

where the normalization has been chosen to make $W(Q,P)$ dimensionless. The Wigner distribution is real and normalized by

$$\int \frac{dQdP}{\hbar} W(Q,P) = 1 , \tag{5.119}$$

but—as we will see later on—there are physical states for which $W(Q,P)$ assumes negative values in some regions of the (Q,P)-plane. For these cases $W(Q,P)$ cannot be interpreted as a probability density like $f(Q,P)$; consequently, the Wigner distribution is called a **quasiprobability** density.

Substituting eqn (2.116) for the density operator into eqn (5.117) leads to the alternative form

$$\begin{aligned}\chi_W(u,v) &= \sum_e \mathcal{P}_e \left\langle \Psi_e \left| e^{-i(u\widehat{p}+v\widehat{q})} \right| \Psi_e \right\rangle \\ &= \sum_e \mathcal{P}_e e^{i\hbar uv/2} \left\langle \Psi_e \left| e^{-iv\widehat{q}} e^{-iu\widehat{p}} \right| \Psi_e \right\rangle ,\end{aligned} \tag{5.120}$$

where the last line follows from the identity (C.67). Since $\exp(-iu\widehat{p})$ is the spatial translation operator, the expectation value can be expressed as

$$\left\langle \Psi_e \left| e^{-iv\widehat{q}} e^{-iu\widehat{p}} \right| \Psi_e \right\rangle = \int dQ' \Psi_e^* (Q') \, e^{-iv(Q'+\hbar u)} \Psi_e (Q' + \hbar u) \, . \tag{5.121}$$

Substituting these results into eqn (5.118) finally leads to

$$W(Q, P) = \sum_e \mathcal{P}_e \int \frac{dX}{\pi} e^{2iXP/\hbar} \Psi_e (Q + X) \, \Psi_e^* (Q - X) \, , \tag{5.122}$$

which is the definition used in Wigner's original paper. Thus the 'momentum' dependence of the Wigner distribution comes from the Fourier transform with respect to the relative coordinate X. Integrating out the momentum dependence yields the marginal distribution in Q:

$$\int \frac{dP}{\hbar} W(Q, P) = \sum_e \mathcal{P}_e \left| \Psi_e (Q) \right|^2 . \tag{5.123}$$

Despite the fact that $W(Q, P)$ can have negative values, the marginal distribution in Q is evidently a genuine probability density.

B *The Wigner distribution for quantum optics*

In the transition to quantum optics the mechanical operators $\widehat{q}$ and $\widehat{p}$ are replaced by the operators q and p for the radiation oscillator. In agreement with our earlier experience, it turns out to be more useful to use the relations (2.66) to rewrite the unitary operator $\exp[-i(up + vq)]$ as $\exp\left[\eta a^\dagger - \eta^* a\right]$, where

$$\eta = \sqrt{\frac{\hbar\omega}{2}} u - i \sqrt{\frac{\hbar}{2\omega}} v \, , \tag{5.124}$$

so that eqn (5.117) is replaced by

$$\chi_W(\eta) = \mathrm{Tr}\left[\rho e^{\eta a^\dagger - \eta^* a}\right] . \tag{5.125}$$

The characteristic function $\chi_W(\eta)$ has the useful properties $\chi_W(0) = 1$ and $\chi_W^*(\eta) = \chi_W(-\eta)$. The Wigner distribution is then defined (Walls and Milburn, 1994, Sec. 4.2.2) as the Fourier transform of $\chi_W(\eta)$:

$$W(\alpha) = \frac{1}{\pi^2} \int d^2\eta e^{\eta^* \alpha - \eta \alpha^*} \chi_W(\eta) \, . \tag{5.126}$$

After verifying the identity

$$\int \frac{d^2\alpha}{\pi^2} e^{\eta^* \alpha - \eta \alpha^*} = \delta_2(\eta) \, , \tag{5.127}$$

where $\delta_2(\eta) \equiv \delta(\mathrm{Re}\,\eta) \, \delta(\mathrm{Im}\,\eta)$, one finds that the Wigner function $W(\alpha)$ is normalized by

$$\int d^2\alpha W(\alpha) = 1\,. \tag{5.128}$$

In order to justify this approach, we next demonstrate that the average, $\operatorname{Tr} \rho X$, of any operator X can be expressed in terms of the moments of the Wigner distribution. The representation (5.127) of the delta function and the identities

$$\alpha^m \alpha^{*n} e^{\eta^*\alpha - \eta\alpha^*} = \left(\frac{\partial}{\partial \eta^*}\right)^m \left(-\frac{\partial}{\partial \eta}\right)^n e^{\eta^*\alpha - \eta\alpha^*} \tag{5.129}$$

allow the moments of $W(\alpha)$ to be evaluated in terms of derivatives of the characteristic function, with the result

$$\int d^2\alpha\; \alpha^m \alpha^{*n} W(\alpha) = \left(-\frac{\partial}{\partial \eta^*}\right)^m \left(\frac{\partial}{\partial \eta}\right)^n \chi_W(\eta, \eta^*)\bigg|_{\eta=0}\,. \tag{5.130}$$

The characteristic function can be cast into a useful form by expanding the exponential in eqn (5.125) and using the operator binomial theorem (C.44) to find

$$\begin{aligned}\chi_W(\eta, \eta^*) &= \sum_{k=0}^{\infty} \frac{1}{k!} \operatorname{Tr}\left[\rho\left(\eta a^\dagger - \eta^* a\right)^k\right] \\ &= \sum_{k=0}^{\infty} \frac{1}{k!} \sum_{j=0}^{k} \eta^{k-j} (-\eta^*)^j \frac{k!}{j!\,(k-j)!} \operatorname{Tr}\left\{\rho \mathcal{S}\left[\left(a^\dagger\right)^{k-j} a^j\right]\right\},\end{aligned} \tag{5.131}$$

where the Weyl—or symmetrical—product $\mathcal{S}\left[\left(a^\dagger\right)^{k-j} a^j\right]$ is the average of all distinct orderings of the operators a and $a^\dagger$. Using this result in eqn (5.130) yields

$$\int d^2\alpha \alpha^m \alpha^{*n} W(\alpha) = \operatorname{Tr}\left\{\rho \mathcal{S}\left[\left(a^\dagger\right)^n a^m\right]\right\}. \tag{5.132}$$

By means of the commutation relations, any operator X that has a power series expansion in a and $a^\dagger$ can be expressed as the sum of Weyl products:

$$X = \sum_{n=0}^{\infty} \sum_{m=0}^{\infty} \mathcal{X}_{nm}^W \mathcal{S}\left[\left(a^\dagger\right)^n a^m\right], \tag{5.133}$$

where the $\mathcal{X}_{nm}^W$s are c-number coefficients. The expectation value of X is then

$$\begin{aligned}\langle X\rangle &= \sum_{n=0}^{\infty} \sum_{m=0}^{\infty} \mathcal{X}_{nm}^W \left\langle \mathcal{S}\left[\left(a^\dagger\right)^n a^m\right]\right\rangle \\ &= \int d^2\alpha \mathcal{X}^W(\alpha)\, W(\alpha)\,,\end{aligned} \tag{5.134}$$

where

$$\mathcal{X}^W(\alpha) = \sum_{n=0}^{\infty} \sum_{m=0}^{\infty} \mathcal{X}_{nm}^W \alpha^m \alpha^{*n}\,. \tag{5.135}$$

Thus the Wigner distribution carries the same physical information as the density operator.

As an example, consider $X = E^2$, where $E = i\sqrt{\hbar\omega_0/2\epsilon_0}\left(a - a^\dagger\right)$ is the electric field amplitude for a single cavity mode. In terms of Weyl products, E^2 is given by

$$E^2 = -\frac{\hbar\omega_0}{2\epsilon_0}\left\{\mathcal{S}\left[a^2\right] - 2\mathcal{S}\left[a^\dagger a\right] + \mathcal{S}\left[a^{\dagger 2}\right]\right\}, \tag{5.136}$$

and substituting this expression into eqn (5.134) yields

$$\left\langle E^2\right\rangle = \frac{\hbar\omega_0}{2\epsilon_0}\int d^2\alpha\left\{2\left|\alpha\right|^2 - \alpha^2 - \alpha^{*2}\right\}W\left(\alpha\right). \tag{5.137}$$

*C Existence of the Wigner distribution**

The general properties of Hilbert space operators, reviewed in Appendix A.3.3, guarantee that the unitary operator $\exp\left(\eta a^\dagger - \eta^* a\right)$ has a complete orthonormal set of (improper) eigenstates $|\Lambda\rangle$, i.e.

$$\exp\left(\eta a^\dagger - \eta^* a\right)|\Lambda\rangle = e^{i\theta_\Lambda(\eta)}\left|\Lambda\right\rangle, \tag{5.138}$$

where $\theta_\Lambda\left(\eta\right)$ is real, $-\infty < \Lambda < \infty$, and $\langle\Lambda'\left|\Lambda\right\rangle = \delta\left(\Lambda - \Lambda'\right)$. Evaluating the trace in the $|\Lambda\rangle$-basis yields

$$\chi_W\left(\eta\right) = \int d\Lambda\left\langle\Lambda\left|\rho\right|\Lambda\right\rangle e^{i\theta_\Lambda(\eta)}. \tag{5.139}$$

This in turn implies that $\chi_W\left(\eta\right)$ is a bounded function, since

$$\left|\chi_W\left(\eta\right)\right| < \int d\Lambda\left|\left\langle\Lambda\left|\rho\right|\Lambda\right\rangle\right| = \int d\eta\left\langle\Lambda\left|\rho\right|\Lambda\right\rangle = \mathrm{Tr}\,\rho = 1, \tag{5.140}$$

where we have used the fact that all diagonal matrix elements of ρ are positive. The Fourier transform of a constant function is a delta function, so the Fourier transform of a bounded function cannot be more singular than a delta function. This establishes the existence of $W\left(\alpha\right)$—at least in the delta function sense—but there is no guarantee that $W\left(\alpha\right)$ is everywhere positive.

D Examples of the Wigner distribution

In some simple cases the Wigner function can be evaluated analytically by means of the characteristic function.

Coherent state. Our first example is the characteristic function for a coherent state, $\rho = |\beta\rangle\langle\beta|$. The calculation of $\chi_W\left(\eta\right)$ in this case can be done more conveniently by applying the identities (C.69) and (C.70) to find

$$e^{\eta a^\dagger - \eta^* a} = e^{-|\eta|^2/2}e^{\eta a^\dagger}e^{-\eta^* a} = e^{|\eta|^2/2}e^{-\eta^* a}e^{\eta a^\dagger}. \tag{5.141}$$

The first of these gives

$$\chi_W\left(\eta\right) = \mathrm{Tr}\left[\rho e^{\eta a^\dagger - \eta^* a}\right] = e^{-|\eta|^2/2}\left\langle\beta\left|e^{\eta a^\dagger}e^{-\eta^* a}\right|\beta\right\rangle = e^{-|\eta|^2/2}e^{\eta\beta^* - \eta^*\beta}. \tag{5.142}$$

This must be inserted into eqn (5.126) to get $W(\alpha)$. These calculations are best done by rewriting the integrals in terms of the real and imaginary parts of the complex integration variables. For the coherent state this yields

$$W(\alpha) = \frac{2}{\pi} e^{-2|\alpha-\beta|^2} . \tag{5.143}$$

The fact that the Wigner function for this case is everywhere positive is not very surprising, since the coherent state is quasiclassical.

Thermal state. The second example is a thermal or chaotic state. In this case, we use the second identity in eqn (5.141) and the cyclic invariance of the trace to write

$$\chi_W(\eta) = e^{|\eta|^2/2} \operatorname{Tr}\left[\rho e^{-\eta^* a} e^{\eta a^\dagger}\right] = e^{|\eta|^2/2} \operatorname{Tr}\left[e^{\eta a^\dagger} \rho e^{-\eta^* a} e^{\eta a^\dagger}\right] . \tag{5.144}$$

Evaluating the trace with the aid of eqn (5.72) leads to the general result

$$\chi_W(\eta) = \int \frac{d^2\alpha}{\pi} e^{\eta\alpha^* - \eta^*\alpha} \langle \alpha | \rho | \alpha \rangle . \tag{5.145}$$

According to eqn (2.178) the density operator for a thermal state is

$$\rho_{\text{th}} = \sum_{n=0}^{\infty} \frac{\overline{n}^n}{(\overline{n}+1)^{n+1}} |n\rangle \langle n| , \tag{5.146}$$

where $\overline{n} = \langle N_{\text{op}} \rangle$ is the average number of photons. The expansion (5.23) of the coherent state yields

$$\langle \alpha | \rho_{\text{th}} | \alpha \rangle = \frac{1}{\overline{n}+1} \exp\left(-\frac{|\alpha|^2}{\overline{n}+1}\right) , \tag{5.147}$$

so that

$$\begin{aligned} \chi_W^{\text{th}}(\eta) &= \frac{1}{\overline{n}+1} \int \frac{d^2\alpha}{\pi} \exp(\eta\alpha^* - \eta^*\alpha) \exp\left(-\frac{|\alpha|^2}{\overline{n}+1}\right) \\ &= \exp\left[-\left(\overline{n}+\frac{1}{2}\right)|\eta|^2\right] . \end{aligned} \tag{5.148}$$

The general relation (5.126) defining the Wigner distribution can be evaluated in the same way, with the result

$$W_{\text{th}}(\alpha) = \frac{1}{\pi} \frac{1}{\overline{n}+1/2} \exp\left(-\frac{|\alpha|^2}{\overline{n}+1/2}\right) , \tag{5.149}$$

which is also everywhere positive.

Number state. For the third example, we choose a pure number state, e.g. $\rho = |1\rangle\langle 1|$, which yields

$$\chi_W(\eta) = \mathrm{Tr}\left[\rho e^{\eta a^\dagger - \eta^* a}\right] = e^{-|\eta|^2/2}\left\langle 1 \left| e^{\eta a^\dagger} e^{-\eta^* a} \right| 1 \right\rangle . \tag{5.150}$$

Expanding the exponential gives

$$e^{-\eta^* a}|1\rangle = |1\rangle - \eta^*|0\rangle , \tag{5.151}$$

so the characteristic function and the Wigner function are respectively

$$\chi_W(\eta) = \left(1 - |\eta|^2\right) e^{-|\eta|^2/2} \tag{5.152}$$

and

$$\begin{aligned} W(\alpha) &= \frac{1}{\pi^2}\int d^2\eta e^{\eta^*\alpha - \eta\alpha^*}\left(1 - |\eta|^2\right) e^{-|\eta|^2/2} \\ &= -\left(1 - 4|\alpha|^2\right)\frac{2}{\pi} e^{-2|\alpha|^2} . \end{aligned} \tag{5.153}$$

In this case $W(\alpha)$ is negative for $|\alpha| < 1/2$, so the Wigner distribution for a number state $|1\rangle\langle 1|$ is a quasiprobability density. A similar calculation for a general number state $|n\rangle$ yields an expression in terms of Laguerre polynomials (Gardiner, 1991, eqn (4.4.91)) which is also a quasiprobability density.

5.6.2 The Q-function

A Antinormal ordering

According to eqn (5.76) ρ is uniquely determined by its diagonal matrix elements in the coherent state basis; therefore, complete knowledge of the **Q-function**,

$$Q(\alpha) = \frac{1}{\pi}\langle\alpha|\rho|\alpha\rangle , \tag{5.154}$$

is equivalent to complete knowledge of ρ. The real function $Q(\alpha)$ satisfies the inequality

$$0 \leqslant Q(\alpha) \leqslant \frac{1}{\pi} , \tag{5.155}$$

and the normalization condition

$$\mathrm{Tr}\,\rho = \int d^2\alpha Q(\alpha) = 1 . \tag{5.156}$$

The argument just given shows that $Q(\alpha)$ contains all the information needed to calculate averages of any operator, but it does not tell us how to extract these results.

The necessary clue is given by eqn (5.78) which expresses any operator X as a sum of antinormally-ordered terms. With this representation for X, the expectation value is

$$\begin{aligned} \langle X \rangle = \text{Tr}\,(\rho X) &= \sum_{m=0}^{\infty}\sum_{n=0}^{\infty} \mathcal{X}_{mn}^{A}\, \text{Tr}\left(\rho a^{m} a^{\dagger n}\right) \\ &= \sum_{m=0}^{\infty}\sum_{n=0}^{\infty} \mathcal{X}_{mn}^{A} \int \frac{d^2\alpha}{\pi} \left\langle \alpha \left| a^{\dagger n} \rho a^{m} \right| \alpha \right\rangle \\ &= \int d^2\alpha Q\left(\alpha\right) \mathcal{X}^{A}\left(\alpha\right), \end{aligned} \tag{5.157}$$

where $\mathcal{X}^{A}(\alpha)$ is defined by eqn (5.80). In other words the expectation value of any physical quantity X can be calculated by writing it in antinormally-ordered form, then replacing the operators a and $a^{\dagger}$ by the complex numbers α and α^* respectively, and finally evaluating the integral in eqn (5.157).

The Q-function, like the Wigner distribution, is difficult to calculate in realistic experimental situations; but there are idealized cases for which a simple expression can be obtained. The easiest is that of a pure coherent state, i.e. $\rho = |\alpha_0\rangle \langle \alpha_0|$, which leads to

$$Q\left(\alpha\right) = \frac{\left|\langle \alpha \left| \alpha_0 \right\rangle\right|^2}{\pi} = \frac{\exp\left(-\left|\alpha - \alpha_0\right|^2\right)}{\pi}\,. \tag{5.158}$$

Despite the fact that this state corresponds to a sharp value of α, the probability distribution has a nonzero spread around the peak at $\alpha = \alpha_0$. This unexpected feature is another consequence of the overcompleteness of the coherent states.

At the other extreme of a pure number state, $\rho = |n\rangle \langle n|$, the expansion of the coherent state in number states yields

$$Q\left(\alpha\right) = \frac{\left|\langle \alpha \left| n \right\rangle\right|^2}{\pi} = \frac{e^{-|\alpha|^2}}{\pi} \frac{|\alpha|^{2n}}{n!}\,, \tag{5.159}$$

which is peaked on the circle of radius $|\alpha| = \sqrt{n}$.

B *Difficulties in computing the Q-function**

For any state of the field, the Q-function is everywhere positive and normalized to unity, so $Q(\alpha)$ is a genuine probability density on the electromagnetic phase space Γ_{em}. The integral in eqn (5.157) is then an average over this distribution. These properties make the Q-function useful for the display and interpretation of experimental data or the results of approximate simulations, but they do not mean that we have found the best of all possible worlds. One difficulty is that there are functions satisfying the inequality (5.155) and the normalization condition (5.156) that do not correspond to any physically realizable density operator, i.e. they are not given by eqn (5.154) for any acceptable ρ. The irreducible quantum fluctuations described by the commutation relation $\left[a, a^{\dagger}\right] = 1$ are the source of this problem. For any density operator ρ,

$$\left\langle a a^{\dagger} \right\rangle = \left\langle a^{\dagger} a + 1 \right\rangle \geqslant 1\,. \tag{5.160}$$

Evaluating the same quantity by means of eqn (5.157) produces the condition

$$\int d^2\alpha Q(\alpha)|\alpha|^2 \geqslant 1 \tag{5.161}$$

on the Q-function. As an example of a spurious Q-function, consider

$$Q(\alpha) = \frac{2}{\pi\sqrt{\pi}\sigma^2}\exp\left(-\frac{|\alpha|^4}{\sigma^4}\right). \tag{5.162}$$

This function satisfies eqns (5.155) and (5.156) for $\sigma^2 > 2/\sqrt{\pi}$, but the integral in eqn (5.161) is

$$\int d^2\alpha Q(\alpha)|\alpha|^2 = \frac{\sigma^2}{\sqrt{\pi}}. \tag{5.163}$$

Thus for $2/\sqrt{\pi} < \sigma^2 < \sqrt{\pi}$, the inequality (5.161) is violated. Finding a Q-function that satisfies this inequality as well is still not good enough, since there are similar inequalities for all higher-order moments $a^2a^{\dagger 2}$, $a^3a^{\dagger 3}$, etc. This poses a serious problem in practice, because of the inevitable approximations involved in the calculation of the Q-function for a nontrivial situation. Any approximation could lead to a violation of one of the infinite set of inequalities and, consequently, to an unphysical prediction for some observable.

The dangers involved in extracting the density operator from an approximate Q-function do not occur in the other direction. Substituting any physically acceptable approximation for the density operator into eqn (5.154) will yield a physically acceptable $Q(\alpha)$. For this reason the results of approximate calculations are often presented in terms of the Q-function. For example, plots of the level lines of $Q(\alpha)$ can provide useful physical insights, since the Q-function is a genuine probability distribution.

5.6.3 The Glauber–Sudarshan $P(\alpha)$-representation

A Normal ordering

We have just seen that the evaluation of the expectation value, $\langle X\rangle$, using the Q-function requires writing out the operator in antinormal-ordered form. This is contrary to our previous practice of writing all observables, e.g. the Hamiltonian, the linear momentum, etc. in normal-ordered form. A more important point is that photon-counting rates are naturally expressed in terms of normally-ordered products, as we will see in Section 9.1.

The commutation relations can be used to express any operator $X\left(a, a^{\dagger}\right)$ in normal-ordered form,

$$X = \sum_{m=0}^{\infty}\sum_{n=0}^{\infty}\mathcal{X}_{nm}^{N}a^{\dagger n}a^{m}, \tag{5.164}$$

so we want a representation of the density operator which is adapted to calculating the averages of normal-ordered products. For this purpose, we apply the coherent state diagonal representation (5.79) to the density operator. This leads to the P-function representation introduced by Glauber (1963) and Sudarshan (1963):

$$\rho = \int d^2\alpha \left|\alpha\right\rangle P\left(\alpha\right)\left\langle\alpha\right| . \tag{5.165}$$

If the coherent states were mutually orthogonal, then $Q(\alpha)$ would be proportional to $P(\alpha)$, but eqn (5.58) for the inner product shows instead that

$$\begin{aligned} Q\left(\alpha\right) &= \int \frac{d^2\beta}{\pi} \left|\left\langle\alpha\left|\beta\right.\right\rangle\right|^2 P\left(\beta\right) \\ &= \int \frac{d^2\beta}{\pi} e^{-\left|\beta\right|^2} P\left(\alpha+\beta\right) . \end{aligned} \tag{5.166}$$

Thus the $Q(\alpha)$ is a Gaussian average of the P-function around the point α.

The average of the generic normal-ordered product $a^{\dagger m}a^n$ is

$$\left\langle a^{\dagger m}a^n\right\rangle = \mathrm{Tr}\left(\rho a^{\dagger m}a^n\right) = \mathrm{Tr}\left(a^n\rho a^{\dagger m}\right) = \int d^2\alpha\alpha^n\alpha^{*m}P\left(\alpha\right), \tag{5.167}$$

which combines with eqn (5.164) to yield

$$\left\langle X\left(a,a^\dagger\right)\right\rangle = \int d^2\alpha\mathcal{X}^N\left(\alpha\right)P\left(\alpha\right), \tag{5.168}$$

where

$$\mathcal{X}^N\left(\alpha\right) = \sum_{m=0}^{\infty}\sum_{n=0}^{\infty}\mathcal{X}^N_{nm}\alpha^{*n}\alpha^m . \tag{5.169}$$

The normalization condition $\mathrm{Tr}\,\rho = 1$ becomes

$$\int d^2\alpha P\left(\alpha\right) = 1 , \tag{5.170}$$

so $P(\alpha)$ is beginning to look like another probability distribution. Indeed, for a pure coherent state, $\rho_{\mathrm{coh}} = \left|\alpha_0\right\rangle\left\langle\alpha_0\right|$, the P-function is

$$P_{\mathrm{coh}}\left(\alpha\right) = \delta_2\left(\alpha-\alpha_0\right), \tag{5.171}$$

where

$$\delta_2\left(\alpha-\alpha_0\right) = \delta\left(\mathrm{Re}\,\alpha - \mathrm{Re}\,\alpha_0\right)\delta\left(\mathrm{Im}\,\alpha - \mathrm{Im}\,\alpha_0\right). \tag{5.172}$$

This is a positive distribution that exactly picks out the coherent state $\left|\alpha_0\right\rangle\left\langle\alpha_0\right|$, so it is more intuitively appealing than the Q-function description of the same state by a Gaussian distribution. Another hopeful result is provided by the P-function for a thermal state. From eqn (2.178) we know that the density operator for a thermal or chaotic state with average number $\overline{n}$ has the diagonal matrix elements

$$\langle n|\rho_{\text{th}}|n\rangle = \frac{\overline{n}^n}{(1+\overline{n})^{n+1}}; \tag{5.173}$$

therefore, the P-function has to satisfy

$$\frac{\overline{n}^n}{(1+\overline{n})^{n+1}} = \int d^2\alpha P_{\text{th}}(\alpha) |\langle n|\alpha\rangle|^2 \tag{5.174}$$

$$= \int d^2\alpha P_{\text{th}}(\alpha) e^{-|\alpha|^2} \frac{|\alpha|^{2n}}{n!}. \tag{5.175}$$

Expressing the remaining integral in polar coordinates suggests that $P(\alpha)$ might be proportional to a Gaussian function of $|\alpha|$, and a little trial and error leads to the result

$$P_{\text{th}}(\alpha) = \frac{1}{\pi\overline{n}} \exp\left[-\frac{|\alpha|^2}{\overline{n}}\right]. \tag{5.176}$$

Thus the P-function acts like a probability distribution for two very different states of light. On the other hand, this is a quantum system, so we should be prepared for surprises.

The interpretation of $P(\alpha)$ as a probability distribution requires $P(\alpha) \geqslant 0$ for all α, and the normalization condition (5.170) implies that $P(\alpha)$ cannot vanish everywhere. The states with nowhere negative $P(\alpha)$ are called **classical states**, and any states for which $P(\alpha) < 0$ in some region of the α-plane are called **nonclassical states**. Multimode states are said to be classical if the function $P(\underline{\alpha})$ in eqn (5.104) satisfies $P(\underline{\alpha}) \geqslant 0$ for all $\underline{\alpha}$.

The meaning of 'classical' intended here is that these are quantum states with the special property that all expectation values can be simulated by averaging over random classical fields with the probability distribution $P(\alpha)$. By virtue of eqn (5.171), all coherent states—including the vacuum state—are classical, and eqn (5.176) shows that thermal states are also classical. The last example shows that classical states need not be quasiclassical,[2] i.e. minimum-uncertainty, states.

Our next objective is to find out what kinds of states are nonclassical. A convenient way to investigate this question is to use eqn (5.165) to calculate the probability that exactly n photons will be detected; this is given by

$$\langle n|\rho|n\rangle = \int d^2\alpha |\langle n|\alpha\rangle|^2 P(\alpha) = \int d^2\alpha e^{-|\alpha|^2} \frac{|\alpha|^{2n}}{n!} P(\alpha). \tag{5.177}$$

If ρ is any classical state—other than the vacuum state—the integrand is non-negative, so the integral must be positive. For the vacuum state, $\rho_{\text{vac}} = |0\rangle\langle 0|$, eqn (5.171) gives $P(\alpha) = \pi\delta_2(\alpha)$, so the integral vanishes for $n \neq 0$ and gives $\langle 0|\rho_{\text{vac}}|0\rangle = 1$ for $n = 0$.

[2]It is too late to do anything about this egregious abuse of language.

Thus for any classical state—other than the vacuum state—the probability for finding n photons cannot vanish for any value of n:

$$\langle n\left|\rho\right|n\rangle \neq 0 \quad \text{for all } n\,. \tag{5.178}$$

Thus a state, $\rho \neq \rho_{\text{vac}}$, such that $\langle n\left|\rho\right|n\rangle = 0$ for some $n > 0$ is nonclassical. The simplest example is the pure number state $\rho = \left|m\right\rangle\left\langle m\right|$, since $\langle n\left|\rho\right|n\rangle = 0$ for $n \neq m$. This can be seen more explicitly by applying eqn (5.177) to the case $\rho = \left|m\right\rangle\left\langle m\right|$, with the result

$$\int d^2\alpha e^{-|\alpha|^2}\frac{\left|\alpha\right|^{2n}}{n!}P\left(\alpha\right) = \begin{cases} 1 & \text{for } n = m\,, \\ 0 & \text{for } n \neq m\,. \end{cases} \tag{5.179}$$

The conditions for $n \neq m$ cannot be satisfied if $P\left(\alpha\right)$ is non-negative; therefore, $P\left(\alpha\right)$ for a pure number state must be negative in some region of the α-plane. A closer examination of this infinite family of equations shows further that $P\left(\alpha\right)$ cannot even be a smooth function; instead it is proportional to the nth derivative of the delta function $\delta_2\left(\alpha\right)$.

*B The normal-ordered characteristic function**

An alternative construction of the $P\left(\alpha\right)$-function can be carried out by using the normally-ordered operator, $e^{\eta a^\dagger}e^{-\eta^* a}$, to define the **normally-ordered characteristic function**

$$\chi_N\left(\eta\right) = \text{Tr}\left(\rho e^{\eta a^\dagger}e^{-\eta^* a}\right). \tag{5.180}$$

The corresponding distribution function, $P\left(\alpha\right)$, is defined by replacing χ_W with χ_N in eqn (5.126) to get

$$P\left(\alpha\right) = \frac{1}{\pi^2}\int d^2\eta e^{\eta^*\alpha - \eta\alpha^*}\chi_N\left(\eta\right). \tag{5.181}$$

The identity (5.141) relates $\chi_N\left(\eta\right)$ and $\chi_W\left(\eta\right)$ by

$$\chi_N\left(\eta\right) = e^{|\eta|^2/2}\chi_W\left(\eta\right), \tag{5.182}$$

so the argument leading to eqn (5.140) yields the much weaker bound $\left|\chi_N\left(\eta\right)\right| < e^{|\eta|^2/2}$ for the normal-ordered characteristic function $\chi_N\left(\eta\right)$. This follows from the fact that $e^{\eta a^\dagger}e^{-\eta^* a}$ is self-adjoint rather than unitary. The eigenvalues are therefore real and need not have unit modulus. This has the important consequence that $P\left(\alpha\right)$ is not guaranteed to exist, even in the delta function sense. In the literature it is often said that $P\left(\alpha\right)$ can be more singular than a delta function.

We already know from eqn (5.171) that $P\left(\alpha\right)$ exists for a pure coherent state, but what about number states? The P-distribution for the number state $\rho = \left|1\right\rangle\left\langle 1\right|$ can be evaluated by combining the general relation (5.182) with the result (5.152) for the Wigner characteristic function of a number state to get

$$P\left(\alpha\right) = \frac{1}{\pi^2}\int d^2\eta e^{\eta^*\alpha - \eta\alpha^*}\left(1 - |\eta|^2\right). \tag{5.183}$$

This can be evaluated by using the identities

$$\eta e^{\eta^*\alpha-\eta\alpha^*} = -\frac{\partial}{\partial\alpha^*}e^{\eta^*\alpha-\eta\alpha^*}\,, \quad \eta^* e^{\eta^*\alpha-\eta\alpha^*} = \frac{\partial}{\partial\alpha}e^{\eta^*\alpha-\eta\alpha^*}\,, \tag{5.184}$$

to find

$$P(\alpha) = \delta_2(\alpha) + \frac{\partial}{\partial\alpha}\frac{\partial}{\partial\alpha^*}\delta_2(\alpha)\,. \tag{5.185}$$

This shows that $P(\alpha)$ is not everywhere positive for a number state. Since $P(\alpha)$ is a generalized function, the meaning of this statement is that there is a real, positive test function $f(\alpha)$ for which

$$\int d^2\alpha P(\alpha)\, f(\alpha) < 0\,, \tag{5.186}$$

e.g. $f(\alpha) = \exp(-2|\alpha|^2)$.

Let ρ be a density operator for which $P(\alpha)$ exists, then in parallel with eqn (5.130) we have

$$\begin{aligned}\int d^2\alpha\; \alpha^{*n}\alpha^m P(\alpha) &= \left(-\frac{\partial}{\partial\eta^*}\right)^m \left(\frac{\partial}{\partial\eta}\right)^n \chi_N(\eta,\eta^*)\Big|_{\eta=0} \\ &= \mathrm{Tr}\left(\rho a^{\dagger n} e^{\eta a^\dagger} a^m e^{-\eta^* a}\right)\Big|_{\eta=0} \\ &= \mathrm{Tr}\left(\rho a^{\dagger n} a^m\right).\end{aligned} \tag{5.187}$$

The case $m = n = 0$ gives the normalization

$$\int d^2\alpha\; P(\alpha) = 1\,, \tag{5.188}$$

and the identity of the averages calculated with $P(\alpha)$ and the averages calculated with ρ shows that the density operator is represented by

$$\rho = \int d^2\alpha\; |\alpha\rangle\, P(\alpha)\, \langle\alpha|\,. \tag{5.189}$$

Thus the definition of $P(\alpha)$ given by eqn (5.181) agrees with the original definition (5.165).

For an operator expressed in normal-ordered form by

$$X\left(a^\dagger, a\right) = \sum_{m=0}^{\infty}\sum_{n=0}^{\infty} \mathcal{X}^N_{nm} a^{\dagger n} a^m\,, \tag{5.190}$$

eqn (5.187) yields

$$\mathrm{Tr}(\rho X) = \int d^2\alpha\; P(\alpha)\, \mathcal{X}^N(\alpha^*, \alpha)\,, \tag{5.191}$$

where

$$\mathcal{X}^N(\alpha^*, \alpha) = \sum_{m=0}^{\infty}\sum_{n=0}^{\infty} \mathcal{X}^N_{nm} \alpha^{*n}\alpha^m\,. \tag{5.192}$$

The P-distribution and the Wigner distribution are related by the following argument. First invert eqn (5.181) to get

$$\chi_N(\eta) = \int d^2\alpha \; e^{\eta\alpha^* - \eta^*\alpha} P(\alpha) . \tag{5.193}$$

Combining this with eqn (5.126) and the relation (5.182) produces

$$\begin{aligned} W(\alpha) &= \frac{1}{\pi^2} \int d^2\eta e^{\eta^*\alpha - \eta\alpha^*} e^{-|\eta|^2/2} \chi_N(\eta) \\ &= \frac{1}{\pi^2} \int d^2\beta P(\beta) \int d^2\eta \; e^{\eta(\beta^* - \alpha^*)} e^{-\eta^*(\beta-\alpha)} e^{-|\eta|^2/2} . \end{aligned} \tag{5.194}$$

The η-integral is readily done by converting to real variables, and the relation between the Wigner distribution and the P-distribution is

$$W(\alpha) = \frac{2}{\pi} \int d^2\beta e^{-2|\beta-\alpha|^2} P(\beta) . \tag{5.195}$$

An interesting consequence of this relation is that a classical state automatically yields a positive Wigner distribution, i.e.

$$P(\alpha) \geqslant 0 \;\text{ implies }\; W(\alpha) \geqslant 0 , \tag{5.196}$$

but the opposite statement is not true:

$$W(\alpha) \geqslant 0 \;\text{ does not imply }\; P(\alpha) \geqslant 0 . \tag{5.197}$$

This is demonstrated by exhibiting a single example—see Exercise 5.7—of a state with a positive Wigner function that is not classical.

It is natural to wonder why $P(\alpha) \geqslant 0$ should be chosen as the definition of a classical state instead of $W(\alpha) \geqslant 0$. The relations (5.196) and (5.197) give one reason, since they show that $P(\alpha) \geqslant 0$ is a stronger condition. A more physical reason is that counting rates are described by expectation values of normal-ordered products, rather than Weyl products. This means that $P(\alpha)$ is more directly related to the relevant experiments than is $W(\alpha)$.

5.6.4 Multimode phase space*

In Section 5.5 we defined multimode coherent states $|\underline{\alpha}\rangle$ by $a_\kappa |\underline{\alpha}\rangle = \alpha_\kappa |\underline{\alpha}\rangle$, where a_κ is the annihilation operator for the mode κ and

$$\underline{\alpha} = (\alpha_1, \alpha_2, \ldots, \alpha_\kappa, \ldots) . \tag{5.198}$$

For states in which only a finite number of modes are occupied, i.e. $a_\kappa |\underline{\alpha}\rangle = 0$ for $\kappa > \kappa'$, the characteristic functions defined previously have the generalizations

$$\chi_W(\underline{\eta}) = \mathrm{Tr}\left(\rho e^{\underline{\eta}\cdot\underline{a}^\dagger - \underline{\eta}^*\underline{a}}\right) , \tag{5.199}$$

$$\chi_N(\underline{\eta}) = \mathrm{Tr}\left(\rho e^{\underline{\eta}\cdot\underline{a}^\dagger} e^{-\underline{\eta}^*\underline{a}}\right) , \tag{5.200}$$

where $\underline{\eta} \equiv (\eta_1, \eta_2, \ldots)$, and

$$\underline{\eta} \cdot \underline{a}^\dagger = \sum_{\kappa \leqslant \kappa'} \eta_\kappa a_\kappa^\dagger . \tag{5.201}$$

The corresponding distributions are defined by multiple Fourier transforms. For example the P-distribution is

$$P(\underline{\alpha}) = \int \left[\prod_{\kappa \leqslant \kappa'} \frac{d^2 \eta_\kappa}{\pi^2} \right] e^{-\underline{\eta} \cdot \underline{\alpha}^* + \underline{\eta}^* \cdot \underline{\alpha}} \chi_N(\underline{\eta}) , \tag{5.202}$$

and the density operator is given by

$$\rho = \int \left[\prod_{\kappa \leqslant \kappa'} d^2 \alpha_\kappa \right] |\underline{\alpha}\rangle P(\underline{\alpha}) \langle \underline{\alpha}| . \tag{5.203}$$

All this is plain sailing as long as κ' remains finite, but some care is required to get the mathematics right when $\kappa' \to \infty$. This has been done in the work of Klauder and Sudarshan (1968), but the $\kappa' \to \infty$ limit is not strictly necessary in practice. The reason is to be found in the alternative characterization of coherent states given by $|\underline{\alpha}\rangle \to |\mathbf{w}\rangle$, where

$$A^{(+)}[\mathbf{v}] |\mathbf{w}\rangle = (\mathbf{v}, \mathbf{w}) |\mathbf{w}\rangle , \tag{5.204}$$

and the wave packets $\mathbf{w}$, $\mathbf{v}$, etc. are expressed as expansions in the chosen modes,

$$\mathbf{w}(\mathbf{r}) = \sum_\kappa \alpha_\kappa \mathbf{w}_\kappa(\mathbf{r}) . \tag{5.205}$$

The vector fields $\mathbf{v}$ and $\mathbf{w}$ belong to the classical phase space Γ_{em} defined in Section 3.5.1, so the expansion coefficients α_κ must go to zero as $\kappa \to \infty$. Thus any real experimental situation can be adequately approximated by a finite number of modes. With this comforting thought in mind, we can express the characteristic and distribution functions as functionals of the wave packets. In this language, the normal-ordered characteristic function and the P-distribution are respectively given by

$$\chi_N(\mathbf{v}) = \text{Tr}\left(\rho \exp\left\{A^{(+)}[\mathbf{v}]\right\} \exp\left\{-A^{(-)}[\mathbf{v}]\right\}\right) \tag{5.206}$$

and

$$P(\mathbf{w}) = \int \mathcal{D}[\mathbf{v}] \exp\{(\mathbf{w}, \mathbf{v}) - (\mathbf{v}, \mathbf{w})\} \chi_N(\mathbf{v}) . \tag{5.207}$$

The symbol $\int \mathcal{D}[\mathbf{v}]$ stands for a (functional) integral over the infinite-dimensional space Γ_{em} of classical wave packets; but, as we have just remarked, it can always be approximated by a finite-dimensional integral over the collection of modes with non-negligible amplitudes.

5.7 Gaussian states*

In classical statistics, the Gaussian (normal) distribution has the useful property that the first two moments determine the values of all other moments (Gardiner, 1985, Sec. 2.8.1). For a Gaussian distribution over N real variables—with the averages of single variables arranged to vanish—all odd moments vanish and the even moments satisfy

$$\langle x_1 \cdots x_{2q} \rangle = \frac{(2q)!}{q!2^q} \left[\langle x_i x_j \rangle \langle x_k x_l \rangle \cdots \langle x_m x_n \rangle \right]_{\text{sym}} , \tag{5.208}$$

where i, j, k, l, m, n range over $1, \ldots, 2q$ and the subscript sym indicates the average over all ways of partitioning the variables into pairs. Two fourth-order examples are

$$\begin{aligned} \langle x_1 \cdots x_4 \rangle &= \frac{4!}{2!2^2} \left[\frac{1}{3} \left\{ \langle x_1 x_2 \rangle \langle x_3 x_4 \rangle + \langle x_1 x_3 \rangle \langle x_2 x_4 \rangle + \langle x_1 x_4 \rangle \langle x_2 x_3 \rangle \right\} \right] \\ &= \langle x_1 x_2 \rangle \langle x_3 x_4 \rangle + \langle x_1 x_3 \rangle \langle x_2 x_4 \rangle + \langle x_1 x_4 \rangle \langle x_2 x_3 \rangle \end{aligned} \tag{5.209}$$

and

$$\langle x_1^4 \rangle = 3 \langle x_1^2 \rangle \langle x_1^2 \rangle . \tag{5.210}$$

This classical property is shared by the coherent states, as can be seen from the general identity

$$\left\langle \alpha \left| a^{\dagger m} a^n \right| \alpha \right\rangle = \alpha^{*m} \alpha^n = \left(\left\langle \alpha \left| a^\dagger \right| \alpha \right\rangle \right)^m \left(\left\langle \alpha \left| a \right| \alpha \right\rangle \right)^n . \tag{5.211}$$

A natural generalization of the classical notion of a Gaussian distribution is to define **Gaussian states** (Gardiner, 1991, Sec. 4.4.5) as those that are described by density operators of the form

$$\rho_G = \mathcal{N} \exp \left[-G \left(a, a^\dagger \right) \right] , \tag{5.212}$$

where

$$G \left(a, a^\dagger \right) = L a^\dagger a + \frac{1}{2} M a^{\dagger 2} + \frac{1}{2} M^* a^2 , \tag{5.213}$$

L and M are free parameters, and the constant $\mathcal{N}$ is fixed by the normalization condition $\mathrm{Tr}\, \rho = 1$.

For the special value $M = 0$, the Gaussian state ρ_G has the form of a thermal state, and we already know (see eqn (5.148)) how to calculate the Wigner characteristic function for this case. We would therefore like to transform the general Gaussian state into this form. If the operators a and $a^\dagger$ were replaced by complex variables α and α^*, this would be easy. The c-number quadratic form $G(\alpha, \alpha^*)$ can always be expressed as a sum of squares by a linear transformation to new variables

$$\begin{aligned} \widetilde{\alpha} &= \mu \alpha + \nu \alpha^* , \\ \widetilde{\alpha}^* &= \mu^* \alpha^* + \nu^* \alpha . \end{aligned} \tag{5.214}$$

What is needed now is the quantum analogue of this transformation, i.e. the new and old operators are related by

$$\widetilde{a} = U a U^\dagger , \tag{5.215}$$

where U is a unitary transformation. We must ensure that eqn (5.215) goes over into eqn (5.214) in the classical limit, and the easiest way to do this is to assume that the unitary transformation has the same form:

$$\widetilde{a} = UaU^{\dagger} = \mu a + \nu a^{\dagger} , \tag{5.216}$$

where μ and ν are c-numbers. The unitary transformation preserves the commutation relations, so the c-number coefficients μ and ν are constrained by

$$|\mu|^2 - |\nu|^2 = 1 . \tag{5.217}$$

Since the overall phase of $\widetilde{a}$ is irrelevant, we can choose μ to be real, and set

$$\mu = \cosh r , \quad \nu = e^{2i\phi} \sinh r . \tag{5.218}$$

The relation between a and $\widetilde{a}$ is an example of the **Bogoliubov transformation** first introduced in low temperature physics (Huang, 1963, Sec. 19.4).

The condition that the transformed Gaussian state is thermal-like is

$$\widetilde{\rho}_G = U\rho_G U^{\dagger} = \mathcal{N} e^{-g_0 a^{\dagger} a} , \tag{5.219}$$

where the constant g_0 is to be determined. The *ansatz* (5.212) shows that this is equivalent to

$$UGU^{\dagger} = g_0 a^{\dagger} a , \tag{5.220}$$

and taking the commutator of both sides of this equation with a produces

$$\left[a, L\widetilde{a}^{\dagger}\widetilde{a} + \frac{1}{2} M \widetilde{a}^{\dagger 2} + \frac{1}{2} M^{*} \widetilde{a}^{2} \right] = g_0 a . \tag{5.221}$$

Evaluating the commutator on the left by means of eqn (5.216) will produce two terms, one proportional to $a^{\dagger}$ and one proportional to a. No $a^{\dagger}$-term can be present if eqn (5.221) is to be satisfied; therefore, the coefficient of $a^{\dagger}$ must be set to zero. A little careful algebra shows that the free parameter ϕ in eqn (5.218) can be chosen to cancel the phase of M. This is equivalent to assuming that M is real and positive to begin with, so that $\phi = 0$. With this simplification, setting the coefficient of $a^{\dagger}$ to zero imposes $\tanh 2r = -M/L$, and using this relation to evaluate the coefficient of the a-term yields in turn $g_0 = \sqrt{L^2 - M^2}$.

We will now show that the Gaussian state has the properties claimed for it by applying the general definition (5.125) to ρ_G, with the result

$$\begin{aligned} \chi_W^G(\eta) &= \mathrm{Tr}\left(\rho_G e^{\eta a^{\dagger} - \eta^{*} a} \right) \\ &= \mathrm{Tr}\left(U\rho_G U^{\dagger} U e^{\eta a^{\dagger} - \eta^{*} a} U^{\dagger} \right) \\ &= \mathcal{N}\, \mathrm{Tr}\left(e^{-g_0 a^{\dagger} a} e^{\eta \widetilde{a}^{\dagger} - \eta^{*} \widetilde{a}} \right) . \end{aligned} \tag{5.222}$$

The remaining $\widetilde{a}$-dependence can be eliminated with the aid of the explicit form (5.216), so that

$$\chi_W^G(\eta) = \mathcal{N}\,\mathrm{Tr}\left(e^{-g_0 a^\dagger a} e^{\zeta a^\dagger - \zeta^* a}\right), \tag{5.223}$$

where

$$\zeta = \eta\mu - \eta^*\nu = \eta\cosh r - \eta^*\sinh r\,. \tag{5.224}$$

The parameter g_0 in eqn (5.219) plays the role of $\hbar\omega/kT$ for the thermal state, so comparison with eqns (2.175)–(2.177) shows that $\mathcal{N} = [1 - \exp(-g_0)]$. An application of eqn (5.148) then yields the Wigner characteristic function

$$\begin{aligned}\chi_W^G(\eta) &= \exp\left[-\left(n_G + \frac{1}{2}\right)|\zeta|^2\right] \\ &= \exp\left[-\left(n_G + \frac{1}{2}\right)|\eta\cosh r - \eta^*\sinh r|^2\right]\end{aligned} \tag{5.225}$$

for the Gaussian state, where $n_G = 1/(e^{g_0} - 1)$ is the analogue of the thermal average number of quanta. The Wigner distribution is given by eqn (5.126), which in the present case becomes

$$W_G(\alpha) = \frac{1}{\pi^2}\int d^2\eta e^{\eta^*\alpha - \eta\alpha^*}\exp\left[-\left(n_G + \frac{1}{2}\right)|\zeta|^2\right]. \tag{5.226}$$

After changing integration variables from η to ζ, this yields

$$W_G(\alpha) = \frac{1}{\pi^2}\int d^2\zeta e^{\zeta^*\beta - \zeta\beta^*}\exp\left[-\left(n_G + \frac{1}{2}\right)|\zeta|^2\right], \tag{5.227}$$

where

$$\beta = \mu\alpha - \nu\alpha^* = \cosh r\;\alpha - \sinh r\;\alpha^*\,. \tag{5.228}$$

According to eqn (5.149), this means that

$$\begin{aligned}W_G(\alpha) &= \frac{1}{\pi}\frac{1}{n_G + 1/2}\exp\left(-\frac{|\beta|^2}{n_G + 1/2}\right) \\ &= \frac{1}{\pi}\frac{1}{n_G + 1/2}\exp\left(-\frac{|\cosh r\;\alpha - \sinh r\;\alpha^*|^2}{n_G + 1/2}\right).\end{aligned} \tag{5.229}$$

It is encouraging to see that the Wigner distribution for a Gaussian state is itself Gaussian, but we previously found that positivity for the Wigner distribution does not guarantee positivity for $P(\alpha)$. In order to satisfy ourselves that $P(\alpha)$ is also Gaussian, we use the relation (5.182) between the normal-ordered and Wigner characteristic functions to carry out a rather long evaluation of $P(\alpha)$ which leads to

$$\begin{aligned}P_G(\alpha) &= \frac{1}{\pi}\frac{1}{\sqrt{(n_G + 1/2)^2 - (n_G + 1/2)\cosh 2r + 1/4}} \\ &\quad\times\exp\left\{-\frac{|\alpha|^2\cosh 2r - \frac{1}{2}\sinh 2r\left(\alpha^2 + \alpha^{*2}\right)}{n_G\cosh^2 r + (n_G + 1)\sinh^2 r}\right\}.\end{aligned} \tag{5.230}$$

Thus all Gaussian states are classical, and both the Wigner function $W_G(\alpha)$ and the $P_G(\alpha)$-function are Gaussian functions of α.

5.8 Exercises

5.1 Are there eigenvalues and eigenstates of $a^\dagger$?

The equation

$$a^\dagger |\phi_\beta\rangle = \beta |\phi_\beta\rangle ,$$

where β is a complex number, is apparently analogous to the eigenvalue problem $a|\alpha\rangle = \alpha|\alpha\rangle$ defining coherent states.

(1) Show that the coordinate-space representation of this equation is

$$\frac{1}{\sqrt{2\hbar\omega}}\left(\omega Q - \hbar\frac{d}{dQ}\right)\phi_\beta(Q) = \beta\phi_\beta(Q) .$$

(2) Find the explicit solution and explain why it does not represent an eigenvector. **Hint**: The solution violates a fundamental principle of quantum mechanics.

5.2 Expectation value of functions of N

Consider the operator-valued function $f(N)$, where $N = a^\dagger a$ and $f(s)$ is a real function of the dimensionless, real argument s.

(1) Show that $f(N)$ is represented by

$$f(N) = \int_{-\infty}^{\infty}\frac{d\theta}{2\pi}f(\theta)\,e^{i\theta N},$$

where $f(\theta)$ is the Fourier transform of $f(s)$.

(2) For any coherent state $|\alpha\rangle$, show that

$$\left\langle\alpha\left|e^{i\theta N}\right|\alpha\right\rangle = \exp\left\{|\alpha|^2\left(e^{i\theta} - 1\right)\right\},$$

and use this to get a representation of $\langle\alpha|f(N)|\alpha\rangle$.

5.3 Approach to orthogonality

By analogy with ordinary vectors, define the angle $\Theta_{\alpha\beta}$ between the two coherent states by $\cos(\Theta_{\alpha\beta}) = |\langle\alpha|\beta\rangle|$. From a plot of $\Theta_{\alpha\beta}$ versus $|\alpha - \beta|$ determine the value at which approximate orthogonality sets in. What is the physical significance of this value?

5.4 Number-phase uncertainty principle

Assume that the quantum fuzzball in Fig. 5.1 is a circle of unit diameter.

(1) What is the physical meaning of this assumption?

(2) Define the *phase uncertainty*, $\Delta\phi$, as the angle subtended by the quantum fuzzball at the origin. In the semiclassical limit $|\alpha_0| \gg 1$, show that $\Delta\phi\Delta n \sim 1$, where Δn is the *rms* deviation of the photon number in the state $|\alpha_0\rangle$.

5.5 Arecchi's experiment

What is the relation of the fourth and second moments of a Poisson distribution? Check this relation for the data given in Fig. 5.4.

5.6 The displacement operator

(1) Show that eqn (5.47) follows from eqn (5.46).

(2) Derive eqn (5.56) and explain why $\Phi(\alpha, \beta)$ has to be real.

(3) Show that $\exp[-i\tau K(\alpha)]$, with $K(\alpha) = i\alpha a^\dagger - i\alpha^* a$, satisfies

$$\frac{\partial}{\partial \tau}\exp[-i\tau K(\alpha)] = \left(\alpha a^\dagger - \alpha^* a\right)\exp[-i\tau K(\alpha)],$$

and that $\exp[-i\tau K(\alpha)] = D(\tau\alpha)$.

(4) Let $\alpha \to \tau\alpha$ and $\beta \to \tau\beta$ in eqn (5.56) and then differentiate both sides with respect to τ. Show that the resulting operator equation reduces to the c-number equation

$$\frac{\partial \Phi(\tau\alpha, \tau\beta)}{\partial \tau} = 2\tau \operatorname{Im}(\alpha\beta^*),$$

and then conclude that $\Phi(\alpha, \beta) = \operatorname{Im}(\alpha\beta^*)$.

5.7 Wigner distribution

(1) Show that the Wigner distribution $W(\alpha)$ for the density operator

$$\rho = \gamma |1\rangle\langle 1| + (1-\gamma)|0\rangle\langle 0|,$$

with $0 < \gamma < 1$, is everywhere positive.

(2) Determine if the state described by ρ is classical.

5.8 The antinormally-ordered characteristic function*

The argument in Section 5.6.1-B begins by replacing the exponential in the classical definition (5.114) by $e^{\eta a^\dagger - \eta^* a}$, but one could just as well start with the classically equivalent form $e^{-\eta^* a}e^{\eta a^\dagger}$, which is antinormally ordered. This leads to the definition

$$\chi_A(\eta) = \operatorname{Tr}\left(\rho e^{-\eta^* a}e^{\eta a^\dagger}\right)$$

of the **antinormally-ordered** characteristic function.

(1) Use eqn (5.72) to show that

$$\chi_A(\eta) = \int d^2\alpha e^{(\eta\alpha^* - \eta^*\alpha)} Q(\alpha).$$

(2) Invert this Fourier integral, e.g. by using eqn (5.127), to find

$$Q(\alpha) = \frac{1}{\pi^2}\int d^2\eta e^{-(\eta\alpha^* - \eta^*\alpha)}\chi_A(\eta).$$

5.9 Classical states

(1) For classical states, with density operators ρ_1 and ρ_2, show that the convex combination $\rho_x = x\rho_1 + (1-x)\,\rho_2$ with $0 < x < 1$ is also a classical state.

(2) Consider the superposition $|\psi\rangle = C\,|\alpha\rangle + C\,|-\alpha\rangle$ of two coherent states, where C and α are both real.

(a) Derive the relation between C and α imposed by the normalization condition $\langle\psi\,|\psi\rangle = 1$.

(b) For the state $\rho = |\psi\rangle\,\langle\psi|$ calculate the probability for observing n photons, and decide whether the state is classical.

5.10 Gaussian states*

Apply the general relation (5.182) to the expression (5.225) for the Wigner characteristic function of a Gaussian state to show that

$$\begin{aligned}P_G(\alpha) = \frac{1}{\pi^2}\int d^2\zeta\; \exp\left[\zeta^*\beta - \zeta\beta^*\right]\exp\left[\left|\cosh r\;\zeta + \sinh r\;\zeta^*\right|^2/2\right] \\ \times\exp\left[-\left(n_G + 1/2\right)|\zeta|^2\right],\end{aligned}$$

where β is given by eqn (5.228). Evaluate the integral to get eqn (5.230).

6

Entangled states

The importance of the quantum phenomenon known as entanglement first became clear in the context of the famous paper by Einstein, Podolsky, and Rosen (EPR) (Einstein *et al.*, 1935), which presented an apparent paradox lying at the foundations of quantum theory. The EPR paradox has been the subject of continuous discussion ever since. In the same year as the EPR paper, Schrödinger responded with several publications (Schrödinger, 1935*a*, 1935*b*[1]) in which he pointed out that the essential feature required for the appearance of the EPR paradox is the application of the all-important superposition principle to the wave functions describing two or more particles that had previously interacted. In these papers Schrödinger coined the name 'entangled states' for the physical situations described by this class of wave functions.

In recent times it has become clear that the importance of this phenomenon extends well beyond esoteric questions about the meaning of quantum theory; indeed, entanglement plays a central role in the modern approach to quantum information processing. The argument for the EPR paradox—which will be presented in Chapter 19—is based on the properties of the EPR states discussed in the following section. After this, we will outline Schrödinger's concept of entanglement, and then continue with a more detailed treatment of the technical issues required for later applications.

6.1 Einstein–Podolsky–Rosen states

As part of an argument intended to show that quantum theory cannot be a complete description of physical reality, Einstein, Podolsky, and Rosen considered two distinguishable spinless particles A and B—constrained to move in a one-dimensional position space—that are initially separated by a distance L and then fly apart like the decay products of a radioactive nucleus. The particular initial state they used is a member of the general family of **EPR states** described by the two-particle wave functions

$$\psi\left(x_A, x_B\right) = \int_{-\infty}^{\infty} \frac{dk}{2\pi} F\left(k\right)\, e^{ik(x_A - x_B)}\,. \tag{6.1}$$

Every function of this form is an eigenstate of the total momentum operator with eigenvalue zero, i.e.

$$\left(\widehat{p}_A + \widehat{p}_B\right)\psi\left(x_A, x_B\right) = 0\,. \tag{6.2}$$

Peculiar phenomena associated with this state appear when we consider a measurement of one of the momenta, say $\widehat{p}_A$. If the result is $\hbar k_0$, then von Neumann's projection

[1]An English translation of this paper is given in Trimmer (1980).

postulate states that the wave function after the measurement is the projection of the initial wave function onto the eigenstate of $\widehat{p}_A$ associated with the eigenvalue $\hbar k_0$. Combining this rule with eqn (6.1) shows that the two-particle wave function after the measurement is reduced to

$$\psi_{\text{red}}(x_A, x_B) \propto F(k_0)\, e^{ik_0(x_A - x_B)} . \tag{6.3}$$

The reduced state is an eigenstate of $\widehat{p}_B$ with eigenvalue $-\hbar k_0$. Since $\widehat{p}_A$ and $\widehat{p}_B$ are constants of the motion for free particles, a measurement of $\widehat{p}_B$ at a later time will always yield the value $-\hbar k_0$. Thus the particular value found in the measurement of $\widehat{p}_A$ uniquely determines the value that would be found in any subsequent measurement of $\widehat{p}_B$.

The true strangeness of this situation appears when we consider the timing of the measurements. Suppose that the first measurement occurs at t_A and the second at $t_B > t_A$. It is remarkable that the prediction of the value $-\hbar k_0$ for the second measurement holds even if $(t_B - t_A) < L/c$. In other words, the result of the measurement of $\widehat{p}_B$ appears to be determined by the measurement of $\widehat{p}_A$ even though the news of the first measurement result could not have reached the position of particle B at the time of the second measurement. This *spooky action-at-a-distance*—which we will study in Chapter 19—was part of the basis for Einstein's conclusion that quantum mechanics is an incomplete theory.

6.2 Schrödinger's concept of entangled states

In order to understand Schrödinger's argument, we first observe that a product wave function,

$$\phi(x_A, x_B) = \eta(x_A)\, \xi(x_B) , \tag{6.4}$$

does not have the peculiar properties of the EPR wave function $\psi(x_A, x_B)$. The joint probability that the position of A is within dx_A of x_{A0} and that the position of B is within dx_B of x_{B0} is the product

$$dp(x_{A0}, x_{B0}) = |\eta(x_{A0})|^2\, dx_A\, |\xi(x_{B0})|^2\, dx_B \tag{6.5}$$

of the individual probabilities, so the positions can be regarded as stochastically independent random variables. The same argument can be applied to the momentum-space wave functions. The joint probability that measurements of $\widehat{p}_A/\hbar$ and $\widehat{p}_B/\hbar$ yield values in the neighborhood dk_A of k_{A0} and dk_B of k_{B0} is the product

$$dp(k_{A0}, k_{B0}) = |\eta(k_{A0})|^2\, dk_A\, |\xi(k_{B0})|^2\, dk_B \tag{6.6}$$

of independent probabilities, analogous to independent coin tosses. Thus a measurement of $\widehat{x}_A$ tells us nothing about the values that may be found in a measurement of $\widehat{x}_B$, and the same holds true for the momentum operators $\widehat{p}_A$ and $\widehat{p}_B$.

One possible response to the conceptual difficulties presented by the EPR states would be to declare them unphysical, but this tactic would violate the superposition principle: *every linear combination of product wave functions also describes a physically possible situation for the two-particle system.* Furthermore, any interaction between the particles will typically cause the wave function for a two-particle system—even if it is initially described by a product function like $\phi(x_A, x_B)$—to evolve into a superposition of product wave functions that is nonfactorizable. Schrödinger called these superpositions entangled states. An example is given by the EPR wave function $\psi(x_A, x_B)$ which is a linear combination of products of plane waves for the two particles. The choice of the name 'entangled' for these states is related to the classical **principle of separability**:

> Complete knowledge of the state of a compound system yields complete knowledge of the individual states of the parts.

This general principle does not require that the constituent parts be spatially separated; however, experimental situations in which there is spatial separation between the parts provide the most striking examples of the failure of classical separability. A classical version of the EPR thought experiment provides a simple demonstration of this principle. We now suppose that the two particles are described by the classical coordinates and momenta (q_A, p_A) and (q_B, p_B), so that the composite system is represented by the four-dimensional phase space (q_A, p_A, q_B, p_B). In classical physics the coordinates and momenta have definite numerical values, so a state of maximum possible information for the two-particle system is a point $(q_{A0}, p_{A0}, q_{B0}, p_{B0})$ in the two-particle phase space. This automatically provides the points (q_{A0}, p_{A0}) and (q_{B0}, p_{B0}) in the individual phase spaces; therefore, the maximum information state for the composite system determines maximum information states for the individual parts. The same argument evidently works for systems with any finite number of degrees of freedom.

In quantum theory, the uncertainty principle implies that the maximum possible information for a physical system is given by a single wave function, rather than a point in phase space. This does not mean, however, that classical separability is necessarily violated. The product function $\phi(x_A, x_B)$ is an example of a maximal information state of the two-particle system, for which the individual wave functions in the product are also maximal information states for the parts. Thus the product function satisfies the classical notion of separability. By contrast, the EPR wave function $\psi(x_A, x_B)$ is another maximal information state, but the individual particles are not described by unique wave functions. Consequently, for an entangled two-particle state we do not possess the maximum possible information for the individual particles; or in Schrödinger's words (Schrödinger, 1935*b*):

> Maximal knowledge of a total system does not necessarily include total knowledge of all its parts, not even when these are fully separated from each other and at the moment are not influencing each other at all.

6.3 Extensions of the notion of entanglement

The EPR states describe two distinguishable particles, e.g. an electron and a proton from an ionized hydrogen atom. Most of the work in the field of quantum information processing has also concentrated on the case of distinguishable particles. We will see later on that particles that are indistinguishable, e.g. two electrons, can be effectively distinguishable under the right conditions; however, it is not always useful—or even possible—to restrict attention to these special circumstances. This has led to a considerable amount of recent work on the meaning of entanglement for indistinguishable particles.

In the present section, we will develop two pieces of theoretical machinery that are needed for the subsequent discussion: the concept of tensor product spaces and the Schmidt decomposition. In the following sections, we will give a definition of entanglement for the general case of two distinguishable quantum objects, and then extend this definition to indistinguishable particles and to the electromagnetic field.

6.3.1 Tensor product spaces

In Section 4.2.1, the Hilbert space $\mathfrak{H}_{\text{QED}}$ for quantum electrodynamics was constructed as the tensor product of the Hilbert space $\mathfrak{H}_{\text{chg}}$ for the atoms and the Fock space $\mathfrak{H}_F$ for the field. This construction only depends on the Born interpretation and the superposition principle; consequently, it works equally well for any pair of distinguishable physical systems A and B described by Hilbert spaces $\mathfrak{H}_A$ and $\mathfrak{H}_B$. Let $\{|\phi_\alpha\rangle\}$ and $\{|\eta_\beta\rangle\}$ be basis sets for $\mathfrak{H}_A$ and $\mathfrak{H}_B$ respectively, then for any pair of vectors $(|\psi\rangle_A, |\vartheta\rangle_B)$ the product vector $|\Lambda\rangle = |\psi\rangle_A |\vartheta\rangle_B$ is defined by the probability amplitudes

$$\langle \phi_\alpha, \eta_\beta | \Lambda \rangle = \langle \phi_\alpha | \psi \rangle \langle \eta_\beta | \vartheta \rangle . \tag{6.7}$$

Since $\{|\phi_\alpha\rangle\}$ and $\{|\eta_\beta\rangle\}$ are complete orthonormal sets of vectors in their respective spaces, the inner product between two such vectors is consistently defined by

$$\begin{aligned}
\langle \Lambda_1 | \Lambda_2 \rangle &= \sum_{\alpha\beta} \langle \Lambda_1 | \phi_\alpha, \eta_\beta \rangle \langle \phi_\alpha, \eta_\beta | \Lambda_2 \rangle \\
&= \sum_{\alpha\beta} \langle \psi_1 | \phi_\alpha \rangle \langle \vartheta_1 | \eta_\beta \rangle \langle \phi_\alpha | \psi_2 \rangle \langle \eta_\beta | \vartheta_2 \rangle \\
&= \langle \psi_1 | \psi_2 \rangle \langle \vartheta_1 | \vartheta_2 \rangle ,
\end{aligned} \tag{6.8}$$

where the inner products $\langle \psi_1 | \psi_2 \rangle$ and $\langle \vartheta_1 | \vartheta_2 \rangle$ refer respectively to $\mathfrak{H}_A$ and $\mathfrak{H}_B$. The linear combination of two product vectors is defined by component-wise addition, i.e. the ket

$$|\Phi\rangle = c_1 |\Lambda_1\rangle + c_2 |\Lambda_2\rangle \tag{6.9}$$

is defined by the probability amplitudes

$$\begin{aligned}
\langle \phi_\alpha, \eta_\beta | \Phi \rangle &= c_1 \langle \phi_\alpha, \eta_\beta | \Lambda_1 \rangle + c_2 \langle \phi_\alpha, \eta_\beta | \Lambda_2 \rangle \\
&= c_1 \langle \phi_\alpha | \psi_1 \rangle \langle \eta_\beta | \vartheta_1 \rangle + c_2 \langle \phi_\alpha | \psi_2 \rangle \langle \eta_\beta | \vartheta_2 \rangle .
\end{aligned} \tag{6.10}$$

The tensor product space $\mathfrak{H}_C = \mathfrak{H}_A \otimes \mathfrak{H}_B$ is the family of all linear combinations of product kets. The family of product kets,

$$\{|\chi_{\alpha\beta}\rangle = |\phi_\alpha, \eta_\beta\rangle = |\phi_\alpha\rangle_A |\eta_\beta\rangle_B\} , \tag{6.11}$$

forms a complete orthonormal set with respect to the inner product (6.8), i.e.

$$\langle \chi_{\alpha'\beta'} \mid \chi_{\alpha\beta} \rangle = \langle \phi_{\alpha'} \mid \phi_\alpha \rangle \langle \eta_{\beta'} | \eta_\beta \rangle = \delta_{\alpha\alpha'} \delta_{\beta\beta'} , \tag{6.12}$$

and a general vector $|\Phi\rangle$ in $\mathfrak{H}_C$ can be expressed as

$$|\Phi\rangle = \sum_{\alpha}\sum_{\beta} \Phi_{\alpha\beta} |\chi_{\alpha\beta}\rangle = \sum_{\alpha}\sum_{\beta} \Phi_{\alpha\beta} |\phi_{\alpha}\rangle_A |\eta_{\beta}\rangle_B . \tag{6.13}$$

The inner product between any two vectors is

$$\langle \Psi | \Phi \rangle = \sum_{\alpha}\sum_{\beta} \Psi^*_{\alpha\beta}\Phi_{\alpha\beta} . \tag{6.14}$$

One can show that choosing new basis sets in $\mathfrak{H}_A$ and $\mathfrak{H}_B$ produces an equivalent basis set for $\mathfrak{H}_C$. This notion can be extended to composite systems composed of N distinguishable subsystems described by Hilbert spaces $\mathfrak{H}_1, \ldots, \mathfrak{H}_N$. The composite system is described by the N-fold tensor product space

$$\mathfrak{H}_C = \mathfrak{H}_1 \otimes \cdots \otimes \mathfrak{H}_N , \tag{6.15}$$

which is defined by repeated use of the two-space definition given above.

It is useful to extend the tensor product construction for vectors to a similar one for operators. Let $\mathfrak{A}$ and $\mathfrak{B}$ be operators acting on $\mathfrak{H}_A$ and $\mathfrak{H}_B$ respectively, then the **operator tensor product**, $\mathfrak{A} \otimes \mathfrak{B}$, is the operator acting on $\mathfrak{H}_C$ defined by

$$(\mathfrak{A} \otimes \mathfrak{B}) |\Phi\rangle = \sum_{\alpha}\sum_{\beta} \Phi_{\alpha\beta}\, \mathfrak{A} |\phi_{\alpha}\rangle_A \mathfrak{B} |\eta_{\beta}\rangle_B . \tag{6.16}$$

This definition immediately yields the rule

$$(\mathfrak{A}_1 \otimes \mathfrak{B}_1)(\mathfrak{A}_2 \otimes \mathfrak{B}_2) = (\mathfrak{A}_1\mathfrak{A}_2) \otimes (\mathfrak{B}_1\mathfrak{B}_2) \tag{6.17}$$

for the product of two such operators. Since the notion of the outer or tensor product of matrices and operators is less familiar than the idea of product wave functions, we sometimes use the explicit $\otimes$ notation for operator tensor products when it is needed for clarity. The definition (6.16) also allows us to treat $\mathfrak{A}$ and $\mathfrak{B}$ as operators acting on the product space $\mathfrak{H}_C$ by means of the identifications

$$\begin{aligned} \mathfrak{A} &\leftrightarrow \mathfrak{A} \otimes I_B , \\ \mathfrak{B} &\leftrightarrow I_A \otimes \mathfrak{B} , \end{aligned} \tag{6.18}$$

where I_A and I_B are respectively the identity operators for $\mathfrak{H}_A$ and $\mathfrak{H}_B$. These relations lead to the rule

$$\mathfrak{A}\mathfrak{B} \leftrightarrow \mathfrak{A} \otimes \mathfrak{B} , \tag{6.19}$$

so we can use either notation as dictated by convenience.

As explained in Section 2.3.2, a mixed state of the composite system is described by a density operator

$$\rho = \sum_{e} \mathcal{P}_e |\Psi_e\rangle \langle \Psi_e| , \tag{6.20}$$

where $\mathcal{P}_e$ is a probability distribution on the ensemble $\{|\Psi_e\rangle\}$ of pure states. The expectation values of observables for the subsystem A are determined by the **reduced density operator**

$$\rho_A = \mathrm{Tr}_B\,(\rho)\,, \tag{6.21}$$

where the **partial trace** over $\mathfrak{H}_B$ of a general operator X acting on $\mathfrak{H}_C$ is the operator on $\mathfrak{H}_A$ with matrix elements

$$\langle \phi_{\alpha'} \left|\mathrm{Tr}_B\,(X)\right| \phi_\alpha\rangle = \sum_\beta \langle \chi_{\alpha'\beta} \left|X\right| \chi_{\alpha\beta}\rangle\,. \tag{6.22}$$

This can be expressed more explicitly by using the fact that every operator on $\mathfrak{H}_C$ can be decomposed into a sum of operator tensor products, i.e.

$$X = \sum_n \mathfrak{A}_n \otimes \mathfrak{B}_n\,. \tag{6.23}$$

Substituting this into the definition (6.22) defines the operator

$$\mathrm{Tr}_B\,(X) = \sum_n \mathfrak{A}_n\, \mathrm{Tr}_B\,(\mathfrak{B}_n) \tag{6.24}$$

acting on $\mathfrak{H}_A$, where the c-number

$$\mathrm{Tr}_B\,(\mathfrak{B}_n) = \sum_\beta \langle \eta_\beta \left|\mathfrak{B}_n\right| \eta_\beta\rangle \tag{6.25}$$

is the trace over $\mathfrak{H}_B$. The average of an observable $\mathfrak{A}$ for the subsystem A is thus given by

$$\mathrm{Tr}\,(\rho\mathfrak{A}) = \mathrm{Tr}_A\,(\rho_A\mathfrak{A})\,. \tag{6.26}$$

In the same way the average of an observable $\mathfrak{B}$ for the subsystem B is

$$\mathrm{Tr}\,(\rho\mathfrak{B}) = \mathrm{Tr}_B\,(\rho_B\mathfrak{B})\,, \tag{6.27}$$

where

$$\rho_B = \mathrm{Tr}_A\,(\rho)\,. \tag{6.28}$$

6.3.2 The Schmidt decomposition

For finite-dimensional spaces, the general expansion (6.13) becomes

$$|\Psi\rangle = \sum_{\alpha=1}^{d_A} \sum_{\beta=1}^{d_B} \Psi_{\alpha\beta} \left|\chi_{\alpha\beta}\right\rangle, \tag{6.29}$$

where $\Psi_{\alpha\beta} = \langle \chi_{\alpha\beta} \left|\Psi\right\rangle$. In the study of entanglement, it is useful to have an alternative representation that is specifically tailored to a particular state vector $|\Psi\rangle$.

For our immediate purposes it is sufficient to explain the geometrical concepts leading to this special expansion; the technical details of the proof are given in Section 6.3.3. The basic idea is illustrated in Fig. 6.1, which shows the original vector, $|\Psi\rangle$, and the normalized product vector, $|\zeta_1\rangle_A\, |\vartheta_1\rangle_B$, that has the largest projection Y_1 onto $|\Psi\rangle$.

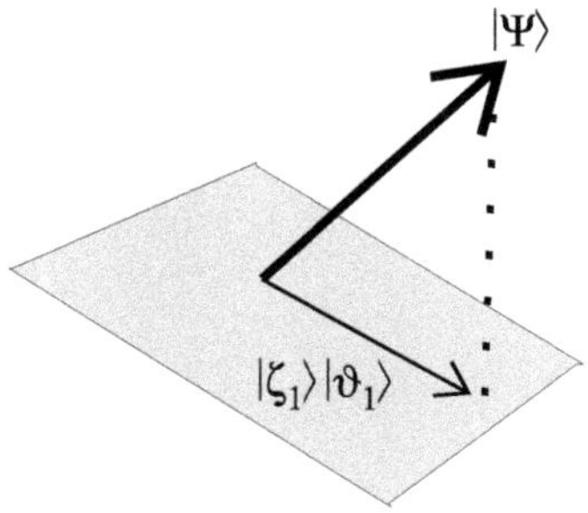

Fig. 6.1 A qualitative sketch of the procedure for deriving the Schmidt decomposition, given by eqn (6.30). The heavy arrow represents the original vector $|\Psi\rangle$ and the plane represents the set of all product vectors $|\zeta\rangle\,|\vartheta\rangle$. The light arrow denotes the projection of $|\Psi\rangle$ onto the plane.

After determining this first product vector, we define a new vector, $|\Psi_1\rangle = |\Psi\rangle - Y_1\,|\zeta_1\rangle_A\,|\vartheta_1\rangle_B$, that is orthogonal to $|\zeta_1\rangle_A\,|\vartheta_1\rangle_B$. The same game can be played with $|\Psi_1\rangle$; that is, we find the normalized product vector $|\zeta_2\rangle_A\,|\vartheta_2\rangle_B$ that has the maximum projection Y_2 onto $|\Psi_1\rangle$ and is orthogonal to $|\zeta_1\rangle_A\,|\vartheta_1\rangle_B$. Since the spaces $\mathfrak{H}_A$ and $\mathfrak{H}_B$ are finite dimensional, this process must terminate after a finite number r of steps, i.e. when $Y_{r+1} = 0$. The orthogonality of the successive product vectors implies that they are linearly independent; therefore, the largest possible number of steps is the smaller of the two dimensions, $\min(d_A, d_B)$. The final result is the **Schmidt decomposition**

$$|\Psi\rangle = \sum_{n=1}^{r} Y_n\,|\zeta_n\rangle_A\,|\vartheta_n\rangle_B\,, \tag{6.30}$$

where the **Schmidt rank** $r \leqslant \min(d_A, d_B)$. The density operator for this pure state is therefore

$$\begin{aligned}\rho &= \sum_{m=1}^{r}\sum_{n=1}^{r} Y_m Y_n^*\,|\zeta_m\rangle_A\,|\vartheta_m\rangle_B\;{}_A\langle\zeta_n|\;{}_B\langle\vartheta_n| \\ &= \sum_{m=1}^{r}\sum_{n=1}^{r} Y_m Y_n^*\left(|\zeta_m\rangle_A\;{}_A\langle\zeta_n|\right)\otimes\left(|\vartheta_m\rangle_B\;{}_B\langle\vartheta_n|\right).\end{aligned} \tag{6.31}$$

The minimum value ($r = 1$) of the Schmidt rank occurs when $|\Psi\rangle$ is a product vector. The product vectors $|\zeta_n\rangle_A\,|\vartheta_n\rangle_B$ are orthonormal by construction, i.e. $\langle\zeta_n\,|\zeta_m\rangle = \langle\vartheta_n\,|\vartheta_m\rangle = \delta_{nm}$, and the coefficients Y_n satisfy the normalization condition

$$\sum_{n=1}^{r} |Y_n|^2 = 1\,. \tag{6.32}$$

In applications of the Schmidt decomposition (6.30), it is important to keep in mind that the basis vectors $|\zeta_n\rangle_A\,|\vartheta_n\rangle_B$ themselves—and not just the coefficients Y_n—are uniquely associated with the vector $|\Psi\rangle$. The Schmidt decomposition for a new vector $|\Phi\rangle$ would require a new set of basis vectors.

6.3.3 Proof of the Schmidt decomposition*

We offer here a proof—modeled on one of the arguments given by Peres (1995, Sec. 5-3)—that the expansion (6.30) exists. For normalized vectors $|\zeta_1\rangle_A$ and $|\vartheta_1\rangle_B$: set

$|\zeta_1, \vartheta_1\rangle = |\zeta_1\rangle_A |\vartheta_1\rangle_B$, and consider the projection operator $P_1 = |\zeta_1, \vartheta_1\rangle \langle \zeta_1, \vartheta_1|$. The identity $|\Psi\rangle = P_1 |\Psi\rangle + (1 - P_1) |\Psi\rangle$ can then be written as $|\Psi\rangle = Y_1 |\zeta_1, \vartheta_1\rangle + |\Psi_1\rangle$, where $Y_1 = \langle \zeta_1, \vartheta_1 | \Psi \rangle$ and the vector $|\Psi_1\rangle = (1 - P_1) |\Psi\rangle$ is orthogonal to $|\zeta_1, \vartheta_1\rangle$. By applying the general expansion (6.29) to the vectors $|\Psi\rangle$ and $|\zeta_1, \vartheta_1\rangle$, one can express $|Y_1|^2$ as

$$|Y_1|^2 = \left| \sum_{\alpha=1}^{d_A} \sum_{\beta=1}^{d_B} \Psi^*_{\alpha\beta} x_\alpha y_\beta \right|^2 \leqslant 1 , \tag{6.33}$$

where $x_\alpha = \langle \phi_\alpha | \zeta_1 \rangle$, $y_\beta = \langle \eta_\beta | \vartheta_1 \rangle$, and the upper bound follows from the normalization of the vectors defining Y_1.

From a geometrical point of view, $|Y_1|$ is the magnitude of the projection of $|\zeta_1, \vartheta_1\rangle$ onto $|\Psi\rangle$. In quantum terms, $|Y_1|^2$ is the probability that a measurement of P_1 will result in the eigenvalue unity and will leave the system in the state $|\zeta_1, \vartheta_1\rangle$. The next step is to choose the product vector $|\zeta_1, \vartheta_1\rangle$—i.e. to find values of x_α and y_β—that maximizes $|Y_1|^2$. This is always possible, since $|Y_1|^2$ is a bounded, continuous function of the finite set of complex variables $(x_1, \ldots, x_{d_A}, y_1, \ldots, y_{d_B})$. The solution is not unique, since the overall phase of $|\zeta_1, \vartheta_1\rangle$ is not determined by the maximization procedure. This is not a real difficulty; the undetermined phases can be chosen so that Y_1 is real. In general, there may be several linearly independent solutions for $|\zeta_1, \vartheta_1\rangle$, but this is also not a serious difficulty. By forming appropriate linear combinations of the degenerate solutions it is always possible to make them mutually orthogonal. We will therefore simplify the discussion by assuming that the maximum is always unique. Note that the maximum value of $|Y_1|^2$ can only be unity if the original vector is itself a product vector.

Now that we have made our choice of $|\zeta_1, \vartheta_1\rangle$, we pick a new product vector $|\zeta_2, \vartheta_2\rangle$—with projection operator $P_2 = |\zeta_2, \vartheta_2\rangle \langle \zeta_2, \vartheta_2|$—and write the identity $|\Psi_1\rangle = P_2 |\Psi_1\rangle + (1 - P_2) |\Psi_1\rangle$ as

$$|\Psi_1\rangle = Y_2 |\zeta_2, \vartheta_2\rangle + |\Psi_2\rangle , \tag{6.34}$$

where $Y_2 = \langle \zeta_2, \vartheta_2 | \Psi_1 \rangle$ and $|\Psi_2\rangle = (1 - P_2) |\Psi_1\rangle$. Since $|\Psi_1\rangle$ is orthogonal to $|\zeta_1, \vartheta_1\rangle$, we can assume that $|\zeta_2, \vartheta_2\rangle$ is also orthogonal to $|\zeta_1, \vartheta_1\rangle$. Now we proceed, as in the first step, by choosing $|\zeta_2, \vartheta_2\rangle$ to maximize $|Y_2|^2$. At this point, we have

$$|\Psi\rangle = Y_1 |\zeta_1, \vartheta_1\rangle + Y_2 |\zeta_2, \vartheta_2\rangle + |\Psi_2\rangle , \tag{6.35}$$

and this procedure can be repeated until the next projection vanishes. The last remark implies that the number of terms is limited by the minimum dimensionality, $\min(d_A, d_B)$; therefore, we arrive at eqn (6.30).

6.4 Entanglement for distinguishable particles

In Section 6.3.1 we saw that the Hilbert space for a composite system formed from any two distinguishable subsystems A and B (which can be atoms, molecules, quantum dots, etc.) is the tensor product $\mathfrak{H}_C = \mathfrak{H}_A \otimes \mathfrak{H}_B$. The current intense interest in quantum information processing has led to the widespread use of the terms *parties* for A and B, and **bipartite system**, for what has traditionally been called a two-particle system. Since our interests in this book are not limited to quantum information processing, we will adhere to the traditional terminology in which the distinguishable objects A and B are called *particles* and the composite system is called a *two-particle* or *two-part* system.

In order to simplify the discussion, we will assume that the two Hilbert spaces have finite dimensions, $d_A, d_B < \infty$. A composite system composed of two distinguishable, spin-1/2 particles—for example, impurity atoms bound to adjacent sites in a crystal lattice—provides a simple example that fits within this framework. In this case, $\mathfrak{H}_A = \mathfrak{H}_B = \mathbb{C}^2$, and all observables can be written as linear combinations of the spin operators, e.g.

$$O^A = C_0 I^A + C_1 \mathbf{n} \cdot \mathbf{S}^A \,, \tag{6.36}$$

where C_0 and C_1 are constants, I^A is the identity operator, $\mathbf{n}$ is a unit vector, $\mathbf{S}^A = \boldsymbol{\sigma}^A/2$, and $\boldsymbol{\sigma} = (\sigma_x, \sigma_y, \sigma_z)$ is the vector of Pauli matrices. A discrete analogue of the EPR wave function is given by the **singlet state**

$$|S=0\rangle_{AB} = \frac{1}{\sqrt{2}} \left\{ |\uparrow\rangle_A |\downarrow\rangle_B - |\downarrow\rangle_A |\uparrow\rangle_B \right\}, \tag{6.37}$$

where the **spin-up** and **spin-down** states are defined by

$$\mathbf{n} \cdot \mathbf{S}^A |\uparrow\rangle_A = +\frac{1}{2} |\uparrow\rangle_A \,, \quad \mathbf{n} \cdot \mathbf{S}^A |\downarrow\rangle_A = -\frac{1}{2} |\downarrow\rangle_A \,, \quad \text{etc.} \tag{6.38}$$

The singlet state has total spin angular momentum zero, so one can show—as in Exercise 6.3—that it has the same expression for every choice of $\mathbf{n}$. If several spin-projections are under consideration, the notation $|\uparrow_{\mathbf{n}}\rangle_A$ and $|\downarrow_{\mathbf{n}}\rangle_A$ can be used to distinguish them.

The most important feature of entanglement for pure states is that the result of one measurement yields information about the probability distribution of a second, independent measurement. For the two-spin system, a measurement of $\mathbf{n} \cdot \mathbf{S}^A$ with the result $\pm 1/2$ guarantees that a subsequent measurement of $\mathbf{n} \cdot \mathbf{S}^B$ will yield the result $\mp 1/2$. A discrete version of the unentangled (separable) state (6.4) is

$$|\phi\rangle = \left\{ c_\uparrow \; |\uparrow\rangle_A + c_\downarrow \; |\downarrow\rangle_A \right\} \left\{ b_\uparrow \; |\uparrow\rangle_B + b_\downarrow \; |\downarrow\rangle_B \right\}. \tag{6.39}$$

In this case, measuring $\mathbf{n} \cdot \mathbf{S}^A$ provides no information at all on the distribution of values for $\mathbf{n} \cdot \mathbf{S}^B$.

6.4.1 Definition of entanglement

We will approach the general idea of entanglement indirectly by first defining separable (unentangled) pure and mixed states, and then defining entangled states as those that are not separable. Since entangled states are the focus of this chapter, this negative procedure may seem a little strange. The explanation is that separable states are simple and entangled states are complicated. We will define separability and entanglement in terms of properties of the state vector or density operator. This is the traditional approach, and it provides a quick entry into the applications of these notions.

A Pure states

The definitions we give here are simply generalizations of the examples presented in Sections 6.1 and 6.2, or rather the finite-dimensional analogues given by eqns (6.37) and (6.39). Thus we say that a pure state $|\Psi\rangle$ of the two-particle system described by the Hilbert space $\mathfrak{H}_C = \mathfrak{H}_A \otimes \mathfrak{H}_B$ is **separable** if it can be expressed as

$$|\Psi\rangle = |\Phi\rangle_A |\Xi\rangle_B \,, \tag{6.40}$$

which is the general version of eqn (6.39), and **entangled** if it is not separable. This awkward negative definition of entanglement as the absence of separability can be avoided by using the Schmidt decomposition (6.30). A little thought shows that the states that cannot be written in the form (6.40) are just the states with $r > 1$. With this in mind, we could define entanglement positively by saying that $|\Psi\rangle$ is entangled if it has Schmidt rank $r > 1$. The discrete analogue (6.37) of the continuous EPR wave function is an example of an entangled state.

The definitions given above imply several properties of the state vector which, conversely, imply the original definitions. Thus the new properties can be used as equivalent definitions of separability and entanglement for pure states. For ease of reference, we present these results as theorems.

Theorem 6.1 *A pure state is separable if and only if the reduced density operators represent pure states, i.e. separable states satisfy the classical separability principle.*

There are two assertions to be proved.

(a) The reduced density operators for a separable pure state $|\Psi\rangle$ represent pure states of A and B.

(b) If the reduced density operators for a pure state $|\Psi\rangle$ describe pure states of A and B, then $|\Psi\rangle$ is separable.

Suggestions for these arguments are given in Exercise 6.1.

Since entanglement is the absence of separability, this result can also be stated as follows.

Theorem 6.2 *A pure state is entangled if and only if the reduced density operators for the subsystems describe mixed states.*

Mixed states are, by definition, not states of maximum information, so this result explicitly demonstrates that possession of maximum information for the total system does not yield maximum information for the constituent parts. However, the statistical properties of the mixed states for the subsystems are closely related. This can be seen by using the Schmidt decomposition (6.31) to evaluate the reduced density operators:

$$\rho_A = \mathrm{Tr}_B(\rho) = \sum_{m=1}^{r} |Y_m|^2 \left(|\zeta_m\rangle \langle\zeta_m|\right) \tag{6.41}$$

and

$$\rho_B = \mathrm{Tr}_A(\rho) = \sum_{m=1}^{r} |Y_m|^2 \left(|\vartheta_m\rangle \langle\vartheta_m|\right). \tag{6.42}$$

Comparing eqns (6.41) and (6.42) shows that the two reduced density operators—although they act in different Hilbert spaces—have the same set of nonzero eigenvalues $\left\{|Y_1|^2, \ldots, |Y_r|^2\right\}$. This implies that the purities of the two reduced states agree,

$$\mathfrak{P}(\rho_A) = \mathrm{Tr}_A\left(\rho_A^2\right) = \sum_{m=1}^{r} |Y_m|^4 = \mathfrak{P}(\rho_B) < 1, \tag{6.43}$$

and that the subsystems have identical von Neumann entropies,

$$\mathfrak{S}(\rho_A) = -\mathrm{Tr}_A\left[\rho_A \ln \rho_A\right] = -\sum_{m=1}^{r} |Y_m|^2 \ln |Y_m|^2 = \mathfrak{S}(\rho_B). \tag{6.44}$$

An entangled pure state is said to be **maximally entangled** if the reduced density operators are maximally mixed according to eqn (2.141), where the number of degenerate nonzero eigenvalues is given by $M = r$. The corresponding values of the purity and von Neumann entropy are respectively $\mathfrak{P}(\rho) = 1/r$ and $\mathfrak{S}(\rho) = \ln r$.

We next turn to results that are more directly related to experiment. For observables $\mathfrak{A}$ and $\mathfrak{B}$ acting on $\mathfrak{H}_A$ and $\mathfrak{H}_B$ respectively and any state $|\Psi\rangle$ in $\mathfrak{H}_C = \mathfrak{H}_A \otimes \mathfrak{H}_B$, we define the averages $\langle\mathfrak{A}\rangle = \langle\Psi|\mathfrak{A}\otimes I_B|\Psi\rangle$ and $\langle\mathfrak{B}\rangle = \langle\Psi|I_A\otimes\mathfrak{B}|\Psi\rangle$ and the **fluctuation operators** $\delta\mathfrak{A} = \mathfrak{A} - \langle\mathfrak{A}\rangle$ and $\delta\mathfrak{B} = \mathfrak{B} - \langle\mathfrak{B}\rangle$. The quantum fluctuations are said to be *uncorrelated* if $\langle\Psi|\delta\mathfrak{A}\,\delta\mathfrak{B}|\Psi\rangle = 0$. With this preparation we can state the following.

Theorem 6.3 *A pure state is separable if and only if the quantum fluctuations of all observables $\mathfrak{A}$ and $\mathfrak{B}$ are uncorrelated.*

See Exercise 6.2 for a suggested proof. Combining this result with the fact that entangled states are not separable leads easily to the following theorem.

Theorem 6.4 *A pure state $|\Psi\rangle$ is entangled if and only if there is at least one pair of observables $\mathfrak{A}$ and $\mathfrak{B}$ with correlated quantum fluctuations.*

Thus the observation of correlations between measured values of $\mathfrak{A}$ and $\mathfrak{B}$ is experimental evidence that the pure state $|\Psi\rangle$ is entangled.

B Mixed states

Since the density operator ρ is simply a convenient description of a probability distribution $\mathcal{P}_e$ over an ensemble, $\{|\Psi_e\rangle\}$, of normalized pure states, the analysis of entanglement for mixed states is based on the previous discussion of entanglement for pure states.

From this point of view, it is natural to define a **separable mixed state** by an ensemble of separable pure states, i.e. $|\Psi_e\rangle = |\zeta_e\rangle |\vartheta_e\rangle$ for all e. The density operator for a separable mixed state is consequently given by a convex linear combination,

$$\rho = \sum_e \mathcal{P}_e |\zeta_e\rangle_A |\vartheta_e\rangle_B \; {}_A\langle\zeta_e| \; {}_B\langle\vartheta_e| , \tag{6.45}$$

of density operators for separable pure states. By writing this in the equivalent form

$$\rho = \sum_e \mathcal{P}_e \left(|\zeta_e\rangle_A \; {}_A\langle\zeta_e|\right) \otimes \left(|\vartheta_e\rangle_B \; {}_B\langle\vartheta_e|\right) , \tag{6.46}$$

we find that the reduced density operators are

$$\rho_A = \mathrm{Tr}_B(\rho) = \sum_e \mathcal{P}_e |\zeta_e\rangle \langle\zeta_e| \tag{6.47}$$

and

$$\rho_B = \mathrm{Tr}_A(\rho) = \sum_e \mathcal{P}_e |\vartheta_e\rangle \langle\vartheta_e| . \tag{6.48}$$

In the special case that both sets of vectors are orthonormal, i.e.

$$\langle\zeta_e | \zeta_f\rangle = \langle\vartheta_e | \vartheta_f\rangle = \delta_{ef} , \tag{6.49}$$

the reduced density operators have the same spectra, so that—just as in the discussion following Theorem 6.2—the two subsystems have the same purity and von Neumann entropy. In the general case that one or both sets of vectors are not orthonormal, the statistical properties can be quite different. An **entangled mixed state** is one that is not separable, i.e. the ensemble contains at least one entangled pure state. Defining useful measures of the degree of entanglement of a mixed state is a difficult problem which is the subject of current research.

The clear experimental tests for separability and entanglement of pure states, presented in Theorems 6.3 and 6.4, are not available for mixed states. To see this, we begin by writing out the correlation function and the averages of the observables $\mathfrak{A}$ and $\mathfrak{B}$ as

$$\begin{aligned} C(\mathfrak{A}, \mathfrak{B}) &= \langle \delta\mathfrak{A}\, \delta\mathfrak{B}\rangle = \mathrm{Tr}\, \rho\delta\mathfrak{A}\, \delta\mathfrak{B} \\ &= \sum_e \mathcal{P}_e \langle\Psi_e | \delta\mathfrak{A}\, \delta\mathfrak{B} | \Psi_e\rangle , \end{aligned} \tag{6.50}$$

and

$$\langle\mathfrak{A}\rangle = \sum_e \mathcal{P}_e \langle\Psi_e | \mathfrak{A} | \Psi_e\rangle , \quad \langle\mathfrak{B}\rangle = \sum_e \mathcal{P}_e \langle\Psi_e | \mathfrak{B} | \Psi_e\rangle . \tag{6.51}$$

We will separate the quantum fluctuations in each pure state from the fluctuations associated with the classical probability distribution, $\mathcal{P}_e$, over the ensemble of pure states, by expressing the fluctuation operator $\delta\mathfrak{A}$ as

$$\delta\mathfrak{A} = \mathfrak{A} - \langle\mathfrak{A}\rangle = \mathfrak{A} - \langle\Psi_e | \mathfrak{A} | \Psi_e\rangle + \langle\Psi_e | \mathfrak{A} | \Psi_e\rangle - \langle\mathfrak{A}\rangle . \tag{6.52}$$

The operator

$$\delta_e \mathfrak{A} = \mathfrak{A} - \langle \Psi_e | \mathfrak{A} | \Psi_e \rangle \tag{6.53}$$

represents the quantum fluctuations of $\mathfrak{A}$ around the average defined by $|\Psi_e\rangle$, and the c-number

$$\delta \langle \mathfrak{A} \rangle_e = \langle \Psi_e | \mathfrak{A} | \Psi_e \rangle - \langle \mathfrak{A} \rangle \tag{6.54}$$

describes the classical fluctuations of the individual quantum averages $\langle \Psi_e | \mathfrak{A} | \Psi_e \rangle$ around the ensemble average $\langle \mathfrak{A} \rangle$. Using eqns (6.52)–(6.54), together with the analogous definitions for $\mathfrak{B}$, in eqn (6.50) leads to

$$C(\mathfrak{A}, \mathfrak{B}) = C_{\text{qu}}(\mathfrak{A}, \mathfrak{B}) + C_{\text{cl}}(\mathfrak{A}, \mathfrak{B}), \tag{6.55}$$

where

$$C_{\text{qu}}(\mathfrak{A}, \mathfrak{B}) = \sum_e \mathcal{P}_e \langle \Psi_e | \delta_e \mathfrak{A}\, \delta_e \mathfrak{B} | \Psi_e \rangle \tag{6.56}$$

represents the quantum part and

$$C_{\text{cl}}(\mathfrak{A}, \mathfrak{B}) = \sum_e \mathcal{P}_e \delta \langle \mathfrak{A} \rangle_e \delta \langle \mathfrak{B} \rangle_e \tag{6.57}$$

represents the classical part.

For a separable mixed state, the quantum correlation functions for each pure state vanish, so that

$$C(\mathfrak{A}, \mathfrak{B}) = C_{\text{cl}}(\mathfrak{A}, \mathfrak{B}) = \sum_e \mathcal{P}_e \delta \langle \mathfrak{A} \rangle_e \delta \langle \mathfrak{B} \rangle_e \,. \tag{6.58}$$

Thus the observables $\mathfrak{A}$ and $\mathfrak{B}$ are correlated in the mixed state, despite the fact that they are uncorrelated for each of the separable pure states. An explicit example of this peculiar situation is presented in Exercise 6.4. As a consequence of this fact, observing correlations between two observables cannot be taken as evidence of entanglement for a mixed state.

6.5 Entanglement for identical particles

6.5.1 Systems of identical particles

In this section, we will be concerned with particles having nonzero rest mass—e.g. electrons, ions, atoms, etc.—described by nonrelativistic quantum mechanics. In quantum theory, particles—as well as more complex systems—are said to be **indistinguishable** or **identical** if all of their intrinsic properties, e.g. mass, charge, spin, etc., are the same. In classical mechanics, this situation poses no special difficulties, since each particle's unique trajectory provides an identifying label, e.g. the position and momentum of the particle at some chosen time. In quantum mechanics, the uncertainty principle removes this possibility, and indistinguishability of particles has radically new consequences.[2]

[2]A more complete discussion of identical particles can be found in any of the excellent texts on quantum mechanics that are currently available, for example Cohen-Tannoudji *et al.* (1977*b*, Chap. XIV) or Bransden and Joachain (1989, Chap. 10).

For identical particles, we will replace the previous labeling A and B by $1, 2, \ldots, N$, for the general case of N identical particles. Since the particles are indistinguishable, the labels have no physical significance; they are merely a bookkeeping device. An N-particle state $|\Psi\rangle$ can be represented by a wave function

$$\Psi(1, 2, \ldots, N) = \langle 1, 2, \ldots, N | \Psi \rangle , \tag{6.59}$$

where the arguments $1, 2, \ldots, N$ stand for a full set of coordinates for each particle. For example, $1 = (\mathbf{r}_1, s_1)$, where $\mathbf{r}_1$ and s_1 are respectively eigenvalues of $\widehat{\mathbf{r}}_1$ and $\widehat{s}_{1z}$.

The permutations on the labels form the **symmetric group** S_N (Hamermesh, 1962, Chap. 7), with group multiplication defined by successive application of permutations. An element P in S_N is defined by its action: $1 \to P(1), 2 \to P(2), \ldots, N \to P(N)$. Each permutation P is represented by an operator Z_P defined by

$$\langle 1, 2, \ldots, N | Z_P | \Psi \rangle = \langle P(1), P(2), \ldots, P(N) | \Psi \rangle , \tag{6.60}$$

or in the more familiar wave function representation,

$$Z_P \Psi(1, 2, \ldots, N) = \Psi(P(1), P(2), \ldots, P(N)) . \tag{6.61}$$

It is easy to show that Z_P is both unitary and hermitian. A transposition is a permutation that interchanges two labels and leaves the rest alone, e.g. $P(1) = 2$, $P(2) = 1$, and $P(j) = j$ for all other values of j. Every permutation P can be expressed as a product of transpositions, and P is said to be even or odd if the number of transpositions is respectively even or odd. These definitions are equally applicable to distinguishable and indistinguishable particles.

One consequence of particle identity is that operators that act on only one of the particles, such as $\mathfrak{A}$ and $\mathfrak{B}$ in Theorems 6.3 and 6.4, are physically meaningless. All physically admissible observables must be unchanged by any permutation of the labels for the particles, i.e. the operator F representing a physically admissible observable must satisfy

$$(Z_P)^\dagger F Z_P = F . \tag{6.62}$$

Suppose, for example, that A is an operator acting in the Hilbert space $\mathfrak{H}^{(1)}$ of one-particle states; then for N particles the physically meaningful **one-particle operator** is

$$A = A(1) + A(2) + \cdots + A(N) , \tag{6.63}$$

where $A(j)$ acts on the coordinates of the particle with the label j.

The restrictions imposed on admissible state vectors by particle identity are a bit more subtle. For systems of identical particles, indistinguishability means that a physical state is unchanged by any permutation of the labels assigned to the particles. For a pure state, this implies that the state vector can at most change by a phase factor under permutation of the labels:

$$Z_P |\Psi\rangle = e^{i\xi_P} |\Psi\rangle . \tag{6.64}$$

By using the special properties of permutations, one can show that the only possibilities are $e^{i\xi_P} = 1$ or $e^{i\xi_P} = (-1)^P$, where $(-1)^P = +1\,(-1)$ for even (odd) permutations.[3] In other words, admissible state vectors must be either completely symmetric or completely antisymmetric under permutation of the particle labels. These two alternatives respectively define orthogonal subspaces $(\mathfrak{H}_C)_{\text{sym}}$ and $(\mathfrak{H}_C)_{\text{asym}}$ of the N-fold tensor product space $\mathfrak{H}_C = \mathfrak{H}^{(1)}(1) \otimes \cdots \otimes \mathfrak{H}^{(1)}(N)$. It is an empirical fact that all elementary particles belong to one of two classes: the **fermions**, described by the antisymmetric states in $(\mathfrak{H}_C)_{\text{asym}}$; and the **bosons**, described by the symmetric states in $(\mathfrak{H}_C)_{\text{sym}}$. As a consequence of the antisymmetry of the state vectors, two fermions cannot occupy the same single-particle state; however the symmetry of bosonic states allows any number of bosons to occupy a single-particle state. For large numbers of particles, these features lead to strikingly different statistical properties for fermions and bosons; the two kinds of particles are said to satisfy **Bose–Einstein** or **Fermi statistics**. This fact has many profound physical consequences, ranging from the Pauli exclusion principle to Bose–Einstein condensation.

In the following discussions, we will often be concerned with the special case of two identical particles. In this situation, a basis for the tensor product space $\mathfrak{H}^{(1)} \otimes \mathfrak{H}^{(1)}$ is provided by the family of product vectors $\{|\chi_{mn}\rangle = |\phi_m\rangle_1 |\phi_n\rangle_2\}$, where $\{|\phi_n\rangle\}$ is a basis for the single-particle space $\mathfrak{H}^{(1)}$. A general state $|\Psi\rangle$ in $\mathfrak{H}^{(1)} \otimes \mathfrak{H}^{(1)}$ can then be expressed as

$$|\Psi\rangle = \sum_m \sum_n \Psi_{mn} |\chi_{mn}\rangle\,, \tag{6.65}$$

where

$$\Psi_{mn} = \langle \chi_{mn} | \Psi \rangle\,. \tag{6.66}$$

The symmetric (bosonic) and antisymmetric (fermionic) subspaces are respectively characterized by the conditions

$$\Psi_{mn} = \Psi_{nm} \tag{6.67}$$

and

$$\Psi_{mn} = -\Psi_{nm}. \tag{6.68}$$

6.5.2 Effective distinguishability

There must be situations in which the indistinguishability of particles makes no difference. If this were not the case, explanations of electron scattering on the Earth would have to take into account the presence of electrons on the Moon. This would create rather serious problems for experimentalists and theorists alike. The key to avoiding this nightmare is the simple observation that experimental devices have a definite position in space and occupy a finite volume. As a concrete example, consider a measuring apparatus that occupies a volume V centered on the point $\mathbf{R}$. Another fact of life is that plane waves are an idealization. Physically meaningful wave functions are always normalizable; consequently, they are localized in some region of space. In many cases, the wave function falls off exponentially, e.g. like $\exp(-|\mathbf{r} - \mathbf{r}_0|/\Lambda)$, or

[3] This is generally true when the particle position space is three dimensional. For systems restricted to two dimensions, continuous values of ξ_P are possible. This leads to the notion of anyons, see for example Leinaas and Myrheim (1977).

$\exp(-|\mathbf{r}-\mathbf{r}_0|^2/\Lambda^2)$, where $\mathbf{r}_0$ is the center of the localization region. In either case, we will say that the wave function is *exponentially small* when $|\mathbf{r}-\mathbf{r}_0| \gg \Lambda$. With this preparation, we will say that an operator F—acting on single-particle wave functions in $\mathfrak{H}^{(1)}$—is a **local observable** in the region V if $F\eta_s(\mathbf{r})$ is exponentially small in V whenever the wave function $\eta_s(\mathbf{r})$ is itself exponentially small in V.

Let us now consider two indistinguishable particles occupying the states $|\phi\rangle$ and $|\eta\rangle$, where $|\phi\rangle$ is localized in the volume V and $|\eta\rangle$ is localized in some distant region—possibly the Moon or just the laboratory next door—so that $\eta_s(\mathbf{r}) = \langle \mathbf{r}s|\eta\rangle$ is exponentially small in V. The state vector for the two bosons or fermions has the form

$$|\Psi\rangle = \frac{1}{\sqrt{2}}\{|\phi\rangle_1 |\eta\rangle_2 \pm |\eta\rangle_1 |\phi\rangle_2\}, \tag{6.69}$$

and a one-particle observable is represented by an operator $F = F(1) + F(2)$. Let Z_{12} be the transposition operator, then $Z_{12}|\Psi\rangle = \pm|\Psi\rangle$ and $Z_{12}F(2)Z_{12} = F(1)$. With these facts in hand it is easy to see that

$$\begin{aligned}\langle\Psi|F|\Psi\rangle &= 2\langle\Psi|F(1)|\Psi\rangle \\ &= \langle\phi|F|\phi\rangle + \langle\eta|F|\eta\rangle \pm \langle\phi|F|\eta\rangle\langle\eta|\phi\rangle \pm \langle\eta|F|\phi\rangle\langle\phi|\eta\rangle. \end{aligned}\tag{6.70}$$

The final two terms in the last equation are negligible because of the small overlap between the one-particle states, but the term $\langle\eta|F|\eta\rangle$ is not small unless the operator F represents a local observable for V. When this is the case, the two-particle expectation value,

$$\langle\Psi|F|\Psi\rangle = \langle\phi|F|\phi\rangle, \tag{6.71}$$

is exactly what one would obtain by assuming that the two particles are distinguishable, and that a measurement is made on the one in V.

The lesson to be drawn from this calculation is that the indistinguishability of two particles can be ignored if the relevant single-particle states are effectively nonoverlapping and only local observables are measured. This does not mean that an electron on the Earth and one on the Moon are in any way different. What we have shown is that the large separation involved makes the indistinguishability of the two electrons irrelevant—for all practical purposes—when analyzing local experiments conducted on the Earth. On the other hand, the measurement of a local observable will be sensitive to the indistinguishability of the particles if the one-particle states have a significant overlap. Consider the situation in which the distant particle is bound to a potential well centered at $\mathbf{r}_0$. Bodily moving the potential well so that the original condition $|\mathbf{r}_0 - \mathbf{R}_A| \gg \Lambda$ is replaced by $|\mathbf{r}_0 - \mathbf{R}_A| \lesssim \Lambda$ restores the effects of indistinguishability.

6.5.3 Definition of entanglement

For identical particles, there are no physically meaningful operators that can single out one particle from the rest; consequently, there is no way to separate a system of two identical particles into distinct subsystems. How then are we to extend the definitions of separability and entanglement given in Section 6.4.1 to systems of identical particles? Since definitions cannot be right or wrong—only more or less useful—it should not be too surprising to learn that this question has been answered in at least two different ways. In the following paragraphs, we will give a traditional answer and compare it to another definition that is preferred by those working in the field of quantum information processing.

For single-particle states $|\zeta\rangle_1$ and $|\eta\rangle_2$, of distinguishable particles 1 and 2, the definition (6.40) tell us that the product vector

$$|\Psi\rangle = |\zeta\rangle_1 |\eta\rangle_2 \tag{6.72}$$

is separable, but if the particles are identical bosons then $|\Psi\rangle$ must be replaced by the symmetrized expression

$$|\Psi\rangle = C\left\{|\zeta\rangle_1 |\eta\rangle_2 + |\eta\rangle_1 |\zeta\rangle_2\right\}, \tag{6.73}$$

where C is a normalization constant. Unless $|\eta\rangle = |\zeta\rangle$, this has the form of an entangled state for distinguishable particles. The traditional approach is to impose the symmetry requirement on the definition of separability used for distinguishable particles; therefore, a state $|\Psi\rangle$ of two identical bosons is said to be **separable** if it can be expressed in the form

$$|\Psi\rangle = |\zeta\rangle_1 |\zeta\rangle_2 . \tag{6.74}$$

In other words, both bosons must occupy the same single-particle state.

It is often useful to employ the definition (6.66) of the expansion coefficients Ψ_{mn} to rewrite the definition of separability as

$$\Psi_{mn} = Z_m Z_n , \tag{6.75}$$

where

$$Z_n = \langle \phi_n | \zeta \rangle . \tag{6.76}$$

Thus separability for bosons is the same as the factorization condition (6.75) for the expansion coefficients. From the original form (6.74) it is clear that eqn (6.75) must hold for all choices of the single-particle basis vectors $|\phi_n\rangle$.

Entangled states are defined as those that are not separable, e.g. the state $|\Psi\rangle$ in eqn (6.73). This seems harmless enough for bosons, but it has a surprising result for fermions. In this case eqn (6.72) must be replaced by

$$|\Psi\rangle = C\left\{|\zeta\rangle_1 |\eta\rangle_2 - |\eta\rangle_1 |\zeta\rangle_2\right\}, \tag{6.77}$$

and setting $|\eta\rangle = |\zeta\rangle$ gives $|\Psi\rangle = 0$, which is simply an expression of the Pauli exclusion principle. Consequently, extending the distinguishable-particle definition of entanglement to fermions leads to the conclusion that every two-fermion state is entangled.

An alternative transition from distinguishable to indistinguishable particles is based on the observation that the symmetrized states

$$|\Psi\rangle = C\left\{|\zeta\rangle_1 |\eta\rangle_2 \pm |\eta\rangle_1 |\zeta\rangle_2\right\} \tag{6.78}$$

for identical particles seem to be the natural analogues of product vectors for distinguishable particles. From this point of view, states that have the minimal form (6.78) imposed by Bose or Fermi symmetry should not be called entangled (Eckert *et al.*,

2002). For those working in the field of quantum information processing, this view is strongly supported by the fact that states of the form (6.78) do not provide a useful *resource*, e.g. for quantum computing. This argument is, however, open to the objection that utility—like beauty—is in the eye of the beholder. We will illustrate this point by way of an example.

A state $|\Psi\rangle$ of two electrons is described by a wave function $\Psi(\mathbf{r}_1, s_1; \mathbf{r}_2, s_2)$ which is antisymmetric with respect to the transposition $(\mathbf{r}_1, s_1) \leftrightarrow (\mathbf{r}_2, s_2)$. For this example, it is convenient to use the wave function representation for the spatial coordinates and to retain the Dirac ket representation for the spins. With this notation, we consider the spin-singlet state

$$|\Psi(\mathbf{r}_1, \mathbf{r}_2)\rangle = \psi(\mathbf{r}_1)\psi(\mathbf{r}_2)\{|\uparrow\rangle_1|\downarrow\rangle_2 - |\downarrow\rangle_1|\uparrow\rangle_2\}, \tag{6.79}$$

which is symmetric in the spatial coordinates and antisymmetric in the spins. If Alice detects a single electron and measures the z-component of its spin to be $s_z = +1/2$, then an electron detected by Bob is guaranteed to have the value $s_z = -1/2$. Thus the state defined in eqn (6.79) displays the most basic feature of entanglement; namely, that the result of one measurement gives information about the possible results of measurements that could be made on another part of the system. This establishes the fundamental utility of the state in eqn (6.79), despite the fact that it does not provide a resource for quantum information processing. A similar example can be constructed for bosons, so we will retain the traditional definition of entanglement for identical particles.

Our preference for extending the traditional definition of entanglement to indistinguishable particles, as opposed to the more restrictive version presented above, does not mean that the latter is not important. On the contrary, the stronger interpretation of entanglement captures an essential physical feature that plays a central role in many applications. In order to distinguish between the two notions of entanglement, we will say that a two-particle state that is entangled in the minimal form (6.78), required by indistinguishability, is **kinematically entangled**, and that an entangled two-particle state is **dynamically entangled** if it cannot be expressed in the form (6.78). The use of the term 'dynamical' is justified by the observation that dynamically entangled states can only be produced by interaction between the indistinguishable particles. For photons, this distinction enters in a natural way in the analysis of the Hong–Ou–Mandel effect in Section 10.2.1. For distinguishable particles, there is no symmetry condition for multiparticle states; consequently, the notion of kinematical entanglement cannot arise and all entangled states are dynamically entangled.

6.6 Entanglement for photons

Since photons are bosons, it seems reasonable to expect that the definition of entanglement introduced in Section 6.5.3 can be applied directly to photons. We will see that this expectation is almost completely satisfied, except for an important reservation arising from the absence of a photon position operator.

The most intuitively satisfactory way to understand entanglement for bosons is in terms of an explicit wave function like

$$\psi_{s_1 s_2}(\mathbf{r}_1, \mathbf{r}_2) = \frac{1}{\sqrt{2}} \left[\zeta_{s_1}(\mathbf{r}_1) \eta_{s_2}(\mathbf{r}_2) + \eta_{s_1}(\mathbf{r}_1) \zeta_{s_2}(\mathbf{r}_2) \right], \tag{6.80}$$

where the subscripts describe internal degrees of freedom such as spin. If we recall that $\zeta_{s_1}(\mathbf{r}_1) = \langle \mathbf{r}_1, s_1 | \zeta \rangle$, where $|\mathbf{r}_1, s_1\rangle$ is an eigenstate of the position operator $\widehat{\mathbf{r}}$ for the particle, then it is clear that the existence of a wave function depends on the existence of a position operator $\widehat{\mathbf{r}}$. For applications to photons, this brings us face to face with the well known absence—discussed in Section 3.6.1—of any acceptable position operator for the photon. In Section 6.6.1 we will show that the absence of position-space wave functions for photons is not a serious obstacle to defining entanglement, and in Section 6.6.2 we will find that the intuitive benefits of the absent wave function can be largely recovered by considering a simple model of photon detection.

6.6.1 Definition of entanglement for photons

In Section 6.5.1 we observed that states of massive bosons belong to the symmetrical subspace $(\mathfrak{H}_C)_{\text{sym}}$ of the tensor product space $\mathfrak{H}_C$ describing a many-particle system. For photons, the definitions of Fock space in Sections 2.1.2-C or 3.1.4 can be understood as a direct construction of $(\mathfrak{H}_C)_{\text{sym}}$ that works for any number of photons. In the example of a two-particle system, the Fock space approach replaces explicitly symmetrized vectors like

$$|\phi_m\rangle_1 |\phi_n\rangle_2 + |\phi_n\rangle_1 |\phi_m\rangle_2 \tag{6.81}$$

by Fock-space vectors,

$$a^\dagger_{\mathbf{k}s} a^\dagger_{\mathbf{k}'s'} |0\rangle , \tag{6.82}$$

generated by applying creation operators to the vacuum. Despite their different appearance, the physical content of the two methods is the same.

We will use box-quantized creation operators to express a general two-photon state as

$$|\Psi\rangle = \frac{1}{\sqrt{2}} \sum_{\mathbf{k}s,\mathbf{k}'s'} C_{\mathbf{k}s,\mathbf{k}'s'} a^\dagger_{\mathbf{k}s} a^\dagger_{\mathbf{k}'s'} |0\rangle , \tag{6.83}$$

where the normalization condition $\langle \Psi | \Psi \rangle = 1$ is

$$\sum_{\mathbf{k}s,\mathbf{k}'s'} |C_{\mathbf{k}s,\mathbf{k}'s'}|^2 = 1 , \tag{6.84}$$

and the expansion (6.83) can be inverted to give

$$C_{\mathbf{k}s,\mathbf{k}'s'} = \frac{\langle 1_{\mathbf{k}s}, 1_{\mathbf{k}'s'} | \Psi \rangle}{\sqrt{2}} . \tag{6.85}$$

By comparing eqns (6.83) and (6.75), we can see that a two-photon state is **separable** if the coefficients in eqn (6.83) factorize:

$$C_{\mathbf{k}s,\mathbf{k}'s'} = \gamma_{\mathbf{k}s} \gamma_{\mathbf{k}'s'} , \tag{6.86}$$

where the $\gamma_{\mathbf{k}s}$s are c-number coefficients. In this case, $|\Psi\rangle$ can be expressed as

$$|\Psi\rangle = \frac{1}{\sqrt{2}} \left(\Gamma^\dagger\right)^2 |0\rangle \,, \tag{6.87}$$

where

$$\Gamma^\dagger = \sum_{\mathbf{k}s} \gamma_{\mathbf{k}s} a^\dagger_{\mathbf{k}s} \,, \tag{6.88}$$

and the normalization condition (6.84) becomes

$$\sum_{\mathbf{k}s} |\gamma_{\mathbf{k}s}|^2 = 1 \,. \tag{6.89}$$

The normalization of the $\gamma_{\mathbf{k}s}$s in turn implies $\left[\Gamma, \Gamma^\dagger\right] = 1$; therefore, $\Gamma^\dagger$ can be interpreted as a creation operator for a photon in the classical wave packet:

$$\boldsymbol{\mathcal{E}}(\mathbf{r}) = \sum_{\mathbf{k}s} \gamma_{\mathbf{k}s} \mathcal{F}_k \mathbf{e}_{\mathbf{k}s} e^{i\mathbf{k}\cdot\mathbf{r}} \,, \tag{6.90}$$

where

$$\mathcal{F}_k = i\sqrt{\frac{\hbar\omega_k}{2\epsilon_0 V}} \,. \tag{6.91}$$

Thus the bosonic character of photons implies that a separable state necessarily contains two photons in the same classical wave packet, in agreement with the definition (6.74) for massive bosons.

A two-photon state that is not separable is said to be **entangled**. This leads in particular to the useful rule

$$|1_{\mathbf{k}s}, 1_{\mathbf{k}'s'}\rangle \text{ is entangled if } \mathbf{k}s \neq \mathbf{k}'s' \,. \tag{6.92}$$

The factorization condition (6.86) provides a definition of separable states and entangled states that works in the absence of position-space wave functions for photons, but the physical meaning of entanglement is not as intuitively clear as it is in ordinary quantum mechanics. The best remedy is to find a substitute for the missing wave function.

6.6.2 The detection amplitude

Let us pretend, for the moment, that the operator $E_s^{(-)}(\mathbf{r}) = \mathbf{e}_s^* \cdot \mathbf{E}^{(-)}(\mathbf{r})$ creates a photon, with polarization $\mathbf{e}_s$, at the point $\mathbf{r}$. If this were true, then the state vector $|\mathbf{r}, s\rangle = E_s^{(-)}(\mathbf{r})|0\rangle$ would describe a situation in which one photon is located at $\mathbf{r}$ with polarization $\mathbf{e}_s$. For a one-photon state $|\Psi\rangle$, this suggests defining a single-photon 'wave function' by

$$\begin{aligned}\Psi(\mathbf{r}, s) &= \langle \mathbf{r}, s | \Psi \rangle \\ &= \left\langle 0 \left| E_s^{(+)}(\mathbf{r}) \right| \Psi \right\rangle \\ &= e_{sj}^* \left\langle 0 \left| E_j^{(+)}(\mathbf{r}) \right| \Psi \right\rangle .\end{aligned} \tag{6.93}$$

Now that our attention has been directed to the appropriate quantity, we can discard this very dubious plausibility argument, and directly investigate the physical significance of $\Psi(\mathbf{r}, s)$. One way to do this is to use eqn (4.74) to evaluate the first-order field correlation function for the one-photon state $|\Psi\rangle$. For equal time arguments, the result is

$$\begin{aligned}
G_{ij}^{(1)}(\mathbf{r}';\mathbf{r}) &= \left\langle \Psi \left| E_i^{(-)}(\mathbf{r}') E_j^{(+)}(\mathbf{r}) \right| \Psi \right\rangle \\
&= \sum_{\underline{n}} \left\langle \Psi \left| E_i^{(-)}(\mathbf{r}') \right| \underline{n} \right\rangle \left\langle \underline{n} \left| E_j^{(+)}(\mathbf{r}) \right| \Psi \right\rangle \\
&= \left\langle \Psi \left| E_i^{(-)}(\mathbf{r}') \right| 0 \right\rangle \left\langle 0 \left| E_j^{(+)}(\mathbf{r}) \right| \Psi \right\rangle ,
\end{aligned} \tag{6.94}$$

where the last line follows from the observation that the vacuum state alone can contribute to the sum over the number states $|\underline{n}\rangle$. By combining these two equations, one finds that

$$\begin{aligned}
G^{(1)}(\mathbf{r}'s';\mathbf{r}s) &= e_{s'i} e_{sj}^* G_{ij}^{(1)}(\mathbf{r}';\mathbf{r}) \\
&= \Psi(\mathbf{r},s)\,\Psi^*(\mathbf{r}',s') .
\end{aligned} \tag{6.95}$$

This result for $G^{(1)}(\mathbf{r}s;\mathbf{r}'s')$ is quite suggestive, since it has the form of the density matrix for a pure state with wave function $\Psi(\mathbf{r},s)$. On the other hand, the usual Born interpretation does not apply to $\Psi(\mathbf{r},s)$, since there is no photon position operator. An important clue pointing to the correct physical interpretation of $\Psi(\mathbf{r},s)$ is provided by the theory of photon detection. In Section 9.1.2-A it is shown that the counting rate for a photon detector—located at $\mathbf{r}$ and equipped with a filter transmitting polarization $\mathbf{e}_s$—is proportional to $G^{(1)}(\mathbf{r}s;\mathbf{r}s)$. According to eqn (6.95), this means that $|\Psi(\mathbf{r},s)|^2$ is the probability that a photon is detected at $\mathbf{r}$, the position of the detector. In view of this fact, we will refer to $\Psi(\mathbf{r},s)$ as the **one-photon detection amplitude**. The important point to keep in mind is that the detector is a classical object which—unlike the photon—has a well-defined location in space. This is what makes the detection amplitude a useful replacement for the missing photon wave function.

We extend this approach to two photons by pretending that $|\mathbf{r}_1,s_1;\mathbf{r}_2,s_2\rangle = E_{s_1}^{(-)}(\mathbf{r}_1)\,E_{s_2}^{(-)}(\mathbf{r}_2)\,|0\rangle$ is a state with one photon at $\mathbf{r}_1$ (with polarization $\mathbf{e}_{s_1}$) and another at $\mathbf{r}_2$ (with polarization $\mathbf{e}_{s_2}$). For a two-photon state $|\Psi\rangle$ this suggests the effective wave function

$$\begin{aligned}
\Psi(\mathbf{r}_1,s_1;\mathbf{r}_2,s_2) &= \langle \mathbf{r}_1,s_1;\mathbf{r}_2,s_2 | \Psi \rangle \\
&= \left\langle 0 \left| E_{s_1}^{(+)}(\mathbf{r}_1)\,E_{s_2}^{(+)}(\mathbf{r}_2) \right| \Psi \right\rangle \\
&= e_{s_1 i}^* e_{s_2 j}^* \Psi_{ij}(\mathbf{r}_1,\mathbf{r}_2) ,
\end{aligned} \tag{6.96}$$

where

$$\Psi_{ij}(\mathbf{r}_1,\mathbf{r}_2) = \left\langle 0 \left| E_i^{(+)}(\mathbf{r}_1)\,E_j^{(+)}(\mathbf{r}_2) \right| \Psi \right\rangle . \tag{6.97}$$

Applying the method used for $G^{(1)}$ to the evaluation of eqn (4.75) for the second-order correlation function (with all time arguments equal) yields

$$\begin{aligned}
G_{klij}^{(2)}(\mathbf{r}_1',\mathbf{r}_2';\mathbf{r}_1,\mathbf{r}_2) &= \left\langle E_k^{(-)}(\mathbf{r}_1')\,E_l^{(-)}(\mathbf{r}_2')\,E_i^{(+)}(\mathbf{r}_1)\,E_j^{(+)}(\mathbf{r}_2) \right\rangle \\
&= \Psi_{ij}(\mathbf{r}_1,\mathbf{r}_2)\,\Psi_{kl}^*(\mathbf{r}_1',\mathbf{r}_2') ,
\end{aligned} \tag{6.98}$$

which has the form of the two-particle density matrix corresponding to the pure two-particle wave function $\Psi_{ij}(\mathbf{r}_1, \mathbf{r}_2)$.

The physical interpretation of $\Psi_{ij}(\mathbf{r}_1, \mathbf{r}_2)$ follows from the discussion of coincidence counting in Section 9.2.4, which shows that the coincidence-counting rate for two fast detectors placed at equal distances from the source of the field is proportional to

$$(\mathbf{e}_{s_1})_k (\mathbf{e}_{s_2})_l (\mathbf{e}_{s_1}^*)_i (\mathbf{e}_{s_2}^*)_j G^{(2)}_{klij}(\mathbf{r}_1, \mathbf{r}_2; \mathbf{r}_1, \mathbf{r}_2) = |\Psi(\mathbf{r}_1, s_1; \mathbf{r}_2, s_2)|^2, \tag{6.99}$$

where $\mathbf{e}_{s_1}$ and $\mathbf{e}_{s_2}$ are the polarizations admitted by the filters associated with the detectors. Since $|\Psi(\mathbf{r}_1, s_1; \mathbf{r}_2, s_2)|^2$ determines the two-photon counting rate, we will refer to $\Psi(\mathbf{r}_1, s_1; \mathbf{r}_2, s_2)$—or $\Psi_{ij}(\mathbf{r}_1, \mathbf{r}_2)$—as the **two-photon detection amplitude**.

6.6.3 Pure state entanglement defined by detection amplitudes

We are now ready to formulate an alternative definition of entanglement, for pure states of photons, that is directly related to observable counting rates. The detection amplitude for the two-photon state $|\Psi\rangle$, defined by eqn (6.83), can be evaluated by using eqns (3.69) and (6.85) in eqn (6.97), with the result:

$$\Psi_{ij}(\mathbf{r}_1, \mathbf{r}_2) = \sqrt{2} \sum_{\mathbf{k}s,\mathbf{k}'s'} C_{\mathbf{k}s,\mathbf{k}'s'} \mathcal{F}_k (\mathbf{e}_{\mathbf{k}s})_i e^{i\mathbf{k}\cdot\mathbf{r}_1} \mathcal{F}_{k'} (\mathbf{e}_{\mathbf{k}'s'})_j e^{i\mathbf{k}'\cdot\mathbf{r}_2}. \tag{6.100}$$

This expansion for the detection amplitude can be inverted, by Fourier transforming with respect to $\mathbf{r}_1$ and $\mathbf{r}_2$ and projecting on the polarization basis, to get

$$C_{\mathbf{k}s,\mathbf{k}'s'} = -\frac{(2\epsilon_0/\hbar)^2}{\sqrt{2}\omega_k\omega_{k'}} \Psi_{\mathbf{k}s,\mathbf{k}'s'}, \tag{6.101}$$

where

$$\Psi_{\mathbf{k}s,\mathbf{k}'s'} = \frac{1}{V} \int d^3r_1 \int d^3r_2 e^{-i\mathbf{k}\cdot\mathbf{r}_1} e^{-i\mathbf{k}'\cdot\mathbf{r}_2} (\mathbf{e}_{\mathbf{k}s}^*)_i (\mathbf{e}_{\mathbf{k}'s'}^*)_j \Psi_{ij}(\mathbf{r}_1, \mathbf{r}_2). \tag{6.102}$$

According to eqns (6.100) and eqn (6.101), the two-photon detection amplitude and the expansion coefficients $C_{\mathbf{k}s,\mathbf{k}'s'}$ provide equivalent descriptions of the two-photon state. From eqn (6.100) we see that factorization of the expansion coefficients, according to eqn (6.86), implies factorization of the detection amplitude, i.e.

$$\Psi_{ij}(\mathbf{r}_1, \mathbf{r}_2) = \phi_i(\mathbf{r}_1)\phi_j(\mathbf{r}_2), \tag{6.103}$$

where

$$\phi_i(\mathbf{r}) = 2^{1/4} \sum_{\mathbf{k}s} \gamma_{\mathbf{k}s} \mathcal{F}_k (\mathbf{e}_{\mathbf{k}s})_i e^{i\mathbf{k}\cdot\mathbf{r}}. \tag{6.104}$$

In other words, the detection amplitude for a separable state factorizes, just as a two-particle wave function does in nonrelativistic quantum mechanics. On the other hand, eqn (6.101) shows that factorization of the detection amplitude implies factorization of the expansion coefficients. Thus we are at liberty to use eqn (6.103) as a definition of a separable state that agrees with the definition (6.86). This approach has the decided advantage that the detection amplitude is closely related to directly observable events, e.g. current pulses emitted by the coincidence counter. The coincidence-counting rate is proportional to the square of the amplitude, so for separable states the coincidence rate is proportional to the product of the singles rates at the two detectors. This means that the random counting events at the two detectors are stochastically independent, i.e. the quantum fluctuations of the electromagnetic field at any pair of detectors are uncorrelated. This is the analogue of Theorem 6.3, which states that a separable state of two distinguishable particles yields uncorrelated quantum fluctuations for any pair of observables.

For $\mathbf{k}s \neq \mathbf{k}'s'$ the state $|\Psi\rangle = |1_{\mathbf{k}s}, 1_{\mathbf{k}'s'}\rangle$ is entangled—according to the traditional definition—and evaluating eqn (6.100) in this case gives

$$\Psi_{ij}(\mathbf{r}_1, \mathbf{r}_2) = \mathcal{F}_k \mathcal{F}_{k'} \left\{ (\mathbf{e}_{\mathbf{k}s})_i e^{i\mathbf{k}\cdot\mathbf{r}_1} (\mathbf{e}_{\mathbf{k}'s'})_j e^{i\mathbf{k}'\cdot\mathbf{r}_2} + (\mathbf{e}_{\mathbf{k}s})_j e^{i\mathbf{k}\cdot\mathbf{r}_2} (\mathbf{e}_{\mathbf{k}'s'})_i e^{i\mathbf{k}'\cdot\mathbf{r}_1} \right\}. \tag{6.105}$$

The definition (6.96) in turn yields

$$\Psi(\mathbf{r}_1, s_1; \mathbf{r}_2, s_2) = \phi_{\mathbf{k}s}(\mathbf{r}_1, s_1)\, \phi_{\mathbf{k}'s'}(\mathbf{r}_2, s_2) + \phi_{\mathbf{k}s}(\mathbf{r}_2, s_2)\, \phi_{\mathbf{k}'s'}(\mathbf{r}_1, s_1), \tag{6.106}$$

where

$$\phi_{\mathbf{k}s}(\mathbf{r}, s_1) = \mathcal{F}_k \mathbf{e}_{s_1}^* \cdot \mathbf{e}_{\mathbf{k}s} e^{i\mathbf{k}\cdot\mathbf{r}}. \tag{6.107}$$

This has the structure of an entangled-state wave function for two bosons—as shown in eqn (6.80)—with similar physical consequences. In particular, if one photon is detected in the mode $\mathbf{k}s$, then a subsequent detection of the remaining photon is guaranteed to find it in the mode $\mathbf{k}'s'$. More generally, quantum fluctuations in the electromagnetic field at the two detectors are correlated. According to the general definition in Section 6.5.3, an entangled two-photon state is dynamically entangled if the detection amplitude cannot be expressed in the minimal form (6.106) required by Bose statistics.

We saw in Section 6.4.1 that reduced density operators, defined by partial traces, are quite useful in the discussion of distinguishable particles, but systems of identical particles—such as photons—cannot be divided into distinguishable subsystems. The key to overcoming this difficulty is found in eqn (6.98) which shows that the second-order correlation function has the form of a density matrix corresponding to the two-photon detection amplitude $\Psi_{ij}(\mathbf{r}_1, \mathbf{r}_2)$. This suggests that the analogue of the reduced density matrix is the first-order correlation function $G_{ij}^{(1)}(\mathbf{r}'; \mathbf{r})$, evaluated for the two-photon state $|\Psi\rangle$.

The first evidence supporting this proposal is provided by considering a separable state defined by eqn (6.87). In this case

$$\begin{aligned} G_{ij}^{(1)}(\mathbf{r}'; \mathbf{r}) &= \left\langle \Psi \left| E_i^{(-)}(\mathbf{r}')\, E_j^{(+)}(\mathbf{r}) \right| \Psi \right\rangle \\ &= \frac{1}{2} \left\langle 0 \left| \Gamma^2 E_i^{(-)}(\mathbf{r}')\, E_j^{(+)}(\mathbf{r})\, \Gamma^{\dagger 2} \right| 0 \right\rangle \\ &= \frac{1}{2} \left\langle 0 \left| \left[\Gamma^2, E_i^{(-)}(\mathbf{r}')\right] \left[E_j^{(+)}(\mathbf{r}), \Gamma^{\dagger 2}\right] \right| 0 \right\rangle, \end{aligned} \tag{6.108}$$

where the last line follows from the identity $E_j^{(+)}(\mathbf{r})|0\rangle = 0$ and its adjoint. The field operators and the operators Γ and $\Gamma^\dagger$ are both linear functions of the creation and annihilation operators, so

$$\left[E_j^{(+)}(\mathbf{r}), \Gamma^{\dagger 2}\right] = 2\left[E_j^{(+)}(\mathbf{r}), \Gamma^\dagger\right]\Gamma^\dagger . \tag{6.109}$$

The remaining commutator is a *c*-number which is evaluated by using the expansions (3.69) and (6.88) to get

$$\left[E_j^{(+)}(\mathbf{r}), \Gamma^\dagger\right] = 2^{-1/4}\phi_j(\mathbf{r}), \tag{6.110}$$

where $\phi_i(\mathbf{r})$ is defined by eqn (6.104). Substituting this result, and the corresponding expression for $\left[\Gamma, E_i^{(-)}(\mathbf{r}')\right]$, into eqn (6.108) yields

$$G_{ij}^{(1)}(\mathbf{r}';\mathbf{r}) = \sqrt{2}\phi_j(\mathbf{r})\phi_i^*(\mathbf{r}') . \tag{6.111}$$

The conclusion is that the first-order correlation function for a separable state factorizes. This is the analogue of Theorem 6.1 for distinguishable particles.

Next let us consider a generic entangled state defined by $|\Psi\rangle = \Gamma^\dagger\Theta^\dagger|0\rangle$, where

$$\Theta^\dagger = \sum_{\mathbf{k}s}\theta_{\mathbf{k}s}a_{\mathbf{k}s}^\dagger \tag{6.112}$$

and

$$\sum_{\mathbf{k}s}|\theta_{\mathbf{k}s}|^2 = 1 . \tag{6.113}$$

For this argument, we can confine attention to operators satisfying $\left[\Gamma, \Theta^\dagger\right] = 0$, which is equivalent to the orthogonality of the classical wave packets:

$$(\theta, \gamma) \equiv \sum_{\mathbf{k}s}\theta_{\mathbf{k}s}^*\gamma_{\mathbf{k}s} = 0 . \tag{6.114}$$

The first-order correlation function for this state is

$$\begin{aligned} G_{ij}^{(1)}(\mathbf{r}';\mathbf{r}) &= \left\langle\Psi\left|E_i^{(-)}(\mathbf{r}')E_j^{(+)}(\mathbf{r})\right|\Psi\right\rangle \\ &= \frac{1}{\sqrt{2}}\left\{\phi_j(\mathbf{r})\phi_i^*(\mathbf{r}') + \eta_j(\mathbf{r})\eta_i^*(\mathbf{r}')\right\}, \end{aligned} \tag{6.115}$$

where $\eta_j(\mathbf{r})$ is defined by replacing $\gamma_{\mathbf{k}s}$ with $\theta_{\mathbf{k}s}$ in eqn (6.104). Thus for the entangled, two-photon state $|\Psi\rangle$, the first-order correlation function (reduced density matrix) has the standard form of the density matrix for a one-particle mixed state. This is the analogue of Theorem 6.2 for distinguishable particles.

6.7 Exercises

6.1 Proof of Theorem 6.1

(1) To prove assertion (a), use the expression for the density operator resulting from eqns (6.40) and (2.81) to evaluate the reduced density operators.

(2) To prove assertion (b), assume that $|\Psi\rangle$ is entangled—so that it has Schmidt rank $r > 1$—and derive a contradiction.

6.2 Proof of Theorem 6.3

(1) For a separable state $|\Psi\rangle$ show that $\langle\Psi|\delta\mathfrak{A}\,\delta\mathfrak{B}|\Psi\rangle = 0$.

(2) Assume that $\langle\Psi|\delta\mathfrak{A}\,\delta\mathfrak{B}|\Psi\rangle = 0$ for all $\mathfrak{A}$ and $\mathfrak{B}$. Apply this to operators that are diagonal in the Schmidt basis for $|\Psi\rangle$ and thus show that $|\Psi\rangle$ must be separable.

6.3 Singlet spin state

(1) Use the standard treatments of the Pauli matrices, given in texts on quantum mechanics, to express the eigenstates of $\mathbf{n}\cdot\boldsymbol{\sigma}$ in the usual basis of eigenstates of σ_z.

(2) Show that the singlet state $|S=0\rangle$, given by eqn (6.37), has the same form for all choices of the quantization axis $\mathbf{n}$.

(3) Show that $\left(\mathbf{S}^A + \mathbf{S}^B\right)^2|S=0\rangle = 0$.

6.4 Correlations in a separable mixed state

Consider a system of two distinguishable spin-1/2 particles described by the ensemble

$$\{|\Psi_1\rangle = |\uparrow\rangle_A\,|\downarrow\rangle_B\,,\quad |\Psi_2\rangle = |\downarrow\rangle_A\,|\uparrow\rangle_B\}$$

of separable states, where the spin states are eigenstates of s_z^A and s_z^B.

(1) Show that the density operator can be written as

$$\rho = p\,|\Psi_1\rangle\langle\Psi_1| + (1-p)\,|\Psi_2\rangle\langle\Psi_2|\,,$$

where $0 \leqslant p \leqslant 1$.

(2) Evaluate the correlation function $\langle\delta s_z^A\,\delta s_z^B\rangle$ and use the result to show that the spins are only uncorrelated for the extreme values $p = 0, 1$.

(3) For intermediate values of p, argue that the correlation is exactly what would be found for a pair of classical stochastic variables taking on the values $\pm 1/2$ with the same assignment of probabilities.

7

Paraxial quantum optics

The generation and manipulation of paraxial beams of light forms the core of experimental practice in quantum optics; therefore, it is important to extend the classical treatment of paraxial optics to situations involving only a few photons, such as the photon pairs produced by spontaneous down-conversion. In addition to the interaction of quantized fields with standard optical elements, the theory of quantum paraxial propagation has applications to fundamental issues such as the generation and control of orbital angular momentum and the meaning of localization for photons.

In geometric optics a beam of light is a bundle of rays making small angles with a central ray directed along a unit vector $\mathbf{u}_0$. The constituent rays of the bundle are said to be **paraxial**. In wave optics, the bundle of rays is replaced by a bundle of unit vectors normal to the wavefront; so a **paraxial wave** is defined by a wavefront that is nearly flat. In this situation it is natural to describe the classical field amplitude, $\boldsymbol{\mathcal{E}}(\mathbf{r}, t)$, as a function of the propagation variable $\zeta = \mathbf{r} \cdot \mathbf{u}_0$, the transverse coordinates $\mathbf{r}_\mathsf{T}$ tangent to the wavefront, and the time t. Paraxial wave optics is more complicated than paraxial ray optics because of diffraction, which couples the $\mathbf{r}_\mathsf{T}$-, ζ-, and t-dependencies of the field. For the most part, we will only consider a single paraxial wave; therefore, we can choose the z-axis along $\mathbf{u}_0$ and set $\zeta = z$.

The definite wavevector associated with the plane wave created by $a_s^\dagger(\mathbf{k})$ makes it possible to recast the geometric-optics picture in terms of photons in plane-wave states. This way of thinking about paraxial optics is useful but—as always—it must be treated with caution. As explained in Section 3.6.1, there is no physically acceptable way to define the position of a photon. This means that the natural tendency to visualize the photons as beads sliding along the rays at speed c must be strictly suppressed. The beads in this naive picture must be replaced by wave packets containing energy $\hbar\omega$ and momentum $\hbar\mathbf{k}$, where $\mathbf{k}$ is directed along the normal to the paraxial wavefront.

In the following section, we begin with a very brief review of classical paraxial wave optics. In succeeding sections we will define a set of paraxial quantum states, and then use them to obtain approximate expressions for the energy, momentum, and photon number operators. This will be followed by the definition of a slowly-varying envelope operator that replaces the classical envelope field $\overline{\boldsymbol{\mathcal{E}}}(\mathbf{r}, t)$. Some more advanced topics—including the general paraxial expansion, angular momentum, and an approximate notion of photon localizability—will be presented in the remaining sections.

7.1 Classical paraxial optics

As explained above, each photon is distributed over a wave packet, with energy $\hbar\omega$ and momentum $\hbar\mathbf{k}$, that propagates along the normal to the wavefront. However, this wave optics description must be approached with equal caution. The standard approach in classical, paraxial wave optics (Saleh and Teich, 1991, Sec. 2.2C) is to set

$$\boldsymbol{\mathcal{E}}(\mathbf{r},t) = \overline{\boldsymbol{\mathcal{E}}}(\mathbf{r},t)\, e^{i(\mathbf{k}_0\cdot\mathbf{r}-\omega_0 t)}\,, \tag{7.1}$$

where ω_0 and $\mathbf{k}_0 = \mathbf{u}_0 n(\omega_0)\,\omega_0/c$ are respectively the carrier frequency and the **carrier wavevector**. The four-dimensional Fourier transform, $\overline{\boldsymbol{\mathcal{E}}}(\mathbf{k},\omega)$, of the slowly-varying envelope is assumed to be concentrated in a neighborhood of $\mathbf{k}=0$, $\omega=0$. The equivalent conditions in the space–time domain are

$$\left|\frac{\partial^2\overline{\boldsymbol{\mathcal{E}}}(\mathbf{r},t)}{\partial t^2}\right| \ll \omega_0\left|\frac{\partial\overline{\boldsymbol{\mathcal{E}}}(\mathbf{r},t)}{\partial t}\right| \ll \omega_0^2\left|\overline{\boldsymbol{\mathcal{E}}}(\mathbf{r},t)\right| \tag{7.2}$$

and

$$\left|\frac{\partial^2\overline{\boldsymbol{\mathcal{E}}}(\mathbf{r},t)}{\partial z^2}\right| \ll k_0\left|\frac{\partial\overline{\boldsymbol{\mathcal{E}}}(\mathbf{r},t)}{\partial z}\right| \ll k_0^2\left|\overline{\boldsymbol{\mathcal{E}}}(\mathbf{r},t)\right|; \tag{7.3}$$

in other words, $\overline{\boldsymbol{\mathcal{E}}}(\mathbf{r},t)$ has negligible variation in time over an optical period and negligible variation in space over an optical wavelength. As we have already seen in the discussion of monochromatic fields, these conditions cannot be applied to the field operator $\mathbf{E}^{(+)}(\mathbf{r},t)$; instead, they must be interpreted as constraints on the allowed states of the field.

7.2 Paraxial states

7.2.1 The paraxial ray bundle

A paraxial beam associated with the carrier wavevector $\mathbf{k}_0$, i.e. a bundle of wavevectors $\mathbf{k}$ clustered around $\mathbf{k}_0$, is conveniently described in terms of relative wavevectors $\mathbf{q} = \mathbf{k}-\mathbf{k}_0$, with $|\mathbf{q}| \ll k_0$. For each $\mathbf{k}=\mathbf{k}_0+\mathbf{q}$ the angle $\vartheta_\mathbf{k}$ between $\mathbf{k}$ and $\mathbf{k}_0$ is given by

$$\sin\vartheta_\mathbf{k} = \frac{|\mathbf{k}_0\times\mathbf{k}|}{k_0 k} = \frac{|\mathbf{k}_0\times\mathbf{q}|}{k_0\,|\mathbf{k}_0+\mathbf{q}|} = \frac{|\mathbf{q}_\mathsf{T}|}{k_0}\left[1+O\left(\frac{q}{k_0}\right)\right], \tag{7.4}$$

where $\mathbf{q}_\mathsf{T} = \mathbf{q} - q_z\widetilde{\mathbf{k}}_0$ and $q_z = \mathbf{q}\cdot\widetilde{\mathbf{k}}_0$. This shows that $\vartheta_\mathbf{k} \simeq |\mathbf{q}_\mathsf{T}|/k_0$, and further suggests defining the small parameter for the paraxial beam as the maximum opening angle,

$$\theta = \frac{\Delta q_\mathsf{T}}{k_0} \ll 1\,, \tag{7.5}$$

where $0 < |\mathbf{q}_\mathsf{T}| < \Delta q_\mathsf{T}$ is the range of the transverse components of $\mathbf{q}$. Variations in the transverse coordinate $\mathbf{r}_\mathsf{T}$ occur over a characteristic distance Λ_T defined by the Fourier transform uncertainty relation $\Lambda_\mathsf{T}\Delta q_\mathsf{T} \sim 1$; consequently, a useful length scale for transverse variations is $\Lambda_\mathsf{T} = 1/\Delta q_\mathsf{T} = 1/(\theta k_0)$.

A natural way to define the characteristic length $\Lambda_\|$ for longitudinal variations is to interpret the transverse length scale Λ_T as the radius of an effective circular aperture. The conventional longitudinal scale is then the distance over which a beam waist, initially equal to $\Lambda_\top$, doubles in size. At this point, a strictly correct argument would bring in classical diffraction theory; but the same end can be achieved—with only a little sleight of hand—with geometric optics. By combining the approximation $\tan\theta \approx \theta$ with elementary trigonometry, it is easy to show that the geometric image of the aperture on a screen at a distance $\Lambda_{\parallel}$ has the radius $\Lambda'_\top = \Lambda_\top + \theta\Lambda_{\parallel}$. The trick is to choose the longitudinal scale length $\Lambda_{\parallel}$ so that $\Lambda'_\top = 2\Lambda_\top$, and this requires

$$\Lambda_{\parallel} = \frac{\Lambda_\top}{\theta} = k_0\Lambda_\top^2 = \frac{1}{\theta^2 k_0}\,. \tag{7.6}$$

We will see in Section 7.4 that $\Lambda_{\parallel} = k_0\Lambda_\top^2$ is twice the Rayleigh range—as usually defined in classical diffraction theory—for the aperture $\Lambda_\top$. Thus our geometric-optics trick has achieved the same result as a proper diffraction theory argument. Since propagation occurs along the direction characterized by $\Lambda_{\parallel}$, the natural time scale is $T = \Lambda_{\parallel}/\left(c/n_0\right) = 1/\left(\theta^2\omega_0\right)$.

The spread, Δq_z, in the longitudinal component of $\mathbf{q}$ satisfies $\Lambda_{\parallel}\Delta q_z \sim 1$, so the longitudinal and transverse widths are related by

$$\frac{\Delta q_z}{k_0} = \left(\frac{\Delta q_\top}{k_0}\right)^2 = \theta^2\,, \tag{7.7}$$

and the $\mathbf{q}$-vectors are effectively confined to a disk-shaped region defined by

$$Q_0 = \left\{\mathbf{q} \text{ satisfying } \left|\mathbf{q}_\top\right| \leqslant \theta k_0\,,\ \ q_z \leqslant \theta^2 k_0\right\}. \tag{7.8}$$

In a dispersive medium with index of refraction $n(\omega)$ the frequency ω_k is a solution of the dispersion relation $ck = \omega_k n(\omega_k)$, and wave packets propagate at the group velocity $v_g(\omega_k) = d\omega_k/dk$. The frequency width is therefore $\Delta\omega = v_{g0}\Delta k$, where v_{g0} is the group velocity at the carrier frequency. The straightforward calculation outlined in Exercise 7.1 yields the estimate

$$\frac{\Delta\omega}{\omega_0} \approx \frac{1}{2}\theta^2 \ll 1\,, \tag{7.9}$$

which is the criterion for a monochromatic field given by eqn (3.107).

7.2.2 The paraxial Hilbert space

The geometric-optics picture of a bundle of rays forming small angles with the central propagation vector $\mathbf{k}_0$ is realized in the quantum theory by a family of states that only contain photons with propagation vectors in the paraxial bundle. In order to satisfy the superposition principle, the family of states must be chosen as the **paraxial space**, $\mathfrak{H}(\mathbf{k}_0,\theta) \subset \mathfrak{H}_F$, spanned by the improper (continuum normalized) number states

$$\left|\{\mathbf{q}s\}_M\right\rangle = \prod_{m=1}^{M} a^\dagger_{0s_m}(\mathbf{q}_m)\left|0\right\rangle, \quad M = 0, 1, \ldots\,, \tag{7.10}$$

where $a_{0s}(\mathbf{q}) = a_s(\mathbf{k}_0+\mathbf{q})$, $\{\mathbf{q}s\}_M \equiv \{\mathbf{q}_1 s_1, \ldots, \mathbf{q}_M s_M\}$, and each relative propagation vector is constrained by the paraxial conditions (7.8). If the paraxial restriction were relaxed, eqn (7.10) would define a continuum basis set for the full Fock space, so the paraxial space is a subspace of $\mathfrak{H}_F$. The states satisfying the paraxiality condition (7.8) also satisfy the monochromaticity condition (3.107); consequently, $\mathfrak{H}(\mathbf{k}_0, \theta)$ is a subspace of the monochromatic space $\mathfrak{H}(\omega_0)$. A state $|\Psi\rangle$ belonging to $\mathfrak{H}(\mathbf{k}_0, \theta)$ is called a **pure paraxial state**, and a density operator ρ describing an ensemble of pure paraxial states is called a **mixed paraxial state**. A useful way to characterize a paraxial state ρ in $\mathfrak{H}(\mathbf{k}_0, \theta)$ is to note that the power spectrum

$$p(\mathbf{k}) = \sum_s \left\langle a_s^\dagger(\mathbf{k})\, a_s(\mathbf{k})\right\rangle = \sum_s \mathrm{Tr}\left[\rho a_s^\dagger(\mathbf{k})\, a_s(\mathbf{k})\right] \tag{7.11}$$

is strongly concentrated near $\mathbf{k} = \mathbf{k}_0$.

In the Schrödinger picture, a general paraxial state $|\Psi(0)\rangle$ has an expansion in the basis $\{|\{\mathbf{q}s\}_M\rangle\}$, and the time evolution is given by

$$|\Psi(t)\rangle = e^{-itH/\hbar}\, |\Psi(0)\rangle\,, \tag{7.12}$$

where H is the total Hamiltonian, including interactions with atoms, etc. It is clear on physical grounds that an initial paraxial state will not in general remain paraxial. For example, a paraxial field injected into a medium containing strong scattering centers will experience large-angle scattering and thus become nonparaxial as it propagates through the medium. In more favorable cases, interaction with matter, e.g. transmission through lenses with moderate focal lengths, will conserve the paraxial property.

The only situation for which it is possible to make a rigorous general statement is free propagation. In this case the basis vectors $|\{\mathbf{q}s\}_M\rangle$ are eigenstates of the total Hamiltonian, $H = H_{\text{em}}$, so that

$$\begin{aligned} |\Psi(t)\rangle = \sum_{M=0}^{\infty} \int \frac{d^3q_1}{(2\pi)^3} \sum_{s_1} \cdots \int \frac{d^3q_M}{(2\pi)^3} \sum_{s_M} F(\{\mathbf{q}s\}_M) \\ \times \exp\left\{-i \sum_{m=1}^{M} \omega(|\mathbf{k}_0 + \mathbf{q}_m|)\, t\right\} |\{\mathbf{q}s\}_M\rangle\,, \end{aligned} \tag{7.13}$$

where $F(\{\mathbf{q}s\}_M) = \langle\{\mathbf{q}s\}_M\,|\Psi(0)\rangle$. Consequently, the state $|\Psi(t)\rangle$ remains in the paraxial space $\mathfrak{H}(\mathbf{k}_0, \theta)$ for all times.

For the sake of simplicity, we have analyzed the case of a single paraxial ray bundle, but in many applications several paraxial beams are simultaneously present. The reasons range from simple reflection by a mirror to complex wave mixing phenomena in nonlinear media. The necessary generalizations can be understood by considering two paraxial bundles with carrier waves $\mathbf{k}_1$ and $\mathbf{k}_2$ and opening angles θ_1 and θ_2. The two beams are said to be **distinct** if the vector $\mathbf{\Delta k} = \mathbf{k}_1 - \mathbf{k}_2$ satisfies

$$|\mathbf{\Delta k}| \gg \max\left[\theta_1 |\mathbf{k}_1|\,, \theta_2 |\mathbf{k}_2|\right], \tag{7.14}$$

i.e. the two bundles of wavevectors do not overlap. The **multiparaxial space**, $\mathfrak{H}(\mathbf{k}_1, \theta_1, \mathbf{k}_2, \theta_2)$, for two distinct paraxial ray bundles is spanned by the basis vectors

$$\prod_{m=1}^{M} a_{1s_m}^{\dagger}(\mathbf{q}_m) \prod_{k=1}^{K} a_{2s_k}^{\dagger}(\mathbf{p}_k) |0\rangle \quad (M, K = 0, 1, \ldots), \tag{7.15}$$

where $a_{\beta s}^{\dagger}(\mathbf{q}) \equiv a_s^{\dagger}(\mathbf{k}_\beta + \mathbf{q})$ $(\beta = 1, 2)$ and the $\mathbf{q}$s and $\mathbf{p}$s are confined to the respective regions Q_1 and Q_2 defined by applying eqn (7.8) to each beam. The argument suggested in Exercise 7.6 shows that the paraxial spaces $\mathfrak{H}(\mathbf{k}_1, \theta_1)$ and $\mathfrak{H}(\mathbf{k}_2, \theta_2)$—which are subspaces of $\mathfrak{H}(\mathbf{k}_1, \theta_1, \mathbf{k}_2, \theta_2)$—may be treated as orthogonal within the paraxial approximation. This description is readily extended to any number of distinct beams.

7.2.3 Photon number, momentum, and energy

The action of the number operator N on the paraxial space $\mathfrak{H}(\mathbf{k}_0, \theta)$ is determined by its action on the basis states in eqn (7.10); consequently, the commutation relation, $\left[N, a_{0s}^{\dagger}(\mathbf{q})\right] = a_{0s}^{\dagger}(\mathbf{q})$, permits the use of the effective form

$$N \simeq N_0 = \int_{Q_0} \frac{d^3q}{(2\pi)^3} \sum_s a_{0s}^{\dagger}(\mathbf{q})\, a_{0s}(\mathbf{q}). \tag{7.16}$$

Applying the same idea to the momentum operator, given by the continuum version of eqn (3.153), leads to $\mathbf{P}_{\text{em}} = \hbar \mathbf{k}_0 N_0 + \mathbf{P}_0$, where

$$\mathbf{P}_0 = \int_{Q_0} \frac{d^3q}{(2\pi)^3} \hbar \mathbf{q} \sum_s a_{0s}^{\dagger}(\mathbf{q})\, a_{0s}(\mathbf{q}) \tag{7.17}$$

is the **paraxial momentum operator**.

The continuum version of eqn (3.150) for the Hamiltonian in a dispersive medium can be approximated by

$$H_{\text{em}} = \int_{Q_0} \frac{d^3q}{(2\pi)^3} \hbar \omega_{|\mathbf{k}_0 + \mathbf{q}|} \sum_s a_{0s}^{\dagger}(\mathbf{q})\, a_{0s}(\mathbf{q}), \tag{7.18}$$

when acting on a paraxial state. The small spread in frequencies across the paraxial bundle, together with the weak dispersion condition (3.120), allows the dispersion relation $\omega_k = ck/n(\omega_k)$ to be approximated by

$$\omega_k = \frac{ck}{n_0 + \left(\frac{dn}{d\omega}\right)_0 (\omega_k - \omega_0)}, \tag{7.19}$$

and a straightforward calculation yields

$$\omega_{|\mathbf{k}_0 + \mathbf{q}|} = \omega_0 + v_{g0} k_0 \left\{ \left| \widetilde{\mathbf{k}}_0 + \frac{\mathbf{q}}{k_0} \right| - 1 \right\} + \cdots. \tag{7.20}$$

The conditions (7.8) allow the expansion

$$\left| \widetilde{\mathbf{k}}_0 + \frac{\mathbf{q}}{k_0} \right| = 1 + \frac{q_z}{k_0} + \frac{q_{\mathsf{T}}^2}{2k_0^2} + O\left(\theta^2\right), \tag{7.21}$$

which in turn leads to the expression $H_{\text{em}} = \hbar\omega_0 N_0 + H_P + O(\theta^2)$, where

$$H_P = \int_{Q_0} \frac{d^3q}{(2\pi)^3} \left\{ v_{g0}\hbar q_z + \frac{\hbar v_{g0} q_{\text{T}}^2}{2k_0} \right\} \sum_s a_{0s}^{\dagger}(\mathbf{q})\, a_{0s}(\mathbf{q}) \tag{7.22}$$

is the **paraxial Hamiltonian** for the space $\mathfrak{H}(\mathbf{k}_0, \theta)$.

The effective orthogonality of distinct paraxial spaces—which corresponds to the distinguishability of distinct paraxial beams—implies that the various global operators are additive. Thus the operators for the total photon number, momentum, and energy for a set of paraxial beams are

$$N = \sum_\beta N_\beta\,, \quad \mathbf{P}_{\text{em}} = \sum_\beta (\hbar\mathbf{k}_\beta N_\beta + \mathbf{P}_\beta)\,, \quad H_{\text{em}} = \sum_\beta (\hbar\omega_\beta N_\beta + H_{P\beta})\,, \tag{7.23}$$

where N_β, $\mathbf{P}_\beta$, and $H_{P\beta}$ are respectively the paraxial number, momentum, and energy operators for the βth beam.

7.3 The slowly-varying envelope operator

We next use the properties of the paraxial space $\mathfrak{H}(\mathbf{k}_0, \theta)$ to justify an approximation for the field operator, $\mathbf{A}^{(+)}(\mathbf{r}, t)$, that replaces eqn (7.1) for the classical field. In order to emphasize the relation to the classical theory, we initially work in the Heisenberg picture. The **slowly-varying envelope operator** $\boldsymbol{\Phi}(\mathbf{r}, t)$ is defined by

$$\mathbf{A}^{(+)}(\mathbf{r}, t) = \sqrt{\frac{\hbar (v_{g0}/c)}{2\epsilon_0 k_0 c}}\, \boldsymbol{\Phi}(\mathbf{r}, t)\, e^{i(\mathbf{k}_0\cdot\mathbf{r} - \omega_0 t)}\,. \tag{7.24}$$

Comparing this definition to the general plane-wave expansion (3.149) shows that

$$\boldsymbol{\Phi}(\mathbf{r}, t) = \int_{Q_0} \frac{d^3q}{(2\pi)^3} f_q \sum_s a_{0s}(\mathbf{q})\, \mathbf{e}_s(\mathbf{k}_0 + \mathbf{q})\, e^{i(\mathbf{q}\cdot\mathbf{r} - \delta_{\mathbf{q}} t)}\,, \tag{7.25}$$

where

$$\delta_{\mathbf{q}} = \omega_{|\mathbf{k}_0 + \mathbf{q}|} - \omega_0 \quad \text{and} \quad f_q = \sqrt{\frac{v_g(|\mathbf{k}_0 + \mathbf{q}|)}{v_{g0}} \frac{k_0}{|\mathbf{k}_0 + \mathbf{q}|}}\,. \tag{7.26}$$

The corresponding expressions in the Schrödinger picture follow from the relation $\mathbf{A}^{(+)}(\mathbf{r}) = \mathbf{A}^{(+)}(\mathbf{r}, t = 0)$.

The envelope operator will only be slowly varying when applied to paraxial states in $\mathfrak{H}(\mathbf{k}_0, \theta)$, so we begin by using eqn (7.10) to evaluate the action of the envelope operator $\boldsymbol{\Phi}(\mathbf{r}) = \boldsymbol{\Phi}(\mathbf{r}, 0)$ on a typical basis vector of $\mathfrak{H}(\mathbf{k}_0, \theta)$:

$$\begin{aligned} \boldsymbol{\Phi}(\mathbf{r}) \left|\{\mathbf{q}s\}_M\right\rangle &= \boldsymbol{\Phi}(\mathbf{r}) \prod_{m=1}^{M} a_{0s_m}^{\dagger}(\mathbf{q}_m) \left|0\right\rangle \\ &= \left[\boldsymbol{\Phi}(\mathbf{r}), \prod_{m=1}^{M} a_{0s_m}^{\dagger}(\mathbf{q}_m)\right] \left|0\right\rangle \\ &= \sum_{m=1}^{M} \left[\boldsymbol{\Phi}(\mathbf{r}), a_{0s_m}^{\dagger}(\mathbf{q}_m)\right] \left\{\prod_{l=1}^{M} (1 - \delta_{lm})\, a_{0s_l}^{\dagger}(\mathbf{q}_l)\right\} \left|0\right\rangle\,, \end{aligned} \tag{7.27}$$

where the last line follows from the identity (C.49). Setting $t = 0$ in eqn (7.25) produces the Schrödinger-picture representation of the envelope operator,

$$\mathbf{\Phi}(\mathbf{r}) = \int_{Q_0} \frac{d^3q}{(2\pi)^3} f_q \sum_s a_{0s}(\mathbf{q})\, \mathbf{e}_s(\mathbf{k}_0 + \mathbf{q})\, e^{i\mathbf{q}\cdot\mathbf{r}}, \tag{7.28}$$

and using this in the calculation of the commutator yields

$$\begin{aligned}\left[\mathbf{\Phi}(\mathbf{r}), a^\dagger_{0s_m}(\mathbf{q}_m)\right] &= f_{q_m}\mathbf{e}_s(\mathbf{k}_0 + \mathbf{q}_m)\, e^{i\mathbf{q}_m\cdot\mathbf{r}} \\ &= \mathbf{e}_s(\mathbf{k}_0)\, e^{i\mathbf{q}_m\cdot\mathbf{r}} + O(\theta).\end{aligned} \tag{7.29}$$

Thus when acting on paraxial states the exact representation (7.28) can be replaced by the approximate form

$$\mathbf{\Phi}(\mathbf{r}) = \sum_s \phi_s(\mathbf{r})\, \mathbf{e}_{0s} + O(\theta), \tag{7.30}$$

where $\mathbf{e}_{0s} = \mathbf{e}_s(\mathbf{k}_0)$, and

$$\phi_s(\mathbf{r}) = \int_{Q_0} \frac{d^3q}{(2\pi)^3} a_{0s}(\mathbf{q})\, e^{i\mathbf{q}\cdot\mathbf{r}}. \tag{7.31}$$

The subscript Q_0 on the integral is to remind us that the integration domain is restricted by eqn (7.8). This representation can only be used when the operator acts on a vector in the paraxial space. It is in this sense that the z-component of the envelope operator is small, i.e.

$$\langle \Psi_1 | \Phi_z(\mathbf{r}) | \Psi_2 \rangle = O(\theta), \tag{7.32}$$

for any pair of normalized vectors $|\Psi_1\rangle$ and $|\Psi_2\rangle$ that both belong to $\mathfrak{H}(k_0, \theta)$. In the leading paraxial approximation, i.e. neglecting $O(\theta)$-terms, the electric field operator is

$$\mathbf{E}^{(+)}(\mathbf{r}, t) = i\sqrt{\frac{\hbar\omega_0 (v_{g0}/c)}{2\epsilon_0 n_0}} \sum_s \mathbf{e}_{0s}\phi_s(\mathbf{r}, t)\, e^{i(\mathbf{k}_0\cdot\mathbf{r} - \omega_0 t)}. \tag{7.33}$$

The commutation relations for the transverse components of the envelope operator have the simple form

$$\left[\Phi_i(\mathbf{r}, t), \Phi_j^\dagger(\mathbf{r}', t)\right] = \delta_{ij}\delta(\mathbf{r} - \mathbf{r}') \quad (i, j = 1, 2), \tag{7.34}$$

which shows that the paraxial electromagnetic field is described by two independent operators $\Phi_1(\mathbf{r})$ and $\Phi_2(\mathbf{r})$ satisfying local commutation relations. This reflects the fact that the paraxial approximation eliminates the nonlocal features exhibited in the exact commutation relations (3.16) by effectively averaging the arguments $\mathbf{r}$ and $\mathbf{r}'$ over volumes large compared to λ_0^3. By the same token, the delta function appearing on the right side of eqn (7.34) is **coarse-grained**, i.e. it only gives correct results when applied to functions that vary slowly on the scale of the carrier wavelength. This feature will be important when we return to the problem of photon localization.

In most applications the operators $\phi_s(\mathbf{r},t)$, corresponding to definite polarization states, are more useful. They satisfy the commutation relations

$$\left[\phi_s(\mathbf{r},t),\phi_{s'}^{\dagger}(\mathbf{r}',t)\right]=\delta_{ss'}\delta(\mathbf{r}-\mathbf{r}')\quad(s,s'=\pm\text{ or }1,2)\,. \tag{7.35}$$

The approximate expansion (7.31) can be inverted to get

$$a_{0s}(\mathbf{q})=\int d^3r\phi_s(\mathbf{r})\,e^{-i\mathbf{q}\cdot\mathbf{r}}=\int d^3r\mathbf{e}_s^*(\mathbf{k}_0)\cdot\mathbf{\Phi}(\mathbf{r})\,e^{-i\mathbf{q}\cdot\mathbf{r}}\,, \tag{7.36}$$

which is valid for $\mathbf{q}$ in the paraxial region Q_0. By using this inversion formula the operators N_0, $\mathbf{P}_0$, and H_P can be expressed in terms of the slowly-varying envelope operator:

$$N_0=\int d^3r\sum_s\phi_s^{\dagger}(\mathbf{r})\,\phi_s(\mathbf{r})\,, \tag{7.37}$$

$$\mathbf{P}_0=\int d^3r\sum_s\phi_s^{\dagger}(\mathbf{r})\,\frac{\hbar}{i}\boldsymbol{\nabla}\phi_s(\mathbf{r})\,, \tag{7.38}$$

$$H_P=\int d^3r\sum_s\phi_s^{\dagger}(\mathbf{r})\left\{v_{g0}\frac{\hbar}{i}\nabla_z-\frac{\hbar v_{g0}\nabla_{\mathsf{T}}^2}{2k_0}\right\}\phi_s(\mathbf{r})\,. \tag{7.39}$$

We can gain a better understanding of the paraxial Hamiltonian by substituting eqns (7.24) and (7.22) into the Heisenberg equation

$$i\hbar\frac{\partial}{\partial t}\mathbf{A}^{(+)}(\mathbf{r},t)=\left[\mathbf{A}^{(+)}(\mathbf{r},t)\,,H_{\text{em}}\right] \tag{7.40}$$

to get

$$\hbar\omega_0\mathbf{\Phi}(\mathbf{r},t)+i\hbar\frac{\partial}{\partial t}\mathbf{\Phi}(\mathbf{r},t)=\hbar\omega_0\left[\mathbf{\Phi}(\mathbf{r},t)\,,N_0\right]+\left[\mathbf{\Phi}(\mathbf{r},t)\,,H_P\right]. \tag{7.41}$$

Since the envelope operator $\mathbf{\Phi}(\mathbf{r},t)$ is a sum of annihilation operators, it satisfies $[\mathbf{\Phi}(\mathbf{r},t)\,,N_0]=\mathbf{\Phi}(\mathbf{r},t)$. Consequently, the term $\hbar\omega_0[\mathbf{\Phi}(\mathbf{r},t)\,,N_0]$ is canceled by the time derivative of the carrier wave. The Heisenberg equation for the envelope field $\mathbf{\Phi}(\mathbf{r},t)$ is therefore

$$i\hbar\frac{\partial}{\partial t}\mathbf{\Phi}(\mathbf{r},t)=\left[\mathbf{\Phi}(\mathbf{r},t)\,,H_P\right]. \tag{7.42}$$

This shows that the paraxial Hamiltonian generates the time translation of the envelope field. By using the explicit form (7.22) of H_P and the commutation relations (7.34), it is simple to see that the Heisenberg equation can be written in the equivalent forms

$$i\left(\nabla_z+\frac{1}{v_{g0}}\frac{\partial}{\partial t}\right)\mathbf{\Phi}(\mathbf{r},t)+\frac{1}{2k_0}\nabla_{\mathsf{T}}^2\mathbf{\Phi}(\mathbf{r},t)=0 \tag{7.43}$$

or

$$i\left(\nabla_z+\frac{1}{v_{g0}}\frac{\partial}{\partial t}\right)\phi_s(\mathbf{r},t)+\frac{1}{2k_0}\nabla_{\mathsf{T}}^2\phi_s(\mathbf{r},t)=0\,. \tag{7.44}$$

Multiplying eqn (7.43) by the normalization factor in eqn (7.24) and passing to the classical limit ($\mathbf{A}^{(+)}(\mathbf{r},t)\to\overline{\boldsymbol{\mathcal{A}}}(\mathbf{r},t)\exp\left[i\left(\mathbf{k}_0\cdot\mathbf{r}-\omega_0 t\right)\right]$) yields the standard paraxial wave equation of the classical theory.

The single-beam argument can be applied to each of the distinct beams to give the Schrödinger-picture representation,

$$\mathbf{A}^{(+)}(\mathbf{r}) = \sum_{\beta s} \sqrt{\frac{\hbar\,(v_{g\beta}/c)}{2\epsilon_0 k_\beta c}}\,\mathbf{e}_{\beta s}\phi_{\beta s}(\mathbf{r})\, e^{i\mathbf{k}_\beta\cdot\mathbf{r}}\,, \tag{7.45}$$

where $\mathbf{e}_{\beta s} = \mathbf{e}_s(\mathbf{k}_\beta)$, $\omega_\beta = \omega(k_\beta) = ck_\beta/n_\beta$, $v_{g\beta}$ is the group velocity for the βth carrier wave,

$$\phi_{\beta s}(\mathbf{r}) = \int_{Q_\beta} \frac{d^3q}{(2\pi)^3} a_{\beta s}(\mathbf{q})\, e^{i\mathbf{q}\cdot\mathbf{r}}\,, \tag{7.46}$$

and

$$\left[\phi_{\beta s}(\mathbf{r}), \phi^\dagger_{\beta' s'}(\mathbf{r}')\right] \approx \delta_{\beta\beta'}\delta_{ss'}\delta(\mathbf{r}-\mathbf{r}') \quad (s, s' = \pm \text{ or } 1, 2)\,. \tag{7.47}$$

The last result—which is established in Exercise 7.3—means that the envelope fields for distinct beams represent independent degrees of freedom.

The corresponding expression for the electric field operator in the paraxial approximation is

$$\mathbf{E}^{(+)}(\mathbf{r}) = \sum_{\beta s} i\sqrt{\frac{\hbar\omega_\beta\,(v_{g\beta}/c)}{2\epsilon_0 n_\beta}}\,\mathbf{e}_{\beta s}\phi_{\beta s}(\mathbf{r})\, e^{i\mathbf{k}_\beta\cdot\mathbf{r}}\,. \tag{7.48}$$

The operators for the photon number N_β, the momentum $\mathbf{P}_\beta$, and the paraxial Hamiltonian $H_{\beta P}$ of the individual beams are obtained by applying eqns (7.37)–(7.39) to each beam.

7.4 Gaussian beams and pulses

It is clear from the relation $\boldsymbol{\mathcal{E}} = -\partial\boldsymbol{\mathcal{A}}/\partial t$ that the electric field also satisfies the paraxial wave equation. For the special case of propagation along the z-axis through vacuum, we find

$$\frac{1}{2k_0}\nabla_{\mathsf{T}}^2\overline{\boldsymbol{\mathcal{E}}} + i\left(\frac{\partial\overline{\boldsymbol{\mathcal{E}}}}{\partial z} + \frac{1}{c}\frac{\partial\overline{\boldsymbol{\mathcal{E}}}}{\partial t}\right) = 0\,. \tag{7.49}$$

For fields with pulse duration much longer than any relevant time scale—or equivalently with spectral width much smaller than any relevant frequency—the time dependence of the slowly-varying envelope function can be neglected; that is, one can set $\partial\overline{\boldsymbol{\mathcal{E}}}/\partial t = 0$ in eqn (7.49). The most useful time-independent solutions of the paraxial equation are those which exhibit minimal diffractive spreading. The fundamental solution with these properties—which is called a **Gaussian beam** or a **Gaussian mode** (Yariv, 1989, Sec. 6.6)—is

$$\overline{\boldsymbol{\mathcal{E}}}(\mathbf{r}, t) = \boldsymbol{\mathcal{E}}_0(\mathbf{r}_\mathsf{T}, z) = \mathcal{E}_0\mathbf{e}_0\frac{w_0\; e^{-i\phi(z)}}{w(z)}\exp\left[ik_0\frac{\rho^2}{2R(z)}\right]\exp\left[-\frac{\rho^2}{w^2(z)}\right], \tag{7.50}$$

where the polarization vector $\mathbf{e}_0$ is in the x–y plane and $\rho = |\mathbf{r}_\mathsf{T}|$. The functions of z on the right side are defined by

$$w(z) = w_0 \sqrt{1 + \left(\frac{z - z_w}{Z_R}\right)^2}, \tag{7.51}$$

$$R(z) = z - z_w + \frac{Z_R^2}{z - z_w}, \tag{7.52}$$

$$\phi(z) = \tan^{-1}\left(\frac{z - z_w}{Z_R}\right), \tag{7.53}$$

where the **Rayleigh range** Z_R is

$$Z_R = \frac{\pi w_0^2}{\lambda_0} > 0\,. \tag{7.54}$$

The function $w(z)$—which defines the width of the transverse Gaussian profile—has the minimum value w_0 (the **spot size**) at $z = z_w$ (the **beam waist**). The solution is completely characterized by $\mathbf{e}_0$, $\mathcal{E}_0$, w_0, and z_w. The function $R(z)$—which represents the radius of curvature of the phase front—is negative for $z < z_w$, and positive for $z > z_w$. The picture is of waves converging from the left and diverging to the right of the focal point at the waist. The definition (7.51) shows that

$$w(z_w + Z_R) = \sqrt{2} w_0\,, \tag{7.55}$$

so the Rayleigh range measures the distance required for diffraction to double the area of the spot. There are also higher-order Gaussian modes that are not invariant under rotations around the beam axis (Yariv, 1989, Sec. 6.9).

The assumption $\partial\overline{\boldsymbol{\mathcal{E}}}/\partial t = 0$ means that the Gaussian beam represents an infinitely long pulse, so we should expect that it is not a normalizable solution. This is readily verified by showing that the normalization integral over the transverse coordinates has the z-independent value

$$\int d^2 r_{\mathsf{T}} \left|\boldsymbol{\mathcal{E}}_0(\mathbf{r}_{\mathsf{T}}, z)\right|^2 = \pi w_0^2 \left|\mathcal{E}_0\right|^2, \tag{7.56}$$

so that the z-integral diverges. A more realistic description is based on the observation that

$$\boldsymbol{\mathcal{E}}_P(\mathbf{r}, t) = \mathcal{F}_P(z - ct)\, \boldsymbol{\mathcal{E}}_0(\mathbf{r}_{\mathsf{T}}, z) \tag{7.57}$$

is a time-dependent solution of eqn (7.49) for any choice of the function $\mathcal{F}_P(z)$. If $\mathcal{F}_P(z)$ is normalizable, then the **Gaussian pulse** (or **Gaussian wave packet**) $\boldsymbol{\mathcal{E}}_P(\mathbf{r}, t)$ is normalizable at all times. The pulse-envelope function is frequently chosen to be Gaussian also, i.e.

$$\mathcal{F}_P(z) = \mathcal{F}_{P0} \exp\left[-\frac{(z - z_0)^2}{L_P^2}\right], \tag{7.58}$$

where L_P is the **pulse length** and $T_P = L_P/c$ is the **pulse duration**.

7.5 The paraxial expansion*

The approach to the quantum paraxial approximation presented above is sufficient for most practical purposes, but it does not provide any obvious way to calculate corrections. A systematic expansion scheme is desirable for at least two reasons.

(1) It is not wise to depend on an approximation in the absence of any method for estimating the errors involved.

(2) There are some questions of principle, e.g. the issue of photon localizability, which require the evaluation of higher-order terms.

We will therefore very briefly outline a systematic expansion in powers of θ (Deutsch and Garrison, 1991*a*) which is an extension of a method developed by Lax *et al.* (1974) for the classical theory. In the interests of simplicity, only propagation in the vacuum will be considered.

In order to construct a consistent expansion in powers of θ, it is first necessary to normalize all physical quantities by using the characteristic lengths introduced in Section 7.2.1. The first step is to define a characteristic volume

$$V_0 = \Lambda_{\mathsf{T}}^2 \Lambda_{\|} = \theta^{-4} \left(\frac{\lambda_0}{2\pi}\right)^3 , \tag{7.59}$$

and a dimensionless wavevector $\overline{\mathbf{q}} = \overline{\mathbf{q}}_{\mathsf{T}} + \overline{q}_z \widetilde{\mathbf{k}}_0$, with $\overline{\mathbf{q}}_{\mathsf{T}} = \mathbf{q}_{\mathsf{T}} \Lambda_{\mathsf{T}}$ and $\overline{q}_z = q_z \Lambda_{\|}$. In terms of the scaled wavevector $\overline{\mathbf{q}}$, the paraxial constraints (7.8) are

$$\overline{Q}_0 = \{\overline{\mathbf{q}} \text{ satisfying } |\overline{\mathbf{q}}_{\mathsf{T}}| \leqslant 1 , \ \overline{q}_z \leqslant 1\} . \tag{7.60}$$

The operators $a_s^{\dagger}(\mathbf{k})$ have dimensions $L^{3/2}$, so the dimensionless operators $\overline{a}_s^{\dagger}(\overline{\mathbf{q}}) = V_0^{-1/2} a_s^{\dagger}(\mathbf{k}_0 + \mathbf{q})$ satisfy the commutation relation

$$\left[\overline{a}_s(\overline{\mathbf{q}}) , \overline{a}_{s'}^{\dagger}(\overline{\mathbf{q}}')\right] = \delta_{ss'} (2\pi)^3 \delta(\overline{\mathbf{q}} - \overline{\mathbf{q}}') . \tag{7.61}$$

In the space–time domain, the operator $\mathbf{\Phi}(\mathbf{r}, t)$ has dimensions $L^{-3/2}$, so it is natural to define a dimensionless envelope field by $\overline{\mathbf{\Phi}}(\overline{\mathbf{r}}, \overline{t}) = \sqrt{V_0}\,\mathbf{\Phi}(\mathbf{r}, t)$, where $\overline{\mathbf{r}} = \overline{\mathbf{r}}_{\mathsf{T}} + \overline{z}\widetilde{\mathbf{k}}_0$ and $\overline{\mathbf{r}}_{\mathsf{T}} = \mathbf{r}_{\mathsf{T}}/\Lambda_{\mathsf{T}}$, $\overline{z} = z/\Lambda_{\|}$. The scaled position-space variables satisfy $\mathbf{q} \cdot \mathbf{r} = \overline{\mathbf{q}} \cdot \overline{\mathbf{r}} = \overline{\mathbf{q}}_{\mathsf{T}} \cdot \overline{\mathbf{r}}_{\mathsf{T}} + \overline{q}_z \overline{z}$. The operator $\overline{\mathbf{\Phi}}(\overline{\mathbf{r}}, \overline{t})$ is related to $\overline{a}_s(\overline{\mathbf{q}})$ by

$$\overline{\mathbf{\Phi}}(\overline{\mathbf{r}}) = \int_{Q_0} \frac{d^3\overline{q}}{(2\pi)^3} \sum_s \overline{a}_s(\overline{\mathbf{q}}) \mathbf{X}_s(\overline{\mathbf{q}}, \theta) e^{i\overline{\mathbf{q}} \cdot \overline{\mathbf{r}}} , \tag{7.62}$$

where $\mathbf{X}_s(\overline{\mathbf{q}}, \theta)$ is the c-number function:

$$\mathbf{X}_s(\overline{\mathbf{q}}, \theta) = \sqrt{\frac{k_0}{|\mathbf{k}_0 + \mathbf{q}|}} \mathbf{e}_s(\mathbf{k}_0 + \mathbf{q}) = \sum_{n=0}^{\infty} \theta^n \mathbf{X}_s^{(n)}(\overline{\mathbf{q}}) . \tag{7.63}$$

Substituting this expansion into eqn (7.62) and exchanging the sum over n with the integral over $\overline{\mathbf{q}}$ yields

$$\overline{\mathbf{\Phi}}(\overline{\mathbf{r}}) = \sum_{n=0}^{\infty} \theta^n \overline{\mathbf{\Phi}}^{(n)}(\overline{\mathbf{r}}), \tag{7.64}$$

where the nth-order coefficient is

$$\overline{\mathbf{\Phi}}^{(n)}(\overline{\mathbf{r}}) = \int' \frac{d^3\overline{q}}{(2\pi)^3} \sum_s \overline{a}_s(\overline{\mathbf{q}}) \mathbf{X}_s^{(n)}(\overline{\mathbf{q}}) e^{i\overline{\mathbf{q}}\cdot\overline{\mathbf{r}}}. \tag{7.65}$$

The zeroth-order relation

$$\overline{\mathbf{\Phi}}^{(0)}(\overline{\mathbf{r}}) = \int' \frac{d^3\overline{q}}{(2\pi)^3} \sum_s \overline{a}_s(\overline{\mathbf{q}}) \mathbf{e}_s(\mathbf{k}_0) e^{i\overline{\mathbf{q}}\cdot\overline{\mathbf{r}}} \tag{7.66}$$

agrees with the previous paraxial approximation (7.31), and it can be inverted to give

$$\overline{a}_s(\overline{\mathbf{q}}) = \int d^3\overline{r}\, \overline{\mathbf{\Phi}}^{(0)}(\overline{\mathbf{r}}) \cdot \mathbf{e}_s^*(\mathbf{k}_0) e^{-i\overline{\mathbf{q}}\cdot\overline{\mathbf{r}}}. \tag{7.67}$$

Carrying out Exercise 7.5 shows that all higher-order coefficients can be expressed in terms of $\overline{\mathbf{\Phi}}_0^{(0)}(\overline{\mathbf{r}})$.

We can justify the operator expansion (7.64) by calculating the action of the exact envelope operator on a typical basis vector in $\mathfrak{H}(\mathbf{k}_0, \theta)$, and showing that the expansion of the resulting vector in θ agrees—order-by-order—with the result of applying the operator expansion. In the same way it can be shown that the operator expansion reproduces the exact commutation relations (Deutsch and Garrison, 1991*a*).

7.6 Paraxial wave packets*

The use of non-normalizable basis states to define the paraxial space can be avoided by employing wave packet creation operators. For this purpose, we restrict the polarization amplitudes, $w_s(\mathbf{k})$, (introduced in Section 3.5.1) to those that have the form $w_s(\mathbf{k}_0 + \mathbf{q}) = V_0^{1/2} \overline{w}_s(\overline{\mathbf{q}})$. Instead of confining the relative wavevectors $\overline{\mathbf{q}}$ to the region $\overline{Q}_0$ described by eqn (7.60), we define a **paraxial wave packet** (with carrier wavevector $\mathbf{k}_0$ and opening angle θ) by the assumption that $\overline{w}_s(\overline{\mathbf{q}})$ vanishes rapidly outside $\overline{Q}_0$, i.e. $\overline{w}_s(\overline{\mathbf{q}})$ belongs to the space

$$\mathcal{P}(\mathbf{k}_0, \theta) = \left\{ \overline{w}_s(\overline{\mathbf{q}}) \quad \text{such that} \quad \lim_{|\overline{\mathbf{q}}|\to\infty} |\overline{\mathbf{q}}|^n |\overline{w}_s(\overline{\mathbf{q}})| = 0 \ \text{ for all } n \geqslant 0 \right\}. \tag{7.68}$$

The inner product for this space of classical wave packets is defined by

$$(w, v) = \int \frac{d^3 q}{(2\pi)^3} \sum_s w_s^*(\mathbf{k}_0 + \mathbf{q}) v_s(\mathbf{k}_0 + \mathbf{q}). \tag{7.69}$$

Since the two wave packets belong to the same space, this can be written in terms of scaled variables as

$$(w, v) = \int \frac{d^3\overline{q}}{(2\pi)^3} \sum_s \overline{w}_s^*(\overline{\mathbf{q}}) \overline{v}_s(\overline{\mathbf{q}}). \tag{7.70}$$

For a paraxial wave packet, we set $\mathbf{k} = \mathbf{k}_0 + \mathbf{q}$ in the general definition (3.191) to get

$$a^{\dagger}[w] = \int \frac{d^3q}{(2\pi)^3} \sum_s a^{\dagger}_{0s}(\mathbf{k}_0 + \mathbf{q})\, w_s(\mathbf{k}_0 + \mathbf{q}) = \int \frac{d^3\overline{q}}{(2\pi)^3} \sum_s \overline{a}^{\dagger}_s(\overline{\mathbf{q}})\, \overline{w}_s(\overline{\mathbf{q}})\,. \quad (7.71)$$

The paraxial space defined by eqn (7.10) can equally well be built up from the vacuum by forming all linear combinations of states of the form

$$|\{w\}_P\rangle = \prod_{p=1}^{P} a^{\dagger}[w_p]\,|0\rangle\,, \quad (7.72)$$

where $\{w\}_P = \{w_1, \ldots, w_P\}$, $P = 0, 1, 2, \ldots$, and the w_ps range over all of $\mathcal{P}(\mathbf{k}_0, \theta)$. The only difference from the construction of the full Fock space is the restriction of the wave packets to the paraxial space $\mathcal{P}(\mathbf{k}_0, \theta) \subset \Gamma_{\text{em}}$, where Γ_{em} is the electromagnetic phase space of classical wave packets defined by eqn (3.189).

The multiparaxial Hilbert spaces introduced in Section 7.2.2 can also be described in wave packet terms. The distinct paraxial beams considered there correspond to the wave packet spaces $\mathcal{P}(\mathbf{k}_1, \theta_1)$ and $\mathcal{P}(\mathbf{k}_2, \theta_2)$. Paraxial wave packets, $w \in \mathcal{P}(\mathbf{k}_1, \theta_1)$ and $v \in \mathcal{P}(\mathbf{k}_2, \theta_2)$, are concentrated around $\mathbf{k}_1$ and $\mathbf{k}_2$ respectively, so it is eminently plausible that w and v are effectively orthogonal. More precisely, it is shown in Exercise 7.6 that

$$\lim_{\theta_2 \to 0} \frac{1}{(\theta_2)^n} |(w, v)| = 0 \quad \text{for all } n \geqslant 1\,, \quad (7.73)$$

i.e. $|(w, v)|$ vanishes faster than any power of θ_2. The symmetry of the inner product guarantees that the same conclusion holds for θ_1; consequently, the wave packet spaces $\mathcal{P}(\mathbf{k}_1, \theta_1)$ and $\mathcal{P}(\mathbf{k}_2, \theta_2)$ can be treated as orthogonal to any finite order in θ_1 or θ_2.

The approximate orthogonality of the wave packets w and v combined with the general rule (3.192) implies

$$\left[a[w], a^{\dagger}[v]\right] = 0 \quad (7.74)$$

whenever w and v belong to distinct paraxial wave packet spaces. From this it is easy to see that the quantum paraxial spaces $\mathfrak{H}(\mathbf{k}_1, \theta_1)$ and $\mathfrak{H}(\mathbf{k}_2, \theta_2)$ are orthogonal to any finite order in the small parameters θ_1 and θ_2. In the paraxial approximation, distinct paraxial wave packets behave as though they were truly orthogonal modes. This means that the multiparaxial Hilbert space describing the situation in which several distinct paraxial beams are present is generated from the vacuum by generalizing eqn (7.72) to

$$\left|\{w_1\}_{P_1}, \{w_2\}_{P_2}, \ldots,\right\rangle = \prod_{\beta} \prod_{p=1}^{P_\beta} a^{\dagger}[w_{\beta p}]\,|0\rangle\,, \quad (7.75)$$

where $P_\beta = 0, 1, \ldots$, and the $w_{\beta p}$s are chosen from $\mathcal{P}(\mathbf{k}_\beta, \theta_\beta)$.

7.7 Angular momentum*

The derivation of the paraxial approximation for the angular momentum $\mathbf{J} = \mathbf{L} + \mathbf{S}$ is complicated by the fact—discussed in Section 3.4—that the operator $\mathbf{L}$ does not have a convenient expression in terms of plane waves. Fortunately, the argument used to show that the energy and the linear momentum are additive also applies to the angular momentum; therefore, we can restrict attention to a single paraxial space. Let us begin by rewriting the expression (3.58) for the helicity operator $\mathbf{S}$ as

$$\mathbf{S} = \hbar \int_P \frac{d^3\overline{q}}{(2\pi)^3} \frac{\widetilde{\mathbf{k}}_0 + \mathbf{q}/k_0}{\left|\widetilde{\mathbf{k}}_0 + \mathbf{q}/k_0\right|} \left[\overline{a}_+^\dagger(\overline{\mathbf{q}})\,\overline{a}_+(\overline{\mathbf{q}}) - \overline{a}_-^\dagger(\overline{\mathbf{q}})\,\overline{a}_-(\overline{\mathbf{q}})\right]. \tag{7.76}$$

The ratio $\mathbf{q}/k_0$ can be expressed as

$$\frac{\mathbf{q}}{k_0} = \frac{\Lambda_\mathsf{T}\mathbf{q}_\mathsf{T}}{\Lambda_\mathsf{T} k_0} + \frac{\Lambda_\| q_z}{\Lambda_\| k_0}\widetilde{\mathbf{k}}_0 = \theta\overline{\mathbf{q}}_\mathsf{T} + \theta^2\overline{q}_z\widetilde{\mathbf{k}}_0\,, \tag{7.77}$$

so expanding in powers of θ gives the simple result

$$\mathbf{S}_0 = \widetilde{\mathbf{k}}_0 S_0 + O(\theta)\,, \tag{7.78}$$

where

$$\begin{aligned} S_0 &= \hbar \int_P \frac{d^3\overline{q}}{(2\pi)^3} \left[\overline{a}_+^\dagger(\overline{\mathbf{q}})\,\overline{a}_+(\overline{\mathbf{q}}) - \overline{a}_-^\dagger(\overline{\mathbf{q}})\,\overline{a}_-(\overline{\mathbf{q}})\right] \\ &= \hbar \int d^3r \left[\phi_+^\dagger(\mathbf{r})\,\phi_+(\mathbf{r}) - \phi_-^\dagger(\mathbf{r})\,\phi_-(\mathbf{r})\right]. \end{aligned} \tag{7.79}$$

Thus, to lowest order, the helicity has only a longitudinal component; the leading transverse component is $O(\theta)$. This is the natural consequence of the fact that each photon has a wavevector close to $\mathbf{k}_0$.

To develop the approximation for $\mathbf{L}$ we substitute the paraxial representation (7.24) and the corresponding expression (7.48) for $\mathbf{E}^{(+)}(\mathbf{r},t)$ into eqn (3.57) to get

$$\begin{aligned} \mathbf{L}_0 &= 2i\epsilon_0 \int d^3r E_j^{(-)} \left(\mathbf{r} \times \frac{1}{i}\boldsymbol{\nabla}\right) A_j^{(+)} \\ &= \int d^3r \Phi_j^\dagger(\mathbf{r},t)\, e^{-i\mathbf{k}_0\cdot\mathbf{r}} \left(\mathbf{r} \times \frac{1}{i}\boldsymbol{\nabla}\right) \Phi_j(\mathbf{r},t)\, e^{i\mathbf{k}_0\cdot\mathbf{r}} \\ &= \int d^3r \Phi_j^\dagger(\mathbf{r},t) \left(\mathbf{r} \times \hbar\mathbf{k}_0 + \mathbf{r} \times \frac{1}{i}\boldsymbol{\nabla}\right) \Phi_j(\mathbf{r},t)\,, \end{aligned} \tag{7.80}$$

where the last line follows from the identity

$$e^{-i\mathbf{k}_0\cdot\mathbf{r}}\boldsymbol{\nabla} e^{i\mathbf{k}_0\cdot\mathbf{r}}\Phi_j(\mathbf{r},t) = (\boldsymbol{\nabla} + i\mathbf{k}_0)\,\Phi_j(\mathbf{r},t)\,. \tag{7.81}$$

This remaining gradient term can be written as

$$\begin{aligned} \mathbf{r} \times \frac{\hbar}{i}\boldsymbol{\nabla} &= \mathbf{r} \times \frac{\hbar}{i}\left(\widetilde{\mathbf{k}}_0\nabla_z + \boldsymbol{\nabla}_\mathsf{T}\right) \\ &= \mathbf{r} \times \widetilde{\mathbf{k}}_0\frac{\hbar}{i}\nabla_z + z\widetilde{\mathbf{k}}_0 \times \frac{\hbar}{i}\boldsymbol{\nabla}_\mathsf{T} + \mathbf{r}_\mathsf{T} \times \frac{\hbar}{i}\boldsymbol{\nabla}_\mathsf{T}, \end{aligned} \tag{7.82}$$

so that

$$\mathbf{L}_0 = \mathbf{L}_{0\mathsf{T}} + \widetilde{\mathbf{k}}_0 L_{0z}\,, \tag{7.83}$$

where the transverse part is given by

$$\mathbf{L}_{0\mathsf{T}} = \int d^3r \Phi_j^{\dagger}(\mathbf{r}) \left(\mathbf{r} \times \hbar \mathbf{k}_0 + \mathbf{r} \times \widetilde{\mathbf{k}}_0 \frac{\hbar}{i} \nabla_z + z \widetilde{\mathbf{k}}_0 \times \frac{\hbar}{i} \boldsymbol{\nabla}_{\mathsf{T}} \right) \Phi_j(\mathbf{r})\,, \tag{7.84}$$

and the longitudinal component is

$$L_{0z} = \int d^3r \Phi_j^{\dagger}(\mathbf{r}) \left\{ r_{\mathsf{T}1} \frac{\hbar}{i} \nabla_{\mathsf{T}2} - r_{\mathsf{T}2} \frac{\hbar}{i} \nabla_{\mathsf{T}1} \right\} \Phi_j(\mathbf{r})\,. \tag{7.85}$$

The transverse part $\mathbf{L}_{0\mathsf{T}}$ is dominated by the term proportional to $\hbar\mathbf{k}_0$. After expressing the integral in terms of the scaled variable $\overline{\mathbf{r}}$ and scaled field $\overline{\boldsymbol{\Phi}}$, one finds that $\mathbf{L}_{0\mathsf{T}} = O(1/\theta)$. The similar terms $\hbar\omega_0 N_0$ and $\hbar\mathbf{k}_0 N_0$ in the momentum and energy are $O(1/\theta^2)$, so they are even larger. This apparently singular behavior is physically harmless; it simply represents the fact that all photons in the wave packet have energies close to $\hbar\omega_0$ and momenta close to $\hbar\mathbf{k}_0$.

For the angular momentum the situation is different. The angular momenta of individual photons in plane-wave modes $\mathbf{k}_0+\mathbf{q}$ must exhibit large fluctuations due to the tight constraints on the polar angle $\vartheta_{\mathbf{k}}$ given by eqn (7.4). These fluctuations are not conjugate to the longitudinal component J_{0z}, since rotations around the z-axis leave $\vartheta_{\mathbf{k}}$ unchanged. On the other hand, the transverse components $\mathbf{L}_{0\mathsf{T}}$ generate rotations around the transverse axes which do change the value of $\vartheta_{\mathbf{k}}$. Thus we should expect large fluctuations in the transverse components of the angular momentum, which are described by the large transverse term $\mathbf{L}_{0\mathsf{T}}$. Thus only the longitudinal component L_{0z} is meaningful for a paraxial state. By combining eqns (7.85) and (7.79), we see that the lowest-order paraxial angular momentum operator is purely longitudinal,

$$\mathbf{J}_0 = \widetilde{\mathbf{k}}_0 \left[L_{0z} + S_0 \right]. \tag{7.86}$$

7.8 Approximate photon localizability*

Mandel's local number operator, defined by eqn (3.204), displays peculiar nonlocal properties. Despite this apparent flaw, Mandel was able to demonstrate that $N(V)$ behaves approximately like a local number operator in the limit $V \gg \lambda_0^3$, where λ_0 is the characteristic wavelength for a monochromatic field state. The important role played by this limit suggests using the paraxial expansion to investigate the alternative definitions of the local number operator in a systematic way. To this end we first introduce a scaled version of the Mandel detection operator by

$$\mathbf{M}(\mathbf{r}) = \frac{1}{\sqrt{V_0}} \overline{\mathbf{M}}(\overline{\mathbf{r}})\, e^{ik_0 z}\,. \tag{7.87}$$

By combining the definition (3.203) with the expansion (7.64), the identity (7.81), and the scaled gradient

$$\frac{\boldsymbol{\nabla}}{k_0} = \frac{1}{k_0}\boldsymbol{\nabla}_\mathsf{T} + \frac{1}{k_0}\mathbf{u}_3\frac{\partial}{\partial z}$$
$$= \theta\overline{\boldsymbol{\nabla}}_\mathsf{T} + \theta^2\mathbf{u}_3\overline{\nabla}_z\,, \tag{7.88}$$

one finds

$$\overline{\mathbf{M}} = \overline{\mathbf{M}}^{(0)} + \theta\overline{\mathbf{M}}^{(1)} + \theta^2\overline{\mathbf{M}}^{(2)} + O\left(\theta^3\right), \tag{7.89}$$

where $\overline{\mathbf{M}}^{(0)} = \overline{\boldsymbol{\Phi}}$, $\overline{\mathbf{M}}^{(1)} = \overline{\boldsymbol{\Phi}}^{(1)}$, and

$$\overline{\mathbf{M}}^{(2)} = \overline{\boldsymbol{\Phi}}^{(2)} - \frac{1}{4}\left(\overline{\nabla}_\mathsf{T}^2 + 2i\overline{\nabla}_z\right)\overline{\boldsymbol{\Phi}}\,. \tag{7.90}$$

The corresponding expansion for $N(V)$ is

$$N(V) = N^{(0)}(V) + \theta^2 N^{(2)}(V) + O\left(\theta^4\right), \tag{7.91}$$

where

$$\begin{aligned} N^{(0)}(V) &= \int d^3\overline{r}\,\overline{\boldsymbol{\Phi}}^{(0)\dagger}(\overline{\mathbf{r}})\cdot\overline{\boldsymbol{\Phi}}^{(0)}(\overline{\mathbf{r}})\,, \\ N^{(2)}(V) &= \int d^3\overline{r}\left\{\overline{\mathbf{M}}^{(1)\dagger}\cdot\overline{\mathbf{M}}^{(1)} + \left[\overline{\mathbf{M}}^{(0)\dagger}\cdot\overline{\mathbf{M}}^{(2)} + \mathrm{HC}\right]\right\}. \end{aligned} \tag{7.92}$$

A simple calculation using the local commutation relations (7.34) for the zeroth-order envelope field yields

$$\left[N^{(0)}(V)\,, N^{(0)}(V')\right] = 0 \tag{7.93}$$

for nonoverlapping volumes, and

$$\left[N^{(0)}(V)\,, \boldsymbol{\Phi}^\dagger(\overline{\mathbf{r}})\right] = \chi_V(\mathbf{r})\,\boldsymbol{\Phi}^\dagger(\overline{\mathbf{r}})\,, \tag{7.94}$$

where the characteristic function $\chi_V(\overline{\mathbf{r}})$ is defined by

$$\chi_V(\mathbf{r}) = \begin{cases} 1 & \text{for } \mathbf{r} \in V\,, \\ 0 & \text{for } \mathbf{r} \notin V\,. \end{cases} \tag{7.95}$$

Thus $N^{(0)}(V)$ acts like a genuine local number operator. The nonlocal features discussed in Section 3.6.2 will only appear in the higher-order terms. It is, however, important to remember that the delta function in the zeroth-order commutation relation (7.34) is really coarse-grained with respect to the carrier wavelength λ_0. For this reason the localization volume V must satisfy $V \gg \lambda_0^3$.

The paraxial expansion of the alternative operator $G(V)$, introduced in eqn (3.210), shows (Deutsch and Garrison, 1991*a*) that the two definitions agree in lowest order, $G^{(0)}(V) = N^{(0)}(V)$, but disagree in second order, $G^{(2)}(V) \neq N^{(2)}(V)$. This disagreement between equally plausible definitions for the local photon number operator is a consequence of the fact that a photon with wavelength λ_0 cannot be localized to a volume of order λ_0^3. Since most experiments are well described by the paraxial approximation, it is usually permissible to think of the photons as localized, provided that the diameter of the localization region is larger than a wavelength.

The negative frequency part $A_i^{(-)}(\mathbf{r})$ is a sum over creation operators, so it is tempting to interpret $A_i^{(-)}(\mathbf{r})$ as creating a photon at the point $\mathbf{r}$. In view of the impossibility of localizing photons, this temptation must be sternly resisted. On the other hand, the cavity operator $a_\kappa^\dagger$ can be interpreted as creating a photon described by the cavity mode $\boldsymbol{\mathcal{E}}_\kappa(\mathbf{r})$, since the mode function extends over the entire cavity. In the same way, the plane-wave operator $a_{\mathbf{k}s}^\dagger$ can be interpreted as creating a photon in the (box-normalized) plane-wave state with wavenumber $\mathbf{k}$ and polarization $\mathbf{e}_{\mathbf{k}s}$. Finally the wave packet operator $a^\dagger[w]$ can be interpreted as creating a photon described by the classical wave packet w, but it would be wrong to think of the photon as strictly localized in the region where $\mathbf{w}(\mathbf{r})$ is large. With this caution in mind, one can regard the pulse-envelope $\mathbf{w}(\mathbf{r})$ as an effective photon wave function, provided that the pulse duration contains many optical periods and the transverse profile is large compared to a wavelength.

There are other aspects of the averaged operators that also require some caution. The operator $N[w] = a^\dagger[w]\,a[w]$ satisfies

$$\left[N[w], a^\dagger[w]\right] = a^\dagger[w], \quad [N[w], a[w]] = -a[w], \tag{7.96}$$

so it serves as a number operator for w-photons, but these number operators are not mutually commutative, since

$$[N[w], N[u]] = (w,u)\left\{a^\dagger[u]\,a[w] - a^\dagger[w]\,a[u]\right\}. \tag{7.97}$$

Thus distinct w photons and u photons cannot be independently counted unless the classical wave packets w and u are orthogonal. This lack of commutativity can be important in situations that require the use of non-orthogonal modes (Deutsch *et al.*, 1991).

7.9 Exercises

7.1 Frequency spread for a paraxial beam

(1) Show that the fractional change in the index of refraction across a paraxial beam is

$$\frac{\Delta n}{n_0} = \frac{\Delta k}{k_0}\frac{\frac{\omega_0}{n_0}\left(\frac{dn}{d\omega}\right)_0}{1+\frac{\omega_0}{n_0}\left(\frac{dn}{d\omega}\right)_0},$$

where $n_0 = n(\omega_0) = \sqrt{\epsilon(\omega_0)/\epsilon_0}$ and $(dn/d\omega)_0$ is evaluated at the carrier frequency.

(2) Combine the relation $k = \sqrt{k_0^2 + |\mathbf{q}_\mathsf{T}|^2 + q_z^2}$ with eqns (7.5) and (7.7) to get

$$\frac{\Delta k}{k_0} = \frac{1}{2}\left(\frac{\Delta q_\mathsf{T}}{k_0}\right)^2 + O\left(\theta^4\right) = \frac{1}{2}\theta^2 + \cdots.$$

(3) Combine this with $\Delta\omega = v_{g0}\Delta k$ to find

$$\frac{\Delta\omega}{\omega_0} = \frac{n_0 v_{g0}}{ck_0}\frac{1}{2}k_0\theta^2 = \frac{n_0}{n_0 + \omega_0\left(\frac{dn}{d\omega}\right)_0}\frac{1}{2}\theta^2 < \frac{1}{2}\theta^2 \,.$$

7.2 Distinct paraxial Hilbert spaces are effectively orthogonal

Consider the paraxial subspaces $\mathfrak{H}(\mathbf{k}_1, \theta_1)$ and $\mathfrak{H}(\mathbf{k}_2, \theta_2)$ discussed in Section 7.2.2.

(1) For a typical basis vector $|\{\mathbf{q}s\}_\kappa\rangle$ in $\mathfrak{H}(\mathbf{k}_1, \theta_1)$ show that $a_s(\mathbf{k})\,|\{\mathbf{q}s\}_\kappa\rangle \approx 0$ whenever $|\mathbf{k} - \mathbf{k}_1| \gg \theta_1 |\mathbf{k}_1|$.

(2) Use this result to argue that each basis vector in $\mathfrak{H}(\mathbf{k}_2, \theta_2)$ is approximately orthogonal to every basis vector in $\mathfrak{H}(\mathbf{k}_1, \theta_1)$.

7.3 Distinct paraxial fields are independent

Combine the definition (7.46) with the definition (7.14) for distinct beams to show that eqn (7.47) is satisfied in the same sense that distinct paraxial spaces are orthogonal.

7.4 An analogy to many-body physics*

Consider a special paraxial state such that the z-dependence of the field $\phi_s(\mathbf{r})$ can be neglected and only one polarization is excited, so that $\phi_s(\mathbf{r}) \to \phi(\mathbf{r}_\mathsf{T})$. Define an effective photon mass M_0 such that the paraxial Hamiltonian H_P for this problem is formally identical to a second quantized description of a two-dimensional, nonrelativistic, many-particle system of bosons with mass M_0 (Huang, 1963, Appendix A.3; Feynman, 1972). This feature leads to interesting analogies between quantum optics and many-body physics (Chiao *et al.*, 1991; Deutsch *et al.*, 1992; Wright *et al.*, 1994).

7.5 Paraxial expansion*

(1) Expand $\mathbf{X}_s(\overline{\mathbf{q}}, \theta)$ through $O(\theta^2)$.

(2) Show that $\overline{\mathbf{\Phi}}^{(1)}(\overline{\mathbf{r}}) = i\widetilde{\mathbf{k}}_0 \mathbf{\nabla}_\mathsf{T} \cdot \mathbf{\Phi}^{(0)}$.

(3) Show that $\overline{\mathbf{\Phi}}^{(2)}(\overline{\mathbf{r}}) = \frac{1}{2}\mathbf{\nabla}_\mathsf{T}\left(\mathbf{\nabla}_\mathsf{T} \cdot \mathbf{\Phi}^{(0)}\right) + \frac{1}{4}\left(\mathbf{\nabla}_\mathsf{T}^2 + 2i\nabla_z\right)\mathbf{\Phi}^{(0)}$.

7.6 Distinct paraxial wave packet spaces are effectively orthogonal*

Consider two paraxial wave packets, $w \in \mathcal{P}(\mathbf{k}_1, \theta_1)$ and $v \in \mathcal{P}(\mathbf{k}_2, \theta_2)$, where $\mathbf{k}_1$ and $\mathbf{k}_2$ satisfy eqn (7.14).

(1) Apply the definitions of $\overline{\mathbf{q}}$ (Section 7.5) and $\overline{w}_s^*(\overline{\mathbf{q}})$ (Section 7.6) to show that

$$(w, v) = \sqrt{\frac{V_2}{V_1}} \int \frac{d^3\overline{q}}{(2\pi)^3} \sum_s \overline{w}_s^*(\overline{\mathbf{q}})\,\overline{v}_s\left(\overline{\mathbf{q}} + \overline{\mathbf{\Delta k}}\right),$$

where $\Delta\mathbf{k} = \mathbf{k}_1 - \mathbf{k}_2$ and the arguments of $\overline{w}_s^*$ and $\overline{v}_s$ are scaled with θ_1 and θ_2 respectively.

(2) Calculate $\overline{\mathbf{\Delta k}}$, explain why $\left|\overline{\mathbf{\Delta k}}\right| \gg |\overline{\mathbf{q}}|$, and combine this with the rapid fall off condition in eqn (7.68) to conclude that $\theta_2^{-n}(w, v) \to 0$ as $\theta_2 \to 0$ for any value of n.

(3) Show that $\theta_2^{-n}\left[a\left[w\right], a^{\dagger}\left[v\right]\right] \rightarrow 0$ as $\theta_2 \rightarrow 0$.

8 Linear optical devices

The manipulation of light beams by passive linear devices, such as lenses, mirrors, stops, and beam splitters, is the backbone of experimental optics. In typical arrangements the individual devices are separated by regions called **propagation segments** in which the light propagates through air or vacuum. The index of refraction is usually piece-wise constant, i.e. it is uniform in each device and in each propagation segment. In most arrangements each device or propagation segment has an axis of symmetry (the **optic axis**), and the angle between the rays composing the beam and the local optic axis is usually small. The light beams are then said to be **piece-wise paraxial**. Under these circumstances, it is useful to treat the interaction of a light beam with a single device as a scattering problem in which the incident and scattered fields both propagate in vacuum. The optical properties of the device determine a linear relation between the complex amplitudes of the incident and scattered classical waves. After a brief review of this classical approach, we will present a phenomenological description of quantized electromagnetic fields interacting with linear optical devices. This approach will show that, at the quantum level, linear optical effects can be viewed—in a qualitative sense—as the propagation of photons guided by classical scattered waves. The scattered waves are a rough analogue of wave functions for particles, so the associated classical rays may be loosely considered as photon trajectories. These classical analogies are useful for visualizing the interaction of photons with linear optical devices but—as is always the case with applications of quantum theory—they must be used with care. A more precise wave-function-like description of quantum propagation through optical systems is given in Section 6.6.2.

8.1 Classical scattering

The general setting for this discussion is a situation in which one or more paraxial beams interact with an optical device to produce several scattered paraxial beams. Both the incident and the scattered beams are assumed to be mutually distinct, in the sense defined by eqn (7.14). Under these circumstances, the paraxial beams will be called **scattering channels**; the **incident** classical fields are input channels and the **scattered** beams are output channels. Since this process is linear in the fields, the initial and final beams can be resolved into plane waves. The conventional classical description of propagation through optical elements pieces together plane-wave solutions of Maxwell's equations by applying the appropriate boundary conditions at the interfaces between media with different indices of refraction, as shown in Fig. 8.1(a). This procedure yields a linear relation between the Fourier coefficients of the incident and

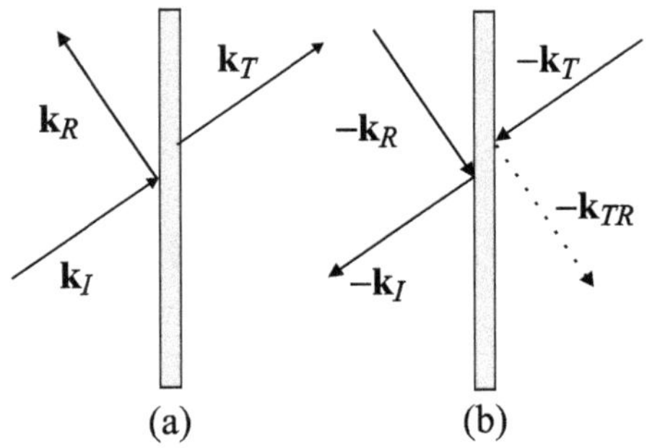

Fig. 8.1 (a) A plane wave $\alpha_{\mathbf{k}_I} \exp(i\mathbf{k}_I \cdot \mathbf{r})$ incident on a dielectric slab. The reflected and transmitted waves are respectively $\alpha_{\mathbf{k}_R} \exp(i\mathbf{k}_R \cdot \mathbf{r})$ and $\alpha_{\mathbf{k}_T} \exp(i\mathbf{k}_T \cdot \mathbf{r})$. (b) The time reversed version of (a). The extra wave at $-\mathbf{k}_{TR}$ is discussed in the text.

scattered waves that is similar to the description of scattering in terms of stationary states in quantum theory (Bransden and Joachain, 1989, Chap. 4). From the viewpoint of scattering theory, the classical piecing procedure is simply a way to construct the **scattering matrix** relating the incident and scattered fields. Before considering the general case, we analyze two simple examples: a propagation segment and a thin slab of dielectric.

For the propagation segment, an incident plane wave $\boldsymbol{\alpha} \exp(i\mathbf{k} \cdot \mathbf{r})$—the input channel—simply acquires the phase kL, where L is the length of the segment along the propagation direction, i.e. the relation between the incident amplitude $\boldsymbol{\alpha}$ and the scattered amplitude $\boldsymbol{\alpha}'$—representing the output channel—is

$$\boldsymbol{\alpha}' = e^{ikL}\boldsymbol{\alpha} = e^{i\omega L/c}\alpha\,. \tag{8.1}$$

In some applications the propagation segment through vacuum is replaced by a length L of dielectric. If the end faces of the dielectric sample are antireflection coated, then the scattering relation is

$$\boldsymbol{\alpha}' = e^{ik(\omega)L}\boldsymbol{\alpha} = e^{in(\omega)\omega L/c}\boldsymbol{\alpha}\,, \tag{8.2}$$

where $n(\omega)$ is the index of refraction for the dielectric. Since the transmitted wave can be expressed as

$$\boldsymbol{\alpha}' e^{ik(z-\omega t)} = \alpha e^{i[kz-\omega(t-\Delta t)]}\,, \tag{8.3}$$

where $\Delta t = n(\omega) L/c$, the dielectric medium is called a **retarder plate**, or sometimes a **phase shifter**.

We next turn to the example of a plane wave incident on a thin dielectric slab—which is not antireflection coated—as shown in Fig. 8.1(a). Ordinary ray tracing, using Snell's law and the law of reflection at each interface between the dielectric and vacuum, determines the directions of the propagation vectors $\mathbf{k}_R$ and $\mathbf{k}_T$ (where R and T stand for the reflected and transmitted waves respectively) relative to the propagation vector $\mathbf{k}_I$ of the incoming wave. Since the transmitted wave crosses the dielectric–vacuum interface twice, we find the familiar result $\mathbf{k}_T = \mathbf{k}_I$, i.e. the incident and transmitted waves are described by the same spatial mode.

The plane of incidence is defined by the vectors $\mathbf{k}_I$ and $\mathbf{n}$, where $\mathbf{n}$ is the unit vector normal to the slab. Every incident electromagnetic plane wave can be resolved into two polarization components: the **TE**- (or **S**-) polarization, with electric vector perpendicular to the plane of incidence, and the **TM**- (or **P**-) polarization, with electric vector in the plane of incidence. For optically isotropic dielectrics, these two polarizations are preserved by reflection and refraction. Since scattering is a linear process, we lose nothing by assuming that the incident wave is either TE- or TM-polarized. This allows us to simplify the vector problem to a scalar problem by suppressing the polarization vectors. The three waves outside the slab are then $\alpha_{\mathbf{k}_I} \exp(i\mathbf{k}_I \cdot \mathbf{r})$, $\alpha'_{\mathbf{k}_R} \exp(i\mathbf{k}_R \cdot \mathbf{r})$, and $\alpha'_{\mathbf{k}_T} \exp(i\mathbf{k}_T \cdot \mathbf{r})$. The solution of Maxwell's equations inside the slab is a linear combination of the transmitted wave at the first interface and the reflected wave from the second interface. Applying the boundary conditions at each interface (Jackson, 1999, Sec. 7.3) yields a set of equations relating the coefficients, and eliminating the coefficients for the interior solution leads to

$$\alpha'_{\mathbf{k}_R} = \mathfrak{r}\, \alpha_{\mathbf{k}_I}\,, \quad \alpha'_{\mathbf{k}_T} = \mathfrak{t}\, \alpha_{\mathbf{k}_I}\,, \tag{8.4}$$

where the complex parameters $\mathfrak{r}$ and $\mathfrak{t}$ are respectively the amplitude reflection and transmission coefficients for the slab. This is the simplest example of the general piecing procedure discussed above.

Important constraints on the coefficients $\mathfrak{r}$ and $\mathfrak{t}$ follow from the time-reversal invariance of Maxwell's equations. What this means is that the time-reversed final field will evolve into the time-reversed initial field. This situation is shown in Fig. 8.1(b), where the incident waves have propagation vectors $-\mathbf{k}_R$ and $-\mathbf{k}_T$ and the scattered waves have $-\mathbf{k}_I$ and $-\mathbf{k}_{TR}$. The amplitudes for this case are written as $\alpha^T_{\mathbf{q}}$, where T stands for time reversal. The usual calculation gives the scattered waves as

$$\begin{aligned} \alpha^T_{-\mathbf{k}_I} &= \mathfrak{r}\, \alpha'^T_{-\mathbf{k}_R} + \mathfrak{t}\, \alpha'^T_{-\mathbf{k}_T}\,, \\ \alpha^T_{-\mathbf{k}_{TR}} &= \mathfrak{t}\, \alpha'^T_{-\mathbf{k}_R} + \mathfrak{r}\, \alpha'^T_{-\mathbf{k}_T}\,. \end{aligned} \tag{8.5}$$

In Appendix B.3.3 it is shown that the linear polarization basis can be chosen so that the time-reversed amplitudes are related to the original amplitudes by eqn (B.80). In the present case, this yields $\alpha^T_{-\mathbf{k}_I} = \alpha^*_{\mathbf{k}_I}$, $\alpha'^T_{-\mathbf{k}_R} = \alpha'^*_{\mathbf{k}_R}$, $\alpha'^T_{-\mathbf{k}_T} = \alpha'^*_{\mathbf{k}_T}$, and $\alpha^T_{-\mathbf{k}_{TR}} = \alpha^*_{\mathbf{k}_{TR}}$. Substituting these relations into eqn (8.5) and taking the complex conjugate gives a second set of relations between the amplitudes $\alpha_{\mathbf{k}_I}$, $\alpha'_{\mathbf{k}_R}$, and $\alpha'_{\mathbf{k}_T}$:

$$\begin{aligned} \alpha_{\mathbf{k}_I} &= \mathfrak{r}^*\, \alpha'_{\mathbf{k}_R} + \mathfrak{t}^*\, \alpha'_{\mathbf{k}_T}\,, \\ \alpha_{\mathbf{k}_{TR}} &= \mathfrak{t}^*\, \alpha'_{\mathbf{k}_R} + \mathfrak{r}^*\, \alpha'_{\mathbf{k}_T}\,. \end{aligned} \tag{8.6}$$

There is an apparent discrepancy here, since the original problem had no wave with propagation vector $\mathbf{k}_{TR}$. Time-reversal invariance for the original problem therefore requires $\alpha_{\mathbf{k}_{TR}} = 0$. Using eqn (8.4) to eliminate $\alpha_{\mathbf{k}_R}$ and $\alpha_{\mathbf{k}_T}$ from eqn (8.6) and imposing $\alpha_{\mathbf{k}_{TR}} = 0$ leads to the constraints

$$\begin{aligned} |\mathfrak{r}|^2 + |\mathfrak{t}|^2 &= 1\,, \\ \mathfrak{r}\, \mathfrak{t}^* + \mathfrak{r}^* \mathfrak{t} &= 0\,. \end{aligned} \tag{8.7}$$

The first relation represents conservation of energy, while the second implies that the transmitted part of $-\mathbf{k}_R$ and the reflected part of $-\mathbf{k}_T$ interfere destructively as required by time-reversal invariance. These relations were originally derived by Stokes (Born and Wolf, 1980, Sec. 1.6).

Setting $\mathfrak{r} = |\mathfrak{r}| \exp(i\theta_{\mathfrak{r}})$ and $\mathfrak{t} = |\mathfrak{t}| \exp(i\theta_{\mathfrak{t}})$ in the second line of eqn (8.7) shows us that time-reversal invariance imposes the relation

$$\theta_{\mathfrak{r}} - \theta_{\mathfrak{t}} = \pm\pi/2\,; \tag{8.8}$$

in other words, the phase of the reflected wave is shifted by $\pm 90°$ relative to the transmitted wave. This phase difference is a measurable quantity; therefore, the $\pm$ sign on the right side of eqn (8.8) is not a matter of convention. In fact, this sign determines whether the reflected wave is retarded or advanced relative to the transmitted wave. In the extreme limit of a perfect mirror, i.e. $|\mathfrak{t}| \to 0$, we can impose the convention $\theta_{\mathfrak{t}} = 0$, so that

$$\theta_{\mathfrak{r}} = \pm\pi/2\,, \quad |\mathfrak{r}| = 1\,. \tag{8.9}$$

For given values of the relevant parameters—the angle of incidence, the index of refraction of the dielectric, and the thickness of the slab—the coefficients $\mathfrak{r}$ and $\mathfrak{t}$ can be exactly calculated (Born and Wolf, 1980, Sec. 1.6.4, eqns (57) and (58)), and the phases $\theta_{\mathfrak{r}}$ and $\theta_{\mathfrak{t}}$ are uniquely determined.

Let us now consider a more general situation in which waves with $\mathbf{k}_I$ and $\mathbf{k}_{TR}$ are both incident. This would be the time-reverse of Fig. 8.1(b), but in this case $\alpha_{\mathbf{k}_{TR}} \neq 0$. The standard calculation then relates $\alpha_{\mathbf{k}_T}$ and $\alpha_{\mathbf{k}_R}$ to $\alpha_{\mathbf{k}_I}$ and $\alpha_{\mathbf{k}_{TR}}$ by

$$\begin{pmatrix} \alpha'_{\mathbf{k}_T} \\ \alpha'_{\mathbf{k}_R} \end{pmatrix} = \begin{bmatrix} \mathfrak{t} & \mathfrak{r} \\ \mathfrak{r} & \mathfrak{t} \end{bmatrix} \begin{pmatrix} \alpha_{\mathbf{k}_I} \\ \alpha_{\mathbf{k}_{TR}} \end{pmatrix}. \tag{8.10}$$

The meaning of the conditions (8.7) is that the 2×2 scattering matrix in this equation is unitary.

Having mastered the simplest possible optical elements, we proceed without hesitation to the general case of linear and nondissipative optical devices. The incident field is to be expressed as an expansion in box-quantized plane waves,

$$\mathbf{f}_{\mathbf{k}s}(\mathbf{r}) = \mathbf{e}_{\mathbf{k}s} \exp(i\mathbf{k} \cdot \mathbf{r}) / \sqrt{V}\,. \tag{8.11}$$

For the single-mode input field $\boldsymbol{\mathcal{E}}_{\text{in}} = \mathbf{f}_{\mathbf{k}s} e^{-i\omega_k t}$, the general piecing procedure yields an output field which we symbolically denote by $(\mathbf{f}_{\mathbf{k}s})_{\text{scat}}$. This field is also expressed as an expansion in box-quantized plane waves. For a given basis function $\mathbf{f}_{\mathbf{k}s}$, we denote the expansion coefficients of the scattered solution by $S_{\mathbf{k}'s',\mathbf{k}s}$, so that

$$(\mathbf{f}_{\mathbf{k}s})_{\text{scat}} = \sum_{\mathbf{k}'s'} \mathbf{f}_{\mathbf{k}'s'} S_{\mathbf{k}'s',\mathbf{k}s}\,. \tag{8.12}$$

Repeating this procedure for all elements of the basis defines the entire scattering matrix $S_{\mathbf{k}'s',\mathbf{k}s}$. The assumption that the device is stationary means that the frequency ω_k associated with the mode $\mathbf{f}_{\mathbf{k}s}$ cannot be changed; therefore the scattering matrix must satisfy

$$S_{\mathbf{k}'s',\mathbf{k}s} = 0 \ \text{ if } \ \omega_{k'} \neq \omega_k\,. \tag{8.13}$$

In general, the sub-matrix connecting plane waves with a common frequency $\omega_k = \omega$ will depend on ω.

The incident classical wave packet is represented by the **in-field**

$$\mathcal{E}_{\text{in}}^{(+)}(\mathbf{r},t)=\sum_{\mathbf{k}s} i\sqrt{\frac{\hbar\omega_k}{2\epsilon_0}}\alpha_{\mathbf{k}s}\mathbf{f}_{\mathbf{k}s}(\mathbf{r})\,e^{-i\omega_k t}\,, \tag{8.14}$$

where the time origin $t=0$ is chosen so that the initial wave packet $\mathcal{E}_{\text{in}}(\mathbf{r},0)$ has not reached the optical element. For $t\,(>0)$ sufficiently large, the scattered wave packet has passed through the optical element, so that it is again freely propagating. The solution after the scattering is completely over is the **out-field**

$$\mathcal{E}_{\text{out}}^{(+)}(\mathbf{r},t)=\sum_{\mathbf{k}'s'} i\sqrt{\frac{\hbar\omega_{k'}}{2\epsilon_0}}\alpha'_{\mathbf{k}'s'}\mathbf{f}_{\mathbf{k}'s'}(\mathbf{r})\,e^{-i\omega_{k'} t}\,, \tag{8.15}$$

where the two sets of expansion coefficients are related by the scattering matrix:

$$\alpha'_{\mathbf{k}'s'}=\sum_{\mathbf{k}s} S_{\mathbf{k}'s',\mathbf{k}s}\alpha_{\mathbf{k}s}\,. \tag{8.16}$$

Time-reversal invariance can be exploited here as well. In the time-reversed problem, the time-reversed output field scatters into the time-reversed input field, so

$$\alpha^T_{-\mathbf{k}s}=\sum_{\mathbf{k}'s'} S_{-\mathbf{k}s,-\mathbf{k}'s'}\,\alpha'^T_{-\mathbf{k}'s'}\,, \tag{8.17}$$

where $-\mathbf{k}s$ is the time reversal of $\mathbf{k}s$. Time-reversal invariance requires

$$S_{-\mathbf{k}s,-\mathbf{k}'s'}=S_{\mathbf{k}'s',\mathbf{k}s}\,, \tag{8.18}$$

where the transposition of the indices reflects the interchange of incoming and outgoing modes. The classical rule (see Appendix B.3.3) for time reversal is

$$\alpha^T_{-\mathbf{k}s}=-\alpha^*_{\mathbf{k}s}\,, \tag{8.19}$$

so using eqn (8.18) in the complex conjugate of eqn (8.17) yields

$$\alpha_{\mathbf{k}s}=\sum_{\mathbf{k}'s'} S^*_{\mathbf{k}'s',\mathbf{k}s}\alpha'_{\mathbf{k}'s'}\,. \tag{8.20}$$

Combining this with eqn (8.16) leads to

$$\alpha_{\mathbf{k}s}=\sum_{\mathbf{k}''s''}\left[\sum_{\mathbf{k}'s'} S^*_{\mathbf{k}'s',\mathbf{k}s}S_{\mathbf{k}'s',\mathbf{k}''s''}\right]\alpha_{\mathbf{k}''s''}\,, \tag{8.21}$$

which must hold for all input fields $\{\alpha_{\mathbf{k}s}\}$. This imposes the constraints

$$\sum_{\mathbf{k}'s'} S^*_{\mathbf{k}'s',\mathbf{k}s}S_{\mathbf{k}'s',\mathbf{k}''s''}=\delta_{\mathbf{k}\mathbf{k}''}\delta_{ss''}\,, \tag{8.22}$$

that are generalizations of eqn (8.7). In matrix form this is $S^\dagger S=SS^\dagger=1$; i.e. every passive linear device is described by a unitary scattering matrix.

8.2 Quantum scattering

We will take a phenomenological approach in which the classical amplitudes are replaced by the Heisenberg-picture operators $a_{\mathbf{k}s}(t)$. Let $t = 0$ be the time at which the Heisenberg and Schrödinger pictures coincide, then according to eqn (3.95) the operator

$$\widetilde{a}_{\mathbf{k}s}(t) = a_{\mathbf{k}s}(t)\, e^{i\omega_k t} \tag{8.23}$$

is independent of time for free propagation. Thus in the scattering problem the time dependence of $\widetilde{a}_{\mathbf{k}'s'}(t)$ comes entirely from the interaction between the field and the optical element. The classical amplitudes $\alpha_{\mathbf{k}s}$ represent the solution prior to scattering, so it is natural to replace them according to the rule

$$\alpha_{\mathbf{k}s} \to \lim_{t\to 0}\left\{a_{\mathbf{k}s}(t)\, e^{i\omega_k t}\right\} = a_{\mathbf{k}s}(0) = a_{\mathbf{k}s}\,. \tag{8.24}$$

Similarly, $\alpha'_{\mathbf{k}'s'}$ represents the solution after scattering, and the corresponding rule,

$$\alpha'_{\mathbf{k}'s'} \to a'_{\mathbf{k}'s'} = \lim_{t\to +\infty}\left\{a_{\mathbf{k}'s'}(t)\, e^{i\omega_{k'} t}\right\} = \lim_{t\to\infty}\left\{\widetilde{a}_{\mathbf{k}'s'}(t)\right\}, \tag{8.25}$$

implies the asymptotic *ansatz*

$$a_{\mathbf{k}'s'}(t) \to a'_{\mathbf{k}'s'} e^{-i\omega_{k'} t}\,. \tag{8.26}$$

At late times the field is propagating in vacuum, so this limit makes sense by virtue of the fact that $\widetilde{a}_{\mathbf{k}'s'}(t)$ is time independent for free propagation.

Thus $a_{\mathbf{k}s}$ and $a'_{\mathbf{k}'s'}$ are respectively the **incident** and **scattered** annihilation operators, and they will be linearly related in the weak-field limit. Furthermore, the correspondence principle tells us that the relation between the operators must reproduce eqn (8.16) in the classical limit $a_{\mathbf{k}s} \to \alpha_{\mathbf{k}s}$. Since both relations are linear, this can only happen if the incident and scattered operators also satisfy

$$a'_{\mathbf{k}'s'} = \sum_{\mathbf{k}s} S_{\mathbf{k}'s',\mathbf{k}s} a_{\mathbf{k}s}\,, \tag{8.27}$$

where $S_{\mathbf{k}'s',\mathbf{k}s}$ is the classical scattering matrix. The **in-field operator $\mathbf{E}_{\text{in}}$** and the **out-field operator $\mathbf{E}_{\text{out}}$** are given by the quantum analogues of eqns (8.14) and (8.15):

$$\mathbf{E}^{(+)}_{\text{in}}(\mathbf{r},t) = \sum_{\mathbf{k}s} i\sqrt{\frac{\hbar\omega_k}{2\epsilon_0}}\, a_{\mathbf{k}s}\mathbf{f}_{\mathbf{k}s}(\mathbf{r})\, e^{-i\omega_k t}\,, \tag{8.28}$$

$$\mathbf{E}^{(+)}_{\text{out}}(\mathbf{r},t) = \sum_{\mathbf{k}'s'} i\sqrt{\frac{\hbar\omega_{k'}}{2\epsilon_0}}\, a'_{\mathbf{k}'s'}\mathbf{f}_{\mathbf{k}'s'}(\mathbf{r})\, e^{-i\omega_{k'} t}\,. \tag{8.29}$$

The operators $\{a_{\mathbf{k}s}\}$ and $\{a'_{\mathbf{k}'s'}\}$ are related by eqn (8.27) and the inverse relation

$$a_{\mathbf{k}s} = \sum_{\mathbf{k}'s'} \left(S^{\dagger}\right)_{\mathbf{k}s,\mathbf{k}'s'} a'_{\mathbf{k}'s'} = \sum_{\mathbf{k}'s'} S^{*}_{\mathbf{k}'s',\mathbf{k}s} a'_{\mathbf{k}'s'}\,. \tag{8.30}$$

The unitarity of the classical scattering matrix guarantees that the scattered operators $\{a'_{\mathbf{k}'s'}\}$ satisfy the canonical commutation relations (3.65), provided that the incident operators $\{a_{\mathbf{k}s}\}$ do so.

The use of the Heisenberg picture nicely illustrates the close relation between the classical and quantum scattering problems, but the Schrödinger-picture description of scattering phenomena is often more useful for the description of experiments. The fixed Heisenberg-picture state vector $|\Psi\rangle$ is the initial state vector in the Schrödinger picture, i.e. $|\Psi(0)\rangle = |\Psi\rangle$, so the time-dependent Schrödinger-picture state vector is

$$|\Psi(t)\rangle = U(t)|\Psi\rangle, \tag{8.31}$$

where $U(t)$ is the unitary evolution operator. Combining the formal solution (3.83) of the Heisenberg operator equations with the *ansatz* (8.26) yields

$$a_{\mathbf{k}s}(t) = U^{\dagger}(t)\, a_{\mathbf{k}s} U(t) \to a'_{\mathbf{k}s} e^{-i\omega_k t} \quad \text{as} \quad t \to \infty, \tag{8.32}$$

which provides some asymptotic information about the evolution operator.

The task at hand is to use this information to find the asymptotic form of $|\Psi(t)\rangle$. Since the scattering medium is linear, it is sufficient to consider a one-photon initial state,

$$|\Psi\rangle = \sum_{\mathbf{k}s} C_{\mathbf{k}s} a^{\dagger}_{\mathbf{k}s} |0\rangle. \tag{8.33}$$

The equivalence between the two pictures implies

$$\langle 0 | a_{\mathbf{k}s} | \Psi(t)\rangle = \langle 0 | a_{\mathbf{k}s}(t) | \Psi\rangle, \tag{8.34}$$

where the left and right sides are evaluated in the Schrödinger and Heisenberg pictures respectively. Since there is neither emission nor absorption in the passive scattering medium, $|\Psi(t)\rangle$ remains a one-photon state at all times, and

$$|\Psi(t)\rangle = \sum_{\mathbf{k}s} \langle 0 | a_{\mathbf{k}s} | \Psi(t)\rangle\, a^{\dagger}_{\mathbf{k}s} |0\rangle. \tag{8.35}$$

The expansion coefficients $\langle 0 | a_{\mathbf{k}s} | \Psi(t)\rangle$ are evaluated by combining eqn (8.34) with the asymptotic rule (8.26) and the scattering law (8.27) to get $\langle 0 | a_{\mathbf{k}s}(t) | \Psi\rangle = e^{-i\omega_k t} C'_{\mathbf{k}s}$, where

$$C'_{\mathbf{k}s} = \sum_{\mathbf{k}'s'} S_{\mathbf{k}s,\mathbf{k}'s'} C_{\mathbf{k}'s'}. \tag{8.36}$$

The evolved state is therefore

$$|\Psi(t)\rangle = \sum_{\mathbf{k}s} e^{-i\omega_k t} C'_{\mathbf{k}s} a^{\dagger}_{\mathbf{k}s} |0\rangle. \tag{8.37}$$

In other words, the prescription for the asymptotic ($t \to \infty$) form of the Schrödinger state vector is simply to replace the initial coefficients $C_{\mathbf{k}s}$ by $e^{-i\omega_k t} C'_{\mathbf{k}s}$, where $C'_{\mathbf{k}s}$ is the transform of the initial coefficient vector by the scattering matrix.

In the standard formulation of scattering theory, the initial state is stationary—i.e. an eigenstate of the free Hamiltonian—in which case all terms in the sum over $\mathbf{k}s$ in eqn (8.33) have the same frequency: $\omega_{\mathbf{k}} = \omega_0$. The energy conservation rule (8.13) guarantees that the same statement is true for the evolved state $|\Psi(t)\rangle$, so the time-dependent exponentials can be taken outside the sum in eqn (8.37) as the overall phase factor $\exp(-i\omega_0 t)$. In this situation the overall phase can be neglected, and the asymptotic evolution law (8.37) can be replaced by the scattering law

$$|\Psi\rangle \to |\Psi\rangle' = \sum_{\mathbf{k}s} C'_{\mathbf{k}s} a^\dagger_{\mathbf{k}s} |0\rangle . \tag{8.38}$$

An equivalent way to describe the asymptotic evolution follows from the observation that the evolved state in eqn (8.37) is obtained from the initial state in eqn (8.33) by the operator transformation

$$a^\dagger_{\mathbf{k}s} \to e^{-i\omega_k t} \sum_{\mathbf{k}'s'} a^\dagger_{\mathbf{k}'s'} S_{\mathbf{k}'s',\mathbf{k}s} . \tag{8.39}$$

When applying this rule to stationary states, the time-dependent exponential can be dropped to get the scattering rule

$$a^\dagger_{\mathbf{k}s} \to a'^\dagger_{\mathbf{k}s} = \sum_{\mathbf{k}'s'} a^\dagger_{\mathbf{k}'s'} S_{\mathbf{k}'s',\mathbf{k}s} . \tag{8.40}$$

For scattering problems involving one- or two-photon initial states, it is often more convenient to use eqn (8.40) directly rather than eqn (8.38). For example, the scattering rule for $|\Psi\rangle = a^\dagger_{\mathbf{k}s} |0\rangle$ is

$$a^\dagger_{\mathbf{k}s} |0\rangle \to a'^\dagger_{\mathbf{k}s} |0\rangle . \tag{8.41}$$

The rule (8.39) also provides a simple derivation of the asymptotic evolution law for multi-photon initial states. For the general n-photon initial state,

$$|\Psi\rangle = \sum_{\mathbf{k}_1 s_1} \cdots \sum_{\mathbf{k}_n s_n} C_{\mathbf{k}_1 s_1,\ldots,\mathbf{k}_n s_n} a^\dagger_{\mathbf{k}_1 s_1} \cdots a^\dagger_{\mathbf{k}_n s_n} |0\rangle , \tag{8.42}$$

applying eqn (8.39) to each creation operator yields

$$|\Psi(t)\rangle = \sum_{\mathbf{k}_1 s_1} \cdots \sum_{\mathbf{k}_n s_n} \exp\left[-i \sum_{m=1}^{n} \omega_{k_m} t\right] C'_{\mathbf{k}_1 s_1,\ldots,\mathbf{k}_n s_n} a^\dagger_{\mathbf{k}_1 s_1} \cdots a^\dagger_{\mathbf{k}_n s_n} |0\rangle , \tag{8.43}$$

where

$$C'_{\mathbf{k}_1 s_1,\ldots,\mathbf{k}_n s_n} = \sum_{\mathbf{p}_1 v_1} \cdots \sum_{\mathbf{p}_n v_n} S_{\mathbf{k}_1 s_1,\mathbf{p}_1 v_1} \cdots S_{\mathbf{k}_n s_n,\mathbf{p}_n v_n} C_{\mathbf{p}_1 \nu_1,\ldots,\mathbf{p}_n \nu_n} . \tag{8.44}$$

For scattering problems the initial state is stationary, so that

$$\sum_{m=1}^{n} \omega_{\mathbf{k}_m} = \omega_0 , \tag{8.45}$$

and the evolution equation (8.43) is replaced by the scattering rule

$$|\Psi\rangle \to |\Psi\rangle' = \sum_{\mathbf{k}_1 s_1} \cdots \sum_{\mathbf{k}_n s_n} C_{\mathbf{k}_1 s_1,\ldots,\mathbf{k}_n s_n} a'^\dagger_{\mathbf{k}_1 s_1} \cdots a'^\dagger_{\mathbf{k}_n s_n} |0\rangle . \tag{8.46}$$

It is important to notice that the scattering matrix in eqn (8.27) has a special property: it relates annihilation operators to annihilation operators only. The scattered annihilation operators do not depend at all on the incident creation operators. This feature follows from the physical assumption that emission and absorption do not occur in passive linear devices. The special form of the scattering matrix has an important consequence for the commutation relations of field operators evaluated at different times. Since all annihilation operators—and therefore all creation operators—commute with one another, eqns (8.28), (8.29), and (8.27) imply

$$\left[E^{(\pm)}_{\mathrm{out},i}\left(\mathbf{r},+\infty\right),E^{(\pm)}_{\mathrm{in},j}\left(\mathbf{r}',-\infty\right)\right]=0 \tag{8.47}$$

for scattering from a passive linear device. In fact, eqn (3.102) guarantees that the positive- (negative-) frequency parts of the field at different finite times commute, as long as the evolution of the field operators is caused by interaction with a passive linear medium. One should keep in mind that commutativity at different times is not generally valid, e.g. if emission and absorption or photon–photon scattering are possible, and further that commutators like $\left[E^{(+)}_i\left(\mathbf{r},t\right),E^{(-)}_j\left(\mathbf{r}',t'\right)\right]$ do not vanish even for free fields or fields evolving in passive linear media. Roughly speaking, this implies that the creation of a photon at $(\mathbf{r}',t')$ and the annihilation of a photon at $(\mathbf{r},t)$ are not independent events.

Putting all this together shows that we can use standard classical methods to calculate the scattering matrix for a given device, and then use eqn (8.27) to relate the annihilation operators for the incident and scattered modes. This apparently simple prescription must be used with care, as we will see in the applications. The utility of this approach arises partly from the fact that each scattering channel in the classical analysis can be associated with a **port**, i.e. a bounding surface through which a well-defined beam of light enters or leaves. Input and output ports are respectively associated with input and output channels. The ports separate the interior of the device from the outside world, and thus allow a black box approach in which the device is completely characterized by an input–output transfer function or scattering matrix. The principle of time-reversal invariance imposes constraints on the number of channels and ports and thus on the structure of the scattering matrix.

The simplest case is a **one-channel** device, i.e. there is one input channel and one output channel. In this case the scattering is described by a 1×1 matrix, as in eqn (8.2). This is more commonly called a **two-port** device, since there is one input port and one output port. As an example, for an antireflection coated thin lens the incident light occupies a single input channel, e.g. a paraxial Gaussian beam, and the transmitted light occupies a single output channel. The lens is therefore a one-channel/two-port device.

8.3 Paraxial optical elements

An optical element that transforms an incident paraxial ray bundle into another paraxial bundle will be called a **paraxial optical element**. The most familiar examples are (ideal) lenses and mirrors. By contrast to the dielectric slab in Fig. 8.1, an ideal lens transmits all of the incident light; no light is reflected or absorbed. Similarly an ideal mirror reflects all of the incident light; no light is transmitted or absorbed. In the non-ideal world inhabited by experimentalists, the conditions defining a paraxial element must be approximated by clever design. The no-reflection limit for a lens is approached by applying a suitable **antireflection coating**. This consists of one or more layers of transparent dielectrics with refractive indices and thicknesses adjusted so that the reflections from the various interfaces interfere destructively (Born and Wolf, 1980, Sec. 1.6). An ideal mirror is essentially the opposite of an antireflection coating; the parameters of the dielectric layers are chosen so that the transmitted waves suffer destructive interference. In both cases the ideal limit can only be approximated for a limited range of wavelengths and angles of incidence. Compound devices made from paraxial elements are automatically paraxial.

For optical elements defined by curved interfaces the calculation of the scattering matrix in the plane-wave basis is rather involved. The classical theory of the interaction of light with lenses and curved mirrors is more naturally described in terms of Gaussian beams, as discussed in Section 7.4. In the absence of this detailed theory it is still possible to derive a useful result by using the general properties of the scattering matrix. We will simplify this discussion by means of an additional approximation. An incident paraxial wave is a superposition of plane waves with wavevectors $\mathbf{k} = \mathbf{k}_0 + \mathbf{q}$, where $|\mathbf{q}| \ll k_0$. According to eqns (7.7) and (7.9), the dispersion in $q_z = \mathbf{q} \cdot \widetilde{\mathbf{k}}_0$ and ω for an incident paraxial wave is small, in the sense that $\Delta\omega / (c\Delta q_{\mathsf{T}}) \sim \Delta q_z / \Delta q_{\mathsf{T}} = O(\theta)$, where $\mathbf{q}_{\mathsf{T}} = \mathbf{q} - (\mathbf{q} \cdot \widetilde{\mathbf{k}}_0) \widetilde{\mathbf{k}}_0$ is the part of $\mathbf{q}$ transverse to $\mathbf{k}_0$ and θ is the opening angle of the beam. This suggests considering an incident classical field that is monochromatic and **planar**, i.e.

$$\boldsymbol{\mathcal{E}}_{\text{in}}^{(+)}(\mathbf{r}, t) = \left\{ \sum_{\mathbf{q}_{\mathsf{T}},\, s} i \sqrt{\frac{\hbar\omega_0}{2\epsilon_0 V}} \alpha_{\mathbf{k}_0 + \mathbf{q}_{\mathsf{T}}, s} \mathbf{e}_{0s} e^{i\mathbf{q}_{\mathsf{T}} \cdot \mathbf{r}_{\mathsf{T}}} \right\} e^{i(k_0 z - \omega_0 t)} . \tag{8.48}$$

In the same spirit the scattering matrix will be approximated by

$$S_{\mathbf{k}s, \mathbf{k}'s'} \approx \delta_{k_z k_0} \delta_{k'_z k_0} \widetilde{S}_{\mathbf{q}_{\mathsf{T}} s, \mathbf{q}'_{\mathsf{T}} s'} , \tag{8.49}$$

with the understanding that the reduced scattering matrix $\widetilde{S}_{\mathbf{q}_{\mathsf{T}} s, \mathbf{q}'_{\mathsf{T}} s'}$ effectively confines $\mathbf{q}_{\mathsf{T}}$ and $\mathbf{q}'_{\mathsf{T}}$ to the paraxial domain defined by eqn (7.8). In this limit, the unitarity condition (8.22) reduces to

$$\sum_{\mathbf{q}''_{\mathsf{T}},\, s''} \widetilde{S}^*_{\mathbf{q}''_{\mathsf{T}} s'', \mathbf{q}_{\mathsf{T}} s} \widetilde{S}_{\mathbf{q}''_{\mathsf{T}} s'', \mathbf{q}'_{\mathsf{T}} s'} = \delta_{\mathbf{q}_{\mathsf{T}} \mathbf{q}'_{\mathsf{T}}} \delta_{ss'} . \tag{8.50}$$

Turning now to the quantum theory, we see that the scattered annihilation operators are given by

$$a'_{\mathbf{k}_0 + \mathbf{q}_{\mathsf{T}}, s} = \sum_{\mathbf{p}_{\mathsf{T}},\, v} \widetilde{S}_{\mathbf{q}_{\mathsf{T}} s, \mathbf{q}'_{\mathsf{T}} s'} a_{\mathbf{k}_0 + \mathbf{q}'_{\mathsf{T}}, s'} . \tag{8.51}$$

Since the eigenvalues of the operator $a^\dagger_{\mathbf{k}s} a_{\mathbf{k}s}$ represent the number of photons in the plane-wave mode $\mathbf{f}_{\mathbf{k}s}$, the operator representing the flux of photons across a transverse plane located to the left ($z < 0$) of the optical element is proportional to

$$F = \sum_{\mathbf{q}_{\mathsf{T}},\, s} a^\dagger_{\mathbf{k}_0 + \mathbf{q}_{\mathsf{T}}, s} a_{\mathbf{k}_0 + \mathbf{q}_{\mathsf{T}}, s} , \tag{8.52}$$

and the operator representing the flux through a plane to the right ($z > 0$) of the optical element is

$$F' = \sum_{\mathbf{q}_T,\, s} a'^{\dagger}_{\mathbf{k}_0+\mathbf{q}_T, s} a'_{\mathbf{k}_0+\mathbf{q}_T, s} \,. \tag{8.53}$$

Combining eqn (8.51) with the unitarity condition (8.50) shows that the incident and scattered flux operators for a transparent optical element are identical, i.e. $F' = F$. This is a strong result, since it implies that all moments of the fluxes are identical,

$$\langle \Psi \left| F'^n \right| \Psi \rangle = \langle \Psi \left| F^n \right| \Psi \rangle \,. \tag{8.54}$$

In other words the overall statistical properties of the light, represented by the set of all moments of the photon flux, are unchanged by passage through a two-port paraxial element, even though the distribution over transverse wavenumbers may be changed by focussing.

8.4 The beam splitter

Beam splitters play an important role in many optical experiments as a method of beam manipulation, and they also exemplify some of the most fundamental issues in quantum optics. The simplest beam splitter is a uniform dielectric slab—such as the one studied in Section 8.1—but in practice beam splitters are usually composed of layered dielectrics, where the index of refraction of each layer is chosen to yield the desired reflection and transmission coefficients $\mathfrak{r}$ and $\mathfrak{t}$. The results of the single-slab analysis are applicable to the layered design, provided that the correct values of $\mathfrak{r}$ and $\mathfrak{t}$ are used. If the surrounding medium is the same on both sides of the device, and the optical properties of the layers are symmetrical around the midplane, then the amplitude reflection and transmission coefficients are the same for light incident from either side. This defines a **symmetrical beam splitter**. In order to simplify the discussion, we will only deal with this case in the text. However, the unsymmetrical beam splitter—which allows for more general phase relations between the incident and scattered waves—is frequently used in practice (Zeilinger, 1981), and an example is studied in Exercise 8.1.

In the typical experimental situation shown in Fig. 8.2, a classical wave, $\alpha_1 \exp(i\mathbf{k}_1 \cdot \mathbf{r})$, which is incident in channel 1, divides at the beam splitter into a

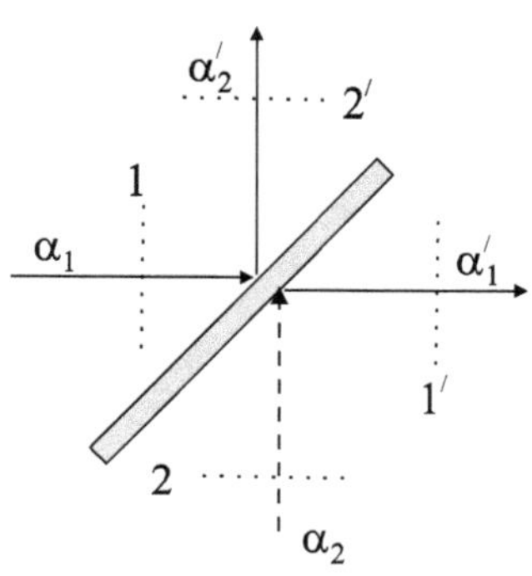

Fig. 8.2 A symmetrical beam splitter. The surfaces 1, 2, 1′, and 2′ are ports and the mode amplitudes α_1, α_2, α'_1, and α'_2 are related by the scattering matrix.

transmitted wave, $\alpha'_1 \exp(i\mathbf{k}_1 \cdot \mathbf{r})$, in channel $1'$ and a reflected wave, $\alpha'_2 \exp(i\mathbf{k}_2 \cdot \mathbf{r})$, in channel $2'$. In the time-reversed version of this event, channel $2'$ is an input channel that scatters into the output channels 1 and 2, where channel 2 is associated with port 2 in the figure. The two output channels in the time-reversed picture correspond to input channels in the original picture; therefore, time-reversal invariance requires that channel 2 be included as an input channel, in addition to the original channel 1. Thus the beam splitter is a **two-channel** device, and the two output channels are related to the two input channels by a 2×2 matrix. The beam splitter can also be described as a **four-port** device, since there are two input ports and two output ports. In the present book we restrict the term 'beam splitter' to devices that are described by the scattering matrix in eqn (8.63), but in the literature this term is often applied to any two-channel/four-port device described by a 2×2 unitary scattering matrix.

In the classical problem, there is no radiation in channel 2, so $\alpha_2 = 0$, and port 2 is said to be an **unused port**. The transmitted and reflected amplitudes are then

$$\alpha'_2 = \mathfrak{r}\,\alpha_1\,, \quad \alpha'_1 = \mathfrak{t}\alpha_1\,. \tag{8.55}$$

The materials composing the beam splitter are chosen to have negligible absorption in the wavelength range of interest, so the reflection and transmission coefficients must satisfy eqn (8.7). Combining eqn (8.7) and eqn (8.55) yields the conservation of energy,

$$|\alpha'_1|^2 + |\alpha'_2|^2 = |\alpha_1|^2\,. \tag{8.56}$$

In many experiments the output fields are measured by square law detectors that are not phase sensitive. In this case the transmission phase $\theta_\mathfrak{t}$ can be eliminated by the redefinition $\alpha_1 \to \alpha_1 \exp(-i\theta_\mathfrak{t})$, and the second line of eqn (8.7) means that we can set $\mathfrak{r} = \pm i\mathfrak{t}$, where $\mathfrak{t}$ is real and positive. The important special case of the **balanced** (50/50) beam splitter is defined by $|\mathfrak{r}| = |\mathfrak{t}| = 1/\sqrt{2}$, and this yields the simple rule

$$\mathfrak{r} = \frac{\pm i}{\sqrt{2}}\,, \quad \mathfrak{t} = \frac{1}{\sqrt{2}}\,. \tag{8.57}$$

Beam splitters are an example of a general class of linear devices called **optical couplers**—or **optical taps**—that split and redirect an input optical signal. In practice optical couplers often consist of one or more waveguides, and the objective is achieved by proper choice of the waveguide geometry. A large variety of optical couplers are in use (Saleh and Teich, 1991, Sec. 7.3), but their fundamental properties are all very similar to those of the beam splitter.

8.4.1 Quantum description of a beam splitter

A loose translation of the argument leading from the classical relation (8.16) to the quantum relation (8.27) might be that classical amplitudes are simply replaced by annihilation operators, according to the rules (8.24) and (8.26). In the present case, this procedure would replace the *c*-number relations (8.55) by the operator relations

$$a'_2 = \mathfrak{r}\,a_1\,, \quad a'_1 = \mathfrak{t}\,a_1\,; \tag{8.58}$$

consequently, the commutation relations for the scattered operators would be

$$\left[a_2', a_2'^\dagger\right] = |\mathfrak{r}|^2, \quad \left[a_1', a_1'^\dagger\right] = |\mathfrak{t}|^2. \tag{8.59}$$

These results are seriously wrong, since they imply a violation of Heisenberg's uncertainty principle for the scattered radiation oscillators. The source of this disaster is the way we have translated the classical statement 'no radiation enters through the unused port 2' to the quantum domain. The condition $\alpha_2 = 0$ is perfectly sensible in the classical problem, but in the quantum theory, eqn (8.59) amounts to claiming that the *operator* a_2 can be set to zero. This is inconsistent with the commutation relation $\left[a_2, a_2^\dagger\right] = 1$, so the classical statement $\alpha_2 = 0$ must instead be interpreted as a condition on the state describing the incident field, i.e.

$$a_2 \left|\Phi_{\text{in}}\right\rangle = 0 \tag{8.60}$$

for a pure state, and

$$a_2 \rho_{\text{in}} = \rho_{\text{in}} a_2^\dagger = 0 \tag{8.61}$$

for a mixed state. It is customary to describe this situation by saying that vacuum fluctuations in the mode $\mathbf{k}_2$ enter through the unused port 2. In other words, the correct quantum calculation resembles a classical problem in which real incident radiation enters through port 1 and mysterious vacuum fluctuations[1] enter through port 2. In this language, the statement 'the operator a_2 cannot be set to zero' is replaced by 'vacuum fluctuations cannot be prevented from entering through the unused port 2.'

Since we cannot impose $a_2 = 0$, it is essential to use the general relation (8.27) which yields

$$\begin{pmatrix} a_1' \\ a_2' \end{pmatrix} = \mathfrak{T} \begin{pmatrix} a_1 \\ a_2 \end{pmatrix}, \tag{8.62}$$

where

$$\mathfrak{T} = \begin{bmatrix} \mathfrak{t} & \mathfrak{r} \\ \mathfrak{r} & \mathfrak{t} \end{bmatrix} \tag{8.63}$$

is the scattering matrix for the beam splitter. The unitarity of $\mathfrak{T}$ guarantees that the scattered operators obey the canonical commutation relations, which in turn guarantee the uncertainty principle.

We can see an immediate consequence of eqns (8.62) and (8.63) by evaluating the number operators $N_2' = a_2'^\dagger a_2'$ and $N_1' = a_1'^\dagger a_1'$. Now

$$\begin{aligned} N_2' &= \left(\mathfrak{r}^* a_1^\dagger + \mathfrak{t}^* a_2^\dagger\right)\left(\mathfrak{r}\, a_1 + \mathfrak{t}\, a_2\right) \\ &= |\mathfrak{r}|^2 N_1 + |\mathfrak{t}|^2 N_2 + \mathfrak{r}^* \mathfrak{t}\, a_1^\dagger a_2 + \mathfrak{r}\, \mathfrak{t}^* a_2^\dagger a_1 . \end{aligned} \tag{8.64}$$

The corresponding formula for N_1' is obtained by interchanging $\mathfrak{r}$ and $\mathfrak{t}$:

$$N_1' = |\mathfrak{t}|^2 N_1 + |\mathfrak{r}|^2 N_2 + \mathfrak{r}\, \mathfrak{t}^* a_1^\dagger a_2 + \mathfrak{r}^*\, \mathfrak{t} a_2^\dagger a_1 , \tag{8.65}$$

and adding the two expressions gives

[1]The universal preference for this language may be regarded as sugar coating for the bitter pill of quantum theory.

$$N_2' + N_1' = N_1 + N_2 + (\mathfrak{r}^* \, \mathfrak{t} + \mathfrak{t} \, \mathfrak{r}^*) \left(a_1^\dagger a_2 + a_2^\dagger a_1 \right) = N_1 + N_2 \,, \tag{8.66}$$

where the Stokes relation (8.7) was used again. This is the operator version of the conservation of energy, which in this case is the same as conservation of the number of photons.

We now turn to the Schrödinger-picture description of scattering from the beam splitter. In accord with the energy-conservation rule (8.13), the operators $\{a_1, a_2, a_1', a_2'\}$ in eqn (8.62) all correspond to modes with a common frequency ω. We therefore begin by considering *single-frequency* problems, i.e. all the incident photons have the same frequency. For the beam splitter, the general operator scattering rule (8.40) reduces to

$$\begin{pmatrix} a_1^\dagger \\ a_2^\dagger \end{pmatrix} \to \mathfrak{T} \begin{pmatrix} a_1^\dagger \\ a_2^\dagger \end{pmatrix} = \begin{pmatrix} \mathfrak{t} \, a_1^\dagger + \mathfrak{r} \, a_2^\dagger \\ \mathfrak{r} \, a_1^\dagger + \mathfrak{t} \, a_2^\dagger \end{pmatrix} , \tag{8.67}$$

and to simplify things further we will only discuss two-photon initial states. With these restrictions, the general input state in eqn (8.42) is replaced by

$$|\Psi\rangle = \sum_{m=1}^{2} \sum_{n=1}^{2} C_{mn} a_m^\dagger a_n^\dagger \, |0\rangle \,. \tag{8.68}$$

Since the creation operators commute with one another, the coefficients satisfy the bosonic symmetry condition $C_{mn} = C_{nm}$.

A simple example—which will prove useful in Section 10.2.1—is a two-photon state in which one photon enters through port 1 and another enters through port 2, i.e.

$$|\Psi\rangle = a_1^\dagger a_2^\dagger \, |0\rangle \,. \tag{8.69}$$

Applying the rule (8.67) to this initial state yields the scattered state

$$|\Psi\rangle' = \mathfrak{r} \, \mathfrak{t} \left(a_1^{\dagger 2} + a_2^{\dagger 2} \right) |0\rangle + \left(\mathfrak{r}^2 + \mathfrak{t}^2 \right) a_1^\dagger a_2^\dagger \, |0\rangle \,. \tag{8.70}$$

Some interesting properties of this solution can be found in Exercise 8.2.

The simplified notation, $a_m = a_{\mathbf{k}_m s_m}$, employed above is useful because the Heisenberg-picture scattering law (8.62) does not couple modes with different frequencies and polarizations. The former property is a consequence of the energy conservation rule (8.13) and the latter follows from the fact that the optically isotropic material of the beam splitter does not change the polarization of the incident light. There are, however, interesting experimental situations with initial states involving several frequencies and more than one polarization state per channel. In these cases the simplified notation is less useful, and it is better to identify the mth input channel solely with the direction of propagation defined by the unit vector $\widetilde{\mathbf{k}}_m$. Photons of either polarization and any frequency can enter and leave through these channels. A notation suited to this situation is

$$a_{ms}(\omega) = a_{\mathbf{q}s} \ \text{ with } \ \mathbf{q} = \frac{\omega}{c} \widetilde{\mathbf{k}}_m \,, \tag{8.71}$$

where $m = 1, 2$ is the channel index and s labels the two possible polarizations. For the following discussion we will use a linear polarization basis $\{\mathbf{e}_h(\widetilde{\mathbf{k}}_m), \mathbf{e}_v(\widetilde{\mathbf{k}}_m)\}$ for each channel, where h and v respectively stand for *horizontal* and *vertical*. The frequency ω can vary continuously, but for the present we will restrict the frequencies to a discrete set. With all this understood, the canonical commutation relations are written as

$$\left[a_{ms}(\omega), a_{nr}^{\dagger}(\omega')\right] = \delta_{mn}\delta_{sr}\delta_{\omega\omega'}\,, \quad \text{with} \quad m,n = 1,2 \quad \text{and} \quad r,s = h,v\,, \tag{8.72}$$

and the operator scattering law (8.67)—which applies to each polarization and frequency separately—becomes

$$\begin{pmatrix} a_{1s}^{\dagger}(\omega) \\ a_{2s}^{\dagger}(\omega) \end{pmatrix} \rightarrow \begin{pmatrix} \mathfrak{t}\, a_{1s}^{\dagger}(\omega) + \mathfrak{r}\, a_{2s}^{\dagger}(\omega) \\ \mathfrak{r}\, a_{1s}^{\dagger}(\omega) + \mathfrak{t}\, a_{2s}^{\dagger}(\omega) \end{pmatrix}. \tag{8.73}$$

Since the coefficients $\mathfrak{t}$ and $\mathfrak{r}$ depend on frequency, they should be written as $\mathfrak{t}(\omega)$ and $\mathfrak{r}(\omega)$, but the simplified notation used in this equation is more commonly found in the literature.

We will only consider two-photon initial states of the form

$$|\Psi\rangle = \sum_{m,n=1}^{2} \sum_{r,s} \sum_{\omega,\omega'} C_{ms,nr}(\omega,\omega')\, a_{ms}^{\dagger}(\omega)\, a_{nr}^{\dagger}(\omega')\, |0\rangle\,, \tag{8.74}$$

where the sums over ω and ω' run over some discrete set of frequencies, and the bosonic symmetry condition is

$$C_{nr,ms}(\omega',\omega) = C_{ms,nr}(\omega,\omega')\,. \tag{8.75}$$

Just as in nonrelativistic quantum mechanics, Bose symmetry applies only to the simultaneous exchange of all the degrees of freedom. Relaxing the simplifying assumption that a single frequency and polarization are associated with all scattering channels opens up many new possibilities.

In the first example—which will be useful in Section 10.2.1-B—the incoming photons have the same polarization, but different frequencies ω_1 and ω_2. In this case the polarization index can be omitted, and the initial state expressed as $|\Psi\rangle = a_1^{\dagger}(\omega_1)\, a_2^{\dagger}(\omega_2)\,|0\rangle$. Applying the scattering law (8.67) to this state yields

$$\begin{aligned} |\Psi\rangle' = &\left\{\mathfrak{t}\,\mathfrak{r}\,\left[a_1^{\dagger}(\omega_1)\, a_1^{\dagger}(\omega_2) + a_2^{\dagger}(\omega_1)\, a_2^{\dagger}(\omega_2)\right]\right\}|0\rangle \\ &+ \left\{\mathfrak{t}^2 a_1^{\dagger}(\omega_1)\, a_2^{\dagger}(\omega_2) + \mathfrak{r}^2 a_2^{\dagger}(\omega_1)\, a_1^{\dagger}(\omega_2)\right\}|0\rangle\,. \end{aligned} \tag{8.76}$$

This solution has a number of interesting features that are explored in Exercise 8.3.

An example of a single-frequency state with two polarizations present is

$$|\Psi\rangle = \frac{1}{\sqrt{2}}\left(a_{1h}^{\dagger} a_{2v}^{\dagger} - a_{1v}^{\dagger} a_{2h}^{\dagger}\right)|0\rangle\,, \tag{8.77}$$

where the frequency argument has been dropped. In this case the expansion coefficients in eqn (8.74) reduce to

$$C_{ms,nr} = \frac{1}{4}\left(\delta_{m1}\delta_{n2} - \delta_{n1}\delta_{m2}\right)\left(\delta_{sh}\delta_{rv} - \delta_{rh}\delta_{sv}\right). \tag{8.78}$$

The antisymmetry in the polarization indices r and s is analogous to the antisymmetric spin wave function for the singlet state of a system composed of two spin-1/2 particles, so $|\Psi\rangle$ is said to have a **singlet-like** character.[2] The overall bosonic symmetry then requires antisymmetry in the spatial degrees of freedom represented by (m, n). More details can be found in Exercise 8.4.

8.4.2 Partition noise

The paraxial, single-channel/two-port devices discussed in Section 8.3 preserve the statistical properties of the incident field. Let us now investigate this question for the beam splitter. Combining the results (8.64) and (8.65) for the number operators of the scattered modes with the condition (8.61) implies

$$\langle N_2' \rangle = \mathrm{Tr}\left(\rho_{\mathrm{in}} N_2'\right) = |\mathfrak{r}|^2 \langle N_1 \rangle , \quad \langle N_1' \rangle = |\mathfrak{t}|^2 \langle N_1 \rangle . \tag{8.79}$$

The intensity for each mode is proportional to the average of the corresponding number operator, so the quantum averages reproduce the classical results, $I_2' = |\mathfrak{r}|^2 I_1$ and $I_1' = |\mathfrak{r}|^2 I_1$. There are no surprises for the average values, so we go on to consider the statistical fluctuations in the incident and transmitted signals. This is done by comparing the normalized variance,

$$\overline{V}\left(N_1'\right) = \frac{V\left(N_1'\right)}{\langle N_1' \rangle^2} = \frac{\langle N_1'^2 \rangle - \langle N_1' \rangle^2}{\langle N_1' \rangle^2} , \tag{8.80}$$

of the transmitted field to the same quantity, $\overline{V}\left(N_1\right)$, for the incident field. The calculation of the transmitted variance involves evaluating $\langle N_1'^2 \rangle$, which can be done by combining eqn (8.65) with eqn (8.61) and using the cyclic invariance property of the trace to get

$$\langle N_1'^2 \rangle = |\mathfrak{t}|^4 \langle N_1^2 \rangle + |\mathfrak{r}|^2 |\mathfrak{t}|^2 \langle N_1 \rangle . \tag{8.81}$$

Substituting this into the definition of the normalized variance leads to

$$\overline{V}\left(N_1'\right) = \overline{V}\left(N_1\right) + \left|\frac{\mathfrak{r}}{\mathfrak{t}}\right|^2 \frac{1}{\langle N_1 \rangle} . \tag{8.82}$$

Thus transmission through the beam splitter—by contrast to transmission through a two-port device—increases the variance in photon number. In other words, the noise in the transmitted field is greater than the noise in the incident field. Since the added noise vanishes for $\mathfrak{r} = 0$, it evidently depends on the partition of the incident field into transmitted and reflected components. It is therefore called **partition noise**.

Partition noise can be blamed on the vacuum fluctuations entering through the unused port 2. This can be seen by temporarily modifying the commutation relation for a_2 to $\left[a_2, a_2^\dagger\right] = \xi_2$, where ξ_2 is a c-number which will eventually be set to unity. This is equivalent to modifying the canonical commutator to $[q_2, p_2] = i\hbar\xi_2$, and this

[2]The spin-statistics connection (Cohen-Tannoudji *et al.*, 1977*b*, Sec. XIV-C) tells us that spin-1/2 particles must be fermions not bosons. This shows that analogies must be handled with care.

in turn yields the uncertainty relation $\Delta q_2 \Delta p_2 \geqslant \xi_2 \hbar/2$. Using this modification in the previous calculation leads to

$$\overline{V}\left(N_1'\right) = \overline{V}\left(N_1\right) + \xi_2 \left|\frac{\mathfrak{r}}{\mathfrak{t}}\right|^2 \frac{1}{\langle N_1 \rangle} \,. \tag{8.83}$$

Thus partition noise can be attributed to the vacuum (zero-point) fluctuations of the mode entering the unused port 2. Additional evidence that partition noise is entirely a quantum effect is provided by the fact that it becomes negligible in the classical limit, $\langle N_1 \rangle \to \infty$. Note that if we consider only the transmitted light, the transparent beam splitter acts as if it were an absorber, i.e. a dissipative element. The increased noise in the transmitted field is then an example of a general relation between dissipation and fluctuation which will be studied later.

8.4.3 Behavior of quasiclassical fields at a beam splitter

We will now analyze an experiment in which a coherent (quasiclassical) state is incident on port 1 of the beam splitter and no light is injected into port 2. The Heisenberg state $|\Phi_{\text{in}}\rangle$ describing this situation satisfies

$$\begin{aligned} a_1 \left|\Phi_{\text{in}}\right\rangle &= \alpha_1 \left|\Phi_{\text{in}}\right\rangle , \\ a_2 \left|\Phi_{\text{in}}\right\rangle &= 0 , \end{aligned} \tag{8.84}$$

where α_1 is the amplitude of the coherent state. The scattering relation (8.62) combines with these conditions to yield

$$\begin{aligned} a_1' \left|\Phi_{\text{in}}\right\rangle &= \left(\mathfrak{r}\, a_2 + \mathfrak{t}\, a_1\right) \left|\Phi_{\text{in}}\right\rangle = \mathfrak{t}\, \alpha_1 \left|\Phi_{\text{in}}\right\rangle , \\ a_2' \left|\Phi_{\text{in}}\right\rangle &= \left(\mathfrak{t}\, a_2 + \mathfrak{r}\, a_1\right) \left|\Phi_{\text{in}}\right\rangle = \mathfrak{r}\, \alpha_1 \left|\Phi_{\text{in}}\right\rangle . \end{aligned} \tag{8.85}$$

In other words, the Heisenberg state vector is also a coherent state with respect to a_1' and a_2', with the respective amplitudes $\mathfrak{t}\, \alpha_1$ and $\mathfrak{r}\, \alpha_1$. This means that the fundamental condition (5.11) for a coherent state is satisfied for both output modes; that is,

$$V\left(a_1'^{\dagger}, a_1'\right) = V\left(a_2'^{\dagger}, a_2'\right) = 0 \,, \tag{8.86}$$

where the variance is calculated for the incident state $|\Phi_{\text{in}}\rangle$. This behavior is exactly parallel to that of a classical field injected into port 1, so it provides further evidence of the nearly classical nature of coherent states.

8.4.4 The polarizing beam splitter

The generic beam splitter considered above consists of a slab of optically isotropic material, but for some purposes it is better to use anisotropic crystals. When light falls on an anisotropic crystal, the two polarizations defined by the crystal axes are refracted at different angles. Devices employing this effect are typically constructed by cementing together two prisms made of uniaxial crystals. The relative orientation of the crystal axes are chosen so that the corresponding polarization components of the incident light are refracted at different angles. Devices of this kind are called **polarizing beam splitters** (PBSs) (Saleh and Teich, 1991, Sec. 6.6). They provide an excellent source for polarized light, and are also used to ensure that the two special polarizations are emitted through different ports of the PBS.

8.5 Y-junctions

In applications to communications, it is often necessary to split the signal so as to send copies down different paths. The beam splitter discussed above can be used for this purpose, but another optical coupler, the **Y-junction**, is often employed instead. A schematic representation of a symmetric Y-junction is shown in Fig. 8.3, where the waveguides denoted by the solid lines are typically realized by optical fibers in the optical domain or conducting walls for microwaves.

The solid arrows in this sketch represent an input beam in channel 1 coupled to output beams in channels 2 and 3. In the time-reversed version, an input beam (the dashed arrow) in channel 3 couples to output beams in channels 1 and 2. Similarly, an input beam in channel 2 couples to output beams in channels 1 and 3. Each output beam in the time-reversed picture corresponds to an input beam in the original picture; therefore, all three channels must be counted as input channels. The three input channels are coupled to three output channels, so the Y-junction is a three-channel device. A strict application of the convention for counting ports introduced above requires us to call this a six-port device, since there are three input ports $(1, 2, 3)$ and three output ports $(1^*, 2^*, 3^*)$. This terminology is logically consistent, but it does not agree with the standard usage, in which the Y-junction is called a three-port device (Kerns and Beatty, 1967, Sec. 2.16). The source of this discrepancy is the fact that—by contrast to the beam splitter—each channel of the Y-junction serves as both input and output channel. In the sketch, the corresponding ports are shown separated for clarity, but it is natural to have them occupy the same spatial location. The standard usage exploits this degeneracy to reduce the port count from six to three.

Applying the argument used for the beam splitter to the Y-junction yields the input–output relation

$$\begin{pmatrix} a_1' \\ a_2' \\ a_3' \end{pmatrix} = Y \begin{pmatrix} a_1 \\ a_2 \\ a_3 \end{pmatrix}, \tag{8.87}$$

where Y is a 3×3 unitary matrix. When the matrix Y is symmetric—$(Y)_{nm} = (Y)_{mn}$— the device is said to be **reciprocal**. In this case, the output at port n from

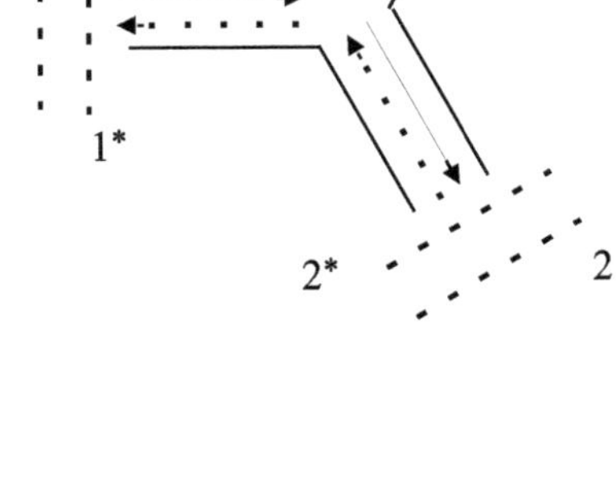

Fig. 8.3 A symmetrical Y-junction. The inward-directed solid arrow denotes a signal injected into channel 1 which is coupled to the output channels 2 and 3 as indicated by the outward-directed solid arrows. The dashed arrows represent the time-reversed process. Ports 1, 2, and 3 are input ports and ports 1^*, 2^*, and 3^* are output ports.

a unit signal injected into port m is the same as the output at port m from a unit signal injected at port n.

For the symmetrical Y-junction considered here, the optical properties of the medium occupying the junction itself and each of the three arms are assumed to exhibit three-fold symmetry. In other words, the properties of the Y-junction are unchanged by any permutation of the channel labels. In particular, this means that the Y-junction is reciprocal. The three-fold symmetry reduces the number of independent elements of Y from nine to two. One can, for example, set

$$Y = \begin{bmatrix} y_{11} & y_{12} & y_{12} \\ y_{12} & y_{11} & y_{12} \\ y_{12} & y_{12} & y_{11} \end{bmatrix}, \tag{8.88}$$

where

$$y_{11} = |y_{11}| \, e^{i\theta_{11}}, \quad y_{12} = |y_{12}| \, e^{i\theta_{12}}. \tag{8.89}$$

The unitarity conditions

$$|y_{11}|^2 + 2\,|y_{12}|^2 = 1, \tag{8.90}$$

$$2\,|y_{11}| \cos(\theta_{11} - \theta_{12}) + |y_{12}| = 0 \tag{8.91}$$

relate the difference between the reflection phase θ_{11} and the transmission phase θ_{12} to the reflection and transmission coefficients $|y_{11}|^2$ and $|y_{12}|^2$. The values of the two real parameters left free, e.g. $|y_{11}|$ and $|y_{12}|$, are determined by the optical properties of the medium at the junction, the optical properties of the arms, and the locations of the degenerate ports $(1, 1^*)$, etc. For the symmetrical Y-junction, the unitarity conditions place strong restrictions on the possible values of $|y_{11}|$ and $|y_{12}|$, as seen in Exercise 8.5.

In common with the beam splitter, the Y-junction exhibits partition noise. For an experiment in which the initial state has photons only in the input channel 1, a calculation similar to the one for the beam splitter sketched in Section 8.4.2—see Exercise 8.6—shows that the noise in the output signal is always greater than the noise in the input signal. In the classical description of this experiment, there are no input signals in channels 2 and 3; consequently, the input ports 2^* and 3^* are said to be unused. Thus the partition noise can again be ascribed to vacuum fluctuations entering through the unused ports.

8.6 Isolators and circulators

In this section we briefly describe two important and closely related devices: the optical isolator and the optical circulator, both of which involve the use of a magnetic field.

8.6.1 Optical isolators

An **optical isolator** is a device that transmits light in only one direction. This property is used to prevent reflected light from traveling upstream in a chain of optical devices. In some applications, this feedback can interfere with the operation of the light source. There are several ways to construct optical isolators (Saleh and Teich,

1991, Sec. 6.6C), but we will only discuss a generally useful scheme that employs Faraday rotation.

The optical properties of a transparent dielectric medium are changed by the presence of a static magnetic field $\mathcal{B}_0$. The source of this change is the response of the atomic electrons to the combined effect of the propagating optical wave and the static field. Since every propagating field can be decomposed into a superposition of plane waves, we will consider a single plane wave. The linearly-polarized electric field $\mathcal{E}$ of the wave is an equal superposition of right- and left-circularly-polarized waves $\mathcal{E}_+$ and $\mathcal{E}_-$; consequently, the electron velocity $\mathbf{v}$—which to lowest order is proportional to $\mathcal{E}$—can be decomposed in the same way. This in turn implies that the velocity components $\mathbf{v}_+$ and $\mathbf{v}_-$ experience different Lorentz forces $e\mathbf{v}_+ \times \mathcal{B}_0$ and $e\mathbf{v}_- \times \mathcal{B}_0$. This effect is largest when $\mathcal{E}$ and $\mathcal{B}_0$ are orthogonal, so we will consider that case. The index of refraction of the medium is determined by the combination of the original wave with the radiation emitted by the oscillating electrons; therefore, the two circular polarizations will have different indices of refraction, n_+ and n_-. For a given polarization s, the change in phase accumulated during propagation through a distance L in the dielectric is $2\pi n_s L/\lambda$, so the phase difference between the two circular polarizations is $\Delta\phi = (2\pi/\lambda)(n_+ - n_-)L$, where λ is the wavelength of the light. The superposition of phase-shifted, right- and left-circularly-polarized waves describes a linearly-polarized field that is rotated through $\Delta\phi$ relative to the incident field.

The rotation of the direction of polarization of linearly-polarized light propagating along the direction of a static magnetic field is called the **Faraday effect** (Landau *et al.*, 1984, Chap. XI, Section 101), and the combination of the dielectric with the magnetic field is called a **Faraday rotator**. Experiments show that the rotation angle $\Delta\phi$ for a single pass through a Faraday rotator of length L is proportional to the strength of the magnetic field and to the length of the sample: $\Delta\phi = VL\mathcal{B}_0$, where V is the **Verdet constant**. Comparing the two expressions for $\Delta\phi$ shows that the Verdet constant is $V = 2\pi(n_+ - n_-)/(\lambda\mathcal{B}_0)$. For a positive Verdet constant the polarization is rotated in the clockwise sense as seen by an observer looking along the propagation direction $\widetilde{\mathbf{k}}$.

The Faraday rotator is made into an optical isolator by placing a linear polarizer at the input face and a second linear polarizer, rotated by $+45°$ with respect to the first, at the output face. When the magnetic field strength is adjusted so that $\Delta\phi = 45°$, the light transmitted through the input polarizer is also transmitted through the output polarizer. On the other hand, light of the same wavelength and polarization propagating in the opposite direction, e.g. the original light reflected from a mirror placed beyond the output polarizer, will undergo a polarization rotation of $-45°$, since $\widetilde{\mathbf{k}}$ has been replaced by $-\widetilde{\mathbf{k}}$. This is a counterclockwise rotation, as seen when looking along the reversed propagation direction $-\widetilde{\mathbf{k}}$, so it is a clockwise rotation as seen from the original propagation direction. Thus the counter-propagating light experiences a further polarization rotation of $+45°$ with respect to the input polarizer. The light reaching the input polarizer is therefore orthogonal to the allowed direction, and it will not be transmitted. This is what makes the device an isolator; it only transmits light propagating in the direction of the external magnetic field. This property has led to the name **optical diodes** for such devices.

Instead of linear polarizers, one could as well use anisotropic, linearly polarizing, single-mode optical fibers placed at the two ends of an isotropic glass fiber. If the polarization axis of the output fiber is rotated by $+45°$ with respect to that of the input fiber and an external magnetic field is applied to the intermediate fiber, then the net effect of this all-fiber device is exactly the same, *viz.* that light will be transmitted in only one direction.

It is instructive to describe the action of the isolator in the language of time reversal. The time-reversal transformations $(\mathbf{k}, s) \to (-\mathbf{k}, s)$ for the wave, and $\mathcal{B}_0 \to -\mathcal{B}_0$ for the magnetic field, combine to yield $\Delta\phi \to \Delta\phi$ for the rotation angle. Thus the time-reversed wave is rotated by $+45°$ clockwise. This is a counterclockwise rotation $(-45°)$ when viewed from the original propagation direction, so it cancels the $+45°$ rotation imposed on the incident field. This guarantees that the polarization of the time-reversed field exactly matches the setting of the input polarizer, so that the wave is transmitted. The transformation $(\mathbf{k}, s) \to (-\mathbf{k}, s)$ occurs automatically upon reflection from a mirror, but the transformation $\mathcal{B}_0 \to -\mathcal{B}_0$ can only be achieved by reversing the currents generating the magnetic field. This is not done in the operation of the isolator, so the time-reversed final state of the field does not evolve into the time-reversed initial state. This situation is described by saying that the external magnetic field violates time-reversal invariance. Alternatively, the presence of the magnetic field in the dielectric is said to create a nonreciprocal medium.

8.6.2 Optical circulators

The beam splitter and the Y-junction can both be used to redirect beams of light, but only at the cost of adding partition noise from the vacuum fluctuations entering through an unused port. We will next study another device—the **optical circulator**, shown in Fig. 8.4(a)—that can redirect and separate beams of light without adding noise. This linear optical device employs the same physical principles as the older microwave waveguide junction circulators discussed in Helszajn (1998, Chap. 1). As shown in Fig. 8.4(a), the circulator has the physical configuration of a symmetric Y-junction, with the addition of a cylindrical resonant cavity in the center of the junction. The central part of the cavity in turn contains an optically transparent ferromagnetic insulator—called a **ferrite pill**—with a magnetization (a permanent internal DC magnetic field $\mathcal{B}_0$) parallel to the cavity axis and thus normal to the plane of the Y-junction. In view of the connection to the microwave case, we will use the conventional terminology in which this is called a three-port device. If the ferrite pill is unmagnetized, this structure is simply a symmetric Y-junction, but we will see that the presence of nonzero magnetization changes it into a nonreciprocal device.

The central resonant cavity supports circulating modes: clockwise $(+)$-modes, in which the field energy flows in a clockwise sense around the cavity, and counterclockwise $(-)$-modes, in which the energy flows in the opposite sense (Jackson, 1999, Sec. 8.7). The $(\pm)$-modes both possess a transverse electric field $\mathcal{E}_{\pm}$, i.e. a field lying in the plane perpendicular to the cavity axis and therefore also perpendicular to the static field $\mathcal{B}_0$. In the Faraday-effect optical isolator the electromagnetic field propagates along the direction of the static magnetic field $\mathcal{B}_0$, which acts on the spin degrees of freedom of the field by rotating the direction of polarization. By contrast, the field in

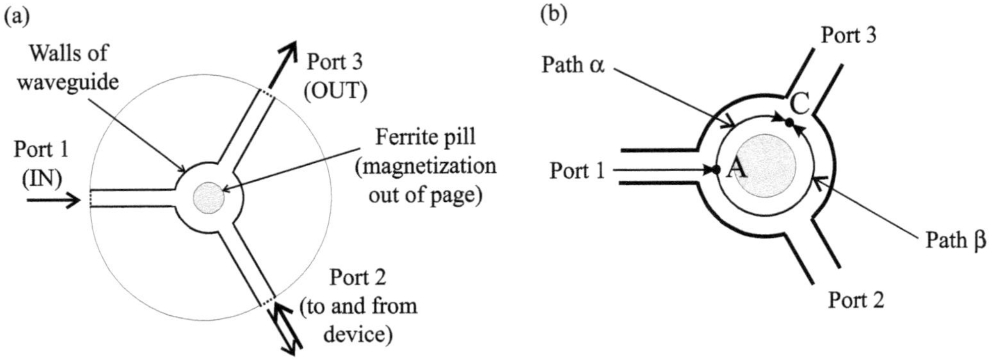

Fig. 8.4 (a) A Y-junction circulator consists of a three-fold symmetric arrangement of three ports with a 'ferrite pill' at the center. All the incoming wave energy is directed solely in an anti-clockwise sense from port 1 to port 2, and all the wave energy coming out of port 2 is directed solely into port 3, etc. (b) Magnified view of central portion of (a). Wave energy can only flow around the ferrite pill in an anti-clockwise sense, since the clockwise energy flow from port 1 to port 3 is forbidden by the destructive interference at point C between paths α and β (see text).

the circulator propagates around the cavity in a plane perpendicular to $\boldsymbol{\mathcal{B}}_0$, and the polarization—i.e. the direction of the electric field—is fixed by the boundary conditions. Despite these differences, the underlying mechanism for the action of the static magnetic field is the same. An electron velocity $\mathbf{v}$ has components $\mathbf{v}_\pm$ proportional to $\boldsymbol{\mathcal{E}}_\pm$, and the corresponding Lorentz forces $\mathbf{v}_+ \times \boldsymbol{\mathcal{B}}_0$ and $\mathbf{v}_- \times \boldsymbol{\mathcal{B}}_0$ are different. This means that the $(+)$- and $(-)$-modes experience different indices of refraction, n_+ and n_-; consequently, they possess different resonant frequencies $\omega_{n,+}$ and $\omega_{n,-}$. In the absence of the static field $\boldsymbol{\mathcal{B}}_0$, time-reversal invariance requires $\omega_{n,+} = \omega_{n,-}$, since the $(+)$- and $(-)$-modes are related by a time-reversal transformation. Thus the presence of the magnetic field in the circulator violates time-reversal invariance, just as it does for the Faraday-effect isolator. There is, however, an important difference between the isolator and the circulator. In the circulator, the static field acts on the spatial mode functions, i.e. on the orbital degrees of freedom of the traveling waves, as opposed to acting on the spin (polarization) degrees of freedom.

The best way to continue this analysis would be to solve for the resonant cavity modes in the presence of the static magnetic field. As a simpler alternative, we offer a wave interference model that is based on the fact that the cavity radius R_c is large compared to the optical wavelength. This argument—which comes close to violating Einstein's rule—begins with the observation that the cavity wall is approximately straight on the wavelength scale, and continues by approximating the circulating mode as a plane wave propagating along the wall. For fixed values of the material properties, the available design parameters are the field strength $\mathcal{B}_0$ and the cavity radius R_c.

Our first task is to impedance match the cavity by ensuring that there are no reflections from port 1, i.e. $y_{11} = 0$. A signal entering port 1 will couple to both of the modes $(+)$ and $(-)$, which will each travel around the full circumference, $L_c = 2\pi R_c$, of the cavity to arrive back at port 1. In our wave interference model this implies $y_{11} \propto e^{i\phi_+} + e^{i\phi_-}$, where $\phi_\pm = n_\pm(\mathcal{B}_0)\,k_0 L_c$ and $k_0 = 2\pi/\lambda_0$. The condition for no reflection is then

$$e^{i\phi_+} + e^{i\phi_-} = 0 \quad \text{or} \quad e^{i\Delta\phi} + 1 = 0\,, \tag{8.92}$$

where

$$\Delta\phi = \phi_+ - \phi_- = \left[n_+(\mathcal{B}_0) - n_-(\mathcal{B}_0)\right] k_0 L_c = \Delta n(\mathcal{B}_0)\, k_0 L_c\,. \tag{8.93}$$

The impedance matching condition (8.92) is imposed by choosing the field strength $\mathcal{B}_0$ and the circumference L_c to satisfy

$$\Delta n(\mathcal{B}_0)\, k_0 L_c = \pm\pi, \pm 3\pi, \dots\,. \tag{8.94}$$

The three-fold symmetry of the circulator geometry then guarantees that $y_{11} = y_{22} = y_{33} = 0$.

The second design step is to guarantee that a signal entering through port 1 will exit entirely through port 2, i.e. that $y_{31} = 0$. For a weak static field, $\Delta n(\mathcal{B}_0)$ is a linear function of $\mathcal{B}_0$ and

$$n_\pm(\mathcal{B}_0) = n_0 \pm \frac{\Delta n(\mathcal{B}_0)}{2}\,, \tag{8.95}$$

where n_0 is the index of refraction at zero field strength. A signal entering through port 1 at the point A will arrive at the point C, leading to port 3, in two ways. In the first way, the $(+)$-mode propagates along path α. In the second way, the $(-)$-mode propagates along the path β. Consequently, the matrix element y_{31} is proportional to $e^{i\phi_\alpha} + e^{i\phi_\beta}$, where

$$\phi_\alpha = n_+(\mathcal{B}_0)\, k_0 \frac{L_c}{3} = n_0 k_0 \frac{L_c}{3} + \frac{\Delta n(\mathcal{B}_0)}{2} k_0 \frac{L_c}{3} \tag{8.96}$$

and

$$\phi_\beta = n_-(\mathcal{B}_0)\, k_0 \frac{2L_c}{3} = n_0 k_0 \frac{2L_c}{3} - \frac{\Delta n(\mathcal{B}_0)}{2} k_0 \frac{2L_c}{3}\,. \tag{8.97}$$

The condition $y_{31} = 0$ is then imposed by requiring $\phi_\beta - \phi_\alpha$ to be an odd multiple of π, i.e.

$$n_0 k_0 \frac{L_c}{3} - \frac{\Delta n(\mathcal{B}_0)}{2} k_0 L_c = \pm\pi, \pm 3\pi, \dots\,. \tag{8.98}$$

The two conditions (8.94) and (8.98) determine the values of L_c and $\mathcal{B}_0$ needed to ensure that the device functions as a circulator. With the convention that the net energy flows along the shortest arc length from one port to the next, this device only allows net energy flow in the counterclockwise sense. Thus a signal entering port 1 can only exit at port 2, a signal entering port 3 can only exit at port 1, and a signal entering through port 2 can only exit at port 3. The scattering matrix

$$C = \begin{pmatrix} 0 & 0 & 1 \\ 1 & 0 & 0 \\ 0 & 1 & 0 \end{pmatrix} \tag{8.99}$$

for the circulator is nonreciprocal but still unitary. By using the input–output relations for this matrix, one can show—as in Exercise 8.7—that the noise in the output signal is the same as the noise in the input signal.

In one important application of the circulator, a wave entering the IN port 1 is entirely transmitted—ideally without any loss—towards an active reflection device, e.g. a reflecting amplifier, that is connected to port 2. The amplified and reflected wave from the active reflection device is entirely transmitted—also without any loss—to the OUT port 3. In this ideal situation the nonreciprocal action of the magnetic field in the ferrite pill ensures that none of the amplified wave from the device connected to port 2 can leak back into port 1. Furthermore, no accidental reflections from detectors connected to port 3 can leak back into the reflection device. The same nonreciprocal action prevents vacuum fluctuations entering the unused port 3 from adding to the noise in channel 2.

In real devices conditions are never perfectly ideal, but the rejection ratio for wave energies traveling in the forbidden direction of the circulator is quite high; for typical optical circulators it is of the order of 30 dB, i.e. a factor of 1000. Moreover, the transparent ferrite pill introduces very little dissipative loss (typically less than tenths of a dB) for the allowed direction of the circulator. This means that the contribution of vacuum fluctuations to the noise can typically be reduced also by a factor of 1000. Fiber versions of optical circulators were first demonstrated by Mizumoto *et al.* (1990), and amplification by optical parametric amplifiers connected to such circulators—where the amplifier noise was reduced well below the standard quantum limit—was demonstrated by Aytur and Kumar (1990).

8.7 Stops

An ancillary—but still important—linear device is a **stop** or **iris**, which is a small, usually circular, aperture (pinhole) in an absorptive or reflective screen. Since the stop only transmits a small portion of the incident beam, it can be used to eliminate aberrations introduced by lenses or mirrors, or to reduce the number of transverse modes in the incident field. This process is called *beam cleanup* or *spatial filtering*.

The problem of transmission through a stop is not as simple as it might appear. The only known exact treatment of diffraction through an aperture is for the case of a thin, perfectly conducting screen (Jackson, 1999, Sec. 10.7). The screen and stop combination is clearly a two-port device, but the strong scattering of the incident field by the screen means that it is not paraxial. It is possible to derive the entire plane-wave scattering matrix from the known solution for the reflected and diffracted fields for a general incident plane wave, but the calculations required are too cumbersome for our present needs. The interesting quantum effects can be demonstrated in a special case that does not require the general classical solution.

In most practical applications the diameter of the stop is large compared to optical wavelengths, so diffraction effects are not important, at least if the distance to the detector is small compared to the Rayleigh range defined by the stop area. By the same token, the polarization of the incident wave will not be appreciably changed by scattering. Thus the transmission through the stop is approximately described by ray optics, and polarization can be ignored. If the coordinate system is chosen so that the screen lies in the (x, y)-plane, then a plane wave propagating from $z < 0$ at normal incidence, e.g. $\alpha_k \exp(ikz)$, with $k > 0$, will scatter according to

$$\begin{gathered}\alpha_k \exp(ikz) \to \alpha'_k \exp(ikz) + \alpha'_{-k} \exp(-ikz)\,, \\ \alpha'_k = \mathfrak{t}\,\alpha_k\,, \quad \alpha'_{-k} = \mathfrak{r}\,\alpha_k\,,\end{gathered} \tag{8.100}$$

where the amplitude transmission coefficient $\mathfrak{t}$ is determined by the area of the stop. This defines the scattering matrix elements $S_{k,k} = \mathfrak{t}$ and $S_{-k,k} = \mathfrak{r}$. Performing this calculation for a plane wave of the same frequency propagating in the opposite direction $(k < 0)$ yields $S_{-k,-k} = \mathfrak{t}$ and $S_{k,-k} = \mathfrak{r}$. In the limit of negligible diffraction, the counter-propagating waves $\exp(\pm ikz)$ can only scatter between themselves, so the scattering matrix for this problem reduces to

$$S = \begin{bmatrix} \mathfrak{t} & \mathfrak{r} \\ \mathfrak{r} & \mathfrak{t} \end{bmatrix}. \tag{8.101}$$

Consequently, the coefficients automatically satisfy the conditions (8.7) which guarantee the unitarity of S. This situation is sketched in Fig. 8.5.

In the classical description, the assumption of a plane wave incident from $z < 0$ is imposed by setting $\alpha_{-k} = 0$, so that P1 and P2 in Fig. 8.5 are respectively the input and output ports. The explicit expression (8.101) and the general relation (8.16) yield the scattered (transmitted and reflected) amplitudes as $\alpha'_k = \mathfrak{t}\,\alpha_k$ and $\alpha'_{-k} = \mathfrak{r}\,\alpha_k$. Warned by our experience with the beam splitter, we know that the no-input condition and the scattering relations of the classical problem cannot be carried over into the quantum theory as they stand. The appropriate translation of the classical assumption $\alpha_{-k} = 0$ is to interpret it as a condition on the quantum field state. As a concrete example, consider a source of light, of frequency $\omega = \omega_k$, placed at the focal point of a converging lens somewhere in the region $z < 0$. The light exits from the lens in the plane-wave mode $\exp(ikz)$, and the most general state of the field for this situation is described by a density matrix of the form

$$\rho_{\text{in}} = \sum_{n_k, m_k = 0}^{\infty} |n;k\rangle\, P_{nm}\, \langle m;k|\,, \tag{8.102}$$

where $|n;k\rangle = (n!)^{-1/2} \left(a_k^\dagger\right)^n |0\rangle$ is a number state for photons in the mode $\exp(ikz)$. The density operator ρ_{in} is evaluated in the Heisenberg picture, so the time-independent coefficients satisfy the hermiticity condition, $P_{nm} = P^*_{mn}$, and the trace condition,

$$\sum_{n=0}^{\infty} P_{nn} = 1\,. \tag{8.103}$$

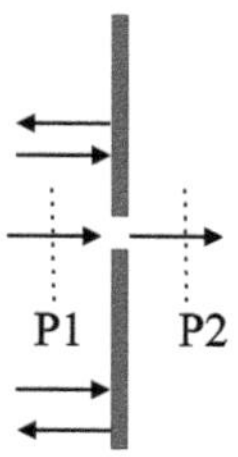

Fig. 8.5 A stop of radius $a \gg \lambda$. The arrows represent a normally incident plane wave together with the reflected and transmitted waves. The surfaces P1 and P2 are ports.

Every one of the number states $|n;k\rangle$ is the vacuum for a_{-k}, therefore the density matrix satisfies

$$a_{-k}\rho_{\text{in}} = \rho_{\text{in}} a_{-k}^{\dagger} = 0\,. \tag{8.104}$$

This is the quantum analogue of the classical condition $\alpha_{-k} = 0$. Since we are not allowed to impose $a_{-k} = 0$, it is essential to use the general relation (8.27) which yields

$$\begin{aligned} a'_k &= \mathfrak{t}\, a_k + \mathfrak{r}\, a_{-k}\,, \\ a'_{-k} &= \mathfrak{t}\, a_{-k} + \mathfrak{r}\, a_k\,. \end{aligned} \tag{8.105}$$

The unitarity of the matrix S in eqn (8.101) guarantees that the scattered operators obey the canonical commutation relations.

Since each incident photon is randomly reflected or transmitted, partition noise is to be expected for stops as well as for beam splitters. Just as for the beam splitter, the additional fluctuation strength in the transmitted field is an example of the general relation between dissipation and fluctuation. In this connection, we should mention that the model of a stop as an aperture in a perfectly conducting, dissipationless screen simplifies the analysis; but it is not a good description of real stops. In practice, stops are usually *black*, i.e. apertures in an absorbing screen. The use of black stops reduces unwanted stray reflections, which are often a source of experimental difficulties. The theory in this case is more complicated, since the absorption of the incident light leads first to excitations in the atoms of the screen. These atomic excitations are coupled in turn to lattice excitations in the solid material. Thus the transmitted field for an absorbing stop will display additional noise, due to the partition between the transmitted light and the excitations of the internal degrees of freedom of the absorbing screen.

8.8 Exercises

8.1 Asymmetric beam splitters

For an asymmetric beam splitter, identify the *upper* (U) and *lower* (L) surfaces as those facing ports 1 and 2 respectively in Fig. 8.2. The general scattering relation is

$$\begin{aligned} a'_1 &= \mathfrak{t}_U\, a_1 + \mathfrak{r}_L\, a_2\,, \\ a'_2 &= \mathfrak{r}_U\, a_1 + \mathfrak{t}_L\, a_2\,. \end{aligned}$$

(1) Derive the conditions on the coefficients guaranteeing that the scattered operators satisfy the canonical commutation relations.

(2) Model an asymmetric beam splitter by coating a symmetric beam splitter (coefficients $\mathfrak{r}$ and $\mathfrak{t}$) with phase shifting materials on each side. Denote the phase shifts for one transit of the coatings by ψ_U and ψ_L and derive the scattering relations. Use your results to express $\mathfrak{t}_U$, $\mathfrak{r}_L$, $\mathfrak{r}_U$, and $\mathfrak{t}_L$ in terms of ψ_U, ψ_L, $\mathfrak{r}$, and $\mathfrak{t}$, and show that the conditions derived in part (1) are satisfied.

(3) Show that the phase shifts can be adjusted so that the scattering relations are

$$\begin{aligned} a'_1 &= \sqrt{1-R}a_1 - \epsilon\sqrt{R}a_2\,, \\ a'_2 &= \epsilon\sqrt{R}a_1 + \sqrt{1-R}a_2\,, \end{aligned}$$

where $R = |\mathfrak{r}|^2$ is the reflectivity and $\epsilon = \pm 1$. This form will prove useful in Section 20.5.3.

8.2 Single-frequency, two-photon state incident on a beam splitter

(1) Treat the coefficients C_{mn} in eqn (8.68) as a symmetric matrix and show that

$$C' = SCS^T ,$$

where S is given by eqn (8.63) and S^T is its transpose.

(2) Evaluate eqn (8.70) for a balanced beam splitter ($\mathfrak{r} = i/\sqrt{2}$, $\mathfrak{t} = 1/\sqrt{2}$). If there are detectors at both output ports, what can you say about the rate of coincidence counting?

(3) Consider the initial state $|\Psi\rangle = N_0 \left[\cos\theta \, a_1^{\dagger 2} + \sin\theta \, a_2^{\dagger 2}\right] |0\rangle$.

(a) Evaluate the normalization constant N_0, calculate the matrices C and C', and then calculate the scattered state $|\Psi\rangle'$.

(b) For a balanced beam splitter, explain why the values $\theta = \pm\pi/4$ are especially interesting.

8.3 Two-frequency state incident on a beam splitter

(1) For the initial state $|\Psi\rangle = a_1^\dagger(\omega_1) \, a_2^\dagger(\omega_2) \, |0\rangle$, calculate the scattered state for the case of a balanced beam splitter, and comment on the difference between this result and the one found in part (2) of Exercise 8.2.

(2) For the initial state $|\Psi\rangle$ no photons of frequency ω_2 are found in channel 1, but they are present in the scattered solution. Where do they come from?

(3) According to the definition in Section 6.5.3, the two states

$$|\Theta_\pm(0)\rangle = \frac{1}{\sqrt{2}} \left[a_1^\dagger(\omega_1) \, a_2^\dagger(\omega_2) \pm a_1^\dagger(\omega_2) \, a_2^\dagger(\omega_1)\right] |0\rangle$$

are dynamically entangled. Evaluate the scattered states for the case of a balanced beam splitter, and compare the different experimental outcomes associated with these examples and with the initial state $|\Psi\rangle$ from part (1).

8.4 Two-polarization state falling on a beam splitter

Consider the initial state $|\Psi\rangle$ defined by eqn (8.77).

(1) Calculate the scattered state for a balanced beam splitter.

(2) Now calculate the scattered state for the alternative initial state

$$|\Psi\rangle = \frac{1}{\sqrt{2}} \left(a_{1h}^\dagger a_{2v}^\dagger + a_{1v}^\dagger a_{2h}^\dagger\right) |0\rangle .$$

Comment on the difference between the results.

8.5 Symmetric Y-junction scattering matrix

Consider the symmetric Y-junction discussed in Section 8.5.

(1) Use the symmetry of the Y-junction to derive eqn (8.88).

(2) Evaluate the upper and lower bounds on $|y_{11}|$ imposed by the unitarity condition on Y.

8.6 Added noise at a Y-junction

Consider the case that photons are incident only in channel 1 of the symmetric Y-junction.

(1) Verify conservation of average photon number, i.e. $\langle N_1' \rangle + \langle N_2' \rangle + \langle N_3' \rangle = \langle N_1 \rangle$.

(2) Evaluate the added noise in output channel 2 by expressing the normalized variance $\overline{V}(N_2')$ in terms of the normalized variance $\overline{V}(N_1)$ in the input channel 1. What is the minimum value of the added noise?

8.7 The optical circulator

For a wave entering port 1 of the circulator depicted in Fig. 8.4(b), paths α and β lead to destructive interference at the mouth of port 3, under the choice of conditions given by eqns (8.94) and (8.98).

(1) What conditions lead to constructive interference at the mouth of port 2?

(2) Show that the scattering matrix given by eqn (8.99) is unitary.

(3) Consider an experimental situation in which a perfect, lossless, retroreflecting mirror terminates port 2. Show that the variance in photon number in the light emitted through port 3 is exactly the same as the variance of the input light entering through port 1.

9

Photon detection

Any experimental measurement sensitive to the discrete nature of photons evidently requires a device that can detect photons one by one. For this purpose a single photon must interact with a system of charged particles to induce a microscopic change, which is subsequently amplified to the macroscopic level. The irreversible amplification stage is needed to raise the quantum event to the classical level, so that it can be recorded. This naturally suggests dividing the treatment of photon detection into several sections. In Section 9.1 we consider the process of primary detection of the incoming photon or photons, and in Section 9.2 we study postdetection signal processing, including the quantum methods of amplification of the primary photon event. Finally in Section 9.3 we study the important techniques of heterodyne and homodyne detection.

9.1 Primary photon detection

In the first section below, we describe six physical mechanisms commonly employed in the primary process of photon detection, and in the second section we present a theoretical analysis of the simplest detection scheme, in which individual atoms are excited by absorption of a single photon. The remaining sections are concerned with the relation of incident photon statistics to the statistics of the ejected photoelectrons, the finite quantum efficiency of detectors, and some general statistical features of the photon distribution.

9.1.1 Photon detection methods

Photon detection is currently based on one of the following physical mechanisms.

(1) **Photoelectric detection**. These detectors fall into two main categories:

   (i) *vacuum tube* devices, in which the incident photon ejects an electron, bound to a photocathode surface, into the vacuum;

   (ii) *solid-state* devices, in which absorption of the incident photon deep within the body of the semiconductor promotes an electron from the valence band to the conduction band (Kittel, 1985).

   In both cases the resulting output signal is proportional to the intensity of the incident light, and thus to the time-averaged square of the electric field strength. This method is, accordingly, also called **square-law detection**.

   There are several classes of vacuum tube devices—for example, the photomultiplier tubes and channeltrons described in Section 9.2.1—but most modern photoelectric detectors are based on semiconductors. The promotion of an electron from the valence band to the conduction band—which is analogous to photoionization of an atom—leaves behind a positively charged hole in the valence band. Both members of the electron–hole pair are free to move through the material.

The energy needed for electron–hole pair production is substantially less than the typical energy—of the order of electron volts—needed to eject a photoelectron into the vacuum outside a metal surface; consequently, semiconductor devices can detect much lower energy photons. Thus the sensitivity of semiconductor detectors extends into the infrared and far-infrared parts of the electromagnetic spectrum. Furthermore, the photon absorption length in the semiconductor material is so small that relatively thin detectors will absorb almost all the incident photons. This means that quantum efficiencies are high (50–90%). Semiconductor detectors are very fast as well as very sensitive, with response times on the scale of nanoseconds. These devices, which are very important for quantum optics, are also called **single-photon counters**.

Solid-state detectors are further divided into two subcategories: **photoconductive** and **photovoltaic**. In photoconductive devices, the photoelectrons are released into a homogeneous semiconducting material, and a uniform internal electric field is applied across the material to accelerate the released photoelectrons. Thus the current in the homogeneous material is proportional to the number of photo-released carriers, and hence to the incident intensity of the light beam falling on the semiconductor. In photovoltaic devices, photons are absorbed and photoelectrons are released in a highly inhomogeneous region inside the semiconductor, where there is a large internal electric field, *viz.*, the depletion range inside a p–n or p–i–n junction. The large internal fields then accelerate the photoelectrons to create a voltage across the junction, which can drive currents in an external circuit. Devices of this type are commonly known as photodiodes (Saleh and Teich, 1991, Chap. 17).

(2) **Rectifying** detection. The oscillating electric field of the electromagnetic wave is rectified, in a diode with a nonlinear I–V characteristic, to produce a direct-current signal which is proportional to the intensity of the wave. The rectification effect arises from a physical asymmetry in the structure of the diode, for example, at the p–n junction of a semiconductor diode device. Such detectors include **Schottky diodes**, consisting of a small metallic contact on the surface of a semiconductor, and biased **superconducting–insulator–superconducting** (SIS) electron tunneling devices. These rectifying detectors are used mainly in the radio and microwave regions of the electromagnetic spectrum, and are commonly called square-law or direct detectors.

(3) **Photothermal** detection. Light is directly converted into heat by absorption, and the resulting temperature rise of the absorber is measured. These detectors are also called **bolometers**. Since thermal response times are relatively long, these detectors are usually slower than many of the others. Nevertheless, they are useful for detection of broad-bandwidth radiation, in experiments allowing long integration times. Thus they are presently being used in the millimeter-wave and far-infrared parts of the electromagnetic spectrum as detectors for astrophysical measurements, including measurements of the anisotropy of the cosmic microwave

background (Richards, 1994).

(4) **Photon beam amplifiers.** The incoming photon beam is coherently amplified by a device such as a maser or a parametric amplifier. These devices are primarily used in the millimeter-wave and microwave region of the electromagnetic spectrum, and play the same role as the electronic pre-amplifiers used at radio frequencies. Rather than providing postdetection amplification, they coherently pre-amplify the incoming electromagnetic wave, by directly providing gain at the carrier frequency. Examples include solid-state masers, which amplify the incoming signal by stimulated emission of radiation (Gordon *et al.*, 1954), and varactor parametric amplifiers (**paramps**), where a pumped, nonlinear, reactive element—such as a nonlinear capacitance of the depletion region in a back-biased p–n junction—can amplify an incoming signal. The nonlinear reactance is modulated by a strong, higher-frequency pump wave which beats with the signal wave to produce an idler wave at the difference frequency between the pump and signal frequencies. The idler wave reacts back via the pump wave to produce more signal wave, etc. This causes a mutual reinforcement, and hence amplification, of both the signal and idler waves, at the expense of power in the pump wave. The idler wave power is dumped into a matched termination.

(5) **Single-microwave-photon counters.** Single microwave photons in a superconducting microwave cavity are detected by using atomic beam techniques to pass individual Rydberg atoms through the cavity. The microwave photon can cause a transition between two high-lying levels (Rydberg levels) of a Rydberg atom, which is subsequently probed by a state-selective field ionization process. The result of this measurement indicates whether a transition has occurred, and therefore provides information about the state of excitation of the microwave cavity (Hulet and Kleppner, 1983; Raushcenbeutal *et al.*, 2000; Varcoe *et al.*, 2000).

(6) **Quantum nondemolition detectors.** The presence of a single photon is detected without destroying it in an absorption process. This detection relies on the phase shift produced by the passage of a single photon through a nonlinear medium, such as a Kerr medium. Such detectors have recently been implemented in the laboratory (Yamamoto *et al.*, 1986).

The last three of these detection schemes, (4) to (6), are especially promising for quantum optics. However, all the basic mechanisms (1) through (3) can be extended, by a number of important auxiliary methods, to provide photon detection at the single-quantum level.

9.1.2 Theory of photoelectric detection

The theory presented here is formulated for the simplest case of excitation of free atoms by the incident light, and it is solely concerned with the primary microscopic detection event. In situations for which photon counting is relevant, the fields are weak; therefore, the response of the atoms can be calculated by first-order perturbation theory. As we will see, the first-order perturbative expression for the counting rate is the product of two factors. The first depends only on the state of the atom, and the second depends only on the state of the field. This clean separation between properties of the detector and properties of the field will hold for any detection scheme that can be described by first-order perturbation theory. Thus the use of the independent atom model does not really restrict the generality of the results. In practice, the sensitivity function describing the detector response is determined empirically, rather than being calculated from first principles.

The primary objective of the theory is therefore to exhibit the information on the state of the field that the counting rate provides. As we will see below, this information is naturally presented in terms of the field–field correlation functions defined in Section 4.7. In a typical experiment, light from an external source, such as a laser, is injected into a sample of some interesting medium and extracted through an output port. The output light is then directed to the detectors by appropriate linear optical elements. An elementary, but nonetheless important, point is that the correlation function associated with a detector signal is necessarily evaluated at the detector, which is typically not located in the interior of the sample being probed. Thus the correlation functions evaluated in the interior of the sample, while of great theoretical interest, are not directly related to the experimental results. Information about the interaction of the light with the sample is effectively stored in the state of the emitted radiation field, which is used in the calculation of the correlation functions at the detectors. Thus for the analysis of photon detection *per se* we only need to consider the interaction of the electromagnetic field with the optical elements and the detectors. The total Hamiltonian for this problem is therefore $H = H_0 + H_{\text{det}}$, where H_{det} represents the interaction with the detectors only. The unperturbed Hamiltonian is $H_0 = H_D + H_{\text{em}} + H_1$, where H_D is the detector Hamiltonian and H_{em} is the field Hamiltonian. The remaining term, H_1, describes the interaction of the field with the passive linear optical devices, e.g. lenses, mirrors, beam splitters, etc., that direct the light to the detectors.

A Single-photon detection

The simplest possible photon detector consists of a single atom interacting with the field. In the interaction picture, $H_{\text{det}} = -\mathbf{d}(t) \cdot \mathbf{E}(\mathbf{r}, t)$ describes the interaction of the field with the detector atom located at $\mathbf{r}$. The initial state is $|\Theta(t_0)\rangle = |\phi_\gamma, \Phi_e\rangle = |\phi_\gamma\rangle |\Phi_e\rangle$, where $|\phi_\gamma\rangle$ is the atomic ground state and $|\Phi_e\rangle$ is the initial state of the radiation field, which is, for the moment, assumed to be pure. According to eqns (4.95) and (4.103) the initial state vector evolves into

$$|\Theta(t)\rangle = |\Theta(t_0)\rangle - \frac{i}{\hbar} \int_{t_0}^{t} dt_1 H_{\text{det}}(t_1) |\Theta(t_0)\rangle + \cdots , \tag{9.1}$$

so the first-order probability amplitude that a joint measurement at time t finds the atom in an excited state $|\phi_\epsilon\rangle$ and the field in the number state $|\underline{n}\rangle$ is

$$\langle \phi_\epsilon, \underline{n} | \Theta(t) \rangle = -\frac{i}{\hbar} \int_{t_0}^{t} dt_1 \langle \phi_\epsilon, \underline{n} | H_{\text{det}}(t_1) | \Theta(t_0) \rangle , \tag{9.2}$$

where $|\phi_\epsilon, \underline{n}\rangle = |\phi_\epsilon\rangle |\underline{n}\rangle$. Only the Rabi operator $\widehat{\Omega}^{(+)}$ in eqn (4.149) can contribute to an absorptive transition, so the matrix element and the probability amplitude are respectively given by

$$\langle \phi_\epsilon, \underline{n} \left| H_{\text{det}}(t_1) \right| \Theta(t_0) \rangle = -e^{i\omega_{\epsilon\gamma} t_1} \mathbf{d}_{\epsilon\gamma} \cdot \left\langle \underline{n} \left| \mathbf{E}^{(+)}(\mathbf{r}, t_1) \right| \Phi_e \right\rangle \tag{9.3}$$

and

$$\langle \phi_\epsilon, \underline{n} \left| \Theta(t) \right\rangle = \frac{i}{\hbar} \int_{t_0}^{t} dt_1 e^{i\omega_{\epsilon\gamma} t_1} \mathbf{d}_{\epsilon\gamma} \cdot \left\langle \underline{n} \left| \mathbf{E}^{(+)}(\mathbf{r}, t_1) \right| \Phi_e \right\rangle, \tag{9.4}$$

where $\mathbf{d}_{\epsilon\gamma} = \left\langle \phi_\epsilon \left| \widehat{\mathbf{d}} \right| \phi_\gamma \right\rangle$ is the dipole matrix element for the transition $\gamma \to \epsilon$.

The conditional probability for finding $|\phi_\epsilon, \underline{n}\rangle$, given $|\phi_\gamma, \Phi_e\rangle$, is therefore

$$\begin{aligned} p(\phi_\epsilon, \underline{n} : \phi_\gamma, \Phi_e) &= \left| \frac{i}{\hbar} \int_{t_0}^{t} dt_1 e^{i\omega_{\epsilon\gamma} t_1} \mathbf{d}_{\epsilon\gamma} \cdot \left\langle \underline{n} \left| \mathbf{E}^{(+)}(\mathbf{r}, t_1) \right| \Phi_e \right\rangle \right|^2 \\ &= \frac{\left(d_{\epsilon\gamma}^*\right)_i \left(d_{\epsilon\gamma}\right)_j}{\hbar^2} \int_{t_0}^{t} dt_1 \int_{t_0}^{t} dt_2 e^{i\omega_{\epsilon\gamma}(t_2 - t_1)} \\ &\quad \times \left\langle \underline{n} \left| E_i^{(+)}(\mathbf{r}, t_1) \right| \Phi_e \right\rangle^* \left\langle \underline{n} \left| E_j^{(+)}(\mathbf{r}, t_2) \right| \Phi_e \right\rangle . \end{aligned} \tag{9.5}$$

The relation $\mathbf{E}^{(-)} = \mathbf{E}^{(+)\dagger}$ implies $\left\langle \underline{n} \middle| E_i^{(+)}(\mathbf{r}, t_1) \middle| \Phi_e \right\rangle^* = \left\langle \Phi_e \middle| E_i^{(-)}(\mathbf{r}, t_1) \middle| \underline{n} \right\rangle$, so that eqn (9.5) can be rewritten as

$$\begin{aligned} p(\phi_\epsilon, \underline{n} : \phi_\gamma, \Phi_e) &= \frac{\left(d_{\epsilon\gamma}^*\right)_i \left(d_{\epsilon\gamma}\right)_j}{\hbar^2} \int_{t_0}^{t} dt_1 \int_{t_0}^{t} dt_2 e^{i\omega_{\epsilon\gamma}(t_2 - t_1)} \\ &\quad \times \left\langle \Phi_e \left| E_i^{(-)}(\mathbf{r}, t_1) \right| \underline{n} \right\rangle \left\langle \underline{n} \left| E_j^{(+)}(\mathbf{r}, t_2) \right| \Phi_e \right\rangle . \end{aligned} \tag{9.6}$$

Since the final state of the radiation field is not usually observed, the relevant quantity is the sum of the conditional probabilities $p(\phi_\epsilon, \underline{n} : \phi_\gamma, \Phi_e)$ over all final field states $|\underline{n}\rangle$:

$$p(\phi_\epsilon : \phi_\gamma, \Phi_e) = \sum_{\underline{n}} p(\phi_\epsilon, \underline{n} : \phi_\gamma, \Phi_e) . \tag{9.7}$$

The completeness identity (3.67) for the number states, combined with eqn (9.6) and eqn (9.7), then yields

$$\begin{aligned} p(\phi_\epsilon : \phi_\gamma, \Phi_e) &= \frac{\left(d_{\epsilon\gamma}^*\right)_i \left(d_{\epsilon\gamma}\right)_j}{\hbar^2} \int_{t_0}^{t} dt_1 \int_{t_0}^{t} dt_2 e^{i\omega_{\epsilon\gamma}(t_2 - t_1)} \\ &\quad \times \left\langle \Phi_e \left| E_i^{(-)}(\mathbf{r}, t_1) E_j^{(+)}(\mathbf{r}, t_2) \right| \Phi_e \right\rangle . \end{aligned} \tag{9.8}$$

This result is valid when the radiation field is known to be initially in the pure state $|\Phi_e\rangle$. In most experiments all that is known is a probability distribution $\mathcal{P}_e$ over an ensemble $\{|\Phi_e\rangle\}$ of pure initial states, so it is necessary to average over this ensemble to get

$$\begin{aligned} p(\phi_\epsilon : \phi_\gamma) &= \sum_e p(\phi_\epsilon : \phi_\gamma, \Phi_e) \mathcal{P}_e \\ &= \frac{\left(d_{\epsilon\gamma}^*\right)_i \left(d_{\epsilon\gamma}\right)_j}{\hbar^2} \int_{t_0}^{t} dt_1 \int_{t_0}^{t} dt_2 e^{i\omega_{\epsilon\gamma}(t_2 - t_1)} \operatorname{Tr}\left[\rho E_i^{(-)}(\mathbf{r}, t_1) E_j^{(+)}(\mathbf{r}, t_2)\right], \end{aligned} \tag{9.9}$$

where

$$\rho = \sum_e \mathcal{P}_e \left|\Phi_e\right\rangle \left\langle\Phi_e\right| \tag{9.10}$$

is the density operator defined by the distribution $\mathcal{P}_e$.

So far it has been assumed that the final atomic state $|\phi_\epsilon\rangle$ can be detected with perfect accuracy, but of course this is never the case. Furthermore, most detection schemes do not depend on a specific transition to a bound level; instead, they involve transitions into excited states lying in the continuum. The atom may be directly ionized, or the absorption of the photon may lead to a bound state that is subject to Stark ionization by a static electric field. The ionized electrons would then be accelerated, and thereby produce further ionization by secondary collisions. All of these complexities are subsumed in the probability $\mathcal{D}(\epsilon)$ that the transition $\gamma \to \epsilon$ occurs and produces a macroscopically observable event, e.g. a current pulse. The overall probability is then

$$p(t) = \sum_\epsilon \mathcal{D}(\epsilon)\, p(\phi_\epsilon : \phi_\gamma) . \tag{9.11}$$

It should be understood that the ϵ-sum is really an integral, and that the factor $\mathcal{D}(\epsilon)$ includes the density of states for the continuum states of the atom. Putting this together with the expression (9.9) leads to

$$p(t) = \int_{t_0}^{t} dt_1 \int_{t_0}^{t} dt_2 \mathfrak{S}_{ji}(t_1 - t_2)\, G_{ij}^{(1)}(\mathbf{r}, t_1; \mathbf{r}, t_2) , \tag{9.12}$$

where the **sensitivity function**

$$\mathfrak{S}_{ji}(t) = \frac{1}{\hbar^2} \sum_\epsilon \mathcal{D}(\epsilon) \left(d^*_{\epsilon\gamma}\right)_i \left(d_{\epsilon\gamma}\right)_j e^{-i\omega_{\epsilon\gamma} t} \tag{9.13}$$

is determined solely by the properties of the atom, and the field–field correlation function

$$G_{ij}^{(1)}(\mathbf{r}, t_1; \mathbf{r}, t_2) = \mathrm{Tr}\left[\rho E_i^{(-)}(\mathbf{r}_1, t_1)\, E_j^{(+)}(\mathbf{r}, t_2)\right] \tag{9.14}$$

is determined solely by the properties of the field.

Since $\mathcal{D}(\epsilon)$ is real and positive, the sensitivity function obeys

$$\mathfrak{S}^*_{ji}(t) = \mathfrak{S}_{ij}(-t) , \tag{9.15}$$

and other useful properties are found by studying the Fourier transform

$$\begin{aligned} \mathfrak{S}_{ji}(\omega) &= \int dt \mathfrak{S}_{ji}(t)\, e^{i\omega t} \\ &= \frac{2\pi}{\hbar^2} \sum_\epsilon \mathcal{D}(\epsilon) \left(d^*_{\epsilon\gamma}\right)_i \left(d_{\epsilon\gamma}\right)_j \delta(\omega - \omega_{\epsilon\gamma}) . \end{aligned} \tag{9.16}$$

The ϵ-sum is really an integral over the continuum of excited states, so $\mathfrak{S}_{ji}(\omega)$ is a smooth function of ω. This explicit expression shows that the 3×3 matrix $\mathfrak{S}(\omega)$, with components $\mathfrak{S}_{ji}(\omega)$, is hermitian—i.e. $\mathfrak{S}_{ji}(\omega) = \mathfrak{S}_{ij}(\omega)^*$—and positive-definite, since

$$v_j^* \mathfrak{S}_{ji}(\omega) v_i = \frac{2\pi}{\hbar^2} \sum_{\epsilon} \mathcal{D}(\epsilon) \left|\mathbf{v}^* \cdot \mathbf{d}_{\epsilon\gamma}\right|^2 \delta(\omega - \omega_{\epsilon\gamma}) > 0 \tag{9.17}$$

for any complex vector $\mathbf{v}$. These properties in turn guarantee that the eigenvalues are real and positive, so the power spectrum,

$$\mathfrak{T}(\omega) = \operatorname{Tr}[\mathfrak{S}(\omega)], \tag{9.18}$$

of the dipole transitions can be used to define averages over frequency by

$$\langle f \rangle_{\mathfrak{T}} = \frac{\int d\omega \mathfrak{T}(\omega) f(\omega)}{\int d\omega \mathfrak{T}(\omega)}. \tag{9.19}$$

The width $\Delta\omega_S$ of the sensitivity function is then defined as the *rms* deviation

$$\Delta\omega_S = \sqrt{\langle \omega^2 \rangle_{\mathfrak{T}} - \langle \omega \rangle_{\mathfrak{T}}^2}. \tag{9.20}$$

The single-photon **counting rate** $w^{(1)}(t)$ is the rate of change of the probability:

$$w^{(1)}(t) = \frac{dp}{dt} = 2\operatorname{Re} \int_{t_0}^{t} dt' \mathfrak{S}_{ji}(t' - t) G_{ij}^{(1)}(\mathbf{r}, t'; \mathbf{r}, t), \tag{9.21}$$

where the final form comes from combining eqn (9.15) with the symmetry property

$$G_{ij}^{(1)*}(\mathbf{r}_1, t_1; \mathbf{r}_2, t_2) = G_{ji}^{(1)}(\mathbf{r}_2, t_2; \mathbf{r}_1, t_1), \tag{9.22}$$

that follows from eqn (9.14). For later use it is better to express the counting rate as

$$w^{(1)}(t) = 2\operatorname{Re} \int \frac{d\omega}{2\pi} \mathfrak{S}_{ji}(\omega) X_{ij}(\omega, t), \tag{9.23}$$

where

$$X_{ij}(\omega, t) = \int_{t_0}^{t} dt' e^{i\omega(t-t')} G_{ij}^{(1)}(\mathbf{r}, t'; \mathbf{r}, t). \tag{9.24}$$

The value of the frequency integral in eqn (9.23) depends on the relative widths of the sensitivity function and $X_{ij}(\omega, t)$, considered as a function of ω with t fixed. One way to get this information is to use eqn (9.24) to evaluate the transform

$$\begin{aligned} X_{ij}(t', t) &= \int \frac{d\omega}{2\pi} e^{i\omega t'} X_{ij}(\omega, t) \\ &= \theta(t') \theta(t - t_0 - t') G_{ij}^{(1)}(\mathbf{r}, t - t'; \mathbf{r}, t). \end{aligned} \tag{9.25}$$

The step functions in this expression guarantee that $X_{ij}(t', t)$ vanishes outside the interval $0 \leqslant t' \leqslant t - t_0$. On the other hand, the correlation function vanishes for $t' \gg T_c$, where T_c is the correlation time. The observation time $t - t_0$ is normally much longer than the correlation time, so the t'-width of $X_{ij}(t', t)$ is approximately T_c. By the uncertainty principle, the ω-width of $X_{ij}(\omega, t)$ is $\Delta\omega_X \sim 1/T_c = \Delta\omega_G$, where $\Delta\omega_G$ is the bandwidth of the correlation function $G_{ij}^{(1)}$.

B Broadband detection

The detector is said to be **broadband** if the bandwidth $\Delta\omega_S$ of the sensitivity function satisfies $\Delta\omega_S \gg \Delta\omega_G = 1/T_c$. For a broadband detector, $X_{ij}(\omega)$ is sharply peaked compared to the sensitivity function; therefore, $\mathfrak{S}_{ji}(\omega)$ can be treated as a constant—$\mathfrak{S}_{ji}(\omega) \approx \mathfrak{S}_{ji}$—and taken outside the integral. This is formally equivalent to setting $\mathfrak{S}_{ji}(t'-t) = \mathfrak{S}_{ji}\delta(t'-t)$ in eqn (9.21), and the result

$$w^{(1)}(t) = \mathfrak{S}_{ji} G_{ij}^{(1)}(\mathbf{r}, t; \mathbf{r}, t) \tag{9.26}$$

is obtained by combining the end-point rule (A.98) for delta functions with the symmetries (9.15) and (9.22). Consequently, the broadband counting rate is proportional to the equal-time correlation function. The argument leading to eqn (9.26) is similar to the derivation of Fermi's golden rule in perturbation theory. In practice, nearly all detectors can be treated as broadband.

The analysis of ideal single-atom detectors can be extended to realistic many-atom detectors when two conditions are satisfied: (1) single-atom absorption is the dominant process; (2) interactions between the atoms can be ignored. These conditions will be satisfied for atoms in a tenuous vapor or in an atomic beam—see item (5) in Section 9.1.1—and they are also satisfied by many solid-state detectors. For atoms located at positions $\mathbf{r}_1, \ldots, \mathbf{r}_N$, the total single-photon counting rate is the average of the counting rates for the individual atoms:

$$w^{(1)}(t) = \frac{1}{N} \sum_{A=1}^{N} \mathfrak{S}_{ji}^{(A)} G_{ij}^{(1)}(\mathbf{r}_A, t; \mathbf{r}_A, t) \,. \tag{9.27}$$

It is often convenient to use a coarse-grained description which replaces the last equation by

$$w^{(1)}(t) = \frac{1}{\overline{n} V_D} \int d^3 r\, n(\mathbf{r})\, \mathfrak{S}_{ji}(\mathbf{r})\, G_{ij}^{(1)}(\mathbf{r}, t; \mathbf{r}, t) \,, \tag{9.28}$$

where $n(\mathbf{r})$ is the density of atoms, $\mathfrak{S}_{ji}(\mathbf{r})$ is the sensitivity function at $\mathbf{r}$, $\overline{n}$ is the mean density of atoms, and V_D is the volume occupied by the detector. A **point detector** is defined by the condition that the correlation function is essentially constant across the volume of the detector. In this case, the counting rate is

$$w^{(1)}(t) = \overline{\mathfrak{S}}_{ji} G_{ij}^{(1)}(\mathbf{r}, t; \mathbf{r}, t) \,, \tag{9.29}$$

where $\overline{\mathfrak{S}}_{ji}$ is the average sensitivity function and $\mathbf{r}$ is the center of mass of the detector. Comparing this to eqn (9.26) shows that a point detector is like a single-atom detector with a modified sensitivity factor.

The sensitivity factor, defined by eqn (9.16), is a 3×3 hermitian matrix which has the useful representation

$$\mathfrak{S}_{ij} = \sum_{a=1}^{3} \mathfrak{S}_a e_{ai} e_{aj}^* \,, \tag{9.30}$$

where the eigenvalues, $\mathfrak{S}_a$, are real and the eigenvectors, $\mathbf{e}_a$, are orthonormal: $\mathbf{e}_b^* \cdot \mathbf{e}_a = \delta_{ab}$. Substituting this representation into eqn (9.26) produces

$$w^{(1)}(t) = \sum_{a=1}^{3} \mathfrak{S}_a G_a^{(1)}(\mathbf{r}, t; \mathbf{r}, t), \tag{9.31}$$

where the new correlation functions,

$$G_a^{(1)}(\mathbf{r}, t; \mathbf{r}, t) = \mathrm{Tr}\left[\rho E_a^{(-)}(\mathbf{r}, t) E_a^{(+)}(\mathbf{r}, t)\right], \tag{9.32}$$

are defined in terms of the scalar field operators $E_a^{(-)}(\mathbf{r}, t) = \mathbf{e}_a \cdot \mathbf{E}^{(-)}(\mathbf{r}, t)$. This form is useful for imposing special conditions on the detector. For example, a detector equipped with a polarization filter is described by the assumption that only one of the eigenvalues, say $\mathfrak{S}_1$, is nonzero. The corresponding eigenvector $\mathbf{e}_1$ is the polarization passed by the filter. In this situation, eqn (9.29) becomes

$$\begin{aligned} w^{(1)}(t) &= \mathfrak{S}\, G^{(1)}(\mathbf{r}, t; \mathbf{r}, t) \\ &= \mathfrak{S}\, \mathrm{Tr}\left[\rho E_1^{(-)}(\mathbf{r}, t) E_1^{(+)}(\mathbf{r}, t)\right], \end{aligned} \tag{9.33}$$

where $E_1^{(+)}(\mathbf{r}, t) = \mathbf{e}^* \cdot \mathbf{E}^{(+)}(\mathbf{r}, t)$, $\mathbf{e}$ is the transmitted polarization, and $\mathfrak{S}$ is the sensitivity factor. As promised, the counting rate is the product of the sensitivity factor $\mathfrak{S}$ and the correlation function $G^{(1)}$. Thus the broadband counting rate provides a direct measurement of the equal-time correlation function $G^{(1)}(\mathbf{r}, t; \mathbf{r}, t)$.

C Narrowband detection

Broadband detectors do not distinguish between photons of different frequencies that may be contained in the incident field, so they do not determine the spectral function of the field. For this purpose, one needs **narrowband detection**, which is usually achieved by passing the light through a **narrowband filter** before it falls on a broadband detector. The filter is a linear device, so its action can be represented mathematically as a linear operation applied to the signal. For a real signal, $X(t) = X^{(+)}(t) + X^{(-)}(t)$, the **filtered signal** at ω—i.e. the part of the signal corresponding to a narrow band of frequencies around ω—is defined by

$$\begin{aligned} X^{(+)}(\omega; t) &= \int_{-\infty}^{\infty} dt' \varpi(t' - t)\, e^{i\omega(t'-t)} X^{(+)}(t') \\ &= \int_{-\infty}^{\infty} dt' \varpi(t')\, e^{i\omega t'} X^{(+)}(t' + t), \end{aligned} \tag{9.34}$$

where the factor $\exp[i\omega(t' - t)]$ serves to pick out the desired frequency. The weighting function $\varpi(t)$ has the following properties.

(1) It is even and positive,

$$\varpi(t) = \varpi(-t) \geqslant 0. \tag{9.35}$$

(2) It is normalized by

$$\int_{-\infty}^{\infty} dt \varpi(t) = 1. \tag{9.36}$$

(3) It is peaked at $t = 0$.

The weighting function is therefore suitable for defining averages, e.g. the **temporal width** ΔT:

$$\Delta T = \left[\int_{-\infty}^{\infty} dt \; \varpi(t) \; t^2 \right]^{1/2} < \infty \,. \tag{9.37}$$

A simple example of an averaging function satisfying eqns (9.35)–(9.37) is

$$\varpi(t) = \begin{cases} \frac{1}{\Delta T} & \text{for } -\frac{\Delta T}{2} \leqslant t \leqslant \frac{\Delta T}{2} \,, \\ 0 & \text{otherwise}\,. \end{cases} \tag{9.38}$$

The meaning of filtering can be clarified by Fourier transforming eqn (9.34) to get

$$X^{(+)}(\omega'; \omega) = F(\omega' - \omega) X^{(+)}(\omega) \,, \tag{9.39}$$

where the **filter function** $F(\omega)$ is the Fourier transform of $\varpi(t)$. Since the normalization condition (9.36) implies $F(0) = 1$, the filtered signal is essentially identical to the original signal in the narrow band defined by the width $\Delta\omega_F \sim 1/\Delta T$ of the filter function; but, it is strongly suppressed outside this band.

The frequency ω selected by the filter varies continuously, so the interesting quantity is the **spectral density** $\mathcal{S}(\omega)$, which is defined as the counting rate per unit frequency interval. Applying the broadband result (9.33) to the filtered field operators yields

$$\begin{aligned} \mathcal{S}(\omega, t) &= \frac{w^{(1)}(\omega, t)}{\Delta\omega_F} \\ &= \frac{\mathfrak{S}}{\Delta\omega_F} \left\langle E_1^{(-)}(\mathbf{r}, t; \omega) E_1^{(+)}(\mathbf{r}, t; \omega) \right\rangle . \end{aligned} \tag{9.40}$$

For the following argument, we choose the simple form (9.38) for the averaging function to calculate the filtered operator:

$$E_1^{(+)}(\mathbf{r}, t; \omega) = \frac{1}{\Delta T} \int_{-\Delta T/2}^{\Delta T/2} dt' e^{i\omega t'} E_1^{(+)}(\mathbf{r}, t' + t) \,. \tag{9.41a}$$

Substituting this result into eqn (9.40) and combining $\Delta\omega_F = 1/\Delta T$ with the definition of the first-order correlation function yields

$$\mathcal{S}(\omega, t) = \frac{\mathfrak{S}}{\Delta T} \int_{-\Delta T/2}^{\Delta T/2} dt_1 \int_{-\Delta T/2}^{\Delta T/2} dt_2 e^{i\omega(t_1 - t_2)} G^{(1)}(\mathbf{r}, t_2 + t; \mathbf{r}, t_1 + t) \,. \tag{9.41b}$$

In almost all applications, we can assume that the correlation function only depends on the difference in the time arguments. This assumption is rigorously valid if the density operator ρ is stationary, and for dissipative systems it is approximately satisfied for large t. Given this property, we set

$$G^{(1)}(\mathbf{r}, t_2 + t; \mathbf{r}, t_1 + t) = \int \frac{d\omega'}{2\pi} G^{(1)}(\mathbf{r}, \omega'; t)\, e^{-i\omega'(t_1 - t_2)}, \tag{9.42}$$

and get

$$\mathcal{S}(\omega) = \mathfrak{S} \int \frac{d\omega'}{2\pi} G^{(1)}(\mathbf{r}, \omega'; t) \frac{\sin^2[(\omega - \omega')\Delta T/2]}{[(\omega - \omega')/2]^2 \Delta T}. \tag{9.43}$$

In this case, the width of the filter is assumed to be very small compared to the width of the correlation function, i.e. $\Delta\omega_S \ll \Delta\omega_G$ ($\Delta T \gg T_c$). By means of the general identity (A.102), one can show that

$$\lim_{\Delta T \to \infty} \frac{\sin^2[\nu \Delta T/2]}{[\nu/2]^2 \Delta T} = \pi\delta(\nu/2) = 2\pi\delta(\nu), \tag{9.44}$$

and substituting this result into eqn (9.43) leads to

$$\mathcal{S}(\omega) = \mathfrak{S} G^{(1)}(\mathbf{r}, \omega; t) = \mathfrak{S} \int d\tau e^{-i\omega\tau} G^{(1)}(\mathbf{r}, \tau + t; \mathbf{r}, t). \tag{9.45}$$

In other words, the spectral density is proportional to the Fourier transform, with respect to the difference of the time arguments, of the two-time correlation function $G^{(1)}(\mathbf{r}, t_2 + t; \mathbf{r}, t_1 + t)$.

It is often useful to have a tunable filter, so that the selected frequency can be swept across the spectral region of interest. The main methods for accomplishing this employ spectrometers to spatially separate the different frequency components. One technique is to use a diffraction grating spectrometer (Hecht, 2002, Sec. 10.2.8) placed on a mount that can be continuously swept in angle, while the input and output slits remain fixed. The spectrometer thus acts as a continuously tunable filter, with bandwidth determined by the width of the slits. Higher resolution can be achieved by using a Fabry–Perot spectrometer (Hecht, 2002, Secs 9.6.1 and 9.7.3) with an adjustable spacing between the plates. A different approach is to use a heterodyne spectrometer, in which the signal is mixed with a local oscillator—usually a laser—which is close to the signal frequency. The beat signal oscillates at an intermediate frequency which is typically in the radio range, so that standard electronics techniques can be used. For example, the radio frequency signal is analyzed by a radio frequency spectrometer or a correlator. The Fourier transform of the correlator output signal yields the radio frequency spectrum of the beat signal.

9.1.3 Photoelectron counting statistics

How does one measure the photon statistics of a light field, such as the Poissonian statistics predicted for the coherent state $|\alpha\rangle$? In practice, these statistics must be inferred from photoelectron counting statistics which, fortunately, often faithfully reproduce the counting statistics of the photons. For example, in the case of light prepared in a coherent state, both the incident photon and the detected photoelectron statistics turn out to be Poissonian.

Consider a light beam—produced, for example, by passing the output of a laser operating far above threshold through an attenuator—that falls on the photocathode surface of a photomultiplier tube. The amplitude of the attenuated coherent state is $\alpha = \exp(-\chi L/2)\alpha_0$, where χ is the absorption coefficient and L is the length of the absorber. The photoelectron probability distribution can be obtained from the probability distribution for the number of incident photons, $p(n)$, by folding it into the Bernoulli distribution function using the standard classical technique (Feller, 1957*a*, Chap. VI). The probability $P(m,\xi)$ of the detection of m photoelectrons found in this way is

$$P(m,\xi) = \sum_{n=m}^{\infty} p(n) \binom{n}{m} \xi^m (1-\xi)^{n-m} , \qquad (9.46)$$

where ξ is the probability that the interaction of a given photon with the atoms in the detector will produce a photoelectron. This quantity—which is called the **quantum efficiency**—is given by

$$\xi = \zeta \left(\frac{c\hbar\omega}{V} \right) T , \qquad (9.47)$$

where ζ (which is proportional to the sensitivity function $\mathfrak{S}$) is the photoelectron ejection probability per unit time per unit light intensity, $(c\hbar\omega/V)$ is the intensity due to a single photon, and T is the integration time of the photon detector. The integration time is usually the RC time constant of the detection system, which in the case of photomultiplier tubes is of the order of nanoseconds. The parameter ζ can be calculated quantum mechanically, but is usually determined empirically. The factors in the summand in eqn (9.46) are: the photon distribution $p(n)$; the binomial coefficient $\binom{n}{m}$ (the number of ways of distributing n photons among m photoelectron ejections); the probability ξ^m that m photons are converted into photoelectrons; and the probability $(1-\xi)^{n-m}$ that the remaining $n-m$ photons are not detected at all. One can show—see Exercise 9.1—that a Poissonian initial photon distribution, with average photon number $\overline{n}$, results in a Poissonian photoelectron distribution,

$$P(m,\xi) = \frac{\overline{m}^m}{m!} e^{-\overline{m}} , \qquad (9.48)$$

where $\overline{m} = \xi\overline{n}$ is the average ejected photoelectron number. In the special case $\xi = 1$, there is a one–one correspondence between an incident photon and a single ejected photoelectron. In this case, the Bernoulli sum in eqn (9.46) consists of only the single term $n = m$, so that the photon and photoelectron distribution functions are identical. Thus the photoelectron statistics will faithfully reproduce the photon statistics in the incident light beam, for example, the Poissonian statistics of the coherent state discussed above. An experiment demonstrating this fact for a helium–neon laser operating far above threshold is described in Section 5.3.2.

9.1.4 Quantum efficiency*

The quantum efficiency ξ introduced in eqn (9.47) is a phenomenological parameter that can represent any of a number of possible failure modes in photon detection: reflection from the front surface of a cathode; a mismatch between the transverse profile of the signal and the aperture of the detector; arrival of the signal during a dead time of the detector; etc. In each case, there is some scattering or absorption channel in addition to the one that yields the current pulse signaling the detection event. We have already seen, in the discussion of beam splitters in Section 8.4, that the presence of additional channels adds partition noise to the signal, due to vacuum fluctuations entering through an unused port. This generic feature allows us to model an imperfect detector as a compound device composed of a beam splitter followed by an ideal detector with 100% quantum efficiency, as shown in Fig. 9.1.

The transmission and reflection coefficients of the fictitious beam splitter must be adjusted to obey the unitarity condition (8.7) and to account for the quantum efficiency of the real detector. These requirements are satisfied by setting

$$\mathfrak{t} = \sqrt{\xi}\,, \quad \mathfrak{r} = i\sqrt{1-\xi}\,. \tag{9.49}$$

The beam splitter is a linear device, so no generality is lost by restricting attention to monochromatic input signals described by a density operator ρ that is the vacuum for all modes other than the signal mode. In this case we can specialize eqn (8.28) for the in-field to

$$E_{\text{in}}^{(+)}(\mathbf{r},t) = i\mathcal{E}_{0s}a_1 e^{ik_s x}e^{-i\omega_s t} + E_{\text{vac,in}}^{(+)}(\mathbf{r},t)\,, \tag{9.50}$$

where we have chosen the x- and y-axes along the $1 \to 1'$ and $2 \to 2'$ arms of the device respectively, $\mathcal{E}_{0s} = \sqrt{\hbar\omega_s/2\epsilon_0 V}$ is the vacuum fluctuation field strength for a plane wave with frequency ω_s, and a_1 is the annihilation operator for the plane-wave mode $\exp\left[i\left(k_s x - \omega_s t\right)\right]$. In principle, the operator $E_{\text{vac,in}}^{(+)}(\mathbf{r},t)$ should be a sum over all modes orthogonal to the signal mode, but the discussion in Section 8.4.1 shows that we need only consider the mode $\exp\left[i\left(k_s y - \omega_s t\right)\right]$ entering through port 2. This leaves us with the simplified in-field

$$E_{\text{in}}^{(+)}(\mathbf{r},t) = i\mathcal{E}_{0s}a_1 e^{ik_s x}e^{-i\omega_s t} + i\mathcal{E}_{0s}a_2 e^{ik_s y}e^{-i\omega_s t}\,. \tag{9.51}$$

An application of eqn (8.63) yields the scattered annihilation operators

$$\begin{aligned} a_1' &= \sqrt{\xi}a_1 + i\sqrt{1-\xi}a_2\,, \\ a_2' &= i\sqrt{1-\xi}a_1 + \sqrt{\xi}a_2\,, \end{aligned} \tag{9.52}$$

and the corresponding out-field

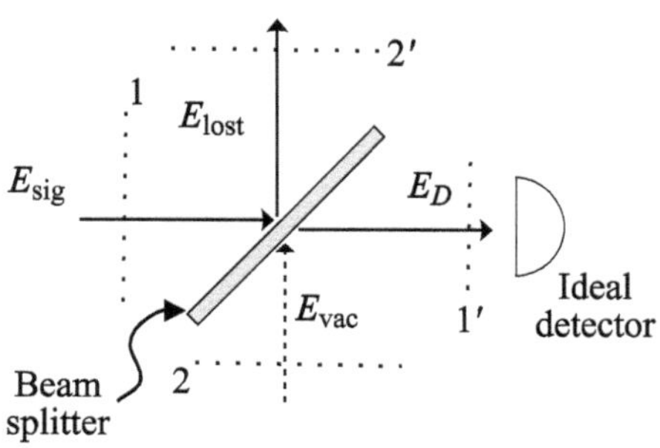

Fig. 9.1 An imperfect detector modeled by combining an ideal detector with a beam splitter. E_{sig} is the signal entering port 1, E_{vac} represents vacuum fluctuations (at the signal frequency) entering port 2, E_D is the effective signal entering the detector, and E_{lost} describes the part of the signal lost due to inefficiencies.

$$E_{\text{out}}^{(+)}(\mathbf{r},t)=E_D^{(+)}(\mathbf{r},t)+E_{\text{lost}}^{(+)}(\mathbf{r},t)\,, \tag{9.53}$$

where

$$E_D^{(+)}(\mathbf{r},t)=i\mathcal{E}_{0s}a_1'e^{ik_sx}e^{-i\omega_st} \tag{9.54}$$

and

$$E_{\text{lost}}^{(+)}(\mathbf{r},t)=i\mathcal{E}_{0s}a_2'e^{ik_sy}e^{-i\omega_st}\,. \tag{9.55}$$

The counting rate of the imperfect detector is by definition the counting rate of the perfect detector viewing port $1'$ of the beam splitter, so—for the simple case of a broadband detector—eqn (9.33) gives

$$\begin{aligned}w^{(1)}(t)&=\mathfrak{S}\left\langle E_D^{(-)}(\mathbf{r}_D,t)\,E_D^{(+)}(\mathbf{r}_D,t)\right\rangle\\&=\mathfrak{S}\,\mathcal{E}_{0s}^2\left\langle a_1'^{\dagger}a_1'\right\rangle\\&=\xi\,\mathfrak{S}\,\mathcal{E}_{0s}^2\left\langle a_1^{\dagger}a_1\right\rangle,\end{aligned} \tag{9.56}$$

where $\langle(\cdots)\rangle=\operatorname{Tr}[\rho(\cdots)]$, $\mathbf{r}_D$ is the location of the detector, and we have used $a_2\rho=\rho a_2^{\dagger}=0$. The operator $E_{\text{lost}}^{(+)}$ represents the part of the signal lost by scattering into the $2\to2'$ channel.

As expected, the counting rate of the imperfect detector is reduced by the quantum efficiency ξ; and the vacuum fluctuations entering through port 2 do not contribute to the average detector output. From our experience with the beam splitter, we know that the vacuum fluctuations will add to the variance of the scattered number operator $N_1'=a_1'^{\dagger}a_1'$. Combining the canonical commutation relations for the creation and annihilation operators with the scattering equation (9.52) and a little algebra gives us

$$V(N_1')=\xi^2V(N_1)+\xi(1-\xi)\langle N_1\rangle\,. \tag{9.57}$$

The first term on the right represents the variance in photon number for the incident field, reduced by the square of the quantum efficiency. The second term is the contribution of the extra partition noise associated with the random response of the imperfect detector, i.e. the arrival of a photon causes a click with probability ξ or no click with probability $1-\xi$.

9.1.5 The Mandel $\mathcal{Q}$-parameter

Most photon detectors are based on the photoelectric effect, and in Section 9.1.2 we have seen that counting rates can be expressed in terms of expectation values of normally-ordered products of electric field operators. In the example of a single mode, this leads to averages of normal-ordered products of the general form $\left\langle a^{\dagger n}a^n\right\rangle$. As seen in Section 5.6.3, the most useful quasi-probability distribution for the description of such measurements is the Glauber–Sudarshan function $P(\alpha)$. If this distribution function is non-negative everywhere on the complex α-plane, then there is a classical model—described by stochastic c-number phasors α with the same $P(\alpha)$ distribution—that reproduces the average values of the quantum theory. It is reasonable to call such light distributions classical, because no measurements based on the photoelectric effect can distinguish between a quantum state and a classical stochastic model that share the same $P(\alpha)$ distribution.

Direct experimental verification of the condition $P(\alpha) \geqslant 0$ requires rather sophisticated methods, which we will study in Chapter 17. A simpler, but still very useful, distinction between classical and nonclassical states of light employs the global statistical properties of the state. Photoelectric counters can measure the moments $\langle N^r \rangle$ $(r = 1, 2, \ldots)$ of the number operator $N = a^\dagger a$, where $\langle (\cdots) \rangle = \mathrm{Tr}[\rho(\cdots)]$, and ρ is the density operator for the state under consideration. We will study the second moment, or rather the variance, $V(N) = \langle N^2 \rangle - \langle N \rangle^2$, which is a measure of the noise in the light. In Section 5.1.3 we found that a coherent state $\rho = |\alpha\rangle\langle\alpha|$ exhibits Poissonian statistics, i.e. for a coherent state the variance in photon number is equal to the average number: $V(N) = \langle N^2 \rangle - \langle N \rangle^2 = \langle N \rangle$, which is the standard quantum limit. Since the rms deviation is $\sqrt{\langle N \rangle}$, this is just another name for the shot noise[1] in the photoelectric detector. The coherent states are constructed to be as classical as possible, so it is useful to compare the variance for a given state ρ with the variance for a coherent state with the same average number of photons. The fractional excess of the variance relative to that of shot noise,

$$\mathcal{Q} \equiv \frac{V(N) - \langle N \rangle}{\langle N \rangle}, \tag{9.58}$$

is called the **Mandel $\mathcal{Q}$ parameter** (Mandel and Wolf, 1995, Sec. 12.10.3). This new usage should not be confused with the Q-function defined by eqn (5.154).

The $\mathcal{Q}$-parameter vanishes for a coherent state, so it can be regarded as a measure of the excess photon-number noise in the light described by the state ρ. Since the operator N is hermitian, the variance $V(N)$ is non-negative, and it only vanishes for number states. Consequently the range of $\mathcal{Q}$-values is

$$-1 \leqslant \mathcal{Q} < \infty . \tag{9.59}$$

A very useful property of the $\mathcal{Q}$-parameter can be derived by first expressing the numerator in eqn (9.58) as

$$\begin{aligned} V(N) - \langle N \rangle &= \langle N^2 \rangle - \langle N \rangle^2 - \langle N \rangle \\ &= \langle a^{\dagger 2} a^2 \rangle - \langle a^\dagger a \rangle^2 , \end{aligned} \tag{9.60}$$

where the last line follows from another application of the commutation relations $[a, a^\dagger] = 1$. Since all the operators are now in normal-ordered form, we may use the P-representation (5.168) to get

$$V(N) - \langle N \rangle = \int \frac{d^2\alpha}{\pi} |\alpha|^4 P(\alpha) - \left(\int \frac{d^2\alpha}{\pi} |\alpha|^2 P(\alpha) \right)^2 . \tag{9.61}$$

By using the fact that $P(\alpha)$ is normalized to unity, the first term can be expressed as a double integral, so that

[1]Shot noise describes the statistics associated with the random arrivals of discrete objects at a detector, e.g. the noise associated with raindrops falling onto a tin rooftop.

$$V(N) - \langle N \rangle = \int \frac{d^2\alpha}{\pi} |\alpha|^4 P(\alpha) \int \frac{d^2\alpha'}{\pi} P(\alpha') - \int \frac{d^2\alpha}{\pi} |\alpha|^2 P(\alpha) \int \frac{d^2\alpha'}{\pi} |\alpha'|^2 P(\alpha') . \qquad (9.62)$$

The final step is to interchange the dummy integration variables α and α' in the first term, and then to average the two equivalent expressions; this yields the final result:

$$V(N) - \langle N \rangle = \frac{1}{2} \int \frac{d^2\alpha}{\pi} \int \frac{d^2\alpha'}{\pi} \left(|\alpha|^2 - |\alpha'|^2\right)^2 P(\alpha) P(\alpha') . \qquad (9.63)$$

The right side is positive for $P(\alpha) \geqslant 0$; therefore, classical states always correspond to non-negative $\mathcal{Q}$ values. An equivalent, but more useful statement, is that negative values of the $\mathcal{Q}$-parameter always correspond to nonclassical states. A point which is often overlooked is that the condition $\mathcal{Q} < 0$ is sufficient but not necessary for a nonclassical state. In other words, there are nonclassical states with $\mathcal{Q} > 0$.

A coherent state has $\mathcal{Q} = 0$ (Poissonian statistics for the vacuum fluctuations), so a state with $\mathcal{Q} < 0$ is said to be **sub-Poissonian**. These states are quieter than coherent states as far as photon number fluctuations are concerned. We will see another example later on in the study of squeezed states. By the same logic, **super-Poissonian** states, with $\mathcal{Q} > 0$, are noisier than coherent states. Thermal states, or more generally chaotic states, are familiar examples of super-Poissonian statistics; and a nonclassical example is presented in Exercise 15.2.

An overall $\mathcal{Q}$-parameter for multimode states can be defined by using the total number operator,

$$N = \sum_M a_M^\dagger a_M , \qquad (9.64)$$

in eqn (9.58). The definition of a classical state is $P(\underline{\alpha}) \geqslant 0$, where $P(\underline{\alpha})$ is the multimode P-function defined by eqn (5.104). A straightforward generalization of the single-mode argument again leads to the conclusion that states with $\mathcal{Q} < 0$ are necessarily nonclassical.

9.2 Postdetection signal processing

In the preceding sections, we discussed several processes for primary photon detection. Now we must study postdetection signal processing, which is absolutely necessary for completing a measurement of the state of a light field. The problem that must be faced in carrying out a measurement on any quantum system is that microscopic processes, such as the events involved in primary photon detection, are inherently reversible. Consider, for example, a photon and a ground-state atom, both trapped in a small cavity with perfectly reflecting walls. The atom can absorb the photon and enter an excited state, but—with equal facility—the excited atom can return to the ground state by emitting the photon. The photon—none the worse for its adventure—can then initiate the process again. We will see in Chapter 12 that this dance can go on indefinitely. In a solid-state photon detector, the cavity is replaced by the crystal lattice, and the ground-state atom is replaced by an electron in the valence band.

The electron can be excited to the conduction band—leaving a hole in the valence band—by absorbing the photon. Just as for the atom, time-reversal invariance assures us that the conduction band electron can return to the valence band by emitting the photon, and so on. This behavior is described by the state vector

$$\begin{aligned}|\text{photon-detector}\rangle &= \alpha(t)\,|\text{photon}\rangle\,|\text{valence-band-electron}\rangle \\ &\qquad + \beta(t)\,|\text{vacuum}\rangle\,|\text{electron–hole-pair}\rangle \\ &= \alpha(t)\,|\text{photon-not-detected}\rangle + \beta(t)\,|\text{photon-detected}\rangle \qquad (9.65)\end{aligned}$$

for the photon-detector system. As long as the situation is described by this entangled state, there is no way to know if the photon was detected or not.

The purpose of a measurement is to put a stop to this quantum dithering by perturbing the system in such a way that it is forced to make a definite choice. An interaction with another physical system having a small number of degrees of freedom clearly will not do, since the reversibility argument could be applied to the enlarged system. Thus the perturbation must involve coupling to a system with a very large number of degrees of freedom, i.e. a macroscopic system. It could be—indeed it has been—argued that this procedure simply produces another entangled state, albeit with many degrees of freedom. While correct in principle, this line of argument brings us back to Schrödinger's diabolical machine and the unfortunate cat. Just as we can be quite certain that looking into this device will reveal a cat that is either definitely dead or definitely alive—and not some spooky superposition of $|\text{dead cat}\rangle$ and $|\text{live cat}\rangle$—we can also be assured that an irreversible interaction with a macroscopic system will yield a definite answer: the photon was detected or it was not detected. In the words of Bohr (1958, p. 88):

> ...every atomic phenomenon is *closed* in the sense that its observation is based on registrations obtained by means of suitable *amplification devices with irreversible functioning* such as, for example, permanent marks on the photographic plate, caused by the penetration of electrons into the emulsion (*emphasis added*).

Thus postdetection signal processing—which bring quantum measurements to a close by processes involving irreversible amplification—is an essential part of photon detection. In the following sections we will discuss several modern postdetection processes: (1) electron multiplication in Markovian avalanche processes, e.g. in vacuum tube photomultipliers, channeltrons, and image intensifiers; (2) solid-state avalanche photodiodes, and solid-state multipliers with noise-free, non-Markovian avalanche electron multiplication. Finally we discuss coincidence detection, which is an important application of postdetection signal processing.

9.2.1 Electron multiplication

We begin with a discussion of electron multiplication processes in photomultipliers, channeltrons, and solid-state avalanche photodiodes. As pointed out above, postdetection gain mechanisms are not only a practical, but also a fundamental, component of all photon detectors. They are necessary for the closing of the quantum process of measurement. As a practical matter, amplification is required to raise the microscopic energy released in the primary photodetection event—$\hbar\omega \sim 10^{-19}$ J for a typical visible photon—to a macroscopic value much larger than the typical thermal noise—$k_B T \sim 10^{-20}$ J—in electronic circuits. From this point on, the signal processing can be easily handled by standard electronics, since the noise in any electronic detection system is determined by the noise in the first-stage electronic amplification process. The typical electron multiplication factor in these postdetection mechanisms is between 10^4 to 10^6.

One amplification mechanism is electron multiplication by secondary impact ionizations occurring at the surfaces of the **dynode** structures of vacuum-tube photomultipliers. A large electric field is applied across successive dynode structures, as shown in Fig. 9.2. The initial photoelectron released from the photocathode is thus accelerated to such high energies that its impact on the surface of the first dynode releases many secondary electrons. By repeated multiplications on successive dynodes, a large electrical signal can be obtained.

In **channeltron** vacuum tubes, which are also called *image intensifiers*, the photoelectrons released from various spots on a single photocathode are collected by a bundle of small, hollow channels, each corresponding to a single pixel. A large electric field applied along the length of each channel induces electron multiplication on the interior surface, which is coated with a thin, conducting film. Repeated multiplications by means of successive impacts of the electrons along the length of each channel produce a large electrical signal, which can be easily handled by standard electronics.

There is a similar postdetection gain mechanism in solid-state **photodiodes**. The primary event is the production of a single electron–hole pair inside the solid-state material, as shown in Fig. 9.3. When a static electric field is applied, the initial electron and hole are accelerated in opposite directions, in the so-called **Geiger mode** of operation. For a sufficiently large field, the electron and hole reach such high energies that secondary pairs are produced. The secondary pairs in turn cause further pair production, so that an **avalanche breakdown** occurs. This process produces a large electrical pulse—like the single click of a Geiger counter—that signals the arrival of a single photon. In this strong-field limit, the secondary emission processes occur so quickly and randomly that all correlations with previous emissions are wiped out. The absence of any dependence on the previous history is the defining characteristic of a **Markov process**.

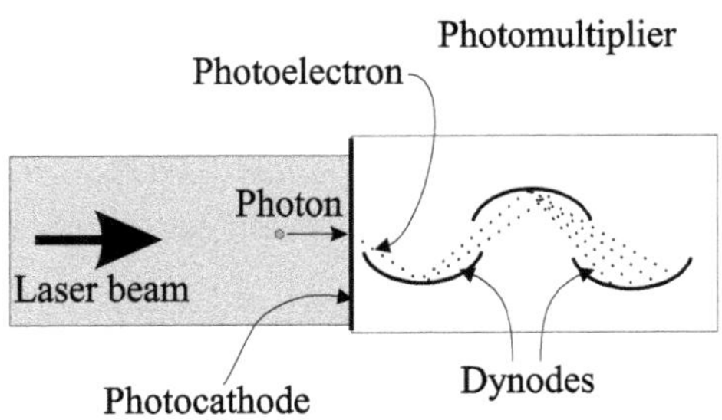

Fig. 9.2 Schematic of a laser beam incident upon a photomultiplier tube.

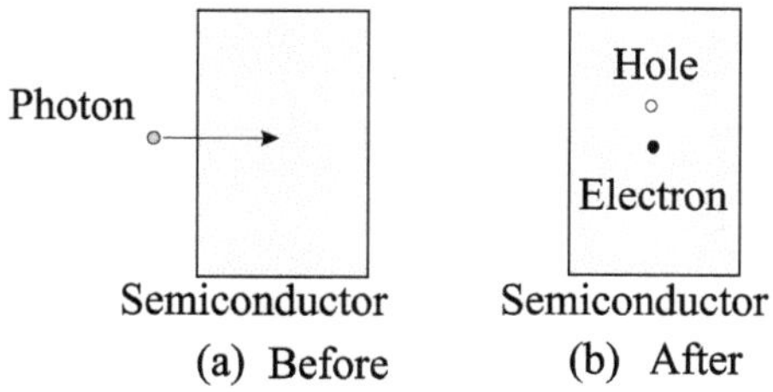

Fig. 9.3 In a semiconductor photodetection device, photoionization occurs inside the body of a semiconductor. In (a) the photon enters the semiconductor. In (b) a photoionization event produces an electron–hole pair inside the semiconductor.

9.2.2 Markovian model for avalanche electron multiplication

We now discuss a simple model (LaViolette and Stapelbroek, 1989) of electron multiplication, such as that of avalanche breakdown in the Geiger mode of silicon solid-state avalanche photon detectors (APDs). This model is based on the Markov approximation; that is, the electron completely forgets all previous scatterings, so that its behavior is solely determined by the initial conditions at each branch point of the avalanche process. The model rests on two underlying assumptions.

(1) The initial photoelectron production always occurs at the same place ($z = 0$), where z is the coordinate along the electric field axis.

(2) Upon impact ionization of an impurity atom, the incoming electron *dies* and two new electrons are *born*. This is the Markov approximation. None of the electrons recombine or otherwise disappear.

The probability that a new carrier is generated in the interval $(z, z + \Delta z)$ is $\alpha(z)\Delta z$, where the gain, $\alpha(z)$, is allowed to vary with z. The probability that n carriers are present at z, given that one carrier is introduced at $z = 0$, is denoted by $p(n, z)$. There are two cases to examine $p(1, z)$ (total failure) and $p(n, z)$ for $n > 1$.

The probability that the incident carrier fails to produce a new carrier in the interval $(z, z + \Delta z)$ is $1 - \alpha(z)\Delta z$. Thus the probability of failure in the next z-interval is

$$p(1, z + \Delta z) = (1 - \alpha(z)\Delta z)\, p(1, z). \tag{9.66}$$

Take the limit $\Delta z \to 0$, or Taylor-series expand the left side, to get the differential equation

$$\frac{\partial p(1, z)}{\partial z} = -\alpha(z)\, p(1, z), \tag{9.67}$$

with the initial condition $p(1, 0) = 1$.

For the successful case that $n > 1$, there are more possibilities, since n carriers at $z + \Delta z$ could come from $n - k$ carriers at z by production of k carriers, where $k = 0, 1, \ldots, n - 1$. Adding up the possible processes gives

$$\begin{aligned} p(n, z + \Delta z) &= (1 - \alpha(z)\Delta z)^n\, p(n, z) + (n - 1)(\alpha(z)\Delta z)\, p(n - 1, z) \\ &\quad + \frac{1}{2}(n - 2)(n - 3)(\alpha(z)\Delta z)^2\, p(n - 2, z) + \cdots . \end{aligned} \tag{9.68}$$

In the limit of small Δz this leads to the differential equation

$$\frac{\partial p(n, z)}{\partial z} = -n\alpha(z)\, p(n, z) + (n - 1)\,\alpha(z)\, p(n - 1, z), \tag{9.69}$$

with the initial condition $p(n, z) = 0$ for $n > 1$.

The solution of eqn (9.67) is easily seen to be

$$p(1, z) = e^{-\zeta(z)}, \quad \text{where } \zeta(z) = \int_0^z dz' \alpha(z'). \tag{9.70}$$

The recursive system of differential equations in eqn (9.69) is a bit more complicated. Perhaps the easiest way is to work out the explicit solutions for $n = 2, 3$ and use the results to guess the general form:

$$p(n, z) = \frac{\left(e^{\zeta(z)} - 1\right)^{n-1}}{e^{n\zeta(z)}}. \tag{9.71}$$

9.2.3 Noise-free, non-Markovian avalanche multiplication

One recent and very important development in postdetection gain mechanisms for photon detectors is noise-free avalanche multiplication in silicon, solid-state photomultipliers (SSPMs) (Kim *et al.*, 1997). Noise-free, postdetection amplification allows the photon detector to distinguish clearly between one and two photons in the primary photodetection event; i.e. the output electronic pulse heights can be cleanly resolved as originating either from a one- or a two-photon primary event. This has led to the direct detection, with high resolution, of the difference between even and odd photon numbers in an incoming beam of light. Applying this photodetection technique to a squeezed state of light shows that there is a pronounced preference for the occupation of even photon numbers; the odd photon numbers are essentially absent. This striking odd–even effect in the photon number distribution is not observed with a coherent state of light, such as that produced by a laser.

A schematic of a noise-free avalanche multiplication device in a SSPM, also known as a visible-light photon counter (VLPC), is shown in Fig. 9.4.

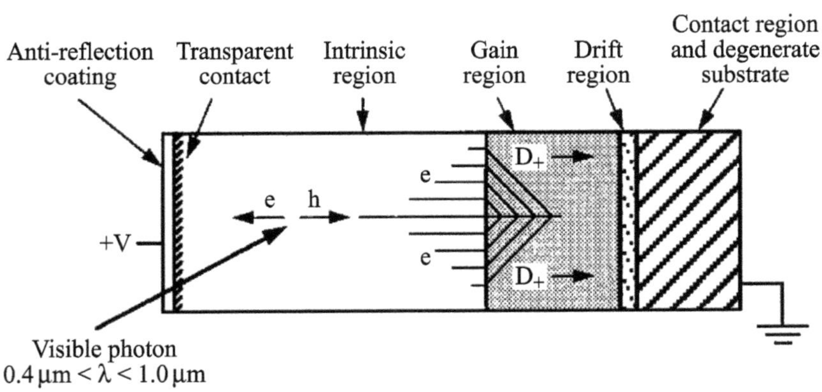

Fig. 9.4 Structure of a solid-state photomultiplier (SSPM) or a visible-light photon counter (VLPC). (Reproduced from Kim *et al.* (1997).)

In contrast to the APD, the SSPM is divided into two separate spatial regions: an *intrinsic* region, inside which the incident photon is converted into a primary electron–hole pair in an intrinsic silicon crystalline material; followed by a *gain* region, consisting of n-doped silicon, inside which well-controlled, noise-free electron multiplication occurs. The electric field in the gain region is larger than in the intrinsic region, due to the difference between the respective dielectric constants. The primary electron and hole, produced by the incoming visible photon, are accelerated in opposite directions by the local electric field in the intrinsic region. The primary electron propagates to the left towards a transparent electrode (the transparent contact) raised to a modest positive potential $+V$. An anti-reflection coating applied to the transparent electrode ensures that the incoming photon is admitted with high efficiency into the interior of the silicon intrinsic region, so that the quantum efficiency of the device can be quite high.

The primary hole propagates to the right and enters the gain region, whereupon the higher electric field present there accelerates it up to the energy (54 meV) required to ionize an arsenic n-type donor atom. The ionization is a single quantum-jump event (a Franck–Hertz-type excitation) in which the hole gives up its entire energy and comes to a complete halt. However, the halted hole is immediately accelerated by the local electric field towards the right, so that the process repeats itself, i.e. the hole again acquires an ionization energy of 54 meV, whereupon it ionizes another local arsenic atom and comes to a complete halt, and so on. In this start-and-stop manner, the hole generates a discrete, deterministic sequence of secondary electrons in a well-controlled manner, as indicated in Fig. 9.4 by the electron vertices inside the gain region. In this way, a sequence of leftwards-propagating secondary electrons is emitted in regular, deterministic manner by the rightwards-propagating hole. Each ionized arsenic atom thus releases a single secondary electron into the conduction band, whereupon it is promptly accelerated to the left towards the interface between the gain and intrinsic region. The secondary electrons enter the intrinsic region, where they are collected, along with the primary electron, by the $+V$ transparent electrode. The result is a noise-free avalanche amplification process, whose gain is given by the number of starts-and-stops of the hole inside the gain region. Measurements of the noise factor, $F \equiv \langle M^2 \rangle / \langle M \rangle^2$, where M is the multiplication factor, show that $F = 1.00 \pm 0.05$ for M between 1×10^4 and 2×10^4 (Kim *et al.*, 1997). This constitutes direct experimental evidence that there is essentially no shot noise in the postdetection electron multiplication process.

Note that this description of the noise-free amplification process depends on the assumption that the motions of the holes and electrons are ballistic, i.e. they propagate freely between collision events. Also, it is assumed that only holes have large enough cross-sections to cause impact ionizations of the arsenic atoms. The resulting process is non-Markovian, in the sense that there is a well-defined, deterministic, nonstochastic delay time between electron multiplication events. Note also that charge conservation requires the number of electrons—collected by the transparent electrode on the left—to be exactly equal to the number of holes—collected on the right by the grounded electrode, labeled as the contact region and degenerate substrate.

9.2.4 Coincidence counting

As we have already seen in Section 1.1.4, one of the most important experimental techniques in quantum optics is **coincidence counting**, in which the output signals of two independent single-photon detectors are sent to a device—the **coincidence counter**—that only emits a signal when the pulses from the two detectors both arrive during a narrow gate window T_{gate}. For simplicity, we will only consider idealized, broadband, point detectors equipped with polarization filters. This means that the detectors can be treated as though they were single atoms, with the understanding that the locations of the 'atoms' are to be treated classically. The detector Hamiltonian is then

$$H_{\text{det}}(t) = \sum_{n=1}^{2} H_{dn}(t), \tag{9.72}$$

$$H_{dn}(t) = -\left(\widehat{\mathbf{d}}_n(t) \cdot \mathbf{e}_n\right) E_n(t), \tag{9.73}$$

where $\mathbf{r}_n$, $\widehat{\mathbf{d}}_n$, $\mathbf{e}_n$, and E_n are respectively the location; the dipole operator; the polarization admitted by the filter; and the corresponding field component

$$E_n(t) = \mathbf{e}_n \cdot \mathbf{E}(\mathbf{r}_n, t) \tag{9.74}$$

for the nth detector. In the following discussion we will show that coincidence counting can be interpreted as a measurement of the second-order correlation function, $G^{(2)}(\mathbf{r}_1, t_1, \mathbf{r}_2, t_2; \mathbf{r}_3, t_3, \mathbf{r}_4, t_4)$, introduced in Section 4.7.

Since a general initial state of the radiation field is described by a density matrix, i.e. an ensemble of pure states, we can begin by assuming that the radiation field is described a pure state $|\Phi_e\rangle$ and that both atoms are in the ground state. The initial state of the total system is then

$$|\Theta_i\rangle = |\phi_\gamma, \phi_\gamma, \Phi_e\rangle = |\phi_\gamma(1)\rangle |\phi_\gamma(2)\rangle |\Phi_e\rangle, \tag{9.75}$$

where $|\phi_\gamma(n)\rangle$ denotes the ground state of the atom located at $\mathbf{r}_n$. For coincidence counting, it is sufficient to consider the final states,

$$|\Theta_f\rangle = |\phi_{\epsilon_1}, \phi_{\epsilon_2}, \underline{n}\rangle = |\phi_{\epsilon_1}(1)\rangle |\phi_{\epsilon_2}(2)\rangle |\underline{n}\rangle, \tag{9.76}$$

where $|\phi_\epsilon(n)\rangle$ denotes a (continuum) excited state of the atom located at $\mathbf{r}_n$ and $|\underline{n}\rangle$ is a general photon number state. The probability amplitude for this transition is

$$\mathfrak{A}_{fi} = \langle \Theta_f | V(t) | \Theta_i \rangle = \delta_{fi} + \left\langle \Theta_f \left| V^{(1)}(t) \right| \Theta_i \right\rangle + \left\langle \Theta_f \left| V^{(2)}(t) \right| \Theta_i \right\rangle + \cdots, \tag{9.77}$$

where the evolution operator $V(t)$ is given by eqn (4.103), with H_{int} replaced by H_{det}. Both atoms must be raised from the ground state to an excited state, so the lowest-order contribution to $\mathfrak{A}_{fi}$ comes from the cross terms in $V^{(2)}(t)$, i.e.

$$\mathfrak{A}_{fi} = \left(-\frac{i}{\hbar}\right)^2 \int_{t_0}^{t} dt_1 \int_{t_0}^{t_1} dt_2 \, \langle \Theta_f | H_{d1}(t_1) H_{d2}(t_2) + H_{d2}(t_1) H_{d1}(t_2) | \Theta_i \rangle. \tag{9.78}$$

The excitation of the two atoms requires the annihilation of two photons; consequently, in evaluating $\mathfrak{A}_{fi}$ the operator $E_n(t)$ in eqn (9.73) can be replaced by the positive-frequency part $E_n^{(+)}(t)$. The detectors are normally located in a passive linear medium, so one can use eqn (3.102) to show that $[H_{d1}(t_1), H_{d2}(t_2)] = 0$ for all (t_1, t_2). This guarantees that the integrand in eqn (9.78) is a symmetrical function of t_1 and t_2, so that eqn (9.78) can be written as

$$\mathfrak{A}_{fi} = \left(-\frac{i}{\hbar}\right)^2 \int_{t_0}^{t} dt_1 \int_{t_0}^{t} dt_2 \left\langle \Theta_f \left| H_{d1}(t_1) H_{d2}(t_2) \right| \Theta_i \right\rangle . \tag{9.79}$$

Finally, substituting the explicit expression (9.73) for the interaction Hamiltonian yields

$$\begin{aligned}\mathfrak{A}_{fi} = {} & \left(-\frac{i}{\hbar}\right)^2 d_{\epsilon_1\gamma} d_{\epsilon_2\gamma} \int_{t_0}^{t} dt_1 \int_{t_0}^{t} dt_2 \exp(i\omega_{\epsilon_1\gamma} t_1) \exp(i\omega_{\epsilon_2\gamma} t_2) \\ & \times \left\langle \underline{n} \left| E_1^{(+)}(t_1) E_2^{(+)}(t_2) \right| \Phi_e \right\rangle ,\end{aligned} \tag{9.80}$$

where we have used the relation between the interaction and Schrödinger pictures to get

$$\left\langle \phi_{\epsilon_n} \left| \widehat{\mathbf{d}}_n(t) \cdot \mathbf{e}_n \right| \phi_\gamma \right\rangle = \exp(i\omega_{\epsilon_1\gamma} t_1) \left\langle \phi_{\epsilon_n} \left| \widehat{\mathbf{d}}_n \cdot \mathbf{e}_n \right| \phi_\gamma \right\rangle = \exp(i\omega_{\epsilon_1\gamma} t_1) \, d_{\epsilon_n\gamma} . \tag{9.81}$$

In a coincidence-counting experiment, the final states of the atoms and the radiation field are not observed; therefore, the transition probability $|\mathfrak{A}_{fi}|^2$ must be summed over ϵ_1, ϵ_2, and $\underline{n}$. This result must then be averaged over the ensemble of pure states defining the initial state ρ of the radiation field. Thus the overall probability, $p(t, t_0)$, that both detectors have clicked during the interval (t_0, t) is

$$p(t, t_0) = \sum_{\epsilon_1} \mathcal{D}_1(\epsilon_1) \sum_{\epsilon_2} \mathcal{D}_2(\epsilon_2) \sum_{\underline{n}} \sum_{e} \mathcal{P}_e \left|\mathfrak{A}_{fi}\right|^2 . \tag{9.82}$$

A calculation similar to the one-photon case shows that $p(t, t_0)$ can be written as

$$\begin{aligned}p(t, t_0) = {} & \int_{t_0}^{t} dt_1' \int_{t_0}^{t} dt_2' \int_{t_0}^{t} dt_1 \int_{t_0}^{t} dt_2 \mathfrak{S}_1(t_1 - t_1') \mathfrak{S}_2(t_2 - t_2') \\ & \times G^{(2)}(\mathbf{r}_1, t_1', \mathbf{r}_2, t_2'; \mathbf{r}_1, t_1, \mathbf{r}_2, t_2) ,\end{aligned} \tag{9.83}$$

where the sensitivity functions are defined by

$$\begin{aligned}\mathfrak{S}_n(t) &= \frac{1}{\hbar^2} \sum_{\epsilon} \mathcal{D}_n(\epsilon) \left|\mathbf{d}_{\epsilon\gamma} \cdot \mathbf{e}_n\right|^2 e^{i\omega_{\epsilon\gamma} t} \quad (n = 1, 2) \\ &= e_{ni}^* e_{nj} \mathfrak{S}_{nij}(t) ,\end{aligned} \tag{9.84}$$

and $G^{(2)}$ is a special case of the scalar second-order correlation function defined by eqn (4.77). The assumption that the detectors are broadband allows us to set $\mathfrak{S}_n(t) = \mathfrak{S}_n \delta(t)$, and thus simplify eqn (9.83) to

$$p(t) = \int_{t_0}^{t} dt_1 \int_{t_0}^{t} dt_2 p^{(2)}(t_1, t_2) , \tag{9.85}$$

where

$$p^{(2)}(t_1, t_2) = \mathfrak{S}_1 \mathfrak{S}_2 G^{(2)}(\mathbf{r}_1, t_1, \mathbf{r}_2, t_2; \mathbf{r}_1, t_1, \mathbf{r}_2, t_2) . \tag{9.86}$$

Since $p(t, t_0)$ is the probability that detections have occurred at $\mathbf{r}_1$ and $\mathbf{r}_2$ sometime during the observation interval (t_0, t), the differential probability that the detections at $\mathbf{r}_1$ and $\mathbf{r}_2$ occur in the subintervals $(t_1, t_1 + dt_1)$ and $(t_2, t_2 + dt_2)$ respectively is $p^{(2)}(t_1, t_2)\, dt_1 dt_2$. The signal pulse from detector n arrives at the coincidence counter at time $t_n + T_n$, where T_n is the signal transit time from the detector to the coincidence counter. The general condition for a coincidence count is

$$|(t_2 + T_2) - (t_1 + T_1)| < T_{\text{gate}} , \tag{9.87}$$

where T_{gate} is the gate width of the coincidence counter. The gate is typically triggered by one of the signals, for example from the detector at $\mathbf{r}_1$. In this case the coincidence condition is

$$t_1 + T_1 < t_2 + T_2 < t_1 + T_1 + T_{\text{gate}} , \tag{9.88}$$

and the coincidence count rate is

$$\begin{aligned} w^{(2)} &= \int_{T_{12}}^{T_{12}+T_{\text{gate}}} d\tau p^{(2)}(t_1, t_1 + \tau) \\ &= \mathfrak{S}_1 \mathfrak{S}_2 \int_{T_{12}}^{T_{12}+T_{\text{gate}}} d\tau G^{(2)}(\mathbf{r}_1, t_1, \mathbf{r}_2, t_1 + \tau; \mathbf{r}_1, t_1, \mathbf{r}_2, t_1 + \tau) , \end{aligned} \tag{9.89}$$

where $T_{12} = T_1 - T_2$ is the offset time for the two detectors. By using delay lines to adjust the signal transit times, coincidence counting can be used to study the correlation function $G^{(2)}(\mathbf{r}_1, t_1, \mathbf{r}_2, t_2; \mathbf{r}_1, t_1, \mathbf{r}_2, t_2)$ for a range of values of $(\mathbf{r}_1, t_1)$ and $(\mathbf{r}_2, t_2)$.

In order to get some practice with the use of the general result (9.89) we will revisit the photon indivisibility experiment discussed in Section 1.4 and preview a two-photon interference experiment that will be treated in Section 10.2.1. The basic arrangement for both experiments is shown in Fig. 9.5.

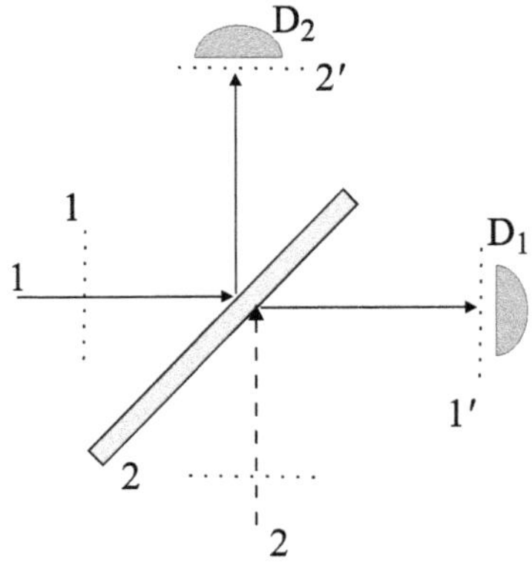

Fig. 9.5 The photon indivisibility and two-photon interference experiments both use this arrangement. The signals from detectors D_1 and D_2 are sent to a coincidence counter.

For the photon indivisibility experiment, we consider a general one-photon input state ρ, i.e. the only condition is $N\rho = \rho N = \rho$, where N is the total number operator. Any one-photon density operator ρ can be expressed in the form

$$\rho = \sum_{\kappa,\lambda} \left|1_\kappa\right\rangle \rho_{\kappa\lambda} \left\langle 1_\lambda\right| , \tag{9.90}$$

where κ and λ are mode labels. The identity $a_\kappa a_\lambda \rho = 0 = \rho a_\lambda^\dagger a_\kappa^\dagger$—which holds for any pair of annihilation operators—implies that

$$\rho E_2^{(-)}\left(\mathbf{r}_2, t_2\right) E_1^{(-)}\left(\mathbf{r}_1, t_1\right) = 0 = E_1^{(+)}\left(\mathbf{r}_1, t_1\right) E_2^{(+)}\left(\mathbf{r}_2, t_2\right) \rho \, . \tag{9.91}$$

The coincidence count rate is determined by the second-order correlation function

$$\begin{aligned} G^{(2)}\left(\mathbf{r}_2, t_2, \mathbf{r}_1, t_1; \mathbf{r}_2, t_2, \mathbf{r}_1, t_1\right) = \operatorname{Tr}\Big[\rho E_2^{(-)}\left(\mathbf{r}_2, t_2\right) E_1^{(-)}\left(\mathbf{r}_1, t_1\right) \\ \times E_1^{(+)}\left(\mathbf{r}_1, t_1\right) E_2^{(+)}\left(\mathbf{r}_2, t_2\right)\Big] , \end{aligned} \tag{9.92}$$

but eqn (9.91) clearly shows that the general second-order correlation function for a one-photon state vanishes everywhere:

$$G^{(2)}\left(\mathbf{r}_1, t_1, \mathbf{r}_2, t_2; \mathbf{r}_1', t_1', \mathbf{r}_2', t_2'\right) \equiv 0 \, . \tag{9.93}$$

The zero coincidence rate in the photon indivisibility experiment is an immediate consequence of this result.

The difference between the photon indivisibility and two-photon interference experiments lies in the choice of the initial state. For the moment, we consider a general incident state which contains at least two photons. This state will be used in the evaluation of the correlation function defined by eqn (9.92). In addition, the original plane-wave modes will be replaced by general wave packets $\mathbf{w}_\kappa(\mathbf{r})$. The field operator produced by scattering from the beam splitter can then be written as

$$\mathbf{E}^{(+)}\left(\mathbf{r}, t\right) = i \sum_\kappa \sqrt{\frac{\hbar\omega_\kappa}{2\epsilon_0}} e^{-i\omega_\kappa t} \mathbf{w}_\kappa\left(\mathbf{r}\right) a_\kappa' \, . \tag{9.94}$$

Substituting this expansion into the general definition (4.75) for $G_{ijkl}^{(2)}$ yields

$$\begin{aligned} G_{ijkl}^{(2)}\left(\{x\};\{x\}\right) = \left(\frac{\hbar}{2\epsilon_0}\right)^2 \sum_{\mu\kappa\lambda\nu} \sqrt{\omega_\mu\omega_\kappa\omega_\lambda\omega_\nu} w_{\mu i}^*\left(\mathbf{r}'\right) w_{\kappa j}^*\left(\mathbf{r}\right) w_{\lambda k}\left(\mathbf{r}\right) w_{\nu l}\left(\mathbf{r}'\right) \\ \times e^{i(\omega_\mu-\omega_\nu)t'} e^{i(\omega_\kappa-\omega_\lambda)t} \operatorname{Tr}\left[\rho a_\mu'^\dagger a_\kappa'^\dagger a_\lambda' a_\nu'\right] , \end{aligned} \tag{9.95}$$

where $\{x\} = \{\mathbf{r}', t', \mathbf{r}, t\}$, but using this in eqn (9.92) would be wrong. The problem is that the last optical element encountered by the field is not the beam splitter, but rather the collimators attached to the detectors. The field scattered from the beam splitter is further scattered, or rather filtered, by the collimators. To be completely precise, we should work out the scattering matrix for the collimator and use eqn (9.94) as the input field. In practice, this is rarely necessary, since the effect of these filters is well approximated by simply omitting the excluded terms when the field is evaluated at a detector location. In this all-or-nothing approximation the explicit use of the collimator scattering matrix is replaced by imposing the following rule at the nth detector:

$$\mathbf{w}_\kappa(\mathbf{r}_n) = 0 \text{ if } \mathbf{w}_\kappa \text{ is blocked by the collimator at detector } n\,. \tag{9.96}$$

We emphasize that this rule is only to be used at the detector locations. For other points, the expression (9.95) must be evaluated without restrictions on the mode functions.

A more realistic description of the incident light leads to essentially the same conclusion. In real experiments, the incident modes are not plane waves but beams (Gaussian wave packets), and the widths of their transverse profiles are usually small compared to the distance from the beam splitter to the detectors. For the two modes pictured in Fig. 9.5, this implies $\mathbf{w}_2(\mathbf{r}_1) \approx 0$ and $\mathbf{w}_1(\mathbf{r}_2) \approx 0$. In other words, the beam $\mathbf{w}_2$ misses detector D_1 and $\mathbf{w}_1$ misses detector D_2. This argument justifies the rule (9.96) even if the collimators are ignored.

For the initial state, $\rho = |\Phi_{\text{in}}\rangle\langle\Phi_{\text{in}}|$, with $|\Phi_{\text{in}}\rangle = a_2^\dagger a_1^\dagger|0\rangle$, each mode sum in eqn (9.95) is restricted to the values $\kappa = 1, 2$. If the rule (9.96) were ignored there would be sixteen terms in eqn (9.95), corresponding to all normal-ordered combinations of $a_1'^\dagger$ and $a_2'^\dagger$ with a_1' and a_2'. Imposing eqn (9.96) reduces this to one term, so that

$$G^{(2)}(\{x\};\{x\}) = \left(\frac{\hbar\omega}{2\epsilon_0}\right)^2 |w_2(\mathbf{r}_2)|^2 |w_1(\mathbf{r}_1)|^2 \left\langle \Phi_{\text{in}} \left| a_2'^\dagger a_1'^\dagger a_1' a_2' \right| \Phi_{\text{in}} \right\rangle, \tag{9.97}$$

where $\omega_2 = \omega_1 = \omega$. Thus the counting rate is proportional to the average of the product of the intensity operators at the two detectors. Combining eqn (9.89) with eqn (8.62) and the relation $\mathfrak{r} = \pm i\,|\mathfrak{t}|$ gives the coincidence-counting rate

$$w^{(2)} = \mathfrak{S}_2\mathfrak{S}_1 T_{\text{gate}} \left(\frac{\hbar\omega}{2\epsilon_0}\right)^2 |w_2(\mathbf{r}_2)|^2 |w_1(\mathbf{r}_1)|^2 \left| |\mathfrak{r}|^2 - |\mathfrak{t}|^2 \right|^2 . \tag{9.98}$$

The combination of eqn (9.95) and eqn (9.96) yields the correct expression for any choice of the incident state. This allows for an explicit calculation of the coincidence rate as a function of the time delay between pulses.

9.3 Heterodyne and homodyne detection

Heterodyne detection is an optical adaptation of a standard method for the detection of weak radio-frequency signals. For almost a century, heterodyne detection in the radio region has been based on square-law detection by diodes, in nonlinear devices known as mixers. After the invention of the laser, this technique was extended to the optical and infrared regions using square-law detectors based on the photoelectric effect. We will first give a brief description of heterodyne detection in classical optics, and then turn to the quantum version. **Homodyne detection** is a special case of heterodyne detection in which the signal and the local oscillator have the same frequency, $\omega_L = \omega_s$. One variant of this scheme (Mandel and Wolf, 1995, Sec. 21.6) uses the heterodyne arrangement shown in Fig. 9.6, but we will describe a different method, called **balanced homodyne detection**, that employs a balanced beam splitter and two identical detectors at the output ports. This technique is especially important at the quantum level, since it is one of the primary tools of measurement for nonclassical states of light, e.g. squeezed states. More generally, it is used in quantum-state tomography—described in Chapter 17—which allows a complete characterization of the quantum state of the light entering the signal port.

9.3.1 Classical analysis of heterodyne detection

Classical heterodyne detection involves a strong monochromatic wave,

$$\mathbf{E}_L(\mathbf{r},t) = \mathcal{E}_L(t)\,\mathbf{w}_L(\mathbf{r})\,e^{-i\omega_L t} + \mathrm{CC}\,, \tag{9.99}$$

called the **local oscillator** (LO), and a weak monochromatic wave,

$$\mathbf{E}_s(\mathbf{r},t) = \mathcal{E}_s(t)\,\mathbf{w}_s(\mathbf{r})\,e^{-i\omega_s t} + \mathrm{CC}\,, \tag{9.100}$$

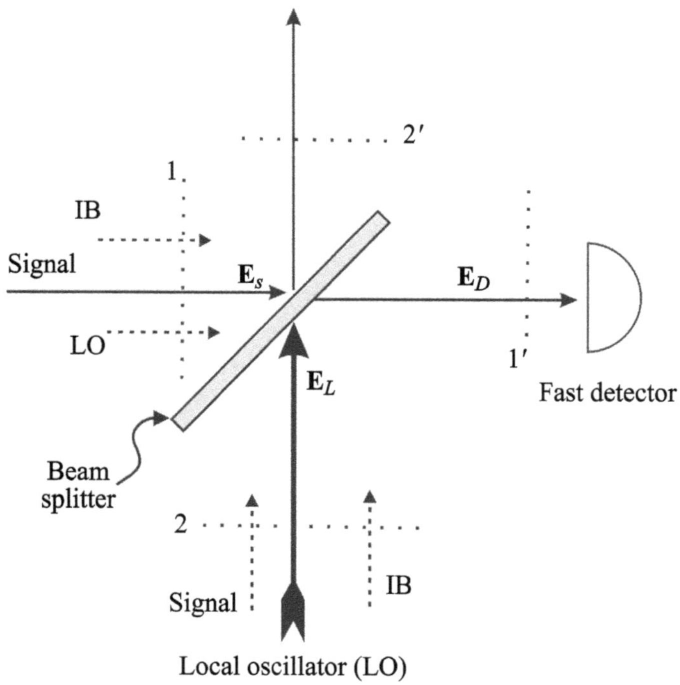

Fig. 9.6 Schematic for heterodyne detection. A strong local oscillator beam (the heavy solid arrow) is combined with a weak signal beam (the light solid arrow) at a beam splitter, and the intensity of the combined beam (light solid arrow) is detected by a fast photodetector. The dashed arrows represent vacuum fluctuations.

called the signal, where $\mathcal{E}_L(t)$ and $\mathcal{E}_s(t)$ are slowly-varying envelope functions. The two waves are mixed at a beam splitter—as shown in Fig. 9.6—so that their combined wavefronts overlap at a fast detector. In a realistic description, the mode functions $\mathbf{w}_L(\mathbf{r})$ and $\mathbf{w}_s(\mathbf{r})$ would be Gaussian wave packets, but in the interests of simplicity we will idealize them as S-polarized plane waves, e.g. $\mathbf{w}_L = \mathbf{e}\exp(ik_L y)/\sqrt{V}$ and $\mathbf{w}_s = \mathbf{e}\exp(ik_s y)/\sqrt{V}$, where V is the quantization volume and $\mathbf{e}$ is the common polarization vector. Since the output fields will also be S-polarized, the polarization vector will be omitted from the following discussion. The two incident waves have different frequencies, so the beam-splitter scattering matrix of eqn (8.63) has to be applied separately to each amplitude. The resulting wave that falls on the detector is $\mathbf{E}_D(\mathbf{r},t) = \boldsymbol{\mathcal{E}}_D(\mathbf{r},t) + \text{CC}$, where

$$\boldsymbol{\mathcal{E}}_D(\mathbf{r},t) = \mathcal{E}'_L(t)\frac{1}{\sqrt{V}}e^{i(k_L x - \omega_L t)} + \mathcal{E}'_s(t)\frac{1}{\sqrt{V}}e^{i(k_s x - \omega_s t)}\,. \tag{9.101}$$

Since the detector surface lies in a plane $x_D = \text{const}$, it is natural to choose coordinates so that $x_D = 0$. The scattered amplitudes are given by $\mathcal{E}'_L(t) = \mathfrak{r}\,\mathcal{E}_L(t)$ and $\mathcal{E}'_s = \mathfrak{t}\,\mathcal{E}_s(t)$, provided that the coefficients $\mathfrak{r}$ and $\mathfrak{t}$ are essentially constant over the frequency bandwidth of the slowly-varying amplitudes $\mathcal{E}_L(t)$ and $\mathcal{E}_s(t)$. Since the signal is weak, it is desirable to lose as little of it as possible. This requires $|\mathfrak{t}| \approx 1$, which in turn implies $|\mathfrak{r}| \ll 1$. The second condition means that only a small fraction of the local oscillator field is reflected into the detector arm, but this loss can be compensated by increasing the incident intensity $|\mathcal{E}_L|^2$. Thus the beam splitter in a heterodyne detector should be highly unbalanced.

The output of the square-law detector is proportional to the average of $|\mathbf{E}_D(\mathbf{r},t)|^2$ over the detector response time T_D, which is always much larger than an optical period. On the other hand, the interference term between the local oscillator and the signal is modulated at the **intermediate frequency**: $\omega_{\text{IF}} \equiv \omega_s - \omega_L$. In optical applications the local oscillator field is usually generated by a laser, with $\omega_L \sim 10^{15}\,\text{Hz}$, but ω_{IF} is typically in the radio-frequency part of the electromagnetic spectrum, around 10^6 to $10^9\,\text{Hz}$. The IF signal is therefore much easier to detect than the incident optical signal. For the remainder of this section we will assume that the bandwidths of both the signal and the local oscillator are small compared to ω_{IF}. This assumption allows us to treat the envelope fields as constants.

In this context, a fast detector is defined by the conditions $1/\omega_L \ll T_D \ll 1/|\omega_{\text{IF}}|$. This inequality, together with the strong-field condition $|\mathcal{E}_L| \gg |\mathcal{E}_s|$, allows the time average over T_D to be approximated by

$$\frac{1}{T_D}\int_{-T_D/2}^{T_D/2} d\tau\, |\boldsymbol{\mathcal{E}}_D(\mathbf{r},t+\tau)|^2 \approx |\mathcal{E}'_L|^2 + 2\,\text{Re}\left[\mathcal{E}'^*_L\mathcal{E}'_s e^{-i\omega_{\text{IF}}t}\right] + \cdots\,. \tag{9.102}$$

The large first term $|\mathcal{E}'_L|^2$ can safely be ignored, since it represents a DC current signal which is easily filtered out by means of a high-pass, radio-frequency filter. The photocurrent from the detector is then dominated by the **heterodyne signal**

$$S_{\text{het}}(t) = 2\,\text{Re}\left[\mathfrak{r}^*\mathfrak{t}\,\mathcal{E}^*_L\mathcal{E}_s e^{-i\omega_{\text{IF}}t}\right], \tag{9.103}$$

which describes the beat signal between the LO and the signal wave at the intermediate frequency ω_{IF}. **Optical heterodyne detection** is the sensitive detection of the heterodyne signal by standard radio-frequency techniques.

Experimentally, it is important to align the directions of the LO and signal beams at the surface of the photon detector, since any misalignment will produce spatial interference fringes over the detector surface. The fringes make both positive and negative contributions to S_{het}; consequently—as can be seen in Exercise 9.4—averaging over the entire surface will wash out the IF signal. Alignment of the two beams can be accomplished by adjusting the tilt of the beam splitter until they overlap interferometrically.

An important advantage of heterodyne detection is that $S_{\mathrm{het}}(t)$ is linear in the local oscillator field $\mathcal{E}_L^*$ and in the signal field $\mathcal{E}_s(t)$. Thus a large value for $|\mathcal{E}_L^*|$ effectively amplifies the contribution of the weak optical signal to the low-frequency heterodyne signal. For instance, doubling the size of $\mathcal{E}_L^*$, doubles the size of the heterodyne signal for a given signal amplitude $\mathcal{E}_s'$. Furthermore, the relative phase between the linear oscillator and the incident signal is faithfully preserved in the heterodyne signal. To make this point more explicit, first rewrite eqn (9.103) as $S_{\mathrm{het}}(t) = \mathcal{F}\cos(\omega_{\mathrm{IF}}t) + \mathcal{G}\sin(\omega_{\mathrm{IF}}t)$, where the Fourier components are given by

$$\mathcal{F} = 2\,\mathrm{Re}\left[\mathfrak{r}^*\mathfrak{t}\;\mathcal{E}_L^*\mathcal{E}_s\right], \quad \mathcal{G} = 2\,\mathrm{Im}\left[\mathfrak{r}^*\mathfrak{t}\;\mathcal{E}_L^*\mathcal{E}_s\right]. \tag{9.104}$$

We use the Stokes relation (8.7), in the form

$$\mathfrak{r}^*\mathfrak{t} = |\mathfrak{r}|\,|\mathfrak{t}|\,e^{\pm i\pi/2}, \tag{9.105}$$

to rewrite eqn (9.104) as

$$\mathcal{F} = \pm 2\,|\mathcal{E}_L^*\mathcal{E}_s|\,|\mathfrak{r}|\,|\mathfrak{t}|\sin(\theta_L - \theta_s), \quad \mathcal{G} = \pm 2\,|\mathcal{E}_L^*\mathcal{E}_s|\,|\mathfrak{r}|\,|\mathfrak{t}|\cos(\theta_L - \theta_s), \tag{9.106}$$

where θ_L and θ_s are respectively the phases of the local oscillator $\mathcal{E}_L$ and the signal $\mathcal{E}_s$.

The quantities $\mathcal{F}$ and $\mathcal{G}$ can be separately measured. For example, $\mathcal{F}$ and $\mathcal{G}$ can be simultaneously determined by means of the apparatus sketched in Fig. 9.7. Note that the insertion of a 90° phase shifter into one of the two local-oscillator arms allows the measurement of both the sine and cosine components of the intermediate-frequency signals at the two photon detectors. Each box labeled 'IF mixer' denotes the combination of a radio-frequency oscillator—conventionally called a *2nd LO*— that operates at the IF frequency, with two local radio-frequency diodes that mix the 2nd LO signal with the two IF signals from the photon detectors. The net result is that these IF mixers produce two DC output signals proportional to the IF amplitudes $\mathcal{F}$ and $\mathcal{G}$. The ratio of $\mathcal{F}$ and $\mathcal{G}$ is a direct measure of the phase difference $\theta_L - \theta_s$ relative to the phase of the 2nd LO, since

$$\frac{\mathcal{F}}{\mathcal{G}} = \tan(\theta_L - \theta_s). \tag{9.107}$$

The heterodyne signal corresponding to $\mathcal{F}$ is maximized when $\theta_L - \theta_s = \pi/2$ and minimized when $\theta_L - \theta_s = 0$, whereas the heterodyne signal corresponding to $\mathcal{G}$ is

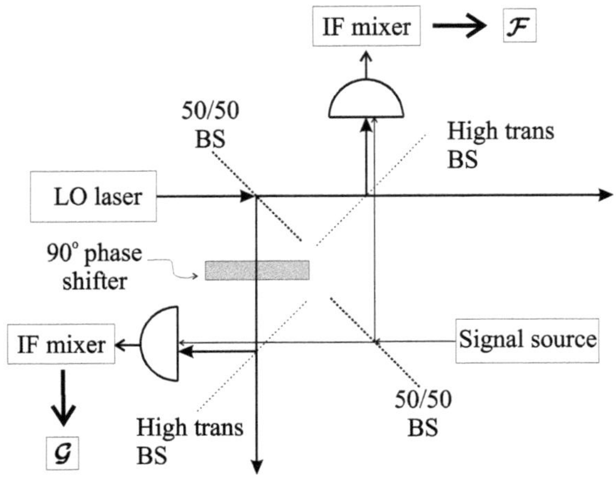

Fig. 9.7 Schematic of an apparatus for two-quadrature heterodyne detection. The beam splitters marked as 'High trans' have $|\mathfrak{t}| \approx 1$.

maximized when $\theta_L - \theta_s = 0$ and minimized when $\theta_L - \theta_s = \pi/2$, where all the phases are defined relative to the 2nd LO phase. The optical phase information in the signal waveform is therefore preserved through the entire heterodyne process, and is stored in the ratio of $\mathcal{F}$ to $\mathcal{G}$. This phase information is valuable for the measurement of small optical time delays corresponding to small differences in the times of arrival of two optical wavefronts; for example, in the difference in the times of arrival at two telescopes of the wavefronts emanating from a single star. Such optical phase information can be used for the measurement of stellar diameters in infrared stellar interferometry with a carbon-dioxide laser as the local oscillator (Hale *et al.*, 2000). This is an extension of the technique of radio-astronomical interferometry to the mid-infrared frequency range.

Examples of important heterodyne systems include: Schottky diode mixers in the radio and microwave regions; superconductor–insulator–superconductor (SIS) mixers, for radio astronomy in the millimeter-wave range; and optical heterodyne mixers, using the carbon-dioxide lasers in combination with semiconductor photoconductors, employed as square-law detectors in infrared stellar interferometry (Kraus, 1986).

9.3.2 Quantum analysis of heterodyne detection

Since the field operators are expressed in terms of classical mode functions and their associated annihilation operators, we can retain the assumptions—i.e. plane waves, S-polarization, etc.—employed in Section 9.3.1. This allows us to use a simplified form of the general expression (8.28) for the in-field operator to replace the classical field (9.101) by the Heisenberg-picture operator

$$E_{\text{in}}^{(+)}(\mathbf{r}, t) = i\mathfrak{e}_L a_{L2} e^{ik_L y} e^{-i\omega_L t} + i\mathfrak{e}_s a_{s1} e^{ik_s x} e^{-i\omega_s t} + E_{\text{vac,in}}^{(+)}(\mathbf{r}, t), \tag{9.108}$$

where $\mathfrak{e}_M = \sqrt{\hbar\omega_M / 2\epsilon_0 V}$ is the vacuum fluctuation field strength for a plane wave with frequency ω_M. This is an extension of the method used in Section 9.1.4 to model imperfect detectors. The annihilation operators a_{L2} and a_{s1} respectively represent the local oscillator field, entering through port 2, and the signal field, entering through port 1; and, we have again assumed that the bandwidths of the signal and local oscillator fields are small compared to ω_{IF}. If this assumption has to be relaxed, then the Schrödinger-picture annihilation operators must be replaced by slowly-varying envelope operators $\overline{a}_{L2}(t)$ and $\overline{a}_{s1}(t)$. In principle, the operator $E^{(+)}_{\text{vac,in}}(\mathbf{r},t)$ includes all modes other than the signal and local oscillator, but most of these terms will not contribute in the subsequent calculations. According to the discussion in Section 8.4.1, each physical input field is necessarily paired with vacuum fluctuations of the same frequency—indicated by the dashed arrows in Fig. 9.6—entering through the other input port. Thus $E^{(+)}_{\text{vac,in}}(\mathbf{r},t)$ must include the operators a_{L1} and a_{s2} describing vacuum fluctuations with frequencies ω_L and ω_s entering through ports 1 and 2 respectively. It should also include any other vacuum fluctuations that could combine with the local oscillator to yield terms at the intermediate frequency, i.e. modes satisfying $\omega_M = \omega_L \pm \omega_{\text{IF}}$. The $+$-choice yields the signal frequency ω_s, which is already included, so the only remaining possibility is $\omega_M = \omega_L - \omega_{\text{IF}}$. Again borrowing terminology from radio engineering, we refer to this mode as the **image band**, and set $M = \text{IB}$ and $\omega_{\text{IB}} = \omega_L - \omega_{\text{IF}}$. The relevant terms in $E^{(+)}_{\text{in}}(\mathbf{r},t)$ are thus

$$\begin{aligned} E^{(+)}_{\text{in}}(\mathbf{r},t) = {} & i\mathfrak{e}_L a_{L2} e^{ik_L y} e^{-i\omega_L t} + i\mathfrak{e}_L a_{L1} e^{ik_L x} e^{-i\omega_L t} \\ & + i\mathfrak{e}_s a_{s1} e^{ik_s x} e^{-i\omega_s t} + i\mathfrak{e}_s a_{s2} e^{ik_s y} e^{-i\omega_s t} \\ & + i\mathfrak{e}_{\text{IB}} a_{\text{IB}2} e^{ik_{\text{IB}} y} e^{-i\omega_{\text{IB}} t} \\ & + i\mathfrak{e}_{\text{IB}} a_{\text{IB}1} e^{ik_{\text{IB}} x} e^{-i\omega_{\text{IB}} t} . \end{aligned} \tag{9.109}$$

A *The heterodyne signal*

The scattered field operator $E^{(+)}_{\text{out}}(\mathbf{r},t)$ is split into two parts, which respectively describe propagation along the $2 \to 2'$ arm and the $1 \to 1'$ arm in Fig. 9.6. The latter part—which we will call $E^{(+)}_{\text{out},D}(\mathbf{r},t)$—is the one driving the detector. The spatial modes in $E^{(+)}_{\text{out},D}(\mathbf{r},t)$ are all of the form $\exp(ikx)$, for various values of k. Since we only need to evaluate the field at the detector location x_D, the calculation is simplified by choosing the coordinates so that $x_D = 0$. In this way we find the expression

$$E^{(+)}_{\text{out},D}(t) = i\mathfrak{e}_L a'_{L1} e^{-i\omega_L t} + i\mathfrak{e}_s a'_{s1} e^{-i\omega_s t} + i\mathfrak{e}_{\text{IB}} a'_{\text{IB}1} e^{-i\omega_{\text{IB}} t} . \tag{9.110}$$

The scattered annihilation operators are obtained by applying the beam-splitter scattering matrix in eqn (8.63) to the incident annihilation operators. This simply amounts to working out how each incident classical mode is scattered into the $1 \to 1'$ arm, with the results

$$a'_{s1} = \mathfrak{t}\, a_{s1} + \mathfrak{r}\, a_{s2} , \quad a'_{L1} = \mathfrak{t}\, a_{L1} + \mathfrak{r}\, a_{L2} , \quad a'_{\text{IB}1} = \mathfrak{t}\, a_{\text{IB}1} + \mathfrak{r}\, a_{\text{IB}2} . \tag{9.111}$$

The finite efficiency of the detector can be taken into account by using the technique discussed in Section 9.1.4 to modify $E^{(+)}_{\text{out},D}(t)$.

Applying eqn (9.33), for the total single-photon counting rate, to this case gives

$$w^{(1)}(t) \propto \left\langle E^{(-)}_{\text{out},D}(t)\, E^{(+)}_{\text{out},D}(t) \right\rangle , \tag{9.112}$$

and the intermediate frequency part of this signal comes from the beat-note terms between the local oscillator part of eqn (9.110)—or rather its conjugate—and the signal and image band parts. This procedure leads to the operator expression

$$S_{\text{het}} = \left[E^{(-)}_{\text{out},D}(t)\, E^{(+)}_{\text{out},D}(t) \right]_{\text{IF}} = F\cos(\omega_{\text{IF}} t) + G\sin(\omega_{\text{IF}} t) , \tag{9.113}$$

where the operators F and G—which correspond to the classical quantities $\mathcal{F}$ and $\mathcal{G}$ respectively—have contributions from both the signal and the image band, i.e.

$$F = F_s + F_{\text{IB}} , \quad G = G_s + G_{\text{IB}} , \tag{9.114}$$

where

$$F_s = \mathfrak{e}_L \mathfrak{e}_s \left(a'^{\dagger}_{L1} a'_{s1} + \text{HC} \right) , \tag{9.115}$$

$$F_{\text{IB}} = \mathfrak{e}_L \mathfrak{e}_{\text{IB}} \left(a'^{\dagger}_{L1} a'_{\text{IB}1} + \text{HC} \right) , \tag{9.116}$$

$$G_s = -i\mathfrak{e}_L \mathfrak{e}_s \left(a'^{\dagger}_{L1} a'_{s1} - \text{HC} \right) , \tag{9.117}$$

and

$$G_{\text{IB}} = -i\mathfrak{e}_L \mathfrak{e}_{\text{IB}} \left(a'^{\dagger}_{L1} a'_{\text{IB}1} - \text{HC} \right) . \tag{9.118}$$

By assumption, the density operator ρ_{in} describing the state of the incident light is the vacuum for all annihilation operators other than a_{L2} and a_{s1}, i.e.

$$a_\Lambda \rho_{\text{in}} = \rho_{\text{in}} a^{\dagger}_\Lambda = 0 , \quad \Lambda = s2, L1, \text{IB}1, \text{IB}2 . \tag{9.119}$$

These conditions immediately yield

$$\left\langle a'^{\dagger}_{L1} a'_{\text{IB}1} \right\rangle = 0 , \tag{9.120}$$

and

$$\left\langle a'^{\dagger}_{L1} a'_{s1} \right\rangle = \mathfrak{r}^* \mathfrak{t} \left\langle a^{\dagger}_{L2} a_{s1} \right\rangle . \tag{9.121}$$

Furthermore, the independently generated signal and local oscillator fields are uncorrelated, so the total density operator can be written as a product

$$\rho_{\text{in}} = \rho_L \rho_s , \tag{9.122}$$

where ρ_L and ρ_s are respectively the density operators for the local oscillator and the signal. This leads to the further simplification

$$\left\langle a^{\dagger}_{L2} a_{s1} \right\rangle = \left\langle a^{\dagger}_{L2} \right\rangle_L \left\langle a_{s1} \right\rangle_s . \tag{9.123}$$

From eqn (9.120) we see that the expectation values of the operators F and G are completely determined by F_s and G_s, and eqn (9.123) allows the final result to be written as

$$\langle F\rangle = \mathfrak{e}_L\mathfrak{e}_s 2\,\mathrm{Re}\left[\mathfrak{r}^*\mathfrak{t}\left\langle a_{L2}^\dagger\right\rangle_L \langle a_{s1}\rangle_s\right], \tag{9.124}$$

$$\langle G\rangle = \mathfrak{e}_L\mathfrak{e}_s 2\,\mathrm{Im}\left[\mathfrak{r}^*\mathfrak{t}\left\langle a_{L2}^\dagger\right\rangle_L \langle a_{s1}\rangle_s\right], \tag{9.125}$$

which suggests defining effective field amplitudes

$$\mathcal{E}_L = \mathfrak{e}_L\langle a_{L2}\rangle_L\,, \quad \mathcal{E}_s = \mathfrak{e}_s\langle a_{s1}\rangle_s\,. \tag{9.126}$$

With this notation, the expectation values of the operators F and G have the same form as the classical quantities $\mathcal{F}$ and $\mathcal{G}$:

$$\begin{aligned}\langle F\rangle &= \pm 2\left|\mathcal{E}_L^*\mathcal{E}_s\right|\left|\mathfrak{r}\right|\left|\mathfrak{t}\right|\sin\left(\theta_L-\theta_s\right),\\ \langle G\rangle &= \pm 2\left|\mathcal{E}_L^*\mathcal{E}_s\right|\left|\mathfrak{r}\right|\left|\mathfrak{t}\right|\cos\left(\theta_L-\theta_s\right).\end{aligned} \tag{9.127}$$

This formal similarity becomes an identity, if both the signal and the local oscillator are described by coherent states, i.e. $a_{L2}\rho_{\text{in}} = \alpha_L\rho_{\text{in}}$ and $a_{s1}\rho_{\text{in}} = \alpha_s\rho_{\text{in}}$.

The result (9.127) is valid for any state, ρ_{in}, that satisfies the factorization rule (9.122). Let us apply this to the extreme quantum situation of the pure number state $\rho_s = |n_s\rangle\langle n_s|$. In this case $\mathcal{E}_s = \mathfrak{e}_s\langle a_{s1}\rangle_s = 0$, and the heterodyne signal vanishes. This reflects the fact that pure number states have no well-defined phase. The same result holds for any density operator, ρ_s, that is diagonal in the number-state basis. On the other hand, for a superposition of number states, e.g.

$$|\psi\rangle = C_0|0\rangle + C_1|1_s\rangle\,, \tag{9.128}$$

the effective field strength for the signal is

$$\mathcal{E}_s = \mathfrak{e}_s\langle\psi|a_{s1}|\psi\rangle = \mathfrak{e}_s C_0^* C_1\,. \tag{9.129}$$

Consequently, a nonvanishing heterodyne signal can be measured even for superpositions of states containing at most one photon.

B *Noise in heterodyne detection*

In the previous section, we carefully included all the relevant vacuum fluctuation terms, only to reach the eminently sensible conclusion that none of them makes any contribution to the average signal. This was not a wasted effort, since we saw in Section 8.4.2 that vacuum fluctuations will add to the noise in the measured signal. We will next investigate the effect of vacuum fluctuations in heterodyne detection by evaluating the variance,

$$V(F) = \left\langle F^2\right\rangle - \langle F\rangle^2\,, \tag{9.130}$$

of the operator F in eqn (9.114).

Since the calculation of fluctuations is substantially more complicated than the calculation of averages, it is a good idea to exploit any simplifications that may turn up. We begin by using eqn (9.114) to write $\langle F^2 \rangle$ as

$$\langle F^2 \rangle = \langle F_s^2 \rangle + \langle F_s F_{\text{IB}} \rangle + \langle F_{\text{IB}} F_s \rangle + \langle F_{\text{IB}}^2 \rangle \,. \tag{9.131}$$

The image band vacuum fluctuations and the signal are completely independent, so there should be no correlations between them, i.e. one should find

$$\langle F_s F_{\text{IB}} \rangle = \langle F_s \rangle \langle F_{\text{IB}} \rangle = \langle F_{\text{IB}} F_s \rangle \,. \tag{9.132}$$

Since the density operator is the vacuum for the image band modes, the absence of correlation further implies

$$\langle F_s F_{\text{IB}} \rangle = \langle F_{\text{IB}} F_s \rangle = 0 \,. \tag{9.133}$$

This result can be verified by a straightforward calculation using eqn (9.119) and the commutativity of operators for different modes.

At this point we have the exact result

$$\begin{aligned} V(F) &= \langle F_s^2 \rangle + \langle F_{\text{IB}}^2 \rangle - \langle F_s \rangle^2 \\ &= V(F_s) + V(F_{\text{IB}}) \,, \end{aligned} \tag{9.134}$$

where we have used $\langle F_{\text{IB}} \rangle = 0$ again to get the final form. A glance at eqns (9.115) and (9.116) shows that this is still rather complicated, but any further simplifications must be paid for with approximations. Since the strong local oscillator field is typically generated by a laser, it is reasonable to model ρ_L as a coherent state,

$$a_{L2} \rho_L = \alpha_L \rho_L \,, \quad \rho_L a_{L2}^{\dagger} = \alpha_L^* \rho_L \,, \tag{9.135}$$

with

$$\alpha_L = |\alpha_L| \, e^{i\theta_L} \,. \tag{9.136}$$

The variance $V(F_{\text{IB}})$ can be obtained from $V(F_s)$ by the simple expedient of replacing the signal quantities $\{a_{s1}, a_{s2}, \mathfrak{e}_s\}$ by the image band equivalents $\{a_{\text{IB}1}, a_{\text{IB}2}, \mathfrak{e}_{\text{IB}}\}$, so we begin by using eqns (9.111), (9.119), and (9.135) to evaluate $V(F_s)$. After a substantial amount of algebra—see Exercise 9.5—one finds

$$\begin{aligned} V(F_s) = &-\mathfrak{e}_s^2 \, |\mathfrak{r}\, \mathfrak{t}|^2 \, |\mathcal{E}_L|^2 \left[e^{-2i\theta_L} V(a_{s1}) + \text{CC} \right] \\ &+ 2\mathfrak{e}_s^2 \, |\mathfrak{r}\, \mathfrak{t}|^2 \, |\mathcal{E}_L|^2 \left[\left\langle a_{s1}^{\dagger} a_{s1} \right\rangle - |\langle a_{s1} \rangle|^2 \right] \\ &+ \mathfrak{e}_s^2 \, |\mathfrak{r}|^2 \, |\mathcal{E}_L|^2 + (\mathfrak{e}_L \mathfrak{e}_s)^2 \, |\mathfrak{t}|^2 \left\langle a_{s1}^{\dagger} a_{s1} \right\rangle , \end{aligned} \tag{9.137}$$

where $|\mathcal{E}_L| = \mathfrak{e}_L \, |\alpha_L|$ is the laser amplitude. We may not appear to be achieving very much in the way of simplification, but it is too soon to give up hope.

The first promising sign comes from the simple result

$$V\left(F_{\text{IB}}\right) = \mathfrak{e}_{\text{IB}}^2 \left|\mathfrak{r}\right|^2 \left|\mathcal{E}_L\right|^2 . \tag{9.138}$$

This represents the amplification—by beating with the local oscillator—of the vacuum fluctuation noise at the image band frequency. With our normalization conventions, the energy density in these vacuum fluctuations is

$$u_{\text{IB}} = 2\epsilon_0 \mathfrak{e}_{\text{IB}}^2 = \frac{\hbar\omega_{\text{IB}}}{V} . \tag{9.139}$$

In Section 1.1.1 we used equipartition of energy to argue that the mean thermal energy for each radiation oscillator is $k_B T$, so the thermal energy density would be $u_T = k_B T/V$. Equating the two energy densities defines an effective **noise temperature**

$$T_{\text{noise}} = \frac{\hbar\omega_{\text{IB}}}{k_B} \approx \frac{\hbar\omega_L}{k_B} . \tag{9.140}$$

This effect will occur for any of the phase-insensitive linear amplifiers studied in Chapter 16, including masers and parametric amplifiers (Shimoda *et al.*, 1957; Caves, 1982).

With this encouragement, we begin to simplify the expression for $V\left(F_s\right)$ by introducing the new creation and annihilation operators

$$b_s^{\dagger}\left(\theta_L\right) = e^{i\theta_L} a_{s1}^{\dagger} , \quad b_s\left(\theta_L\right) = e^{-i\theta_L} a_{s1} . \tag{9.141}$$

This eliminates the explicit dependence on θ_L from eqn (9.137), but the new operators are still non-hermitian. The next step is to consider the observable quantities represented by the hermitian **quadrature operators**

$$X\left(\theta_L\right) = \frac{b_s\left(\theta_L\right) + b_s^{\dagger}\left(\theta_L\right)}{2} = \frac{e^{-i\theta_L} a_{s1} + e^{i\theta_L} a_{s1}^{\dagger}}{2} \tag{9.142}$$

and

$$Y\left(\theta_L\right) = \frac{b_s\left(\theta_L\right) - b_s^{\dagger}\left(\theta_L\right)}{2i} = \frac{e^{-i\theta_L} a_{s1} - e^{i\theta_L} a_{s1}^{\dagger}}{2i} . \tag{9.143}$$

These operators are the hermitian and anti-hermitian parts of the annihilation operator:

$$b_s\left(\theta_L\right) = X\left(\theta_L\right) + iY\left(\theta_L\right) , \tag{9.144}$$

and the canonical commutation relations imply

$$\left[X\left(\theta_L\right), Y\left(\theta_L\right)\right] = \frac{i}{2} . \tag{9.145}$$

By writing the defining equations (9.142) and (9.143) as

$$\begin{aligned} X\left(\theta_L\right) &= X\left(0\right)\cos\theta_L + Y\left(0\right)\sin\theta_L , \\ Y\left(\theta_L\right) &= X\left(0\right)\sin\theta_L - Y\left(0\right)\cos\theta_L , \end{aligned} \tag{9.146}$$

the quadrature operators can be interpreted as a rotation of the phase plane through the angle θ_L, given by the phase of the local oscillator field. In the calculations to follow we will shorten the notation by $X\left(\theta_L\right) \to X$, etc.

After substituting eqns (9.141) and (9.144), into eqn (9.134), we arrive at

$$V(F) = 4|\mathfrak{r}\,\mathfrak{t}|^2 |\mathcal{E}_L|^2 \mathfrak{e}_s^2 \left[V(Y) - \frac{1}{4}\right] + |\mathfrak{r}|^2 |\mathcal{E}_L|^2 \left[\mathfrak{e}_s^2 + \mathfrak{e}_{\mathrm{IB}}^2\right] + |\mathfrak{t}|^2 \mathfrak{e}_s^2 \left\langle a_{s1}^\dagger a_{s1} \right\rangle \mathfrak{e}_L^2 . \tag{9.147}$$

The combination $V(Y) - 1/4$ vanishes for any coherent state, in particular for the vacuum, so it represents the **excess noise** in the signal. It is important to realize that the excess noise can be either positive or negative, as we will see in the discussion of squeezed states in Section 15.1.2. The first term on the right of eqn (9.147) represents the amplification of the excess signal noise by beating with the strong local oscillator field. The second term represents the amplification of the vacuum noise at the signal and the image band frequencies. Finally, the third term describes amplification—by beating against the signal—of the vacuum noise at the local oscillator frequency. The strong local oscillator assumption can be stated as $|\mathfrak{r}|^2 |\alpha_L|^2 \gg |\mathfrak{t}|^2$, so the third term is negligible. Neglecting it allows us to treat the local oscillator as an effectively classical field.

The noise terms discussed above are fundamental, in the sense that they arise directly from the uncertainty principle for the radiation oscillators. In practice, experimentalists must also deal with additional noise sources, which are called *technical* in order to distinguish them from fundamental noise. In the present context the primary technical noise arises from various disturbances—e.g. thermal fluctuations in the laser cavity dimensions, Johnson noise in the electronics, etc.—affecting the laser providing the local oscillator field. By contrast to the fundamental vacuum noise, the technical noise is—at least to some degree—subject to experimental control. Standard practice is therefore to drive the local oscillator by a master oscillator which is as well controlled as possible.

9.3.3 Balanced homodyne detection

This technique combines heterodyne detection with the properties of the ideal balanced beam splitter discussed in Section 8.4. A strong quasiclassical field (the LO) is injected into port 2, and a weak signal with the same frequency is injected into port 1 of a balanced beam splitter, as shown in Fig. 9.8. In practice, it is convenient to generate both fields from a single master oscillator. Note, however, that the signal and local oscillator *mode functions* are orthogonal, because the plane-wave propagation vectors are orthogonal. If the beam splitter is balanced, and the rest of the system is designed to be as bilaterally symmetric as possible, this device is called a **balanced homodyne detector**. In particular, the detectors placed at the output ports 1′ and 2′ are required to be identical within close tolerances. In practice, this is made possible by the high reproducibility of semiconductor-based photon detectors fabricated on the same homogeneous, single-crystal wafer using large-scale integration techniques.

The difference between the outputs of the two identical detectors is generated by means of a balanced, differential electronic amplifier. Since the two input transistors of the differential amplifier—whose noise figure dominates that of the entire postdetection electronics—are themselves semiconductor devices fabricated on the same wafer, they can also be made identical within close tolerances. The symmetry achieved in this way guarantees that the technical noise in the laser source—from which both the signal

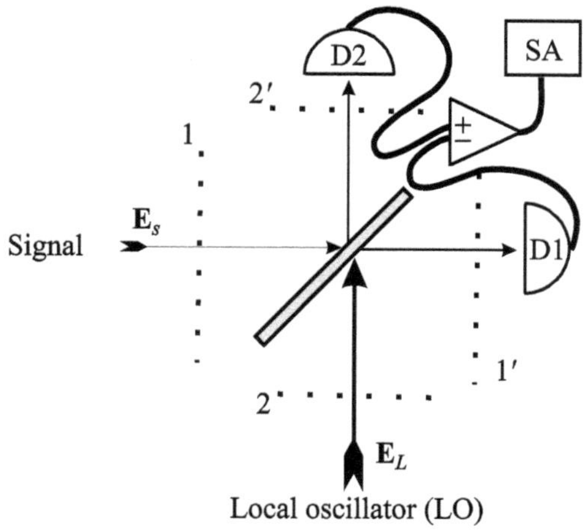

Fig. 9.8 Schematic of a balanced homodyne detector. Detectors D1 and D2 respectively collect the output of ports 1′ and 2′. The outputs of D2 and D1 are respectively fed into the non-inverting input (+) and the inverting input (−) of a differential amplifier. The output of the differential amplifier, i.e. the difference between the two detected signals, is then fed into a radio-frequency spectrum analyzer SA.

and the local oscillator are derived—will produce essentially identical fluctuations in the outputs of detectors D1 and D2. These common-mode noise waveforms will cancel out upon subtraction in the differential amplifier. This technique can, therefore, lead to almost ideal detection of purely quantum statistical properties of the signal. We will encounter this method of detection later in connection with experiments on squeezed states of light.

A Classical analysis of homodyne detection

It is instructive to begin with a classical analysis for general values of the reflection and transmission coefficients $\mathfrak{r}$ and $\mathfrak{t}$ before specializing to the balanced case. The classical amplitudes at detectors D1 and D2 are related to the input fields by

$$\begin{aligned} \mathcal{E}_{\mathrm{D1}} &= \mathfrak{r}\,\mathcal{E}_L + \mathfrak{t}\,\mathcal{E}_s\,, \\ \mathcal{E}_{\mathrm{D2}} &= \mathfrak{t}\,\mathcal{E}_L + \mathfrak{r}\,\mathcal{E}_s\,, \end{aligned} \tag{9.148}$$

and the difference in the outputs of the square-law detectors is proportional to the difference in the intensities, so the homodyne signal is

$$\begin{aligned} S_{\mathrm{hom}} &= \left|\mathcal{E}_{\mathrm{D2}}\right|^2 - \left|\mathcal{E}_{\mathrm{D1}}\right|^2 \\ &= \left(1 - 2\left|\mathfrak{r}\right|^2\right)\left|\mathcal{E}_L\right|^2 - \left(1 - 2\left|\mathfrak{r}\right|^2\right)\left|\mathcal{E}_s\right|^2 + 4\left|\mathfrak{t}\,\mathfrak{r}\right|\operatorname{Im}\left[\mathcal{E}_L^*\mathcal{E}_s\right], \end{aligned} \tag{9.149}$$

where we have used the Stokes relations (8.7) and set $\mathfrak{r}^*\mathfrak{t} = i\left|\mathfrak{r}\,\mathfrak{t}\right|$ (this is the +-sign in eqn (9.105)) to simplify the result. The first term on the right side is not sensitive to the phase θ_L of the local oscillator, so it merely provides a constant background for measurements of the homodyne signal as a function of θ_L. By design, the signal intensity is small compared to the local oscillator intensity, so the $|\mathcal{E}_s|^2$-term can be neglected altogether. As mentioned in Section 9.3.2, the local oscillator amplitude is subject to technical fluctuations $\delta\mathcal{E}_L$—e.g. variations in the laser power due to acoustical-noise-induced changes in the laser cavity dimensions—which in turn produce phase-sensitive fluctuations in the output,

$$\delta S_{\text{hom}} = -\left(1 - 2|\mathfrak{r}|^2\right) 2\,\text{Re}\left[\mathcal{E}_L^* \delta\mathcal{E}_L\right] + 4\,|\mathfrak{t}\;\mathfrak{r}|\,\text{Im}\left[\delta\mathcal{E}_L^* \mathcal{E}_s\right]. \tag{9.150}$$

The fluctuations associated with the direct detection signal, $\left(1 - 2|\mathfrak{r}|^2\right)|\mathcal{E}_L|^2$, for the local oscillator are negligible compared to the fluctuations in the $\mathcal{E}_s$ contribution if

$$\left(1 - 2|\mathfrak{r}|^2\right) \ll \frac{|\mathcal{E}_s|}{|\mathcal{E}_L|}, \tag{9.151}$$

and this is certainly satisfied for an ideal balanced beam splitter, for which $|\mathfrak{r}|^2 = |\mathfrak{t}|^2 = 1/2$, and

$$S_{\text{hom}} = 2\,\text{Im}\left[\mathcal{E}_L^* \mathcal{E}_s\right]. \tag{9.152}$$

B Quantum analysis of homodyne detection

We turn now to the quantum analysis of homodyne detection, which is simplified by the fact that the local oscillator and the signal have the same frequency. The complications associated with the image band modes are therefore absent, and the in-field is simply

$$E_{\text{in}}^{(+)}(\mathbf{r}, t) = i\mathfrak{e}_s a_L e^{ik_s y} e^{-i\omega_s t} + i\mathfrak{e}_s a_s e^{ik_s x} e^{-i\omega_s t}. \tag{9.153}$$

In this case all relevant vacuum fluctuations are dealt with by the operators a_L and a_s, so the operator $E_{\text{vac,in}}^{(+)}(\mathbf{r}, t)$ will not contribute to either the signal or the noise.

The homodyne signal. The out-field is

$$E_{\text{out}}^{(+)}(\mathbf{r}, t) = E_{\text{D1}}^{(+)}(\mathbf{r}, t) + E_{\text{D2}}^{(+)}(\mathbf{r}, t), \tag{9.154}$$

where the fields

$$E_{\text{D1}}^{(+)}(\mathbf{r}, t) = i\mathfrak{e}_s a_s' e^{ik_s x} e^{-i\omega_s t} \tag{9.155}$$

and

$$E_{\text{D2}}^{(+)}(\mathbf{r}, t) = i\mathfrak{e}_s a_L' e^{ik_s y} e^{-i\omega_s t} \tag{9.156}$$

drive the detectors D1 and D2 respectively, and the scattered annihilation operators satisfy the operator analogue of (9.148):

$$\begin{aligned} a_L' &= \mathfrak{t}\,a_L + \mathfrak{r}\,a_s\,, \\ a_s' &= \mathfrak{r}\,a_L + \mathfrak{t}\,a_s\,. \end{aligned} \tag{9.157}$$

The difference in the two counting rates is proportional to

$$S_{\text{hom}} = \left\langle E_{\text{D2}}^{(-)}(\mathbf{r},t)\, E_{\text{D2}}^{(+)}(\mathbf{r},t) - E_{\text{D1}}^{(-)}(\mathbf{r},t)\, E_{\text{D1}}^{(+)}(\mathbf{r},t) \right\rangle$$
$$= \mathfrak{e}_s^2 \langle N'_{21} \rangle\,, \tag{9.158}$$

where

$$N'_{21} = a_L'^{\dagger} a'_L - a_s'^{\dagger} a'_s$$
$$= \left(1 - 2|\mathfrak{r}|^2\right) a_L^{\dagger} a_L - \left(1 - 2|\mathfrak{r}|^2\right) a_s^{\dagger} a_s - 2i\,|\mathfrak{r}\ \mathfrak{t}| \left(a_L^{\dagger} a_s - a_s^{\dagger} a_L\right) \tag{9.159}$$

is the quantum analogue of the classical result (9.149). For a balanced beam splitter, this simplifies to

$$N'_{21} = -i\left(a_L^{\dagger} a_s - a_s^{\dagger} a_L\right); \tag{9.160}$$

consequently, the balanced homodyne signal is

$$S_{\text{hom}} = 2\mathfrak{e}_s^2\, \text{Im}\left\langle a_L^{\dagger} a_s \right\rangle. \tag{9.161}$$

If we again assume that the signal and local oscillator are statistically independent, then $\langle a_L^{\dagger} a_s \rangle = \langle a_L^{\dagger} \rangle \langle a_s \rangle$, and

$$S_{\text{hom}} = 2\,\text{Im}\left(\mathcal{E}_L^{*} \mathcal{E}_s\right), \tag{9.162}$$

where the effective field amplitudes are again defined by

$$\mathcal{E}_L = \mathfrak{e}_s \langle a_L \rangle = \mathfrak{e}_s\, |\langle a_L \rangle|\, e^{i\theta_L}\,, \tag{9.163}$$

and

$$\mathcal{E}_s = \mathfrak{e}_s \langle a_s \rangle\,. \tag{9.164}$$

Just as for heterodyne detection, the phase sensitivity of homodyne detection guarantees that the detection rate vanishes for signal states described by density operators that are diagonal in photon number. Alternatively, for the calculation of the signal we can replace the difference of number operators by

$$a_L'^{\dagger} a'_L - a_s'^{\dagger} a'_s \to -i\left(\langle a_L \rangle^{*} a_s - a_s^{\dagger} \langle a_L \rangle\right) = 2\,|\langle a_L \rangle|\, Y\,, \tag{9.165}$$

where Y is the quadrature operator defined by eqn (9.143). This gives the equivalent result

$$S_{\text{hom}} = 2\,|\mathcal{E}_L|\, \mathfrak{e}_s \langle Y \rangle \tag{9.166}$$

for the homodyne signal.

Noise in homodyne detection. Just as in the classical analysis, the first term in the expression (9.159) for N'_{21} would produce a phase-insensitive background, but for $|\mathfrak{r}|^2$ significantly different from the balanced value 1/2, the variance in the homodyne output associated with technical noise in the local oscillator could seriously degrade the signal-to-noise ratio. This danger is eliminated by using a balanced system, so that N'_{21} is given by eqn (9.160). The calculation of the variance $V(N'_{21})$ is considerably simplified by the assumption that the local oscillator is approximately described by a coherent state with $\alpha_L = |\alpha_L| \exp(i\theta_L)$. In this case one finds

$$V(N'_{21}) = |\alpha_L|^2 + \langle a_s^\dagger a_s \rangle + 2|\alpha_L|^2 V(a_s^\dagger, a_s) - |\alpha_L|^2 V(e^{-i\theta_L} a_s) - |\alpha_L|^2 V(e^{-i\theta_L} a_s^\dagger). \tag{9.167}$$

Expressing this in terms of the quadrature operator Y gives the simpler result

$$V(N'_{21}) = 4|\alpha_L|^2 V(Y) + \langle a_s^\dagger a_s \rangle \simeq 4|\alpha_L|^2 V(Y), \tag{9.168}$$

where the last form is valid in the usual case that the input signal flux is negligible compared to the local oscillator flux.

C Corrections for finite detector efficiency*

So far we have treated the detectors as though they were 100% efficient, but perfect detectors are very hard to find. We can improve the argument given above by using the model for imperfect detectors described in Section 9.1.4. Applying this model to detector D1 requires us to replace the operator a'_s—describing the signal transmitted through the beam splitter in Fig. 9.8—by

$$a''_s = \sqrt{\xi} a'_s + i\sqrt{1-\xi} c'_s, \tag{9.169}$$

where the annihilation operator c'_s is associated with the mode $\exp[i(k_s y - \omega_s t)]$ entering through port 2 of the imperfect-detector model shown in Fig. 9.1. A glance at Fig. 9.8 shows that this is also the mode associated with a_L. Since the quantization rules assign a unique annihilation operator to each mode, things are getting a bit confusing. This difficulty stems from a violation of Einstein's rule caused by an uncritical use of plane-wave modes. For example, the local oscillator entering port 2 of the homodyne detector, as shown in Fig. 9.8, should be described by a Gaussian wave packet $\mathbf{w}_L$ with a transverse profile that is approximately planar at the beam splitter and effectively zero at the detector D1. Correspondingly, the operator c'_s, representing the vacuum fluctuations blamed for the detector noise, should be associated with a wave packet that is approximately planar at the fictitious beam splitter of the imperfect-detector model and effectively zero at the real beam splitter in Fig. 9.8. In other words, the noise in detector D1 does not enter the beam splitter. All of this can be done precisely by using the wave packet quantization methods developed in Section 3.5.2, but this is not necessary as long as we keep our wits about us. Thus we impose $c'_s \rho = 0$, $a_L \rho \neq 0$, and $\left[a_L^\dagger, c'_s\right] = 0$, even though—in the oversimplified plane-wave picture—both operators c'_s and a_L are associated with the same plane-wave mode. In the same way, the noise in detector D2 is simulated by replacing the transmitted LO-field a'_L with

$$a''_L = \sqrt{\xi} a'_L + i\sqrt{1-\xi} c'_L, \tag{9.170}$$

where $c'_L \rho = 0$, and $\left[c'_L, a_s^\dagger\right] = 0$.

Continuing in this vein, the difference operator N'_{21} is replaced by

$$\begin{aligned} N''_{21} &= a''^\dagger_L a''_L - a''^\dagger_s a''_s \\ &= \xi N'_{21} + \delta N''_{21}. \end{aligned} \tag{9.171}$$

Each term in $\delta N''_{21}$ contains at least one creation or annihilation operator for the vacuum modes discussed above. Since the vacuum operators commute with the operators for the signal and local oscillator, the expectation value of $\delta N''_{21}$ vanishes, and the homodyne signal is

$$S_{\text{hom}} = \mathfrak{e}_s^2 \langle N''_{21} \rangle = \xi \mathfrak{e}_s^2 \langle N'_{21} \rangle = 2\xi \, \text{Im} \left(\mathcal{E}_L^* \mathcal{E}_s \right). \tag{9.172}$$

As expected, the signal from the imperfect detector is just the perfect detector result reduced by the quantum efficiency.

We next turn to the noise in the homodyne signal, which is proportional to the variance $V(N''_{21})$. It is not immediately obvious how the extra partition noise in each detector will contribute to the overall noise, so we first use eqn (9.171) again to get

$$\left\langle \left(N''_{21}\right)^2 \right\rangle = \xi^2 \left\langle \left(N'_{21}\right)^2 \right\rangle + \xi \left\langle N'_{21} \delta N''_{21} \right\rangle + \xi \left\langle \delta N''_{21} N'_{21} \right\rangle + \left\langle \left(\delta N''_{21}\right)^2 \right\rangle . \tag{9.173}$$

There are no correlations between the vacuum fields c'_L and c'_s entering the imperfect detector and the signal and local oscillator fields, so we should expect to find that the second and third terms on the right side of eqn (9.173) vanish. An explicit calculation shows that this is indeed the case. Evaluating the fourth term in the same way leads to the result

$$V\left(N''_{21}\right) = \xi^2 V\left(N'_{21}\right) + \xi\left(1-\xi\right) \left[\left\langle a'^{\dagger}_L a'_L \right\rangle + \left\langle a'^{\dagger}_s a'_s \right\rangle \right]. \tag{9.174}$$

Comparing this to the single-detector result (9.57) shows that the partition noises at the two detectors add, despite the fact that N''_{21} represents the difference in the photon counts at the two detectors. After substituting eqn (9.168) for $V(N'_{21})$; using the scattering relations (9.157); and neglecting the small signal flux, we get the final result

$$V\left(N''_{21}\right) = \xi^2 4 \left|\alpha_L\right|^2 V(Y) + \xi\left(1-\xi\right) \left|\alpha_L\right|^2 . \tag{9.175}$$

9.4 Exercises

9.1 Poissonian statistics are reproduced

Use the Poisson distribution $p(n) = (n!)^{-1} \overline{n}^n \exp(-\overline{n})$ for the incident photons in eqn (9.46) to derive eqn (9.48).

9.2 m-fold coincidence counting

Generalize the two-detector version of coincidence counting to any number m. Show that the m-photon coincidence rate is

$$w^{(m)} = \left(\frac{1}{m!}\right)^2 \left(\prod_{n=1}^{m} \mathfrak{S}_n\right) \int_{T_{12}}^{T_{12}+T_{\text{gate}}} d\tau_2 \cdots \int_{T_{1m}}^{T_{1m}+T_{\text{gate}}} d\tau_m$$
$$G^{(m)}\left(\mathbf{r}_1, t_1, \ldots, \mathbf{r}_m, t_m + \tau_m; \mathbf{r}_1, t_1, \ldots, \mathbf{r}_m, t_m + \tau_m\right),$$

where the signal from the first detector is used to gate the coincidence counter and $T_{1n} = T_1 - T_n$.

9.3 Sub-Poissonian statistics

Consider the state $|\Psi\rangle = \alpha |n\rangle + \beta |n+1\rangle$, with $|\alpha|^2 + |\beta|^2 = 1$. Show that $|\Psi\rangle$ is a nonclassical state that exhibits sub-Poissonian statistics.

9.4 Alignment in heterodyne detection

For the heterodyne scheme shown in Fig. 9.6, assume that the reflected LO beam has the wavevector $\mathbf{k}_L = k_L \cos\varphi \mathbf{u}_x + k_L \sin\varphi \mathbf{u}_y$. Rederive the expression for S_{het} and show that averaging over the detector surface wipes out the heterodyne signal.

9.5 Noise in heterodyne detection

Use eqn (9.111), eqn (9.119), and eqn (9.135) to derive eqn (9.137).

10 Experiments in linear optics

In this chapter we will study a collection of significant experiments which were carried out with the aid of the linear optical devices described in Chapter 8 and the detection techniques discussed in Chapter 9.

10.1 Single-photon interference

The essential features of quantum interference between alternative Feynman paths are illustrated by the familiar Young's arrangement—sketched in Fig. 10.1—in which there are two pinholes in a perfectly reflecting screen. The screen is illuminated by a plane-wave mode occupied by a single photon with energy $\hbar\omega$, and after many successive photons have passed through the pinholes the detection events—e.g. spots on a photographic plate—build up the pattern observed in classical interference experiments.

An elementary quantum mechanical explanation of the single-photon interference pattern can be constructed by applying **Feynman's rules of interference** (Feynman *et al.*, 1965, Chaps 1–7).

(1) The probability of an event in an ideal experiment is given by the square of the absolute value of a complex number $\mathcal{A}$ which is called the probability amplitude:

$$\begin{aligned} P &= \text{probability}\,, \\ \mathcal{A} &= \text{probability amplitude}\,, \\ P &= |\mathcal{A}|^2\,. \end{aligned} \tag{10.1}$$

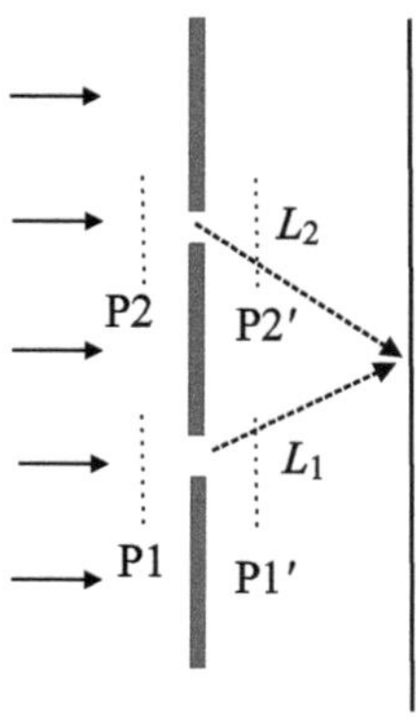

Fig. 10.1 A two-pinhole interferometer. The arrows represent an incident plane wave. The four ports are defined by the surfaces P1, P1′, P2, P2′, and the path lengths from the pinholes 1 and 2—bracketed by the ports (P1, P1′) and (P2, P2′) respectively—to the interference point are L_1 and L_2.

(2) When an event can occur in several alternative ways, the probability amplitude for the event is the sum of the probability amplitudes for each way considered separately; i.e. there is interference between the alternatives:

$$\begin{aligned} \mathcal{A} &= \mathcal{A}_1 + \mathcal{A}_2\,, \\ P &= \left|\mathcal{A}_1 + \mathcal{A}_2\right|^2 . \end{aligned} \tag{10.2}$$

(3) If an experiment is performed which is capable of determining whether one or another alternative is actually taken, the probability of the event is the sum of the probabilities for each alternative. In this case,

$$P = P_1 + P_2\,, \tag{10.3}$$

and there is no interference.

In applying rule (2) it is essential to be sure that the situation described in rule (3) is excluded. This means that the experimental arrangement must be such that it is impossible—even in principle—to determine which of the alternatives actually occurs. In the literature—and in the present book—it is customary to refer to the alternative ways of reaching the final event as **Feynman processes** or **Feynman paths**.

In the two-pinhole experiment, the two alternative processes are passage of the photon through the lower pinhole 1 or the upper pinhole 2 to arrive at the final event: detection at the same point on the screen. In the absence of any experimental procedure for determining which process actually occurs, the amplitudes for the two alternatives must be added. Let $\mathcal{A}_{\text{in}}$ be the quantum amplitude for the incoming wave; then the amplitudes for the two processes are $\mathcal{A}_1 = \mathcal{A}_{\text{in}} \exp(ikL_1)$ and $\mathcal{A}_2 = \mathcal{A}_{\text{in}} \exp(ikL_2)$, where $k = \omega/c$. The probability of detection at the point on the screen (determined by the values of L_1 and L_2) is therefore

$$\left|\mathcal{A}_1 + \mathcal{A}_2\right|^2 = 2\left|\mathcal{A}_{\text{in}}\right|^2 + 2\left|\mathcal{A}_{\text{in}}\right|^2 \cos\left[k\left(L_2 - L_1\right)\right], \tag{10.4}$$

which has the same form as the interference pattern in the classical theory.

This thought experiment provides one of the simplest examples of wave–particle duality. The presence of the interference term in eqn (10.4) exhibits the wave-aspect of the photon, while the detection of the photon at a point on the screen displays its particle-aspect. Arguments based on the uncertainty principle (Cohen-Tannoudji *et al.*, 1977*a*, Complement D1; Bransden and Joachain, 1989, Sec. 2.5) show that any experimental procedure that actually determines which pinhole the photon passed through—this is called **which-path** information—will destroy the interference pattern. These arguments typically involve an interaction with the particle—in this case a photon—which introduces uncontrollable fluctuations in physical properties, such as the momentum. The arguments based on the uncertainty principle show that which-path information obtained by disturbing the particle destroys the interference pattern, but this is not the only kind of experiment that can provide which-path information. In Section 10.3 we will describe an experiment demonstrating that single-photon interference is destroyed by an experimental arrangement that merely makes it *possible* to obtain which-path information, even if none of the required measurements are actually made and there is no interaction with the particle.

The description of the two-pinhole experiment presented above provides a simple physical model which helps us to understand single-photon interference, but a more detailed analysis requires the use of the scattering theory methods developed in Sections 8.1 and 8.2. For the two-pinhole problem, the effects of diffraction cannot be ignored, so it will not be possible to confine attention to a small number of plane waves, as in the analysis of the beam splitter and the stop. Instead, we will use the general relations (8.29) and (8.27) to guide a calculation of the field operator in position space. This is equivalent to using the classical Green function defined by this boundary value problem to describe the propagation of the field operator through the pinhole.

In the plane-wave basis the positive frequency part of the out-field is given by

$$\mathbf{E}_{\text{out}}^{(+)}(\mathbf{r},t)=\sqrt{\frac{\hbar}{2\epsilon_0 cV}}\sum_{\mathbf{k}s} i\omega_{\mathbf{k}} a'_{\mathbf{k}s}\mathbf{e}_s e^{i(\mathbf{k}\cdot\mathbf{r}-\omega_{\mathbf{k}}t)}\,, \tag{10.5}$$

where the scattered annihilation operators obey

$$a'_{\mathbf{k}s}=\sum_{\mathbf{k}'s'} S_{\mathbf{k}s,\mathbf{k}'s'}a_{\mathbf{k}'s'}\,. \tag{10.6}$$

If the source of the incident field is on the left ($z<0$), then the problem is to calculate the transmitted field on the right ($z>0$). The field will be observed at points $\mathbf{r}$ lying on a *detection plane* at $z=L$. The plane waves that impinge on a detector at $\mathbf{r}$ must have $k_z>0$, and the terms in eqn (10.6) can be split into those with $k'_z>0$ (forward waves) and $k'_z<0$ (backwards waves). The contribution of the forward waves to eqn (10.5) represents the part of the incident field transmitted through the pinholes, while the backward waves—vacuum fluctuations in this case—scatter into forward waves by reflection from the screen. The total field in the region $z>0$ is then the sum of three terms:

$$\mathbf{E}_{\text{out}}^{(+)}(\mathbf{r},t)=\mathbf{E}_1^{(+)}(\mathbf{r},t)+\mathbf{E}_2^{(+)}(\mathbf{r},t)+\mathbf{E}_3^{(+)}(\mathbf{r},t)\,, \tag{10.7}$$

where $\mathbf{E}_1^{(+)}$ and $\mathbf{E}_2^{(+)}$ are the fields coming from pinholes 1 and 2 respectively, and the field resulting from reflections of backwards waves at the screen is

$$\mathbf{E}_3^{(+)}(\mathbf{r},t)=\sqrt{\frac{\hbar}{2\epsilon_0 cV}}\sum_{\mathbf{k}s,k_z>0} i\omega_{\mathbf{k}} a'^{<}_{\mathbf{k}s}\mathbf{e}_s e^{i(\mathbf{k}\cdot\mathbf{r}-\omega_{\mathbf{k}}t)}\,, \tag{10.8}$$

where

$$a'^{<}_{\mathbf{k}s}=\sum_{\mathbf{k}'s',k'_z<0} S_{\mathbf{k}s,\mathbf{k}'s'}a_{\mathbf{k}'s'}\,. \tag{10.9}$$

In the absence of the reflected vacuum fluctuations, $\mathbf{E}_3^{(+)}$, the total field $\mathbf{E}_{\text{out}}^{(+)}$ would not satisfy the commutation relation (3.17), and this would lead to violations of the uncertainty principle, as shown in Exercise 10.1.

If the distance to the observation point $\mathbf{r}$ is large compared to the sizes of the pinholes and to the distance between them—this is called *Fraunhofer diffraction* or the *far-field approximation*—the fields due to the two pinholes are given by

$$\mathbf{E}_p^{(+)}(\mathbf{r},t) = iD_p\mathbf{E}^{(+)}(\mathbf{r}_p, t - L_p/c) \quad (p = 1, 2), \tag{10.10}$$

where L_p is the distance from the pth pinhole to the observation point $\mathbf{r}$, and D_p is a real coefficient that depends on the pinhole geometry. For simplicity we will assume that the pinholes are identical, $D_1 = D_2 = D$, and that the incident radiation is monochromatic. If the direction of the incident beam and the vectors $\mathbf{r} - \mathbf{r}_1$ and $\mathbf{r} - \mathbf{r}_2$ are approximately orthogonal to the screen, then $D_p \approx \sigma/(\lambda_0 L)$, where σ is the common area of the pinholes and λ_0 is the average wavelength in the incident field (Born and Wolf, 1980, Sec. 8.3). This is the standard classical expression, except for replacing the classical field in the pinhole by the quantum field operator. The average intensity in a definite polarization $\mathbf{e}$ at a detection point $\mathbf{r}$ is proportional to

$$\begin{aligned} I_{\text{tot}} &= \left\langle E_{\text{out}}^{(-)}(\mathbf{r},t)\, E_{\text{out}}^{(+)}(\mathbf{r},t)\right\rangle \\ &= \sum_{q=1}^{3}\sum_{p=1}^{3}\left\langle E_q^{(-)}(\mathbf{r},t)\, E_p^{(+)}(\mathbf{r},t)\right\rangle, \end{aligned} \tag{10.11}$$

where $E_{\text{out}}^{(-)} = \mathbf{e}\cdot\mathbf{E}_{\text{out}}^{(-)}$, $E_q^{(-)} = \mathbf{e}\cdot\mathbf{E}_q^{(-)}$, and the indices p and q represent the three terms in eqn (10.7). The density operator, ρ, that defines the ensemble average, $\langle\cdots\rangle$, contains no backwards waves, since it represents the field generated by a source to the left of the screen. According to eqn (10.8) and eqn (10.9) the operator $E_3^{(+)}$ is a linear combination of annihilation operators for backwards waves, therefore

$$E_3^{(+)}\rho = 0 = \rho E_3^{(-)}. \tag{10.12}$$

By using this fact, plus the cyclic invariance of the trace, it is easy to show that eqn (10.11) reduces to

$$\begin{aligned} I_{\text{tot}} &= \sum_{q=1}^{2}\sum_{p=1}^{2}\left\langle E_q^{(-)}(\mathbf{r},t)\, E_p^{(+)}(\mathbf{r},t)\right\rangle \\ &= I_1 + I_2 + I_{12}, \end{aligned} \tag{10.13}$$

where I_p is the intensity due to the pth pinhole alone,

$$I_p = |D|^2\left\langle E^{(-)}(\mathbf{r}_p, t - L_p/c)\, E^{(+)}(\mathbf{r}_p, t - L_p/c)\right\rangle \quad (p = 1, 2), \tag{10.14}$$

I_{12} is the interference term,

$$\begin{aligned} I_{12} &= 2\,\text{Re}\left\langle E_1^{(-)}(\mathbf{r},t)\, E_2^{(+)}(\mathbf{r},t)\right\rangle \\ &= 2D^2\,\text{Re}\left\langle E^{(-)}(\mathbf{r}_1, t - L_1/c)\, E^{(+)}(\mathbf{r}_2, t - L_2/c)\right\rangle, \end{aligned} \tag{10.15}$$

and $E^{(-)}(\mathbf{r},t) = \mathbf{e}\cdot\mathbf{E}^{(-)}(\mathbf{r},t)$.

The expectation values appearing in these expressions are special cases of the first-order field correlation function $G^{(1)}$ defined by eqn (4.76). In this notation, the results are

$$I_p = |D|^2 G^{(1)}(\mathbf{r}_p, t - L_p/c; \mathbf{r}_p, t - L_p/c) \quad (p = 1, 2), \tag{10.16}$$

and

$$I_{12} = 2D^2 \operatorname{Re} G^{(1)}(\mathbf{r}_1, t - L_1/c; \mathbf{r}_2, t - L_2/c). \tag{10.17}$$

From the classical theory of two-pinhole interference we know that high visibility interference patterns are obtained with monochromatic light. In quantum theory this means that the power spectrum $\langle a^\dagger_{\mathbf{k}s} a_{\mathbf{k}s}\rangle$ is strongly peaked at $|\mathbf{k}| = k_0 = \omega_0/c$. If the density operator ρ satisfies this condition, then the plane-wave expansion for $E^{(+)}$ implies that the temporal Fourier transform of $\operatorname{Tr}\left[E^{(+)}(\mathbf{r}, t)\rho\right]$ is strongly peaked at ω_0. This means that the envelope operator $\overline{E}^{(+)}$ defined by

$$\overline{E}^{(+)}(\mathbf{r}, t) = E^{(+)}(\mathbf{r}, t)\, e^{i\omega_0 t} \tag{10.18}$$

can be treated as slowly varying—on the time scale $1/\omega_0$—provided that it is applied to the monochromatic density matrix ρ. In this case, the correlation functions can be written as

$$\begin{aligned} G^{(1)}(\mathbf{r}_1, t_1; \mathbf{r}_2, t_2) &= \operatorname{Tr}\left[\rho \overline{E}^{(-)}(\mathbf{r}_1, t_1)\, \overline{E}^{(+)}(\mathbf{r}_2, t_2)\right] e^{-i\omega_0(t_2 - t_1)} \\ &\equiv \overline{G}^{(1)}(\mathbf{r}_1, t_1; \mathbf{r}_2, t_2)\, e^{-i\omega_0(t_2 - t_1)}, \end{aligned} \tag{10.19}$$

where $\overline{G}^{(1)}(\mathbf{r}_1, t_1; \mathbf{r}_2, t_2)$ is a slowly-varying function of t_1 and t_2. For sufficiently long pulses, the incident radiation is approximately stationary, so the correlation functions are unchanged by a time translation $t_p \to t_p + \tau$. In other words they only depend on the time difference $t_1 - t_2$, so the direct terms become

$$I_p = D^2 \overline{G}^{(1)}(\mathbf{r}_p, 0; \mathbf{r}_p, 0) \quad (p = 1, 2), \tag{10.20}$$

while the interference term reduces to

$$I_{12} = 2D^2 \operatorname{Re} \overline{G}^{(1)}(\mathbf{r}_1, \tau; \mathbf{r}_2, 0)\, e^{i\omega_0 \tau}, \tag{10.21}$$

where $\tau = (L_2 - L_1)/c$ is the difference in the light travel time for the two pinholes. All three terms are independent of the time t. The direct terms only depend on the average intensities at the pinholes, but the factor

$$e^{i\omega_0 \tau} = e^{ik_0(L_2 - L_1)} \tag{10.22}$$

in the interference term produces rapid oscillations along the detection plane. This is explicitly exhibited by expressing $\overline{G}^{(1)}$ in terms of its amplitude and phase:

$$\overline{G}^{(1)}(\mathbf{r}_1, t_1; \mathbf{r}_2, t_2) = \left|\overline{G}^{(1)}(\mathbf{r}_1, t_1; \mathbf{r}_2, t_2)\right| e^{i\Phi(\mathbf{r}_1, t_1; \mathbf{r}_2, t_2)}, \tag{10.23}$$

so that I_{12} is given by

$$I_{12} = 2D^2 \left| \overline{G}^{(1)} (\mathbf{r}_1, \tau; \mathbf{r}_2, 0) \right| \cos \left[\Phi (\mathbf{r}_1, \tau; \mathbf{r}_2, 0) + \omega_0 \tau \right]. \tag{10.24}$$

The interference pattern is modulated by slow variations in the amplitude and phase of $\overline{G}^{(1)}$ due to the finite length of the pulse. When these modulations are ignored, the interference maxima occur at the path length differences

$$L_2 - L_1 = c\tau = n\lambda_0 - \frac{\Phi \lambda_0}{2\pi}, \quad n = 0, \pm 1, \pm 2, \dots. \tag{10.25}$$

The interference pattern calculated from the first-order quantum correlation function is identical to the classical interference pattern. Since this is true even if the field state contains only one photon, first-order interference is also called one-photon interference.

An important quantity for interference experiments is the **fringe visibility**

$$\mathcal{V} \equiv \frac{\langle I \rangle_{\max} - \langle I \rangle_{\min}}{\langle I \rangle_{\max} + \langle I \rangle_{\min}}, \tag{10.26}$$

where $\langle I \rangle_{\max}$ and $\langle I \rangle_{\min}$ are respectively the maximum and minimum values of the total intensity on the detection plane. If the slow variations in $\overline{G}^{(1)}$ are neglected, then one finds

$$\langle I \rangle_{\max} = D^2 \left\{ \overline{G}^{(1)} (\mathbf{r}_1, 0; \mathbf{r}_1, 0) + \overline{G}^{(1)} (\mathbf{r}_2, 0; \mathbf{r}_2, 0) + 2 \left| \overline{G}^{(1)} (\mathbf{r}_1, \tau; \mathbf{r}_2, 0) \right| \right\}, \tag{10.27}$$

$$\langle I \rangle_{\min} = D^2 \left\{ \overline{G}^{(1)} (\mathbf{r}_1, 0; \mathbf{r}_1, 0) + \overline{G}^{(1)} (\mathbf{r}_2, 0; \mathbf{r}_2, 0) - 2 \left| \overline{G}^{(1)} (\mathbf{r}_1, \tau; \mathbf{r}_2, 0) \right| \right\}, \tag{10.28}$$

so the visibility is

$$\mathcal{V} = \frac{2 \left| \overline{G}^{(1)} (\mathbf{r}_1, \tau; \mathbf{r}_2, 0) \right|}{\overline{G}^{(1)} (\mathbf{r}_1, 0; \mathbf{r}_1, 0) + \overline{G}^{(1)} (\mathbf{r}_2, 0; \mathbf{r}_2, 0)}. \tag{10.29}$$

The field–field correlation function $\overline{G}^{(1)} (\mathbf{r}_1, \tau; \mathbf{r}_2, 0)$ is therefore a measure of the coherence of the signals from the two pinholes. There are no fringes ($\mathcal{V} = 0$) if the correlation function vanishes. On the other hand, the inequality (4.85) shows that the visibility is bounded by

$$\mathcal{V} \leqslant \frac{2\sqrt{\overline{G}^{(1)} (\mathbf{r}_1, 0; \mathbf{r}_1, 0) \, \overline{G}^{(1)} (\mathbf{r}_2, 0; \mathbf{r}_2, 0)}}{\overline{G}^{(1)} (\mathbf{r}_1, 0; \mathbf{r}_1, 0) + \overline{G}^{(1)} (\mathbf{r}_2, 0; \mathbf{r}_2, 0)} \leqslant 1, \tag{10.30}$$

where the maximum value of unity occurs when the intensities at the two pinholes are equal. This suggests introducing a normalized correlation function, the **mutual coherence function**,

$$g^{(1)} (x; x') = \frac{\overline{G}^{(1)} (x; x')}{\sqrt{\overline{G}^{(1)} (x; x) \, \overline{G}^{(1)} (x'; x')}}, \tag{10.31}$$

which satisfies $\left|g^{(1)}(x;x')\right| \leqslant 1$. In these terms, perfect coherence corresponds to $\left|g^{(1)}(x;x')\right| = 1$, and the fringe visibility is

$$\mathcal{V} = \frac{2\sqrt{\overline{G}^{(1)}(\mathbf{r}_1,0;\mathbf{r}_1,0)\,\overline{G}^{(1)}(\mathbf{r}_2,0;\mathbf{r}_2,0)}\,\left|g^{(1)}(\mathbf{r}_1,\tau;\mathbf{r}_2,0)\right|}{\overline{G}^{(1)}(\mathbf{r}_1,0;\mathbf{r}_1,0)+\overline{G}^{(1)}(\mathbf{r}_2,0;\mathbf{r}_2,0)}\,. \tag{10.32}$$

Thus measurements of the intensity at each pinhole, the fringe visibility, and the fringe spacing completely determine the complex mutual coherence function $g^{(1)}(x;x')$. This means that the correlation function $G^{(1)}(x;x')$ or $g^{(1)}(x;x')$ can always be interpreted in terms of a Young's-style interference experiment.

10.1.1 Hanbury Brown–Twiss effect

We have just seen that first-order interference, e.g. in Young's experiment or in the Michelson interferometer, is described by the first-order field correlation function $G^{(1)}$. The Hanbury Brown–Twiss effect (Hanbury Brown, 1974) was one of the earliest observations that demonstrated optical interference in the intensity–intensity correlation function $G^{(2)}$. This observation was interpreted as a measurement of photon–photon correlation, so it eventually led to the founding of the field of quantum optics. The effect was originally discovered in a simple laboratory experiment in which light from a mercury arc lamp passes through an interference filter that singles out a strong green line of the mercury atom at a wavelength of 546.1 nm. The spectrally pure green light is split by means of a balanced beam splitter into two beams, which are detected by square-law detectors placed at the output ports of the beam splitter. The experimental arrangement is shown in Fig. 10.2. The output current $I(t)$ from each detector is a measure of the intensity in that arm of the beam splitter. The intensities are slowly varying on the optical scale, with typical Fourier components in the radio range. The outputs of the two detectors are fed into a radio-frequency mixer that accumulates the time integral of the product of the two signals. By sending the signal from one of the detectors through a variable delay line the intensity–intensity correlation,

$$f(\tau) = \int_{-\infty}^{\infty} I(t)I(t-\tau)dt\,, \tag{10.33}$$

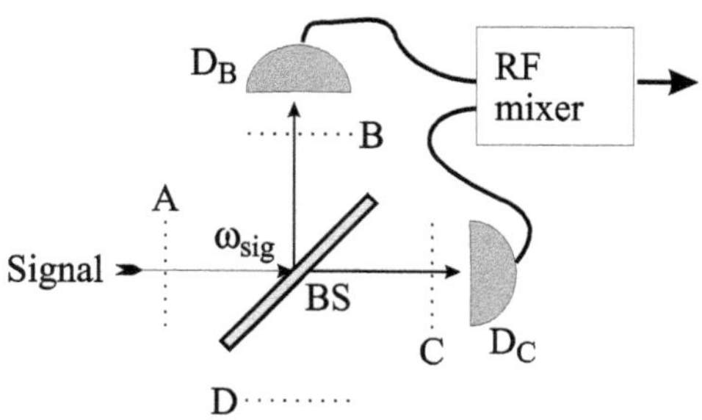

Fig. 10.2 Experimental arrangement for observing the Hanbury Brown–Twiss effect. The signal is split by a 50/50 beam splitter and the split fields enter detectors at B and C. The output of the detectors is fed into a radio-frequency (RF) mixer which integrates the product of the two signals.

is measured as a function of the delay time τ. The data (Hanbury Brown and Twiss, 1957) show a peak in the intensity–intensity correlation function $f(\tau)$ near $\tau = 0$. Hanbury Brown and Twiss interpreted this as a **photon-bunching** effect explained by the fact that the Bose character of photons enhances the probability that two photons will arrive simultaneously at the two detectors. However, Glauber showed that classical intensity fluctuations in the thermal light emitted by the mercury arc lamp yield a completely satisfactory description, so that there is no need to invoke the Bose statistics of photons.

The experimental technique for measuring the intensity–intensity correlation was later changed from simple square-law detection to coincidence detection based on a photoelectron counting technique using photomultipliers. Since this technique can register clicks associated with the arrival of individual photons, it would seem to be closer to a measurement of a photon–photon correlation function.

For the thermal light source which was used in this experiment, this hope is unjustified, because we can explain the results on the basis of classical-field notions by using the semiclassical theory of the photoelectric effect. A quantum description of this experiment, to be presented later on, employs an expansion of the density operator in the basis of coherent states. We will see that the radiation emitted by the thermal source is described by a completely positive quasi-probability distribution function $P(\alpha)$, which is consistent with a semiclassical explanation in terms of fluctuations in the intensity of the classical electromagnetic field.

On the other hand, for a pure coherent state the Hanbury Brown–Twiss effect *per se* does not exist. Thus if we were to replace the mercury arc lamp by a laser operating far above threshold, the photon arrivals would be described by a pure Poissonian random process, with no photon-bunching effect.

This intensity–intensity correlation method was applied to astrophysical stellar interferometry to measure stellar diameters (Hanbury Brown and Twiss, 1956). Stellar interferometry depends on the difference in path lengths to the telescope from points on opposite limbs of the star. For example, Michelson stellar interferometry (Born and Wolf, 1980, Sec. 7.3.6) is based on first-order interference—i.e. on the field–field correlation function—so the optical path lengths must be equalized to high precision. This is done by adjusting the positions of the interferometer mirrors attached to the telescope so that all wavelengths of light interfere constructively in the field of view. Under these conditions, white light entering the telescope will result in a bright **white-light fringe**. The white-light fringe condition must be met before attempting to measure a stellar diameter by this method.

By contrast, the beauty of the intensity stellar interferometer is that one can compensate for the delays corresponding to the difference in path lengths in the radio-wavelength region after detection, rather than in the optical-wavelength region before detection. Compensating the optical delay by an electronic delay produces a maximum in the intensity–intensity correlation function of the optical signals.

Furthermore, the optical quality of the telescope surfaces for the intensity interferometer can be much lower than that required for Michelson stellar interferometry, so that one can use the reflectors of searchlights as *light buckets*, rather than astronomical telescopes with optically perfect surfaces. However, the disadvantage of the intensity interferometer is that it requires higher intensity sources than the Michelson stellar interferometer. Thus intensity interferometry can only be used to measure the diameters of the brightest stars.

10.2 Two-photon interference

The results in Section 10.1 provide support for Dirac's dictum that each photon interferes with itself, but he went on to say (Dirac, 1958, Sec. I.3)

> Each photon then interferes only with itself. Interference between two different photons never occurs.

This is one of the very few instances in which Dirac was wrong. Further experimental progress in the generation of states containing exactly two photons has led to the realization that different photons can indeed interfere. These phenomena involve the second-order correlation function $G^{(2)}$, defined in Section 4.7, so they are sometimes called **second-order interference**. Another terminology calls them **fourth-order interference**, since $G^{(2)}$ is an average over the product of four electric field operators.

We will study two important examples of two-photon interference: the Hong–Ou–Mandel interferometer, in which interference between two photons occurs locally at a single beam splitter, and the Franson interferometer, where the interference occurs between two photons falling on spatially-separated beam splitters.

10.2.1 The Hong–Ou–Mandel interferometer

The quantum property of photon indivisibility was demonstrated by allowing a single photon to enter through one port of a beam splitter. In an experiment performed by Hong, Ou, and Mandel (Hong *et al.*, 1987), interference between two Feynman processes was demonstrated by illuminating a beam splitter with a two-photon state produced by pumping a crystal of potassium dihydrogen phosphate (**KDP**) with an ultraviolet laser beam, as shown in Fig. 10.3. In a process known as spontaneous down-conversion—which will be discussed in Section 13.3.2—a pump photon with frequency ω_p splits into a pair of lower frequency photons, traditionally called the

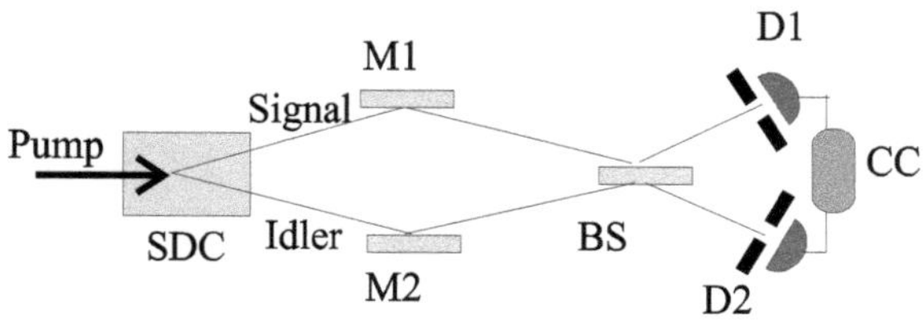

Fig. 10.3 The Hong–Ou–Mandel interferometer illuminated by a two-photon state, produced by spontaneous down-conversion in the crystal labeled SDC. The two photon wave packets are reflected from mirrors M1 and M2 so that they meet at the beam splitter BS. The outputs of detectors D1 and D2 are fed to the coincidence counter CC. (Adapted from Hong *et al.* (1987).)

signal and idler.[1] Since photons are indistinguishable, they cannot be assigned labels; therefore, the traditional language must be used carefully and sparingly. The words 'signal photon' or 'idler photon' simply mean that a photon occupies the signal mode or the idler mode. It is the modes, rather than the photons, that are distinguishable. Prior to their arrival at the beam splitter, e.g. at the mirrors M1 and M2, the diffraction patterns of the signal and idler modes do not overlap.

In the following discussion, the production process can be treated as a black box; we only need to know that one pump photon enters the crystal and that two (down-converted) photons are produced simultaneously and leave the crystal as wave packets with widths of the order of 15 fs. In the notation used in Fig. 8.2, the signal mode $(\mathbf{k}_{\text{sig}}, s_{\text{sig}})$ enters through port 1 and the idler mode $(\mathbf{k}_{\text{idl}}, s_{\text{idl}})$ enters through port 2 of the beam splitter BS.

A *Degenerate plane-wave model*

It is instructive to analyze this situation in terms of interference between Feynman processes. We begin with the idealized case of plane-wave modes—propagating from the beam splitter to the detectors—with degenerate frequencies: $\omega_{\text{idl}} = \omega_{\text{sig}} = \omega_0 = \omega_p/2$. The experimental feature of interest is the coincidence-counting rate. Since a given photon can only be counted once, the events leading to coincidence counts are those in which each detector receives one photon.

There are, consequently, two processes leading to coincidence events.

(1) The reflection–reflection ($\mathfrak{rr}$) process: both wave packets are reflected from the beam splitter towards the two detectors.

(2) The transmission–transmission ($\mathfrak{tt}$) process: both wave packets are transmitted through the beam splitter towards the two detectors.

In the absence of which-path information these processes are indistinguishable, since they both lead to the same final state: one scattered photon is in the idler mode and the other is in the signal mode. This results in simultaneous clicks in the two detectors, and one cannot know, even in principle, which of the two processes actually occurred. According to the Feynman rules of interference we must add the probability amplitudes for the two processes, and then calculate the absolute square of the sum to find the total probability. If the incident amplitude is set to one, the amplitudes of the two processes are $\mathcal{A}_{\mathfrak{rr}} = \mathfrak{r}^2$ and $\mathcal{A}_{\mathfrak{tt}} = \mathfrak{t}^2$, where $\mathfrak{r}$ and $\mathfrak{t}$ are respectively the complex reflection and transmission coefficients for the beam splitter; therefore, the coincidence amplitude is

$$\mathcal{A}_{\text{coinc}} = \mathcal{A}_{\mathfrak{rr}} + \mathcal{A}_{\mathfrak{tt}} = \mathfrak{r}^2 + \mathfrak{t}^2 \,. \tag{10.34}$$

According to eqn (8.8), $\mathfrak{r}$ and $\mathfrak{t}$ are $\pi/2$ out of phase; therefore the coincidence probability is

$$P_{\text{coinc}} = \left|\mathcal{A}_{\text{coinc}}\right|^2 = \left(\left|\mathfrak{r}\right|^2 - \left|\mathfrak{t}\right|^2\right)^2 , \tag{10.35}$$

[1] These names are borrowed from radio engineering, which in turn borrowed the 'idler' from the mechanical term 'idler gear'.

which, happily, agrees with the result (9.98) for the coincidence-counting rate. The partial destructive interference between the $\mathfrak{rr}$- and $\mathfrak{tt}$-processes, demonstrated by the expression for P_{coinc}, becomes total interference for the special case of a balanced beam splitter, i.e. the coincidence probability vanishes. We will refer to this as the **Hong–Ou–Mandel (HOM) effect**. This is a strictly quantum interference effect which cannot be explained by any semiclassical theory.

Another way of describing this phenomenon is that two photons, in the appropriate initial state, impinging simultaneously onto a balanced beam splitter will pair off and leave together through one of the two exit ports, i.e. both photons occupy one of the output modes, $(\mathbf{k}_{\text{sig}}, s_{\text{sig}})$ or $(\mathbf{k}_{\text{idl}}, s_{\text{idl}})$. This behavior is permitted for photons, which are bosons, but it would be forbidden by the Pauli principle for electrons, which are fermions. As a result of this pairing effect, detectors placed at the two exit ports of a balanced beam splitter will never register a coincidence count. The exit port used by the photon pair varies randomly from one incident pair to the next.

The argument based on the Feynman rules very effectively highlights the fundamental principles involved in two-photon interference, but it is helpful to derive the result by using a Schrödinger-picture scattering analysis. The Schrödinger-picture state produced by degenerate, spontaneous down-conversion is $a_{\text{sig}}^{\dagger} a_{\text{idl}}^{\dagger} |0\rangle$, but the initial state for the beam splitter scattering calculation is modified by the further propagation from the twin-photon source to the beam splitter. According to eqn (8.1) the scattering matrix S for propagation through vacuum is simply multiplication by $\exp(ikL)$, where k is the wavenumber and L is the propagation distance; therefore, the general rule (8.44) shows that the state incident on the beam splitter is

$$|\Phi_{\text{in}}\rangle = e^{ik_0 L_{\text{sig}}} e^{ik_0 L_{\text{idl}}} a_{\text{sig}}^{\dagger} a_{\text{idl}}^{\dagger} |0\rangle , \tag{10.36}$$

where L_{idl} and L_{sig} are respectively the distances along the idler and signal arms from the point of creation of the photon pair to the beam splitter. For the present calculation this phase factor is not important; however, it will play a significant role in Section 10.2.1-B. According to eqn (6.92), $|\Phi_{\text{in}}\rangle$ is an entangled state, and the final state

$$\begin{aligned} |\Phi_{\text{fin}}\rangle &= \mathfrak{r}\,\mathfrak{t}\, e^{-2i\omega_0 t} e^{ik_0 L_{\text{sig}}} e^{ik_0 L_{\text{idl}}} \left\{ \left(a_{\text{idl}}^{\dagger}\right)^2 + \left(a_{\text{sig}}^{\dagger}\right)^2 \right\} |0\rangle \\ &\quad + e^{ik_0 L_{\text{sig}}} e^{ik_0 L_{\text{idl}}} \left(\mathfrak{r}^2 + \mathfrak{t}^2\right) a_{\text{idl}}^{\dagger} a_{\text{sig}}^{\dagger} |0\rangle , \end{aligned} \tag{10.37}$$

obtained by using eqn (8.43), is also entangled. For a balanced beam splitter this reduces to

$$|\Phi_{\text{fin}}\rangle = \frac{i}{2} e^{-2i\omega_0 t} e^{ik_0 L_{\text{sig}}} e^{ik_0 L_{\text{idl}}} \left\{ \left(a_{\text{idl}}^{\dagger}\right)^2 + \left(a_{\text{sig}}^{\dagger}\right)^2 \right\} |0\rangle , \tag{10.38}$$

which explicitly exhibits the final state as a superposition of paired-photon states. Once again the conclusion is that the coincidence rate vanishes for a balanced beam splitter.

The quantum nature of this result can be demonstrated by considering a semiclassical model in which the signal and idler beams are represented by c-number amplitudes α_{sig} and α_{idl}. The classical version of the beam splitter equation (8.62) is

$$\begin{aligned} \alpha'_{\text{sig}} &= \mathfrak{t}\,\alpha_{\text{sig}} + \mathfrak{r}\,\alpha_{\text{idl}}\,, \\ \alpha'_{\text{idl}} &= \mathfrak{r}\,\alpha_{\text{sig}} + \mathfrak{t}\,\alpha_{\text{idl}}\,, \end{aligned} \tag{10.39}$$

and the singles counting rates at detectors D1 and D2 are respectively proportional to $|\alpha'_{\text{idl}}|^2$ and $|\alpha'_{\text{sig}}|^2$. The coincidence-counting rate is proportional to $|\alpha'_{\text{idl}}|^2\,|\alpha'_{\text{sig}}|^2 = |\alpha'_{\text{idl}}\alpha'_{\text{sig}}|^2$, and eqn (10.39) yields

$$\begin{aligned} \alpha'_{\text{idl}}\alpha'_{\text{sig}} &= \mathfrak{r}\,\mathfrak{t}\,\left(\alpha^2_{\text{sig}} + \alpha^2_{\text{idl}}\right) + \left(\mathfrak{r}^2 + \mathfrak{t}^2\right)\alpha_{\text{sig}}\alpha_{\text{idl}} \\ &\to \frac{i}{2}\left(\alpha^2_{\text{sig}} + \alpha^2_{\text{idl}}\right), \end{aligned} \tag{10.40}$$

where the last line is the result for a balanced beam splitter. This classical result resembles eqn (10.38), but now the coincidence rate cannot vanish unless one of the singles rates does. A more satisfactory model can be constructed along the lines of the argument used for the discussion of photon indivisibility in Section 1.4. Spontaneous emission is a real transition, while the down-conversion process depends on the virtual excitation of the quantum states of the atoms in the crystal; nevertheless, spontaneous down-conversion is a quantum event. A semiclassical model can be constructed by assuming that the quantum down-conversion event produces classical fields that vary randomly from one coincidence gate to the next. With this model one can show, as in Exercise 10.2, that

$$\frac{p_{\text{coinc}}}{p_{\text{sig}}p_{\text{idl}}} > \frac{1}{2}\,, \tag{10.41}$$

where p_{coinc} is the probability for a coincidence count, and p_{sig} and p_{idl} are the probabilities for singles counts—all averaged over many counting windows. This semiclassical model limits the visibility of the interference minimum to 50%; the essentially perfect null seen in the experimental data can only be predicted by using the complete destructive interference between probability amplitudes allowed by the full quantum theory. Thus the *HOM null* provides further evidence for the indivisibility of photons.

*B Nondegenerate wave packet analysis**

The simplified model used above suffices to explain the physical basis of the Hong–Ou–Mandel interferometer, but it is inadequate for describing some interesting applications to precise timing, such as the measurement of the propagation velocity of single-photon wave packets in a dielectric, and the nonclassical dispersion cancelation effect, discussed in Sections 10.2.2 and 10.2.3 respectively. These applications exploit the fact that the signal and idler modes produced in the experiment are not plane waves; instead, they are described by wave packets with temporal widths $T \sim 15\,\text{fs}$. In order to deal with this situation, it is necessary to allow continuous variation of the frequencies and to relax the degeneracy condition $\omega_{\text{idl}} = \omega_{\text{sig}}$, while retaining the simple geometry of the scattering problem. To this end, we first use eqn (3.64) to replace the box-normalized operator $a_{\mathbf{k}s}$ by the continuum operator $a_s(\mathbf{k})$, which obeys the canonical commutation relations (3.26). In polar coordinates the propagation vectors are described by $\mathbf{k} = (k, \theta, \phi)$, so the propagation directions of the modes $(\mathbf{k}_{\text{sig}}, s_{\text{sig}})$ and $(\mathbf{k}_{\text{idl}}, s_{\text{idl}})$ are given by $(\theta_\sigma, \phi_\sigma)$, where $\sigma = \text{sig}, \text{idl}$ is the channel index. The assumption of frequency degeneracy can be eliminated, while maintaining the scattering geometry, by considering wave packets corresponding to narrow cones of propagation directions. The wave packets are described by real averaging functions $f_\sigma(\theta, \phi)$ that are strongly peaked at $(\theta, \phi) = (\theta_\sigma, \phi_\sigma)$ and normalized by

$$\int d\Omega \, |f_\sigma(\theta, \phi)|^2 = 1\,, \tag{10.42}$$

where $d\Omega = d(\cos\theta)\, d\phi$. In practice the widths of the averaging functions can be made so small that

$$\int d\Omega f_\sigma(\theta, \phi)\, f_\rho(\theta, \phi) \approx \delta_{\sigma\rho}\,. \tag{10.43}$$

With this preparation, we define wave packet operators

$$a_\sigma^\dagger(\omega) \equiv \frac{\omega}{c^{3/2}} \int \frac{d\Omega}{2\pi} f_\sigma(\theta, \phi)\, a_{s_\sigma}^\dagger(\mathbf{k})\,, \tag{10.44}$$

that satisfy

$$\begin{aligned} \left[a_\sigma(\omega), a_\rho^\dagger(\omega')\right] &= \delta_{\sigma\rho} 2\pi\delta(\omega - \omega')\,, \\ \left[a_\sigma(\omega), a_\rho(\omega')\right] &= 0\,. \end{aligned} \tag{10.45}$$

For a given value of the channel index σ, the operator $a_\sigma^\dagger(\omega)$ creates photons in a wave packet with propagation unit vectors clustered near the channel value $\widetilde{\mathbf{k}}_\sigma = \mathbf{k}_\sigma / k_\sigma$, and polarization s_σ; however, the frequency ω can vary continuously. These operators are the continuum generalization of the operators $a_{ms}(\omega)$ defined in eqn (8.71).

With this machinery in place, we next look for the appropriate generalization of the incident state in eqn (10.36). Since the frequencies of the emitted photons are not fixed, we assume that the source generates a state

$$\int \frac{d\omega}{2\pi} \int \frac{d\omega'}{2\pi} C(\omega, \omega')\, a_{\text{sig}}^\dagger(\omega)\, a_{\text{idl}}^\dagger(\omega')\, |0\rangle\,, \tag{10.46}$$

describing a pair of photons, with one in the signal channel and the other in the idler channel. As discussed above, propagation from the source to the beam splitter multiplies the state $a_{\text{sig}}^\dagger(\omega)\, a_{\text{idl}}^\dagger(\omega')\, |0\rangle$ by the phase factor $\exp(ikL_{\text{sig}}) \exp(ik'L_{\text{idl}})$. It is more convenient to express this as

$$e^{ikL_{\text{sig}}} e^{ik'L_{\text{idl}}} = e^{i(k+k')L_{\text{idl}}} e^{ik\Delta L}\,, \tag{10.47}$$

where $\Delta L = L_{\text{sig}} - L_{\text{idl}}$ is the difference in path lengths. Consequently, the initial state for scattering from the beam splitter has the general form

$$|\Phi_{\text{in}}\rangle = \int \frac{d\omega}{2\pi} \int \frac{d\omega'}{2\pi} C(\omega, \omega')\, e^{ik\Delta L} a_{\text{sig}}^\dagger(\omega)\, a_{\text{idl}}^\dagger(\omega')\, |0\rangle\,, \tag{10.48}$$

where we have absorbed the symmetrical phase factor $\exp\left[i(k + k')\, L_{\text{idl}}\right]$ into the coefficient $C(\omega, \omega')$.

By virtue of the commutation relations (10.45), every two-photon state $a_{\text{sig}}^{\dagger}(\omega)\, a_{\text{idl}}^{\dagger}(\omega')\,|0\rangle$ satisfies Bose symmetry; consequently, the two-photon wave packet state $|\Phi_{\text{in}}\rangle$ satisfies Bose symmetry for any choice of $C(\omega,\omega')$. However, not all states of this form will exhibit the two-photon interference effect. To see what further restrictions are needed, we consider the balanced case $\Delta L = 0$, and examine the effects of the alternative processes on $|\Phi_{\text{in}}\rangle$. In the transmission–transmission process the directions of propagation are preserved, but in the reflection–reflection process the directions of propagation are interchanged. Thus the actions on the incident state are respectively given by

$$|\Phi_{\text{in}}\rangle \xrightarrow{\mathfrak{tt}} |\Phi_{\text{in}}\rangle_{\mathfrak{tt}} = \frac{1}{2}\int\frac{d\omega}{2\pi}\int\frac{d\omega'}{2\pi}C(\omega,\omega')\,a_{\text{sig}}^{\dagger}(\omega)\,a_{\text{idl}}^{\dagger}(\omega')\,|0\rangle\,, \tag{10.49}$$

and

$$\begin{aligned}|\Phi_{\text{in}}\rangle \xrightarrow{\mathfrak{rr}} |\Phi_{\text{in}}\rangle_{\mathfrak{rr}} &= -\frac{1}{2}\int\frac{d\omega}{2\pi}\int\frac{d\omega'}{2\pi}C(\omega,\omega')\,a_{\text{idl}}^{\dagger}(\omega)\,a_{\text{sig}}^{\dagger}(\omega')\,|0\rangle \\ &= -\frac{1}{2}\int\frac{d\omega}{2\pi}\int\frac{d\omega'}{2\pi}C(\omega',\omega)\,a_{\text{sig}}^{\dagger}(\omega)\,a_{\text{idl}}^{\dagger}(\omega')\,|0\rangle\,.\end{aligned} \tag{10.50}$$

For interference to take place, the final states $|\Phi_{\text{in}}\rangle_{\mathfrak{tt}}$ and $|\Phi_{\text{in}}\rangle_{\mathfrak{rr}}$ must agree up to a phase factor, i.e. $|\Phi_{\text{in}}\rangle_{\mathfrak{tt}} = \exp(i\Lambda)\,|\Phi_{\text{in}}\rangle_{\mathfrak{rr}}$. This in turn implies $C(\omega,\omega') = -\exp(i\Lambda)\,C(\omega',\omega)$, and a second use of this relation shows that $\exp(2i\Lambda) = 1$. Consequently the condition for interference is

$$C(\omega,\omega') = \pm C(\omega',\omega)\,. \tag{10.51}$$

We will see below that the (+)-version of this condition leads to the photon pairing effect as in the degenerate case. The (−)-version is a new feature which is possible only in the nondegenerate case. As shown in Exercise 10.5, it leads to destructive interference for the emission of photon pairs.

In order to see what happens when the interference condition is violated, consider the function

$$C(\omega,\omega') = (2\pi)^2\,C_0\,\delta(\omega-\omega_1)\,\delta(\omega'-\omega_2) \tag{10.52}$$

describing the input state $a_{\text{sig}}^{\dagger}(\omega_1)\,a_{\text{idl}}^{\dagger}(\omega_2)\,|0\rangle$, where $\omega_1 \neq \omega_2$. In this situation photons entering through port 1 always have frequency ω_1 and photons entering through port 2 always have frequency ω_2; therefore, a measurement of the photon energy at either detector would provide which-path information by determining the path followed by the photon through the beam splitter. This leads to a very striking conclusion: even if no energy determination is actually made, the mere possibility that it could be made is enough to destroy the interference effect.

The input state defined by eqn (10.52) is entangled, but this is evidently not enough to ensure the HOM effect. Let us therefore consider the symmetrized function

$$C(\omega,\omega') = (2\pi)^2\,C_0\left[\delta(\omega-\omega_1)\,\delta(\omega'-\omega_2) + \delta(\omega'-\omega_1)\,\delta(\omega-\omega_2)\right], \tag{10.53}$$

which does satisfy the interference condition. The corresponding state

$$|\Phi_{\text{in}}\rangle = C_0 \left\{ a^{\dagger}_{\text{sig}}(\omega_1)\, a^{\dagger}_{\text{idl}}(\omega_2)\, |0\rangle + a^{\dagger}_{\text{sig}}(\omega_2)\, a^{\dagger}_{\text{idl}}(\omega_1)\, |0\rangle \right\} \tag{10.54}$$

is not just entangled, it is *dynamically entangled*, according to the definition in Section 6.5.3. Thus dynamical entanglement is a necessary condition for the photon pairing or antipairing effect associated with the $\pm$ sign in eqn (10.51). This feature plays an important role in quantum information processing with photons.

In the experiments to be discussed below, the two-photon state is generated by the spontaneous down-conversion process in which momentum and energy are conserved:

$$\begin{aligned} \hbar\omega_p &= \hbar\omega + \hbar\omega' , \\ \hbar\mathbf{k}_p &= \hbar\mathbf{k} + \hbar\mathbf{k}' , \end{aligned} \tag{10.55}$$

where $(\hbar\omega_p, \hbar\mathbf{k}_p)$ is the energy–momentum four-vector of the parent ultraviolet photon, and $(\hbar\omega, \hbar\mathbf{k})$ and $(\hbar\omega', \hbar\mathbf{k}')$ are the energy–momentum four-vectors for the daughter photons. The energy conservation law allows $C(\omega, \omega')$ to be written as

$$C(\omega, \omega') = 2\pi\delta(\omega + \omega' - \omega_p)\, g(\nu) , \tag{10.56}$$

where

$$\nu = \frac{\omega - \omega'}{2} , \quad \omega = \omega_0 + \nu , \quad \omega' = \omega_0 - \nu . \tag{10.57}$$

The interference condition (10.51), which ensures that the two Feynman processes lead to the same final state, becomes $g(\nu) = \pm g(-\nu)$.

The conservation rule (10.55) tells us that the down-converted photons are anti-correlated in energy. A *bluer* photon ($\omega > \omega_0$) is always associated with a *redder* photon ($\omega' < \omega_0$). Furthermore, the photons are produced with equal amplitudes on either side of the degeneracy value, $\omega = \omega_0 = \omega_p/2$, i.e. $g(\nu) = g(-\nu)$. Thus the coefficient function $C(\omega, \omega')$ for down-conversion satisfies the $(+)$-version of eqn (10.51). The width, $\Delta\nu$, of the power spectrum $|g(\nu)|^2$ is jointly determined by the properties of the KDP crystal and the filters that select out a particular pair of conjugate photons. The two-photon coherence time corresponding to $\Delta\nu$ is

$$\tau_2 \sim \frac{1}{\Delta\nu} . \tag{10.58}$$

We are now ready to carry out a more realistic analysis of the Hong–Ou–Mandel experiment in terms of the interference between the $\mathfrak{tt}$- and $\mathfrak{rr}$-processes. For a given value of $\nu = (\omega - \omega')/2$, the amplitudes are

$$\mathcal{A}_{\mathfrak{tt}}(\nu) = \mathfrak{t}^2 g(\nu)\, e^{i\Phi_{\mathfrak{tt}}(\nu)} \to \frac{1}{2} g(\nu)\, e^{i\Phi_{\mathfrak{tt}}(\nu)} \tag{10.59}$$

and

$$\mathcal{A}_{\mathfrak{rr}}(\nu) = \mathfrak{r}^2 g(\nu)\, e^{i\Phi_{\mathfrak{rr}}(\nu)} \to -\frac{1}{2} g(\nu)\, e^{i\Phi_{\mathfrak{rr}}(\nu)} , \tag{10.60}$$

where the final forms hold for a balanced beam splitter and $\Phi_{\mathfrak{tt}}(\nu)$ and $\Phi_{\mathfrak{rr}}(\nu)$ are the phase shifts for the $\mathfrak{rr}$- and $\mathfrak{tt}$-processes respectively. The total coincidence probability is therefore

$$P_{\text{coinc}} = \int d\nu \left|\mathcal{A}_{\mathfrak{tt}}(\nu) + \mathcal{A}_{\mathfrak{rr}}(\nu)\right|^2 = \int d\nu \left|g(\nu)\right|^2 \sin^2\left(\frac{\Delta\Phi(\nu)}{2}\right), \tag{10.61}$$

where

$$\Delta\Phi(\nu) = \Phi_{\mathfrak{tt}}(\nu) - \Phi_{\mathfrak{rr}}(\nu). \tag{10.62}$$

The phase changes $\Phi_{\mathfrak{tt}}(\nu)$ and $\Phi_{\mathfrak{rr}}(\nu)$ depend on the frequencies of the two photons and the geometrical distances involved. The distances traveled by the idler and signal wave packets in the $\mathfrak{tt}$-process are

$$\begin{aligned} L_{\text{idl}}^{\mathfrak{tt}} &= L_{\text{idl}} + L_1, \\ L_{\text{sig}}^{\mathfrak{tt}} &= L_{\text{sig}} + L_2, \end{aligned} \tag{10.63}$$

where L_1 (L_2) is the distance from the beam splitter to the detector D1 (D2). The corresponding distances for the $\mathfrak{rr}$-process are

$$\begin{aligned} L_{\text{idl}}^{\mathfrak{rr}} &= L_{\text{idl}} + L_2, \\ L_{\text{sig}}^{\mathfrak{rr}} &= L_{\text{sig}} + L_1. \end{aligned} \tag{10.64}$$

In the $\mathfrak{tt}$-process the idler (signal) wave packet enters detector D1 (D2), so the phase change is

$$\Phi_{\mathfrak{tt}}(\nu) = \frac{\omega}{c} L_{\text{idl}}^{\mathfrak{tt}} + \frac{\omega'}{c} L_{\text{sig}}^{\mathfrak{tt}}. \tag{10.65}$$

According to eqn (10.50), ω and ω' switch roles in the $\mathfrak{rr}$-process; consequently,

$$\Phi_{\mathfrak{rr}}(\nu) = \frac{\omega}{c} L_{\text{idl}}^{\mathfrak{rr}} + \frac{\omega'}{c} L_{\text{sig}}^{\mathfrak{rr}}. \tag{10.66}$$

Substituting eqns (10.63)–(10.66) into eqn (10.62) leads to the simple result

$$\Delta\Phi(\nu) = 2\nu \frac{\Delta L}{c}. \tag{10.67}$$

Since the two photons are created simultaneously, the difference in arrival times of the signal and idler wave packets is

$$\Delta t = \frac{\Delta L}{c}. \tag{10.68}$$

The resulting form for the coincidence probability,

$$P_{\text{coinc}}(\Delta t) = \int d\nu \left|g(\nu)\right|^2 \sin^2(\nu \Delta t), \tag{10.69}$$

has a width determined by $\left|g(\nu)\right|^2$ and a null at $\Delta t = 0$, as shown in Exercise 10.3. As expected, the null occurs for the balanced case,

$$L_{\text{sig}} = L_{\text{idl}} = L_0. \tag{10.70}$$

In this argument, we have replaced the plane waves of Section 10.2.1-A with Gaussian pulses. Each pulse is characterized by two parameters, the pulse width, T_σ, and the arrival time, t_σ, of the pulse peak at the beam splitter. If the absolute difference in arrival times, $|\Delta t| = |L_{\text{sig}} - L_{\text{idl}}|/c$, is larger than the sum of the pulse widths ($|\Delta t| > T_{\text{sig}} + T_{\text{idl}}$) the pulses are nonoverlapping, and the destructive interference effect will not occur. This case simply represents two repetitions of the photon indivisibility experiment with a single photon. What happens in this situation depends on the width, T_{gate}, of the acceptance window for the coincidence counter. If $T_{\text{gate}} < |\Delta t|$ no coincidence count will occur, but in the opposite situation, $T_{\text{gate}} > |\Delta t|$, coincidence counts will be recorded with probability 1/2. For $\Delta t = 0$ the wave packets overlap, and interference between the alternative Feynman paths prevents any coincidence counts. In order to increase the contrast between the overlapping and nonoverlapping cases, one should choose $T_{\text{gate}} > \Delta t_{\max}$, where $\Delta t_{\max}$ is the largest value of the absolute time delay. The result is an extremely narrow dip—the **HOM dip**—in the coincidence count rate as a function of Δt, as seen in Fig. 10.4.

The alternative analysis using the Schrödinger-picture scattering technique is also instructive. For this purpose, we substitute the special form (10.56) for $C(\omega, \omega')$ into eqn (10.48) to find the initial state for scattering by the beam splitter:

$$|\Phi_{\text{in}}\rangle = e^{i\omega_0 \Delta t} \int \frac{d\nu}{2\pi} g(\nu)\, e^{i\nu\Delta t} a_{\text{sig}}^\dagger(\omega_0 + \nu)\, a_{\text{idl}}^\dagger(\omega_0 - \nu)\, |0\rangle\,. \tag{10.71}$$

Applying eqn (8.76) to each term in this superposition yields

$$|\Phi_{\text{fin}}\rangle = |\Phi_{\text{pair}}\rangle + |\Phi_{\text{coinc}}\rangle\,, \tag{10.72}$$

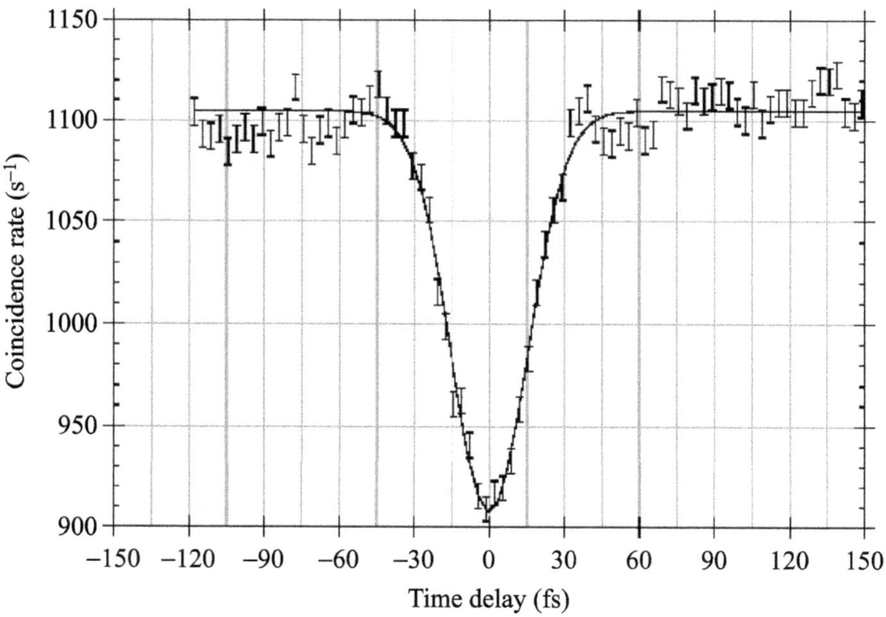

Fig. 10.4 Coincidence rate as a function of the relative optical time delay in the interferometer. The solid line is a Gaussian fit, with an rms width of 15.3 fs. This profile serves as a map of the overlapping photon wave packets. (Reproduced from Steinberg *et al.* (1992).)

where

$$|\Phi_{\text{pair}}\rangle = ie^{i\omega_p t}e^{i\omega_0 \Delta t}\frac{1}{2}\int \frac{d\nu}{2\pi} g(\nu)\cos(\nu\Delta t)$$
$$\times \left[a_{\text{sig}}^{\dagger}(\omega_0+\nu)\, a_{\text{sig}}^{\dagger}(\omega_0-\nu)\,|0\rangle + a_{\text{idl}}^{\dagger}(\omega_0+\nu)\, a_{\text{idl}}^{\dagger}(\omega_0-\nu)\,|0\rangle\right] \quad (10.73)$$

describes the pairing behavior, and

$$|\Phi_{\text{coinc}}\rangle = ie^{i\omega_p t}e^{i\omega_0 \Delta t}\frac{1}{2}\int \frac{d\nu}{2\pi} g(\nu)\sin(\nu\Delta t)$$
$$\times \left[a_{\text{sig}}^{\dagger}(\omega_0+\nu)\, a_{\text{idl}}^{\dagger}(\omega_0-\nu)\,|0\rangle - a_{\text{idl}}^{\dagger}(\omega_0+\nu)\, a_{\text{sig}}^{\dagger}(\omega_0-\nu)\,|0\rangle\right] \quad (10.74)$$

represents the state leading to coincidence counts.

10.2.2 The single-photon propagation velocity in a dielectric*

The down-converted photons are *twins*, i.e. they are born at precisely the same instant inside the nonlinear crystal. On the other hand, the strict conservation laws in eqn (10.55) are only valid if $(\hbar\omega_p, \hbar\mathbf{k}_p)$ is sharply defined. In practice this means that the incident pulse length must be long compared to any other relevant time scale, i.e. the pump laser is operated in continuous-wave (cw) mode. Thus the twin photons are born at the same time, but this time is fundamentally unknowable because of the energy–time uncertainty principle.

These properties allow a given pair of photons to be used, in conjunction with the Hong–Ou–Mandel interferometer, to measure the speed with which an individual photon traverses a transparent dielectric medium. This allows us to investigate the following question: Does an individual photon wave packet move at the group velocity through the medium, just as an electromagnetic wave packet does in classical electrodynamics? The answer is yes, if the single-photon state is monochromatic and the medium is highly transparent. This agrees with the simple theory of the quantized electromagnetic field in a transparent dielectric, which leads to the expectation that an electromagnetic wave packet containing a single photon propagates with the classical group velocity through a dispersive and nondissipative dielectric medium.

A schematic of an experiment (Steinberg *et al.*, 1992) which demonstrates that individual photons do indeed travel at the group velocity is shown in Fig. 10.5. In this arrangement an argon-ion UV laser beam, operating at wavelength of 351 nm, enters a KDP crystal, where entangled pairs of photons are produced. Degenerate red photons at a wavelength of 702 nm are selected out for detection by means of two irises, I1 and I2, placed in front of detectors D1 and D2, which are single-photon counting modules (silicon avalanche photodiodes). The signal wave packet, which follows the upper path of the interferometer, traverses a glass sample of length L, and subsequently enters an optical-delay mechanism, consisting of a right-angle **trombone prism** mounted on a computer-controlled translation stage. This prism retroreflects the signal wave packet onto one input port of the final beam splitter, with a variable time delay. Consequently, the location of the trombone prism can be chosen so that the signal wave packet will overlap with the idler wave packet.

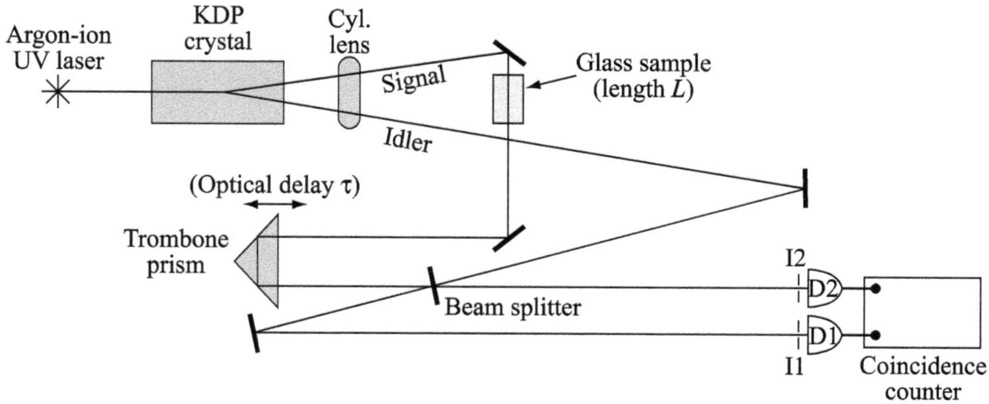

Fig. 10.5 Apparatus to measure photon propagation times. (Reproduced from Steinberg *et al.* (1992).)

Meanwhile, the idler wave packet has been traveling along the lower path of the interferometer, which is empty of all optical elements, apart from a single mirror which reflects the idler wave packet onto the other input port of the beam splitter. If the optical path length difference between the upper and lower paths of the interferometer is adjusted to be zero, then the signal and idler wave packets will meet at the same instant at the final beam splitter. For this to happen, the longitudinal position of the trombone prism must be adjusted so as to exactly compensate for the delay—relative to the idler wave packets transit time through vacuum—experienced by the signal wave packet, due to its propagation through the glass sample at the group velocity, $v_g < c$.

As explained in Section 10.2.1, the bosonic character of photons allows a pair of photons meeting at a balanced beam splitter to pair off, so that they both go towards the same detector. The essential condition is that the initial two-photon state contains no which-path hints. When this condition is satisfied, there is a minimum (a perfect null under ideal circumstances) of the coincidence-counting signal. The overlap of the signal and idler wave packets at the beam splitter must be as complete as possible, in order to produce the Hong–Ou–Mandel minimum in the coincidence count rate. As the time delay produced by the trombone prism is varied, the result is an inverted Gaussian profile, similar to the one pictured in Fig. 10.4, near the minimum in the coincidence rate. As can be readily seen from the first line in Table 10.1, a compensating delay of $35\,219 \pm 1$ fs must be introduced by the trombone prism in order to produce the Hong–Ou–Mandel minimum in the coincidence rate. This delay is very close to what one expects for a classical electromagnetic wave packet propagating at the group velocity through a 1/2 inch length of SF11 glass.

This experiment was repeated for several samples of glass in various configurations. From Table 10.1, we see that the theoretical predictions, based on the assumption that single-photon wave packets travel at the group velocity, agree very well with experimental measurements. The predictions based on the alternative supposition that

Glass	L (μm)	τ_t (expt) (fs)	τ_g (theory) (fs)	τ_p (theory) (fs)
SF11 ($\frac{1}{2}''$)	12687 ± 13	35219 ± 1	35181 ± 35	32642 ± 33
SF11 ($\frac{1}{4}''$)	-6337 ± 13	-17559.6 ± 1	-17572 ± 35	-16304 ± 33
SF11 ($\frac{1}{2}''$ & $\frac{1}{4}''$)	19033 ± 0.5	52782.4 ± 1	52778.6 ± 1.4	48949 ± 46
BK7 ($\frac{1}{2}''$ & $\frac{1}{4}''$)	18894 ± 18	33513 ± 1	33480 ± 33	32314 ± 32
All BK7 & SF11	n/a*	-19264 ± 1	-19269 ± 1.4	-16635 ± 56
BK7 ($\frac{1}{2}''$)	12595 ± 13	22349.5 ± 1	22318 ± 22	21541 ± 21

*This measurement involved both pieces of BK7 in one arm and both pieces of SF11 in the other, so no individual length measurement is meaningful.

Table 10.1 Measured delay times compared to theoretical values computed using the group and phase velocities. (Reproduced from Steinberg *et al.* (1992).)

the photon travels at the phase velocity seriously disagree with experiment.

10.2.3 The dispersion cancelation effect*

In addition to providing evidence that single photons propagate at the group velocity, the experiment reported above displays a feature that is surprising from a classical point of view. For the experimental run with the 1/2 in glass sample inserted in the signal arm, Fig. 10.6 shows that the HOM dip has essentially the same width as the vacuum-only case shown in Fig. 10.4. This is surprising, because a classical wave packet passing through the glass sample experiences dispersive broadening, due to the fact that plane waves with different frequencies propagate at different phase velocities. This raises the question: Why is the width of the coincidence-count dip not changed by the broadening of the signal wave packet? One could also ask the more fundamental question: How is it that the presence of the glass sample in the signal arm does not altogether destroy the delicate interference phenomena responsible for the null in the coincidence count?

To answer these questions, we first recall that the existence of the HOM null depends on starting with an initial state such that the $\mathfrak{rr}$- and $\mathfrak{tt}$-processes lead to the same final state. When this condition for interference is satisfied, it is impossible—even in principle—to determine which photon passed through the glass sample. This means that each of the twin photons traverses both the $\mathfrak{rr}$- and the $\mathfrak{tt}$-paths—just as each photon in a Young's interference experiment passes through both pinholes. In this way, each photon experiences two different values of the frequency-dependent index of refraction—one in glass, the other in vacuum—and this fact is the basis for a quantitative demonstration that the two-photon interference effect also takes place in the unbalanced HOM interferometer.

The only difference between this experiment and the original Hong–Ou–Mandel experiment discussed in Section 10.2.1-B is the presence of the glass sample in the signal arm of the apparatus; therefore, we only need to recalculate the phase difference $\Delta\Phi(\nu)$ between the two paths. The new phase shifts for each path are obtained from the old phase shifts by adding the difference in phase shift between the length L of

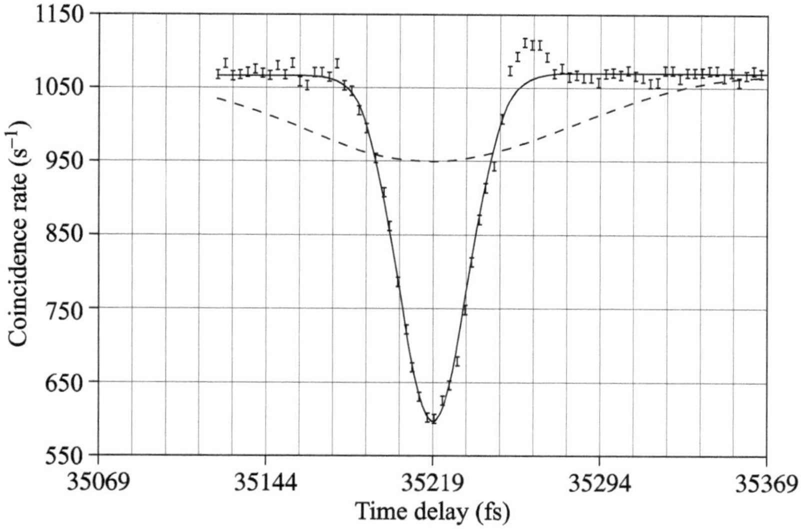

Fig. 10.6 Coincidence profile after a 1/2 in piece of SF11 glass is inserted in the signal arm of the interferometer. The location of the minimum is shifted by 35 219 fs from the corresponding vacuum result, but the width is essentially unchanged. For comparison the dashed curve shows a classically broadened 15 fs pulse. (Reproduced from Steinberg *et al.* (1992).)

the glass sample and the same length of vacuum; therefore

$$\Phi_{\mathrm{tt}}(\nu) = \Phi_{\mathrm{tt}}^{(0)}(\nu) + \left[k(\omega) - \frac{\omega}{c}\right] L \tag{10.75}$$

and

$$\Phi_{\mathrm{rr}}(\nu) = \Phi_{\mathrm{rr}}^{(0)}(\nu) + \left[k(\omega') - \frac{\omega'}{c}\right] L , \tag{10.76}$$

where $\Phi_{\mathrm{tt}}^{(0)}(\nu)$ and $\Phi_{\mathrm{rr}}^{(0)}(\nu)$ are respectively given by eqns (10.65) and (10.66). The new phase difference is

$$\Delta\Phi(\nu) = \Delta\Phi^{(0)}(\nu) + \left\{\left[k(\omega) - \frac{\omega}{c}\right] - \left[k(\omega') - \frac{\omega'}{c}\right]\right\} L , \tag{10.77}$$

so using eqn (10.67) for $\Delta\Phi^{(0)}(\nu)$ yields

$$\Delta\Phi(\nu) = \frac{2\nu}{c}(\Delta L - L) + \left[k(\omega_0 + \nu) - k(\omega_0 - \nu)\right] L , \tag{10.78}$$

where $\omega_0 = (\omega + \omega')/2 = \omega_p/2$. The difference $k(\omega_0 + \nu) - k(\omega_0 - \nu)$ represents the fact that both of the anti-correlated twin photons pass through the glass sample.

As a consequence of dispersion, the difference between the wavevectors is not in general a linear function of ν; therefore, it is not possible to choose a single value of ΔL that ensures $\Delta\Phi(\nu) = 0$ for all values of ν. Fortunately, the limited range of values for ν allowed by the sharply-peaked function $|g(\nu)|^2$ in eqn (10.69) justifies a Taylor series expansion,

$$k(\omega_0 \pm \nu) = k(\omega_0) + \left(\frac{dk}{d\omega}\right)_0 (\pm\nu) + \frac{1}{2}\left(\frac{d^2k}{d\omega^2}\right)_0 (\pm\nu)^2 + O(\nu^3), \tag{10.79}$$

around the degeneracy value $\nu = 0$ $(\omega = \omega' = \omega_0)$. When this expansion is substituted into eqn (10.78) all even powers of ν cancel out; we call this the **dispersion cancelation effect**. In this approximation, the phase difference is

$$\begin{aligned} \Delta\Phi(\nu) &= \frac{2\nu}{c}(\Delta L - L) + 2\left(\frac{dk}{d\omega}\right)_0 \nu L + O(\nu^3) \\ &= \frac{2\nu}{c}(\Delta L - L) + \frac{2\nu}{v_{g0}} L + O(\nu^3), \end{aligned} \tag{10.80}$$

where the last line follows from the definition (3.142) of the group velocity. If the third-order dispersive terms are neglected, the null condition $\Delta\Phi(\nu) = 0$ is satisfied for all ν by setting

$$\Delta L = \left(1 - \frac{c}{v_{g0}}\right) L < 0, \tag{10.81}$$

where the inequality holds for normal dispersion, i.e. $v_{g0} < c$. Thus the signal path length must be shortened, in order to compensate for slower passage of photons through the glass sample.

The second-order term in the expansion (10.79) defines the group velocity dispersion coefficient β:

$$\beta = \frac{1}{2}\left.\frac{d^2k}{d\omega^2}\right|_{\omega=\omega_0} = -\frac{1}{2}\frac{1}{v_{g0}^2}\left(\frac{dv_g}{d\omega}\right)_0. \tag{10.82}$$

Since β cancels out in the calculation of $\Delta\Phi(\nu)$, it does not affect the width of the Hong–Ou–Mandel interference minimum.

10.2.4 The Franson interferometer*

The striking phenomena discussed in Sections 10.2.1–10.2.3 are the result of a quantum interference effect that occurs when twin photons—which are produced simultaneously at a single point in the KDP crystal—are reunited at a single beam splitter. In an even more remarkable interference effect, first predicted by Franson (1989), the two photons never meet again. Instead, they only interact with spatially-separated interferometers, that we will label as *nearby* and *distant*. The final beam splitter in each interferometer has two output ports: the one positioned between the beam splitter and the detector is called the *detector* port, since photons emerging from this port fall on the detector; the other is called the *exit* port, since photons emitted from this port leave the apparatus. At the final beam splitter in each interferometer the photon randomly passes through the detector or the exit port. Speaking anthropomorphically, the *choice* made by each photon at its final beam splitter is completely random, but the two—apparently independent—choices are in fact correlated. For certain settings of the interferometers, when one photon chooses the detector port, so does the other, i.e. the random choices of the two photons are perfectly correlated. This happens despite the fact that the photons have never interacted since their joint production in the KDP crystal. Even more remarkably, an experimenter can force a change, from perfectly correlated choices to perfectly anti-correlated choices, by altering the setting of only one of the interferometers, e.g. the nearby one.

This situation is so radically nonclassical that it is difficult to think about it clearly. A common mistake made in this connection is to conclude that altering the setting at the nearby interferometer is somehow *causing* an instantaneous change in the choices made by the photon in the distant interferometer. In order to see why this is wrong, it is useful to imagine that there are two experimenters: Alice, who adjusts the nearby interferometer and observes the choices made by photons at its final beam splitter; and Bob, who observes the choices made by successive photons at the final beam splitter in the distant interferometer, but makes no adjustments. An important part of the experimental arrangement is a secret classical channel through which Alice is informed—without Bob's knowledge—of the results of Bob's measurements. Let us now consider two experimental runs involving many successive pairs of photons. In the first, Alice uses her secret information to set her interferometer so that the choices of the two photons are perfectly correlated. In the meantime, Bob—who is kept in the dark regarding Alice's machinations—accumulates a record of the detection-exit choices at his beam splitter. In the second run, Alice alters the settings so that the photon choices are perfectly anti-correlated, and Bob innocently continues to acquire data. Since the individual quantum events occurring at Bob's beam splitter are perfectly random, it is clear that his two sets of data will be statistically indistinguishable. In other words, Bob's local observations at the distant interferometer—made without benefit of a secret channel—cannot detect the changes made by Alice in the settings of the nearby interferometer. The same could be said of any local observations made by Alice, if she were deprived of her secret channel. The difference between the two experiments is not revealed until the two sets of data are brought together—via the classical communication channel—and compared. Alice's manipulations do not cause events through instantaneous action at a distance; instead, her actions cause a change in the correlation between distant events that are individually random as far as local observations are concerned.

The peculiar phenomena sketched above can be better understood by describing a Franson interferometer that was used in an experiment with down-converted pairs (Kwiat *et al.*, 1993). In this arrangement, shown schematically in Fig. 10.7, each photon passes through one interferometer.

An examination of Fig. 10.7 shows that each interferometer I_j (defined by the components Mj, B1_j, and B2_j, with $j = 1, 2$) contains two paths, from the initial to the final beam splitter, that send the photon to the associated detector: a long path with length L_j and a short path with length S_j. This arrangement is called an *unbalanced Mach–Zehnder interferometer*. The difference $\Delta L_j = L_j - S_j$ in path lengths serves as an optical delay line that can be adjusted by means of the trombone prism. We will label the signal and idler wave packets with 1 and 2 according to the interferometer that is involved.

A photon traversing an interferometer does not split at the beam splitters, but the

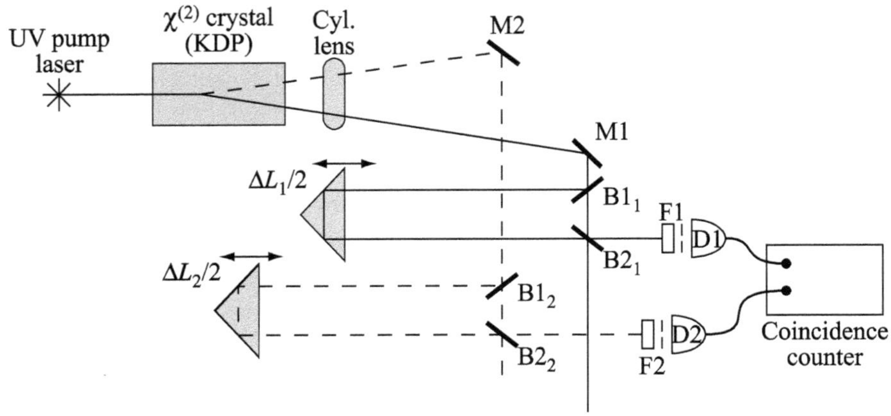

Fig. 10.7 Experimental configuration for a Franson interferometer. (Reproduced from Kwiat *et al.* (1993).)

probability amplitude defining the wave packet does; consequently—just as in Young's two-pinhole experiment—the two paths available to the photon could produce single-photon interference. In the present case, the interference would appear as a temporal oscillation of the intensity emitted from the final beam splitter. We will abuse the terminology slightly by also referring to these oscillations as interference fringes. This effect can be prevented by choosing the optical delay $\Delta L_j/c$ to be much greater than the typical coherence time τ_1 of a single-photon wave packet:

$$\frac{\Delta L_j}{c} \gg \tau_1 \,. \tag{10.83}$$

When this is the case, the two partial wave packets—one following the long path and the other following the short path through the interferometer—completely miss each other at the final beam splitter, so there is no single-photon interference.

The motivation for eliminating single-photon interference is that the oscillation of the singles rates at one or both detectors would confuse the measurement of the coincidence rate, which is the signal for two-photon interference. Further examination of Fig. 10.7 shows that there are four paths that can result in the detection of both photons: l–l (each wave packet follows its long path); l–s (wave packet 1 follows its long path and wave packet 2 follows its short path); s–l (wave packet 1 follows its short path and wave packet 2 follows its long path); and s–s (each wave packet follows its short path).

According to Feynman's rules, two paths leading to distinct final states cannot interfere, so we need to determine which pairs of paths lead to different final states. The first step in this task is to calculate the arrival time of the wave packets at their respective detectors. For interferometer I_j, let T_j be the propagation time to the first beam splitter plus the propagation time from the final beam splitter to the detector; then the arrival times at the detector via the long or short path are

$$t_{jl} = T_j + L_j/c \tag{10.84}$$

and

$$t_{js} = T_j + S_j/c\,, \tag{10.85}$$

respectively. This experiment uses a cw pump to produce the photon pairs; therefore, only the differences in arrival times at the detectors are meaningful. The four processes yield the time differences

$$\begin{aligned}
\Delta t_{ls} &= t_{1l} - t_{2s} = T_1 - T_2 + \frac{L_1 - S_2}{c}\,, &(10.86)\\
\Delta t_{sl} &= t_{1s} - t_{2l} = T_1 - T_2 - \frac{L_2 - S_1}{c}\,, &(10.87)\\
\Delta t_{ll} &= t_{1l} - t_{2l} = T_1 - T_2 + \frac{L_1 - L_2}{c}\,, &(10.88)\\
\Delta t_{ss} &= t_{1s} - t_{2s} = T_1 - T_2 + \frac{S_1 - S_2}{c}\,, &(10.89)
\end{aligned}$$

and two processes will not interfere if the difference between their Δts is larger than the two-photon coherence time τ_2 defined by eqn (10.58). For example, eqns (10.86) and (10.87) yield the difference

$$\Delta t_{ls} - \Delta t_{sl} = \frac{\Delta L_1 + \Delta L_2}{c} \gg \tau_2\,, \tag{10.90}$$

where the final inequality follows from the condition (10.83) and the fact that $\tau_1 \sim \tau_2$. The conclusion is that the processes l–s and s–l cannot interfere, since they lead to different final states. Similar calculations show that l–s and s–l are distinguishable from l–l and s–s; therefore, the only remaining possibility is interference between l–l and s–s. In this case the difference is

$$\Delta t_{ll} - \Delta t_{ss} = \frac{\Delta L_1 - \Delta L_2}{c}\,, \tag{10.91}$$

so that interference between these two processes can occur if the condition

$$\frac{|\Delta L_1 - \Delta L_2|}{c} \ll \tau_2 \tag{10.92}$$

is satisfied. The practical effect of these conditions is that the interferometers must be almost identical, and this is a source of experimental difficulty.

When the condition (10.92) is satisfied, the final states reached by the short–short and long–long paths are indistinguishable, so the corresponding amplitudes must be added in order to calculate the coincidence probability, i.e.

$$P_{12} = \left|\mathcal{A}_{ll} + \mathcal{A}_{ss}\right|^2. \tag{10.93}$$

The amplitudes for the two paths are

$$\begin{aligned}
\mathcal{A}_{ll} &= \mathfrak{r}_1 \mathfrak{t}_1' \mathfrak{r}_2 \mathfrak{t}_2' e^{i\Phi_{ll}}\,,\\
\mathcal{A}_{ss} &= \mathfrak{r}_1' \mathfrak{t}_1 \mathfrak{r}_2' \mathfrak{t}_2 e^{i\Phi_{ss}}\,,
\end{aligned} \tag{10.94}$$

where $(\mathfrak{r}_j, \mathfrak{t}_j)$ and $(\mathfrak{r}_j', \mathfrak{t}_j')$ are respectively the reflection and transmission coefficients for the first and second beam splitter in the jth interferometer, and the phases Φ_{ll}

and Φ_{ss} are the sums of the one-photon phases for each path. We will simplify this calculation by assuming that all beam splitters are balanced and that the photon frequencies are degenerate, i.e. $\omega_1 = \omega_2 = \omega_0 = \omega_p/2$. In this case the phases are

$$\begin{aligned} \Phi_{ll} &= \omega_0 \left(t_{1l} + t_{2l}\right) = \omega_0 \left(T_1 + T_2\right) + \frac{\omega_0}{c} \left(L_1 + L_2\right), \\ \Phi_{ss} &= \omega_0 \left(t_{1s} + t_{2s}\right) = \omega_0 \left(T_1 + T_2\right) + \frac{\omega_0}{c} \left(S_1 + S_2\right), \end{aligned} \tag{10.95}$$

and the coincidence probability is

$$P_{12} = \cos^2\left(\frac{\Delta\Phi}{2}\right), \tag{10.96}$$

where

$$\Delta\Phi = \Phi_{ll} - \Phi_{ss} = \frac{\omega_0}{c}\left(\Delta L_1 + \Delta L_2\right). \tag{10.97}$$

Now suppose that Bob and Alice initially choose the same optical delay for their respective interferometers, i.e. they set $\Delta L_1 = \Delta L_2 = \Delta L$, then

$$\frac{\Delta\Phi}{2} = \frac{\omega_0}{c}\Delta L = 2\pi\frac{\Delta L}{\lambda_0}, \tag{10.98}$$

where $\lambda_0 = 2\pi c/\omega_0$ is the common wavelength of the two photons. If the delay ΔL is arranged to be an integer number m of wavelengths, then $\Delta\Phi/2 = 2\pi m$ and P_{12} achieves the maximum value of unity. In other words, with these settings the behavior of the photons at the final beam splitters are perfectly correlated, due to constructive interference between the two probability amplitudes.

Next consider the situation in which Bob keeps his settings fixed, while Alice alters her settings to $\Delta L_1 = \Delta L + \delta L$, so that

$$\frac{\Delta\Phi}{2} = 2\pi m + \pi\frac{\delta L}{\lambda_0}, \tag{10.99}$$

and

$$P_{12} = \cos^2\left(\pi\frac{\delta L}{\lambda_0}\right). \tag{10.100}$$

For the special choice $\delta L = \lambda_0/2$, the coincidence probability vanishes, and the behavior of the photons at the final beam splitters are anti-correlated, due to complete destructive interference of the probability amplitudes. This drastic change is brought about by a very small adjustment of the optical delay in only one of the interferometers. We should stress the fact that macroscopic physical events—the firing of the detectors—that are spatially separated by a large distance behave in a correlated or anti-correlated way, by virtue of the settings made by Alice in only one of the interferometers.

In Chapter 19 we will see that these correlations-at-a-distance violate the Bell inequalities that are satisfied by any so-called local realistic theory. We recall that a theory is said to be local if no signals can propagate faster than light, and it is said to be realistic if physical objects can be assumed to have definite properties in the absence of observation. Since the results of experiments with the Franson interferometer violate Bell's inequalities—while agreeing with the predictions of quantum theory—we can conclude that the quantum theory of light is not a local realistic theory.

10.3 Single-photon interference revisited*

The experimental techniques required for the Hong–Ou–Mandel demonstration of two-photon interference—creation of entangled photon pairs by spontaneous down-conversion (SDC), mixing at beam splitters, and coincidence detection—can also be used in a beautiful demonstration of a remarkable property of single-photon interference. In our discussion of Young's two-pinhole interference in Section 10.1, we have already remarked that any attempt to obtain which-path information destroys the interference pattern. The usual thought experiments used to demonstrate this for the two-pinhole configuration involve an actual interaction of the photon—either with some piece of apparatus or with another particle—that can determine which pinhole was used. The experiment to be described below goes even further, since the mere possibility of making such a determination destroys the interference pattern, even if the measurements are not actually carried out and no direct interaction with the photons occurs. This is a real experimental demonstration of Feynman's rule that interference can only occur between alternative processes if there is no way—even in principle—to distinguish between them. In this situation, the complex amplitudes for the alternative processes must first be added to produce the total probability amplitude, and only then is the probability for the final event calculated by taking the absolute square of the total amplitude.

10.3.1 Mandel's two-crystal experiment

In the two-crystal experiment of Mandel and his co-workers (Zou *et al.*, 1991), shown in Fig. 10.8, the beam from an argon laser, operating at an ultraviolet wavelength, falls on the beam splitter BS_p. This yields two coherent, parallel pump beams that enter into two staggered nonlinear crystals, NL1 and NL2, where they can undergo spontaneous down-conversion. The rate of production of photon pairs in the two crystals is so low that at most a single photon pair exists inside the apparatus at any given instant. In

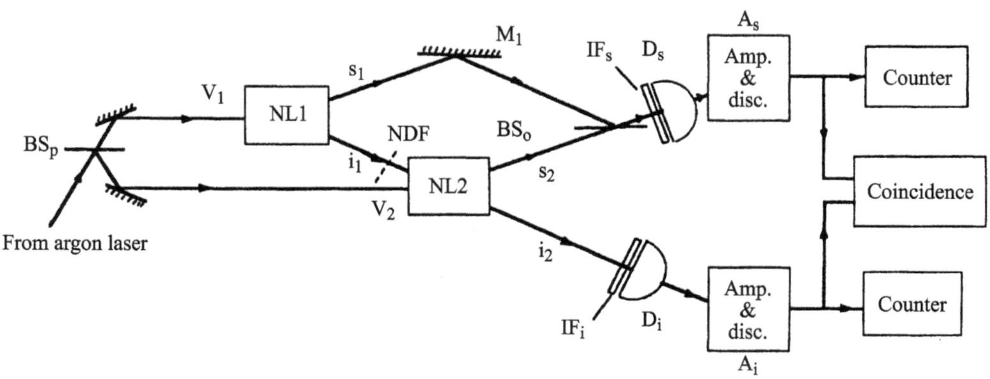

Fig. 10.8 Spontaneous down-conversion (SDC) occurs in two crystals NL1 and NL2. The two idler modes i_1 and i_2 from these two crystals are carefully aligned so that they coincide on the face of detector D_i. The dashed line in beam path i_1 in front of crystal NL2 indicates a possible position of a beam block, e.g. an opaque card. (Reproduced from Zou *et al.* (1991).)

other words, we can assume that the simultaneous emission of two photon pairs, one from each crystal, is so rare that it can be neglected.

The idler beams i_1 and i_2, emitted from the crystals NL1 and NL2 respectively, are carefully aligned so that their transverse Gaussian-mode beam profiles overlap as exactly as possible on the face of the idler detector D_i. Thus, when a click occurs in D_i, it is impossible—even in principle—to know whether the detected photon originated from the first or the second crystal. It therefore follows that it is also impossible—even in principle—to know whether the twin signal wave packet, produced together with the idler wave packet describing the detected photon, originated from the first crystal as a signal wave packet in beam s_1, or from the second crystal as a signal wave packet in beam s_2. The two processes resulting in the appearance of s_1 or s_2 are, therefore, indistinguishable; and their amplitudes must be added before calculating the final probability of a click at detector D_s.

10.3.2 Analysis of the experiment

The two indistinguishable Feynman processes are as follows. The first is the emission of the signal wave packet by the first crystal into beam s_1, reflection by the mirror M_1, reflection at the output beam splitter BS_o, and detection by the detector D_s. This is accompanied by the emission of a photon in the idler mode i_1 that traverses the crystal NL2—which is transparent at the idler wavelength—and falls on the detector D_i. The second process is the emission by the second crystal of a photon in the signal wave packet s_2, transmission through the output beam splitter BS_o, and detection by the same detector D_s, accompanied by emission of a photon into the idler mode i_2 which falls on D_i. This experiment can be analyzed in two apparently different ways that we consider below.

A Second-order interference

Let us suppose that the photon detections at D_s are registered in coincidence with the photon detections at D_i, and that the two idler beams are perfectly aligned. If a click were to occur in D_s in coincidence with a click in D_i, it would be impossible to determine whether the signal–idler pair came from the first or the second crystal. In this situation Feynman's interference rule tells us that the probability amplitude $\mathcal{A}_1$ that the photon pair originates in crystal NL1 and the amplitude $\mathcal{A}_2$ of pair emission by NL2 must be added to get the probability

$$|\mathcal{A}_1 + \mathcal{A}_2|^2 \tag{10.101}$$

for a coincidence count. When the beam splitter BS_o is slowly scanned by small translations in its transverse position, the signal path length of the first process is changed relative to the signal path length of the second process. This in turn leads to a change in the phase difference between $\mathcal{A}_1$ and $\mathcal{A}_2$; therefore, the coincidence count rate would exhibit interference fringes.

From Section 9.2.4 we know that the coincidence-counting rate for this experiment is proportional to the second-order correlation function

$$G^{(2)}(x_s, x_i; x_s, x_i) = \mathrm{Tr}\left[\rho_{\mathrm{in}} E_s^{(-)}(x_s) E_i^{(-)}(x_i) E_i^{(+)}(x_i) E_s^{(+)}(x_s)\right], \tag{10.102}$$

where ρ_{in} is the density operator describing the initial state of the photon pair produced by down-conversion. The subscripts s and i respectively denote the polarizations of the signal and idler modes. The variables x_s and x_i are defined as $x_s = (\mathbf{r}_s, t_s)$ and $x_i = (\mathbf{r}_i, t_i)$, where $\mathbf{r}_s$ and $\mathbf{r}_i$ are respectively the locations of the detectors $\mathrm{D_s}$ and $\mathrm{D_i}$, while t_s and t_i are the arrival times of the photons at the detectors. This description of the experiment as a second-order interference effect should not be confused with the two-photon interference studied in Section 10.2.1. In the present experiment at most one photon is incident on the beam splitter $\mathrm{BS_o}$ during a coincidence-counting window; therefore, the pairing phenomena associated with Bose statistics for two photons in the same mode cannot occur.

B First-order interference

Since the state ρ_{in} involves two photons—the signal and the idler—the description in terms of $G^{(2)}$ offered in the previous section seems very natural. On the other hand, in the ideal case in which there are no absorptive or scattering losses and the classical modes for the two idler beams $\mathrm{i_1}$ and $\mathrm{i_2}$ are perfectly aligned, an idler wave packet will fall on $\mathrm{D_i}$ whenever a signal wave packet falls on $\mathrm{D_s}$. In this situation, the detector $\mathrm{D_i}$ is actually superfluous; the counting rate of detector $\mathrm{D_s}$ will exhibit interference whether or not coincidence detection is actually employed. In this case the amplitudes $\mathcal{A}_1$ and $\mathcal{A}_2$ refer to the processes in which the signal wave packet originates in the first or the second crystal. The counting rate $|\mathcal{A}_1 + \mathcal{A}_2|^2$ at detector $\mathrm{D_s}$ will therefore exhibit the same interference fringes as in the coincidence-counting experiment, even if the clicks of detector $\mathrm{D_i}$ are not recorded. In this case the interference can be characterized solely by the first-order correlation function

$$G^{(1)}(x_s; x_s) = \mathrm{Tr}\left[\rho_{\text{in}} E_s^{(-)}(x_s)\, E_s^{(+)}(x_s)\right]. \tag{10.103}$$

In the actual experiment, no coincidence detection was employed during the collection of the data. The first-order interference pattern shown as trace A in Fig. 10.9 was obtained from the signal counter $\mathrm{D_s}$ alone. In fact, the detector $\mathrm{D_i}$ and the entire coincidence-counting circuitry could have been removed from the apparatus without altering the experimental results.

10.3.3 Bizarre aspects

The interference effect displayed in Fig. 10.9 may appear strange at first sight, since the signal wave packets $\mathrm{s_1}$ and $\mathrm{s_2}$ are emitted spontaneously and at random by two spatially well-separated crystals. In other words, they appear to come from independent sources. Under these circumstances one might expect that photons emitted into the two modes $\mathrm{s_1}$ and $\mathrm{s_2}$ should have nothing to do with each other. Why then should they produce interference effects at all? The explanation is that the presence of at most one photon in a signal wave packet during a given counting window, combined with the perfect alignment of the two idler beams $\mathrm{i_1}$ and $\mathrm{i_2}$, makes it impossible—even in principle—to determine which crystal actually emitted the detected photon in the signal mode. This is precisely the situation in which the Feynman rule (10.2) applies; consequently, the amplitudes for the processes involving signal photons $\mathrm{s_1}$ or $\mathrm{s_2}$ must be added, and interference is to be expected.

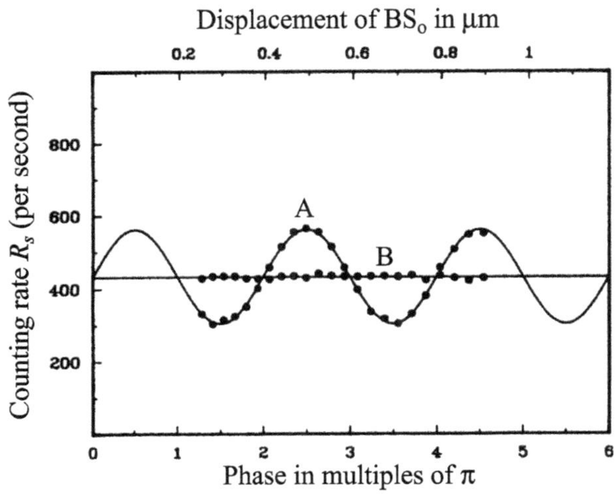

Fig. 10.9 Interference fringes of the signal photons detected by D_s, as the transverse position of the final splitter BS_o is scanned (see Fig. 10.8). Trace A is taken with a neutral 91% transmission density filter placed between the two crystals. Trace B is taken with the beam path i_1 blocked by an opaque card (i.e. a 'beam block'). (Reproduced from Zou *et al.* (1991).)

Now let us examine what happens if the experimental configuration is altered in such a way that which-path information becomes available in principle. For this purpose we assign Alice to control the position of the beam splitter BS_o and record the counting rate at detector D_s, while Bob is put in charge of the entire idler arm, including the detector D_i. As part of an investigation of possible future modifications of the experiment, Bob inserts a neutral density filter (an ideal absorber with amplitude transmission coefficient $\mathfrak{t}$ independent of frequency) between NL1 and NL2, as shown by the line NDF in Fig. 10.8. Since the filter interacts with the idler photons, but does not interact with the signal photons in any way, Bob expects that he can carry out this modification without any effect on Alice's measurements. In the extreme limit $\mathfrak{t} \approx 0$—i.e. the idler photon i_1 is completely blocked, so that it will never arrive at D_i—Bob is surprised when Alice excitedly reports that the interference pattern at D_s has completely disappeared, as shown in trace B of Fig. 10.9.

Alice and Bob eventually arrive at an explanation of this truly bizarre result by a strict application of the Feynman interference rules (10.1)–(10.3). They reason as follows. With the i_1-beam block in place, suppose that there is a click at D_s but not at D_i. Under the assumption that both D_s and D_i are ideal (100% effective) detectors, it then follows with certainty that no idler photon was emitted by NL2. Since the signal and idler photons are emitted in pairs from the same crystal, it also follows that the signal photon must have been emitted by NL1. Under the same circumstances, if there are simultaneous clicks at D_s and D_i, then it is equally certain that the signal photon must have come from NL2. This means that Bob and Alice could obtain which-path information by monitoring both counters. Therefore, in the new experimental configuration, it is in principle possible to determine which of the alternative processes actually occurred. This is precisely the situation covered by rule (10.3), so the probability of a count at D_s is the sum of the probabilities for the two processes considered separately; there is no interference. A truly amazing aspect of this situation is that the interference pattern disappears even if the detector D_i is not present. In fact—just as before—the detector D_i and the entire coincidence-counting circuitry could have been removed from the apparatus without altering the experimental results. Thus the mere *possibility* that which-path information could be gathered by inserting a beam block is sufficient to eliminate the interference effect.

The phenomenon discussed above provides another example of the nonlocal character of quantum physics. Bob's insertion or withdrawal of the beam blocker leads to very different observations by Alice, who could be located at any distance from Bob. This situation is an illustration of a typically Delphic remark made by Bohr in the course of his dispute with Einstein (Bohr, 1935):

> But even at this stage there is essentially the question of *an influence on the very conditions which define the possible types of predictions regarding the future behavior of the system.*

With this hint, we can understand the effect of Bob's actions as setting the overall conditions of the experiment, which produce the nonlocal effects.

An interesting question which has not been addressed experimentally is the following: How soon after a sudden blocking of beam path i_1 does the interference pattern disappear for the signal photons? Similarly, how soon after a sudden unblocking of beam path i_1 does the interference pattern reappear for the signal photons?

10.4 Tunneling time measurements*

Soon after its discovery, it was noticed that the Schrödinger equation possessed real, exponentially damped solutions in classically forbidden regions of space, such as the interior of a rectangular potential barrier for a particle with energy below the top of the barrier. This phenomenon—which is called **tunneling**—is mathematically similar to evanescent waves in classical electromagnetism.

The first observation of tunneling quickly led to the further discoveries of important early examples, such as the field emission of electrons from the tips of cold, sharp metallic needles, and Gamow's explanation of the emission of alpha particles (helium nuclei) from radioactive nuclei undergoing α decay.

Recent examples of the applications of tunneling include the Esaki tunnel diode (which allows the generation of high-frequency radio waves), Josephson tunneling between two superconductors separated by a thin oxide barrier (which allows the sensitive detection of magnetic fields in a *S*uperconducting *QU*antum *I*nterference *D*evice (SQUID)), and the scanning tunneling microscope (which allows the observation of individual atoms on surfaces).

In spite of numerous useful applications and technological advances based on tunneling, there remained for many decades after its early discovery a basic, unresolved physics problem. How fast does a particle traverse the barrier during the tunneling process? In the case of quantum optics, we can rephrase this question as follows: How quickly does a photon pass through a tunnel barrier in order to reach the far side?

First of all, it is essential to understand that this question is physically meaningless in the absence of a concrete description of the method of measuring the transit time. This principle of operationalism is an essential part of the scientific method, but it is especially crucial in the studies of phenomena in quantum mechanics, which are far removed from everyday experience. A definition of the operational procedure starts with a careful description of an idealized thought experiment. Thought experiments were especially important in the early days of quantum mechanics, and they are still very important today as an aid for formulating physically meaningful questions. Many of these thought experiments can then be turned into real experiments, as measurements of the tunneling time illustrate.

Let us therefore first consider a thought experiment for measuring the tunneling time of a photon. In Fig. 10.10, we show an experimental method which uses twin photons γ_1 and γ_2, born simultaneously by spontaneous down-conversion. Placing two Geiger counters at equal distances from the crystal would lead—in the absence of any tunnel barrier—to a pair of simultaneous clicks. Now suppose that a tunnel barrier is inserted into the path of the upper photon γ_1. One might expect that this would impede the propagation of γ_1, so that the click of the upper Geiger counter—placed behind the barrier—would occur later than the click of the lower Geiger counter. The surprising result of an experiment to be described below is that exactly the opposite happens. The arrival of the tunneling photon γ_1 is registered by a click of the upper Geiger counter that occurs *before* the click signaling the arrival of the nontunneling photon γ_2. In other words, the tunneling photon seems to have traversed the barrier superluminally. However, for reasons to be given below, we shall see that there is no operational way to use this superluminal tunneling phenomenon to send true signals faster than the speed of light.

This particular thought experiment is not practical, since it would require the use of Geiger counters with extremely fast response times, comparable to the femtosecond time scales typical of tunneling. However, as we have seen earlier, the Hong–Ou–Mandel two-photon interference effect allows one to resolve the relative times of arrival of two photons at a beam splitter to within fractions of a femtosecond. Hence, the

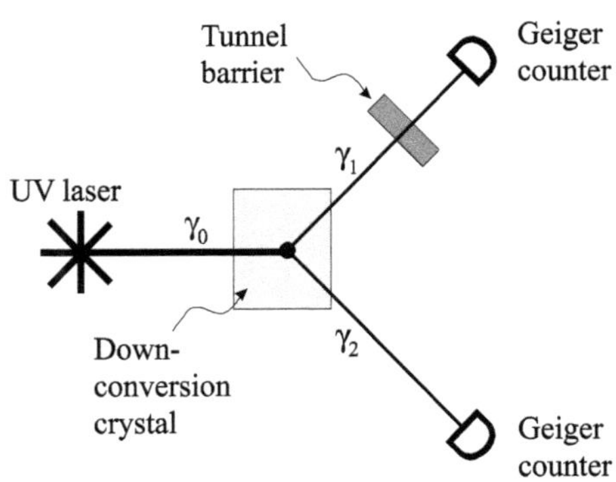

Fig. 10.10 Schematic of a thought experiment to measure the tunneling time of the photon. Spontaneous down-conversion generates twin photons γ_1 and γ_2 by absorption of a photon from a UV pump laser. In the absence of a tunnel barrier, the two photons travel the same distance to two Geiger counters placed equidistantly from the crystal, and two simultaneous clicks occur. A tunnel barrier (shaded rectangle) is now inserted into the path of photon γ_1. The tunneling time is given by the time difference between the clicks of the two Geiger counters.

impractical thought experiment can be turned into a realistic experiment by inserting a tunnel barrier into one arm of a Hong–Ou–Mandel interferometer (Steinberg and Chiao, 1995), as shown in Fig. 10.11.

The two arms of the interferometer are initially made equal in path length (perfectly balanced), so that there is a minimum—a Hong–Ou–Mandel (HOM) dip—in the coincidence count rate. After the insertion of the tunnel barrier into the upper arm of the interferometer, the mirror M_1 must be slightly displaced in order to recover the HOM dip. This procedure compensates for the extra delay—which can be either positive or negative—introduced by the tunnel barrier. Measurements show that the delay due to the tunnel barrier is negative in sign; the mirror M_1 has to be moved away from the barrier in order to recover the HOM dip. This is contrary to the normal expectation that all such delays should be positive in sign. For example, one would expect a positive sign if the tunnel barrier were an ordinary piece of glass, in which case the mirror would have to be moved towards the barrier to recover the HOM dip. Thus the sign of the necessary displacement of mirror M_1 determines whether tunneling is superluminal or subluminal in character.

The tunnel barrier used in this experiment—which was first performed at Berkeley in 1993 (Steinberg *et al.*, 1993; Steinberg and Chiao, 1995)—is a dielectric mirror formed by an alternating stack of high- and low-index coatings, each a quarter wavelength thick. The multiple Bragg reflections from the successive interfaces of the dielectric coatings give rise to constructive interference in the backwards direction of propagation for the photon and destructive interference in the forward direction. The result is an exponential decay in the envelope of the electric field amplitude as a function of propagation distance into the periodic structure, i.e. an evanescent wave. This constitutes a **photonic bandgap**, that is, a range of classical wavelengths—equivalent to energies for photons—for which propagation is forbidden. This is similar to the ex-

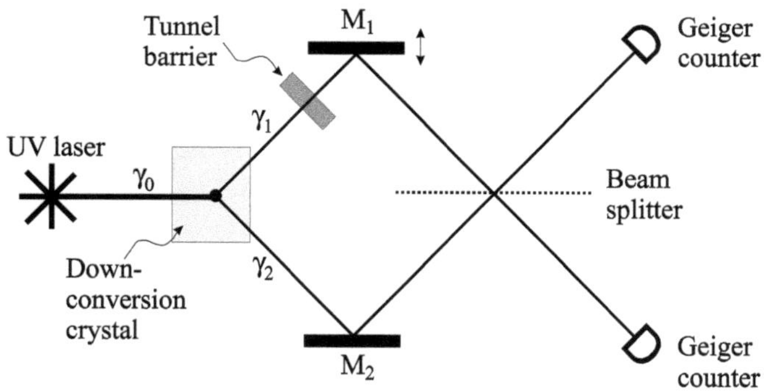

Fig. 10.11 Schematic of a realistic tunneling-time experiment, such as that performed in Berkeley (Steinberg *et al.*, 1993; Steinberg and Chiao, 1995), to measure the tunneling time of a photon by means of Hong–Ou–Mandel two-photon interference. The double-headed arrow to the right of mirror M_1 indicates that it can be displaced so as to compensate for the tunneling time delay introduced by the tunnel barrier. The sign of this displacement indicates whether the tunneling time is superluminal or subluminal.

ponential decay of the electron wave function inside the classically forbidden region of a tunnel barrier.

In this experiment, the photonic bandgap stretched from a wavelength of 600 nm to 800 nm, with a center at 700 nm, the wavelength of the photon pairs used in the Hong–Ou–Mandel interferometer. The exponential decay of the photon probability amplitude with propagation distance is completely analogous to the exponential decay of the probability amplitude of an electron inside a periodic crystal lattice, when its energy lies at the center of the electronic bandgap. The tunneling probability of the photon through the photonic tunnel barrier was measured to be around 1%, and was spectrally flat over the typical 10 nm-wide bandwidths of the down-conversion photon wave packets. This is much narrower than the 200 nm total spectral width of the photonic bandgap. The carrier wavelength of the single-photon wave packets was chosen to coincide with the center of the bandgap. After the tunneling process was completed, the transmitted photon wave packets suffered a 99% reduction in intensity, but the distortion from the initial Gaussian shape was observed to be completely negligible.

In Fig. 10.12, the data for the tunneling time obtained using the Hong–Ou–Mandel

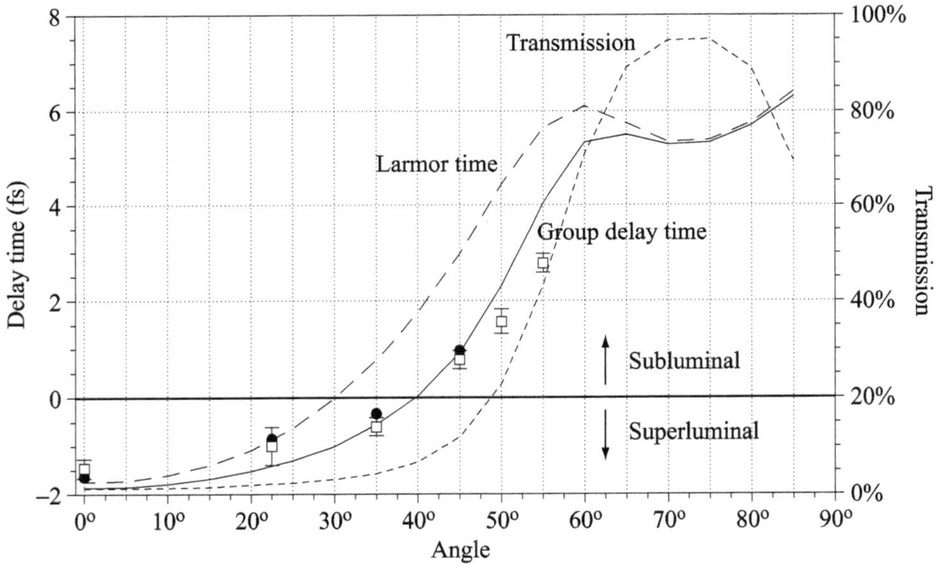

Fig. 10.12 Summary of tunneling time data taken using the Hong–Ou–Mandel interferometer, shown schematically in Fig. 10.11, as the tunnel barrier sample was tilted: starting from normal incidence at 0° towards 60° for p-polarized down-converted photons. As the sample was tilted towards Brewster's angle (around 60°), the tunneling time changed sign from a *negative* relative delay, indicating a *superluminal* tunneling time, to a *positive* relative delay, indicating a *subluminal* tunneling time. Note that the sign reversal occurs at a tilt angle of 40°. Two different samples used as barriers are represented respectively by the circles and the squares. (Reproduced from Steinberg and Chiao (1995).)

interferometer are shown as a function of the tilt angle of the tunnel barrier sample relative to normal incidence, with the plane of polarization of the incident photon lying in the plane of incidence (this is called *p*-polarization). As the tilt angle is increased towards Brewster's angle (around 60°), the reflectivity of the successive interfaces between the dielectric layers tends to zero. In this limit the destructive interference in the forward direction disappears, so the photonic bandgap, along with its associated tunnel barrier, is eliminated.

Thus as one tilts the tunnel barrier towards Brewster's angle, it effectively behaves more and more like an ordinary glass sample. One then expects to obtain a positive delay for the passage of the photon γ_1 through the barrier, corresponding to a subluminal tunneling delay time. Indeed, for the three data points taken at the large tilt angles of 45°, 50°, and 55° (near Brewster's angle) the mirror M_1 had to be moved towards the sample, as one would normally expect for the compensation of positive delays. However, for the three data points at the small tilt angles of 0°, 22°, and 35°, the data show that the tunneling delay of photon γ_1 is negative relative to photon γ_2. In other words, for incidence angles near normal the mirror M_1 had to be moved in the counterintuitive direction, away from the tunnel barrier. The change in sign of the effect implies a superluminal tunneling time for these small angles of incidence. The displacement of mirror M_1 required to recover the HOM dip changed from positive to negative at 40°, corresponding to a smooth transition from subluminal to superluminal tunneling times. From these data, one concludes that, near normal incidence, the tunneling wave packet γ_1 passes through the barrier superluminally (i.e. effectively faster than c) relative to wave packet γ_2. The interpretation of this seemingly paradoxical result evidently requires some care.

We first note that the existence of apparently superluminal propagation of classical electromagnetic waves is well understood. An example, that shares many features with tunneling, is propagation of a Gaussian pulse with carrier frequency in a region of anomalous dispersion. The fact that this would lead to superluminal propagation of a greatly reduced pulse was first predicted by Garrett and McCumber (1969) and later experimentally demonstrated by Chu and Wong (1982). The classical explanation of this phenomenon is that the pulse is reshaped during its propagation through the medium. The locus of maximum constructive interference—the pulse peak—is shifted forward toward the leading edge of the pulse, so that the peak of a small replica of the original pulse arrives before the peak of a similar pulse propagating through vacuum. Another way of saying this is that the trailing edge of the pulse is more strongly absorbed than the leading edge. The resulting movement of the peak is described by the group velocity, which can be greater than c or even negative. These phenomena are actually quite general; in particular, they will also occur in an amplifying medium (Bolda *et al.*, 1993). In this case it is possible for a Gaussian pulse with carrier frequency detuned from a gain line to propagate—with little change in amplitude and shape—with a group velocity greater than c or negative (Chiao, 1993; Steinberg and Chiao, 1994).

The method used above to explain classical superluminal propagation is mathematically similar to Wigner's theory of tunneling in quantum mechanics (Wigner, 1955). This theory of the tunneling time was based on the idea, roughly speaking, that the peak of the tunneling wave packet would be delayed with respect to the peak of a nontunneling wave packet by an amount determined by the maximum constructive interference of different energy components, which defines the peak of the tunneling wave packet. The method of stationary phase then leads to the expression

$$\tau_{\text{Wigner}} = \hbar \left. \frac{d \arg T(E)}{dE} \right|_{E_0} \tag{10.104}$$

for the **group-delay** tunneling time, where E_0 is the most probable energy of the tunneling particle's wave packet, and $T(E)$ is the particle's tunneling probability amplitude as a function of its energy E. Wigner's theory predicts that the tunneling delay becomes superluminal because—for sufficiently thick barriers—the time τ_{Wigner} depends only on the tunneling particle's energy, and not on the thickness of the barrier. Since the Wigner tunneling time saturates at a finite value for thick barriers, this produces a seeming violation of relativistic causality when $\tau_{\text{Wigner}} < d/c$, where d is the thickness of the barrier.

Wigner's theory was not originally intended to apply to photons, but we have already seen in Section 7.8 that a classical envelope satisfying the paraxial approximation can be regarded as an effective probability amplitude for the photon. This allows us to use the classical wave calculations to apply Wigner's result to photons. From this point of view, the rare occasions when a tunneling photon penetrates through the barrier—approximately 1% of the photons appear on the far side—is a result of the small probability amplitude that is transmitted. This in turn corresponds to the 1% transmission coefficient of the sample at 0° tilt. It is only for these lucky photons that the click of the upper Geiger counter occurs earlier than a click of the lower Geiger counter announcing the arrival of the nontunneling photon γ_2. The average of all data runs at normal incidence shows that the peak of the tunneling wave packet γ_1 arrived 1.47 ± 0.21 fs earlier than the peak of the wave packet γ_2 that traveled through the air. This is in reasonable agreement (within two standard deviations) with the prediction of 1.9 fs based on eqn (10.104).

Some caveats need to be made here, however. The first is this: the observation of a superluminal tunneling time does not imply the possibility of sending a true signal faster than the vacuum speed of light, in violation of special relativity. By 'true signal' we mean a signal which connects a cause to its effect; for example, a signal sent by closing a switch at one end of a transmission-wire circuit that causes an explosion to occur at the other end. Such *causal* signals are characterized by discontinuous fronts—produced by the closing of the switch, for example—and these fronts are prohibited by relativity from ever traveling faster than c. However, it should be stressed that it is perfectly permissible, and indeed, under certain circumstances—arising from the principle of relativistic causality itself—absolutely necessary, for the group velocity of a wave packet to exceed the vacuum speed of light (Bolda *et al.*, 1993; Chiao and Steinberg, 1997). From a quantum mechanical point of view, this kind of superluminal behavior is not surprising in the case of the tunneling phenomenon considered here. Since this phenomenon is fundamentally probabilistic in nature, there is no deterministic way of controlling whether any given tunneling event will occur or not. Hence there is no possibility of sending a controllable signal faster than c by means of any tunneling particle, including the photon.

It may seem paradoxical that a particle of light can, in some sense, travel faster than light, but we must remember that it is not logically impossible for a particle of light in a medium to travel faster than a particle of light in the vacuum. Nevertheless, it behooves us to discuss the fundamental questions raised by these kinds of counterintuitive superluminal phenomena concerning the meaning of causality in quantum optics. This will be done in more detail below.

The second caveat is this: it would seem that the above data would rule out all theories of the tunneling time other than Wigner's, but this is not so. One can only say that for the specific operational method used to obtain the data shown in Fig. 10.12, Wigner's theory is singled out as the closest to being correct. However, by using a different operational method which employs different experimental conditions to measure a physical quantity—such as the time of interaction of a tunneling particle with a modulated barrier, as was suggested by Büttiker and Landauer (1982)—one will obtain a different result from Wigner's. One striking difference between the predictions of these two particular theories of tunneling times is that in Wigner's theory, the group-delay tunneling time is predicted to be independent of barrier thickness in the case of thick barriers, whereas in Büttiker and Landauer's theory, their interaction tunneling time is predicted to be linearly dependent upon barrier thickness. A linear dependence upon the thickness of a tunnel barrier has indeed been measured for one of the two tunneling times observed by Balcou and Dutriaux (1997), who used a 2D tunnel barrier based on the phenomenon of frustrated total internal reflection between two closely spaced glass prisms. Thus in Balcou and Dutriaux's experiment, the existence of Büttiker and Landauer's interaction tunneling time has in fact been established. For a more detailed review of these and yet other tunneling times, wave propagation speeds, and superluminal effects, see Chiao and Steinberg (1997).

The conflicts between the predictions of the various tunneling-time theories discussed above illustrate the fact that the interpretation of measurements in quantum theory may depend sensitively upon the exact operational conditions used in a given experiment, as was emphasized early on by Bohr. Hence it should not surprise us that the operationalism principle introduced at the beginning of this chapter must always be carefully taken into account in any treatment of these problems. More concretely, the phrase 'the tunneling time' is meaningless unless it is accompanied by a precise operational description of the measurement to be performed.

10.5 The meaning of causality in quantum optics*

The appearance of counterintuitive, superluminal tunneling times in the above experiments necessitates a careful re-examination of what is meant by causality in the context of quantum optics. We begin by reviewing the notion of causality in classical electromagnetic theory. In Section 8.1, we have seen that the interaction of a classical electromagnetic wave with any linear optical device—including a tunnel barrier—can be described by a scattering matrix. We will simplify the discussion by only considering planar waves, e.g. superpositions of plane waves with all propagation vectors directed along the z-axis. An incident classical, planar wave $\mathcal{E}_{\text{in}}(z,t)$ propagating in vacuum is a function of the retarded time $t_r = t - z/c$ only; therefore we replace $\mathcal{E}_{\text{in}}(z,t)$ by $\mathcal{E}_{\text{in}}(t_r)$. This allows the incident field to be expressed as a one-dimensional Fourier integral transform:

$$\mathcal{E}_{\text{in}}(t_r) = \int_{-\infty}^{\infty} \frac{d\omega}{2\pi} \mathcal{E}_{\text{in}}(\omega) e^{-i\omega t_r} . \tag{10.105}$$

The output wave, also propagating in vacuum, is described in the same way by a function $\mathcal{E}_{\text{out}}(\omega)$ that is related to $\mathcal{E}_{\text{in}}(\omega)$ by

$$\mathcal{E}_{\text{out}}(\omega) = S(\omega)\mathcal{E}_{\text{in}}(\omega) , \tag{10.106}$$

where $S(\omega)$ is the scattering matrix—or transfer function—for the device in question. The transfer function $S(\omega)$ describes the reshaping of the input wave packet to produce the output wave packet. By means of the convolution theorem, we can transform the frequency-domain relation (10.106) into the time-domain relation

$$\mathcal{E}_{\text{out}}(t_r) = \int_{-\infty}^{+\infty} S(\tau)\mathcal{E}_{\text{in}}(t_r - \tau)d\tau , \tag{10.107}$$

where

$$S(\tau) = \int_{-\infty}^{\infty} \frac{d\omega}{2\pi} S(\omega)e^{-i\omega\tau} . \tag{10.108}$$

The fundamental principle of causality states that no effect can ever precede its cause. This implies that the transfer function must strictly vanish for all negative delays, i.e.

$$S(\tau) = 0 \quad \text{for all } \tau < 0 . \tag{10.109}$$

Therefore, the range of integration in eqn (10.107) is restricted to positive values, so that

$$\mathcal{E}_{\text{out}}(t_r) = \int_{0}^{\infty} S(\tau)\mathcal{E}_{\text{in}}(t_r - \tau)d\tau . \tag{10.110}$$

Thus we reach the intuitively appealing conclusion that the output field at time t_r can only depend on values of the input field in the past. In particular, if the input signal has a **front** at $t_r = 0$, that is

$$\mathcal{E}_{\text{in}}(t_r) = 0 \quad \text{for all } t_r < 0 \text{ (or equivalently } z > ct) , \tag{10.111}$$

then it follows from eqn (10.110) that

$$\mathcal{E}_{\text{out}}(t_r) = 0 \quad \text{for all } t_r < 0 . \tag{10.112}$$

Thus the classical meaning of causality for linear optical systems is that the reshaping, by whatever mechanism, of the input wave packet to produce the output wave packet cannot produce a nonvanishing output signal before the arrival of the input signal front at the output face.

In the quantum theory, one replaces the classical electric field amplitudes by time-dependent, positive-frequency electric field operators in the Heisenberg picture. By virtue of the correspondence principle, the linear relation between the classical input and output fields must also hold for the field operators, so that

$$\mathbf{E}_{\text{out}}^{(+)}(t_r) = \int_0^{+\infty} S(\tau)\mathbf{E}_{\text{in}}^{(+)}(t_r - \tau)d\tau . \tag{10.113}$$

One new feature in the quantum version is that the frequency ω in $S(\omega)$ is now interpreted in terms of the Einstein relation $E = \hbar\omega$ for the photon energy. Another important change is in the definition of a signal front. We have already learnt that field operators cannot be set to zero; consequently, the statement that the input signal has a front must be reinterpreted as an assumption about the quantum state of the field. The quantum version of eqn (10.111) is, therefore,

$$\mathbf{E}_{\text{in}}^{(+)}(t_r)\rho = 0 \quad \text{for all } t_r < 0 , \tag{10.114}$$

where ρ is the time-independent density operator describing the state of the system in the Heisenberg picture. It therefore follows from eqn (10.113) that

$$\mathbf{E}_{\text{out}}^{(+)}(t_r)\rho = 0 \quad \text{for all } t_r < 0 . \tag{10.115}$$

The physics behind this statement is that if the system starts off in the vacuum state at $t = 0$ at the input, nothing that the optical system can do to it can promote it out of the vacuum state at the output, *before* the arrival of the front. Therefore, causality has essentially similar meanings at the classical and the quantum levels of description of linear optical systems.

10.6 Interaction-free measurements*

A familiar procedure for determining if an object is present in a given location is to illuminate the region with a beam of light. By observing scattering or absorption of the light by the object, one can detect its presence or determine its absence; consequently, the first step in locating an object in a dark room is to turn on the light. Thus in classical optics, the *interaction* of light with the object would seem to be necessary for its observation. One of the strange features of quantum optics is that it is sometimes possible to determine an object's presence or absence without interacting with the object. The idea of interaction-free measurements was first suggested by Elitzur and Vaidman (1993), and it was later dubbed 'quantum seeing in the dark' (Kwiat *et al.*, 1996). A useful way to think about this phenomenon is to realize that **null events**—e.g. a detector does *not* click during a given time window—can convey information just as much as the positive events in which a click does occur.

When it is certain that there is one and only one photon inside an interferometer, some very counterintuitive nonlocal quantum effects—including interaction-free measurements—are possible. In an experiment performed in 1995 (Kwiat *et al.*, 1995*a*), this aim was achieved by pumping a lithium-iodate crystal with a 351 nm wavelength ultraviolet laser, in order to produce entangled photon pairs by spontaneous down-conversion. As shown in Fig. 10.13, one member of the pair, the *gate* photon, is directed to a silicon avalanche photodiode T, and the signal from this detector is used to

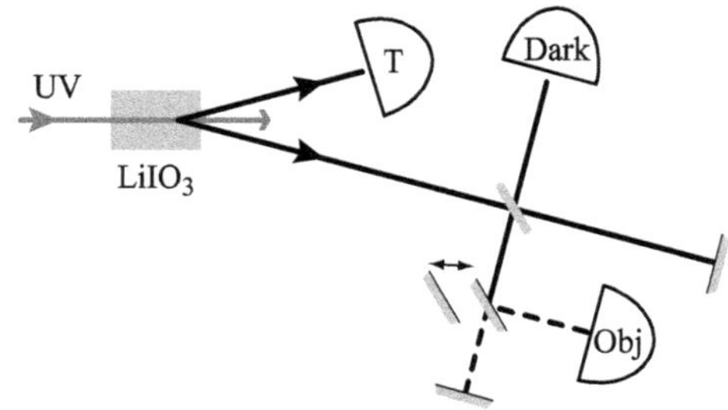

Fig. 10.13 Schematic of an experiment using a down-conversion source to demonstrate one form of interaction-free measurement. The object to be detected is represented by a translatable 100% mirror, with translation denoted by the double-arrow symbol ↔. (Reproduced from Kwiat *et al.* (1995*a*).)

open the gate for the other detectors. The other member of the pair, the *test* photon, is injected into a Michelson interferometer, which is prepared in a dark fringe near the equal-path length, white-light fringe condition; see Exercise 10.6. Thus the detector *Dark* at the output port of the Michelson is a *dark fringe* detector. It will never register any counts at all, if both arms of the interferometer are unblocked. However, the presence of an absorbing or nontransmitting object in the lower arm of the Michelson completely changes the possible outcomes by destroying the destructive interference leading to the dark fringe.

In the real experimental protocol, the unknown object is represented by a translatable, 100% reflectivity mirror. In the original Elitzur–Vaidman thought experiment, this role is played by a 100%-sensitivity detector that triggers a bomb. This raises the stakes,[2] but does not alter the physical principles involved. When the mirror blocks the lower arm of the interferometer in the real experiment, it completely deflects the test photon to the detector *Obj*. A click in *Obj* is the signal that the blocking object is present. When the mirror is translated out of the lower arm, the destructive interference condition is restored, and the test photon never shows up at the *Dark* detector.

For a central Michelson beam splitter with (intensity) reflectivity R and transmissivity $T = 1 - R$ (neglecting losses), an incident test photon will be sent into the lower arm with probability R. If the translatable mirror is present in the lower arm, the photon is deflected into the detector *Obj* with unit probability; therefore, the probability of absorption is

$$P\,(\text{absorption}) = P(\text{failure}) = R\,. \tag{10.116}$$

This is not as catastrophic as the exploding bomb, but it still represents an unsuccessful outcome of the interaction-free measurement attempt. However, there is also a mutually exclusive possibility that the test photon will be transmitted by the central beam splitter, with probability T, and—upon its return—reflected by the beam splitter, with probability R, to the *Dark* detector. Thus clicks at the *Dark* port occur with probability RT. When a *Dark* click occurs there is no possibility that the test photon was absorbed by the object—the bomb did not go off—since there was only a single photon in the system at the time. Hence, the probability of a successful interaction-free measurement of the presence of the object is

[2]One of the virtues of thought experiments is that they are not subject to health and safety inspections.

$$P(\text{detection}) = P(\text{success}) = RT\,. \tag{10.117}$$

For a lossless Michelson interferometer, the fraction η of successful interaction-free measurements is therefore

$$\begin{aligned}\eta &\equiv \frac{P\,(\text{success})}{P\,(\text{success}) + P\,(\text{failure})} = \frac{P\,(\text{detection})}{P\,(\text{detection}) + P\,(\text{absorption})} \\ &= \frac{RT}{RT+R} = \frac{1-R}{2-R}\,,\end{aligned} \tag{10.118}$$

which tends to an upper limit of 50% as R approaches zero.

This quantum effect is called an **interaction-free measurement**, because the single photon injected into the interferometer did not interact at all—either by absorption or by scattering—with the object, and yet we can infer its presence by means of the absence of any interaction with it. Furthermore, the inference of presence or absence can be made with complete certainty based on the principle of the indivisibility of the photon, since the same photon could not both have been absorbed by the object and later caused the click in the dark detector. Actually, it is Bohr's wave–particle complementarity principle that plays a central role in this kind of measurement. In the absence of the object, it is the wave-like nature of light that ensures—through destructive interference—that the photon never exits through the dark port. In the presence of the object, it is the particle-like nature of the light—more precisely the indivisibility of the quantum of light—which enforces the mutual exclusivity of a click at the dark port or absorption by the object.

Thus a null event—here the absence of a click at *Obj*—constitutes just as much of a measurement in quantum mechanics as the observation of a click. This feature of quantum theory was already emphasized by Renninger (1960) and by Dicke (1981), but its implementation in quantum interference was first pointed out by Elitzur and Vaidman. Note that this effect is nonlocal, since one can determine remotely the presence or the absence of the unknown object, by means of an arbitrarily remote dark detector. The fact that the entire interferometer configuration must be set up ahead of time in order to see this nonlocal effect is another example of the general principle in Bohr's Delphic remark quoted in Section 10.3.3.

The data in Fig. 10.14 show that the fraction of successful measurements is nearly 50%, in agreement with the theoretical prediction given by eqn (10.118). By technical refinements of the interferometer, the probability of a successful interaction-free measurement could, in principle, be increased to as close to 100% as desired (Kwiat *et al.*, 1995*a*). A success rate of $\eta = 73\%$ has already been demonstrated (Kwiat *et al.*, 1999*a*). In the 100% success-rate limit, one could determine the presence or absence of an object with minimal absorption of photons.

This possibility may have important practical applications. In an extension of this interaction-free measurement method to 2D imaging, one could use an array of these devices to map out the silhouette of an unknown object, while restricting the number of absorbed photons to as small a value as desired. In conjunction with X-ray interferometers—such as the Bonse–Hart type—this would, for example, allow X-ray pictures of the bones of a hand to be taken with an arbitrarily low X-ray dosage.

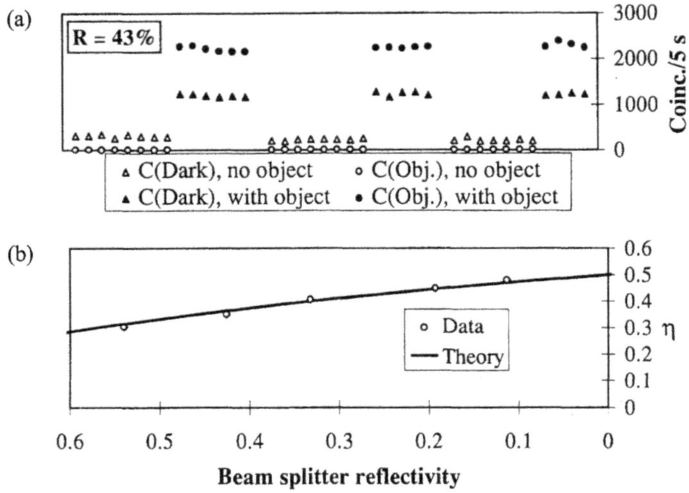

Fig. 10.14 (a) Data demonstrating interaction-free measurement. The Michelson beam splitter reflectivity for the upper set of data was 43%. (b) Data and theoretical fit for the figure of merit η as a function of beam splitter reflectivity. (Reproduced from Kwiat *et al.* (1995*a*).)

10.7 Exercises

10.1 Vacuum fluctuations

Drop the term $\mathbf{E}_3^{(+)}(\mathbf{r},t)$ from the expression (10.7) for $\mathbf{E}_{\text{out}}^{(+)}(\mathbf{r},t)$ and evaluate the equal-time commutator $\left[\mathbf{E}_{\text{out},i}^{(+)}(\mathbf{r},t), \mathbf{E}_{\text{out},j}^{(-)}(\mathbf{r}',t)\right]$. Compare this to the correct form in eqn (3.17) and show that restoring $\mathbf{E}_3^{(+)}(\mathbf{r},t)$ will repair the flaw.

10.2 Classical model for two-photon interference

Construct a semiclassical model for two-photon interference, along the lines of Section 1.4, by assuming: the down-conversion mechanism produces classical amplitudes $\alpha_{\sigma n} = \sqrt{I_{\sigma n}}\exp(i\theta_{\sigma n})$, where $\sigma = \text{sig, idl}$ is the channel index and the gate windows are labeled by $n = 1, 2, \ldots$; the phases $\theta_{\sigma n}$ vary randomly over $(0, 2\pi)$; the phases and intensities $I_{\sigma n}$ are statistically independent; the intensities $I_{\sigma n}$ for the two channels have the same average and *rms* deviation.

Evaluate the coincidence-count probability p_{coinc} and the singles probabilities p_{sig} and p_{idl}, and thus derive the inequality (10.41).

10.3 The HOM dip*

Assume that the function $|g(\nu)|^2$ in eqn (10.69) is a Gaussian:

$$|g(\nu)|^2 = \left(\tau_2/\sqrt{\pi}\right)\exp\left(-\tau_2^2\nu^2\right).$$

Evaluate and plot $P_{\text{coinc}}(\Delta t)$.

10.4 HOM by scattering theory*

(1) Apply eqn (8.76) to eqn (10.71) to derive eqn (10.72).

(2) Use the definition (6.96) to obtain a formal expression for the coincidence-counting detection amplitude, and then use the rule (9.96) to show that $|\Phi_{\text{pair}}\rangle$ will not contribute to the coincidence-count rate.

10.5 Anti-HOM*

Consider the two-photon state given by eqn (10.48), where $C(\omega, \omega')$ satisfies the $(-)$-version of eqn (10.51).

(1) Why does $C(\omega, \omega') = -C(\omega', \omega)$ not violate Bose symmetry?

(2) Assume that $C(\omega, \omega')$ satisfies eqn (10.56) and the $(-)$-version of eqn (10.51). Use eqns (10.71)–(10.74) to conclude that the photons in this case behave like fermions, i.e. the pairing behavior seen in the HOM interferometer is forbidden.

10.6 Interaction-free measurements*

(1) Work out the relation between the lengths of the arms of the Michelson interferometer required to ensure that a dark fringe occurs at the output port.

(2) Explain why the probabilities $P\,(\text{failure})$ and $P\,(\text{success})$, respectively defined by eqns (10.116) and (10.117), do not sum to one.

11

Coherent interaction of light with atoms

In Chapter 4 we used perturbation theory to describe the interaction between light and matter. In addition to the assumption of weak fields—i.e. the interaction energy is small compared to individual photon energies—perturbation theory is only valid for times in the interval $1/\omega_0 \ll t \ll 1/W$, where ω_0 and W are respectively the unperturbed frequency and the perturbative transition rate for the system under study. When ω_0 is an optical frequency, the lower bound is easily satisfied, but the upper bound can be violated. Let ρ be a stationary density matrix for the field; then the field–field correlation function, for a fixed spatial point $\mathbf{r}$ but two different times, will typically decay exponentially:

$$G_{ij}^{(1)}(\mathbf{r}, t_1; \mathbf{r}, t_2) = \mathrm{Tr}\left[\rho E_i^{(-)}(\mathbf{r}, t_1) E_j^{(+)}(\mathbf{r}, t_2)\right] \sim \exp\left(-|t_1 - t_2|/T_c\right), \tag{11.1}$$

where T_c is the **coherence time** for the state ρ. For some states, e.g. the Planck distribution, the coherence time is short, in the sense that $T_c \ll 1/W$. Perturbation theory is applicable to these states, but there are many situations—in particular for laser fields—in which $T_c > 1/W$. Even though the field is weak, perturbation theory cannot be used in these cases; therefore, we need to develop nonperturbative methods that are applicable to weak fields with long coherence times.

11.1 Resonant wave approximation

The phenomenon of resonance is ubiquitous in physics and it plays a central role in the interaction of light with atoms. Resonance will occur if there is an allowed atomic transition $q \to p$ with transition frequency $\omega_{qp} = (\varepsilon_q - \varepsilon_p)/\hbar$ and a matching optical frequency $\omega \approx \omega_{qp}$. In Section 4.9.2 we saw that the weak-field condition can be expressed as $\Omega \ll \omega_0$, where Ω is the characteristic Rabi frequency defined by eqn (4.147). In the interaction picture, the state vector satisfies the Schrödinger equation (4.94), in which the full Hamiltonian is replaced by the interaction Hamiltonian; consequently,

$$i\frac{\partial}{\partial t}|\Psi(t)\rangle \sim \Omega |\Psi(t)\rangle\,. \tag{11.2}$$

Thus the weak-field condition tells us that the changes in the interaction-picture state vector occur on the time scale $1/\Omega \gg 1/\omega_0$. Consequently, the state vector does not change appreciably over an optical period. This disparity in time scales is the basis for a nonperturbative approximation scheme. In the interests of clarity, we will first develop this method for a simple model called the two-level atom.

11.1.1 Two-level atoms

The spectra of real atoms and the corresponding sets of stationary states display a daunting complexity, but there are situations of theoretical and practical interest in which this complexity can be ignored. In the simplest case, the atomic state vector is a superposition of only two of the stationary states. Truncated models of this kind are called **two-level atoms**. This simplification can occur when the atom interacts with a narrow band of radiation that is only resonant with a transition between two specific energy levels. In this situation, the two atomic states involved in the transition are the only dynamically active degrees of freedom, and the probability amplitudes for all the other stationary states are negligible.

In the semiclassical approximation, the Feynman–Vernon–Hellwarth theorem (Feynman *et al.*, 1957) shows that the dynamical equations for a two-level atom are isomorphic to the equations for a spin-1/2 particle in an external magnetic field. This provides a geometrical picture which is useful for visualizing the solutions. The general zeroth-order Hamiltonian for the fictitious spin system is $H_0 = -\mu \boldsymbol{\mathcal{B}} \cdot \boldsymbol{\sigma}$, and we will choose the fictitious $\boldsymbol{\mathcal{B}}$-field as $\boldsymbol{\mathcal{B}} = -B\mathbf{u}_3$, so that the spin-up state is higher in energy than the spin-down state.

To connect this model to the two-level atom, let the two resonantly connected atomic states be $|\varepsilon_1\rangle$ and $|\varepsilon_2\rangle$, with $\varepsilon_1 < \varepsilon_2$. The atomic Hilbert space is effectively truncated to the two-dimensional space spanned by $|\varepsilon_1\rangle$ and $|\varepsilon_2\rangle$, so the atomic Hamiltonian and the atomic dipole operator $\widehat{\mathbf{d}}$ are represented by 2×2 matrices. Every 2×2 matrix can be expressed in terms of the standard Pauli matrices; in particular, the truncated atomic Hamiltonian is

$$H_{\text{at}} = \begin{bmatrix} \varepsilon_2 & 0 \\ 0 & \varepsilon_1 \end{bmatrix} = \frac{\varepsilon_2 + \varepsilon_1}{2} I_2 + \frac{\hbar\omega_{21}}{2}\sigma_z \,, \tag{11.3}$$

where I_2 is the 2×2 identity matrix and $\hbar\omega_{21} = \varepsilon_2 - \varepsilon_1$. The term proportional to I_2 can be eliminated by choosing the zero of energy so that $\varepsilon_2 + \varepsilon_1 = 0$. This enforces the relation $\mu\mathcal{B} \leftrightarrow \hbar\omega_{21}/2$ between the two-level atom and the fictitious spin.

When the very small effects of weak interactions are ignored, atomic states have definite parity; therefore, the odd-parity operator $\widehat{\mathbf{d}}$ has no diagonal matrix elements. For the two-level atom, this implies $\widehat{\mathbf{d}} = \mathbf{d}^*\sigma_- + \mathbf{d}\,\sigma_+$, where $\mathbf{d} = \langle\varepsilon_2|\widehat{\mathbf{d}}|\varepsilon_1\rangle$, σ_+ is the spin-raising operator, and σ_- is the spin-lowering operator. Combining this with the decomposition $\mathbf{E} = \mathbf{E}^{(+)} + \mathbf{E}^{(-)}$ and the plane-wave expansion (3.69) for $\mathbf{E}^{(+)}$ leads to

$$H_{\text{int}} = H_{\text{int}}^{(\text{r})} + H_{\text{int}}^{(\text{ar})} \,, \tag{11.4}$$

$$\begin{aligned} H_{\text{int}}^{(\text{r})} &= -\mathbf{d}\cdot\mathbf{E}^{(+)}\sigma_+ - \mathbf{d}^*\cdot\mathbf{E}^{(-)}\sigma_- \\ &= -i\hbar\sum_{\mathbf{k}s}\left[\sqrt{\frac{\hbar\omega_k}{2\epsilon_0 V}}\frac{\mathbf{d}\cdot\mathbf{e}_{\mathbf{k}s}}{\hbar}\right] a_{\mathbf{k}s}\sigma_+ + \text{HC}\,, \end{aligned} \tag{11.5}$$

$$\begin{aligned} H_{\text{int}}^{(\text{ar})} &= -\mathbf{d}\cdot\mathbf{E}^{(-)}\sigma_+ - \mathbf{d}^*\cdot\mathbf{E}^{(+)}\sigma_- \\ &= -i\hbar\sum_{\mathbf{k}s}\left[\sqrt{\frac{\hbar\omega_k}{2\epsilon_0 V}}\frac{\mathbf{d}^*\cdot\mathbf{e}_{\mathbf{k}s}}{\hbar}\right] a_{\mathbf{k}s}\sigma_- + \text{HC}\,. \end{aligned} \tag{11.6}$$

In $H_{\text{int}}^{(\text{r})}$ the annihilation (creation) operator $a_{\mathbf{k}s}\left(a_{\mathbf{k}s}^{\dagger}\right)$ is paired with the energy-raising (-lowering) operator $\sigma_{+}\left(\sigma_{-}\right)$, while $H_{\text{int}}^{(\text{ar})}$ has the opposite pairings. In the perturbative calculations of Section 4.9.3 the emission (absorption) of a photon is associated with lowering (raising) the energy of the atom, subject to the resonance condition $\omega_k = \omega_{21}$, so $H_{\text{int}}^{(\text{r})}$ and $H_{\text{int}}^{(\text{ar})}$ are respectively called the **resonant** and **antiresonant** Hamiltonians.

The full Hamiltonian in the Schrödinger picture is

$$H = H_0 + H_{\text{int}}\,, \tag{11.7}$$

where

$$H_0 = \sum_{\mathbf{k}s} \hbar\omega_k a_{\mathbf{k}s}^{\dagger} a_{\mathbf{k}s} + \frac{\hbar\omega_{21}}{2}\sigma_z\,. \tag{11.8}$$

In the interaction picture, the operators satisfy the uncoupled equations of motion

$$i\hbar\frac{\partial}{\partial t}a_{\mathbf{k}s}(t) = \left[a_{\mathbf{k}s}(t)\,, H_0\right] = \hbar\omega_k a_{\mathbf{k}s}(t)\,, \tag{11.9}$$

$$i\hbar\frac{\partial}{\partial t}\sigma_z(t) = \left[\sigma_z(t)\,, H_0\right] = 0\,, \tag{11.10}$$

$$i\hbar\frac{\partial}{\partial t}\sigma_{\pm}(t) = \left[\sigma_{\pm}(t)\,, H_0\right] = \mp\frac{\hbar\omega_{21}}{2}\sigma_{\pm}(t)\,, \tag{11.11}$$

with the solution

$$a_{\mathbf{k}s}(t) = a_{\mathbf{k}s}e^{-i\omega_k t}\,,\quad \sigma_z(t) = \sigma_z\,,\quad \sigma_{\pm}(t) = e^{\pm i\omega_{21}t}\sigma_{\pm}\,, \tag{11.12}$$

where $a_{\mathbf{k}s}$, σ_z, and $\sigma_{\pm}$ are the Schrödinger-picture operators. Thus the time dependence of the operators is explicitly expressed in terms of the atomic transition frequency ω_{21} and the optical frequencies ω_k. This is a great advantage for the calculations to follow.

The interaction-picture state vector $|\Theta(t)\rangle$ satisfies the Schrödinger equation

$$i\hbar\frac{\partial}{\partial t}\left|\Theta(t)\right\rangle = H_{\text{int}}(t)\left|\Theta(t)\right\rangle\,, \tag{11.13}$$

where

$$H_{\text{int}}(t) = H_{\text{int}}^{(\text{r})}(t) + H_{\text{int}}^{(\text{ar})}(t)\,, \tag{11.14}$$

$$H_{\text{int}}^{(\text{r})}(t) = -i\hbar\sum_{\mathbf{k}s}\left[\sqrt{\frac{\hbar\omega_k}{2\epsilon_0 V}}\frac{\mathbf{d}\cdot\mathbf{e}_{\mathbf{k}s}}{\hbar}\right]e^{i(\omega_{21}-\omega_k)t}a_{\mathbf{k}s}\sigma_{+} + \text{HC} \tag{11.15}$$

and

$$H_{\text{int}}^{(\text{ar})}(t) = -i\hbar\sum_{\mathbf{k}s}\left[\sqrt{\frac{\hbar\omega_k}{2\epsilon_0 V}}\frac{\mathbf{d}^*\cdot\mathbf{e}_{\mathbf{k}s}}{\hbar}\right]e^{-i(\omega_{21}+\omega_k)t}a_{\mathbf{k}s}\sigma_{-} + \text{HC} \tag{11.16}$$

are obtained by replacing the operators in eqns (11.5) and (11.6) by the explicit solutions in eqn (11.12).

11.1.2 Time averaging

The slow and fast time scales can be separated explicitly by means of a temporal filtering operation, like the one introduced in Section 9.1.2-C to describe narrowband detection. We use an averaging function, $\varpi(t)$, satisfying eqns (9.35)–(9.37), to define running averages by

$$\overline{f}(t) \equiv \int_{-\infty}^{\infty} dt' \varpi(t-t') f(t') = \int_{-\infty}^{\infty} dt' \varpi(t') f(t+t') . \tag{11.17}$$

The temporal width ΔT defined by eqn (9.37) will now be renamed the **memory interval** T_{mem}. The idea behind this new language is that the temporally coarse-grained picture imposed by averaging over the time scale T_{mem} causes *amnesia*, i.e. averaged operators at time t will not be correlated with averaged operators at an earlier time, $t' < t - T_{\text{mem}}$. The average in eqn (11.17) washes out oscillations with periods smaller than T_{mem}, and the average of the derivative is the derivative of the average:

$$\overline{\frac{df}{dt}}(t) = \frac{d}{dt}\overline{f}(t) . \tag{11.18}$$

The separation of the two time scales is enforced by imposing the condition

$$\frac{1}{\omega_{21}} \ll T_{\text{mem}} \ll \frac{1}{\Omega} \tag{11.19}$$

on T_{mem}. A function $g(t)$ that varies on the time scale $1/\Omega$ is essentially constant over the averaging interval, so that

$$\overline{g}(t) \equiv \int_{-\infty}^{\infty} dt' \varpi(t-t') g(t') \approx g(t) . \tag{11.20}$$

The combination of this feature with the normalization condition (9.36) leads to the following rule:

$$\varpi(t-t') \approx \delta(t-t') \quad \text{when applied to slowly-varying functions} . \tag{11.21}$$

It is also instructive to describe the averaging procedure in the frequency domain. We would normally denote the Fourier transform of $\varpi(t)$ by $\varpi(\omega)$, but this particular function plays such an important role in the theory that we will honor it with a special name:

$$K(\omega) = \int_{-\infty}^{\infty} dt \; \varpi(t) \; e^{i\omega t} . \tag{11.22}$$

The properties of $\varpi(t)$ guarantee that $K(\omega)$ is real and even, $K^*(\omega) = K(-\omega) = K(\omega)$, and that it has a finite width, w_K, related to the averaging interval by $w_K \sim 1/T_{\text{mem}}$. The frequency-domain conditions corresponding to eqn (11.19) are

$$\Omega \ll w_K \ll \omega_{21} , \tag{11.23}$$

and the time-domain normalization condition (9.36) implies $K(0) = 1$. Performing the Fourier transform of eqn (11.17) gives the frequency-domain description of the averaging procedure as $\overline{f}(\omega) = K(\omega) f(\omega)$. Thus for small frequencies, $\omega \ll w_K$, the original function $f(\omega)$ is essentially unchanged, but frequencies larger than the width w_K are strongly suppressed. For this reason $K(\omega)$ is called the **cut-off function.**[1]

11.1.3 Time-averaged Schrödinger equation

Since $|\Theta(t)\rangle$ only varies on the slow time scale, the rule (11.21) tells us that it is effectively unchanged by the running average, i.e. $|\Theta(t)\rangle \approx \overline{|\Theta(t)\rangle}$. Consequently, averaging the Schrödinger equation (11.13), with the help of eqn (11.18), yields the approximate equation

$$i\hbar \frac{\partial}{\partial t} |\Theta(t)\rangle = \overline{H}_{\text{int}}(t) |\Theta(t)\rangle . \tag{11.24}$$

According to eqn (11.16), all terms in $H_{\text{int}}^{(\text{ar})}(t)$ are rapidly oscillating; therefore, we expect that $\overline{H}_{\text{int}}^{(\text{ar})}(t) \approx 0$. This expectation is justified by the explicit calculation in Exercise 11.1, which shows that the cut-off function in each term of $\overline{H}_{\text{int}}^{(\text{ar})}(t)$ is evaluated with its argument on the optical scale. In the **resonant wave approximation** (RWA), the antiresonant part is discarded, i.e. the full interaction Hamiltonian $H_{\text{int}}(t)$ is replaced by the resonant part $\overline{H}_{\text{int}}^{(\text{r})}(t)$. The traditional name, *rotating wave approximation*, is suggested by the mathematical similarity between the two-level atom and a spin-1/2 particle precessing in a magnetic field (Yariv, 1989, Chap. 15).

Turning next to the expression (11.15) for $\overline{H}_{\text{int}}^{(\text{r})}(t)$, we see that the exponentials involve the detuning $\Delta_k = \omega_k - \omega_{21}$ which will be small near resonance; therefore, the average of $H_{\text{int}}^{(\text{r})}(t)$ will not vanish. The explicit calculation gives

$$H_{\text{rwa}}(t) \equiv \overline{H}_{\text{int}}^{(\text{r})}(t) = -i\hbar \sum_{\mathbf{k}s} \mathfrak{g}_{\mathbf{k}s} e^{-i\Delta_k t} \sigma_+ a_{\mathbf{k}s} + \text{HC} , \tag{11.25}$$

where

$$\mathfrak{g}_{\mathbf{k}s} = K(\Delta_k) \left[\sqrt{\frac{\hbar \omega_k}{2\epsilon_0 V}} \frac{\mathbf{d} \cdot \mathbf{e}_{\mathbf{k}s}}{\hbar} \right] , \tag{11.26}$$

and we have introduced the new notation $H_{\text{rwa}}(t)$ as a reminder of the approximation in use. The cut-off function in the definition of the coupling constant guarantees that only terms satisfying the **resonance condition** $|\omega_{21} - \omega_k| < w_K$ will contribute to H_{rwa}.

With the resonant wave approximation in force, we can transform to the Schrödinger picture by the simple expedient of omitting the time-dependent exponentials in eqn (11.25). Thus the RWA Hamiltonian in the Schrödinger picture is

$$\begin{aligned} H_{\text{rwa}} &= H_0 - \mathbf{d} \cdot \mathbf{E}^{(+)} \sigma_+ - \mathbf{d}^* \cdot \mathbf{E}^{(-)} \sigma_- \\ &= H_0 - i\hbar \sum_{\mathbf{k}s} \mathfrak{g}_{\mathbf{k}s} a_{\mathbf{k}s} \sigma_+ + i\hbar \sum_{\mathbf{k}s} \mathfrak{g}_{\mathbf{k}s}^* a_{\mathbf{k}s}^\dagger \sigma_- , \end{aligned} \tag{11.27}$$

where H_0 is given by eqn (11.8). This observation provides the following general scheme for defining the resonant wave approximation directly in the Schrödinger picture.

[1]This is physics jargon. An engineer would probably call $K(\omega)$ a low-pass filter.

(1) Discard all terms in H_{int} that do not conserve energy in a first-order perturbation calculation.

(2) Multiply the coupling constants in the remaining terms by the cut-off function $K(\Delta_k)$.

It is also useful to note that this rule mandates that each term in H_{rwa} is the product of an energy-raising (-lowering) operator for the atom with an energy-lowering (-raising) operator for the field. We emphasize that the discarded part, $H^{(\text{ar})}$, is not unphysical; it simply does not contribute to the first-order transition amplitude. The antiresonant Hamiltonian $H^{(\text{ar})}$ can and does contribute in higher orders of perturbation theory, but the time averaging argument shows that H_{rwa} is the dominant part of the Hamiltonian for long-term evolution under the influence of weak fields.

11.1.4 Multilevel atoms

Our object is this section is to introduce a family of operators that play the role of the Pauli matrices for an atom with more than two active levels. We will only consider the interaction of the field with a single atom, since the generalization to the many-atom case is straightforward. The **atomic transition operators** S_{qp} are defined by

$$S_{qp} = |\varepsilon_q\rangle\langle\varepsilon_p|\,, \tag{11.28}$$

where $|\varepsilon_q\rangle$ and $|\varepsilon_p\rangle$ are eigenstates of H_{at}. As explained in Appendix C.1.2, this notation means that the operator S_{qp} projects any atomic state $|\Psi\rangle$ onto $|\varepsilon_q\rangle$ with coefficient $\langle\varepsilon_p|\Psi\rangle$, i.e.

$$S_{qp}|\Psi\rangle = |\varepsilon_q\rangle\langle\varepsilon_p|\Psi\rangle\,. \tag{11.29}$$

When this definition is applied to the two-level case, it is easy to see that $S_{21} = \sigma_+$, $S_{12} = \sigma_-$, and $S_{22} - S_{11} = \sigma_z$. The energy eigenvalue equation for the states, $H_{\text{at}}|\varepsilon_q\rangle = \varepsilon_q|\varepsilon_q\rangle$, implies the operator eigenvalue equation $[S_{qp}, H_{\text{at}}] = -\hbar\omega_{qp}S_{qp}$ for S_{qp}, so the transition operators are sometimes called **eigenoperators**.

The eigenstates $|\varepsilon_q\rangle$ of H_{at} satisfy the completeness relation

$$\sum_q |\varepsilon_q\rangle\langle\varepsilon_q| = I_A\,, \tag{11.30}$$

where I_A is the identity operator in $\mathfrak{H}_A$; therefore,

$$O \equiv I_A O I_A = \sum_q \sum_p \langle\varepsilon_q|O|p\rangle\, S_{qp}\,. \tag{11.31}$$

Thus the S_{qp}s form a complete set for the expansion of any atomic operator, just as every 2×2 matrix can be expressed as a linear combination of Pauli matrices.

The algebraic properties

$$S_{qp} = S_{pq}^{\dagger}\,, \tag{11.32}$$

$$S_{qp}S_{q'p'} = \delta_{pq'}S_{qp'}\,, \tag{11.33}$$

$$[S_{qp}, S_{q'p'}] = \{\delta_{pq'}S_{qp'} - \delta_{p'q}S_{q'p}\} \tag{11.34}$$

are readily derived by using the orthogonality of the eigenstates. The special case $q = p$ and $q' = p'$ of eqn (11.33) shows that the S_{qq}s are a set of orthogonal projection operators for the atom. For any atomic state $|\Psi\rangle$, eqn (11.29) yields $S_{qq}|\Psi\rangle = |\varepsilon_q\rangle\langle\varepsilon_q|\Psi\rangle$, i.e. S_{qq} projects out the $|\varepsilon_q\rangle$ component of $|\Psi\rangle$. The S_{qq}s are called **population** operators, since the expectation value,

$$\langle \Psi | S_{qq} | \Psi \rangle = |\langle \varepsilon_q | \Psi \rangle|^2 , \tag{11.35}$$

is the probability for finding the value ε_q, and the corresponding eigenstate $|\varepsilon_q\rangle$, in a measurement of the energy of an atom prepared in the state $|\Psi\rangle$. Because of the convention that $q > p$ implies $\varepsilon_q > \varepsilon_p$, the operator S_{qp} for $q > p$ is called a **raising operator**. It is analogous to the angular momentum raising operator, or to the creation operator $a^\dagger_{\mathbf{k}s}$ for a photon. By the same token, $S_{pq} = S^\dagger_{qp}$ is a **lowering operator**, analogous to the lowering operator for angular momentum, or to the photon annihilation operator $a_{\mathbf{k}s}$.

In this representation the atomic Hamiltonian in the Schrödinger picture has the simple form

$$H_{\text{at}} = \sum_q \varepsilon_q S_{qq} , \tag{11.36}$$

and the interaction Hamiltonian is given by

$$H_{\text{int}} = -\sum_{q,p} S_{qp} \mathbf{d}_{qp} \cdot \mathbf{E}(0) , \tag{11.37}$$

where $\mathbf{d}_{qp} = \langle \varepsilon_q | \widehat{\mathbf{d}} | \varepsilon_p \rangle$. Since $\mathbf{d}_{qq} = 0$, the sum over q and p splits into two parts with $q > p$ and $p > q$. Combining this with $\mathbf{E} = \mathbf{E}^{(+)} + \mathbf{E}^{(-)}$ leads to an expression involving four sums. After interchanging the names of the summation indices in the $q < p$ sums, the result can be arranged as follows:

$$H_{\text{int}} = H^{(\text{r})}_{\text{int}} + H^{(\text{ar})}_{\text{int}} , \tag{11.38}$$

$$\begin{aligned} H^{(\text{r})}_{\text{int}} &= -\sum_{q>p} S_{qp} \mathbf{d}_{qp} \cdot \mathbf{E}^{(+)}(\mathbf{0}) + \text{HC} , \\ H^{(\text{ar})}_{\text{int}} &= -\sum_{q>p} S_{qp} \mathbf{d}_{qp} \cdot \mathbf{E}^{(-)}(\mathbf{0}) + \text{HC} . \end{aligned} \tag{11.39}$$

In $H^{(\text{r})}_{\text{int}}$ the raising (lowering) operator S_{qp} (S_{pq}) is associated with the annihilation (creation) operator $\mathbf{E}^{(+)}$ ($\mathbf{E}^{(-)}$), while the opposite pairing appears in $H^{(\text{ar})}_{\text{int}}$. It is not necessary to carry out the explicit time averaging procedure; the results of the two-level problem have already provided us with a general rule for writing down the RWA Hamiltonian. Since all antiresonant terms are to be discarded, we can dispense with $H^{(\text{ar})}_{\text{int}}$ and set

$$H_{\text{rwa}} = -\sum_{q>p} S_{qp} \mathbf{d}_{qp} \cdot \mathbf{E}^{(+)}(\mathbf{0}) + \text{HC} . \tag{11.40}$$

Expanding the field operator in plane waves yields the equivalent form

$$H_{\text{rwa}} = -i\hbar \sum_{\mathbf{k}s} \sum_{q>p} \mathfrak{g}_{qp,\mathbf{k}s} S_{qp} a_{\mathbf{k}s} + \text{HC} , \tag{11.41}$$

where the coupling frequencies,

$$\mathfrak{g}_{qp,\mathbf{k}s} = \sqrt{\frac{\hbar\omega_k}{2\epsilon_0 V}} K\left(\omega_{qp} - \omega_k\right) \frac{\mathbf{d}_{qp} \cdot \mathbf{e}_{\mathbf{k}s}}{\hbar}, \tag{11.42}$$

include the cut-off function, so that only those terms satisfying a resonance condition $|\omega_{qp} - \omega_k| < w_K$ will contribute to the RWA interaction Hamiltonian. The Schrödinger-picture form in eqn (11.41) becomes

$$H_{\text{rwa}}(t) = -i\hbar \sum_{\mathbf{k}s} \sum_{q>p} \mathfrak{g}_{qp,\mathbf{k}s} e^{i(\omega_{qp}-\omega_k)t} S_{qp} a_{\mathbf{k}s} + \text{HC} \tag{11.43}$$

in the interaction picture.

11.2 Spontaneous emission II

11.2.1 Propagation of spontaneous emission

The discussion of spontaneous emission in Section 4.9.3 is concerned with the calculation of the rate of quantum jumps associated with the emission of a photon. This approach does not readily lend itself to answering other kinds of questions. For example, if an atom at the origin is prepared in its excited state at $t = 0$, what is the earliest time at which a detector located at a distance r can register the arrival of a photon? Questions of this kind are best answered by using the Heisenberg picture.

Since the Heisenberg, Schrödinger, and interaction pictures all coincide at $t = 0$, the interaction Hamiltonian in the Heisenberg picture can be inferred from eqn (11.25) by setting $t = 0$ in the exponentials. The total Hamiltonian in the resonant wave approximation is therefore

$$H = H_{\text{at}} + H_{\text{em}} + H_{\text{rwa}}, \tag{11.44}$$

$$H_{\text{at}} = \frac{\hbar\omega_{21}}{2}\sigma_z(t), \quad H_{\text{em}} = \sum_{\mathbf{k}s} \hbar\omega_k a^\dagger_{\mathbf{k}s}(t) a_{\mathbf{k}s}(t), \tag{11.45}$$

$$H_{\text{rwa}} = -i\hbar \sum_{\mathbf{k}s} \left(\mathfrak{g}_{\mathbf{k}s}\sigma_+(t) a_{\mathbf{k}s}(t) - \mathfrak{g}^*_{\mathbf{k}s}\sigma_-(t) a^\dagger_{\mathbf{k}s}(t) \right), \tag{11.46}$$

where the operators are all evaluated in the Heisenberg picture. The Heisenberg equations of motion,

$$\frac{d}{dt}\sigma_z(t) = -2\sum_{\mathbf{k}s} \left(\mathfrak{g}_{\mathbf{k}s}\sigma_+(t) a_{\mathbf{k}s}(t) + \mathfrak{g}^*_{\mathbf{k}s}\sigma_-(t) a^\dagger_{\mathbf{k}s}(t) \right), \tag{11.47}$$

$$\frac{d}{dt}\sigma_-(t) = -i\omega_{21}\sigma_-(t) + \sum_{\mathbf{k}s} \mathfrak{g}_{\mathbf{k}s} a_{\mathbf{k}s}(t)\sigma_z(t), \tag{11.48}$$

$$\frac{d}{dt}a_{\mathbf{k}s}(t) = -i\omega_k a_{\mathbf{k}s}(t) + \mathfrak{g}^*_{\mathbf{k}s}\sigma_-(t), \tag{11.49}$$

show that the field operators $a_{\mathbf{k}s}(t)$ and the atomic operators $\boldsymbol{\sigma}(t)$, which are independent at $t = 0$, are coupled at all later times. For this reason, it is usually impossible to obtain closed-form solutions.

Let us study the time dependence of the field emitted by an initially excited atom. In the Heisenberg picture, the plane-wave expansion (3.69) for the positive-frequency part of the field is

$$\mathbf{E}^{(+)}(\mathbf{r},t)=\sum_{\mathbf{k}s} i\sqrt{\frac{\hbar\omega_k}{2\epsilon_0 V}}a_{\mathbf{k}s}(t)\,\mathbf{e}_{\mathbf{k}s}e^{i\mathbf{k}\cdot\mathbf{r}}\,, \tag{11.50}$$

so we begin by using the standard integrating factor method to get the formal solution,

$$a_{\mathbf{k}s}(t)=a_{\mathbf{k}s}(0)\,e^{-i\omega_k t}+\mathfrak{g}^*_{\mathbf{k}s}\int_0^t dt' e^{-i\omega_k(t-t')}\sigma_-(t')\,, \tag{11.51}$$

of eqn (11.49). Substituting this into eqn (11.50) gives $\mathbf{E}^{(+)}(\mathbf{r},t)$ as the sum of two terms:

$$\mathbf{E}^{(+)}(\mathbf{r},t)=\mathbf{E}^{(+)}_{\text{vac}}(\mathbf{r},t)+\mathbf{E}^{(+)}_{\text{rad}}(\mathbf{r},t)\,, \tag{11.52}$$

where

$$\mathbf{E}^{(+)}_{\text{vac}}(\mathbf{r},t)=\sum_{\mathbf{k}s} i\sqrt{\frac{\hbar\omega_k}{2\epsilon_0 V}}a_{\mathbf{k}s}(0)\,\mathbf{e}_{\mathbf{k}s}e^{i(\mathbf{k}\cdot\mathbf{r}-\omega_k t)} \tag{11.53}$$

describes vacuum fluctuations and

$$\mathbf{E}^{(+)}_{\text{rad}}(\mathbf{r},t)=\sum_{\mathbf{k}s} i\sqrt{\frac{\hbar\omega_k}{2\epsilon_0 V}}\mathfrak{g}^*_{\mathbf{k}s}\mathbf{e}_{\mathbf{k}s}e^{i\mathbf{k}\cdot\mathbf{r}}\int_0^t dt' e^{-i\omega_k(t-t')}\sigma_-(t') \tag{11.54}$$

represents the field radiated by the atom. The state vector,

$$|\text{in}\rangle=|\varepsilon_2,0\rangle=|\varepsilon_2\rangle\,|0\rangle\,, \tag{11.55}$$

describes the situation with the atom in the excited state and no photons in the field. In Section 9.1 we saw that the counting rate for a detector located at $\mathbf{r}$ is proportional to $\langle\text{in}|\mathbf{E}^{(-)}(\mathbf{r},t)\cdot\mathbf{E}^{(+)}(\mathbf{r},t)|\text{in}\rangle$. Since $|\text{in}\rangle$ is the vacuum for photons, $\mathbf{E}^{(+)}_{\text{vac}}(\mathbf{r},t)$ will not contribute, and the counting rate is proportional to $\langle\text{in}|\mathbf{E}^{(-)}_{\text{rad}}(\mathbf{r},t)\cdot\mathbf{E}^{(+)}_{\text{rad}}(\mathbf{r},t)|\text{in}\rangle$.

Calculating the atomic radiation operator $\mathbf{E}^{(+)}_{\text{rad}}(\mathbf{r},t)$ from eqn (11.54) requires an evaluation of the sum over polarizations, followed by the conversion of the $\mathbf{k}$-sum to an integral, as outlined in Exercise 11.3. After carrying out the integral over the directions of $\mathbf{k}$, the result is

$$\begin{aligned}\mathbf{E}^{(+)}_{\text{rad}}(\mathbf{r},t)=i\int\frac{k^2dk}{(2\pi)^3}\frac{\omega_k K(\omega_k-\omega_{21})}{2\epsilon_0}\left[\mathbf{d}^*+\frac{(\mathbf{d}^*\cdot\boldsymbol{\nabla})\boldsymbol{\nabla}}{k^2}\right]\frac{4\pi\sin(kr)}{kr}\\ \times\int_0^t dt' e^{-i\omega_k(t-t')}\sigma_-(t')\,.\end{aligned} \tag{11.56}$$

The cut-off function $K(\omega_k-\omega_{21})$ imposes $k\approx k_{21}=\omega_{21}/c$, so we can define the **radiation zone** by $kr\approx k_{21}r\gg 1$. For a detector in the radiation zone,

$$\left[\mathbf{d}^*+\frac{1}{k^2}(\mathbf{d}^*\cdot\boldsymbol{\nabla})\boldsymbol{\nabla}\right]\frac{4\pi\sin(kr)}{kr}=\frac{4\pi\sin(kr)}{kr}\mathbf{d}^*_{\top}+O\left(\frac{1}{k^2r^2}\right), \tag{11.57}$$

where

$$\mathbf{d}_{\top}^{*} = \mathbf{d}^{*} - (\widetilde{\mathbf{r}} \cdot \mathbf{d}^{*})\,\widetilde{\mathbf{r}} = (\mathbf{d}^{*} \times \widetilde{\mathbf{r}}) \times \widetilde{\mathbf{r}} \tag{11.58}$$

is the component of $\mathbf{d}^*$ transverse to the vector $\mathbf{r}$ linking the atom to the detector. This is the same as the rule for the polarization of radiation emitted by a classical dipole (Jackson, 1999, Sec. 9.2). After changing the integration variable from k to $\omega = \omega_k = ck$, we find

$$\begin{aligned} \mathbf{E}_{\mathrm{rad}}^{(+)}(\mathbf{r},t) = {} & \frac{i}{4\pi^2 c^2 \epsilon_0} \frac{\mathbf{d}_{\top}^{*}}{r} \int_0^{\infty} d\omega \omega^2 K(\omega - \omega_{21}) \sin\left(\frac{\omega r}{c}\right) \\ & \times \int_0^t dt' e^{-i\omega(t-t')} \sigma_-(t') \,. \end{aligned} \tag{11.59}$$

Approximating the slowly-varying factor ω^2 by ω_{21}^2, and unpacking $\sin(kr)$, yields the expression

$$\mathbf{E}_{\mathrm{rad}}^{(+)}(\mathbf{r},t) = \frac{k_{21}^2}{8\pi^2 \epsilon_0} \frac{\mathbf{d}_{\top}^{*}}{r} \left[I(r) - I(-r) \right] \tag{11.60}$$

for the field, where

$$\begin{aligned} I(r) &= \int_0^t dt' \left[\int_0^{\infty} d\omega K(\omega - \omega_{21})\, e^{i\omega r/c} e^{-i\omega(t-t')} \right] \sigma_-(t') \\ &= e^{ik_{21}r} e^{-i\omega_{21}t} \int_0^t dt' \left[\int_{-\omega_{21}}^{\infty} d\omega K(\omega)\, e^{i\omega[r/c-(t-t')]} \right] e^{i\omega_{21}t'} \sigma_-(t') \,. \end{aligned} \tag{11.61}$$

The condition $w_K \ll \omega_{21}$ allows us to extend the lower limit of the ω-integral to $-\infty$ with negligible error, so

$$\begin{aligned} \int_{-\omega_{21}}^{\infty} d\omega K(\omega)\, e^{i\omega\tau} &\approx 2\pi \int_{-\infty}^{\infty} \frac{d\omega}{2\pi} K(\omega)\, e^{i\omega\tau} \\ &= 2\pi \varpi(\tau) \,, \end{aligned} \tag{11.62}$$

where $\varpi(\tau)$ is the averaging function introduced in eqn (11.17). The results derived in Exercise 11.4 include the fact that

$$\overline{\sigma}_-(t') = e^{i\omega_{21}t'} \sigma_-(t') \tag{11.63}$$

is a slowly-varying envelope operator, i.e. it varies on the time scale set by $|\mathfrak{g}_{\mathbf{k}s}|$. Combining these observations with the approximate delta function rule (11.21) leads to

$$\begin{aligned} I(r) &= 2\pi e^{ik_{21}r} e^{-i\omega_{21}t} \int_0^t dt' \delta\left(r/c - (t-t')\right) \overline{\sigma}_-(t') \\ &= 2\pi e^{ik_{21}r} e^{-i\omega_{21}t} \overline{\sigma}_-(t - r/c) \,, \end{aligned} \tag{11.64}$$

and

$$I(-r) = 2\pi e^{-ik_{21}r} e^{-i\omega_{21}t} \int_0^t dt' \delta\left(-r/c - (t-t')\right) \overline{\sigma}_-(t') = 0 \,. \tag{11.65}$$

The final result for the radiated field is

$$\mathbf{E}^{(+)}_{\text{rad}}(\mathbf{r},t) = \frac{k_{21}^2 \mathbf{d}_{\top}^*}{4\pi\epsilon_0}\frac{e^{ik_{21}r}}{r}e^{-i\omega_{21}t}\overline{\sigma}_-(t - r/c)\,. \tag{11.66}$$

Thus the field operator behaves as an expanding spherical wave with source given by the atomic dipole operator at the retarded time $t - r/c$. Just as in the classical theory, the detector will not fire before the first arrival time $t = r/c$. We should emphasize that this fundamental result does not depend on the resonant wave approximation and the other simplifications made here. A rigorous calculation leading to the same conclusion has been given by Milonni (1994).

11.2.2 The Weisskopf–Wigner method

The perturbative calculation of the spontaneous emission rate can apparently be improved by including higher-order terms from eqn (4.103). Since the initial and final states are fixed, these terms must describe virtual emission and absorption of photons. In other words, the higher-order terms—called **radiative corrections**—involve vacuum fluctuations. We know, from Section 2.5, that the contributions from vacuum fluctuations are infinite, so it will not come as a surprise to learn that all of the integrals defining the higher-order contributions are divergent.

A possible remedy would be to include the cut-off function $K(\Delta_k)$, in the coupling frequencies, i.e. to replace $\mathfrak{G}_{\mathbf{k}s}$ by $\mathfrak{g}_{\mathbf{k}s}$. This will cure the divergent integrals, but it must then be proved that the results do not depend on the detailed shape of $K(\Delta_k)$. This can be done, but only at the expense of importing the machinery of renormalization theory from quantum electrodynamics (Greiner and Reinhardt, 1994).

A more important drawback of the perturbative approach is that it is only valid in the limited time interval $t \ll 1/|\mathfrak{g}_{\mathbf{k}s}| \approx \tau_{sp} = 1/A_{2\to 1}$. Thus perturbation theory cannot be used to follow the evolution of the system for times comparable to the spontaneous decay time. We will use the RWA to pursue a nonperturbative approach (see Cohen-Tannoudji *et al.* (1977*b*, Complement D-XIII), or the original paper Weisskopf and Wigner (1930)) which can describe the behavior of the atom–field system for long times, $t > \tau_{sp}$.

The key to this nonperturbative method is the following simple observation. In the resonant wave approximation, the atom–field state $|\varepsilon_2; 0\rangle$, in which the atom is in the excited state and there are no photons, can only make transitions to one of the states $|\varepsilon_1; 1_{\mathbf{k}s}\rangle$, in which the atom is in the ground state and there is exactly one photon present. Conversely, the state $|\varepsilon_1; 1_{\mathbf{k}s}\rangle$ can only make a transition into the state $|\varepsilon_2; 0\rangle$. This is demonstrated more explicitly by using eqn (11.25) for H_{rwa} to find

$$H_{\text{rwa}}(t)|\varepsilon_2; 0\rangle = i\hbar\sum_{\mathbf{k}s}\mathfrak{g}_{\mathbf{k}s}^* e^{i\Delta_k t}|\varepsilon_1; 1_{\mathbf{k}s}\rangle\,, \tag{11.67}$$

and

$$H_{\text{rwa}}(t)|\varepsilon_1; 1_{\mathbf{k}s}\rangle = -i\hbar\mathfrak{g}_{\mathbf{k}s}e^{-i\Delta_k t}|\varepsilon_2; 0\rangle\,. \tag{11.68}$$

Consequently, the **spontaneous emission subspace**

$$\mathfrak{H}_{\text{se}} = \text{span}\{|\varepsilon_2; 0\rangle\,, |\varepsilon_1; 1_{\mathbf{k}s}\rangle \text{ for all } \mathbf{k}s\} \tag{11.69}$$

is sent into itself by the action of the RWA Hamiltonian: $H_{\text{rwa}}(t)\,\mathfrak{H}_{\text{se}} \to \mathfrak{H}_{\text{se}}$. This means that an initial state in $\mathfrak{H}_{\text{se}}$ will evolve into another state in $\mathfrak{H}_{\text{se}}$. The time-dependent state can therefore be expressed as

$$|\Theta(t)\rangle = C_2(t)\,|\varepsilon_2;0\rangle + \sum_{\mathbf{k}s} C_{1\mathbf{k}s}(t)\,e^{-i\Delta_k t}\,|\varepsilon_1;1_{\mathbf{k}s}\rangle\,, \tag{11.70}$$

where the exponential in the second term is included to balance the explicit time dependence of the interaction-picture Hamiltonian. Substituting this into the Schrödinger equation (11.13) produces equations for the coefficients:

$$\frac{dC_2(t)}{dt} = -\sum_{\mathbf{k}s} \mathfrak{g}_{\mathbf{k}s} C_{1\mathbf{k}s}(t)\,, \tag{11.71}$$

$$\left(\frac{d}{dt} + i\Delta_k\right) C_{1\mathbf{k}s}(t) = \mathfrak{g}^*_{\mathbf{k}s} C_2(t)\,. \tag{11.72}$$

For the discussion of spontaneous emission, it is natural to assume that the atom is initially in the excited state and no photons are present, i.e.

$$C_2(0) = 1\,, \quad C_{1\mathbf{k}s}(0) = 0\,. \tag{11.73}$$

Inserting the formal solution,

$$C_{1\mathbf{k}s}(t) = \int_0^t dt'\,\mathfrak{g}^*_{\mathbf{k}s} e^{-i\Delta_k(t-t')} C_2(t')\,, \tag{11.74}$$

of eqn (11.72) into eqn (11.71) leads to the integro-differential equation

$$\frac{dC_2(t)}{dt} = -\int_0^t dt' \left\{\sum_{\mathbf{k}s} |\mathfrak{g}_{\mathbf{k}s}|^2\, e^{-i\Delta_k(t-t')}\right\} C_2(t') \tag{11.75}$$

for C_2. This presents us with a difficult problem, since the evolution of $C_2(t)$ now depends on its past history. The way out is to argue that the function in curly brackets decays rapidly as $t - t'$ increases, so that it is a good approximation to set $C_2(t') = C_2(t)$. This allows us to replace eqn (11.75) by

$$\frac{dC_2(t)}{dt} = -\left\{\int_0^t dt' \sum_{\mathbf{k}s} |\mathfrak{g}_{\mathbf{k}s}|^2\, e^{-i\Delta_k t'}\right\} C_2(t)\,, \tag{11.76}$$

which has the desirable feature that $C_2(t+\Delta t)$ only depends on $C_2(t)$, rather than $C_2(t')$ for all $t' < t$. As we already noted in Section 9.2.1, evolutions with this property are called Markov processes, and the transition from eqn (11.75) to eqn (11.76) is called the Markov approximation. In the following paragraphs we will justify the assumptions underlying the Markov approximation by a Laplace transform method that is also useful in related problems.

The differential equations for $C_1(t)$ and $C_2(t)$ define a linear initial value problem that can be solved by the Laplace transform method reviewed in Appendix A.5. Applying the general scheme in eqns (A.73)–(A.75) to the initial conditions (11.73) and the differential equations (11.71) and (11.72) produces the algebraic equations

$$\zeta\,\widetilde{C}_2(\zeta) = 1 - \sum_{\mathbf{k}s} \mathfrak{g}_{\mathbf{k}s}\widetilde{C}_{1\mathbf{k}s}(\zeta)\,, \tag{11.77}$$

$$(\zeta + i\Delta_k)\,\widetilde{C}_{1\mathbf{k}s}(\zeta) = \mathfrak{g}^*_{\mathbf{k}s}\widetilde{C}_2(\zeta)\,. \tag{11.78}$$

Substituting the solution of the second of these equations into the first leads to

$$\widetilde{C}_2(\zeta) = \frac{1}{\zeta + D(\zeta)}\,, \tag{11.79}$$

where

$$D(\zeta) = \sum_{\mathbf{k}s} \frac{|\mathfrak{g}_{\mathbf{k}s}|^2}{\zeta + i\Delta_k}\,. \tag{11.80}$$

In order to carry out the limit $V \to \infty$, we introduce

$$\mathfrak{g}^2(\mathbf{k}) = V\sum_{s} |\mathfrak{g}_{\mathbf{k}s}|^2\,, \tag{11.81}$$

which allows $D(\zeta)$ to be expressed as

$$D(\zeta) = \frac{1}{V}\sum_{\mathbf{k}} \frac{\mathfrak{g}^2(\mathbf{k})}{\zeta + i\Delta_k} \to \int \frac{d^3k}{(2\pi)^3}\frac{\mathfrak{g}^2(\mathbf{k})}{\zeta + i\Delta_k}\,. \tag{11.82}$$

According to eqn (4.160),

$$\mathfrak{g}^2(\mathbf{k}) = \frac{\omega_k\,|K(\Delta_k)|^2}{2\epsilon_0\hbar}\left[|\mathbf{d}|^2 - \left|\mathbf{d}\cdot\widetilde{\mathbf{k}}\right|^2\right], \tag{11.83}$$

and the integral over the directions of $\mathbf{k}$ in eqn (11.82) can be carried out by the method used in eqn (4.161). The relation $|\mathbf{k}| = \omega_k/c$ is then used to change the integration variable from $|\mathbf{k}|$ to $\Delta = \omega_k - \omega_{21}$. The lower limit of the Δ-integral is $\Delta = -\omega_{21}$, but the width of the cut-off function is small compared to the transition frequency ($w_K \ll \omega_{21}$); therefore, there is negligible error in extending the integral to $\Delta = -\infty$ to get

$$D(\zeta) = \frac{w_{21}}{2\pi}\int_{-\infty}^{\infty} d\Delta \frac{\left(1 + \frac{\Delta}{\omega_{21}}\right)^3 |K(\Delta)|^2}{\zeta + i\Delta}\,, \tag{11.84}$$

where

$$w_{21} = \frac{|\mathbf{d}|^2\,\omega_{21}^3}{3\pi\epsilon_0\hbar c^3} = A_{2\to 1} \tag{11.85}$$

is the spontaneous decay rate previously found in Section 4.9.3.

The time dependence of $C_2(t)$ is determined by the location of the poles in $\widetilde{C}_2(\zeta)$, which are in turn determined by the roots of

$$\zeta + D(\zeta) = 0\,. \tag{11.86}$$

A peculiar feature of this approach is that it is absolutely essential to solve this equation without knowing the function $D(\zeta)$ exactly. The reason is that an exact evaluation of $D(\zeta)$ would require an explicit model for $|K(\Delta)|^2$, but no physically meaningful results can depend on the detailed behavior of the cut-off function. What is needed is an approximate evaluation of $D(\zeta)$ which is as insensitive as possible to the shape of $|K(\Delta)|^2$. The key to this approximation is found by combining eqn (11.86) with eqn (11.84) to conclude that the relevant values of ζ are small compared to the width of the cut-off function, i.e.

$$\zeta = O(w_{21}) \ll w_K\,. \tag{11.87}$$

This is the step that will justify the Markov approximation (11.76). In the time domain, the function $C_2(t)$ varies significantly over an interval of width $\Delta t \sim 1/w_{21}$; consequently, the condition (11.87) is equivalent to $T_{\text{mem}} \ll \Delta t$; that is, the memory of the averaging function is short compared to the time scale on which the function $C_2(t)$ varies. The physical source of this feature is the continuous phase space of final states available to the emitted photon. Summing over this continuum of final photon states effectively erases the memory of the atomic state that led to the emission of the photon.

For values of ζ satisfying eqn (11.87), $D(\zeta)$ can be approximated by combining the normalization condition $K(0) = 1$ with the identity

$$\lim_{\zeta \to 0} \frac{1}{\zeta + i\Delta} = \pi\delta(\Delta) - iP\frac{1}{\Delta}\,, \tag{11.88}$$

where P denotes the Cauchy principal value—see eqn (A.93). The result is

$$D(\zeta) = \frac{w_{21}}{2} + i\delta\omega_{21}\,, \tag{11.89}$$

where the imaginary part,

$$\delta\omega_{21} = -\frac{w_{21}}{2\pi} P\int_{-\infty}^{\infty} d\Delta \left(1 + \frac{\Delta}{\omega_{21}}\right)^3 \frac{|K(\Delta)|^2}{\Delta}\,, \tag{11.90}$$

is the **frequency shift**. It is customary to compare $\delta\omega_{21}$ to the Lamb shift (Cohen-Tannoudji *et al.*, 1992, Sec. II-E.1), but this is somewhat misleading. The result for $\operatorname{Re} D(\zeta)$ is robust, in the sense that it is independent of the details of the cut-off function, but the result for $\operatorname{Im} D(\zeta)$ is not robust, since it depends on the shape of $|K(\Delta)|^2$. In Exercise 11.2, eqn (11.90) is used to get the estimate, $\delta\omega_{21}/w_{21} = O(w_K/\omega_{21}) \ll 1$, for the size of the frequency shift. This is comforting, since it tells us that $\delta\omega_{21}$ is at least very small, even if its exact numerical value has no physical significance. The experimental fact that measured shifts are small compared to the line widths is even more comforting. A strictly consistent application of the RWA neglects all terms of the order w_K/ω_{21}; therefore, we will set $\delta\omega_{21} = 0$.

Substituting $D(\zeta)$ from eqn (11.89) into eqn (11.79) gives the simple result

$$\widetilde{C}_2(\zeta) = \frac{1}{\zeta + w_{21}/2}, \tag{11.91}$$

and evaluating the inverse transform (A.72) by the rule (A.80) produces the corresponding time-domain result

$$C_2(t) = e^{-w_{21}t/2}. \tag{11.92}$$

Thus the nonperturbative Weisskopf–Wigner method displays an irreversible decay,

$$|C_2(t)|^2 = e^{-w_{21}t}, \tag{11.93}$$

of the upper-level occupation probability. This conclusion depends crucially on the coupling of the discrete atomic states to the broad distribution of electromagnetic modes available in the infinite volume limit. In the time domain, we can say that the atom forgets the emission event before there is time for reabsorption. We will see later on that the irreversibility of the decay does not hold for atoms in a cavity with dimensions comparable to a wavelength.

In addition to following the decay of the upper-level occupation probability, we can study the probability that the atom emits a photon into the mode $\mathbf{k}s$. According to eqn (11.78),

$$\widetilde{C}_{1\mathbf{k}s}(\zeta) = \frac{\mathfrak{g}^*_{\mathbf{k}s}}{(\zeta + i\Delta_k)(\zeta + w_{21}/2)}. \tag{11.94}$$

The probability amplitude for a photon with wavevector $\mathbf{k}$ and polarization $\mathbf{e}_{\mathbf{k}s}$ is $C_{1\mathbf{k}s}(t)\, e^{i\Delta_k t}$, so another application of eqn (A.80) yields

$$C_{1\mathbf{k}s}(t) = i\mathfrak{g}^*_{\mathbf{k}s}\frac{e^{-i\Delta_k t} - e^{-w_{21}t/2}}{\Delta_k + iw_{21}/2}. \tag{11.95}$$

After many decay times ($w_{21}t \gg 1$), the probability for emission is

$$\begin{aligned} p_{\mathbf{k}s} &= \lim_{t\to\infty}\left|C_{1\mathbf{k}s}(t)\, e^{i\Delta_k t}\right|^2 \\ &= \frac{|\mathfrak{g}_{\mathbf{k}s}|^2}{(\Delta_k)^2 + \left(\frac{w_{21}}{2}\right)^2} \\ &= \left[\frac{\omega_k |K(\Delta_k)|^2}{2\epsilon_0\hbar V}\right]\frac{|\mathbf{d}\cdot\mathbf{e}_{\mathbf{k}s}|^2}{(\Delta_k)^2 + \left(\frac{w_{21}}{2}\right)^2}. \end{aligned} \tag{11.96}$$

The denominator of the second factor effectively constrains Δ_k by $|\Delta_k| < w_{21}$, so it is permissible to set $|K(\Delta_k)| = 1$ in the following calculations.

As explained in Section 3.1.4, physically meaningful results are found by passing to the limit of infinite quantization volume. In the present case, this is done by using the rule $1/V \to d^3k/(2\pi)^3$, which yields

$$dp_s(\mathbf{k}) = \left[\frac{\omega_k}{2\epsilon_0\hbar}\right]\frac{|\mathbf{d}\cdot\mathbf{e}_{\mathbf{k}s}|^2}{(\Delta_k)^2 + \left(\frac{w_{21}}{2}\right)^2}\frac{d^3k}{(2\pi)^3} \tag{11.97}$$

for the probability of emitting a photon with polarization $\mathbf{e}_{\mathbf{k}s}$ into the momentum-space volume element d^3k. Summing over polarizations and integrating over the angles of $\mathbf{k}$, by the methods used in Section 4.9.3, gives the probability for emission of a photon in the frequency interval $(\omega, \omega + d\omega)$:

$$dp(\omega) = \frac{\frac{w_{21}}{2}}{(\omega - \omega_{21})^2 + \left(\frac{w_{21}}{2}\right)^2}\frac{d\omega}{\pi}. \tag{11.98}$$

This has the form of the **Lorentzian line shape**

$$\mathcal{L}(\nu) = \frac{\gamma}{\nu^2 + \gamma^2}, \tag{11.99}$$

where ν is the detuning from the resonance frequency, γ is the half-width-at-half-maximum (HWHM), and the normalization condition is

$$\int_{-\infty}^{\infty}\frac{d\nu}{\pi}\mathcal{L}(\nu) = 1. \tag{11.100}$$

From eqn (11.98) we see that the **line width** w_{21} is the full-width-at-half-maximum, but also that the normalization condition is not exactly satisfied. The trouble is that $\omega = \omega_k$ is required to be positive, so the integral over all physical frequencies is

$$\int_{-\omega_{21}}^{\infty}\frac{d\nu}{\pi}\frac{\frac{w_{21}}{2}}{\nu^2 + \left(\frac{w_{21}}{2}\right)^2} < 1. \tag{11.101}$$

This is not a serious problem since $\omega_{21} \gg w_{21}$, i.e. the optical transition frequency is much larger than the line width. Thus the lower limit of the integral can be extended to $-\infty$ with small error. The spectrum of spontaneous emission is therefore well represented by a Lorentzian line shape.

11.2.3 Two-photon cascade*

The photon indivisibility experiment of Grangier, Roget, and Aspect, discussed in Section 1.4, used a two-photon cascade transition as the source of an entangled two-photon state. The simplest model for this process is a three-level atom, as shown in Fig. 11.1.

This concrete example will illustrate the use of the general techniques discussed in the previous section. The one-photon detunings, $\Delta_{32,k} = ck - \omega_{32}$ and $\Delta_{21,k'} = ck' - \omega_{21}$, are related to the two-photon detuning, $\Delta_{31,kk'} = ck + ck' - \omega_{31}$, by

$$\Delta_{31,kk'} = \Delta_{32,k} + \Delta_{21,k'} = \Delta_{32,k'} + \Delta_{21,k}. \tag{11.102}$$

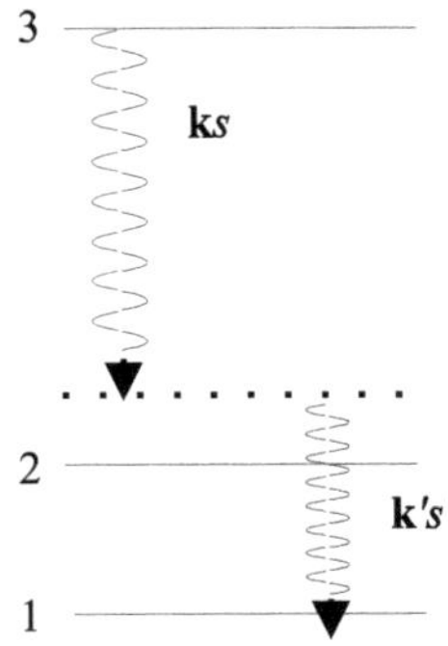

Fig. 11.1 Two-photon cascade emission from a three-level atom. The frequencies are assumed to satisfy $\omega = ck \approx \omega_{32}$, $\omega' = ck' \approx \omega_{21}$, and $\omega_{32} \gg \omega_{21}$.

According to the general result (11.43), the RWA Hamiltonian is

$$H_{\text{rwa}}(t) = -i\hbar \sum_{\mathbf{k}s} \left[\mathfrak{g}_{32,\mathbf{k}s} e^{-i\Delta_{32,k}t} S_{32} a_{\mathbf{k}s} + \mathfrak{g}_{21,\mathbf{k}s} e^{-i\Delta_{21,k}t} S_{21} a_{\mathbf{k}s} \right] + \text{HC}\,, \tag{11.103}$$

where the coupling constants are

$$\begin{aligned} \mathfrak{g}_{32,\mathbf{k}s} &= \sqrt{\frac{\hbar\omega_k}{2\epsilon_0 V}} \frac{\mathbf{d}_{32} \cdot \mathbf{e}_{\mathbf{k}s}}{\hbar} K(\Delta_{32,k})\,, \\ \mathfrak{g}_{21,\mathbf{k}s} &= \sqrt{\frac{\hbar\omega_k}{2\epsilon_0 V}} \frac{\mathbf{d}_{21} \cdot \mathbf{e}_{\mathbf{k}s}}{\hbar} K(\Delta_{21,k})\,. \end{aligned} \tag{11.104}$$

Initially the atom is in the uppermost excited state $|\varepsilon_3\rangle$ and the field is in the vacuum state $|0\rangle$, so the combined system is described by the product state $|\varepsilon_3; 0\rangle = |\varepsilon_3\rangle |0\rangle$. The excited atom can decay to the intermediate state $|\varepsilon_2\rangle$ with the emission of a photon, and then subsequently emit a second photon while making the final transition to the ground state $|\varepsilon_1\rangle$. It may seem natural to think that the $3 \to 2$ photon must be emitted first and the $2 \to 1$ photon second, but the order could be reversed. The reason is that we are not considering a sequence of *completed* spontaneous emissions, each described by an Einstein A coefficient, but instead a coherent process in which the atom emits two photons during the overall transition $3 \to 1$. Since the final states are the same, the processes ($3 \to 2$ followed by $2 \to 1$) and ($2 \to 1$ followed by $3 \to 2$) are indistinguishable. Feynman's rules then tell us that the two amplitudes must be coherently added before squaring to get the transition probability. If the level spacings were nearly equal, both processes would be equally important. In the situation we are considering, $\omega_{32} \gg \omega_{21}$, the process ($2 \to 1$ followed by $3 \to 2$) would be far off resonance; therefore, we can safely neglect it. This approximation is formally justified by the estimate

$$\mathfrak{g}_{32,\mathbf{k}s}\mathfrak{g}_{21,\mathbf{k}s} \approx 0\,, \tag{11.105}$$

which is a consequence of the fact that the cut-off functions $|K(\Delta_{32,k})|$ and $|K(\Delta_{21,k})|$ do not overlap.

The states $|\varepsilon_2; 1_{\mathbf{k}s}\rangle = |\varepsilon_2\rangle |1_{\mathbf{k}s}\rangle$ and $|\varepsilon_1; 1_{\mathbf{k}s}, 1_{\mathbf{k}'s'}\rangle = |\varepsilon_1\rangle |1_{\mathbf{k}s}, 1_{\mathbf{k}'s'}\rangle$ will appear as the state vector $|\Theta(t)\rangle$ evolves. It is straightforward to show that applying the

Hamiltonian to each of these states results in a linear combination of the same three states. The standard terminology for this situation is that the subspace spanned by $|\varepsilon_3; 0\rangle$, $|\varepsilon_2; 1_{\mathbf{k}s}\rangle$, and $|\varepsilon_1; 1_{\mathbf{k}s}, 1_{\mathbf{k}'s'}\rangle$ is *invariant* under the action of the Hamiltonian. We have already met with a case like this in Section 11.2.2, and we can use the ideas of the Weisskopf–Wigner model to analyze the present problem. To this end, we make the following *ansatz* for the state vector:

$$|\Theta(t)\rangle = Z(t)\,|\varepsilon_3; 0\rangle + \sum_{\mathbf{k}s} Y_{\mathbf{k}s}(t)\, e^{i\Delta_{32,k}t}\,|\varepsilon_2; 1_{\mathbf{k}s}\rangle + \sum_{\mathbf{k}s}\sum_{\mathbf{k}'s'} X_{\mathbf{k}s,\mathbf{k}'s'}(t)\, e^{i\Delta_{31,kk'}t}\,|\varepsilon_1; 1_{\mathbf{k}s}, 1_{\mathbf{k}'s'}\rangle\,, \tag{11.106}$$

where the time-dependent exponentials have been introduced to cancel the time dependence of $H_{\text{rwa}}(t)$. Note that the coefficient $X_{\mathbf{k}s,\mathbf{k}'s'}$ is necessarily symmetric under $\mathbf{k}s \leftrightarrow \mathbf{k}'s'$.

Substituting this expansion into the Schrödinger equation—see Exercise 11.5—leads to a set of linear differential equations for the coefficients. We will solve these equations by the Laplace transform technique, just as in Section 11.2.2. The initial conditions are $Z(0) = 1$ and $Y_{\mathbf{k}s}(0) = X_{\mathbf{k}s,\mathbf{k}'s'}(0) = 0$, so the differential equations are replaced by the algebraic equations

$$\zeta\widetilde{Z}(\zeta) = 1 - \sum_{\mathbf{k}s} \mathfrak{g}_{32,\mathbf{k}s}\widetilde{Y}_{\mathbf{k}s}(\zeta)\,, \tag{11.107}$$

$$[\zeta + i\Delta_{32,k}]\,\widetilde{Y}_{\mathbf{k}s}(\zeta) = \mathfrak{g}^*_{32,\mathbf{k}s}\widetilde{Z}(\zeta) - 2\sum_{\mathbf{k}'s'} \mathfrak{g}_{21,\mathbf{k}'s'}\widetilde{X}_{\mathbf{k}s,\mathbf{k}'s'}(\zeta)\,, \tag{11.108}$$

$$[\zeta + i\Delta_{31,kk'}]\,\widetilde{X}_{\mathbf{k}s,\mathbf{k}'s'}(\zeta) = \frac{1}{2}\left[\mathfrak{g}^*_{21,\mathbf{k}s}\widetilde{Y}_{\mathbf{k}'s'}(\zeta) + \mathfrak{g}^*_{21,\mathbf{k}'s'}\widetilde{Y}_{\mathbf{k}s}(\zeta)\right]. \tag{11.109}$$

Solving the final equation for $\widetilde{X}_{\mathbf{k}s,\mathbf{k}'s'}$ and substituting the result into eqn (11.108) produces

$$[\zeta + i\Delta_{32,k} + D_k(\zeta)]\,\widetilde{Y}_{\mathbf{k}s}(\zeta) = \mathfrak{g}^*_{32,\mathbf{k}s}\widetilde{Z}(\zeta) - \sum_{\mathbf{k}'s'} \frac{\mathfrak{g}_{21,\mathbf{k}'s'}\mathfrak{g}^*_{21,\mathbf{k}s}}{\zeta + i\Delta_{31,kk'}}\widetilde{Y}_{\mathbf{k}'s'}(\zeta)\,, \tag{11.110}$$

where

$$D_k(\zeta) = \sum_{\mathbf{k}'s'} \frac{|\mathfrak{g}_{21,\mathbf{k}'s'}|^2}{\zeta + i\Delta_{32,k} + i\Delta_{21,k'}}\,. \tag{11.111}$$

As far as the **k**-dependence is concerned, eqn (11.110) is an integral equation for $\widetilde{Y}_{\mathbf{k}s}(\zeta)$, but there is an approximation that simplifies matters. The first-order term on the right side shows that $\widetilde{Y}_{\mathbf{k}s} \sim \mathfrak{g}^*_{32,\mathbf{k}s}$, but this implies that the $\mathbf{k}'$-sum in the second term includes the product $\mathfrak{g}_{21,\mathbf{k}'s'}\mathfrak{g}^*_{32,\mathbf{k}'s'}$, which can be neglected by virtue of eqn (11.105). Thus the second term can be dropped, and an approximate solution to eqn (11.110) is given by

$$\widetilde{Y}_{\mathbf{k}s}(\zeta) = \frac{\mathfrak{g}^*_{32,\mathbf{k}s}\widetilde{Z}(\zeta)}{\zeta + i\Delta_{32,k} + D_k(\zeta)}. \tag{11.112}$$

Calculations similar to those in Section 11.2.2 allow us to carry out the limit $V \to \infty$ and express $D_k(\zeta)$ as

$$D_k(\zeta) = \frac{w_{21}}{2\pi}\int d\Delta' \frac{|K(\Delta')|^2}{\zeta + i\Delta_{32,k} + i\Delta'}, \tag{11.113}$$

where w_{21}, the decay rate for the $2 \to 1$ transition, is given by eqn (11.85).

The poles of $\widetilde{Y}_{\mathbf{k}s}(\zeta)$ are partly determined by the zeroes of $\zeta + i\Delta_{32,k} + D_k(\zeta)$, so the relevant values of ζ satisfy

$$\zeta + i\Delta_{32,k} = O(w_{21}). \tag{11.114}$$

Another application of the argument used in Section 11.2.2 yields $D_k \approx w_{21}/2$, so the expression for $\widetilde{Y}_{\mathbf{k}s}(\zeta)$ simplifies to

$$\widetilde{Y}_{\mathbf{k}s}(\zeta) = \frac{\mathfrak{g}^*_{32,\mathbf{k}s}\widetilde{Z}(\zeta)}{\zeta + i\Delta_{32,k} + \frac{w_{21}}{2}}. \tag{11.115}$$

Substituting this into eqn (11.107) gives

$$\widetilde{Z}(\zeta) = \frac{1}{\zeta + F(\zeta)}, \tag{11.116}$$

where

$$F(\zeta) = \frac{w_{32}}{2\pi}\int d\Delta \frac{|K(\Delta)|^2}{\zeta + \frac{w_{21}}{2} + i\Delta}, \tag{11.117}$$

and w_{32} is the decay rate for the $3 \to 2$ transition. In this case $\zeta = O(w_{32})$, so $\zeta + w_{21}/2$ is also small compared to the width w_K of the cut-off function. A third application of the same argument yields $F(\zeta) = w_{32}/2$, so the Laplace transforms of the expansion coefficients are given by

$$\widetilde{Z}(\zeta) = \frac{1}{\zeta + \frac{w_{32}}{2}}, \tag{11.118}$$

$$\widetilde{Y}_{\mathbf{k}s}(\zeta) = \frac{\mathfrak{g}^*_{32,\mathbf{k}s}}{\left[\zeta + i\Delta_{32,k} + \frac{w_{21}}{2}\right]\left(\zeta + \frac{w_{32}}{2}\right)}, \tag{11.119}$$

$$\widetilde{X}_{\mathbf{k}s,\mathbf{k}'s'}(\zeta) = \frac{1}{2}\frac{\mathfrak{g}^*_{32,\mathbf{k}s}\mathfrak{g}^*_{21,\mathbf{k}'s'}}{\left[\zeta + i\Delta_{31,kk'}\right]\left(\zeta + \frac{w_{32}}{2}\right)\left[\zeta + i\Delta_{32,k} + \frac{w_{21}}{2}\right]} + (\mathbf{k}s \leftrightarrow \mathbf{k}'s'). \tag{11.120}$$

The rule (A.80) shows that the inverse Laplace transform of eqn (11.120) has the form

$$X_{\mathbf{k}s,\mathbf{k}'s'}(t) = G_1 \exp\left(-\frac{w_{32}t}{2}\right) + G_2 \exp\left(-\frac{w_{21}t}{2}\right) \exp\left[-i\Delta_{32,k}t\right] + G_3 \exp\left[-i\Delta_{31,kk'}t\right] + (\mathbf{k}s \leftrightarrow \mathbf{k}'s') . \tag{11.121}$$

In the limit of long times, i.e. $w_{32}t \gg 1$ and $w_{21}t \gg 1$, only the third term survives. Evaluating the residue for the pole at $\zeta = -i\Delta_{31,kk'}$ provides the explicit expression for G_3 and thus the long-time probability amplitude for the state $|\varepsilon_3; 1_{\mathbf{k}s}, 1_{\mathbf{k}'s'}\rangle$:

$$X^{\infty}_{\mathbf{k}s,\mathbf{k}'s'} = -\frac{1}{2} \frac{\mathfrak{g}^*_{32,\mathbf{k}s}\mathfrak{g}^*_{21,\mathbf{k}'s'}}{\left[\Delta_{31,kk'} + \frac{i}{2}w_{32}\right]\left[\Delta_{21,k'} + \frac{i}{2}w_{21}\right]} + (\mathbf{k}s \leftrightarrow \mathbf{k}'s') . \tag{11.122}$$

Since the two one-photon resonances are nonoverlapping, only one of these two terms will contribute for a given $(\mathbf{k}s, \mathbf{k}'s')$-pair. In order to pass to the infinite volume limit, we introduce $\mathfrak{g}_{32,s}(\mathbf{k}) = \sqrt{V}\mathfrak{g}_{32,\mathbf{k}s}$ and $\mathfrak{g}_{21,s'}(\mathbf{k}') = \sqrt{V}\mathfrak{g}_{21,\mathbf{k}'s'}$ and use the argument leading to eqn (11.97) to get the differential probability

$$dp(\mathbf{k}s, \mathbf{k}'s') = \frac{1}{4} \frac{|\mathfrak{g}_{32,s}(\mathbf{k})|^2}{\left\{[\Delta_{13,kk'}]^2 + \frac{1}{4}w_{32}^2\right\}} \frac{|\mathfrak{g}_{21,s'}(\mathbf{k}')|^2}{\left\{[\Delta_{21,k'}]^2 + \frac{1}{4}w_{21}^2\right\}} \frac{d^3k}{(2\pi)^3} \frac{d^3k'}{(2\pi)^3} . \tag{11.123}$$

For early times, i.e. $w_{32}t < 1$, $w_{21}t < 1$, the full solution in eqn (11.121) must be used, and the expansion (11.106) shows that the atom and the field are described by an entangled state. At late times, the irreversible decay of the upper-level occupation probabilities destroys the necessary coherence, and the system is described by the product state $|\varepsilon_3; 1_{\mathbf{k}s}, 1_{\mathbf{k}'s'}\rangle = |\varepsilon_3\rangle |1_{\mathbf{k}s}, 1_{\mathbf{k}'s'}\rangle$. Thus the atom is no longer entangled with the field, but the two photons remain entangled with one another, as described by the state $|1_{\mathbf{k}s}, 1_{\mathbf{k}'s'}\rangle$. The entanglement of the photons in the final state is the essential feature of the design of the photon indivisibility experiment.

11.3 The semiclassical limit

Since we have a fully quantum treatment of the electromagnetic field, it should be possible to derive the semiclassical approximation—which was simply assumed in Section 4.1—and combine it with the quantized description of spontaneous emission. This is an essential step, since there are many applications in which an effectively classical field, e.g. the single-mode output of a laser, interacts with atoms that can also undergo spontaneous emission into other modes. Of course, the entire electromagnetic field could be treated by the quantized theory, but this would unnecessarily complicate the description of the interesting applications. The final result—which is eminently plausible on physical grounds—can be stated as the following rule.

In the presence of an external classical field $\boldsymbol{\mathcal{E}}(\mathbf{r}, t) = -\partial\boldsymbol{\mathcal{A}}(\mathbf{r}, t)/\partial t$, the total Schrödinger-picture Hamiltonian is

$$H = H^{\mathrm{sc}}_{\mathrm{chg}}(t) + H_{\mathrm{em}} + H_{\mathrm{int}} , \tag{11.124}$$

where

$$H_{\mathrm{em}} = \sum_f \hbar\omega_f a^\dagger_f a_f \tag{11.125}$$

is the Hamiltonian for the quantized radiation field, and

$$H_{\text{int}} = -\int d^3r\, \widehat{\mathbf{j}}(\mathbf{r}) \cdot \mathbf{A}^{(+)}(\mathbf{r}) \tag{11.126}$$

is the interaction Hamiltonian between the quantized field and the charges. The remaining term,

$$H_{\text{chg}}^{\text{sc}}(t) = \sum_{n=1}^{N} \frac{\widehat{\mathbf{p}}_n^2}{2M_n} + \frac{1}{4\pi\epsilon_0} \sum_{n \neq l} \frac{q_n q_l}{|\widehat{\mathbf{r}}_n - \widehat{\mathbf{r}}_l|} - \sum_{n=1}^{N} \frac{q_n}{M_n} \boldsymbol{\mathcal{A}}(\widehat{\mathbf{r}}_n, t) \cdot \widehat{\mathbf{p}}_n\,, \tag{11.127}$$

includes the mutual Coulomb interaction between the charges and the interaction of the charges with the external classical field.

The rule (11.124) is derived in Section 11.3.1—where some subtleties concerning the separation of the quantized radiation field and the classical field are explained—and applied to the treatment of Rabi oscillations and the optical Bloch equation in the following sections.

11.3.1 The semiclassical Hamiltonian*

In the presence of a classical source current $\boldsymbol{\mathcal{J}}(\mathbf{r}, t)$, the complete Schrödinger-picture Hamiltonian is the sum of the microscopic Hamiltonian, given by eqn (4.29), and the hemiclassical interaction term given by eqn (5.36):

$$H = H_{\text{em}} + H_{\text{chg}} + H_{\text{int}} + H_{\mathcal{J}}(t)\,, \tag{11.128}$$

where H_{em}, H_{chg}, H_{int}, and $H_{\mathcal{J}}$ are given by eqns (5.29), (4.31), (5.27), and (5.36) respectively. The description of the internal states of atoms, etc. is contained in this Hamiltonian, since H_{chg} includes all Coulomb interactions between the charges. The hemiclassical interaction Hamiltonian is an explicit function of time—by virtue of the presence of the prescribed external current—which is conveniently expressed as

$$H_{\mathcal{J}}(t) = -\sum_{\kappa} \left[G_{\kappa}(t)\, a_{\kappa}^{\dagger} + G_{\kappa}^{*}(t)\, a_{\kappa} \right], \tag{11.129}$$

where

$$G_{\kappa}(t) = \sqrt{\frac{\hbar}{2\epsilon_0\omega_{\kappa}}} \int d^3r\, \boldsymbol{\mathcal{J}}(\mathbf{r}, t) \cdot \boldsymbol{\mathcal{E}}_{\kappa}^{*}(\mathbf{r}) \tag{11.130}$$

is the multimode generalization of the coefficients introduced in eqn (5.39).

The familiar semiclassical approximation involves a prescribed classical field, rather than a classical current, so our immediate objective is to show how to replace the current by the field. For this purpose, it is useful to transform to the Heisenberg picture, i.e. to replace the time-independent, Schrödinger-picture operators by their time-dependent, Heisenberg forms:

$$\left\{ a_{\kappa}, a_{\kappa}^{\dagger}, \widehat{\mathbf{r}}_n, \widehat{\mathbf{p}}_n \right\} \to \left\{ a_{\kappa}(t)\,, a_{\kappa}^{\dagger}(t)\,, \widehat{\mathbf{r}}_n(t)\,, \widehat{\mathbf{p}}_n(t) \right\}. \tag{11.131}$$

The c-number current $\boldsymbol{\mathcal{J}}(\mathbf{r}, t)$ is unchanged, so the full Hamiltonian in the Heisenberg picture is still an explicit function of time. The advantage of this transformation is that we can apply familiar methods for treating first-order, ordinary differential equations to the Heisenberg equations of motion for the quantum operators.

By using the equal-time commutation relations to evaluate $[a_\kappa(t), H(t)]$, one finds the Heisenberg equation for the annihilation operator $a_\kappa(t)$:

$$i\hbar\frac{d}{dt}a_\kappa(t) = \hbar\omega_\kappa a_\kappa(t) - G_\kappa(t) + [a_\kappa(t), H_{\text{int}}]. \tag{11.132}$$

The general solution of this linear, inhomogeneous differential equation for $a_\kappa(t)$ is the sum of the general solution of the homogeneous equation and any special solution of the inhomogeneous equation. The result (5.40) for the single-mode problem suggests the choice of the special solution $\alpha_\kappa(t)$, where $\alpha_\kappa(t)$ is a c-number function satisfying

$$i\hbar\frac{d}{dt}\alpha_\kappa(t) = \hbar\omega_\kappa\alpha_\kappa(t) - G_\kappa(t). \tag{11.133}$$

The *ansatz*

$$a_\kappa(t) = \alpha_\kappa(t) + a_\kappa^{\text{rad}}(t) \tag{11.134}$$

for the general solution defines a new operator, $a_\kappa^{\text{rad}}(t)$, that satisfies the canonical, equal-time commutation relations

$$\left[a_\kappa^{\text{rad}}(t), a_\lambda^{\text{rad}\dagger}(t)\right] = \delta_{\kappa\lambda}. \tag{11.135}$$

Substituting eqn (11.134) into eqn (11.132) produces the homogeneous differential equation

$$i\hbar\frac{d}{dt}a_\kappa^{\text{rad}}(t) = \hbar\omega_\kappa a_\kappa^{\text{rad}}(t) + \left[a_\kappa^{\text{rad}}(t), H_{\text{int}}\right]. \tag{11.136}$$

In order to express H_{int} in terms of the new operators $a_\kappa^{\text{rad}}(t)$, we substitute eqn (11.134) into the Heisenberg-picture version of the expansion (5.28) to get

$$\mathbf{A}^{(+)}(\mathbf{r}, t) = \boldsymbol{\mathcal{A}}^{(+)}(\mathbf{r}, t) + \mathbf{A}^{\text{rad}(+)}(\mathbf{r}, t). \tag{11.137}$$

The operator part,

$$\mathbf{A}^{\text{rad}(+)}(\mathbf{r}, t) = \sum_\kappa \sqrt{\frac{\hbar}{2\epsilon_0\omega_\kappa}}\, a_\kappa^{\text{rad}}(t)\, \boldsymbol{\mathcal{E}}_\kappa(\mathbf{r}), \tag{11.138}$$

is defined in terms of the new annihilation operators $a_\kappa^{\text{rad}}(t)$. The c-number part,

$$\boldsymbol{\mathcal{A}}^{(+)}(\mathbf{r}, t) = \sum_\kappa \sqrt{\frac{\hbar}{2\epsilon_0\omega_\kappa}}\, \alpha_\kappa(t)\, \boldsymbol{\mathcal{E}}_\kappa(\mathbf{r}) = \left\langle \underline{\alpha} \middle| \mathbf{A}^{(+)}(\mathbf{r}, t) \middle| \underline{\alpha} \right\rangle, \tag{11.139}$$

is the positive-frequency part of the classical field $\boldsymbol{\mathcal{A}}$ defined by the coherent state, $|\underline{\alpha}\rangle$, that is emitted by the classical current $\boldsymbol{\mathcal{J}}$. Substitution of eqn (11.137) into eqn (5.27) yields

$$H_{\text{int}} = H_{\text{int}}^{\text{sc}} + H_{\text{int}}^{\text{rad}}, \tag{11.140}$$

where

$$H_{\mathrm{int}}^{\mathrm{sc}} = -\int d^3r \, \widehat{\mathbf{j}}(\mathbf{r},t) \cdot \boldsymbol{\mathcal{A}}(\mathbf{r},t) \tag{11.141}$$

and

$$H_{\mathrm{int}}^{\mathrm{rad}} = -\int d^3r \, \widehat{\mathbf{j}}(\mathbf{r},t) \cdot \mathbf{A}^{\mathrm{rad}}(\mathbf{r},t) \tag{11.142}$$

respectively describe the interaction of the charges with the classical field, $\boldsymbol{\mathcal{A}}(\mathbf{r},t)$, and the quantized radiation field $\mathbf{A}^{\mathrm{rad}}(\mathbf{r},t)$. Since $a_{\kappa}^{\mathrm{rad}}(t)$ commutes with $H_{\mathrm{int}}^{\mathrm{sc}}$, the Heisenberg equation for $a_{\kappa}^{\mathrm{rad}}(t)$ is

$$i\hbar\frac{d}{dt}a_{\kappa}^{\mathrm{rad}}(t) = \hbar\omega_{\kappa}a_{\kappa}^{\mathrm{rad}}(t) + \left[a_{\kappa}^{\mathrm{rad}}(t), H_{\mathrm{int}}^{\mathrm{rad}}\right]. \tag{11.143}$$

The operators $\widehat{\mathbf{r}}_n(t)$ and $\widehat{\mathbf{p}}_n(t)$ for the charges commute with $H_{\mathcal{J}}(t)$, so their Heisenberg equations are

$$\begin{aligned} i\hbar\frac{d}{dt}\widehat{\mathbf{r}}_n(t) &= \left[\widehat{\mathbf{r}}_n(t), H_{\mathrm{chg}}\right] + \left[\widehat{\mathbf{r}}_n(t), H_{\mathrm{int}}^{\mathrm{sc}}\right] + \left[\widehat{\mathbf{r}}_n(t), H_{\mathrm{int}}^{\mathrm{rad}}\right], \\ i\hbar\frac{d}{dt}\widehat{\mathbf{p}}_n(t) &= \left[\widehat{\mathbf{p}}_n(t), H_{\mathrm{chg}}\right] + \left[\widehat{\mathbf{p}}_n(t), H_{\mathrm{int}}^{\mathrm{sc}}\right] + \left[\widehat{\mathbf{p}}_n(t), H_{\mathrm{int}}^{\mathrm{rad}}\right], \end{aligned} \tag{11.144}$$

where H_{chg} is given by eqn (4.31).

The complete Heisenberg equations, (11.143) and (11.144), follow from the new form,

$$H = H_{\mathrm{chg}}^{\mathrm{sc}} + H_{\mathrm{em}}^{\mathrm{rad}} + H_{\mathrm{int}}^{\mathrm{rad}}, \tag{11.145}$$

of the Hamiltonian, where

$$H_{\mathrm{chg}}^{\mathrm{sc}} = H_{\mathrm{chg}} + H_{\mathrm{int}}^{\mathrm{sc}} \tag{11.146}$$

and

$$H_{\mathrm{em}}^{\mathrm{rad}} = \sum_{\kappa}\hbar\omega_{\kappa}a_{\kappa}^{\mathrm{rad}\dagger}(t)\,a_{\kappa}^{\mathrm{rad}}(t). \tag{11.147}$$

We have, therefore, succeeded in replacing the classical current $\boldsymbol{\mathcal{J}}$ by the classical field $\boldsymbol{\mathcal{A}}$.

The definition (5.26) of the current operator and the explicit expression (4.31) for H_{chg} yield

$$H_{\mathrm{chg}}^{\mathrm{sc}} = \sum_{n=1}^{N}\frac{\widehat{\mathbf{p}}_n^2(t)}{2M_n} + \frac{1}{4\pi\epsilon_0}\sum_{n\neq l}\frac{q_n q_l}{\left|\widehat{\mathbf{r}}_n(t) - \widehat{\mathbf{r}}_l(t)\right|} - \sum_{n=1}^{N}\frac{q_n}{M_n}\boldsymbol{\mathcal{A}}\left(\widehat{\mathbf{r}}_n(t), t\right)\cdot\widehat{\mathbf{p}}_n(t), \tag{11.148}$$

which agrees with the semiclassical Hamiltonian in eqn (4.3), in the approximation that the $\boldsymbol{\mathcal{A}}^2$-terms are neglected. The explicit time dependence of the Schrödinger-picture form for the Hamiltonian—which is obtained by inverting the replacement rule (11.131)—now comes from the appearance of the classical field $\boldsymbol{\mathcal{A}}(\mathbf{r},t)$, rather than the classical current $\boldsymbol{\mathcal{J}}(\mathbf{r},t)$.

The replacement of a_κ by a_κ^{rad} is not quite as straightforward as it appears to be. The equal-time canonical commutation relation (11.135) guarantees the existence of a vacuum state $\left|0^{\text{rad}}\right\rangle$ for the a_κ^{rad}s, i.e.

$$a_\kappa^{\text{rad}}(t)\left|0^{\text{rad}}\right\rangle = 0 \quad \text{for all modes}\,, \tag{11.149}$$

but the physical interpretation of $\left|0^{\text{rad}}\right\rangle$ requires some care. The meaning of the new vacuum state becomes clear if one uses eqn (11.134) to express eqn (11.149) as

$$a_\kappa(t)\left|0^{\text{rad}}\right\rangle = \alpha_\kappa(t)\left|0^{\text{rad}}\right\rangle. \tag{11.150}$$

This shows that the Heisenberg-picture 'vacuum' for $a_\kappa^{\text{rad}}(t)$ is in fact the coherent state $|\underline{\alpha}\rangle$ generated by the classical current. In the Schrödinger picture this becomes

$$a_\kappa\left|0^{\text{rad}}(t)\right\rangle = \alpha_\kappa(t)\left|0^{\text{rad}}(t)\right\rangle, \tag{11.151}$$

which means that the modified vacuum state is even time dependent. In either picture, the excitations created by $a_\kappa^{\text{rad}\,\dagger}$ represent vacuum fluctuations relative to the coherent state $|\underline{\alpha}\rangle$. These subtleties are not very important in practice, since the classical field is typically confined to a single mode or a narrow band of modes. For other modes, i.e. those modes for which $\alpha_\kappa(t)$ vanishes at all times, the modified vacuum is the true vacuum. For this reason the superscript 'rad' in a_κ^{rad}, etc. will be omitted in the applications, and we arrive at eqn (11.124).

11.3.2 Rabi oscillations

The resonant wave approximation is also useful for describing the interaction of a two-level atom with a classical field having a long coherence time T_c, e.g. the field of a laser. From Section 4.8.2, we know that perturbation theory cannot be used if $T_c > 1/A$, where A is the Einstein A coefficient, but the RWA provides a nonperturbative approach. We will assume that there is only one mode, with frequency ω_0, which is nearly resonant with the atomic transition. In this case the interaction-picture state vector $|\Theta(t)\rangle$ satisfies

$$i\hbar\frac{\partial}{\partial t}|\Theta(t)\rangle = H_{\text{rwa}}(t)|\Theta(t)\rangle\,, \tag{11.152}$$

and specializing eqn (11.25) to the single mode $(\mathbf{k}_0, s_0)$ gives

$$H_{\text{rwa}}(t) = -i\hbar\mathfrak{g}_0 e^{-i\delta t}\sigma_+ a_0 + i\hbar\mathfrak{g}_0^* e^{i\delta t}\sigma_- a_0^\dagger\,, \tag{11.153}$$

where $\delta = \omega_0 - \omega_{21}$ is the detuning.

In Chapter 12 we will study the full quantum dynamics associated with this Hamiltonian (also known as the Jaynes–Cummings Hamiltonian), but for our immediate purposes we will assume that the combined system of field and atom is initially described by the state

$$|\Theta(0)\rangle = |\Psi(0)\rangle\,|\alpha\rangle\,, \tag{11.154}$$

where $|\Psi(0)\rangle$ is the initial state vector for the atom and $|\alpha\rangle$ is a coherent state for a_0, i.e.

$$a_0|\alpha\rangle = \alpha|\alpha\rangle\,. \tag{11.155}$$

This is a simple model for the output of a laser. As explained above, the operator $a_0^{\text{rad}} = a_0 - \alpha$ represents vacuum fluctuations around the coherent state, so replacing

a_0 by α in eqn (11.153) amounts to neglecting all vacuum fluctuations, including spontaneous emission from the upper level. This approximation defines the semiclassical Hamiltonian:

$$\begin{aligned} H_{\rm sc}(t) &= -i\hbar\mathfrak{g}e^{-i\delta t}\sigma_+ + i\hbar\mathfrak{g}^* e^{i\delta t}\sigma_- \\ &= \hbar\Omega_L e^{-i\delta t}\sigma_+ + \hbar\Omega_L^* e^{i\delta t}\sigma_- \,, \end{aligned} \tag{11.156}$$

where

$$\Omega_L = -i\mathfrak{g}_0\alpha = -\frac{\mathbf{d}\cdot\boldsymbol{\mathcal{E}}_L}{\hbar}\,, \tag{11.157}$$

and $\boldsymbol{\mathcal{E}}_L$ is the classical field amplitude corresponding to α. With the conventions adopted in Section 11.1.1, the atomic state is described by

$$|\Psi\rangle \to \begin{pmatrix} \Psi_2 \\ \Psi_1 \end{pmatrix}, \tag{11.158}$$

where $\Psi_2\,(\Psi_1)$ is the amplitude for the excited (ground) state. In this basis the Schrödinger equation becomes

$$i\frac{d}{dt}\begin{pmatrix} \Psi_2 \\ \Psi_1 \end{pmatrix} = \begin{bmatrix} 0 & \Omega_L e^{-i\delta t} \\ \Omega_L^* e^{i\delta t} & 0 \end{bmatrix} \begin{pmatrix} \Psi_2 \\ \Psi_1 \end{pmatrix}. \tag{11.159}$$

The transformation $\Psi_2 = \exp(-i\delta t/2)\,C_2$ and $\Psi_1 = \exp(i\delta t/2)\,C_1$ produces an equation with constant coefficients,

$$i\frac{d}{dt}\begin{pmatrix} C_2 \\ C_1 \end{pmatrix} = \begin{bmatrix} -\frac{\delta}{2} & \Omega_L \\ \Omega_L^* & \frac{\delta}{2} \end{bmatrix} \begin{pmatrix} C_2 \\ C_1 \end{pmatrix}. \tag{11.160}$$

The eigenvalues of the 2×2 matrix on the right are $\pm\Omega_R$, where

$$\Omega_R = \sqrt{\frac{\delta^2}{4} + |\Omega_L|^2} \tag{11.161}$$

is the Rabi frequency. The general solution is

$$\begin{pmatrix} C_2(t) \\ C_2(t) \end{pmatrix} = C_+\xi_+ \exp(-i\Omega_R t) + C_-\xi_- \exp(i\Omega_R t)\,, \tag{11.162}$$

where ξ_+ and ξ_- are the eigenvectors corresponding to $\pm\Omega_R$ and the constants $C_\pm$ are determined by the initial conditions. For exact resonance ($\delta = 0$) and an atom initially in the ground state, the occupation probabilities are

$$|\Psi_1(t)|^2 = \cos^2(\Omega_R t)\,, \tag{11.163}$$

$$|\Psi_2(t)|^2 = \sin^2(\Omega_R t)\,. \tag{11.164}$$

The oscillation between the ground and excited states is also known as **Rabi flopping**.

11.3.3 The Bloch equation

The pure-state description of an atom employed in the previous section is not usually valid, so the Schrödinger equation must be replaced by the quantum Liouville equation introduced in Section 2.3.2-A. In the interaction picture, eqn (2.119) becomes

$$i\hbar\frac{\partial}{\partial t}\rho\left(t\right)=\left[H_{\text{int}}\left(t\right),\rho\left(t\right)\right], \tag{11.165}$$

where $\rho(t)$ is the density operator for the system under study. We now consider a two-level atom interacting with a monochromatic classical field defined by the positive-frequency part,

$$\boldsymbol{\mathcal{E}}^{(+)}\left(\mathbf{r},t\right)=\overline{\boldsymbol{\mathcal{E}}}\left(\mathbf{r},t\right)e^{-i\omega_0 t}, \tag{11.166}$$

where ω_0 is the carrier frequency and $\boldsymbol{\mathcal{E}}(\mathbf{r},t)$ is the slowly-varying envelope. The RWA interaction Hamiltonian is then

$$\begin{aligned}H_{\text{rwa}}\left(t\right)&=-\mathbf{d}\cdot\boldsymbol{\mathcal{E}}^{(+)}\left(t\right)\sigma_{+}\left(t\right)-\mathbf{d}^{*}\cdot\boldsymbol{\mathcal{E}}^{(-)}\left(t\right)\sigma_{-}\left(t\right)\\&=-\mathbf{d}\cdot\overline{\boldsymbol{\mathcal{E}}}\left(t\right)e^{-i\delta t}\sigma_{+}-\mathbf{d}^{*}\cdot\overline{\boldsymbol{\mathcal{E}}}^{*}\left(t\right)e^{i\delta t}\sigma_{-},\end{aligned} \tag{11.167}$$

where $\overline{\boldsymbol{\mathcal{E}}}(t)=\overline{\boldsymbol{\mathcal{E}}}(\mathbf{R},t)$ is the slowly-varying envelope evaluated at the position $\mathbf{R}$ of the atom. The explicit time dependence of the atomic operators has been displayed by using eqn (11.12). In this special case, the quantum Liouville equation has the form

$$i\frac{d}{dt}\rho\left(t\right)=-\Omega\left(t\right)e^{-i\delta t}\left[\sigma_{+},\rho\left(t\right)\right]-\Omega^{*}\left(t\right)e^{i\delta t}\left[\sigma_{-},\rho\left(t\right)\right], \tag{11.168}$$

where the complex, time-dependent Rabi frequency is defined by

$$\Omega\left(t\right)=\frac{\mathbf{d}\cdot\overline{\boldsymbol{\mathcal{E}}}\left(t\right)}{\hbar}. \tag{11.169}$$

Combining the notation $\rho_{qp}(t)=\langle\varepsilon_q|\rho(t)|\varepsilon_p\rangle$ with the hermiticity condition $\rho_{12}(t)=\rho_{21}^{*}(t)$ allows eqn (11.168) to be written out explicitly as

$$i\frac{d}{dt}\rho_{11}\left(t\right)=-\Omega\left(t\right)e^{-i\delta t}\rho_{21}\left(t\right)+\Omega^{*}\left(t\right)e^{i\delta t}\rho_{12}\left(t\right), \tag{11.170}$$

$$i\frac{d}{dt}\rho_{22}\left(t\right)=\Omega\left(t\right)e^{-i\delta t}\rho_{21}\left(t\right)-\Omega^{*}\left(t\right)e^{i\delta t}\rho_{12}\left(t\right), \tag{11.171}$$

$$i\frac{d}{dt}\rho_{12}\left(t\right)=-\Omega\left(t\right)e^{-i\delta t}\left[\rho_{22}\left(t\right)-\rho_{11}\left(t\right)\right], \tag{11.172}$$

where ρ_{11} and ρ_{22} are the occupation probabilities for the two levels and the off-diagonal term ρ_{12} is called the **atomic coherence**. For most applications, it is better to eliminate the explicit exponentials by setting

$$\rho_{12}\left(t\right)=e^{-i\delta t}\overline{\rho}_{12}\left(t\right),\quad\rho_{22}\left(t\right)=\overline{\rho}_{22}\left(t\right),\quad\rho_{11}\left(t\right)=\overline{\rho}_{11}\left(t\right), \tag{11.173}$$

to get

$$\frac{d}{dt}\overline{\rho}_{22}(t) = i\left[\Omega(t)\overline{\rho}_{12}(t) - \Omega^*(t)\overline{\rho}_{21}(t)\right], \tag{11.174}$$
$$\frac{d}{dt}\overline{\rho}_{11}(t) = -i\left[\Omega(t)\overline{\rho}_{12}(t) - \Omega^*(t)\overline{\rho}_{21}(t)\right], \tag{11.175}$$
$$\frac{d}{dt}\overline{\rho}_{21}(t) = i\delta\overline{\rho}_{21}(t) + i\Omega(t)\left(\overline{\rho}_{11}(t) - \overline{\rho}_{22}(t)\right). \tag{11.176}$$

The sum of eqns (11.174) and (11.175) conveys the reassuring news that the total occupation probability, $\overline{\rho}_{11}(t) + \overline{\rho}_{22}(t)$, is conserved.

For a strictly monochromatic field, $\Omega(t) = \Omega$, these equations can be solved to obtain a generalized description of Rabi flopping, but there is a more pressing question to be addressed. This is the neglect of the decay of the upper level by spontaneous emission. We have seen in Section 11.2.2 that the upper-level amplitude $C_2(t) \sim \exp(-w_{21}t/2)$, so in the absence of the external field the occupation probability $\overline{\rho}_{22}$ of the upper level and the coherence $\overline{\rho}_{12}(t)$ should behave as

$$\begin{aligned}\overline{\rho}_{22}(t) &\sim C_2(t)\,C_2^*(t) \sim e^{-w_{21}t},\\ \overline{\rho}_{21}(t) &\sim C_2(t)\,C_1^*(t) \sim e^{-w_{21}t/2}.\end{aligned} \tag{11.177}$$

An equivalent statement is that the terms $-w_{21}\overline{\rho}_{22}(t)$ and $-w_{21}\overline{\rho}_{21}(t)/2$ should appear on the right sides of eqns (11.174) and (11.175) respectively. This would be the end of the story if spontaneous emission were the only thing that has been left out, but there are other effects to consider. In atomic vapors, elastic scattering from other atoms will disturb the coherence $\overline{\rho}_{12}(t)$ and cause an additional decay rate, and in crystals similar effects arise due to lattice vibrations and local field fluctuations.

The general description of dissipative effects will be studied Chapter 14, but for the present we will adopt a phenomenological approach in which eqns (11.174)–(11.176) are replaced by the **Bloch equations**:

$$\frac{d}{dt}\overline{\rho}_{22}(t) = -w_{21}\overline{\rho}_{22}(t) + i\left[\Omega(t)\overline{\rho}_{12}(t) - \Omega^*(t)\overline{\rho}_{21}(t)\right], \tag{11.178}$$
$$\frac{d}{dt}\overline{\rho}_{11}(t) = w_{21}\overline{\rho}_{22}(t) - i\left[\Omega(t)\overline{\rho}_{12}(t) - \Omega^*(t)\overline{\rho}_{21}(t)\right], \tag{11.179}$$
$$\frac{d}{dt}\overline{\rho}_{21}(t) = \left(i\delta - \Gamma_{21}\right)\overline{\rho}_{21}(t) + i\Omega(t)\left(\overline{\rho}_{11}(t) - \overline{\rho}_{22}(t)\right), \tag{11.180}$$

where the **decay rate** w_{21} and the **dephasing rate** Γ_{21} are parameters to be determined from experiment. In this simple two-level model the lower level is the ground state, so the term $w_{21}\overline{\rho}_{22}$ in eqn (11.179) is required in order to guarantee conservation of the total occupation probability. This allows eqns (11.179) and (11.180) to be replaced by

$$\overline{\rho}_{11}(t) + \overline{\rho}_{22}(t) = 1, \tag{11.181}$$

$$\frac{d}{dt}\left[\overline{\rho}_{22}(t) - \overline{\rho}_{11}(t)\right] = -w_{21} - w_{21}\left[\overline{\rho}_{22}(t) - \overline{\rho}_{11}(t)\right] + 2i\left[\Omega(t)\overline{\rho}_{12}(t) - \Omega^*(t)\overline{\rho}_{21}(t)\right], \tag{11.182}$$

where $\overline{\rho}_{22}(t) - \overline{\rho}_{11}(t)$ is the **population inversion**. In the literature, the parameters w_{21} and Γ_{21} are often represented as

$$w_{21} = \frac{1}{T_1}\,, \quad \Gamma_{21} = \frac{1}{T_2}\,, \tag{11.183}$$

where T_1 and T_2 are respectively called the **longitudinal** and **transverse relaxation times**. This terminology is another allusion to the analogy with a spin-1/2 system precessing in an external magnetic field. Another common usage is to call T_1 and T_2 respectively the *on-diagonal* and *off-diagonal* relaxation times.

In the frequency domain, the slow time variation of the field envelope $\overline{\boldsymbol{\mathcal{E}}}(t)$ is represented by the condition $\Delta\omega_0 \ll \omega_0$, where $\Delta\omega_0$ is the spectral width of $\overline{\boldsymbol{\mathcal{E}}}(\omega)$. The detuning and the dephasing rate are also small compared to the carrier frequency, but either or both can be large compared to $\Delta\omega_0$. This limit can be investigated by means of the formal solution,

$$\overline{\rho}_{21}(t) = \overline{\rho}_{21}(t_0)\, e^{(i\delta - \Gamma_{21})(t-t_0)} - i \int_{t_0}^{t} dt'\Omega(t')\left[\overline{\rho}_{22}(t') - \overline{\rho}_{11}(t')\right] e^{(i\delta-\Gamma_{21})(t-t')}\,, \tag{11.184}$$

of eqn (11.180). Since $\Gamma_{21} \geqslant w_{21}/2 > 0$, the formal solution has the $t_0 \to -\infty$ limit

$$\overline{\rho}_{21}(t) = -i \int_{-\infty}^{t} dt'\Omega(t')\left[\overline{\rho}_{22}(t') - \overline{\rho}_{11}(t')\right] e^{(i\delta-\Gamma_{21})(t-t')}\,. \tag{11.185}$$

The exponential factor $\exp\left[-\Gamma_{21}(t-t')\right]$ implies that the main contribution to the integral comes from the interval $t - 1/\Gamma_{21} < t' < t$, while the rapidly oscillating exponential $\exp\left[i\delta(t-t')\right]$ similarly restricts contributions to the interval $t - 1/|\delta| < t' < t$. Thus if either of the conditions $\Gamma_{21} \gg \max(\Delta\omega_0, w_{21})$ or $|\delta| \gg \max(\Delta\omega_0, w_{21})$ is satisfied, the main contribution to the integral comes from a small interval $t - \Delta t < t' < t$. In this interval, the remaining terms in the integrand are effectively constant; consequently, they can be evaluated at the upper limit to find:

$$\overline{\rho}_{21}(t) = \frac{\Omega(t)\left[\overline{\rho}_{22}(t) - \overline{\rho}_{11}(t)\right]}{\delta + i\Gamma_{21}}\,. \tag{11.186}$$

The approximation of the atomic coherence by this limiting form is called **adiabatic elimination**, by analogy to the behavior of thermodynamic systems. A thermodynamic parameter, such as the pressure of a gas, will change in step with slow changes in a control parameter, e.g. the temperature. The analogous behavior is seen in eqn (11.186) which shows that the atomic coherence $\overline{\rho}_{21}(t)$ follows the slower changes in the populations. For a large dephasing rate, exponential decay drives $\overline{\rho}_{21}(t)$ to the equilibrium value given by eqn (11.186). In the case of large detuning, the deviation from the equilibrium value oscillates so rapidly that its contribution averages to zero.

Once the mechanism of adiabatic elimination is understood, its application reduces to the following simple rule.

(a) If $|\Gamma_{qp} + i\Delta_{qp}|$ is large, set $d\overline{\rho}_{qp}/dt = 0$.

(b) Use the resulting algebraic relations to eliminate as many ρ_{qp}s as possible.

(11.187)

Substituting $\overline{\rho}_{21}(t)$ from eqn (11.186) into eqn (11.182) leads to

$$\frac{d}{dt}\left[\overline{\rho}_{22}(t)-\overline{\rho}_{11}(t)\right]=-w_{21}-\left\{w_{21}+\frac{4\left|\Omega(t)\right|^{2}\Gamma_{21}}{\delta^{2}+\Gamma_{21}^{2}}\right\}\left[\overline{\rho}_{22}(t)-\overline{\rho}_{11}(t)\right], \qquad (11.188)$$

which shows that the adiabatic elimination of the atomic coherence does not necessarily imply the adiabatic elimination of the population inversion. The solution of this differential equation also shows that no pumping scheme for a strictly two-level atom can change the population inversion from negative to positive. Since laser amplification requires a positive inversion, this implies that laser action can only be described by atoms with at least three active levels.

If $w_{21}=O(\Delta\omega_0)$ the population inversion and the external field change on the same time scale. Adiabatic elimination of the population inversion will only occur for $w_{21}\gg\Delta\omega_0$. In this limit the adiabatic elimination rule yields

$$\overline{\rho}_{22}(t)-\overline{\rho}_{11}(t)=-\frac{w_{21}}{w_{21}+\frac{4|\Omega(t)|^{2}\Gamma_{21}}{\delta^{2}+\Gamma_{21}^{2}}}<0\,. \qquad (11.189)$$

When adiabatic elimination is possible for both the atomic coherence and the population inversion, the atomic density matrix appears to react instantaneously to changes in the external field. What this really means is that transient effects are either suppressed by rapid damping ($w_{21}\gg\Delta\omega_0$, and $\Gamma_{21}\gg\Delta\omega_0$) or average to zero due to rapid oscillations ($|\delta|\gg\Delta\omega_0$). The apparently instantaneous response of the two-level atom is also displayed by multilevel atoms when the corresponding conditions are satisfied.

For later applications it is more useful to substitute the adiabatic form (11.186) into the original equations (11.179) and (11.178) to get a pair of equations for the occupation probabilities $P_q=\overline{\rho}_{qq}$. In the strictly monochromatic case, one finds

$$\begin{aligned}\frac{dP_2}{dt}&=W_{12}P_1-\left(w_{21}+W_{12}\right)P_2\,,\\ \frac{dP_1}{dt}&=-W_{12}P_1+\left(w_{21}+W_{12}\right)P_2\,,\end{aligned} \qquad (11.190)$$

where

$$W_{12}=\frac{2\left|\Omega\right|^{2}\Gamma_{21}}{\delta^{2}+\Gamma_{21}^{2}} \qquad (11.191)$$

is the rate of $1\rightarrow2$ transitions (absorptions) driven by the field. By virtue of the equality $B_{1\rightarrow2}=B_{2\rightarrow1}$, explained in Section 1.2.2, this is equal to the rate of $2\rightarrow1$ transitions (stimulated emissions) driven by the field. Equations (11.190) are called **rate equations** and their use is called the **rate equation approximation**. The occupation probability of $|\varepsilon_2\rangle$ is increased by absorption from $|\varepsilon_1\rangle$ and decreased by the combination of spontaneous and stimulated emission to $|\varepsilon_1\rangle$. The inverse transitions determine the rate of change of P_1, in such a way that probability is conserved. The rate equations can be generalized to atoms with three or more levels by adding up all of the (incoherent) processes feeding and depleting the occupation probability of each level.

11.4 Exercises

11.1 The antiresonant Hamiltonian

Apply the definition (11.17) of the running average to $H_{\text{int}}^{(\text{ar})}(t)$ to find:

$$\overline{H}_{\text{int}}^{(\text{ar})}(t) = -i\hbar \sum_{\mathbf{k}s} \left[\sqrt{\frac{\hbar\omega_k}{2\epsilon_0 V}} \frac{\mathbf{d}^* \cdot \mathbf{e}_{\mathbf{k}s}}{\hbar} \right] e^{-i(\omega_{21}+\omega_k)t} K(\omega_{21} + \omega_k) \, a_{\mathbf{k}s} \sigma_- + \text{HC}\,.$$

Use the properties of the cut-off function and the conventions $\omega_{21} > 0$ and $\omega_k > 0$ to explain why dropping $\overline{H}_{\text{int}}^{(\text{ar})}(t)$ is a good approximation.

11.2 The Weisskopf–Wigner method

(1) Fill in the steps needed to go from eqn (11.80) to eqn (11.84).

(2) Assume that $|K(\Delta)|^2$ is an even function of Δ and show that

$$\frac{\delta\omega_{21}}{w_{21}} = -\frac{3}{2\pi\omega_{21}} \int_{-\infty}^{\infty} d\Delta \; |K(\Delta)|^2 - \frac{1}{2\pi\omega_{21}^3} \int_{-\infty}^{\infty} d\Delta \; \Delta^2 \, |K(\Delta)|^2 \,.$$

Use this to derive the estimate $\delta\omega_{21}/w_{21} = O(w_K/\omega_{21}) \ll 1$.

11.3 Atomic radiation field

(1) Use the eqns (11.26) and (B.48) to show that

$$\sum_s \sqrt{\frac{\hbar\omega_k}{2\epsilon_0 V}} \mathfrak{g}_{\mathbf{k}s}^* \mathbf{e}_{\mathbf{k}s} e^{i\mathbf{k}\cdot\mathbf{r}} = \frac{\omega_k K(\Delta_k)}{2\epsilon_0 V} \left[\mathbf{d}^* + \frac{(\mathbf{d}^* \cdot \boldsymbol{\nabla}) \boldsymbol{\nabla}}{k^2} \right] e^{i\mathbf{k}\cdot\mathbf{r}}\,.$$

(2) With the aid of this result, convert the **k**-sum in eqn (11.54) to an integral. Show that

$$\int d\Omega_{\mathbf{k}} e^{i\mathbf{k}\cdot\mathbf{r}} = \frac{4\pi \sin(kr)}{kr}\,,$$

and then derive eqn (11.56).

11.4 Slowly-varying envelope operators

Define envelope operators $\overline{\sigma}_-(t) = \exp(i\omega_{21}t)\,\sigma_-(t)$, $\overline{\sigma}_z(t) = \sigma_z(t)$, and $\overline{a}_{\mathbf{k}s}(t) = \exp(i\omega_k t)\, a_{\mathbf{k}s}(t)$.

(1) Use eqns (11.47)–(11.49) to derive the equations satisfied by the envelope operators.

(2) From these equations argue that the envelope operators are slowly varying, i.e. essentially constant over an optical period.

11.5 Two-photon cascade*

(1) Substitute the *ansatz* (11.106) into the Schrödinger equation for the Hamiltonian (11.103) and obtain the differential equations for the coefficients.

(2) Use the given initial conditions to derive eqns (11.107)–(11.109).

(3) Carry out the steps needed to arrive at eqn (11.113).

(4) Starting with the normalization $|K(0)| = 1$ and the fact that $|K(\Delta')|^2$ is an even function, use an argument similar to the derivation of eqn (11.89) to show that $D_k \approx w_{21}/2$.

(5) Evaluate the residue for the poles of $\widetilde{X}_{\mathbf{k}s,\mathbf{k}'s'}(\zeta)$ to find the coefficients G_1, G_2, and G_3, and then derive eqn (11.122).

12

Cavity quantum electrodynamics

In Section 4.9 we studied spontaneous emission in free space and also in the modified geometry of a planar cavity. The large dimensions in both cases—three for free space and two for the planar cavity—provide the densely packed energy levels that are essential for the validity of the Fermi golden rule calculation of the emission rate.

Cavity quantum electrodynamics is concerned with the very different situation of an atom trapped in a cavity with all three dimensions comparable to the wavelength of the emitted radiation. In this case the radiation modes are discrete, and the Fermi golden rule cannot be used. Instead of disappearing into the blackness of infinite space, the emitted radiation is reflected from the nearby cavity walls, and soon absorbed again by the atom. The re-excitation of the atom results in a cycle of emissions and absorptions, rather than irreversible decay. In the limit of strong fields, i.e. many photons in a single mode, this cyclic behavior is described in Section 11.3.2 as Rabi flopping. The exact periodicity of Rabi flopping is, however, an artifact of the semiclassical approximation, in which the discrete nature of photons is ignored. In the limit of weak fields, the grainy nature of light makes itself felt in the nonclassical features of collapse and revival of the probability for atomic excitation.

There are several possible experimental realizations of cavity quantum electrodynamics, but the essential physical features of all of them are included in the Jaynes–Cummings model discussed in Section 12.1. In Section 12.2 we will use this model to describe the intrinsically quantum phenomena of collapse and revival of the radiation field in the cavity. A particular experimental realization is presented in Section 12.3.

12.1 The Jaynes–Cummings model

12.1.1 Definition of the model

In its simplest form, the Jaynes–Cummings model consists of a single two-level atom located in an ideal cavity. For the two-level atom we will use the treatment given in Section 11.1.1, in which the two atomic eigenstates are $|\epsilon_1\rangle$ and $|\epsilon_2\rangle$ with $\epsilon_1 < \epsilon_2$. The Hamiltonian is then

$$H_{\text{at}} = \frac{\hbar\omega_0}{2}\sigma_z\,, \tag{12.1}$$

where we have chosen the zero of energy so that $\epsilon_2 + \epsilon_1 = 0$, and set $\omega_0 \equiv (\epsilon_2 - \epsilon_1)/\hbar$. For the electromagnetic field, we use the formulation in Section 2.1, so that

$$H_{\text{em}} = \sum_{\kappa} \hbar\omega_\kappa a_\kappa^\dagger a_\kappa \tag{12.2}$$

is the Hamiltonian, and

$$\mathbf{E}^{(+)}(\mathbf{r}) = i\sum_{\kappa}\sqrt{\frac{\hbar\omega_\kappa}{2\epsilon_0}}a_\kappa\boldsymbol{\mathcal{E}}_\kappa(\mathbf{r}) \tag{12.3}$$

is the positive-frequency part of the electric field (in the Schrödinger picture).

Adapting the general result (11.27) to the cavity problem gives the RWA interaction Hamiltonian

$$\begin{aligned} H_{\text{rwa}} &= -\mathbf{d}\cdot\mathbf{E}^{(+)}\sigma_+ - \mathbf{d}^*\cdot\mathbf{E}^{(-)}\sigma_- \\ &= -i\hbar\sum_{\kappa} g_\kappa a_\kappa\sigma_+ + i\hbar\sum_{\kappa} g_\kappa^* a_\kappa^\dagger\sigma_- \, ; \end{aligned} \tag{12.4}$$

where $\mathbf{d} = \left\langle\epsilon_1\middle|\widehat{\mathbf{d}}\middle|\epsilon_2\right\rangle$ is the dipole matrix element; the coupling frequencies are

$$g_\kappa = K(\omega_0 - \omega_\kappa)\sqrt{\frac{\hbar\omega_\kappa}{2\epsilon_0}}\frac{\mathbf{d}\cdot\boldsymbol{\mathcal{E}}_\kappa(\mathbf{R})}{\hbar}\, ; \tag{12.5}$$

$K(\omega_0 - \omega_\kappa)$ is the RWA cut-off function; and $\mathbf{R}$ is the position of the atom.

We will now drastically simplify this model in two ways. The first is to assume that the center-of-mass motion of the atom can be treated classically. This means that ω_0 should be interpreted as the Doppler-shifted resonance frequency. In many cases the Doppler effect is not important; for example, for microwave transitions in Rydberg atoms passing through a resonant cavity, or single atoms confined in a trap. The second simplification is enforced by choosing the cavity parameters so that the lowest (fundamental) mode frequency is nearly resonant with the atomic transition, while all higher frequency modes are well out of resonance. This guarantees that only the lowest mode contributes to the resonant Hamiltonian; consequently, the family of annihilation operators a_κ can be reduced to the single operator a for the fundamental mode. From now on, we will call the fundamental frequency the **cavity frequency** ω_C and the corresponding mode function $\boldsymbol{\mathcal{E}}_C(\mathbf{R})$ the **cavity mode**.

The total Hamiltonian for the Jaynes–Cummings model is therefore $H_{\text{JC}} = H_0 + H_{\text{int}}$, where

$$H_0 = \hbar\omega_C a^\dagger a + (\hbar\omega_0/2)\,\sigma_z\,, \tag{12.6}$$

$$H_{\text{int}} = -i\hbar g a\sigma_+ + i\hbar g a^\dagger\sigma_-\,, \tag{12.7}$$

and

$$g = \sqrt{\frac{\hbar\omega_C}{2\epsilon_0}}\frac{\mathbf{d}\cdot\boldsymbol{\mathcal{E}}_C(\mathbf{R})}{\hbar}\,. \tag{12.8}$$

By appropriate choice of the phases in the atomic eigenstates $|\epsilon_1\rangle$ and $|\epsilon_2\rangle$, we can always arrange that g is real.

12.1.2 Dressed states

The interaction Hamiltonian in eqn (12.7) has the same general form as the interaction Hamiltonian (11.25) for the Weisskopf–Wigner model of Section 11.2.2, but it is greatly simplified by the fact that only one mode of the radiation field is active. In the Weisskopf–Wigner case, the infinite-dimensional subspaces $\mathfrak{H}_{\text{se}}$ are left invariant (mapped into themselves) under the action of the Hamiltonian. Since the Hamiltonians have the same structure, a similar behavior is expected in the present case.

The product states,

$$|\epsilon_j, n\rangle^{(0)} = |\epsilon_j\rangle |n\rangle \quad (n = 0, 1, \ldots)\,, \tag{12.9}$$

where the $|\epsilon_j\rangle$s $(j = 1, 2)$ are the atomic eigenstates and the $|n\rangle$s are number states for the cavity mode, provide a natural basis for the Hilbert space $\mathfrak{H}_{\text{JC}}$ of the Jaynes–Cummings model. The $|\epsilon_j, n\rangle^{(0)}$s are called **bare states**, since they are eigenstates of the non-interacting Hamiltonian H_0:

$$H_0 |\epsilon_j, n\rangle^{(0)} = (\epsilon_j + n\hbar\omega_C) |\epsilon_j, n\rangle^{(0)}. \tag{12.10}$$

Turning next to H_{int}, a straightforward calculation shows that

$$H_{\text{int}} |\epsilon_1, 0\rangle^{(0)} = 0\,, \tag{12.11}$$

which means that spontaneous absorption from the bare vacuum is forbidden in the resonant wave approximation. Consequently, the ground-state energy and state vector for the atom–field system are, respectively,

$$\varepsilon_G = \epsilon_1 = -\frac{\hbar\omega_0}{2} \quad \text{and} \quad |G\rangle = |\epsilon_1, 0\rangle^{(0)}. \tag{12.12}$$

Furthermore, for each photon number n the pairs of bare states $|\epsilon_2, n\rangle^{(0)}$ and $|\epsilon_1, n+1\rangle^{(0)}$ satisfy

$$\begin{aligned} H_{\text{int}} |\epsilon_2, n\rangle^{(0)} &= i\hbar g\sqrt{n+1}\, |\epsilon_1, n+1\rangle^{(0)}, \\ H_{\text{int}} |\epsilon_1, n+1\rangle^{(0)} &= -i\hbar g\sqrt{n+1}\, |\epsilon_2, n\rangle^{(0)}. \end{aligned} \tag{12.13}$$

Consequently, each two-dimensional subspace

$$\mathfrak{H}_n = \text{span}\left\{|\epsilon_2, n\rangle^{(0)}, |\epsilon_1, n+1\rangle^{(0)}\right\} \quad (n = 0, 1, \ldots) \tag{12.14}$$

is left invariant by the total Hamiltonian. This leads to the natural decomposition of $\mathfrak{H}_{\text{JC}}$ as

$$\mathfrak{H}_{\text{JC}} = \mathfrak{H}_G \oplus \mathfrak{H}_0 \oplus \mathfrak{H}_1 \oplus \cdots\,, \tag{12.15}$$

where $\mathfrak{H}_G = \text{span}\left\{|\epsilon_1, 0\rangle^{(0)}\right\}$ is the one-dimensional space spanned by the ground state. In the subspace $\mathfrak{H}_n$ the Hamiltonian is represented by a 2×2 matrix

$$H_{\text{JC},n} = \left(n + \frac{1}{2}\right)\hbar\omega_C \begin{bmatrix} 1 & 0 \\ 0 & 1 \end{bmatrix} + \frac{\hbar}{2}\begin{bmatrix} \delta & -2ig\sqrt{n+1} \\ 2ig\sqrt{n+1} & -\delta \end{bmatrix}, \tag{12.16}$$

where $\delta = \omega_0 - \omega_C$ is the detuning. This construction allows us to reduce the solution of the full Schrödinger equation, $H_{\text{JC}} |\Phi\rangle = \varepsilon |\Phi\rangle$, to the diagonalization of the 2×2-matrix $H_{\text{JC},n}$ for each n. The details are worked out in Exercise 12.1. For each subspace

$\mathfrak{H}_n$, the exact eigenvalues and eigenvectors, which will be denoted by $\varepsilon_{j,n}$ and $|j,n\rangle$ $(j = 1,2)$, respectively, are

$$\varepsilon_{1,n} = \left(n + \frac{1}{2}\right)\hbar\omega_C + \frac{\hbar\Omega_n}{2}\,, \tag{12.17}$$

$$|1,n\rangle = \sin\theta_n\,|\epsilon_2, n\rangle^{(0)} + \cos\theta_n\,|\epsilon_1, n+1\rangle^{(0)}, \tag{12.18}$$

$$\varepsilon_{2,n} = \left(n + \frac{1}{2}\right)\hbar\omega_C - \frac{\hbar\Omega_n}{2}\,, \tag{12.19}$$

$$|2,n\rangle = \cos\theta_n\,|\epsilon_2, n\rangle^{(0)} - \sin\theta_n\,|\epsilon_1, n+1\rangle^{(0)}, \tag{12.20}$$

where

$$\Omega_n = \sqrt{\delta^2 + 4g^2\,(n+1)} \tag{12.21}$$

is the Rabi frequency for oscillations between the two bare states in $\mathfrak{H}_n$. The probability amplitudes for the bare states are given by

$$\begin{aligned}\cos\theta_n &= \frac{\Omega_n - \delta}{\sqrt{(\Omega_n - \delta)^2 + 4g^2\,(n+1)}}\,,\\ \sin\theta_n &= \frac{2g\sqrt{n+1}}{\sqrt{(\Omega_n - \delta)^2 + 4g^2\,(n+1)}}\,.\end{aligned} \tag{12.22}$$

The bare $(g = 0)$ eigenvalues

$$\begin{aligned}\varepsilon_{1,n}^{(0)} &= (n + 1/2)\,\hbar\omega_C + \hbar\delta/2\,,\\ \varepsilon_{2,n}^{(0)} &= (n + 1/2)\,\hbar\omega_C - \hbar\delta/2\end{aligned} \tag{12.23}$$

are degenerate at resonance $(\delta = 0)$, but the exact eigenvalues satisfy

$$\varepsilon_{1,n} - \varepsilon_{2,n} = \hbar\Omega_n \geqslant 2\hbar g\sqrt{n+1}\,. \tag{12.24}$$

This is an example of the ubiquitous phenomenon of **avoided crossing** (or **level repulsion**) which occurs whenever two states are coupled by a perturbation.

The eigenstates $|1,n\rangle$ and $|2,n\rangle$ of the full Jaynes–Cummings Hamiltonian H_{JC} are called **dressed states**, since the interaction between the atom and the field is treated exactly. By virtue of this interaction, the dressed states are entangled states of the atom and the field.

12.2 Collapses and revivals

With the dressed eigenstates of H_{JC} in hand, we can write the general solution of the time-dependent Schrödinger equation as

$$|\Psi(t)\rangle = e^{-i\varepsilon_G t/\hbar} C_G\,|G\rangle + \sum_{n=0}^{\infty}\sum_{j=1}^{2} C_{j,n} e^{-i\varepsilon_{j,n} t/\hbar}\,|j,n\rangle\,, \tag{12.25}$$

where the expansion coefficients are determined by the initial state vector according to $C_G = \langle G\,|\Psi(0)\rangle$ and $C_{j,n} = \langle j,n\,|\Psi(0)\rangle$ $(j = 1,2)$ $(n = 0,1,\ldots)$. If the atom is initially in the excited state $|\epsilon_2\rangle$ and exactly m cavity photons are present, i.e. $|\Psi(0)\rangle = |\epsilon_2, m\rangle^{(0)}$, the general solution (12.25) specializes to $|\Psi(t)\rangle = |\epsilon_2, m; t\rangle$, where

$$
\begin{aligned}
|\epsilon_2, n; t\rangle \equiv e^{-i(n+1/2)\omega_C t} \left[\cos\left(\frac{\Omega_n t}{2}\right) + i\cos(2\theta_n)\sin\left(\frac{\Omega_n t}{2}\right)\right] |\epsilon_2, n\rangle^{(0)} \\
- ie^{-i(n+1/2)\omega_C t}\sin(2\theta_n)\sin\left(\frac{\Omega_n t}{2}\right)|\epsilon_1, n+1\rangle^{(0)}.
\end{aligned}
\tag{12.26}
$$

At resonance, the probabilities for the states $|\epsilon_2, m\rangle^{(0)}$ and $|\epsilon_1, m+1\rangle^{(0)}$ are

$$
\begin{aligned}
P_{2,m}(t) &= \left| {}^{(0)}\langle \epsilon_2, m\,|\Psi(t)\rangle \right|^2 = \cos^2\left(g\sqrt{m+1}t\right), \\
P_{1,m+1}(t) &= \left| {}^{(0)}\langle \epsilon_1, m+1\,|\Psi(t)\rangle \right|^2 = \sin^2\left(g\sqrt{m+1}t\right),
\end{aligned}
\tag{12.27}
$$

so—as expected—the system oscillates between the two atomic states by emission and absorption of a single photon. The exact periodicity displayed here is a consequence of the special choice of an initial state with a definite number of photons. For $m > 0$, this is analogous to the semiclassical problem of Rabi flopping driven by a field with definite amplitude and phase. The analogy to the classical case fails for $m = 0$, i.e. an excited atom with no photons present. The classical analogue of this case would be a vanishing field, so that no Rabi flopping would occur. The occupation probabilities $P_{2,0}(t) = \cos^2(gt)$ and $P_{1,1}(t) = \sin^2(gt)$ describe **vacuum Rabi flopping**, which is a consequence of the purely quantum phenomenon of spontaneous emission, followed by absorption, etc.

For initial states that are superpositions of several photon number states, exact periodicity is replaced by more complex behavior which we will now study. A superposition,

$$
|\Psi(0)\rangle = \sum_{n=0}^{\infty} K_n |\epsilon_2, n\rangle^{(0)},
\tag{12.28}
$$

of the initial states $|\epsilon_2, n\rangle^{(0)}$ that individually lead to Rabi flopping evolves into

$$
|\Psi(t)\rangle = \sum_{n=0}^{\infty} K_n |\epsilon_2, n; t\rangle,
\tag{12.29}
$$

so the probability to find the atom in the upper state, without regard to the number of photons, is

$$
P_2(t) = \sum_{n=0}^{\infty} \left| {}^{(0)}\langle \epsilon_2, n\,|\Psi(t)\rangle \right|^2 = \sum_{n=0}^{\infty} |K_n|^2 \left| {}^{(0)}\langle \epsilon_2, n\,|\epsilon_2, n; t\rangle \right|^2.
\tag{12.30}
$$

At resonance, eqn (12.27) allows this to be written as

$$
P_2(t) = \frac{1}{2} + \frac{1}{2}\sum_{n=0}^{\infty} |K_n|^2 \cos\left(2\sqrt{n+1}gt\right).
\tag{12.31}
$$

If more than one of the coefficients K_n is nonvanishing, this function is a sum of oscillatory terms with incommensurate frequencies. Thus true periodicity is only found for the special case $|K_n| = \delta_{nm}$ for some fixed value of m. For any choice of the K_ns the time average of the upper-level population is $\langle P_2(t)\rangle = 1/2$.

In order to study the behavior of $P_2(t)$, we need to make an explicit choice for the K_ns. Let us suppose, for example, that the initial state is $|\Psi(0)\rangle = |\epsilon_2\rangle |\alpha\rangle$, where $|\alpha\rangle$ is a coherent state for the cavity mode. The coefficients are then $|K_n|^2 = e^{-|\alpha|^2} |\alpha|^{2n}/n!$, and

$$P_2(t) = \frac{1}{2} + \frac{e^{-|\alpha|^2}}{2} \sum_{n=0}^{\infty} \frac{|\alpha|^{2n}}{n!} \cos\left(2\sqrt{n+1}gt\right). \tag{12.32}$$

Photon numbers for the coherent state follow a Poisson distribution, so the main contribution to the sum over n will come from the range $(\overline{n} - \Delta n, \overline{n} + \Delta n)$, where $\overline{n} = |\alpha|^2$ is the mean photon number and $\Delta n = |\alpha|$ is the variance. For large $\overline{n}$, the corresponding spread in Rabi frequencies is $\Delta\Omega \sim 2g$. At very early times, $t \ll 1/g$, the arguments of the cosines are essentially in phase, and $P_2(t)$ will execute an almost coherent oscillation. At later times, the variation of the Rabi frequencies with photon number will lead to an effectively random distribution of phases and destructive interference. This effect can be estimated analytically by replacing the sum over n with an integral and evaluating the integral in the stationary-phase approximation. The result,

$$P_2(t) = \frac{1}{2} + \frac{e^{-|gt|^2}}{2} \cos(2|\alpha| gt) \quad \text{for } gt \lesssim 1, \tag{12.33}$$

describes the **collapse** of the upper-level population to the time-averaged value of 1/2. This decay in the oscillations is neither surprising nor particularly quantal in character. A superposition of Rabi oscillations due to classical fields with random field strengths would produce a similar decay.

What is surprising is the behavior of the upper-level population at still later times. A numerical evaluation of eqn (12.32) reveals that the oscillations reappear after a *rephasing* time $t_{\text{rp}} \sim 4\pi|\alpha|/g$. This **revival**—with $P_2(t) = O(1)$—is a specifically quantum effect, explained by photon indivisibility. The revival is in turn followed by another collapse. The first collapse and revival are shown in Fig. 12.1.

The classical nature of the collapse is illustrated by the dashed curve in the same figure, which is calculated by replacing the discrete sum in eqn (12.32) by an integral. The two curves are indistinguishable in the initial collapse phase, but the classical (dashed) curve remains flat at the value 1/2 during the quantum revival. Thus the experimental observation of a revival provides further evidence for the indivisibility of photons. After a few collapse–revival cycles, the revivals begin to overlap and—as shown in Exercise 12.2—$P_2(t)$ becomes irregular.

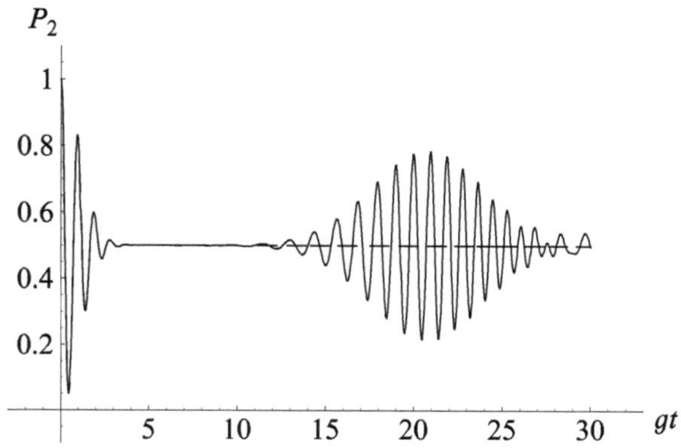

Fig. 12.1 The solid curve shows the probability $P_2(t)$ versus gt, where the upper-level population $P_2(t)$ is given by eqn (12.32), and the average photon number is $\overline{n} = |\alpha|^2 = 10$. The dashed curve is the corresponding classical result obtained by replacing the discrete sum over photon number by an integral.

12.3 The micromaser

The interaction of a Rydberg atom with the fundamental mode of a microwave cavity provides an excellent realization of the Jaynes–Cummings model. The configuration sketched in Fig. 12.2 is called a **micromaser** (Walther, 2003). It is designed so that—with high probability—at most one atom is present in the cavity at any given time. A velocity-selected beam of alkali atoms from an oven is sent into a laser excitation region, where the atoms are promoted to highly excited Rydberg states. The size of a Rydberg atom is characterized by the radius, $a_{\text{Ryd}} = n_p^2\hbar^2/me^2$, of its Bohr orbit, where n_p is

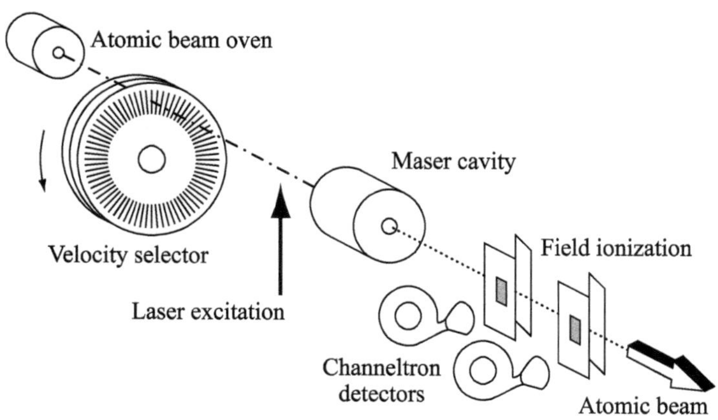

Fig. 12.2 Rubidium Rydberg atoms from an oven pass successively through a velocity selector, a laser excitation region, and a superconducting microwave cavity. After emerging from the cavity, they are detected—in a state-selective manner—by field ionization, followed by channeltron detectors. (Reproduced from Rempe *et al.* (1990).)

the principal quantum number, and $\hbar^2/me^2$ is the Bohr radius for the ground state of the hydrogen atom. These atoms are truly macroscopic in size; for example, the radius of a Rydberg atom with $n_p \simeq 100$ is on the order of microns, instead of nanometers. The dipole matrix element $\mathbf{d} = \langle n_p | e\widehat{\mathbf{r}} | n_p + 1 \rangle$ for a transition between two adjacent Rydberg states $n_p + 1 \to n_p$ is proportional to the diameter of the atom, so it scales as n_p^2. On the other hand, for transitions between high angular momentum (circular) states the frequency scales as $\omega \propto 1/n_p^3$, which is in the microwave range. According to eqn (4.162) the Einstein A coefficient scales like $A \propto |\mathbf{d}|^2 \omega^3 \propto 1/n_p^5$. Thus the lifetime $\tau = 1/A \propto n_p^5$ of the upper level is very long, and the neglect of spontaneous emission is a very good approximation.

The opposite conclusion follows for absorption and stimulated emission, since the relation (4.166) between the A and B coefficients shows that $B \propto n_p^4$. For the same applied field, the absorption rate for a Rydberg atom with $n_p \simeq 100$ is typically 10^8 times larger than the absorption rate at the Lyman transition between the $2p$ and $1s$ states of the hydrogen atom. Since stimulated emission is also described by the Einstein B coefficient, stimulated emission from the Rydberg atom can occur when there are only a few photons inside a microwave cavity.

As indicated in Fig. 12.2, a single Rydberg atom enters and leaves a superconducting microwave cavity through small holes drilled on opposite sides. During the transit time of the atom across the cavity the photons already present can stimulate emission of a single photon into the fundamental cavity mode; conversely, the atom can sometimes reabsorb a single photon. The interaction of the atom with a single mode of the cavity is described by the Jaynes–Cummings Hamiltonian in eqn (12.7). By monitoring whether or not the Rydberg atom has made a transition, $n_p + 1 \to n_p$, between the adjacent Rydberg states, one can infer indirectly whether or not a single microwave photon has been deposited in the cavity. This is possible because of the entangled nature of the dressed states in eqns (12.18) and (12.20). A measurement of the state of the atom, with the outcome $|\epsilon_2\rangle$, forces a reduction of the total state vector of the atom–radiation system, with the result that the radiation field is definitely in the state $|n\rangle$. In other words, the number of photons in the cavity has not changed. Conversely, a measurement with the outcome $|\epsilon_1\rangle$ guarantees that the field is in the state $|n + 1\rangle$, i.e. a photon has been added to the cavity.

The discrimination between the two Rydberg states is easily accomplished, since the ionization of the Rydberg atom by a DC electric field depends very sensitively on its principal quantum number n_p. The higher number $n_p + 1$ corresponds to a larger, more easily ionized atom, and the lower number n_p corresponds to a smaller, less easily ionized atom. The electric field in the first ionization region—shown in Fig. 12.2—is strong enough to ionize all $(n_p + 1)$-atoms, but too weak to ionize any n_p-atoms. Thus an atom that remains in the excited state is detected in the first region. If the atom has made a transition to the lower state, then it will be ionized by the stronger field in the second region. In this way, it is possible sensitively to identify the state of the Rydberg atom. If the atom is in the appropriate state, it will be ionized and release a single electron into the corresponding ionizing field region. The free electron is accelerated by the ionizing field and enters into an electron-multiplication region of a channeltron detector. As explained in Section 9.2.1, the channeltron detector can enormously multiply the single electron released by the Rydberg atom, and this provides an indirect method for continuously monitoring the photon-number state of the cavity.

A frequency-doubled dye laser ($\lambda = 297\,\text{nm}$) is used to excite rubidium (^{85}Rb) atoms to the $n_p = 63$, $P_{3/2}$ state from the $n_p = 5$, $S_{1/2}$ ($F = 3$) state. The cavity is tuned to the 21.456 GHz transition from the upper maser level in the $(n_p = 63, P_{3/2})$ state to the $(n_p = 61,\ D_{5/2})$ lower maser state. For this experiment a superconducting cavity with a Q-value of 3×10^8 was used, corresponding to a photon lifetime inside the cavity of 2 ms. The transit time of the Rydberg atom through the cavity is controlled by changing the atomic velocity with the velocity selector. On the average, only a small fraction of an atom is inside the cavity at any given time. In order to reduce the number of thermally excited photons in the cavity, a liquid helium environment reduces the temperature of the superconducting niobium microwave cavity to 2.5 K, corresponding to the average photon number $\overline{n} \approx 2$.

If the transit time of the atom is larger than the collapse time but smaller than the time of the first revival, then the solution (12.32) tells us that the atom will come into equilibrium with the cavity field, as seen in Fig. 12.1. In this situation the atom leaving the chamber is found in the upper or lower state with equal probability, i.e. $P_2 = 1/2$. When the transit time is increased to a value comparable to the first revival time, the probability for the excited state becomes larger than 1/2. The data in Fig. 12.3 show a quantum revival of the population of atoms in the upper maser state that occurs after a transit time of around 150 μs. Such a revival would be impossible in any semiclassical picture of the atom–field interaction; it is prima-facie evidence for the quantized nature of the electromagnetic field.

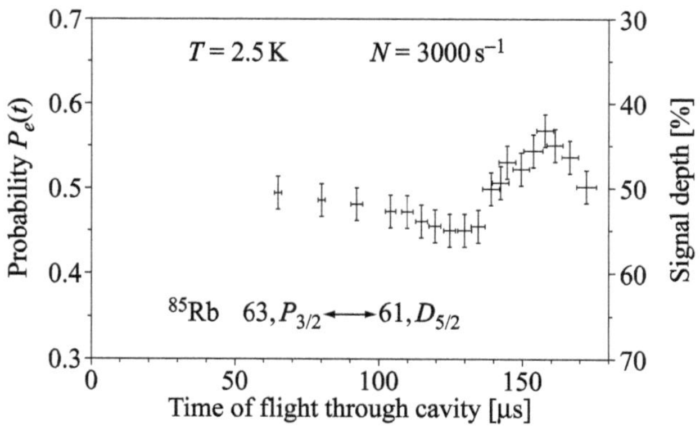

Fig. 12.3 Probability of finding the atom in the upper maser level as a function of the time of flight of a Rydberg atom through a superconducting cavity. The flux of atoms was around 3000 atoms per second. Note the revival of upper state atoms which occurs at around 150 μs. (Reproduced from Rempe *et al.* (1987).)

12.4 Exercises

12.1 Dressed states

(1) Verify eqns (12.10)–(12.16).

(2) Solve the eigenvalue problem for eqn (12.16) and thus derive eqns (12.17)–(12.22).

(3) Display level repulsion by plotting the (normalized) bare eigenvalues $\varepsilon_{1,n}^{(0)}/\hbar\omega_C$ and $\varepsilon_{2,n}^{(0)}/\hbar\omega_C$, and dressed eigenvalues $\varepsilon_{1,n}/\hbar\omega_C$ and $\varepsilon_{2,n}/\hbar\omega_C$ as functions of the detuning δ/ω_C.

12.2 Collapse and revival for pure initial states

(1) For the initial state $|\Psi(0)\rangle = |\epsilon_2, m\rangle^{(0)}$, verify the solution (12.26).

(2) Carry out the steps required to derive eqn (12.31).

(3) Write a program to evaluate eqn (12.32), and use it to study the behavior of $P_2(t)$ at times following the first revival.

12.3 Collapse and revival for a mixed initial state*

Replace the pure initial state of the previous problem with the mixed state

$$\rho = \sum_{n=0}^{\infty} p_n \, |\epsilon_2, n\rangle^{(0)} \, {}^{(0)}\langle\epsilon_2, n| \, .$$

(1) Show that this state evolves into

$$\rho = \sum_{n=0}^{\infty} p_n \, |\epsilon_2, n; t\rangle^{(0)} \, {}^{(0)}\langle\epsilon_2, n; t| \, .$$

(2) Derive the expression for $P_2(t)$.

(3) Assume that p_n is the thermal distribution for a given average photon number $\overline{n}$. Evaluate and plot $P_2(t)$ numerically for the value of $\overline{n}$ used in Fig. 12.1. Comment on the comparison between the two plots.

13 Nonlinear quantum optics

The interaction of light beams with linear optical devices is adequately described by the quantum theory of light propagation explained in Section 3.3, Chapter 7, and Chapter 8, but some of the most important applications involve modification of the incident light by interactions with nonlinear media, e.g. by frequency doubling, spontaneous down-conversion, four-wave mixing, etc. These phenomena are the province of nonlinear optics. Classical nonlinear optics deals with fields that are strong enough to cause appreciable change in the optical properties of the medium, so that the weak-field condition of Section 3.3.1 is violated. A Bloch equation that includes dissipative effects, such as scattering from other atoms and spontaneous emission, describes the response of the atomic density operator to the classical field.

For the present, we do not need the details of the Bloch equation. All we need to know is that there is a characteristic response time, T_{med}, for the medium. The classical envelope field evolves on the time scale $T_{\text{fld}} \sim 1/\Omega$, where Ω is the characteristic Rabi frequency. If $T_{\text{med}} \approx T_{\text{fld}}$ the coupled equations for the atoms and the field must be solved together. This situation arises, for example, in the phenomenon of self-induced transparency and in the theory of free-electron lasers (Yariv, 1989, Chaps 13, 15).

In many applications of interest for nonlinear optics, the incident radiation is detuned from the atomic resonances in order to avoid absorption. As shown in Section 11.3.3, this justifies the evaluation of the atomic density matrix by adiabatic elimination. In this approximation, the atoms appear to follow the envelope field instantaneously; they are said to be *slaved* to the field. Even with this simplification, the Bloch equation cannot be solved exactly, so the atomic density operator is evaluated by using time-dependent perturbation theory in the atom–field coupling. In this calculation, excited states of an atom only appear as virtual intermediate states; the atom is always returned to its original state. This means that both spontaneous emission and absorption are neglected.

13.1 The atomic polarization

Substituting the perturbative expression for the atomic density matrix into the source terms for Maxwell's equations results in the apparent disappearance, via adiabatic elimination, of the atomic degrees of freedom. This in turn produces an expansion of the medium polarization in powers of the field, which is schematically represented by

$$\mathcal{P}_i = \epsilon_0 \left[\chi_{ij}^{(1)} \mathcal{E}_j + \chi_{ijk}^{(2)} \mathcal{E}_j \mathcal{E}_k + \chi_{ijkl}^{(3)} \mathcal{E}_j \mathcal{E}_k \mathcal{E}_l + \cdots \right], \tag{13.1}$$

where the $\chi^{(n)}$s are the tensor **nonlinear susceptibilities** required for dealing with anisotropic materials and $\boldsymbol{\mathcal{E}}$ is the classical electric field. The term $\chi^{(2)}_{ijk}\mathcal{E}_j\mathcal{E}_k$ describes the combination of two waves to provide the source for a third, so it is said to describe **three-wave mixing**. In the same way $\chi^{(3)}_{ijkl}\mathcal{E}_j\mathcal{E}_k\mathcal{E}_l$ is associated with **four-wave mixing**.

A substance is called **weakly nonlinear** if the dielectric response is accurately represented by a small number of terms in the expansion (13.1). This approximation is the basis for most of nonlinear optics,[1] but there are nonlinear optical effects that cannot be described in this way, e.g. saturation in lasers (Yariv, 1989, Sec. 8.7). The higher-order terms in the polarization lead to nonlinear terms in Maxwell's equations that represent self-coupling of individual modes as well as coupling between different modes. These terms describe **self-actions** of the electromagnetic field that are mediated by the interaction of the field with the medium.

Quantum nonlinear optics is concerned with situations in which there are a small number of photons in some or all of the field modes. In this case the quantized field theory is required, but the correspondence principle assures us that the effects arising in classical nonlinear optics must also be present in the quantum theory. Thus the classical three- and four-wave mixing terms correspond to three- and four-photon interactions. Since the quantum fields are typically weak, these nonlinear phenomena are often unobservably small. There are, however, at least two situations in which this is not the case. According to eqn (2.188), the vacuum fluctuation field strength in a physical cavity of volume V is $\mathfrak{e}_f = \sqrt{\hbar\omega_f/2\epsilon_0 V}$. This shows that substantial field strengths can be achieved, even for a single photon, in a small enough cavity. A second exception depends on the fact that the frequency-dependent nonlinear susceptibilities display resonant behavior. If the detuning from resonance is made as small as possible—i.e. without violating the conditions required for adiabatic elimination—the nonlinear couplings are said to be **resonantly enhanced**.

When both of these conditions are met, the interaction between the medium and the field can be so strong that the electromagnetic field will interact with itself, even when there are only a few quanta present. This happens, for example, when microwave photons inside a cavity interact with each other via a medium composed of Rydberg atoms excited near resonance. In this case the interacting microwave photons can even form a photon fluid.

In addition to these practical issues, there are situations in which the use of quantum theory is mandatory. In the phenomenon of spontaneous down-conversion, a nonlinear optical process couples vacuum fluctuations of the electromagnetic field to an incident beam of ultraviolet light so that an ultraviolet photon decays into a pair of lower-energy photons. Effects of this kind cannot be described by the semiclassical theory.

In Section 13.2 we will briefly review some features of classical nonlinear optics and introduce the corresponding quantum description. In the following two sections we will discuss examples of three- and four-photon coupling. In each case the quantum theory

[1]For a selection of recent texts on nonlinear optics, see Shen (1984), Schubert and Wilhelmi (1986), Butcher and Cotter (1990), Boyd (1992), and Newell and Moloney (1992).

will be developed in a phenomenological way, i.e. it will be based on a conjectured form for the Hamiltonian. This is in fact the standard way of formulating a quantum theory. The choice of the Hamiltonian must ultimately be justified by comparing the results of calculations with experiment, as there will always be ambiguities—such as in operator ordering, coordinate choices (e.g. Cartesian versus spherical), etc.—which cannot be settled by theoretical arguments alone. Quantum theory is richer than classical theory; consequently, there is no unique way of deriving the quantum Hamiltonian from the classical energy.

13.2 Weakly nonlinear media

13.2.1 Classical theory

A *Plane waves in crystals*

Many applications of nonlinear optics involve the interaction of light with crystals, so we briefly review the form of the fundamental plane waves in a crystal. As explained in Appendix B.5.3, the field can be expressed as

$$\boldsymbol{\mathcal{E}}^{(+)}(\mathbf{r},t)=i\frac{1}{\sqrt{V}}\sum_{\mathbf{k}s}\mathfrak{F}_{ks}\alpha_{\mathbf{k}s}\boldsymbol{\varepsilon}_{\mathbf{k}s}e^{i(\mathbf{k}\cdot\mathbf{r}-\omega_{ks}t)}\,, \tag{13.2}$$

where $\boldsymbol{\varepsilon}_{\mathbf{k}s}$ is a crystal eigenpolarization, the polarization-dependent frequency ω_{ks} is a solution of the dispersion relation

$$c^2k^2=\omega^2n_s^2(\omega)\,, \tag{13.3}$$

and $n_s(\omega)$ is the index of refraction associated with the eigenpolarization $\boldsymbol{\varepsilon}_{\mathbf{k}s}$. The normalization constant,

$$\mathfrak{F}_{ks}=\sqrt{\frac{\hbar\omega_{ks}v_g(\omega_{ks})}{2\epsilon_0 n_s(\omega_{ks})\,c}}\,, \tag{13.4}$$

has been chosen to smooth the path toward quantization, and $v_g(\omega_{ks})=d\omega_{ks}/dk$ is the group velocity. For a polychromatic field, the expression (3.116) for the envelope $\overline{\boldsymbol{\mathcal{E}}}_\beta^{(+)}$ is replaced by

$$\overline{\boldsymbol{\mathcal{E}}}_\beta^{(+)}(\mathbf{r},t)=\frac{1}{\sqrt{V}}\sum_{\mathbf{k}s}{}'\,\mathfrak{F}_{ks}\alpha_{\mathbf{k}s}\boldsymbol{\varepsilon}_{\mathbf{k}s}e^{i(\mathbf{k}\cdot\mathbf{r}-\Delta_{\beta k}t)}\,, \tag{13.5}$$

where the prime on the **k**-sum indicates that it is restricted to **k**-values such that the detuning, $\Delta_{\beta ks}=\omega_{ks}-\omega_\beta$, is small compared to the minimum spacing between carrier frequencies, i.e. $|\Delta_{\beta ks}|\ll\min\{|\omega_\alpha-\omega_\beta|\,,\ \alpha\neq\beta\}$.

B *Nonlinear susceptibilities*

Symmetry, or lack of symmetry, with respect to spatial inversion is a fundamental distinction between different materials. A medium is said to have a **center of symmetry**, or to be **centrosymmetric**, if there is a spatial point (which is conventionally chosen as the origin of coordinates) with the property that the inversion transformation $\mathbf{r} \to -\mathbf{r}$ leaves the medium invariant. When this is true, the polarization must behave as a polar vector, i.e. $\boldsymbol{\mathcal{P}} \to -\boldsymbol{\mathcal{P}}$. The electric field is also a polar vector, so eqn (13.1) implies that all even-order susceptibilities—in particular $\chi^{(2)}_{ijk}$—vanish for centrosymmetric media. Vapors, liquids, amorphous solids, and some crystals are centrosymmetric. The absence of a center of symmetry defines a **non-centrosymmetric** crystal. This is the only case in which it is possible to obtain a nonvanishing $\chi^{(2)}_{ijk}$. There is no such general restriction on $\chi^{(3)}_{ijkl}$—or any odd-order susceptibility—since the third-order polarization, $\mathcal{P}^{(3)}_i = \chi^{(3)}_{ijkl}\mathcal{E}_j\mathcal{E}_k\mathcal{E}_l$, is odd under $\boldsymbol{\mathcal{E}} \to -\boldsymbol{\mathcal{E}}$.

The schematic expansion (13.1) does not explicitly account for dispersion, so we now turn to the exact constitutive relation

$$\begin{aligned}\mathcal{P}^{(n)}_i(\mathbf{r},t) = \epsilon_0 \int dt_1 \cdots \int dt_n \chi^{(n)}_{ij_1j_2\cdots j_n}(t-t_1, t-t_2, \ldots, t-t_n)\\ \times \mathcal{E}_{j_1}(\mathbf{r},t_1)\cdots\mathcal{E}_{j_n}(\mathbf{r},t_n)\end{aligned} \tag{13.6}$$

for the nth-order polarization, which is treated in greater detail in Appendix B.5.4. This time-domain form explicitly displays the history dependence of the polarization—previously encountered in Section 3.3.1-B—but the equivalent frequency-domain form

$$\begin{aligned}\mathcal{P}^{(n)}_i(\mathbf{r},\nu) = \epsilon_0 \int \frac{d\nu_1}{2\pi} \cdots \int \frac{d\nu_n}{2\pi} 2\pi\delta\left(\nu - \sum_{p=1}^{n}\nu_p\right)\chi^{(n)}_{ij_1j_2\cdots j_n}(\nu_1, \ldots, \nu_n)\\ \times \mathcal{E}_{j_1}(\mathbf{r},\nu_1)\cdots\mathcal{E}_{j_n}(\mathbf{r},\nu_n)\end{aligned} \tag{13.7}$$

is more useful in practice.

C Effective electromagnetic energy

The derivation in Section 3.3.1-B of the effective electromagnetic energy for a linear, dispersive dielectric can be restated in the following simplified form.

(1) Start with the expression for the energy in a static field.

(2) Replace the static field by a time-dependent field.

(3) Perform a running time-average—as in eqn (3.136)—on the resulting expression.

For a nonlinear dielectric, we carry out step (1) by using the result

$$\begin{aligned}\mathcal{U}_{\text{es}} &= \int_{V_c} d^3r \int_0^{\boldsymbol{\mathcal{D}}} \boldsymbol{\mathcal{E}}(\mathbf{r}) \cdot d(\boldsymbol{\mathcal{D}}(\mathbf{r}))\\ &= \frac{\epsilon_0}{2}\int_{V_c} d^3r \boldsymbol{\mathcal{E}}^2(\mathbf{r}) + \int_{V_c} d^3r \int_0^{\boldsymbol{\mathcal{P}}} \boldsymbol{\mathcal{E}}(\mathbf{r}) \cdot d(\boldsymbol{\mathcal{P}}(\mathbf{r}))\end{aligned} \tag{13.8}$$

for the energy of a static field in a dielectric occupying the volume V_c (Jackson, 1999, Sec. 4.7). Substituting eqn (13.1) into this expression leads to an expansion of the energy in powers of the field amplitude:

$$\mathcal{U}_{\text{es}} = \mathcal{U}^{(2)}_{\text{es}} + \mathcal{U}^{(3)}_{\text{es}} + \mathcal{U}^{(4)}_{\text{es}} + \cdots . \tag{13.9}$$

The first term on the right is discussed in Section 3.3.1-B, so we can concentrate on the higher-order ($n \geqslant 3$) terms:

$$\mathcal{U}_{\text{es}}^{(n)} = \frac{1}{n} \int_{V_c} d^3 r \mathcal{E}_i (\mathbf{r}) \mathcal{P}_i^{(n-1)} (\mathbf{r}) .$$

In steps (2) and (3), we replace the static energy by the effective energy,

$$\mathcal{U}_{\text{es}}^{(n)} \to \mathcal{U}_{\text{em}}^{(n)} (t) = \frac{1}{n} \int_{V_c} d^3 r \left\langle \mathcal{E}_i (\mathbf{r}, t) \mathcal{P}_i^{(n-1)} (\mathbf{r}, t) \right\rangle \quad \text{for } n \geqslant 3 , \tag{13.10}$$

and use eqn (13.6) to evaluate the nth-order polarization. Our experience with the quadratic term, $\mathcal{U}_{\text{em}}^{(2)} (t)$, tells us that eqn (13.10) will only be useful for polychromatic fields; therefore, we impose the condition $1/\omega_{\min} \ll T \ll 1/\Delta\omega_{\max}$ on the averaging time, where $\omega_{\min}$ is the smallest carrier frequency and $\Delta\omega_{\max}$ is the largest spectral width for the polychromatic field. This time-averaging eliminates all rapidly-varying terms, while leaving the slowly-varying envelope fields unchanged.

The lowest-order energy associated with the nonlinear polarizations is

$$\mathcal{U}_{\text{em}}^{(3)} (t) = \frac{1}{3} \int_{V_c} d^3 r \left\langle \mathcal{E}_i (\mathbf{r}, t) \mathcal{P}_i^{(2)} (\mathbf{r}, t) \right\rangle , \tag{13.11}$$

so the next task is to evaluate $\mathcal{P}_i^{(2)} (\mathbf{r}, t)$ for a polychromatic field. This is done by applying the exact relation (13.7) for $n = 2$, and using the expansion (3.119) for a polychromatic field to find:

$$\begin{aligned} \mathcal{P}_i^{(2)} (\mathbf{r}, \nu) = \epsilon_0 \sum_{\beta,\gamma} \sum_{\sigma',\sigma''=\pm} \int \frac{d\nu_1}{2\pi} \int \frac{d\nu_2}{2\pi} 2\pi\delta (\nu - \nu_1 - \nu_2) \chi_{ijk}^{(2)} (\nu_1, \nu_2) \\ \times \overline{\mathcal{E}}_{\beta j}^{(\sigma')} (\mathbf{r}, \nu_1 - \sigma'\omega_\beta) \overline{\mathcal{E}}_{\gamma k}^{(\sigma'')} (\mathbf{r}, \nu_2 - \sigma''\omega_\gamma) . \end{aligned} \tag{13.12}$$

Weak dispersion means that the susceptibility is essentially constant across the spectral width of each sharply-peaked envelope function, $\overline{\mathcal{E}}_{\beta j}^{(\pm)} (\mathbf{r}, \nu)$; therefore, $\mathcal{P}_i^{(2)} (\mathbf{r}, \nu)$ can be approximated by

$$\begin{aligned} \mathcal{P}_i^{(2)} (\mathbf{r}, \nu) = \epsilon_0 \sum_{\beta,\gamma} \sum_{\sigma',\sigma''=\pm} \int \frac{d\nu_1}{2\pi} \int \frac{d\nu_2}{2\pi} 2\pi\delta (\nu - \nu_1 - \nu_2) \chi_{ijk}^{(2)} (\sigma'\omega_\beta, \sigma''\omega_\gamma) \\ \times \overline{\mathcal{E}}_{\beta j}^{(\sigma')} (\mathbf{r}, \nu_1 - \sigma'\omega_\beta) \overline{\mathcal{E}}_{\gamma k}^{(\sigma'')} (\mathbf{r}, \nu_2 - \sigma''\omega_\gamma) . \end{aligned} \tag{13.13}$$

Carrying out an inverse Fourier transform yields the time-domain relation,

$$\begin{aligned} \mathcal{P}_i^{(2)} (\mathbf{r}, t) = \epsilon_0 \sum_{\beta,\gamma} \sum_{\sigma',\sigma''=\pm} \chi_{ijk}^{(2)} (\sigma'\omega_\beta, \sigma''\omega_\gamma) \\ \times \overline{\mathcal{E}}_{\beta j}^{(\sigma')} (\mathbf{r}, t) \overline{\mathcal{E}}_{\gamma k}^{(\sigma'')} (\mathbf{r}, t) e^{-i(\sigma'\omega_\beta + \sigma''\omega_\gamma)t} , \end{aligned} \tag{13.14}$$

which shows that the time-averaging has eliminated the history dependence of the polarization.

Using eqn (13.14) to evaluate the expression (13.11) for $\mathcal{U}_{\text{em}}^{(3)}(t)$ is simplified by the observation that the slowly-varying envelope fields can be taken outside the time average, so that

$$\mathcal{U}_{\text{em}}^{(3)}(t) = \frac{1}{3}\int_{V_c} d^3r \sum_{\alpha,\beta,\gamma}\sum_{\sigma,\sigma',\sigma''} \chi_{ijk}^{(2)}(\sigma'\omega_\beta, \sigma''\omega_\gamma)\,\overline{\mathcal{E}}_{\alpha i}^{(\sigma)}(\mathbf{r},t)\,\overline{\mathcal{E}}_{\beta j}^{(\sigma')}(\mathbf{r},t)$$
$$\times\, \overline{\mathcal{E}}_{\gamma k}^{(\sigma'')}(\mathbf{r},t)\left\langle e^{-i(\sigma\omega_\alpha+\sigma'\omega_\beta+\sigma''\omega_\gamma)t}\right\rangle . \tag{13.15}$$

The frequencies in the exponential all satisfy $\omega T \gg 1$, so the remaining time-average,

$$\left\langle e^{-i(\sigma\omega_\alpha+\sigma'\omega_\beta+\sigma''\omega_\gamma)t}\right\rangle = \frac{1}{T}\int_{-T/2}^{T/2} d\tau e^{-i(\sigma\omega_\alpha+\sigma'\omega_\beta+\sigma''\omega_\gamma)(t+\tau)} ,$$

vanishes unless

$$\sigma\omega_\alpha + \sigma'\omega_\beta + \sigma''\omega_\gamma = 0 . \tag{13.16}$$

This is called **phase matching**. By convention, the carrier frequencies are positive; consequently, phase matching in eqn (13.15) always imposes conditions of the form

$$\omega_\alpha = \omega_\beta + \omega_\gamma . \tag{13.17}$$

This in turn means that only terms of the form $\overline{\mathcal{E}}^{(+)}\overline{\mathcal{E}}^{(+)}\overline{\mathcal{E}}^{(-)}$ or $\overline{\mathcal{E}}^{(-)}\overline{\mathcal{E}}^{(-)}\overline{\mathcal{E}}^{(+)}$ will contribute. By making use of the symmetry properties of the susceptibility, reviewed in Appendix B.5.4, one finds the explicit result

$$\mathcal{U}_{\text{em}}^{(3)}(t) = \epsilon_0 \sum_{\alpha,\beta,\gamma} \chi_{ijk}^{(2)}(\omega_\beta, \omega_\gamma)\,\delta_{\omega_\alpha,\omega_\beta+\omega_\gamma}$$
$$\times \int_{V_c} d^3r \left[\overline{\mathcal{E}}_{\alpha i}^{(-)}(\mathbf{r},t)\,\overline{\mathcal{E}}_{\beta j}^{(+)}(\mathbf{r},t)\,\overline{\mathcal{E}}_{\gamma k}^{(+)}(\mathbf{r},t) + \text{CC}\right] . \tag{13.18}$$

In many applications, the envelope fields will be expressed by an expansion in some appropriate set of basis functions. For example, if the nonlinear medium is placed in a resonant cavity, then the carrier frequencies can be identified with the frequencies of the cavity modes, and each envelope field is proportional to the corresponding mode function. More generally, the field can be represented by the plane-wave expansion (13.2), provided that the power spectrum $|\alpha_{\mathbf{k}s}|^2$ exhibits well-resolved peaks at $\omega_{ks} = \omega_\alpha$, where ω_α ranges over the distinct monochromatic carrier frequencies. With this restriction held firmly in mind, the explicit sums over the distinct monochromatic waves can be replaced by sums over the plane-wave modes, so that

$$\mathcal{U}_{\text{em}}^{(3)} = \frac{i}{V^{3/2}} \sum_{\mathbf{k}_0 s_0, \mathbf{k}_1 s_1, \mathbf{k}_2 s_2} g_{s_0 s_1 s_2}^{(3)}(\omega_1, \omega_2)\left[\alpha_0\alpha_1^*\alpha_2^* - \text{CC}\right]$$
$$\times\, \mathcal{C}(\mathbf{k}_0 - \mathbf{k}_1 - \mathbf{k}_2)\,\delta_{\omega_0,\omega_1+\omega_2} , \tag{13.19}$$

where $\alpha_0 = \alpha_{\mathbf{k}_0 s_0}$, etc., and

$$\mathcal{C}(\mathbf{k}) = \int_{V_c} d^3 r e^{i\mathbf{k}\cdot\mathbf{r}} \tag{13.20}$$

is the spatial cut-off function for the crystal. The three-wave coupling strength is related to the second-order susceptibility by

$$g^{(3)}_{s_0 s_1 s_2}(\omega_1, \omega_2) = \epsilon_0 \mathfrak{F}_0 \mathfrak{F}_1 \mathfrak{F}_2 (\boldsymbol{\varepsilon}_{\mathbf{k}_0 s_0})_i (\boldsymbol{\varepsilon}_{\mathbf{k}_1 s_1})_j (\boldsymbol{\varepsilon}_{\mathbf{k}_2 s_2})_k \chi^{(2)}_{ijk}(\omega_1, \omega_2), \tag{13.21}$$

where $\omega_p = \omega_{k_p s_p}$ and $\mathfrak{F}_p = \mathfrak{F}_{k_p s_p}$ $(p = 0, 1, 2)$.

In the limit of a large crystal, i.e. when all dimensions are large compared to optical wavelengths,

$$\mathcal{C}(\mathbf{k}) \sim V_c \delta_{\mathbf{k},0} \to (2\pi)^3 \delta(\mathbf{k}). \tag{13.22}$$

This tells us that for large crystals the only terms that contribute to $\mathcal{U}^{(3)}_{\text{em}}$ are those satisfying the complete phase-matching conditions

$$\mathbf{k}_0 = \mathbf{k}_1 + \mathbf{k}_2, \quad \omega_0 = \omega_1 + \omega_2. \tag{13.23}$$

The same kind of analysis for $\mathcal{U}^{(4)}_{\text{em}}$ reveals two possible phase-matching conditions:

$$\mathbf{k}_0 = \mathbf{k}_1 + \mathbf{k}_2 + \mathbf{k}_3, \quad \omega_0 = \omega_1 + \omega_2 + \omega_3, \tag{13.24}$$

corresponding to terms of the form $\alpha_0^* \alpha_1 \alpha_2 \alpha_3 + \text{CC}$, and

$$\mathbf{k}_0 + \mathbf{k}_1 = \mathbf{k}_2 + \mathbf{k}_3, \quad \omega_0 + \omega_1 = \omega_2 + \omega_3, \tag{13.25}$$

corresponding to terms like $\alpha_0^* \alpha_1^* \alpha_2 \alpha_3 + \text{CC}$. As shown in Exercise 13.1, the coupling constants associated with these processes are related to the third-order susceptibility, $\chi^{(3)}$.

The definition (13.21) relates the nonlinear coupling term to a fundamental property of the medium, but this relation is not of great practical value. The first-principles evaluation of the susceptibilities is an important problem in condensed matter physics, but such a priori calculations typically involve other approximations. With the exception of hydrogen, the unperturbed atomic wave functions for single atoms are not known exactly; therefore, various approximations—such as the atomic shell model—must be used. In the important case of crystalline materials, corrections due to local field effects are also difficult to calculate (Boyd, 1992, Sec. 3.8). In practice, approximate calculations of the susceptibilities can readily incorporate the symmetry properties of the medium, but otherwise they are primarily useful as a rough guide to the feasibility of a proposed experiment. Fortunately, the analysis of experiments does not require the full solution of these difficult problems. An alternative procedure is to use symmetry arguments to determine the form of expressions, such as (13.19), for the energy. The coupling constants, which in principle depend on the nonlinear susceptibilities, can then be determined by ancillary experiments.

13.2.2 Quantum theory

The approximate quantization scheme for an isotropic dielectric given in Section 3.3.2 can be applied to crystals by the simple expedient of replacing the classical amplitude $\alpha_{\mathbf{k}s}$ in eqn (13.5) by the annihilation operator $a_{\mathbf{k}s}$, i.e.

$$\overline{\boldsymbol{\mathcal{E}}}_{\beta}^{(+)}(\mathbf{r},0) \rightarrow \overline{\mathbf{E}}_{\beta}^{(+)}(\mathbf{r}) = \frac{i}{\sqrt{V}} \sum_{\mathbf{k}s}{}' \mathfrak{F}_{ks} a_{\mathbf{k}s} \mathbf{e}_{\mathbf{k}s} e^{i\mathbf{k}\cdot\mathbf{r}} . \tag{13.26}$$

In the linear approximation, the electromagnetic Hamiltonian in a crystal—which we will now treat as the zeroth-order Hamiltonian, $H_{\text{em}}^{(0)}$—is obtained from eqn (3.150) by using the polarization-dependent frequency ω_{ks} in place of ω_k:

$$H_{\text{em}}^{(0)} = \sum_{\mathbf{k}s} \hbar\omega_{ks} a_{\mathbf{k}s}^{\dagger} a_{\mathbf{k}s} . \tag{13.27}$$

The assumption that the classical power spectrum $|\alpha_{\mathbf{k}s}|^2$ is peaked at the carrier frequencies is replaced by the rule that the expressions (13.26) and (13.27) are only valid when the operators act on a polychromatic space $\mathfrak{H}(\{\omega_\beta\})$, as defined in Section 3.3.4.

In a weakly nonlinear medium, we will employ a phenomenological approach in which the total electromagnetic Hamiltonian is given by

$$H_{\text{em}} = H_{\text{em}}^{(0)} + H_{\text{em}}^{\text{NL}} . \tag{13.28}$$

The higher-order terms comprising $H_{\text{em}}^{\text{NL}}$ can be constructed from classical energy expressions, such as (13.19), by applying the quantization rule (13.26) and putting all the terms into normal order. An alternative procedure is to use the correspondence principle and symmetry arguments to determine the form of the Hamiltonian. In this approach, the weak-field condition is realized by assuming that the terms in the $H_{\text{em}}^{\text{NL}}$ are given by low-order polynomials in the field operators. Since the field interacts with itself through the medium, the coupling constants must transform appropriately under the symmetry group for the medium. The coupling constants must, therefore, have the same symmetry properties as the classical susceptibilities. The Hamiltonian must also be invariant with respect to time translations, and—for large crystals—spatial translations. The general rules of quantum theory (Bransden and Joachain, 1989, Sec. 5.9) tell us that these invariances are respectively equivalent to the conservation of energy and momentum. Applying these conservation laws to the individual terms in the Hamiltonian yields—after dividing through by $\hbar$—the classical phase-matching conditions (13.23)–(13.25).

The expansion (13.9) for the classical energy is replaced by

$$H_{\text{em}}^{\text{NL}} = H_{\text{em}}^{(3)} + H_{\text{em}}^{(4)} + \cdots , \tag{13.29}$$

where the symmetry considerations mentioned above lead to expressions of the form

$$\begin{aligned} H_{\text{em}}^{(3)} = \frac{i}{V^{3/2}} & \sum_{\mathbf{k}_0 s_0, \mathbf{k}_1 s_1, \mathbf{k}_2 s_2} \mathcal{C}(\mathbf{k}_0 - \mathbf{k}_1 - \mathbf{k}_2)\, \delta_{\omega_0, \omega_1+\omega_2} \\ & \times g_{s_0 s_1 s_2}^{(3)}(\omega_1, \omega_2) \left[a_{\mathbf{k}_1 s_1}^{\dagger} a_{\mathbf{k}_2 s_2}^{\dagger} a_{\mathbf{k}_0 s_0} - \text{HC} \right] \end{aligned} \tag{13.30}$$

and

$$H_{\text{em}}^{(4)} = \frac{1}{V^2} \sum_{\mathbf{k}_0 s_0, \ldots, \mathbf{k}_3 s_3} \mathcal{C}\left(\mathbf{k}_0 - \mathbf{k}_1 - \mathbf{k}_2 - \mathbf{k}_3\right) \delta_{\omega_0, \omega_1 + \omega_2 + \omega_3}$$
$$\times g_{s_0 s_1 s_2 s_3}^{(4)}\left(\omega_1, \omega_2, \omega_3\right) \left[a_{\mathbf{k}_0 s_0}^{\dagger} a_{\mathbf{k}_1 s_1} a_{\mathbf{k}_2 s_2} a_{\mathbf{k}_3 s_3} + \text{HC}\right]$$
$$+ \frac{1}{V^2} \sum_{\mathbf{k}_0 s_0, \ldots, \mathbf{k}_3 s_3} \mathcal{C}\left(\mathbf{k}_0 + \mathbf{k}_1 - \mathbf{k}_2 - \mathbf{k}_3\right) \delta_{\omega_0 + \omega_1, \omega_2 + \omega_3}$$
$$\times f_{s_0 s_1 s_2 s_3}^{(4)}\left(\omega_1, \omega_2, \omega_3\right) \left[a_{\mathbf{k}_2 s_2}^{\dagger} a_{\mathbf{k}_3 s_3}^{\dagger} a_{\mathbf{k}_0 s_0} a_{\mathbf{k}_1 s_1} + \text{HC}\right]. \tag{13.31}$$

Another important feature follows from the observation that the susceptibilities are necessarily proportional to the density of atoms. When combined with the assumption that the susceptibilities are uniform over the medium, this implies that the operators $H_{\text{em}}^{(3)}$ and $H_{\text{em}}^{(4)}$ represent the coherent interaction of the field with the entire material sample. First-order transition amplitudes are thus proportional to N_{at}, and the corresponding transition rates are proportional to N_{at}^2. In contrast to this, scattering of the light from individual atoms adds incoherently, so that the transition rate is proportional to N_{at} rather than N_{at}^2.

The Hamiltonian obtained in this way contains many terms describing a variety of nonlinear processes allowed by the symmetry properties of the medium. For a given experiment, only one of these processes is usually relevant, so a model Hamiltonian is constructed by neglecting the other terms. The relevant coupling constants must then be determined experimentally.

13.3 Three-photon interactions

The mutual interaction of three photons corresponds to classical three-wave mixing, which can only occur in a crystal with nonvanishing $\chi^{(2)}$, e.g. lithium niobate, or ammonium dihydrogen phosphate (ADP). A familiar classical example is **up-conversion** (Yariv, 1989, Sec. 17.6), which is also called **sum-frequency generation** (Boyd, 1992, Sec. 2.4). In this process, waves $\boldsymbol{\mathcal{E}}_1$ and $\boldsymbol{\mathcal{E}}_2$, with frequencies ω_1 and ω_2, mix in a noncentrosymmetric $\left(\chi^{(2)}\right)$ crystal to produce a wave $\boldsymbol{\mathcal{E}}_0$ with frequency $\omega_0 = \omega_1 + \omega_2$. The traditional applications for this process involve strong fields that can be treated classically, but we are interested in a quantum approach. To this end we replace classical wave mixing by a microscopic process in which photons with energy and momentum $(\hbar\mathbf{k}_1, \hbar\omega_1)$ and $(\hbar\mathbf{k}_2, \hbar\omega_2)$ are absorbed and a photon with energy and momentum $(\hbar\mathbf{k}_0, \hbar\omega_0)$ is emitted. The phase-matching conditions (13.23) are then interpreted as conservation of energy and momentum in each microscopic interaction.

As a result of crystal anisotropy, phase matching can only be achieved by an appropriate choice of polarizations for the three photons. The uniaxial crystals usually employed in these experiments—which are described in Appendix B.5.3-A—have a principal axis of symmetry, so they exhibit birefringence. This means that there are two refractive indices for each frequency: the ordinary index $n_o(\omega)$ and the extraordinary index $n_e(\omega, \theta)$. The ordinary index $n_o(\omega)$ is independent of the direction of propagation, but the extraordinary index $n_e(\omega, \theta)$ depends on the angle θ between the propagation vector and the principal axis. The crystal is said to be negative (positive) when $n_e < n_o$ ($n_e > n_o$). For typical crystals, the refractive indices exhibit a large amount of dispersion between the lower frequencies of the input beams and the higher frequency of the output beam; therefore, it is necessary to exploit the birefringence of the crystal in order to satisfy the phase-matching conditions.

In **type I phase matching**, for negative uniaxial crystals, the incident beams have parallel polarizations as ordinary rays inside the crystal, while the output beam propagates in the crystal as an extraordinary ray. Thus the input photons obey

$$k_1 = \frac{\omega_1 n_o(\omega_1)}{c}, \quad k_2 = \frac{\omega_2 n_o(\omega_2)}{c}, \tag{13.32}$$

while the output photon satisfies the dispersion relation

$$k_0 = \frac{\omega_0 n_e(\omega_0, \theta_0)}{c}, \tag{13.33}$$

where θ_0 is the angle between the output direction and the optic axis. In **type II phase matching**, for negative uniaxial crystals, the linear polarizations of the input beams are orthogonal, so that one is an ordinary ray, and the other an extraordinary ray, e.g.

$$k_1 = \frac{\omega_1 n_o(\omega_1)}{c}, \quad k_2 = \frac{\omega_2 n_e(\omega_2, \theta_2)}{c}. \tag{13.34}$$

In this case the output beam also propagates in the crystal as an extraordinary ray. For positive uniaxial crystals the roles of ordinary and extraordinary rays are reversed (Boyd, 1992).

With an appropriate choice of the angle θ_0, which can be achieved either by suitably cutting the crystal face or by adjusting the directions of the input beams with respect to the crystal axis, it is always possible to find a pair of input frequencies for which all three photons have parallel propagation vectors. This is called **collinear** phase matching.

From Appendix B.3.3 and Section 4.4, we know that the classical and quantum theories of light are both invariant under time reversal; consequently, the time-reversed process—in which an incident high-frequency field $\mathcal{E}_0$ generates the low-frequency output fields $\boldsymbol{\mathcal{E}}_1$ and $\boldsymbol{\mathcal{E}}_2$—must also be possible. This process is called **down-conversion.** In the classical case, one of the down-converted fields, say $\boldsymbol{\mathcal{E}}_1$, must be initially present; and the growth of the field $\boldsymbol{\mathcal{E}}_2$ is called **parametric amplification** (Boyd, 1992, Sec. 2.5). The situation is quite different in quantum theory, since the initial state need not contain either of the down-converted photons. For this reason the time-reversed quantum process is called **spontaneous down-conversion** (SDC). Spontaneous down-conversion plays a central role in modern quantum optics. For somewhat obscure historical reasons, this process is frequently called *spontaneous parametric down-conversion* or else *parametric fluorescence.* In this context 'parametric' simply means that the optical medium is unchanged, i.e. each atom returns to its initial state.

13.3.1 The three-photon Hamiltonian

We will simplify the notation by imposing the convention that the polarization index is understood to accompany the wavevector. The three modes are thus represented by $(\mathbf{k}_0, \omega_0)$, $(\mathbf{k}_1, \omega_1)$, and $(\mathbf{k}_2, \omega_2)$ respectively. The fundamental interaction processes are shown in Fig. 13.1, where the Feynman diagram (b) describes down-conversion, while diagram (a) describes the time-reversed process of sum-frequency generation. Strictly speaking, Feynman diagrams represent scattering amplitudes; but they are frequently used to describe terms in the interaction Hamiltonian. The excuse is that the first-order perturbation result for the scattering amplitude is proportional to the matrix element of the interaction Hamiltonian between the initial and final states.

Since the nonlinear process is the main point of interest, we will simplify the problem by assuming that the entire quantization volume V is filled with a medium having the same linear index of refraction as the nonlinear crystal. This is called **index matching**. The simplified version of eqn (13.30) is then

$$H_{\text{em}}^{(3)} = \frac{1}{V^{3/2}} \sum_{\mathbf{k}_0} \sum_{\mathbf{k}_2} \sum_{\mathbf{k}_3} g^{(3)} \mathcal{C}\left(\mathbf{k}_0 - \mathbf{k}_1 - \mathbf{k}_2\right) a_{\mathbf{k}_1}^{\dagger} a_{\mathbf{k}_2}^{\dagger} a_{\mathbf{k}_0} + \text{HC} . \tag{13.35}$$

This is the relevant Hamiltonian for detection in the far field of the crystal, i.e. when the distance to the detector is large compared to the size of the crystal, since all atoms can then contribute to the generation of the down-converted photons.

The two terms in $H_{\text{em}}^{(3)}$ describe down-conversion and sum-frequency generation respectively. Note that both terms must be present in order to ensure the Hermiticity of the Hamiltonian. The down-conversion process is analogous to a radioactive decay in which a single parent particle (the ultraviolet photon) decays into two daughter particles, while sum-frequency generation is an analogue of particle–antiparticle annihilation.

13.3.2 Spontaneous down-conversion

Spontaneous down-conversion is the preferred light source for many recent experiments in quantum optics, e.g. single-photon number-state production, entanglement phenomena (such as the Einstein–Podolsky–Rosen effect and Franson two-photon interference), and tunneling time measurements. One reason for the popularity of this light source is that it is highly directional, whereas the atomic cascade sources discussed in Sections 1.4 and 11.2.3 emit light in all directions. In SDC, correlated photon pairs are emitted into narrow cones in the form of a rainbow surrounding the pump beam direction. The two photons of a pair are always emitted on opposite sides of the rainbow axis. Since the photon pairs are emitted within a few degrees of the pump

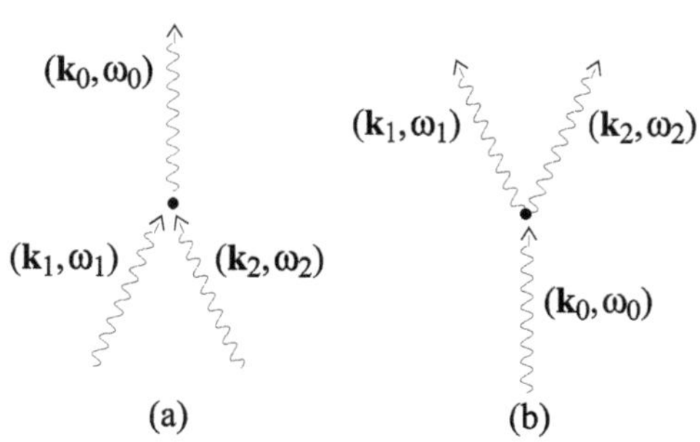

Fig. 13.1 Three-photon interactions (time flows upward in the diagrams): (a) represents sum-frequency generation, and (b) represents the time-reversed process of down-conversion.

beam direction, detection of the output within small solid angles is relatively straightforward. Another practical reason for the choice of SDC is that it is much easier to implement experimentally, since the heart of the light source is a nonlinear crystal. This method eliminates the vacuum technology required by the use of atomic beams in a cascade emission source.

A Generation of entangled photon pairs

In spontaneous down-conversion the incident field is called the **pump beam**, and the down-converted fields are traditionally called the signal and idler. To accommodate this terminology we change the notation $(\boldsymbol{\mathcal{E}}_0, \mathbf{k}_0, \omega_0)$ for the input field to $(\boldsymbol{\mathcal{E}}_P, \mathbf{p}, \omega_P)$. There is no physical distinction between the signal and idler, so we will continue to use the previous notation for the conjugate modes in the down-converted light. The emission angles and frequencies of the down-converted photons vary continuously, but they are subject to overall conservation of energy and momentum in the down-conversion process.

The interaction Hamiltonian (13.35) is more general than is required in practice, since it is valid for any distribution in the pump photon momenta. In typical experiments, the pump photons are supplied by a continuous wave (cw) ultraviolet laser, so the pump field is well approximated by a classical plane-wave mode with amplitude $\mathcal{E}_P$. A suitable quantum model is given by a Heisenberg-picture state satisfying

$$a_{\mathbf{k}}(t)\left|\alpha_{\mathbf{p}}\right\rangle = \delta_{\mathbf{k},\mathbf{p}}\alpha_{\mathbf{p}}e^{-i\omega_P t}\left|\alpha_{\mathbf{p}}\right\rangle . \tag{13.36}$$

In other words $|\alpha_{\mathbf{p}}\rangle$ is a coherent state built up from pump photons that are all in the mode $\mathbf{p}$. The coherent-state parameter $\alpha_{\mathbf{p}}$ is related to the classical field amplitude $\mathcal{E}_P$ by

$$\mathcal{E}_P \equiv e^{-i\mathbf{p}\cdot\mathbf{r}}\left\langle \alpha_{\mathbf{p}}\left|\mathbf{e}_{\mathbf{p}}\cdot\mathbf{E}^{(+)}(\mathbf{r})\right|\alpha_{\mathbf{p}}\right\rangle = i\mathfrak{F}_p\frac{\alpha_{\mathbf{p}}}{\sqrt{V}} , \tag{13.37}$$

where the expansion (13.26) was used to get the final result. Since the number of pump photons is large, the loss of one pump photon in each down-conversion event can be neglected. This **undepleted pump approximation** allows the semiclassical limit described in Section 11.3 to be applied. Thus we replace the Heisenberg-picture operator $a_{\mathbf{p}}(t)$ for the pump mode by $\alpha_{\mathbf{p}}\exp(-i\omega_P t) + \delta a_{\mathbf{p}}(t)$, and then neglect the terms involving the vacuum fluctuation operators $\delta a_{\mathbf{p}}(t)$.

Since the pump mode is treated classically and the coherent state $|\alpha_{\mathbf{p}}\rangle$ is the vacuum for the down-converted modes, we replace the notation $|\alpha_{\mathbf{p}}\rangle$ by $|0\rangle$. The classical amplitude, $\alpha_{\mathbf{p}}\exp(-i\omega_P t)$, is unchanged by the transformation from the Heisenberg picture to the Schrödinger picture; therefore, the semiclassical Hamiltonian in the Schrödinger picture is

$$H = H_0 + H_{\text{em}}^{(3)}(t) , \tag{13.38}$$

$$H_0 = \hbar\omega_P\left|\alpha_{\mathbf{p}}\right|^2 + \sum_{\mathbf{q}}\hbar\omega_{\mathbf{q}}a_{\mathbf{q}}^{\dagger}a_{\mathbf{q}} , \tag{13.39}$$

$$H_{\text{em}}^{(3)}(t) = -\frac{i}{V}\sum_{\mathbf{k}_1,\mathbf{k}_2} G^{(3)}e^{-i\omega_P t}\mathcal{C}(\mathbf{p}-\mathbf{k}_1-\mathbf{k}_2)\,a_{\mathbf{k}_1}^{\dagger}a_{\mathbf{k}_2}^{\dagger} + \text{HC} , \tag{13.40}$$

where the pump-enhanced coupling constant is $G^{(3)} = \mathcal{E}_P g^{(3)}/\mathfrak{F}_p$. The explicit time dependence of the Schrödinger-picture Hamiltonian is a result of treating the pump beam as an external classical field. The c-number term, $\hbar\omega_P \left|\alpha_{\mathbf{p}}\right|^2$, in the unperturbed Hamiltonian can be dropped, since it shifts all unperturbed energy levels by the same amount.

We will eventually need the limit of infinite quantization volume, so we use the rules (3.64) to express the (Schrödinger-picture) Hamiltonian as

$$H = H_0 + H_{\text{em}}^{(3)}(t) \,, \tag{13.41}$$

$$H_0 = \int \frac{d^3q}{(2\pi)^3} \hbar\omega_{\mathbf{q}} a^\dagger(\mathbf{q}) a(\mathbf{q}) \,, \tag{13.42}$$

$$H_{\text{em}}^{(3)}(t) = -i \int \frac{d^3k_1}{(2\pi)^3} \int \frac{d^3k_2}{(2\pi)^3} G^{(3)} e^{-i\omega_P t} \mathcal{C}(\mathbf{p} - \mathbf{k}_1 - \mathbf{k}_2) a^\dagger(\mathbf{k}_1) a^\dagger(\mathbf{k}_2) + \text{HC} \,. \tag{13.43}$$

The Hamiltonian has the same form in the Heisenberg picture, with $a^\dagger(\mathbf{k}_1)$ replaced by $a^\dagger(\mathbf{k}_1, t)$, etc. Let

$$N(\mathbf{k}_1, t) = a^\dagger(\mathbf{k}_1, t) a(\mathbf{k}_1, t) \tag{13.44}$$

denote the (Heisenberg-picture) number operator for the $\mathbf{k}_1$-mode, then a straightforward calculation using eqn (3.26) yields

$$[N(\mathbf{k}_1, t), H] = -2ie^{-i\omega_P t} \int \frac{d^3k_2}{(2\pi)^3} G^{(3)} \mathcal{C}(\mathbf{p} - \mathbf{k}_1 - \mathbf{k}_2) a^\dagger(\mathbf{k}_1, t) a^\dagger(\mathbf{k}_2, t) - \text{HC} \,. \tag{13.45}$$

The illuminated volume of the crystal is typically large on the scale of optical wavelengths, so the approximation (13.22) can be used to simplify this result to

$$[N(\mathbf{k}_1, t), H] = -2ie^{-i\omega_P t} G^{(3)} a^\dagger(\mathbf{k}_1, t) a^\dagger(\mathbf{p} - \mathbf{k}_1, t) \,. \tag{13.46}$$

In this approximation we see that

$$[N(\mathbf{k}_1, t) - N(\mathbf{p} - \mathbf{k}_1, t), H] = 0 \,, \tag{13.47}$$

i.e. the difference between the population operators for signal and idler photons is a constant of the motion. An experimental test of this prediction is to measure the expectation values $n(\mathbf{k}_1, t) = \langle N(\mathbf{k}_1, t) \rangle$ and $n(\mathbf{p} - \mathbf{k}_1, t) = \langle N(\mathbf{p} - \mathbf{k}_1, t) \rangle$. This can be done by placing detectors behind each of a pair of stops that select out a particular signal–idler pair $(\mathbf{k}_1, \mathbf{p} - \mathbf{k}_1)$. According to eqn (13.47), the expectation values satisfy

$$\begin{aligned} n(\mathbf{k}_1, t) - n(\mathbf{p} - \mathbf{k}_1, t) &= \langle N(\mathbf{k}_1, t) - N(\mathbf{p} - \mathbf{k}_1, t) \rangle \\ &= \langle N(\mathbf{k}_1, 0) - N(\mathbf{p} - \mathbf{k}_1, 0) \rangle \\ &= 0 \,, \end{aligned} \tag{13.48}$$

which provides experimental evidence that the conjugate photons are created at the same time.

B Entangled state of the signal and idler photons

Even with pump enhancement, the coupling parameter $G^{(3)}(\mathbf{k}_1, \mathbf{k}_2)$ is small, so the interaction-picture state vector, $|\Psi(t)\rangle$, for the field can be evaluated by first-order perturbation theory. These calculations are simplified by returning to the box-quantized form (13.40). In this notation, the interaction Hamiltonian is

$$H_{\text{em}}^{(3)}(t) = -i\frac{1}{V}\sum_{\mathbf{k}_1,\mathbf{k}_2} G^{(3)}\mathcal{C}(\mathbf{p} - \mathbf{k}_1 - \mathbf{k}_2)\, e^{-i\Delta t} a_{\mathbf{k}_1}^{\dagger} a_{\mathbf{k}_2}^{\dagger} + \text{HC}, \tag{13.49}$$

where we have transformed to the interaction picture by using the rule (4.98), and introduced the detuning, $\Delta = \omega_P - \omega_2 - \omega_1$, for the down-conversion transition. Applying the perturbation series (4.103) for the state vector leads to

$$\begin{aligned} |\Psi(t)\rangle &= |0\rangle + \left|\Psi^{(1)}(t)\right\rangle + \cdots, \\ \left|\Psi^{(1)}(t)\right\rangle &= -\frac{1}{V}\sum_{\mathbf{k}_1,\mathbf{k}_2} \frac{2G^{(3)}}{\hbar}\mathcal{C}(\mathbf{p} - \mathbf{k}_1 - \mathbf{k}_2)\, e^{-i\Delta t/2}\frac{\sin[\Delta t/2]}{\Delta} a_{\mathbf{k}_1}^{\dagger} a_{\mathbf{k}_2}^{\dagger} |0\rangle . \end{aligned} \tag{13.50}$$

According to the discussion in Chapter 6, each term in the $\mathbf{k}_1, \mathbf{k}_2$-sum (with the exception of the degenerate case $\mathbf{k}_1 = \mathbf{k}_2$) describes an entangled state of the signal and idler photons. Combining the limit, $V \to \infty$, of infinite quantization volume with the large-crystal approximation (13.22) for $\mathcal{C}$ yields

$$\begin{aligned} |\Psi(t)\rangle = |0\rangle &- \int \frac{d^3k_1}{(2\pi)^3} \int \frac{d^3k_2}{(2\pi)^3} \frac{2G^{(3)}}{\hbar} (2\pi)^3 \delta(\mathbf{p} - \mathbf{k}_1 - \mathbf{k}_2) \\ &\times e^{-i\Delta t/2}\frac{\sin[\Delta t/2]}{\Delta} a^{\dagger}(\mathbf{k}_1) a^{\dagger}(\mathbf{k}_2) |0\rangle . \end{aligned} \tag{13.51}$$

The limit $t \to \infty$ is relevant for cw pumping, so we can use the identity

$$\lim_{t\to\infty} e^{-i\Delta t/2}\frac{\sin(\Delta t/2)}{\Delta} = \frac{\pi}{2}\delta(\Delta), \tag{13.52}$$

which is a special case of eqn (A.102), to find

$$\begin{aligned} |\Psi(\infty)\rangle = |0\rangle &- \int \frac{d^3k_1}{(2\pi)^3} \int \frac{d^3k_2}{(2\pi)^3} \left(\frac{1}{2}\frac{G^{(3)}}{\hbar}\right) \\ &\times (2\pi)^3 \delta(\mathbf{p} - \mathbf{k}_1 - \mathbf{k}_2)(2\pi)\delta(\omega_P - \omega_1 - \omega_2) \\ &\times a^{\dagger}(\mathbf{k}_1) a^{\dagger}(\mathbf{k}_2) |0\rangle , \end{aligned} \tag{13.53}$$

where $\omega_1 = \omega_{\mathbf{k}_1}$ and $\omega_2 = \omega_{\mathbf{k}_2}$.

The conclusion is that down-conversion produces a superposition of states that are dynamically entangled in energy as well as momentum. The entanglement in energy, which is imposed by the phase-matching condition, $\omega_1 + \omega_2 = \omega_P$, provides an explanation for the observation that the two photons are created almost simultaneously. A strictly correct proof would involve the second-order correlation function

$G^{(2)}(\mathbf{r}_1, t_1, \mathbf{r}_1, t_1; \mathbf{r}_2, t_2, \mathbf{r}_2, t_2)$, but the same end is served by a simple uncertainty principle argument. If we interpret t_1 and t_2 as the creation times of the two photons, then the average time, $t_P = (t_1 + t_2)/2$, can be interpreted as the pair creation time, and the time interval between the two individual photon creation events is $\tau = t_1 - t_2$. The respective conjugate frequencies are $\Omega = \omega_1 + \omega_2$ and $\nu = (\omega_1 - \omega_2)/2$. The uncertainty in the pair creation time, $\Delta t_P \sim 1/\Delta\Omega$, is large by virtue of the tight phase-matching condition, $\Omega \simeq \omega_P$. On the other hand, the individual frequencies have large spectral bandwidths, so that $\Delta\nu$ is large and $\tau \sim 1/\Delta\nu$ is small. Consequently, the absolute time at which the pair is created is undetermined, but the time interval between the creations of the two photons is small.

13.3.3 Experimental techniques and results

Spontaneous down-conversion in a lithium niobate crystal was first observed by Harris *et al.* (1967). Shortly thereafter, it was observed in an ammonium dihydrogen phosphate (ADP) crystal by Magde and Mahr (1967). A sketch of the apparatus used by Harris *et al.* is shown in Fig. 13.2. The beam from an argon-ion laser, operating at a wavelength of 488 nm, impinges on a lithium niobate crystal oriented so that collinear, type I phase matching is achieved. The laser beam enters the crystal polarized as an extraordinary ray. Temperature tuning of the index of refraction allows the adjustment of the wavelength of the down-converted, collinear signal and idler beams, which are ordinary rays produced inside the crystal. These beams are spectrally analyzed by means of a prism monochromator, and then detected. In the Magde and Mahr experiment, a pulsed 347 nm beam is produced by means of second-harmonic generation pumped by a pulsed ruby laser beam. The peak pulse power in the ultraviolet beam is 1 MW, with a pulse duration of 20 ns. Spontaneous down-conversion occurs when the pulsed 347 nm beam of light enters the ADP crystal. Instead of temperature tuning, angle tuning is used to produce collinearly phase-matched signal and idler beams of various wavelengths.

Zel'dovich and Klyshko (1969) were the first to notice that phased-matched, down-converted photons should be observable in coincidence detection. Burnham and Weinberg (1970) performed the first experiment to observe these predicted coincidences, and in the same experiment they were also the first to produce a pair of non-collinear signal and idler beams in SDC. Their apparatus, sketched in Fig. 13.3, uses a 9 mW,

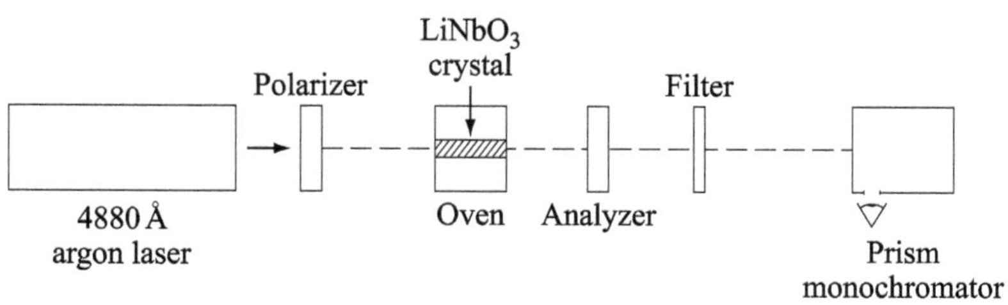

Fig. 13.2 Apparatus used to observe spontaneous down-conversion in 1967 by Harris, Oshman, and Byer. (Reproduced from Harris *et al.* (1967).)

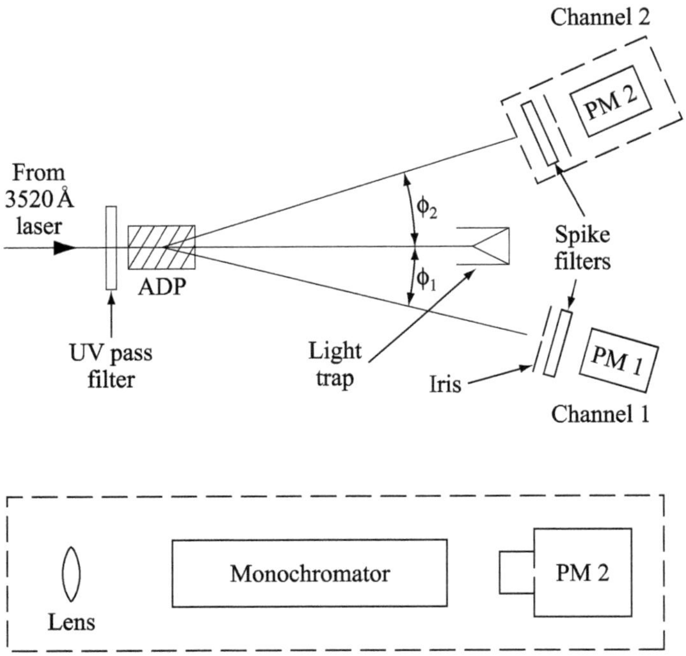

Fig. 13.3 Apparatus used by Burnham and Weinberg (1970) to observe the simultaneity of photodetection of the photon pairs generated in spontaneous down-conversion in an ammonium dihydrogen phosphate (ADP) crystal. Coincidence-counting electronics (not shown) is used to register coincidences between pulses in the outputs of the two photomultipliers PM1 and PM2. These detectors are placed at angles ϕ_1 and ϕ_2 such that phase matching is satisfied inside the crystal for the two members (i.e. signal and idler) of a given photon pair. (Reproduced from Burnham and Weinberg (1970).)

continuous-wave, helium–cadmium, ultraviolet laser—operating at a wavelength of 325 nm—as the pump beam to produce SDC in an ADP crystal. The crystal is cut so as to produce conical rainbow emissions of the signal and idler photon pairs around the pump beam direction. The ultraviolet (UV) laser beam enters an inch-long ADP crystal, and pairs of phase-matched signal ($\lambda_1 = 633\,\text{nm}$) and idler ($\lambda_2 = 668\,\text{nm}$) photons emerge from the crystal at the respective angles of $\phi_1 = 52\,\text{mrad}$ and $\phi_2 = 55\,\text{mrad}$, with respect to the pump beam. After passing through the crystal, the pump beam enters a beam dump which eliminates any background due to scattering of the UV photons. After passing through narrowband filters—actually a combination of interference filter and monochromator in the case of the idler photon—with 4 nm and 1.5 nm passbands centered on the signal and idler wavelengths respectively, the individual signal and idler photons are detected by photomultipliers with near-infrared-sensitive S20 photocathodes. Pinholes with effective diameters of 2 mm are used to define precisely the angles of emission of the detected photons around the phase-matching directions. Most importantly, Burnham and Weinberg were also the first to use coincidence detection to demonstrate that the phase-matched signal and idler photons are produced essentially simultaneously inside the crystal, within a narrow coincidence window of ±20 ns, that is limited only by the response time of the electronic circuit.

In more modern versions of the Burnham–Weinberg experiment, vacuum photomultipliers are replaced by solid-state silicon avalanche photodiodes (single-photon counting modules), which function exactly like a Geiger counter, except that—by means of an internal discriminator—the output consists of standardized TTL (transistor–transistor logic), five-volt level square pulses with subnanosecond rise times for each detected photon. This makes the coincidence detection of single photons much easier.

13.3.4 Absolute measurement of the quantum efficiency of detectors

In Section 13.3.2 we have seen that the process of spontaneous down-conversion provides a source of entangled pairs of photons. Burnham and Weinberg (1970) used coincidence-counting techniques—originally developed in nuclear and elementary particle physics—to observe the extremely tight correlation between the emission times of the two photons. As they pointed out, this correlation allows a direct measurement of the absolute quantum efficiency of a photon counter. Migdall (2001) subsequently developed this suggestion into a measurement protocol. The idea behind this technique is as follows: when a click occurs in one photon counter (the trigger detector), we are then certain that there must have been another photon emitted in the conjugate direction, defined by momentum and energy conservation. Thus we know precisely the direction of emission of the conjugate photon, and also its time of arrival—within a very narrow time window relative to the trigger photon—at any point along its direction of propagation.

As shown in Fig. 13.4, the procedure is to place the detector under test (DUT) and the trigger detector so that the coincidence counter can only be triggered by signals from a single entangled pair. For a long series of measurements, the respective quantum efficiencies η_1 and η_2 of the trigger detector and the DUT are defined by

$$N_1 = \eta_1 N \tag{13.54}$$

and

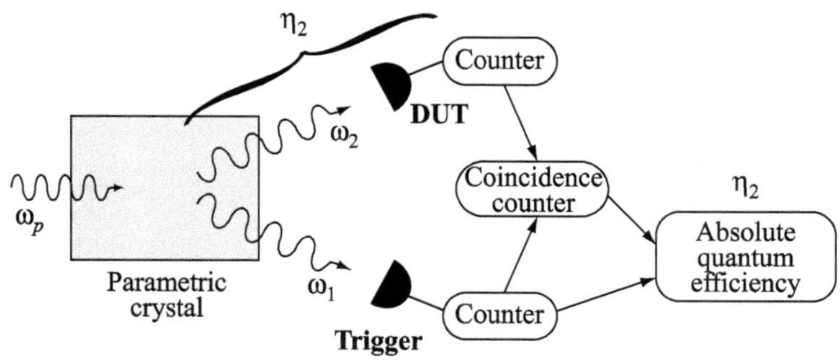

Fig. 13.4 Scheme for absolute measurement of quantum efficiency. A pair of entangled photons originating in the crystal head toward the 'trigger' detector and the 'detector under test' (DUT). The parameter η_2 is the quantum efficiency for the entire path from the point of emission to the DUT. (Reproduced from Migdall (2001).)

$$N_2 = \eta_2 N\,, \tag{13.55}$$

where N is the total number of conjugate photon pairs emitted by the crystal into the directions of the two detectors, N_1 is the number of counts registered by the trigger detector, and N_2 is the number of counts registered by the DUT. We may safely assume that the clicks at the two detectors are uncorrelated, so the probability of a coincidence count is $\eta_{\text{coinc}} = \eta_1\eta_2$. Thus the number of coincidence counts is

$$N_{\text{coinc}} = \eta_1\eta_2 N\,, \tag{13.56}$$

and combing this with eqn (13.54) shows that the absolute quantum efficiency η_2 of the DUT is the ratio

$$\eta_2 = \frac{N_{\text{coinc}}}{N_1} \tag{13.57}$$

of two measurable quantities. The beauty of this scheme is that this result is independent of the quantum efficiency, η_1, of the trigger detector.

Systematic errors, however, must be carefully taken into account. Any losses along the optical path—from the point of emission of the twin photons inside the crystals all the way to the point of detection in the DUT—will contribute to a systematic error in the measurement. Thus the exit face of the crystal must be carefully anti-reflection coated, and measured. Care also must be taken to use a large enough iris in the collection optics for the conjugate photon. This will minimize absorption, by the iris, of photons which should have impinged on the DUT. Furthermore, this iris must be carefully aligned, so that it passes all photons propagating in the conjugate direction determined by phase matching with the trigger photon. This ensures that no conjugate photons are missed due to misalignment. This alignment error can, however, be minimized by maximizing the detected signal as a function of small transverse motions of the test detector.

However, the most serious systematic error arises in the electronic, rather than the optical, part of the system. The electronic gate window used in the coincidence counter is usually not a perfectly rectangular pulse shape; typically, it has small tails of lesser counting efficiency, due to which some coincidence counts can be missed. These tails can, however, be calibrated out in separate electronic measurements of the coincidence circuitry.

13.3.5 Two-crystal source of hyperentangled photon pairs

For many applications of quantum optics, e.g. quantum cryptography, quantum dense coding, quantum entanglement-swapping, quantum teleportation, and quantum computation, it is very convenient—and often necessary—to employ an intense source of **hyperentangled** pairs of photons, i.e. photons that are entangled in two or more degrees of freedom. A particularly simple, and yet powerful, light source which yields photon pairs entangled in polarization and momentum was demonstrated by Kwiat *et al.* (1999*b*).

A schematic of the apparatus used for generating hyperentangled photon pairs with high intensity is shown in Fig. 13.5. The heart of this photon-pair light source

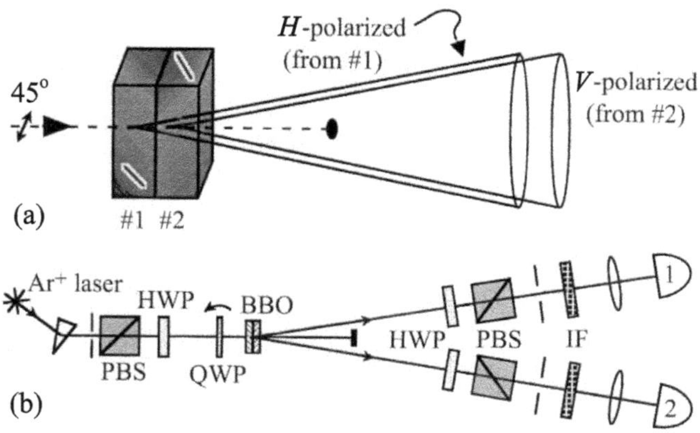

Fig. 13.5 (a) High-intensity spontaneous down-conversion light source: two identical, thin, highly nonlinear crystals are stacked in a 'crossed' configuration, i.e. the crystal axes lie in perpendicular planes, as indicated by the diagonal markings on the sides. The crystals are so thin that it is not possible to tell if a given photon pair emitted by the stack comes from the first or from the second crystal. Hence the crossed stack produces polarization-entangled pairs of photons. (b) Schematic of apparatus to produce and to characterize this photon-pair light source. (Reproduced from Kwiat *et al.* (1999*b*).)

consists of two identically cut, thin (0.59 mm), type I down-conversion crystals—β barium borate (BBO)—that are stacked in a *crossed* configuration, i.e. with their optic axes lying in perpendicular planes. What we will call the vertical plane is defined by the optic axis of the first crystal and the direction of the pump beam, while the horizontal plane is defined by the optic axis of the second crystal and the pump beam. The crystals are sufficiently thin so that the waist of the pump beam—a continuous-wave, ultraviolet (wavelength 351 nm), argon-ion laser—overlaps both. Since these are birefringent (type I) crystals, the ultraviolet pump enters as an extraordinary ray, and the pair of red, down-converted photon beams leave as ordinary rays. The two crystals are identically cut with their optic axes oriented at 33.9° with respect to the normal to the input face. The phase-matching conditions guarantee that two degenerate-frequency photons at 702 nm wavelength are emitted into a cone with a half-opening angle of 3.0°.

Under certain conditions, this arrangement allows one to determine the crystal of origin of the twin photons. For example, if the pump laser is V-polarized (i.e. linearly-polarized in the vertical plane), then type I down-conversion would only occur in the first crystal, which would produce H-polarized (i.e. linearly-polarized in the horizontal plane) twin photons. Similarly, if the pump laser were H-polarized, then type I down-conversion would only occur in the second crystal, which would produce V-polarized twin photons. However, suppose that the pump laser polarization is neither horizontal nor vertical, but instead makes an angle of 45° with respect to the vertical axis. This state is a coherent superposition, with equal amplitudes, of horizontal and vertical polarizations. Thus when this 45°-polarized pump beam is incident on the two-crystal stack, a down-conversion event can occur, with equal probability, either in the first or in the second crystal. If the photon pair originates in the first crystal, both photons would be H-polarized, whereas if the photon pair originates in the second crystal, both photons would be V-polarized.

The thickness of each crystal is much smaller than the Rayleigh range (a few centimeters) of the pump beam, and diffraction ensures that the spatial modes—i.e. the cones of emission in Fig. 13.5(a)—from the two crystals overlap in the far field, where the photons are detected. This situation provides the guiding principle behind this light source: for a 45°-polarized pump beam, it is impossible—even in principle—to know whether a given photon pair originated in the first or in the second crystal. We must therefore apply Feynman's superposition rule to obtain the state at the output of the pair of tandem crystals. If the crystals are identical in thickness and the pump is normally incident on the crystal face, the result is the entangled state

$$\left|\Phi^+\right\rangle = \frac{1}{\sqrt{2}}\left|1_{\mathbf{k}_1 H}, 1_{\mathbf{k}_2 H}\right\rangle + \frac{1}{\sqrt{2}}\left|1_{\mathbf{k}_1 V}, 1_{\mathbf{k}_2 V}\right\rangle . \tag{13.58}$$

The notation $1_{\mathbf{k}_1 H}$ denotes the horizontal polarization state of one member of the photon pair—originating in the first crystal—and $1_{\mathbf{k}_2 H}$ denotes the horizontal polarization state of the conjugate member, also originating in the first crystal. Similarly, $1_{\mathbf{k}_1 V}$ denotes the vertical polarization state of one member of the photon pair—originating in the second crystal—and $1_{\mathbf{k}_2 V}$ denotes the vertical polarization state of the conjugate member, also originating in the second crystal. The phase-matching conditions ensure that the down-converted photon pairs are emitted into azimuthally conjugate directions along rainbow-like cones, so that they are entangled both in momentum and in polarization. Hence this light source produces hyperentangled photon pairs.

The entangled state $|\Phi^+\rangle$ is one of the four **Bell states** defined by

$$\left|\Phi^+\right\rangle \equiv \frac{1}{\sqrt{2}}\left|1_{\mathbf{k}_1 H}, 1_{\mathbf{k}_2 H}\right\rangle + \frac{1}{\sqrt{2}}\left|1_{\mathbf{k}_1 V}, 1_{\mathbf{k}_2 V}\right\rangle , \tag{13.59}$$

$$\left|\Phi^-\right\rangle \equiv \frac{1}{\sqrt{2}}\left|1_{\mathbf{k}_1 H}, 1_{\mathbf{k}_2 H}\right\rangle - \frac{1}{\sqrt{2}}\left|1_{\mathbf{k}_1 V}, 1_{\mathbf{k}_2 V}\right\rangle , \tag{13.60}$$

$$\left|\Psi^+\right\rangle \equiv \frac{1}{\sqrt{2}}\left|1_{\mathbf{k}_1 H}, 1_{\mathbf{k}_2 V}\right\rangle + \frac{1}{\sqrt{2}}\left|1_{\mathbf{k}_1 V}, 1_{\mathbf{k}_2 H}\right\rangle , \tag{13.61}$$

$$\left|\Psi^-\right\rangle \equiv \frac{1}{\sqrt{2}}\left|1_{\mathbf{k}_1 H}, 1_{\mathbf{k}_2 V}\right\rangle - \frac{1}{\sqrt{2}}\left|1_{\mathbf{k}_1 V}, 1_{\mathbf{k}_2 H}\right\rangle . \tag{13.62}$$

These are maximally entangled states that form a basis set for the polarization states of pairs of entangled photons with wavevectors $\mathbf{k}_1$ and $\mathbf{k}_2$. The states $|\Phi^+\rangle$ and $|\Phi^-\rangle$ can be generated by two crossed type I crystals, and the states $|\Psi^+\rangle$ and $|\Psi^-\rangle$ can be generated by a pair of crossed type II crystals.

More generally, the two crystals could be tilted away from normal incidence around an axis perpendicular to the direction of the pump laser beam. This would result in phase changes which lead to the output entangled state

$$\left|\Phi^+;\xi\right\rangle = \frac{1}{\sqrt{2}}\left|1_{\mathbf{k}_1 H}, 1_{\mathbf{k}_2 H}\right\rangle + \frac{e^{i\xi}}{\sqrt{2}}\left|1_{\mathbf{k}_1 V}, 1_{\mathbf{k}_2 V}\right\rangle , \tag{13.63}$$

where the phase ξ depends on the tilt angle. Instead of tilting the two tandem crystals, it is more convenient to tilt a quarter-wave plate placed in front of them, so that an elliptically-polarized pump beam emerges from the quarter-wave plate with the major axis of the ellipse oriented at 45° with respect to the vertical. Then the down-converted photon pair emerges from the tandem crystals in the entangled state $|\Phi^+;\xi\rangle$, with a nonvanishing phase difference ξ between the H–H and V–V polarization-product states. The phase of the entanglement parameter ξ can be easily adjusted by changing the relative phase between the horizontal and vertical polarization components of the pump light, i.e. by changing the ellipticity of the ultraviolet laser beam polarization.

In the actual experiment, schematically shown Fig. 13.5(b), a combination of a prism and an iris acts as a filter to separate out the ultraviolet laser pump beam from the unwanted fluorescence of the argon-ion discharge tube. A polarizing beam splitter (PBS) acts as a prefilter to select a linear polarization of the laser beam. Following this, a half-wave plate (HWP) allows the selected linear polarization to be rotated around the laser beam axis. The beam then enters a quarter-wave plate (QWP)—whose tilt angle allows the adjustment of the relative phase ξ of the entangled state in eqn (13.63)—placed in front of the tandem crystals (BBO).

Separate half-wave plates (HWP) and polarizing beam splitters (PBS) provide polarization analyzers, placed in front of detectors 1 and 2, that allow independent variations of the two angles of linear polarization, θ_1 and θ_2, of the photons detected by Geiger counters 1 and 2, respectively. The irises in front of these detectors were around 2 mm in size, and the interference filters (IF) had typical bandwidths of 5 nm in wavelength. The iris sizes and interference-filter bandwidths were determined by the criterion that the detection should occur in the far field of the crystals, and by phase-matching considerations.

Under these conditions, with a 150 mW incident pump beam and a 10% solid-angle collection efficiency—arising from the finite sizes of the irises placed in front of the detectors—the hyperentangled pair production rate was around 20,000 coincidences per second. Standard coincidence detection of the correlated photon pairs in this experiment was accomplished by means of solid-state Geiger counters (silicon avalanche photodiodes with around 70% quantum efficiency, operated in the Geiger mode), in conjunction with a time-to-amplitude converter and a single-channel analyzer, with a coincidence time window of 7 ns. The polarization states of the individual photons were analyzed by means of rotatable linear polarizers, with the analyzer angle for detector 2 being rotated relative to that of detector 1 (whose analyzer angle was kept fixed at $-45°$).

Typical data are shown in Fig. 13.6. The singles rate (the output of an individual Geiger counter) shows no dependence on the relative angle of the two analyzers, indicating that the photons were individually unpolarized. On the other hand, coincidence measurements showed that the relative polarization of one photon in a given entangled pair with respect to the conjugate photon was very high (with a visibility of $99.6 \pm 0.3\%$). This means that an extremely pure two-photon entangled state has been produced with a high degree of polarization entanglement. Such a high visibility in the two-photon coincidence fringes indicates a violation of Bell's inequalities—see eqn (19.38)—by over 200 standard deviations, for data collected in about 3 minutes.

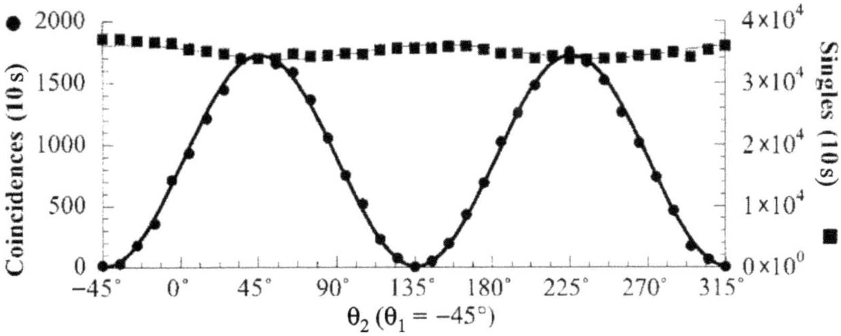

Fig. 13.6 Coincidence rates (indicated by circles, with values on the left axis) and singles rates—the outputs of the *individual* Geiger counters—(indicated by squares, with values on the right axis) versus the relative angle $\theta_2 - \theta_1$ between the two linear analyzers (i.e. polarizing beam splitters, PBSs) placed in front of detectors 1 and 2 in Fig. 13.5(b). These data were taken by varying θ_2 with θ_1 kept fixed at $-45°$. (Reproduced from Kwiat *et al.* (1999*b*).)

A further experiment demonstrated that it is possible to tune the entanglement phase ξ continuously over a range from 0 to 5.5π by tilting the quarter-wave plate, placed in front of the tandem crystals, from 0° to 30°.

13.4 Four-photon interactions

Four-photon processes correspond to classical four-wave mixing, so they involve the third-order susceptibility $\chi^{(3)}$. The parity argument shows that $\chi^{(3)}$ can be nonzero for an isolated atom, therefore four-photon processes can take place in any medium, including a vapor. In Section 13.4.2-B we will describe experimental observation of photon–photon scattering in a rubidium vapor cell.

13.4.1 Frequency tripling and down-conversion

The four-photon analogue of sum-frequency generation is **frequency tripling** or **third harmonic generation** in which three photons are absorbed to produce a single final photon. The energy and momentum conservation (phase matching) rules are then

$$\hbar\omega_0 = \hbar\omega_1 + \hbar\omega_2 + \hbar\omega_3 \,, \tag{13.64}$$

$$\hbar\mathbf{k}_0 = \hbar\mathbf{k}_1 + \hbar\mathbf{k}_2 + \hbar\mathbf{k}_3 \,, \tag{13.65}$$

and the Feynman diagram is shown in Fig. 13.7(a). In the degenerate case $\omega_1 = \omega_2 = \omega_3 = \omega$, energy conservation requires $\omega_0 = 3\omega$. This effect was first observed in the early 1960s by Maker *et al.* (1963).

The time-reversed process, which describes down-conversion of one photon into three, is shown in Fig. 13.7(b). In the photon indivisibility experiment described in Section 1.4, one of the two entangled photons is used to trigger the counters. This guaranteed that a genuine one-photon state would be incident on the beam splitter. In nondegenerate three-photon down-conversion, the three final photons are all entangled.

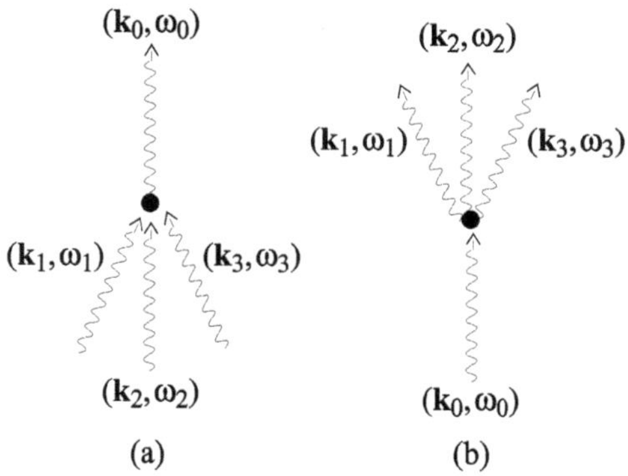

Fig. 13.7 (a) Sum-frequency generation with three photons. (b) Down-conversion of one photon into three.

It would therefore be possible to use one photon to trigger the counters, and thus guarantee that a genuine entangled state of two photons is incident on another part of the apparatus.

13.4.2 Photon–photon scattering

In three-photon coupling, the phase-matching conditions (13.23) are the only possibility, but with four photons there are two arrangements for conserving energy and momentum: namely, eqns (13.64) and (13.65), and

$$\hbar\omega_0 + \hbar\omega_1 = \hbar\omega_2 + \hbar\omega_3 \,, \tag{13.66}$$

$$\hbar\mathbf{k}_0 + \hbar\mathbf{k}_1 = \hbar\mathbf{k}_2 + \hbar\mathbf{k}_3 \,. \tag{13.67}$$

The corresponding Feynman diagram, shown in Fig. 13.8, describes photon–photon scattering. In quantum electrodynamics, this process depends on the virtual production of electron–positron pairs in the vacuum. This scattering cross-section is so small that it cannot be observed with currently available techniques (Schweber, 1961, Chap. 16a). The situation in a nonlinear medium is quite different, since the incident photons can excite an atom near resonance and thus produce an enormously enhanced photon–photon cross-section.

A The phenomenological Hamiltonian

We will restrict our attention to a vapor, since this is the simplest medium allowing four-photon processes. In this case there are no preferred directions, so the coupling

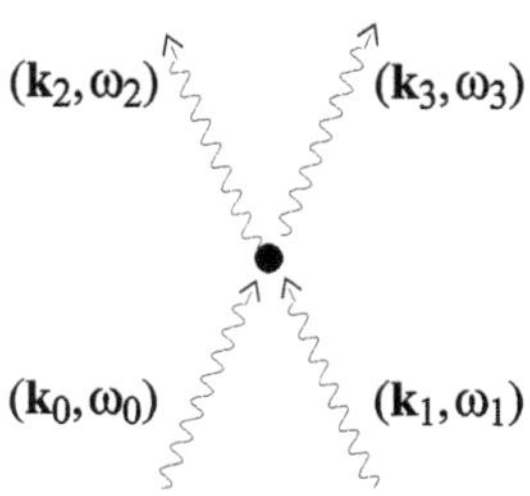

Fig. 13.8 Photon–photon scattering mediated by interaction with atoms in the medium.

between modes can only depend on the inner products of the polarization basis vectors. These geometrical factors are readily calculated for any given process, so we will simplify the notation by suppressing the polarization indices. From this point on, the argument parallels the one used for the three-photon Hamiltonian, so the simplest interaction Hamiltonian that yields Fig. 13.8 in lowest order is

$$H_{\text{int}} = \frac{1}{4}\frac{1}{V^2}\sum_{\mathbf{k}_0,\mathbf{k}_1,\mathbf{k}_2,\mathbf{k}_3} n_{\text{at}}\gamma\left(\mathbf{k}_2,\mathbf{k}_3;\mathbf{k}_0,\mathbf{k}_1\right)\mathcal{C}\left(\mathbf{k}_2+\mathbf{k}_3-\mathbf{k}_1-\mathbf{k}_0\right)a^{\dagger}_{\mathbf{k}_3}a^{\dagger}_{\mathbf{k}_2}a_{\mathbf{k}_1}a_{\mathbf{k}_0}\,, \tag{13.68}$$

where the coupling constants satisfy $\gamma\left(\mathbf{k}_2,\mathbf{k}_3;\mathbf{k}_0,\mathbf{k}_1\right) = \gamma^*\left(\mathbf{k}_0,\mathbf{k}_1;\mathbf{k}_2,\mathbf{k}_3\right)$, and $\mathcal{C}\left(\mathbf{k}\right)$ is defined by eqn (13.20).

B *Experimental observation of photon–photon scattering*

An experiment has been performed to observe head-on photon–photon collisions—mediated by the atoms in a rubidium vapor cell—leading to 90° scattering. In the experiment the rubidium atoms are excited close enough to resonance to get resonant enhancement, but far enough from resonance to eliminate photon absorption and resonance fluorescence. The resonant enhancement of the coupling is what makes this experiment possible, by contrast to the observation of photon–photon scattering in the vacuum.

The detailed theoretical analysis of this experiment is rather complicated (Mitchell *et al.*, 2000), but the model Hamiltonian of eqn (13.68) suffices for a qualitative treatment. In particular, one would expect coincidence detections for pairs of photons scattered in opposite directions—in the center-of-mass frame of a pair of incident photons—as if the two incident photons had undergone an elastic hard-sphere scattering in a head-on collision.

As shown in Fig. 13.9, a diode laser beam at 780 nm wavelength passes through two isolators (this prevents the retroreflected beam from a mirror placed behind the cell from re-entering the laser, and thus interfering with its operation). In order to minimize absorption and resonance fluorescence, the frequency of the laser beam is detuned from the nearest rubidium-atom absorption line by 1.3 GHz, which is somewhat larger than the atomic Doppler line width at room temperature. The incident diode laser beam passes through a single-mode, polarization-maintaining fiber that spatially filters it. This produces a single-transverse (TEM_{00}) mode beam that is incident onto a square, glass rubidium vapor cell. This cell is identical in shape and size to the standard cuvettes used in Beckmann spectrophotometers. Two vertically-polarized photons, one from the incident beam direction, and one from the retroreflecting mirror, thus could collide head-on—inside a beam waist of area $(0.026\,\text{cm})^2$—in the interior of the vapor cell. The atomic density of rubidium atoms inside the cell is around $1.6\times10^{10}\,\text{atoms/cm}^3$.

The two colliding photons—like two hard spheres—will sometimes scatter off each other at right angles to the incident laser beam direction. The scattered photons would be produced simultaneously, much like the twin photons in spontaneous down-conversion. They could therefore be detected by means of coincidence counters, e.g. two silicon avalanche photodiode Geiger counters, or single-photon counting modules

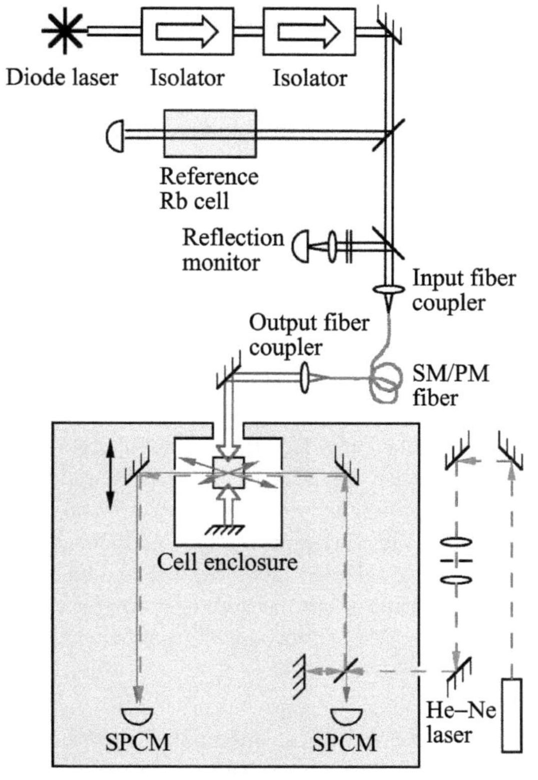

Fig. 13.9 The apparatus used for observing photon–photon scattering mediated by rubidium atoms excited off resonance. (Reproduced from Mitchell *et al.* (2000).)

(SPCM). The reference rubidium cell is used to monitor how close to atomic resonance the diode laser is tuned, and an auxiliary helium–neon laser is used to align the optics of the scattered-light detection system.

In Fig. 13.10, we show experimental data for the coincidence-counting signal as a function of the time delay between coincidence-counting pulses. The coincidence-counting electronic circuitry was used to scan the time delay from negative to positive values. By inspection, there is a peak in coincidence counts around zero time delay, which is consistent with the coincidence-detection window of 1 ns. This is evidence for photon–photon collisions mediated by the atoms. As a control experiment, the same scan of coincidence counts was made after a deliberate misalignment of the two detectors by 0.14 rad with respect to the exact back-to-back scattering direction. This misalignment was large enough to violate the momentum-conservation condition (13.67). As expected, the coincidence peak disappeared.

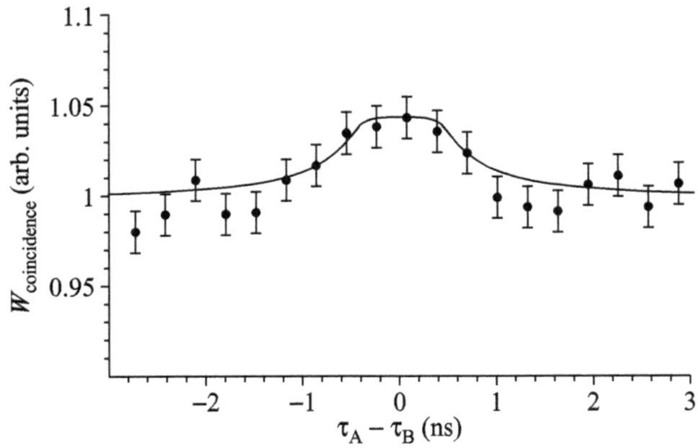

Fig. 13.10 Observed coincidence rates for 90° photon–photon scattering mediated by rubidium atoms excited off resonance. See Fig. 13.9 for the setup of the apparatus. Error bars indicate statistical errors on the data acquired with detectors aligned to collect back-to-back scattering products. The observed maximum in the coincidence rate disappears when the two detectors are deliberately misaligned from the back-to-back scattering direction. The solid curve is a theoretical fit using three measured parameters: the beam shape, the finite detection time, and the detector area. (Reproduced from Mitchell *et al.* (2000).)

13.4.3 Kerr media

For vapors and liquids the second-order susceptibility vanishes, and the absence of any preferred direction implies that the third-order polarization envelope for a single monochromatic wave, $\boldsymbol{\mathcal{E}}(t)\exp(-i\omega_0 t)$, is given by

$$\boldsymbol{\mathcal{P}}^{(3)} = \widetilde{\chi}^{(3)} \left|\boldsymbol{\mathcal{E}}\right|^2 \boldsymbol{\mathcal{E}}\,. \tag{13.69}$$

This is also valid for some centrosymmetric crystals, e.g. those with cubic symmetry. In these cases the lowest-order optical response of the medium is given by the linear index of refraction $n = \sqrt{1+\chi^{(1)}}$. The nonlinear optical response is conveniently described in terms of a field-dependent index $n(\mathcal{E})$ defined as

$$n^2(\mathcal{E}) = 1 + \chi = n^2 + \widetilde{\chi}^{(3)} \left|\boldsymbol{\mathcal{E}}\right|^2 + \cdots\,. \tag{13.70}$$

Since $\widetilde{\chi}^{(3)} \left|\boldsymbol{\mathcal{E}}\right|^2$ is small, this can be approximated by

$$\begin{aligned} n(\mathcal{E}) &= n + \widetilde{n}_2 \left|\boldsymbol{\mathcal{E}}\right|^2 + \cdots\,, \\ \widetilde{n}_2 &= \frac{1}{2}\frac{\widetilde{\chi}^{(3)}}{n}\,. \end{aligned} \tag{13.71}$$

This is more often expressed in terms of the intensity $\mathcal{I}$ as

$$\begin{aligned} n(\mathcal{E}) &= n + n_2 \mathcal{I} + \cdots\,, \\ n_2 &= \sqrt{\frac{\mu_0}{\epsilon_0}}\frac{\widetilde{\chi}^{(3)}}{n^2}\,. \end{aligned} \tag{13.72}$$

The dependence of the atomic polarization, or equivalently the index of refraction, on the intensity of the field is called the **optical Kerr effect**. Media with non-negligible values of n_2/n are called **Kerr media**. In a Kerr medium, the phase of a classical plane wave traversing a distance L increases by $\varphi = kL = n(\mathcal{E})\,L/c$, and the increment in phase due to the intensity-dependent term is

$$\Delta\varphi = n_2 I \frac{\omega}{c} L = \frac{2\pi n_2 I}{\lambda_0} L\,. \tag{13.73}$$

This dependence of the phase on the intensity is called **self-phase modulation**. The intensity dependence of the index of refraction also leads to the phenomenon of **self-focussing** (Saleh and Teich, 1991, Sec. 19.3).

In the quantum description of the Kerr effect, the interaction Hamiltonian is given by the general expression (13.68); but substantial simplifications occur in real applications. We consider an experimental configuration in which the Kerr medium is enclosed in a resonant cavity with discrete modes. In this case, one mode is typically dominant. In principle, the quantization scheme should be carried out from the beginning using the cavity modes as a basis, but the result would have the same form as obtained from the degenerate case $\mathbf{k}_0 = \mathbf{k}_1 = \mathbf{k}_2 = \mathbf{k}_3$ of Fig. 13.8. The model Hamiltonian is then

$$H = \hbar\omega_0 a^\dagger a + \frac{1}{2}\hbar g a^{\dagger 2} a^2\,, \tag{13.74}$$

where the coupling constant g is proportional to $\widetilde{\chi}^{(3)}$and a is the annihilation operator for the favored mode. By means of the canonical commutation relations, the Hamiltonian can be expressed as

$$H = \hbar\omega_0 N + \frac{1}{2}\hbar g\left(N^2 - N\right), \tag{13.75}$$

where $N = a^\dagger a$. In the Heisenberg picture, this form makes it clear that $N(t)$ is a constant of the motion: $N(t) = N(0) = N$. This corresponds to the classical result that the intensity is fixed and only the phase changes.

The evolution of the quantum amplitude is given by the Heisenberg equation for the annihilation operator:

$$\begin{aligned}\frac{da(t)}{dt} &= -i\omega_0 a(t) - ig a^\dagger(t)\, a^2(t) \\ &= -i\left(\omega_0 + gN\right) a(t)\,.\end{aligned} \tag{13.76}$$

Since the number operator is independent of time, the solution is

$$a(t) = e^{-i(\omega_0 + gN)t} a\,, \tag{13.77}$$

and the matrix elements of the annihilation operator in the number-state basis are

$$\langle m\,|a(t)|\,m'\rangle = e^{-i(\omega_0+gm)t}\langle m\,|a|\,m'\rangle = \delta_{m,m'-1}\sqrt{m+1}e^{-i(\omega_0+gm)t}\,. \tag{13.78}$$

Thus the modulus of the matrix element is constant, and the term mgt in the phase represents the quantum analogue of the classical phase shift $\Delta\varphi$.

It is also useful to consider situations in which the classical field is the sum of two monochromatic fields with different carrier frequencies:

$$\boldsymbol{\mathcal{E}}(t) = \boldsymbol{\mathcal{E}}_1(t)\exp(-i\omega_1 t) + \boldsymbol{\mathcal{E}}_2(t)\exp(-i\omega_2 t)\,. \tag{13.79}$$

The polarization will then have contributions of the form $|\boldsymbol{\mathcal{E}}_1|^2\boldsymbol{\mathcal{E}}_1$ and $|\boldsymbol{\mathcal{E}}_2|^2\boldsymbol{\mathcal{E}}_2$—describing self-phase modulation—and also terms proportional to $|\boldsymbol{\mathcal{E}}_1|^2\boldsymbol{\mathcal{E}}_2$ and $|\boldsymbol{\mathcal{E}}_2|^2\boldsymbol{\mathcal{E}}_1$—describing **cross-phase modulation.** This is called a **cross-Kerr medium**, and the Hamiltonian is

$$H = \hbar\omega_1 a_1^\dagger a_1 + \hbar\omega_2 a_2^\dagger a_2 + \frac{\hbar g_1}{2} a_1^{\dagger 2} a_1^2 + \frac{\hbar g_2}{2} a_2^{\dagger 2} a_2^2 + \hbar g_{12}\, a_1^\dagger a_2^\dagger a_1 a_2\,. \tag{13.80}$$

The coupling frequencies g_1, g_2, and g_{12} are all proportional to components of the $\chi^{(3)}$-tensor. For isotropic media, the three coupling frequencies are identical; but for crystals it is possible to have $g_1 = g_2 = 0$, while $g_{12} \neq 0$. This situation represents pure cross-phase modulation.

13.5 Exercises

13.1 The fourth-order classical energy

Apply the line of argument used to derive the effective energy expression (13.18) for $\mathcal{U}_{\text{em}}^{(3)}$ to show that the fourth-order effective energy is

$$\begin{aligned}\mathcal{U}_{\text{em}}^{(4)} = \frac{1}{V^2} &\sum_{\mathbf{k}_0 s_0,\ldots,\mathbf{k}_3 s_3} g_{s_0 s_1 s_2 s_3}^{(4)}(\omega_1,\omega_2,\omega_3)\left[\alpha_0^*\alpha_1\alpha_2\alpha_3 + \text{CC}\right] \\ &\times \delta_{\omega_0,\omega_1+\omega_2+\omega_3}\,\mathcal{C}(\mathbf{k}_0 - \mathbf{k}_1 - \mathbf{k}_2 - \mathbf{k}_3) \\ &+ \frac{1}{V^2}\sum_{\mathbf{k}_0 s_0,\ldots,\mathbf{k}_3 s_3} f_{s_0 s_1 s_2 s_3}^{(4)}(\omega_1,\omega_2,\omega_3)\left[\alpha_0\alpha_1\alpha_2^*\alpha_3^* + \text{CC}\right] \\ &\times \delta_{\omega_0+\omega_1,\omega_2+\omega_3}\,\mathcal{C}(\mathbf{k}_0 + \mathbf{k}_1 - \mathbf{k}_2 - \mathbf{k}_3)\,,\end{aligned}$$

where

$$\begin{aligned} g_{s_0 s_1 s_2 s_3}^{(4)}(\omega_1,\omega_2,\omega_3) &= -\epsilon_0\mathfrak{F}_0\mathfrak{F}_1\mathfrak{F}_2\mathfrak{F}_3\chi_{s_0 s_1 s_2 s_3}^{(3)}(\omega_1,\omega_2,\omega_3)\,, \\ f_{s_0 s_1 s_2 s_3}^{(4)}(\omega_1,\omega_2,\omega_3) &= \frac{3}{4}\epsilon_0\mathfrak{F}_0\mathfrak{F}_1\mathfrak{F}_2\mathfrak{F}_3\chi_{s_0 s_1 s_2 s_3}^{(3)}(\omega_1,-\omega_2,-\omega_3)\,,\end{aligned}$$

and

$$\chi_{s_0 s_1 s_2 s_3}^{(3)}(\omega_1,\omega_2,\omega_3) = (\boldsymbol{\varepsilon}_{\mathbf{k}_0 s_0})_i(\boldsymbol{\varepsilon}_{\mathbf{k}_1 s_1})_j(\boldsymbol{\varepsilon}_{\mathbf{k}_2 s_2})_k(\boldsymbol{\varepsilon}_{\mathbf{k}_3 s_3})_l\,\chi_{ijkl}^{(2)}(\omega_1,\omega_2,\omega_3)\,.$$

13.2 Kerr medium

Consider a Kerr medium with the Hamiltonian given by eqn (13.74).

(1) For a coherent state $|\alpha\rangle$, use the result of part (2) of Exercise 5.2 to show that

$$\langle\alpha\,|a(t)|\,\alpha\rangle = e^{-i\omega_0 t}\alpha\exp\left\{\left(e^{-igt}-1\right)|\alpha|^2\right\}.$$

(2) For a nearly classical state, i.e. $|\alpha|^2 \gg 1$, one might intuitively expect that the number operator N in eqn (13.77) could be replaced by $|\alpha|^2$ in the evaluation of $\langle \alpha | a(t) | \alpha \rangle$. Write down the resulting expression and compare it to the exact result given above to determine the range of values of t for which the conjectured expression is valid. What is the behavior of the correct expression for $\langle \alpha | a(t) | \alpha \rangle$ as $|\alpha| \to \infty$?

(3) Using the form (13.75) of the Hamiltonian, exhibit the solution of the Schrödinger equation $i\hbar\partial/\partial t |\psi(t)\rangle = H |\psi(t)\rangle$—with initial condition $|\psi(0)\rangle = |\alpha\rangle$—as an expansion in number states. Use this solution to explain the counterintuitive results of part (2) and to decide if $|\psi(t)\rangle$ remains a nearly coherent state for all times t.

13.3 Cross-Kerr medium

Consider a cross-Kerr medium described by the Hamiltonian in eqn (13.80).

(1) Derive the Heisenberg equations of motion for the annihilation operators and show that the number operators $N_1(t) = a_1^\dagger(t)\,a_1(t)$ and $N_2(t) = a_2^\dagger(t)\,a_2(t)$ are constants of the motion.

(2) For the two-mode coherent state $|\alpha_1, \alpha_2\rangle$, evaluate $\langle \alpha_1, \alpha_2 | a_1(t) | \alpha_1, \alpha_2 \rangle$.

(3) For a pure cross-Kerr medium, expand the interaction-picture state vector $|\Psi(t)\rangle$ in the number-state basis $\{|n_1, n_2\rangle\}$ and show that the exact solution of the interaction-picture Schrödinger equation is

$$|\Psi(t)\rangle = \sum_{n_1 n_2} |n_1, n_2\rangle \langle n_1 n_2 | \Psi(0)\rangle\, e^{-ig_{12} n_1 n_2 t}\,.$$

13.4 The cross-Kerr medium as a QND*

In a *quantum nondemolition* (QND) measurement (Braginsky and Khalili, 1996; Grangier *et al.*, 1998) the *quantum back actions* of normal measurements—e.g. the randomization of the momentum of a free particle induced by a measurement of its position—are partially avoided by forming an entangled state of the signal with a second system, called the *meter*. For the pure cross-Kerr medium in Section 13.4.3, identify a_1 and a_2 as the signal and meter operators respectively. Assume that the (interaction-picture) input state is $|\Psi(0)\rangle = |n_1, \alpha_2\rangle$, i.e. a number state for the signal and a coherent state for the meter.

(1) Use the results of Exercise 13.3 to show that $|\Psi(t)\rangle = \left|n_1, \alpha_2 e^{-i\gamma n_1}\right\rangle$, where $\gamma = g_{12} t$.

(2) Devise a homodyne measurement scheme that can distinguish between the phase shifts experienced by the meter beam for different values of n_1, e.g. $n_1 = 0$ and $n_1 = 1$. For example, measure the quadrature $X_2 = \left(a_2 \exp[-i\varphi] + a_2^\dagger \exp[i\varphi]\right)/2$, where φ is the phase of the local oscillator in the homodyne apparatus.

14
Quantum noise and dissipation

In the majority of the applications considered so far—e.g. photons in an ideal cavity, photons passing through passive linear media, atoms coupled to the radiation field, etc.—we have neglected all dissipative effects, such as absorption and scattering. In terms of the fundamental microscopic theory, this means that all interactions between the system under study and the external world have been ignored. When this assumption is in force, the system is said to be **closed**. The evolution of a closed system is completely determined by its Hamiltonian. A pure state of a closed system is described by a state vector obeying the Schrödinger equation (2.108), and a mixed state is represented by a density operator obeying the quantum Liouville equation (2.119). With the possible exception of the entire universe, the assumption that a system is closed is always an approximation. Every experimentally relevant physical system is unavoidably coupled to other physical systems in its vicinity, and usually very little is known about the neighboring systems or about the coupling mechanisms. If interactions with the external world cannot be neglected, the system is said to be **open**. In this chapter, we begin the study of open systems.

14.1 The world as sample and environment

For the discussion of open systems, we will divide the world into two parts: the **sample**[1]—the physical objects of experimental interest—and the **environment**—everything else. Deciding which degrees of freedom should be assigned to the sample and which to the environment requires some care, as we will shortly see.

In fact, we have already studied three open systems in previous chapters. In the discussion of blackbody radiation in Section 2.4.2, the radiation field is assumed to be in thermal equilibrium with the cavity walls. In this case the sample is the radiation field in the cavity, and some coupling to the cavity walls (the environment) is required to enforce thermal equilibrium. In line with standard practice in statistical mechanics, we simply assume the existence of a weak coupling that imposes equilibrium, but otherwise plays no role. In the discussion of the Weisskopf–Wigner method in Section 11.2.2 the sample is a two-level atom, and the modes of the radiation field are assigned to the environment. In this case, an approximate treatment of the coupling to the environment leads to a derivation of the irreversible decay of the excited atom. A purely phenomenological treatment of other dissipative terms in the Bloch equation for the two-level atom can be found in Section 11.3.3.

[1]Overuse has leached almost all meaning from the word 'system', so we have replaced it with 'sample' for this discussion.

As an illustration of the choices involved in separating the world into sample and environment, we begin by revisiting the problem of transmission through a stop. In Section 8.7 the radiation field is treated as a closed system by assuming that the screen is a perfect reflector, and by including both the incident and the reflected modes in the sample. Let us now look at this problem in a different way, by assigning the reflected modes—i.e. the modes propagating from right to left in Fig. 8.5—to the environment. The newly defined sample consists of the modes propagating from left to right. It is clearly an open system, since the right-going modes of the sample scatter into left-going modes that belong to the environment. The loss of photons from the sample represents dissipation, and the result (8.82) shows that this dissipation is accompanied by an increase in fluctuations of photon number in the transmitted field. This is a simple example of a general principle which is often called the fluctuation dissipation theorem.

14.1.1 Reservoir model for the environment

Our next task is to work out a more systematic way of dealing with open systems. This effort would be doomed from the start if it required a detailed description of the environment, but there are many experimentally interesting situations for which such knowledge is not necessary. These favorable cases are characterized by generalizations of the conditions required for the Weisskopf–Wigner (WW) treatment of spontaneous emission.

(1) The modes of the environment (the radiation field for WW) have a continuous spectrum.

(2) The sample (the two-level atom for WW) has—to a good approximation—the following properties.
   - (a) The sample Hamiltonian has a discrete spectrum. This is guaranteed if the sample (like the atom) has a finite number of degrees of freedom. If the sample has an infinite number of degrees of freedom (like the radiation field) a discrete spectrum is guaranteed by confinement to a finite region of space, e.g. a cavity.
   - (b) The sample is weakly coupled to a broad spectral range of environmental modes.

In the Weisskopf–Wigner model these features justify the Markov approximation. Applying the general rule (11.23) of the resonant wave approximation to the WW model provides the condition

$$|\Omega_{\mathbf{k}s}| \ll \Delta_K \ll \omega_{21}\,, \tag{14.1}$$

where $|\Omega_{\mathbf{k}s}|$ is the one-photon Rabi frequency defined in eqn (4.153), and Δ_K is the width of the cut-off function for the RWA. This inequality defines what is meant by coupling to a broad spectral range of the radiation field.

Turning now to the general problem, we assume the environmental degrees of freedom that couple to the sample have continuous spectra, and that the coupling is weak. Expressing the characteristic coupling strength as $\hbar\Omega_S$ defines a characteristic response frequency Ω_S, and the condition of weak coupling to a broad range of environmental excitations is

$$\Omega_S \ll \Delta_E \ll \omega_S\,. \tag{14.2}$$

Here Δ_E is the spectral width of the environmental modes that are coupled to the sample, and ω_S is a characteristic mode frequency for the unperturbed sample.

In the Weisskopf–Wigner model, the environment is the radiation field, and we have a detailed theory for this example. This luxury is missing in the general case, so we will instead devise a generic model that is based on the assumption of weak interaction between the sample and the environment. An important consequence of this assumption is that the sample can only excite low energy modes of the environment. As we have previously remarked, the low-lying modes of many systems can be approximated by harmonic oscillators. For example, suppose that the environment includes some solid material, e.g. the walls of a cavity, and that interaction with the sample excites vibrations in the crystal lattice of the solid. In the quantum theory of solids, these lattice vibrations are called **phonons** (Cohen-Tannoudji *et al.* 1977*a*, Complement JV, p. 586; Kittel 1985, Chap. 2). The νth phonon mode—which is an analogue of the $\mathbf{k}s$-mode of the radiation field—is represented by a harmonic oscillator with fundamental frequency Ω_ν, analogous to $\omega_{\mathbf{k}s}$. For macroscopic bodies, the discrete index ν becomes effectively continuous, so this environment has a continuous spectrum. Generalizing from this example suggests modeling the environment by one or more families of harmonic oscillators with continuous spectra. Each family of oscillators is called a **reservoir**. Weak coupling to the reservoir implies that the amplitudes of the oscillator displacements and momenta will be small; therefore, we will make the crucial assumption that the interaction Hamiltonian H_{SE} is linear in the creation and annihilation operators for the reservoir modes.

Within this schematic model of the **world**—the combined system of sample and environment—the reservoirs can be grouped into two classes, according to their uses. A reservoir which is not itself subjected to any experimental measurements will be called a **noise reservoir**. In this case, the reservoir model simply serves as a useful theoretical device for describing dissipative effects. This is the most common situation, but there are important applications in which the primary experimental signal is carried by the modes of one of the reservoirs. In these cases, we will call the reservoir under observation a **signal reservoir**. In the optical experiments discussed below, the signal reservoir excitations are—naturally enough—photons.

For noise reservoirs, the objective is to carry out an approximate elimination of the reservoir degrees of freedom, in order to arrive at a description of the sample as an open system. The two principal methods used for this purpose are the quantum Langevin equations for the field operator and atomic operator (which are formulated in the Heisenberg picture) and the master equation for the density operator (which is expressed in the interaction picture). The Langevin approach is, in some ways, more intuitive and technically simpler. It is particularly useful for problems that have simple analytical solutions or are amenable to perturbation theory, but it produces equations of motion for sample operators that do not lend themselves to the numerical simulations required for more complex problems. For such cases, the approach through the master equation is essential. We will explain the Langevin method in the present chapter, and introduce the master equation in Chapter 18.

In the case of a signal reservoir—which, after all, carries the experimental information—it would evidently be foolish to eliminate the reservoir degrees of freedom.

Instead, the objective is to determine the effect of the sample on the reservoir modes to be observed. Despite this difference in aim, the theoretical techniques developed for dealing with noise reservoirs can also be applied to signal reservoirs. The principal reason for this happy outcome is the assumption that both kinds of reservoirs are coupled to the sample by an interaction Hamiltonian that is linear in the reservoir operators. This approach to signal reservoirs, which is usually called the input–output method, is described in Section 14.3.

A The world Hamiltonian

The division of the world into sample and environment implies that the Hilbert space for the world is the tensor product,

$$\mathfrak{H}_W = \mathfrak{H}_S \otimes \mathfrak{H}_E \,, \tag{14.3}$$

of the sample and environment spaces. For most applications, it is necessary to model the environment by means of several independent reservoirs; therefore, the space $\mathfrak{H}_E$ is itself a tensor product,

$$\mathfrak{H}_E = \mathfrak{H}_1 \otimes \mathfrak{H}_2 \otimes \cdots \otimes \mathfrak{H}_{N_{\text{res}}} \,, \tag{14.4}$$

of the Hilbert spaces for the N_{res} independent reservoirs that define the environment. Pure states, $|\chi\rangle$, in $\mathfrak{H}_W$ are linear combinations of product states:

$$|\chi\rangle = C_1 |\Psi_1\rangle |\Lambda_1\rangle + C_2 |\Psi_2\rangle |\Lambda_2\rangle + \cdots \,, \tag{14.5}$$

where $|\Psi_j\rangle$ and $|\Lambda_j\rangle$ belong respectively to $\mathfrak{H}_S$ and $\mathfrak{H}_E$. In most situations, however, both the sample and the reservoirs must be described by mixed states.

In general, the sample may be acted on by time-dependent external classical fields or currents, and its constituent parts may interact with each other. Thus the total Schrödinger-picture Hamiltonian for the sample is

$$H_S(t) = H_{S0} + H_{S1}(t) \,, \tag{14.6}$$

where H_{S0} is the noninteracting part of the sample Hamiltonian. The interaction term $H_{S1}(t)$ is

$$H_{S1}(t) = H_{\text{SS}} + V_S(t) \,, \tag{14.7}$$

where H_{SS} describes the internal sample interactions and $V_S(t)$ represents any interactions with external classical fields or currents. The time dependence of the external fields is the source of the explicit time dependence of $V_S(t)$ in the Schrödinger picture. In typical cases, $V_S(t)$ is a linear function of the sample operators. The Hamiltonian for the isolated sample is

$$H_S = H_{S0} + H_{\text{SS}} \,. \tag{14.8}$$

The total Schrödinger-picture Hamiltonian for the world is then

$$H_W = H_S(t) + H_E + H_{\text{SE}} \,, \tag{14.9}$$

where

$$H_E = \sum_{J=1}^{N_{\text{res}}} H_J \tag{14.10}$$

is the free Hamiltonian for the environment, H_J is the Hamiltonian for the Jth reservoir,

$$H_{\text{SE}} = \sum_{J=1}^{N_{\text{res}}} H_{\text{SE}}^{(J)} \tag{14.11}$$

is the total interaction Hamiltonian between the sample and the environment, and $H_{\text{SE}}^{(J)}$ is the interaction Hamiltonian of the sample with the Jth reservoir. The world is, by definition, a closed system.

We will initially use a box-quantization description of the reservoir oscillators that parallels the treatment of the radiation field in Section 3.1.4, i.e. each family of oscillators will be labeled by a discrete index ν. The free Hamiltonian for reservoir J is therefore given by

$$H_J = \sum_{\nu} \hbar\Omega_\nu b_{J\nu}^\dagger b_{J\nu} \,, \tag{14.12}$$

where $b_{J\nu}$ is the annihilation operator for the νth mode of the Jth reservoir. We have simplified the model by assuming that each reservoir has the same set of fundamental frequencies $\{\Omega_\nu\}$, rather than a different set $\{\Omega_{J\nu}\}$ for each reservoir. This is not a serious restriction, since in the continuum limit each $\Omega_{J\nu}$ is replaced by a continuous variable Ω. The kinematical independence of the reservoirs is imposed by the commutation relations

$$[b_{J\nu}, b_{K\mu}] = 0 \,, \quad \left[b_{J\nu}, b_{K\mu}^\dagger\right] = \delta_{JK}\delta_{\nu\mu} \,.$$

In typical applications, the sample is coupled to the environment through sample operators, O_J, that can be chosen to satisfy

$$[O_J, H_{S0}] \approx \hbar\omega_J O_J \,, \tag{14.13}$$

where $\omega_J \geqslant 0$. For $\omega_J > 0$, this means that O_J is an approximate energy-lowering operator for the unperturbed sample Hamiltonian H_{S0}. We will also need the limiting case $\omega_J = 0$, which means that O_J is an approximate constant of the motion.

In the resonant wave approximation, the sample–environment interaction can be written as

$$H_{\text{SE}}^{(J)} = i\hbar \sum_{\nu} v_J(\Omega_\nu) \left(O_J^\dagger b_{J\nu} - b_{J\nu}^\dagger O_J\right) , \tag{14.14}$$

where $v_J(\Omega_\nu)$ is a real, positive coupling frequency. This *ansatz* incorporates the assumption that each sample–reservoir interaction Hamiltonian is a linear function of the reservoir operators. The restriction to real coupling frequencies is not significant, as shown in Exercise 14.1. Each coupling frequency is a candidate for the characteristic, sample-response frequency Ω_S, so it must satisfy the condition

$$v_J(\Omega_\nu) \ll \Delta_E \ll \omega_S \,. \tag{14.15}$$

The choice of the sample operator O_J is determined by the physical damping mechanism associated with the Jth reservoir.

B The world density operator

The probability distributions relevant to experiments are determined by the Schrödinger-picture density operator, $\rho_W^S(t)$, that describes the state of the world. We must, therefore, begin by choosing an initial form, $\rho_W^S(t_0)$, for the density operator. The natural assumption is that the sample and the reservoirs are uncorrelated for a sufficiently early time t_0. Since the time-independent, Heisenberg-picture density operator, ρ_W^H, satisfies $\rho_W^H = \rho_W^S(t_0)$, this is equivalent to assuming that

$$\rho_W = \rho_S\,\rho_E\,, \tag{14.16}$$

where ρ_S acts on $\mathfrak{H}_S$, and ρ_E acts on $\mathfrak{H}_E$. We have dropped the superscript H, since the remaining argument is conducted entirely in the Heisenberg picture. Furthermore, it is equally natural to assume that the various reservoirs are mutually uncorrelated at the initial time, so that

$$\rho_E = \rho_1\,\rho_2\,\cdots\,\rho_{N_{\text{res}}}\,, \tag{14.17}$$

where ρ_J acts on $\mathfrak{H}_J$ for $J = 1, 2, \ldots, N_{\text{res}}$. One or more of the density operators ρ_J is often assumed to describe a thermal equilibrium state, in which case the corresponding reservoir is called a **heat bath.**

The average value of any observable O is given by

$$\langle O\rangle = \text{Tr}_W\left(\rho_W O\right), \tag{14.18}$$

where Tr_W is defined by the sum over a basis set for $\mathfrak{H}_W = \mathfrak{H}_S \otimes \mathfrak{H}_E$. By using the definition of partial traces in Section 6.3.1, it is straightforward to show that

$$\text{Tr}_W\left(SR\right) = \left(\text{Tr}_S\, S\right)\left(\text{Tr}_E\, R\right), \tag{14.19}$$

if S acts only on $\mathfrak{H}_S$ and R acts only on $\mathfrak{H}_E$. The average of an operator product, SR, with respect to the world density operator $\rho_W = \rho_S\rho_E$ is then

$$\begin{aligned}\langle SR\rangle &= \text{Tr}_W\left[\left(\rho_S\,\rho_E\right)\left(SR\right)\right]\\ &= \text{Tr}_W\left[\rho_S S\;\rho_E R\right]\\ &= \left[\text{Tr}_S\left(\rho_S S\right)\right]\left[\text{Tr}_E\left(\rho_E R\right)\right]\\ &= \langle S\rangle\,\langle R\rangle\,,\end{aligned} \tag{14.20}$$

where the identities (6.17) and (14.19) were used to get the second and third lines. Applying this relation to $S = 1$ (more precisely, $S = I_S$, where I_S is the identity operator for $\mathfrak{H}_S$), and $R = R_J R_K$, where R_J acts on $\mathfrak{H}_J$, R_K acts on $\mathfrak{H}_K$, and $J \neq K$, yields

$$\langle R_J R_K\rangle = \langle R_J\rangle\,\langle R_K\rangle\,. \tag{14.21}$$

In other words, distinct reservoirs are statistically independent.

C Noise statistics

The statistical independence of the various reservoirs allows them to be treated individually, so we drop the reservoir index in the present section. For most experimental arrangements, the reservoir is not subjected to any special preparation; therefore, we will assume that distinct reservoir modes are uncorrelated, i.e. the reservoir density operator is **factorizable**:

$$\rho = \bigotimes_{\nu} \rho_{\nu} \,, \tag{14.22}$$

where ρ_ν is the density operator for the νth mode. For operators F_ν and G_μ that are respectively functions of b_ν, $b_\nu^\dagger$ and b_μ, $b_\mu^\dagger$, this assumption implies $\langle F_\nu G_\mu \rangle = \langle F_\nu \rangle \langle G_\mu \rangle$ for $\mu \neq \nu$.

For the discussion of quantum noise, only fluctuations around mean values are of interest. We will say that a factorizable density operator ρ is a **noise distribution** if the natural oscillator variables b_ν and $b_\nu^\dagger$ satisfy

$$\left\langle b_\nu^\dagger \right\rangle = \langle b_\nu \rangle = 0 \quad \text{for all } \nu \,. \tag{14.23}$$

These conditions can always be achieved by using the fluctuation operator $\delta b_\nu = b_\nu - \langle b_\nu \rangle$ in place of b_ν. By means of suitable choices of the operators F_ν and G_μ, the combination of eqns (14.23) and (14.22) can be used to derive restrictions on the moments of a noise distribution ρ. For example, the results

$$\left\langle b_\nu^\dagger b_\mu \right\rangle = \left\langle b_\mu b_\nu^\dagger \right\rangle = \left\langle b_\nu^\dagger \right\rangle \langle b_\mu \rangle = 0 \quad \text{and} \quad \langle b_\nu b_\mu \rangle = \langle b_\nu \rangle \langle b_\mu \rangle = 0 \quad \text{for } \mu \neq \nu \tag{14.24}$$

lead to the useful rules

$$\left\langle b_\nu^\dagger b_\mu \right\rangle = \delta_{\nu\mu} \left\langle b_\nu^\dagger b_\nu \right\rangle , \quad \left\langle b_\mu b_\nu^\dagger \right\rangle = \delta_{\nu\mu} \left\langle b_\nu b_\nu^\dagger \right\rangle , \tag{14.25}$$

$$\langle b_\nu b_\mu \rangle = \delta_{\nu\mu} \left\langle b_\nu^2 \right\rangle \tag{14.26}$$

for the fundamental second-order moments of a noise distribution ρ. For some applications it is more convenient to employ symmetrically-ordered moments, e.g.

$$\frac{1}{2} \left\langle b_\nu^\dagger b_\mu + b_\mu b_\nu^\dagger \right\rangle = \mathfrak{N}_\mu \delta_{\mu\nu} \,, \tag{14.27}$$

where

$$\mathfrak{N}_\mu = \left\langle b_\mu^\dagger b_\mu \right\rangle + \frac{1}{2} \tag{14.28}$$

is the **noise strength**. One virtue of this choice is that the lower bound in the inequality $\mathfrak{N}_\mu \geqslant 1/2$ represents the presence of vacuum fluctuations.

If we neglect the weak reservoir–sample interaction, the time-domain analogue of these relations can be expressed in terms of the Heisenberg-picture noise operator, $\xi(t)$, defined as a solution of the Heisenberg equation,

$$i\hbar \frac{d}{dt} \xi(t) = [\xi(t), H_{\text{res}}] \,, \tag{14.29}$$

where H_{res} is given by eqn (14.12). The value of $\xi(t)$ at the initial time $t = t_0$—when the Schrödinger and Heisenberg pictures coincide—is taken to be

$$\xi(t_0) = \sum_\nu C_\nu b_\nu \,, \tag{14.30}$$

where C_ν is a c-number coefficient. The explicit solution,

$$\xi(t) = \sum_\nu C_\nu b_\nu e^{-i\Omega_\nu(t-t_0)}, \tag{14.31}$$

leads to the results

$$G(t,t') = \langle \xi^\dagger(t)\,\xi(t')\rangle = \sum_\nu |C_\nu|^2 \langle b_\nu^\dagger b_\nu\rangle e^{i\Omega_\nu(t-t')} \tag{14.32}$$

and

$$F(t,t') = \langle \xi(t)\,\xi(t')\rangle = \sum_\nu C_\nu^2 \langle b_\nu^2\rangle e^{-i\Omega_\nu(t+t'-2t_0)} \tag{14.33}$$

for the second-order correlation functions $G(t,t')$ and $F(t,t')$.

The factorizability assumption (14.22) alone is sufficient to show that $G(t,t')$ is invariant under the uniform time translation ($t \to t+\tau$, $t' \to t'+\tau$) for any set of coefficients C_ν, but the same cannot be said for $F(t,t')$. The only way to ensure time-translation invariance of $F(t,t')$ is to impose

$$\langle b_\nu^2\rangle = 0, \tag{14.34}$$

which in turn implies $F(t,t') \equiv 0$. A distribution satisfying eqns (14.27) and (14.34) is said to represent **phase-insensitive noise**. It is possible to discuss many noise properties using only the second-order correlation functions F and G (Caves, 1982), but for our purposes it is simpler to impose the stronger assumption that the distribution ρ is stationary. From the general discussion in Section 4.5, we know that a stationary density operator commutes with the Hamiltonian. The simple form (14.12) of H in turn implies that each ρ_ν commutes with the mode number operator N_ν; consequently, ρ_ν is diagonal in the number-state basis. This very strong feature subsumes eqn (14.34) in the general result

$$\left\langle \left(b_\nu^\dagger\right)^n (b_\nu)^m \right\rangle = \delta_{nm} \left\langle \left(b_\nu^\dagger\right)^n (b_\nu)^n \right\rangle, \tag{14.35}$$

which guarantees time-translation invariance for correlation functions of all orders.

14.1.2 Adiabatic elimination of the reservoir operators

In the Schrödinger picture, the reservoir and sample operators act in different spaces, so $[b_{J\nu}, O] = 0$ for any sample operator, O. Since the Schrödinger and Heisenberg pictures are connected by a time-dependent unitary transformation, the equal-time commutators vanish at all times,

$$[O(t), b_{J\nu}(t)] = \left[O(t), b_{J\nu}^\dagger(t)\right] = 0. \tag{14.36}$$

With this fact in mind, it is straightforward to use the explicit form of H_W to find the Heisenberg equations for the reservoir operators:

$$\frac{\partial b_{J\nu}(t)}{\partial t} = -i\Omega_\nu b_{J\nu}(t) - v_J(\Omega_\nu)\, O_J(t). \tag{14.37}$$

Each of these equations has the formal solution

$$b_{J\nu}(t) = b_{J\nu}(t_0)\, e^{-i\Omega_\nu(t-t_0)} - v_J(\Omega_\nu) \int_{t_0}^{t} dt' e^{-i\Omega_\nu(t-t')} O_J(t'), \tag{14.38}$$

where t_0 is the initial time at which the Schrödinger and Heisenberg pictures coincide. This convention allows the identification of $b_{J\nu}(t_0)$ with the Schrödinger-picture operator $b_{J\nu}$. The first term on the right side of this equation describes free evolution of the reservoir, and the second term represents radiation reaction, i.e. the emission and absorption of reservoir excitations by the sample.

The Heisenberg equation for a sample operator O_K is

$$\frac{\partial O_K(t)}{\partial t} = \frac{1}{i\hbar}\left[O_K(t), H_W(t)\right] = \frac{1}{i\hbar}\left[O_K(t), H_S(t)\right] + \frac{1}{i\hbar}\left[O_K(t), H_{\text{SE}}(t)\right]. \tag{14.39}$$

The explicit form (14.14) for $H_{\text{SE}}(t)$, together with the equal-time commutation relations, allow us to express the final term in eqn (14.39) as

$$\frac{1}{i\hbar}\left[O_K(t), H_{\text{SE}}(t)\right] = \sum_\nu v_J(\Omega_\nu)\left\{\left[O_K(t), O_J^\dagger(t)\right] b_{J\nu}(t) - b_{J\nu}^\dagger(t)\left[O_K(t), O_J(t)\right]\right\}. \tag{14.40}$$

The equal-time commutation relations (14.36) guarantee that the products of sample and reservoir operators in this equation can be written in any order without changing the result, but the individual terms in the formal solution (14.38) for the reservoir operators do not commute with the sample operators. Consequently, it is essential to decide on a definite ordering before substituting the formal solution for the reservoir operators into eqn (14.40), and this ordering must be strictly enforced throughout the subsequent calculation. The final physical predictions are independent of the original order chosen, but the interpretation of intermediate results may vary. This is another example of ordering ambiguities like those that allow one to have the zero-point energy by choosing symmetrical ordering, or to eliminate it by using normal ordering. We have chosen to write eqn (14.40) in normal order with respect to the reservoir operators.

Substituting the formal solution (14.38) into eqn (14.40) yields two kinds of terms. One depends explicitly on the initial reservoir operators $b_{J\nu}(t_0)$ and the other arises from the radiation-reaction term. We can now proceed to eliminate the reservoir degrees of freedom—in parallel with the elimination of the radiation field in the Weisskopf–Wigner model—but the necessary calculations depend on the details of the sample–environment interaction. Consequently, we will carry out the adiabatic elimination process in several illustrative examples.

14.2 Photons in a lossy cavity

In this example, the sample consists of the discrete modes of the radiation field in an ideal physical cavity, and the environment consists of one or more reservoirs which schematically describe the mechanism for the loss of electromagnetic energy. For an enclosed cavity—such as the microcavities discussed in Chapter 12—a single reservoir representing the exchange of energy between the radiation field and the cavity walls will suffice. For the commonly encountered four-port devices—e.g. a resonant cavity capped by mirrors—it is necessary to invoke two reservoirs representing the vacuum modes entering and leaving the cavity through each port. In the present section we will concentrate on the simpler case of the enclosed cavity; the four-port devices will be discussed in Section 14.3.

In order for the discrete cavity modes to retain their identity, the characteristic interaction energy, $\hbar\Omega_S$, between the sample and the reservoir must be small compared to the minimum energy difference, $\hbar\Delta\omega$, between adjacent modes, i.e.

$$\Omega_S \ll \Delta\omega \,. \tag{14.41}$$

For example, a rectangular cavity with dimensions L_1, L_2, and L_3 satisfying $L_1 \leqslant L_2 \leqslant L_3$ has $\Delta\omega = 2\pi c/L_3$. When eqn (14.41) is satisfied the radiation modes are weakly coupled through their interaction with the reservoir modes, and—to a good approximation—we may treat each radiation mode separately.

We may, therefore, consider a reduced sample consisting of a single mode of the field, with frequency ω_0, and drop the mode index. The unperturbed sample Hamiltonian is then

$$H_{S0} = \hbar\omega_0 a^\dagger a \,, \tag{14.42}$$

and we will initially allow for the presence of an interaction term $H_{S1}(t)$. In this case there is only one sample operator and one reservoir, so the general expression (14.14) reduces to

$$H_{\text{SE}} = i\hbar \sum_\nu v(\Omega_\nu)\left(a^\dagger b_\nu - b_\nu^\dagger a\right) . \tag{14.43}$$

The coupling constant $v(\Omega_\nu)$ is proportional to the RWA cut-off function defined by eqn (11.22):

$$v(\Omega_\nu) = v_0(\Omega_\nu) K(\Omega_\nu - \omega_0) \,. \tag{14.44}$$

This is an explicit realization of the assumption that the sample is coupled to a broad spectrum of reservoir excitations.

In this connection, we note that the interaction Hamiltonian H_{SE} is similar to the RWA interaction Hamiltonian H_{rwa}, in eqn (11.46), that describes spontaneous emission by a two-level atom. In the present case, the annihilation operator a for the discrete cavity mode plays the role of the atomic lowering operator σ_- and the modes of the radiation field are replaced by the reservoir excitation modes. The mathematical similarity between H_{SE} and H_{rwa} allows similar physical conclusions to be drawn. In particular, a reservoir excitation—which carries positive energy—will never be reabsorbed once it is emitted. The implication that the interaction between the sample and a physically realistic reservoir is inherently dissipative is supported by the explicit calculations shown below.

This argument apparently rules out any description of an amplifying medium in terms of coupling to a reservoir. There is a formal way around this difficulty, but it requires the introduction of an **inverted-oscillator reservoir** which has distinctly unphysical properties. In this model, all reservoir excitations have negative energy; therefore, emitting a reservoir excitation would increase the energy of the sample.

Since the emission is irreversible, the result would be an amplification of the cavity mode. For more details, see Gardiner (1991, Chap. 7.2.1) and Exercise 14.5.

14.2.1 The Langevin equation for the field

The Heisenberg equation for $a(t)$ is

$$\frac{d}{dt}a(t) = -i\omega_0 a(t) + \sum_\nu v(\Omega_\nu) b_\nu(t) + \frac{1}{i\hbar}\left[a(t), H_{S1}(t)\right], \tag{14.45}$$

while the formal solution (14.38) for this case is

$$b_\nu(t) = b_\nu(t_0)e^{-i\Omega_\nu(t-t_0)} - v(\Omega_\nu)\int_{t_0}^t dt' e^{-i\Omega_\nu(t-t')} a(t'). \tag{14.46}$$

The general rule (14.2) requires $\omega_0 \gg |v(\Omega_\nu)|$, and we will also assume that H_{S1} is weak compared to H_{S0}. Thus the first term on the right side of eqn (14.45) describes oscillations that are much faster than those due to the remaining terms. This suggests the introduction of **slowly-varying envelope operators**,

$$\overline{a}(t) = a(t)e^{i\omega_0 t}, \quad \overline{b}_\nu(t) = b_\nu(t)e^{i\omega_0 t}, \tag{14.47}$$

that satisfy

$$\frac{d}{dt}\overline{a}(t) = \sum_\nu v(\Omega_\nu)\overline{b}_\nu(t) + \frac{1}{i\hbar}\left[\overline{a}(t), H_{S1}(t)\right], \tag{14.48}$$

and

$$\overline{b}_\nu(t) = \overline{b}_\nu(t_0)e^{-i(\Omega_\nu-\omega_0)(t-t_0)} - v(\Omega_\nu)\int_{t_0}^t dt' e^{-i(\Omega_\nu-\omega_0)(t-t')}\overline{a}(t'). \tag{14.49}$$

The envelope operator $\overline{a}(t)$ varies on the time scale $T_S = 1/\Omega_S$, so it is the operator version of the slowly-varying classical envelope introduced in Section 3.3.1.

We are now ready to carry out the elimination of the reservoir degrees of freedom, by substituting eqn (14.49) into eqn (14.48). The H_{S1}-term plays no role in this argument, so we will simplify the intermediate calculation by omitting it. The simplified equation for $\overline{a}(t)$ is

$$\frac{d}{dt}\overline{a}(t) = -\int_{t_0}^t dt' \mathcal{K}(t-t')\overline{a}(t') + \xi(t), \tag{14.50}$$

where

$$\mathcal{K}(t-t') = \sum_\nu |v(\Omega_\nu)|^2 e^{-i(\Omega_\nu-\omega_0)(t-t')}, \tag{14.51}$$

and

$$\xi(t) = \sum_\nu v(\Omega_\nu)\overline{b}_\nu(t_0)e^{-i(\Omega_\nu-\omega_0)(t-t_0)}. \tag{14.52}$$

At this stage, the passage to the continuum limit is essential; therefore, we change the sum over the discrete modes to an integral according to the rule

$$\sum_{\nu} f_\nu \to \int_0^\infty d\Omega \mathcal{D}(\Omega) f(\Omega), \tag{14.53}$$

where $\mathcal{D}(\Omega)$ is the density of states for the reservoir modes. The exact form of $\mathcal{D}(\Omega)$ depends on the particular model chosen for the reservoir. For example, if the reservoir is defined by modes of the radiation field, then $\mathcal{D}(\Omega)$ is given by eqn (4.158). In practice these details are not important, since they will be absorbed into a phenomenological decay constant. Applying the rule (14.53) to $\mathcal{K}(t-t')$ and using eqn (14.44) leads to the useful representation

$$\mathcal{K}(t-t') = \int_0^\infty d\Omega \mathcal{D}(\Omega) |v_0(\Omega)|^2 |K(\Omega-\omega_0)|^2 e^{-i(\Omega-\omega_0)(t-t')}. \tag{14.54}$$

The frequency width of the Fourier transform $\mathcal{K}(\Omega)$ of $\mathcal{K}(t-t')$ is well approximated by the width Δ_K of the cut-off function. According to the uncertainty principle for Fourier transforms, the temporal width of $\mathcal{K}(t-t')$ is therefore of the order of $1/\Delta_K$. Since $\mathcal{K}(t-t')$ decays to zero for $|t-t'| > 1/\Delta_K$, we use the terminology introduced in Section 11.1.2 to call $T_{\text{mem}} = 1/\Delta_K$ the memory interval for the reservoir. The general rule (14.2) for cut-off functions, which in the present case is

$$\Omega_S = \max|v(\Omega)| \ll \Delta_K \ll \omega_0, \tag{14.55}$$

imposes the relation $T_{\text{mem}} \ll T_S$. In other words, the assumption of a broad spectral range for the sample–reservoir interaction is equivalent to the statement that the reservoir has a short memory. This assumption effectively restricts the integral in eqn (14.50) to the interval $t - T_{\text{mem}} < t' < t$, in which $\overline{a}(t')$ is essentially constant. The short memory of the reservoir justifies the Markov approximation, $\overline{a}(t') \approx \overline{a}(t)$, and this allows us to replace the integro-differential equation (14.50) by the ordinary differential equation

$$\frac{d}{dt}\overline{a}(t) = -\frac{\Lambda(t)}{2}\overline{a}(t) + \xi(t), \tag{14.56}$$

where

$$\Lambda(t) = 2\int_{t_0}^t dt' \mathcal{K}(t-t'). \tag{14.57}$$

Substituting the explicit form for $\mathcal{K}(t-t')$ gives

$$\Lambda(t) = 2\int_0^{t-t_0} d\tau \int_0^\infty d\Omega \mathcal{D}(\Omega) |v_0(\Omega)|^2 |K(\Omega-\omega_0)|^2 e^{-i(\Omega-\omega_0)\tau}. \tag{14.58}$$

We can assume that the cut-off function $|K(\Omega-\omega_0)|^2$ is sharply peaked with respect to the prefactor in the Ω-integrand, so that $\mathcal{D}(\Omega)|v_0(\Omega)|^2$ can be removed from the Ω-integral to get

$$\Lambda(t) = 2\mathcal{D}(\omega_0) |v_0(\omega_0)|^2 \int_0^{t-t_0} d\tau \int_{-\omega_0}^\infty d\Omega |K(\Omega)|^2 e^{-i\Omega\tau}. \tag{14.59}$$

The width Δ_K of the cut-off function satisfies $\Delta_K \ll \omega_0$, so the lower limit of the Ω-integral can be replaced by $-\infty$ with negligible error. This approximation ensures that $\Lambda(t)$ is real. After interchanging the Ω- and τ-integrals and noting that $|K(\Omega)|^2$ is an even function of Ω, one finds that

$$\Lambda(t) = 2\mathcal{D}(\omega_0)|v_0(\omega_0)|^2 \int_{-\infty}^{\infty} d\Omega\, |K(\Omega)|^2 \frac{1}{2} \int_{-(t-t_0)}^{t-t_0} d\tau e^{-i\Omega\tau} . \tag{14.60}$$

The definition (14.57) shows that $\Lambda(t_0) = 0$, but we are only concerned with much later times such that $t - t_0 > T_S \gg T_{\text{mem}}$, where $T_S = 1/\Omega_S$ is the response time for the slowly-varying envelope operator. In this limit, i.e. after several memory times have passed, eqn (14.56) can be replaced by

$$\frac{d}{dt}\overline{a}(t) = -\frac{\kappa}{2}\overline{a}(t) + \xi(t) , \tag{14.61}$$

where

$$\kappa = \lim_{t_0 \to -\infty} \Lambda(t) = 2\pi\mathcal{D}(\omega_0)|v(\omega_0)|^2 . \tag{14.62}$$

If we had not extended the lower integration limit in eqn (14.59) to $-\infty$, the constant κ would have a small imaginary part. This is reminiscent of the Weisskopf–Wigner model, in which the decay constant for the upper level of an atom is found to have a small imaginary part nominally related to the Lamb shift. In Section 11.2.2, we showed that a consistent application of the resonant wave approximation requires one to drop the imaginary part. Applying this idea to the present case implies that extending the lower limit to $-\infty$ is required for consistency with the resonant wave approximation.

The Fermi-golden-rule result, eqn (14.62), demonstrates that κ is positive for every initial state of the reservoir. This agrees with the expectation—expressed at the beginning of Section 14.2—that the interaction of the cavity mode and the reservoir is necessarily dissipative. From now on we will call κ the decay rate for the cavity mode. One can easily verify that the H_{S1}-contribution could have been carried along throughout this calculation, to get the complete equation

$$\frac{d}{dt}\overline{a}(t) = -\frac{\kappa}{2}\overline{a}(t) + \frac{1}{i\hbar}\left[\overline{a}(t), H_{S1}(t)\right] + \xi(t) . \tag{14.63}$$

The last vestiges of the reservoir degrees of freedom are in the operator $\xi(t)$. This is conventionally called a **noise operator**, since eqn (14.61) is the operator analogue of the Langevin equations describing the evolution of a classical oscillator subjected to a random driving force. The most famous application for these equations is the analysis of Brownian motion (Chandler, 1987, Sec. 8.8). This formal similarity has led to the name **operator Langevin equation** for eqn (14.61). This language is extended to eqn (14.63), even when an internal interaction H_{SS} contributes nonlinear terms.

According to eqn (14.52), $\xi(t)$ is a linear function of the initial reservoir operators $b_\nu(t_0)$ alone; it does not depend on the field operators. Noise operators of this kind are said to be **additive**, but not all noise operators have this property. In Section 14.4 we will see that the noise operators for atoms involve products of reservoir operators and atomic operators. Noise operators of this kind are said to represent **multiplicative noise**. An example of multiplicative noise for the radiation field is given in Exercise 14.2.

The additivity property of the noise operator $\xi(t)$ implies that the initial sample operators, $\overline{a}(t_0)$ and $\overline{a}^\dagger(t_0)$, commute with $\xi(t)$ for any t. On the other hand, the sample operators at later times depend on the operators $b_\nu(t_0)$ and $b_\nu^\dagger(t_0)$; therefore, they will not in general commute with $\xi(t)$ or $\xi^\dagger(t)$. This is an example of the general ordering problem discussed in Section 14.1.2; it is solved by strictly adhering to the original ordering of factors.

At first glance, the noise operator $\xi(t)$ may appear to be merely another nuisance—like the zero-point energy—but this is not true. To illustrate the importance of $\xi(t)$, let us drop the noise operator from eqn (14.61). The solution is then $\overline{a}(t) = e^{-\kappa(t-t_0)/2}\overline{a}(t_0)$, which in turn gives the equal-time commutator

$$\left[a(t), a^\dagger(t)\right] = \left[\overline{a}(t), \overline{a}^\dagger(t)\right] = e^{-\kappa(t-t_0)}\left[a, a^\dagger\right] = e^{-\kappa(t-t_0)}. \tag{14.64}$$

This is disastrously wrong! Unitary time evolution preserves the commutation relations, so we should find $\left[a(t), a^\dagger(t)\right] = 1$ at all times. This contradiction shows that the noise operator is essential for preserving the canonical commutation relations and, consequently, the uncertainty principle. In this example—with no H_{S1}-term—the Langevin equation is so simple that one can immediately write down the solution

$$\overline{a}(t) = e^{-\kappa(t-t_0)/2}a + \int_{t_0}^{t} dt' e^{-\kappa(t-t')/2}\xi(t'), \tag{14.65}$$

and then calculate the equal-time commutator explicitly:

$$\left[a(t), a^\dagger(t)\right] = e^{-\kappa(t-t_0)} + \int_{t_0}^{t} dt' \int_{t_0}^{t} dt'' e^{-\kappa(t-t')/2}e^{-\kappa(t-t'')/2}\left[\xi(t'), \xi^\dagger(t'')\right]. \tag{14.66}$$

The definition (14.52) leads to

$$\left[\xi(t'), \xi^\dagger(t'')\right] = \sum_\nu \left|v(\Omega_\nu)\right|^2 e^{-i(\Omega_\nu-\omega_0)(t'-t'')}. \tag{14.67}$$

In the continuum limit, the arguments used to get from eqn (14.58) to eqn (14.60) can be applied to get

$$\left[\xi(t'), \xi^\dagger(t'')\right] = \kappa\delta(t'-t''). \tag{14.68}$$

It should be understood that this result is valid only when applied to functions that vary slowly on the time scale T_{mem} of the reservoir. Substituting eqn (14.68) into eqn (14.66) shows that indeed $\left[a(t), a^\dagger(t)\right] = 1$ at all times t.

14.2.2 Noise correlation functions

We next apply the general results in Section 14.1.1-C to study the properties of the noise operator. According to the definition (14.52) of $\xi(t)$ and the convention (14.23), the average of $\xi(t)$ vanishes, i.e.

$$\langle\xi(t)\rangle = \mathrm{Tr}_E\left[\rho_E\xi(t)\right] = 0. \tag{14.69}$$

This is of course what one should expect of a sensible noise source. Turning next to the correlation function, we know—from previous experience with vacuum fluctuations—that we should proceed cautiously by evaluating $\langle \xi^\dagger(t)\,\xi(t')\rangle$ for $t' \neq t$. Since $\xi^\dagger(t)\,\xi(t')$ only acts on the reservoir degrees of freedom, an application of eqn (14.19) gives

$$\langle \xi^\dagger(t)\,\xi(t')\rangle = \mathrm{Tr}_E\left[\rho_E \xi^\dagger(t)\,\xi(t')\right]. \tag{14.70}$$

Substituting the explicit definition (14.52) of the noise operator yields

$$\begin{aligned}\langle \xi^\dagger(t')\,\xi(t)\rangle = \sum_\nu \sum_\mu v(\Omega_\nu)\, v(\Omega_\mu) \left\langle \overline{b}_\nu^\dagger(t_0)\, \overline{b}_\mu(t_0)\right\rangle_E \\ \times e^{i(\Omega_\nu-\omega_0)(t'-t_0)} e^{-i(\Omega_\nu-\omega_0)(t-t_0)},\end{aligned} \tag{14.71}$$

and the assumption of uncorrelated reservoir modes simplifies this to

$$\langle \xi^\dagger(t')\,\xi(t)\rangle = \sum_\nu |v(\Omega_\nu)|^2\, n_\nu\, e^{i(\Omega_\nu-\omega_0)(t'-t)}, \tag{14.72}$$

where

$$n_\nu = \left\langle b_\nu^\dagger b_\nu\right\rangle \tag{14.73}$$

is the average occupation number of the νth mode of the reservoir. Taking the continuum limit and applying the Markov approximation yields the normal-ordered correlation function,

$$\begin{aligned}\langle \xi^\dagger(t')\,\xi(t)\rangle &= \int d\Omega \mathcal{D}(\Omega)\, |v(\Omega)|^2\, |K(\Omega-\omega_0)|^2\, n(\Omega)\, e^{i(\Omega-\omega_0)(t'-t)} \\ &\approx n_0 \kappa \delta(t-t'),\end{aligned} \tag{14.74}$$

where $n_0 = n(\omega_0)$. A similar calculation yields the antinormal-ordered correlation function

$$\langle \xi(t)\,\xi^\dagger(t')\rangle = (n_0+1)\,\kappa\delta(t-t'). \tag{14.75}$$

The noise operator is said to be **delta correlated**, because of the factor $\delta(t-t')$. Since this is an effect of the short memory of the reservoir, the delta function only makes sense when applied to functions that vary slowly on the time scale T_{mem}. The noise strength is given by the power spectrum, i.e. the Fourier transform of the correlation function. For delta-correlated noise operators the spectrum is said to be **white noise**, because the power spectrum has the same value, $n_0\kappa$ (or $(n_0+1)\,\kappa$), for all frequencies. This relation between the noise strength and the dissipation rate is another example of the fluctuation dissipation theorem.

The delta correlation of the noise operator is the source of other useful properties of the solutions of the linear Langevin equation (14.61). By using the formal solution (14.65), one finds that

$$\begin{aligned}\langle \xi^\dagger(t+\tau)\,\overline{a}(t)\rangle = \langle \xi^\dagger(t+\tau)\,\overline{a}(t_0)\rangle\, e^{-\kappa(t-t_0)/2} \\ + \int_{t_0}^{t} dt_1 e^{-\kappa(t-t_1)/2} \langle \xi^\dagger(t+\tau)\,\xi(t_1)\rangle.\end{aligned} \tag{14.76}$$

The first term on the right side vanishes, by virtue of the assumption that the field and the reservoir are initially uncorrelated. The second term also vanishes, because the delta function from eqn (14.74) vanishes for $0 \leqslant t_1 \leqslant t$ and $\tau > 0$. Thus the operator $\overline{a}(t)$ satisfies

$$\left\langle \xi^{\dagger}(t+\tau)\,\overline{a}(t)\right\rangle = 0 \quad \text{for } \tau > 0\,, \tag{14.77}$$

and is consequently said to be **nonanticipating** with respect to the noise operator (Gardiner, 1985, Sec. 4.2.4). In anthropomorphic language, the field at time t cannot know what the randomly fluctuating noise term will do in the future.

14.3 The input–output method

In Section 14.2 our attention was focussed on the interaction of cavity modes with a noise reservoir, but there are important applications in which the excitation of reservoir modes themselves is the experimentally observable signal. In these situations some of the reservoirs are not noise reservoirs; consequently, averages like $\langle b_\nu \rangle$ need not vanish. Consider—as shown in Fig. 14.1—an open-sided cavity formed by two mirrors M1 and M2 that match the curvature of a particular Gaussian mode. Analysis of this classical wave problem shows that the mode is effectively confined to the resonator (Yariv, 1989, Chap. 7), so that the main loss mechanism is transmission through the end mirrors.

The geometry of the cavity might lead one to believe that it is a two-port device, but this would be a mistake. The reason is that radiation can both enter and leave through each of the mirrors. We have indicated this feature by drawing the input and output ports separately in Fig. 14.1. The labeling conventions are modeled after the beam splitter in Fig. 8.2, but in this case the radiation is normally incident to the partially transmitting mirror. Thus the resonant cavity is a four-port device.

If we only consider the fundamental cavity mode with frequency ω_0, the sample Hamiltonian is

$$H_S = H_{S0} + H_{S1}(t)\,, \tag{14.78}$$

where

$$H_{S0} = \hbar\omega_0 a^{\dagger} a\,, \tag{14.79}$$

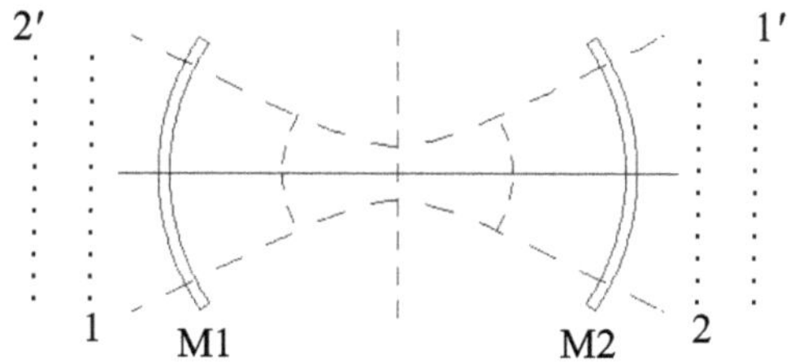

Fig. 14.1 A Gaussian mode in a resonant cavity. The upper and lower dashed curves represent lines of constant intensity for a Gaussian solution given by eqn (7.50), and the left and right dashed curves represent the local curvature of the wavefront. The curvature of mirrors M1 and M2 are chosen to match the wavefront curvature at their locations. Under these conditions the Gaussian mode is confined to the cavity. Ports 1 and 2 are input ports for photons entering from the left and right respectively. Ports 1′ and 2′ are output ports for photons exiting to the right and left respectively. The cavity is therefore a four-port device like the beam splitter.

and a is the mode annihilation operator. The internal sample interaction Hamiltonian $H_{S1}(t)$ can depend explicitly on time in the presence of external classical fields, and a model of this sort will be used later on to describe nonlinear coupling between cavity modes induced by spontaneous down-conversion.

Losses through the end mirrors are described by two reservoirs consisting of vacuum modes of the field propagating in space to the left and right of the cavity. We could treat these reservoirs by using the exact theory of vacuum propagation, but the simpler description in terms of the generic reservoir operators $b_{J\nu}$ introduced in Section 14.1.1 is sufficient. For this application it is better to go to the continuum limit from the beginning, as opposed to the end, of the analysis. For this purpose, we construct a simplified reservoir model by imposing periodic boundary conditions on a one-dimensional (1D) cavity of length L. The index ν then runs over the integers, and the corresponding wavevectors are $k = 2\pi\nu/L$. In the limit $L \to \infty$, the operators $b_{J\nu}$ are replaced by new operators $b_{Jk} = \sqrt{L} b_{J\nu}$ satisfying

$$\left[b_{Jk}, b_{Jk'}^{\dagger}\right] = 2\pi\delta\left(k - k'\right), \tag{14.80}$$

and the environment Hamiltonian is

$$H_E = \sum_{J=1}^{2} \int_{-\infty}^{\infty} \frac{dk}{2\pi}\, \hbar\Omega_k\, b_{J,k}^{\dagger} b_{J,k}\,. \tag{14.81}$$

The standard approach to in- and out-fields (Gardiner, 1991, Sec. 5.3) employs creation and annihilation operators for modes of definite frequency Ω, rather than definite wavenumber k. In the 1D model this can be achieved by assuming that the mode frequency Ω_k is a monotone-increasing function of the continuous label k. This assumption justifies the change of variables $k \to \Omega$ in eqn (14.81), with the result

$$H_E = \int_{0}^{\infty} \frac{d\Omega}{2\pi}\, \hbar\Omega \sum_{J=1}^{2} b_{J,\Omega}^{\dagger} b_{J,\Omega}\,, \tag{14.82}$$

where

$$b_{J,\Omega} = \frac{1}{\sqrt{\left|d\Omega_k/dk\right|}} b_{J,k}\,. \tag{14.83}$$

Using this definition in eqn (14.80) leads to the Heisenberg-picture, equal-time commutation relations

$$\left[b_{J,\Omega}(t), b_{K,\Omega'}^{\dagger}(t)\right] = 2\pi\delta_{JK}\delta\left(\Omega - \Omega'\right), \quad J, K = 1, 2\,, \tag{14.84}$$

$$\left[b_{J,\Omega}(t), a^{\dagger}(t)\right] = 0\,, \quad J = 1, 2\,. \tag{14.85}$$

It should be kept in mind that Ω simply replaces the mode label k; it is not a Fourier transform variable.

We should also mention that the usual presentation of this theory extends the Ω-integral in eqn (14.82) to $-\infty$, and thus introduces unphysical negative-energy modes. In expert hands, this formal device simplifies the mathematics without really violating any physical principles, but it clearly defies Einstein's rule. Furthermore, the restriction to the physically allowed, positive-energy modes clarifies the physical significance of the approximations to be imposed below.

In our approach, the generic sample–environment Hamiltonian, given by eqn (14.43), is

$$H_{\mathrm{SE}} = i\hbar \sum_{J=1}^{2} \int_0^\infty \frac{d\Omega}{2\pi} \sqrt{L \frac{dk}{d\Omega}}\, v_J(\Omega) \left\{ a^\dagger b_{J,\Omega} - b_{J,\Omega}^\dagger a \right\}. \tag{14.86}$$

The looming disaster of the uncompensated factor $\sqrt{L}$ is an illusion. In the finite cavity, the unit cell for wavenumbers is $2\pi/L$; therefore, the density of states $\mathcal{D}(\Omega)$ satisfies

$$\mathcal{D}(\Omega)\, d\Omega = \frac{dk}{2\pi/L}. \tag{14.87}$$

This observation allows the dangerous-looking result for H_{SE} to be replaced by

$$H_{\mathrm{SE}} = i\hbar \sum_{J=1}^{2} \int_0^\infty d\Omega \sqrt{\frac{\mathcal{D}(\Omega)}{2\pi}}\, v_J(\Omega) \left\{ a^\dagger b_{J,\Omega} - b_{J,\Omega}^\dagger a \right\}. \tag{14.88}$$

The terms in eqn (14.88) have simple interpretations; for example, $b_{2,\Omega}^\dagger a$ represents the disappearance of a cavity photon balanced by the emission of a photon into the environment through the mirror M2.

The slowly-varying envelope operators—$\overline{a}(t) = a(t)\exp(i\omega_0 t)$ and $\overline{b}_{J,\Omega}(t) = b_{J,\Omega}(t)\exp(i\omega_0 t)$ $(J = 1, 2)$—obey the Heisenberg equations of motion:

$$\frac{d}{dt}\overline{b}_{J,\Omega}(t) = -i(\Omega - \omega_0)\,\overline{b}_{J,\Omega}(t) - \sqrt{2\pi\mathcal{D}(\Omega)}\, v_J(\Omega)\,\overline{a}(t) \quad (J = 1, 2), \tag{14.89}$$

$$\frac{d}{dt}\overline{a}(t) = \frac{1}{i\hbar}\left[\overline{a}(t), H_{S1}(t)\right] + \sum_{J=1}^{2} \int_0^\infty d\Omega \sqrt{\frac{\mathcal{D}(\Omega)}{2\pi}}\, v_J(\Omega)\,\overline{b}_{J,\Omega}(t). \tag{14.90}$$

14.3.1 In-fields

We begin by choosing a time t_0 earlier than any time at which interactions occur. A formal solution of eqn (14.89) is given by

$$\overline{b}_{J,\Omega}(t) = \overline{b}_{J,\Omega}(t_0)\, e^{-i(\Omega-\omega_0)(t-t_0)} - \sqrt{2\pi\mathcal{D}(\Omega)}\, v_J(\Omega) \int_{t_0}^{t} dt' e^{-i(\Omega-\omega_0)(t-t')}\,\overline{a}(t'), \tag{14.91}$$

and substituting this into eqn (14.90) yields

$$\begin{aligned}\frac{d}{dt}\overline{a}(t) = {} & \frac{1}{i\hbar}\left[\overline{a}(t), H_{S1}(t)\right] \\ & + \sum_{J=1}^{2} \int_0^\infty d\Omega \sqrt{\frac{\mathcal{D}(\Omega)}{2\pi}}\, v_J(\Omega)\,\overline{b}_{J,\Omega}(t_0)\, e^{-i(\Omega-\omega_0)(t-t_0)} \\ & - \sum_{J=1}^{2} \int_{-\omega_0}^\infty d\Omega \mathcal{D}(\Omega+\omega_0)\, \left|v_J(\Omega+\omega_0)\right|^2 \int_{t_0}^{t} dt' e^{-i\Omega(t-t')}\,\overline{a}(t'),\end{aligned} \tag{14.92}$$

where the integration variable Ω has been shifted by $\Omega \to \Omega + \omega_0$ in the final term. Since the operator $\overline{a}(t')$ is slowly varying, the t'-integral in this term defines a function of Ω that is sharply peaked at $\Omega = 0$; in particular, the width of this function is small compared to ω_0. This implies that the lower limit of the Ω-integral can be extended to $-\infty$ with negligible error. In addition, we impose the Markov approximation by the *ansatz*:

$$2\pi\mathcal{D}(\Omega)|v_J(\Omega)|^2 \simeq \kappa_J = 2\pi\mathcal{D}(\omega_0)|v_J(\omega_0)|^2 \quad (J = 1, 2), \tag{14.93}$$

representing the assumption that the sample interacts with a broad spectrum of reservoir excitations. Note that this replaces eqn (14.91) by

$$\overline{b}_{J,\Omega}(t) = \overline{b}_{J,\Omega}(t_0)e^{-i(\Omega-\omega_0)(t-t_0)} - \sqrt{\kappa_J}\int_{t_0}^{t} dt' e^{-i(\Omega-\omega_0)(t-t')}\overline{a}(t'). \tag{14.94}$$

When the approximation (14.93) is used in eqn (14.92), the extended Ω-integral in the third term produces $2\pi\delta(t-t')$. Evaluating the t'-integral, with the aid of the end-point rule (A.98), then leads to the Langevin equation,

$$\frac{d}{dt}\overline{a}(t) = \frac{1}{i\hbar}[\overline{a}(t), H_{S1}(t)] - \frac{\kappa_C}{2}\overline{a}(t) + \xi_C(t), \tag{14.95}$$

where

$$\kappa_C = \kappa_1 + \kappa_2 \tag{14.96}$$

is the total cavity damping rate. The cavity noise-operator,

$$\xi_C(t) = \sum_{J=1}^{2}\sqrt{\kappa_J}\,\overline{b}_{J,\text{in}}(t), \tag{14.97}$$

is expressed in terms of the **in-fields**

$$\overline{b}_{J,in}(t) = \int_0^{\infty}\frac{d\Omega}{2\pi}\overline{b}_{J,\Omega}(t_0)e^{-i(\Omega-\omega_0)(t-t_0)}. \tag{14.98}$$

For later use it is convenient to write out the Langevin equation as

$$\frac{d}{dt}\overline{a}(t) = \frac{1}{i\hbar}[\overline{a}(t), H_{S1}(t)] + \sqrt{\kappa_1}\,\overline{b}_{1,\text{in}}(t) + \sqrt{\kappa_2}\,\overline{b}_{2,\text{in}}(t) - \frac{\kappa_C}{2}\overline{a}(t). \tag{14.99}$$

The operator $\overline{a}(t)$ depends on the initial reservoir operators through the in-fields, so eqn (14.99) is called the **retarded Langevin equation**. Since t_0 precedes any interactions, the reservoir fields and the sample fields are uncorrelated at $t = t_0$.

The in-fields have an unexpected algebraic property. Combining the equal-time commutation relations (14.84) with the definition (14.98) leads to

$$\left[\overline{b}_{J,\text{in}}(t), \overline{b}^{\dagger}_{K,\text{in}}(t')\right] = \delta_{JK}\int_{-\omega_0}^{\infty}\frac{d\Omega}{2\pi}e^{-i\Omega(t-t')}. \tag{14.100}$$

The correct interpretation of the ambiguous expression on the right side involves both mathematics and physics. The mathematical part of the argument is to interpret the

Ω-integral as a generalized function of $t-t'$. According to Appendix A.6.2, this is done by applying the generalized function to a *good* function $f(t')$ to find:

$$\int_{-\infty}^{\infty} dt' \left[\int_{-\omega_0}^{\infty} \frac{d\Omega}{2\pi} e^{-i\Omega(t-t')}\right] f(t') = \int_{-\omega_0}^{\infty} \frac{d\Omega}{2\pi} e^{-i\Omega t} f(\Omega) . \tag{14.101}$$

The physical part of the argument is that only slowly-varying good functions are relevant. In the frequency domain, this means that $f(\Omega)$ is peaked at $\Omega = 0$ and has a width that is small compared to ω_0. Thus, just as in the argument following eqn (14.92), the lower limit can be extended to $-\infty$ with negligible error. This last step replaces the right side of eqn (14.101) by $f(t)$, and this in turn implies the unequal-time commutation relations:

$$\left.\begin{aligned} \left[\overline{b}_{J,\text{in}}(t), \overline{b}^{\dagger}_{K,\text{in}}(t')\right] &= \delta_{JK}\delta(t-t') \\ \left[\overline{b}_{J,\text{in}}(t), \overline{b}_{K,\text{in}}(t')\right] &= 0 \end{aligned}\right\} \quad (J, K = 1, 2) , \tag{14.102}$$

for the in-fields.

If the environment density operator represents the vacuum, i.e.

$$\overline{b}_{J,\Omega}(t_0)\,\rho_E = \rho_E \overline{b}^{\dagger}_{J,\Omega}(t_0) = 0 \quad (J = 1, 2) , \tag{14.103}$$

and $\langle[\overline{a}, H_{\text{ss}}]\rangle = 0$, then one can show that

$$\frac{d}{dt}\left\langle \overline{a}^{\dagger}(t)\,\overline{a}(t)\right\rangle = -2\frac{\kappa_C}{2}\left\langle \overline{a}^{\dagger}(t)\,\overline{a}(t)\right\rangle = -(\kappa_1+\kappa_2)\left\langle \overline{a}^{\dagger}(t)\,\overline{a}(t)\right\rangle . \tag{14.104}$$

This justifies the interpretation of κ_1 and κ_2 as the rate of loss of cavity photons through mirrors M1 and M2 respectively.

14.3.2 Out-fields

In most applications, only the emitted fields are experimentally accessible; thus we will be interested in the reservoir fields at late times, after all interactions inside the cavity have occurred. For this purpose, we choose a late time t_1 and write a formal solution of eqn (14.89) as

$$\overline{b}_{J,\Omega}(t) = \overline{b}_{J,\Omega}(t_1)\, e^{-i(\Omega-\omega_0)(t-t_1)} + \sqrt{\kappa_J}\int_{t}^{t_1} dt' e^{-i(\Omega-\omega_0)(t-t')}\overline{a}(t') \quad (J = 1, 2) . \tag{14.105}$$

After substituting this into eqn (14.90), we find the **advanced Langevin** equation

$$\frac{d}{dt}\overline{a}(t) = \frac{1}{i\hbar}\left[\overline{a}(t), H_{S1}(t)\right] + \sqrt{\kappa_2}\,\overline{b}_{2,\text{out}}(t) + \sqrt{\kappa_1}\,\overline{b}_{1,\text{out}}(t) + \frac{\kappa_C}{2}\overline{a}(t) , \tag{14.106}$$

where the **out-fields** $\overline{b}_{J,\text{out}}(t)$ are defined by

$$\overline{b}_{J,\text{out}}(t) = \int_{0}^{\infty} \frac{d\Omega}{2\pi}\overline{b}_{J,\Omega}(t_1)\, e^{-i(\Omega-\omega_0)(t-t_1)} . \tag{14.107}$$

The sign difference between the final terms of eqns (14.106) and (14.99) can be traced back to the minus sign in the second term of eqn (14.105). This in turn reflects the free evolution of $\overline{b}_{1,\text{out}}(t)$ and $\overline{b}_{2,\text{out}}(t)$ toward the *future* values $\overline{b}_{1,\Omega}(t_1)$ and $\overline{b}_{2,\Omega}(t_1)$. Another important difference from the retarded case is that the operators $\overline{b}_{1,\Omega}(t_1)$ and $\overline{b}_{2,\Omega}(t_1)$ are necessarily correlated with the sample operator $\overline{a}(t_1)$, since the time t_1 follows all interactions inside the cavity.

A relation between the in- and out-fields—similar to the scattering relations discussed in Section 8.2—follows from equating the alternate expressions (14.94) and (14.105) for $\overline{b}_{J,\Omega}(t)$ to get

$$\overline{b}_{J,\Omega}(t_1)\, e^{-i(\Omega-\omega_0)(t-t_1)} = \overline{b}_{J,\Omega}(t_0)\, e^{-i(\Omega-\omega_0)(t-t_0)} - \sqrt{\kappa_J}\int_{t_0}^{t_1} dt' e^{-i(\Omega-\omega_0)(t-t')}\overline{a}(t') . \tag{14.108}$$

The left side of this equation is the integrand of the expression (14.107) defining $\overline{b}_{J,\text{out}}(t)$, so we take the hint and integrate over Ω to find the **input–output** equation:

$$\overline{b}_{J,\text{out}}(t) = \overline{b}_{J,\text{in}}(t) - \sqrt{\kappa_J}\,\overline{a}(t) . \tag{14.109}$$

14.3.3 The empty cavity

In order to get some insight into the meaning of all this formalism, we consider the case of an empty cavity, i.e. $H_{S1} = 0$. In this case, the equation of motion (14.99) for the intracavity field is a linear differential equation with constant coefficients,

$$\frac{d}{dt}\overline{a}(t) = \sqrt{\kappa_2}\,\overline{b}_{2,\text{in}}(t) + \sqrt{\kappa_1}\,\overline{b}_{1,\text{in}}(t) - \frac{\kappa_C}{2}\,\overline{a}(t) . \tag{14.110}$$

Equations of this type are commonly solved by introducing the Fourier transform pairs

$$\overline{F}(\omega) = \int_{-\infty}^{\infty} dt e^{i\omega t}\overline{F}(t) , \tag{14.111}$$

$$\overline{F}(t) = \int_{-\infty}^{\infty} \frac{d\omega}{2\pi} e^{-i\omega t}\overline{F}(\omega) . \tag{14.112}$$

In the present case, $\overline{F}(t)$ stands for any of the envelope operators $\overline{a}(t)$, $\overline{b}_{1,\text{in}}(t)$, and $\overline{b}_{2,\text{in}}(t)$. Since these operators are not hermitian, a convention regarding adjoints is needed. We choose to use the same convention in the time and frequency domains:

$$\overline{F}^{\dagger}(t) = \left[\overline{F}(t)\right]^{\dagger} , \quad \overline{F}^{\dagger}(\omega) = \left[\overline{F}(\omega)\right]^{\dagger} . \tag{14.113}$$

With this convention in force, the adjoint of eqn (14.112) yields

$$\overline{F}^{\dagger}(t) = \int_{-\infty}^{\infty} \frac{d\omega}{2\pi} e^{i\omega t}\overline{F}^{\dagger}(\omega) = \int_{-\infty}^{\infty} \frac{d\omega}{2\pi} e^{-i\omega t}\overline{F}^{\dagger}(-\omega) . \tag{14.114}$$

Substituting the expansions (14.112) and (14.114) into eqn (14.102) produces the frequency-domain commutation relations

$$\begin{aligned} \left[\overline{b}_{J,\text{in}}(\omega), \overline{b}^{\dagger}_{K,\text{in}}(\omega')\right] &= 2\pi\delta_{JK}\delta(\omega-\omega') , \\ \left[\overline{b}_{J,\text{in}}(\omega), \overline{b}_{K,\text{in}}(\omega')\right] &= 0 . \end{aligned} \tag{14.115}$$

In general, it is not correct to think of eqn (14.112) as a mode expansion for $\overline{F}(t)$. For example, $\overline{a}(t)$ is the Heisenberg-picture annihilation operator associated with a particular cavity mode; this is as far as mode expansions go. Consequently the application of eqn (14.112) to $\overline{a}(t)$ cannot be regarded as a further mode expansion. The in-fields are a special case in this regard, since Fourier transforming the definition (14.98) yields

$$\overline{b}_{J,\text{in}}(\omega) = \overline{b}_{J,\omega+\omega_0}(t_0)\, e^{i\omega t_0} \quad (J = 1, 2). \tag{14.116}$$

This close relation between the Fourier transform and the mode expansion is a result of the explicit definition of the in-field as a superposition of freely propagated annihilation operators for the individual modes.

We can now proceed by Fourier transforming the differential equation (14.110) for $\overline{a}(t)$, to get the algebraic equation

$$-i\omega\overline{a}(t) = \sqrt{\kappa_2}\,\overline{b}_{2,\text{in}}(\omega) + \sqrt{\kappa_1}\,\overline{b}_{1,\text{in}}(\omega) - \frac{\kappa_C}{2}\,\overline{a}(\omega), \tag{14.117}$$

with the solution

$$\overline{a}(\omega) = \frac{\sqrt{\kappa_2}\,\overline{b}_{2,\text{in}}(\omega) + \sqrt{\kappa_1}\,\overline{b}_{1,\text{in}}(\omega)}{\kappa_C/2 - i\omega}. \tag{14.118}$$

In the frequency domain, the unmodified operators and the slowly-varying envelope operators are related by the translation rule

$$\overline{F}(\omega) = F(\omega + \omega_0). \tag{14.119}$$

This kind of rule is often expressed by saying that ω is replaced by $\omega + \omega_0$, but this is a bit misleading. The translation rule really means that the *argument* of the function is translated; for example, $\overline{F}(-\omega)$ is replaced by $F(-\omega + \omega_0)$, not $F(-\omega - \omega_0)$. Thus the argument in $\overline{F}(\omega)$ represents the displacement, either positive or negative, from the carrier frequency ω_0. Applying the translation rule to eqn (14.116) and to the expression for $\overline{a}(\omega)$ yields

$$b_{J,\text{in}}(\omega) = b_{J,\omega}(t_0)\, e^{i\omega t_0} \quad (J = 1, 2), \tag{14.120}$$

and

$$a(\omega) = \frac{\sqrt{\kappa_2}\, b_{2,\text{in}}(\omega) + \sqrt{\kappa_1}\, b_{1,\text{in}}(\omega)}{\kappa_C/2 - i(\omega - \omega_0)}. \tag{14.121}$$

The frequency-domain version of the scattering equation (14.109) for $b_{1,\text{out}}(\omega)$, where $b_{1,\text{out}}(\omega) = \overline{b}_{1,\text{out}}(\omega - \omega_0)$, combines with the explicit solution (14.121) to yield the input–output equation

$$b_{1,\text{out}}(\omega) = \frac{[(\kappa_2 - \kappa_1)/2 - i(\omega - \omega_0)]\, b_{1,\text{in}}(\omega) - \sqrt{\kappa_1\kappa_2}\, b_{2,\text{in}}(\omega)}{\kappa_C/2 - i(\omega - \omega_0)}. \tag{14.122}$$

For far off-resonance radiation, i.e. $|\omega - \omega_0| \gg \kappa_C/2$, this relation reduces to

$$b_{1,\text{out}}(\omega) \approx b_{1,\text{in}}(\omega), \tag{14.123}$$

which corresponds to complete reflection of the radiation incident on M1. For a symmetrical resonator, i.e. $\kappa_1 = \kappa_2 = \kappa_C/2$, the input–output relation simplifies to

$$b_{1,\mathrm{out}}(\omega) = \frac{i(\omega-\omega_0)\, b_{1,\mathrm{in}}(\omega) + (\kappa_C/2)\, b_{2,\mathrm{in}}(\omega)}{i(\omega-\omega_0) - \kappa_C/2}, \tag{14.124}$$

and for nearly resonant radiation, $\omega \approx \omega_0$, this becomes

$$b_{1,\mathrm{out}}(\omega) = \frac{(\kappa_C/2)\, b_{2,\mathrm{in}}(\omega)}{i(\omega-\omega_0) - \kappa_C/2}. \tag{14.125}$$

In this limit, the output field from mirror M1 is simply proportional to the input field at mirror M2, i.e. there is essentially no reflection of radiation incident on mirror M2. In this situation the cavity is called a Lorentzian filter, since the output intensity,

$$\left\langle b_{1,\mathrm{out}}^{\dagger}(\omega)\, b_{1,\mathrm{out}}(\omega)\right\rangle = \frac{(\kappa_C/2)^2}{(\kappa_C/2)^2 + (\omega-\omega_0)^2} \left\langle b_{2,\omega}^{\dagger}(t_0)\, b_{2,\omega}(t_0)\right\rangle, \tag{14.126}$$

has a typical Lorentzian line shape.

14.4 Noise and dissipation for atoms

In Section 11.3.3 we obtained a dissipative form of the Bloch equation for a two-level atom by adding phenomenological damping terms to the quantum Liouville equation for the atomic density operator. The Liouville equation is defined in the Schrödinger picture, or sometimes in the interaction picture; consequently, the Bloch equation does not immediately fit into the Heisenberg-picture formulation of the sample–reservoir model employed above. In order to make the connection, we first recall that an N-level atom is completely described by the transition operators, $S_{qp} = |\varepsilon_q\rangle\langle\varepsilon_p|$, defined in Section 11.1.4. In particular, the matrix elements $\rho_{pq}(t) = \langle\varepsilon_p|\rho(t)|\varepsilon_q\rangle$ of the density operator are given by

$$\rho_{pq}(t) = \mathrm{Tr}\, \rho(t)\, S_{qp}\,. \tag{14.127}$$

The trace is invariant under unitary transformations, so this result can equally well be written as

$$\rho_{pq}(t) = \mathrm{Tr}\, \rho S_{qp}(t)\,, \tag{14.128}$$

where ρ and $S_{qp}(t)$ are both expressed in the Heisenberg picture. Since the Heisenberg-picture density operator ρ is time independent, the Bloch equation for the matrix elements of the density operator is an immediate consequence of the Heisenberg equations of motion for the transition operators. For this reason, we will sometimes use the name **operator Bloch equation** for these particular Heisenberg equations.

14.4.1 Two-level atoms

In order to avoid unnecessary complications, we will restrict the detailed discussion to the simplest case of two-level atoms. With these results in hand, the generalization to

N-level atoms is straightforward. For a sample consisting of a single two-level atom, the sample Hamiltonian is $H_S = H_{S0} + H_{S1}(t)$, where

$$H_{S0} = \frac{\hbar\omega_{21}}{2}\left[S_{22} - S_{11}\right]. \qquad (14.129)$$

The terms in the Heisenberg equations of motion contributed by $H_{S1}(t)$ play no role in the following discussion, so we will omit them from the intermediate calculations and restore them at the end to get the final form of the Langevin equations.

A Noise reservoirs

There are now two forms of dissipation to be considered: spontaneous emission (sp) and **phase-changing perturbations** (pc). We already have the complete theory for spontaneous emission, but in the present context it is more instructive to use the schematic approach of Section 14.1.2. The creation and annihilation operators for the reservoir excitations (photons) that are emitted and absorbed in the $2 \leftrightarrow 1$ transition are denoted by $b_\Omega^\dagger$ and b_Ω. The second form of dissipation is associated with the decay of the atomic dipole, due to perturbations that do not cause real transitions between the two levels. In the simplest case, the atom is excited from an initial state to a virtual intermediate state and then returned to the original state. In a vapor, this effect arises primarily from collisions with other atoms. In a solid, phase-changing perturbations are often caused by local field fluctuations. The phase-changing perturbations of the two levels may arise from different mechanisms, so we need a reservoir for each level, with creation and annihilation operators $c_{q\Omega}^\dagger$ and $c_{q\Omega}$ $(q = 1, 2)$.

The environment Hamiltonian is therefore

$$H_E = \int_0^\infty \frac{d\Omega}{2\pi}\hbar\Omega b_\Omega^\dagger b_\Omega + \sum_{q=1}^{2}\int_0^\infty \frac{d\Omega}{2\pi}\hbar\Omega c_{q\Omega}^\dagger c_{q\Omega}\,, \qquad (14.130)$$

and the sample–environment interaction Hamiltonian $H_{\rm SE}$ is

$$H_{\rm SE} = H_{\rm sp} + H_{\rm pc}\,, \qquad (14.131)$$

where $H_{\rm sp}$ and $H_{\rm pc}$ are responsible for spontaneous emission and phase-changing perturbations respectively. The spontaneous emission Hamiltonian,

$$H_{\rm sp} = i\hbar\int_0^\infty d\Omega\sqrt{\frac{\mathcal{D}(\Omega)}{2\pi}}v(\Omega)\left(b_\Omega^\dagger S_{12} - S_{21}b_\Omega\right), \qquad (14.132)$$

is modeled directly on the RWA Hamiltonian of eqn (11.25), with the coupling constant $v(\Omega)$ playing the role of the dipole matrix element.

The simplest phase-changing perturbation is a second-order process in which the atom starts and ends in the same state. The transition from an initial state $|\varepsilon_q\rangle$ to an intermediate state $|\varepsilon_p\rangle$ is represented by the operator S_{pq}, and the return to the original state is described by S_{qp}; consequently, the complete transition is described by the product $S_{qp}S_{pq} = S_{qq}$. Since there is no overall change in energy, the resonance for this transition occurs at zero frequency. We model the phase-changing mechanism by coupling the atom to two reservoirs according to

$$H_{\text{pc}} = i\hbar \sum_{q=1}^{2} \int_0^\infty d\Omega \sqrt{\frac{\mathcal{D}(\Omega)}{2\pi}} u_q(\Omega) \left(c_{q\Omega}^\dagger S_{qq} - S_{qq} c_{q\Omega} \right). \tag{14.133}$$

Coupling to the zero-frequency resonance is enforced by assuming that the coupling constant $u_q(\Omega)$ is proportional to the cut-off function centered at zero frequency.

B Langevin equations

Since the sample and environment operators commute at equal times, the terms in the total Hamiltonian can be written in any desired order. We chose to put them in normal order with respect to the environment operators, so that the Heisenberg equations

$$\frac{d}{dt} S_{qp}(t) = \frac{1}{i\hbar} \left[S_{qp}(t), H_E + H_S + H_{\text{sp}} + H_{\text{pc}} \right] \tag{14.134}$$

are also normally ordered. The resonance frequencies for the interaction of the sample with the spontaneous-emission and phase-changing reservoirs are $\omega = \omega_{21}$ and $\omega = 0$ respectively; therefore, we express eqn (14.134) in terms of the envelope operators

$$\begin{aligned} \overline{S}_{12}(t) &= S_{12}(t)\, e^{i\omega_{21}t}, \quad \overline{S}_{qq}(t) = S_{qq}(t), \\ \overline{b}_\Omega(t) &= b_\Omega(t)\, e^{i\omega_{21}t}, \quad \overline{c}_{q\Omega}(t) = c_{q\Omega}(t), \end{aligned} \tag{14.135}$$

to find

$$\begin{aligned} \frac{d}{dt}\overline{S}_{12}(t) = {} & \left\{ \overline{S}_{22}(t) - \overline{S}_{11}(t) \right\} \beta(t) + \left\{ \gamma_2^\dagger(t) - \gamma_1^\dagger(t) \right\} \overline{S}_{12}(t) \\ & - \overline{S}_{12}(t) \left\{ \gamma_2(t) - \gamma_1(t) \right\}, \end{aligned} \tag{14.136}$$

$$\frac{d}{dt}\overline{S}_{22}(t) = -\beta^\dagger(t)\, \overline{S}_{12}(t) - \overline{S}_{21}(t)\, \beta(t), \tag{14.137}$$

$$\frac{d}{dt}\overline{b}_\Omega(t) = -i(\Omega - \omega_{21})\, \overline{b}_\Omega(t) + \sqrt{2\pi\mathcal{D}(\Omega)}\, v(\Omega)\, \overline{S}_{12}(t), \tag{14.138}$$

$$\frac{d}{dt}\overline{c}_{q\Omega}(t) = -i\Omega\overline{c}_{q\Omega}(t) + \sqrt{2\pi\mathcal{D}(\Omega)}\, u_q(\Omega)\, \overline{S}_{qq}(t), \tag{14.139}$$

where

$$\beta(t) = \int_0^\infty d\Omega \sqrt{\frac{\mathcal{D}(\Omega)}{2\pi}}\, v(\Omega)\, \overline{b}_\Omega(t) \tag{14.140}$$

and

$$\gamma_q(t) = \int_0^\infty d\Omega \sqrt{\frac{\mathcal{D}(\Omega)}{2\pi}}\, u_q(\Omega)\, \overline{c}_{q\Omega}(t) \quad (q = 1, 2). \tag{14.141}$$

The equation for $\overline{S}_{11}(t)$ has been omitted, by virtue of the identity $\overline{S}_{11}(t) + \overline{S}_{22}(t) = 1$.

The Langevin equations for the atomic transition operators are derived by an argument similar to the one employed in Section 14.2.1. The formal solutions of eqns

(14.138) and (14.139) for the reservoir operators are combined with the Markov conditions

$$2\pi\mathcal{D}(\Omega)\left|v(\Omega)\right|^{2} \simeq w_{21} = 2\pi\mathcal{D}(\omega_{21})\left|v(\omega_{21})\right|^{2} \tag{14.142}$$

and

$$2\pi\mathcal{D}(\Omega)\left|u_{q}(\Omega)\right|^{2} \simeq w_{qq} = 2\pi\mathcal{D}(0)\left|u_{q}(0)\right|^{2}, \tag{14.143}$$

to get

$$\beta(t) = \sqrt{w_{21}}\, b_{\text{in}}(t) + \frac{w_{21}}{2}\overline{S}_{12}(t) \tag{14.144}$$

and

$$\gamma_{q}(t) = \sqrt{w_{qq}}\, c_{q,\text{in}}(t) + \frac{w_{qq}}{2}\overline{S}_{qq}(t) \quad (q = 1, 2). \tag{14.145}$$

The in-fields for the reservoirs are given by

$$b_{\text{in}}(t) = \int_{0}^{\infty}\frac{d\Omega}{2\pi}b_{\Omega}(t_{0})\,e^{-i(\Omega-\omega_{21})(t-t_{0})} \tag{14.146}$$

and

$$c_{q,\text{in}}(t) = \int_{0}^{\infty}\frac{d\Omega}{2\pi}c_{q\Omega}(t_{0})\,e^{-i\Omega(t-t_{0})}. \tag{14.147}$$

Substituting these results into eqns (14.136) and (14.137) yields the Langevin equations for the transition operators:

$$\frac{d}{dt}S_{12}(t) = \left[i\omega_{12} - \Gamma_{12}\right]S_{12}(t) + \frac{1}{i\hbar}\left[S_{12}(t), H_{S1}(t)\right] + \xi_{12}(t), \tag{14.148}$$

$$\frac{d}{dt}S_{22}(t) = -w_{21}S_{22}(t) + \frac{1}{i\hbar}\left[S_{22}(t), H_{S1}(t)\right] + \xi_{22}(t), \tag{14.149}$$

$$\frac{d}{dt}S_{11}(t) = w_{21}S_{22}(t) + \frac{1}{i\hbar}\left[S_{11}(t), H_{S1}(t)\right] + \xi_{11}(t), \tag{14.150}$$

where w_{21} is the spontaneous decay rate for the $2 \to 1$ transition, w_{11} and w_{22} are the rates of the phase-changing perturbations, and

$$\Gamma_{12} = \frac{1}{2}\left(w_{21} + w_{22} + w_{11}\right) \tag{14.151}$$

is the dephasing rate for the atomic dipole. We have restored the $H_{S1}(t)$-terms and also imposed $\xi_{11}(t) = -\xi_{22}(t)$ in accord with the conservation of population.

The operators $\xi_{12}(t)$ and $\xi_{22}(t)$ represent multiplicative noise, since they involve products of sample and reservoir operators. This raises a new difficulty, because there is no general argument proving that multiplicative noise operators are delta correlated. Even in the special cases for which a proof can be given—e.g. those considered in Exercise 14.2—the calculations are quite involved. In this situation, the only general procedure available is to include the delta-correlation assumption as part of the Markov approximation. For the problem at hand the *ansatz* is

$$\left\langle \xi_{qp}(t)\,\xi^{\dagger}_{q'p'}(t')\right\rangle = C_{qp,q'p'}\,\delta(t-t'). \tag{14.152}$$

The coefficients $C_{qp,q'p'}$ can be evaluated, at least partially, by the general methods described in Section 14.6.

We will see, in the following section, that the use of atomic transition operators is a great advantage for the generalization from two-level to N-level atoms, but for applications to two-level atoms themselves, it is often easier to work in terms of the familiar Pauli matrices. The relations

$$S_{22} = \frac{1}{2}(1+\sigma_z), \quad S_{11} = \frac{1}{2}(1-\sigma_z), \quad S_{12} = \sigma_-, \quad S_{21} = \sigma_+ \tag{14.153}$$

lead to the equivalent Langevin equations

$$\frac{d}{dt}\overline{\sigma}_-(t) = -\Gamma_{12}\overline{\sigma}_-(t) + \frac{1}{i\hbar}\left[\overline{\sigma}_-(t), H_{S1}(t)\right] + \xi_-(t), \tag{14.154}$$

$$\frac{d}{dt}\overline{\sigma}_z(t) = -w_{21}\left[1+\overline{\sigma}_z(t)\right] + \frac{1}{i\hbar}\left[\overline{\sigma}_z(t), H_{S1}(t)\right] + \xi_z(t), \tag{14.155}$$

where $\xi_-(t) = \xi_{12}(t)$ and $\xi_z(t) = 2\xi_{22}(t)$.

14.4.2 N-level atoms

The derivation of the Langevin equations for atoms with N levels could be carried out by applying the approach followed for the two-level atom, but this would require assigning a reservoir for every real decay and another reservoir for each level subjected to phase-changing perturbations. One can escape burial under this avalanche of reservoirs by paying careful attention to the structure of eqns (14.148)–(14.151) for the two-level atom. If we assume that the dissipative effects involve transitions between pairs of atomic levels or phase-changing perturbations of single levels, then a little thought shows that the N-level Langevin equations must have the general form

$$\frac{d}{dt}S_{qp}(t) = (i\omega_{qp} - \Gamma_{qp})S_{qp}(t) + \frac{1}{i\hbar}\left[S_{qp}(t), H_{S1}(t)\right] + \xi_{qp}(t) \quad \text{for } q \neq p, \tag{14.156}$$

$$\frac{d}{dt}S_{qq}(t) = \sum_p w_{pq}S_{pp}(t) - \sum_p w_{qp}S_{qq}(t) + \frac{1}{i\hbar}\left[S_{qq}(t), H_{S1}(t)\right] + \xi_{qq}(t). \tag{14.157}$$

The envelope operators are defined by generalizing eqn (14.135) to

$$S_{qp}(t) = \overline{S}_{qp}(t)\, e^{i\omega_{qp}t} e^{i[\theta_q(t)-\theta_p(t)]}, \tag{14.158}$$

where each $\theta_q(t)$ is a real function. The reason for including the θ_qs in this definition is that—in favorable cases—they can be chosen to eliminate explicit time dependencies due to $\left[\overline{S}_{qp}(t), H_{S1}(t)\right]$. Substituting eqn (14.158) into eqns (14.156) and (14.157) leads to the envelope equations

$$\frac{d}{dt}\overline{S}_{qp}(t) = \left[-i\left(\dot{\theta}_q - \dot{\theta}_p\right) - \Gamma_{qp}\right]\overline{S}_{qp}(t) + \frac{1}{i\hbar}\left[\overline{S}_{qp}(t), H_{S1}(t)\right] + \xi_{qp}(t) \quad \text{for } q \neq p, \tag{14.159}$$

$$\frac{d}{dt}\overline{S}_{qq}(t) = \sum_p w_{pq}\overline{S}_{pp}(t) - \sum_p w_{qp}\overline{S}_{qq}(t) + \frac{1}{i\hbar}\left[\overline{S}_{qq}(t), H_{S1}(t)\right] + \xi_{qq}(t), \tag{14.160}$$

where

$$w_{pq} = \begin{cases} \text{transition rate for } p \to q \text{ if } \varepsilon_p > \varepsilon_q \,, \\ 0 \text{ if } \varepsilon_p < \varepsilon_q \,, \\ \text{the phase-changing rate for the } q\text{th level when } q = p \,. \end{cases} \tag{14.161}$$

$$\text{For } q \neq p, \; \Gamma_{qp} = \frac{1}{2} \sum_r \left(w_{qr} + w_{pr} \right) \; \text{ is the dephasing rate for } \overline{S}_{qp}(t) \,. \tag{14.162}$$

Strictly speaking, one should also define envelope noise operators,

$$\overline{\xi}_{qp}(t) = e^{-i\omega_{qp}t} e^{-i[\theta_q(t) - \theta_p(t)]} \xi_{qp}(t) \,, \tag{14.163}$$

but the assumption that the original operators $\xi_{qp}(t)$ are delta correlated implies that the envelope noise operators would have the same correlation functions. Since the correlation functions are all that matters for noise operators, it is safe to ignore the distinction between $\xi_{qp}(t)$ and $\overline{\xi}_{qp}(t)$.

14.5 Incoherent pumping

Incoherent pumping processes—which raise rather than lower the energy of an atom—are used to produce population inversion; consequently, they play a central role in laser physics. As we have seen in Section 14.4, the interaction of an atom with a short-memory reservoir is necessarily dissipative. This raises the following question: Can incoherent pumping be described by a reservoir model? This feat has been accomplished, but only at the cost of introducing an unphysical reservoir (Gardiner, 1991, Sec. 7.2.1). The idea is to describe pumping by coupling the atom to a reservoir composed of oscillators with an inverted energy spectrum, $\varepsilon_\Omega = -\hbar\Omega$, as in Exercise 14.5. Emitting an excitation into this reservoir lowers the reservoir energy and therefore raises the energy of the atom. We have previously mentioned the formal use of unphysical negative-energy modes in the discussion of the input–output method in Section 14.3, but in that situation the probability for exciting the unphysical modes is negligible. This cannot be the case for the inverted-oscillator reservoir; otherwise, there would be no pumping. Since this model violates Einstein's rule, we must accept some added complexity.

The interaction between an atom and a classical field, with rapid fluctuations in phase, provides a physically acceptable model for incoherent pumping. Unfortunately, building such a model for the simplest case of a two-level atom is pointless, since the discussion in Section 11.3.3 shows that no pumping scheme for a two-level atom can produce an inverted population. We will, therefore, grudgingly admit that real atoms have more than two levels and add a third. The added complexity will be offset by ignoring phase-changing perturbations.

The sample is a collection of three-level atoms, with the energy-level diagram shown in Fig. 14.2. The $3 \leftrightarrow 1$ transition is driven by a strong, classical pump field

$$\boldsymbol{\mathcal{E}}_P(t) = \mathbf{e}_P \mathcal{E}_{P0} e^{i\vartheta_P(t)} e^{-i\omega_P t} \,, \tag{14.164}$$

where $\omega_P \approx \omega_{31}$ and $\vartheta_P(t)$ is a rapidly fluctuating phase. Since there is no coupling between the atoms, we can restrict our attention to a single atom located at $\mathbf{r} = 0$. For this reduced sample, the interaction Hamiltonian is $H_{S1} = V_P(t) + H'_{S1}$, where

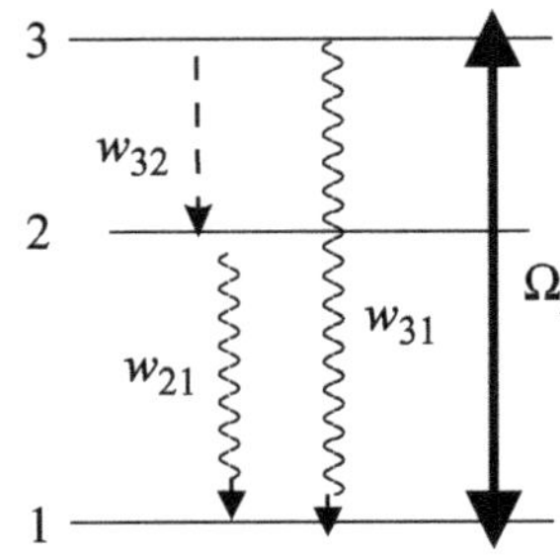

Fig. 14.2 A three-level atom with dipole allowed transitions $1 \leftrightarrow 3$ and $1 \leftrightarrow 2$. The spontaneous emission rates are w_{31} and w_{21} respectively. The $1 \leftrightarrow 3$ transition is also driven by a classical field with Rabi frequency Ω_P. A non-radiative decay $3 \to 2$, with rate w_{32}, is indicated by the dashed arrow. The wavy arrows denote the spontaneous emissions.

$$V_P(t) = \hbar \left[\Omega_P e^{i\vartheta_P(t)} e^{-i\omega_P t} e^{i\omega_{31} t} e^{i[\theta_3(t) - \theta_1(t)]} \overline{S}_{31} + \text{HC} \right], \tag{14.165}$$

$\overline{S}_{31}$ is the envelope operator defined by eqn (14.158), Ω_P is the Rabi frequency associated with the constant amplitude $\mathcal{E}_{P0}$, and H'_{S1} includes any other interactions with external fields as well as any sample–sample interactions. The remaining interaction term H'_{S1} influences some of the choices to be made, but the terms contributed by H'_{S1} to the equations of motion play no direct role in the following argument. We will therefore omit them from the intermediate steps and restore them at the end. In addition to the spontaneous emissions, $2 \to 1$ and $3 \to 1$, we assume that there is a non-radiative decay channel: $3 \to 2$.

The Langevin equations for this problem are derived in Exercise 14.6 by dropping the phase-changing terms from the ($N = 3$)-case of eqns (14.159) and (14.160). It is also useful to impose $\theta_3(t) - \theta_1(t) = \Delta_P t - \vartheta_P(t)$—where $\Delta_P = \omega_P - \omega_{31}$ is the pump detuning—in order to eliminate the explicit time dependence of $V_P(t)$. The remaining phase differences $\theta_1 - \theta_2$ and $\theta_2 - \theta_3$ are related by

$$\begin{aligned}
\theta_2 - \theta_3 &= (\theta_1 - \theta_3) - (\theta_1 - \theta_2) \\
&= \Delta_P t - \vartheta_P(t) - (\theta_1 - \theta_2),
\end{aligned} \tag{14.166}$$

so we can only impose one more condition on the phases. The choice of this condition depends on H'_{S1}. In the problem at hand, we have assumed that the transition $2 \leftrightarrow 1$ is dipole allowed, but the transition $3 \leftrightarrow 2$ is not. Thus only the transition $2 \leftrightarrow 1$ can be dipole-coupled to the electromagnetic field. We therefore reserve $\theta_1 - \theta_2$ to deal with any such coupling, and use eqn (14.166) as the definition of $\theta_2 - \theta_3$. For the sake of simplicity, we will assume that $\Delta_{21} = \dot{\theta}_2 - \dot{\theta}_1$ is a constant; this assumption is valid in most applications.

The central idea of this approach is that the envelope operators are effectively independent of the randomly fluctuating pump phase $\vartheta_P(t)$. This means that $\langle \overline{S}_{qp} \rangle_P \simeq \overline{S}_{qp}$, where $\langle \cdots \rangle_P$ denotes averaging over the distribution of pump phases. This allows the rapid fluctuations in the phase to be exploited by a variant of the adiabatic elimination argument. As an illustration of this approach, we start with the Langevin equation,

$$\frac{d\overline{S}_{23}(t)}{dt} = -i\left(\dot{\theta}_2 - \dot{\theta}_3\right) \overline{S}_{23}(t) - i\Omega_P \overline{S}_{21}(t) - \Gamma_{23} \overline{S}_{23}(t) + \xi_{23}(t), \tag{14.167}$$

for the atomic coherence operator $\overline{S}_{23}(t)$, and impose the phase choice (14.166). Writing out the formal solution and averaging it over the phase distribution of the pump then leads to

$$\begin{aligned}\overline{S}_{23}(t) = \overline{S}_{23}(t_0)\, e^{(i\Delta_P - i\Delta_{21} - \Gamma_{23})(t-t_0)} C_P(t, t_0) \\ - i\Omega_P \int_{t_0}^{t} dt' e^{(i\Delta_P - i\Delta_{21} - \Gamma_{23})(t-t')} C_P(t, t')\, \overline{S}_{21}(t') \\ + \int_{t_0}^{t} dt' e^{(i\Delta_P - i\Delta_{21} - \Gamma_{23})(t-t')} C_P(t, t')\, \langle \xi_{23}(t')\rangle_P\,,\end{aligned} \tag{14.168}$$

where

$$C_P(t, t') \equiv \left\langle e^{-i\vartheta_P(t)} e^{i\vartheta_P(t')} \right\rangle_P . \tag{14.169}$$

For a time-stationary distribution of pump phase, $C_P(t, t')$ only depends on the time difference $t - t'$; and it decays rapidly for $|t - t'|$ larger than the pump correlation time. For the function $C_P(t, t_0)$, this means that transient effects, associated with turning on the pump, will fade away for $t - t_0$ larger than the pump correlation time. This is mathematically equivalent to the limit $t_0 \to -\infty$, so that $C_P(t, t_0) \to 0$. In the remaining terms, the rapid decay of $C_P(t, t')$ justifies evaluating the other functions in the t'-integrals at $t' = t$. The result is

$$\overline{S}_{23}(t) = -i\Omega_P T_P \overline{S}_{21}(t) + T_P \langle \xi_{23}(t)\rangle_P\,, \tag{14.170}$$

where

$$T_P = \lim_{t_0 \to -\infty} \int_{t_0}^{t} dt' \left\langle e^{-i[\vartheta_P(t) - \vartheta_P(t')]} \right\rangle_P \tag{14.171}$$

is a measure of the correlation time for the incoherent pump. The same procedure applied to $\overline{S}_{13}(t)$ yields

$$\overline{S}_{13}(t) = -i\Omega_P T_P \left\{\overline{S}_{11}(t) - \overline{S}_{33}(t)\right\} + T_P \langle \xi_{13}(t)\rangle_P\,. \tag{14.172}$$

The strengths of the noise operators $\langle \xi_{qp}(t)\rangle_P$ are determined by the atomic transition rates, which we can assume are small compared to Ω_P. This justifies neglecting the noise terms in eqns (14.170) and (14.172) to get

$$\begin{aligned}\overline{S}_{23}(t) &= -i\Omega_P T_P \overline{S}_{21}(t)\,, \\ \overline{S}_{13}(t) &= -i\Omega_P T_P \left[\overline{S}_{11}(t) - \overline{S}_{33}(t)\right].\end{aligned} \tag{14.173}$$

Substituting these results in the remaining Langevin equations and restoring the contributions from H'_{S1} produces the reduced equations:

$$\frac{d\overline{S}_{11}(t)}{dt} = -R_P \overline{S}_{11}(t) + w_{21}\overline{S}_{22}(t) + (w_{31} + R_P)\,\overline{S}_{33}(t) + \frac{1}{i\hbar}\left[\overline{S}_{11}(t), H'_{S1}\right] + \xi_{11}(t)\,, \tag{14.174}$$

$$\frac{d\overline{S}_{22}(t)}{dt} = w_{32}\overline{S}_{33}(t) - w_{21}\overline{S}_{22}(t) + \frac{1}{i\hbar}\left[\overline{S}_{22}(t), H'_{S1}\right] + \xi_{22}(t)\,, \tag{14.175}$$

$$\frac{d\overline{S}_{33}(t)}{dt} = R_P \overline{S}_{11}(t) - (w_{31} + R_P + w_{32}) \overline{S}_{33}(t) + \frac{1}{i\hbar} \left[\overline{S}_{33}(t), H'_{S1} \right] + \xi_{33}(t), \quad (14.176)$$

$$\frac{d\overline{S}_{12}(t)}{dt} = \left[i\Delta_{21} - \frac{1}{2}(w_{21} + R_P) \right] \overline{S}_{12}(t) + \frac{1}{i\hbar} \left[\overline{S}_{12}(t), H'_{S1} \right] + \xi_{12}(t), \quad (14.177)$$

where $R_P = 2\Omega_P^2 T_P$ is the incoherent pumping rate. The more familiar c-number Bloch equations describing incoherent pumping are derived in Exercise 14.7 by averaging these equations with the initial density operator ρ. The correlation functions for the remaining noise operators can be calculated by means of the Einstein relation discussed in Section 14.6.2 and Exercise 14.8.

In eqn (14.177) we have explicitly exhibited the dephasing rate $(w_{21} + R_P)/2$, in order to show that the pumping rate, R_P, contributes to the dephasing rate in exactly the same way as the decay rate w_{21}. This suggests that we modify the general definition (14.162) for Γ_{pq} to include the effects of any pumping transitions that may be present. This is done by replacing the decay rates w_{qp} with $w_{qp} + R_{qp}$, where $R_{qp} = R_{pq}$ is the rate for an incoherent pump driving $q \leftrightarrow p$.

14.6 The fluctuation dissipation theorem*

Now that we have seen several examples of the fluctuation dissipation theorem, it is time to seek a more general result. In the examples considered above, the O_Js satisfy commutation relations of the general form

$$[O_J, O_K] = \sum_I \Lambda^I_{JK} O_I \quad (14.178)$$

(e.g. the operators $\{1, \overline{a}, \overline{a}^\dagger\}$ or $\{\overline{S}_{qp}\}$), and in some cases product relations

$$O_J O_K = \sum_I \Phi^I_{JK} O_I \quad (14.179)$$

(e.g. the transition operators $\{\overline{S}_{qp}\}$), where the Λ^I_{JK}s and Φ^I_{JK}s are c-number coefficients. The O_Js in the previous examples also satisfy

$$[O_J, H_{S0}] = \hbar\omega_J O_J\,. \quad (14.180)$$

The last property permits the definition of slowly-varying envelope operators $\overline{O}_J(t)$ by

$$\overline{O}_J(t) = O_J(t)\exp(i\omega_J t)\,. \quad (14.181)$$

In practice these features are quite typical; they are not restricted to the specific examples in Sections 14.2 and 14.4. For a given sample, it is usually easy to pick out these operators by inspection.

A potentially significant weakness of the discussions in Sections 14.2 and 14.4 is their neglect of the effects of internal sample interactions or interactions with external classical fields. In particular, the proof of the important nonanticipating property in eqn (14.77) uses the explicit solution (14.65) of the linear Langevin equation (14.61), which is only correct for $H_{S1} = 0$. This is an example of the following general feature of the theory of noise and dissipation. If the Heisenberg equations for the sample operators are linear, then results that are needed for subsequent applications—such as the nonanticipating property—can be proved by fairly simple arguments. Since the internal interaction H_{SS} describes coupling between different degrees of freedom of the sample, it will necessarily produce nonlinear terms in the Heisenberg equations for the sample operators. In order to avoid these complications as much as possible, we will make two assumptions. The first is that the internal interactions can be neglected when considering dissipative effects, i.e. $H_{SS} \sim 0$. The second is that any external interactions produce linear terms in the Heisenberg equation, i.e.

$$\frac{1}{i\hbar}\left[\overline{O}_J(t), V_S(t)\right] = i\sum_K \Omega_{JK}(t)\,\overline{O}_K(t), \tag{14.182}$$

where the $\Omega_{JK}(t)$s are c-number functions. The plausibility of these assumptions depends on the following points.

(1) The effect of H_{SS} and $V_S(t)$ is to cause additional unitary—and thus non-dissipative—evolution of the sample.

(2) By convention, H_{SS} is weak compared to H_{S0}.

(3) In typical cases—e.g. atoms interacting with a laser or field modes excited by a classical current—$V_S(t)$ is linear in the sample operators, and they satisfy the commutation relations (14.178).

With these facts in mind, it is quite plausible that ignoring H_{SS} and imposing eqn (14.182) on $V_S(t)$ will not cause any serious errors in the treatment of dissipation and noise. A more sophisticated argument that dispenses with these simplifying assumptions is briefly sketched in Exercise 14.9.

14.6.1 Generic Langevin equations

The argument just given allows us to replace the general Heisenberg equation (14.39) for the O_Js by the equation of motion

$$i\hbar\frac{d}{dt}\overline{O}_J(t) = \left[\overline{O}_J(t), H_{SE}(t)\right] + \left[\overline{O}_J(t), V_S(t)\right] \tag{14.183}$$

for the slowly-varying envelope operators. We can then substitute the formal solutions (14.38) for the reservoir operators into this equation, and impose the Markov approximation, i.e. the assumption that the reservoir memory T_{mem} is much shorter than any dynamical time scale for the sample. The resulting Langevin equations take the general form

$$\frac{d}{dt}\overline{O}_J(t) = D_J(t) + \xi_J(t), \tag{14.184}$$

where

$$D_J(t) = \sum_K Z_{JK}(t)\,\overline{O}_K(t) \tag{14.185}$$

is the (generalized) **drift term**, and the noise operators are defined so that

$$\langle \xi_J(t) \rangle = 0 \,. \tag{14.186}$$

The complex coefficients $Z_{JK}(t)$ are given by

$$Z_{JK}(t) = -\Gamma_{JK} + i\Omega_{JK}(t) \,, \tag{14.187}$$

where the real, positive constants Γ_{JK} arise from the elimination of the reservoir variables—combined with the Markov approximation—and the real functions $\Omega_{JK}(t)$ are defined by eqn (14.182). The decay constants Γ_{JK} can be expressed as functions of the coupling strengths $v_J(\Omega_\nu)$, but in practice they are treated as phenomenological parameters. The Markov approximation includes the assumption that the noise operators $\xi_J(t)$ are delta correlated,

$$\left\langle \xi_J(t)\, \xi_K^\dagger(t') \right\rangle = C_{JK}\,\delta(t - t') \,. \tag{14.188}$$

The coefficients C_{JK} define the **correlation matrix** for the noise operators, and $C_{JK}/2$ is also known as the diffusion matrix. The names 'drift term' and 'diffusion matrix' arise in connection with the master equation approach, which will be discussed in Chapter 18.

14.6.2 The Einstein relations

The direct calculation of the correlation matrix C_{JK} is very difficult, except in the case of additive noise. Fortunately, yet another consequence of the Markov approximation can be used to express the C_{JK}s in terms of sample correlation functions. We first show that the sample operators are nonanticipating with respect to the noise operators. For this purpose we can use eqns (14.184) and (14.188) to find the equations of motion for the correlation functions $\langle \xi_K^\dagger(t')\, \overline{O}_J(t) \rangle$:

$$\frac{\partial}{\partial t} \left\langle \xi_K^\dagger(t')\, \overline{O}_J(t) \right\rangle = \left\langle \xi_K^\dagger(t')\, D_J(t) \right\rangle + C_{KJ}\,\delta(t - t') \,. \tag{14.189}$$

For $t' > t$ the delta function term vanishes, and we find a set of linear, homogeneous differential equations

$$\frac{\partial}{\partial t} \left\langle \xi_K^\dagger(t')\, \overline{O}_J(t) \right\rangle = \sum_I Z_{JI} \left\langle \xi_K^\dagger(t')\, \overline{O}_I(t) \right\rangle \tag{14.190}$$

for the set of correlation functions $\langle \xi_K^\dagger(t')\, \overline{O}_J(t) \rangle$. The assumption that the sample and the reservoirs are uncorrelated at $t = t_0$ ensures that all the correlation functions vanish at $t = t_0$,

$$\left\langle \xi_K^\dagger(t')\, \overline{O}_I(t_0) \right\rangle = 0 \,; \tag{14.191}$$

therefore, we can conclude that

$$\left\langle \xi_K^\dagger(t')\, \overline{O}_J(t) \right\rangle = 0 \quad \text{for } t' > t \,. \tag{14.192}$$

Similar arguments show that $\langle \overline{O}_J(t)\, \xi_K^\dagger(t') \rangle = 0$ for $t' > t$, etc.

To use this fact, we start with the identity (Meystre and Sargent, 1990, Sec. 14-4)

$$\begin{aligned}\overline{O}_J(t) &= \overline{O}_J(t-\Delta t) + \int_{t-\Delta t}^{t} dt' \frac{d\overline{O}_J(t')}{dt'} \\ &= \overline{O}_J(t-\Delta t) + \int_{t-\Delta t}^{t} dt'' \left\{D_J(t'') + \xi_J(t'')\right\}, \end{aligned} \tag{14.193}$$

which in turn implies

$$\begin{aligned}\left\langle \overline{O}_J(t)\,\xi_K^\dagger(t)\right\rangle &= \left\langle \overline{O}_J(t-\Delta t)\,\xi_K^\dagger(t)\right\rangle + \int_{t-\Delta t}^{t} dt' \left\langle D_J(t')\,\xi_K^\dagger(t)\right\rangle \\ &\quad + \int_{t-\Delta t}^{t} dt'' \left\langle \xi_J(t')\,\xi_K^\dagger(t)\right\rangle . \end{aligned} \tag{14.194}$$

The nonanticipating property guarantees that the first term vanishes and that the integrand of the second term also vanishes, except possibly at the end point $t' = t$. Thus the integral must vanish unless the correlation function $\left\langle D_J(t')\,\xi_K^\dagger(t)\right\rangle$ is proportional to $\delta(t-t')$. This cannot be the case, since the drift term is slowly varying compared to the noise term. Thus only the third term contributes, and

$$\begin{aligned}\left\langle \overline{O}_J(t)\,\xi_K^\dagger(t)\right\rangle &= \int_{t-\Delta t}^{t} dt'' \left\langle \xi_J(t')\,\xi_K^\dagger(t)\right\rangle \\ &= \int_{t-\Delta t}^{t} dt'' C_{JK}\,\delta(t-t') = \frac{1}{2}C_{JK} . \end{aligned} \tag{14.195}$$

A similar calculation shows that

$$\left\langle \xi_J(t)\,\overline{O}_K^\dagger(t)\right\rangle = \frac{1}{2}C_{JK} . \tag{14.196}$$

We will now use these results to investigate the equation of motion of the equal-time correlation function $\left\langle \overline{O}_J(t)\,\overline{O}_K^\dagger(t)\right\rangle$. The Langevin equation (14.184) combines with eqns (14.195) and (14.196) to yield

$$\begin{aligned}\frac{d}{dt}\left\langle \overline{O}_J(t)\,\overline{O}_K^\dagger(t)\right\rangle &= \left\langle \left\{D_J(t) + \xi_J(t)\right\}\overline{O}_K^\dagger(t)\right\rangle + \left\langle \overline{O}_J(t)\left\{D_K^\dagger(t) + \xi_K^\dagger(t)\right\}\right\rangle \\ &= \left\langle D_J(t)\,\overline{O}_K^\dagger(t)\right\rangle + \left\langle \overline{O}_J(t)\,D_K^\dagger(t)\right\rangle \\ &\quad + \left\langle \overline{O}_J(t)\,\xi_K^\dagger(t)\right\rangle + \left\langle \xi_J(t)\,\overline{O}_K^\dagger(t)\right\rangle \\ &= \left\langle D_J(t)\,\overline{O}_K^\dagger(t)\right\rangle + \left\langle \overline{O}_J(t)\,D_K^\dagger(t)\right\rangle + C_{JK} . \end{aligned} \tag{14.197}$$

We turn this around to obtain the **Einstein relation**,

$$C_{JK} = \frac{d}{dt}\left\langle \overline{O}_J(t)\,\overline{O}_K^\dagger(t)\right\rangle - \left\langle D_J(t)\,\overline{O}_K^\dagger(t)\right\rangle - \left\langle \overline{O}_J(t)\,D_K^\dagger(t)\right\rangle , \tag{14.198}$$

that expresses the noise correlation matrix in terms of equal-time sample correlation functions. The sample correlation functions depend on the decay constants, so this is the general form of the fluctuation dissipation theorem. The calculation of the noise correlation matrix is thereby reduced to obtaining the values of the equal-time correlation functions $\langle \overline{O}_I(t) O_K^\dagger(t) \rangle$. In general the sample correlation functions must be independently calculated—e.g. by means of the master equation discussed in Chapter 18—but approximate estimates are often sufficient.

For an illustration of the use of eqn (14.198), we turn to the incoherently pumped three-level atom of Section 14.5. The index J now runs over the nine pairs (q,p), with $q,p = 1,2,3$. Let us, for example, calculate the correlation coefficient $C_{12,12}$ appearing in

$$\left\langle \xi_{12}(t)\, \xi_{12}^\dagger(t') \right\rangle = C_{12,12}\,\delta(t-t') \,. \tag{14.199}$$

For the case of pure pumping, i.e. $H'_{S1} = 0$, the Langevin equation (14.177) tells us that the drift term $D_{12} = -\Gamma_{12}\overline{S}_{12}$. Applying eqn (14.198) yields

$$\begin{aligned} C_{12,12} &= \frac{d}{dt}\left\langle \overline{S}_{12}\overline{S}_{12}^\dagger \right\rangle - \left\langle D_{12}\overline{S}_{12}^\dagger \right\rangle - \left\langle \overline{S}_{12}D_{12}^\dagger \right\rangle \\ &= \frac{d}{dt}\left\langle \overline{S}_{11} \right\rangle + 2\Gamma_{12}\left\langle \overline{S}_{12}\overline{S}_{12}^\dagger \right\rangle \\ &= -R_P N_1(t) + w_{21}N_2(t) + (w_{31} + R_P)\, N_3(t) + 2\Gamma_{12}N_1(t)\,, \end{aligned} \tag{14.200}$$

where $N_q(t) = \langle \overline{S}_{qq}(t) \rangle$. At long times (i.e. for $t_0 \to -\infty$) the populations are given by the steady-state solution of the c-number Bloch equations obtained by averaging eqns (14.174)–(14.177). One then finds

$$C_{12,12} = \frac{2\Gamma_{12}w_{21}\left(R_P + w_{31} + w_{32}\right)}{R_P\left(2w_{21} + w_{32}\right) + w_{21}\left(w_{31} + w_{32}\right)} \,. \tag{14.201}$$

Note that $C_{12,12}$, which represents the strength of the noise operator ξ_{12}, vanishes for $w_{21} = 0$. This justifies the interpretation of ξ_{12} as the noise due to the spontaneous emission $2 \to 1$. A similar calculation yields

$$C_{21,21} = \frac{2\Gamma_{12}R_P w_{32}}{R_P\left(2w_{21} + w_{32}\right) + w_{21}\left(w_{31} + w_{32}\right)} \,, \tag{14.202}$$

which implies

$$\left\langle \xi_{12}^\dagger(t)\, \xi_{12}(t') \right\rangle = \left\langle \xi_{21}(t)\, \xi_{21}^\dagger(t') \right\rangle = C_{21,21}\,\delta(t-t') \,. \tag{14.203}$$

14.7 Quantum regression*

All experimentally relevant numerical information is contained in the expectation values of functions of the sample operators, so we begin by observing that the expectation values $\langle \overline{O}_J(t) \rangle$ obey the averaged form of the Langevin equations (14.184):

$$\frac{d}{dt}\langle \overline{O}_J(t) \rangle = \sum_K Z_{JK}(t)\,\langle \overline{O}_K(t) \rangle \,. \tag{14.204}$$

A standard method for solving sets of linear first-order equations like (14.204) is to define a Green function $G_{JK}(t,t')$ by

$$\begin{aligned}\frac{d}{dt}G_{JK}(t,t') &= \sum_I Z_{JI}(t)\,G_{IK}(t,t')\,,\\ G_{JK}(t',t') &= \delta_{JK}\,,\end{aligned} \tag{14.205}$$

which allows the solution of eqn (14.204) to be written as

$$\left\langle \overline{O}_J(t)\right\rangle = \sum_K G_{JK}(t,t')\left\langle \overline{O}_K(t')\right\rangle . \tag{14.206}$$

In classical statistics, the relation (14.206) between the averages of the stochastically-dependent variables $\overline{O}_J(t)$ and $\overline{O}_K(t')$ is called a **linear regression**. This solution for the time dependence of the averages of the sample operators is moderately useful, but the correlation functions $\left\langle \overline{O}_J(t)\overline{O}_K(t')\right\rangle$ are of much greater interest, since their Fourier transforms describe the spectral response functions measured in experiments. Using the Langevin equation for $\overline{O}_J(t)$ to evaluate the time derivative of the correlation function leads to

$$\frac{d}{dt}\left\langle \overline{O}_J(t)\overline{O}_K(t')\right\rangle = -\sum_I Z_{JI}(t)\left\langle \overline{O}_I(t)\overline{O}_K(t')\right\rangle + \left\langle \xi_J(t)\overline{O}_K(t')\right\rangle . \tag{14.207}$$

For $t' < t$ the nonanticipating property (14.192) imposes $\left\langle \xi_J(t)\overline{O}_K(t')\right\rangle = 0$, and the correlation function satisfies

$$\frac{d}{dt}\left\langle \overline{O}_J(t)\overline{O}_K(t')\right\rangle = -\sum_I Z_{JI}(t)\left\langle \overline{O}_I(t)\overline{O}_K(t')\right\rangle . \tag{14.208}$$

Since this has the same form as eqn (14.204), the solution is obtained by using the same Green function:

$$\left\langle \overline{O}_J(t)\overline{O}_K(t')\right\rangle = \sum_I G_{JI}(t,t')\left\langle \overline{O}_I(t')\overline{O}_K(t')\right\rangle . \tag{14.209}$$

In other words, the two-time correlation function $\left\langle \overline{O}_J(t)\overline{O}_K(t')\right\rangle$ is related to the equal-time correlation functions $\left\langle \overline{O}_I(t')\overline{O}_K(t')\right\rangle$ by the same regression law that relates the single-time averages $\left\langle \overline{O}_J(t)\right\rangle$ at time t to the averages $\left\langle \overline{O}_I(t')\right\rangle$ at the earlier time t'. A little thought shows that a similar derivation gives the more general result

$$\left\langle X(t')\overline{O}_J(t)Y(t')\right\rangle = \sum_K G_{JK}(t,t')\left\langle X(t')\overline{O}_K(t')Y(t')\right\rangle , \tag{14.210}$$

where $X(t')$ and $Y(t')$ are sample operators that depend on $\left\{\overline{O}_J(t'')\right\}$ for $t'' < t' < t$. Equations (14.209) and (14.210) are special cases of the **quantum regression theorem** first proved by Lax (1963). We will study the general version in Chapter 18.

14.8 Photon bunching*

We mentioned in Section 10.1.1 that the Hanbury Brown–Twiss effect can be measured by coincidence counting. As explained in Section 9.2.4, the coincidence-count rate is proportional to the second-order correlation function

$$G^{(2)}\left(\mathbf{r}',t',\mathbf{r},t;\mathbf{r}',t',\mathbf{r},t\right)=\left\langle E^{(-)}\left(\mathbf{r}',t'\right)E^{(-)}\left(\mathbf{r},t\right)E^{(+)}\left(\mathbf{r},t\right)E^{(+)}\left(\mathbf{r}',t'\right)\right\rangle, \quad (14.211)$$

where $\mathbf{r}'$ and $\mathbf{r}$ are the locations of the detectors, $t' = t + \tau$, and the fields are all projected on a common polarization vector. By placing suitable filters in front of the detectors, we can confine our attention to a single mode, so that $G^{(2)}$ is proportional to the correlation function

$$\mathcal{C}\left(\tau\right)=\left\langle a^{\dagger}\left(t+\tau\right)a^{\dagger}\left(t\right)a\left(t\right)a\left(t+\tau\right)\right\rangle=\left\langle a^{\dagger}\left(t\right)N\left(t+\tau\right)a\left(t\right)\right\rangle, \quad (14.212)$$

where $N\left(t\right) = a^{\dagger}\left(t\right)a\left(t\right)$ is the mode number operator in the Heisenberg picture.

The quantum regression theorem can be applied to the evaluation of $C\left(\tau\right)$ by using the Langevin equation for $a\left(t\right)$ to derive the differential equation

$$\frac{d\left\langle N\left(t\right)\right\rangle}{dt}=-\kappa\left\langle N\left(t\right)\right\rangle+\left\langle \xi^{\dagger}\left(t\right)\overline{a}\left(t\right)+\overline{a}^{\dagger}\left(t\right)\xi\left(t\right)\right\rangle \quad (14.213)$$

for the average photon number. It is shown in Exercise 14.2 that

$$\left\langle \xi^{\dagger}\left(t\right)\overline{a}\left(t\right)+\overline{a}^{\dagger}\left(t\right)\xi\left(t\right)\right\rangle=n_{0}\kappa\,, \quad (14.214)$$

so that the equation for $\left\langle N\left(t\right)\right\rangle$ can be rewritten as

$$\frac{d\left\langle \delta N\left(t\right)\right\rangle}{dt}=-\kappa\left\langle \delta N\left(t\right)\right\rangle, \quad (14.215)$$

where $\delta N\left(t\right) = N\left(t\right) - n_0$. The solution,

$$\left\langle \delta N\left(t\right)\right\rangle=e^{-\kappa\left(t-t_{0}\right)}\left\langle \delta N\left(t_{0}\right)\right\rangle, \quad (14.216)$$

of this equation is a special case of the linear regression equation (14.206), with the Green function $G\left(\tau\right) = \exp\left(-\kappa\tau\right)$. According to the quantum regression theorem (14.210), the correlation function $\left\langle a^{\dagger}\left(t\right)\delta N\left(t+\tau\right)a\left(t\right)\right\rangle$ obeys the same regression law, so

$$\left\langle a^{\dagger}\left(t\right)\delta N\left(t+\tau\right)a\left(t\right)\right\rangle=e^{-\kappa\tau}\left\langle a^{\dagger}\left(t\right)\delta N\left(t\right)a\left(t\right)\right\rangle, \quad (14.217)$$

and

$$\mathcal{C}\left(\tau\right)=e^{-\kappa\tau}\left\langle a^{\dagger 2}\left(t\right)a^{2}\left(t\right)\right\rangle+\left(1-e^{-\kappa\tau}\right)n_{0}\left\langle N\left(t\right)\right\rangle. \quad (14.218)$$

For large times, $\kappa\left(t - t_0\right) \gg 1$, eqn (14.216) shows that $\left\langle N\left(t\right)\right\rangle \approx n_0$. The remaining expectation value $\left\langle a^{\dagger 2}\left(t\right)a^{2}\left(t\right)\right\rangle$ can be calculated by using the solution (14.65) for

$a(t)$. In the same large-time limit, the initial-value term in eqn (14.65) can be dropped to get the asymptotic result

$$\left\langle a^{\dagger 2}(t)\, a^{2}(t)\right\rangle = \int_{t_0}^{t} dt_1 \cdots \int_{t_0}^{t} dt_4 \exp\left[-\frac{\kappa}{2}\sum_{j=1}^{4}(t-t_j)\right]\left\langle \xi^{\dagger}(t_1)\,\xi^{\dagger}(t_2)\,\xi(t_3)\,\xi(t_4)\right\rangle . \tag{14.219}$$

For a thermal noise distribution,

$$\rho_E = \exp\left[-\beta\sum_{\nu}\hbar\Omega_\nu N_\nu\right], \tag{14.220}$$

the discussion in Section 14.2.2 shows that

$$\left\langle \xi^{\dagger}(t_1)\,\xi^{\dagger}(t_2)\,\xi(t_3)\,\xi(t_4)\right\rangle = (n_0\kappa)^2\left\{\delta(t_1-t_3)\,\delta(t_2-t_4)+\delta(t_1-t_4)\,\delta(t_2-t_3)\right\}. \tag{14.221}$$

Substituting this result into eqn (14.219) and carrying out the integrals yields

$$\left\langle a^{\dagger 2}(t)\, a^{2}(t)\right\rangle = 2n_0^2 \quad \text{for } \kappa(t-t_0) \gg 1\,. \tag{14.222}$$

The correlation function $\mathcal{C}(\tau)$ is then given by

$$\mathcal{C}(\tau) = n_0^2\left(1+e^{-\kappa\tau}\right), \tag{14.223}$$

which shows that the coincidence rate is largest at $\tau = 0$. In other words, photon detections are more likely to occur at small rather than large time separations, as shown explicitly by eqn (14.223) which yields

$$\mathcal{C}(0) = 2\mathcal{C}(\infty)\,. \tag{14.224}$$

This effect is called **photon bunching**; it represents the quantum aspect of the Hanbury Brown–Twiss effect. For a contrasting situation, consider an experiment in which the thermal light is replaced by light from a laser operated well above threshold. There are no cavity walls and consequently no external reservoir, so the operator $a(t)$ evolves freely as $a\exp(-i\omega_0 t)$. The density operator for the field is a coherent state $|\alpha\rangle\langle\alpha|$, so that

$$\mathcal{C}(\tau) = \left\langle \alpha\left|a^{\dagger 2}a^{2}\right|\alpha\right\rangle = |\alpha|^4\,. \tag{14.225}$$

In this case, the coincidence rate is independent of the delay time τ; photon bunching is completely absent.

14.9 Resonance fluorescence*

When an atom is exposed to a strong, plane-wave field that is nearly resonant with an atomic transition, some of the incident light will be inelastically scattered into all directions. This phenomenon, which is called **resonance fluorescence**, has been studied experimentally and theoretically for over a century. Early experiments (Wood,

1904, 1912; Dunoyer, 1912) provided support for Bohr's model of the atom, and after the advent of a quantum theory for light the effects were explained theoretically (Weisskopf, 1931).

In the ideal case of scattering from an isolated atom at rest, the theory predicts (Mollow, 1969) a three-peaked spectrum (the **Mollow triplet**) for the scattered radiation. After the invention of the laser and the development of atomic beam techniques, it became possible to approximate this ideal situation. The first experimental verifications of the Mollow triplet were obtained by crossing an atomic beam with a laser beam at right angles, and observing the resulting fluorescent emission (Schuda *et al.*, 1974; Wu *et al.*, 1975; Hartig *et al.*, 1976). This experimental technique was later refined by reducing the atomic beam current—so that at most one atom is in the interaction region at any given time—and by employing counter-propagating laser beams to reduce the Doppler broadening due to atomic motion transverse to the beam direction. These improvements cannot, however, eliminate the **transit broadening** $\Delta\omega_{\text{tran}} \sim 1/T_{\text{tran}}$ caused by the finite transit time T_{tran} for an atom crossing the laser beam. In more recent experiments (Schubert *et al.*, 1995; Stalgies *et al.*, 1996) the ideal case is almost exactly realized by observing resonance fluorescence from a laser-cooled ion in an electrodynamic trap.

In the interests of simplicity, we will only consider the case of resonance fluorescence from a two-level atom. The previous discussion of Rabi oscillations, in Section 11.3.2, neglected spontaneous emission, but a theory of resonance fluorescence must include both the classical driving field and the quantized radiation field. This can be done by using the result— obtained in Section 11.3.1—that the effective Hamiltonian is the sum of the semiclassical Hamiltonian for the atom in the presence of the laser field and the radiation Hamiltonian describing the interaction with the quantized radiation field. In the present case, this yields the effective Schrödinger-picture Hamiltonian

$$H_W = H_{S0} + V_S(t) + H_E + H_{\text{SE}}\,, \tag{14.226}$$

where

$$H_{S0} = \frac{\hbar\omega_{21}}{2}\sigma_z\,, \tag{14.227}$$

$$V_S(t) = \hbar\Omega_L e^{-i\omega_L t}\sigma_+ + \hbar\Omega_L^* e^{i\omega_L t}\sigma_-\,, \tag{14.228}$$

$$H_E = \sum_{\mathbf{k}s} \hbar\omega_k a_{\mathbf{k}s}^\dagger a_{\mathbf{k}s}\,, \tag{14.229}$$

and

$$H_{\text{SE}} = i\hbar\sum_{\mathbf{k}s} v_{\mathbf{k}s}\left(\sigma_- a_{\mathbf{k}s}^\dagger - \sigma_+ a_{\mathbf{k}s}\right). \tag{14.230}$$

The explicit time dependence of $V_S(t)$ comes from the semiclassical treatment of the laser field. Since we are dealing with a single atom, the location of the atom can be chosen as the origin of coordinates.

The quantity to be measured is the counting rate for photons of polarization $\mathbf{e}$ at a detector located at $\mathbf{r}$. According to eqn (9.33),

$$\begin{aligned} w^{(1)}(t) &= \mathfrak{S}\, G^{(1)}(\mathbf{r},t;\mathbf{r},t) \\ &= \mathfrak{S}\,\mathrm{Tr}\left[\rho\mathbf{e}^{*}\cdot\mathbf{E}^{(-)}(\mathbf{r},t)\,\mathbf{e}\cdot\mathbf{E}^{(+)}(\mathbf{r},t)\right], \end{aligned} \tag{14.231}$$

where $\mathfrak{S}$ is the sensitivity factor for the detector, and the Heisenberg-picture density operator,

$$\rho_W = \rho_{\text{atom}}\,|\alpha_0\rangle\langle\alpha_0|\,, \tag{14.232}$$

is the product of the density operator for the coherent state $|\alpha_0\rangle$ describing the laser field and the initial density operator ρ_{atom} for the atom. Our first objective is to show that the counting rate can be expressed in terms of atomic correlation functions.

14.9.1 The counting rate

The discussion in Section 11.3, in particular eqn (11.149), shows that the density operator ρ in eqn (14.232) is the vacuum state for the fluorescent modes; consequently, the only difference between the problem at hand and the spontaneous emission calculation in Section 11.2.1 is the effect of the laser field on the atom. Furthermore, the operator $a_{\mathbf{k}s}(t)$ commutes with $H_{\text{sc}}(t)$, so the atom–laser coupling does not change the form of the Heisenberg equation for $a_{\mathbf{k}s}(t)$. Consequently, we can still use the formal solution (11.51) and the argument contained in eqns (11.52)–(11.66). The new feature is that the definition (11.63) of the slowly-varying envelope operators for the atom must be replaced by

$$\overline{\sigma}_{-}(t) = e^{i\omega_L t}\sigma_{-}(t)\,, \tag{14.233}$$

in order to eliminate the explicit time dependence in $V_S(t)$. This is permissible, because of the near-resonance assumption $|\delta| \ll \omega_{21}$, where $\delta = \omega_{21} - \omega_L$ is the detuning. For a detector in the radiation zone, the counting rate is therefore given by

$$w^{(1)}(t) = \mathfrak{S}\,\mathrm{Tr}_W\left[\rho_W\mathbf{e}\cdot\mathbf{E}_{\text{rad}}^{(-)}(\mathbf{r},t)\,\mathbf{e}^{*}\cdot\mathbf{E}_{\text{rad}}^{(+)}(\mathbf{r},t)\right], \tag{14.234}$$

where

$$\mathbf{E}_{\text{rad}}^{(+)}(\mathbf{r},t) = \frac{k_L^2\left[(\mathbf{d}^{*}\times\widetilde{\mathbf{r}})\times\widetilde{\mathbf{r}}\right]}{4\pi\epsilon_0}\frac{e^{ik_L r}}{r}e^{-i\omega_L t}\overline{\sigma}_{-}(t-r/c)\,, \tag{14.235}$$

and $k_L = \omega_L/c$. Combining the last two equations gives us the desired result

$$w^{(1)}(t) = R\left\langle\overline{\sigma}_{+}(t-r/c)\,\overline{\sigma}_{-}(t-r/c)\right\rangle_S\,, \tag{14.236}$$

where $\langle X\rangle_S = \mathrm{Tr}_S(\rho_{\text{atom}}X)$, and the rate

$$R = \frac{\mathfrak{S}}{r^2}\left(\frac{k_L^2}{4\pi\epsilon_0}\right)^2\left|(\mathbf{d}^{*}\times\widetilde{\mathbf{r}})\times\widetilde{\mathbf{r}}\right|^2 \tag{14.237}$$

carries all the information on the angular distribution of the radiation.

14.9.2 Langevin equations for the atom

The result (14.236) has eliminated any direct reference to the radiation field; therefore, we are free to treat the fluorescent field modes as a reservoir and the atom—under the influence of the laser field—as the sample. Elimination of the field operators by means of the formal solution (11.51) and the Markov approximation yields the Langevin equations

$$\frac{d\overline{\sigma}_+(t)}{dt} = -(\Gamma - i\delta)\,\overline{\sigma}_+(t) - i\Omega_L^*\overline{\sigma}_z(t) + \xi_+ \,, \tag{14.238}$$

$$\frac{d\overline{\sigma}_z(t)}{dt} = -w\left[1 + \overline{\sigma}_z(t)\right] + 2i\Omega_L^*\overline{\sigma}_-(t) - 2i\Omega_L\overline{\sigma}_+(t) + \xi_z \,, \tag{14.239}$$

where $\Gamma = \Gamma_{12}$ is the dipole dephasing rate, $w = w_{21}$ is the spontaneous decay rate, and the noise operators are defined in Section 14.4.

We begin with the averaged Langevin equations,

$$\frac{d\left\langle\overline{\sigma}_+(t)\right\rangle}{dt} = -(\Gamma - i\delta)\left\langle\overline{\sigma}_+(t)\right\rangle - i\Omega_L^*\left\langle\overline{\sigma}_z(t)\right\rangle, \tag{14.240}$$

$$\frac{d\left\langle\overline{\sigma}_z(t)\right\rangle}{dt} = -w\left[1 + \left\langle\overline{\sigma}_z(t)\right\rangle\right] + 2i\Omega_L^*\left\langle\overline{\sigma}_-(t)\right\rangle - 2i\Omega_L\left\langle\overline{\sigma}_+(t)\right\rangle, \tag{14.241}$$

and note that the averaged atomic operators approach steady-state values, $\left\langle\overline{\sigma}_+\right\rangle_{\text{ss}}$ and $\left\langle\overline{\sigma}_z\right\rangle_{\text{ss}}$, for times $t \gg \max(1/\Gamma, 1/w)$. These values are determined by setting the time derivatives to zero and solving the resulting algebraic equations, to get

$$\left\langle\overline{\sigma}_z\right\rangle_{\text{ss}} = -\frac{1}{1 + \left|\Omega_L\right|^2/\Omega_{\text{sat}}^2}\,, \tag{14.242}$$

$$\left\langle\overline{\sigma}_+\right\rangle_{\text{ss}} = -i\frac{\Omega_L^*}{\Gamma - i\delta}\left\langle\overline{\sigma}_z\right\rangle_{\text{ss}}\,, \tag{14.243}$$

where

$$\Omega_{\text{sat}} = \sqrt{\frac{w\left(\Gamma^2 + \delta^2\right)}{4\Gamma}} \tag{14.244}$$

is the **saturation value** for the Rabi frequency. For $\left|\Omega_L\right| \gg \Omega_{\text{sat}}$, $\left\langle\overline{\sigma}_z\right\rangle_{\text{ss}} \approx 0$, which means that the two levels are equally populated. In the same limit, one finds

$$\left\langle\overline{\sigma}_+\right\rangle_{\text{ss}} \to 0\,, \tag{14.245}$$

i.e. the average dipole moment goes to zero for large laser intensities. This effect is called **bleaching**. The ratio $\left|\Omega_L\right|^2/\Omega_{\text{sat}}^2$ is often expressed as

$$\frac{\left|\Omega_L\right|^2}{\Omega_{\text{sat}}^2} = \frac{I_L}{I_{\text{sat}}}\,, \tag{14.246}$$

where I_L is the laser intensity and

$$I_{\text{sat}} = \frac{3\hbar^2\epsilon_0 c w\left(\delta^2 + \Gamma^2\right)}{8\Gamma\left|\mathbf{d}\right|^2} \tag{14.247}$$

is the **saturation intensity**.

The fact that the population difference $\langle\overline{\sigma}_z\rangle_{ss}$ and the dipole moment $\langle\overline{\sigma}_+\rangle_{ss}$ are independent of time raises a question: What happened to the Rabi oscillations of the atom? The answer is that they are still present, but concealed by the ensemble average defined by the initial density operator. This can be seen more explicitly by applying the long-time averaging procedure

$$\langle\overline{\sigma}_\lambda\rangle_\infty = \lim_{T\to\infty}\frac{1}{T}\int_0^T dt\,\langle\overline{\sigma}_\lambda(t)\rangle \quad (\lambda = +, z, -) \tag{14.248}$$

to eqns (14.240) and (14.241). It is easy to show that the average of the left side vanishes in both equations, so that the time averages $\langle\overline{\sigma}_\lambda\rangle_\infty$ satisfy the same equations as the steady-state solutions $\langle\overline{\sigma}_\lambda\rangle_{ss}$. Thus the steady-state solutions are equivalent to a long-time average over the Rabi oscillations. This result is conceptually similar to the famous ergodic theorem in statistical mechanics (Chandler, 1987, Chap. 3).

Since the distance r to the detector is fixed, we can use the retarded time $t_r = t - r/c$ instead of t. With this understanding, the total number of counts in the interval $(t_{r0}, t_{r0}+T)$ is

$$N(T) = R\int_{t_{r0}}^{t_{r0}+T} dt_r\,\langle\overline{\sigma}_+(t_r)\,\overline{\sigma}_-(t_r)\rangle\,, \tag{14.249}$$

and the Pauli-matrix identity,

$$\overline{\sigma}_+(t_r)\,\overline{\sigma}_-(t_r) = \frac{1}{2}\left[1+\overline{\sigma}_z(t_r)\right] = S_{22}(t_r)\,, \tag{14.250}$$

allows this to be written in the equivalent form

$$N(T) = R\int_{t_{r0}}^{t_{r0}+T} dt_r\,\langle S_{22}(t_r)\rangle\,. \tag{14.251}$$

For sufficiently large t_{r0} the average in eqn (14.251) can be replaced by the stationary value, so that

$$N(T) = RT\,\langle S_{22}\rangle_{ss} = \frac{RT}{2}\frac{|\Omega_L|^2/\Omega_{sat}^2}{1+|\Omega_L|^2/\Omega_{sat}^2}\,. \tag{14.252}$$

This result tells us the total number of counts, but it does not distinguish between the coherent contribution due to Rabi oscillations of the atomic dipole and the incoherent contribution arising from quantum noise, i.e. spontaneous emission. In order to bring out this feature, we introduce the fluctuation operators

$$\delta\overline{\sigma}_z(t_r) = \overline{\sigma}_z(t_r) - \langle\overline{\sigma}_z(t_r)\rangle\,,\quad \delta\overline{\sigma}_\pm(t_r) = \overline{\sigma}_\pm(t_r) - \langle\overline{\sigma}_\pm(t_r)\rangle\,, \tag{14.253}$$

and rewrite eqn (14.249) as

$$N(T) = N_{coh}(T) + N_{inc}(T)\,, \tag{14.254}$$

with

$$N_{coh}(T) = R\int_{t_{r0}}^{t_{r0}+T} dt_r\,\langle\overline{\sigma}_+(t_r)\rangle\,\langle\overline{\sigma}_-(t_r)\rangle \tag{14.255}$$

and

$$N_{\text{inc}}(T) = R\int_{t_{r0}}^{t_{r0}+T} dt_r \left\langle \delta\overline{\sigma}_+(t_r)\,\delta\overline{\sigma}_-(t_r)\right\rangle . \tag{14.256}$$

The coherent contribution is what one would predict from forced oscillations of a classical dipole with magnitude $|\langle\overline{\sigma}_-(t_r)\rangle|$, and the incoherent contribution depends on the strength of the quantum fluctuation operators $\delta\overline{\sigma}_+(t_r)$ and $\delta\overline{\sigma}_-(t_r)$. In the limit of large t_{r0} the coherent contribution is obtained by substituting the asymptotic result (14.243) into eqn (14.256), with the result

$$N_{\text{coh}}(T) = RT\frac{w}{4\Gamma}\frac{|\Omega_L|^2/\Omega_{\text{sat}}^2}{\left(1+|\Omega_L|^2/\Omega_{\text{sat}}^2\right)^2} . \tag{14.257}$$

The incoherent contribution can be evaluated directly from eqn (14.256), but it is easier to use eqns (14.252) and (14.254) to get

$$N_{\text{inc}}(T) = \frac{RT}{2}\frac{I_L}{I_{\text{sat}}}\frac{1-(w/2\Gamma)+\left(|\Omega_L|^2/\Omega_{\text{sat}}^2\right)}{\left(1+|\Omega_L|^2/\Omega_{\text{sat}}^2\right)^2} . \tag{14.258}$$

In the high intensity limit, the laser field should become more classical, and one might expect that the coherent contribution would dominate the counting rate. Examination of the results shows exactly the opposite; $N_{\text{coh}}(T) \to 0$ and $N_{\text{inc}}(T) \to RT/2$. This apparent paradox is resolved by the bleaching of the average dipole moment—shown in eqn (14.245)—and the fact that half the atoms are in the excited state and consequently available for spontaneous emission.

14.9.3 The fluorescence spectrum

Spectral data for fluorescent emission are acquired by using one of the narrowband counting techniques discussed in Section 9.1.2-C. It is safe to assume that the field correlation functions approximately satisfy time-translation invariance for times t_r much larger than the decay times for the sample; therefore, we can immediately use the result (9.45) for the spectral density to get

$$\mathcal{S}(\omega, t_r) = \mathfrak{S}G^{(1)}(\mathbf{r},\omega) = \mathfrak{S}\int d\tau e^{-i\omega\tau}G^{(1)}(\mathbf{r},\tau+t_r;\mathbf{r},t_r) . \tag{14.259}$$

Substituting the solution (14.235) for the radiation field into this expression yields

$$\mathcal{S}(\omega, t_r) = R\int d\tau e^{i(\omega_L-\omega)\tau}\left\langle \overline{\sigma}_+(\tau+t_r)\,\overline{\sigma}_-(t_r)\right\rangle . \tag{14.260}$$

Once again, we can use the fluctuation operators defined by eqn (14.253) to split the spectral density into a coherent contribution, due to oscillations driven by the external laser field, and an incoherent contribution, due to quantum noise. Thus $\mathcal{S}(\omega, t_r) = \mathcal{S}_{\text{coh}}(\omega, t_r) + \mathcal{S}_{\text{inc}}(\omega, t_r)$, where

$$\mathcal{S}_{\text{coh}}(\omega, t_r) = R\int d\tau e^{i(\omega_L-\omega)\tau}\langle\overline{\sigma}_+(\tau+t_r)\rangle\langle\overline{\sigma}_-(t_r)\rangle \tag{14.261}$$

and

$$\mathcal{S}_{\text{inc}}(\omega, t_r) = R\int d\tau e^{i(\omega_L-\omega)\tau}\langle\delta\overline{\sigma}_+(\tau+t_r)\,\delta\overline{\sigma}_-(t_r)\rangle\,. \tag{14.262}$$

The assumption that t_r is much larger than the atomic decay times means that $\langle\overline{\sigma}_+(\tau+t_r)\rangle$ and $\langle\overline{\sigma}_-(t_r)\rangle$ are respectively given by the asymptotic steady-state values $\langle\overline{\sigma}_+\rangle_{\text{ss}}$ and $\langle\overline{\sigma}_+\rangle_{\text{ss}}^*$ from eqn (14.243); consequently, the coherent contribution is

$$\mathcal{S}_{\text{coh}}(\omega, t_r) = R\left|\langle\overline{\sigma}_+\rangle_{\text{ss}}\right|^2\int d\tau e^{i(\omega_L-\omega)\tau} = 2\pi R\left|\langle\overline{\sigma}_+\rangle_{\text{ss}}\right|^2\delta(\omega-\omega_L)\,. \tag{14.263}$$

The first step in the calculation of the incoherent contribution is to write eqn (14.262) as

$$\begin{aligned}\mathcal{S}_{\text{inc}}(\omega, t_r) &= R\int_0^\infty d\tau e^{i\Delta\tau}\langle\delta\overline{\sigma}_+(\tau+t_r)\,\delta\overline{\sigma}_-(t_r)\rangle \\ &\quad + R\int_{-\infty}^0 d\tau e^{i\Delta\tau}\langle\delta\overline{\sigma}_+(\tau+t_r)\,\delta\overline{\sigma}_-(t_r)\rangle\,,\end{aligned} \tag{14.264}$$

where $\Delta = \omega_L - \omega$. In the second integral, one can change $\tau \to -\tau$ and use time-translation invariance to get

$$\begin{aligned}\langle\delta\overline{\sigma}_+(-\tau+t_r)\,\delta\overline{\sigma}_-(t_r)\rangle &= \langle\delta\overline{\sigma}_+(t_r)\,\delta\overline{\sigma}_-(\tau+t_r)\rangle \\ &= \langle\delta\overline{\sigma}_+(\tau+t_r)\,\delta\overline{\sigma}_-(t_r)\rangle^*\,,\end{aligned} \tag{14.265}$$

so that

$$\mathcal{S}_{\text{inc}}(\omega, t_r) = 2R\,\text{Re}\int_0^\infty d\tau e^{i\Delta\tau}\langle\delta\overline{\sigma}_+(\tau+t_r)\,\delta\overline{\sigma}_-(t_r)\rangle\,. \tag{14.266}$$

The correlation function in the integrand is one component of the matrix

$$F_{\lambda\mu}(\tau, t_r) = \langle\delta\overline{\sigma}_\lambda(\tau+t_r)\,\delta\overline{\sigma}_\mu(t_r)\rangle \quad (\lambda, \mu = +, z, -)\,, \tag{14.267}$$

so

$$\begin{aligned}\mathcal{S}_{\text{inc}}(\omega, t_r) &= 2R\,\text{Re}\int_0^\infty d\tau e^{i\Delta\tau}F_{+-}(\tau, t_r) \\ &= 2R\lim_{\epsilon\to 0+}\text{Re}\int_0^\infty d\tau e^{-(\epsilon-i\Delta)\tau}F_{+-}(\tau, t_r) \\ &= 2R\lim_{\epsilon\to 0+}\text{Re}\,\widetilde{F}_{+-}(\epsilon-i\Delta, t_r)\,,\end{aligned} \tag{14.268}$$

where $\widetilde{F}_{+-}(\zeta, t_r)$ is the Laplace transform of $\widetilde{F}_{+-}(\tau, t_r)$ with respect to τ.

The evaluation of the Laplace transform is accomplished with the techniques used to prove the quantum regression theorem. We begin by subtracting eqns (14.240) and (14.241) from eqns (14.238) and (14.239), to get the equations of motion for the fluctuation operators. By including the equation for $\delta\overline{\sigma}_-(t)$—the conjugate of eqn (14.238)—the equations can be written in matrix form as

$$\frac{d}{dt}\delta\overline{\sigma}_\lambda(t) = \sum_\mu V_{\lambda\mu}\delta\overline{\sigma}_\mu(t) + \xi_\lambda(t), \tag{14.269}$$

where

$$V = \begin{bmatrix} -(\Gamma - i\delta) & -i\Omega_L^* & 0 \\ -2i\Omega_L & -w & 2i\Omega_L^* \\ 0 & i\Omega_L & -(\Gamma + i\delta) \end{bmatrix}. \tag{14.270}$$

After differentiating eqn (14.267) with respect to τ, with t_r fixed, and using eqn (14.267) one finds

$$\frac{\partial}{\partial\tau}F_{\lambda\mu}(\tau, t_r) = \sum_\nu V_{\lambda\nu}F_{\nu\mu}(\tau, t_r), \tag{14.271}$$

where we have used the nonanticipating property $\langle\xi_\lambda(\tau + t_r)\delta\overline{\sigma}_\mu(t_r)\rangle = 0$ for $\tau > 0$. The Laplace transform technique for initial value problems—explained in Appendix A.5—turns these differential equations into the algebraic equations

$$\zeta\widetilde{F}_{\lambda\mu}(\zeta, t_r) - \sum_\nu V_{\lambda\nu}\widetilde{F}_{\nu\mu}(\zeta, t_r) = F_{\lambda\mu}(0, t_r), \tag{14.272}$$

which determine the matrix $\widetilde{F}_{\lambda\mu}(\zeta, t_r)$. Since t_r is large, the initial values $F_{\lambda\mu}(0, t_r)$ defined by eqn (14.267) are given by the steady-state average

$$\begin{aligned} F_{\lambda\mu}(0, t_r) &= \langle\delta\overline{\sigma}_\lambda\delta\overline{\sigma}_\mu\rangle_{\text{ss}} \\ &= \langle\overline{\sigma}_\lambda\overline{\sigma}_\mu\rangle_{\text{ss}} - \langle\overline{\sigma}_\lambda\rangle_{\text{ss}}\langle\overline{\sigma}_\mu\rangle_{\text{ss}}. \end{aligned} \tag{14.273}$$

The product of two Pauli matrices can always be reduced to an expression linear in the Pauli matrices, so the initial values are determined by eqns (14.242) and (14.243).

The evaluation of the incoherent part of the spectral density by eqn (14.268) only requires $\widetilde{F}_{+-}(-i\Delta, t_r)$, which is readily obtained by applying Cramers rule to eqn (14.272) to find the on-resonance ($\delta = 0$) result

$$\widetilde{F}_{+-}(-i\Delta, t_r) = \frac{N_{+-}(\Delta)}{D(\Delta)} \equiv \digamma(\Delta). \tag{14.274}$$

The numerator is a linear function of the initial values:

$$\begin{aligned} N_{+-}(\Delta) = &-2i\Omega_L^{*2}F_{--}(0, t_r) + i\Omega_L^*(\Delta + i\Gamma)F_{z-}(0, t_r) \\ &+ i\left[\Delta^2 - 2|\Omega_L|^2 - \Gamma w + i(\Gamma + w)\Delta\right]F_{+-}(0, t_r), \end{aligned} \tag{14.275}$$

and the denominator is the product of three factors: $D(\Delta) = D_0(\Delta)D_+(\Delta)D_-(\Delta)$, where $D_0(\Delta) = \Delta + i\Gamma$,

$$D_{\pm}(\Delta) = \Delta \pm 2\Omega'_L + i\left(\frac{\Gamma + w}{2}\right), \tag{14.276}$$

and

$$\Omega'_L = \sqrt{|\Omega_L|^2 - \left(\frac{\Gamma - w}{4}\right)^2}\,. \tag{14.277}$$

The factorization of the denominator suggests using the method of partial fractions to express $\mathcal{F}(\Delta)$ as

$$\mathcal{F}(\Delta) = \frac{C(\Delta)}{D_0(\Delta)} + \frac{C(\Delta)}{D_+(\Delta)} + \frac{C(\Delta)}{D_-(\Delta)}, \tag{14.278}$$

with

$$C(\Delta) = \frac{N_{+-}(\Delta)}{D_0(\Delta)\left[D_+(\Delta) + D_-(\Delta)\right] + D_+(\Delta)\, D_-(\Delta)}\,. \tag{14.279}$$

The functions $D_0(\Delta)$ and $D_{\pm}(\Delta)$ have zeroes at $\Delta_0 = -i\Gamma$ and $\Delta_{\pm} = \mp 2\Omega'_L - i(\Gamma + w)/2$ respectively, so $\mathcal{F}(\Delta)$ has three poles in the lower-half Δ-plane. If the laser field is weak, in the sense that

$$|\Omega_L|^2 < \left(\frac{\Gamma - w}{4}\right)^2, \tag{14.280}$$

then eqn (14.277) shows that Ω'_L is pure imaginary. All three poles are then located on the negative imaginary axis, so that $\operatorname{Re}\mathcal{F}(\Delta)$ will have a single peak at $\Delta = 0$, on the real Δ-axis. For a strong laser, Ω'_L is real, and the poles at $\Delta_{\pm}$ are displaced away from the imaginary axis. In this case, $\operatorname{Re}\mathcal{F}(\Delta)$ will exhibit three peaks on the real Δ-axis, at $\Delta_+ = -2\Omega'_L$, $\Delta_0 = 0$, and $\Delta_- = 2\Omega'_L$.

An explicit evaluation of eqn (14.278) can be carried out in the general case, but the resulting expressions are too cumbersome to be of much use. One then has the choice of studying the behavior of the spectral density numerically, or making simplifications to produce a manageable analytic result. We will leave the numerical study to the exercises and impose three simplifying assumptions. The first is that the laser is exactly on resonance with the atomic transition ($\delta = 0$), and the second is that the laser field is strong ($|\Omega_L| \gg \Gamma, w$). The third simplification is to evaluate the numerator $C(\Delta)$ at the location of the pole in each of the three terms. This procedure will give an accurate picture of the behavior of the function $\mathcal{S}_{\text{inc}}(\omega, t_r)$ in the vicinity of the peaks, but will be slightly in error in the regions between them. With these assumptions in place, one finds

$$\mathcal{S}_{\text{inc}}(\omega, t_r) = \mathcal{S}_{\text{inc}}^{(+)}(\omega, t_r) + \mathcal{S}_{\text{inc}}^{(0)}(\omega, t_r) + \mathcal{S}_{\text{inc}}^{(-)}(\omega, t_r), \tag{14.281}$$

$$\mathcal{S}_{\text{inc}}^{(0)}(\omega, t_r) = \frac{R}{2}\frac{\Gamma}{(\omega - \omega_L)^2 + \Gamma^2}, \tag{14.282}$$

$$\mathcal{S}_{\text{inc}}^{(\pm)}(\omega, t_r) = \frac{R}{8}\frac{\Gamma + w}{(\omega - \omega_L \mp 2|\Omega_L|)^2 + (\Gamma + w)^2/4}\,. \tag{14.283}$$

This clearly displays the three peaks of the Mollow triplet. The presence of the side peaks is evidence of persistent Rabi oscillations that modulate the primary resonance at $\omega = \omega_L$. The heights and widths of the peaks are related by

$$\frac{\text{central peak height}}{\text{side peak height}} = 1 + \frac{w}{\Gamma} \; (= 3 \;\text{ for the pure radiative case}) \, , \tag{14.284}$$

$$\frac{\text{side peak width}}{\text{central peak width}} = \frac{1}{2}\left(1 + \frac{w}{\Gamma}\right) \left(= \frac{3}{2} \;\text{ for the pure radiative case}\right) , \tag{14.285}$$

where the pure radiative case occurs when spontaneous emission is the only decay mechanism. In this situation eqn (14.151) yields $w = 2\Gamma$. These features have been experimentally demonstrated.

14.10 Exercises

14.1 Sample–environment coupling

Consider a single reservoir, so that the index J in eqn (14.14) can be suppressed. The general *ansatz* for an interaction, H_{SE}, that is linear in both reservoir and sample operators is

$$H_{\text{SE}} = i\hbar \sum_{\nu} \left(v\left(\Omega_{\nu}\right) O^{\dagger} b_{\nu} - v^{*}\left(\Omega_{\nu}\right) b_{\nu}^{\dagger} O \right) ,$$

where $v\left(\Omega_{\nu}\right)$ is a complex coupling constant. Show that there is a simple unitary transformation, $b_{\nu} \to b_{\nu}'$, that allows the complex $v\left(\Omega_{\nu}\right)$ to be replaced by $\left|v\left(\Omega_{\nu}\right)\right|$.

14.2 Multiplicative noise for the radiation field*

(1) Derive the evolution equation

$$\frac{dN\left(t\right)}{dt} = -\kappa N\left(t\right) + \chi\left(t\right)$$

for the number operator, where $\chi\left(t\right) = \xi^{\dagger}\left(t\right) \overline{a}\left(t\right) + \overline{a}^{\dagger}\left(t\right) \xi\left(t\right)$ is a multiplicative noise operator.

(2) Combine the nonanticipating property (14.77), the delta correlation property (14.74), and the end-point rule (A.98) for delta functions to find $\left\langle \chi\left(t\right)\right\rangle = n_0 \kappa$. Is this result consistent with interpreting the evolution equation as a Langevin equation?

(3) Rewrite the equation for $N\left(t\right)$ in terms of the new noise operator $\xi_N\left(t\right) = \chi\left(t\right) - \left\langle \chi\left(t\right)\right\rangle$, and then derive the result

$$\left\langle N\left(t\right)\right\rangle = \left\langle N\left(t_0\right)\right\rangle e^{-\kappa\left(t-t_0\right)} + n_0 \left[1 - e^{-\kappa\left(t-t_0\right)}\right]$$

describing the relaxation of the average photon number to the equilibrium value n_0.

(4) Use the explicit solution (14.65) for $\overline{a}\left(t\right)$ to show that

$$\left\langle \xi_N\left(t\right) \xi_N\left(t'\right)\right\rangle = C_{NN}\left(t\right) \delta\left(t - t'\right) ,$$

where $C_{NN}\left(t\right)$ approaches a constant value for $\kappa t \gg 1$.

14.3 Approach to thermal equilibrium

The constant κ in eqn (14.61) represents the rate at which field energy is lost to the walls, so it should be possible to recover the blackbody distribution for radiation in a cavity with walls at temperature T. For this purpose, enlarge the sample to include all the modes ($\mathbf{k}s$) of the radiation field; but keep things simple by assuming that all modes are coupled to a single reservoir with the same value of κ.

(1) Generalize the single-mode treatment by writing down the Langevin equation for $\overline{a}_{\mathbf{k}s}$. Give the expression for the noise operator, $\xi_{\mathbf{k}}(t)$, and show that

$$\left\langle \xi_{\mathbf{k}}^{\dagger}(t)\,\xi_{\mathbf{k}}(t')\right\rangle = \kappa n(\omega_{\mathbf{k}})\,\delta(t-t'),$$

where $n(\omega_{\mathbf{k}})$ is the average number of reservoir excitations at the mode frequency $\omega_{\mathbf{k}}$.

(2) Apply the result in part (3) of Exercise 14.2 to find $\lim_{t\to\infty}\langle N_{\mathbf{k}s}(t)\rangle = n(\omega_{\mathbf{k}})$. What is the physical meaning of the limit $t\to\infty$?

(3) Finally, use the discussion following eqn (2.177) to argue that the photon distribution in the cavity asymptotically relaxes to a blackbody distribution.

14.4 Noise operators for the two-level atom

By following the derivation of the Langevin equations (14.148)–(14.150) show that the noise operators are

$$\xi_{22}(t) = -\sqrt{w_{21}}\left[b_{\text{in}}^{\dagger}(t)\,\overline{S}_{12}(t) + \text{HC}\right] = -\xi_{11}(t),$$

$$\begin{aligned}\xi_{12}(t) = {} & \left\{\overline{S}_{22}(t) - \overline{S}_{11}(t)\right\}\sqrt{w_{21}}\,b_{\text{in}}(t) + \left\{\sqrt{w_{22}}\,c_{2,\text{in}}^{\dagger}(t) - \sqrt{w_{11}}\,c_{1,\text{in}}^{\dagger}(t)\right\}\overline{S}_{12}(t)\\ & - \overline{S}_{12}(t)\left\{\sqrt{w_{22}}\,c_{2,\text{in}}(t) - \sqrt{w_{11}}\,c_{1,\text{in}}(t)\right\}.\end{aligned}$$

14.5 Inverted-oscillator reservoir*

A gain medium enclosed in a resonant cavity has been modeled (Gardiner, 1991, Sec. 7.2.1) by the interaction of the cavity mode $a(t)$ of Section 14.3 with an inverted-oscillator reservoir described by the Hamiltonian

$$H_{\text{IO}} = -\int_0^{\infty}\frac{d\Omega}{2\pi}\hbar\Omega c_{\Omega}^{\dagger}c_{\Omega},$$

where $\left[c_{\Omega}, c_{\Omega'}^{\dagger}\right] = 2\pi\delta(\Omega-\Omega')$.

(1) Express the energy-raising and energy-lowering operators for the reservoir in terms of c_{Ω} and $c_{\Omega}^{\dagger}$.

(2) In addition to the two terms in eqn (14.88), the interaction Hamiltonian H_{SE} now has a third term, $H_{S,\text{IO}}$ describing the interaction with the inverted oscillators. In the resonant wave approximation, show that $H_{S,\text{IO}}$ must have the form

$$H_{S,\text{IO}} = i\hbar\int_0^{\infty}\frac{d\Omega}{2\pi}\chi(\Omega)\left(c_{\Omega}a - a^{\dagger}c_{\Omega}^{\dagger}\right).$$

(3) Using the discussion in Section 14.3 as a guide, derive the Langevin equation

$$\frac{d}{dt}\overline{a}(t) = \frac{1}{2}(g - \kappa_C)\,\overline{a}(t) + \xi(t),$$

and give expressions for the gain g and the noise operator $\xi(t)$.

14.6 Langevin equations for incoherent pumping

Use the ($N = 3$)-case of eqns (14.159) and (14.160), without the phase-changing terms, to derive the full set of Langevin equations for the three-level atom of Fig. 14.2.

14.7 Bloch equations for incoherent pumping

Consider the case of pure pumping, i.e. $H'_{S1} = 0$.

(1) Derive the c-number Bloch equations by averaging eqns (14.174)–(14.177).

(2) Find the steady-state solutions for the populations.

14.8 Noise correlation coefficients

Consider the reduced Langevin equations (14.174)–(14.177), with $H'_{S1} = 0$.

(1) How many independent coefficients $C_{qp,lk}$ ($q, p, k, l = 1, 2, 3$) are there?

(2) Use the Einstein relation and the steady-state populations to calculate the independent coefficients in the limit $w_{32} \to \infty$.

14.9 Generalized transition operators*

The two important simplifying assumptions $H_{SS} \sim 0$ and eqn (14.182) were made for the sole purpose of ensuring the linearity of the Heisenberg equations of motion, which is essential for the relatively simple arguments establishing the nonanticipating property (14.192) and the quantum regression theorem (14.209). Both of these assumptions can be eliminated by a special choice of the sample operators. To this end, define the stationary states, $|\Phi_A\rangle$, of the full sample Hamiltonian $H_S = H_{S0} + H_{SS}$ by $H_S |\Phi_A\rangle = \varepsilon_A |\Phi_A\rangle$, and for simplicity's sake assume that A is a discrete label.

(1) Explain why the use of the $|\Phi_A\rangle$s renders the assumption $H_{SS} \sim 0$ unnecessary.

(2) Show that the **generalized transition operators** $S_{AB} = |\Phi_A\rangle\langle\Phi_B|$ satisfy the following:

   (a) $[S_{AB}, H_S] = -\hbar\omega_{AB} S_{AB}$, with $\omega_{AB} = (\varepsilon_A - \varepsilon_B)/\hbar$;

   (b) $S_{AB} S_{CD} = \delta_{BC} S_{AD}$;

   (c) $[S_{AB}, S_{CD}] = \delta_{BC} S_{AD} - \delta_{AD} S_{CB}$;

   (d) $X = \sum_A \sum_B \langle\Phi_A |X| \Phi_B\rangle S_{AB}$, for any sample operator X.

(3) For an external field acting on the sample through $V_S(t)$, derive eqn (14.182) by showing that

$$\frac{1}{i\hbar}\left[\overline{S}_{AB}(t), V_S(t)\right] = i \sum_{CD} \Omega_{AB,CD}(t)\, \overline{S}_{CD}(t)\,.$$

Give the explicit expression for $\Omega_{AB,CD}(t)$ in terms of the matrix elements of $V_S(t)$.

14.10 Mollow triplet*

Use eqn (14.268) for a numerical evaluation of $\mathcal{S}_{\text{inc}}/R$ as a function of Δ/Γ. Assume resonance ($\delta = 0$) and pure radiative decay ($w = 2\Gamma$), and consider two cases: $|\Omega_L| = 5\Gamma$ and $|\Omega_L| = \Gamma/\sqrt{2}$. In each case, plot the numerical evaluation of eqn (14.278) and the numerical evaluation of eqn (14.281) against Δ/Γ.

15

Nonclassical states of light

In Section 5.6.3 we defined a classical state for a single mode of the electromagnetic field by the requirement that the Glauber–Sudarshan $P(\alpha)$-function is everywhere non-negative. When this condition is satisfied $P(\alpha)$ may be regarded as a probability distribution for the classical field amplitude α. Advances in experimental techniques have resulted in the controlled generation of nonclassical states of the field, for which $P(\alpha)$ is not a true probability density. In this chapter, we study the nonclassical states that have received the most attention: squeezed states and number states.

15.1 Squeezed states

In the correspondence-principle limit, a coherent state of light approaches a noiseless classical electromagnetic field as closely as allowed by the uncertainty principle for the radiation oscillators. This might lead one to expect that a coherent state would describe a light beam with the minimum possible quantum noise. On theoretical grounds, it has long been known that this is not the case, and in recent years states with noise levels below the standard quantum limit—known as **squeezed states**—have been demonstrated experimentally.

15.1.1 Squeezed states for a radiation oscillator

As an introduction to the ideas involved, let us begin by considering a single field mode which is described by the operators q and p for the corresponding radiation oscillator. In Section 5.1 we saw that the coherent states are minimum-uncertainty states, with

$$\Delta q_0 = \sqrt{\hbar/2\omega}\,, \quad \Delta p_0 = \sqrt{\hbar\omega/2}\,, \quad \Delta q_0 \Delta p_0 = \hbar/2\,. \tag{15.1}$$

The simplest example is the vacuum state, which is described, in the momentum representation, by

$$\Phi_0(P) = \left(2\pi\Delta p_0^2\right)^{-1/4} \exp\left(-\frac{P^2}{4\Delta p_0^2}\right). \tag{15.2}$$

Suppose that the radiation oscillator is prepared in the initial state,

$$\psi(P,0) = \left(2\pi\Delta p^2\right)^{-1/4} \exp\left(-\frac{P^2}{4\Delta p^2}\right), \tag{15.3}$$

which is called a **squeezed vacuum state** if $\Delta p < \Delta p_0$. This wave function cannot be a stationary state of the oscillator; instead, it is a superposition over the whole family of energy eigenstates:

$$\psi(P,0)=\sum_{n=0}^{\infty}C_n\Phi_n(P)\,, \tag{15.4}$$

where Φ_n is the nth excited state ($H\Phi_n=n\hbar\omega\Phi_n$), and we have, as usual, subtracted the zero-point energy. The excited state $\Phi_n(P)$ is an n-photon state, so we have reached the paradoxical sounding conclusion that the squeezed vacuum contains many photons.

The energy eigenvalues are $n\hbar\omega$, so the initial state $\psi(P,0)$ evolves into

$$\psi(P,t)=\sum_{n=0}^{\infty}C_n\Phi_n(P)\,e^{-in\omega t}\,. \tag{15.5}$$

By virtue of the equal spacing of the energy levels—a unique property of the harmonic oscillator—the wave function is periodic, with period $T=2\pi/\omega$. This in turn implies that the time-dependent width,

$$\Delta p(t)=\sqrt{\langle\psi(t)|P^2|\psi(t)\rangle-\langle\psi(t)|P|\psi(t)\rangle^2}\,, \tag{15.6}$$

will exhibit the same periodicity. In other words, $\psi(P,t)$ is a breathing Gaussian wave packet which expands in size—as measured by $\Delta p(t)$—from its minimum initial value to a maximum size half a period later, and then contracts back to its initial size. This periodic cycling from minimum to maximum spread repeats indefinitely. We recall from eqns (2.99) and (2.100) that the operators p and q respectively correspond to the electric and magnetic fields. According to Section 2.5 this means that the variance in the electric field for the squeezed vacuum state (15.3) is smaller than the vacuum-fluctuation variance.

The Hamiltonian for a radiation oscillator is unchanged by the (unitary) parity transformation $p\to-p$, $q\to-q$ on the operators p and q; therefore the energy eigenstates, e.g. the momentum-space eigenfunctions $\Phi_n(P)$, are also eigenstates of parity:

$$\Phi_n(P)\to(-1)^n\,\Phi_n(P)\quad\text{for }P\to-P\,.$$

An immediate consequence of this fact is that an initial state having definite parity, i.e. a superposition of eigenstates which all have the same parity, will evolve into a state with the same parity at all times. Inspection of eqn (15.3) shows that this initial Gaussian state is an even function of P; consequently, the coefficients C_n in the expansion (15.5) must vanish for all odd integers n. In other words, the evolution of the squeezed vacuum state can only involve even-parity eigenfunctions for the radiation oscillators. Since these eigenfunctions represent number states, an equivalent statement is that only even integer number states can be involved in the production and the time evolution of a squeezed vacuum state. Thus we arrive at the important conclusion that the simplest elementary process leading to such a state is photon pair production.

For production of photons in pairs one needs to look to nonlinear optical interactions, such as those provided by $\chi^{(2)}$ and $\chi^{(3)}$ media. The first experiment demonstrating a squeezed state of light was performed by Slusher *et al.* (1985), who used four-wave mixing in an atomic-vapor medium with a $\chi^{(3)}$ nonlinearity. More strongly squeezed states of light were subsequently generated in $\chi^{(2)}$ crystals by Kimble and coworkers (Wu *et al.*, 1986). In both cases the internal interaction in the sample induced by the external classical field has the form

$$H_{\mathrm{SS}} = i\Omega_P \left(a^{\dagger 2} - \mathrm{HC} \right) , \tag{15.7}$$

for some c-number, phenomenological coupling constant Ω_P. Long before these experiments were performed, squeezed states were discovered theoretically by Stoler (1970), in a study of minimum-uncertainty wave packets that are unitarily equivalent to coherent states. Yuen (1976) introduced squeezed states into quantum optics through the notion of **two-photon coherent states.** He also made the important observation that squeezed states would lead to the possibility of quantum noise reduction. Caves (1981) studied squeezed states in the context of possible improvements in the fundamental sensitivity of gravitational-wave detectors based on optical interferometers that use squeezed light.

But how are squeezed states of light to be detected? If there is a synchronous experimental method to measure $p(t)$, i.e. the electric field, just at the integer multiples of the period—when the p-noise, $\Delta p(t)$, is at a minimum—it is plausible that one can observe p-noise that is less than the standard quantum limit. The price we pay for reduced p-noise at integer multiples of the period ($t = 0, T, 2T, \ldots$) is an increased p-noise at odd multiples of a half-period ($t = T/2, 3T/2, 5T/2, \ldots$). This increase must be such that the product of the alternating deviations, e.g. $\Delta p(T)\, \Delta p(3T/2)$, remains larger than $\hbar/2$. An equivalent argument is based on the fact that $\dot{q} = p$, so that the deviation in displacement, $\Delta q(t)$, is 90° out of phase (in quadrature) with $\Delta p(t)$. Consequently $\Delta q(t)$ is a maximum when $\Delta p(t)$ is a minimum, and the uncertainty relation is maintained at all times. A synchronous measurement method is provided by balanced homodyne detection, as discussed in Section 9.3.3. This kind of detection scheme has blind spots precisely at those times when the p-noise is at a maximum, and sensitive spots at the intermediate times when the p-noise is at a minimum. In this way, the signal-to-noise ratio of a synchronous measurement scheme for the electric field can, in principle, be increased over the prediction of the standard quantum limit associated with a coherent state.

The theory required to describe the generation of squeezed states is significantly more complex than the discussion showing that coherent states are generated by classical currents. For this reason, we will follow the historical sequence outlined above, by first studying the formal properties of squeezed states. This background is quite useful for the analysis of experiments, even in the absence of a detailed model of the source. In subsequent sections we will present the theory of squeezed-light generation, and finally describe an actual experiment.

15.1.2 General properties of squeezed states

A *Quadrature operators*

In place of eqn (15.3) we could equally well consider a squeezed vacuum for which the deviation in the magnetic field (i.e. $\Delta q(t)$) periodically achieves minimum values less than the vacuum fluctuation value $\mathcal{B}_0$. This can be done by using the coordinate representation, and replacing P by Q everywhere in the discussion. More generally, there is no reason to restrict attention to purely electric or purely magnetic fluctuations; we could, instead, decide to measure any linear combination of the two. For this discussion, let us first introduce the dimensionless canonical operators

$$\begin{aligned} X_0 &\equiv \frac{1}{2}\left(a^\dagger + a\right) = \frac{q}{2\Delta q_0} = \sqrt{\frac{\omega}{2\hbar}}q\,, \\ Y_0 &\equiv \frac{i}{2}\left(a^\dagger - a\right) = \frac{p}{2\Delta p_0} = \sqrt{\frac{1}{2\hbar\omega}}p\,, \end{aligned} \tag{15.8}$$

which satisfy the commutation relation

$$[X_0, Y_0] = \frac{i}{2}\,. \tag{15.9}$$

Comparing this to the canonical relations $[q,p] = i\hbar$ and $\Delta q \Delta p \geqslant \hbar/2$ shows that the corresponding uncertainty product is

$$\Delta X_0 \Delta Y_0 \geqslant \frac{1}{4}\,. \tag{15.10}$$

The solution $a\,(t) = a \exp\left(-i\omega t\right)$ of the free-field Heisenberg equations yields the time evolution of X_0 and Y_0:

$$\begin{aligned} X_0\,(t) &= X_0 \cos\left(\omega t\right) + Y_0 \sin\left(\omega t\right), \\ Y_0\,(t) &= -X_0 \sin\left(\omega t\right) + Y_0 \cos\left(\omega t\right), \end{aligned} \tag{15.11}$$

which describes a rotation in the phase plane. It is often useful to generalize the conventional choice, $t = 0$, of the reference time to $t = t_0$, so that the annihilation operator is given by

$$a\,(t) = ae^{-i\omega(t-t_0)} = ae^{-i\beta}e^{-i\omega t}\,, \tag{15.12}$$

where $\beta = -\omega t_0$. In a mechanical context, choosing t_0 amounts to setting a clock; but in the optical context, the preceding equation shows that choosing the reference time t_0 is equivalent to choosing the reference phase β. In the homodyne detection experiments to be described later on, the phase β can be controlled by means of changes in the relative phase between a local oscillator beam and the squeezed light which is being measured. With this choice of reference phase, the time evolution of the magnetic and electric fields is given by

$$\begin{aligned} X_0\,(t) &= X \cos\left(\omega t\right) + Y \sin\left(\omega t\right), \\ Y_0\,(t) &= -X \sin\left(\omega t\right) + Y \cos\left(\omega t\right), \end{aligned} \tag{15.13}$$

where

$$\begin{aligned} X &= \frac{1}{2}\left(ae^{-i\beta} + a^\dagger e^{i\beta}\right) = X_0 \cos\left(\beta\right) + Y_0 \sin\left(\beta\right), \\ Y &= \frac{1}{2i}\left(ae^{-i\beta} - a^\dagger e^{i\beta}\right) = -X_0 \sin\left(\beta\right) + Y_0 \cos\left(\beta\right). \end{aligned} \tag{15.14}$$

These are the same quadrature operators introduced in the analysis of heterodyne and homodyne detection in Section 9.3; they are related to the canonical operators by a rotation through the angle β in the phase plane. The cases considered previously correspond to $\beta = -\pi/2$ and $\beta = 0$ for the electric and magnetic fields respectively.

For any value of β, the quadrature operators satisfy eqns (15.9) and (15.10). Consequently, for any coherent state $|\alpha\rangle$—in particular for the vacuum state—the variances of the quadrature operators are

$$V(X) = V(Y) = \frac{1}{4}, \tag{15.15}$$

and the uncertainty product $\Delta X \Delta Y = 1/4$ has the minimum possible value at all times. Fig. 5.1 shows that the phase space portrait of the coherent state in the dimensionless variables (X_0, Y_0) consists of a circular **quantum fuzzball**, which surrounds the tip of the coherent-state phasor α. The rotation to X and Y amounts to choosing the X-axis along the phasor. The isotropic quantum fuzzball corresponds to a quasi-probability distribution which has the form of an isotropic Gaussian in phase space.

A state ρ is said to be **squeezed along the quadrature** X, if the variance $V(X) = \langle X^2 \rangle - \langle X \rangle^2$ satisfies $V(X) < 1/4$, where $\langle Z \rangle = \mathrm{Tr}(\rho Z)$, for any operator Z. This condition can be expressed more conveniently in terms of the **normal-ordered variance** $V_N(X) \equiv \langle : X^2 : \rangle - \langle : X : \rangle^2$, where $: Z :$ is the normal-ordering operation defined by eqn (2.107). Since X is a linear function of the creation and annihilation operators, $: X : = X$, but $: X^2 : \neq X^2$. An explicit calculation leads to the relation

$$V_N(X) = V(X) - \frac{1}{4}. \tag{15.16}$$

With this notation, the squeezing condition becomes $V_N(X) < 0$ and perfect squeezing, i.e. $V(X) = 0$, corresponds to $V_N(X) = -1/4$.

The straightforward calculation suggested in Exercise 15.1 establishes the relations

$$V_N(X) = \frac{1}{2}\mathrm{Re}\left[e^{-2i\beta} V(a)\right] + \frac{1}{2} V\left(a^\dagger, a\right), \tag{15.17}$$

$$V_N(Y) = -\frac{1}{2}\mathrm{Re}\left[e^{-2i\beta} V(a)\right] + \frac{1}{2} V\left(a^\dagger, a\right), \tag{15.18}$$

between the normal quadrature variances and variances of the annihilation operators. The quantity $V\left(a^\dagger, a\right) = \left\langle a^\dagger a \right\rangle - \left\langle a^\dagger \right\rangle \langle a \rangle$ is an example of the joint variance, $V(F, G) = \langle FG \rangle - \langle F \rangle \langle G \rangle$, introduced in Section 5.1.1. It is easy to see that $V\left(a^\dagger, a\right) \geqslant 0$; therefore, necessary conditions for squeezing along X or Y are

$$\mathrm{Re}\left[e^{-2i\beta} V(a)\right] < 0 \tag{15.19}$$

and

$$\mathrm{Re}\left[e^{-2i\beta} V(a)\right] > 0 \tag{15.20}$$

respectively. Thus a state for which $V(a) = 0$ is not squeezed along any quadrature. This fact excludes both number states and coherent states from the category of squeezed states.

B The squeezing operator

As an aid to understanding how single-mode squeezing is generated by the interaction Hamiltonian (15.7), let us recall the argument used in Section 5.4.1 to guess the form of the displacement operator that generates coherent states from the vacuum. The Hamiltonian H_{int} describing the interaction of a classical current with a single mode of the radiation field is linear in the creation and annihilation operators. For the mode exactly in resonance with a purely sinusoidal current, the time evolution of the state vector in the interaction picture is represented by the unitary operator $\exp\left(-itH_{\text{int}}/\hbar\right)$, which leads to the form $D(\alpha) = \exp\left(\alpha a^{\dagger} - \alpha^{*} a\right)$ for the displacement operator.

By analogy with this argument, the quadratic interaction Hamiltonian (15.7) suggests that the **squeezing operator** should be defined by

$$S(\zeta) = e^{\frac{1}{2}\left(\zeta^{*} a^{2} - \zeta a^{\dagger 2}\right)}, \tag{15.21}$$

where the c-number $\zeta = r\exp(2i\phi)$ is called the **complex squeeze parameter**. The modulus $r = |\zeta|$ describes the amount of squeezing, and the phase 2ϕ determines the angle of the squeezing axis in phase space.

The unitary squeezing operator applied to a pure state $|\Psi\rangle$ defines the **squeezing transformation**,

$$|\Psi(\zeta)\rangle = S(\zeta)|\Psi\rangle, \tag{15.22}$$

for states. It is also useful to define squeezed operators by

$$X(\zeta) = S(\zeta) X S^{\dagger}(\zeta), \tag{15.23}$$

so that expectation values are preserved, i.e.

$$\langle\Psi(\zeta)|X(\zeta)|\Psi(\zeta)\rangle = \langle\Psi|X|\Psi\rangle. \tag{15.24}$$

Applying eqn (15.23) to the density operator describing a mixed state, as well as to the observable X, shows that mixed-state expectation values are also preserved:

$$\operatorname{Tr}\left[\rho(\zeta) X(\zeta)\right] = \operatorname{Tr}\left[\rho X\right]. \tag{15.25}$$

The first example is the squeezed vacuum state $\psi(P,0)$ in eqn (15.3). With the correct choice for ζ this can be expressed as

$$\psi(P,0) = \langle P|S(\zeta)|0\rangle. \tag{15.26}$$

In the limit of weak squeezing, i.e. $|\zeta| \ll 1$, the operator in eqn (15.22) can be expanded to get

$$\begin{aligned} S(\zeta)|0\rangle &= |0\rangle + \frac{1}{2}\left(\zeta^{*} a^{2} - \zeta a^{\dagger 2}\right)|0\rangle + \cdots \\ &= |0\rangle - \frac{1}{2}\zeta a^{\dagger 2}|0\rangle + \cdots. \end{aligned} \tag{15.27}$$

The first-order term on the right side is the output state for the degenerate case of the down-conversion process discussed in Section 13.3.2. Thus down-conversion represents incipient single-mode squeezing. The transformation of a single pump photon of frequency ω_P into a pair of photons, each with frequency $\omega_0 = \omega_P/2$, is the source of the photons in the squeezed vacuum $S(\zeta)|0\rangle$. The general case of nondegenerate down-conversion can similarly serve as the source of a two-mode squeezed state. In this case, the nonlocal phenomena associated with entangled states would play an important role.

For general squeezed states, the features of experimental interest are expressible in terms of variances of the quadrature operators or other observables, such as the number operator. For example, the variance, $V(X)$, of X in the squeezed state is

$$V(X) = \mathrm{Tr}\left[\rho(\zeta) X^2\right] - \left(\mathrm{Tr}\left[\rho(\zeta) X\right]\right)^2 . \tag{15.28}$$

The easiest way to evaluate these expressions is to use the relation (15.23) between the original operators X and their squeezed versions $X(\zeta)$. Since all observables can be expressed in terms of the creation and annihilation operators it is sufficient to consider

$$a(\zeta) = S(\zeta)\, a S^{\dagger}(\zeta) . \tag{15.29}$$

The first step in evaluating the right side of this equation is to define the **squeezing generator** $K(\zeta)$ by

$$K(\zeta) = -\frac{i}{2}\left(\zeta^* a^2 - \zeta a^{\dagger 2}\right) , \tag{15.30}$$

so that $S(\zeta) = \exp\left[iK(\zeta)\right]$. The second step is to imitate eqn (5.49) by introducing the interpolating operators

$$c(\tau) = e^{i\tau K(\zeta)} a e^{-i\tau K(\zeta)} , \tag{15.31}$$

where τ is a real variable in the interval $(0, 1)$. The interpolation formula has the form of a time evolution with Hamiltonian K, so the interpolating operators satisfy the Heisenberg-like equations

$$i\frac{d}{d\tau} c(\tau) = \left[c(\tau), \widetilde{K}\right] , \tag{15.32}$$

where

$$\widetilde{K} = -\frac{i}{2}\left[\zeta^* c^2(\tau) - \zeta c^{\dagger 2}(\tau)\right] . \tag{15.33}$$

If we identify ζ with $-2i\Omega_P$, then K has the general form (15.7). This means that we will be able to use the results obtained here to treat the model for squeezed-state generation to be given in Section 15.2.

The explicit form (15.33), together with the canonical commutation relation $\left[c(\tau), c^{\dagger}(\tau)\right] = 1$, yields a pair of first-order equations for c and $c^{\dagger}$:

$$\frac{dc}{d\tau} = \zeta c^{\dagger} , \quad \frac{dc^{\dagger}}{d\tau} = \zeta^* c , \tag{15.34}$$

and eliminating $c^{\dagger}$ produces a single second-order equation:

$$\frac{d^2 c}{d\tau^2} = |\zeta|^2 c . \tag{15.35}$$

Since $|\zeta|^2 = r^2$ is real and positive the fundamental solutions are $e^{\pm r\tau}$ and the general solution is $c(\tau) = C_+ e^{r\tau} + C_- e^{-r\tau}$. Substituting this form into either of the first-order equations yields one relation between C_+ and C_-, and the initial condition $c(0) = a$ gives another. The solution of this pair of algebraic equations provides the expression $a(\zeta) = \mu a + \nu a^\dagger$ for the squeezed annihilation operator, where the coefficients

$$\mu = \cosh(r) \, , \quad \nu = e^{2i\phi} \sinh(r) \, , \tag{15.36}$$

satisfy the identity $\mu^2 - |\nu|^2 = 1$. The relation between a and $a(\zeta)$ is another example of the Bogoliubov transformation. The inverse transformation,

$$a = \mu a(\zeta) - \nu a^\dagger(\zeta) \, , \tag{15.37}$$

will be useful in subsequent calculations.

Let us first apply eqn (15.37) to express the quadrature operators, defined by eqn (15.14), as

$$\begin{aligned} X &= \frac{1}{2}\left[\cosh(r) - e^{-2i(\phi-\beta)}\sinh(r)\right] a(\zeta) e^{-i\beta} + \text{HC} \, , \\ Y &= \frac{i}{2}\left\{\left[\cosh(r) + e^{-2i(\phi-\beta)}\sinh(r)\right] a(\zeta) e^{-i\beta} - \text{HC}\right\} . \end{aligned} \tag{15.38}$$

For the quadrature angle $\beta = \phi$ this simplifies to $X = e^{-r} X(\zeta)$ and $Y = e^{r} Y(\zeta)$, so that

$$\begin{aligned} V(X) &= V\left(e^{-r} X(\zeta)\right) = e^{-2r} V(X(\zeta)) = e^{-2r} V_0(X) \, , \\ V(Y) &= V\left(e^{r} Y(\zeta)\right) = e^{2r} V(Y(\zeta)) = e^{2r} V_0(Y) \, , \end{aligned} \tag{15.39}$$

i.e. the X-quadrature is squeezed and the Y-quadrature is stretched, relative to the variances V_0 in the original state. The alternative choice $\beta = \phi - \pi/2$ reverses the roles of X and Y. For either choice, the deviations in the squeezed state satisfy

$$\Delta X = \sqrt{V(X)} = e^{\pm r} \Delta_0 X \, , \quad \Delta Y = \sqrt{V(Y)} = e^{\mp r} \Delta_0 Y \, , \tag{15.40}$$

which shows that the uncertainty product is unchanged by squeezing. In particular, if $|\Psi\rangle$ is a minimum-uncertainty state, then so is the squeezed state $|\Psi(\zeta)\rangle$, i.e.

$$\Delta X \Delta Y = \Delta_0 X \Delta_0 Y = \frac{1}{4} \, . \tag{15.41}$$

We now turn to the question of the classical versus nonclassical nature of squeezed states. Suppose that $\rho(\zeta)$ is squeezed along X. The P-representation (5.168) can be used to express the variance as

$$\begin{aligned} V(X) &= \int \frac{d^2\alpha}{\pi} P(\alpha) \left\langle \alpha \left| (X - \langle X \rangle)^2 \right| \alpha \right\rangle \\ &= \int \frac{d^2\alpha}{\pi} P(\alpha) \left\{ \langle \alpha | X^2 | \alpha \rangle - 2 \langle X \rangle \langle \alpha | X | \alpha \rangle + \langle X \rangle^2 \right\} , \end{aligned} \tag{15.42}$$

where $P(\alpha)$ is the P-function representing the squeezed state $\rho(\zeta)$ and $\langle X \rangle = \text{Tr}\left[\rho(\zeta) X\right]$. The coherent-state expectation values can be evaluated by first using eqn

(15.14) and the commutation relations to express X^2 in normal-ordered form. After a little further algebra one finds that the normal-ordered variance is

$$V_N(X) = \int \frac{d^2\alpha}{\pi} P(\alpha) \left(\frac{\alpha e^{-i\beta} + \alpha^* e^{i\beta}}{2} - \langle X \rangle \right)^2 . \tag{15.43}$$

Now let us suppose that the squeezed state $\rho(\zeta)$ is classical, i.e. $P(\alpha) \geqslant 0$, then the last result shows that $V_N(X) > 0$. Since this contradicts the assumption that $\rho(\zeta)$ is squeezed along X, we conclude that all squeezed states are nonclassical.

15.1.3 Multimode squeezed states*

A description of multimode squeezed states can be constructed by imitating the treatment of multimode coherent states in Section 5.5.1. The single-mode squeezing operator can be applied to any member of a complete set of modes, e.g. the plane waves of a box-quantized description; consequently, the simplest definition of a multimode squeezed state is

$$|\Psi(\underline{\zeta})\rangle = S(\underline{\zeta})\,|\Psi\rangle , \tag{15.44}$$

where

$$S(\underline{\zeta}) = \prod_{\mathbf{k}s} \exp\left[\frac{1}{2}\left(\zeta^*_{\mathbf{k}s} a^2_{\mathbf{k}s} - \zeta_{\mathbf{k}s} a^{\dagger 2}_{\mathbf{k}s}\right)\right] \tag{15.45}$$

is the multimode squeezing operator. Since the individual squeezing generators commute, the definition of $S(\underline{\zeta})$ can also be expressed as

$$S(\underline{\zeta}) = \exp\left[\frac{1}{2}\sum_{\mathbf{k}s}\left(\zeta^*_{\mathbf{k}s} a^2_{\mathbf{k}s} - \zeta_{\mathbf{k}s} a^{\dagger 2}_{\mathbf{k}s}\right)\right] . \tag{15.46}$$

15.1.4 Special squeezed states*

Coherent states are minimum-uncertainty states, so eqn (15.41) implies that the **squeezed coherent states**,

$$|\zeta;\alpha\rangle \equiv S(\zeta)\,|\alpha\rangle = S(\zeta)\,D(\alpha)\,|0\rangle , \tag{15.47}$$

are also minimum-uncertainty states. In this notation, the squeezed vacuum state discussed previously is denoted by $|\zeta;0\rangle$. The squeezed vacuum is generated by injecting pump radiation into a nonlinear medium with an effective interaction given by eqn (15.7), and the more general squeezed coherent state can be obtained by simultaneously injecting the pump beam and the output of a laser matching the squeezed mode. Furthermore, the squeezed coherent states are eigenstates of the transformed operator $a(\zeta)$, since

$$a(\zeta)\,|\zeta;\alpha\rangle = S(\zeta)\,a\,|\alpha\rangle = \alpha\,|\zeta;\alpha\rangle . \tag{15.48}$$

The state $|\zeta;\alpha\rangle$ is therefore an analogue of the coherent state $|\alpha\rangle$, but it is generated by creating and annihilating pairs of photons. The squeezed coherent states are therefore the two-photon coherent states introduced by Yuen.

For a fixed value of the squeezing parameter ζ, the squeezed coherent states have the same orthogonality and completeness properties as the coherent states. The orthogonality property follows from the unitary relation (15.47), which shows that the inner product of two squeezed coherent states is

$$\langle \zeta ; \beta | \zeta ; \alpha \rangle = \langle \beta \left| S^{\dagger} (\zeta) \, S (\zeta) \right| \alpha \rangle = \langle \beta | \alpha \rangle . \tag{15.49}$$

The resolution of the identity follows in the same way, since combining eqn (15.47) with eqn (5.69) gives us

$$\int \frac{d^2 \alpha}{\pi} |\zeta ; \alpha\rangle \langle \zeta ; \alpha| = S(\zeta) \left\{ \int \frac{d^2 \alpha}{\pi} |\alpha\rangle \langle \alpha| \right\} S^{\dagger} (\zeta) = 1 . \tag{15.50}$$

An alternative family of states is defined by the **displaced squeezed states**

$$|\alpha ; \zeta\rangle \equiv D(\alpha) |\zeta\rangle = D(\alpha) \, S(\zeta) |0\rangle , \tag{15.51}$$

which are constructed by displacing a squeezed vacuum state. An idealized physical model for this is to inject the output of a squeezed vacuum generator into a laser amplifier for the squeezed mode. The squeezed vacuum is the simplest example of a squeezed state, so the displaced squeezed states are also called **ideal squeezed states** (Caves, 1981).

The states $|\zeta ; \alpha\rangle$ and $|\alpha ; \zeta\rangle$ are quite different, since the operators $S(\zeta)$ and $D(\alpha)$ do not commute. For this reason it is important to remember that ζ is the squeezing parameter and α is the displacement parameter. Despite their differences, these two states are both normalized, so there must be a unitary transformation connecting them. Indeed it is not difficult to show that they are related by

$$|\zeta ; \alpha\rangle = |\alpha_- ; \zeta\rangle \tag{15.52}$$

and

$$|\alpha ; \zeta\rangle = |\zeta ; \alpha_+\rangle , \tag{15.53}$$

where

$$\begin{aligned} \alpha_{\pm} &= \mu\alpha \pm \nu\alpha^* \\ &= \alpha \cosh r \pm \alpha^* e^{2i\phi} \sinh r . \end{aligned} \tag{15.54}$$

According to eqn (15.53) the displaced squeezed state $|\alpha ; \zeta\rangle$ is also an eigenvector of $a(\zeta)$,

$$a(\zeta) |\alpha ; \zeta\rangle = \alpha_+ |\alpha ; \zeta\rangle , \tag{15.55}$$

but the eigenvalue is α_+ rather than α.

The relation (15.53) allows us to transfer the orthogonality and completeness relations for squeezed coherent states to the displaced squeezed states. Applying eqn (15.53) to eqns (15.49) and (15.50) yields

$$\langle \beta ; \zeta | \alpha ; \zeta \rangle = \langle \zeta ; \beta_+ | \zeta ; \alpha_+ \rangle = \langle \beta_+ | \alpha_+ \rangle , \tag{15.56}$$

and

$$\int \frac{d^2\beta}{\pi} |\beta_-;\zeta\rangle \langle\beta_-;\zeta| = \int \frac{d^2\beta}{\pi} |\zeta;\beta\rangle \langle\zeta;\beta| = 1 . \tag{15.57}$$

The general result (15.39) shows that squeezing any minimum-uncertainty state produces the quadrature variances $V(X) = e^{-r}/4$ and $V(Y) = e^{r}/4$. For the case of the squeezed coherent state $|\zeta;\alpha\rangle$, with $\alpha = |\alpha| e^{i\theta}$, the quadrature averages are given by

$$\begin{aligned} \langle\zeta;\alpha|X|\zeta;\alpha\rangle &= |\alpha| e^{-r} \cos(\theta - \phi) , \\ \langle\zeta;\alpha|Y|\zeta;\alpha\rangle &= |\alpha| e^{r} \sin(\theta - \phi) . \end{aligned} \tag{15.58}$$

For the special choice $\theta = \phi$ one finds $\langle\zeta;\alpha|Y|\zeta;\alpha\rangle = 0$ and

$$\langle\zeta;\alpha|X|\zeta;\alpha\rangle = |\alpha| e^{-r} , \tag{15.59}$$

so the squeezed quadrature X represents the amplitude of the coherent state. Consequently this process is called **amplitude squeezing.** This example has led to the frequent use of the names **amplitude quadrature** and **phase quadrature** for X and Y respectively.

Of course, the roles of X and Y can always be changed by making a different phase choice. If we choose $\theta - \phi = \pi/2$, then $\langle\zeta;\alpha|X|\zeta;\alpha\rangle = 0$ and $\langle\zeta;\alpha|Y|\zeta;\alpha\rangle = |\alpha| e^{r}$. The amplitude of the coherent state is now carried by the stretched quadrature Y, and the squeezed quadrature X is conjugate to Y. Roughly speaking, the operator conjugate to the amplitude is related to the phase; consequently, this process is called **phase squeezing.**

15.1.5 Photon-counting statistics*

The variances and averages of the quadrature operators were used in the interpretation of the homodyne detection scheme discussed in Section 9.3.3, but photon-counting experiments are related to the average and variance of the photon number operator. For the special squeezed states defined by eqns (15.47) and (15.51), the most direct way to calculate these quantities is first to use eqn (15.37) to express the operators N and N^2 in terms of the transformed operators $a(\zeta)$ and $a^\dagger(\zeta)$, and then to rearrange these expressions in normal-ordered form with respect to $a(\zeta)$ and $a^\dagger(\zeta)$. Finally, the eigenvalue equations (15.48) and (15.55), together with their adjoints, can be used to get the expectation values of N and N^2 as explicit functions of ζ and α.

By virtue of the relation (15.52), it is enough to consider the expectation values for the displaced squeezed state $|\alpha;\zeta\rangle$. Using eqn (15.37) produces the expression

$$\begin{aligned} N = a^\dagger a &= \left(\mu a^\dagger(\zeta) - \nu^* a(\zeta)\right)\left(\mu a(\zeta) - \nu a^\dagger(\zeta)\right) \\ &= \left(\mu^2 + |\nu|^2\right) a^\dagger(\zeta) a(\zeta) - \nu^*\mu a(\zeta)^2 - \nu\mu a^\dagger(\zeta) + |\nu|^2 \end{aligned} \tag{15.60}$$

for the number operator, so eqn (15.55) and its adjoint yield

$$\begin{aligned} \langle N\rangle = \langle\alpha;\zeta|N|\alpha;\zeta\rangle &= \left(\mu\alpha_+^* - \nu^*\alpha_+\right)\left(\mu\alpha_+ - \nu\alpha_+^*\right) + |\nu|^2 \\ &= |\alpha|^2 + |\nu|^2 = |\alpha|^2 + \sinh^2(r) . \end{aligned} \tag{15.61}$$

To get the final result we have used the solution $\alpha = \mu\alpha_+ - \nu\alpha_+^*$ of eqn (15.54).

For the calculation of $\langle N^2 \rangle$, we first use the commutation relations to establish the identity $N^2 = a^{\dagger 2} a^2 + N$, which leads to

$$\langle N^2 \rangle = \langle a^{\dagger 2} a^2 \rangle + \langle N \rangle . \tag{15.62}$$

The next step is to use eqn (15.37) to derive the normal-ordered expression—with respect to the squeezed operators $a(\zeta)$ and $a^\dagger(\zeta)$—for a^2:

$$a^2 = \mu^2 a(\zeta)^2 - 2\mu\nu a^\dagger(\zeta) a(\zeta) + \nu^2 a^\dagger(\zeta) - \mu\nu . \tag{15.63}$$

This can be used in turn to derive the normal-ordered form for $a^{\dagger 2} a^2$ and thus to evaluate $\langle a^{\dagger 2} a^2 \rangle$ in the same way as $\langle N \rangle$. This calculation is straightforward but rather lengthy. A somewhat more compact method is to use the completeness relation (15.57) to get

$$\begin{aligned} \langle a^{\dagger 2} a^2 \rangle &= \langle \alpha; \zeta | a^{\dagger 2} a^2 | \alpha; \zeta \rangle \\ &= \langle \alpha; \zeta | a^{\dagger 2} \left(\int \frac{d^2\beta}{\pi} |\beta_-; \zeta\rangle \langle \beta_-; \zeta| \right) a^2 |\alpha; \zeta\rangle \\ &= \int \frac{d^2\beta}{\pi} \left| \langle \beta_-; \zeta | a^2 | \alpha; \zeta \rangle \right|^2 . \end{aligned} \tag{15.64}$$

Applying the eigenvalue equation (15.55) to $|\alpha; \zeta\rangle$ and the adjoint equation to $\langle \beta_-; \zeta|$ produces $a(\zeta)|\alpha; \zeta\rangle = \alpha_+ |\alpha; \zeta\rangle$ and $\langle \beta_-; \zeta| a^\dagger(\zeta) = \langle \beta_-; \zeta| \beta^*$, so the matrix element in the integrand is given by

$$\langle \beta_-; \zeta | a^2 | \alpha; \zeta \rangle = f(\beta^*) \langle \beta_-; \zeta | \alpha; \zeta \rangle = f(\beta^*) \langle \beta | \alpha_+ \rangle , \tag{15.65}$$

where

$$f(\beta^*) = \mu^2 \alpha_+^2 - 2\mu\nu\beta^*\alpha_+ + \nu^2 \beta^{*2} - \mu\nu . \tag{15.66}$$

Substituting this result in eqn (15.64) and using the explicit formula (5.58) for the inner product leaves us with

$$\begin{aligned} \langle a^{\dagger 2} a^2 \rangle &= \int \frac{d^2\beta}{\pi} |f(\beta^*)|^2 e^{-|\beta - \alpha_+|^2} \\ &= \int \frac{d^2\beta}{\pi} \left| f(\beta^* + \alpha_+^*) \right|^2 e^{-|\beta|^2} , \end{aligned} \tag{15.67}$$

where the last line was obtained by the change of integration variables $\beta \to \beta + \alpha_+$.

This rather elaborate preparation would be useless if the remaining integrals could not be easily evaluated. Fortunately, the integrals can be readily done in polar coordinates, $\beta = b \exp(i\vartheta)$, as can be seen in Exercise 15.4. After a certain amount of algebra, one finds

$$\langle a^{\dagger 2} a^2 \rangle = |\alpha|^4 + \mu^2 |\nu|^2 - \mu \left(\alpha^2 \nu^* + \text{CC} \right) + 4 |\alpha|^2 |\nu|^2 + 2 |\nu|^4 . \tag{15.68}$$

Combining this result with eqns (15.36), (15.62), and (9.58) leads to the general expression

$$\mathcal{Q} = \frac{\sinh^2 r \ \cosh 2r + 2 |\alpha|^2 \sinh r \left[\sinh r - \cosh r \cos(\theta - \phi) \right]}{|\alpha|^2 + \sinh^2 r} \tag{15.69}$$

for the Mandel $\mathcal{Q}$ parameter.

The $\mathcal{Q}$ parameter is positive (super-Poissonian statistics) for $\cos(\theta - \phi) \leqslant 0$, but it can be negative (sub-Poissonian statistics) if $\cos(\theta - \phi) > 0$. In the case $\theta = \phi$ we have amplitude squeezing (see eqn (15.59) for the squeezed quadrature X), so the general result becomes

$$\mathcal{Q} = \frac{\sinh^2 r \, \cosh 2r - |\alpha|^2 \left[1 - e^{-2r}\right]}{|\alpha|^2 + \sinh^2 r} . \tag{15.70}$$

In the strong-field limit $|\alpha| \gg \exp(4r)$, $\mathcal{Q}$ becomes

$$\mathcal{Q} \approx -\left[1 - e^{-2r}\right] . \tag{15.71}$$

If we also assume strong squeezing ($r \gg 1$), then $\mathcal{Q} \approx -1$, i.e. there is negligible noise in photon number. Consequently, amplitude squeezed states are also called **number squeezed states**. This terminology is rather misleading, since eqn (15.19) shows that a squeezed state can never be a number state.

15.1.6 Are squeezed states robust?*

In Section 8.4.3 we saw that a coherent state $|\alpha_1\rangle$ incident on a beam splitter is scattered into a two-mode coherent state $|\alpha_1', \alpha_2'\rangle$, where $\alpha_1' = \mathfrak{t}\,\alpha_1$ and $\alpha_2' = \mathfrak{r}\,\alpha_1$. A similar result would be found for any passive, linear optical element. An even more impressive feature appears in Section 18.5.2, where it is shown that an initial coherent state $|\alpha_0\rangle$ coupled to a zero-temperature reservoir evolves into the coherent state $\left|\alpha_0 e^{-\Gamma t/2} e^{-i\omega_0 t}\right\rangle$. In other words, the defining statistical property, $V\left(a^\dagger, a\right) = 0$, of the coherent state is unchanged by this form of dissipation. Only the amplitude of the parameter α_0 is reduced. For these reasons the coherent state is regarded as **robust**. The situation for squeezed states turns out to be a bit more subtle.

Let us first consider an experiment in which light in a squeezed state enters through port 1 of a beam splitter, as shown in Fig. 8.2. The input state $|\Psi\rangle$ is the vacuum for the mode entering through the unused port 2, i.e.

$$a_2 |\Psi\rangle = 0 , \tag{15.72}$$

but it is squeezed along a quadrature

$$X_1 = \frac{1}{2}\left[a_1 e^{-i\beta} + a_1^\dagger e^{i\beta}\right] \tag{15.73}$$

of the incident mode 1, i.e. $V_N(X_1) < 0$. According to eqn (8.62) the scattered operators a_1' and a_2' are related to the incident operators a_1 and a_2 by

$$\begin{aligned} a_2' &= \mathfrak{r}\, a_1 + \mathfrak{t}\, a_2 , \\ a_1' &= \mathfrak{t}\, a_1 + \mathfrak{r}\, a_2 , \end{aligned} \tag{15.74}$$

where $|\mathfrak{t}|^2 + |\mathfrak{r}|^2 = 1$. We choose the phases of $\mathfrak{r}$ and $\mathfrak{t}$ so that the transmission coefficient $\mathfrak{t}$ is real and the reflection coefficient $\mathfrak{r}$ is purely imaginary.

The question to be investigated is whether there is squeezing along any output quadrature. We begin by examining general quadratures

$$X_1' = \frac{1}{2}\left[a_1' e^{-i\beta_1} + a_1'^{\dagger} e^{i\beta_1}\right] \tag{15.75}$$

and

$$X_2' = \frac{1}{2}\left[a_2' e^{-i\beta_2} + a_2'^{\dagger} e^{i\beta_2}\right] \tag{15.76}$$

for the transmitted and reflected modes respectively. Applying eqns (15.17), (15.72), and (15.74) to the X_1'-quadrature leads to

$$\begin{aligned} V_N\left(X_1'\right) &= \frac{1}{2}\operatorname{Re}\left[V\left(a_1' e^{-i\beta_1}\right)\right] + \frac{1}{2}V\left(a_1'^{\dagger}, a_1'\right) \\ &= \frac{1}{2}\operatorname{Re}\left[\mathfrak{t}^2 V\left(a_1 e^{-i\beta_1}\right)\right] + \frac{1}{2}\left|\mathfrak{t}\right|^2 V\left(a_1^{\dagger}, a_1\right) \\ &= \mathfrak{t}^2 V_N\left(X_1\right) + \frac{\mathfrak{t}^2}{2}\operatorname{Re}\left[\left(e^{i\varphi} - 1\right) V\left(a_1 e^{-i\beta}\right)\right], \end{aligned} \tag{15.77}$$

where $\varphi = 2\left(\beta - \beta_1\right)$. Squeezing along X_1 means that $V_N\left(X_1\right) < 0$, but the second term depends on the value of β_1. The simplest choice—$\beta_1 = \beta$—leads to

$$V_N\left(X_1'\right) = \mathfrak{t}^2 V_N\left(X_1\right), \tag{15.78}$$

which shows that squeezing along X_1 implies squeezing along X_1' for the quadrature angle $\beta_1 = \beta$. As might be expected, the inescapable partition noise at the beam splitter reduces the amount of squeezing by the intensity transmission coefficient $\mathfrak{t}^2 < 1$. This particular choice of output quadrature does answer the squeezing question, but it does not necessarily yield the largest degree of squeezing.

A similar argument applied to X_2' begins with

$$V_N\left(X_2'\right) = \frac{1}{2}\operatorname{Re}\left[V\left(a_2' e^{-i\beta_2}\right)\right] + \frac{1}{2}V\left(a_2'^{\dagger}, a_2'\right), \tag{15.79}$$

but the relation $\mathfrak{r}^2 = -\left|\mathfrak{r}\right|^2$ produces

$$\operatorname{Re}\left[V\left(a_2' e^{-i\beta_2}\right)\right] = \operatorname{Re}\left[\mathfrak{r}^2 V\left(a_1 e^{-i\beta_2}\right)\right] = -\left|\mathfrak{r}\right|^2 \operatorname{Re}\left[V\left(a_1 e^{-i\beta_2}\right)\right]. \tag{15.80}$$

The final result in this case is

$$V\left(X_2'\right) = \left|\mathfrak{r}\right|^2 V_N\left(X_1\right) - \frac{\left|\mathfrak{r}\right|^2}{2}\operatorname{Re}\left[\left(e^{i\varphi} + 1\right) V\left(a_1 e^{-i\beta}\right)\right], \tag{15.81}$$

where $\varphi = 2\left(\beta - \beta_2\right)$. For the reflected mode, the choice $\beta_2 = \beta - \pi/2$ $(\varphi = \pi)$ shows reduced squeezing along X_2'. Alternatively, we can use the relation

$$X_2'\big|_{\beta_2 = \beta - \pi/2} = -\left.Y_2'\right|_{\beta_2 = \beta} \tag{15.82}$$

to say that squeezing occurs along the conjugate quadrature Y_2' for $\beta_2 = \beta$.

We next consider the evolution of a squeezed state coupled to a zero-temperature reservoir. For the quadrature

$$X_\beta = \frac{1}{2}\left(ae^{-i\beta} + a^\dagger e^{i\beta}\right), \tag{15.83}$$

eqn (15.43) gives us

$$V_N\left(X_\beta; t\right) = \int \frac{d^2\alpha}{\pi} P\left(\alpha, \alpha^*; t\right) \left(\frac{\alpha e^{-i\beta} + \alpha^* e^{i\beta}}{2} - \left\langle X_\beta; t\right\rangle\right)^2, \tag{15.84}$$

where

$$\left\langle X_\beta; t\right\rangle = \int \frac{d^2\alpha}{\pi} P\left(\alpha, \alpha^*; t\right) \frac{\alpha e^{-i\beta} + \alpha^* e^{i\beta}}{2}. \tag{15.85}$$

The assumption that the state is initially squeezed along X_β means that

$$V_N\left(X_\beta; 0\right) = \int \frac{d^2\alpha}{\pi} P_0\left(\alpha, \alpha^*\right) \left(\frac{\alpha e^{-i\beta} + \alpha^* e^{i\beta}}{2} - \left\langle X_\beta\right\rangle_0\right)^2 < 0, \tag{15.86}$$

where $P_0\left(\alpha, \alpha^*\right) = P\left(\alpha, \alpha^*; t = 0\right)$. Anticipating the general solution (18.88) for dissipation by interaction with a zero-temperature reservoir leads to

$$\begin{aligned} V_N\left(X_\beta; t\right) = &\int \frac{d^2\alpha}{\pi} P_0\left(e^{(\Gamma/2 + i\omega_0)t}\alpha, e^{(\Gamma/2 - i\omega_0)t}\alpha^*\right) e^{\Gamma t} \\ &\times \left(\frac{\alpha e^{-i\beta} + \alpha^* e^{i\beta}}{2} - \left\langle X_\beta; t\right\rangle\right)^2, \end{aligned} \tag{15.87}$$

and

$$\begin{aligned} \left\langle X_\beta; t\right\rangle = &\int \frac{d^2\alpha}{\pi} P_0\left(e^{(\Gamma/2 + i\omega_0)t}\alpha, e^{(\Gamma/2 - i\omega_0)t}\alpha^*\right) e^{\Gamma t} \\ &\times \left(\frac{\alpha e^{-i\beta} + \alpha^* e^{i\beta}}{2}\right). \end{aligned} \tag{15.88}$$

Our next step is to make the change of integration variables $\alpha \to \alpha \exp\left[-\left(\Gamma/2 + i\omega_0\right)t\right]$ in the last two equations. For eqn (15.88) the result is

$$\begin{aligned} \left\langle X_\beta; t\right\rangle &= e^{-\Gamma t/2} \int \frac{d^2\alpha}{\pi} P_0\left(\alpha, \alpha^*\right) \left(\frac{\alpha e^{-i(\beta + \omega_0 t)} + \alpha^* e^{-i(\beta + \omega_0 t)}}{2}\right) \\ &= e^{-\Gamma t/2} \left\langle X_{\beta + \omega_0 t}\right\rangle_0, \end{aligned} \tag{15.89}$$

and a similar calculation starting with eqn (15.87) yields

$$V_N\left(X_\beta; t\right) = e^{-\Gamma t} V_N\left(X_{\beta + \omega_0 t}; 0\right). \tag{15.90}$$

Just as in the case of the beam splitter, we are free to choose new quadratures to investigate, in this case at different times. At time t we take advantage of this freedom to let $\beta \to \beta - \omega_0 t$, so that

$$V_N\left(X_{\beta - \omega_0 t}; t\right) = e^{-\Gamma t} V_N\left(X_\beta; 0\right) < 0. \tag{15.91}$$

Thus at any time t, there is a squeezed quadrature—with the amount of squeezing reduced by $\exp\left(-\Gamma t\right)$—but the required quadrature angle rotates with frequency ω_0.

With the results (15.78) and (15.91) in hand, we can now judge the robustness of squeezed states. Let us begin by recalling that coherent states are regarded as robust because the defining property, $V\left(a^{\dagger}, a\right) = 0$, is strictly conserved by dissipative scattering—i.e. coupling to a zero-temperature reservoir—as well as by passage through passive, linear devices. By contrast, dissipative scattering degrades the *degree* of squeezing as well as the overall intensity of the squeezed input light, so that

$$|V_N(X:t)| \to 0 \quad \text{as}\ t \to \infty\,. \tag{15.92}$$

Even this result depends on the detection of a quadrature that is rotating at the optical frequency ω_0. Detector response times are large compared to optical periods, so even the reduced squeezing shown by eqn (15.91) would be extremely difficult to detect. Passage through a linear optical device also degrades the degree of squeezing, as shown by eqn (15.78). This combination of properties is the basis for the general opinion that squeezed states are not robust.

15.2 Theory of squeezed-light generation*

The method used by Kimble and co-workers (Wu *et al.*, 1986) to generate squeezed states relies on the microscopic process responsible for the spontaneous down-conversion effect discussed in Section 13.3.2; but two important changes in the experimental arrangement are shown in Fig. 15.1. The first is that the $\chi^{(2)}$ crystal is cut so as to produce collinear phase matching with degenerate pairs ($\omega_1 = \omega_2 = \omega_0 = \omega_P/2$) of photons, and also anti-reflection coated for both the first- and the second-harmonic frequencies ω_0 and $\omega_P = 2\omega_0$. In this configuration the down-converted photons have identical frequencies and propagate in the same direction as the pump photons; in other words, this is time-reversed second harmonic generation. The second change is that the crystal is enclosed by a resonant cavity that is tuned to the degenerate frequency $\omega_0 = \omega_P/2$ and, therefore, also to the pump frequency ω_P.

The degeneracy conditions between the down-converted photons and the cavity resonance frequency are maintained by a combination of temperature tuning for the crystal and a servo control of the optical resonator length. This arrangement strongly

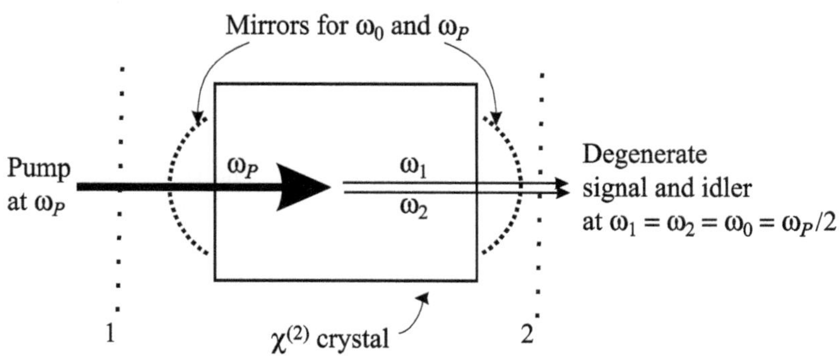

Fig. 15.1 A simplified schematic for the squeezed state generator employed in the experiment of Kimble and co-workers (Wu *et al.*, 1986).

favors the degenerate pairs over all other pairs of photons that are produced by down-conversion. In this way, the crystal—pumped by the strong laser beam at the second-harmonic frequency $2\omega_0$—becomes an **optical parametric amplifier**[1] (OPA) for the degenerate photon pairs at the first-harmonic frequency ω_0.

This device can be understood at the classical level in the following way. The $\chi^{(2)}$ nonlinearity couples the two weak down-converted light beams to the strong-pump laser beam, so that the weak light signals can be amplified by drawing energy from the pump. The basic process is analogous to that of a child pumping a swing by standing and squatting twice per period of the swing, thus increasing the amplitude of the motion. This kind of amplification process depends on the timing (phase) of the pumping motion relative to the timing (phase) of the swinging motion.

In the case of light beams, the mechanism for the transfer of energy from the pump to the degenerate weak beams is the mixing of the strong-pump beam,

$$E_P = \mathrm{Re}\left(\mathcal{E}_P e^{-i\omega_P t}\right) = \mathrm{Re}\left(|\mathcal{E}_P|\, e^{i\theta_P} e^{-i\omega_P t}\right), \tag{15.93}$$

with the two weak beams via the $\chi^{(2)}$ nonlinearity. This leads to a mutual reinforcement of the weak beams at the expense of the pump beam. If the depletion of the strong-pump beam by the parametric amplification process is ignored, the mutual-reinforcement mechanism leads to an exponential growth of both of the weak beams. With sufficient feedback from the mirrors surrounding the crystal, this amplifier—like that of a laser—can begin to oscillate, and thereby become an **optical parametric oscillator** (OPO). When operated just below the threshold of oscillation, the optical parametric amplifier emits strongly squeezed states of light.

The resonant enhancement at the degenerate signal and idler frequencies justifies the use of the phenomenological model Hamiltonian,

$$H_S = H_{S0} + H_{\mathrm{SS}}\,, \tag{15.94}$$

$$H_{S0} = \hbar\omega_0 a^\dagger a\,, \tag{15.95}$$

$$H_{\mathrm{SS}} = \frac{i\hbar}{2}\left[\Omega_P e^{-i\omega_P t} a^{\dagger 2} - \Omega_P^* e^{i\omega_P t} a^2\right], \tag{15.96}$$

for the sample shown in Fig. 15.1. The resonant mode associated with the annihilation operator a is jointly defined by the collinear phase-matching condition for the nonlinear crystal and by the boundary conditions at the two mirrors forming the optical resonator.

Note that H_{SS} has exactly the form of the squeezing generator defined by eqn (15.30). The coupling frequency Ω_P, which is proportional to the product $\chi^{(2)}\mathcal{E}_P$, characterizes the strength of the nonlinear interaction. The term $\Omega_P a^{\dagger 2}$ describes the down-conversion process in which a pump photon is converted into the degenerate signal and idler photons. It is important to keep in mind that the complex coupling parameter Ω_P is proportional to $\mathcal{E}_P = |\mathcal{E}_P|\exp(i\theta_P)$, so that the parametric gain depends on the phase of the pump wave. The consequences of this phase dependence will be examined in the following sections.

[1]The term ‘parametric’ amplifier was originally introduced in microwave engineering. The ‘parameter’ in the optical case is the pump wave amplitude, which is assumed to be unchanged by the nonlinear interaction.

A The Langevin equations

The experimental signal in this case is provided by photons that escape the cavity, e.g. through the mirror M2. In Section 14.3 this situation was described by means of in- and out-fields for a general interaction H_{SS}. In the present application, H_{SS} is given by eqn (15.96), and an explicit evaluation of the interaction term $[a, H_{SS}]/i\hbar$ gives us

$$\frac{d}{dt}\overline{a}(t) = -\frac{\kappa_C}{2}\,\overline{a}(t) + \Omega_P\, e^{i(2\omega_0-\omega_P)t}\,\overline{a}^\dagger(t) + \xi_C(t), \tag{15.97}$$

where $\overline{a}(t) = a(t)\exp(i\omega_0 t)$ is the slowly-varying envelope operator, $\kappa_C = \kappa_1 + \kappa_2$ is the cavity damping rate, and $\xi_C(t)$ is the cavity noise operator defined by eqn (14.97). The explicit time dependence on the right side is eliminated by imposing the resonance condition $\omega_P = 2\omega_0$ on the cavity. The equation for the adjoint envelope operator $\overline{a}^\dagger(t)$ is then

$$\frac{d}{dt}\overline{a}^\dagger(t) = -\frac{\kappa_C}{2}\,\overline{a}^\dagger(t) + \Omega_P^*\,\overline{a}(t) + \xi_C^\dagger(t). \tag{15.98}$$

Before considering the solution of the operator equations, it is instructive to write the ensemble-averaged equations in matrix form:

$$\frac{d}{dt}\begin{pmatrix}\langle\overline{a}(t)\rangle\\ \langle\overline{a}^\dagger(t)\rangle\end{pmatrix} = \begin{bmatrix}-\kappa_C/2 & \Omega_P\\ \Omega_P^* & -\kappa_C/2\end{bmatrix}\begin{pmatrix}\langle\overline{a}(t)\rangle\\ \langle\overline{a}^\dagger(t)\rangle\end{pmatrix}, \tag{15.99}$$

where we have used $\langle\xi_C(t)\rangle = 0$. The 2×2 matrix on the right side has eigenvalues $\Lambda_\pm = -\kappa_C/2 \pm |\Omega_P|$, so the general solution is a linear combination of special solutions varying as $\exp[\Lambda_\pm(t-t_0)]$. Since $\kappa_C > 0$, the eigenvalue Λ_- always describes an exponentially decaying solution. On the other hand, the eigenvalue Λ_+ can describe an exponentially growing solution if $|\Omega_P| > \kappa_C/2$.

At the threshold value $|\Omega_P| = \kappa_C/2$, the average $\langle\overline{a}(t)\rangle$ of the slowly-varying envelope operator approaches a constant for times $t - t_0 \gg \kappa_C$, so that $\langle a(t)\rangle \sim \exp(-i\omega_0 t)$ is oscillatory at large times. This describes the transition from optical parametric amplification to optical parametric oscillation. Operation above the oscillation threshold would produce an exponentially rapid build-up of the intracavity field that would quickly lead to a violation of the weak-field assumptions justifying the model Hamiltonian H_{SS} in eqn (15.96). Dealing with pump fields exceeding the threshold value requires the inclusion of nonlinear effects that would lead to gain saturation and thus prevent runaway amplification. We avoid these complications by imposing the condition $|\Omega_P| < \kappa_C/2$. On the other hand, we will see presently that the largest squeezing occurs for pump fields just below the threshold value.

The coupled equations (15.97) and (15.98) for $\overline{a}(t)$ and $\overline{a}^\dagger(t)$ are a consequence of the special form of the down-conversion Hamiltonian. Since the differential equations are linear, they can be solved by a variant of the Fourier transform technique used for the empty-cavity problem in Section 14.3.3. In the frequency domain the differential equations are transformed into algebraic equations:

$$-i\omega\overline{a}(\omega) = -\frac{\kappa_C}{2}\,\overline{a}(\omega) + \Omega_P\overline{a}^\dagger(-\omega) + \xi_C(\omega), \tag{15.100}$$

$$-i\omega\overline{a}^\dagger(-\omega) = -\frac{\kappa_C}{2}\,\overline{a}^\dagger(\omega) + \Omega_P^*\overline{a}(\omega) + \xi_C^\dagger(-\omega), \tag{15.101}$$

which have the solution

$$\begin{aligned}\overline{a}(\omega) &= \frac{(\kappa_C/2 - i\omega)\,\xi_C(\omega) + \Omega_P \xi_C^\dagger(-\omega)}{(\kappa_C/2 - i\omega)^2 - |\Omega_P|^2}\,,\\ \overline{a}^\dagger(-\omega) &= \frac{(\kappa_C/2 - i\omega)\,\xi_C^\dagger(-\omega) + \Omega_P^* \xi_C(\omega)}{(\kappa_C/2 - i\omega)^2 - |\Omega_P|^2}\,.\end{aligned} \tag{15.102}$$

Combining the definition (14.97) with the result (14.116) for the in-fields in the frequency domain gives

$$\xi_C(\omega) = \left[\sqrt{\kappa_1}\,\overline{b}_{1,\omega+\omega_0}(t_0) + \sqrt{\kappa_2}\,\overline{b}_{2,\omega+\omega_0}(t_0)\right] e^{i\omega t_0}\,. \tag{15.103}$$

This shows that $\overline{a}(\omega)$ and $\overline{a}^\dagger(\omega)$ are entirely expressed in terms of the reservoir operators at the initial time. The correlation functions of the intracavity field $\overline{a}(t)$ are therefore expressible in terms of the known statistical properties of the reservoirs.

Before turning to these calculations, we note that operator $\overline{a}(\omega)$ has two poles—determined by the roots of the denominator in eqn (15.102)—located at

$$\omega = \omega_\pm = -i\left[\frac{\kappa_C}{2} \pm |\Omega_P|\right]. \tag{15.104}$$

Since κ_C is positive, the pole at ω_+ always remains in the lower half plane—corresponding to the exponentially damped solution of eqn (15.99)—but when the coupling frequency exceeds the threshold value, $|\Omega_P|_{\text{crit}} = \kappa_C/2$, the pole at ω_- infiltrates into the upper half plane—corresponding to the exponentially growing solution of eqn (15.99). Thus the OPA–OPO transition occurs at the same value for the operator solution and the ensemble-averaged solution.

B *Squeezing of the intracavity field*

As explained in Section 15.1.2, the properties of squeezed states are best exhibited in terms of the normal-ordered variances $V_N(X)$ and $V_N(Y)$ of conjugate pairs of quadrature operators. According to eqns (15.17) and (15.18), these quantities can be evaluated in terms of the joint variance $V\left(a^\dagger(t), a(t)\right)$ and the variance $V(a(t))$, which can in turn be expressed in terms of the Fourier transforms $a^\dagger(\omega)$ and $a(\omega)$. For example, eqns (14.112) and (14.114) lead to

$$V\left(a^\dagger(t), a(t)\right) = \int \frac{d\omega}{2\pi} \int \frac{d\omega'}{2\pi} e^{-i(\omega'+\omega)t} V\left(a^\dagger(-\omega'), a(\omega)\right). \tag{15.105}$$

Applying the relations

$$a(\omega) = \overline{a}(\omega - \omega_0)\,, \quad a^\dagger(-\omega') = \overline{a}^\dagger(-\omega' - \omega_0) \tag{15.106}$$

that follow from eqn (14.119), and the change of variables $\omega \to \omega + \omega_0$, $\omega' \to \omega' - \omega_0$ allows this to be expressed in terms of the slowly-varying operators $\overline{a}(\omega)$:

$$V\left(a^\dagger(t), a(t)\right) = \int \frac{d\omega}{2\pi} \int \frac{d\omega'}{2\pi} e^{-i(\omega'+\omega)t} V\left(\overline{a}^\dagger(-\omega'), \overline{a}(\omega)\right). \tag{15.107}$$

The solution (15.102) gives $\overline{a}(\omega)$ and $\overline{a}^\dagger(\omega)$ as linear combinations of the initial reservoir creation and annihilation operators. In the experiment under consideration, there is no injected signal at the resonance frequency ω_0, and the incident pump field at ω_P is treated classically. The Heisenberg-picture density operator can therefore be treated as the vacuum for the initial reservoir fields, i.e. $\rho_E = |0\rangle \langle 0|$, where

$$b_{1,\Omega}(t_0)|0\rangle = b_{2,\Omega}(t_0)|0\rangle = 0\,. \tag{15.108}$$

This means that only antinormally-ordered products of the reservoir operators will contribute to the right side of eqn (15.107). The fact that the variance is defined with respect to the reservoir vacuum greatly simplifies the calculation. To begin with, first calculating $\overline{a}(\omega)|0\rangle$ provides the happy result that many terms vanish. Once this is done, the commutation relations (14.115) lead to

$$V\left(a^{\dagger}(t), a(t)\right) = \frac{1}{2}\frac{|\Omega_P|^2}{(\kappa_C/2)^2 - |\Omega_P|^2}\,. \tag{15.109}$$

In the same way, the crucial variance $V(a(t))$ is found to be

$$V(a(t)) = \frac{1}{4}\frac{\Omega_P \kappa_C}{(\kappa_C/2)^2 - |\Omega_P|^2}\,, \tag{15.110}$$

so that

$$V_N(X_\beta) = \frac{1}{8}\frac{\kappa_C}{(\kappa_C/2)^2 - |\Omega_P|^2}\,\mathrm{Re}\left[e^{-2i\beta}\Omega_P\right] + \frac{1}{4}\frac{|\Omega_P|^2}{(\kappa_C/2)^2 - |\Omega_P|^2}\,. \tag{15.111}$$

The minimum value of $V_N(X)$ is attained at the quadrature phase

$$\beta = \frac{\theta_P}{2} - \frac{\pi}{2}\,, \tag{15.112}$$

where θ_P is the phase of Ω_P. For this choice of β,

$$V_N(X) = -\frac{1}{4}\frac{|\Omega_P|}{\kappa_C/2 + |\Omega_P|} \tag{15.113}$$

and

$$V_N(Y) = \frac{1}{4}\frac{|\Omega_P|}{\kappa_C/2 - |\Omega_P|}\,. \tag{15.114}$$

Keeping in mind the necessity of staying below the oscillation threshold, i.e. $|\Omega_P| < \kappa_C/2$, we see that $V_N(X) > -1/8$. The relation (15.16) then yields

$$\frac{1}{8} < V(X) < \frac{1}{4}\,; \tag{15.115}$$

in other words, the cavity field cannot be squeezed by more than 50%. In this connection, it is important to note that these results only depend on the symmetrical combination $\kappa_C = \kappa_1 + \kappa_2$ and not on κ_1 or κ_2 separately. This feature reflects the fact that the mode associated with $a(t)$ is a standing wave that is jointly determined by the boundary conditions at the two mirrors.

C Squeezing of the emitted light

The limits on cavity field squeezing are not the end of the story, since only the *output* of the OPA—i.e. the field emitted through one of the mirrors—can be experimentally studied. We therefore consider a time $t_1 \gg t_0$ when the light emitted—say through mirror M2—reaches a detector. The detected signal is represented by the out-field operator $b_{2,\text{out}}(t)$ introduced in Section 14.3. We reproduce the definition,

$$b_{2,\text{out}}(t) = \int_{-\infty}^{\infty} \frac{d\Omega}{2\pi} b_{2,\Omega}(t_1) e^{-i\Omega(t-t_1)}, \tag{15.116}$$

here, in order to emphasize the dependence of the output signal on the *final* value $b_{2,\Omega}(t_1)$ of the reservoir operator.

Combining the Fourier transforms of the scattering relations (14.109) with eqn (15.102) produces the following relations between the in- and out-fields:

$$\overline{b}_{J,\text{out}}(\omega) = \sum_{L=1}^{2} P_{JL}(\omega) \overline{b}_{L,\text{in}}(\omega) + \sum_{L=1}^{2} C_{JL}(\omega) \overline{b}_{L,\text{in}}^{\dagger}(-\omega), \tag{15.117}$$

with

$$P_{JL}(\omega) = \delta_{JL} - \sqrt{\kappa_J \kappa_L} \frac{[\kappa_C/2 - i\omega]}{[\kappa_C/2 - i\omega]^2 - |\Omega_P|^2}, \tag{15.118}$$

$$C_{JL}(\omega) = -\sqrt{\kappa_J \kappa_L} \frac{\Omega_P}{[\kappa_C/2 - i\omega]^2 - |\Omega_P|^2}. \tag{15.119}$$

The M2-output quadratures are defined by replacing $a(t)$ with $b_{2,\text{out}}(t)$ in eqn (15.14) to get

$$\begin{aligned} X_{\text{out}}(t) &= \frac{1}{2}\left(b_{2,\text{out}}(t) e^{-i\beta} + b_{2,\text{out}}^{\dagger}(t) e^{i\beta}\right), \\ Y_{\text{out}}(t) &= \frac{1}{2i}\left(b_{2,\text{out}}(t) e^{-i\beta} - b_{2,\text{out}}^{\dagger}(t) e^{i\beta}\right) \end{aligned} \tag{15.120}$$

in the time domain, or

$$\begin{aligned} X_{\text{out}}(\omega) &= \frac{1}{2}\left(b_{2,\text{out}}(\omega) e^{-i\beta} + b_{2,\text{out}}^{\dagger}(-\omega) e^{i\beta}\right), \\ Y_{\text{out}}(\omega) &= \frac{1}{2i}\left(b_{2,\text{out}}(\omega) e^{-i\beta} - b_{2,\text{out}}^{\dagger}(-\omega) e^{i\beta}\right) \end{aligned} \tag{15.121}$$

in the frequency domain. The parameter β is again chosen to satisfy eqn (15.112). The normal-ordered variances for the output quadratures are

$$V_N(X_{\text{out}}(t)) = \int \frac{d\omega''}{2\pi} \int \frac{d\omega'}{2\pi} e^{-i(\omega''+\omega')t} V_N(X_{\text{out}}(\omega''), X_{\text{out}}(\omega')), \tag{15.122}$$

$$V_N(Y_{\text{out}}(t)) = \int \frac{d\omega''}{2\pi} \int \frac{d\omega'}{2\pi} e^{-i(\omega''+\omega')t} V_N(Y_{\text{out}}(\omega''), Y_{\text{out}}(\omega')), \tag{15.123}$$

where

$$V_N(F,G) = \langle : FG : \rangle - \langle : F : \rangle \langle : G : \rangle \tag{15.124}$$

is the **joint normal-ordered variance**. Calculations very similar to those for the cavity quadratures lead to

$$V_N\left(X_{\text{out}}(\omega''), X_{\text{out}}(\omega')\right) = -\frac{1}{2}\frac{|\Omega_P|\,\kappa_2}{\left[\kappa_C/2 + |\Omega_P|\right]^2 + (\omega'' - \omega_0)^2} 2\pi\delta\left(\omega'' + \omega' - 2\omega_0\right), \tag{15.125}$$

$$V_N\left(Y_{\text{out}}(\omega''), Y_{\text{out}}(\omega')\right) = \frac{1}{2}\frac{|\Omega_P|\,\kappa_2}{\left[\kappa_C/2 - |\Omega_P|\right]^2 + (\omega'' - \omega_0)^2} 2\pi\delta\left(\omega'' + \omega' - 2\omega_0\right). \tag{15.126}$$

The delta functions in the last two equations reflect the fact that the output field $b_{2,\text{out}}(t)$—by contrast to the discrete cavity mode described by $a(t)$—lies in a continuum of reservoir modes. In this situation, it is necessary to measure the time-dependent correlation function $V_N\left(X_{\text{out}}(t), X_{\text{out}}(0)\right)$, or rather the corresponding spectral function,

$$\begin{aligned} S_N(\omega) &= \int dt e^{i\omega t} V_N\left(X_{\text{out}}(t), X_{\text{out}}(0)\right), \\ &= \int \frac{d\omega'}{2\pi} V_N\left(X_{\text{out}}(\omega), X_{\text{out}}(\omega')\right). \end{aligned} \tag{15.127}$$

Using eqn (15.125) to carry out the remaining integral produces

$$S_N(\omega) = -\frac{1}{2}\frac{|\Omega_P|\,\kappa_2}{\left[\kappa_C/2 + |\Omega_P|\right]^2 + (\omega - \omega_0)^2}, \tag{15.128}$$

which has its minimum value for $|\Omega_P| = \kappa_C/2 = (\kappa_1 + \kappa_2)/2$ and $\omega = \omega_0$, i.e.

$$S_N(\omega) > -\frac{1}{4}\frac{\kappa_2}{\kappa_1 + \kappa_2}. \tag{15.129}$$

For a symmetrical cavity—i.e. $\kappa_1 = \kappa_2$—the degree of squeezing is bounded by

$$S_N(\omega) > -\frac{1}{8}; \tag{15.130}$$

therefore, the output field can at best be squeezed by 50%, just as for the intracavity field. However, the degree of squeezing for the output field is not a symmetrical function of κ_1 and κ_2. For an extremely unsymmetrical cavity—e.g. $\kappa_1 \ll \kappa_2$—we see that

$$S_N(\omega) \gtrapprox -\frac{1}{4}; \tag{15.131}$$

in other words, the output light can be squeezed by almost 100%.

The surprising result that the emitted light can be more squeezed than the light in the cavity demands some additional discussion. The first point to be noted is that the intracavity mode associated with the operator $a(t)$ is a standing wave. Thus photons generated in the nonlinear crystal are emitted into an equal superposition of left- and right-propagating waves. The left-propagating component of the intracavity mode is partially reflected from the mirror M1 and then partially transmitted through the mirror M2, together with the right-propagating component. Reflection from the ideal mirror M1 does not introduce any phase jitter between the two waves; therefore, interference is possible between the two right-propagating waves emitted from the mirror M2. This makes it possible to achieve squeezing in one quadrature of the emitted light.

In estimating the degree of squeezing that can be achieved, it is essential to account for the vacuum fluctuations in the M1 reservoir that are partially transmitted through the mirror M1 into the cavity. Interference between these fluctuations and the right-propagating component of the intracavity mode is impossible, since the phases are statistically independent. For a symmetrical cavity, $\kappa_1 = \kappa_2$, the result is that the squeezing of the output light can be no greater than the squeezing of the intracavity light. On the other hand, if the mirror M1 is a perfect reflector at ω_0, i.e. $\kappa_1 = 0$, then the vacuum fluctuations in the M1 reservoir cannot enter the cavity. In this case it is possible to approach 100% squeezing in the light emitted through the mirror M2.

15.3 Experimental squeezed-light generation

In Fig. 15.2, an experiment by Kimble and co-workers (Wu *et al.*, 1986) to generate squeezed light is sketched. The light source for this experiment is a ring laser contain-

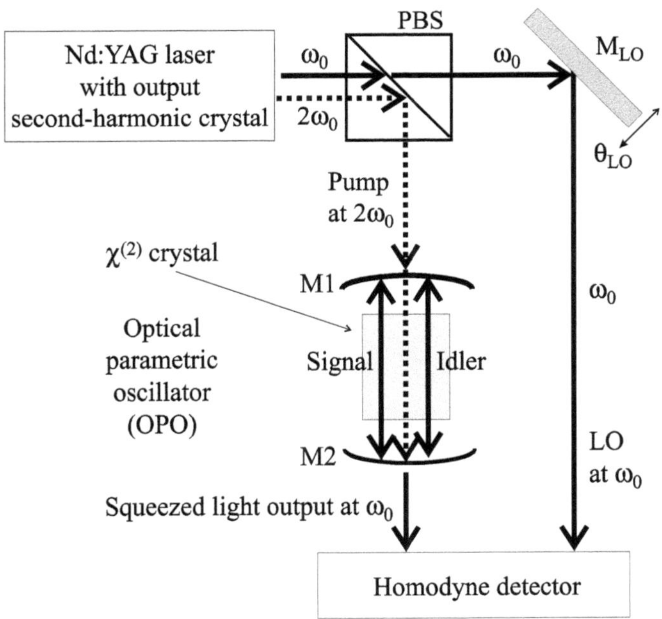

Fig. 15.2 Simplified schematic of an experiment to generate squeezed light. 'PBS' stands for 'polarizing beam splitter' and 'M_{LO}' is a mirror for the local oscillator (LO) beam at ω_{LO}. (Adapted from Wu *et al.* (1986).)

ing a diode-laser-pumped, neodymium-doped, yttrium aluminum garnet (Nd:YAG) crystal—which produces an intense laser beam at the first-harmonic frequency ω_0—and an intracavity, second-harmonic crystal (barium sodium niobate), which produces a strong beam at the second-harmonic frequency $2\omega_0$. The solid lines represent beams at the first harmonic, corresponding to a wavelength of $1.06\,\mu$m, and the dashed lines represent beams at the second harmonic, corresponding to a wavelength of $0.53\,\mu$m.

The two outputs of the ring laser source are each linearly polarized along orthogonal axes, so that the polarizing beam splitter (PBS) can easily separate them into two beams. The first-harmonic beam is transmitted through the polarizing beam splitter and then directed downward by the mirror M_{LO}. This beam serves as the local oscillator (LO) for the homodyne detector, and the mirror M_{LO} is mounted on a translation stage so as to be able to adjust the LO phase θ_{LO}. The second-harmonic beam is directed downward by the polarizing beam splitter, and it provides the pump beam of the optical parametric oscillator (OPO).

The heart of the experiment is the OPO system, which is operated just below the threshold of oscillation, where a maximum of squeezed-light generation occurs. The OPO consists of a $\chi^{(2)}$ crystal (lithium niobate doped with magnesium oxide), surrounded by the two confocal mirrors M1 and M2. The crystal is cut so that the signal and idler modes have the same frequency, ω_0, and are also collinear. The entrance mirror M1 has an extremely high reflectivity at the first-harmonic frequency ω_0, but only a moderately high reflectivity at the second-harmonic frequency $2\omega_0$. Thus M1 allows the second-harmonic, pump light to enter the OPO, while also serving as one of the reflecting surfaces defining a resonant cavity for both the first- and second-harmonic frequencies. This arrangement enhances the pump intensity inside the crystal.

By contrast, the exit mirror M2 has an extremely high reflectivity for the second-harmonic frequency, but only a moderately high reflectivity at the first-harmonic frequency. Thus the mirrors M1 and M2 form a resonator—for both the first- and second-harmonic frequencies—but at the same time M2 allows the degenerate signal and idler beams—at the first-harmonic frequency ω_0—to escape toward the homodyne detector.

In Fig. 15.2, the left and right ports of the box indicating the homodyne detector correspond to two ports of a central balanced beam splitter which respectively emit the signal and local oscillator beams. The output ports of the beam splitter are followed by two balanced photodetectors, and the detected outputs of the photodetectors are then subtracted by means of a balanced differential amplifier. Finally, the output of the differential amplifier is fed into a spectrum analyzer, as explained in Section 9.3.3.

It is important to emphasize that the extremely high reflectivity, for frequency ω_0, of the entrance mirror, M1, blocks out vacuum fluctuations from entering the system, thereby preventing them from contributing unwanted vacuum fluctuation noise at this frequency. As explained in Section 15.2-C, the asymmetry in the reflectivities of the mirrors M1 and M2 at the first-harmonic frequency ω_0 allows more squeezing of the light to occur outside than inside the cavity.

The resulting data is shown in Fig. 15.3, where the output noise voltage, $V(\theta)$, of the spectrum analyzer associated with the homodyne detector is plotted versus the local oscillator phase $\theta = \theta_{LO}$, for a fixed intermediate frequency of 1.8 MHz.

The crucial comparison of this noise output is with the noise from the standard

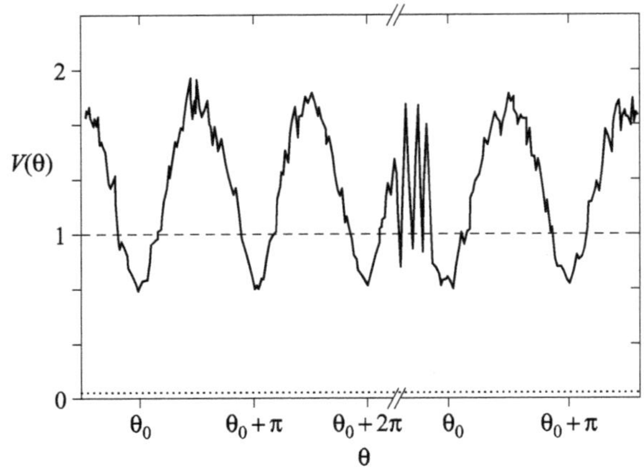

Fig. 15.3 Homodyne-detector, spectrum-analyzer output noise voltage (i.e. the rms noise voltage at an intermediate frequency of 1.8 MHz) versus the local oscillator phase. (Reproduced from Wu *et al.* (1986).)

quantum limit (SQL), which is determined either by blocking the output of the OPO, or by changing the temperature of the lithium niobate crystal so that the signal and idler modes are detuned away from the cavity resonance. The SQL level—which represents the noise from vacuum fluctuations—is indicated by the dashed line in this figure. By inspection of these and similar data, the authors concluded that, in the absence of linear attenuation, the light output from the OPO would have been squeezed by a factor $e^{2r} > 10$. This means the semiminor axis of the noise ellipse of the Gaussian Wigner function in phase space would be more than ten times the semimajor axis.

Strictly speaking, this experiment in squeezed state generation and detection did not involve exactly degenerate photon pairs, since the detected photons were symmetrically displaced from exact degeneracy by 1.8 MHz (within a bandwidth of 100 kHz). The exact conservation of energy in parametric down-conversion guarantees that the shifts in the two frequencies are anti-correlated, i.e. $\omega_i' = \omega_0 + \Delta\omega_i$ and $\omega_s' = \omega_0 + \Delta\omega_s$, with $\Delta\omega_i = -\Delta\omega_s$. Thus the beat notes produced by interference of the upper and lower sidebands with the local oscillator are exactly the same. Both sidebands are detected in the balanced homodyne detector, but their phases are correlated in just such a way that for one particular phase θ_{LO} of the local oscillator—which can be adjusted by the piezoelectric translator that controls the location of the mirror M_{LO}—the sensitive spots of homodyne detection coincide with the least noisy quadrature of the squeezed light. This is true in spite of the fact that the two conjugate photons may not be exactly degenerate in frequency, as long as they are inside the gain-narrowed line width of the optical parametric amplification profile just below threshold. The noise analysis for this case of slightly nondegenerate parametric down-conversion can be found in Kimble (1992).

15.4 Number states

We have seen in Section 2.1.2 that the number states provide a natural basis for the Fock space of a single mode of the radiation field. Any state, whether pure or mixed, can be expressed in terms of number states. By definition, the variance of the number operator vanishes for a number state $|n\rangle$; so evaluating eqn (9.58) for the Mandel $\mathcal{Q}$-parameter of the number state $|n\rangle$ gives

$$\mathcal{Q}\left(|n\rangle\right) \equiv \frac{V\left(N\right) - \langle N\rangle}{\langle N\rangle} = -1\,, \tag{15.132}$$

where $\langle X\rangle = \langle n\left|X\right|n\rangle$. Thus the number states saturate the general inequality $\mathcal{Q}\left(|\Psi\rangle\right) \geqslant -1$. Furthermore, every state with negative $\mathcal{Q}$ is nonclassical; consequently, a pure number state is as nonclassical as it can be. Since this is true no matter how large n is, the classical limit cannot be identified with the large-n limit. Further evidence of the nonclassical nature of number states is provided by eqn (5.153), which shows that the Wigner distribution $W\left(\alpha\right)$ for the single-photon number state $|1\rangle$ is negative in a neighborhood of the origin in phase space.

15.4.1 Single-photon wave packets from SDC

States containing exactly one photon in a classical traveling-wave mode, e.g. a Gaussian wave packet, are of particular interest in contemporary quantum optics. In the approximate sense discussed in Section 7.8 the photon is localized within the wave packet. With almost complete certainty, such a single-photon wave packet state would register a single click when it falls on an ideal photodetector with unit quantum efficiency.

The first experiment demonstrating the existence of single-photon wave packet states was performed by Hong and Mandel (1986). The single-photon state is formed by one of the pair of photons emitted in spontaneous down-conversion, using the apparatus shown in Fig. 15.4. An argon-ion UV laser beam at a wavelength of $\lambda = 351$ nm enters a crystal—potassium dihydrogen phosphate (KDP)—with a $\chi^{(2)}$ nonlinearity.

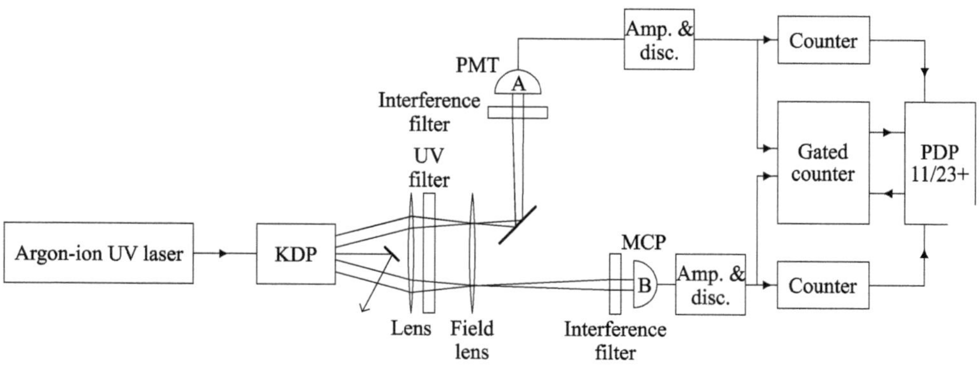

Fig. 15.4 Schematic of Hong and Mandel's experiment to generate and detect single-photon wave packets. (Reproduced from Hong and Mandel (1986).)

Conjugate down-converted photon pairs are generated on opposite sides of the UV beam wavelength at the signal and idler wavelengths of 746 nm and 659 nm, respectively, and enter the photon counters A and B. Counter B is gated by the pulse derived from counter A, for a counting time interval of 20 ns.

Whenever a click is registered by counter A—and the less-than-unity quantum efficiency of counter B is accounted for—there is one and only one click at counter B. This is shown in Fig. 15.5, in which the derived probability $p(n)$ for a count at counter B—conditioned on the detection of a signal photon at counter A—is plotted versus the photon number n.

The data show that within small uncertainties (indicated by the cross-hatched regions),

$$p(n) = \delta_{n,1}\,; \tag{15.133}$$

that is, the idler photons detected by B have been prepared in the single-photon number state $|n = 1\rangle$. In other words, the moment that the click goes off in counter A, one can, with almost complete certainty, predict that there is one and only one photon within a well-defined wave packet propagating in the idler channel. The Mandel $\mathcal{Q}$-parameter derived from these data, $\mathcal{Q} = -1.06 \pm 0.11$, indicates that this state of light is maximally nonclassical, as expected for a number state.

15.4.2 Single photons on demand

The spontaneous down-conversion events that yield the single-photon wave packet states occur randomly, so there is no way to control the time of emission of the wave packet from the nonlinear crystal. Recently, work has been done on a controlled production process in which the time of emission of a well-defined single-photon wave packet is closely determined. Such a deterministic emission process for an individual photon wave packet is called **single photons on demand** or a **photon gun**. One such method involves quantum dots placed inside a high-Q cavity. When a single electron is controllably injected into the quantum dot—via the Coulomb blockade mechanism—the resonant enhancement of the rate of spontaneous emission by the high-Q cavity produces an almost immediate emission of a single photon. Deterministic production of single-photon states can be useful for quantum information processing and quantum

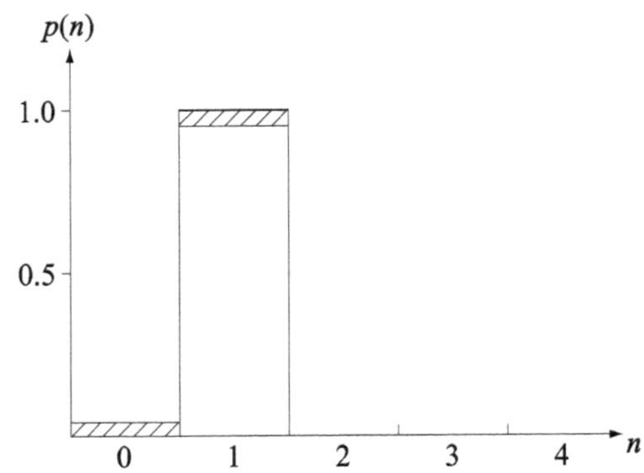

Fig. 15.5 The derived probability $p(n)$ for the detection of n idler photons conditioned on the detection of a single signal photon in the 1986 experiment of Hong and Mandel. The cross-hatched regions indicate the uncertainties of $p(n)$. (Reproduced from Hong and Mandel (1986).)

computation, since often the photons must be synchronized with the computer cycles in a controllable manner.

15.4.3 Number states in a micromaser

Number states have been produced in a standing-wave mode inside a cavity, as opposed to the traveling-wave packet described above. In the microwave region, number states inside a microwave cavity have been produced by means of the micromaser described in Section 12.3. This is accomplished by two methods described below.

In the first method, a completed measurement of the final state of the atom after it exits the cavity allows the experimenter to know—with certainty—whether the atom has made a downwards transition inside the cavity. Combining this knowledge with the conservation of energy determines—again with certainty—the number state of the cavity field.

In the second method, an exact integer number of photons is maintained inside the cavity by means of a **trapping state** (Walther, 2003). According to eqn (12.21), the effective Rabi frequency for an on-resonance, n-photon state is $\Omega_n = 2g\sqrt{(n+1)}$, where g is the coupling constant of the two-level atom with the cavity mode. The Rabi period is therefore $T_n = 2\pi/\Omega_n = \pi/\left(g\sqrt{n+1}\right)$. If the interaction time T_{int} of the atom with the field satisfies $T_{\text{int}} = kT_n$, where k is an integer, then an atom that enters the cavity in an excited state will leave in an excited state. Thus the number of photons in the cavity will be unchanged—i.e. *trapped*—if the condition

$$\sqrt{n+1}gT_{\text{int}} = k\pi \tag{15.134}$$

is satisfied.

Trapping states are characterized by the number of photons remaining in the cavity, and the number of Rabi cycles occurring during the passage of an atom through the cavity. Thus the trapping state $(n, k) = (1, 1)$ denotes a state in which a one-photon, one-Rabi-oscillation trapped field state is maintained by a continuous stream of Rydberg atoms prepared in the upper level. Experiments show that, under steady-state excitation conditions, the one-photon cavity state is stable. Although this technique produces number states of microwave photons in a beautifully simple and clean way, it is difficult to extract them from the high-Q superconducting cavity for use in external experiments.

15.5 Exercises

15.1 Quadrature variances

(1) Use eqn (15.14) and the canonical commutation relations to calculate $:X^2:$ and to derive eqns (15.17) and (15.18).

(2) Are the conditions (15.19) and (15.20) sufficient, as well as necessary? If not, what are the sufficient conditions?

(3) Explain why number states and coherent states are not squeezed states.

(4) Is the state $|\psi\rangle = \cos\theta\ |0\rangle + \sin\theta\ |1\rangle$ squeezed for any value of θ? In other words, for a given θ, is there a quadrature X with $V_N(X) < 0$?

15.2 Squeezed number state

Number states are not squeezed, but it is possible to squeeze a number state. Consider $|\zeta, n\rangle = S(\zeta)|n\rangle$.

(1) Evaluate the Mandel $\mathcal{Q}$-parameter for this state and comment on the result.

(2) What quadrature exhibits maximum squeezing?

15.3 Displaced squeezed states and squeezed coherent states*

Use the properties of $S(\zeta)$ and $D(\alpha)$ to derive the relations (15.52)–(15.54).

15.4 Photon statistics for the displaced squeezed state*

Carry out the integral in eqn (15.67) using polar coordinates and combine this with the other results to get eqn (15.69).

16
Linear optical amplifiers*

Generally speaking, an optical amplifier is any device that converts a set of input modes into a set of output modes with increased intensity. The only absolutely necessary condition is that the creation and annihilation operators for the two sets of modes must be connected by a unitary transformation. Paradoxically, this level of generality makes it impossible to draw any general conclusions; consequently, further progress requires some restriction on the family of amplifiers to be studied.

To this end, we consider the special class of unitary input–output transformations that can be expressed as follows. The annihilation operator for each output mode is a linear combination, with c-number coefficients, of the creation and annihilation operators for the input modes. Devices of this kind are called **linear amplifiers**. We note in passing that linear amplifiers are quite different from laser oscillator-amplifiers, which typically display the highly nonlinear phenomenon of saturation (Siegman, 1986, Sec. 4.5).

For typical applications of linear, optical amplifiers—e.g. optical communication or the generation of nonclassical states of light—it is desirable to minimize the noise added to the input signal by the amplifier. The first source of noise is the imperfect coupling of the incident signal into the amplifier. Some part of the incident radiation will be scattered or absorbed, and this loss inevitably adds partition noise to the transmitted signal. In the literature, this is called **insertion-loss noise**, and it is gathered together with other effects—such as noise in the associated electronic circuits—into the category of **technical noise**. Since these effects vary from device to device, we will concentrate on the intrinsic quantum noise associated with the act of amplification itself.

In the present chapter we first discuss the general properties of linear amplifiers and then describe several illustrative examples. In the final sections we present a simplified version of a general theory of linear amplifiers due to Caves (1982), which is an extension of the earlier work of Haus and Mullen (1962).

16.1 General properties of linear amplifiers

The degenerate optical parametric amplifier (OPA) studied in Section 15.2 is a linear device, by virtue of the assumption that depletion of the pump field can be neglected. In the application to squeezing, the input consists of vacuum fluctuations—represented by $b_{2,\text{in}}(t)$—entering the mirror M2, and the corresponding output is the squeezed state—represented by $b_{2,\text{out}}(t)$—emitted from M2. Both the input and the output have the carrier frequency ω_0. Rather than extending this model to a general theory of linear amplifiers that allows for multiple inputs and outputs and frequency shifts between them, we choose to explain the basic ideas in the simplest possible context: linear amplifiers with a single input field and a single output field—denoted by $b_{\text{in}}(t)$ and $b_{\text{out}}(t)$ respectively—having a common carrier frequency.

We will also assume that the characteristic response frequency of the amplifier and the bandwidth of the input field are both small compared to the carrier frequency. This narrowband assumption justifies the use of the slowly-varying amplitude operators introduced in Chapter 14, but it should be remembered that both the input and the output are reservoir modes that do not have sharply defined frequencies. Just as in the calculation of the squeezing of the emitted light in Section 15.2, the input and output are described by continuum modes.

All other modes involved in the analysis are called **internal modes** of the amplifier. In the sample–reservoir language, the internal modes consist of the sample modes and any reservoir modes other than the input and output. A peculiarity of this jargon is that some of the 'internal' modes are field modes, e.g. vacuum fluctuations, that exist in the space outside the physical amplifier.

The definition of the amplifier is completed by specifying the Heisenberg-picture density operator ρ that describes the state of both the input field and the internal modes of the amplifier. This is the same thing as specifying the initial value of the Schrödinger-picture density operator. Since we want to use the amplifier for a broad range of input fields, it is natural to require that the operating state of the amplifier is independent of the incident field state. This condition is imposed by the factorizability assumption

$$\rho = \rho_{\text{in}}\rho_{\text{amp}}\,, \tag{16.1}$$

where ρ_{in} and ρ_{amp} respectively describe the states of the input field and the amplifier.

In the generic states of interest for communications, the expectation value of the input field does not vanish identically:

$$\langle b_{\text{in}}(t)\rangle = \text{Tr}\left[\rho_{\text{in}} b_{\text{in}}(t)\right] \neq 0\,. \tag{16.2}$$

Situations for which $\langle b_{\text{in}}(t)\rangle = 0$ for all t—e.g. injecting the vacuum state or a squeezed-vacuum state into the amplifier—are to be treated as special cases.

The identification of the measured values of the input and output fields with the expectation values $\langle \overline{b}_{\text{in}}(t)\rangle$ and $\langle \overline{b}_{\text{out}}(t)\rangle$ runs into the apparent difficulty that the annihilation operators $\overline{b}_{\text{in}}(t)$ and $\overline{b}_{\text{out}}(t)$ do not represent measurable quantities. To see why this is not really a problem, we recall the discussion in Section 9.3, which showed that both heterodyne and homodyne detection schemes effectively measure a hermitian quadrature operator. For example, it is possible to measure one member of the conjugate pair $(X_{\beta,\text{in}}(t)\,, Y_{\beta,\text{in}}(t))$, where

$$\begin{aligned} X_{\beta,\text{in}}(t) &= \frac{1}{2}\left[e^{-i\beta}\overline{b}_{\text{in}}(t) + e^{i\beta}\overline{b}^{\dagger}_{\text{in}}(t)\right], \\ Y_{\beta,\text{in}}(t) &= \frac{1}{2i}\left[e^{-i\beta}\overline{b}_{\text{in}}(t) - e^{i\beta}\overline{b}^{\dagger}_{\text{in}}(t)\right]. \end{aligned} \tag{16.3}$$

The quadrature angle β is determined by the relative phase between the input signal and the local oscillator employed in the detection scheme. The operational significance of the complex expectation value $\langle \overline{b}_{\text{in}}(t) \rangle$ is established by carrying out measurements of $X_{\beta,\text{in}}(t)$ for several quadrature angles and using the relation

$$\langle X_{\beta,\text{in}}(t) \rangle = \frac{1}{2}\left[e^{-i\beta}\langle \overline{b}_{\text{in}}(t)\rangle + e^{i\beta}\left\langle \overline{b}_{\text{in}}^{\dagger}(t)\right\rangle\right] = \text{Re}\left[e^{-i\beta}\langle \overline{b}_{\text{in}}(t)\rangle\right]. \tag{16.4}$$

With this reassuring thought in mind, we are free to use the algebraically simpler approach based on the annihilation operators. An important example is provided by the **phase transformation**,

$$\overline{b}_{\text{in}}(t) \to \overline{b}'_{\text{in}}(t) = e^{-i\theta}\overline{b}_{\text{in}}(t), \tag{16.5}$$

of the annihilation operator. The corresponding transformation for the quadratures,

$$X_{\beta,\text{in}}(t) \to X'_{\text{in}}(t) = X_{\beta,\text{in}}(t)\cos\theta + Y_{\beta,\text{in}}(t)\sin\theta, \tag{16.6}$$

$$Y_{\beta,\text{in}}(t) \to Y'_{\text{in}}(t) = Y_{\beta,\text{in}}(t)\cos\theta - X_{\beta,\text{in}}(t)\sin\theta, \tag{16.7}$$

represents a rotation through the angle θ in the (X, Y)-plane. As explained in Section 8.1, these transformations are experimentally realized by the use of phase shifters.

16.1.1 Phase properties of linear amplifiers

From Section 14.1.1-C, we know that the noise properties of the input/output fields are described by the correlation functions of the fluctuation operators, $\delta\overline{b}_{\gamma}(\omega) \equiv \overline{b}_{\gamma}(\omega) - \langle \overline{b}_{\gamma}(\omega)\rangle$, where $\gamma = \text{in}, \text{out}$. Thus the input/output noise correlation functions are defined by

$$K_{\gamma}(\omega, \omega') = \frac{1}{2}\left\langle \delta\overline{b}_{\gamma}(\omega)\,\delta\overline{b}_{\gamma}^{\dagger}(\omega') + \delta\overline{b}_{\gamma}^{\dagger}(\omega')\,\delta\overline{b}_{\gamma}(\omega)\right\rangle \quad (\gamma = \text{in}, \text{out}). \tag{16.8}$$

The definitions (14.98) and (14.107) relating the input/output fields to the reservoir operators allow us to apply the conditions (14.27) and (14.34) for phase-insensitive noise. The input/output noise reservoir is phase insensitive if the following conditions are satisfied.

(1) The noise in different frequencies is uncorrelated, i.e.

$$K_{\gamma}(\omega, \omega') = \mathfrak{N}_{\gamma}(\omega)\,2\pi\delta(\omega - \omega'), \tag{16.9}$$

where

$$\mathfrak{N}_{\gamma}(\omega) = \left\langle \delta\overline{b}_{\gamma}^{\dagger}(\omega)\,\delta\overline{b}_{\gamma}(\omega)\right\rangle + \frac{1}{2} \tag{16.10}$$

is the noise strength.

(2) The phases of the fluctuation operators are randomly distributed, so that

$$\langle \delta\overline{b}_{\gamma}(\omega)\,\delta\overline{b}_{\gamma}(\omega')\rangle = 0. \tag{16.11}$$

With this preparation, we are now ready to introduce an important division of the family of linear amplifiers into two classes. A **phase-insensitive amplifier** is defined by the following conditions.

(i) The output field strength, $\left|\left\langle\overline{b}_{\text{out}}(\omega)\right\rangle\right|^2$, is invariant under phase transformations of the input field.

(ii) If the input noise is phase insensitive, so is the output noise.

Condition (i) means that the only effect of a phase shift in the input field—i.e. a rotation of the quadratures—is to produce a corresponding phase shift in the output field. Condition (ii) means that the noise added by the amplifier is randomly distributed in phase. An amplifier is said to be **phase sensitive** if it fails to satisfy either one of these conditions.

In addition to the categories of phase sensitive and phase-insensitive, amplifiers can also be classified according to their physical configuration. In the degenerate OPA the gain medium is enclosed in a resonant cavity, and the input field is coupled into one of the cavity modes. The cavity mode in turn couples to an output mode to produce the amplified signal. This configuration is called a **regenerative amplifier**, which is yet another term borrowed from radio engineering. One way to understand the regenerative amplifier is to visualize the cavity mode as a traveling wave bouncing back and forth between the two mirrors. These waves make many passes through the gain medium before exiting through the output port.

The advantage of greater overall gain, due to multiple passes through the gain medium, is balanced by the disadvantage that the useful gain bandwidth is restricted to the bandwidth of the cavity. This restriction on the bandwidth can be avoided by the simple expedient of removing the mirrors. In this configuration, there are no reflected waves—and therefore no multiple passes through the gain medium—so these devices are called **traveling-wave amplifiers**.

16.2 Regenerative amplifiers

In this section we take advantage of the remarkable versatility of the spontaneous down-conversion process to describe three regenerative amplifiers, two phase insensitive and one phase sensitive.

16.2.1 Phase-insensitive amplifiers

A modification of the degenerate OPA design of Section 15.2 provides two examples of phase-insensitive amplifiers. In the modified design, shown in Fig. 16.1, the signal and idler modes are frequency degenerate, but not copropagating. In the absence of the mirrors M1 and M2, down-conversion of the pump radiation would produce symmetrical cones of light around the pump direction, but this azimuthal symmetry is broken by the presence of the cavity axis joining the two mirrors. This arrangement picks out a single pair of conjugate modes: the idler and the signal.

The boundary conditions at the mirrors define a set of discrete cavity modes, and the fundamental cavity mode—which we will call the idler—is chosen to satisfy the phase-matching condition $\omega_0 = \omega_P/2$. The discrete idler mode is represented by a single operator $a(t)$. On the other hand, the signal mode is a traveling wave with propagation direction determined by the phase-matching conditions in the nonlinear crystal. Thus the signal mode is represented by a continuous family of operators $b_{\text{sig},\Omega}(t)$.

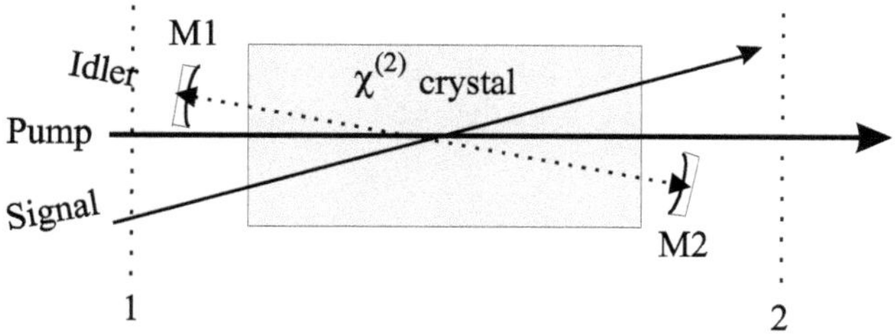

Fig. 16.1 Two examples of phase-insensitive optical amplifiers based on down-conversion in a $\chi^{(2)}$ crystal: (a) taking the signal-mode in- and out-fields as the input and output of the amplifier defines a phase-preserving amplifier; (b) taking the signal-mode in-field as the input and the out-field through mirror M2 as the output defines a phase-conjugating amplifier.

The first step in dividing the world into sample and reservoirs is to identify the sample. From the experimental point of view, the sample in this case evidently consists of the atoms in the nonlinear crystal, combined with the idler mode in the cavity. The theoretical description is a bit simpler, since—as we have seen in Chapter 13—the atoms in the crystal are only virtually excited. This means that the effect of the atoms is completely accounted for by the signal–idler coupling constant; consequently, the sample can be taken to consist of the idler mode alone. There are then three environmental reservoirs: the signal reservoir represented by the operators $b_{\text{sig},\Omega}(t)$ and two noise reservoirs represented by the operators $b_{1,\Omega}(t)$ and $b_{2,\Omega}(t)$ describing radiation entering and leaving the cavity through the mirrors.

Analyzing this model requires a slight modification of the method of in- and out-fields described in Section 14.3. The new feature requiring the modification is the form of the coupling between the idler (sample) mode and the signal (reservoir) mode. This term in the interaction Hamiltonian H_{SE} does not have the generic form of eqn (14.88); instead, it is described by eqn (15.7). In a notation suited to the present discussion:

$$H_{\text{SE}}^{\text{sig-idl}} = i\hbar \int_0^\infty d\Omega \sqrt{\frac{\mathcal{D}(\omega)}{2\pi}} \left\{ v_P(\Omega)\, e^{-i\omega_P t} b_{\text{sig},\Omega}^\dagger a^\dagger - v_P^*(\Omega)\, e^{i\omega_P t} a b_{\text{sig},\Omega} \right\}, \quad (16.12)$$

where $v_P(\Omega)$ is the strength of the coupling—induced by the nonlinear crystal—between the signal mode, the idler mode, and the pump field. The presence of the products $b_{\text{sig},\Omega}^\dagger a^\dagger$ and $a b_{\text{sig},\Omega}$ represents the fact that the signal and idler photons are created and annihilated in pairs in down-conversion.

After including this new term in H_{SE}, the procedures explained in Section 14.3 can be applied to the present problem. The interaction term in eqn (16.12) leads to the modified Heisenberg equations

$$\frac{d}{dt}\overline{a}(t) = \int_0^\infty d\Omega \sqrt{\frac{\mathcal{D}(\omega)}{2\pi}} v_P(\Omega)\, \overline{b}_{\text{sig},\Omega}^\dagger(t) + \sum_{m=1}^{2} \int_0^\infty d\Omega \sqrt{\frac{\mathcal{D}(\omega)}{2\pi}} v_m(\Omega)\, \overline{b}_{m,\Omega}(t), \quad (16.13)$$

$$\frac{d}{dt}\overline{b}_{\text{sig},\Omega}(t) = -i(\Omega - \omega_0)\, \overline{b}_{\text{sig},\Omega}(t) + \sqrt{\frac{\mathcal{D}(\omega)}{2\pi}} v_P(\Omega)\, \overline{a}^\dagger(t), \quad (16.14)$$

where $v_m(\Omega)$ describes the coupling of the idler to the noise modes, and $\overline{a}(t) = a(t)\exp(i\omega_0 t)$, etc. The equations for the noise reservoir operators $b_{m,\Omega}(t)$ have the generic form of eqn (14.89). The retarded and advanced solutions of eqn (16.14) for the signal mode are respectively

$$\overline{b}_{\text{sig},\Omega}(t) = \overline{b}_{\text{sig},\Omega}(t_0)\, e^{-i(\Omega-\omega_0)(t-t_0)} + v_P(\Omega)\int_{t_0}^{t} dt' \overline{a}^{\dagger}(t')\, e^{-i(\Omega-\omega_0)(t-t')} \tag{16.15}$$

and

$$\overline{b}_{\text{sig},\Omega}(t) = \overline{b}_{\text{sig},\Omega}(t_1)\, e^{-i(\Omega-\omega_0)(t-t_1)} - v_P(\Omega)\int_{t}^{t_1} dt' \overline{a}^{\dagger}(t')\, e^{-i(\Omega-\omega_0)(t-t')}. \tag{16.16}$$

The corresponding results for the noise reservoir operators, $\overline{b}_{m,\Omega}(t)$, are given by eqns (14.94) and (14.105).

After substituting the retarded solutions for $\overline{b}_{\text{sig},\Omega}(t)$ and $\overline{b}_{m,\Omega}(t)$ into the equation of motion (16.13), we impose the Markov approximation by assuming that the idler mode is coupled to a broad band of excitations in the two mirror reservoirs and in the signal reservoir. The general discussion in Section 14.3 yields the broadband rule $v_m(\Omega) \sim \sqrt{\kappa_m}$ for the noise modes. The signal mode must be treated differently, since $v_P(\Omega)$ is proportional to the classical pump field, which has a well-defined phase θ_P. In this case the broadband rule is $v_P(\Omega) \sim \sqrt{g_P}\exp(i\theta_P)$, where g_P is positive.

The contributions from the noise reservoirs yield the expected loss term $-\kappa_C \overline{a}(t)/2$, but the contribution from the signal reservoir instead produces a gain term $+g_P \overline{a}(t)/2$. This new feature is another consequence of the fact that the down-conversion mechanism creates and annihilates the signal and idler photons in pairs. Emission of a photon into the continuum signal reservoir can never be reversed; therefore, the associated idler photon can also never be lost. On the other hand, the inverse process—in which a signal–idler pair is annihilated to create a pump photon—does not contribute in the approximation of constant pump strength. Consequently, in the linear approximation the coupling of the signal and idler modes through down-conversion leads to an increase in the strength of both signal and idler fields at the expense of the (undepleted) classical pump field.

After carrying out these calculations, one finds the retarded Langevin equation for the idler mode:

$$\frac{d}{dt}\overline{a}(t) = -\frac{1}{2}(\kappa_C - g_P)\,\overline{a}(t) + \sqrt{g_P}e^{i\theta_P}\overline{b}^{\dagger}_{\text{sig,in}}(t) + \sqrt{\kappa_1}\,\overline{b}_{1,\text{in}}(t) + \sqrt{\kappa_2}\,\overline{b}_{2,\text{in}}(t)\,, \tag{16.17}$$

where

$$\overline{b}_{\text{sig,in}}(t) = \int_{-\infty}^{\infty}\frac{d\Omega}{2\pi}\overline{b}_{\text{sig},\Omega}(t_0)\, e^{-i(\Omega-\omega_0)(t-t_0)} \tag{16.18}$$

is the signal in-field, and the in-fields for the mirrors are given by eqn (14.98). For $g_P > \kappa_C$, eqn (16.17) predicts an exponential growth of the idler field that would violate the weak-field assumptions required for the model. Consequently—just as in the treatment of squeezing in Section 15.2-A—the pump field must be kept below the threshold value ($g_P < \kappa_C$).

We now imitate the empty-cavity analysis of Section 14.3.3 by transforming eqn (16.17) to the frequency domain and solving for $\overline{a}(\omega)$, with the result

$$\overline{a}(\omega) = \frac{e^{i\theta_P}\sqrt{g_P}\,\overline{b}^{\dagger}_{\text{sig,in}}(-\omega) + \sqrt{\kappa_1}\,\overline{b}_{1,\text{in}}(\omega) + \sqrt{\kappa_2}\,\overline{b}_{2,\text{in}}(\omega)}{\frac{1}{2}(\kappa_C - g_P) - i\omega}. \tag{16.19}$$

The input–output relation for the signal mode is obtained by equating the right sides of eqns (16.15) and (16.16) and integrating over Ω to get

$$\overline{b}_{\text{sig,out}}(t) = \overline{b}_{\text{sig,in}}(t) + \sqrt{g_P}\,e^{i\theta_P}\overline{a}^{\dagger}(t) \tag{16.20}$$

in the time domain, or

$$\overline{b}_{\text{sig,out}}(\omega) = \overline{b}_{\text{sig,in}}(\omega) + \sqrt{g_P}\,e^{i\theta_P}\overline{a}^{\dagger}(-\omega) \tag{16.21}$$

in the frequency domain. The input–output relations for the mirror reservoirs are given by the frequency-domain form of eqn (14.109):

$$\overline{b}_{1,\text{out}}(\omega) = \overline{b}_{1,\text{in}}(\omega) - \sqrt{\kappa_1}\,\overline{a}(\omega), \tag{16.22}$$

$$\overline{b}_{2,\text{out}}(\omega) = \overline{b}_{2,\text{in}}(\omega) - \sqrt{\kappa_2}\,\overline{a}(\omega). \tag{16.23}$$

A Phase-transmitting OPA

The first step in defining an amplifier is to decide on the identity of the input and output fields. In other words: What is to be measured? For the first example, we choose the in-field and out-field of the signal mode as the input and output fields, i.e. $\overline{b}_{\text{in}}(\omega) = \overline{b}_{\text{sig,in}}(\omega)$ and $\overline{b}_{\text{out}}(\omega) = \overline{b}_{\text{sig,out}}(\omega)$. The idler field and the two mirror reservoir in-fields are then internal modes of the amplifier. Substituting these identifications and the solution (16.19) into eqn (16.21) yields the amplifier input–output equation

$$\overline{b}_{\text{out}}(\omega) = P(\omega)\,\overline{b}_{\text{in}}(\omega) + \eta(\omega), \tag{16.24}$$

where the coefficient

$$P(\omega) = \frac{\frac{1}{2}(\kappa_C + g_P) - i\omega}{\frac{1}{2}(\kappa_C - g_P) - i\omega} \tag{16.25}$$

has the symmetry property

$$P(\omega) = P^{*}(-\omega), \tag{16.26}$$

and the operator

$$\begin{aligned}\eta(\omega) &= \frac{\sqrt{g_P}\,e^{i\theta_P}\xi^{\dagger}_C(-\omega)}{\frac{1}{2}(\kappa_C - g_P) - i\omega} \\ &= \frac{\sqrt{g_P}\,e^{i\theta_P}}{\frac{1}{2}(\kappa_C - g_P) - i\omega}\left[\sqrt{\kappa_1}\,\overline{b}^{\dagger}_{1,\text{in}}(-\omega) + \sqrt{\kappa_2}\,\overline{b}^{\dagger}_{2,\text{in}}(-\omega)\right]\end{aligned} \tag{16.27}$$

is called the **amplifier noise**.

This result shows that the noise added by the amplifier is entirely due to the noise reservoirs associated with the mirrors. The absence of noise added by the atoms in the nonlinear crystal is a consequence of the fact that the excitations of the atoms are purely virtual. In most applications, only vacuum fluctuations enter through M1 and M2, but the following calculations are valid in the more general situation that both mirrors are coupled to any phase-insensitive noise reservoirs. In particular, the vanishing ensemble average of the noise operator $\eta(\omega)$ implies that the input–output equation for the average field is

$$\left\langle \overline{b}_{\text{out}}(\omega)\right\rangle = P(\omega)\left\langle \overline{b}_{\text{in}}(\omega)\right\rangle . \tag{16.28}$$

Subtracting this equation from eqn (16.24) yields the input–output equation

$$\delta\overline{b}_{\text{out}}(\omega) = P(\omega)\,\delta\overline{b}_{\text{in}}(\omega) + \eta(\omega) \tag{16.29}$$

for the fluctuation operators.

The first step in the proof that this amplifier is phase insensitive is to use eqn (16.28) to show that the effect of a phase transformation applied to the input field is

$$\left\langle \overline{b}'_{\text{out}}(\omega)\right\rangle = P(\omega)\left\langle \overline{b}'_{\text{in}}(\omega)\right\rangle = e^{i\theta}\left\langle \overline{b}_{\text{out}}(\omega)\right\rangle . \tag{16.30}$$

In other words, changes in the phase of the input signal are simply passed through the amplifier. Amplifiers with this property are said to be **phase transmitting**. The field strength $\left|\left\langle \overline{b}_{\text{out}}(\omega)\right\rangle\right|^2$ is evidently unchanged by a phase transformation; therefore the amplifier satisfies condition (i) of Section 16.1.1.

Turning next to condition (ii), we note that the operators $\delta b_{\text{in}}(\omega)$ and $\eta(\omega)$ are linear functions of the uncorrelated reservoir operators $b_{\text{sig},\Omega}(t_0)$ and $b_{m,\Omega}(t_0)$ $(m = 1, 2)$. This feature combines with eqn (16.29) to give

$$K_{\text{out}}(\omega,\omega') = P(\omega)\,P^*(\omega')\,K_{\text{in}}(\omega,\omega') + K_{\text{amp}}(\omega,\omega'), \tag{16.31}$$

where

$$K_{\text{amp}}(\omega,\omega') = \frac{1}{2}\left\langle \eta(\omega)\,\eta^\dagger(\omega') + \eta^\dagger(\omega')\,\eta(\omega)\right\rangle \tag{16.32}$$

is the amplifier–noise correlation function. Since $\eta(\omega)$ is a linear combination of the mirror noise operators, the assumption that the mirror noise is phase insensitive guarantees that

$$K_{\text{amp}}(\omega,\omega') = \mathfrak{N}_{\text{amp}}(\omega)\,2\pi\delta(\omega-\omega'), \tag{16.33}$$

where $\mathfrak{N}_{\text{amp}}(\omega)$ is the amplifier noise strength. If the correlation function $K_{\text{in}}(\omega,\omega')$ satisfies eqn (16.9), then eqns (16.31) and (16.33) guarantee that $K_{\text{out}}(\omega,\omega')$ does also. The output noise strength is then given by

$$\mathfrak{N}_{\text{out}}(\omega) = |P(\omega)|^2\,\mathfrak{N}_{\text{in}}(\omega) + \mathfrak{N}_{\text{amp}}(\omega). \tag{16.34}$$

It is also necessary to verify that the output noise satisfies eqn (16.11), when the input noise does. This is an immediate consequence of the phase insensitivity of the amplifier noise and the input–output equation (16.29), which together yield

$$\langle \delta b_{\text{out}}(\omega)\,\delta b_{\text{out}}(\omega')\rangle = P(\omega)\,P(\omega')\,\langle \delta b_{\text{in}}(\omega)\,\delta b_{\text{in}}(\omega')\rangle . \tag{16.35}$$

Putting all this together shows that the amplifier is phase insensitive, since it satisfies conditions (i) and (ii) from Section 16.1.1.

For this amplifier, it is reasonable to define the gain as the ratio of the output field strength to the input field strength:

$$G(\omega) = \frac{\left|\left\langle \overline{b}_{\text{out}}(\omega)\right\rangle\right|^2}{\left|\left\langle \overline{b}_{\text{in}}(\omega)\right\rangle\right|^2} = \frac{\left|\left\langle b_{\text{out}}(\omega+\omega_0)\right\rangle\right|^2}{\left|\left\langle b_{\text{in}}(\omega+\omega_0)\right\rangle\right|^2}\,. \tag{16.36}$$

Using eqn (16.28) yields the explicit expression

$$G(\omega-\omega_0) = \frac{\left(\kappa_C+g_P\right)^2/4+\left(\omega-\omega_0\right)^2}{\left(\kappa_C-g_P\right)^2/4+\left(\omega-\omega_0\right)^2}\,, \tag{16.37}$$

which displays the expected peak in the gain at the resonance frequency ω_0. An alternative procedure is to define the gain in terms of the quadrature operators, and then to show—see Exercise 16.1—that the gain is the same for all quadratures.

B Phase-conjugating OPA

The crucial importance of the choice of input and output fields is illustrated by using the apparatus shown in Fig. 16.1 to define a quite different amplifier. In this version the input field is still the signal-mode in-field $\overline{b}_{\text{sig,in}}(\omega)$, but the output field is the out-field $\overline{b}_{2,\text{out}}(\omega)$ for the mirror M2. The internal modes are the same as before. The input–output equation for this amplifier—which is derived from eqn (16.23) by using the solution (16.19) and the identifications $\overline{b}_{\text{in}}(\omega) = \overline{b}_{\text{sig,in}}(\omega)$ and $\overline{b}_{\text{out}}(\omega) = \overline{b}_{2,\text{out}}(\omega)$—has the form

$$\overline{b}_{\text{out}}(\omega) = C(\omega)\,e^{i\theta_P}\,\overline{b}_{\text{in}}^{\dagger}(-\omega) + \eta(\omega)\,. \tag{16.38}$$

The coefficient $C(\omega)$ and the amplifier noise operator are respectively given by

$$C(\omega) = -\frac{\sqrt{\kappa_2 g_P}}{\frac{1}{2}\left(\kappa_C-g_P\right)-i\omega} \tag{16.39}$$

and

$$\eta(\omega) = \frac{\frac{1}{2}\left(\kappa_1-\kappa_2-g_P\right)-i\omega}{\frac{1}{2}\left(\kappa_C-g_P\right)-i\omega}\,\overline{b}_{2,\text{in}}(\omega) - \frac{\sqrt{\kappa_1\kappa_2}}{\frac{1}{2}\left(\kappa_C-g_P\right)-i\omega}\,\overline{b}_{1,\text{in}}(\omega)\,. \tag{16.40}$$

The important difference from eqn (16.24) is that the output field depends on the adjoint of the input field. Note that $C(\omega)$ has the same symmetry as $P(\omega)$:

$$C(\omega) = C^*(-\omega)\,. \tag{16.41}$$

The ensemble average of eqn (16.38) is

$$\left\langle \overline{b}_{\text{out}}(\omega)\right\rangle = C(\omega)\left\langle \overline{b}_{\text{in}}^{\dagger}(-\omega)\right\rangle, \tag{16.42}$$

so the phase transformation $\overline{b}_{\text{in}}(\omega) \to \overline{b}'_{\text{in}}(\omega) = \exp(i\theta)\,\overline{b}_{\text{in}}(\omega)$ results in

$$\left\langle \overline{b}'_{\text{out}}(\omega)\right\rangle = e^{-i\theta}C(\omega)\left\langle \overline{b}_{\text{in}}^{\dagger}(-\omega)\right\rangle = e^{-i\theta}\left\langle \overline{b}_{\text{out}}(\omega)\right\rangle. \tag{16.43}$$

Instead of being passed through the amplifier unchanged, the phasor $\exp(i\theta)$ is replaced by its conjugate. A device with this property is called a **phase-conjugating amplifier**.

This amplifier nevertheless satisfies condition (i) of Section 16.1.1, since

$$\left|\left\langle \overline{b}'_{\text{out}}(\omega)\right\rangle\right|^2 = \left|\left\langle \overline{b}_{\text{out}}(\omega)\right\rangle\right|^2 . \tag{16.44}$$

The argument used in Section 16.2.1-A to establish condition (ii) works equally well here; therefore, the alternative design also defines a phase-insensitive amplifier. The form of the input–output relation in this case suggests that the gain should be defined as

$$G(\omega) = \frac{\left|\left\langle \overline{b}_{\text{out}}(\omega)\right\rangle\right|^2}{\left|\left\langle \overline{b}^{\dagger}_{\text{in}}(-\omega)\right\rangle\right|^2} = |C(\omega)|^2 = \frac{\kappa_2 g_P}{\frac{1}{4}(\kappa_C - g_P)^2 + \omega^2} . \tag{16.45}$$

16.2.2 Phase-sensitive OPA

In the design shown in Fig. 16.2 the fields entering and leaving the cavity through the mirror M1 are designated as the input and output fields respectively, i.e. $\overline{b}_{\text{in}}(t) = \overline{b}_{1,\text{in}}(t)$ and $\overline{b}_{\text{out}}(t) = \overline{b}_{1,\text{out}}(t)$. The degenerate signal and idler modes of the cavity and the input field $\overline{b}_{2,\text{in}}(t)$ for the mirror M2 are the internal modes of the amplifier. The input–output relation is obtained from eqn (15.117) by applying this identification of the input and output fields:

$$\overline{b}_{\text{out}}(\omega) = P(\omega)\,\overline{b}_{\text{in}}(\omega) + C(\omega)\,e^{i\theta_P}\,\overline{b}^{\dagger}_{\text{in}}(-\omega) + \eta(\omega) . \tag{16.46}$$

The *phase-transmitting* and *phase-conjugating* coefficients are respectively

$$P(\omega) = 1 - \frac{\kappa_1(\kappa_C/2 - i\omega)}{(\kappa_C/2 - i\omega)^2 - |\Omega_P|^2} \tag{16.47}$$

and

$$C(\omega) = -\frac{|\Omega_P|\,\kappa_1}{(\kappa_C/2 - i\omega)^2 - |\Omega_P|^2} . \tag{16.48}$$

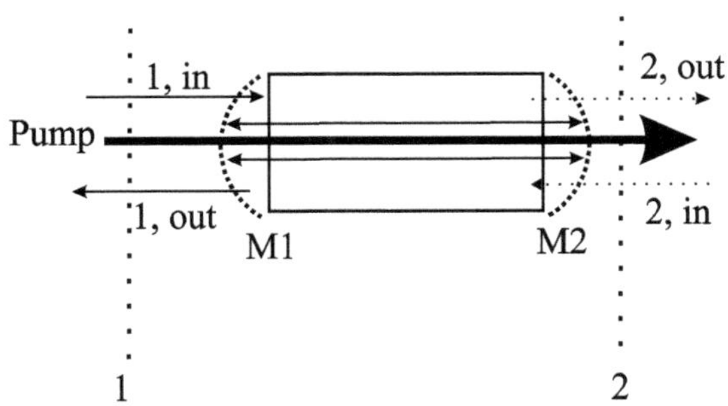

Fig. 16.2 A phase-sensitive amplifier based on the degenerate OPA. The heavy solid arrow represents the classical pump; the thin solid arrows represent the input and output modes for the mirror M1; and the dashed arrows represent the input and output for the mirror M2.

The functions $P(\omega)$ and $C(\omega)$ satisfy eqns (16.26) and (16.41) respectively. The amplifier noise operator,

$$\eta(\omega) = -\frac{\sqrt{\kappa_1\kappa_2}}{(\kappa_C/2 - i\omega)^2 - |\Omega_P|^2}\left[(\kappa_C/2 - i\omega)\,\overline{b}_{2,\mathrm{in}}(\omega) + \Omega_P \overline{b}^{\dagger}_{2,\mathrm{in}}(-\omega)\right], \tag{16.49}$$

only depends on the reservoir operators associated with the mirror M2, so the amplifier noise is entirely caused by vacuum fluctuations passing through the unused port at M2.

According to eqn (16.46), the output field strength is

$$\begin{aligned}\left|\langle \overline{b}_{\mathrm{out}}(\omega)\rangle\right|^2 &= |P(\omega)|^2\left|\langle \overline{b}_{\mathrm{in}}(\omega)\rangle\right|^2 + |C(\omega)|^2\left|\left\langle \overline{b}^{\dagger}_{\mathrm{in}}(-\omega)\right\rangle\right|^2 \\ &\quad + 2\,\mathrm{Re}\left[P(\omega)\,C^*(\omega)\,\langle \overline{b}_{\mathrm{in}}(\omega)\rangle \left\langle \overline{b}^{\dagger}_{\mathrm{in}}(-\omega)\right\rangle^*\right].\end{aligned} \tag{16.50}$$

We first test condition (i) of Section 16.1.1, by applying the phase transformation (16.5) to the input field and evaluating the difference between the transformed and the original output intensities to get

$$\begin{aligned}\delta\left|\langle \overline{b}_{\mathrm{out}}(\omega)\rangle\right|^2 &= \left|\left\langle \overline{b}'_{\mathrm{out}}(\omega)\right\rangle\right|^2 - \left|\langle \overline{b}_{\mathrm{out}}(\omega)\rangle\right|^2 \\ &= 2\,\mathrm{Re}\left[\left(e^{2i\theta} - 1\right)P(\omega)\,C^*(\omega)\,\langle \overline{b}_{\mathrm{in}}(\omega)\rangle \left\langle \overline{b}^{\dagger}_{\mathrm{in}}(-\omega)\right\rangle^*\right].\end{aligned} \tag{16.51}$$

Satisfying condition (i) would require the right side of this equation to vanish as an identity in θ. The generic assumption (16.2) combined with the explicit forms of the functions $P(\omega)$ and $C(\omega)$ makes this impossible; therefore, the amplifier is phase sensitive.

This feature is a consequence of the fact that $P(\omega)$ and $C(\omega)$ are both nonzero, so that the right side of eqn (16.46) depends jointly on $\overline{b}_{\mathrm{in}}(\omega)$ and $\overline{b}^{\dagger}_{\mathrm{in}}(-\omega)$. A straightforward calculation shows that condition (ii) of Section 16.1.1 is also violated, even for the simple case that the reservoir for the mirror M2 is the vacuum. Choosing an appropriate definition of the gain for a phase-sensitive amplifier is a bit trickier than for the phase-insensitive cases, so this step will be postponed to the general treatment in Section 16.4.

The alert reader will have noticed that the amplified signal is propagating backwards toward the source of the input signal. Devices of this kind are sometimes called **reflection amplifiers**. This is not a useful feature for communications applications; therefore, it is necessary to reverse the direction of the amplifier output so that it propagates in the same direction as the input signal. Mirrors will not do for this task, since they would interfere with the input. One solution is to redirect the amplifier output by using an optical circulator, as described in Section 8.6. This device will redirect the output signal, but it will not interfere with the input signal or add further noise.

16.3 Traveling-wave amplifiers

The regenerative amplifiers discussed above enhance the nonlinear interaction for a relatively weak cw pump beam by means of the resonant cavity formed by the mirrors M1 and M2. This approach has the disadvantage of restricting the useful bandwidth to that of the cavity. An alternative method is to remove the mirrors M1 and M2 to get the configuration shown in Fig. 16.3, but this experimental simplification inevitably comes at the expense of some theoretical complication.

The mirrors in the regenerative amplifiers perform two closely related functions. The first is to guarantee that the field inside the cavity is a superposition of a discrete set of cavity modes. In practice, the design parameters are chosen so that only one cavity mode is excited. The position dependence of the field is then entirely given by the corresponding mode function; in effect, the cavity is a zero-dimensional system. The second function—which follows from the first—is to justify the sample–reservoir model that treats the discrete modes inside the cavity and the continuum of reservoir modes outside the cavity as kinematically-independent degrees of freedom.

Removing the mirrors eliminates both of these conceptual simplifications. Since there are no discrete cavity modes, each of the continuum of external modes propagates through the amplifier and interacts with the gain medium. Thus all field modes are reservoir modes, and the sample consists of the atoms in the gain medium.

The interaction of the field with the gain medium could be treated by generalizing the scattering description of passive, linear devices developed in Section 8.2, but this approach would be quite complicated in the present application. The fact that the sample occupies a fixed interval, say $0 \leqslant z \leqslant L_S$, along the propagation (z) axis violates translation invariance and therefore conservation of momentum. Consequently, the scattering matrix for the amplifier connects each incident plane wave, $\exp(ikz)$, to a continuum of scattered waves $\exp(ik'z)$.

We will avoid this complication by employing a position–space approach that closely resembles the classical theory of parametric amplification (Yariv, 1989, Chap. 17). This technique can also be regarded as the Heisenberg-picture version of a method developed to treat squeezing in a traveling-wave configuration (Deutsch and Garrison, 1991*b*).

16.3.1 Laser amplifier

As a concrete example, we consider a sample composed of a collection of three-level atoms—with the level structure displayed in Fig. 16.4—which is made into a gain

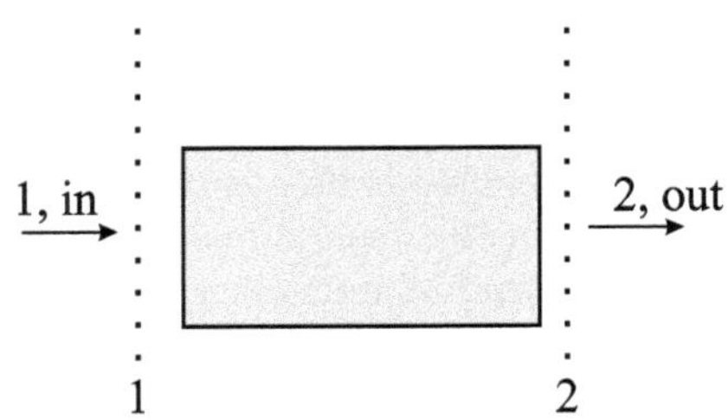

Fig. 16.3 A *black box* schematic of a traveling-wave amplifier. The shaded box indicates the gain medium and the fields at the two ports are the input and output values of the signal. The vacuum fluctuations entering port 2 are not indicated, since they do not couple to the signal.

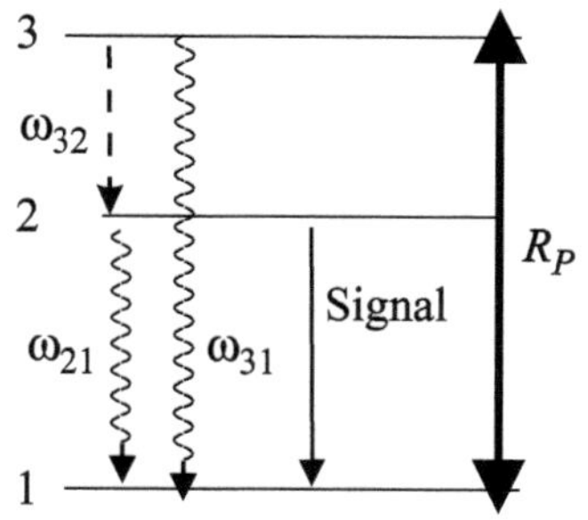

Fig. 16.4 A three-level atom with a population inversion between levels 1 and 2, maintained by an incoherent pump (dark double arrow) with rate R_P. The solid arrow, the dashed arrow, and the wavy arrows respectively represent the amplified signal transition, a nonradiative decay, and spontaneous emission.

medium by maintaining a population inversion between levels 1 and 2 through the use of the incoherent pumping mechanism described in Section 14.5. By virtue of the cylindrical shape of the gain medium, the end-fire modes—i.e. field modes with frequencies $\omega \simeq \omega_{21}$ and propagation vectors, $\mathbf{k}$, lying in a narrow cone around the axis of the cylinder—will be preferentially amplified.

This new feature requires a modification of the reservoir assignment used for the pumping calculation. The noise reservoir previously associated with the spontaneous emission $2 \to 1$ is replaced by two reservoirs: (1) a noise reservoir associated with spontaneous emission into modes with propagation vectors outside the end-fire cone; and (2) a signal reservoir associated with the end-fire modes.

In the undepleted pump approximation, the back action of the atoms on the pump field can be ignored. This certainly cannot be done for the interaction with the signal reservoir; after all, the action of the gain medium on the signal is the whole purpose of the device. Thus the coupling of the entire collection of atoms to the signal reservoir must be included by using the interaction Hamiltonian

$$H'_{S1} = -\sum_{n} \left[S_{21}^{(n)}(t)\, \mathbf{d}_{21} \cdot \mathbf{E}^{(+)}(\mathbf{r}_n, t) + \mathrm{HC} \right], \tag{16.52}$$

where the sum runs over the atoms in the sample and the coordinate, $\mathbf{r}_n$, of the nth atom is treated classically.

The description of the signal reservoir given above amounts to the assumption that the Heisenberg-picture density operator for the input signal is a paraxial state with respect to the z-axis; consequently, the contribution of the end-fire modes to the field operator can be represented in terms of the slowly-varying envelope operators $\phi_s(\mathbf{r}, t)$ appearing in eqn (7.33). We will assume that the amplifier has been designed so that only one polarization will be amplified; consequently, only one operator $\phi(\mathbf{r}, t)$ will be needed.

Turning next to the input signal, we recall that a paraxial state is characterized by transverse and longitudinal length scales $\Lambda_{\top} = 1/(\theta k_0)$ and $\Lambda_{\parallel} = 1/\left(\theta^2 k_0\right)$ respectively, where θ is the opening angle of the paraxial ray bundle. The scale lengths $\Lambda_{\top}$ and $\Lambda_{\parallel}$ correspond respectively to the spot size and Rayleigh range of a classical Gaussian beam. We choose θ so that $\Lambda_{\top} > 2R_S$ and $\Lambda_{\parallel} \gg L_S$, where R_S and L_S are respectively the radius and length of the cylinder. This allows a further simplification in which diffraction is ignored and the envelope operator is approximated by

$$\phi(\mathbf{r},t) = \frac{1}{\sqrt{\sigma}}\phi(z,t), \tag{16.53}$$

where $\sigma = \pi R_S^2$. In this 1D approximation, the field expansion (7.33) and the commutation relation (7.35) are respectively replaced by

$$\mathbf{E}^{(+)}(\mathbf{r},t) = i\sqrt{\frac{\hbar\omega_0 (v_{g0}/c)}{2\epsilon_0 n_0 \sigma}}\mathbf{e}_0\phi(z,t)\, e^{i(k_0 z - \omega_0 t)} \tag{16.54}$$

and

$$\left[\phi(z,t), \phi^\dagger(z',t)\right] = \delta(z-z'). \tag{16.55}$$

The discretely distributed atoms and the continuous field are placed on a more even footing by introducing the spatially coarse-grained operator density

$$S_{qp}(z,t) = \frac{1}{\Delta z}\sum_n S_{qp}^{(n)}(t)\,\chi(z-z_n). \tag{16.56}$$

The averaging interval Δz is chosen to satisfy the following two conditions. (1) A slab with volume $\sigma\Delta z$ contains many atoms. (2) The envelope operator $\phi(z,t)$ is essentially constant over an interval of length Δz. The function

$$\chi(z-z_n) = \theta(\Delta z/2 - z + z_n)\,\theta(z - z_n + \Delta z/2) \tag{16.57}$$

serves to confine the n-sum to the atoms in a slab of thickness Δz centered at z. The atomic envelope operators are defined by

$$S_{qp}(z,t) = \overline{S}_{qp}(z,t)\, e^{i\omega_{qp}t} e^{i[\psi_q(z,t) - \psi_p(z,t)]}, \tag{16.58}$$

where the phases satisfy

$$\psi_2(z,t) - \psi_1(z,t) = \Delta_0 t - k_0 z. \tag{16.59}$$

Using this notation, together with eqn (16.54), allows us to rewrite eqn (16.52) as

$$H'_{S1} = -i\hbar\int_0^{L_S} dz\left[\mathfrak{f}\,\overline{S}_{21}(z,t)\,\phi(z,t) - \text{HC}\right], \tag{16.60}$$

where

$$\mathfrak{f} \equiv \sqrt{\frac{\hbar(v_{g0}/c)\,\omega_0}{2\epsilon_0\sigma}}\frac{\mathbf{d}_{21}\cdot\mathbf{e}_0}{\hbar} \tag{16.61}$$

is the coupling constant.

The total electromagnetic part of the Hamiltonian for this 1D model is, therefore,

$$H_{\text{em}} = \int_{-\infty}^{\infty} dz\phi^\dagger(z,t)\, v_{g0}\frac{\hbar}{i}\nabla_z\phi(z,t) + H'_{S1}. \tag{16.62}$$

This leads to the corresponding Heisenberg equation

$$\left(\frac{\partial}{\partial t} + v_{g0}\frac{\partial}{\partial z}\right)\phi(z,t) = \mathfrak{f}^*\overline{S}_{12}(z,t) \quad \text{for } 0 \leqslant z \leqslant L_S, \tag{16.63}$$

$$\left(\frac{\partial}{\partial t} + v_{g0}\frac{\partial}{\partial z}\right)\phi(z,t) = 0 \quad \text{for } z < 0 \text{ or } z > L_S \tag{16.64}$$

for the field.

The atomic operators are coupled to the reservoirs describing the incoherent pump and spontaneous emission into off-axis modes; therefore, we insert eqn (16.60) into the coarse-grained version of eqn (14.177) to find

$$\frac{d}{dt}\overline{S}_{12}(z,t) = [i\Delta_0 - \Gamma_{12}]\,\overline{S}_{12}(z,t) - \mathfrak{f}\left\{\overline{S}_{11}(z,t) - \overline{S}_{22}(z,t)\right\}\phi(z,t) + \xi_{12}(z,t)\,. \tag{16.65}$$

The coarse-grained noise operator

$$\xi_{12}(z,t) = \frac{1}{\Delta z}\sum_n \xi_{12}^{(n)}(t)\,\chi(z - z_n) \tag{16.66}$$

has the correlation function

$$\left\langle \xi_{12}(z,t)\,\xi_{12}^{\dagger}(z',t')\right\rangle = n_{\mathrm{at}}\sigma C_{12,12}\delta(t-t')\,\delta(z-z')\,, \tag{16.67}$$

where $\delta(z-z')$ is a coarse-grained delta function, n_{at} is the density of atoms, and $C_{12,12}$ is an element of the noise correlation matrix discussed in Section 14.6.2.

In the strong-pump limit, the dephasing rate $\Gamma_{12} = (w_{21} + R_P)/2$ is large compared to the other terms in eqn (16.65); therefore, applying the adiabatic elimination rule (11.187) provides the approximate solution

$$\overline{S}_{12}(z,t) = \mathfrak{f}\,\frac{\overline{S}_{22}(z,t) - \overline{S}_{11}(z,t)}{\Gamma_{12} - i\Delta_0}\phi(z,t) + \frac{\xi_{12}(z,t)}{\Gamma_{12} - i\Delta_0}\,. \tag{16.68}$$

We have to warn the reader that this procedure is something of a swindle, since $\xi_{12}(z,t)$ is not a slowly-varying function of t. Fortunately, the result can be justified—see Exercise 16.3—by interpreting $\delta(t-t')$ in eqn (16.67) as an even coarser-grained delta function, that only acts on test functions that vary slowly on the dephasing time scale $T_{12} = 1/\Gamma_{12}$.

In the linear approximation for eqn (16.68), the operator $\overline{S}_{22}(z,t) - \overline{S}_{11}(z,t)$ can be simplified in two ways. The first is to neglect the small quantum fluctuations, i.e. to replace the operator by its average $\langle \overline{S}_{22}(z,t) - \overline{S}_{11}(z,t)\rangle$. The next step is to solve the averaged form of the operator Bloch equations (14.174)–(14.177), with the approximation that $H'_{S1} = 0$. The result is

$$\overline{S}_{22}(z,t) - \overline{S}_{11}(z,t) \approx \langle \overline{S}_{22}(z,t) - \overline{S}_{11}(z,t)\rangle = n_{\mathrm{at}}\sigma D\,, \tag{16.69}$$

where D is the steady-state inversion for a single atom. With these approximations, the propagation equation (16.63) becomes

$$\left(\frac{\partial}{\partial t} + v_{g0}\frac{\partial}{\partial z}\right)\phi(z,t) = \frac{|\mathfrak{f}|^2 n_{\mathrm{at}}\sigma D}{\Gamma_{12} - i\Delta_0}\phi(z,t) + \frac{\mathfrak{f}^*}{\Gamma_{12} - i\Delta_0}\xi_{12}(z,t)\,. \tag{16.70}$$

This equation is readily solved by transforming to the **wave coordinates**:

$$\begin{aligned} \tau &= t - z/v_{g0} \quad \text{(the retarded time for the signal wave)}\,, \\ \mathfrak{z} &= z\,, \end{aligned} \tag{16.71}$$

to get

$$\frac{\partial}{\partial \mathfrak{z}}\phi(\mathfrak{z},\tau) = g\phi(\mathfrak{z},\tau) + \frac{f^*}{v_{g0}(\Gamma_{12} - i\Delta_0)}\xi_{12}(\mathfrak{z},\tau), \tag{16.72}$$

where

$$g = \frac{|\mathfrak{f}|^2}{v_{g0}}\frac{n_{\text{at}}\sigma D}{[\Gamma_{12} - i\Delta_0]} = \frac{k_0 |\mathbf{d}_{21}\cdot\mathbf{e}_0|^2 n_{\text{at}} D}{2\epsilon_0\hbar[\Gamma_{12} - i\Delta_0]} \tag{16.73}$$

is the (complex) small-signal gain. The retarded time τ can be treated as a parameter in eqn (16.72), so the solution is

$$\phi(\mathfrak{z},\tau) = \phi(0,\tau)e^{g\mathfrak{z}} + \frac{f^*}{v_{g0}(\Gamma_{12} - i\Delta_0)}\int_0^{\mathfrak{z}} d\mathfrak{z}_1 e^{g(\mathfrak{z}-\mathfrak{z}_1)}\xi_{12}(\mathfrak{z}_1,\tau), \tag{16.74}$$

which has the form

$$\phi(z,t) = \phi(0,t - z/v_{g0})e^{gz} + \frac{f^*}{v_{g0}(\Gamma_{12} - i\Delta_0)}\int_0^z dz_1 e^{g(z-z_1)}\xi_{12}\left(z_1, t - \frac{z - z_1}{v_{g0}}\right) \tag{16.75}$$

in the laboratory coordinates (z, t).

Setting $z = L_S$ and letting $t \to t + L_S/v_{g0}$ gives the field value at the output face:

$$\phi(L_S, t + L_S/v_{g0}) = \phi(0,t)e^{gL_S} + \frac{f^*}{v_{g0}(\Gamma_{12} - i\Delta_0)}\int_0^{L_S} dz_1 e^{g(L_S-z_1)}\xi_{12}\left(z_1, t + \frac{z_1}{v_{g0}}\right). \tag{16.76}$$

In order to recover the standard form for input–output relations we introduce the representations

$$\phi(z,t) = \frac{1}{\sqrt{v_{g0}}}\int\frac{d\omega}{2\pi}b_{\text{in}}(\omega)e^{-i\omega(t-z/v_{g0})} \quad \text{for } z < 0 \tag{16.77}$$

and

$$\phi(z,t) = \frac{1}{\sqrt{v_{g0}}}\int\frac{d\omega}{2\pi}b_{\text{out}}(\omega)e^{-i\omega(t-z/v_{g0})} \quad \text{for } z > L_S, \tag{16.78}$$

for the solutions of eqn (16.64) outside the crystal. The factor $1/\sqrt{v_{g0}}$ is inserted to guarantee that the commutation relation (16.55) for $\phi(z,t)$ and the standard input–output commutation relations

$$\left[b_\gamma(\omega), b_\gamma^\dagger(\omega')\right] = 2\pi\delta(\omega - \omega') \quad (\gamma = \text{in}, \text{out}) \tag{16.79}$$

are both satisfied. Substituting eqns (16.77) and (16.78) into eqn (16.76) and carrying out a Fourier transform produces the input–output relation

$$b_{\text{out}}(\omega) = e^{gL_S}b_{\text{in}}(\omega) + \eta(\omega), \tag{16.80}$$

where

$$\eta(\omega) = \frac{f^*}{\sqrt{v_{g0}}(\Gamma_{12} - i\Delta_0)}\int_0^{L_S} dz_1 e^{g(L_S-z_1)}e^{-i\omega z_1/v_{g0}}\xi_{12}(z_1,\omega) \tag{16.81}$$

is the amplifier noise operator. By using the frequency-domain form of eqn (16.67), one can show that the noise correlation function is

$$\begin{aligned} K_{\mathrm{amp}}(\omega, \omega') &= \frac{1}{2}\left\langle \eta(\omega)\eta^{\dagger}(\omega') + \eta^{\dagger}(\omega)\eta(\omega')\right\rangle \\ &= \mathfrak{N}_{\mathrm{amp}} 2\pi\delta(\omega - \omega'), \end{aligned} \tag{16.82}$$

where the noise strength is

$$\mathfrak{N}_{\mathrm{amp}} = \frac{n_{\mathrm{at}} k_0 \left|\mathbf{d}_{21} \cdot \mathbf{e}_0\right|^2}{2\epsilon_0 \hbar \left(\Gamma_{12}^2 + \Delta_0^2\right)} \frac{e^{2gL_S} - 1}{2g} \frac{1}{2}\left(C_{12,12} + C_{21,21}\right). \tag{16.83}$$

Comparing eqn (16.80) to eqn (16.46) shows that $P(\omega) = e^{gL_S}$ and $C(\omega) = 0$. Consequently, this amplifier is phase insensitive and phase transmitting.

16.3.2 Traveling-wave OPA

For this example, we return to the down-conversion technique by removing the mirrors from the phase-sensitive design shown in Fig. 16.2. Even without the mirrors, appropriately cutting the ends of the crystal will guarantee that the pump beam and the degenerate signal and idler beams copropagate along the length of the crystal. We assume a Gaussian pump beam with spot size w_0 and Rayleigh range Z_R focussed on a nonlinear crystal with radius R_S and length L_S.

If $w_0 > 2R_S$ and $Z_R \gg L_S$, the effects of diffraction are negligible; consequently, the problem is effectively one-dimensional. In this limit, the classical pump field can be expressed as

$$\boldsymbol{\mathcal{E}}_P(\mathbf{r}, t) = \mathbf{e}_P \left|\mathcal{E}_{P0}\right| e^{i\theta_P} f_P\left(t - z/v_{gP}\right) e^{i(k_P z - \omega_P t)}, \tag{16.84}$$

where we have assumed that the medium outside the crystal is linearly index matched. The temporal shape of the pump pulse is described by the function $f_P(\tau)$, with maximum value $f_P(0) = 1$ and pulse duration τ_P. In the long-pulse limit, $\tau_P \to \infty$, the problem is further simplified by setting $f_P\left(t - z/v_{gP}\right) = 1$.

In the 1D limit the signal–idler mode is described by a paraxial state, so the field can again be represented by eqn (16.54), with $\omega_0 = \omega_P/2$. The polarization of the signal–idler mode is fixed, relative to that of the pump, by the phase-matching conditions in the nonlinear crystal. Applying the 1D approximation and the long-pulse limit to the expressions (7.39) and (13.30) yields the effective field Hamiltonian

$$H_{\mathrm{em}} = \int_{-\infty}^{\infty} dz \phi^{\dagger}(z, t) v_{g0} \frac{\hbar}{i} \nabla_z \phi(z, t) + \frac{\hbar}{2} g^{(3)} \int_0^{L_S} dz \left\{e^{-i\theta_P} \phi^2(z, t) + \mathrm{HC}\right\}. \tag{16.85}$$

The special form of the interaction Hamiltonian—which represents the pair-production aspect of down-conversion—produces a propagation equation

$$\left(\frac{\partial}{\partial t} + v_{g0}\frac{\partial}{\partial z}\right)\phi(z, t) = -ig^{(3)} e^{i\theta_P} \phi^{\dagger}(z, t) \tag{16.86}$$

that couples the field $\phi(z,t)$ to its adjoint $\phi^\dagger(z,t)$. This means that the propagation equation and its adjoint must be solved together. In the wave coordinates defined by eqn (16.71) the equations to be solved are

$$\frac{\partial}{\partial \mathfrak{z}}\phi(\mathfrak{z},\tau) = -ige^{i\theta_P}\phi^\dagger(\mathfrak{z},\tau), \tag{16.87}$$

$$\frac{\partial}{\partial \mathfrak{z}}\phi^\dagger(\mathfrak{z},\tau) = ige^{-i\theta_P}\phi(\mathfrak{z},\tau), \tag{16.88}$$

where $g = g^{(3)}/v_{g0}$ is the weak signal gain. Since the retarded time τ only appears as a parameter, these equations can be solved by standard techniques to find

$$\phi(z,t) = \phi(0,t-z/v_{g0})\cosh(gz) - ie^{i\theta_P}\phi^\dagger(0,t-z/v_{g0})\sinh(gz), \tag{16.89}$$

where we have reverted to the original (z,t)-variables. Thus the solution at z,t is expressed in terms of the field operators evaluated at the input face, $z=0$, and the retarded time $\tau = t - z/v_{g0}$.

The time-domain, input–output relation for the traveling-wave amplifier is obtained by evaluating this solution at the output face, $z = L_S$, and letting $t \to t + L_S/v_{g0}$:

$$\phi(L_S, t + L_S/v_{g0}) = \phi(0,t)\cosh(gL_S) - ie^{i\theta_P}\phi^\dagger(0,t)\sinh(gL_S). \tag{16.90}$$

Fourier transforming this equation yields

$$e^{-iL_S/v_{g0}}\phi(L_S,\omega) = \phi(0,\omega)\cosh(gL_S) - ie^{i\theta_P}\phi^\dagger(0,-\omega)\sinh(gL_S), \tag{16.91}$$

which can be brought into the standard form for input–output relations by using the representations (16.77) and (16.78) to find

$$b_{\text{out}}(\omega) = Pb_{\text{in}}(\omega) - ie^{i\theta_P}Cb_{\text{in}}^\dagger(0,-\omega), \tag{16.92}$$

with

$$P = \cosh(gL_S) \quad \text{and} \quad C = \sinh(gL_S). \tag{16.93}$$

Comparing this to eqn (16.46) reveals two things: (1) the amplifier is phase sensitive; and (2) the noise operator is missing! In other words, the degenerate, traveling-wave, parametric amplifier is intrinsically noiseless. This does not mean that the right-to-left propagating vacuum fluctuations entering port 2 have been magically eliminated; rather, they do not contribute to the output noise because they are not scattered into the left-to-right propagating signal–idler mode.

16.4 General description of linear amplifiers

We now turn from the examples considered above to a general description of the class of single-input, single-output linear amplifiers introduced at the beginning of this chapter. This will be a black box treatment with no explicit assumptions about the internal structure of the amplifier.

At a given time t, $b_{\text{in}}(t)$ and $b_{\text{out}}(t)$ are annihilation operators for photons in the input and output modes respectively. The basic assumption for linear amplifiers is that $b_{\text{out}}(t)$ can be expressed as a linear combination of the input-mode creation and annihilation operators $b_{\text{in}}^{\dagger}(t')$ and $b_{\text{in}}(t')$, for times $t' < t$, plus an operator representing additional noise contributed by the amplifier. The mathematical statement of this physical assumption is

$$b_{\text{out}}(t) = \int dt' P(t-t')\, b_{\text{in}}(t') + \int dt' C(t-t')\, b_{\text{in}}^{\dagger}(t') + \eta(t) . \tag{16.94}$$

Carrying out a Fourier transform, and combining the convolution theorem (A.55) with the representation (14.114) leads to the frequency-domain form

$$\overline{b}_{\text{out}}(\omega) = P(\omega)\,\overline{b}_{\text{in}}(\omega) + C(\omega)\,\overline{b}_{\text{in}}^{\dagger}(-\omega) + \eta(\omega) . \tag{16.95}$$

Since the right side involves both $\overline{b}_{\text{in}}(\omega)$ and $\overline{b}_{\text{in}}^{\dagger}(-\omega)$, the adjoint equation

$$\overline{b}_{\text{out}}^{\dagger}(-\omega) = C^{*}(-\omega)\,\overline{b}_{\text{in}}(\omega) + P^{*}(-\omega)\,\overline{b}_{\text{in}}^{\dagger}(-\omega) + \eta^{\dagger}(-\omega) \tag{16.96}$$

is also required. This construction guarantees that the amplifier noise operator $\eta(\omega)$ only depends on the internal modes of the amplifier.

The input–output relations (16.46) for the phase-sensitive amplifier described in Section 16.2.2 have the form of eqns (16.95) and (16.96), except for the explicit phase factor $\exp(i\theta_P)$ associated with the particular pumping mechanism for that example. This blotch can be eliminated by carrying out the uniform phase transformation

$$\overline{b}_{\text{out}}(\omega) = \overline{b}'_{\text{out}}(\omega)\, e^{i\theta_P/2} , \quad \overline{b}_{\text{in}}(\omega) = \overline{b}'_{\text{in}}(\omega)\, e^{i\theta_P/2} , \quad \overline{b}_{2,\text{in}}(\omega) = \overline{b}'_{2,\text{in}}(\omega)\, e^{i\theta_P/2} . \tag{16.97}$$

When expressed in terms of the transformed (primed) operators the input–output relation (16.46) is scrubbed clean of the offending phase factor.

This kind of maneuver is usually expressed in a condensed form something like this: let $\overline{b}_{\text{out}}(\omega) \to \overline{b}_{\text{out}}(\omega)\exp(i\theta_P/2)$, etc. This is all very well, except for the following puzzle: What has happened to the reference phase that was supposed to be provided by the pump? The answer is that the input–output relation is only half the story. The rest is provided by the density operator $\rho = \rho_{\text{in}}\rho_{\text{amp}}$. For the amplifier of Section 16.2.2, ρ_{amp} is assumed to be a phase-insensitive noise reservoir, so the phase transformation of $\overline{b}_{2,\text{in}}(\omega)$ is not a problem. On the other hand, the input signal state ρ_{in} is not—one hopes—pure noise; therefore, more care is needed.

To illustrate this point, consider the opposite extreme $\rho_{\text{in}} = |\underline{\beta}\rangle\langle\underline{\beta}|$, where $|\underline{\beta}\rangle$ is a multimode coherent state defined in Section 5.5.1. In the present case, this means

$$\overline{b}_{\Omega}(t_0)\,|\underline{\beta}\rangle = \beta_{\Omega}\,|\underline{\beta}\rangle , \tag{16.98}$$

which in turn yields

$$\overline{b}_{\text{in}}(t)\,|\underline{\beta}\rangle = \beta_{\text{in}}(t)\,|\underline{\beta}\rangle , \tag{16.99}$$

where

$$\beta_{\text{in}}(t) = \int_{-\infty}^{\infty} \frac{d\Omega}{2\pi} \beta_{\Omega} e^{-i(\Omega-\omega_0)(t-t_0)} \,. \tag{16.100}$$

This coherent state is defined with respect to the original in-field operators; consequently, the action of the transformed operators is given by

$$\overline{b}'_{\text{in}}(\omega) \left|\underline{\beta}\right\rangle = e^{-i\theta_P/2} \beta_{\text{in}}(t) \left|\underline{\beta}\right\rangle . \tag{16.101}$$

Thus the pump phase removed from the input–output relation is not lost; it reappears in the calculation of the ensemble averages that are to be compared to experimental results.

The same trick works for the examples of phase-insensitive amplifiers in Sections 16.2.1-A and 16.2.1-B. With this reassurance, we can assume that the most general input–output equation can be written in the form of eqns (16.95) and (16.96).

16.4.1 The input–output equation

The linearity assumption embodied in eqns (16.95) and (16.96) does not in itself impose any additional conditions on the coefficients $P(\omega)$ and $C(\omega)$, but in all the three of the examples given above the explicit expressions for these functions satisfy the useful symmetry condition

$$P^*(-\omega) = P(\omega) \,, \;\; C^*(-\omega) = C(\omega) \,. \tag{16.102}$$

It is worthwhile to devote some effort to finding out the source of this property. The first step is to recall that the Langevin equations for the sample and reservoir modes are derived from the Heisenberg equations for the fields. In the three examples considered above, $\overline{a}(t)$ is the only sample operator; and the internal sample interaction Hamiltonian H_{SS} is a quadratic function of $\overline{a}(t)$ and $\overline{a}^\dagger(t)$. The Heisenberg equation for $\overline{a}(t)$ is therefore linear. The equations for the reservoir variables are also linear by virtue of the general assumption, made in Section 14.1.1-A, that the interaction Hamiltonian is linear in the reservoir operators.

In all three examples, these properties allow the Langevin equations for $\overline{a}(t)$ and $\overline{a}^\dagger(t)$ to be written in the form

$$\frac{d}{dt}\varphi_S(t) = -W\varphi_S(t) + \mathcal{F}(t) \,, \tag{16.103}$$

where

$$\varphi_S(t) = \begin{pmatrix} \overline{a}(t) \\ \overline{a}^\dagger(t) \end{pmatrix}, \;\; \mathcal{F}(t) = \begin{pmatrix} \xi(t) \\ \xi^\dagger(t) \end{pmatrix}, \tag{16.104}$$

W is a 2×2 hermitian matrix, and the noise operator $\xi(t)$ is a linear combination of reservoir operators. Solving eqn (16.103) via a Fourier transform leads to

$$\varphi_S(\omega) = V(\omega)\mathcal{F}(\omega) \,, \tag{16.105}$$

where the 2×2 matrix

$$V(\omega) = (W - i\omega)^{-1} \tag{16.106}$$

satisfies

$$V^{\dagger}(-\omega) = V(\omega) . \tag{16.107}$$

Substituting this solution into an input–output relation, such as eqn (14.109), produces coefficients that have the symmetry property (16.102).

This analysis raises the following question: How restrictive is the assumption that the sample operators satisfy linear equations of motion? To address this question, let us assume that H_{SS} contains terms that are more than quadratic in the sample operators, so that the equations of motion are nonlinear. The solution would then express the sample operators as nonlinear functions of the noise operators. This situation raises two further questions, one physical and the other mathematical.

The physical question concerns the size of the higher-order terms in H_{SS}. If they are small, then H_{SS} can be approximated by a quadratic form, and the linear model is regained. If the higher-order terms cannot be neglected, then the sample must be experiencing large amplitude excitations. Under these circumstances it is difficult to see how the overall amplification process could be linear.

The mathematical issue is that nonlinear differential equations for the sample operators cannot readily be solved by the Fourier transform method. This makes it hard to see how a frequency-domain relation like (16.95) could be derived.

These arguments are far from conclusive, but they do suggest that imposing the assumption of weak sample excitations will not cause a significant loss of generality. We will therefore extend the definition of linear amplifiers to include the assumption that the internal modes all satisfy linear equations of motion. This in turn implies that the symmetry property (16.102) can be applied in general.

The necessity of working with the pair of input–output equations (16.95) and (16.96) suggests that a matrix notation would be useful. The input–output equations can be written as

$$\varphi_{\text{out}}(\omega) = \mathcal{R}(\omega)\,\varphi_{\text{in}}(\omega) + \zeta(\omega) , \tag{16.108}$$

where

$$\varphi_{\gamma}(\omega) = \begin{pmatrix} \overline{b}_{\gamma}(\omega) \\ \overline{b}_{\gamma}^{\dagger}(-\omega) \end{pmatrix} \quad (\gamma = \text{in}, \text{out}) , \quad \zeta(\omega) = \begin{pmatrix} \eta(\omega) \\ \eta^{\dagger}(-\omega) \end{pmatrix} , \tag{16.109}$$

and

$$\mathcal{R}(\omega) = \begin{bmatrix} P(\omega) & C(\omega) \\ C(\omega) & P(\omega) \end{bmatrix} \tag{16.110}$$

is the **input–output** matrix. In this notation the symmetry condition (16.102) is

$$\mathcal{R}^{\dagger}(-\omega) = \mathcal{R}(\omega) . \tag{16.111}$$

The matrix $\mathcal{R}(\omega)$ is neither hermitian nor unitary, but it does commute with its adjoint, i.e. $\mathcal{R}^{\dagger}(\omega)\,\mathcal{R}(\omega) = \mathcal{R}(\omega)\,\mathcal{R}^{\dagger}(\omega)$. Matrices with this property are called *normal*, and all normal matrices have a complete, orthonormal set of eigenvectors. An explicit calculation yields the eigenvalue–eigenvector pairs

$$\begin{aligned} z_1(\omega) &= P(\omega) + C(\omega), \quad \Theta_1 = \frac{1}{\sqrt{2}}\begin{pmatrix}1\\1\end{pmatrix}, \\ z_2(\omega) &= P(\omega) - C(\omega), \quad \Theta_2 = \frac{i}{\sqrt{2}}\begin{pmatrix}1\\-1\end{pmatrix}. \end{aligned} \tag{16.112}$$

It is instructive to express the input–output equation in the basis $\{\Theta_1, \Theta_2\}$. By writing the expansion for the in-operator φ_{in} as

$$\varphi_{\text{in}}(\omega) = \sqrt{2}X_{\text{in}}(\omega)\Theta_1 + \sqrt{2}Y_{\text{in}}(\omega)\Theta_2, \tag{16.113}$$

one finds the operator-valued coefficients to be

$$\begin{aligned} X_{\text{in}}(\omega) &= \frac{1}{\sqrt{2}}\Theta_1^\dagger\varphi_{\text{in}}(\omega) = \frac{1}{2}\left[\overline{b}_{\text{in}}(\omega) + \overline{b}_{\text{in}}^\dagger(-\omega)\right] = X_{\beta=0,\text{in}}(\omega), \\ Y_{\text{in}}(\omega) &= \frac{1}{\sqrt{2}}\Theta_2^\dagger\varphi_{\text{in}}(\omega) = \frac{1}{2i}\left[\overline{b}_{\text{in}}(\omega) - \overline{b}_{\text{in}}^\dagger(-\omega)\right] = Y_{\beta=0,\text{in}}(\omega). \end{aligned} \tag{16.114}$$

The special value, $\beta = 0$, of the quadrature angle is an artefact of the phase transformation trick—explained at the beginning of Section 16.4—used to ensure the absence of explicit phase factors in the general input–output equation (16.95).

In this basis the input–output relations have the diagonal form

$$\begin{aligned} X_{\text{out}}(\omega) &= [P(\omega) + C(\omega)]\, X_{\text{in}}(\omega) + \zeta_1(\omega), \\ Y_{\text{out}}(\omega) &= [P(\omega) - C(\omega)]\, Y_{\text{in}}(\omega) + \zeta_2(\omega), \end{aligned} \tag{16.115}$$

where

$$\begin{aligned} \zeta_1(\omega) &= \frac{1}{\sqrt{2}}\Theta_1^\dagger\zeta(\omega) = \frac{1}{2}\left[\eta(\omega) + \eta^\dagger(-\omega)\right] = \zeta_1^\dagger(-\omega), \\ \zeta_2(\omega) &= \frac{1}{\sqrt{2}}\Theta_2^\dagger\zeta(\omega) = \frac{1}{2i}\left[\eta(\omega) - \eta^\dagger(-\omega)\right] = \zeta_2^\dagger(-\omega). \end{aligned} \tag{16.116}$$

We will refer to $X_{\text{out}}(\omega)$ and $Y_{\text{out}}(\omega)$ as the **principal quadratures**.

The ensemble average of eqn (16.108) is

$$\Phi_{\text{out}}(\omega) = \mathcal{R}(\omega)\,\Phi_{\text{in}}(\omega), \tag{16.117}$$

where

$$\Phi_\gamma(\omega) = \langle\varphi_\gamma(\omega)\rangle = \begin{pmatrix}\langle\overline{b}_\gamma(\omega)\rangle \\ \left\langle\overline{b}_\gamma^\dagger(-\omega)\right\rangle\end{pmatrix} = \begin{pmatrix}\langle\overline{b}_\gamma(\omega)\rangle \\ \langle\overline{b}_\gamma(-\omega)\rangle^*\end{pmatrix} \quad (\gamma = \text{in}, \text{out}). \tag{16.118}$$

Subtracting eqn (16.117) from eqn (16.108) produces the input–output equation

$$\delta\varphi_{\text{out}}(\omega) = \mathcal{R}(\omega)\,\delta\varphi_{\text{in}}(\omega) + \zeta(\omega), \tag{16.119}$$

where $\delta\varphi_{\text{in}}(\omega) = \varphi_{\text{in}}(\omega) - \Phi_{\text{in}}(\omega)$ and $\delta\varphi_{\text{out}}(\omega) = \varphi_{\text{out}}(\omega) - \Phi_{\text{out}}(\omega)$ are respectively the fluctuation operators for the input and output. In the principal quadrature basis this becomes

$$\begin{aligned} \delta X_{\text{out}}(\omega) &= [P(\omega) + C(\omega)]\, \delta X_{\text{in}}(\omega) + \zeta_1(\omega), \\ \delta Y_{\text{out}}(\omega) &= [P(\omega) - C(\omega)]\, \delta Y_{\text{in}}(\omega) + \zeta_2(\omega). \end{aligned} \tag{16.120}$$

The diagonalized form (16.115) of the input–output relation suggests two natural definitions for gain in a general linear amplifier. These are the **principal gains** defined by

$$G_1(\omega) = \frac{|\langle X_{\text{out}}(\omega)\rangle|^2}{|\langle X_{\text{in}}(\omega)\rangle|^2} = |P(\omega) + C(\omega)|^2, \tag{16.121}$$

$$G_2(\omega) = \frac{|\langle Y_{\text{out}}(\omega)\rangle|^2}{|\langle Y_{\text{in}}(\omega)\rangle|^2} = |P(\omega) - C(\omega)|^2. \tag{16.122}$$

The principal gains can also be defined as the eigenvalues of the **gain matrix**

$$\mathcal{G}(\omega) = \mathcal{R}^{\dagger}(\omega)\,\mathcal{R}(\omega), \tag{16.123}$$

which has the same eigenvectors as the input–output matrix. For phase-insensitive amplifiers, the gain matrix is diagonal, and the principal gains are the same: $G_1(\omega) = G_2(\omega)$.

The complex functions $P(\omega) \pm C(\omega)$ that appear in eqn (16.115) are expressed in terms of the principal gains as

$$P(\omega) + C(\omega) = \sqrt{G_1(\omega)}\, e^{i\vartheta_1(\omega)}, \tag{16.124}$$

$$P(\omega) - C(\omega) = \sqrt{G_2(\omega)}\, e^{i\vartheta_2(\omega)}, \tag{16.125}$$

so that the symmetry condition (16.102) becomes

$$\left.\begin{aligned} G_j(\omega) &= G_j(-\omega), \\ \vartheta_j(\omega) &= -\vartheta_j(-\omega) \mod 2\pi \end{aligned}\right\} \quad (j = 1, 2). \tag{16.126}$$

With this notation eqn (16.115) is replaced by

$$\begin{aligned} X_{\text{out}}(\omega) &= \sqrt{G_1(\omega)}\, e^{i\vartheta_1(\omega)} X_{\text{in}}(\omega) + \zeta_1(\omega), \\ Y_{\text{out}}(\omega) &= \sqrt{G_2(\omega)}\, e^{i\vartheta_2(\omega)} Y_{\text{in}}(\omega) + \zeta_2(\omega). \end{aligned} \tag{16.127}$$

16.4.2 Conditions for phase insensitivity

According to eqn (16.51), imposing condition (i) of Section 16.1.1 requires

$$2\,\text{Re}\left[\left(e^{2i\theta} - 1\right) P(\omega)\, C^*(\omega) \left\langle \overline{b}_{\text{in}}(\omega)\right\rangle \left\langle \overline{b}_{\text{in}}(-\omega)\right\rangle\right] = 0. \tag{16.128}$$

This is supposed to hold as an identity in θ for all input values $\left\langle \overline{b}_{\text{in}}(\omega)\right\rangle$; consequently, the coefficients must satisfy

$$P(\omega)\, C^*(\omega) = 0. \tag{16.129}$$

Thus all phase-insensitive amplifiers fall into one of the two classes illustrated in Section 16.2.1: (1) phase-transmitting amplifiers, with $C(\omega) = 0$ and $P(\omega) \neq 0$; or (2) phase-conjugating amplifiers, with $P(\omega) = 0$ and $C(\omega) \neq 0$.

Turning next to condition (ii), we recall that the fluctuation operators satisfy the input–output relation (16.95), and that the amplifier noise operator is not correlated with the input fields. Combining these observations with the condition (16.129) yields

$$K_{\text{out}}(\omega,\omega') = P(\omega)\,P^*(\omega')\,K_{\text{in}}(\omega,\omega') + C(\omega)\,C^*(\omega')\,K_{\text{in}}(-\omega',-\omega) + K_{\text{amp}}(\omega,\omega')\,. \tag{16.130}$$

Imposing eqn (16.9) on both $K_{\text{out}}(\omega,\omega')$ and $K_{\text{in}}(\omega,\omega')$ then implies that

$$K_{\text{amp}}(\omega,\omega') = \mathfrak{N}_{\text{amp}}(\omega)\,2\pi\delta(\omega-\omega')\,, \tag{16.131}$$

where

$$\mathfrak{N}_{\text{amp}}(\omega) = \mathfrak{N}_{\text{out}}(\omega) - G(\omega)\,\mathfrak{N}_{\text{in}}(\omega)\,, \tag{16.132}$$

and

$$G(\omega) = \begin{cases} |P(\omega)|^2 & \text{(phase-transmitting amplifier)}\,, \\ |C(\omega)|^2 & \text{(phase-conjugating amplifier)} \end{cases} \tag{16.133}$$

is the gain for the phase-insensitive amplifier. A similar calculation yields

$$\begin{aligned}\left\langle \delta\overline{b}_{\text{out}}(\omega)\,\delta\overline{b}_{\text{out}}(\omega')\right\rangle &= P(\omega)\,P(\omega')\left\langle \delta\overline{b}_{\text{in}}(\omega)\,\delta\overline{b}_{\text{in}}(\omega')\right\rangle \\ &\quad + C(\omega)\,C(\omega')\left\langle \delta\overline{b}_{\text{in}}^{\dagger}(-\omega)\,\delta\overline{b}_{\text{in}}^{\dagger}(-\omega')\right\rangle \\ &\quad + \langle \eta(\omega)\,\eta(\omega')\rangle\,.\end{aligned} \tag{16.134}$$

Imposing eqn (16.11) on the input and output fields implies

$$\langle \eta(\omega)\,\eta(\omega')\rangle = 0\,. \tag{16.135}$$

In other words, the amplifier noise is itself phase insensitive, since it satisfies eqns (16.9) and (16.11).

16.4.3 Unitarity constraints

The derivation of the Langevin equation from the linear Heisenberg equations of motion imposes the symmetry condition (16.102) on the coefficients $P(\omega)$ and $C(\omega)$, but the sole constraint on the amplifier noise is that it can only depend on the internal modes of the amplifier. Additional constraints follow from the requirement that the out-field operators are related to the in-field operators by a unitary transformation.

An immediate consequence is that the out-field operators and the in-field operators satisfy the same canonical commutation relations:

$$\left.\begin{aligned}\left[\overline{b}_{\gamma}(\omega)\,,\overline{b}_{\gamma}^{\dagger}(\omega')\right] &= 2\pi\delta(\omega-\omega')\,, \\ \left[\overline{b}_{\gamma}(\omega)\,,\overline{b}_{\gamma}(\omega')\right] &= 0\end{aligned}\right\}\quad (\gamma = \text{in, out})\,. \tag{16.136}$$

Substituting eqns (16.95) and (16.96) into eqn (16.136), with $\gamma = \text{out}$, imposes conditions on the amplifier noise operator. Once again, we recall that $\overline{b}_{\text{in}}(\omega)$ and $\overline{b}_{\text{in}}^{\dagger}(\omega)$ are linear functions of the input field operators evaluated at the initial time $t = t_0$. The amplifier noise operator depends on the noise reservoir operators evaluated at the same time t_0, e.g. see eqn (16.49). The equal-time commutators between the internal mode operators and the input operators all vanish; therefore, the in-field operators $\overline{b}_{\text{in}}(\omega)$ and $\overline{b}_{\text{in}}^{\dagger}(\omega)$ commute with the amplifier noise operators $\eta(\omega)$ and $\eta^{\dagger}(\omega)$.

With this simplification in mind, eqn (16.136) imposes two relations between the amplifier noise operator and the c-number coefficients:

$$[\eta(\omega), \eta(\omega')] = i\sqrt{G_1(\omega) G_2(\omega)} \sin[\vartheta_{12}(\omega)] 2\pi\delta(\omega + \omega'), \tag{16.137}$$

and

$$\left[\eta(\omega), \eta^{\dagger}(\omega')\right] = \left\{1 - \sqrt{G_1(\omega) G_2(\omega)} \cos[\vartheta_{12}(\omega)]\right\} 2\pi\delta(\omega - \omega'), \tag{16.138}$$

where $\vartheta_{12}(\omega) = \vartheta_1(\omega) - \vartheta_2(\omega)$. The two kinds of phase-insensitive amplifiers correspond to the values $\vartheta_{12}(\omega) = 0$—the phase-transmitting amplifiers—and $\vartheta_{12}(\omega) = \pi$—the phase-conjugating amplifiers.

Combining the expression (16.114) for the input quadratures with eqn (16.136) and the identities

$$X_{\beta,\text{in}}^{\dagger}(\omega) = X_{\beta,\text{in}}(-\omega), \quad Y_{\beta,\text{in}}^{\dagger}(\omega) = Y_{\beta,\text{in}}(-\omega) \tag{16.139}$$

yields the commutation relations

$$[X_{\text{in}}(\omega), X_{\text{in}}(\omega')] = [Y_{\text{in}}(\omega), Y_{\text{in}}(\omega')] = 0, \tag{16.140}$$

and

$$\left[X_{\text{in}}(\omega), Y_{\text{in}}^{\dagger}(\omega')\right] = \frac{i}{2} 2\pi\delta(\omega - \omega'). \tag{16.141}$$

The unitary connection between the in- and out-fields requires the output quadratures to satisfy the same relations. Substituting the input–output equation (16.115) into eqns (16.140) and (16.141) yields an equivalent form of the unitarity conditions:

$$[\zeta_j(\omega), \zeta_j(\omega')] = 0 \quad (j = 1, 2), \tag{16.142}$$

$$\left[\zeta_1(\omega), \zeta_2^{\dagger}(\omega')\right] = \frac{i}{2}\left\{1 - \sqrt{G_1(\omega) G_2(\omega)}\, e^{i\vartheta_{12}(\omega)}\right\} 2\pi\delta(\omega - \omega'). \tag{16.143}$$

16.5 Noise limits for linear amplifiers

The familiar uncertainty relations of quantum mechanics can be derived from the canonical commutation relations by specializing the general result in Appendix C.3.7. By a similar argument, the unitarity constraints on the noise operators impose lower bounds on the noise added by an amplifier.

16.5.1 Phase-insensitive amplifiers

For phase-insensitive amplifiers, the commutation relations (16.137) and (16.138) respectively reduce to

$$[\eta(\omega), \eta(\omega')] = 0, \tag{16.144}$$

and

$$\left[\eta(\omega), \eta^{\dagger}(\omega')\right] = \{1 \mp G(\omega)\} 2\pi\delta(\omega - \omega'), \tag{16.145}$$

where $G(\omega)$ is the gain. The upper and lower signs correspond respectively to phase-transmitting and phase-conjugating amplifiers. In both cases, the amplifier noise is phase insensitive, so $K_{\text{amp}}(\omega,\omega')$ satisfies eqn (16.131). Substituting this form into the definition (16.32) then leads to

$$\begin{aligned}\mathfrak{N}_{\text{amp}}(\omega)\,2\pi\delta(\omega-\omega') &= \frac{1}{2}\left\langle \eta(\omega)\,\eta^{\dagger}(\omega') + \eta^{\dagger}(\omega')\,\eta(\omega)\right\rangle \\ &= \left\langle \eta^{\dagger}(\omega')\,\eta(\omega)\right\rangle + \frac{1}{2}\left\langle \left[\eta(\omega),\eta^{\dagger}(\omega')\right]\right\rangle \\ &= \left\langle \eta^{\dagger}(\omega')\,\eta(\omega)\right\rangle + \frac{1}{2}\left\{1 \mp G(\omega)\right\}2\pi\delta(\omega-\omega').\end{aligned} \tag{16.146}$$

Since $\left\langle \eta^{\dagger}(\omega')\,\eta(\omega)\right\rangle$ is a positive-definite integral kernel, we see that

$$\mathfrak{N}_{\text{amp}}(\omega) \geqslant \frac{1}{2}\left|1 \mp G(\omega)\right|. \tag{16.147}$$

Thus a phase-insensitive amplifier with $G(\omega) > 1$ necessarily adds noise to any input signal.

For phase-insensitive input noise, the output noise is also phase insensitive; and eqn (16.132) can be rewritten as

$$\mathfrak{N}_{\text{out}}(\omega) = G(\omega)\,\mathfrak{N}_{\text{in}}(\omega) + \mathfrak{N}_{\text{amp}}(\omega). \tag{16.148}$$

For some purposes it is useful to treat the amplifier noise as though it were due to the amplification of a fictitious input noise $\mathfrak{A}(\omega)$. This additional input noise—which is called the **amplifier noise number**—is defined by

$$\mathfrak{A}(\omega) = \frac{\mathfrak{N}_{\text{amp}}(\omega)}{G(\omega)}. \tag{16.149}$$

With this notation, the relation (16.148) and the inequality (16.147) are respectively replaced by

$$\mathfrak{N}_{\text{out}}(\omega) = G(\omega)\left[\mathfrak{N}_{\text{in}}(\omega) + \mathfrak{A}(\omega)\right] \tag{16.150}$$

and

$$\mathfrak{A}(\omega) \geqslant \frac{1}{2}\left|1 \mp \frac{1}{G(\omega)}\right|. \tag{16.151}$$

Applying this inequality to eqn (16.150) yields a lower bound for the output noise:

$$\mathfrak{N}_{\text{out}}(\omega) \geqslant G(\omega)\left\{\mathfrak{N}_{\text{in}}(\omega) + \frac{1}{2}\left|1 \mp \frac{1}{G(\omega)}\right|\right\}. \tag{16.152}$$

If the input noise is due to a heat bath at temperature T, the continuum versions of eqns (14.28) and (2.177) combine to give the noise strength,

$$\begin{aligned}\mathfrak{N}_{\text{in}}(\omega) &= \frac{1}{\exp\left[\beta\hbar(\omega_0+\omega)\right]-1} + \frac{1}{2} \\ &= \frac{1}{2}\coth\left[\frac{\hbar(\omega_0+\omega)}{2k_BT}\right].\end{aligned} \tag{16.153}$$

This result suggests a more precise definition of the effective noise temperature, first discussed in Section 9.3.2-B. The idea is to ask what increase in temperature ($T \to T+$

T_{amp}) is required to blame the total pre-amplification noise on a fictitious thermal reservoir. A direct application of this idea leads to

$$\frac{1}{2}\coth\left[\frac{\hbar(\omega_0+\omega)}{2k_B(T+T_{\text{amp}})}\right]=\frac{1}{2}\coth\left[\frac{\hbar(\omega_0+\omega)}{2k_BT}\right]+\mathfrak{A}(\omega)\,, \tag{16.154}$$

but this would make T_{amp} depend on the input-noise temperature T and on the frequency ω. A natural way to get something that can be regarded as a property of the amplifier alone is to impose eqn (16.154) only for the case $T=0$ and for the resonance frequency $\omega=0$. This yields the **amplifier noise temperature**

$$k_BT_{\text{amp}}=\frac{\hbar\omega_0}{\ln\left(1+1/\mathfrak{A}(0)\right)}\,. \tag{16.155}$$

For $G(0)=G(\omega_0)>1$, eqns (16.155) and (16.151) provide the lower bound

$$k_BT_{\text{amp}}\geqslant\frac{\hbar\omega_0}{\ln\left(\frac{3G(\omega_0)\mp1}{G(\omega_0)\mp1}\right)}\to\frac{\hbar\omega_0}{\ln(3)} \tag{16.156}$$

on the noise temperature. The final form is the limiting value for high gains, i.e. $G(\omega_0)\gg1$.

16.5.2 Phase-sensitive amplifiers

The definition of a phase-sensitive amplifier is purely negative. An amplifier is phase sensitive if it is *not* phase insensitive. One consequence of this broad definition is that explicit constraints—such as the special form imposed on the noise correlation function by eqn (16.131)—are not available for phase-sensitive amplifiers. In the general case, e.g. when considering broadband amplifiers, further restrictions on the family of amplifiers are used to make up for the absence of constraints (Caves, 1982). For the narrowband amplifiers we are studying, an alternative approach will be described below. It is precisely the presence of the constraint (16.131) which makes the alternative approach unnecessary for the noise analysis of phase-insensitive amplifiers.

The basic idea of the alternative approach is to treat narrow frequency bands of the input and output as though they were discrete modes. For this purpose, let $\Delta\omega$ be a frequency interval that is small compared to the characteristic widths of the functions $G_j(\omega)$ and $\vartheta_j(\omega)$—or $P(\omega)$ and $C(\omega)$—and define coarse-grained quadratures and noise operators by

$$F^c(\omega)=\int_{\omega-\Delta\omega/2}^{\omega+\Delta\omega/2}\frac{d\omega_1}{\sqrt{2\pi\Delta\omega}}F(\omega_1)\,, \tag{16.157}$$

where F stands for any of the operators in the set

$$\mathbb{F}=\{X_{\text{in}}(\omega)\,,Y_{\text{in}}(\omega)\,,X_{\text{out}}(\omega)\,,Y_{\text{out}}(\omega)\,,\zeta_1(\omega)\,,\zeta_2(\omega)\}\,. \tag{16.158}$$

All of these operators satisfy $F^\dagger(\omega)=F(-\omega)$, and this property is inherited by the coarse-grained versions: $F^{c\dagger}(\omega)=F^c(-\omega)$. From the experimental point of view, the coarse-graining operation is roughly equivalent to the use of a narrowband-pass filter.

The noise strength for the non-hermitian operator $F^c(\omega)$ is

$$[\Delta F^c(\omega)]^2 = \frac{1}{2}\left\langle \delta F^c(\omega)\,\delta F^{c\dagger}(\omega) + \delta F^{c\dagger}(\omega)\,\delta F^c(\omega)\right\rangle , \tag{16.159}$$

but this general result can be simplified by using the special properties of the operators in $\mathbb{F}$. The commutation relations (16.140) and (16.142) guarantee that all the operators in $\mathbb{F}$ satisfy $[F(\omega), F(\omega')] = 0$, and the property $F^\dagger(\omega) = F(-\omega)$ shows that this is equivalent to $\left[F(\omega), F^\dagger(\omega')\right] = 0$. Averaging ω and ω' over the interval $(\omega - \Delta\omega/2, \omega + \Delta\omega/2)$ in the latter form yields the coarse-grained relation

$$\left[F^c(\omega), F^{c\dagger}(\omega)\right] = 0 , \tag{16.160}$$

and this allows eqn (16.159) to be replaced by

$$[\Delta F^c(\omega)]^2 = \left\langle \delta F^{c\dagger}(\omega)\,\delta F^c(\omega)\right\rangle = \left\langle \delta F^c(\omega)\,\delta F^{c\dagger}(\omega)\right\rangle . \tag{16.161}$$

The output noise strength can be related to the input noise strength and the amplifier noise by means of the coarse-grained input–output equations:

$$\begin{aligned} X^c_{\text{out}}(\omega) &= \sqrt{G_1(\omega)}\, e^{i\vartheta_1(\omega)} X^c_{\text{in}}(\omega) + \zeta^c_1(\omega) , \\ Y^c_{\text{out}}(\omega) &= \sqrt{G_2(\omega)}\, e^{i\vartheta_2(\omega)} Y^c_{\text{in}}(\omega) + \zeta^c_2(\omega) . \end{aligned} \tag{16.162}$$

These relations are obtained by applying the averaging procedure (16.157) to eqn (16.127), and using the assumption that the gain functions are essentially constant over the interval $(\omega - \Delta\omega/2, \omega + \Delta\omega/2)$. The lack of correlation between the in-fields and the amplifier noise implies that the output noise strength in each principal quadrature is the sum of the amplified input noise and the amplifier noise in that quadrature:

$$\begin{aligned} [\Delta X^c_{\text{out}}(\omega)]^2 &= G_1(\omega)\,[\Delta X^c_{\text{in}}(\omega)]^2 + [\Delta\zeta^c_1(\omega)]^2 , \\ [\Delta Y^c_{\text{out}}(\omega)]^2 &= G_2(\omega)\,[\Delta Y^c_{\text{in}}(\omega)]^2 + [\Delta\zeta^c_2(\omega)]^2 . \end{aligned} \tag{16.163}$$

In this situation there is an amplifier noise number for each principal quadrature:

$$\mathfrak{A}_j(\omega) = \frac{\left[\Delta\zeta^c_j(\omega)\right]^2}{G_j(\omega)} \quad (j = 1, 2) , \tag{16.164}$$

so that eqn (16.163) can be written as

$$\begin{aligned} [\Delta X^c_{\text{out}}(\omega)]^2 &= G_1(\omega)\left\{[\Delta X^c_{\text{in}}(\omega)]^2 + \mathfrak{A}_1(\omega)\right\} , \\ [\Delta Y^c_{\text{out}}(\omega)]^2 &= G_2(\omega)\left\{[\Delta Y^c_{\text{in}}(\omega)]^2 + \mathfrak{A}_2(\omega)\right\} . \end{aligned} \tag{16.165}$$

The **signal-to-noise** ratios for the principal quadratures are defined by

$$\begin{aligned} [\text{SNR}(X)]_\gamma &= \frac{\left|\langle X^c_\gamma(\omega)\rangle\right|^2}{\left[\Delta X^c_\gamma(\omega)\right]^2} \quad (\gamma = \text{in}, \text{out}) , \\ [\text{SNR}(Y)]_\gamma &= \frac{\left|\langle Y^c_\gamma(\omega)\rangle\right|^2}{\left[\Delta Y^c_\gamma(\omega)\right]^2} \quad (\gamma = \text{in}, \text{out}) . \end{aligned} \tag{16.166}$$

Input–output relations for the signal-to-noise ratios follow by combining the ensemble average of the operator input–output equation, eqn (16.162), with eqn (16.165) to get

$$
\begin{aligned}
[\mathrm{SNR}\,(X)]_{\mathrm{out}} &= \frac{[\mathrm{SNR}\,(X)]_{\mathrm{in}}}{1+\mathfrak{A}_1\,(\omega)\,/\,[\Delta X^c_{\mathrm{in}}\,(\omega)]^2}\,, \\
[\mathrm{SNR}\,(Y)]_{\mathrm{out}} &= \frac{[\mathrm{SNR}\,(Y)]_{\mathrm{in}}}{1+\mathfrak{A}_2\,(\omega)\,/\,[\Delta Y^c_{\mathrm{in}}\,(\omega)]^2}\,.
\end{aligned}
\tag{16.167}
$$

Lower bounds on the amplifier noise strengths $\Delta\zeta_1^c\,(\omega)$ and $\Delta\zeta_2^c\,(\omega)$ can be derived by applying the coarse-graining operation to the commutation relation (16.143) to get

$$
\left[\zeta_1^c\,(\omega)\,,\zeta_2^{c\dagger}\,(\omega)\right] = \left\{1-\sqrt{G_1\,(\omega)\,G_2\,(\omega)}e^{i\vartheta_{12}(\omega)}\right\}\frac{i}{2}\,. \tag{16.168}
$$

This looks like the commutation relations between a canonical pair, except for the fact that the operators $\zeta_1^c\,(\omega)$ and $\zeta_2^{c\dagger}\,(\omega)$ are not hermitian. This flaw can be circumvented by applying the generalized uncertainty relation, $2\Delta C\Delta D \geqslant |\langle[C,D]\rangle|$, that is derived in Appendix C.3.7. This result is usually quoted only for hermitian operators, but it is actually valid for any pair of normal operators C and D, i.e. operators satisfying $\left[C,C^\dagger\right] = \left[D,D^\dagger\right] = 0$. By virtue of eqn (16.160), $\zeta_1^c\,(\omega)$ and $\zeta_2^{c\dagger}\,(\omega)$ are both normal operators; therefore, the product of the noise strengths in the principal quadratures satisfies the **amplifier uncertainty principle**:

$$
\Delta\zeta_1^c\,(\omega)\,\Delta\zeta_2^c\,(\omega) \geqslant \frac{1}{4}\left|1-\sqrt{G_1\,(\omega)\,G_2\,(\omega)}\;e^{i\vartheta_{12}(\omega)}\right|. \tag{16.169}
$$

This can be expressed in terms of the amplifier noise numbers as

$$
\sqrt{\mathfrak{A}_1\,(\omega)\,\mathfrak{A}_2\,(\omega)} \geqslant \frac{1}{4}\left|1-\frac{1}{\sqrt{G_1\,(\omega)\,G_2\,(\omega)}}e^{-i\vartheta_{12}(\omega)}\right|. \tag{16.170}
$$

At the carrier frequency, $\omega = 0$, the symmetry condition (16.126) only allows the values $\vartheta_{12}\,(0) = 0,\pi$, and the general amplifier uncertainty principle is replaced by

$$
\sqrt{\mathfrak{A}_1\,(0)\,\mathfrak{A}_2\,(0)} \geqslant \frac{1}{4}\left|1\mp\frac{1}{\sqrt{G_1\,(0)\,G_2\,(0)}}\right|, \tag{16.171}
$$

where the upper and lower signs correspond to $\vartheta_{12}\,(0) = 0$ and $\vartheta_{12}\,(0) = \pi$ respectively.

16.6 Exercises

16.1 Quadrature gain

(1) Show that the frequency-domain form of eqn (16.3) is

$$
\begin{aligned}
X_{\beta,\mathrm{in}}\,(\omega) &= \frac{1}{2}\left[e^{-i\beta}\overline{b}_{\mathrm{in}}\,(\omega)+e^{i\beta}\overline{b}^{\dagger}_{\mathrm{in}}\,(-\omega)\right], \\
Y_{\beta,\mathrm{in}}\,(\omega) &= \frac{1}{2i}\left[e^{-i\beta}\overline{b}_{\mathrm{in}}\,(\omega)-e^{i\beta}\overline{b}^{\dagger}_{\mathrm{in}}\,(-\omega)\right].
\end{aligned}
$$

(2) Show that the frequency-domain operators satisfy $X_{\beta,\text{in}}^{\dagger}(\omega) = X_{\beta,\text{in}}(-\omega)$, $Y_{\beta,\text{in}}^{\dagger}(\omega) = Y_{\beta,\text{in}}(-\omega)$, and

$$\left[X_{\beta,\text{in}}(\omega), X_{\beta,\text{in}}^{\dagger}(\omega')\right] = \left[X_{\beta,\text{in}}(\omega), X_{\beta,\text{in}}(-\omega')\right] = 0\,,$$
$$\left[Y_{\beta,\text{in}}(\omega), Y_{\beta,\text{in}}^{\dagger}(\omega')\right] = \left[Y_{\beta,\text{in}}(\omega), Y_{\beta,\text{in}}(-\omega')\right] = 0\,.$$

(3) Use the input–output relation (16.24) and its adjoint to conclude that the output quadrature is related to the input quadrature by

$$X_{\beta,\text{out}}(\omega) = P(\omega) X_{\beta,\text{in}}(\omega) + \frac{1}{2}\left[e^{-i\beta}\eta(\omega) + e^{i\beta}\eta^{\dagger}(-\omega)\right].$$

(4) Define the gain for this quadrature by

$$G_{\beta}(\omega) = \frac{\left|\left\langle X_{\beta,\text{out}}(\omega)\right\rangle\right|^2}{\left|\left\langle X_{\beta,\text{in}}(\omega)\right\rangle\right|^2}\,,$$

and show that the gain is the same for all quadratures.

16.2 Phase-insensitive traveling-wave amplifier

(1) Work out the coarse-grained version of eqns (14.174)–(14.177), and then use eqn (16.60) for H'_{S1} to derive the reduced Langevin equations for the amplifier.

(2) Use the properties of $\xi_{12}^{(n)}(t)$ to derive eqn (16.67).

(3) Show that

$$\left[\overline{S}_{qp}(z,t), \overline{S}_{kl}(z',t)\right] = \left\{\delta_{pk}\overline{S}_{ql}(z,t) - \delta_{lq}\overline{S}_{kp}(z,t)\right\}\delta(z-z')\,.$$

(4) Show that $\left\langle \overline{S}_{22}(z,t) - \overline{S}_{11}(z,t)\right\rangle \approx n_{\text{at}}\sigma D$.

(5) Derive eqns (16.82) and (16.83).

16.3 Colored noise

Reconsider the use of adiabatic elimination to solve eqn (16.65).

(1) Use the formal solution of eqn (16.65) to conclude that the noise term on the right side of eqn (16.68) should be replaced by

$$\zeta_{12}(z,t) = \int_{t_0}^{t} dt_1 e^{(i\Delta_0 - \Gamma_{12})(t-t_1)}\xi_{12}(z,t_1)\,.$$

(2) Use the properties of $\xi_{12}(z,t_1)$ to show that

$$\left\langle \zeta_{12}(z,t)\,\zeta_{12}^{\dagger}(z',t')\right\rangle = n_{\text{at}}\sigma C_{12,12}\delta(z-z')\,e^{i\Delta_0(t-t')}\frac{e^{-\Gamma_{12}|t-t'|}}{2\Gamma_{12}}\,.$$

(3) Justify eqn (16.68) by evaluating

$$\int dt' \left\langle \zeta_{12}(z,t)\,\zeta_{12}^{\dagger}(z',t')\right\rangle f(t')\,,$$

where $f(t')$ is slowly varying on the scale $T_{12} = 1/\Gamma_{12}$.

17
Quantum tomography

Classical tomography is an experimental method for examining the interior of a physical object by scanning a penetrating beam of radiation, for example, X-rays, through its interior. In medicine, the density profile of the interior of the body is reconstructed by using the method of CAT scans (computer-assisted tomographic scans). This procedure allows a high-resolution image of an interior section of the human body to be formed, and is therefore very useful for diagnostics.

In quantum tomography, the subject of interest is not the density distribution inside a physical object, but rather the Wigner distribution describing a quantum state. By exploiting the mathematical similarity between a physical density distribution and the quasiprobability distribution $W(\alpha)$, the methods of tomography can be applied to perform a high-resolution determination of a quantum state of light. We begin with a review of the mathematical techniques used in classical tomography, and then proceed to the application of these methods to the Wigner function and the description of a representative set of experiments.

17.1 Classical tomography

Classical tomography consists of a sequence of measurements, called *scans*, of the detected intensity of an X-ray beam at the end of a given path through the object. The fraction of the intensity absorbed in a small interval Δs is $\kappa\rho\Delta s$, where κ is the opacity and ρ the density of the material. For the usual case of uniform opacity, the ratio of the detected intensity to the source intensity is proportional to the line integral of the density along the path. After a scan of lateral displacements through the object is completed, the angle of the X-ray beam is changed, and a new sequence of lateral scans is performed. When these lateral scans are completed, the angle is then again incremented, etc. Thus a complete set of data for X-ray absorption can be obtained by translations and rotations of the path of the X-ray beam through the object. The density profile is then recovered from these data by the mathematical technique described below.

The medical motivation for this procedure is the desire to locate a single lump of matter—such as a tumor which possesses a density differing from that of normal tissue—in the interior of a body. The source and the detector straddle the body in such a way that the line of sight connecting them can be stepped through lateral displacements, and then stepped through different angles with respect to the body. One can thereby determine—in fact, overdetermine—the location of the lump by observing which of the translational and rotational data sets yield the maximum absorption.

17.1.1 Procedure for classical tomography

Consider an object whose density profile $\rho(x, y, z)$ we wish to map by probing its interior with a thin beam of X-rays directed from the source S to the detector D, as shown in Fig. 17.1. We place the origin O of coordinates near the center of the object, and choose a plane containing the source and the detector as the (x, y)-plane, i.e. the $(z = 0)$-plane. The line $\overline{\mathrm{SD}}$ that joins the source to the detector is traditionally called the **line of sight**. For a given line of sight, we introduce a rotated coordinate system (x', y'), where the y'-axis is parallel to the line of sight, the x'-axis is perpendicular to it, and θ is the rotation angle between the x'- and x-axes. Two lines of sight that differ only by interchanging the source and detector are redundant, since they provide the same information; consequently, the rotation angle θ can be restricted to the range $0 < \theta < \pi$.

The intensity ratio measured by passing the X-ray beam along the line of sight $\overline{\mathrm{SD}}$ is proportional to the line integral

$$P_\theta(x') = \int_{\overline{\mathrm{SD}}} \rho(x, y, 0)\, ds\,, \tag{17.1}$$

where s is a coordinate measured along the line of sight. We will call this line integral the *projection* of the density along the $\overline{\mathrm{SD}}$ direction. It is also commonly called a *line-out* of the density. Incrementing the x'-value, while keeping the line of sight parallel to the y'-axis, generates a set of data which yields information about the integrated column density of the object as a function of x'. After a sequence of scans at different x'-values has been completed, a new set of line-outs can be generated by incrementing the rotation angle θ. For applications of classical tomography to real three-dimensional objects, data for slices at $z \neq 0$ can be obtained by translating the source–detector system in the z-direction, and then repeating the steps listed above. This part of the procedure will not be relevant for the application to quantum tomography, so from now on we only consider $z = 0$ and replace $\rho(x, y, z = 0)$ by $\rho(x, y)$.

From the above considerations, we formulate the following (not necessarily optimal) procedure for collecting tomographic data.

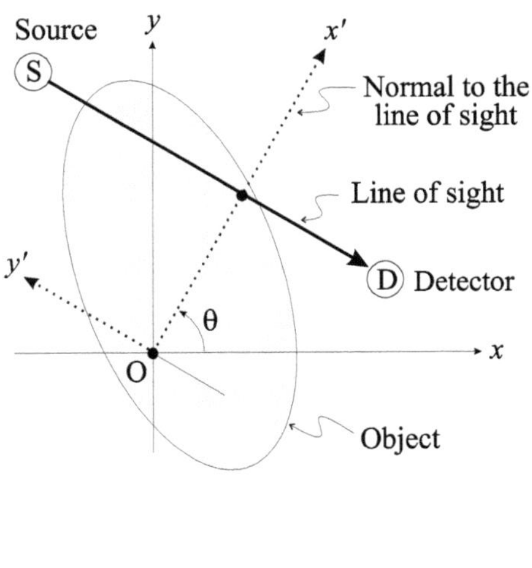

Fig. 17.1 Coordinate system used in tomography.

(1) Collect the projections for lines of sight at a fixed angle θ, while scanning the coordinate x' from one side of O to the other.

(2) Repeat this procedure after incrementing the angle θ by a small amount.

(3) Repeat steps (1) and (2), collecting data for $P_\theta(x')$ for $-\infty < x' < \infty$ and $0 < \theta < \pi$.

(4) Determine the original density $\rho(x, y)$ by means of the inverse Radon transform described below.

17.1.2 The Radon transform

The rotated coordinates (x', y') are related to the fixed coordinates by

$$x' = x \cos\theta + y \sin\theta\,, \quad y' = -x \sin\theta + y \cos\theta\,, \tag{17.2}$$

and the inverse relation is

$$x = x' \cos\theta - y' \sin\theta\,, \quad y = x' \sin\theta + y' \cos\theta\,. \tag{17.3}$$

The projection $P_\theta(x')$ defines the **forward Radon transform**:

$$P_\theta(x') = \int_{\overline{\mathrm{SD}}} \rho(x, y)\, ds \tag{17.4}$$

$$= \int_{-\infty}^{+\infty} \rho(x' \cos\theta - y' \sin\theta, x' \sin\theta + y' \cos\theta)\, dy'\,. \tag{17.5}$$

For the application at hand, the convention for Fourier transforms used in the other parts of this book can lead to confusion; therefore, we revert to the usual notation in which $\widetilde{f}$ denotes the Fourier transform of f. Let us then consider the one-dimensional Fourier transform of the projection $P_\theta(x')$,

$$\widetilde{P}_\theta(k) \equiv \int_{-\infty}^{+\infty} dx'\, P_\theta(x') e^{-ikx'}\,, \tag{17.6}$$

and the two-dimensional Fourier transform of the density,

$$\widetilde{\rho}(u, v) \equiv \int_{-\infty}^{+\infty} dx \int_{-\infty}^{+\infty} dy\, \rho(x, y) e^{-i(xu+yv)}\,. \tag{17.7}$$

The **Fourier slice theorem** states that

$$\widetilde{P}_\theta(k) = \widetilde{\rho}(k \cos\theta, k \sin\theta)\,. \tag{17.8}$$

The proof proceeds as follows. Inspection of Fig. 17.1 shows that the two-dimensional wavevector

$$\mathbf{k} = (k \cos\theta, k \sin\theta) \tag{17.9}$$

is directed along the line $\overline{\mathrm{OP}}$, i.e. the x'-axis. For any point on the line of sight, with coordinates $\mathbf{r} = (x, y)$, one finds $\mathbf{k} \cdot \mathbf{r} = kx \cos\theta + ky \sin\theta = kx'$. Substituting this relation and the definition (17.5) of the forward Radon transform into eqn (17.6) then leads to

$$\begin{aligned}\widetilde{P}_{\theta}(k) &= \int_{-\infty}^{+\infty} dx' \int_{-\infty}^{+\infty} dy'\, e^{-i\mathbf{k}\cdot\mathbf{r}} \rho(x(x',y'), y(x',y')) \\ &= \int_{-\infty}^{+\infty} dx \int_{-\infty}^{+\infty} dy\, e^{-i\mathbf{k}\cdot\mathbf{r}} \rho(x,y)\,, \end{aligned} \tag{17.10}$$

where the last form follows by changing integration variables and using the fact that the transformation linking (x',y') to (x,y) has unit Jacobian. This result is just the definition of the Fourier transform of the density, so we arrive at eqn (17.8).

For the final step, we first express the density in physical space, $\rho(x,y)$, as the inverse Fourier transform of the density in reciprocal space:

$$\rho(x,y) = \frac{1}{4\pi^2} \int_{-\infty}^{\infty} du \int_{-\infty}^{\infty} dv\, \widetilde{\rho}(u,v) e^{i(xu+yv)}\,. \tag{17.11}$$

In order to use the Fourier slice theorem, we identify (u,v) with the two-dimensional vector $\mathbf{k}$, defined in eqn (17.9), so that $u = k\cos\theta$ and $v = k\sin\theta$. This resembles the familiar transformation to polar coordinates, but one result of Exercise 17.1 is that the restriction $0 < \theta < \pi$ requires k to take on negative as well as positive values, i.e. $-\infty < k < \infty$. This transformation implies that $du\, dv = dk\, |k|\, d\theta$, so that eqn (17.11) becomes

$$\rho(x,y) = \frac{1}{4\pi^2} \int_{-\infty}^{\infty} |k|\, dk \int_{0}^{\pi} d\theta\, \widetilde{\rho}\,(k\cos\theta, k\sin\theta)\, e^{ik(x\cos\theta + y\sin\theta)}\,, \tag{17.12}$$

and the Fourier slice theorem allows this to be expressed as

$$\rho(x,y) = \frac{1}{4\pi^2} \int_{-\infty}^{\infty} |k|\, dk \int_{0}^{\pi} d\theta\, \widetilde{P}_{\theta}(k) e^{ik(x\cos\theta + y\sin\theta)}\,. \tag{17.13}$$

Substituting eqn (17.6) in this relation yields the *inverse Radon transform*:

$$\rho(x,y) = \frac{1}{4\pi^2} \int_{-\infty}^{\infty} |k|\, dk \int_{0}^{\pi} d\theta \int_{-\infty}^{\infty} dx'\, P_{\theta}(x') e^{ik(x\cos\theta + y\sin\theta - x')}\,. \tag{17.14}$$

This result reconstructs the density distribution $\rho(x,y)$ from the measured data set $P_{\theta}(x')$.

17.2 Optical homodyne tomography

In eqn (5.126) we introduced a version of the Wigner distribution, $W(\alpha)$, that is particularly well suited to quantum optics. The complex argument α, which is the amplitude defining a coherent state, is equivalent to the pair of real variables $x = \operatorname{Re}\alpha$ and $y = \operatorname{Im}\alpha$; consequently, $W(\alpha)$ can equally well be regarded as a function of x and y, as in Exercise 17.2. Expressing the Wigner distribution in this form suggests that $W(x,y)$ is an analogue of the density function $\rho(x,y)$. With this interpretation, the mathematical analysis used for classical tomography can be applied to recover $W(x, y)$ from an appropriate set of measurements. The objection that the quasiprobability $W(x, y)$ can be negative—as shown by the number-state example in eqn (5.153)—does not pose a serious difficulty, since negative absorption in the classical problem would simply correspond to amplification.

In order to apply the inverse Radon transform (17.14) to quantum optics, we must first understand the physical significance of the projection $P_\theta(x')$. In this context, the parameter θ is not a geometrical angle; instead, it is the phase of the local oscillator field in a homodyne measurement scheme. As explained in Section 9.3, this parameter labels the natural quadratures,

$$X_\theta = X_0 \cos\theta + Y_0 \sin\theta\,, \quad Y_\theta = X_0 \sin\theta - Y_0 \cos\theta\,, \tag{17.15}$$

for homodyne measurement. Generalizing eqn (5.123) tells us that integrating the Wigner distribution over one of the conjugate variables generates the marginal probability distribution for the other; so applying the forward Radon transform (17.5) to the Wigner distribution leads to the conclusion that the projection,

$$P_\theta(x') = \int_{-\infty}^{+\infty} W(x' \cos\theta - y' \sin\theta, x' \sin\theta + y' \cos\theta) dy'\,, \tag{17.16}$$

is the probability distribution for measured values x' of the operator X_θ.

The difference between the physical interpretations of $P_\theta(x')$ in classical and quantum tomography requires corresponding changes in the experimental protocol. Setting the phase of the local oscillator in a homodyne measurement scheme is analogous to setting the angle θ of the X-ray beam, but there is no analogue for setting the lateral position x'. In the quantum optics application, the variable x' is not under experimental control. Instead, it represents the possible values of the quadrature X_θ, which are subject to quantum fluctuations.

In this situation, the procedure is to set a value of θ and then carry out many homodyne measurements of X_θ. A histogram of the results determines the fraction of the values falling in the interval x' to $x' + \Delta x'$, and thus the probability distribution $P_\theta(x')$. This is easier said than done, and it represents a substantial advance beyond previous experiments, that simply measured the average and variance of the quadrature. Once $P_\theta(x')$ has been experimentally determined, the inverse Radon transform yields the Wigner function as

$$W(x, y) = \frac{1}{4\pi^2} \int_{-\infty}^{\infty} |k|\, dk \int_0^{\pi} d\theta \int_{-\infty}^{\infty} dx'\, P_\theta(x') e^{ik(x\cos\theta + y\sin\theta - x')}\,. \tag{17.17}$$

As shown in Section 5.6.1, the Wigner distribution permits the evaluation of the average of any observable; consequently, this reconstruction of the Wigner distribution provides a complete description of the quantum state of the light.

17.3 Experiments in optical homodyne tomography

The method of optical homodyne tomography sketched above is one example from a general field variously called **quantum-state tomography** (Raymer and Funk, 2000) or **quantum-state reconstruction** (Altepeter *et al.*, 2005). Techniques for recovering the density matrix from measured values have been applied to atoms (Ashburn *et al.*, 1990), molecules (Dunn *et al.*, 1995), and Bose–Einstein condensates (Bolda *et al.*, 1998). In the domain of quantum optics, Raymer and co-workers (Smithey *et al.*, 1993) studied the properties of squeezed states by using pulsed light for the signal and the local oscillator. This is an important technique for obtaining time-resolved data for various processes (Raymer *et al.*, 1995), but the simple theory presented above is more suitable for describing experiments with continuous-wave (cw) beams.

17.3.1 Optical tomography for squeezed states

Following Raymer's pulsed-light, quantum-state tomography experiments, Mlynek and his co-workers (Breitenbach *et al.*, 1997) performed experiments in which they generated and then analyzed squeezed states. The description of the experiment is therefore naturally divided into the generation and measurement steps.

A *Squeezed state generation*

The light used in this experiment is provided by an Nd:YAG (neodymium-doped, yttrium–aluminum garnet) laser (1064 nm and 500 mW) operated in cw mode. As shown in Fig. 17.2, the laser beam, at frequency ω, first passes through a **mode cleaning cavity** (a high finesse Fabry–Perot resonator with a 170 kHz bandwidth) in order to reduce technical noise arising from relaxation oscillations in the laser. The filtered beam is then split into three parts: the upper part is sent into a **second-harmonic generator** (SHG); the middle part is sent into an **electro-optic modulator** (EOM)

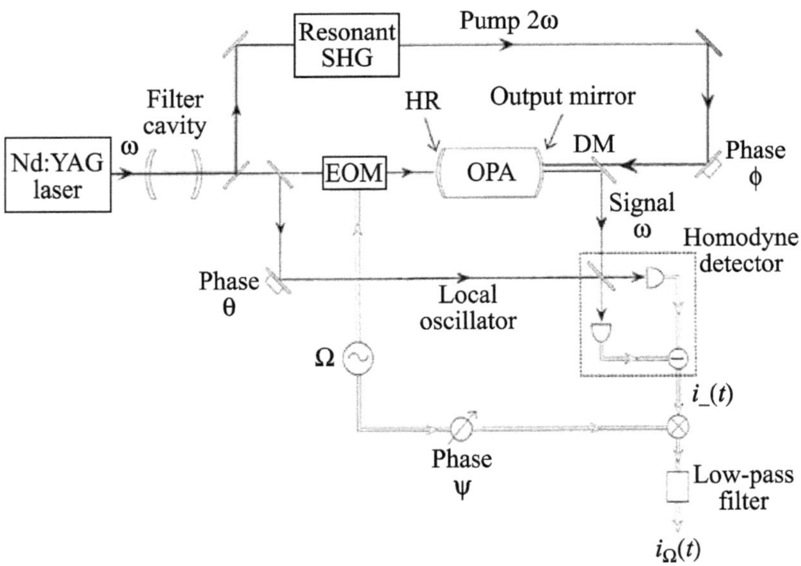

Fig. 17.2 Experimental setup used for generating and detecting squeezed light in a tomographic scheme. (Reproduced from Mlynek *et al.* (1998).)

(Saleh and Teich, 1991, Sec. 18.1-B); and the lower part serves as the local oscillator for the homodyne detector.

The resonant SHG—a $\chi^{(2)}$ crystal placed inside a 2ω-resonator—produces a second-harmonic pump beam that enters the OPA through the right-hand mirror. This mirror also serves as the output port for the squeezed light near frequency ω. The OPA consists of a $\chi^{(2)}$ crystal coated on the left end with a mirror (HR) that is highly reflective at both ω and 2ω and on the right end with the output mirror. The two mirrors define a cavity that is resonant at both the first and second harmonics. For an unmodulated input, e.g. vacuum fluctuations, this is a degenerate OPA configuration.

The down-converted photons in each pair share the same spatial mode, polarization, and frequency. For a sufficiently high transmission coefficient of the output mirror at frequency ω, the OPA produces an intense, squeezed-light output signal in the vicinity of ω. The parametric gain of the OPA at the pump frequency is maximized by adjusting the temperature of the crystal. The **dichroic** mirror (DM)—located to the right of the output mirror—transmits the incoming 2ω-pump beam toward the OPA, but deflects the outgoing squeezed-light beam into the homodyne detector.

The EOM voltage is modulated at frequency Ω, where $\Omega/2\pi = 1.5$ or $2.5\,\text{MHz}$. This adds two side bands to the coherent middle beam, at frequencies $\omega \pm \Omega$ that are well within the cavity bandwidth, $\Gamma/2\pi = 17\,\text{MHz}$. The OPA is operated in a *dual port* configuration, i.e. the pump beam enters through the output mirror on the right and the coherent signal is injected through the mirror HR on the left. The OPA cavity is also highly asymmetric; the transmission coefficient at frequency ω is less than 0.1% for the mirror HR, but about 2.1% for the output mirror.

Due to this high asymmetry, the transmitted sidebands and their quantum fluctuations are strongly attenuated, as shown in Exercise 17.3, so that the squeezed output comes primarily from the vacuum fluctuations at ω, entering through the output coupler. The output of the OPA then consists of squeezed vacuum at ω together with bright sidebands at $\omega \pm \Omega$. If the output from the EOM is blocked, the OPA emits a pure squeezed vacuum state. If the output from the SHG is blocked, the OPA emits a coherent state.

B Tomographic measurements

The output of the OPA is sent into the homodyne detector, but this is a new way of using homodyne methods. The usual approach, presented in Section 9.3, assumes that the detectors are only sensitive to the overall energy flux of the light; consequently, the homodyne signal is defined by averaging over the field state: $S_{\text{hom}} \propto \langle N'_{21} \rangle$, where $\langle N'_{21} \rangle$ represents the difference in the firing rates of the two detectors.

For **photoemissive** detectors, i.e. those with frequency-independent quantum efficiency (Raymer *et al.*, 1995), the quantum fluctuations represented by the operator N'_{21} are visible as fluctuations in the difference between the output currents of the detectors. In the present case, $N'_{21} = -i\left(\alpha_L^* b_{\text{out}} - b_{\text{out}}^\dagger \alpha_L\right)$, where $\alpha_L = |\alpha_L| \exp(-i\theta)$ is the classical amplitude of the local oscillator and b_{out} describes the output field of the OPA. Expressing N'_{21} in terms of quadrature operators as

$$N'_{21} \propto X \cos\theta + Y \sin\theta = X_\theta \tag{17.18}$$

shows that observations of the current fluctuations represent measurements of the quadrature X_θ.

The data (about half a million points per trace) for the current i_Ω were taken with a high-speed 12 bit analog-to-digital converter, as the phase of the local oscillator was swept by 360° in approximately 200 ms. Time traces of i_Ω for coherent states and for squeezed states are shown in the left-most column of Fig. 17.3.

The top trace represents the coherent state output, which is obtained by blocking the second-harmonic pump beam. This characterizes and calibrates the laser system used for the local oscillator and the first-harmonic input into the resonant, second-harmonic generator crystal. The next three traces represent squeezed coherent states. The second trace is the waveform for a phase-squeezed state, where the noise is minimum at the zero-crossings of the waveform. The third trace represents a state squeezed along the $\phi = 48°$ quadrature, where ϕ is the relative phase between the pump wave and the coherent input wave. The fourth trace represents the waveform for amplitude-

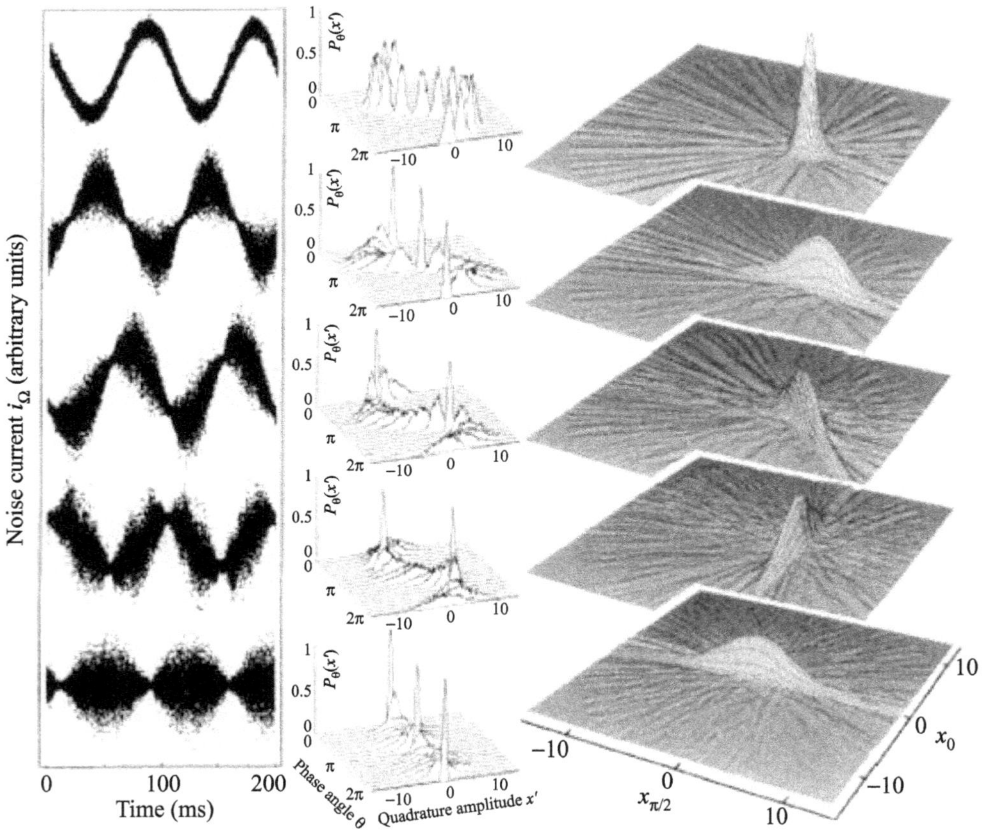

Fig. 17.3 Data showing noise waveforms for a coherent state (top trace) and various kinds of squeezed states (lower traces), along with their phase space tomographic portraits on the right. (Reproduced from Mlynek *et al.* (1998).)

squeezed light, where the noise is minimum at the maxima of the waveform. Finally, the fifth trace represents the squeezed vacuum state, where the coherent state input to the parametric amplifier has been completely blocked, so that only vacuum fluctuations are admitted into the OPA. Ones sees that the noise vanishes periodically at the zero-crossings of the noise envelope.

The middle column of Fig. 17.3 depicts the tomographic projections $P_\theta(x')$, which are substituted into the inverse Radon transform (17.17) to generate the portraits of Wigner functions depicted in the third column of the figure. Numerical analysis of the distributions for the second through the fifth traces shows that they all have the Gaussian shape predicted for squeezed coherent states.

17.4 Exercises

17.1 Radon transform

(1) For the transformation $u = k\cos\theta$, $v = k\sin\theta$, with the restriction $0 < \theta < \pi$, work out the inverse transformation expressing k and θ as functions of u and v, and thus show that k must have negative as well as positive values.

(2) Derive the relation $du\, dv = |k|\, dk\, d\theta$ by evaluating the Jacobian or else by just drawing the appropriate picture.

17.2 Wigner distribution

Starting from the definition (5.126), show that $W(\alpha) = W(x, y)$ can be written in the form

$$W(x, y) = \int \frac{d^2k}{(2\pi)^2} e^{i\mathbf{k}\cdot\mathbf{r}} \chi_W(\mathbf{k}),$$

where $\mathbf{k} = (k_1, k_2)$ and $\mathbf{r} = (x, y)$. Derive the explicit form of the Wigner characteristic function in terms of the density operator ρ and the quadrature operators X_0 and Y_0. What normalization condition does $W(x, y)$ satisfy?

17.3 Dual port OPA*

Model the dual port OPA discussed in Section 17.3.1 by identifying the input and output fields as $b_{\text{out}} = b_{1,\text{out}}$ and $b_{\text{in}} = b_{2,\text{in}}$, where the notation is taken from Fig. 16.2.

(1) Use eqn (15.117) to work out the coefficients P and C in the input–output relations for this amplifier.

(2) Explicitly evaluate the amplifier noise operator.

(3) For the unbalanced case $\kappa_2 \ll \kappa_1$ show that the incident field is strongly attenuated and that the primary source of the squeezed output is the vacuum fluctuations entering through the mirror M1.

18
The master equation

In this chapter we will study the time evolution of an open system—the *sample* discussed in Chapter 14—by means of the quantum Liouville equation for the world density operator. This approach, which employs the interaction-picture description of the density operator, is complementary to the Heisenberg-picture treatment presented in Chapter 14. The physical ideas involved in the two methods are, however, the same.

The equation of motion of the reduced sample density operator is derived by an approximate elimination of the environment degrees of freedom that depends crucially on the Markov approximation. This approximate equation of motion for the sample density operator is mainly used to derive c-number equations that can be solved by numerical methods. In this connection, we will discuss the Fokker–Planck equation in the P-representation and the method of quantum Monte Carlo wave functions.

18.1 Reduced density operators

As explained in Section 14.1.1, the world—the composite system of the sample and the environment—is described by a density operator ρ_W acting on the Hilbert space $\mathfrak{H}_W = \mathfrak{H}_S \otimes \mathfrak{H}_E$. The application of the general definition (6.21) of the reduced density operator to ρ_W produces two reduced density operators: $\varrho_S = \mathrm{Tr}_E(\rho_W)$ and $\varrho_E = \mathrm{Tr}_S(\rho_W)$, that describe the sample and the environment respectively. For example, the rule (6.26) for partial traces shows that the average of a sample operator Q,

$$\langle Q \rangle = \mathrm{Tr}_W \left[\rho_W \left(Q \otimes I_E \right) \right] = \mathrm{Tr}_S \left(\varrho_S Q \right), \tag{18.1}$$

is entirely determined by the reduced density operator for the sample.

For an open system, the reduced density operator ϱ_S will always describe a mixed state. According to Theorem 6.1 in Section 6.4.1, the reduced density operators for the sample and the environment can only describe pure states if the world density operator describes a separable pure state, i.e. $\rho_W = |\Omega_W\rangle \langle \Omega_W|$, and $|\Omega_W\rangle = |\Psi_S\rangle |\Phi_E\rangle$. Even if this were initially the case, the interaction between the sample and the environment would inevitably turn the separable pure state $|\Omega_W\rangle$ into an entangled pure state. The reduced density operators for an entangled state necessarily describe mixed states, so the sample state will always evolve into a mixed state.

18.2 The environment picture

In the Schrödinger picture, the world density operator satisfies the quantum Liouville equation

$$i\hbar\frac{\partial}{\partial t}\rho_W(t) = [H_W(t), \rho_W(t)]$$
$$= [H_S(t) + H_E + H_{\text{SE}}, \rho_W(t)], \tag{18.2}$$

where the terms in H_W are defined in eqns (14.6)–(14.11). Since the sample–environment interaction is assumed to be weak, it is natural to regard $H_{W0}(t) = H_S(t) + H_E$ as the zeroth-order part, and $H_{\text{SE}}(t)$ as the perturbation. This allows us to introduce an interaction picture, through the unitary transformation,

$$|\Psi^{\text{env}}(t)\rangle = U_{W0}^{\dagger}(t)|\Psi(t)\rangle, \tag{18.3}$$

where $U_{W0}(t)$ satisfies

$$i\hbar\frac{\partial}{\partial t}U_{W0}(t) = H_{W0}(t)U_{W0}(t), \quad U_{W0}(0) = 1. \tag{18.4}$$

We will call this interaction picture the **environment picture**, since it plays a special role in the theory.

The differential equation (18.4) has the same form as eqn (4.90), but now the ordering of the operators $U_{W0}(t)$ and $H_{W0}(t)$ is important, since the time-dependent Hamiltonian $H_{W0}(t)$ will not in general commute with $U_{W0}(t)$. If this warning is kept firmly in mind, the formal procedure described in Section 4.8 can be used again to find the Schrödinger equation

$$i\hbar\frac{\partial}{\partial t}|\Psi^{\text{env}}(t)\rangle = H_{\text{SE}}^{\text{env}}(t)|\Psi^{\text{env}}(t)\rangle, \tag{18.5}$$

and the quantum Liouville equation

$$i\hbar\frac{\partial}{\partial t}\rho_W^{\text{env}}(t) = [H_{\text{SE}}^{\text{env}}(t), \rho_W^{\text{env}}(t)] \tag{18.6}$$

in the environment picture. The transformed operators,

$$O^{\text{env}}(t) = U_{W0}^{\dagger}(t)\,O\,U_{W0}(t), \tag{18.7}$$

satisfy the equations of motion

$$i\hbar\frac{\partial}{\partial t}O^{\text{env}}(t) = [O^{\text{env}}(t), H_{W0}^{\text{env}}(t)]. \tag{18.8}$$

One should keep in mind that $H_{W0}^{\text{env}}(t) = H_S^{\text{env}}(t) + H_E^{\text{env}}(t)$, and that the sample Hamiltonian, $H_S^{\text{env}}(t)$, still contains all interaction terms between different degrees of freedom in the sample. Only the sample–environment interaction is treated as a perturbation. Thus the environment-picture sample operators obey the full Heisenberg equations for the sample.

18.3 Averaging over the environment

In line with our general convention, we will now drop the identifying superscript 'env', and replace it by the understanding that states and operators in the following discussion are normally expressed in the environment picture. Exceptions to this rule will be explicitly identified. Our immediate task is to derive an equation of motion for the reduced density operator ρ_S. In pursuing this goal, we will generally follow Gardiner's treatment (Gardiner, 1991, Chap. 5).

The first part of the argument corresponds to the formal elimination of the reservoir operators in Section 14.1.2, and we begin in a similar way by incorporating the quantum Liouville equation (18.6) and the initial density operator $\rho_W(0)$ into the equivalent integral equation,

$$\rho_W(t) = \rho_W(0) - \frac{i}{\hbar}\int_0^t dt_1 \left[H_{\mathrm{SE}}(t_1), \rho_W(t_1)\right]. \tag{18.9}$$

The assumption that H_{SE} is weak—compared to H_S and H_E—suggests solving this equation by perturbation theory, but a perturbation expansion would only be valid for a very short time. From Chapter 14, we know that typical sample correlation functions decay exponentially:

$$\langle Q_1(t+\tau) Q_2(t)\rangle \sim e^{-\gamma\tau}, \tag{18.10}$$

with a decay rate, $\gamma \sim g^2$, where g is the sample–environment coupling constant. This exponential decay can not be recovered by an expansion of $\rho_W(t)$ to any finite order in g.

As the first step toward finding a better approach, we iterate the integral equation (18.9) twice—this is suggested by the fact that γ is second order in g—to find

$$\begin{aligned}\rho_W(t) = \rho_W(0) &- \frac{i}{\hbar}\int_0^t dt_1 \left[H_{\mathrm{SE}}(t_1), \rho_W(0)\right] \\ &+ \left(-\frac{i}{\hbar}\right)^2 \int_0^t dt_1 \int_0^{t_1} dt_2 \left[H_{\mathrm{SE}}(t_1), \left[H_{\mathrm{SE}}(t_2), \rho_W(t_2)\right]\right].\end{aligned} \tag{18.11}$$

Tracing over the environment then produces the exact equation

$$\begin{aligned}\rho_S(t) = \rho_S(0) &- \frac{i}{\hbar}\int_0^t dt_1 \,\mathrm{Tr}_E\left(\left[H_{\mathrm{SE}}(t_1), \rho_W(0)\right]\right) \\ &+ \left(-\frac{i}{\hbar}\right)^2 \int_0^t dt_1 \int_0^{t_1} dt_2 \,\mathrm{Tr}_E\left(\left[H_{\mathrm{SE}}(t_1), \left[H_{\mathrm{SE}}(t_2), \rho_W(t_2)\right]\right]\right)\end{aligned} \tag{18.12}$$

for the reduced density operator. Since our objective is an equation of motion for $\rho_S(t)$, the next step is to differentiate with respect to t to find

$$\begin{aligned}\frac{\partial}{\partial t}\rho_S(t) = &-\frac{i}{\hbar}\mathrm{Tr}_E\left(\left[H_{\mathrm{SE}}(t), \rho_W(0)\right]\right) \\ &- \frac{1}{\hbar^2}\int_0^t dt' \,\mathrm{Tr}_E\left(\left[H_{\mathrm{SE}}(t), \left[H_{\mathrm{SE}}(t'), \rho_W(t')\right]\right]\right).\end{aligned} \tag{18.13}$$

This equation is exact, but it is useless as it stands, since the unknown world density operator $\rho_W(t')$ appears on the right side. Further progress depends on finding approximations that will lead to a manageable equation for $\rho_S(t)$ alone. The first simplifying assumption is that the sample and the environment are initially uncorrelated: $\rho_W(0) = \rho_S(0)\,\rho_E(0)$. By combining the generic expression,

$$H_{\mathrm{SE}} = i\hbar \sum_r \sum_\nu \left(g^*_{r\nu} b^\dagger_{r\nu} Q_r - g_{r\nu} Q^\dagger_r b_{r\nu} \right), \tag{18.14}$$

for the sample–environment interaction with the conventional assumption, $\langle b_{r\nu}\rangle_E = 0$, and the initial factorization condition, it is straightforward to show that

$$\mathrm{Tr}_E\left(\left[H_{\mathrm{SE}}(t), \rho_W(0) \right] \right) = 0\,. \tag{18.15}$$

If one or more of the reservoirs has $\langle b_{r\nu}\rangle_E \neq 0$, one can still get this result by writing H_{SE} in terms of the fluctuation operators $\delta b_{r\nu} = b_{r\nu} - \langle b_{r\nu}\rangle_E$, and absorbing the extra terms by suitably redefining H_S and H_E, as in Exercise 18.1.

Thus the initial factorization assumption always allows eqn (18.13) to be replaced by the simpler form

$$\frac{\partial}{\partial t}\rho_S(t) = -\frac{1}{\hbar^2}\int_0^t dt'\, \mathrm{Tr}_E\left\{ \left[H_{\mathrm{SE}}(t), \left[H_{\mathrm{SE}}(t'), \rho_W(t') \right] \right] \right\}. \tag{18.16}$$

Replacing $\rho_W(t')$ by $\rho_W(0) = \rho_S(0)\,\rho_E(0)$ in eqn (18.16) would provide a perturbative solution that is correct to second order, but—as we have just seen—this would not correctly describe the asymptotic time dependence of the correlation functions.

The key to finding a better approximation is to exploit the extreme asymmetry between the sample and the environment. The environment is very much larger than the sample; indeed, it includes the rest of the universe. It is therefore physically reasonable to assume that the fractional change in the sample, caused by interaction with the environment, is much larger than the fractional change in the environment, caused by interaction with the sample. If this is the case, there will be no reciprocal correlation between the sample and the environment, and the density operator $\rho_W(t')$ will be approximately factorizable at all times.

This argument suggests the *ansatz*

$$\rho_W(t') \approx \rho_S(t')\,\rho_E(0)\,, \tag{18.17}$$

and using this in eqn (18.16) produces the **master equation**:

$$\frac{\partial}{\partial t}\rho_S(t) = -\frac{1}{\hbar^2}\int_0^t dt'\, \mathrm{Tr}_E\left\{ \left[H_{\mathrm{SE}}(t), \left[H_{\mathrm{SE}}(t'), \rho_S(t')\,\rho_E(0) \right] \right] \right\}. \tag{18.18}$$

The double commutator in eqn (18.18) can be rewritten in a more convenient way by exploiting the fact that typical interactions have the form

$$H_{\mathrm{SE}}(t) = \hbar\left\{ \mathfrak{F}(t) + \mathfrak{F}^\dagger(t) \right\}. \tag{18.19}$$

This in turn allows the double commutator to be written as

$$\mathfrak{C}_2(t,t') = \frac{1}{\hbar^2}\left[H_{\text{SE}}(t), \left[H_{\text{SE}}(t'), \rho_S(t')\,\rho_E(0)\right]\right]$$
$$= \left\{\left[\mathfrak{F}(t), \mathfrak{G}(t')\right] + \text{HC}\right\} + \left\{\left[\mathfrak{F}^\dagger(t), \mathfrak{G}(t')\right] + \text{HC}\right\}, \tag{18.20}$$

where

$$\mathfrak{G}(t') = \left[\mathfrak{F}(t'), \rho_S(t')\,\rho_E(0)\right]. \tag{18.21}$$

18.4 Examples of the master equation

In order to go on, it is necessary to assume an explicit form for H_{SE}. For this purpose, we will consider the two concrete examples that were studied in Chapter 14: the single cavity mode and the two-level atom.

18.4.1 Single cavity mode

In the environment picture, the definition (14.43) of system–reservoir interaction for a single cavity mode becomes

$$H_{\text{SE}} = i\hbar\sum_\nu v(\Omega_\nu)\left\{a^\dagger(t)\,b_\nu(t) - \text{HC}\right\}. \tag{18.22}$$

Since $a(t)$ and $b_\nu(t)$ are evaluated in the environment picture, they satisfy the Heisenberg equations

$$\frac{\partial}{\partial t}a(t) = \frac{1}{i\hbar}\left[a(t), H_S(t)\right]$$
$$= -i\omega_0 a(t) + \frac{1}{i\hbar}\left[a(t), H_{S1}(t)\right], \tag{18.23}$$

and

$$\frac{\partial}{\partial t}b_\nu(t) = -i\Omega_\nu b_\nu(t). \tag{18.24}$$

By introducing the slowly-varying envelope operators $\overline{a}(t) = a(t)\exp(i\omega_0 t)$ and $\overline{b}_\nu(t) = b_\nu(t)\exp(i\omega_0 t)$, we can express $H_{\text{SE}}(t)$ in the form (18.19), with

$$\mathfrak{F}(t) = -i\xi^\dagger(t)\,\overline{a}(t), \tag{18.25}$$

where we recognize

$$\xi(t) = \sum_\nu v(\Omega_\nu)\,\overline{b}_\nu(t_0)\,e^{-i(\Omega_\nu-\omega_0)(t-t_0)} \tag{18.26}$$

as the noise operator defined in eqn (14.52).

The terms in $[\mathfrak{F}(t), \mathfrak{G}(t')]$ contain products of $\xi^\dagger(t)$, $\xi^\dagger(t')$, and $\rho_E(0)$ in various orders. When the partial trace over the environment states in eqn (18.18) is performed, the cyclic invariance of the trace can be exploited to show that all terms are proportional to

$$\left\langle \xi^\dagger(t)\,\xi^\dagger(t')\right\rangle_E = \text{Tr}_E\left[\rho_E(0)\,\xi^\dagger(t)\,\xi^\dagger(t')\right]. \tag{18.27}$$

Just as in Section 14.2, we will assume that $\rho_E(0)$ is diagonal in the reservoir oscillator occupation number—this amounts to assuming that $\rho_E(0)$ is a stationary distribution—so that the correlation functions in eqn (18.27) all vanish. This assumption is convenient, but it is not strictly necessary. A more general treatment—that includes, for example, a reservoir described by a squeezed state—is given in Walls and Milburn (1994, Sec. 6.1).

When $\rho_E(0)$ is stationary, $[\mathfrak{F}(t), \mathfrak{G}(t')]$ and its adjoint will not contribute to the master equation. By contrast, the commutator $\left[\mathfrak{F}^\dagger(t), \mathfrak{G}(t')\right]$ is a sum of terms containing products of $\xi(t)$, $\xi^\dagger(t')$, and $\rho_E(0)$ in various orders. In this case, the cyclic invariance of the partial trace produces two kinds of terms, proportional respectively to

$$\left\langle \xi^\dagger(t)\,\xi(t')\right\rangle_E = n_{\text{cav}}\kappa\delta(t-t') \tag{18.28}$$

and

$$\left\langle \xi(t)\,\xi^\dagger(t')\right\rangle = (n_{\text{cav}}+1)\,\kappa\delta(t-t')\,, \tag{18.29}$$

where n_{cav} is the average number of reservoir oscillators at the cavity-mode frequency ω_0.

The explicit expressions on the right sides of these equations come from eqns (14.74) and (14.75), which were derived by using the Markov approximation. Thus the master equation also depends on the Markov approximation, in particular on the assumption that the envelope operator $\overline{a}(t)$ is essentially constant over the memory interval $(t - T_{\text{mem}}/2, t + T_{\text{mem}}/2)$.

After evaluating the partial trace of the double commutator $\mathfrak{C}_2(t,t')$—see Exercise 18.2—the environment-picture form of the master equation for the field is found to be

$$\begin{aligned}\frac{\partial}{\partial t}\rho_S(t) = &-\frac{\kappa}{2}(n_{\text{cav}}+1)\left\{\overline{a}^\dagger(t)\,\overline{a}(t)\,\rho_S(t) + \rho_S(t)\,\overline{a}^\dagger(t)\,\overline{a}(t) - 2\overline{a}(t)\,\rho_S(t)\,\overline{a}^\dagger(t)\right\}\\ &-\frac{\kappa}{2}n_{\text{cav}}\left\{\overline{a}(t)\,\overline{a}^\dagger(t)\,\rho_S(t) + \rho_S(t)\,\overline{a}(t)\,\overline{a}^\dagger(t) - 2\overline{a}^\dagger(t)\,\rho_S(t)\,\overline{a}(t)\right\}.\end{aligned} \tag{18.30}$$

The slowly-varying envelope operators $\overline{a}(t)$ and $\overline{a}^\dagger(t)$ are paired in every term; consequently, they can be replaced by the original environment-picture operators $a(t)$ and $a^\dagger(t)$ without changing the form of the equation. The right side of the equation of motion is therefore entirely expressed in terms of environment-picture operators, so we can easily transform back to the Schrödinger picture to find

$$\frac{\partial}{\partial t}\rho_S(t) = \mathcal{L}_S\rho_S(t) + \mathcal{L}_{\text{dis}}\rho_S(t)\,. \tag{18.31}$$

The **Liouville operators** $\mathcal{L}_S$—describing the free Hamiltonian evolution of the sample—and $\mathcal{L}_{\text{dis}}$—describing the dissipative effects arising from coupling to the environment—are respectively given by

$$\mathcal{L}_S\rho_S(t) = \frac{1}{i\hbar}\left[\hbar\omega_0 a^\dagger a + H_{S1}(t), \rho_S(t)\right] \tag{18.32}$$

and

$$\begin{aligned}\mathcal{L}_{\text{dis}}\rho_S(t) = &-\frac{\kappa}{2}(n_{\text{cav}}+1)\left\{\left[a^\dagger, a\rho_S(t)\right] + \left[\rho_S(t)\,a^\dagger, a\right]\right\}\\ &-\frac{\kappa}{2}n_{\text{cav}}\left\{\left[a, a^\dagger\rho_S(t)\right] + \left[\rho_S(t)\,a, a^\dagger\right]\right\}.\end{aligned} \tag{18.33}$$

The operators we are used to, such as the Hamiltonian or the creation and annihilation operators, send one Hilbert-space vector to another. By contrast, the Liouville operators send one operator to another operator. For this reason they are sometimes called **super operators**.

A Thermal equilibrium again

In Exercise 14.2 it is demonstrated that the average photon number asymptotically approaches the Planck distribution. With the aid of the master equation, we can study this limit in more detail. In this case, $H_{S1}(t) = 0$, so we can expect the density operator to be diagonal in photon number. The diagonal matrix elements of eqn (18.31) in the number-state basis yield

$$\begin{aligned}\frac{dp_n(t)}{dt} &= -\kappa\left\{(n_{\text{cav}}+1)\,n + n_{\text{cav}}(n+1)\right\}p_n(t) \\ &\quad + \kappa(n_{\text{cav}}+1)(n+1)\,p_{n+1}(t) + \kappa n_{\text{cav}} n\, p_{n-1}(t)\,,\end{aligned} \tag{18.34}$$

where $p_n(t) = \langle n|\rho(t)|n\rangle$. The first term on the right represents the rate of flow of probability from the n-photon state to all other states, while the second and third terms represent the flow of probability into the n-photon state from the $(n+1)$-photon state and the $(n-1)$-photon state respectively.

In order to study the approach to equilibrium, we write the equation as

$$\frac{dp_n(t)}{dt} = Z_{n+1}(t) - Z_n(t)\,, \tag{18.35}$$

where

$$Z_n(t) = n\kappa\left\{(n_{\text{cav}}+1)\,p_n(t) - n_{\text{cav}}p_{n-1}(t)\right\}. \tag{18.36}$$

The equilibrium condition is $Z_{n+1}(\infty) = Z_n(\infty)$, but this is the same as $Z_n(\infty) = 0$, since $Z_0(t) \equiv 0$. Thus equilibrium imposes the recursion relations

$$(n_{\text{cav}}+1)\,p_n(\infty) = n_{\text{cav}}p_{n-1}(\infty)\,. \tag{18.37}$$

This is an example of the principle of detailed balance; the rate of probability flow from the n-photon state to the $(n-1)$-photon state is the same as the rate of probability flow of the $(n-1)$-photon state to the n-photon state. The solution of this recursion relation, subject to the normalization condition

$$\sum_{n=0}^{\infty} p_n(\infty) = 1\,, \tag{18.38}$$

is the Bose–Einstein distribution

$$p_n(\infty) = \frac{(n_{\text{cav}})^n}{(n_{\text{cav}}+1)^{n+1}}\,. \tag{18.39}$$

18.4.2 Two-level atom

For the two-level atom, the sample–reservoir interaction Hamiltonian H_{SE} is given by eqns (14.131)–(14.133). In this case, the operator $\mathfrak{F}(t)$ in eqn (18.19) is the sum of two terms: $\mathfrak{F}(t) = \mathfrak{F}_{\mathrm{sp}}(t) + \mathfrak{F}_{\mathrm{pc}}(t)$, that are respectively given by

$$\mathfrak{F}_{\mathrm{sp}}(t) = i\sqrt{w_{21}}b_{\mathrm{in}}^{\dagger}(t)\,\overline{S}_{12}(t) = i\sqrt{w_{21}}b_{\mathrm{in}}^{\dagger}(t)\,\overline{\sigma}_{-}(t) \tag{18.40}$$

and

$$\begin{aligned}\mathfrak{F}_{\mathrm{pc}}(t) &= i\left\{\sqrt{w_{11}}c_{1,\mathrm{in}}^{\dagger}(t)\,\overline{S}_{11}(t) + \sqrt{w_{22}}c_{2,\mathrm{in}}^{\dagger}(t)\,\overline{S}_{22}(t)\right\}\\ &= i\left\{\sqrt{w_{11}}c_{1,\mathrm{in}}^{\dagger}(t)\left(\frac{1-\overline{\sigma}_z(t)}{2}\right) + \sqrt{w_{22}}c_{2,\mathrm{in}}^{\dagger}(t)\left(\frac{1+\overline{\sigma}_z(t)}{2}\right)\right\}.\end{aligned} \tag{18.41}$$

The envelope operators $\overline{\sigma}_{-}(t)$ and $\overline{\sigma}_z(t)$ are related to the environment-picture forms by $\overline{\sigma}_{-}(t) = \sigma_{-}(t)\exp(i\omega_{21}t)$ and $\overline{\sigma}_z(t) = \sigma_z(t)$, while the operators $b_{\mathrm{in}}^{\dagger}(t)$ and $c_{q,\mathrm{in}}^{\dagger}(t)$ are the in-fields defined by eqns (14.146) and (14.147) respectively.

We will assume that the reservoirs are uncorrelated, i.e. $\rho_E(0) = \rho_{\mathrm{sp}}(0)\,\rho_{\mathrm{pc}}(0)$, and that the individual reservoirs are stationary. These assumptions guarantee that most of the possible terms in the double commutator will vanish when the partial trace over the environment is carried out.

After performing the invigorating algebra suggested in Exercise 18.4.2, the surviving terms yield the Schrödinger-picture master equation

$$\frac{\partial}{\partial t}\rho_S(t) = \mathcal{L}_S\rho_S(t) + \mathcal{L}_{\mathrm{dis}}\rho_S(t), \tag{18.42}$$

where the Hamiltonian part,

$$\mathcal{L}_S\rho_S(t) = \frac{1}{i\hbar}\left[\frac{\hbar\omega_{21}}{2}\sigma_z + H_{S1}(t), \rho_S(t)\right], \tag{18.43}$$

includes $H_{S1}(t)$. The dissipative part is

$$\begin{aligned}\mathcal{L}_{\mathrm{dis}}\rho_S(t) = &-\frac{w_{21}}{2}(n_{\mathrm{sp}}+1)\left\{[\sigma_{+},\sigma_{-}\rho_S] + [\rho_S\sigma_{+},\sigma_{-}]\right\}\\ &-\frac{w_{21}}{2}n_{\mathrm{sp}}\left\{[\sigma_{-},\sigma_{+}\rho_S] + [\rho_S\sigma_{-},\sigma_{+}]\right\}\\ &-\frac{w_{\mathrm{pc}}}{2}[\rho_S\sigma_z,\sigma_z],\end{aligned} \tag{18.44}$$

where n_{sp} is the average number of reservoir excitations (photons) at the transition frequency ω_{21}. The phase-changing rate in the last term is

$$w_{\mathrm{pc}} = \frac{1}{2}\sum_{q=1}^{2}(2n_{\mathrm{pc},q}+1)\,w_{qq}, \tag{18.45}$$

where $n_{\mathrm{pc},q}$ is the average number of excitations in the phase-perturbing reservoir coupled to the atomic state $|\varepsilon_q\rangle$.

18.5 Phase space methods

In Section 5.6.3 we have seen that the density operator for a single cavity mode can be described in the Glauber–Sudarshan $P(\alpha)$ representation (5.165). As we will show below, this representation provides a natural way to express the operator master equation (18.31) as a differential equation for the $P(\alpha)$-function. In single-mode applications—and also in some more complex situations—this equation is mathematically identical to the Fokker–Planck equation studied in classical statistics (Risken, 1989, Sec. 4.7).

By defining an atomic version of the P-function—as the Fourier transform of a properly chosen quantum characteristic function—it is possible to apply the same techniques to the master equation for atoms, but we will restrict ourselves to the simpler case of a single mode of the radiation field. The application to atomic master equations can be found, for example, in Haken (1984, Sec. IX.2) or Walls and Milburn (1994, Chap. 13).

For the discussion of the master equation in terms of $P(\alpha)$, it is better to use the alternate convention in which $P(\alpha)$ is regarded as a function, $P(\alpha,\alpha^*)$, of the independent variables α and α^*. In this notation the P-representation is

$$\rho_S(t) = \int \frac{d^2\alpha}{\pi} \left|\alpha\right\rangle P(\alpha,\alpha^*;t) \left\langle\alpha\right| . \tag{18.46}$$

The function $P(\alpha,\alpha^*;t)$ is real and satisfies the normalization condition

$$\int \frac{d^2\alpha}{\pi} P(\alpha,\alpha^*;t) = 1 , \tag{18.47}$$

but it cannot always be interpreted as a probability distribution. The trouble is that, for nonclassical states, $P(\alpha,\alpha^*;t)$ must take on negative values in some region of the α-plane.

18.5.1 The Fokker–Planck equation

In order to use the P-representation in the master equation, we must translate the products of Fock space operators, e.g. a, $a^\dagger$, and ρ, in the master equation into the action of differential operators on the c-number function $P(\alpha,\alpha^*;t)$. For this purpose it is useful to write the coherent state $|\alpha\rangle$ as

$$\left|\alpha\right\rangle = e^{-|\alpha|^2/2} \left|\alpha;B\right\rangle , \tag{18.48}$$

where the **Bargmann state** $|\alpha;B\rangle$ is

$$\left|\alpha;B\right\rangle = \sum_{n=0}^{\infty} \frac{\alpha^n}{\sqrt{n!}} \left|n\right\rangle . \tag{18.49}$$

The virtue of the Bargmann states is that they are analytic functions of α. More precisely, for any fixed state $|\Psi\rangle$ the c-number function

$$\left\langle\Psi\right.\left|\alpha;B\right\rangle = \sum_{n=0}^{\infty} \frac{\alpha^n}{\sqrt{n!}} \left\langle\Psi\right.\left|n\right\rangle \tag{18.50}$$

is analytic in α. In the same sense, $\langle\alpha;B|$ is an analytic function of α^*, so it is independent of α.

Since $|\alpha; B\rangle$ is proportional to $|\alpha\rangle$, the action of a on the Bargmann states is just

$$a\,|\alpha; B\rangle = \alpha\,|\alpha; B\rangle\,. \tag{18.51}$$

The action of $a^\dagger$ is found by using eqn (18.49) to get

$$\begin{aligned} a^\dagger\,|\alpha; B\rangle &= \sum_{n=0}^{\infty} \frac{\alpha^n}{\sqrt{n!}} \sqrt{n+1}\,|n+1\rangle \\ &= \frac{\partial}{\partial\alpha}\,|\alpha; B\rangle\,. \end{aligned} \tag{18.52}$$

The adjoint of this rule is

$$\langle\alpha; B|\,a = \frac{\partial}{\partial\alpha^*}\,\langle\alpha; B|\,. \tag{18.53}$$

In the Bargmann notation, the P-representation of the density operator is

$$\rho_S\,(t) = \int \frac{d^2\alpha}{\pi} P\,(\alpha,\alpha^*;t)\,e^{-|\alpha|^2}\,|\alpha; B\rangle\,\langle\alpha; B|\,. \tag{18.54}$$

The rule (18.51) then gives

$$\begin{aligned} a\rho_S\,(t) &= \int \frac{d^2\alpha}{\pi} P\,(\alpha,\alpha^*;t)\,e^{-|\alpha|^2}\alpha\,|\alpha; B\rangle\,\langle\alpha; B| \\ &= \int \frac{d^2\alpha}{\pi} \alpha P\,(\alpha,\alpha^*;t)\,|\alpha\rangle\,\langle\alpha|\,. \end{aligned} \tag{18.55}$$

Applying the rule (18.52) yields

$$a^\dagger\rho_S\,(t) = \int \frac{d^2\alpha}{\pi} P\,(\alpha,\alpha^*;t)\,e^{-|\alpha|^2}\left\{\frac{\partial}{\partial\alpha}\,|\alpha; B\rangle\right\}\langle\alpha; B|\,, \tag{18.56}$$

but this is not expressed in terms of a differential operator acting on $P\,(\alpha,\alpha^*;t)$. Integrating by parts on α leads to the desired form:

$$\begin{aligned} a^\dagger\rho_S\,(t) &= -\int \frac{d^2\alpha}{\pi}\frac{\partial}{\partial\alpha}\left\{P\,(\alpha,\alpha^*;t)\,e^{-\alpha\alpha^*}\right\}|\alpha; B\rangle\,\langle\alpha; B| \\ &= \int \frac{d^2\alpha}{\pi}\left\{\left(\alpha^* - \frac{\partial}{\partial\alpha}\right)P\,(\alpha,\alpha^*;t)\right\}|\alpha\rangle\,\langle\alpha|\,. \end{aligned} \tag{18.57}$$

This result depends on the fact that the normalization condition requires $P\,(\alpha,\alpha^*;t)$ to vanish as $|\alpha| \to \infty$.

Combining eqns (18.55) and (18.57) with their adjoints gives us the translation table

$$\begin{aligned} &a\rho_S\,(t) \leftrightarrow \alpha P\,(\alpha,\alpha^*;t) && \rho_S\,(t)\,a \leftrightarrow (\alpha - \partial/\partial\alpha^*)\,P\,(\alpha,\alpha^*;t) \\ &a^\dagger\rho_S\,(t) \leftrightarrow (\alpha^* - \partial/\partial\alpha)\,P\,(\alpha,\alpha^*;t) && \rho_S\,(t)\,a^\dagger \leftrightarrow \alpha^* P\,(\alpha,\alpha^*;t)\,. \end{aligned} \tag{18.58}$$

Applying the rules in eqn (18.58) to eqn (18.32)—for the simple case with $H_{S1} = 0$—and to eqn (18.33) yields the translations

$$\mathcal{L}_S \rho_S(t) = \frac{1}{i\hbar}\left[\hbar\omega_0 a^\dagger a, \rho_S(t)\right] \leftrightarrow i\omega_0\left(\frac{\partial}{\partial\alpha}\alpha - \frac{\partial}{\partial\alpha^*}\alpha^*\right) P(\alpha,\alpha^*;t) \tag{18.59}$$

and

$$\mathcal{L}_{\text{dis}}\rho_S(t) \leftrightarrow \frac{\kappa}{2}\left\{\frac{\partial}{\partial\alpha}\left[\alpha P(\alpha,\alpha^*;t)\right] + \frac{\partial}{\partial\alpha^*}\left[\alpha^* P(\alpha,\alpha^*;t)\right]\right\} + \kappa n_{\text{cav}}\frac{\partial^2}{\partial\alpha\partial\alpha^*}P(\alpha,\alpha^*;t)\,, \tag{18.60}$$

for the Hamiltonian and dissipative Liouville operators respectively.

In the course of carrying out these calculations, it is easy to get confused about the correct order of operations. The reason is that products like $a^\dagger a\rho$—with operators standing on the left of ρ—and products like $\rho a^\dagger a$—with operators standing to the right of ρ—are both translated into differential operators acting from the left on the function $P(\alpha,\alpha^*;t)$.

Studying a simple example, e.g. carrying out a direct derivation of both $a^\dagger a\rho$ and $\rho a^\dagger a$, shows that the order of the differential operators is reversed from the order of the Fock space operators when the Fock space operators stand to the right of ρ. Another way of saying this is that one should work from the inside to the outside; the first differential operator acting on P corresponds to the Hilbert space operator closest to ρ. This rule gives the correct result for Fock space operators to the left or to the right of ρ.

The master equation for an otherwise unperturbed cavity mode is, therefore, represented by

$$\frac{\partial}{\partial t}P(\alpha,\alpha^*;t) = \frac{\partial}{\partial\alpha}\left[Z(\alpha)P(\alpha,\alpha^*;t)\right] + \frac{\partial}{\partial\alpha^*}\left[Z^*(\alpha)P(\alpha,\alpha^*;t)\right] + \kappa n_{\text{cav}}\frac{\partial^2}{\partial\alpha\partial\alpha^*}P(\alpha,\alpha^*;t)\,, \tag{18.61}$$

where

$$Z(\alpha) = \left(\frac{\kappa}{2} + i\omega_0\right)\alpha\,. \tag{18.62}$$

We can achieve a firmer grip on the meaning of this equation by changing variables from (α,α^*) to $\mathbf{u} = (u_1,u_2)$, where $u_1 = \operatorname{Re}\alpha$ and $u_2 = \operatorname{Im}\alpha$. In these variables, $P(\alpha,\alpha^*;t) = P(\mathbf{u};t)$, and the α-derivative is

$$\frac{\partial}{\partial\alpha} = \frac{1}{2}\left(\frac{\partial}{\partial u_1} - i\frac{\partial}{\partial u_2}\right). \tag{18.63}$$

In this notation, the master equation takes the form of a classical Fokker–Planck equation in two dimensions:

$$\frac{\partial}{\partial t}P(\mathbf{u};t) = -\boldsymbol{\nabla}\cdot\left[\mathbf{F}(\mathbf{u})P(\mathbf{u};t)\right] + \frac{D_0}{2}\boldsymbol{\nabla}^2 P(\mathbf{u};t)\,, \tag{18.64}$$

where

$$D_0 = \kappa n_{\text{cav}}/2 \tag{18.65}$$

is the **diffusion constant**, and we have introduced the following shorthand notation:

$$\begin{aligned} \nabla \cdot \mathbf{X} &= \frac{\partial X_1}{\partial u_1} + \frac{\partial X_2}{\partial u_2}, \\ \mathbf{F}(\mathbf{u}) &= (-\operatorname{Re} Z, -\operatorname{Im} Z) = \left(\omega_0 u_2 - \frac{\kappa}{2} u_1, -\omega_0 u_1 - \frac{\kappa}{2} u_2\right), \\ \nabla^2 &= \left[\left(\frac{\partial}{\partial u_1}\right)^2 + \left(\frac{\partial}{\partial u_2}\right)^2\right]. \end{aligned} \tag{18.66}$$

The first- and second-order differential operators in eqn (18.64) are respectively called the **drift term** and the **diffusion term.**

A Classical Langevin equations

The Fokker–Planck equation (18.64) is a special case of a general family of equations of the form

$$\frac{\partial}{\partial t} P(\underline{u}; t) = -\underline{\nabla} \cdot [\underline{F}(\underline{u}, t) P(\underline{u}; t)] + \frac{1}{2} \sum_{m=1}^{N} \sum_{n=1}^{N} \frac{\partial}{\partial u_m} \frac{\partial}{\partial u_n} D_{mn}(\underline{u}, t) P(\underline{u}; t), \tag{18.67}$$

where $\underline{u} = (u_1, \ldots, u_N)$, $\underline{F} = (F_1, \ldots, F_N)$, D_{mn} is the **diffusion matrix**, and

$$\underline{X} \cdot \underline{Y} = X_1 Y_1 + \cdots + X_N Y_N. \tag{18.68}$$

For the two-component case, given by eqn (18.64), the diffusion matrix is diagonal, $D_{mn} = D_0 \delta_{mn}$, so it has a single eigenvalue $D_0 > 0$. The corresponding condition in the general N-component case is that all eigenvalues of the diffusion matrix D are positive, i.e. D is a positive-definite matrix. In this case D has a square root matrix B that satisfies $D = BB^T$.

When D is positive definite, then eqn (18.67) is exactly equivalent to the set of classical Langevin equations (Gardiner, 1985, Sec. 4.3.5)

$$\frac{du_n(t)}{dt} = C_n(\underline{u}, t) + \sum_{m=1}^{N} B_{nm}(\underline{u}, t) w_m(t), \tag{18.69}$$

where the u_ns are stochastic variables and the w_ms are independent white noise sources of unit strength, i.e. $\langle w_m(t) \rangle = 0$ and

$$\langle w_m(t) w_n(t') \rangle = \delta_{mn} \delta(t - t'). \tag{18.70}$$

In particular, the Langevin equations corresponding to eqn (18.64) are

$$\frac{d\underline{u}(t)}{dt} = \underline{F}(\underline{u}) + \sqrt{D_0}\, \underline{w}(t). \tag{18.71}$$

These real Langevin equations are essential for numerical simulations, but for analytical work it is useful to write them in complex form. This is done by combining $\alpha = u_1 + iu_2$ with eqns (18.62) and (18.66) to get

$$\frac{d\alpha(t)}{dt} = -Z(\alpha(t)) + \sqrt{2D_0}\eta(t)\,, \tag{18.72}$$

where $\alpha(t)$ is a complex stochastic variable, and

$$\eta(t) = \frac{1}{\sqrt{2}}\left[w_1(t) + iw_2(t)\right] \tag{18.73}$$

is a complex white noise source satisfying

$$\langle \eta(t) \rangle = 0\,, \quad \langle \eta(t)\,\eta(t') \rangle = 0\,, \quad \langle \eta^*(t)\,\eta(t') \rangle = \delta(t - t')\,. \tag{18.74}$$

The equivalence of the Fokker–Planck equation and the classical Langevin equations for a positive-definite diffusion matrix is important in practice, since the numerical simulation of the Langevin equations is usually much easier than the direct numerical solution of the Fokker–Planck equation itself.

For some problems—e.g. when the appropriate reservoir is described by a squeezed state—the diffusion matrix derived from the Glauber–Sudarshan P-function is not positive definite, so the Fokker–Planck equation is not equivalent to a set of classical Langevin equations. In such cases, another representation of the density operator may be more useful (Walls and Milburn, 1994, Sec. 6.3.1).

18.5.2 Applications of the Fokker–Planck equation

A Coherent states are robust

Let us begin with a simple example in which $n_{\text{cav}} = 0$, so that the diffusion term in eqn (18.64) vanishes. If we interpret the reservoir oscillators as phonons in the cavity walls, then this model describes the idealized situation of material walls at absolute zero. Alternatively, the reservoir could be defined by other modes of the electromagnetic field, into which the particular mode of interest is scattered by a gas of nonresonant atoms. In this case, it is natural to assume that the initial reservoir state is the vacuum. In other words, the universe is big and dark and cold.

The terms remaining after setting $n_{\text{cav}} = 0$ can be rearranged to produce

$$\frac{\partial}{\partial t}P(\mathbf{u};t) + \mathbf{F}(\mathbf{u})\cdot\boldsymbol{\nabla}P(\mathbf{u};t) = \kappa P(\mathbf{u};t)\,. \tag{18.75}$$

Let us study the evolution of a field state initially defined by $P(\mathbf{u};0) = P_0(\mathbf{u})$. The general technique for solving linear, first-order, partial differential equations like eqn (18.75) is the method of characteristics (Zauderer, 1983, Sec. 2.2), but we will employ an equivalent method that is well suited to the problem at hand.

The first step is to introduce an integrating factor, by setting

$$P(\mathbf{u};t) = \overline{P}(\mathbf{u};t)\,e^{\kappa t}\,, \tag{18.76}$$

so that

$$\frac{\partial}{\partial t}\overline{P}(\mathbf{u};t)+\mathbf{F}(\mathbf{u})\cdot\boldsymbol{\nabla}\overline{P}(\mathbf{u};t)=0\,. \tag{18.77}$$

The second step is to transform to new variables $(\mathbf{u}', t')$ by

$$\mathbf{u}'=\mathbf{V}(\mathbf{u},t)\,,\quad t'=t\,, \tag{18.78}$$

where we require $\mathbf{u}'=\mathbf{u}$ at $t=0$, and also assume that the function $\mathbf{V}(\mathbf{u},t)$ is linear in $\mathbf{u}$, i.e.

$$V_n(\mathbf{u},t)=\sum_{m=1}^{2}G_{nm}(t)\,u_m\,. \tag{18.79}$$

The reason for trying a linear transformation is that the coefficient vector,

$$F_j(\mathbf{u})=\sum_l W_{jl}u_l\,,\quad \text{where } W=\begin{bmatrix}-\kappa/2 & \omega_0\\ -\omega_0 & -\kappa/2\end{bmatrix}, \tag{18.80}$$

is itself linear in $\mathbf{u}$.

The chain rule calculation in Exercise 18.5 yields expressions for the operators $\partial/\partial t$ and $\partial/\partial u_l$ in terms of the new variables, so that eqn (18.77) becomes

$$\frac{\partial}{\partial t'}\overline{P}(\mathbf{u}';t')+\sum_n\left\{\sum_l\left[G(t)\,W+\frac{dG(t)}{dt}\right]_{nl}u_l\right\}\frac{\partial}{\partial u'_n}\overline{P}(\mathbf{u}';t)=0\,. \tag{18.81}$$

Choosing the matrix $G(t)$ to satisfy

$$\frac{dG(t)}{dt}+G(t)\,W=0 \tag{18.82}$$

ensures that the coefficient of $\partial/\partial u'_n$ vanishes identically in $\mathbf{u}$, and this in turn simplifies the equation for $\overline{P}(\mathbf{u}';t')$ to

$$\frac{\partial}{\partial t'}\overline{P}(\mathbf{u}';t')=0\,. \tag{18.83}$$

Thus $\overline{P}(\mathbf{u}';t')=\overline{P}(\mathbf{u}';0)$, but $t'=0$ is the same as $t=0$, so $\overline{P}(\mathbf{u}';t')=P_0(\mathbf{u}')$.

In this way the solution to the original problem is found to be

$$P(\mathbf{u},t)=e^{\kappa t}P_0(\mathbf{V}(\mathbf{u},t))\,, \tag{18.84}$$

and the only remaining problem is to evaluate $\mathbf{V}(\mathbf{u},t)$. This is most easily done by writing $G(t)$ as

$$G(t)=\begin{bmatrix}b_1(t) & b_2(t)\\ c_1(t) & c_2(t)\end{bmatrix}, \tag{18.85}$$

and substituting this form into eqn (18.82). This yields simple differential equations for the vectors $\mathbf{b}(t)$ and $\mathbf{c}(t)$, with initial conditions $\mathbf{b}(0)=(1,0)$ and $\mathbf{c}(0)=(0,1)$. The solution of these auxiliary equations gives

$$G(t) = e^{\kappa t/2} R(t) = e^{\kappa t/2} \begin{bmatrix} \cos(\omega_0 t) & -\sin(\omega_0 t) \\ \sin(\omega_0 t) & \cos(\omega_0 t) \end{bmatrix}, \tag{18.86}$$

so that

$$P(\mathbf{u}; t) = P_0 \left(e^{\kappa t/2} R(t)\, \mathbf{u} \right) e^{\kappa t}. \tag{18.87}$$

Thus, in the absence of the diffusive term, the shape of the distribution is unchanged; the argument $\mathbf{u}$ is simply scaled by $\exp(\kappa t/2)$ and rotated by the angle $\omega_0 t$. In the complex-α description the solution is given by

$$P(\alpha, \alpha^*; t) = P_0 \left(e^{(\kappa/2 + i\omega_0)t} \alpha, e^{(\kappa/2 - i\omega_0)t} \alpha^* \right) e^{\kappa t}. \tag{18.88}$$

This result is particularly interesting if the field is initially in a coherent state $|\alpha_0\rangle$, i.e. the initial P-function is $P_0(\mathbf{u}) = \delta_2(\mathbf{u} - \mathbf{u}_0)$. In this case, the standard properties of the delta function lead to

$$P(\mathbf{u}; t) = e^{\kappa t} \delta_2 \left(e^{\kappa t/2} R(t)\, \mathbf{u} - \mathbf{u}_0 \right) = \delta_2 (\mathbf{u} - \mathbf{u}(t)), \tag{18.89}$$

where

$$\mathbf{u}(t) = e^{-\kappa t/2} [R(t)]^{-1} \mathbf{u}_0. \tag{18.90}$$

The conclusion is that a coherent state interacting with a zero-temperature reservoir will remain a coherent state, with a decaying amplitude

$$\alpha(t) = \alpha_0 e^{-\kappa t/2} e^{-i\omega_0 t}. \tag{18.91}$$

Consequently, the time-dependent joint variance of $a^\dagger$ and a vanishes at all times:

$$V(a^\dagger, a\,; t) = \left\langle \alpha(t) \left| a^\dagger a \right| \alpha(t) \right\rangle - \left\langle \alpha(t) \left| a^\dagger \right| \alpha(t) \right\rangle \left\langle \alpha(t) \left| a \right| \alpha(t) \right\rangle = 0. \tag{18.92}$$

In other words, coherent states are *robust*: scattering and absorption will not destroy the coherence properties, as long as the environment is at zero temperature.

This apparently satisfactory result raises several puzzling questions. The first is that the initially pure state remains pure, even after interaction with the environment. This seems to contradict the general conclusion, established in Section 18.1, that interaction with a reservoir inevitably produces a mixed state for the sample.

The resolution of this discrepancy is that the general argument is true for the exact theory, while the master equation is derived with the aid of the approximation—see eqn (18.17)—that back-action of the sample on the reservoir can be neglected. This means that the robustness property of the coherent states is only as strong as the approximations leading to the master equation. Furthermore, we will see in Section 18.6 that coherent states are the only pure states that can take advantage of this loophole in the general argument of Section 18.1.

The second difficulty with the robustness of coherent states is that it seems to violate the fluctuation dissipation theorem. The field suffers dissipation, but there is no added noise. Consequently, it is a relief to realize that the strength of the noise term in the equivalent classical Langevin equations (18.71) vanishes for $n_{cav} = 0$. Further reassurance comes from the operator Langevin approach, in particular eqn (14.74), which shows that the strength of the Langevin noise operator also vanishes for $n_{cav} = 0$.

B Thermalization of an initial coherent state

The coordinates defined by eqn (18.78) are also useful for solving eqn (18.64), the Fokker–Planck equation with diffusion. According to eqn (18.86), the transformation from $\mathbf{u}$ to $\mathbf{u}'$ is a rotation followed by scaling with $\exp(\kappa t/2)$. The operator $\boldsymbol{\nabla}^2$ on the right side of eqn (18.64) is invariant under rotations, so $\boldsymbol{\nabla}^2 = e^{\kappa t}\boldsymbol{\nabla}'^2$ and the Fokker–Planck equation becomes

$$\frac{\partial}{\partial t'}\overline{P}(\mathbf{u}';t') = \frac{D_0}{2}e^{\kappa t}\boldsymbol{\nabla}'^2\overline{P}(\mathbf{u}';t'), \tag{18.93}$$

which is the diffusion equation with a time-dependent diffusion coefficient. The Fourier transform,

$$\overline{P}(\mathbf{q}';t') = \int d^2u'\overline{P}(\mathbf{u}';t')\, e^{-i\mathbf{q}'\cdot\mathbf{u}'}, \tag{18.94}$$

then satisfies the ordinary differential equation

$$\frac{d}{dt'}\overline{P}(\mathbf{q}';t') = \frac{D_0}{2}e^{\kappa t}\mathbf{q}'^2\overline{P}(\mathbf{q}';t'), \tag{18.95}$$

which has the solution

$$\overline{P}(\mathbf{q}';t') = \exp\left[\frac{D_0}{2\kappa}\left(e^{\kappa t}-1\right)\mathbf{q}'^2\right]\overline{P}_0(\mathbf{q}'). \tag{18.96}$$

For the initial coherent state, $P_0(\mathbf{u}) = \delta_2(\mathbf{u}-\mathbf{u}_0)$, one finds

$$\overline{P}_0(\mathbf{q}') = \exp\left[-i\mathbf{q}'\cdot\mathbf{u}_0\right], \tag{18.97}$$

and the inverse transform can be explicitly evaluated to yield

$$P(\mathbf{u};t) = \frac{1}{\pi w(t)}\exp\left[-\frac{(\mathbf{u}-\mathbf{u}(t))^2}{w(t)}\right], \tag{18.98}$$

where $\mathbf{u}(t)$ is given by eqn (18.90) and

$$w(t) = n_{\text{cav}}\left(1-e^{-\kappa t}\right). \tag{18.99}$$

For long times ($t \gg 1/\kappa$) $\mathbf{u}(t) \to 0$, and the P-function approaches the thermal distribution given by eqn (5.176); in other words, the field comes into equilibrium with the cavity walls as expected. At short times, $t \ll 1/\kappa$, we see that $w(t) \sim n_{\text{cav}}\kappa t \ll 1$ and the initial delta function is recovered.

C A driven mode in a lossy cavity

In Section 5.2 we presented a simple model for generating a coherent state of a single mode in a lossless cavity. We can be sure that losses will be present in any real experiment, so we turn to the Fokker–Planck equation for a more realistic treatment.

The off-resonant term in the Heisenberg equation (5.38) defining our model can safely be neglected, so the situation is adequately represented by the simplified Hamiltonian

$$H_S(t) = \hbar\omega_0 a^\dagger a - \hbar W e^{-i\Omega t} a^\dagger - \hbar W^* e^{i\Omega t} a\,, \tag{18.100}$$

that leads to the Liouville operator

$$\mathcal{L}_S \rho_S(t) = \frac{1}{i}\left[\omega_0 a^\dagger a, \rho_S(t)\right] - \frac{1}{i}\left[W e^{-i\Omega t} a^\dagger + W^* e^{i\Omega t} a, \rho_S(t)\right]. \tag{18.101}$$

After including the new terms in the master equation and applying the rules (18.58), one finds an equation of the same form as eqn (18.61), except that the $Z(\alpha)$ function is replaced by

$$Z(\alpha) = \left(\frac{\kappa}{2} + i\omega_0\right)\alpha - iWe^{-i\Omega t}. \tag{18.102}$$

Instead of directly solving the Fokker–Planck equation, it is instructive to use the equivalent set of classical Langevin equations. Substituting the new $Z(\alpha)$ function into the general result (18.72) yields

$$\frac{d\alpha(t)}{dt} = -\left(\frac{\kappa}{2} + i\omega_0\right)\alpha(t) + iWe^{-i\Omega t} + \sqrt{2D_0}\,\eta(t)\,, \tag{18.103}$$

which has the solution

$$\alpha(t) = \alpha(0)\,e^{-(i\omega_0+\kappa/2)t} + \alpha_{\text{coh}}(t) + \sqrt{2D_0}\,\vartheta(t)\,, \tag{18.104}$$

where

$$\alpha_{\text{coh}}(t) = \frac{iW}{i\Delta + \kappa/2} e^{-(i\omega_0+\kappa/2)t}\left[e^{(i\Delta+\kappa/2)t} - 1\right] \tag{18.105}$$

is a definite (i.e. nonrandom) function, and

$$\vartheta(t) = \int_0^t dt_1 e^{-(i\omega_0+\kappa/2)(t-t_1)}\eta(t_1)\,. \tag{18.106}$$

The initial value, $\alpha(0)$, is a complex random variable, not a definite complex number.

The average of any function $f(\alpha(0))$ is given by

$$\langle f(\alpha(0))\rangle = \int d^2\alpha(0)\, P_0(\alpha(0), \alpha^*(0))\, f(\alpha(0))\,, \tag{18.107}$$

but special problems arise if the initial state is not classical. For a classical state—i.e. $P_0(\alpha,\alpha^*) \geqslant 0$—standard methods can be used to draw $\alpha(0)$ randomly from the distribution, but these methods fail when $P_0(\alpha,\alpha^*)$ is negative. For these nonclassical states, the c-number Langevin equations are of doubtful utility for numerical simulations.

For the problem at hand, the initial state is the vacuum, with the positive distribution $P_0(\alpha(0), \alpha^*(0)) = \delta(\alpha(0))$. The initial value $\alpha(0)$ and the noise term $\vartheta(t)$ both have vanishing averages, so the average value of $\alpha(t)$ is given by

$$\langle \alpha(t) \rangle = \alpha_{\text{coh}}(t) . \tag{18.108}$$

For the nondissipative case, $\kappa = D_0 = 0$, the average agrees with eqn (5.41); but, when dissipation is present, the long time ($t \gg 1/\kappa$) solution approaches

$$\langle \alpha(t) \rangle = \frac{iW}{i\Delta + \kappa/2} e^{-i\Omega t} . \tag{18.109}$$

Thus the decay of the average field due to dissipation—shown by eqn (18.91)—is balanced by radiation from the classical current, and the average field amplitude has a definite phase determined by the phase of the classical current. This would also be true if the sample were described by the coherent state $\rho_{\text{coh}}(t) = |\alpha_{\text{coh}}(t)\rangle \langle \alpha_{\text{coh}}(t)|$, so it will be necessary to evaluate second-order moments in order to see if eqn (18.104) corresponds to a true coherent state.

We will first investigate the coherence properties of the state by using the explicit solution (18.104) to get

$$\begin{aligned} \langle \alpha^2(t) \rangle &= \left\langle \left[\alpha(0) e^{-(i\omega_0 + \kappa/2)t} + \alpha_{\text{coh}}(t) + \sqrt{2D_0}\vartheta(t) \right]^2 \right\rangle \\ &= \alpha_{\text{coh}}^2(t) + 2D_0 \langle \vartheta^2(t) \rangle . \end{aligned} \tag{18.110}$$

The simple form of the second line depends on two facts: (i) $\alpha_{\text{coh}}(t)$ is a definite function; and (ii) the distribution of $\alpha(0)$ is concentrated at $\alpha(0) = 0$. A further simplification comes from using eqn (18.105) to evaluate $\langle \vartheta^2(t) \rangle$. The result is a double integral with integrand proportional to $\langle \eta(t_1) \eta(t_2) \rangle$, but eqn (18.74) shows that this average vanishes for all values of t_1 and t_2. The final result is then

$$\langle \alpha^2(t) \rangle = \alpha_{\text{coh}}^2(t) = \langle \alpha(t) \rangle^2 , \tag{18.111}$$

which also agrees with the prediction for a true coherent state.

Before proclaiming that we have generated a true coherent state in a lossy cavity, we must check the remaining second-order moment, $\langle |\alpha(t)|^2 \rangle$, which represents the average of the number operator $a^\dagger a$. Since $\alpha(t)$ is concentrated at the origin, we can simplify the calculation by setting $\alpha(0) = 0$ at the outset. This gives us

$$\left\langle |\alpha(t)|^2 \right\rangle = |\alpha_{\text{coh}}(t)|^2 + 2D_0 \left\langle |\vartheta(t)|^2 \right\rangle . \tag{18.112}$$

Combining eqns (18.106) and (18.74) leads to

$$\begin{aligned} \left\langle |\vartheta(t)|^2 \right\rangle &= \int_0^t dt_1 \int_0^t dt_2 e^{-(-i\omega_0 + \kappa/2)(t - t_1)} e^{-(i\omega_0 + \kappa/2)(t - t_2)} \langle \eta^*(t_1) \eta(t_2) \rangle \\ &= \int_0^t dt_1 e^{-\kappa(t - t_1)} = \frac{1 - e^{-\kappa t}}{\kappa} , \end{aligned} \tag{18.113}$$

so that

$$\left\langle \left| \alpha \left(t \right) \right|^2 \right\rangle - \left| \alpha_{\text{coh}} \left(t \right) \right|^2 = \frac{2D_0}{\kappa} \left(1 - e^{-\kappa t} \right) = n_{\text{cav}} \left(1 - e^{-\kappa t} \right) , \tag{18.114}$$

where we used eqn (18.65) to get the final result. The left side of this equation would vanish for a true coherent state, so the state generated in a lossy cavity is only coherent if $n_{\text{cav}} = 0$, i.e. if the cavity walls are at zero temperature.

18.6 The Lindblad form of the master equation*

The master equations (18.31) and (18.42) share three important properties.

(a) The trace condition, $\text{Tr} \left[\rho_S \left(t \right) \right] = 1$, is conserved.

(b) The positivity of ρ_S is conserved, i.e. $\left\langle \Psi \left| \rho_S \left(t \right) \right| \Psi \right\rangle \geqslant 0$ for all states $\left| \Psi \right\rangle$ and all times t.

(c) The equations are derivable from a model of the sample interacting with a collection of reservoirs.

The most general linear, dissipative time evolution that satisfies (a), (b), and (c) is given by

$$\frac{\partial \rho_S}{\partial t} = \mathfrak{L}_S \rho_S + \mathfrak{L}_{\text{dis}} \rho_S , \tag{18.115}$$

where

$$\mathfrak{L}_S \rho_S = \frac{1}{i\hbar} \left[H_S \left(t \right) , \rho_S \right] \tag{18.116}$$

describes the Hamiltonian evolution of the sample, and the dissipative term has the **Lindblad form** (Lindblad, 1976)

$$\mathfrak{L}_{\text{dis}} \rho_S = -\frac{1}{2} \sum_{k=1}^{K} \left\{ C_k^\dagger C_k \rho_S + \rho_S C_k^\dagger C_k - 2 C_k \rho_S C_k^\dagger \right\} . \tag{18.117}$$

Each of operators $C_1, C_2, \ldots, C_K$ acts on the sample space $\mathfrak{H}_S$ and there can be a finite or infinite number of them, depending on the sample under study.

One can see by inspection that there are two Lindblad operators, i.e. $K = 2$, for the single-mode master equation (18.31):

$$C_1 = \sqrt{\kappa \left(n_{\text{cav}} + 1 \right)} a , \quad C_2 = \sqrt{\kappa n_{\text{cav}}} a^\dagger . \tag{18.118}$$

A slightly longer calculation—see Exercise 18.6—shows that there are three operators for the master equation (18.42) describing the two-level atom.

The Lindblad form (18.117) for the dissipative operator can be used to investigate a variety of questions. For example, in Section 2.3.4 we introduced a quantitative measure of the degree of mixing by defining the purity of the state ρ_S as $\mathfrak{P} \left(t \right) = \text{Tr} \left[\rho_S^2 \left(t \right) \right] \leqslant 1$. One can show from eqn (18.115) that the time derivative of the purity is

$$\frac{d}{dt} \mathfrak{P} \left(t \right) = -2 \sum_{k=1}^{K} \text{Tr} \left\{ \rho_S \left(t \right) C_k^\dagger C_k \rho_S \left(t \right) - \rho_S \left(t \right) C_k \rho_S \left(t \right) C_k^\dagger \right\} . \tag{18.119}$$

At first glance, it may seem natural to assume that interaction with the environment can only cause further mixing of the sample state, so one might expect that the time derivative of the purity is always negative. If $\rho_S(0)$ is a mixed state this need not be true. For example, the purity of a thermal state would be increased by interaction with a colder reservoir, as seen in Exercise 18.7.

On the other hand, for a pure state there is no way to go but down; therefore, the intuitive expectation of declining purity should be satisfied. In order to check this, we evaluate eqn (18.119) for an initially pure state $\rho_S(0) = |\Psi\rangle\langle\Psi|$, to find

$$\begin{aligned}\left.\frac{d}{dt}\mathfrak{P}(t)\right|_{t=0} &= -2\sum_{k=1}^{K}\left\{\left\langle\Psi\left|C_k^{\dagger}C_k\right|\Psi\right\rangle - \left\langle\Psi\left|C_k^{\dagger}\right|\Psi\right\rangle\langle\Psi|C_k|\Psi\rangle\right\} \\ &= -2\sum_{k=1}^{K}\left\langle\Psi\left|\delta C_k^{\dagger}\delta C_k\right|\Psi\right\rangle \leqslant 0\,,\end{aligned} \tag{18.120}$$

where

$$\delta C_k = C_k - \langle\Psi|C_k|\Psi\rangle\,. \tag{18.121}$$

Thus the Lindblad form guarantees the physically essential result that initially pure states cannot increase in purity (Gallis, 1996).

The appearance of an inequality like eqn (18.120) prompts the following question: Are there any physical samples possessing states that saturate the inequality? We can answer this question in one instance by studying the master equation (18.31) with a zero-temperature reservoir. In this case eqn (18.118) gives us $C_2 = 0$ and $C_1 = \sqrt{\kappa}a$, so that eqn (18.120) becomes

$$\left.\frac{d}{dt}\mathfrak{P}(t)\right|_{t=0} = -2\kappa\left\langle\Psi\left|(a-\alpha)^{\dagger}(a-\alpha)\right|\Psi\right\rangle \leqslant 0\,, \tag{18.122}$$

where $\alpha = \langle\Psi|a|\Psi\rangle$. The inequality can only be saturated if $|\Psi\rangle$ satisfies

$$a|\Psi\rangle = \alpha|\Psi\rangle\,, \tag{18.123}$$

i.e. when $|\Psi\rangle$ is a coherent state. For all other pure states, interaction with a zero-temperature reservoir will decrease the purity, i.e. the state becomes mixed.

18.7 Quantum jumps

18.7.1 An elementary description of quantum jumps

The notion of quantum jumps was a fundamental part of the earliest versions of the quantum theory, but for most of the twentieth century it was assumed that the phenomenon itself would always be unobservable, since there were no experimental methods available for isolating and observing individual atoms, ions, or photons.

This situation began to change in the 1980s with Dehmelt's proposal (Dehmelt, 1982) for an improvement in frequency standards based on observations of a single ion, and the subsequent development of electromagnetic traps (Paul, 1990) that made such observations possible.

The following years have seen a considerable improvement in both experimental and theoretical techniques. The improved experimental methods have made possible the direct observation of the quantum jumps postulated by the founders of quantum theory.

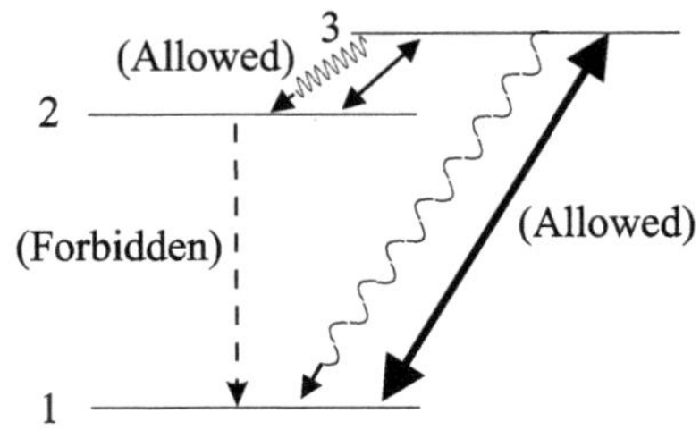

Fig. 18.1 A three-level ion with dipole-allowed transitions $3 \leftrightarrow 2$ and $3 \leftrightarrow 1$, indicated by wavy arrows, and a dipole-forbidden transition $2 \to 1$, indicated by the light dashed arrow. The heavy double arrows denote strong incoherent couplings on the $3 \leftrightarrow 1$ and $3 \to 2$ transitions.

A A three-level model

It is always good to have a simple, concrete example in mind, so we will study a single, trapped, three-level ion, with the level structure shown in Fig. 18.1. The dipole-allowed transitions, $3 \to 1$ and $3 \to 2$, have Einstein-A coefficients κ_{31} and κ_{32} respectively, so the total decay rate of level 3 is $\kappa_3 = \kappa_{31} + \kappa_{32}$. Since the dipole-forbidden transition, $2 \to 1$, has a unique final state, it is described by a single decay rate κ_2, which is small compared to both κ_{31} and κ_{32}.

In addition to the spontaneous emission processes, we assume that an incoherent radiation source, at the frequency ω_{31}, drives the ion between levels 1 and 3 by absorption and spontaneous emission. As explained in Section 1.2.2, both of these processes occur with the rate $W_{31} = B_{31}\rho(\omega_{31})$, where $\rho(\omega_{31})$ is the energy density of the external field and B_{31} is the Einstein-B coefficient.

When level 3 is occupied, the ion can isotropically emit fluorescent radiation, i.e. radiation at frequency ω_{31}. Another way of saying this is that the ion scatters the pump light in all directions. Consequently, observing the fluorescent intensity—say at right angles to the direction of the pump radiation—effectively measures the population of level 3.

We will further assume that the levels 2 and 3 are closely spaced in energy, compared to their separation from level 1, so that $\omega_{32} \ll \omega_{31}$. From eqn (4.162) we know that the Einstein-A coefficient is proportional to the cube of the transition frequency; therefore, the transition rate for $3 \to 2$ will be small compared to the transition rate for $3 \to 1$, i.e. $\kappa_{32} \ll \kappa_{31}$. In some cases, the small size of κ_{32} may cause an excessive delay in the transition from 3 to 2, so we also allow for an incoherent driving field on the $3 \leftrightarrow 2$ transition such that

$$\kappa_{32} \ll W_{32} = B_{32}\rho(\omega_{32}) \ll \kappa_{31} \,. \tag{18.124}$$

Under these conditions, the ion will spend most of its time shuttling between levels 1 and 3, with infrequent transitions from 3 to the intermediate level 2. The forbidden transition $2 \to 1$ occurs very slowly compared to $3 \to 1$ and $3 \to 2$, so level 2 effectively traps the occupation probability for a relatively long time. When this happens the fluorescent signal will turn off, and it will not turn on again until the ion decays back to level 1. We will refer to these transitions as quantum jumps.[1]

[1]It would be equally correct—but not nearly as exciting—to refer to this phenomenon as 'interrupted fluorescence'.

During the dark interval the ion is said to be *shelved* and $|\varepsilon_2\rangle$ is called a *shelving state*. The shelving effect is emphasized when the $1 \leftrightarrow 3$ transition is strongly saturated and the state $|\varepsilon_2\rangle$ is long-lived compared to $|\varepsilon_3\rangle$, i.e. when

$$W_{31} \gg \kappa_3 \gg \kappa_2 \,. \tag{18.125}$$

During the bright periods when fluorescence is observed, the state vector $|\Psi_{\text{ion}}\rangle$ will be a linear combination of $|\varepsilon_1\rangle$ and $|\varepsilon_3\rangle$; in other words, $|\Psi_{\text{ion}}\rangle$ is in the subspace $\mathfrak{H}_{13}$.

B A possible experimental realization

As a possible experimental realization of the three-level model, consider the intermittent resonance fluorescence of the strong Lyman-alpha line, emitted by a singly-ionized helium ion (He^+) in a Paul trap. One advantage of this choice is that the spectrum is hydrogenic, so that it can be calculated exactly.

The complementary relation between theory and experiment guarantees the presence of several real-world features that complicate the situation. The level diagram in part (a) of Fig. 18.2 shows not one, but two intermediate states, $2S_{1/2}$ and $2P_{1/2}$, that are separated in energy by the celebrated Lamb shift, $\Delta E_L/\hbar = 14.043\,\text{GHz}$. The $2S_{1/2}$-level is a candidate for a shelving state, since there is no dipole-allowed transition to the $1S_{1/2}$ ground state, but the $2P_{1/2}$-level does have a dipole-allowed transition, $2P_{1/2} \to 1S_{1/2}$. This adds unwanted complexity.

An additional theoretical difficulty is caused by the fact that the dominant mechanism for the transition $2S_{1/2} \to 1S_{1/2}$ is a two-photon decay. This is a problem, because the reservoir model introduced in Section 14.1.1 is built on the emission or absorption of single reservoir quanta; consequently, the standard reservoir model would not apply directly to this case.

Fortunately, these complications can be exploited to achieve a closer match to our simple model. The first step is to apply a weak DC electric field $\boldsymbol{\mathcal{E}}_0$ to the ion. In this

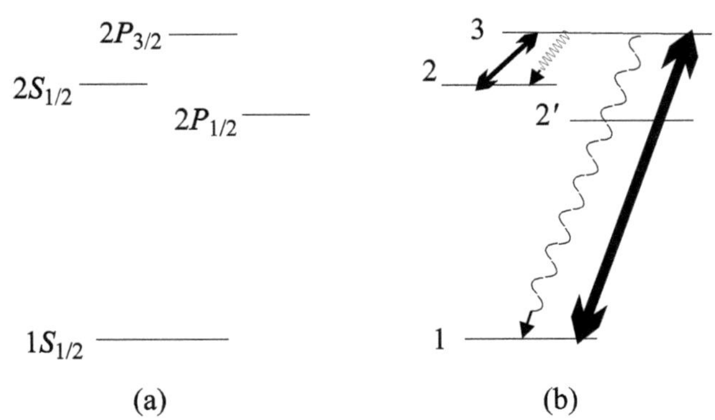

Fig. 18.2 (a) Level diagram for the He^+ ion. The spacing between the $2S_{1/2}$ and $2P_{1/2}$ levels is exaggerated for clarity. (b) The unperturbed $|2S_{1/2}\rangle$ and $|2P_{1/2}\rangle$ states are replaced by the Stark-mixed states $|\varepsilon_2\rangle$ and $|\varepsilon_2'\rangle$. The wavy arrows indicate dipole-allowed decays, while the solid arrows indicate incoherent driving fields. A dipole-allowed decay $2' \to 1$ is not shown, since $2'$ is effectively isolated by the method explained in the text.

application 'weak' means that the energy-level shift caused by the static field is small compared to the Lamb shift, i.e.

$$\left|\left\langle 2S_{1/2}\left|\boldsymbol{\mathcal{E}}_0 \cdot \mathbf{d}\right| 2P_{1/2}\right\rangle\right| \ll \Delta E_L \,. \tag{18.126}$$

In this case there will be no first-order Stark shift, and the second-order Stark effect (Bethe and Salpeter, 1977) mixes the $2S_{1/2}$ and $2P_{1/2}$ states to produce two new states,

$$|\varepsilon_2\rangle = C_S \left|2S_{1/2}\right\rangle + C_P \left|2P_{1/2}\right\rangle, \tag{18.127}$$

$$|\varepsilon_2'\rangle = C_S' \left|2S_{1/2}\right\rangle + C_P' \left|2P_{1/2}\right\rangle, \tag{18.128}$$

as illustrated in part (b) of Fig. 18.2.

A second-order perturbation calculation—using the Stark interaction $H_{\text{Stark}} = -\mathbf{d} \cdot \boldsymbol{\mathcal{E}}_0$—shows that $|C_S| \gg |C_P|$, i.e. $|\varepsilon_2\rangle$ is dominantly like $\left|2S_{1/2}\right\rangle$, while $|\varepsilon_2'\rangle$ is mainly like $\left|2P_{1/2}\right\rangle$. The states $|\varepsilon_1\rangle$ and $|\varepsilon_3\rangle$ of the simple model pictured in Fig. 18.1 are identified with $\left|1S_{1/2}\right\rangle$ and $\left|2P_{3/2}\right\rangle$ respectively.

Since neither $|\varepsilon_2\rangle$ nor $|\varepsilon_2'\rangle$ has definite parity, the dipole selection rules now allow single-photon transitions $2 \to 1$ and $2' \to 1$. The rate for $2 \to 1$ is proportional to $|C_P|^2$, so a proper choice of $|\boldsymbol{\mathcal{E}}_0|$ will guarantee that the single-photon process dominates the two-photon process, while still being slow compared to the rate κ_{31} for the Lyman-alpha transition.

By the same token, there are dipole-allowed transitions from 3 to both 2 and $2'$. The unwanted level $2'$ can be effectively eliminated by applying a microwave field resonant with the $3 \to 2$ transition, but not with the $3 \to 2'$ transition. The strength of this field can be adjusted so that the stimulated emission rate for $3 \to 2$ is large compared to the spontaneous rates for $3 \to 2'$ and $3 \to 2$, but small compared to the stimulated and spontaneous rates for the $3 \to 1$ transition. These settings ensure that the population of $|\varepsilon_2'\rangle$ will remain small at all times and that $|\varepsilon_2\rangle$ is an effective shelving state.

The main practical difficulty for this experiment is that the pump would have to operate at the vacuum-UV wavelength, 30.38 nm, of the Lyman-alpha line of the He^+ ion. One possible way around this difficulty is to use the radiation from a synchrotron light source.

The transition $3 \to 2$ is primarily due to the microwave-frequency transition $2P_{3/2} \to 2S_{1/2}$, which occurs at 44 GHz. Our assumption that the spontaneous emission rate for this transition is small compared to the transition rate for the Lyman-alpha transition is justified by the rough estimate

$$\frac{A\,(\text{Lyman alpha})}{A\,(2P_{3/2} \to 2S_{1/2})} \sim \left(\frac{\nu\,(\text{Lyman alpha})}{\nu\,(2P_{3/2} \to 2S_{1/2})}\right)^3 \sim 10^{16} \tag{18.129}$$

(Bethe and Salpeter, 1977), which uses the values 2.47×10^{15} Hz and 11 GHz for the Lyman-alpha transition and the $2P_{3/2} \to 2S_{1/2}$ microwave transition in hydrogen respectively.

The combination of the low rate for the $3 \to 2$ transition and the long lifetime of the shelving level 2 will permit easy observation of interrupted resonance fluorescence at the helium ion Lyman-alpha line, i.e. quantum jumps.

For hydrogen, the lifetime of the $2P_{3/2}$ state is 1.595 ns, so the estimate (18.129) tells us that the lifetime for the microwave transition is approximately 2×10^7 s, i.e. of the order of a year. The lifetime of the same transition for a hydrogenic ion scales as Z^{-4}, so for $Z = 2$ the microwave transition lifetime is 1.5×10^6 seconds, which is about a month.

This is still a rather long time to wait for a quantum jump. The solution is to adjust the strength of the resonant microwave field driving the $3 \leftrightarrow 2$ transition to bring this lifetime within the limits of the experimentalist's patience.

C Rate equation analysis

The assumption that the driving field is incoherent allows us to extend the rate equation approximation (11.190) for two-level atoms to our simple model to get

$$\frac{dP_3}{dt} = -\left(\kappa_3 + W_{31} + W_{32}\right) P_3 + W_{31} P_1 + W_{32} P_2 \,, \tag{18.130}$$

$$\frac{dP_2}{dt} = -\left(\kappa_2 + W_{32}\right) P_2 + \left(\kappa_{32} + W_{32}\right) P_3 \,, \tag{18.131}$$

$$\frac{dP_1}{dt} = -W_{31} P_1 + \left(W_{31} + \kappa_{31}\right) P_3 + \kappa_2 P_2 \,. \tag{18.132}$$

Adding the equations shows that the sum of the three probabilities is constant:

$$P_1 + P_2 + P_3 = 1 \,. \tag{18.133}$$

The inequalities (18.125) suggest that the adiabatic elimination rule (11.187) can be applied to the rate equations (18.130)–(18.132). To see how the rule works in this case, it is useful to express the rate equations in terms of the probability $P_{31} = P_3 + P_1$ that the ionic state is in $\mathfrak{H}_{13}$, and the inversion $D_{31} = P_3 - P_1$. The new form of the rate equations is

$$\begin{aligned} \frac{d}{dt} D_{31} = &-\left(2W_{31} + \kappa_{31} + \frac{1}{2}\kappa_{32} + \frac{1}{2}W_{32}\right) D_{31} \\ &-\left(\kappa_{31} + \frac{1}{2}\kappa_{32} + \frac{1}{2}W_{32}\right) P_{31} + \left(W_{32} - \kappa_2\right) P_2 \,, \end{aligned} \tag{18.134}$$

$$\frac{d}{dt} P_{31} = -\frac{1}{2}\left(\kappa_{32} + W_{32}\right) P_{31} - \frac{1}{2}\left(\kappa_{32} + W_{32}\right) D_{31} + \left(\kappa_2 + W_{32}\right) P_2 \,, \tag{18.135}$$

$$\frac{d}{dt} P_2 = -\left(\kappa_2 + W_{32}\right) P_2 + \frac{1}{2}\left(\kappa_2 + W_{32}\right) P_{31} + \frac{1}{2}\left(\kappa_2 + W_{32}\right) D_{31} \,. \tag{18.136}$$

The rate multiplying D_{31} on the right side of eqn (18.134) is much larger than any other rate in the equations; therefore, $D_{31}(t)$ will rapidly decay to the steady-state solution of eqn (18.134), i.e.

$$D_{31} = -\frac{\kappa_{31} + \frac{1}{2}\kappa_{32} + \frac{1}{2}W_{32}}{2W_{31} + \kappa_{31} + \frac{1}{2}\kappa_{32} + \frac{1}{2}W_{32}} P_{31} + \frac{W_{32} - \kappa_2}{2W_{31} + \kappa_{31} + \frac{1}{2}\kappa_{32} + \frac{1}{2}W_{32}} P_2 \,. \tag{18.137}$$

The coefficients of the probabilities P_{31} and P_2 are very small, so we can set $D_{31} \simeq 0$ in the rest of the calculation.

With this approximation, the remaining rate equations are

$$\frac{dP_2}{dt} = -R_{\text{on}}P_2 + R_{\text{off}}P_{31} \tag{18.138}$$

and

$$\frac{dP_{31}}{dt} = R_{\text{on}}P_2 - R_{\text{off}}P_{31}\,, \tag{18.139}$$

where $R_{\text{on}} = \kappa_2 + W_{32}$ is the rate at which the fluorescence turns on, and $R_{\text{off}} = (\kappa_{32} + W_{32})/2$ is the rate at which fluorescence turns off. Solving eqns (18.138) and (18.139) for $P_{31}(t)$ yields

$$P_{31}(t) = P_{31}(0)\,e^{-(R_{\text{on}}+R_{\text{off}})t} + \frac{R_{\text{on}}}{R_{\text{on}}+R_{\text{off}}}\left[1 - e^{-(R_{\text{on}}+R_{\text{off}})t}\right]. \tag{18.140}$$

The fluorescent intensity $I_F(t)$ is proportional to $P_{31}(t)$, so $I_F(t)$ evolves smoothly from its initial value $I_F(0)$ to the steady-state value

$$I_F \propto \frac{R_{\text{on}}}{R_{\text{on}}+R_{\text{off}}}\,. \tag{18.141}$$

This result is completely at odds with the flickering on-and-off behavior predicted above. The source of this discrepancy is the fact that the quantities P_1, P_2, and P_3 in the rate equations (18.138) and (18.139) are unconditional probabilities. This means that P_1, for example, is the probability that the ion is in level 1 without regard to its past history or any other conditions. Another way of saying this is that P_1 refers to an ensemble of ions which have reached level 1 in all possible ways.

Before the development of single-ion traps, resonance fluorescence experiments dealt with dilute atomic gases, and the total fluorescence signal would be correctly described by eqn (18.140). In this case, the on-and-off behavior of the individual atoms would be washed out by averaging over the random fluorescence of the atoms in the gas. For a single trapped ion, the smooth behavior in eqn (18.140) can only be recovered by averaging over many observations, all starting with the ion in the same state, e.g. the ground state.

In addition to the inability of the rate equations to predict quantum jumps, it is also the case that statistical properties—such as the distribution of waiting times between jumps—are beyond their reach. Thus any improvement must involve putting in some additional information; that is, reducing the size of the ensemble.

The first step in this direction was taken by Cook and Kimble (1985) who introduced the conditional probability $P_{31,n}(t, t+T)$ that the ion is in $\mathfrak{H}_{13}$ after making n transitions between $\mathfrak{H}_{13}$ and $|\varepsilon_2\rangle$ during the interval $(t, t+T)$. The number of transitions defines a subensemble of ions with this history. The complementary object $P_{2,n}(t, t+T)$ is the probability that the ion is in level $|\varepsilon_2\rangle$ after n transitions between $\mathfrak{H}_{13}$ and $|\varepsilon_2\rangle$ during the interval $(t, t+T)$.

By using the approximations leading to eqns (18.138) and (18.139), it is possible to derive an infinite set of coupled rate-like equations for $P_{31,n}(t, t+T)$ and $P_{2,n}(t, t+T)$, with $n = 0, 1, \ldots$. This approach permits the calculation of various statistical features of the quantum jumps, but it is not easy to connect it with the more refined quantum-jump theories to be developed later on.

D A stochastic model

We will now consider a simple on-and-off model which is qualitatively similar to the more sophisticated quantum-jump theories. In this approach, the analytical treatment based on conditional probabilities is replaced by an equivalent stochastic simulation.

We first assume that the fluorescent intensity can only have the values $I = 0$ (off) or $I = I_F$ (on). If the signal is on at time t, then the probability that it will turn off in the interval $(t, t + \Delta t)$ is $\Delta p_{\text{off}} = R_{\text{off}} \Delta t$. Conversely, if the signal is off at time t, then the probability that it will turn on in the interval $(t, t + \Delta t)$ is $\Delta p_{\text{on}} = R_{\text{on}} \Delta t$. For sufficiently small Δt, we can assume that only one of these events occurs.

The fluorescent intensities I_n at the discrete times $t_n = (n - 1) \Delta t$ can then be calculated by the following algorithm.

$$\begin{aligned}
&\text{For } I_n = 0 \text{ choose a random number } r \text{ in } (0,1)\,; \\
&\text{then set } I_{n+1} = 0 \text{ if } \Delta p_{\text{on}} < r \text{ or } I_{n+1} = I_F \text{ if } \Delta p_{\text{on}} > r\,. \\
&\text{For } I_n = I_F \text{ choose a random number } r \text{ in } (0,1)\,; \\
&\text{then set } I_{n+1} = I_F \text{ if } \Delta p_{\text{off}} < r \text{ or } I_{n+1} = 0 \text{ if } \Delta p_{\text{off}} > r\,.
\end{aligned} \tag{18.142}$$

The random choices in this algorithm are a special case of the rejection method (Press *et al.*, 1992, Sec. 7.3) for choosing random variables from a known distribution.

From a physical point of view, the algorithm is an approximate embodiment of the collapse postulate for measurements in quantum theory. The value I_n is the outcome of a measurement of the fluorescent intensity at $t = t_n$, so it corresponds to a collapse of the state vector of the ion into the state with the value I_n. If $I_{n+1} = I_n$ the subsequent collapse at $t = t_{n+1}$ is into the same state as at $t = t_n$. For $I_{n+1} \neq I_n$ the collapse at t_{n+1} is into the other state, so we see a quantum jump.

A typical[2] sequence of quantum jumps is shown in Fig. 18.3. Random sequences of binary choices (*dots* and *dashes*) of this kind are called **random telegraph signals**. This plot exhibits the expected on-and-off behavior for a single ion, but the smooth fluorescence curve predicted by the rate equations is nowhere to be seen.

In order to recover an approximation to eqn (18.140), we consider M experiments, all starting with $I_1 = I_F$, and define the average fluorescent intensity at time t_n by

$$I_{n,\text{av}} = \frac{1}{M} \sum_{j=1}^{M} I_{n,j}\,, \tag{18.143}$$

where $I_{n,j}$ is the fluorescent intensity at time t_n for the jth run. A comparison of $I_{n,\text{av}}$ with the values predicted by eqn (18.140) is shown in Fig. 18.4, for $M = 100$.

E Experimental evidence

We have demonstrated a simple model displaying quantum jumps and a plausible experimental realization for it, but the question remains if any such phenomena have

[2]The stochastic algorithm (18.142) gives a different plot for each run with the same input parameters. The 'typical' plot shown here was chosen to illustrate the effect most convincingly. This kind of data selection is not unknown in experimental practice.

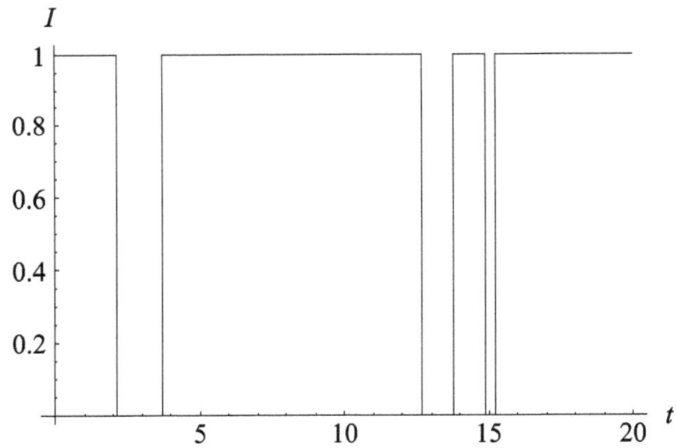

Fig. 18.3 Normalized fluorescent intensity I/I_F versus time (in units of the radiative lifetime $1/\kappa_b$ of the shelving state). In these units, $R_{\text{on}} = 1.6$, $R_{\text{off}} = 0.3$, and $\Delta t = 0.1$. The initial intensity is $I(0) = I_F$.

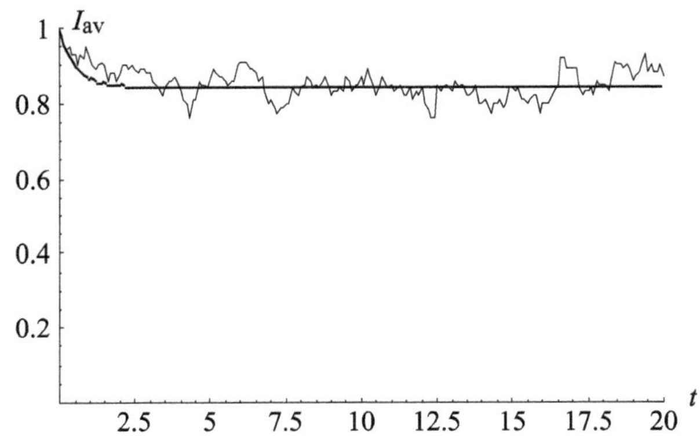

Fig. 18.4 Fluorescent intensity I/I_F (averaged over 100 runs) *vs* time (in units of the radiative lifetime of the shelving state). The smooth and jagged curves correspond respectively to eqns (18.140) and (18.143). Parameter values from Fig. 18.3.

been seen in reality. For this evidence we turn to an experiment in which intermittent fluorescence was observed from a single, laser-cooled Ba^+ ion in a radio frequency trap (Nagourney *et al.*, 1986).

The complementary relation between theory and experiment is in full play in this case, as seen by comparing the level diagram for this experiment—shown in Fig. 18.5—with Fig. 18.1. Fortunately, the complications involved in the real experiment do not change the essential nature of the effect, which is seen in Fig. 18.6.

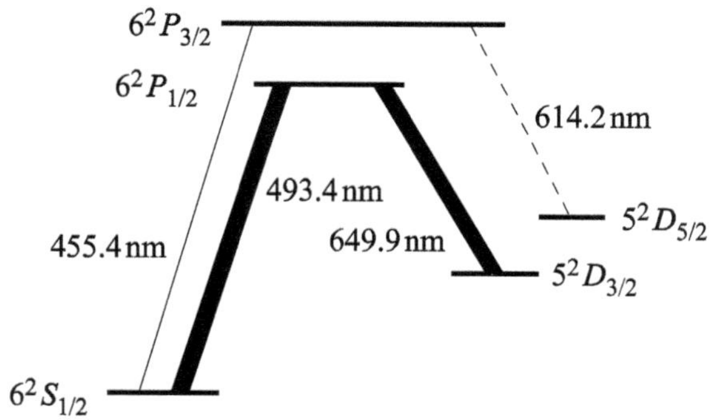

Fig. 18.5 Level structure of Ba^+. The states in the simple three-level model discussed in the text are $|\varepsilon_1\rangle = |6^2S_{1/2}\rangle$, $|\varepsilon_2\rangle = |6^2P_{3/2}\rangle$, and $|\varepsilon_3\rangle = |5^2D_{5/2}\rangle$, which is the shelf state. The remaining states are only involved in the laser cooling process indicated by the heavy solid lines. The $1 \leftrightarrow 2$ transition is driven by an incoherent source (a lamp) indicated by the light solid line. (Reproduced from Nagourney *et al.* (1986).)

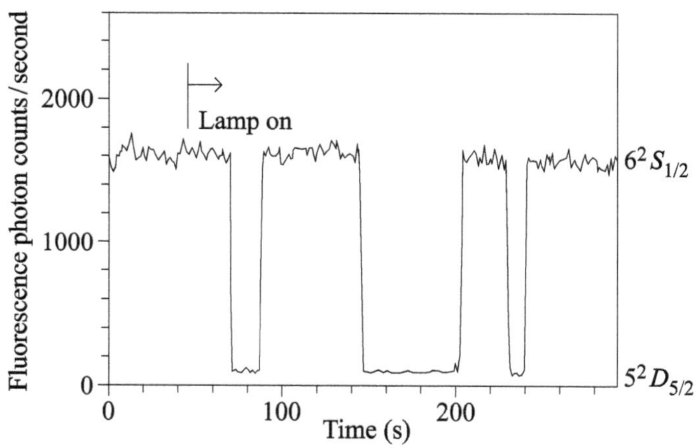

Fig. 18.6 A typical trace of the 493 nm fluorescence from the $6^2P_{1/2}$-level showing the quantum jumps after the hollow cathode lamp is turned on. The atom is definitely known to be in the shelf level during the low fluorescence periods. (Reproduced from Nagourney *et al.* (1986).)

18.7.2 Quantum jumps and the master equation*

Many features of quantum-jump experiments are well described—at least semi-quantitatively—by the rate equation approximation for the conditional probabilities, e.g. $P_{31,n}$, or by the equivalent stochastic simulation; but the rate equation model has definite limitations. The most important of these is the restriction to incoherent excitation of the atomic states. Many experiments employ laser excitation, which is inherently coherent in character.

The effort required to incorporate coherence effects eventually led to the creation of several closely related approaches to the problem of quantum jumps. These techniques are known by names like the *Monte Carlo wave function method*, *quantum trajectories*, and *quantum state diffusion*. Sorting out the relations between them is a complicated story, which we will not attempt to tell in detail. For an authoritative account, we recommend the excellent review article of Plenio and Knight (1998) which carries the history up to 1999.

We will present a brief account of the Monte Carlo wave function technique for the solution of the master equation. The other approaches mentioned above are technically similar; but they differ in the original motivations leading to them, in their physical interpretations, and in the kinds of experimental situations they can address.

There are two complementary views of these theoretical approaches. One may regard them simply as algorithms for the solution of the master equation, or as conceptually distinct views of quantum theory. The discussion therefore involves both computational and fundamental physics issues. We will first consider the computational aspects of the Monte Carlo wave function technique, and then turn to the conceptual relations between this method and the approaches based on quantum trajectories or quantum state diffusion.

The master equation (18.115) is a differential equation describing the time evolution of the sample density operator. Except in highly idealized situations—for which analytical solutions are known—the solution of the master equation requires numerical methods. Even for the apparently simple case of a single cavity mode, the sample Hilbert space $\mathfrak{H}_S$ is infinite dimensional, so the annihilation operator a is represented by an infinite matrix. A direct numerical attack would therefore require replacing $\mathfrak{H}_S$ by a finite-dimensional space, e.g. the subspace spanned by the number states $|0\rangle, \ldots, |M-1\rangle$. This would entail representing the creation and annihilation operators and the density operator by $M \times M$ matrices.

In some situations, such as those discussed in Section 18.5.2, an alternative approach is to replace the infinite-dimensional space $\mathfrak{H}_S$ by the two-dimensional quantum phase space, and to use—for a restricted class of problems—the Fokker–Planck equation (18.61) or the equivalent classical Langevin equation (18.72). In general, this method will fail if the diffusion matrix D is not positive definite.

The master equation for an atom can also be represented by a Fokker–Planck equation on a finite-dimensional phase space, but the collection of problems amenable to this treatment is restricted by the same kind of considerations, e.g. a positive-definite diffusion kernel, that apply to the radiation field. In many cases the center-of-mass motion of the atom can be neglected—or at least treated classically—so the sample Hilbert space is finite dimensional. In this situation the master equation for a two-level atom is simply a differential equation for a 2×2 hermitian matrix. This is equivalent to a set of four coupled ordinary differential equations, so it is not computationally onerous.

Unfortunately, in the real world of experimental physics, atoms often have more than two relevant levels, or it may be necessary to consider more than one atom at a time. In either case the computational difficulty grows rapidly with the dimensionality of the sample Hilbert space.

In general, a numerical simulation will take place in a sample Hilbert space with some dimension M. The master equation is then an equation for an $M \times M$ matrix, and the computational cost for solving the problem scales as M^2. This is an important consideration, since increasing the accuracy of the simulation typically requires enlarging the Hilbert space. On the other hand, if one could work with a state vector instead of the density operator, the cost of a solution would only scale as M. This gain alone justifies the development of the Monte Carlo wave function technique described below.

18.7.3 The Monte Carlo wave function method*

According to eqn (18.115), the change in the density operator over a time step Δt is

$$\rho_S(t+\Delta t) = \rho_S(t) + \frac{\Delta t}{i\hbar}[H_S, \rho_S] + \Delta t \mathcal{L}_{\text{dis}}\rho_S + O\left(\Delta t^2\right). \tag{18.144}$$

By combining the first two terms in eqn (18.117) for $\mathcal{L}_{\text{dis}}$ with the Hamiltonian term, this can be rewritten as

$$\rho_S(t+\Delta t) = \rho_S(t) - \frac{i\Delta t}{\hbar}H_{\text{dis}}\rho_S(t) + \frac{i\Delta t}{\hbar}\rho_S(t)H_{\text{dis}}^\dagger + \Delta t\sum_{k=1}^{K} C_k\rho_S(t)C_k^\dagger, \tag{18.145}$$

where the *dissipative* Hamiltonian is

$$H_{\text{dis}} = H_S - \frac{i\hbar}{2}\sum_{k=1}^{K} C_k^\dagger C_k. \tag{18.146}$$

This suggests defining a dissipative, nonunitary time translation operator,

$$U_{\text{dis}}(\Delta t) = e^{-i\Delta t H_{\text{dis}}/\hbar} = 1 - \frac{i\Delta t}{\hbar}H_{\text{dis}} + O\left(\Delta t^2\right), \tag{18.147}$$

and then using it to rewrite eqn (18.145) as

$$\rho_S(t+\Delta t) = U_{\text{dis}}(\Delta t)\rho_S(t)U_{\text{dis}}^\dagger(\Delta t) + \Delta t\sum_{k=1}^{K} C_k\rho_S(t)C_k^\dagger, \tag{18.148}$$

correct to $O(\Delta t)$.

The ensemble definition (2.116) of the density operator shows that this is equivalent to

$$\begin{aligned}\sum_e |\Psi_e(t+\Delta t)\rangle \mathcal{P}_e \langle\Psi_e(t+\Delta t)| &= \sum_e \mathcal{P}_e U_{\text{dis}}(\Delta t)|\Psi_e(t)\rangle\langle\Psi_e(t)|U_{\text{dis}}^\dagger(\Delta t) \\ &\quad + \sum_e \mathcal{P}_e \Delta t\sum_{k=1}^{K} C_k|\Psi_e(t)\rangle\langle\Psi_e(t)|C_k^\dagger,\end{aligned} \tag{18.149}$$

where the $\mathcal{P}_e$s are the probabilities defining the initial state, and $|\Psi_e(0)\rangle = |\Theta_e\rangle$. The first term on the right side of this equation evidently represents the dissipative evolution of each state in the ensemble. This is closely related to the Weisskopf–Wigner approach to perturbation theory, which we used in Section 11.2.2 to derive the decay of an excited atomic state by spontaneous emission.

This is all very well, but what is the meaning of the second term on the right side of eqn (18.149)? One way to answer this question is to fix attention on a single state in the ensemble, say $|\Psi_e(t)\rangle$, and to define the normalized states

$$|\phi_{ek}(t)\rangle = \frac{C_k |\Psi_e(t)\rangle}{\sqrt{\left\langle \Psi_e(t) \left| C_k^\dagger C_k \right| \Psi_e(t) \right\rangle}}, \quad k = 1, \dots, K. \tag{18.150}$$

With this notation, the contribution of $|\Psi_e(t)\rangle$ to the second term in eqn (18.149) is $(\kappa_e(t)\,\Delta t)\,\rho^e_{\text{meas}}(t)$, where

$$\rho^e_{\text{meas}}(t) = \sum_{k=1}^{K} \mathcal{P}^e_k |\phi_{ek}(t)\rangle \langle \phi_{ek}(t)|, \tag{18.151}$$

$$\mathcal{P}^e_k = \frac{\left\langle \Psi_e(t) \left| C_k^\dagger C_k \right| \Psi_e(t) \right\rangle}{\kappa_e(t)}, \tag{18.152}$$

and

$$\kappa_e(t) = \sum_{k=1}^{K} \left\langle \Psi_e(t) \left| C_k^\dagger C_k \right| \Psi_e(t) \right\rangle \tag{18.153}$$

is the total transition (quantum-jump) rate of $|\Psi_e(t)\rangle$ into the collection of normalized states defined by eqn (18.150). Since the coefficients $\mathcal{P}^e_k$ satisfy $0 \leqslant \mathcal{P}^e_k \leqslant 1$ and

$$\sum_{k=1}^{K} \mathcal{P}^e_k = 1, \tag{18.154}$$

they can be treated as probabilities.

With this interpretation, ρ^e_{meas} has the form (2.127) of the mixed state describing the sample after a measurement has been performed, but before the particular outcome is known. This suggests that we interpret the second term on the right side of eqn (18.149) as a wave packet reduction resulting from a measurement-like interaction with the reservoir.

After summing over the ensemble, eqn (18.148) becomes

$$\rho_S(t + \Delta t) = U_{\text{dis}}(\Delta t)\,\rho_S(t)\,U^\dagger_{\text{dis}}(\Delta t) + (\overline{\kappa}(t)\,\Delta t)\,\rho_{\text{meas}}(t), \tag{18.155}$$

where

$$\rho_{\text{meas}}(t) = \sum_e \mathcal{P}'_e \rho^e_{\text{meas}}(t), \tag{18.156}$$

$$\mathcal{P}'_e = \frac{\mathcal{P}_e \kappa_e(t)}{\overline{\kappa}(t)}, \tag{18.157}$$

and

$$\overline{\kappa}(t) = \sum_{e} \mathcal{P}_e \kappa_e(t) \tag{18.158}$$

is the ensemble-averaged transition rate.

A The Monte Carlo wave function algorithm

In quantum theory, a system evolves smoothly by the Schrödinger equation until a measurement event forces a discontinuous change. This feature is the basis for the procedure described here.

It is plausible to expect that only one of the two terms in eqn (18.155)—dissipative evolution or wave packet reduction—will operate during a sufficiently small time step. We will first describe the Monte Carlo wave function algorithm (MCWFA) that follows from this assumption, and then show that the density operator calculated in this way is an approximate solution of the master equation (18.115).

In order to simplify the presentation we assume that the initial ensemble is defined by

$$\begin{aligned} &\text{states } \{|\Theta_1\rangle, \ldots, |\Theta_M\rangle\}, \\ &\text{probabilities } \{\mathcal{P}_1, \ldots, \mathcal{P}_M\}, \end{aligned} \tag{18.159}$$

so that the index $e = 1, 2, \ldots, M$.

In each time step, a choice between dissipative evolution and wave packet reduction—i.e. a quantum jump—has to be made. For this purpose, we note that the probability of a quantum jump during the interval $(t, t + \Delta t)$ is $\Delta P_e(t) = \kappa_e(t)\Delta t$, where $\kappa_e(t)$ is the total transition rate defined by eqn (18.153). The discrete scheme will only be accurate if the jump probability during a time step is small, i.e. $\Delta P_e(t) \ll 1$. Consequently, the time step Δt must satisfy $\kappa_e(t)\Delta t \ll 1$.

With this preparation, we are now ready to state the algorithm for integrating the master equation in the interval $(0, T)$.

(1) Set $e = 1$ and define the discrete times $t_n = (n-1)\Delta t$, where $1 \leqslant n \leqslant N$ and $(N-1)\Delta t = T$.

(2) At the initial time $t = 0$, set $|\Psi(0)\rangle = |\Psi_e(0)\rangle = |\Theta_e\rangle$.

(3) For $n = 2, \ldots, N$ choose a random number r in the interval $(0, 1)$. If $\Delta P_e(t_{n-1}) < r$ go to (a), and if $\Delta P_e(t_{n-1}) > r$ go to (b). Since we have imposed $\Delta P_e(t) \ll 1$, this procedure guarantees that quantum jumps are relatively rare interruptions of continuous evolution.

   (a) In this case there is no quantum jump, and the state vector is advanced from t_{n-1} to t_n by dissipative evolution followed by normalization:

$$\begin{aligned} |\Psi_e(t_n)\rangle &= \frac{U_{\text{dis}}(\Delta t)|\Psi_e(t_{n-1})\rangle}{\sqrt{\left\langle \Psi_e(t_{n-1}) \middle| U_{\text{dis}}^{\dagger}(\Delta t) U_{\text{dis}}(\Delta t) \middle| \Psi_e(t_{n-1}) \right\rangle}} \\ &= \frac{\left(1 - \frac{i\Delta t}{\hbar} H_{\text{dis}}\right)|\Psi_e(t_{n-1})\rangle}{\sqrt{1 - \Delta P_e(t_{n-1})}}, \end{aligned} \tag{18.160}$$

   where the last line follows from the definition (18.147) of $U_{\text{dis}}(\Delta t)$.

(b) In this case there is a quantum jump, and the new state vector is defined by choosing k randomly from $\{1, 2, \ldots, K\}$—conditioned by the probability distribution $\mathcal{P}_k^e$ defined in eqn (18.152)—and setting

$$|\Psi_e(t_n)\rangle = |\phi_{ek}(t_{n-1})\rangle\,, \tag{18.161}$$

i.e. $|\Psi_e(t_n)\rangle$ jumps to one of the states permitted by the second term in eqn (18.148).

(4) Repeat step (3) N_{traj} times to get N_{traj} discrete representations

$$\{|\Psi_{ej}(t_n)\rangle\,, 1 \leqslant n \leqslant N\}\,, \quad j = 1, \ldots, N_{\text{traj}} \tag{18.162}$$

of the state vector. These representations are distinct, due to the random choices made in each time step. The density operator that evolves from the original pure state $|\Theta_e\rangle$ is then given by

$$\rho_e(t_n) = \frac{1}{N_{\text{traj}}} \sum_{j=1}^{N_{\text{traj}}} |\Psi_{ej}(t_n)\rangle \langle \Psi_{ej}(t_n)|\,. \tag{18.163}$$

(5) Replace e by $e+1$. If $e+1 \leqslant M$ go to step (2). If $e+1 > M$ go to step (6).

(6) The density operator $\rho(t)$ that evolves from the initial density operator $\rho(0)$—defined by the ensemble (18.159)—is given by

$$\rho(t_n) = \sum_{e=1}^{M} \mathcal{P}_e \rho_e(t_n)\,. \tag{18.164}$$

The computational cost of this method scales as $N_{\text{traj}}N$, where N is the dimensionality of the sample Hilbert space $\mathfrak{H}_S$. Consequently, the MCWFA would not be very useful as a technique for solving the master equation, if the required number of trials is itself of order N. Fortunately, there are applications with large N for which one can get good statistics with $N_{\text{traj}} \ll N$.

B *Proof that the MCWFA generates a solution*

If each of the density operators $\rho_e(t)$ satisfies the master equation, then so will the overall density operator defined by eqn (18.164); therefore, it is sufficient to give the proof for a single $\rho_e(t)$. For a sufficiently large number of trials, the evolution of the pure state operators,

$$\rho_{ej}(t_n) = |\Psi_{ej}(t_n)\rangle \langle \Psi_{ej}(t_n)|\,, \tag{18.165}$$

is effectively given by step (2a) with probability $1 - \Delta P_e(t_{n-1})$ and by step (2b) with probability $\Delta P_e(t_{n-1})$. In other words,

$$\begin{aligned}\rho_{ej}(t_n) = {} & (1 - \Delta P_e(t_{n-1})) \left|\Psi_{ej}^{\text{dis}}(t_n)\right\rangle \left\langle \Psi_{ej}^{\text{dis}}(t_n)\right| \\ & + \Delta P_e(t_{n-1}) \sum_k \mathcal{P}_k^e(t_{n-1}) |\phi_{ek}(t_{n-1})\rangle \langle \phi_{ek}(t_{n-1})|\,,\end{aligned} \tag{18.166}$$

where

$$\left|\Psi_{ej}^{\text{dis}}\left(t_n\right)\right\rangle = \frac{\left(1 - \frac{i\Delta t}{\hbar} H_{\text{dis}}\right)\left|\Psi_{ej}\left(t_{n-1}\right)\right\rangle}{\sqrt{1 - \Delta P_e\left(t_{n-1}\right)}} . \tag{18.167}$$

The $\left|\phi_{ek}\left(t_{n-1}\right)\right\rangle$s are defined by substituting $\left|\Psi_{ej}\left(t_{n-1}\right)\right\rangle$ for $\left|\Psi_e\left(t_{n-1}\right)\right\rangle$ in eqn (18.150). Using the definitions of ΔP_e, $\mathcal{P}_k^e$, and H_{dis} in this equation and neglecting $O\left(\Delta t^2\right)$-terms leads to

$$\frac{\rho_{ej}\left(t_n\right) - \rho_{ej}\left(t_{n-1}\right)}{\Delta t} = -\frac{i}{\hbar}\left[H_S, \rho_{ej}\left(t_{n-1}\right)\right] + \mathcal{L}_{\text{dis}}\rho_{ej}\left(t_{n-1}\right). \tag{18.168}$$

Averaging this result over the trials, according to eqn (18.163), and taking the limit $\Delta t \to 0$ shows that $\rho_e(t)$ satisfies the master equation (18.115).

18.7.4 Laser-induced fluorescence*

For a concrete application of the MCWFA, we return to the trapped three-level ion considered in Section 18.7.1. For this example, however, we replace the incoherent source driving $3 \leftrightarrow 1$ by a coherent laser field $\mathcal{E}_L e^{-i\omega_L t}$ that is close to resonance, i.e. $\left|\omega_L - \omega_{31}\right| \ll \omega_L$. In the interests of simplicity, we also drop the field driving $3 \leftrightarrow 2$. The semiclassical approximation for the laser is applied by substituting $\mathbf{E}^{(+)} \to \mathcal{E}_L e^{-i\omega_L t}$ into eqn (11.40).

In the resonant wave approximation, the Schrödinger-picture Hamiltonian is $H_S = H_{S0} + H_{S1}$, where

$$H_{S0} = \sum_q \epsilon_q S_{qq} , \tag{18.169}$$

$$H_{S1} = \hbar\Omega_L S_{31} e^{-i\omega_L t} + \text{HC} , \tag{18.170}$$

and $\Omega_L = -\mathbf{d}_{31} \cdot \mathcal{E}_L/\hbar$ is the Rabi frequency for the laser driving the $1 \leftrightarrow 3$ transition. The S_{qp}s are the atomic transition operators defined in Section 11.1.4, and the labels q and p range over the values $1, 2, 3$.

The form of the dissipative operator $\mathcal{L}_{\text{dis}}$ for the three-level ion can be inferred from the result (18.44) for the two-level atom, by identifying each pair of levels connected by a decay channel with a two-level atom. For example, the lowering operator σ_- in eqn (18.44) will be replaced by S_{13} for the $3 \to 1$ decay channel, and the remaining transitions are treated in the same way.

There are two important simplifications in the present case. The first is that the phase-changing collision term in eqn (18.44) is absent for an isolated ion. The second simplification is the assumption that the reservoirs coupled to the three transitions—i.e. the modes of the radiation field near resonance—are at zero temperature. This approximation is generally accurate at optical frequencies, since $k_B T \ll \omega_{\text{opt}}$ for any reasonable temperature.

One can use these features to show that $\mathcal{L}_{\text{dis}}$ is defined by

$$\begin{aligned} \mathcal{L}_{\text{dis}}\rho_S = &-\frac{\kappa_{31}}{2}\left(S_{31}S_{13}\rho_S + \rho_S S_{31}S_{13} - 2S_{13}\rho_S S_{31}\right) \\ &-\frac{\kappa_{32}}{2}\left(S_{32}S_{23}\rho_S + \rho_S S_{32}S_{23} - 2S_{23}\rho_S S_{32}\right) \\ &-\frac{\kappa_2}{2}\left(S_{21}S_{12}\rho_S + \rho_S S_{21}S_{12} - 2S_{12}\rho_S S_{21}\right). \end{aligned} \tag{18.171}$$

This expression for $\mathcal{L}_{\text{dis}}$ can be cast into the general Lindblad form (18.117) by setting $K = 3$ and defining the operators

$$C_1 = \sqrt{\kappa_{31}} S_{13}\,, \quad C_2 = \sqrt{\kappa_{32}} S_{23}\,, \quad C_3 = \sqrt{\kappa_2} S_{12}\,, \tag{18.172}$$

corresponding respectively to the decay channels $3 \to 1$, $3 \to 2$, and $2 \to 1$.

The Rabi frequency Ω_L is small compared to the laser frequency ω_L, so the Schrödinger-picture master equation,

$$i\hbar \frac{\partial}{\partial t} \rho_S(t) = [H_S, \rho_S(t)] + \mathcal{L}_{\text{dis}} \rho_S(t)\,, \tag{18.173}$$

involves two very different time scales, $1/\omega_L \ll 1/\Omega_L$. Differential equations with this feature are said to be **stiff**, and it is usually very difficult to obtain accurate numerical solutions for them (Press *et al.*, 1992, Sec. 16.6). In the case at hand, this difficulty can be avoided by transforming to the interaction picture.

The general results in Section 4.8 yield the transformed master equation

$$i\hbar \frac{\partial}{\partial t} \rho_S^I(t) = \left[H_{S1}^I, \rho_S^I(t)\right] + U_0^\dagger(t)\, \mathcal{L}_{\text{dis}} \rho_S(t)\, U_0(t)\,, \tag{18.174}$$

where $U_0(t) = \exp(-iH_{S0}t/\hbar)$ and the transform of any operator X is $X^I(t) = U_0^\dagger(t)\, X U_0(t)$. Applying this rule to the transition operators gives

$$S_{qp}^I(t) = U_0^\dagger(t)\, S_{qp} U_0(t) = e^{i\omega_{qp}t} S_{qp}\,, \tag{18.175}$$

and this in turn leads to

$$U_0^\dagger(t)\, \mathcal{L}_{\text{dis}} \rho_S(t)\, U_0(t) = \mathcal{L}_{\text{dis}} \rho_S^I(t)\,. \tag{18.176}$$

Thus we arrive at the useful conclusion that $\mathcal{L}_{\text{dis}}$ has the same form in both pictures.

The transformed interaction Hamiltonian is

$$H_{S1}^I = \hbar\Omega_L S_{31} e^{-i\delta t} + \text{HC}\,, \tag{18.177}$$

where $\delta = \omega_L - \omega_{31}$. The interaction-picture master equation (18.174) is not stiff, but it still has time-dependent coefficients. This annoyance can be eliminated by a further transformation

$$\overline{\rho}_S(t) = e^{itF} \rho_S^I(t)\, e^{-itF}\,, \tag{18.178}$$

where

$$F = \sum_q f_q S_{qq}\,. \tag{18.179}$$

The algebra involved here is essentially identical to the original transformation to the interaction picture, and it is not difficult to show that the equation of motion for $\overline{\rho}(t)$ will have constant coefficients provided that the parameters f_q are chosen to satisfy

$$f_3 - f_1 = \delta\,. \tag{18.180}$$

The simple solution $f_1 = f_2 = 0$, and $f_3 = \delta$, leads to

$$i\hbar\frac{\partial}{\partial t}\overline{\rho}_S(t) = \left[\overline{H}_{S1}, \overline{\rho}_S(t)\right] + \mathcal{L}_{\text{dis}}\overline{\rho}_S(t)\,, \tag{18.181}$$

where the transformed interaction Hamiltonian is

$$\overline{H}_{S1} = \hbar\begin{bmatrix} 0 & 0 & \Omega_L^* \\ 0 & 0 & 0 \\ \Omega_L & 0 & -\delta \end{bmatrix}. \tag{18.182}$$

We are now in a position to calculate all the bits and pieces that are needed for the direct solution of the master equation (18.181), or the application of the MCWFA. We leave the algebra as an exercise for the reader and proceed directly to the numerical solution of the master equation. The density operator for this problem is represented by a 3×3 hermitian matrix which is determined by nine real numbers. Thus the master equation in this case consists of nine linear, ordinary differential equations with constant coefficients. There are many packaged programs that can be used to solve this problem.

Of course, this means that we do not really need the MCWFA, but it is still useful to have a solvable problem as a check on the method. In Fig. 18.7 we compare the direct solution to the average over 48 trials of the MCWFA. The match between the averaged results and the direct solution can be further improved by using more trials in the average, but it should already be clear that the MCWFA is converging on a solution of the master equation.

Following the general practice in physics, we assume—on the basis of this special case—that the MCWFA can be confidently applied in all cases. In particular, this includes those applications for which the dimension of the relevant Hilbert space is large compared to the number of trials needed.

18.7.5 Quantum trajectories*

The results displayed in Fig. 18.7 show that the full-blown master equation—whether solved directly or by averaging over repeated trials of the MCWFA—does no better than the rate equations of Section 18.7.1 in describing the phenomenon of interrupted fluorescence. This should not be a surprise, since the master equation describes the evolution of the entire ensemble of state vectors for the ion.

What is needed for the description of quantum jumps (interrupted fluorescence) is an improved version of the simple on-and-off model used to derive the random telegraph signal in Fig. 18.3. This is where single trials of the MCWFA come into play. Each trial yields a sequence of state vectors

$$|\Psi(t_1)\rangle\,, |\Psi(t_2)\rangle\,, \ldots, |\Psi(t_N)\rangle\,, \tag{18.183}$$

which is a discrete sampling of a continuous function $|\Psi(t)\rangle$. This has led to the use of the name **discrete quantum trajectory** for each individual trial of the MCWFA.

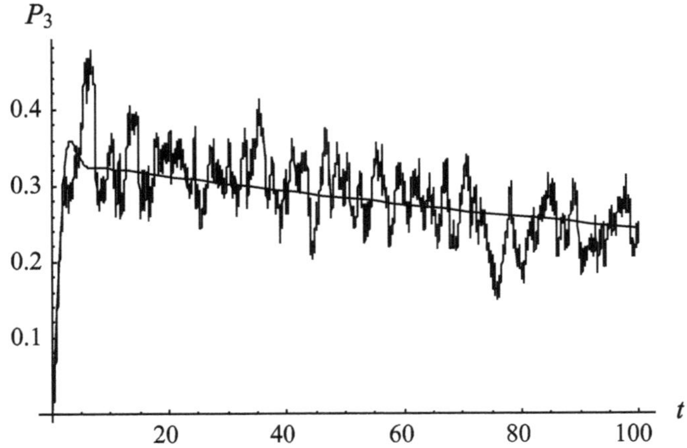

Fig. 18.7 The population of $|\varepsilon_3\rangle$ as a function of time. The smooth curve represents the direct solution of eqn (18.181) and the jagged curve is the result of averaging over 48 trials of the Monte Carlo wave function algorithm. Time is measured in units of the decay time $1/\kappa_{31}$ for the $3 \to 1$ transition. In these units $\Omega_L = 0.5$, $\delta = 0$, $\kappa_{32} = 0.01$, and $\kappa_{21} = 0.001$.

An example of the upper-level population P_3 obtained from a single quantum trajectory is shown in Fig. 18.8. Once again, a judicious choice from the results for several trajectories nicely exhibits the random telegraph signal characterizing interrupted fluorescence.

According to the standard rules of quantum theory, the information from a completed measurement—in particular, the collapse of the state vector—should be taken into account immediately. In the algorithm presented in Section 18.7.3 the new infor-

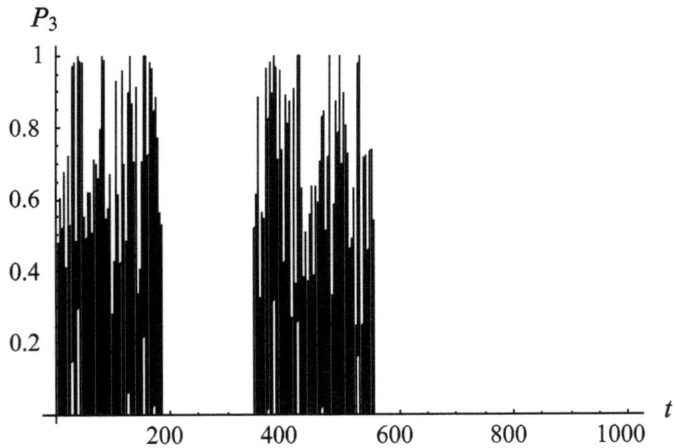

Fig. 18.8 The population of $|\varepsilon_3\rangle$ as a function of time for a single quantum trajectory. The parameter values are the same as in Fig. 18.7.

mation is not used until the next time step at $t_n + \Delta t$, so single trials of the Monte Carlo wave function method are approximations to the true quantum trajectory.

A more refined treatment involves allowing for the projection or collapse event to occur one or more times during the interval Δt, and using the dissipative Hamiltonian to propagate the state vector in the subintervals between collapses. With this kind of analysis, it can be shown that the Monte Carlo method is accurate to order Δt. Increasing the accuracy to order Δt^2 requires the inclusion of jumps at both ends of the interval and also the possibility that two jumps can occur in succession (Plenio and Knight, 1998).

Results like that shown in Fig. 18.8 might tempt one to believe that the Monte Carlo technique—or the more refined quantum trajectory method—provides a description of single quantum events in isolated microscopic samples. Any such conclusion would be completely false. A large sample of trials for the Monte Carlo technique will resemble a corresponding set of experimental runs, but the relation between the two sets is purely statistical. Both will yield the same expectation values, correlation functions, etc. In other words, the Monte Carlo or quantum trajectory methods are still based on ensembles. The difference between these methods and the full master equation is that the ensembles are conditioned, i.e. reduced, by taking experimental results into account.

18.7.6 Quantum state diffusion*

As explained above, the standard formulations of quantum theory do not apply to individual microscopic samples, but rather to ensembles of identically prepared samples. Several of the founders of the quantum theory, including Einstein (Einstein *et al.*, 1935) and Schrödinger (Schrödinger, 1935*b*), were not at all satisfied with this feature, and there have been many subsequent efforts to reformulate the theory so that it applies to individual microscopic objects. One approach, which has attracted a great deal of attention, is to replace the Schrödinger equation for an ensemble by a stochastic equation—e.g. a diffusion equation in the Hilbert space of quantum states—for an individual system.

The universal empirical success of conventional quantum theory evidently requires that the new stochastic equation should agree with the Schrödinger equation when applied to ensembles. Many such equations are possible, but symmetry considerations—see Gisin and Percival (1992) and references contained therein—have led to an essentially unique form.

For a sample described by the Lindblad master equation (18.115) the stochastic equation for the state vector can be written as

$$\begin{aligned}\frac{d}{dt}\left|\Psi(t)\right\rangle &= \frac{1}{i\hbar}H_{\text{dis}}\left|\Psi(t)\right\rangle + \sum_k \left\langle C_k^\dagger(t)\right\rangle_\Psi \left[C_k(t) - \frac{1}{2}\left\langle C_k(t)\right\rangle_\Psi\right]\left|\Psi(t)\right\rangle \\ &\quad + \sum_k \left[C_k(t) - \left\langle C_k(t)\right\rangle_\Psi\right]\left|\Psi(t)\right\rangle \zeta_k(t)\,, \end{aligned} \tag{18.184}$$

where H_{dis} is the dissipative Hamiltonian defined by eqn (18.146), and

$$\langle X\rangle_\Psi = \langle\Psi\,|X|\,\Psi\rangle \tag{18.185}$$

is the expectation value in the state. The c-numbers $\zeta_k(t)$ are delta-correlated random variables, i.e.

$$\langle \zeta_k^* (t) \zeta_{k'} (t') \rangle_P = \delta_{kk'} \delta (t - t') , \tag{18.186}$$

where the average $\langle \cdots \rangle_P$ is defined by the probability distribution P for the random variables ζ_k.

We have chosen to write the stochastic equation for the state vector so that it resembles the operator Langevin equations discussed in Chapter 14, but most authors prefer to use the more mathematically respectable Ito form (Gardiner, 1991). The presence of the averages $\langle C_k (t) \rangle_\Psi$ makes this equation nonlinear, so that analytical solutions are hard to come by.

In this approach, quantum jumps appear as smooth transitions between discrete quantum states. The transitions occur on a short time scale, that is determined by the equation itself. Physical interactions describing measurements of an observable lead to irreversible diffusion toward one of the eigenstates of the observable, so that no separate collapse postulate is required. In applications, the numerical solution of eqn (18.184) has the same kind of advantage over the direct solution of the master equation as the Monte Carlo wave function method.

Given the close relation between the master equation, quantum jumps, and quantum state diffusion, it is not very surprising to learn that quantum state diffusion can be derived as a limiting case of the quantum-jump method. The limiting case is that of infinitely many jumps, where each jump causes an infinitesimal change in the state vector. This mathematical procedure is related to the experimental technique of balanced heterodyne detection discussed in Section 9.3. Thus the quantum state diffusion method can be regarded as a new conceptual approach to quantum theory, or as a particular method for solving the master equation.

18.8 Exercises

18.1 Averaging over the environment

(1) Combine $\rho_W (0) = \rho_S (0) \rho_E (0)$ and the assumption $\langle b_{r\nu} \rangle_E = 0$ with eqn (18.14) to derive eqn (18.15).

(2) Drop the assumption $\langle b_{r\nu} \rangle_E = 0$, and introduce the fluctuation operators $\delta b_{r\nu} = b_{r\nu} - \langle b_{r\nu} \rangle_E$. Show how to redefine H_S and H_E, so that eqn (18.15) will still be valid.

18.2 Master equation for a cavity mode

(1) Use the discussion in Section 18.4.1 to argue that the general expression (18.20) for the double commutator $\mathfrak{C}_2 (t, t')$ can be replaced by $\mathfrak{C}_2 (t, t') = \{ [\mathfrak{F}^\dagger (t) , \mathfrak{G} (t')] + \text{HC} \}$.

(2) Use the expression (18.25) for $\mathfrak{F}$ to show that $\text{Tr}_E \left[\mathfrak{F}^\dagger (t) , \mathfrak{G} (t')\right]$ can be expressed in terms of the correlation functions in eqns (18.28) and (18.29).

(3) Put everything together to derive eqn (18.30). Do not forget the end-point rule.

(4) Transform back to the Schrödinger picture to derive eqns (18.31)–(18.33).

18.3 Master equation for a two-level atom

(1) Use the Markov assumptions (14.142) and (14.143) to verify eqns (18.40) and (18.41).

(2) Use these expressions to evaluate the double commutator $\mathfrak{G}_2$.

(3) Given the assumptions made in Section 18.4.2, find out which terms in $\mathfrak{G}_2$ have vanishing traces over the environment.

(4) Evaluate the traces of the surviving terms and thus derive the master equation in the environment picture.

(5) Transform back to the Schrödinger picture to derive eqns (18.42)–(18.44).

18.4 Thermal equilibrium for a cavity mode

(1) Derive eqn (18.34) from eqn (18.31).

(2) Solve the recursion relation (18.37), subject to eqn (18.38), to find eqn (18.39).

18.5 Fokker–Planck equation

(1) Carry out the chain rule calculation needed to derive eqn (18.81).

(2) Derive and solve the differential equations for the functions introduced in eqn (18.85).

(3) Derive eqn (18.93).

18.6 Lindblad form for the two-level atom*

Determine the three operators C_1, C_2, and C_3 for the two-level atom.

18.7 Evolution of the purity of a general state*

(1) Use the cyclic invariance of the trace operation to deduce eqn (18.119) from eqn (18.115).

(2) Suppose that a single cavity mode is in thermal equilibrium with the cavity walls at temperature T. At $t = 0$ the cavity walls are suddenly cooled to zero temperature. Calculate the initial rate of change of the purity.

19

Bell's theorem and its optical tests

Since this is a book on quantum optics, we have assumed throughout that quantum theory is correct in its entirety, including all its strange and counterintuitive predictions. As far as we know, all of these predictions—even the most counterintuitive ones—have been borne out by experiment. Einstein accepted the experimentally verified predictions of quantum theory, but he did not believe that quantum mechanics could be the entire story. His position was that there must be some underlying, more fundamental theory, which satisfied the principles of locality and realism.

According to the principle of **locality**, a measurement occurring in a finite volume of space in a given time interval could not possibly influence—or be influenced by—measurements in a distant volume of space at a time before any light signal could connect the two localities. In the language of special relativity, two such localities are said to be **space-like** separated.

The principle of **realism** contains two ideas. The first is that the physical properties of objects exist independently of any measurements or observations. This point of view was summed up in his rhetorical question to Abraham Pais, while they were walking one moonless night together on a path in Princeton: 'Is the Moon there when nobody looks?' The second is the condition of **spatial separability**: the physical properties of spatially-separated systems are mutually independent.

The combination of the principles of locality and realism with the EPR thought experiment convinced Einstein that quantum theory must be an incomplete description of physical reality.

For many years after the EPR paper, this discussion appeared to be more concerned with philosophy than physics. The situation changed dramatically when Bell (1964) showed that every **local realistic** theory—i.e. a theory satisfying a plausible interpretation of the metaphysical principles of locality and realism favored by Einstein—predicts that a certain linear combination of correlations is uniformly bounded. Bell further showed that this inequality is violated by the predictions of quantum mechanics.

Subsequent work has led to various generalizations and reformulations of Bell's original approach, but the common theme continues to be an inequality satisfied by some linear combination of correlations. We will refer to these inequalities generically as **Bell inequalities**. Most importantly, two-photon, coincidence-counting experiments have shown that a particular Bell inequality is, in fact, violated by nature. One must therefore give up one or the other—or possibly even both—of the principles of locality and realism (Chiao and Garrison, 1999).

Bell thereby successfully transformed what seemed to be an essentially philosophical problem into experimentally testable physical propositions. This resulted in what Shimony has aptly called *experimental metaphysics.* The first experiment to test Bell's theorem was performed by Freedman and Clauser (1972). This early experiment already indicated that there must be something wrong with Einstein's fundamental principles.

One of the most intriguing developments in recent years is that the Bell inequalities —which began as part of an investigation into the conceptual foundations of quantum theory—have turned out to have quite practical applications to fields like quantum cryptography and quantum computing.

Quantum optics is an important tool for investigating the phenomenon of quantum nonlocality connected with EPR states and the EPR paradox. Although Einstein, Podolsky, and Rosen formulated their argument in the language of nonrelativistic quantum mechanics, the problem they posed also arises in the case of two relativistic particles flying off in different directions, for example, the two photons emitted in spontaneous down-conversion.

19.1 The Einstein–Podolsky–Rosen paradox

The Einstein–Podolsky–Rosen paper (Einstein *et al.*, 1935) adds two further ideas to the principles of locality and realism presented above. The first is the definition of an **element of physical reality**:

> If, without in any way disturbing a system, we can predict with certainty (i.e. with probability equal to unity) the value of a physical quantity, then there exists an element of physical reality corresponding to this physical quantity.

The second is a criterion of completeness for a physical theory:

> ...every element of the physical reality must have a counterpart in the physical theory.

The argument in the EPR paper was formulated in terms of the entangled two-body wave function

$$\psi\left(x_A, x_B\right) = \int_{-\infty}^{\infty} \frac{dk}{2\pi} e^{ik(x_A - x_B - L)}, \tag{19.1}$$

which is a special case of the EPR states defined by eqn (6.1), but we will use a simpler example due to Bohm (1951, Chap. 22), which more closely resembles the actual experimental situations that we will study. Hints for carrying out the original argument can be found in Exercise 19.1.

Bohm's example is modeled on the decay of a spin-zero particle into two distinguishable spin-1/2 particles, and it—like the original EPR argument—is expressed in the language of nonrelativistic quantum mechanics. In the rest frame of the parent particle, conservation of the total linear momentum implies that the daughter particles are emitted in opposite directions, and conservation of spin angular momentum implies that the total spin must vanish.

In this situation, the decay channel in which the particles travel along the z-axis, with momenta $\hbar k_0$ and $-\hbar k_0$, is described by the two-body state

$$|\Psi\rangle_{AB} = e^{ik_0(z_A - z_B)}\,|\Phi\rangle_{AB}\,, \tag{19.2}$$

where the spins σ_A and σ_B are described by the **Bohm singlet state**

$$|\Phi\rangle_{AB} = \frac{1}{\sqrt{2}}\left\{|\uparrow\rangle_A\,|\downarrow\rangle_B - |\downarrow\rangle_A\,|\uparrow\rangle_B\right\}, \tag{19.3}$$

which is expressed in the notation introduced in eqns (6.37) and (6.38).

The choice of the quantization axis $\mathbf{n}$ is left open, since—as seen in Exercise 6.3—the spherical symmetry of the Bohm singlet state guarantees that it has the same form for any choice of $\mathbf{n}$. Since only spin measurements will be considered, the following discussion will be carried out entirely in terms of the spin part $|\Phi\rangle_{AB}$ of the two-body state vector.

The spins of the daughter particles can be measured separately by means of two Stern–Gerlach magnets placed to intercept them, as shown in Fig. 19.1. Correlations between the spatially well-separated spin measurements can then be determined by means of coincidence-counting circuitry connecting the four counters.

Let us first suppose that the magnetic fields—and consequently the spatial quantization axes—of the two Stern–Gerlach magnets are directed along the x-axis, i.e. $\mathbf{n} = \mathbf{u}_x$. A measurement of the spin component S_x^A with the result $+1/2$ is signalled by a click in the upper Geiger counter of the Stern–Gerlach apparatus A. Applying von Neumann's projection postulate to the Bohm singlet state yields the reduced state

$$|\Phi\rangle_{AB}^{x} = |\uparrow_x\rangle_A\,|\downarrow_x\rangle_B\,, \tag{19.4}$$

where $|\uparrow_x\rangle_A$ is an eigenstate of S_x^A with eigenvalue $+1/2$, etc.

The reduced state is also an eigenstate of S_x^B with eigenvalue $-1/2$; therefore, any measurement of S_x^B would certainly yield the value $-1/2$, corresponding to a click in the lower counter of apparatus B. Since this prediction of a definite value for S_x^B

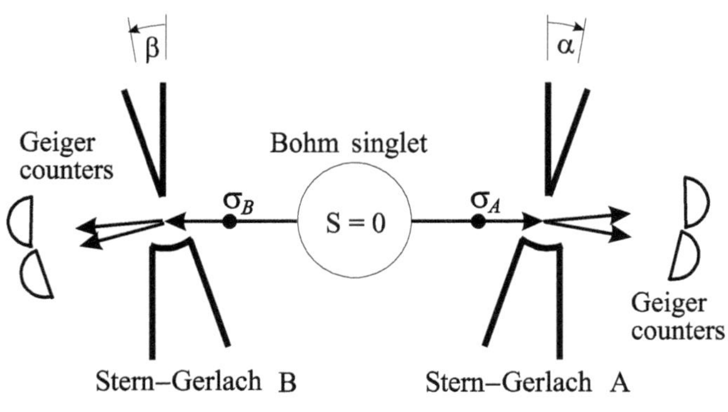

Fig. 19.1 The Bohm singlet version of the EPR experiment. σ_A and σ_B are spin-1/2 particles in a singlet state, and α and β are the angles of orientation of the two Stern–Gerlach magnets.

does not require any measurement at all, the system is not disturbed in any way. Consequently, S_x^B is an element of physical reality at B.

Now consider the alternative scenario in which the quantization axes are directed along y. In this case, a measurement of S_y^A with the outcome $+1/2$ leaves the system in the reduced state

$$|\Phi\rangle_{AB}^{y} = |\uparrow_y\rangle_A |\downarrow_y\rangle_B , \tag{19.5}$$

and this in turn implies that the value of S_y^B is certainly $-1/2$. This prediction is also possible without disturbing the system; therefore, S_y^B is also an element of physical reality at B.

From the local-realistic point of view, a believer in quantum theory now faces a dilemma. The spin components S_x^B and S_y^B are represented by noncommuting operators:

$$\left[S_x^B, S_y^B\right] = i\hbar S_z^B \neq 0 , \tag{19.6}$$

so they cannot be simultaneously predicted or measured. This leaves two alternatives.

(1) If S_x^B and S_y^B are both elements of physical reality, then quantum theory—which cannot predict values for both of them—is incomplete.

(2) Two physical quantities, like S_x^B and S_y^B, that are associated with noncommuting operators cannot be simultaneously real.

The latter alternative implies a more restrictive definition of physical reality in which, for example, two quantities cannot be simultaneously real unless they can be simultaneously measured or predicted. This would, however, mean that the physical reality of S_x^B or S_y^B at B depends on which measurement was carried out at the distant apparatus A.

The state reductions in eqns (19.4) or (19.5), i.e. the replacement of the original state $|\Phi\rangle_{AB}$ by $|\Phi\rangle_{AB}^{x}$ or $|\Phi\rangle_{AB}^{y}$ respectively, occur as soon as the measurement at A is completed. This is true no matter where apparatus B is located; in particular, when the light transit time from A to B is larger than the time required to complete the measurement at A. Thus the global change in the state vector occurs before any signal could travel from A to B. This evidently violates local realism.

In the words of Einstein, Podolsky, and Rosen, 'No reasonable definition of reality could be expected to permit this.' On this basis, they concluded that quantum theory is incomplete. In this connection, it is interesting to quote Einstein's reaction to Schrödinger's introduction of the notion of entangled states. In a letter to Born, written in 1948, Einstein wrote the following (Einstein, 1971):

> There seems to me no doubt that those physicists who regard the descriptive methods of quantum mechanics as definitive in principle would react to this line of thought in the following way: they would drop the requirement for *the independent existence of the physical reality present in different parts of space*; they would be justified in pointing out that the quantum theory nowhere makes explicit use of this requirement. [*Emphasis added*]

19.2 The nature of randomness in the quantum world

If the EPR claim that quantum theory is incomplete is accepted, then the next step would be to find some way to complete it. One advantage of such a construction would be that the randomness of quantum phenomena, e.g. in radioactive decay, might be explained by a mechanism similar to ordinary statistical mechanics.

In other words, there may exist some set of **hidden variables** within the radioactive nucleus that evolve in a deterministic way. The apparent randomness of radioactive decay would then be merely the result of our ignorance of the initial values of the hidden variables. From this point of view, there is no such thing as an uncaused random event, and the characteristic randomness of the quantum world originates at the very beginning of each microscopic event.

This should be contrasted with the quantum description, in which the state vector evolves in a perfectly deterministic way from its initial value, and randomness enters only at the time of measurement.

A simple example of a hidden variable theory is shown in Fig. 19.2. Imagine a box containing many small, hard spheres that bounce elastically from the walls of the box, and also scatter elastically from each other. The properties of such a system of particles can be described by classical statistical mechanics.

Cutting a small hole into one of the walls of the box will result in an exponential decay law for the number of particles remaining in the box as a function of time. In this model for a nucleus undergoing radioactive decay, the apparent randomness is ascribed to the observers ignorance of the initial conditions of the balls, which obey completely deterministic laws of motion. The unknown initial conditions are the hidden variables responsible for the observed phenomenon of randomness.

For an alternative model, we jump from the nineteenth to the twentieth century, and imagine that the box is equipped with a computer running a program generating random numbers, which are used to decide whether or not a particle is emitted in a given time interval. In this case the apparently random behavior is generated by a deterministic algorithm, and the hidden variables are concealed in the program code and the seed value used to begin it.

Let us next consider a series of random events occurring in a time interval $(t - \Delta t/2, t + \Delta t/2)$ at two distant points $\mathbf{r}_1$ and $\mathbf{r}_2$. If the two sets of events are space-like separated, i.e. $|\mathbf{r}_1 - \mathbf{r}_2| > c\Delta t$, then the principle of local realism requires that correlations between the random series can only occur as a result of an earlier, common cause. We will call this the principle of **statistical separability**.

In the absence of a common cause, the separated random events are like independent coin tosses, located at $\mathbf{r}_1$ and $\mathbf{r}_2$, so it would seem that they must obey a common-sense factorization condition. For example, the joint probability of the outcomes heads-at-$\mathbf{r}_1$ and heads-at-$\mathbf{r}_2$ should be the product of the independent probabilities for heads at each location.

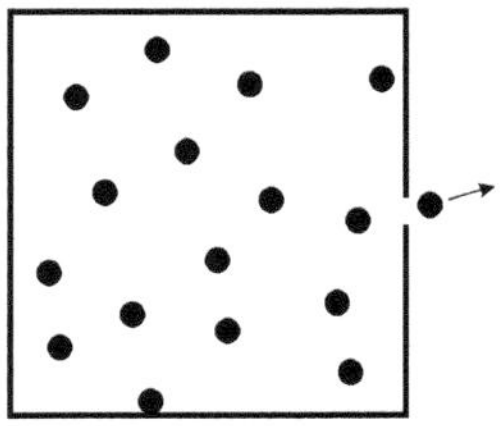

Fig. 19.2 A simple model for radioactive decay, consisting of small balls inside a large box with a small hole cut into one of the walls. Einstein's 'hidden variables' would be the unknown initial conditions of these balls.

In quantum mechanics, the factorizability of joint probabilities implies the factorizability of joint probability amplitudes (up to a phase factor); for example, a situation in which measurements at $\mathbf{r}_1$ and $\mathbf{r}_2$ are statistically independent is described by a separable two-body wave function, i.e. the product of a wave function of $\mathbf{r}_1$ and a wave function of $\mathbf{r}_2$. Conversely, the absolute square of a product wave function is the product of two separate probabilities, just as for two independent coin tosses at $\mathbf{r}_1$ and $\mathbf{r}_2$.

By contrast, an entangled state of two particles, e.g. a superposition of two product wave functions, is not factorizable. The result is that the probability distribution defined by an entangled state does not satisfy the principle of statistical separability, even when the parts are far apart in space.

The EPR argument emphasizes the importance of these disparities between the classical and quantum descriptions of the world, but it does not point the way to an experimental method for deciding between the two views. Bell realized that the key is the fact that the nonfactorizability of entangled states in quantum mechanics violates the common-sense, independent-coin-toss rule for joint probabilities.

He then formulated the statistical separability condition in terms of a factorizability condition on the joint probability for correlations between measurements on two distant particles. Bell's analysis applies completely generally to all local realistic theories, in a sense to be explained in the next section.

19.3 Local realism

Converting the qualitative disparities between the classical and quantum approaches into experimentally testable differences requires a quantitative formulation of local realism that does not depend on quantum theory. We will follow Shimony's version (Shimony, 1990) of Bell's solution for this problem. This analysis can be presented in a very general way, but it is easier to understand when it is described in terms of a concrete experiment. For this purpose, we first sketch an optical version of the Bohm singlet experiment.

19.3.1 Optical Bohm singlet experiment

As shown in Fig. 19.3, the entangled pair of spin-1/2 particles in Fig. 19.1 is replaced by a pair of photons emitted back-to-back in an entangled state, and the Stern–Gerlach magnets are replaced by calcite prisms that act as polarization analyzers. The beam of unpolarized right-going photons γ_A is split by the calcite prism A into an extraordinary ray e and an ordinary ray o. Similarly, the beam of left-going photons γ_B is split by calcite prism B into e and o rays.

The ordinary-ray and extraordinary-ray output ports of the calcite prisms are monitored by four counters. The two calcite prisms A and B can be independently rotated around the common decay axis by the azimuthal angles α and β respectively. The values of α and β—which determine the division of the incident wave into e- and o-waves—correspond to the direction of the magnetic field in a Stern–Gerlach apparatus.

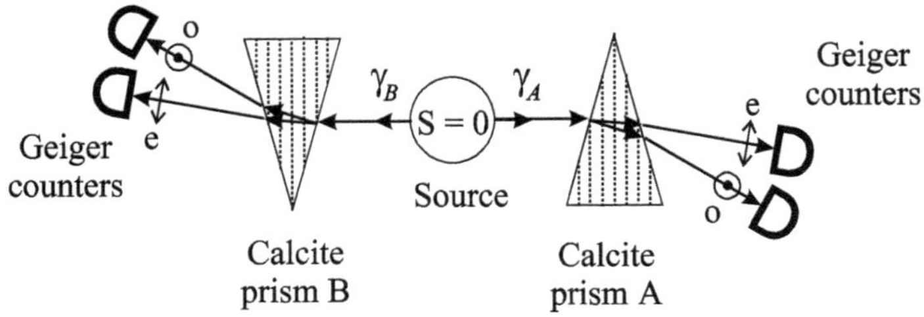

Fig. 19.3 An optical implementation of the EPR experiment. Calcite prisms replace the Stern–Gerlach magnets shown in Fig. 19.1. The source emits an entangled state of two oppositely-directed photons, such as the Bell state $|\Psi^-\rangle$. The birefringent prisms split the light into ordinary 'o' and extraordinary 'e' rays. The vertical dotted lines inside the prisms indicate the optic axes of the calcite crystals. Coincidence-counting circuitry connecting the Geiger counters is not shown.

The counters on each side of the apparatus are mounted rigidly with respect to the calcite prisms, so that they corotate with the prisms. Thus the four counters will constantly monitor the o and e outputs of the calcite prisms for all values of α and β.

The azimuthal angles α and β are examples of what are called **parameter settings**, or simply parameters, of the EPR experiment. The experimentalist on the right side of the apparatus, Alice, is free to choose the parameter setting α (the azimuthal angle of rotation of calcite prism A) as she pleases. Likewise, the experimentalist on the left side, Bob, is free to choose the parameter setting β (the azimuthal angle of rotation of calcite prism B) as he pleases, independently of Alice's choice.

19.3.2 Conditions defining locality and realism

Bell's seminal paper has inspired many proposals for realizations of the metaphysical notions of realism and locality, including both deterministic and stochastic forms of hidden variables theories. In this section we present a general class of realizations by specifying the conditions that a theory must satisfy in order to be called local and realistic.

We will say that a theory is **realistic** if it describes all required elements of physical reality for a system by means of a space, Λ, of *completely specified* states λ—i.e. the states of maximum information—satisfying the following two conditions.

Objective reality

$$\Lambda \text{ is defined without reference to any measurements.} \tag{19.7}$$

Spatial separability

$$\text{The state spaces } \Lambda_A \text{ and } \Lambda_B \text{ for the spatially-separated systems A and B are independently defined.} \tag{19.8}$$

The only other condition imposed on Λ is that it must support probability distributions $\rho(\lambda)$ in order to describe situations in which maximum information is not available.

The only conditions imposed on an admissible distribution $\rho(\lambda)$ are that it be positive definite, i.e. $\rho(\lambda) \geqslant 0$, normalized to unity,

$$\int d\lambda \rho(\lambda) = 1\,, \tag{19.9}$$

and independent of the parameter values α and β. The last condition incorporates the intuitive idea that the states λ are determined at the source S, before any encounters with the measuring devices at A and B.

One possible example for Λ would be the classical phase space involved in the simple model of radioactive decay presented above. In this case, the completely specified states λ are simply points in the phase space, and a probability distribution $\rho(\lambda)$ would be the usual phase space distribution.

A much more surprising example comes from a *disentangled* version of quantum theory, which is defined by excluding all entangled states of spatially-separated systems. This mutilated theory violates the superposition principle, but by doing so it allows us to identify Λ with the Hilbert space $\mathfrak{H}$ for the local system. An individual state λ is thereby identified with a pure state $|\psi\rangle$.

According to the standard interpretation of quantum theory, this choice of λ gives a complete description of the state of an isolated system. In this case $\rho(\lambda)$ is just the distribution defining a mixed state. The fact that the disentangled version of quantum theory is realistic illustrates the central role played by entanglement in differentiating the quantum view from the local realistic view.

We next turn to the task of developing a quantitative realization of locality. For this purpose, we need a language for describing measurements at the spatially-separated stations A and B, shown in Fig. 19.3. For the sake of simplicity, it is best to consider experiments that have a discrete set of possible outcomes $\{\mathfrak{A}_m,\ m = 1, \ldots, M\}$ and $\{\mathfrak{B}_n,\ n = 1, \ldots, N\}$ at the stations A and B respectively, e.g. $\mathfrak{A}_1$ could describe a detector firing at station A during a certain time interval. With each outcome $\mathfrak{A}_m$, we associate a numerical value, $\mathcal{A}_m$, called an **outcome parameter**. The definition of the output parameters is at our disposal, so they can be chosen to satisfy the following convenient conditions:

$$-1 \leqslant \mathcal{A}_m \leqslant +1 \quad \text{and} \quad -1 \leqslant \mathcal{B}_n \leqslant +1\,. \tag{19.10}$$

For the two-calcite-prism experiment, sketched in Fig. 19.3, the indices m and n can only assume the values o and e, corresponding respectively to the ordinary and the extraordinary rays emerging from a given prism. The source S emits a pair of photons prepared at birth in some state λ. The experimental signals in this case are clicks in one of the counters, so one useful definition of the outcome parameters is

$$\begin{aligned}
\mathcal{A}_e &= 1 \quad \text{for outcome } \mathfrak{A}_e \quad \text{(Alice's e-counter clicks)}\,,\\
\mathcal{A}_o &= -1 \quad \text{for outcome } \mathfrak{A}_o \quad \text{(Alice's o-counter clicks)}\,,\\
\mathcal{B}_e &= 1 \quad \text{for outcome } \mathfrak{B}_e \quad \text{(Bob's e-counter clicks)}\,,\\
\mathcal{B}_o &= -1 \quad \text{for outcome } \mathfrak{B}_o \quad \text{(Bob's o-counter clicks)}\,.
\end{aligned} \tag{19.11}$$

The outcome $\mathfrak{A}_e$ occurs when a rightwards-propagating photon from the source S is deflected through the e port of the calcite prism A, and subsequently registered by

Alice's e-counter, etc. In this thought experiment we imagine that all counters have 100% sensitivity; consequently, if an e-counter does not click, we can be sure that the corresponding o-counter will click.

The following conditional probabilities will be useful.

$$p(\mathfrak{A}_m|\lambda,\alpha,\beta) \equiv \text{probability of outcome } \mathfrak{A}_m, \text{ given the system state } \lambda \text{ and parameter settings } \alpha,\, \beta\,. \tag{19.12}$$

$$p(\mathfrak{B}_n|\lambda,\alpha,\beta) \equiv \text{probability of outcome } \mathfrak{B}_n, \text{ given the system state } \lambda \text{ and parameter settings } \alpha,\, \beta\,. \tag{19.13}$$

$$p(\mathfrak{A}_m|\lambda,\alpha,\beta,\mathfrak{B}_n) \equiv \text{probability of outcome } \mathfrak{A}_m, \text{ given the system state } \lambda, \text{ parameter settings } \alpha,\, \beta\,, \text{ and outcome } \mathfrak{B}_n\,. \tag{19.14}$$

$$p(\mathfrak{B}_n|\lambda,\alpha,\beta,\mathfrak{A}_m) \equiv \text{probability of outcome } \mathfrak{B}_n, \text{ given the system state } \lambda, \text{ parameter settings } \alpha,\, \beta\,, \text{ and outcome } \mathfrak{A}_m\,. \tag{19.15}$$

$$p(\mathfrak{A}_m,\mathfrak{B}_n|\lambda,\alpha,\beta) \equiv \text{joint probability of outcomes } \mathfrak{A}_m \text{ and } \mathfrak{B}_n\,, \text{ given the system state } \lambda \text{ and the parameter settings } \alpha,\, \beta\,. \tag{19.16}$$

Following the work of Jarrett (1984), as presented by Shimony (1990), we will say that a theory is **local** if it satisfies the following conditions.

Parameter independence

$$p(\mathfrak{A}_m|\lambda,\alpha,\beta) = p(\mathfrak{A}_m|\lambda,\alpha)\,, \tag{19.17}$$

$$p(\mathfrak{B}_n|\lambda,\alpha,\beta) = p(\mathfrak{B}_n|\lambda,\beta)\,. \tag{19.18}$$

Outcome independence

$$p(\mathfrak{A}_m|\lambda,\alpha,\beta,\mathfrak{B}_n) = p(\mathfrak{A}_m|\lambda,\alpha,\beta)\,, \tag{19.19}$$

$$p(\mathfrak{B}_n|\lambda,\alpha,\beta,\mathfrak{A}_m) = p(\mathfrak{B}_n|\lambda,\alpha,\beta)\,. \tag{19.20}$$

Parameter independence states that the parameter settings chosen by one observer have no effect on the outcomes seen by the other. For example, eqn (19.17) tells us that the probability distribution of the outcomes observed by Alice at A does not depend on the parameter settings chosen by Bob at B.

This apparently innocuous statement is, in fact, extremely important. If parameter independence were violated, then Bob—who might well be space-like separated from Alice—could send her an instantaneous message by merely changing β, e.g. twisting his calcite crystal. Such a possibility would violate the relativistic prohibition against sending signals faster than light. Likewise, eqn (19.18) prohibits Alice from sending instantaneous messages to Bob.

The principle of outcome independence states that the probability of outcomes seen by one observer does not depend on which outcomes are actually seen by the other. This is what one would expect for two independent coin tosses—since the outcome of one coin toss is clearly independent of the outcome of the other—but eqns (19.19) and (19.20) also seem to prohibit correlations due to a common cause, e.g. in the source S.

This incorrect interpretation stems from overlooking the assumption that λ is a *complete* description of the state, including any secret mechanism that builds in correlations at the source (Bub, 1997, Chap. 2). With this in mind, the conditions (19.19) and (19.20) simply reflect the fact that the actual outcomes $\mathfrak{B}_n$ or $\mathfrak{A}_m$ are superfluous, if λ is given as part of the conditions. We will return to the issue of correlations after deriving Bell's strong-separability condition.

It is also important to realize that the individual events at A and B can be truly random, even if they are correlated. This situation is exhibited in the experiment sketched in Fig. 19.3. When the polarizations of photons γ_A and γ_B, in the Bell state $|\Psi^-\rangle$, are measured separately—i.e. without coincidence counting—they are randomly polarized; that is, the individual sequences of e- or o-counts at A and B are each as random as two independent sequences of coin tosses.

Finally, we note that a violation of outcome independence does not imply any violations of relativity. The conditional probability $p(\mathfrak{A}_m|\lambda, \alpha, \beta, \mathfrak{B}_n)$ describes a situation in which Bob has already performed a measurement and transmitted the result to Alice by a respectably subluminal channel. Thus protecting the world from superluminal messages and the accompanying causal anomalies is the responsibility of parameter independence alone.

19.3.3 Strong separability

Bell's theorem is concerned with the strength of correlations between the random outcomes at A and B, so the first step is to find the constraints imposed by the combined effects of realism and locality—in the form of parameter and outcome independence—on the joint probability $p(\mathfrak{A}_m, \mathfrak{B}_n|\lambda, \alpha, \beta)$ defined by eqn (19.16).

We begin by applying the compound probability rule (A.114) to find

$$p(\mathfrak{A}_m, \mathfrak{B}_n|\lambda, \alpha, \beta) = p(\mathfrak{A}_m|\lambda, \alpha, \beta, \mathfrak{B}_n)p(\mathfrak{B}_n|\lambda, \alpha, \beta)\,. \tag{19.21}$$

In other words, the joint probability for outcome $\mathfrak{A}_m$ and outcome $\mathfrak{B}_n$ is the product of the probability for outcome $\mathfrak{A}_m$ (conditioned on the occurrence of the outcome $\mathfrak{B}_n$) with the probability that outcome $\mathfrak{B}_n$ actually occurred. All three probabilities are conditioned by the assumption that the state of the system was λ and the parameter settings were α and β. The situation is symmetrical in A and B, so we also find

$$p(\mathfrak{A}_m, \mathfrak{B}_n|\lambda, \alpha, \beta) = p(\mathfrak{B}_n|\lambda, \alpha, \beta, \mathfrak{A}_m)p(\mathfrak{A}_m|\lambda, \alpha, \beta)\,. \tag{19.22}$$

Applying outcome independence, eqn (19.19), to the right side of eqn (19.21) yields

$$p(\mathfrak{A}_m, \mathfrak{B}_n|\lambda, \alpha, \beta) = p(\mathfrak{A}_m|\lambda, \alpha, \beta)p(\mathfrak{B}_n|\lambda, \alpha, \beta)\,, \tag{19.23}$$

and applying parameter independence to both terms on the right side of this equation results in the **strong-separability condition**:

$$p(\mathfrak{A}_m, \mathfrak{B}_n|\lambda, \alpha, \beta) = p(\mathfrak{A}_m|\lambda, \alpha)p(\mathfrak{B}_n|\lambda, \beta)\,. \tag{19.24}$$

This is the mathematical expression of the following, seemingly common-sense, statement: for a given specification, λ, of the state, whatever Alice does or observes must be independent of whatever Bob does or observes, since they could reside in space-like separated regions.

Before using the strong-separability condition to prove Bell's theorem, we return to the question of correlations that might be imposed by a common cause. In typical experiments, the complete specification of the state represented by λ is not available—for example, the values of the hidden variables cannot be determined—so the strong-separability condition must be averaged over a distribution $\rho(\lambda)$ that represents the experimental information that is available.

The result is

$$p(\mathfrak{A}_m, \mathfrak{B}_n|\alpha, \beta) = \int d\lambda \rho(\lambda)\, p(\mathfrak{A}_m|\lambda, \alpha) p(\mathfrak{B}_n|\lambda, \beta)\,, \tag{19.25}$$

where

$$p(\mathfrak{A}_m, \mathfrak{B}_n|\alpha, \beta) = \int d\lambda \rho(\lambda)\, p(\mathfrak{A}_m, \mathfrak{B}_n|\lambda, \alpha, \beta)\,. \tag{19.26}$$

The corresponding averaged probabilities for single outcomes are

$$\begin{aligned} p(\mathfrak{A}_m|\alpha) &= \int d\lambda \rho(\lambda)\, p(\mathfrak{A}_m|\lambda, \alpha)\,, \\ p(\mathfrak{B}_n|\beta) &= \int d\lambda \rho(\lambda)\, p(\mathfrak{B}_n|\lambda, \beta)\,; \end{aligned} \tag{19.27}$$

consequently, the condition for statistical independence,

$$p(\mathfrak{A}_m, \mathfrak{B}_n|\alpha, \beta) = p(\mathfrak{A}_m|\alpha) p(\mathfrak{B}_n|\beta)\,, \tag{19.28}$$

can only be satisfied—for general choices of $\mathfrak{A}_m$ and $\mathfrak{B}_n$—when $\rho(\lambda) = \delta(\lambda - \lambda_0)$.

A closer connection with experiment is afforded by defining **Bell's expectation values.**

(1) The expectation value of outcomes seen by Alice is

$$E(\lambda, \alpha) = \sum_m p(\mathfrak{A}_m|\lambda, \alpha)\mathcal{A}_m\,. \tag{19.29}$$

(2) The expectation value of outcomes seen by Bob is

$$E(\lambda, \beta) = \sum_n p(\mathfrak{B}_n|\lambda, \beta)\mathcal{B}_n\,. \tag{19.30}$$

(3) The expectation value of joint outcomes seen by both Alice and Bob is

$$E(\lambda, \alpha, \beta) = \sum_{m,n} p(\mathfrak{A}_m, \mathfrak{B}_n|\lambda, \alpha, \beta)\mathcal{A}_m\mathcal{B}_n\,. \tag{19.31}$$

The quantity $E(\lambda, \alpha, \beta)$ is the average value of joint outcomes as measured, for example, in a coincidence-counting experiment. The bounds $|\mathcal{A}_m| \leqslant 1$ and $|\mathcal{B}_n| \leqslant 1$, together with the normalization of the probabilities, imply that the absolute values of all these expectation values are bounded by unity.

From Bell's strong-separability condition, it follows that the joint expectation value—for a given complete state λ—also factorizes:

$$E(\lambda, \alpha, \beta) = E(\lambda, \alpha)E(\lambda, \beta)\,, \tag{19.32}$$

but in the absence of complete state information, the relevant expectation values are

$$E\left(\alpha\right) \equiv \int d\lambda \rho\left(\lambda\right) E(\lambda, \alpha) = \sum_m p(\mathfrak{A}_m|\alpha)\mathcal{A}_m\,, \tag{19.33}$$

etc. Thus the correlation function

$$C\left(\alpha, \beta\right) = E(\alpha, \beta) - E\left(\alpha\right) E\left(\beta\right) \tag{19.34}$$

can only vanish in the extreme case, $\rho\left(\lambda\right) = \delta\left(\lambda - \lambda_0\right)$, of perfect information.

19.4 Bell's theorem

An evaluation of any one of Bell's expectation values, e.g. $E(\lambda, \alpha)$, would depend on the details of the particular local realistic theory under consideration. One of the consequences of Bell's original work (Bell, 1964) has been the discovery of various linear combinations of expectation values, which have the useful property that upper and lower bounds can be derived for the entire class of local realistic theories defined above. We follow Shimony (1990), by considering the particular sum

$$S\left(\lambda\right) \equiv E(\lambda, \alpha_1, \beta_1) + E(\lambda, \alpha_1, \beta_2) + E(\lambda, \alpha_2, \beta_1) - E(\lambda, \alpha_2, \beta_2)\,, \tag{19.35}$$

which was first suggested by Clauser *et al.* (1969). With a fixed value, λ, of the hidden variables, the four combinations (α_1, β_1), (α_1, β_2), (α_2, β_1), and (α_2, β_2) represent independent choices α_1 or α_2 by Alice and β_1 or β_2 by Bob, as shown in Fig. 19.4.

For the typical situation in which the complete state λ is not known, $S\left(\lambda\right)$ should be replaced by the experimentally relevant quantity:

$$S \equiv E(\alpha_1, \beta_1) + E(\alpha_1, \beta_2) + E(\alpha_2, \beta_1) - E(\alpha_2, \beta_2)\,. \tag{19.36}$$

Bell's theorem is then stated as follows.

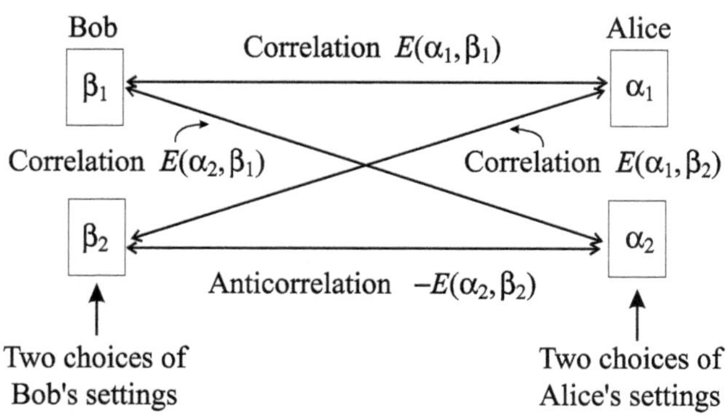

Fig. 19.4 The four terms in the sum S defined in eqn (19.35). The dependence of the expectation values $E(\lambda, \alpha, \beta)$ on the system state λ has been suppressed in this figure.

Theorem 19.1 *For all local realistic theories,*

$$-2 \leqslant E(\lambda, \alpha_1, \beta_1) + E(\lambda, \alpha_1, \beta_2) + E(\lambda, \alpha_2, \beta_1) - E(\lambda, \alpha_2, \beta_2) \leqslant +2\,. \tag{19.37}$$

Averaging over the distribution of states produces the Bell inequality:

$$-2 \leqslant E(\alpha_1, \beta_1) + E(\alpha_1, \beta_2) + E(\alpha_2, \beta_1) - E(\alpha_2, \beta_2) \leqslant +2\,. \tag{19.38}$$

This result limits the total amount of correlation, as measured by S, that is allowed for a local realistic theory. Experiments using coincidence-detection measurements performed on two-photon decays have shown that this bound can be violated.

19.4.1 Mermin's lemma

In order to prove Bell's theorem, we first prove the following lemma due to Mermin.

Lemma 19.2 *If x_1, x_2, y_1, y_2 are real numbers in the interval $[-1, +1]$, then the sum $S \equiv x_1 y_1 + x_1 y_2 + x_2 y_1 - x_2 y_2$ lies in the interval $[-2, +2]$, i.e. $|S| \leqslant 2$.*

Proof Since S is a linear function of each of the four variables x_1, x_2, y_1, y_2, it must take on its extreme values when the arguments of the function themselves are extrema, i.e. when $(x_1, x_2, y_1, y_2) = (\pm 1, \pm 1, \pm 1, \pm 1)$, where the four $\pm$s are independent. There are four terms in S, and each term is bounded between -1 and $+1$; consequently, $|S| \leqslant 4$. However, we can also rewrite S as

$$S = (x_1 + x_2)(y_1 + y_2) - 2x_2 y_2\,. \tag{19.39}$$

The extrema of $x_1 + x_2$ are 0 or ± 2, and similarly for $y_1 + y_2$. Therefore the extrema of the product $(x_1 + x_2)(y_1 + y_2)$ are 0 or ± 4. The extrema for $2x_2 y_2$ are ± 2. Hence the extrema for S are ± 2 or ± 6. The latter possibility is ruled out by the previously determined limit $|S| \leqslant 4$; therefore, the extrema of S are ± 2, i.e. $|S| \leqslant 2$. ■

19.4.2 Proof of Bell's theorem

Proof Bell's theorem now follows as a corollary of Mermin's lemma. With the identifications

$$\begin{aligned} x_1 &= E(\lambda, \alpha_1)\,, \quad \text{where } |E(\lambda, \alpha_1)| \leqslant 1\,, \\ x_2 &= E(\lambda, \alpha_2)\,, \quad \text{where } |E(\lambda, \alpha_2)| \leqslant 1\,, \\ y_1 &= E(\lambda, \beta_1)\,, \quad \text{where } |E(\lambda, \beta_1)| \leqslant 1\,, \\ y_2 &= E(\lambda, \beta_2)\,, \quad \text{where } |E(\lambda, \beta_2)| \leqslant 1\,, \end{aligned} \tag{19.40}$$

Lemma 19.2 implies

$$|E(\lambda, \alpha_1)E(\lambda, \beta_1) + E(\lambda, \alpha_1)E(\lambda, \beta_2) + E(\lambda, \alpha_2)E(\lambda, \beta_1) - E(\lambda, \alpha_2)E(\lambda, \beta_2)| \leqslant 2\,. \tag{19.41}$$

Using the strong-separability condition (19.32) for each term, i.e. $E(\lambda, \alpha, \beta) = E(\lambda, \alpha)E(\lambda, \beta)$, we now arrive at

$$-2 \leqslant E(\lambda, \alpha_1, \beta_1) + E(\lambda, \alpha_1, \beta_2) + E(\lambda, \alpha_2, \beta_1) - E(\lambda, \alpha_2, \beta_2) \leqslant +2\,, \tag{19.42}$$

and averaging over λ yields eqn (19.38). ■

19.5 Quantum theory versus local realism

As a prelude to the experimental tests of local realism, we first support our previous claim that quantum theory violates outcome independence and satisfies parameter independence. In addition, we give an explicit example for which the quantum prediction of the correlations violates Bell's theorem.

19.5.1 Quantum theory is not local

The issues of parameter independence and outcome independence will be studied by considering an experiment simpler than the one presented in Section 19.3.1. In this arrangement, shown in Fig. 19.5, pairs of polarization-entangled photons are produced by down-conversion, and Alice and Bob are supplied with linear polarization filters and a single counter apiece. This reduces the outcomes for Alice to: $\mathfrak{A}_{\text{yes}}$ (Alice's detector clicks) and $\mathfrak{A}_{\text{no}}$ (there is no click). The corresponding outcome parameters are $\mathcal{A}_{\text{yes}} = 1$ and $\mathcal{A}_{\text{no}} = 0$. Bob's outcomes and outcome parameters are defined in the same way.

We begin by assuming that the source produces the entangled state

$$|\chi\rangle = F\,|h_A, v_B\rangle + G\,|v_A, h_B\rangle\,, \tag{19.43}$$

where

$$|h_A, v_B\rangle \equiv a^{\dagger}_{\mathbf{k}_A h} a^{\dagger}_{\mathbf{k}_B v}\,|0\rangle\,, \quad |v_A, h_B\rangle \equiv a^{\dagger}_{\mathbf{k}_A v} a^{\dagger}_{\mathbf{k}_B h}\,|0\rangle\,, \tag{19.44}$$

$\mathbf{k}_A$ and $\mathbf{k}_B$ are directed toward Alice and Bob respectively, and h and v label orthogonal polarizations: $\mathbf{e}_h$ (*horizontal*) and $\mathbf{e}_v$ (*vertical*). The parameters are the angles α and β defining the linear polarizations $\mathbf{e}_\alpha$ and $\mathbf{e}_\beta$ transmitted by the polarizers.

Since $a_{\mathbf{k}_A h} \propto \mathbf{e}_{\mathbf{k}_A h} \cdot E^{(+)}$, etc., the annihilation operators in the (h, v)-basis are related to the annihilation operators in the $(\alpha, \overline{\alpha} = \pi/2 - \alpha)$-basis by

$$\begin{pmatrix} a_{\mathbf{k}_A \alpha} \\ a_{\mathbf{k}_A \overline{\alpha}} \end{pmatrix} = \begin{bmatrix} \cos\alpha & \sin\alpha \\ -\sin\alpha & \cos\alpha \end{bmatrix} \begin{pmatrix} a_{\mathbf{k}_A h} \\ a_{\mathbf{k}_A v} \end{pmatrix}. \tag{19.45}$$

The corresponding relation for Bob follows by letting $\alpha \to \beta$ and $\mathbf{k}_A \to \mathbf{k}_B$.

A Parameter independence

For this experiment, the role of $p(\mathfrak{A}_m|\lambda, \alpha, \beta)$ in eqn (19.17) is played by $p(\mathfrak{A}_{\text{yes}}|\chi, \alpha, \beta)$, the probability that Alice's detector clicks for the given state and parameter settings. This is proportional to the detection rate for $\mathbf{e}_\alpha$-polarized photons, i.e.

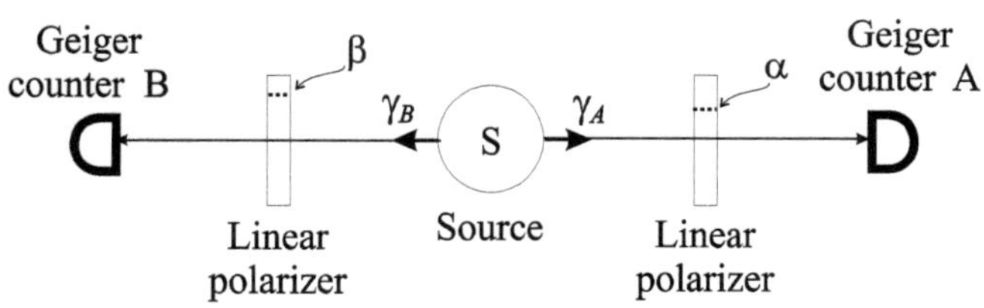

Fig. 19.5 Schematic of an apparatus to measure the polarization correlations of the entangled photon pair γ_A and γ_B emitted back-to-back from the source S. The coincidence-counting circuitry connecting the two Geiger counters is not shown.

$$p(\mathfrak{A}_{\text{yes}}|\chi,\alpha,\beta) \propto G_{\alpha}^{(1)}(\mathbf{r}_A,t_A;\mathbf{r}_A,t_A) \propto \left\langle \chi \left| a_{\mathbf{k}_A\alpha}^{\dagger} a_{\mathbf{k}_A\alpha} \right| \chi \right\rangle . \tag{19.46}$$

A calculation—see Exercise 19.2 —using eqns (19.43)–(19.45) yields

$$p(\mathfrak{A}_{\text{yes}}|\chi,\alpha,\beta) \propto |F|^2 \cos^2\alpha + |G|^2 \sin^2\alpha \,. \tag{19.47}$$

Thus the quantum result for the probability of a click of Alice's detector is independent of the setting β of Bob's polarizer, although it can depend on her own polarizer setting α. In other words, quantum theory—at least in this example—satisfies parameter independence. The symmetry of the experimental arrangement guarantees that the probability, $p(\mathfrak{B}_{\text{yes}}|\chi,\alpha,\beta)$, seen by Bob is independent of α.

This single example does not constitute a general proof that quantum theory satisfies parameter independence, but the features of the calculation provide guidance for crafting such a proof. In general, the calculation of outcome probabilities for Alice take the same form as in the example, i.e. the expectation value of an operator—which may well depend on Alice's parameter settings—is evaluated by using the state vector determined by the source. Neither Alice's operator nor the state vector depend on Bob's parameter settings; therefore, parameter independence is guaranteed for quantum theory.

For the special values $F = -G = 1/\sqrt{2}$, the entangled state $|\chi\rangle$ becomes the singlet-like Bell state

$$\left|\Psi^-\right\rangle = \frac{1}{\sqrt{2}}\left\{|h_A,v_B\rangle - |v_A,h_B\rangle\right\}, \tag{19.48}$$

first defined in Section 13.3.5. In this case, $p(\mathfrak{A}_{\text{yes}}|\chi,\alpha,\beta)$ is independent of α as well as β, so that Alice's singles-counting measurements are the same as expected from an unpolarized beam. This supports our previous claim that the individual measurements can be as random as coin tosses.

B *Outcome independence*

Checking outcome independence requires the evaluation of the conditional probability $p(\mathfrak{A}_{\text{yes}}|\lambda,\alpha,\beta,\mathfrak{B}_{\text{rslt}})$ that Alice hears a click, given that Bob has observed the outcome $\mathfrak{B}_{\text{rslt}}$, where rslt $=$ yes, no. In this case, we will simplify the calculation by setting $|\chi\rangle = \left|\Psi^-\right\rangle$ at the beginning.

With the usual assumption of 100% detector sensitivity, both possible outcomes for Bob—$\mathfrak{B}_{\text{yes}}$ (click) or $\mathfrak{B}_{\text{no}}$ (no click)—constitute a measurement. According to von Neumann's projection postulate, we must then replace the original state $\left|\Psi^-\right\rangle$ by the reduced state $\left|\Psi^-\right\rangle_{\text{rslt}}$, to find

$$p(\mathfrak{A}_{\text{yes}}|\lambda,\alpha,\beta,\mathfrak{B}_{\text{rslt}}) \propto {}_{\text{rslt}}\left\langle \Psi^- \left| a_{\mathbf{k}_A\alpha}^{\dagger} a_{\mathbf{k}_A\alpha} \right| \Psi^- \right\rangle_{\text{rslt}} . \tag{19.49}$$

The reduced state for either of Bob's outcomes can be constructed by inverting Bob's version of eqn (19.45) to express the creation operators in the (h,v)-basis in terms of the creation operators in the $(\beta,\overline{\beta})$-basis:

$$\begin{pmatrix} a^{\dagger}_{\mathbf{k}_B h} \\ a^{\dagger}_{\mathbf{k}_B v} \end{pmatrix} = \begin{bmatrix} \cos\beta & -\sin\beta \\ \sin\beta & \cos\beta \end{bmatrix} \begin{pmatrix} a^{\dagger}_{\mathbf{k}_B \beta} \\ a^{\dagger}_{\mathbf{k}_B \overline{\beta}} \end{pmatrix}. \tag{19.50}$$

Using this in the definition (19.44) exhibits the original states as superpositions of states containing β-polarized photons and states containing $\overline{\beta}$-polarized photons.

For the outcome $\mathfrak{B}_{\text{yes}}$—Bob heard a click—the projection postulate instructs us to drop the states containing the $\overline{\beta}$-polarized photons, since they are blocked by the polarizer. This produces the reduced state

$$\left|\Psi^{-}\right\rangle_{\text{yes}} = \frac{1}{\sqrt{2}}\left\{\sin\beta\ \left|h_A, \beta_B\right\rangle - \cos\beta\ \left|v_A, \beta_B\right\rangle\right\}, \tag{19.51}$$

where $\left|\beta_B\right\rangle = a^{\dagger}_{\mathbf{k}_B \beta}\left|0\right\rangle$. Substituting this into eqn (19.49) leads—by way of the calculation in Exercise 19.3—to the simple result

$$p(\mathfrak{A}_{\text{yes}}|\lambda, \alpha, \beta, \mathfrak{B}_{\text{yes}}) \propto \sin^2(\alpha - \beta). \tag{19.52}$$

For the opposite outcome, $\mathfrak{B}_{\text{no}}$, the projection postulate tells us to drop the states containing β-polarized photon states instead, and the result is

$$p(\mathfrak{A}_{\text{yes}}|\lambda, \alpha, \beta, \mathfrak{B}_{\text{no}}) \propto \cos^2(\alpha - \beta). \tag{19.53}$$

The conclusion is that quantum theory violates outcome independence, since the probability that Alice hears a click depends on the outcome of Bob's previous measurement. The fact that Alice's probabilities only depend on the difference in polarizer settings follows from the assumption that the source produces the special state $\left|\Psi^{-}\right\rangle$, which is invariant under rotations around the common propagation axis.

The violation of outcome independence implies that the two sets of experimental outcomes must be correlated. The probability that both detectors click is proportional to the coincidence-count rate, which—as we learnt in Section 9.2.4—is determined by the second-order Glauber correlation function; consequently,

$$\begin{aligned} p\left(\mathfrak{A}_{\text{yes}}, \mathfrak{B}_{\text{yes}}\left|\Psi^{-}, \alpha, \beta\right.\right) &\propto G^{(2)}_{\alpha\beta}(\mathbf{r}_1 t_1; \mathbf{r}_2 t_2) \\ &\propto \left\langle\Psi^{-}\right| a^{\dagger}_{\mathbf{k}_A \alpha} a^{\dagger}_{\mathbf{k}_B \beta} a_{\mathbf{k}_B \beta} a_{\mathbf{k}_A \alpha}\left|\Psi^{-}\right\rangle. \end{aligned} \tag{19.54}$$

The techniques used above give

$$\begin{aligned} p\left(\mathfrak{A}_{\text{yes}}, \mathfrak{B}_{\text{yes}}\left|\Psi^{-}, \alpha, \beta\right.\right) &= \sin^2(\alpha - \beta) \\ &= \frac{1}{2} - \frac{1}{2}\cos(2\alpha - 2\beta), \end{aligned} \tag{19.55}$$

which describes an interference pattern, e.g. if β is held fixed while α is varied. Furthermore, this pattern has 100% visibility, since perfect nulls occur for the values $\alpha = \beta, \beta + \pi, \beta + 2\pi, \ldots$, at which the planes of polarization of the two photons are parallel. The surprise is that an interference pattern with 100% visibility occurs in the second-order correlation function $G^{(2)}_{\alpha\beta}$ while the first-order functions $G^{(1)}_{\alpha}$ and $G^{(1)}_{\beta}$ display zero visibility, i.e. no interference at all.

19.5.2 Quantum theory violates Bell's theorem

The results (19.52), (19.53), and (19.55) show that quantum theory violates outcome independence and the strong-separability principle; consequently, quantum theory does not satisfy the hypothesis of Bell's theorem. Nevertheless, it is still logically possible that quantum theory could satisfy the conclusion of Bell's theorem, i.e. the inequality (19.37). We will now dash this last, faint hope by exhibiting a specific example in which the quantum prediction violates the Bell inequality (19.38).

For the experiment depicted in Fig. 19.3, let us now calculate what quantum theory predicts for $S(\lambda)$ when λ is represented by the Bell state $|\Psi^-\rangle$. For general parameter settings α and β, the definition (19.31) for Bell's joint expectation value can be written as

$$E(\alpha,\beta) = p_{ee}(\alpha,\beta)\mathcal{A}_e\mathcal{B}_e + p_{eo}(\alpha,\beta)\mathcal{A}_e\mathcal{B}_o + p_{oe}(\alpha,\beta)\mathcal{A}_o\mathcal{B}_e + p_{oo}(\alpha,\beta)\mathcal{A}_o\mathcal{B}_o\,, \quad (19.56)$$

where we have omitted the λ-dependence of the expectation value, and adopted the simplified notation

$$p_{mn}(\alpha,\beta) \equiv p(\mathcal{A}_m, \mathfrak{B}_n|\lambda,\alpha,\beta) \quad (19.57)$$

for the joint probabilities.

In Exercise 19.4, the calculation of the probabilities is done by using the techniques leading to eqn (19.55), with the result

$$p_{ee}(\alpha,\beta) = p_{oo}(\alpha,\beta) = \frac{1}{2}\sin^2(\alpha-\beta)\,, \quad (19.58)$$

$$p_{eo}(\alpha,\beta) = p_{oe}(\alpha,\beta) = \frac{1}{2}\cos^2(\alpha-\beta)\,. \quad (19.59)$$

After combining these expressions for the probabilities with the definition (19.11) for the outcome parameters, Bell's joint expectation value (19.56) becomes

$$E(\alpha,\beta) = \sin^2(\alpha-\beta) - \cos^2(\alpha-\beta) = -\cos(2\alpha-2\beta)\,. \quad (19.60)$$

Our objective is to choose values $(\alpha_1,\beta_1,\alpha_2,\beta_2)$ such that S violates the inequality $|S| \leqslant 2$. A set of values that accomplishes this,

$$\alpha_1 = 0^\circ\,,\quad \alpha_2 = 45^\circ\,,\quad \beta_1 = 22.5^\circ\,,\quad \beta_2 = -22.5^\circ\,, \quad (19.61)$$

is illustrated in Fig. 19.6.

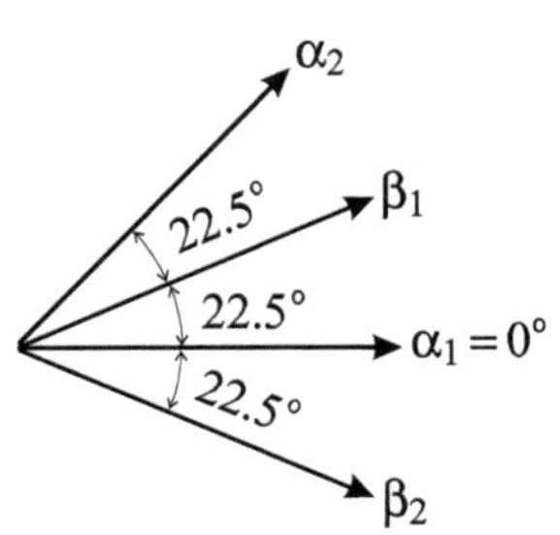

Fig. 19.6 A choice of angular settings $\alpha_1, \alpha_2, \beta_1, \beta_2$ in the calcite-prism-pair experiment (see Fig. 19.3) that maximizes the violation of Bell's bounds (19.42) by the quantum theory.

For these settings, the expectation values are given by

$$
\begin{aligned}
E(\alpha_1 = 0, \beta_1 = 22.5^\circ) &= -\cos(45^\circ) = -\frac{1}{\sqrt{2}}, \\
E(\alpha_1 = 0, \beta_2 = -22.5^\circ) &= -\cos(-45^\circ) = -\frac{1}{\sqrt{2}}, \\
E(\alpha_2 = 45^\circ, \beta_1 = 22.5^\circ) &= -\cos(-45^\circ) = -\frac{1}{\sqrt{2}}, \\
E(\alpha_2 = 45^\circ, \beta_2 = -22.5^\circ) &= -\cos(-135^\circ) = +\frac{1}{\sqrt{2}},
\end{aligned}
\tag{19.62}
$$

so that $S = -2\sqrt{2}$. This violation of the bound $|S| \leqslant 2$ by a factor of $\sqrt{2}$ shows that quantum theory violates the Bell inequality (19.38) by a comfortable margin.

19.5.3 Motivation for the definition of the sum S

What motivates the choice of four terms and the signs $(+,+,+,-)$ in eqn (19.35)? The answers to this question now becomes clear in light of the above calculation. The independent observers, Alice and Bob, need to make two independent choices in their respective parameter settings α and β, in order to observe changes in the correlations between the polarizations of the photons γ_A and γ_B. This explains the four pairs of parameter settings appearing in the definition of S, and pictured in Fig. 19.4.

The motivation for the choice of signs $(+,+,+,-)$ in S can be explained by reference to Fig. 19.6. Alice and Bob are free to choose the first three pairs of parameters settings, (α_1, β_1), (α_1, β_2), and (α_2, β_1), so that all three pairs have the same setting difference, 22.5°, and negative correlations. In the quantum theory calculation of S for the Bell state $|\Psi^-\rangle$, these choices yield the same negative correlation, $-1/\sqrt{2}$, since the expectation values only depend on the difference in the polarizer settings.

By contrast, the fourth pair of settings, (α_2, β_2), describes the two angles that are the farthest away from each other in Fig. 19.6, and it yields a positive expectation value $E(\alpha_2, \beta_2) = +1/\sqrt{2}$. This arises from the fact that, for this particular pair of angles ($\alpha_2 = 45^\circ$, $\beta_2 = -22.5^\circ$), the relative orientations of the planes of polarization of the back-to-back photons γ_A and γ_B are almost orthogonal. The opposite sign of this expectation value compared to the first three can be exploited by deliberately choosing the opposite sign for this term in eqn (19.35). This stratagem ensures that all four terms contribute with the same sign, and this gives the best chance of violating the inequality.

It should be emphasized that the violation of this Bell inequality by quantum theory is not restricted to this particular example. However, it turns out that this special choice of angular settings defines an extremum for S in the important case of maximally entangled states. Consequently, these parameter settings maximize the quantum theory violation of the Bell inequality (Su and Wódkiewicz, 1991).

19.6 Comparisons with experiments

19.6.1 Visibility of second-order interference fringes

For comparison with experiments with two counters, such as the one sketched in Fig. 19.5, the visibility of the second-order interference fringes observed in coincidence detection can be defined—by analogy to eqn (10.26)—as

$$\mathcal{V} \equiv \frac{G^{(2)}_{\alpha\beta}\Big|_{\max} - G^{(2)}_{\alpha\beta}\Big|_{\min}}{G^{(2)}_{\alpha\beta}\Big|_{\max} + G^{(2)}_{\alpha\beta}\Big|_{\min}}, \tag{19.63}$$

where $G^{(2)}_{\alpha\beta}\big|_{\max}$ and $G^{(2)}_{\alpha\beta}\big|_{\min}$ are respectively the maximum and minimum, with respect to the angles α and β, of the second-order Glauber correlation function. Let us assume that data analysis shows that an empirical fit to the second-order interference fringes has the form

$$G^{(2)}_{\alpha\beta} \propto 1 - \eta \cos(2\alpha - 2\beta), \tag{19.64}$$

for some value of the fitting parameter η. Given appropriate assumptions about the curve-fitting technique, one can show that

$$\eta = \mathcal{V}. \tag{19.65}$$

The physical meaning of a high, but imperfect ($\mathcal{V} < 1$), visibility is that decoherence of some sort has occurred between the two photons γ_A and γ_B during their propagation from the source to Alice and Bob. Thus the entangled pure state emitted by the source changes, for either fundamental or technical reasons, into a slightly mixed state before arriving at the detectors.

Next, let us consider experiments with four counters, such as the one sketched in Fig. 19.3. Again, using data analysis that assumes a finite-visibility fitting parameter η, the joint probabilities (19.58) and (19.59) have the following modified forms:

$$p_{ee}(\alpha, \beta) = p_{oo}(\alpha, \beta) = \frac{1}{2} - \frac{1}{2}\eta \cos(2\alpha - 2\beta), \tag{19.66}$$

$$p_{eo}(\alpha, \beta) = p_{oe}(\alpha, \beta) = \frac{1}{2} + \frac{1}{2}\eta \cos(2\alpha - 2\beta), \tag{19.67}$$

so that Bell's joint expectation value becomes

$$E(\alpha, \beta) = -\eta \cos(2\alpha - 2\beta). \tag{19.68}$$

For the special settings in eqn (19.61), one finds

$$S = -\frac{4}{\sqrt{2}}\eta = -2\sqrt{2}\eta. \tag{19.69}$$

This implies that the maximum amount of visibility $\mathcal{V}_{\max}$ permitted by Bell's inequality $|S| \leqslant 2$ is

$$\mathcal{V}_{\max} = \eta_{\max} = \frac{1}{\sqrt{2}} = 70.7\%. \tag{19.70}$$

19.6.2 Data from the tandem-crystal experiment violates the Bell inequality $|S| \leqslant 2$

For comparison with experiment, we show once again the data from the tandem two-crystal experiment discussed in Section 13.3.5, but this time we superpose a finite-visibility, sinusoidal interference-fringe pattern, of the form (19.64), with the maximum visibility $\mathcal{V}_{\max} = 70.7\%$ permitted by Bell's theorem. This is shown as a light, dotted curve in Fig. 19.7.

One can see by inspection that the data violate the Bell inequality (19.38) by many standard deviations. Indeed, detailed statistical analysis shows that these data violate the constraint $|S| \leqslant 2$ by 242 standard deviations. However, this data exhibits a high signal-to-noise ratio, so that systematic errors will dominate random errors in the data analysis.

19.6.3 Possible experimental loopholes

A The detection loophole

Since the quantum efficiencies of photon counters are never unity, there is a possible experimental loophole, called the **detection loophole**, in most quantum optical tests of Bell's theorem. If the quantum efficiency is less than 100%, then some of the photons will not be counted. This could be important, if the ensemble of photons generated by the source is not homogeneous. For example, it is conceivable—although far-fetched—that the photons that were not counted just happen to have different correlations than the ones that were counted. For example, the second-order interference fringes for the undetected photons might have a visibility that is less than the maximum allowable amount $\mathcal{V}_{\max} = 70.7\%$. Averaging the visibility of the undetected photons with the visibility of the detected photons, which do have a measured visibility greater than 70.7%, might produce a total distribution which just barely manages to satisfy the inequality (19.38).

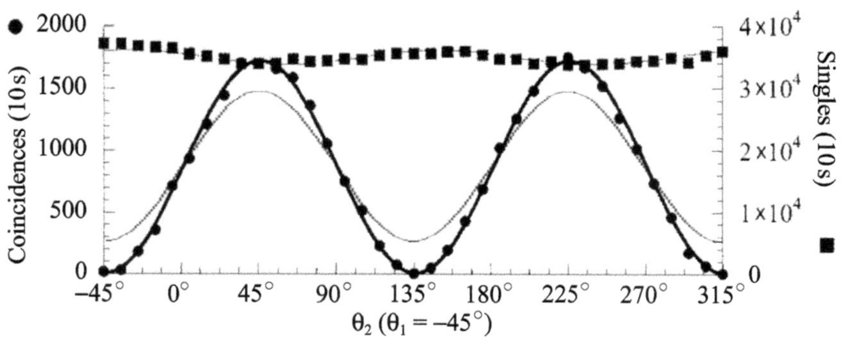

Fig. 19.7 Data from the tandem-crystals experiment (see Fig. 13.6) compared to maximum-visibility sinusoidal interference fringes with $\mathcal{V}_{\max} = 70.7\%$ (light, dotted curve), which is the maximum visibility permitted by Bell's theorem. (Adapted from Kwiat *et al.* (1999*b*).)

This scenario is ruled out if one adopts the entirely reasonable, **fair-sampling** assumption that the detected photons represent a fair sample of the undetected photons. In this case, the undetected photons would not have substantially distorted the observed interference fringes if they had been included in the data analysis. Nevertheless, the fair-sampling assumption is difficult to prove or disprove by experiment.

One way out of this difficulty is to repeat the quantum optical tests of Bell's theorem with extremely high quantum efficiency photon counters, such as solid-state photomultipliers (Kwiat *et al.*, 1994). This would minimize the chance of missing any appreciable fraction of the photons in the total ensemble of photon pairs from the source. To close the detection loophole, a quantum efficiency of greater than 83% is required for maximally entangled photons, but this requirement can be reduced to 67% by the use of nonmaximally entangled photons (Eberhard, 1993).

Replacing photons by ions allows much higher quantum efficiencies of detection, since ions can be detected much more efficiently than photons. In practice, nearly all ions can be counted, so that almost none will be missed. An experiment using entangled ions has been performed (Rowe *et al.*, 2001). With the detection loophole closed, the experimenters observed an 8 standard deviation violation of the Clauser–Horne–Shimony–Holt inequality (Clauser *et al.*, 1969)

$$|E(\alpha_1, \beta_1) + E(\alpha_2, \beta_1)| + |E(\alpha_1, \beta_2) - E(\alpha_2, \beta_2)| \leqslant 2\,. \qquad (19.71)$$

This is one of several experimentally useful Bell inequalities that are equivalent in physical content to the condition $|S| \leqslant 2$ discussed above.

B *The locality loophole*

Another possible loophole—which is conceptually much more important than the question of detector efficiency—is the locality loophole. Closing this loophole is especially vital in light of the incorporation of the extremely important Einsteinian principle of locality into Bell's theorem.

Since photons travel at the speed of light, they are much better suited than atoms or ions for closing the locality loophole. Using photons, it is easy to ensure that Alice's and Bob's decisions for the settings of their parameters α and β are space-like separated, and therefore truly independent.

For example, Alice and Bob could randomly and quickly reset α and β during the time interval after emission from the source and before arrival of the photons at their respective calcite prisms. There would then be no way for any secret machinery at the source to know beforehand what values of α and β Alice or Bob would eventually decide upon for their measurements. Therefore, properties of the photons that were predetermined at the source could not possibly influence the outcomes of the measurements that Alice and Bob were about to perform.

The first attempt to close the locality loophole was an experiment with a separation of 12 m between Alice and Bob. Rapidly varying the settings of α and β, by means of two acousto-optical switches (Aspect *et al.*, 1982), produced a violation of the Clauser–Horne–Shimony–Holt inequality (19.71) by 6 standard deviations.

However, the time variation of the two polarizing elements in this experiment was periodic and deterministic, so that the settings of α and β at the time of arrival of the photons could, in principle, be predicted. This would still allow the properties of the photons that led to the observed outcomes of measurements to be predetermined at the source.

A more satisfactory experiment *vis-à-vis* closing the locality loophole was performed with a separation of 400 m between the two polarizers. Two separate, ultrafast electro-optic modulators, driven by two local, independent random number generators, rapidly varied the settings of α and β in a completely random fashion. The result was a violation of the Clauser–Horne–Shimony–Holt inequality (19.71) by 30 standard deviations.

The two random number generators operated at the very high toggle frequency of 500 MHz. After accounting for various extraneous time delays, the experimenters concluded that no given setting of α or β could have been influenced by any event that occurred more than $0.1\,\mu$s earlier, which is much shorter than the $1.3\,\mu$s light transit time across 400 meters.

Hence the locality loophole was firmly closed. However, the detection loophole was far from being closed in this experiment, since only 5% of all the photon pairs were detected. Thus a heavy reliance on the fair-sampling assumption was required in the data analysis.

19.6.4 Relativistic issues

An experiment with a very large separation, of 10.9 km, between Alice and Bob has been performed using optical fiber technology, in conjunction with a spontaneous down-conversion light source (Tittel *et al.*, 1998). A violation of Bell's inequalities by 16 standard deviations was observed in this experiment.

Relativistic issues, such as putting limits on the so-called *speed of collapse* of the two-photon wave function, could then be examined experimentally using this type of apparatus. Depending on assumptions about the detection process and about which inertial frame is used, the speed of collapse was shown to be at least 10^4c to 10^7c (Zbinden *et al.*, 2001). Further experiments with rapidly rotating absorbers ruled out an alternative theory of nonlocal collapse (Suarez and Scarani, 1997).

19.6.5 Greenberger–Horne–Zeilinger states

The previous discussion of experiments testing Bell's theorem was based on constraints on the total amount of correlation between random events observable in two-particle coincidence experiments. These constraints are fundamentally statistical in nature. Greenberger, Horne, and Zeilinger (GHZ) (Kafatos, 1989, pp. 69–72) showed that using three particles, as opposed to two, in a maximally entangled state such as

$$|\psi_{\text{GHZ}}\rangle \propto |a, b, c\rangle - |a', b', c'\rangle \,, \tag{19.72}$$

allows a test of the combined principles of locality and realism by observing, or failing to observe, a single triple-coincidence click. Thus, in principle, the use of statistical correlations is unnecessary for testing local realistic theories. However, in practice, the detectors with quantum efficiencies less than 100% used in real experiments again required the use of inequalities. Violations of these inequalities have been observed in experiments involving nonmaximally entangled states generated by spontaneous down-conversion (Torgerson *et al.*, 1995; White *et al.*, 1999). Once again, the results contradict all local realistic theories.

For a review of these and other quantum optical tests of the foundations of physics, see Steinberg *et al.* (2005).

19.7 Exercises

19.1 The original EPR argument

(1) Show that the EPR wave function, given by eqn (19.1), is an eigenfunction of the total momentum $\widehat{p}_A + \widehat{p}_B$, with eigenvalue 0, and also an eigenfunction of the operator $\widehat{x}_A - \widehat{x}_B$, with eigenvalue L.

(2) Calculate the commutator $[\widehat{p}_A + \widehat{p}_B, \widehat{x}_A - \widehat{x}_B]$ and use the result to explain why (1) does not violate the uncertainty principle.

(3) If $\widehat{p}_A$ is measured, show that $\widehat{p}_B$ has a definite value. Alternatively, if $\widehat{x}_A$ is measured, show that $\widehat{x}_B$ has a definite value.

(4) Argue from the previous results that both $\widehat{x}_B$ and $\widehat{p}_B$ are elements of physical reality, and explain why this leads to the EPR paradox.

19.2 Parameter independence for quantum theory

(1) Use eqns (19.43)–(19.45) to derive eqn (19.47).

(2) Verify parameter independence when $|\chi\rangle$ is replaced by any of the four Bell states $\{|\Psi^{\pm}\rangle, |\Phi^{\pm}\rangle\}$ defined by eqns (13.59)–(13.62).

19.3 Violation of outcome independence

(1) Use eqn (19.50) to expand $|h_A, v_B\rangle$ and $|v_A, h_B\rangle$ in terms of $|h_A, \beta_B\rangle$ and $|v_A, \overline{\beta}_B\rangle$.

(2) Evaluate the reduced states $|\Psi^-\rangle_{\text{yes}}$ and $|\Psi^-\rangle_{\text{no}}$.

(3) Calculate the conditional probabilities $p(\mathfrak{A}_{\text{yes}}|\lambda, \alpha, \beta, \mathfrak{B}_{\text{yes}})$ and $p(\mathfrak{A}_{\text{yes}}|\lambda, \alpha, \beta, \mathfrak{B}_{\text{no}})$.

(4) Calculate the joint probability $p(\mathfrak{A}_{\text{yes}}, \mathfrak{B}_{\text{yes}}|\Psi^-, \alpha, \beta)$.

(5) If $|\Psi^-\rangle$ is replaced by $|\phi\rangle = |h_A, v_B\rangle$, is outcome independence still violated?

19.4 Violation of Bell's inequality

(1) Carry out the calculations needed to derive eqns (19.58) and (19.59).

(2) If $|\Psi^-\rangle$ is replaced by $|\phi\rangle = |h_A, v_B\rangle$, is the Bell inequality still violated?

20 Quantum information

Quantum optics began in the early years of the twentieth century, but its applications to communications, cryptography, and computation are of much more recent vintage. The progress of communications technology has made quantum effects a matter of practical interest, as evidenced in the discussion of noise control in optical transmission lines in Section 20.1. The issue of inescapable quantum noise is also related to the difficulty—discussed in Section 20.2—of copying or cloning quantum states.

Other experimental and technological advances are opening up new directions for development in which the quantum properties of light are a resource, rather than a problem. Streams of single photons with randomly chosen polarizations have already been demonstrated as a means for the secure transmission of cryptographic keys, as discussed in Section 20.3. Multiphoton states offer additional options that depend on quantum entanglement, as shown by the descriptions of quantum dense coding and quantum teleportation in Section 20.4. This set of ideas plays a central role in the closely related field of quantum computing, which is briefly reviewed in Section 20.5.

20.1 Telecommunications

Optical methods of communication—e.g. signal fires, heliographs, Aldis lamps, etc.—have been in use for a very long time, but high-speed optical telecommunications are a relatively recent development. The appearance of low-loss optical fibers and semiconductor lasers in the 1960s and 1970s provided the technologies that made new forms of optical communication a practical possibility.

The subsequent increases in bandwidth to 10^4 GHz and transmission rates to the multiterabit range have led—under the lash of Moore's law—to substantial decreases in the energy per bit and the size of the physical components involved in switching and amplification of signals. An inevitable consequence of this technologically driven development is that phenomena at the quantum level are rapidly becoming important for real-world applications.

Long-haul optical transmission lines require repeater stations that amplify the signal in order to compensate for attenuation. This process typically adds noise to the signal; for example, erbium doped fiber amplifiers (EDFA) degrade the signal-to-noise ratio by about 4 dB. Only 1 dB arises from technical losses in the components; the remaining 3 dB loss is due to intrinsic quantum noise.

Thus quantum noise is dominant, even for apparently classical signals containing a very large number of photons. Similar effects arise when the signal is divided by a passive device such as an optical coupler. Future technological developments can be expected to increase the importance of quantum noise; therefore, we devote Sections 20.1.2 and 20.1.3 to the problem of quantum noise management.

20.1.1 Optical transmission lines*

Let us consider an optical transmission line in which the repeater stations employ phase-insensitive amplifiers. For phase-insensitive input noise, the input and output signal-to-noise ratios are defined by

$$[\mathrm{SNR}]_\gamma = \frac{\left|\left\langle \overline{b}_\gamma(\omega)\right\rangle\right|^2}{\mathfrak{N}_\gamma(\omega)} \quad (\gamma = \mathrm{in}, \mathrm{out})\,, \tag{20.1}$$

where $\mathfrak{N}_{\mathrm{in}}$ and $\mathfrak{N}_{\mathrm{out}}$ are the noise in the input and output respectively. The relation between the input and output signal-to-noise ratios is obtained by combining eqns (16.36) and (16.150) to get

$$[\mathrm{SNR}]_{\mathrm{out}} = \frac{\left|\left\langle \overline{b}_{\mathrm{out}}(\omega)\right\rangle\right|^2}{\mathfrak{N}_{\mathrm{out}}(\omega)} = \frac{\left|\left\langle \overline{b}_{\mathrm{in}}(\omega)\right\rangle\right|^2}{\mathfrak{N}_{\mathrm{in}}(\omega) + \mathfrak{A}(\omega)} = \frac{[\mathrm{SNR}]_{\mathrm{in}}}{1 + \mathfrak{A}(\omega)/\mathfrak{N}_{\mathrm{in}}(\omega)}\,. \tag{20.2}$$

The most favorable situation occurs when the input noise strength has the standard quantum limit value 1/2. In this case one finds

$$\frac{[\mathrm{SNR}]_{\mathrm{out}}}{[\mathrm{SNR}]_{\mathrm{in}}} = \frac{1}{1 + 2\mathfrak{A}(\omega)} \leqslant \frac{1}{2 \mp 1/G(\omega)} \to \frac{1}{2}\,. \tag{20.3}$$

The inequality follows from eqn (16.151) and the final result represents the high-gain limit. The decibel difference between the signal-to-noise ratios is therefore bounded by

$$d = 10\log\left[\frac{[\mathrm{SNR}]_{\mathrm{out}}}{[\mathrm{SNR}]_{\mathrm{in}}}\right] \leqslant -10\log 2 \approx -3\,. \tag{20.4}$$

In other words, the quantum noise added by a high-gain, phase-insensitive amplifier degrades the signal-to-noise ratio by at least three decibels. This result holds even for strong input fields containing many photons. For example, if the input is described by the multi-mode coherent state defined by eqns (16.98)–(16.100), then the input noise strength is $\mathfrak{N}_{\mathrm{in}}(\omega) = 1/2$. In this case the inequality (20.4) is valid for any value of the effective classical intensity $|\beta_{\mathrm{in}}(t)|^2$, no matter how large.

This result demonstrates that high-gain, phase-insensitive amplifiers are intrinsically noisy. This noise is generated by fundamental quantum processes that are at work even in the absence of the technical noise—e.g. insertion-loss noise and Johnson noise in the associated electronic circuits—always encountered in real devices.

20.1.2 Reduction of amplifier noise*

In the discussion of squeezing in Section 15.1 we have seen that quantum noise can be unequally shared between different field quadratures by using nonlinear optical effects. This approach—which can yield essentially noise-free amplification for one quadrature by dumping the unwanted noise in the conjugate quadrature—is presented in the present section.

There is an alternative scheme, based on the special features of cavity quantum electrodynamics, in which the signal propagates through a **photonic bandgap**. This is a three-dimensional structure in which periodic variations of the refractive index produce a dispersion relation that does not allow propagating solutions in one or more frequency bands—the *bandgaps*—so that vacuum fluctuations and the associated noise are forbidden at those frequencies (Abram and Grangier, 2003).

In the discussion of linear optical amplifiers in Chapter 16, we derived the inequality (16.147) which shows that the amplifier noise for a phase-conjugating amplifier is always larger than the vacuum noise, i.e. $\mathfrak{N}_{\text{amp}} > 1/2$. On the other hand, the noise added by a phase-transmitting amplifier can be made as small as desired by allowing $G(\omega)$ to approach unity. Thus noise reduction can be achieved with a phase-transmitting amplifier, provided that we are willing to give up any significant amplification.

Achieving noise reduction by giving up amplification scarcely recommends itself as a useful strategy for long-haul communications, so we turn next to phase-sensitive amplifiers. In this case, the lower bound (16.147) on amplifier noise is replaced by the amplifier uncertainty principle (16.169). The resemblance between eqn (16.169) and the standard uncertainty principle for canonically conjugate variables is promising, since the latter is known to allow squeezing.

Furthermore, the amplifier uncertainty principle has the additional advantage that the lower bound is itself adjustable; indeed, it can be set to zero. Even when this is not possible, the noise in one quadrature can be reduced at the expense of increasing the noise in the conjugate quadrature. We first demonstrate two examples in which the amplifier noise actually vanishes, and then discuss what can be achieved in less favorable situations.

The phase-sensitive, traveling-wave amplifier described in Section 16.3.2 is intrinsically noiseless, so the lower bound of the amplifier uncertainty principle automatically vanishes. For applications requiring the generally larger gains possible for regenerative amplifiers, the phase-sensitive OPA presented in Section 16.2.2 can be modified to provide noise-free amplification.

For the phase-sensitive OPA, the amplifier noise comes from vacuum fluctuations entering the cavity through the mirror M2, as shown in Fig. 16.2. Thus the amplifier noise would be eliminated by preventing the vacuum fluctuations from entering the cavity. In an ideal world, this can be accomplished by making M2 a perfect reflector, i.e. setting $\kappa_2 = 0$ in eqns (16.47)–(16.49). Under these circumstances, eqn (16.49) reduces to $\eta(\omega) = 0$, so that the amplifier is noiseless.

In these examples, the noise vanishes for both of the principal quadratures, i.e. $\mathfrak{A}_1(\omega) = \mathfrak{A}_2(\omega) = 0$. According to eqn (16.167), this means that the signal-to-noise ratio is preserved by amplification. This is only possible if the lower bound in eqn (16.170) vanishes, and this in turn requires $G_1(\omega)\, G_2(\omega) = 1$. Consequently, the price for noise-free amplification is that one quadrature is attenuated while the other is amplified.

In the real world—where traveling-wave amplifiers may not provide sufficient gain and there are no perfect mirrors—other options must be considered. The general idea is to achieve high gain and low noise for the same quadrature. For this purpose, the signal should be carried by modulation of either the amplitude or the phase of the chosen quadrature, e.g. $X_{\text{in}}^{c}(\omega)$; and the input noise should be small, i.e. $\Delta X_{\text{in}}^{c}(\omega) \ll 1/2$.

In the high-gain limit, the lower bound in eqn (16.169) is proportional to $\sqrt{G_1(\omega)\, G_2(\omega)}$; consequently, the amplifier noise in the conjugate quadrature is necessarily large. This is not a problem as long as the noisy quadrature is strongly rejected by the detectors in use. The degree to which these objectives can be attained depends on the details of the overall design.

20.1.3 Reduction of branching noise

The information encoded in an optical signal is often intended for more than one recipient, so that it is necessary to split the signal into two or more identical parts, usually by means of a directional coupler. These junction points—which are often called **optical taps**—may also be used to split off a small part of the signal for measurement purposes.

Whatever the motive for the tap, it is in effect a measurement of the radiation field. A measurement of any quantum system perturbs it in an uncontrollable fashion; consequently, the optical tap must add noise to the signal. A succession of taps will therefore degrade the signal, even if there is no associated amplifier noise.

In fact, we have already met with this effect, in the guise of the partition noise at a beam splitter. The explanation that partition noise arises from vacuum fluctuations entering through the unused port of the beam splitter suggests that injecting a squeezed vacuum state into the unused port might help with the noise problem. This idea was initially proposed in 1980 (Shapiro, 1980) and experimentally realized in 1997 (Bruckmeier *et al.*, 1997). We discuss below a simple model that illustrates this approach.

The idea is to add two elements, shown in Fig. 20.1, to the simple beam splitter described in Section 8.4: (1) a squeezed-light generator (SQLG); and (2) a pair of variable retarder plates (see Exercise 20.1). The SQLG, which is the essential part of the modified beam splitter, injects squeezed light into the previously unused input port 2. The function of the variable retarder plates, which are placed at the input port 2 and the output port 2′, is to simplify the overall scattering matrix.

The phase transformations, $a_2 \to e^{i\theta} a_2$ and $a_2' \to e^{i\theta'} a_2'$, imposed by the retarder plates are more usefully described as rotations of the input and output quadratures through the angles θ and θ'. Combining the phase transformations with eqn (8.63)—as outlined in Exercise 20.2—yields the scattering matrix

$$S = \begin{bmatrix} \sqrt{T} & i\sqrt{R}e^{i\theta} \\ i\sqrt{R}e^{i\theta'} & \sqrt{T}e^{i(\theta+\theta')} \end{bmatrix} . \tag{20.5}$$

The phases of the beam splitter coefficients have been chosen so that $\mathfrak{t} = \sqrt{T}$ is real and $\mathfrak{r} = i\sqrt{R}$ is pure imaginary, where T and R are respectively the intensity transmission and reflection coefficients.

The SQLG is designed to emit a squeezed state for the quadrature

$$X_2 = \frac{1}{2}\left(e^{-i\beta} a_2 + e^{i\beta} a_2^{\dagger}\right), \tag{20.6}$$

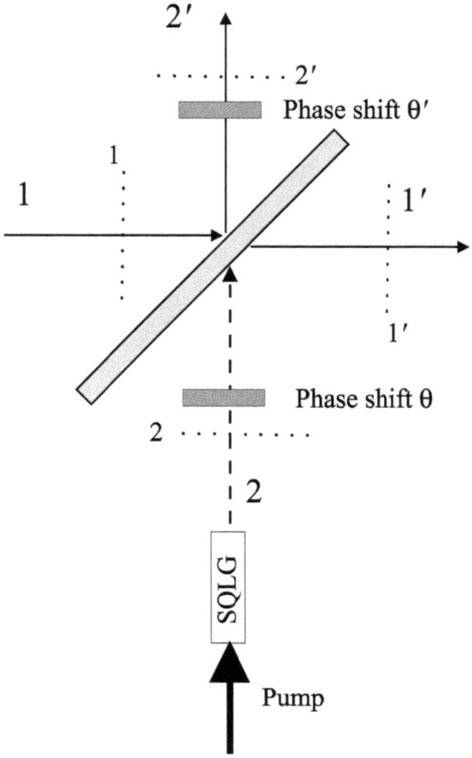

Fig. 20.1 Modified beam splitter for noiseless branching. The OPA injects squeezed light into port 2 and the phase plates are used to obtain a convenient form for the scattering matrix.

so an application of eqn (15.39) produces the variances

$$V(X_2) = \frac{e^{-2r}}{4}, \quad V(Y_2) = \frac{e^{2r}}{4}, \tag{20.7}$$

where Y_2 is the conjugate quadrature and r is the magnitude of the squeezing parameter. For the special values $\theta = -\pi/2$ and $\theta' = \pi/2$ the input–output relations for the amplitude quadratures are

$$\begin{pmatrix} X_1' \\ X_2' \end{pmatrix} = \begin{bmatrix} \sqrt{T} & \sqrt{R} \\ -\sqrt{R} & \sqrt{T} \end{bmatrix} \begin{pmatrix} X_1 \\ X_2 \end{pmatrix}, \tag{20.8}$$

where the quadratures for the input channel 1, and the output channels 1′ and 2′ are defined by the angle β used in eqn (20.6).

Let us now specialize to a balanced beam splitter, and assume that the signal is carried by X_1. The squeezed state satisfies $\langle X_2 \rangle = 0$; consequently, the two output signals have the same average magnitude:

$$\langle X_1' \rangle = -\langle X_2' \rangle = \frac{1}{\sqrt{2}} \langle X_1 \rangle . \tag{20.9}$$

Since the input signal X_1 and the SQLG output are uncorrelated, the variances of the output signals X_1' and X_2' are also identical:

$$V(X_1') = V(X_2') = \frac{1}{2}V(X_1) + \frac{e^{-r}}{8}. \tag{20.10}$$

The 50% reduction of the output variances compared to the input variance does not mean that the output signals are quieter; it merely reflects the reduction of the amplitudes by the factor $1/\sqrt{2}$. This can be seen by defining the signal-to-noise ratios,

$$\mathrm{SNR}(X_m) = \frac{|\langle X_m \rangle|^2}{V(X_m)}, \quad \mathrm{SNR}(X_m') = \frac{|\langle X_m' \rangle|^2}{V(X_m')} \quad (m = 1, 2), \tag{20.11}$$

and using the previous results to find

$$\mathrm{SNR}(X_1') = \mathrm{SNR}(X_2') = \frac{\mathrm{SNR}(X_1)}{1 + e^{-r}/[4V(X_1)]}. \tag{20.12}$$

In the limit of strong squeezing, this coupler almost exactly preserves the signal-to-noise ratio of the input signal. Consequently, the output signals are faithful copies of the input signal down to the level of the quantum fluctuations. The injection of the squeezed light into port 2 has effectively diverted almost all of the partition noise into the unobserved output quadratures Y_1' and Y_2'.

This scheme succeeds in splitting the signal without adding any noise, but at the cost of reducing the intensity of the output signals by 50%. This drawback can be overcome by inserting a noiseless amplifier, e.g. the traveling-wave OPA described in Section 16.3.2, prior to port 1 of the beam splitter. The gain of the amplifier can be adjusted so that each of the split signals has the same strength and the same signal-to-noise ratio as the original signal.

20.2 Quantum cloning

At first glance, it may seem that the noiseless beam splitter of Section 20.1.3 produces a perfect copy or *clone* of the input signal. This impression is misleading, since only the expectation values and variances of the particular input quadrature X_1 are faithfully copied; indeed, the variance of the output conjugate quadrature Y_1' is much larger than the variance of Y_1.

This observation suggests a general question: To what extent does quantum theory allow cloning? In the following section, we will review the famous no-cloning theorem (Dieks, 1982; Wootters and Zurek, 1982), which outlaws perfect cloning of an unknown quantum state.

We should note that this work was not done to answer the question we have just raised. It was a response to a proposal by Herbert (1982) for a superluminal communications scheme employing EPR correlations. The connection between no-cloning and no-superluminal-signaling is a recurring theme in later work (Ghirardi and Weber, 1983; Bussey, 1987).

The no-cloning theorem quickly became an important physical principle which was, for example, used to argue for the security of quantum cryptography (Bennett and Brassard, 1984). The final step in the initial development was the extension of the result from pure to mixed states (Barnum *et al.*, 1996).

This was immediately followed by the work of Bužek and Hillery (1996) who began the investigation of imperfect cloning. We will study the degree of cloning allowed by quantum theory in Section 20.2.2.

In the study of quantum information, the systems of interest are usually described by states in a finite-dimensional Hilbert space $\mathfrak{H}_{\text{sys}}$. For the special case of two-state systems—e.g. a two-level atom, a spin-1/2 particle, or the two polarizations of a photon—$\mathfrak{H}_{\text{sys}}$ is two-dimensional, and a vector $|\gamma\rangle$ in $\mathfrak{H}_{\text{sys}}$ is called a **qubit**. The generic description of qubits employs the so-called **computational basis** $\{|0\rangle, |1\rangle\}$ defined by

$$\sigma_z |0\rangle = |0\rangle, \quad \sigma_z |1\rangle = -|1\rangle. \tag{20.13}$$

In this notation a general qubit is represented by $|\gamma\rangle = \gamma_0 |0\rangle + \gamma_1 |1\rangle$.

In the more general case, $\dim(\mathfrak{H}_{\text{sys}}) = d > 2$, the state is called a **qudit**. We will follow the usual convention by referring to the systems under study as qubits, but it should be kept in mind that many of the results also hold in the general finite-dimensional case. In the interests of simplicity, we will only treat closed systems undergoing unitary time evolution.

For the applications considered below, it is often necessary to consider one or more **ancillary** (helper) systems in addition to the system of interest. The reservoirs used in the treatment of dissipation in Chapter 14 are an example of ancillary systems or *ancillas*. In that case the unitary evolution of the closed sample–reservoir system was used to derive the dissipative equations by tracing over the ancilla degrees of freedom.

Another common theme in this field is the assumption that the total system consists of a family of distinguishable qubits. The Hilbert space $\mathfrak{H}$ for the system is $\mathfrak{H} = \mathfrak{H}_Q \otimes \mathfrak{H}_{\text{anc}}$, where $\mathfrak{H}_{\text{anc}}$ and $\mathfrak{H}_Q$ are respectively the state spaces for the ancillas and the family of qubits. This abstract approach has the great advantage that the results do not depend on the specific details of particular physical realizations, but there are, nevertheless, some implicit physical assumptions involved.

If the qubits are particles, then—as we learnt in Section 6.5.1—the Hilbert space $\mathfrak{H}_Q$ for two qubits is

$$\mathfrak{H}_Q = \begin{cases} \left(\mathfrak{H}_{\text{sys}} \otimes \mathfrak{H}_{\text{sys}}\right)_{\text{sym}} & \text{for bosons}, \\ \left(\mathfrak{H}_{\text{sys}} \otimes \mathfrak{H}_{\text{sys}}\right)_{\text{asym}} & \text{for fermions}. \end{cases} \tag{20.14}$$

For massive particles—e.g. atoms, molecules, quantum dots, etc.—a way around this complication is to choose an experimental arrangement in which each particle's center-of-mass position can be treated classically. In these circumstances, as we saw in Section 6.5.2, the symmetrization or antisymmetrization normally required for identical particles can be ignored. In this model, a qubit located at $\mathbf{r}_a$ is described by a copy of $\mathfrak{H}_{\text{sys}}$, called $\mathfrak{H}_a$. The vectors $|\gamma\rangle_a$ in $\mathfrak{H}_a$ represent the *internal states* of the qubit.

For a family of two qubits, located at $\mathbf{r}_a$ and $\mathbf{r}_b$, the space $\mathfrak{H}_Q$ is the unsymmetrized tensor product: $\mathfrak{H}_Q = \mathfrak{H}_a \otimes \mathfrak{H}_b$. The Bell states, first defined in Section 13.3.5 for photons, are represented by

$$\left|\Psi^{\pm}\right\rangle_{ab} = \frac{1}{\sqrt{2}}\left\{\left|0,1\right\rangle_{ab} \pm \left|1,0\right\rangle_{ab}\right\},$$
$$\left|\Phi^{\pm}\right\rangle_{ab} = \frac{1}{\sqrt{2}}\left\{\left|0,0\right\rangle_{ab} \pm \left|1,1\right\rangle_{ab}\right\}, \tag{20.15}$$

where

$$\left|u,v\right\rangle_{ab} \equiv \left|u\right\rangle_{a}\left|v\right\rangle_{b}. \tag{20.16}$$

More features of the Bell states can be found in Exercise 20.3.

In the general case of qubits located at $\mathbf{r}_1, \ldots, \mathbf{r}_N$ the qubit space is

$$\mathfrak{H}_Q = \bigotimes_{a=1}^{N} \mathfrak{H}_a\,, \tag{20.17}$$

and a generic state is denoted by $\left|u_1, \ldots, u_N\right\rangle_{12\cdots N}$. When no confusion will result, the notation is simplified by omitting the subscripts on the kets, e.g. $\left|u,v\right\rangle_{ab} \to \left|u,v\right\rangle$. The application of these ideas to photons requires a bit more care, as we will see below.

20.2.1 The no-cloning theorem

For closed systems, we can assume that every physically permitted operation is described by a unitary transformation U acting on the Hilbert space $\mathfrak{H}$ describing the qubits and the ancillas. To set the scene for the cloning discussion, we assume that there is a set of qubits, $\left|\mathbb{B}\right\rangle_b$, all in the same (internal) *blank* state $\left|\mathbb{B}\right\rangle$, and a cloning device which is initially in the *ready* state $\left|\mathbb{R}\right\rangle_{\text{anc}} \in \mathfrak{H}_{\text{anc}}$.

If we only want to make one copy—this is called $1 \to 2$ cloning—the total initial state is

$$\left|\gamma, \mathbb{B}, \mathbb{R}\right\rangle \equiv \left|\gamma\right\rangle_a \otimes \left|\mathbb{B}\right\rangle_b \otimes \left|\mathbb{R}\right\rangle_{\text{anc}} = \left|\gamma\right\rangle_a \left|\mathbb{B}\right\rangle_b \left|\mathbb{R}\right\rangle_{\text{anc}}. \tag{20.18}$$

The cloning assumption is that there is a unitary operator U such that

$$U\left|\gamma, \mathbb{B}, \mathbb{R}\right\rangle = \left|\gamma, \gamma, \mathbb{R}_\gamma\right\rangle = \left|\gamma\right\rangle_a \left|\gamma\right\rangle_b \left|\mathbb{R}_\gamma\right\rangle_{\text{anc}}, \tag{20.19}$$

where $\left|\mathbb{R}_\gamma\right\rangle_{\text{anc}}$ is the state of the cloner after it has cloned the state $\left|\gamma\right\rangle_a$. In this approach, cloning is not the creation of a new particle, but instead the imposition of a specified internal state on an existing particle.

After this preparation, the **no-cloning theorem** can be stated as follows (Scarani *et al.*, 2005).

Theorem 20.1 *There is no quantum operation that can perfectly duplicate an unknown quantum state.*

We will use a proof given by Peres (1995, Sec. 9-4) that exhibits a contradiction following from the assumption that a cloning operation does exist, i.e. that there is a unitary operator satisfying eqn (20.19).

Since the cloning device is supposed to work in the absence of any knowledge of the initial state, it must be possible to use U to clone a different state $|\zeta\rangle$, so that

$$U\,|\zeta,\mathbb{B},\mathbb{R}\rangle = |\zeta,\zeta,\mathbb{R}_\zeta\rangle\,. \tag{20.20}$$

A direct use of the unitarity of U yields

$$\langle\gamma,\gamma,\mathbb{R}_\gamma|\,\zeta,\zeta,\mathbb{R}_\zeta\rangle = \langle\gamma,\mathbb{B},\mathbb{R}|\,\zeta,\mathbb{B},\mathbb{R}\rangle = \langle\gamma\,|\zeta\rangle\,, \tag{20.21}$$

where we have imposed the convention that the initial states $|\mathbb{R}\rangle_{\text{anc}}$, $|\gamma\rangle_a$, $|\mathbb{B}\rangle_b$, and $|\zeta\rangle_a$ are all normalized and that the inner product between internal states does not depend on the location of the qubit.

Using the explicit tensor products in eqns (20.19) and (20.20) produces the alternative form

$$\langle\gamma,\gamma,\mathbb{R}_\gamma|\,\zeta,\zeta,\mathbb{R}_\zeta\rangle = \langle\mathbb{R}_\gamma\,|\mathbb{R}_\zeta\,\rangle\,\langle\gamma\,|\zeta\,\rangle^2\,. \tag{20.22}$$

For non-orthogonal qubits, $|\gamma\rangle$ and $|\zeta\rangle$, equating the two results leads to

$$\langle\mathbb{R}_\gamma\,|\mathbb{R}_\zeta\,\rangle\,\langle\gamma\,|\zeta\,\rangle = 1\,. \tag{20.23}$$

The inner product $\langle\mathbb{R}_\gamma\,|\mathbb{R}_\zeta\,\rangle$ automatically satisfies $|\langle\mathbb{R}_\gamma\,|\mathbb{R}_\zeta\,\rangle| \leqslant 1$, and we can always choose $|\gamma\rangle$ and $|\zeta\rangle$ so that $|\langle\gamma\,|\zeta\,\rangle| < 1$; therefore, there are states $|\gamma\rangle$ and $|\zeta\rangle$ for which eqn (20.23) cannot be satisfied. This contradiction proves the theorem.

This elegant proof shows that the impossibility of perfect cloning of unknown, and hence arbitrary, states is a fundamental feature of quantum theory; indeed, the only requirement is that quantum operations are represented by unitary transformations. In this respect it is similar to the Heisenberg uncertainty principle, for which the sole requirement is the canonical commutation relation $[\widehat{q},\widehat{p}] = i\hbar$.

We should emphasize, however, that this argument only excludes **universal cloning machines**, i.e. those that can clone any given state. This leaves open the possibility that specific states could be cloned. In fact the argument does not prohibit the cloning of each member of a known set of mutually orthogonal states.

The application of this theorem in the context of quantum optics raises some problems. The proof rests on the assumption that the qubits are distinguishable and localizable, but photons are indistinguishable, massless bosons that cannot be precisely localized and are easily created and destroyed. Thus it is not immediately obvious that the proof of the no-cloning theorem given above applies to photons.

A second problem arises from the observation that stimulated emission—which produces new photons with the same wavenumber and polarization as the incident photons—would seem to provide a ready-made copying mechanism. Why is it that stimulated emission is not a counterexample to the no-cloning theorem? In the following paragraphs we will address these questions in turn.

Since photons are indistinguishable bosons, we cannot add any identifying subscript to a photonic qubit $|\gamma\rangle$, and the two-qubit space is the two-photon Fock space, $\mathfrak{H}^{(2)}$. The simplest way to define a photonic qubit is to choose a specific wavevector $\mathbf{k}$, and set $|\gamma\rangle = \Gamma^\dagger\,|0\rangle$, where

$$\Gamma^\dagger = \sum_s \gamma_{\mathbf{k}s} a^\dagger_{\mathbf{k}s}\,. \tag{20.24}$$

Since s takes on two values, the state $|\gamma\rangle$ qualifies as a qubit.

Cloning this qubit can only mean that a second photon is added in the same mode; therefore, the cloning transformation (20.19) for this case would be

$$U\Gamma^{\dagger}|0\rangle|\mathbb{R}\rangle_{\text{anc}} = \frac{1}{\sqrt{2}}\Gamma^{\dagger 2}|0\rangle|\mathbb{R}_{\gamma}\rangle_{\text{anc}} \,. \tag{20.25}$$

By contrast to the distinguishable qubit model, the polarization state is not imposed on an existing photon in a blank state; instead, a new photon is created with the same polarization as the original. Despite this significant physical difference, a similar proof of the no-cloning theorem can be constructed by following the hints in Exercise 20.4.

The proof of the no-cloning theorem—either the standard version starting with eqn (20.19) or the photonic version treated in Exercise 20.4—does not suggest any specific mechanism that prevents cloning. Finding a mechanism of this sort for photons turns out to be related to the second problem noted above. Could stimulated emission provide a cloning method?

The discussion of stimulated emission starts with a photon incident on an atom in an excited state. In this case, the nonzero ratio $A/B = \hbar k^3/\pi^2 \neq 0$ of the Einstein A and B coefficients provides the essential clue: stimulated emission is unavoidably accompanied by spontaneous emission. Since the spontaneously emitted photons have random directions and polarizations, they will violate the cloning assumptions (20.25).

This argument eliminates cloning machines based on excited atoms, but what about parametric amplifiers, such as the traveling-wave OPA in Section 16.3.2, in which there are no population inversions and, consequently, no excited atoms? This possible loophole was closed by the work of Milonni and Hardies (1982), in which it is shown that stimulated emission is necessarily accompanied by spontaneous emission, even in the absence of inverted atoms.

In the context of quantum optics, the impossibility of perfect, universal cloning can therefore be understood as a consequence of the unavoidable pairing of stimulated and spontaneous emission.

The no-cloning theorem does not exclude devices that can clone each member of a known set of orthogonal states. For example, two orthogonal polarization states can be cloned by exploiting stimulated emission. For this purpose, suppose that the sum over polarizations in eqn (20.24) refers to the linear polarization vectors $\mathbf{e}_h$ (*horizontal*) and $\mathbf{e}_v$ (*vertical*).

The cloning device consists of a trap containing a single excited atom, followed by a polarizing beam splitter. The PBS is oriented so that h- and v-polarized photons are sent through ports 1 and 2 respectively. For an initial state $|1_{\mathbf{k}h}\rangle$, the first-order perturbation calculation suggested in Exercise 20.5 shows that the combination of stimulated and spontaneous emission produces an output state proportional to

$$\sqrt{2}\,|2_{\mathbf{k}h}\rangle + |1_{\mathbf{k}h},\ 1_{\mathbf{k}v}\rangle \,. \tag{20.26}$$

Since the PBS sends the unwanted v-polarized photon through port 2, the only two-photon state emitted through port 1 is the desired cloned state $|2_{\mathbf{k}h}\rangle$. The argument is symmetrical under the simultaneous exchange of h with v and port 1 with port 2; therefore, the device is equally good at cloning v-polarized photons.

This design produces perfect clones of each state in the basis, but only if the basis is known in advance, so that the PBS can be properly oriented. As usual, the experimental realization is a different matter. This idea depends on having detectors that can reliably distinguish between one and two photons in a given mode, but such detectors are—to say the least—very hard to find.

Since classical theory is an approximation to quantum theory, we are left with a final puzzle: How is it that the no-cloning theorem does not prohibit the everyday practice of amplifying and copying classical signals? To understand this, we observe that for an incident state with n_i photons, the total emission probability for the amplifier is proportional to n_i+1, where n_i and 1 respectively correspond to stimulated and spontaneous emission.

If $n_i = 1$, the two processes are equally probable, but if $n_i \gg 1$, then stimulated emission dominates the output signal. Thus the classical copying process can achieve its aim, despite the fact that it cannot create a perfect clone of the input.

20.2.2 Quantum cloning machines*

The ideal cloning operation in eqn (20.19) would—if only it were possible—produce an exact copy of a qubit without damaging the original. In their seminal paper on imperfect cloning, Hillery and Bužek posed two questions: (1) How close can one come to perfect cloning? (2) What happens to the original qubit in the process?

Attempts to answer these questions have generated a large and rapidly developing field of research. In the remainder of this section, we will give a very brief outline of the basic notions, and discuss one optical implementation. For those interested in a more detailed account, the best strategy is to consult a recent review article, e.g. Scarani *et al.* (2005) or Fan (2006).

A Cloning distinguishable qubits*

The unattainable ideal of perfect cloning is replaced by the idea of a **quantum cloning machine** (QCM), which consists of a chosen ancillary state $|\mathbb{R}\rangle_{\text{anc}}$ in $\mathfrak{H}_{\text{anc}}$ and a unitary transformation U acting on $\mathfrak{H} = \mathfrak{H}_Q \otimes \mathfrak{H}_{\text{anc}}$. We will only discuss the simplest case of $1 \to 2$ cloning, for which the action of U on the initial state $|\gamma, \mathbb{B}, \mathbb{R}\rangle$ defines the *cloned state*

$$|\gamma; \gamma\rangle \equiv U |\gamma, \mathbb{B}, \mathbb{R}\rangle . \tag{20.27}$$

In general, the vector $|\gamma; \gamma\rangle$ represents an entangled state of the ancilla and the two qubits, so the state of the qubits alone is described by the reduced density operator

$$\rho_{ab} = \text{Tr}_{\text{anc}} |\gamma; \gamma\rangle \langle \gamma; \gamma| , \tag{20.28}$$

where the trace is defined by summing over a basis for the ancillary space $\mathfrak{H}_{\text{anc}}$. The states of the individual qubits are in turn represented by the reduced density operators

$$\rho_a = \text{Tr}_b \, \rho_{ab} \quad \text{and} \quad \rho_b = \text{Tr}_a \, \rho_{ab} \, . \tag{20.29}$$

The task is to choose $|\mathbb{R}\rangle_{\text{anc}}$ and U to achieve the best possible result, as opposed to imposing the form of $|\gamma; \gamma\rangle$ a priori. This effort clearly depends on defining what is meant by 'best possible'.

Of the many available measures of success, the most commonly used is the **fidelity**:

$$F_a(\gamma) = {}_a\langle\gamma|\rho_a|\gamma\rangle_a\,, \quad F_b(\gamma) = {}_b\langle\gamma|\rho_b|\gamma\rangle_b\,, \tag{20.30}$$

which measures the overlap between the mixed state produced by the cloning operation and the original pure state. A QCM is said to be a **universal QCM** if the fidelities are independent of $|\gamma\rangle$, i.e. the machine does equally well at cloning every state.

A nonuniversal QCM is called a **state-dependent QCM**. The QCM is a **symmetric QCM** if the fidelities of the output states are equal, i.e. $F_a(\gamma) = F_b(\gamma)$, and it is an **optimal QCM** if the fidelities are as large as quantum theory allows.

The unitary operator U for a QCM is linear, so its action on the general input state $|\gamma, \mathbb{B}, \mathbb{R}\rangle$ is completely determined by its action on the special states $|0, \mathbb{B}, \mathbb{R}\rangle$ and $|1, \mathbb{B}, \mathbb{R}\rangle$, where 0 and 1 label the computational basis vectors defined by eqn (20.13). For the Bužek–Hillery QCM, the ancilla consists of a single qubit, $|\mathbb{R}\rangle_{\text{anc}} = R_0|0\rangle_{\text{anc}} + R_1|1\rangle_{\text{anc}}$, and the transformation U is defined by

$$U|0, \mathbb{B}, \mathbb{R}\rangle = \sqrt{\frac{2}{3}}|0\rangle_a|0\rangle_b|1\rangle_{\text{anc}} - \sqrt{\frac{1}{3}}|\Psi^+\rangle_{ab}|0\rangle_{\text{anc}}\,, \tag{20.31}$$

$$U|1, \mathbb{B}, \mathbb{R}\rangle = -\sqrt{\frac{2}{3}}|1\rangle_a|1\rangle_b|0\rangle_{\text{anc}} + \sqrt{\frac{1}{3}}|\Psi^+\rangle_{ab}|1\rangle_{\text{anc}}\,. \tag{20.32}$$

The Bell state $|\Psi^+\rangle_{ab}$ is defined in eqn (20.15). In Exercise 20.6, these explicit expressions are used to evaluate the reduced density operators ρ_a and ρ_b which yield the fidelities $F_a(\gamma) = F_b(\gamma) = 5/6$. Thus the Bužek–Hillery QCM is universal and symmetric. It has also been shown—see the references given in Scarani *et al.* (2005)—that it is optimal.

B Cloning photons*

In order to carry out an actual experiment, the abstractions of the preceding discussion must be replaced by real hardware. Furthermore, the application of these ideas in quantum optics also requires a more careful use of the theory. Both of these considerations are illustrated by an experimental demonstration of a cloning machine for photons (Lamas-Linares *et al.*, 2002).

The basic idea, as shown in Fig. 20.2, is to use stimulated emission in a type II down-conversion crystal, which is adjusted so that the down-converted photons propagating along certain directions are entangled in polarization (Kwiat *et al.*, 1995*b*).

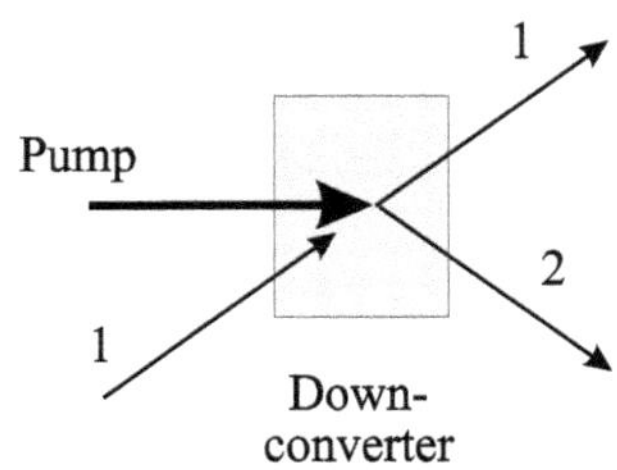

Fig. 20.2 Schematic for a photon cloning machine. The type II down-converter produces nondegenerate signal and idler modes with wavevectors $\mathbf{k}_1$ (mode 1) and $\mathbf{k}_2$ (mode 2). The photons are entangled in polarization.

The pump beam and the single photon to be injected into the crystal are both derived from a Ti–sapphire laser producing 120 fs pulses. The pump is created by frequency-doubling the laser beam, and the single-photon state is generated by splitting off a small part of the beam, which is then attenuated below the single-photon level.

With this method, there is still a small probability that two photons could be injected. If no down-conversion occurs, the transmitted two-photon state will appear as a false count for cloning. These false counts can be avoided by triggering the detectors for the $\mathbf{k}_1$-photons with the detection of the conjugate $\mathbf{k}_2$-photon, which is a signature of down-conversion.

To model this situation, we first pick a pair of orthogonal linear polarizations, $\mathbf{e}_h$ and $\mathbf{e}_v$, for each of the wavevectors. The production of polarization-entangled signal and idler modes is then described by the interaction Hamiltonian

$$H_{\mathrm{SS}} = \hbar\Omega_P \left(a^\dagger_{\mathbf{k}_1 v} a^\dagger_{\mathbf{k}_2 h} - a^\dagger_{\mathbf{k}_1 h} a^\dagger_{\mathbf{k}_2 v}\right) + \mathrm{HC}\,. \tag{20.33}$$

By following the hints in Exercise 20.7, one can show that this Hamiltonian is invariant under joint and identical rotations of the two polarization bases around their respective wavevectors.

The cloning effect is consequently independent of the polarization of the input photon; that is, this should be a universal QCM. It is therefore sufficient to consider a particular input state, say $|1_{\mathbf{k}_1 v}\rangle = a^\dagger_{\mathbf{k}_1 v}|0\rangle$, which evolves into $|\varphi(t)\rangle = \exp(-iH_{\mathrm{SS}}t/\hbar)|1_{\mathbf{k}_1 v}\rangle$. The relevant time t is limited by the pulse duration of the pump, which satisfies $\Omega_P t_P \ll 1$; therefore, the action of the evolution operator can be approximated by a Taylor series expansion of the exponential in powers of $\Omega_P t$:

$$\begin{aligned} |\varphi(t)\rangle &\approx \{1 - iH_{\mathrm{SS}}t/\hbar + \cdots\}|1_{\mathbf{k}_1 v}\rangle \\ &= |1_{\mathbf{k}_1 v}\rangle - i\Omega_P t\left\{\sqrt{2}\,|2_{\mathbf{k}_1 v}, 1_{\mathbf{k}_2 h}\rangle - |1_{\mathbf{k}_1 v}, 1_{\mathbf{k}_1 h}, 1_{\mathbf{k}_2 v}\rangle + \cdots\right\}. \end{aligned} \tag{20.34}$$

This result for $|\varphi(t)\rangle$ displays the probabilistic character of this QCM; the most likely outcome is that the injected photon passes through the crystal without producing a clone. The cloning effect occurs with a probability determined by the first-order term in the expansion. The factor $\sqrt{2}$ in the first part of this expression represents the enhancement due to stimulated emission.

According to von Neumann's projection rule, the detection of a trigger photon with wavevector $\mathbf{k}_2$ and either polarization leaves the system in the state

$$|\varphi(t)\rangle_{\mathrm{red}} = \frac{P_2|\varphi(t)\rangle}{\sqrt{\langle\varphi(t)|P_2|\varphi(t)\rangle}}\,, \tag{20.35}$$

where

$$P_2 = |1_{\mathbf{k}_2 v}\rangle\langle 1_{\mathbf{k}_2 v}| + |1_{\mathbf{k}_2 h}\rangle\langle 1_{\mathbf{k}_2 h}| \tag{20.36}$$

is the projection operator describing the reduction of the state associated with this measurement. Combining eqns (20.34) and (20.36) yields

$$|\varphi(t)\rangle_{\mathrm{red}} = -i\left\{\sqrt{\frac{2}{3}}\,|2_{\mathbf{k}_1 v}, 1_{\mathbf{k}_2 h}\rangle - \sqrt{\frac{1}{3}}\,|1_{\mathbf{k}_1 v}, 1_{\mathbf{k}_1 h}, 1_{\mathbf{k}_2 v}\rangle\right\}. \tag{20.37}$$

The probability of detecting two photons in the mode $\mathbf{k}_1 v$ is $2/3$ and the probability of detecting one photon in each of the modes $\mathbf{k}_1 v$ and $\mathbf{k}_1 h$ is $1/3$. The factor of two between the probabilities is also a consequence of stimulated emission.

The indistinguishability of the photons guarantees that the QCM is symmetric, but it also prevents the definition of reduced density operators like those in eqn (20.29). In this situation, the cloning fidelity can be defined as the probability that an output photon with wavevector $\mathbf{k}_1$ has the same polarization as the input photon. This happens with unit probability for the first term in $|\varphi(t)\rangle_{\text{red}}$ and with probability $1/2$ in the second term; therefore, the fidelity is

$$F = \left(\frac{2}{3}\right) \times (1) + \left(\frac{1}{3}\right) \times \left(\frac{1}{2}\right) = \frac{5}{6}\,. \tag{20.38}$$

The theoretical model for this QCM therefore predicts that it is universal, symmetric, and optimal.

In the experiment, the incident photon first passes through an adjustable optical delay line, which is used to control the time lapse ΔT between its arrival and that of the laser pulse that generates the down-converted photons. Stimulated emission should only occur when the photon wave packet and the pump pulse overlap. The results of the experiment, which are shown in Fig. 20.3, support this prediction.

The number of counts, $N(2,0)$, with two photons in the mode $\mathbf{k}_1 v$ and no photon in the mode $\mathbf{k}_1 h$ is shown in Fig. 20.3 as a function of the distance $c\Delta T$, for three different polarization states—curves (a)–(c)—of the injected photon. As expected, there is a pronounced peak at zero distance. The corresponding plots (d)–(f) of $N(1,1)$—the number of counts with one photon in each polarization mode—show no such effect.

The experimental fidelity can be derived from the ratio

$$R = \frac{N_{\text{peak}}(2,0)}{N_{\text{base}}(2,0)} \tag{20.39}$$

between the peak value and the base value of the $N(2,0)$ curve. At maximum overlap between the incident single-photon wave packet and the pump pulse ($c\Delta T = 0$), the probability of the $(2,0)$-configuration is

$$\mathcal{P}(2,0) = \frac{N_{\text{peak}}(2,0)}{N_{\text{peak}}(2,0) + N_{\text{peak}}(1,1)}\,, \tag{20.40}$$

which becomes

$$\mathcal{P}(2,0) = \frac{R}{R + N_{\text{peak}}(1,1)/N_{\text{base}}(2,0)} \tag{20.41}$$

when expressed in terms of R.

The base values $N_{\text{base}}(1,1)$ and $N_{\text{base}}(2,0)$ represent the situation in which there is no overlap between the single-photon wave packet and the pump pulse. In this case, the detection of the original photon and a down-converted photon in the spatial mode $\mathbf{k}_1$ are independent events. Down-conversion produces $\mathbf{k}_1 v$ and $\mathbf{k}_1 h$ photons with equal

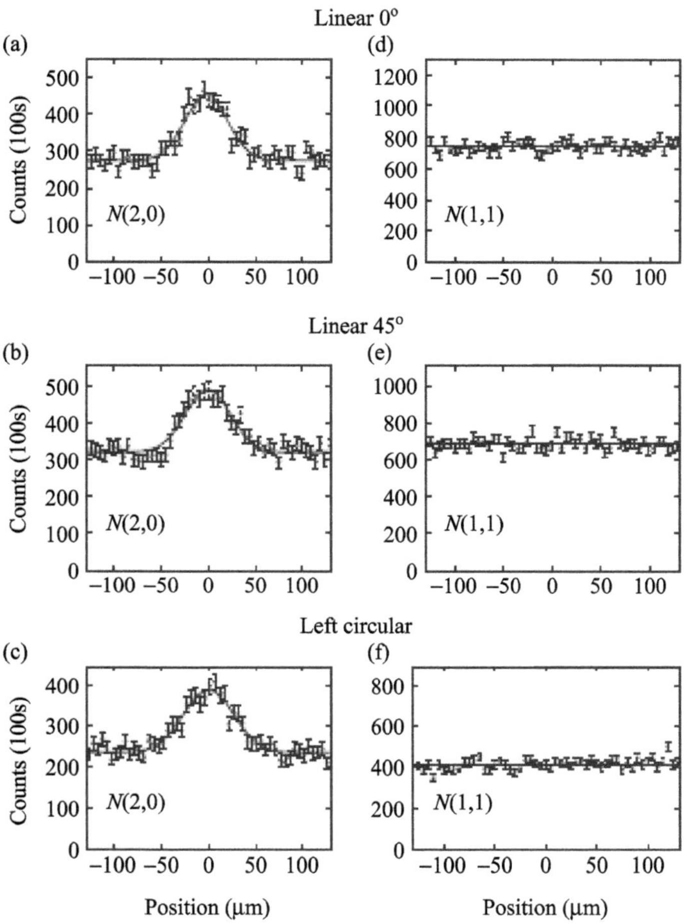

Fig. 20.3 Plots (a)–(c) show $N(2,0)$ as a function of $c\Delta T$ for linear at 0° (vertical), linear at 45°, and left circular polarizations respectively. Plots (d)–(f) show $N(1,1)$ for the same polarizations. (Reproduced from Lamas-Linares *et al.* (2002).)

probability; therefore, the probability that the polarizations of the two $\mathbf{k}_1$-photons are the same is 1/2. This implies that

$$N_{\text{base}}(1,1) = N_{\text{base}}(2,0)\,. \tag{20.42}$$

The apparent disagreement between eqn (20.42) and the data in the plot pairs (a) and (d), (b) and (e), and (c) and (f) is an artefact of the detection method used to count the (2, 0)-configurations; see Exercise 20.7.

The data show that $N_{\text{peak}}(1,1) = N_{\text{base}}(1,1)$; therefore

$$\mathcal{P}(2,0) = \frac{R}{R+1}\,, \tag{20.43}$$

and

$$\mathcal{P}(1,1) = \frac{1}{R+1}\,. \tag{20.44}$$

Applying the argument used to derive eqn (20.38) leads to

$$F = \left(\frac{R}{R+1}\right) \times 1 + \left(\frac{1}{R+1}\right) \times \frac{1}{2} = \frac{R+1/2}{R+1}\,. \tag{20.45}$$

The data yield essentially the same fidelity, $F = 0.81 \pm 0.01$, for all polarizations. This is close to the optimal value $F = 5/6 \simeq 0.833$; consequently, this QCM is very nearly universal and optimal.

20.3 Quantum cryptography

The history of **cryptography**—the art of secure communication through the use of secret writing or codes—can be traced back at least two thousand years (Singh, 1999), and the importance of this subject continues to increase. In current practice, the message is expressed as a string of binary digits M, and then combined with a second string, known as the *key*, by an algorithm or *cipher*. The critical issue is the possibility that the encrypted message could be read by an unauthorized person.

For most applications, it is sufficient to make this task so difficult that the message remains confidential for as long as the information has value. The commonly employed method of public key cryptography enforces this condition by requiring the solution of a computationally difficult problem, e.g. factoring a very large integer. This kind of encryption is not provably secure, since it is subject to attack by *cryptanalysis*, e.g. through the use of better factorization algorithms or faster computers.

In classical cryptography, the only provably secure method is the **one-time pad**, i.e. the key is only used once (Gisin *et al.*, 2002). In one version of this scheme, the key shared by Alice and Bob is a randomly generated number K which must have a binary representation at least as long as the message. Since the binary digits of K are random, the key itself contains no information. Alice encrypts her message as the signal $S = M \oplus K$, where $\oplus$ indicates bit-wise addition without carry, i.e. addition modulo 2. This means that corresponding bits are added according to the rules $0 + 0 = 0$, $0 + 1 = 1$, and $1 + 1 = 0$.

The bits of S are as random as those of K, so the signal carries no information for Eve, the lurking eavesdropper. On the other hand, Bob can decipher the message by bit-wise subtraction of K from S to recover M. The security of the messages is weakened by repeated use of the key. For example, if two messages $M1$ and $M2$ are sent, then the identity $K \oplus K = 0$ implies

$$\begin{aligned} S1 \oplus S2 &= M1 \oplus K \oplus M2 \oplus K \\ &= M1 \oplus M2 \oplus K \oplus K \\ &= M1 \oplus M2\,. \end{aligned} \tag{20.46}$$

The bits in $M1$ and $M2$ are not random; therefore, Eve gains some information about the messages themselves. With enough messages, the encryption system could be broken.

The use of a one-time pad solves the problem of secure communication, only to raise a new problem. How is the key itself to be safely transmitted through a potentially insecure channel? If Alice and Bob have to meet for this purpose, she might as well deliver the message itself.

One of the most intriguing discoveries in recent years (Wiesner, 1983; Bennett and Brassard, 1984, 1985) is that the peculiar features of quantum theory offer a solution to the problem of secure transmission of cryptographic keys. Once this is done, the message itself can be sent as a string of classical bits. Thus quantum cryptography really reduces to the secure transmission of keys, i.e. **quantum key distribution**.

A quantum method for distributing a key evidently involves encoding the key in the quantum states of some microscopic system. Since the electromagnetic field provides the most useful classical communication channel, it is natural to use a property of photons, e.g. polarization, to carry the information in a quantum channel.

As a concrete illustration, consider orthogonal linear polarizations $\mathbf{e}_h(\mathbf{k})$ and $\mathbf{e}_v(\mathbf{k})$ that define the basis of single-photon states:

$$\mathfrak{B} = \left\{ |h\rangle = a^{\dagger}_{\mathbf{k}h}|0\rangle, |v\rangle = a^{\dagger}_{\mathbf{k}v}|0\rangle \right\}. \tag{20.47}$$

One can then encode 0 as $|h\rangle$ and 1 as $|v\rangle$. We will see below that a scheme based on $\mathfrak{B}$ alone is too simple to foil Eve, so we add a second basis

$$\overline{\mathfrak{B}} = \left\{ |\overline{h}\rangle = a^{\dagger}_{\mathbf{k}\overline{h}}|0\rangle, |\overline{v}\rangle = a^{\dagger}_{\mathbf{k}\overline{v}}|0\rangle \right\}, \tag{20.48}$$

where the new polarization basis

$$\begin{aligned}\mathbf{e}_{\overline{h}}(\mathbf{k}) &= \frac{1}{\sqrt{2}}\left[\mathbf{e}_h(\mathbf{k}) + \mathbf{e}_v(\mathbf{k})\right], \\ \mathbf{e}_{\overline{v}}(\mathbf{k}) &= \frac{1}{\sqrt{2}}\left[\mathbf{e}_v(\mathbf{k}) - \mathbf{e}_h(\mathbf{k})\right]\end{aligned} \tag{20.49}$$

is the first polarization basis rotated through 45°. The creation operators and the single-photon basis states transform just like the polarization vectors. The corresponding encoding for $\overline{\mathfrak{B}}$ is: $0 \leftrightarrow |\overline{h}\rangle$ and $1 \leftrightarrow |\overline{v}\rangle$.

The two basis sets have the essential property that no member of one basis is orthogonal to either member of the other. The bases are also as different as possible, in the sense that $|\langle \overline{s}|s\rangle|^2 = 1/2$ for $\overline{s} = \overline{h}, \overline{v}$ and $s = h, v$. Pairs of bases related in this way are said to be **mutually unbiased**, and they are a feature of many quantum key distribution schemes.

20.3.1 The BB84 protocol

We now consider the **BB84 protocol**, named after Bennett and Brassard and the year they proposed the scheme (Bennett and Brassard, 1984). In the initial step, Alice sends a string of photons to Bob. For each photon, she uses a random number generator to choose a polarization from the four possibilities in $\mathfrak{B}$ and $\overline{\mathfrak{B}}$. At this stage, the only restriction is that Bob and Alice must be able to establish a one–one correspondence between the transmitted and received photons.

Bob, who is equipped with an independent random number generator, chooses one of the basis sets, $\mathfrak{B}$ or $\overline{\mathfrak{B}}$, in which to measure each incoming photon. If Alice sends $|h\rangle$ or $|v\rangle$ and Bob happens to choose $\mathfrak{B}$, his measurement will pick out the correct state, and his bit assignment will exactly match the one Alice sent. If, on the other hand, Bob chooses $\overline{\mathfrak{B}}$, then a measurement on $|h\rangle$ will yield $|\overline{h}\rangle$ or $|\overline{v}\rangle$ with equal probability. Thus if Alice sent 0, Bob will assign 1 half the time. Since Bob will make the wrong choice of basis about half the time, his average error rate will be 25%. The bit string resulting from this procedure is called the **raw key**.

An error rate of 25% would overwhelm any standard error correction scheme, but the BB84 protocol provides another option. For each bit, Bob announces—through the insecure public channel—his choice of measurement basis, but not the result of his measurement. Alice replies by stating whether or not the encoding basis and the measurement basis agree for that bit. If their bases agree, the bit is kept; otherwise, it is discarded. The remaining bit string, which is about half the length of the raw string, is called the **sifted key**.

The first experimental demonstration of this scheme was a table top experiment in which the signals from Alice to Bob were carried by faint pulses of light containing less than one photon on average (Bennett *et al.*, 1992). The distance between sender and receiver in this experiment was only 30 cm, but within a few years quantum key distribution was demonstrated (Muller *et al.*, 1995, 1996) over a distance of 23 km with signals carried by a commercial optical fiber network.

In order to understand the quantum basis for the security of the BB84 protocol, let us first imagine an alternative in which the bits are encoded in classical pulses of polarized light. If Eve intercepts a particular pulse, so that it does not arrive at Bob's detector, then Alice and Bob can agree to discard that bit from the string. This lowers the bit rate for transmitting the key, but Eve gains no information.

Thus it is not enough for Eve to detect the pulse; she must also make a copy for herself and send the original on to Bob. This tactic would provide information about the key without alerting Alice and Bob. In the classical case, this procedure is—at least in principle—always possible. For example, Eve could split off a small part of each pulse by means of a strongly unbalanced beam splitter, and record the polarization. The remaining pulse could then be amplified to match the original, and sent on to Bob.

Eve faces the same problem for the quantum BB84 protocol. She must make a copy of each single-photon state sent by Alice, and then send the original on to Bob. Furthermore, she must be able to do this for photons described by either of the bases $\mathfrak{B}$ or $\overline{\mathfrak{B}}$. Since the basis vectors in $\mathfrak{B}$ are not orthogonal to the basis vectors in $\overline{\mathfrak{B}}$, this is precisely what the no-cloning theorem says cannot be done.

Furthermore, when Eve intercepts a signal and sends a new signal on to Bob, she is bound—again according to the no-cloning theorem—to make a certain number of errors on average. If she carries out this strategy too often, Alice and Bob will become aware of her activity. According to this ideal description, the BB84 protocol is invulnerable to attack.

In practice—as one might expect—things are more complicated. Transmission of the key will be degraded by technical imperfections as well as Eve's machinations. It is also possible for Eve to gain some knowledge of the key by means of the imperfect cloning methods discussed in Section 20.2.2, without necessarily revealing her presence to Alice and Bob. The techniques for countering such attacks are primarily classical in nature (Gisin *et al.*, 2002), so we will not pursue them further.

Thus the no-cloning theorem—which was originally introduced as a purely negative statement about quantum theory—is the conceptual basis for the security of quantum key distribution protocols. In this connection, it is important to realize that the classical proof of the absolute security of the one-time pad depends on the assumption that the bits of K are truly random. For this reason, the choices made by Alice and Bob must be equally random.

This turns out to be a rather delicate issue. The standard random number generators for computers are deterministic programs of finite length; consequently, their output cannot be truly random. The ultimate security of BB84, or any other quantum key distribution protocol, therefore depends on generating a truly random sequence of numbers by some physical means. The behavior of a single photon at a beam splitter provides a natural way to satisfy this need. A single photon incident on an ideal balanced beam splitter with 100% detectors at each output port will—according to quantum theory—generate a perfectly random sequence of firings in the detectors. Associating 0 with one detector and 1 with the other defines a perfect coin flip.

As always, reality is more complicated; for example, the dead time of real detectors can impose a strong anti-correlation between successive bits. This effect limits the bit rate of quantum random number generation to a few megahertz (Gisin *et al.*, 2002). Leaving these practical issues aside, we see that the security of quantum key distribution is guaranteed by the perfectly random nature of individual quantum events. This is a historically unique situation; the security of quantum cryptography ultimately depends on the validity of quantum theory itself.

20.4 Entanglement as a quantum resource

The quantum effects on communications studied in the previous sections are primarily a source of difficulties. The use of phase-sensitive amplifiers to eliminate the quantum noise added by amplification, and the injection of squeezed light to minimize branching noise at an optical coupler are responses to these difficulties.

The role of the no-cloning theorem in providing a basis for the secure transmission of a cryptographic key is usually presented in a positive light, but this is a partisan view. For the frustrated Eve, the no-cloning theorem is still a negative result.

In these applications, quantum theory may provide new options, but it does not provide any new resources. For example, the qubits used by Alice and Bob in the key distribution protocol each carry only one classical bit, sometimes called a **cbit**.

It is the fundamental quantum property of entanglement that provides a novel communications resource. In the present section, we will consider two examples, quantum dense coding and quantum teleportation, which employ this resource. In both cases the ancilla is an entangled qubit pair provided by an external source, and Alice and Bob are each provided with one qubit of the pair. Local operations carried out by Alice and Bob on their respective qubits change the entangled state in a nonlocal way, and detection of these changes can be used to transfer information.

Before considering the specific applications, we must discuss some special features arising from the use of photons to carry the qubits. The abstract language used above implicitly assumes that the qubits are distinguishable quantum systems with definite locations. Since photons are indistinguishable bosons that cannot be precisely localized, there appears to be a conceptual problem.

The first point to note is that the indistinguishability of photons renders statements like 'Bob carries out a local operation on his photon' meaningless. The correct statement is 'Bob carries out a local operation on a photon.' This brings us to the second point: the word 'local' in 'local operation' applies to the hardware that realizes the theoretical manipulation, not to the photon.

We made this remark for detectors in Section 6.6.2, but it applies equally to retarder plates, beam splitters, etc. These classical devices—unlike photons—are both distinguishable and localizable. On the other hand, the physical operations they perform are represented by unitary operators that apply to the entire state of the electromagnetic field. By virtue of the peculiar properties of entangled states, this means that local operations can have nonlocal effects.

In the experiments we will discuss, the photons in the pair are ideally described by plane waves, with wavevectors $\mathbf{k}_A$ (directed toward Alice) and $\mathbf{k}_B$ (directed toward Bob), and equal frequencies, $\omega_A = \omega_B$. An example is shown in Fig. 20.4. The polarization-entangled, two-photon state emitted by the source is therefore a superposition of the states $|1_{\mathbf{k}_A s}, 1_{\mathbf{k}_B s'}\rangle$, where $s, s' = h, v$.

We will only consider situations with fixed directions for the wavevectors, so the shorthand notation

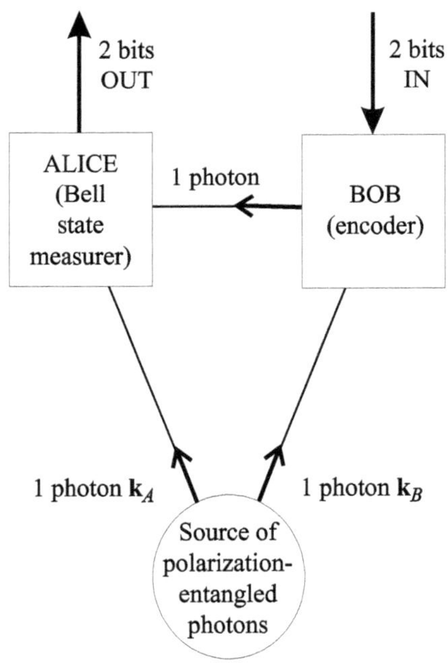

Fig. 20.4 Quantum dense coding: a source of polarization-entangled photons provides a communications resource. Bob's local operations on a photon alter the nonlocal entangled state, so that a single photon sent from Bob to Alice allows her to receive two bits of information.

$$\begin{aligned} |s_\gamma, s'_{\gamma'}\rangle &\equiv a^\dagger_{\mathbf{k}_\gamma s} a^\dagger_{\mathbf{k}_{\gamma'} s'} |0\rangle \text{ for } (\gamma, s) \neq (\gamma', s'), \\ |s_\gamma, s_\gamma\rangle &\equiv \frac{1}{\sqrt{2}} a^{\dagger 2}_{\mathbf{k}_\gamma s} |0\rangle, \end{aligned} \tag{20.50}$$

with $\gamma, \gamma' \in \{A, B\}$, $s, s' \in \{h, v\}$, is adequate.

A third point related to local operations is that these plane waves are idealizations of Gaussian wave packets with finite transverse widths. This means that the realistic $\mathbf{k}_A$-mode is effectively zero at Bob's location, and the $\mathbf{k}_B$-mode is effectively zero at Alice's location. The mathematical consequence is that Bob's local manipulations are represented by unitary operators that only act on the $\mathbf{k}_B$-mode, i.e. on the second argument of the two-photon state $|s_A, s'_B\rangle$. By the same token, Alice's operations only act on the first argument. This is formally similar to performing operations on distinguishable qubits, but we emphasize that it is the modes that are distinguishable, not the photons.

20.4.1 Quantum dense coding

The common currency for classical digital communication and computation is the *bit*, i.e. the binary digits 0 and 1, which are physically represented by classical two-state systems. For storage, e.g. in a magnetic storage device, 0 and 1 can be respectively represented by a spin-down state (a downwards-pointing net magnetization), and a spin-up state of a magnetic resolution element. For transmission, 0 and 1 are typically represented by two resolvable voltages $\mathcal{V}_0$ and $\mathcal{V}_1$.

In either case, the two states of a macroscopic system encode the binary choice between 0 and 1; that is, one bit of information is carried by a classical two-state system. Conversely, the one-to-one relation between the two states of the classical system and the two logical states 0 and 1 assures us that a classical, two-state system can carry at most one bit of information.

For a two-state quantum system the outcome is quite different. A surprising result of quantum theory is that two bits of information can be transmitted by sending a single qubit. This apparent doubling of the transmission rate is called **quantum dense coding**.

A *A generic model for quantum dense coding*

A thought experiment (Bennett and Wiesner, 1992) to implement quantum dense coding is sketched in Fig. 20.4. In this scenario, Bob has received two bits of classical information through his input port IN, and he wants to communicate this news to Alice. Since there are four possible two-bit messages, an encoding scheme with four alternatives is needed. The resource Bob will use is the pair of entangled qubits provided by the source.

Bob can carry out local operations to change the original two-qubit state into any one of the four Bell states, chosen according to a prearranged mapping of the four possible messages onto the four Bell states. Once this is done, Bob sends the qubit in his apparatus to Alice, so that she has the entire entangled state at her disposal. Alice then performs a **Bell state measurement**, i.e. an observation that determines which of the four Bell states describes the two-qubit state. By means of this measurement Alice acquires the two bits of information sent by Bob.

The fact that Alice obtains the message after receiving the qubit sent by Bob suggests that the two classical bits were somehow packed into this single qubit. This is an essentially classical point of view that does not really fit the present case. Alice receives two qubits, one from the original source of the entangled state and one sent by Bob. The qubit from the original source may well have been sent long before Bob's actions, so it seems eminently reasonable to assume that it carries no information.

On the other hand, Bob's qubit by itself also carries no information. For example, if the ever resourceful Eve manages to intercept Bob's qubit, she will learn absolutely nothing. Furthermore, if Alice's qubit from the source does not arrive, then she also will learn nothing from receiving Bob's qubit. This should make it clear that the information is carried, nonlocally, by the entangled state itself.

The real advantage of this scheme is that Bob can send two bits with a single operation. This is twice the rate possible for a classical channel; consequently, quantum dense coding might better be called *quantum rapid coding.*

B Quantum dense coding with photons

In an experimental demonstration of quantum dense coding (Mattle *et al.*, 1996), a polarization-entangled, two-photon state is generated by means of down-conversion in a type II crystal, as shown for example in Fig. 13.5. The two down-converted photons have the same frequency, but different propagation directions, selected by means of irises. The source is adjusted so that it emits the state

$$|\Theta\rangle = \frac{i}{\sqrt{2}}|h_A, v_B\rangle + \frac{1}{\sqrt{2}}|v_A, h_B\rangle . \tag{20.51}$$

Bob allows the input photon in the $\mathbf{k}_B$-mode to pass successively through a half-wave and a quarter-wave retarder. These devices are reviewed in Exercise 20.8. The experimentally adjustable parameter for each retarder is the angle ϑ between the fast axis and the horizontal polarization vector $\mathbf{e}_h$. The unitary operations needed to generate the four Bell states,

$$|\Phi^{\pm}\rangle \equiv \frac{1}{\sqrt{2}}|h_A, h_B\rangle \pm \frac{1}{\sqrt{2}}|v_A, v_B\rangle , \tag{20.52}$$

$$|\Psi^{\pm}\rangle \equiv \frac{1}{\sqrt{2}}|h_A, v_B\rangle \pm \frac{1}{\sqrt{2}}|v_A, h_B\rangle , \tag{20.53}$$

correspond to different settings of the retarder angles, $\vartheta_{\lambda/2}$ and $\vartheta_{\lambda/4}$.

The source of entangled pairs has been arranged so that the emitted state $|\Theta\rangle$ scatters into the Bell state $|\Psi^{+}\rangle$, for the settings $\vartheta_{\lambda/2} = \vartheta_{\lambda/4} = 0$. Using the operations discussed in Exercise 20.9, Bob encodes his two bits by choosing the two angles $\vartheta_{\lambda/2}$ and $\vartheta_{\lambda/4}$, and then sends the photon to Alice. Bob's local operations have changed the entangled state, but Alice can only detect these changes by a Bell state measurement that requires both photons.

This means that Alice cannot begin to decode the message before she receives the photon sent by Bob, as well as the photon from the source. In common with all other communication schemes, the time required for transmission of information by quantum dense coding is restricted by the speed of light.

The next step is for Alice to decode the message, which turns out to be quite a bit more difficult than encoding it. Linear optical techniques are constrained by a *no-go* theorem, which states that the four Bell states cannot be distinguished with a probability greater than 50% (Calsamiglia and Lutkenhaus, 2001). Indeed, the Bell state analysis used in the particular experiment discussed above could not distinguish between the states $|\Phi^+\rangle$ and $|\Phi^-\rangle$.

However, for entangled photon pairs produced by down-conversion, there is a way around this prohibition. The proof of the no-go theorem involves the assumption that the Bell states are not entangled in any degrees of freedom other than the polarization; consequently, the no-go theorem can be circumvented by the use of hyperentangled states (Kwiat and Weinfurter, 1998). The example discussed in Section 13.3.5—in which the photons are entangled in both polarization and momentum—is one candidate.

An alternative, and experimentally easier, scheme exploits the fact that down-conversion automatically produces photon pairs that are entangled in both energy and polarization. As we have seen in Section 13.3.2-B, energy entanglement implies that the two photons are produced at essentially the same time.

This feature is the basis for a complete Bell state analysis. In addition to its intrinsic interest, this scheme illustrates the application of various theoretical and experimental techniques; therefore, we will discuss it in some detail. A schematic diagram illustrating the idea for this measurement is shown in Fig. 20.5.

As one can see from Exercise 20.10, the Bell state $|\Psi^-\rangle$ has the curious property that it is unchanged by scattering from a balanced beam splitter, i.e. $|\Psi^-\rangle' = |\Psi^-\rangle$. This implies that the photons exhibit **anti-pairing**, i.e. one photon exits through each of the two output ports. The other Bell states display the opposite behavior; whenever $|\Psi^+\rangle$ or $|\Phi^\pm\rangle$ are incident, the photons are paired, as discussed in Section 10.2.1. In other words, both scattered photons are emitted through one or the other of the two output ports.

This difference allows $|\Psi^-\rangle$ to be distinguished from the remaining Bell states: when $|\Psi^-\rangle$ is incident, detectors in the A and B arms of the apparatus will both fire so that a coincidence count is registered. For the other Bell states, only the detectors in one arm will fire, so there will be no coincidence counts between the two arms. This effect only depends on the behavior at the beam splitter, so it would work even if the photons were not hyperentangled.

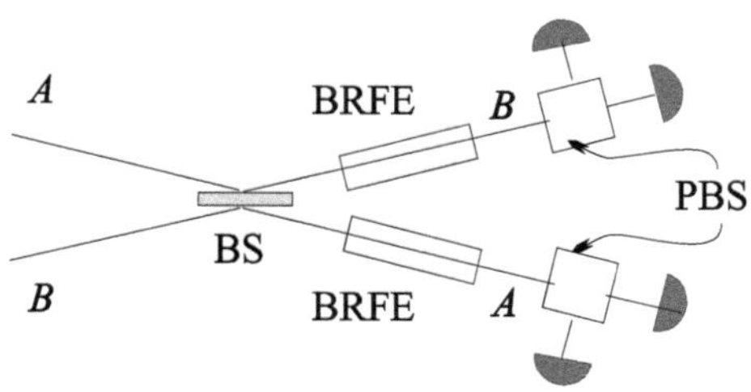

Fig. 20.5 Schematic of an experiment for a complete Bell state analysis using hyperentangled photons. (1) The beam splitter (BS) identifies $|\Psi^-\rangle$. (2) The birefringent elements (BRFEs) identify $|\Psi^+\rangle$. (3) The polarizing beam splitters (PBSs) distinguish $|\Phi^-\rangle$ from $|\Phi^+\rangle$. (Adapted from Kwiat and Weinfurter (1998).)

Next we turn to the task of distinguishing $|\Psi^{+}\rangle$ from $|\Phi^{\pm}\rangle$. This is accomplished by means of the two birefringent elements, which have optic axes aligned along the h- and v-polarizations. The two down-converted photons are emitted simultaneously in matched wave packets with widths of the order of 15 fs, but the h- and v-components experience different group velocities due to the difference between the indices of refraction for the two polarizations.

The resulting separation between the two wave packets means that the detections of the two photons will also be separated in time. In principle, it is only necessary to separate the two packets by an amount greater than their widths, but in practice the delay must be larger than the resolution time—of the order of 1 ns—of the detectors. The detection events for $|\Phi^{\pm}\rangle$ are expected to be simultaneous, since $|\Phi^{\pm}\rangle$ is a superposition of states with pairs of photons having the same polarization.

The final task of separating $|\Phi^{+}\rangle$ and $|\Phi^{-}\rangle$ begins with the action of the beam splitter:

$$|\Phi^{\pm}\rangle \to |\Phi^{\pm}\rangle' = \frac{i}{2}\left\{|h_A, h_A\rangle \pm |v_A, v_A\rangle\right\} + (A \leftrightarrow B)\,. \tag{20.54}$$

Applying eqn (8.2) to each polarization produces the scattering matrix for a birefringent element of length L:

$$S_{\mathbf{k}s,\mathbf{k}'s'} = e^{i\phi_s}\delta_{\mathbf{k}\mathbf{k}'}\delta_{ss'}\,, \tag{20.55}$$

where $\phi_s = n_s(\omega)\,L/c$ is the phase shift for the s-polarization. Propagation through the birefringent elements therefore produces

$$|\Phi^{\pm}\rangle'' = \frac{ie^{2i\phi_0}}{2}\left\{e^{i\delta}|h_A, h_A\rangle \pm e^{-i\delta}|v_A, v_A\rangle\right\} + (A \leftrightarrow B)\,, \tag{20.56}$$

where $\phi_0 = (\phi_h + \phi_v)/2$, and $\delta = \phi_h - \phi_v$.

For both $|\Phi^{+}\rangle''$ and $|\Phi^{-}\rangle''$ two photons will strike a single detector, so the two states are still not distinguished. The last trick is to send the light into a polarizing beam splitter oriented along the 45°-rotated basis $\overline{\mathfrak{B}}$ defined in eqn (20.48). In Exercise 20.11, it is shown that expressing $|\Phi^{\pm}\rangle''$ in the new basis yields

$$|\Phi^{+}\rangle'' = \frac{i}{2}e^{2i\phi_0}\left\{\cos\delta\;\left[|\overline{h}_A, \overline{h}_A\rangle + |\overline{v}_A, \overline{v}_A\rangle\right] - \sqrt{2}i\sin\delta\;|\overline{h}_A, \overline{v}_A\rangle\right\} + (A \leftrightarrow B)\,, \tag{20.57}$$

$$|\Phi^{-}\rangle'' = \frac{i}{2}e^{2i\phi_0}\left\{i\sin\delta\;\left[|\overline{h}_A, \overline{h}_A\rangle + |\overline{v}_A, \overline{v}_A\rangle\right] - \sqrt{2}\cos\delta\;|\overline{h}_A, \overline{v}_A\rangle\right\} + (A \leftrightarrow B)\,. \tag{20.58}$$

Coincidence counts between the detectors at the output ports of the PBS will arise from $|\overline{h}_A, \overline{v}_A\rangle$, but not from $|\overline{h}_A, \overline{h}_A\rangle$ and $|\overline{v}_A, \overline{v}_A\rangle$. Since the coefficients depend on the phase difference δ, the two outcomes—coincidence counts or counts in one detector only—can be separated by choosing δ to achieve destructive interference for one of the terms. For example, adjusting L so that

$$\delta = \frac{(n_h - n_v)\,\omega}{c}L = n\pi \tag{20.59}$$

leads to the greatly simplified states

$$|\Phi^+\rangle'' = \frac{i}{2}e^{2i\phi_0}(-)^n \left[|\overline{h}_A, \overline{h}_A\rangle + |\overline{v}_A, \overline{v}_A\rangle\right] + (A \leftrightarrow B) \tag{20.60}$$

and

$$|\Phi^-\rangle'' = \frac{i}{\sqrt{2}}e^{2i\phi_0}(-)^{n+1} |\overline{h}_A, \overline{v}_A\rangle + (A \leftrightarrow B)\,. \tag{20.61}$$

In this case $|\Phi^-\rangle''$ produces coincidence counts between the h- and v-counters, while $|\Phi^+\rangle''$ leads to two-photon counts in one or the other of the detectors.

The procedure outlined above constitutes a complete Bell measurement, but the two photons must be hyperentangled. This Bell state analysis also makes substantial demands on the photon counters. A demonstration experiment based on this scheme has recently been carried out (Schuck *et al.*, 2006). The result was that the four Bell states could be identified with a probability in the range of 81%–89%. This is already substantially greater than the 50% bound imposed by the no-go theorem for linear optics, and further improvements of the experimental technique are to be expected.

20.4.2 Quantum teleportation

In quantum dense coding, the apparently arcane and counterintuitive property of entanglement is precisely what allows Bob to transmit two classical bits of information by means of local operations carried out on a single qubit. We next consider an even more remarkable demonstration of the power of entanglement. In this scenario, Alice has received a qubit in an unknown state $|\gamma\rangle_T \in \mathfrak{H}_T$—where $\mathfrak{H}_T$ is the internal state space of the qubit—and she wants to transmit this quantum information to Bob by sending him two classical bits. This is the inverse of the quantum dense coding problem, and the method used to accomplish this magic feat is called **quantum teleportation** (Bennett *et al.*, 1993).

If Alice were sent an unknown classical signal, she could simply make a copy and send it to Bob, but the no-cloning theorem prohibits this action for an unknown quantum signal. What, then, is Alice to do in the quantum case? Let us first consider what can be done without the aid of any ancilla. In this situation, the only available option is to measure the value of some observable $O_T = \mathbf{n} \cdot \boldsymbol{\sigma}_T$, where $\mathbf{n}$ is a unit vector. Alice can measure O_T and then tell Bob the components of $\mathbf{n}$ and the result, $\epsilon\,(= \pm 1)$, of the measurement.

Bob's task is to generate an approximation to the unknown state by using this information. The only thing Bob knows is that the state $|\gamma\rangle_T$ has a nonvanishing projection on the eigenstate $|\epsilon\rangle_T$ of O_T, so the best he can do is to prepare a qubit in the mixed state

$$\rho = (1 + \epsilon\mathbf{n} \cdot \boldsymbol{\sigma}_B)\,/2\,; \tag{20.62}$$

see Ralph (2006) and Exercise 20.12. Under these circumstances, the average fidelity is 2/3. Since the attempt to send classical instructions for replicating $|\gamma\rangle_T$ does not seem to be very promising, we next turn to the situation shown in Fig. 20.6, in which Alice and Bob are supplied with an ancilla.

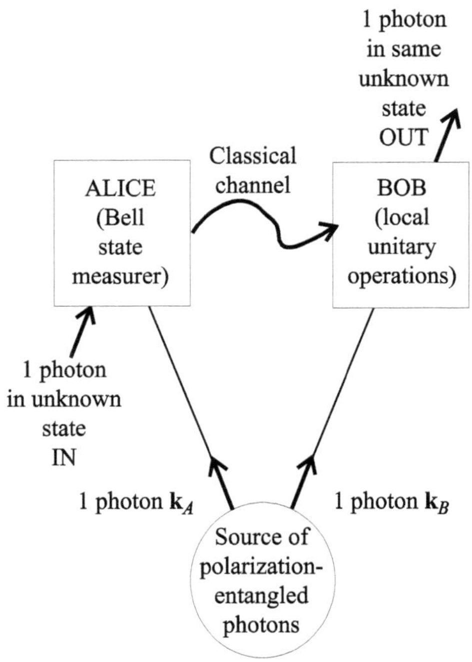

Fig. 20.6 Schematic for quantum teleportation, in which an unknown polarization state of a photon entering Alice's IN port is teleported to become the same unknown polarization state for the photon leaving Bob's OUT port.

A A generic teleportation model

In order to emphasize that the remarkable results of the following discussion apply to all quantum systems, not just to photons, we will use the generic computational basis defined in eqn (20.13). In this notation, one example of an ancilla is provided by the Bell state $|\Psi^-\rangle_{AB}$ defined in eqn (20.15).

The complete three-particle system is described by the state

$$\begin{aligned}|\Theta\rangle_{ABT} &= |\Psi^-\rangle_{AB}\,|\gamma\rangle_T \\ &= \frac{1}{\sqrt{2}}\left[|0\rangle_A\,|1\rangle_B\,|\gamma\rangle_T - |1\rangle_A\,|0\rangle_B\,|\gamma\rangle_T\right]. \end{aligned} \tag{20.63}$$

The order of the Hilbert-space vectors in the tensor product has no physical significance, so the three-particle state is equally well represented by

$$|\Theta\rangle_{ATB} = \frac{1}{\sqrt{2}}\,|0\rangle_A\,|\gamma\rangle_T\,|1\rangle_B - \frac{1}{\sqrt{2}}\,|1\rangle_A\,|\gamma\rangle_T\,|0\rangle_B \in \mathfrak{H}_A\otimes\mathfrak{H}_T\otimes\mathfrak{H}_B\,. \tag{20.64}$$

The tensor products $|u\rangle_A\,|\gamma\rangle_T$ $(u = 0, 1)$ are given by

$$|u\rangle_A\,|\gamma\rangle_T = \gamma_0\,|u, 0\rangle_{AT} + \gamma_1\,|u, 1\rangle_{AT}\,, \tag{20.65}$$

and the vectors $|u, v\rangle_{AT}$ are linear combinations of the Bell states spanning $\mathfrak{H}_A\otimes\mathfrak{H}_T$; consequently—as one can show in Exercise 20.13—eqn (20.64) can be rewritten as

$$\begin{aligned}|\Theta\rangle_{ATB} = & \frac{1}{2}\left|\Psi^-\right\rangle_{AT}\{\gamma_0|0\rangle_B + \gamma_1|1\rangle_B\} \\ & + \frac{1}{2}\left|\Psi^+\right\rangle_{AT}\{-\gamma_0|0\rangle_B + \gamma_1|1\rangle_B\} \\ & + \frac{1}{2}\left|\Phi^-\right\rangle_{AT}\{\gamma_1|0\rangle_B + \gamma_0|1\rangle_B\} \\ & + \frac{1}{2}\left|\Phi^+\right\rangle_{AT}\{-\gamma_1|0\rangle_B + \gamma_0|1\rangle_B\}.\end{aligned} \tag{20.66}$$

Having mastered this theory, Alice now performs a Bell measurement on her two qubits. According to von Neumann, the result will be to project $|\Theta\rangle_{ATB}$ onto one of the four Bell states of $\mathfrak{H}_A \otimes \mathfrak{H}_T$. Alice then sends Bob a message—of length two bits—informing him which of the four possible outcomes actually occurred.

Bob, who has also learnt the theory, then knows that his qubit is in one of the states shown in the four lines of eqn (20.66). For example, if Alice found $\left|\Psi^-\right\rangle_{AT}$, then Bob knows that his qubit is guaranteed to be in the original unknown state $|\gamma\rangle_T$. The other three states are related to the original state in one of three ways: (1) a **phase-flip** (changing the relative phase of $|0\rangle_B$ and $|1\rangle_B$ by 180°); (2) a **bit-flip** (interchanging $|0\rangle_B$ and $|1\rangle_B$); and (3) a combined phase- and bit-flip. In each of these cases, there is a unitary operator—U_{pf} for the phase-flip, U_{bf} for the bit-flip, and $U_{\text{pf}}U_{\text{bf}}$ for the combination—that transforms the corresponding state into the state $|\gamma\rangle_B$.

In an optical experiment, the unitary operators are realized by appropriate combinations of beam splitters and phase shifters (Reck *et al.*, 1994). By sending the photon in his apparatus through the optical elements corresponding to the appropriate unitary transformation, Bob can be sure that the qubit emitted from his OUT port is an exact replica of the qubit given to Alice.

In this process, the only physical objects transferred from Alice to Bob are the carriers of the two bits delivered through the classical channel. Consequently, the teleportation process is limited by the speed of light, and it does not violate any conservation laws.

This result raises several puzzles. The first is: What happened to the no-cloning theorem? After all, we have just claimed that the procedure ends with Bob in possession of a perfect copy of the qubit sent to Alice. The answer is that the original qubit no longer exists, so that the no-cloning theorem is not violated.

For any outcome of Alice's Bell state measurement, the T-qubit is described by the corresponding Bell state of $\mathfrak{H}_A \otimes \mathfrak{H}_T$; no information about the original state $|\gamma\rangle_T$ is left in the A–T subsystem. In fact, any attempt on Alice's part to find out something about $|\gamma\rangle_T$, before performing the Bell state measurement, would frustrate the teleportation process. This is analogous to the destruction of the interference pattern by any attempt to determine which pinhole a photon passes through in a Young's-type experiment. This leads to the very strange conclusion that neither Alice nor Bob has any information about the mystery qubit $|\gamma\rangle_T$, despite the fact that Bob can be certain that he has a perfect copy.

An equally puzzling issue is the apparent discrepancy between the amount of information that is needed to specify $|\gamma\rangle_T$ and the two bits actually sent by Alice. To see this explicitly, let us write

$$|\gamma\rangle_T = \cos\left(\frac{\theta_T}{2}\right)|0\rangle_T + e^{i\phi_T}\sin\left(\frac{\theta_T}{2}\right)|1\rangle_T\,, \tag{20.67}$$

so that the state is represented by the point (θ_T, ϕ_T) on the Poincaré sphere. Precisely specifying this point would require an infinite number of bits, and even a crude approximation would require many more than two bits. Thus it would seem that Alice is getting an infinite return on her two bit investment.

The key to understanding this situation is that quantum results require careful interpretation. In the present instance, the apparently infinite information carried by $|\gamma\rangle_T$ is only potentially available. Measuring an observable $O_B = \mathbf{n}\cdot\boldsymbol{\sigma}_B$ will provide exactly one bit of information: the binary choice between the eigenvalues $+1$ and -1. This is, nevertheless, an amazing result. A potentially infinite number of bits have been delivered by combining the entanglement resource with just two classical bits of information.

Finally, there is a conceptual issue arising from the use of the word 'teleportation'. The question is: What has actually been transported? For this discussion, it is better to replace the abstract formulation used above by a concrete example. Suppose that the mystery qubit $|\gamma\rangle_T$ is a superposition of the states of a two-level atom, and that the ancilla is an entangled state of a photon (sent to Alice) and an electron (sent to Bob).

At the end of the process, Bob's particle is described by the same superposition as the one supplied to Alice, but the physical substrate is the two spin states of the electron, not another two-level atom. For this example, one could argue that the term *quantum faxing* might be more appropriate. It is true that quantum faxing—unlike classical faxing—requires the destruction of the original information, but that is simply the price that must be paid for working in the quantum domain.

A sceptically inclined onlooker might conclude that 'teleportation' is simply another example of the irrationally exuberant terminology sometimes found in the field of quantum information, but this would not be quite fair. Let us now consider a different example in which all three particles are photons. In this case, the photon in Bob's possession at the end is physically indistinguishable—at the most fundamental level—from the original photon supplied to Alice; consequently, using the evocative term 'teleportation' seems entirely reasonable.

B *Teleportation of photons*

Since this is a book on quantum optics, we will now concentrate on the three-photon case. The only formal change in the theory is that the tensor products of states used above are replaced by products of creation operators acting on the vacuum. Thus the initial three-photon state is

$$|\Theta\rangle_{ABT} = a_T^\dagger[\gamma]\,|\Psi^-\rangle_{AB}\,, \tag{20.68}$$

where $a_T^\dagger[\gamma] = \gamma_h a_{Th}^\dagger + \gamma_v a_{Tv}^\dagger$ creates the unknown photon state in the T-channel, and the ancilla shared by Alice and Bob is given by the Bell state

$$|\Psi^-\rangle_{AB} = \frac{1}{\sqrt{2}}\{|h_A, v_B\rangle - |v_A, h_B\rangle\}$$

$$= \frac{1}{\sqrt{2}} \left\{ a^{\dagger}_{\mathbf{k}_A h} a^{\dagger}_{\mathbf{k}_B v} - a^{\dagger}_{\mathbf{k}_A v} a^{\dagger}_{\mathbf{k}_B h} \right\} |0\rangle \, . \tag{20.69}$$

The tensor product algebra used in the generic discussion is exactly mirrored by algebraic manipulations of the products of creation operators, so the theoretical argument, as seen in Exercise 20.14, goes through as before.

The first laboratory demonstration of quantum teleportation for photons was carried out by Bouwmeester *et al.* (1997). In this experiment a pulse of UV light produces the ancillary photons in the A- and B-channels by down-conversion. The pulse is then retroreflected to pass through the nonlinear crystal again, and thus produce another pair of photons in the T- and T'-channels. The T-channel photon is prepared in the polarization state $\gamma = (\gamma_h, \gamma_v)$, and detection of the T' photon signals that the mystery photon is on the way.

In this proof-of-principle experiment the full Bell state analysis was replaced by a simpler procedure in which the A–T pair is allowed to fall on the two input ports of a beam splitter. The experimental arrangement can be extracted from Fig. 20.5 by changing B to T and omitting the birefringent elements and the polarizing beam splitters.

The necessary two-photon interference effects at the beam splitter will only occur if the two wave packets overlap. In other words, it must not be possible to distinguish the A- and T-wave packets by their arrival times. For this purpose, both photons were sent through frequency filters that narrowed their frequency spread and therefore broadened their temporal spread. Of course, the filters also cut down substantially on the count rate, but this sort of trade-off is a common feature of optical experiments.

As we have already seen, coincidence counts in the detectors in the A and B arms of the apparatus signal that the Bell state $|\Psi^-\rangle_{AT}$ has been detected. Alice relays this information to Bob, who then knows that the photon in the B-channel is in the same polarization state as the photon that was sent to Alice. This will happen only one time out of four, so the success rate for teleportation is less than 25%. In a later version of this experiment (Pan *et al.*, 2003) fidelity in the successful cases exceeded 80%.

It should now be clear that Alice's Bell state measurement poses substantial experimental difficulties. In Section 20.4.1-B we presented a complete Bell state analysis due to Kwiat and Weinfurter (1998), but their method avoids the no-go theorem by relying on the hyperentanglement of down-converted photon pairs.

In a teleportation experiment, the photon state to be teleported and the two ancilla photons are generated by independent sources; consequently, the photon in the T-channel is only entangled with the ancilla photons in the A- and B-channels to the minimal extent required by Bose statistics. Thus the no-go theorem limits any linear optical scheme for discriminating between the photonic Bell states $\{|\Psi^{\pm}\rangle_{AT}, |\Phi^{\pm}\rangle_{AT}\}$ in a teleportation experiment to a 50% success rate.

This limitation on the success rate does not, however, mean that only one Bell state can be detected. A three-Bell-state analyzer (van Houwelingen *et al.*, 2006)—employing only linear optics and no additional ancillary photons—and a four-Bell-state analyzer (Walther and Zeilinger, 2005)—depending on additional ancillary photons—have both been experimentally demonstrated.

The obstacles presented by the no-go theorem for linear optics suggest exploiting nonlinear optical effects. An experiment of this kind has been performed (Kim *et al.*, 2001) by using sum-frequency generation (SFG)—the inverse of down-conversion—in type-I and type-II crystals. This technique permits a full Bell state analysis, but the efficiency is strongly limited by the weakness of the SFG effect and the necessity of ensuring a good overlap between the spatial modes. The observed fidelity of $F = 0.83$ is a convincing demonstration of quantum teleportation, but the low count rate means that this method is not yet useful for quantum communication protocols.

20.5 Quantum computing

The first proposals for quantum computing were independently made in 1982 by Benioff (1982) and Feynman (1982). Benioff presented a quantum version of a Turing machine that would operate without dissipation of energy, while Feynman was interested in the possible use of a quantum computer to simulate the behavior of other quantum systems.

These papers excited a substantial amount of interest at the time, but the rapid growth in this field was first stimulated by the work of Deutsch and Jozsa (1992), Grover (1997), and Shor (1997).

Deutsch and Jozsa demonstrated a quantum algorithm for a certain decision problem that is guaranteed to be exponentially faster than any classical algorithm.

Grover showed that a quantum computer could search a database of length N in a time—i.e. a number of steps—proportional to $\sqrt{N}$. The optimum time for a classical search strategy is proportional to N, so Grover's work constitutes a rigorous demonstration of a problem of practical interest for which a quantum computer is superior to any classical computer.

Shor's work concerned the problem of finding the prime factors of an integer N. The most efficient known classical algorithm, the **number field sieve**, requires a time $t \sim \exp\left[2\left(\ln N\right)^{1/3}\left(\ln\ln N\right)^{2/3}\right]$ to find the factors. This time grows faster than any power of $\ln N$, and it is firmly believed—but not proven—that all classical factorization algorithms share this property. Shor demonstrated a quantum algorithm with a factorization time $t \sim (\ln N)^3$, i.e. it is only polynomial in $\ln N$. The appearance of a quantum computer would therefore be very bad news for those using trapdoor codes that depend on the difficulty of factoring large integers.

The Grover and Shor algorithms are quite complicated, and in any case are beyond the purview of this book. For the general topic of quantum computing, we will restrict ourselves to a very brief discussion of the prevailing generic model. More detailed descriptions can be found in several texts, e.g. Nielsen and Chuang (2000). This introduction will be followed by a brief discussion of a proposed all-optical scheme. For topics like this that are the subject of current investigations the best strategy is to consult recent review articles, e.g. Ralph (2006).

20.5.1 A generic model for quantum computers

Feynman's original proposal was motivated by the extreme computational demands of quantum theory. Consider, for example, a very simple classical system composed of N bits. In this case there are 2^N possible states, each labeled by an N-digit binary number.

By contrast, the states of a quantum system consisting of N qubits occupy a Hilbert space of dimension 2^N. The number of basis vectors is the same as the number of classical states, but the superposition principle requires the inclusion of all possible linear combinations of the basis vectors.

As we have seen in Section 18.7.2, the density matrix for this system has $O\left(2^{2N}\right)$ elements. For a system of modest size, e.g. $N = 100$, the dimension of the quantum state space is $O\left(10^{30}\right)$. Simulating this system on a classical computer is possible in principle, but the memory and running time needed make it impossible in practice. This prompted Feynman to consider replacing the classical computer by a quantum computer.

Generally speaking, a *quantum computer* is any device that employs specifically quantum effects, such as entanglement, to accomplish a computational task. The standard conceptual model currently in use includes a collection of N qubits called a **quantum register**, which is initially in some state $|\Lambda_{\text{in}}\rangle$, and a unitary transformation U_{alg} that implements the algorithm.

Since unitary transformations are invertible, this scheme represents a **reversible quantum computer**. The unitary transformation is expressed as the product of a set of standard transformations, called **quantum gates**, that operate on a few qubits at a time. The result of the computation is read out by performing measurements on some or all of the qubits. The corresponding theoretical operation is the projection of the output state $U_{\text{alg}}|\Lambda_{\text{in}}\rangle$ onto the basis vector describing the measurement outcome.

A Quantum parallelism

The procedure outlined above has two crucial features related to the unitary transformation and the measurement step respectively. The unitary transformation is invertible, so it preserves the enormous amount of information in the state vector. This property, which is called **quantum parallelism**, offers the possibility of converting the high dimension of the Hilbert space from a difficulty into an advantage.

The measurement step renders the outcome probabilistic; there is no way of predicting which of the possible measurement outcomes will occur. Running the algorithm twice will in general produce different results. Furthermore, the reduction of the state vector accompanying the measurement destroys all the information associated with the measurement outcomes that did not occur.

Successful quantum algorithms—such as those of Grover and Shor—are cleverly contrived to achieve good results in spite of the evident tension between the unitary algorithm and the reductive measurement. For example, Shor's algorithm does not always result in factorization, but it does succeed with high probability.

A simple example illustrating quantum parallelism is provided by the following toy problem which employs a variant of the Deutsch–Jozsa algorithm. Consider a function, $f(x)$, where x ranges over $\{0, 1\}$ and $f(x)$ can only have the values 0 or 1. There are exactly four such functions, so a classical algorithm for $f(x)$ must be provided with two bits of data to specify which function is to be evaluated.

The computer and the algorithm are shrouded in secrecy inside a black box, but we are allowed to submit values of x in order to get $f(x)$. If we want to know both

$f(0)$ and $f(1)$, then we must either run the algorithm twice—once for each input—or else run two identically programmed computers in parallel.

As an alternative, suppose there is a hidden quantum computer with a two-qubit register. In this situation, programming the computer to yield a given set of values $f = (f(0), f(1))$ is the same as the quantum dense coding problem. In Section 20.4.1 we saw that it is always possible to devise a set of unitary operations that convert a known initial state into one of the Bell states. We may as well simplify this part of the problem by assuming that the initial state of the quantum register is itself a Bell state, e.g. the initial state $|\Theta\rangle$ of the dense coding discussion is replaced by $|\Phi^+\rangle$.

In accord with the usual conventions in the field of quantum information processing, we will also assume that the unitary operators act on the first, rather than the second, qubit. If we associate the possible functions f with operators U^f according to the encoding scheme

$$\begin{aligned} U^{(0,0)} &= \begin{bmatrix} 1 & 0 \\ 0 & 1 \end{bmatrix}, \quad U^{(1,1)} = \begin{bmatrix} 1 & 0 \\ 0 & -1 \end{bmatrix}, \\ U^{(0,1)} &= \begin{bmatrix} 0 & 1 \\ 1 & 0 \end{bmatrix}, \quad U^{(1,0)} = \begin{bmatrix} 0 & 1 \\ -1 & 0 \end{bmatrix}, \end{aligned} \tag{20.70}$$

then it is easy to verify that

$$U^{(1,1)} |\Phi^+\rangle = |\Phi^-\rangle, \quad U^{(0,1)} |\Phi^+\rangle = |\Psi^+\rangle, \quad U^{(1,0)} |\Phi^+\rangle = |\Psi^-\rangle. \tag{20.71}$$

After the programmer supplies the two bits needed to choose the operator U^f—i.e. the one that gives the same output as the classical computer—the output of the computation is obtained by performing a Bell state measurement. If the result is $|\Psi^+\rangle$, then $f = (0, 1)$, etc. The important point is that it is only necessary to run the quantum algorithm once to get both values $f(0)$ and $f(1)$. Thus quantum parallelism gives the same result as classical parallelism, but the work of the two classical computers is done by one quantum computer.

B *Quantum logic gates**

The description of the simple quantum computer given in the last section fits conveniently with the discussion of quantum dense coding in Section 20.4.1, but it does not have the form commonly used in the quantum computing literature. The usual procedure is to express the operator U_{alg} as the product of a standard set of unitary operators, called **quantum logic gates**, that typically act on one or two qubits out of the N qubits in the register. Since the output of each gate serves as input to the next, the collection of gates can be visualized as a **quantum circuit**.

Classical computers employ operations on single bits and pairs of bits, and it has been shown that the most general computation can be performed by means of a single kind of two-bit gate combined with a collection of single-bit gates. An analogous result holds for quantum computers, so we only need to consider a single kind of two-qubit gate.

A one-qubit logic gate is completely specified by its action on the basis vectors $|0\rangle$ and $|1\rangle$; for example, the **X gate** is defined by $\mathbb{X}|0\rangle = |1\rangle$, and $\mathbb{X}|1\rangle = |0\rangle$. This is analogous to the classical **NOT gate** that interchanges 0 (false) and 1 (true). There are also useful one-qubit gates that do not have classical analogues, such as the **Z gate**: $\mathbb{Z}\left|0\right\rangle = \left|0\right\rangle$, $\mathbb{Z}\left|1\right\rangle = -\left|1\right\rangle$, and the **Hadamard gate**:

$$\mathbb{H}\left|0\right\rangle = \frac{1}{\sqrt{2}}\left\{\left|0\right\rangle + \left|1\right\rangle\right\}, \quad \mathbb{H}\left|1\right\rangle = \frac{1}{\sqrt{2}}\left\{\left|0\right\rangle - \left|1\right\rangle\right\}.$$

These gates can all be expressed as 2×2 matrices, and—as seen in Exercises 20.15 and 20.16—they are also related to rotations on the Poincaré sphere.

An important two-qubit gate is the **controlled-NOT** (C-NOT) gate, defined by

$$\mathbb{C}_{\text{NOT}}\left|a,b\right\rangle = \left|a, b \oplus a\right\rangle \quad (a, b = 0, 1), \tag{20.72}$$

where $\oplus$ represents addition modulo 2. The first and second qubits in the two-qubit state $\left|a,b\right\rangle = \left|a\right\rangle\left|b\right\rangle$ are conventionally called the **control** qubit and the **target** qubit respectively.

Thus the C-NOT gate has the following effects. (1) The control qubit is left unchanged. (2) The target qubit is flipped if the control qubit is 1, and left alone if the control qubit is 0. A convenient graphical notation for these standard gates is shown in Fig 20.7.

Another useful two-qubit gate is the **controlled-sign** or *controlled-phase* gate defined by

$$\mathbb{C}_S\left|a,b\right\rangle = (-1)^{ab}\left|a,b\right\rangle \quad (a, b = 0, 1). \tag{20.73}$$

This operation does nothing unless both the control and target qubits are $\left|1\right\rangle$, in which case it multiplies the two-qubit state by -1.

*C Quantum circuits**

In Section 20.5.1-A we flouted the convention that the register always begins in a standard state, e.g. $\left|\Lambda_{\text{in}}\right\rangle = \left|0,0\right\rangle$. It is easy to verify that $\left|\Phi^+\right\rangle = \mathbb{C}_{\text{NOT}}\mathbb{H}\left|0,0\right\rangle$, i.e. the initial state used in the previous discussion is built up from the standard state by applying a Hadamard gate followed by a controlled-NOT gate.

Inspection of eqn (20.70) shows that the operator $U^{(0,1)}$ leading to the outcome $\left|\Psi^+\right\rangle$ is an X gate, so the result $f = (0,1)$ is achieved by the unitary transformation $\left|\Psi^+\right\rangle = U^{(0,1)}\left|\Phi^+\right\rangle = \mathbb{X}\mathbb{C}_{\text{NOT}}\mathbb{H}\left|0,0\right\rangle$. The corresponding quantum circuit diagram, shown in Fig. 20.8, is to be read from left to right. Other examples are considered in Exercise 20.17.

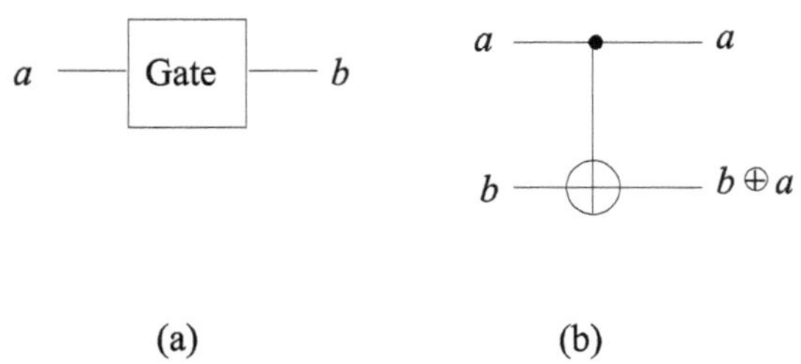

Fig. 20.7 Graphical representations of quantum logic gates: (a) a generic one-qubit gate, and (b) a controlled-NOT gate, with control qubit $\left|a\right\rangle$ and target qubit $\left|b\right\rangle$.

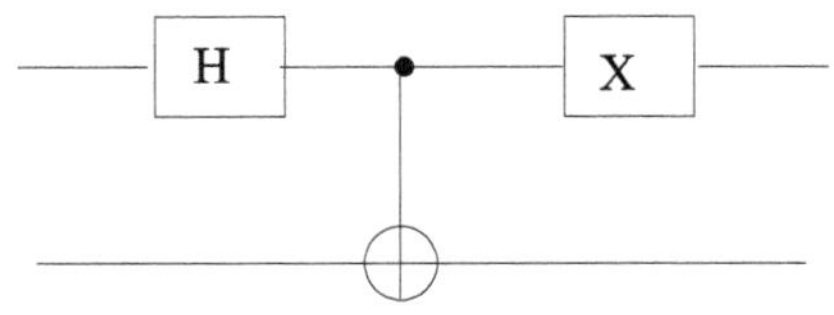

Fig. 20.8 Quantum circuit diagram for the program implemented by the sequence: Hadamard gate, controlled-NOT gate, and X gate.

20.5.2 Quantum computing experiments*

Experimental realizations of the idealized devices discussed above must overcome a number of very serious difficulties. To begin with, the qubits must be controllable to one part in 10^4 by means of analog pulses (Berggren, 2004). This is an especially acute problem if the qubits are carried by photons. The dissipative interaction of the qubits with the environment poses a still more daunting obstacle, since the resulting decoherence will destroy the entangled state.

Decoherence can be reduced by clever design, but it is impossible to eliminate it altogether. This fact has necessitated the introduction of error-correction protocols, first by Shor (1995), and later by Bennett *et al.* (1996) and Knill and Laflamme (1997). A common feature of these schemes is the use of a large number of ancillary qubits to guarantee the accuracy of the computation.

The necessity of error correction is a strong contributor to estimates that something like 10^6 qubits would be needed for a computation of practical interest (Berggren, 2004). Experiments performed to date only involve a few qubits, but **scalability**, i.e. the potential for extending a scheme to a very large number of qubits, is a primary concern.

The first experimental demonstrations of quantum computing (Chuang *et al.*, 1998; Vandersypen *et al.*, 2001) used the method of *bulk quantum computation* (Knill *et al.*, 1998), in which a large number of qubits—provided by spin-1/2 nuclei in molecules—are manipulated in parallel by nuclear magnetic resonance (NMR) techniques. This approach is adequate for proof-of-principle demonstrations but cannot be used for register sizes much greater than ten.

In order to achieve scalability, subsequent proposals have concentrated on various solid-state systems, e.g. nuclear spins of donor atoms in Si (Kane, 1998), electron spins in quantum dots (Loss and DiVincenzo, 1998; Petta *et al.*, 2005), qubits formed by counter-circulating persistent currents in Josephson junction circuits (Mooij *et al.*, 1999), electron-spin-resonance transistors (Vrijen *et al.*, 2000), and electron spins bound to deep donor states in Si (Stoneham *et al.*, 2003).

The physical system of greatest interest for us—the photon—is conspicuously absent from this list of candidates for quantum computers. The reason is that a two-qubit logic gate, such as the C-NOT gate discussed in Section 20.5.1-B, can only produce an entangled state—in our terminology a *dynamically* entangled state—of two photons by means of photon–photon coupling, i.e. an optical nonlinearity.

As suggested by Milburn (1989), one way to do this would be to induce a cross-Kerr coupling—see Section 13.4.3—between two optical modes. Unfortunately, the materials provided by nature have $\chi^{(3)}$s that are orders-of-magnitude too small to accomplish the desired effects. Increasing the length of the nonlinear region does not help, because the accompanying linear absorption will defeat the purpose of the device.

Another possibility is to trap an atom in a very small, high finesse cavity, but this approach has not yet been successful. This situation led to the general feeling that large-scale quantum computing by optical means is not a practical possibility.

20.5.3 Two-photon logic gates with linear optics*

The consensus view that optical methods are not suitable for quantum computing was challenged by the work of Knill, Laflamme, and Milburn (KLM) (Knill *et al.*, 2001), who showed that quantum algorithms could be implemented by combining single-photon sources, photon detectors, and passive linear optical elements.

Their scheme eliminated the need for strong optical nonlinearities in the manipulation of photonic qubits. This is a complex and rapidly evolving subject, so we will only sketch the first step in its development. More details can be found in recent review articles, e.g. Ralph (2006).

One possible design—adapted from the work of Hofmann and Takeuchi (2002)—for a two-photon logic gate utilizing only linear optics and photon detection is shown in Fig. 20.9.

This is a four channel/eight port device; the four input ports are the Control-in port, the Target-in port, and the unused ports of beam splitters 1 and 3, that

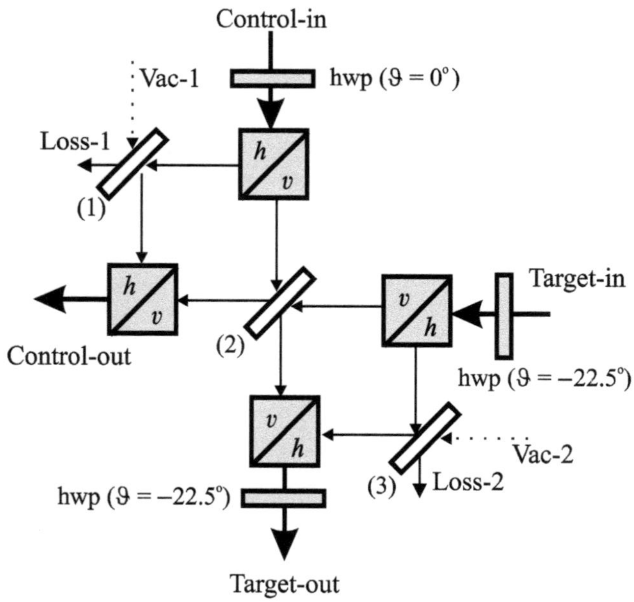

Fig. 20.9 Schematic for a nondeterministic control-NOT gate. The polarizing beam splitters transmit v-polarized light and reflect h-polarized light at 90°. The half-wave plate (hwp) at the control input is aligned at $\vartheta = 0$, while the hwps at the target input and output ports are aligned at $\vartheta = -22.5°$, where ϑ is the angle between the h-polarization and the fast axis; see Exercise 20.8. The beam splitters are asymmetric.

communicate with the vacuum channels Vac-1 and Vac-2. The beam splitters are asymmetric, with scattering laws of the general form

$$\begin{aligned} a_1' &= \sqrt{1-R}a_1 - \epsilon\sqrt{R}a_2 \,, \\ a_2' &= \epsilon\sqrt{R}a_1 + \sqrt{1-R}a_2 \,, \end{aligned} \tag{20.74}$$

worked out in Exercise 8.1. All three beam splitters are assumed to have the same reflectivity R, and the sign factors $\epsilon = \pm 1$ are chosen to accomplish the design objectives. The half-wave plate at the control input is a Z gate, and the half-wave plates at the target ports are Hadamard gates.

The use of passive optical elements ensures that photon number is conserved, so for two incident photons—one in the control channel and one in the target channel—we can be sure that exactly two photons will be emitted. However, the mixing occurring at the beam splitters implies that the output state will be a superposition of all possible two-photon states in the output channels: Control-out, Target-out, Loss-1, and Loss-2.

The central beam splitter is particularly important in this regard, since photons are incident from both sides. As we have seen in Section 10.2.1, this is precisely the situation required for the strictly quantum interference effects associated with different Feynman paths having the same end point.

The key to the operation of this gate is **postselection**, i.e. discarding all outcomes that do not satisfy a chosen criterion. In the present case, the first part of the criterion is that detectors in the Control-out and Target-out channels should eventually register a coincidence count. The states that can contribute to such a coincidence event are superpositions of the **coincidence basis** states $\{|h_C, h_T\rangle, |h_C, v_T\rangle, |v_C, h_T\rangle, |v_C, v_T\rangle\}$.

Satisfying the condition (20.72) for a control-NOT gate further requires that exactly one member of the coincidence basis occurs in the output state for each of the four possible input states. A rather lengthy calculation, outlined in Exercise 20.18, shows that this goal can only be reached for the value $R = 1/3$ and asymmetry parameters satisfying $\epsilon_2 = -\epsilon_3 = \epsilon_1$.

With these values, the operation of the gate is given by

$$\begin{aligned} \mathbb{C}_{\mathrm{NOT}} |h_C, h_T\rangle &= \frac{1}{3} |h_C, h_T\rangle + \cdots \,, \quad \mathbb{C}_{\mathrm{NOT}} |h_C, v_T\rangle = \frac{1}{3} |h_C, v_T\rangle + \cdots \,, \\ \mathbb{C}_{\mathrm{NOT}} |v_C, h_T\rangle &= \frac{1}{3} |v_C, v_T\rangle + \cdots \,, \quad \mathbb{C}_{\mathrm{NOT}} |v_C, v_T\rangle = \frac{1}{3} |v_C, h_T\rangle + \cdots \,, \end{aligned} \tag{20.75}$$

where '$\cdots$' contains the terms that are not in the subspace spanned by the coincidence basis. The target photon polarization is unchanged if the control photon is h-polarized but flipped ($h \leftrightarrow v$), when the control photon is v-polarized. With the identification $h \leftrightarrow \mathtt{0}$ and $v \leftrightarrow \mathtt{1}$, this is the photonic version of eqn (20.72).

A simple modification of the design in Fig. 20.9 yields the gate action

$$\begin{aligned} \mathbb{C}_S |h_C, h_T\rangle &= \frac{1}{3} |h_C, h_T\rangle + \cdots \,, \quad \mathbb{C}_S |h_C, v_T\rangle = \frac{1}{3} |h_C, v_T\rangle + \cdots \,, \\ \mathbb{C}_S |v_C, h_T\rangle &= \frac{1}{3} |v_C, h_T\rangle + \cdots \,, \quad \mathbb{C}_S |v_C, v_T\rangle = -\frac{1}{3} |v_C, v_T\rangle + \cdots \,, \end{aligned} \tag{20.76}$$

which satisfies the definition (20.73) of a controlled-sign gate.

The postselection criterion picks out the appropriate outcomes, but the probability for successful operation is $(1/3)^2 = 1/9$. Eight times out of nine, the two photons are emitted into the wrong channels, e.g. one photon into Loss-2 and one into Control-1, or two photons into Control-out, etc. The success or failure of the gate is **heralded**—i.e. the outcome is known—by the presence or absence of a coincidence between the Control-out and Target-out channels. Additional checks could be made by detecting photons emitted into the loss channels, or by discriminating between one- or two-photon events in the control and target channels.

This gate has been experimentally realized (O'Brien *et al.*, 2003) by using down-conversion to produce the input photons and quantum state tomography to verify that the output states agreed with the theoretical model. In this approach, the necessity for dynamically coupling the two photons has been avoided by incorporating the coincidence measurement as part of the action of the device.

20.5.4 Linear optical quantum computing*

The quantum logic gate discussed above does provide a nontrivial two-qubit operation, but it fails the scalability test. A device containing many such gates, each with a success probability of 1/9, would almost never work. In their general scheme, KLM answered this objection by making use of the ideas involved in quantum teleportation.

The use of quantum teleportation to carry out general quantum computations was first suggested by Gottesman and Chuang (1999), and KLM showed that a so-called **teleportation gate** could be realized with a high probability of success, given a sufficiently complex entangled state.

The KLM approach avoids the failure mode associated with a vanishingly small success probability, but the resources required are too large for practical scalability. For example, the number of **Bell pairs**—i.e. pairs of photons described by a Bell state—needed to implement a single controlled-sign gate with success probability of 95% is of the order 10 000 (Ralph, 2006).

This resource cost can be greatly reduced by using parity-state encoding (Hayes *et al.*, 2004). Single-qubit **parity states** are the alternative basis states,

$$|\pm\rangle = (|0\rangle \pm |1\rangle)/\sqrt{2}, \tag{20.77}$$

that satisfy $\mathbb{X}|\pm\rangle = \pm|\pm\rangle$. In parity-state encoding, the logical 0 and 1 are represented by n-qubit states:

$$\begin{aligned} |0\rangle^{(n)} &= \frac{1}{\sqrt{2}}\left[\otimes_{j=1}^{n}|+\rangle_j + \otimes_{j=1}^{n}|-\rangle_j\right], \\ |1\rangle^{(n)} &= \frac{1}{\sqrt{2}}\left[\otimes_{j=1}^{n}|+\rangle_j - \otimes_{j=1}^{n}|-\rangle_j\right]. \end{aligned} \tag{20.78}$$

A clever application of this encoding scheme reduces the overhead cost to the order of 100 Bell pairs per gate.

An alternative scheme—which actually amounts to a fundamentally different model for quantum computing—grew out of a theoretical proposal by Raussendorf and Briegel (2001). In the standard model sketched in Section 20.5.1, an algorithm is represented as a sequence of unitary operators that are physically realized by quantum logic gates.

The logic gates produce a sequence of entangled states that ends in the desired final state, which is measured to produce the computational result.

In the new model, the entanglement resource is prepared beforehand, in the form of a highly entangled, multi-qubit, initial state. The nature of this state is most easily understood by visualizing the qubits as spin-1/2 particles attached to the sites of a lattice and interacting through nearest-neighbor coupling. A *cluster* is a collection of occupied lattice points such that each pair of sites is connected by jumps across nearest-neighbor links. Each qubit is initially prepared in the parity state $|+\rangle$ and a **cluster state** is generated by pair-wise entanglement of the initial qubits.

As an example, consider a one-dimensional lattice with three occupied sites, 1, 2, 3, so that the initial state is $|+\rangle_1 |+\rangle_2 |+\rangle_3$. The corresponding cluster state can be generated by successive application of controlled-sign gates as follows:

$$|\Phi_{\text{lin3}}\rangle = \mathbb{C}_S^{(1,2)} |+\rangle_1 \left\{ \mathbb{C}_S^{(2,3)} |+\rangle_2 |+\rangle_3 \right\}, \tag{20.79}$$

where $\mathbb{C}_S^{(i,j)}$ acts on two-qubit states $|a\rangle_i |b\rangle_j$. Carrying out the gate operations, with the aid of the definitions (20.73) and (20.77), leads to the explicit expression

$$|\Phi_{\text{lin3}}\rangle = \frac{1}{\sqrt{2}} \left[|+\rangle_1 |\mathsf{0}\rangle_2 |+\rangle_3 + |-\rangle_1 |\mathsf{1}\rangle_2 |-\rangle_3 \right] \tag{20.80}$$

for the cluster state. The cluster states needed for nontrivial calculations generally involve clusters on two-dimensional lattices and many more qubits.

The cluster state provides the essential substrate for the computation, but the algorithm itself is defined by combining two further elements: (1) a sequence of **local measurements** (von Neumann measurements on individual qubits); and (2) **classical feedforward**. The latter term means that the result of one measurement in the sequence can be used to determine the choice of the measurement basis used in a subsequent measurement.

These two elements can replace any of the operations considered in Section 20.5.1. For example, any unitary operation on a single qubit can be simulated by means of a four-qubit cluster state and three measurements. In general, one-qubit measurements are used to imprint the initial data onto the cluster state, and then process it to yield the final result.

The use of irreversible measurements as an integral part of the algorithm, rather than just the final readout step, has led to the name **one-way quantum computing** for this approach. For a sufficiently large cluster state, it has been shown that these elements are sufficient to implement a universal quantum computer. The different structures of reversible and one-way computing make comparisons a bit difficult, but the current estimate is that one-way computing requires roughly 60 Bell pairs per two-qubit gate operation.

Highly entangled states of many atoms have been experimentally produced by precise control of the interactions between neutral atoms bound by dipole forces to the sites of an optical lattice (Mandel *et al.*, 2003), but we are more interested in optical realizations of cluster states. Walther *et al.* (2005) demonstrated a one-way version of a simple example of Grover's search algorithm.

In their experiment, a four-photon cluster state was directly produced by down-conversion techniques. Four-qubit cluster states have also been produced by entangling EPR pairs with a controlled-sign gate (Kiesel *et al.*, 2005), and by a technique called type-I qubit fusion (Zhang *et al.*, 2006), which combines Bell states by mixing at a beam splitter and postselection. One-way quantum computing may, therefore, be a promising application of quantum optics to quantum computing.

20.6 Exercises

20.1 Variable retarder plate

Design a variable retarder plate by joining two identical, thin, right-angle prisms along their hypotenuses. Sketch the appropriate arrangement and carry out the following.

(1) Assuming that the light passes through the central part of the retarder, show how the optical path length can be adjusted by sliding the prisms along their common hypotenuse.

(2) Express the optical path length in terms of the index of refraction and the geometrical parameters of the device. Assign numerical values for a practical design.

(3) Calculate the optical path lengths required to obtain the phase shifts $\theta = -\pi/2$ and $\theta' = \pi/2$.

20.2 Modified beam splitter

Consider the modified beam splitter pictured in Fig. 20.1.

(1) Derive eqn (20.5) for the scattering matrix.

(2) For general values of θ and θ', use the scattering matrix to express the output quadratures X_1', Y_1', X_2', Y_2' in terms of the input quadratures X_1, Y_1, X_2, Y_2. Calculate the variances of the output quadratures. Explain why the values $\theta = -\pi/2$ and $\theta' = \pi/2$ are particularly useful.

(3) If no variable retarder plates are available, i.e. $\theta = \theta' = 0$, how can the design of the SQLG be changed to achieve the same noiseless splitting of the input signal X_1.

20.3 Bell states

Consider the Bell states defined in eqn (20.15).

(1) Show that the Bell states are mutually orthogonal and all normalized to unity.

(2) Explain—ideally without any further algebra—why the Bell states form a basis for $\mathfrak{H}_a \otimes \mathfrak{H}_b$.

20.4 No-cloning theorem for photons

Consider cloning the one-photon states $|\gamma\rangle = \Gamma^\dagger |0\rangle$ and $|\zeta\rangle = Z^\dagger |0\rangle$, where

$$Z^\dagger = \sum_s \zeta_{\mathbf{k}s} a_{\mathbf{k}s}^\dagger \,.$$

(1) Derive the commutator

$$\left[Z,\Gamma^{\dagger}\right]=\sum_{\mathbf{k}s}\zeta^{*}_{\mathbf{k}s}\gamma_{\mathbf{k}s}\equiv(\zeta,\gamma)\,.$$

(2) Adapt the general proof for the no-cloning theorem to show that the cloning assumption (20.25), and the corresponding assumption for $|\zeta\rangle$, cannot be satisfied for all choices of the operators $\Gamma^{\dagger}$ and $Z^{\dagger}$.

20.5 Cloning a known state

For the device in Section 20.2.1 that clones a known state, assume the model interaction $H_{\text{int}}=g\sigma_{-}\left(a^{\dagger}_{\mathbf{k}h}+a^{\dagger}_{\mathbf{k}v}\right)+\text{HC}$, between the two-level atom and the field. For the initial state $|1_{\mathbf{k}h}\rangle$, use first-order, time-dependent perturbation theory to calculate the change in the initial state vector and thus derive eqn (20.26).

20.6 Bužek–Hillery QCM*

Use the explicit expressions (20.31) and (20.32) to evaluate the reduced qubit density operators ρ_{ab}, ρ_{a}, and ρ_{b}. Use the results to calculate the fidelities for the clones a and b.

20.7 Photon cloning machine*

Consider the photon cloning machine described in Section 20.2.2-B.

(1) Denote the polarization basis for the $\mathbf{k}_n$-mode $(n=1,2)$ by $\{\mathbf{e}_h(\mathbf{k}_n),\mathbf{e}_v(\mathbf{k}_n)\}$. For a rotation of each basis around $\mathbf{k}_n$ by the angle θ, i.e.

$$\begin{aligned}\mathbf{e}'_h(\mathbf{k}_n)&=\cos\theta\,\mathbf{e}_h(\mathbf{k}_n)+\sin\theta\,\mathbf{e}_v(\mathbf{k}_n)\,,\\ \mathbf{e}'_v(\mathbf{k}_n)&=-\sin\theta\,\mathbf{e}_h(\mathbf{k}_n)+\cos\theta\,\mathbf{e}_v(\mathbf{k}_n)\,,\end{aligned}$$

derive the corresponding transformation of the creation operators $a^{\dagger}_{\mathbf{k}_n h}$, $a^{\dagger}_{\mathbf{k}_n v}$ and show that the Hamiltonian in eqn (20.33) has the same form in the new basis.

(2) The $(2,0)$-events in which two photons are present in the $\mathbf{k}_1 v$-mode are counted by letting the output fall on a beam splitter with detectors at each output port. A coincidence count shows that two photons were present. For an ideal balanced beam splitter and 100% detectors, show that the probability of a coincidence count is 1/2. Use this to explain the discrepancy between eqn (20.42) and the baseline data in Fig. 20.3.

20.8 Wave plates

A polarization-dependent retarder plate (**wave plate**) is made from an anisotropic crystal, with indices of refraction n_F and n_S for light polarized along the *fast axis* $\mathbf{e}_F$ and the *slow axis* $\mathbf{e}_S$ respectively (Saleh and Teich, 1991, Sec. 6.1-B).

Consider a classical field with amplitude $\boldsymbol{\mathcal{E}}=\mathcal{E}_h\mathbf{e}_h+\mathcal{E}_v\mathbf{e}_v$ propagating in the z-direction, that falls on a retarder plate of thickness Δz lying in the (x,y)-plane.

(1) By discarding an overall phase factor show that the output field $\mathcal{E}' = \mathcal{E}_h'\mathbf{e}_h + \mathcal{E}_v'\mathbf{e}_v$ is related to the input field by $\mathrm{col}\left(\mathcal{E}_h',\mathcal{E}_v'\right) = T_\xi\left(\vartheta\right)\mathrm{col}\left(\mathcal{E}_h,\mathcal{E}_v\right)$, where the **Jones matrix** $T_\xi\left(\vartheta\right)$ is given by

$$T_\xi\left(\vartheta\right) = \begin{bmatrix} \cos^2\vartheta + \sin^2\vartheta e^{i\xi} & -\sin\vartheta\cos\vartheta\left(1 - e^{i\xi}\right) \\ -\sin\vartheta\cos\vartheta\left(1 - e^{i\xi}\right) & \sin^2\vartheta + \cos^2\vartheta e^{i\xi} \end{bmatrix},$$

ϑ is the angle between $\mathbf{e}_h$ and $\mathbf{e}_F$, and $\xi = \left(n_S - n_F\right)\omega\Delta z/c$.

(2) Evaluate the Jones matrix for $\xi = \pi/2$ (the **quarter-wave plate**) and $\xi = \pi$ (the **half-wave plate**).

(3) For $\vartheta = 0$ and a 45°-polarized input, i.e. $\mathcal{E}_h = \mathcal{E}_v$, what is the output polarization state? Answer the same question if $\vartheta = \pi/4$ and the input field is h-polarized.

20.9 Quantum dense coding

The unitary operators used by Bob for quantum dense coding are defined by $U\left(\vartheta_{\lambda/4}, \vartheta_{\lambda/2}\right) = T_{\pi/2}\left(\vartheta_{\lambda/4}\right)T_\pi\left(\vartheta_{\lambda/2}\right)$, where $T_\xi\left(\vartheta\right)$ is given by the result of the previous exercise. As explained in the text, this operator only acts on the second argument of $\left|s_A, s_B'\right\rangle$.

(1) For the general state

$$\left|\Theta\right\rangle = c_{hh}\left|h_A, h_B\right\rangle + c_{hv}\left|h_A, v_B\right\rangle + c_{vh}\left|v_A, h_B\right\rangle + c_{vv}\left|v_A, v_B\right\rangle$$

determine the expansion coefficients for which $U\left(0,0\right)\left|\Theta\right\rangle = \left|\Psi^+\right\rangle$.

(2) For $\left|\Theta\right\rangle$ from part (1), find $\left(\vartheta_{\lambda/4}, \vartheta_{\lambda/2}\right)$ such that $U\left(\vartheta_{\lambda/4}, \vartheta_{\lambda/2}\right)\left|\Theta\right\rangle$ is equal (up to a phase factor) to the remaining Bell states.

20.10 Bell states incident on a balanced beam splitter

For the Bell states in eqns (20.52) and (20.53) use the method described in Section 8.4.1 to show that the scattered states produced by a balanced beam splitter are

$$\begin{aligned} \left|\Psi^-\right\rangle' &= \left|\Psi^-\right\rangle, \\ \left|\Psi^+\right\rangle' &= \frac{i}{\sqrt{2}}\left|h_A, v_A\right\rangle + \left(A \leftrightarrow B\right), \\ \left|\Phi^\pm\right\rangle' &= \frac{i}{2}\left\{\left|h_A, h_A\right\rangle \pm \left|v_A, v_A\right\rangle\right\} + \left(A \leftrightarrow B\right). \end{aligned}$$

20.11 Rotated polarization basis

Consider the 45°-rotated polarization basis defined by eqn (20.48).

(1) Derive

$$a_{\gamma h}^\dagger = \left(a_{\gamma\bar{h}}^\dagger - a_{\gamma\bar{v}}^\dagger\right)/\sqrt{2}, \quad a_{\gamma v}^\dagger = \left(a_{\gamma\bar{h}}^\dagger + a_{\gamma\bar{v}}^\dagger\right)/\sqrt{2},$$

where $\gamma \in \{A, B\}$.

(2) Show that

$$\begin{aligned}|h_A, h_A\rangle &= \frac{1}{2}\left\{\left|\overline{h}_A, \overline{h}_A\right\rangle + \left|\overline{v}_A, \overline{v}_A\right\rangle\right\} - \frac{1}{\sqrt{2}}\left|\overline{h}_A, \overline{v}_A\right\rangle, \\ |v_A, v_A\rangle &= \frac{1}{2}\left\{\left|\overline{h}_A, \overline{h}_A\right\rangle + \left|\overline{v}_A, \overline{v}_A\right\rangle\right\} + \frac{1}{\sqrt{2}}\left|\overline{h}_A, \overline{v}_A\right\rangle.\end{aligned}$$

(3) Starting with eqn (20.56), derive eqns (20.57) and (20.58).

20.12 Insufficient information

Consider Alice's attempt to give Bob instructions for making an approximate copy of her unknown qubit $|\gamma\rangle$.

(1) Given the unit vector $\mathbf{n}$ and the eigenvalue ϵ of $\mathbf{n}\cdot\boldsymbol{\sigma}$, explain why Bob's best estimate for the unknown state $|\gamma\rangle$ is given by eqn (20.62).

(2) Why cannot Alice get more information for Bob by making further measurements?

(3) Suppose that the sender of Alice's qubit, who does know the state $|\gamma\rangle$, is willing to send her an endless stream of qubits, all prepared in the same state. Alice's research budget, however, limits her to a finite number of measurements. Can Alice supply Bob with enough information to permit an exact reproduction (up to an overall phase) of $|\gamma\rangle$?

20.13 Teleportation of qubits

(1) Express the basis states $|u, v\rangle_{AT}$ $(u, v = 0, 1)$ as linear combinations of the Bell states, and then derive eqn (20.66).

(2) Show that the Pauli matrices are unitary as well as hermitian, and use this fact to construct unitary operators for the phase-flip and the bit-flip.

(3) Suppose that Alice does her Bell state measurement, but that Eve intercepts the message to Bob. Calculate the reduced density operator ρ_B that Bob must use in this circumstance, and comment on the result.

(4) Now suppose that Alice misunderstands the theory, and thinks that she should make a measurement that projects onto the basis vectors $|u, v\rangle_{AT}$. After Alice tells Bob which of the four possibilities occurred, what information does Bob have about his qubit?

20.14 Teleportation of photons

Consider the application of the teleportation protocol to photons.

(1) Write out the explicit expressions for the Bell states in the A–T subsystem.

(2) Derive the photonic version of eqn (20.66).

(3) Give explicit forms for the action of the unitary transformations U_{pf} (phase-flip) and U_{bf} (bit-flip) on the creation operators.

20.15 Quantum logic gates*

(1) Show that the X, Z, and Hadamard gates are unitary operators.

(2) Use the representation $|\gamma\rangle = \gamma_0 |0\rangle + \gamma_1 |1\rangle$ of a general qubit to express all three gates as 2×2 matrices. Explain the names for the X and Z gates by relating them to Pauli matrices.

(3) For a spin-1/2 particle, the operator for a rotation through the angle α around the axis directed along the unit vector $\mathbf{u}$ is (Bransden and Joachain, 1989, Sec. 6.9)

$$R_{\mathbf{u}}(\alpha) = \cos\left(\frac{\alpha}{2}\right) - i\sin\left(\frac{\alpha}{2}\right)\mathbf{u}\cdot\boldsymbol{\sigma}\,.$$

Combine this with the Poincaré-sphere representation

$$|\gamma\rangle = \cos\left(\frac{\theta}{2}\right)|0\rangle + e^{i\phi}\sin\left(\frac{\theta}{2}\right)|1\rangle$$

for qubits to show that the X, Z, and Hadamard gates are respectively given by $iR_{\mathbf{u}_x}(\pi)$, $iR_{\mathbf{u}_z}(\pi)$, and $iR_{\mathbf{h}}(\pi)$, where $\mathbf{u}_x$, $\mathbf{u}_y$, $\mathbf{u}_z$ are the coordinate unit vectors and $\mathbf{h} = \mathbf{u}_x/\sqrt{2} + \mathbf{u}_z/\sqrt{2}$.

(4) Show that the control-NOT operator $\mathbb{C}_{\text{NOT}}$, defined by eqn (20.72), is unitary. Use the basis $\{|0,0\rangle, |0,1\rangle, |1,0\rangle, |1,1\rangle\}$ to express $\mathbb{C}_{\text{NOT}}$ as a 4×4 matrix.

20.16 Single-photon gates*

Identify the polarization states of a single photon with the logical states by $|h\rangle \leftrightarrow |0\rangle$ and $|v\rangle \leftrightarrow |1\rangle$. Use the results of Exercise 20.8 to show that the Z and Hadamard gates can be realized by means of half-wave plates.

20.17 Quantum circuits*

Work out the gates required for the outcomes $|\Phi^-\rangle$ and $|\Psi^-\rangle$ in the computation discussed in Section 20.5.1-A and draw the corresponding quantum circuit diagrams.

20.18 Controlled-NOT gate*

For the nondeterministic control-NOT gate sketched in Section 20.5.3, use the notation a_{Ch}, a_{Cv}, a_{Th}, a_{Tv} for the control and target modes and b_{1h}, b_{2h} for h-polarized vacuum fluctuations in the Vac-1 and Vac-2 channels. Devise a suitable notation for the operators associated with the internal lines in Fig. 20.9, and carry out the following steps.

(1) Write out the scattering relations for each of the optical elements in the gate. For this purpose it is useful to impose a consistent convention for assigning the $\pm\epsilon$s to the asymmetric beam splitters, e.g. assign $-\epsilon\sqrt{R}$ for reflection from the lower surface of a beam splitter.

(2) Explain why the vacuum v-polarizations b_{1v}, b_{2v} can be omitted.

(3) Use the scattering relations to eliminate the internal variables and thus find the overall scattering relations $(a_{Ch}, a_{Cv}, \ldots) \to (a'_{Ch}, a'_{Cv}, \ldots)$ which define the elements of the scattering matrix for the gate.

(4) Employ the general result (8.40) to determine the action of the gate on each input state in the coincidence basis, and thus show that

$$|h_C, h_T\rangle \to \frac{1}{2}\epsilon_1 (\epsilon_2 - \epsilon_3) R |h_C, h_T\rangle - \frac{1}{2}\epsilon_1 (\epsilon_2 + \epsilon_3) R |h_C, v_T\rangle + \cdots ,$$

$$|h_C, v_T\rangle \to -\frac{1}{2}\epsilon_1 (\epsilon_2 + \epsilon_3) R |h_C, h_T\rangle + \frac{1}{2}\epsilon_1 (\epsilon_2 - \epsilon_3) R |h_C, v_T\rangle + \cdots ,$$

$$|v_C, h_T\rangle \to \frac{1}{2} [(2 - \epsilon_2\epsilon_3) R - 1] |v_C, h_T\rangle + \frac{1}{2} [1 - (2 + \epsilon_2\epsilon_3) R] |v_C, v_T\rangle + \cdots ,$$

$$|v_C, v_T\rangle \to \frac{1}{2} [1 - (2 + \epsilon_2\epsilon_3) R] |v_C, h_T\rangle + \frac{1}{2} [(2 - \epsilon_2\epsilon_3) R - 1] |v_C, v_T\rangle + \cdots .$$

Determine the value of R and the assignment of the ϵs needed to define a control-NOT gate.

Appendix A
Mathematics

A.1 Vector analysis

Our conventions for elementary vector analysis are as follows. The **unit vectors** corresponding to the Cartesian coordinates x, y, z are $\mathbf{u}_x$, $\mathbf{u}_y$, $\mathbf{u}_z$. For a general vector $\mathbf{v}$, we denote the unit vector in the direction of $\mathbf{v}$ by $\widetilde{\mathbf{v}} = \mathbf{v}/|\mathbf{v}|$.

The scalar product of two vectors is $\boldsymbol{a} \cdot \boldsymbol{b} = a_x b_x + a_y b_y + a_z b_z$, or

$$\boldsymbol{a} \cdot \boldsymbol{b} = \sum_{i=1}^{3} a_i b_i \,, \tag{A.1}$$

where $(a_1, a_2, a_3) = (a_x, a_y, a_z)$, etc. Since expressions like this occur frequently, we will use the Einstein **summation convention**: repeated vector indices are to be summed over; that is, the expression $a_i b_i$ is understood to imply the sum in eqn (A.1). *The summation convention will only be employed for three-dimensional vector indices.* The cross product is

$$(\boldsymbol{a} \times \boldsymbol{b})_i = \epsilon_{ijk} a_j b_k \,, \tag{A.2}$$

where the **alternating tensor** ϵ_{ijk} is defined by

$$\epsilon_{ijk} = \begin{cases} 1 & ijk \text{ is an even permutation of } 123\,, \\ -1 & ijk \text{ is an odd permutation of } 123\,, \\ 0 & \text{otherwise}\,. \end{cases} \tag{A.3}$$

A.2 General vector spaces

A **complex vector space** is a set $\mathfrak{H}$ on which the following two operations are defined.

(1) *Multiplication by scalars.* For every pair (α, ψ), where α is a *scalar*, i.e. a complex number, and $\psi \in \mathfrak{H}$, there is a unique element of $\mathfrak{H}$ that is denoted by $\alpha\psi$.

(2) *Vector addition.* For every pair ψ, ϕ of vectors in $\mathfrak{H}$ there is a unique element of $\mathfrak{H}$ denoted by $\psi + \phi$.

The two operations satisfy (a) $\alpha(\beta\psi) = (\alpha\beta)\,\psi$, and (b) $\alpha\,(\psi + \phi) = \alpha\psi + \alpha\phi$. It is assumed that there is a special **null** vector, usually denoted by 0, such that $\alpha 0 = 0$ and $\psi + 0 = \psi$. If the scalars are restricted to real numbers these conditions define a **real vector space**.

Ordinary displacement vectors, $\mathbf{r}$, belong to a real vector space denoted by $\mathbb{R}^3$. The set $\mathbb{C}^n$ of n-tuplets $\psi = (\psi_1, \ldots, \psi_n)$, where each component ψ_i is a complex number, defines a complex vector space with component-wise operations:

$$\begin{aligned} \alpha\psi &= (\alpha\psi_1, \ldots, \alpha\psi_n) , \\ \psi + \phi &= (\psi_1 + \phi_1, \ldots, \psi_n + \phi_n) . \end{aligned} \tag{A.4}$$

Each vector in $\mathbb{R}^3$ or $\mathbb{C}^n$ is specified by a finite number of components, so these spaces are said to be **finite dimensional**.

The set of complex functions, $C(\mathbb{R})$, of a single real variable defines a vector space with point-wise operations:

$$(\alpha\psi)(x) = \alpha\psi(x) , \tag{A.5}$$

$$(\psi + \phi)(x) = \psi(x) + \phi(x) , \tag{A.6}$$

where α is a scalar, and $\psi(x)$ and $\phi(x)$ are members of $C(\mathbb{R})$. This space is said to be **infinite dimensional**, since a general function is not determined by any finite set of values.

For any subset $\mathcal{U} \subset \mathfrak{H}$, the set of all linear combinations of vectors in $\mathcal{U}$ is called the **span** of $\mathcal{U}$, written as $\text{span}(\mathcal{U})$. A family $\mathcal{B} \subset \mathfrak{H}$ is a **basis** for $\mathfrak{H}$ if $\mathfrak{H} = \text{span}(\mathcal{B})$, i.e. every vector in $\mathfrak{H}$ can be expressed as a linear combination of vectors in $\mathcal{B}$. In this situation $\mathfrak{H}$ is said to be **spanned** by $\mathcal{B}$.

A **linear operator** is a rule that assigns a new vector $M\psi$ to each vector $\psi \in \mathfrak{H}$, such that

$$M(\alpha\psi + \beta\phi) = \alpha M\psi + \beta M\phi \tag{A.7}$$

for any pair of vectors ψ and ϕ, and any scalars α and β. The action of a linear operator M on $\mathfrak{H}$ is completely determined by its action on the vectors of a basis $\mathcal{B}$.

A.3 Hilbert spaces

A.3.1 Definition

An **inner product** on a vector space $\mathfrak{H}$ is a rule that assigns a complex number, denoted by (ϕ, ψ), to every pair of elements ϕ and $\psi \in \mathfrak{H}$, with the following properties:

$$(\phi, \alpha\psi + \beta\chi) = \alpha(\phi, \psi) + \beta(\phi, \chi) , \tag{A.8a}$$

$$(\phi, \psi) = (\psi, \phi)^* , \tag{A.8b}$$

$$0 \leqslant (\phi, \phi) < \infty , \tag{A.8c}$$

$$(\phi, \phi) = 0 \text{ if and only if } \phi = 0 . \tag{A.8d}$$

An **inner product space** is a vector space equipped with an inner product. The inner product satisfies the **Cauchy–Schwarz inequality**:

$$|(\phi, \psi)|^2 \leqslant (\phi, \phi)(\psi, \psi) . \tag{A.9}$$

Two vectors are **orthogonal** if $(\phi, \psi) = 0$. If $\mathfrak{F}$ is a subspace of $\mathfrak{H}$, then the **orthogonal complement** of $\mathfrak{F}$ is the subspace $\mathfrak{F}^\perp$ of vectors orthogonal to every vector in $\mathfrak{F}$.

The **norm** $\|\psi\|$ of ψ is defined as $\|\psi\| = \sqrt{(\psi, \psi)}$, so that $\|\psi\| = 0$ implies $\psi = 0$. Vectors with $\|\psi\| = 1$ are said to be **normalized**. A set of vectors is **complete** if the only vector orthogonal to every vector in the set is the null vector. Each complete set contains a basis for the space. A vector space with a countable basis set, $\mathcal{B} = \{\phi^{(1)}, \phi^{(2)}, \ldots\}$, is said to be **separable**. The vector spaces relevant to quantum theory are all separable. A basis for which $\left(\phi^{(n)}, \phi^{(m)}\right) = \delta_{nm}$ holds is called **orthonormal**. Every vector in $\mathfrak{H}$ can be uniquely expanded in an orthonormal basis, e.g.

$$\psi = \sum_{n=1}^{\infty} \psi_n \phi^{(n)} , \tag{A.10}$$

where the expansion coefficients are $\psi_n = \left(\phi^{(n)}, \psi\right)$.

A sequence $\psi^1, \psi^2, \ldots, \psi^k, \ldots$ of vectors in $\mathfrak{H}$ is **convergent** if

$$\left\|\psi^k - \psi^j\right\| \to 0 \quad \text{as } k, j \to \infty . \tag{A.11}$$

A vector ψ is a **limit** of the sequence if

$$\left\|\psi^k - \psi\right\| \to 0 \quad \text{as } k \to \infty . \tag{A.12}$$

A **Hilbert space** is an inner product space that contains the limits of all convergent sequences.

A.3.2 Examples

The finite-dimensional spaces $\mathbb{R}^3$ and $\mathbb{C}^N$ are both Hilbert spaces. The inner product for $\mathbb{R}^3$ is the familiar dot product, and for $\mathbb{C}^N$ it is

$$(\psi, \phi) = \sum_{n=1}^{N} \psi_n^* \phi_n . \tag{A.13}$$

If we constrain the complex functions $\psi(x)$ by the normalizability condition

$$\int_{-\infty}^{\infty} dx \left|\psi(x)\right|^2 < \infty , \tag{A.14}$$

then the Cauchy–Schwarz inequality for integrals,

$$\left|\int_{-\infty}^{\infty} dx \psi^*(x) \phi(x)\right|^2 \leqslant \int_{-\infty}^{\infty} dx \left|\psi(x)\right|^2 \int_{-\infty}^{\infty} dx \left|\phi(x)\right|^2 , \tag{A.15}$$

is sufficient to guarantee that the inner product defined by

$$(\psi, \phi) = \int_{-\infty}^{\infty} dx \psi^*(x) \phi(x) \tag{A.16}$$

makes the vector space of complex functions into a Hilbert space, which is called $L_2(\mathbb{R})$.

A.3.3 Linear operators

Let A be a linear operator acting on $\mathfrak{H}$; then the **domain** of A, called $\mathcal{D}(A)$, is the subspace of vectors $\psi \in \mathfrak{H}$ such that $\|A\psi\| < \infty$. An operator A is **positive definite** if $(\psi, A\psi) \geqslant 0$ for all $\psi \in \mathcal{D}(A)$, and it is **bounded** if $\|A\psi\| < b\|\psi\|$, where b is a constant independent of ψ. The **norm** of an operator is defined by

$$\|A\| = \max \frac{\|A\psi\|}{\|\psi\|} \quad \text{for } \psi \neq 0\,, \tag{A.17}$$

so a bounded operator is one with finite norm.

If $A\psi = \lambda\psi$, where λ is a complex number and ψ is a vector in the Hilbert space, then λ is an **eigenvalue** and ψ is an **eigenvector** of A. In this case λ is said to belong to the **point spectrum** of A. The eigenvalue λ is **nondegenerate** if the eigenvector ψ is unique (up to a multiplicative factor). If ψ is not unique, then λ is **degenerate.** The linearly-independent solutions of $A\psi = \lambda\psi$ form a subspace called the **eigenspace** for λ, and the dimension of the eigenspace is the **degree of degeneracy** for λ. The **continuous spectrum** of A is the set of complex numbers λ such that: (1) λ is not an eigenvalue, and (2) the operator $\lambda - A$ does not have an inverse.

The **adjoint (hermitian conjugate)** $A^\dagger$ of A is defined by

$$\left(\psi, A^\dagger \phi\right) = (\phi, A\psi)^*\,, \tag{A.18}$$

and A is **self-adjoint (hermitian)** if $\mathcal{D}\left(A^\dagger\right) = \mathcal{D}(A)$ and $(\phi, A\psi) = (A\phi, \psi)$. Bounded self-adjoint operators have real eigenvalues and a complete orthonormal set of eigenvectors. For unbounded self-adjoint operators, the point and continuous spectra are subsets of the real numbers. Note that $\left(\psi, A^\dagger A\psi\right) = (\phi, \phi)$, where $\phi = A\psi$, so that

$$\left(\psi, A^\dagger A\psi\right) \geqslant 0\,, \tag{A.19}$$

i.e. $A^\dagger A$ is positive definite.

A self-adjoint operator, P, satisfying

$$P^2 = P \tag{A.20}$$

is called a **projection operator**; it has only a point spectrum consisting of $\{0, 1\}$. Consider the set of vectors $P\mathfrak{H}$, consisting of all vectors of the form $P\psi$ as ψ ranges over $\mathfrak{H}$. This is a subspace of $\mathfrak{H}$, since

$$\alpha P\phi + \beta P\chi = P(\alpha\phi + \beta\chi) \tag{A.21}$$

shows that every linear combination of vectors in $P\mathfrak{H}$ is also in $P\mathfrak{H}$. Conversely, let $\mathfrak{S}$ be a subspace of $\mathfrak{H}$ and $\left\{\phi^{(n)}\right\}$ an orthonormal basis for $\mathfrak{S}$. The operator P, defined by

$$P\psi = \sum_n \left(\phi^{(n)}, \psi\right)\phi^{(n)}\,, \tag{A.22}$$

is a projection operator, since

$$P^2\psi = \sum_n \left(\phi^{(n)}, \psi\right) P\phi^{(n)} = \sum_n \left(\phi^{(n)}, \psi\right)\phi^{(n)} = P\psi\,. \tag{A.23}$$

Thus there is a one-to-one correspondence between projection operators and subspaces of $\mathfrak{H}$. Let P and Q be projection operators and suppose that the vectors in $P\mathfrak{H}$ are orthogonal to the vectors in $Q\mathfrak{H}$; then $PQ = QP = 0$ and P and Q are said to be **orthogonal projections**. In the extreme case that $\mathfrak{S} = \mathfrak{H}$, the expansion (A.10) shows that P is the identity operator, $P\psi = \psi$.

A self-adjoint operator with pure point spectrum $\{\lambda_1, \lambda_2, \ldots\}$ has the **spectral resolution**

$$A = \sum_n \lambda_n P_n \,, \tag{A.24}$$

where P_n is the projection operator onto the subspace of eigenvectors with eigenvalue λ_n. The spectral resolution for a self-adjoint operator A with a continuous spectrum is

$$A = \int \lambda \; d\mu\,(\lambda)\,, \tag{A.25}$$

where $d\mu\,(\lambda)$ is an **operator-valued measure** defined by the following statement: for each subset Δ of the real line,

$$P\,(\Delta) = \int_\Delta d\mu\,(\lambda) \tag{A.26}$$

is the projection operator onto the subspace of vectors ψ such that $\left\|(\lambda - A)^{-1}\,\psi\right\| < \infty$ for all $\lambda \notin \Delta$ (Riesz and Sz.-Nagy, 1955, Chap. VIII, Sec. 120).

A linear operator U is **unitary** if it preserves inner products, i.e.

$$(U\psi, U\phi) = (\psi, \phi) \tag{A.27}$$

for any pair of vectors ψ, ϕ in the Hilbert space. A necessary and sufficient condition for unitarity is that the operator is **norm preserving**, i.e.

$$(U\psi, U\psi) = (\psi, \psi) \quad \text{for all } \psi \text{ if and only if } U \text{ is unitary}\,. \tag{A.28}$$

The spectral resolution for a unitary operator with a pure point spectrum is

$$U = \sum_n e^{i\theta_n} P_n \,, \quad \theta_n \text{ real}\,, \tag{A.29}$$

and for a continuous spectrum

$$U = \int e^{i\theta} \; d\mu\,(\theta)\,, \quad \theta \;\text{ real}\,. \tag{A.30}$$

A linear operator N is said to be a **normal operator** if

$$\left[N, N^\dagger\right] = 0\,. \tag{A.31}$$

The hermitian and unitary operators are both normal. The hermitian operators $N_1 = \left(N + N^\dagger\right)/2$ and $N_2 = \left(N - N^\dagger\right)/2i$ satisfy $N = N_1 + iN_2$ and $[N_1, N_2] = 0$. Normal operators therefore have the spectral resolutions

$$N = \sum_n \left(x_n P_{1n} + i y_n P_{2n}\right), \quad [P_{1n}, P_{2m}] = 0 \tag{A.32}$$

for a point spectrum, and

$$N = \int x \, d\mu_1(x) + i \int y \, d\mu_2(y), \quad \left[\int_{\Delta_1} d\mu_1(x), \int_{\Delta_1} d\mu_2(y) \right] = 0 \tag{A.33}$$

for a continuous spectrum.

A.3.4 Matrices

A linear operator X acting on an N-dimensional Hilbert space, with basis $\{f^{(1)}, \ldots, f^{(N)}\}$, is represented by the $N \times N$ *matrix*

$$X_{mn} = \left(f^{(m)}, X f^{(n)} \right). \tag{A.34}$$

The operator and its matrix are both called X. The matrix for the product XY of two operators is the **matrix product**

$$(XY)_{mn} = \sum_{k=1}^{N} X_{mk} Y_{kn}. \tag{A.35}$$

The **determinant** of X is defined as

$$\det(X) = \sum_{n_1 \cdots n_N} \epsilon_{n_1 \cdots n_N} X_{1n_1} \cdots X_{Nn_N}, \tag{A.36}$$

where the generalized alternating tensor is

$$\sum_{n_1 \cdots n_N} \epsilon_{n_1 \cdots n_N} = \begin{cases} 1 & n_1 \cdots n_N \text{ is an even permutation of } 12 \cdots N, \\ -1 & n_1 \cdots n_N \text{ is an odd permutation of } 12 \cdots N, \\ 0 & \text{otherwise}. \end{cases} \tag{A.37}$$

The **trace of** X is

$$\operatorname{Tr} X = \sum_{n=1}^{N} X_{nn}. \tag{A.38}$$

The **transpose matrix** X^T is defined by $X^T_{nm} = X_{mn}$. The **adjoint matrix** $X^\dagger$ is the complex conjugate of the transpose: $X^\dagger_{nm} = X^*_{mn}$. A matrix X is **symmetric** if $X = X^T$, self-adjoint or hermitian if $X^\dagger = X$, and unitary if $X^\dagger X = XX^\dagger = I$, where I is the $N \times N$ identity matrix. Unitary transformations preserve the inner product. The hermitian and unitary matrices both belong to the larger class of **normal matrices** defined by $X^\dagger X = XX^\dagger$.

A matrix X is **positive definite** if all of its eigenvalues are real and non-negative. This immediately implies that the determinant and trace of the matrix are both non-negative. An equivalent definition is that X is positive definite if

$$\phi^\dagger X \phi \geqslant 0 \tag{A.39}$$

for all vectors ϕ. For a positive-definite matrix X, there is a matrix Y such that $X = YY^\dagger$.

The normal matrices have the following important properties (Mac Lane and Birkhoff, 1967, Sec. XI-10).

Theorem A.1 *(i) If f is an eigenvector of the normal matrix Z with eigenvalue z, then f is an eigenvector of $Z^\dagger$ with eigenvalue z^*, i.e. $Zf = zf \Rightarrow Z^\dagger f = z^* f$.*
(ii) Every normal matrix has a complete, orthonormal set of eigenvectors.

Thus hermitian matrices have real eigenvalues and unitary matrices have eigenvalues of modulus 1.

A.4 Fourier transforms

A.4.1 Continuous transforms

In the mathematical literature it is conventional to denote the **Fourier (integral) transform** of a function $f(x)$ of a single, real variable by

$$\widetilde{f}(k) = \int_{-\infty}^{\infty} dx f(x) e^{-ikx}, \tag{A.40}$$

so that the **inverse Fourier transform** is

$$f(x) = \int_{-\infty}^{\infty} \frac{dk}{2\pi} \widetilde{f}(k) e^{ikx}. \tag{A.41}$$

The virtue of this notation is that it reminds us that the two functions are, generally, drastically different, e.g. if $f(x) = 1$, then $\widetilde{f}(k) = 2\pi\delta(k)$.

On the other hand, the $\tilde{}$ is a typographical nuisance in any discussion involving many uses of the Fourier transform. For this reason, we will sacrifice precision for convenience. In our convention, the Fourier transform is indicated by the same letter, and the distinction between the *functions* is maintained by paying attention to the *arguments.*

The Fourier transform pair is accordingly written as

$$f(k) = \int_{-\infty}^{\infty} dx f(x) e^{-ikx}, \tag{A.42}$$

$$f(x) = \int_{-\infty}^{\infty} \frac{dk}{2\pi} f(k) e^{ikx}. \tag{A.43}$$

This is analogous to the familiar idea that the meaning of a vector $\mathbf{V}$ is independent of the coordinate system used, despite the fact that the components (V_x, V_y, V_z) of $\mathbf{V}$ are changed by transforming to a new coordinate system. From this point of view, the functions $f(x)$ and $f(k)$ are simply different representations of the same physical quantity. Confusion is readily avoided by paying attention to the physical significance of the arguments, e.g. x denotes a point in **position space**, while k denotes a point in the **reciprocal space** or k-space.

If the position-space function $f(x)$ is real, then the Fourier transform satisfies

$$f^*(k) = [f(k)]^* = f(-k). \tag{A.44}$$

When the position variable x is replaced by the time t, it is customary in physics to use the opposite sign convention:

$$f(\omega) = \int_{-\infty}^{\infty} dx f(x) e^{i\omega t}, \tag{A.45}$$

$$f(t) = \int_{-\infty}^{\infty} \frac{d\omega}{2\pi} f(k) e^{-i\omega t}. \tag{A.46}$$

Fourier transforms of functions of several variables, typically $f(\mathbf{r})$, are defined similarly:

$$f(\mathbf{k}) = \int d^3 r f(\mathbf{r}) e^{-i\mathbf{k}\cdot\mathbf{r}}, \tag{A.47}$$

$$f(\mathbf{r}) = \int \frac{d^3 k}{(2\pi)^3} f(\mathbf{k}) e^{i\mathbf{k}\cdot\mathbf{r}}, \tag{A.48}$$

where the integrals are over position space and reciprocal space ($\mathbf{k}$-space) respectively. If $f(\mathbf{r})$ is real then

$$f^*(\mathbf{k}) = f(-\mathbf{k}). \tag{A.49}$$

Combining these conventions for a space–time function $f(\mathbf{r}, t)$ yields the transform pair

$$f(\mathbf{k}, \omega) = \int d^3 r \int dt f(\mathbf{r}, t) e^{-i(\mathbf{k}\cdot\mathbf{r} - \omega t)}, \tag{A.50}$$

$$f(\mathbf{r}, t) = \int \frac{d^3 k}{(2\pi)^3} \int \frac{d\omega}{2\pi} f(\mathbf{k}, \omega) e^{i(\mathbf{k}\cdot\mathbf{r} - \omega t)}. \tag{A.51}$$

The last result is simply the plane-wave expansion of $f(\mathbf{r}, t)$. If $f(\mathbf{r}, t)$ is real, then the Fourier transform satisfies

$$f^*(\mathbf{k}, \omega) = f(-\mathbf{k}, -\omega). \tag{A.52}$$

Two related and important results on Fourier transforms—which we quote for the one- and three-dimensional cases—are **Parseval's** theorem:

$$\int dt f^*(t) g(t) = \int \frac{d\omega}{2\pi} f^*(\omega) g(\omega), \tag{A.53}$$

$$\int d^3 r f^*(\mathbf{r}) g(\mathbf{r}) = \int \frac{d^3 k}{(2\pi)^3} f^*(\mathbf{k}) g(\mathbf{k}), \tag{A.54}$$

and the **convolution** theorem:

$$h(t) = \int dt' f(t - t') g(t') \quad \text{if and only if} \quad h(\omega) = f(\omega) g(\omega), \tag{A.55}$$

$$h(\omega) = \int \frac{d\omega'}{2\pi} f(\omega - \omega') g(\omega') \quad \text{if and only if} \quad h(t) = f(t) g(t), \tag{A.56}$$

$$h(\mathbf{r}) = \int d^3 r' f(\mathbf{r} - \mathbf{r}') g(\mathbf{r}') \quad \text{if and only if} \quad h(\mathbf{k}) = f(\mathbf{k}) g(\mathbf{k}), \tag{A.57}$$

$$h(\mathbf{k}) = \int \frac{d^3 k'}{(2\pi)^3} f(\mathbf{k} - \mathbf{k}') g(\mathbf{k}') \quad \text{if and only if} \quad h(\mathbf{r}) = f(\mathbf{r}) g(\mathbf{r}). \tag{A.58}$$

These results are readily derived by using the delta function identities (A.95) and (A.96).

A.4.2 Fourier series

It is often useful to simplify the mathematics of the one-dimensional continuous transform by considering the functions to be defined on a finite interval $(-L/2, L/2)$ and imposing **periodic boundary conditions.** The basis vectors are still of the form $u_k(x) = C\exp(ikx)$, but the periodicity condition, $u_k(-L/2) = u_k(L/2)$, restricts k to the discrete values

$$k = \frac{2\pi n}{L} \quad (n = 0, \pm 1, \pm 2, \ldots) . \tag{A.59}$$

Normalization requires $C = 1/\sqrt{L}$, so the transform is

$$f_k = \frac{1}{\sqrt{L}} \int_{-L/2}^{L/2} dx f(x) e^{-ikx} , \tag{A.60}$$

and the inverse transform $f(x)$ is

$$f(x) = \frac{1}{\sqrt{L}} \sum_k f_k e^{ikx} . \tag{A.61}$$

The continuous transform is recovered in the limit $L \to \infty$ by first using eqn (A.60) to conclude that

$$\sqrt{L} f_k \to f(k) \quad \text{as } L \to \infty , \tag{A.62}$$

and writing the inverse transform as

$$f(x) = \frac{1}{L} \sum_k \sqrt{L} f_k e^{ikx} . \tag{A.63}$$

The difference between neighboring k-values is $\Delta k = 2\pi/L$, so this equation can be recast as

$$f(x) = \sum_k \frac{\Delta k}{2\pi} \sqrt{L} f_k e^{ikx} \to \int \frac{dk}{2\pi} f(k) e^{ikx} . \tag{A.64}$$

In Cartesian coordinates the three-dimensional discrete transform is defined on a rectangular parallelepiped with dimensions L_x, L_y, L_z. The one-dimensional results then imply

$$f_{\mathbf{k}} = \frac{1}{\sqrt{V}} \int_V d^3 r f(\mathbf{r}) e^{-i\mathbf{k}\cdot\mathbf{r}} , \tag{A.65}$$

where the $\mathbf{k}$-vector is restricted to

$$\mathbf{k} = \frac{2\pi n_x}{L_x} \mathbf{u}_x + \frac{2\pi n_y}{L_y} \mathbf{u}_y + \frac{2\pi n_z}{L_z} \mathbf{u}_z , \tag{A.66}$$

and $V = L_x L_y L_z$. The inverse transform is

$$f(\mathbf{r}) = \frac{1}{\sqrt{V}} \sum_{\mathbf{k}} f_{\mathbf{k}} e^{i\mathbf{k}\cdot\mathbf{r}} , \tag{A.67}$$

and the integral transform is recovered by

$$\sqrt{V} f_{\mathbf{k}} \to f(\mathbf{k}) \quad \text{as } V \to \infty\,. \tag{A.68}$$

The sum and integral over $\mathbf{k}$ are related by

$$\frac{1}{V}\sum_{\mathbf{k}} \to \int \frac{d^3k}{(2\pi)^3}\,, \tag{A.69}$$

which in turn implies

$$V\delta_{\mathbf{k},\mathbf{k}'} \to (2\pi)^3\,\delta(\mathbf{k}-\mathbf{k}')\,. \tag{A.70}$$

A.5 Laplace transforms

Another useful idea—which is closely related to the one-dimensional Fourier transform—is the **Laplace transform** defined by

$$\widetilde{f}(\zeta) = \int_0^\infty dt\; e^{-\zeta t} f(t)\,. \tag{A.71}$$

In this case, we will use the standard mathematical notation $\widetilde{f}(\zeta)$, since we do not use Laplace transforms as frequently as Fourier transforms. The inverse transform is

$$f(t) = \int_{\zeta_0 - i\infty}^{\zeta_0 + i\infty} \frac{d\zeta}{2\pi i}\, e^{\zeta t}\widetilde{f}(\zeta)\,. \tag{A.72}$$

The line $(\zeta_0 - i\infty, \zeta_0 + i\infty)$ in the complex ζ-plane must lie to the right of any poles in the transform function $\widetilde{f}(\zeta)$.

The identity

$$\widetilde{\left(\frac{df}{dt}\right)}(\zeta) = \zeta\widetilde{f}(\zeta) - f(0) \tag{A.73}$$

is useful in treating initial value problems for sets of linear, differential equations. Thus to solve the equations

$$\frac{df_n}{dt} = \sum_m V_{nm} f_m\,, \tag{A.74}$$

with a constant matrix V, and initial data $f_n(0)$, one takes the Laplace transform to get

$$\zeta\widetilde{f}_n(\zeta) - \sum_m V_{nm}\widetilde{f}_m(\zeta) = f_n(0)\,. \tag{A.75}$$

This set of algebraic equations can be solved to express $\widetilde{f}_n(\zeta)$ in terms of $f_n(0)$. Inverting the Laplace transform yields the solution in the time domain.

The convolution theorem for Laplace transforms is

$$\int_0^t dt'\, g(t-t')\, f(t') = \int_{\zeta_0 - i\infty}^{\zeta_0 + i\infty} \frac{d\zeta}{2\pi i}\,\widetilde{g}(\zeta)\,\widetilde{f}(\zeta)\, e^{\zeta t}\,, \tag{A.76}$$

where the integration contour is to the right of any poles of both $\widetilde{g}(\zeta)$ and $\widetilde{f}(\zeta)$.

An important point for applications to physics is that poles in the Laplace transform correspond to exponential time dependence. For example, the function $f(t) = \exp(zt)$ has the transform

$$\widetilde{f}(\zeta) = \frac{1}{\zeta - z}\,. \tag{A.77}$$

More generally, consider a function $\widetilde{f}(\zeta)$ with N simple poles in ζ:

$$\widetilde{f}(\zeta) = \frac{1}{(\zeta - z_1)\cdots(\zeta - z_N)}\,, \tag{A.78}$$

where the complex numbers $z_1, \ldots, z_N$ are all distinct. The inverse transform is

$$f(t) = \int_{\zeta_0 - i\infty}^{\zeta_0 + i\infty} \frac{d\zeta}{2\pi i} \frac{e^{\zeta t}}{(\zeta - z_1)\cdots(\zeta - z_N)}\,, \tag{A.79}$$

where $\zeta_0 > \max[\operatorname{Re} z_1, \ldots, \operatorname{Re} z_N]$. The contour can be closed by a large semicircle in the left half plane, and for $N > 1$ the contribution from the semicircle can be neglected. The integral is therefore given by the sum of the residues,

$$f(t) = \sum_{n=1}^{N} e^{z_n t} \prod_{j \neq n} \frac{1}{z_n - z_j}\,, \tag{A.80}$$

which explicitly exhibits $f(t)$ as a sum of exponentials.

A.6 Functional analysis

A.6.1 Linear functionals

In normal usage, a function, e.g. $f(x)$, is a rule assigning a unique value to each value of its argument. The argument is typically a point in some finite-dimensional space, e.g. the real numbers $\mathbb{R}$, the complex numbers $\mathbb{C}$, three-dimensional space $\mathbb{R}^3$, etc. The values of the function are also points in a finite-dimensional space. For example, the classical electric field is represented by a function $\boldsymbol{\mathcal{E}}(\mathbf{r})$ that assigns a vector—a point in $\mathbb{R}^3$—to each position $\mathbf{r}$ in $\mathbb{R}^3$.

A rule, X, assigning a value to each point f in an infinite-dimensional space $\mathfrak{M}$ (which is usually a space of functions) is called a **functional** and written as $X[f]$. The square brackets surrounding the argument are intended to distinguish functionals from functions of a finite number of variables.

If $\mathfrak{M}$ is a vector space, e.g. a Hilbert space, then a functional $Y[f]$ that obeys

$$Y[\alpha f + \beta g] = \alpha Y[f] + \beta Y[g]\,, \tag{A.81}$$

for all scalars α, β and all functions $f, g \in \mathfrak{M}$, is called a **linear functional**. The family, $\mathfrak{M}'$, of linear functionals on $\mathfrak{M}$ is called the **dual space of** $\mathfrak{M}$. The dual space is also a vector space, with linear combinations of its elements defined by

$$(\alpha X + \beta Y)[f] = \alpha X[f] + \beta Y[f] \tag{A.82}$$

for all $f \in \mathfrak{M}$.

A.6.2 Generalized functions

In Section 3.1.2 the definition (3.18) and the rule (3.21) are presented with the cavalier disregard for mathematical niceties that is customary in physics. There are however some situations in which more care is required. For these contingencies we briefly outline a more respectable treatment. The chief difficulty is the existence of the integrals defining the operators $s\left(-\nabla^2\right)$. This problem can be overcome by restricting the functions $\varphi(\mathbf{r})$ in eqn (3.18) to **good** functions (Lighthill, 1964, Chap. 2), i.e. infinitely-differentiable functions that fall off faster than any power of $|\mathbf{r}|$. The Fourier transform of a good function is also a good function, so all of the relevant integrals exist, as long as $s(|\mathbf{k}|)$ does not grow exponentially at large $|\mathbf{k}|$. The examples we need are all of the form $|\mathbf{k}|^\alpha$, where $-1 \leqslant \alpha \leqslant 1$, so eqns (3.18) and (3.21) are justified. For physical applications the really important assumption is that all functions can be approximated by good functions.

A **generalized function** is a linear functional, say $G[\varphi]$, defined on the good functions, i.e.

$$G[\varphi] \text{ is a complex number and } G[\alpha\varphi + \beta\psi] = \alpha G[\varphi] + \beta G[\varphi] \tag{A.83}$$

for any scalars α, β and any good functions φ, ψ. A familiar example is the delta function. The rule

$$\int d^3r\delta(\mathbf{r}-\mathbf{R})\varphi(\mathbf{r}) = \varphi(\mathbf{R}) \tag{A.84}$$

maps the function $\varphi(\mathbf{r})$ into the single number $\varphi(\mathbf{R})$. In this language, the transverse delta function $\Delta_{ij}^{\perp}(\mathbf{r}-\mathbf{r}')$ is also a generalized function. An alternative terminology, often found in the mathematical literature, labels good functions as **test** functions and generalized functions as **distributions**.

In quantum field theory, the notion of a generalized function is extended to linear functionals sending good functions to operators, i.e. for each good function φ,

$$X[\varphi] \text{ is an operator and } X[\alpha\varphi + \beta\psi] = \alpha X[\varphi] + \beta X[\varphi]\,. \tag{A.85}$$

Such functionals are called **operator-valued generalized functions**. For any density operator ρ describing a physical state, $X[\varphi]$ defines an ordinary (c-number) generalized function $X_\rho[\varphi]$ by

$$X_\rho[\varphi] = \operatorname{Tr}(\rho X[\varphi])\,. \tag{A.86}$$

A.7 Improper functions

A.7.1 The Heaviside step function

The **step function** $\theta(x)$ is defined by

$$\theta(x) = \begin{cases} 1 & \text{for } x > 0\,, \\ 0 & \text{for } x < 0\,, \end{cases} \tag{A.87}$$

and it has the useful representation

$$\theta(x) = -\lim_{\epsilon\to 0}\int_{-\infty}^{\infty}\frac{ds}{2\pi i}\frac{e^{-isx}}{s+i\epsilon}, \tag{A.88}$$

which is proved using contour integration.

A.7.2 The Dirac delta function

A Standard properties

(1) If the function $f(x)$ has isolated, simple zeros at the points $x^1, x^2, \ldots$ then

$$\delta(f(x)) = \sum_i \frac{1}{\left|\left(\frac{df}{dx}\right)_{x=x^i}\right|}\delta\left(x - x^i\right). \tag{A.89}$$

The multidimensional generalization of this rule is

$$\delta\left(\underline{f}(\underline{x})\right) = \sum_i \frac{1}{\left|\det\left(\frac{\partial \underline{f}}{\partial \underline{x}}\right)_{\underline{x}=\underline{x}^i}\right|}\delta\left(\underline{x} - \underline{x}^i\right), \tag{A.90}$$

where $\underline{x} = (x_1, x_2, \ldots, x_N)$, $\underline{f}(\underline{x}) = (f_1(\underline{x}), f_2(\underline{x}), \ldots, f_N(\underline{x}))$,

$$\begin{aligned}\delta\left(\underline{f}(\underline{x})\right) &= \delta\left(f_1(\underline{x})\right)\cdots\delta\left(f_N(\underline{x})\right),\\ \delta\left(\underline{x} - \underline{x}^i\right) &= \delta\left(x_1 - x_1^i\right)\cdots\delta\left(x_N - x_N^i\right),\end{aligned} \tag{A.91}$$

the Jacobian $\partial\underline{f}/\partial\underline{x}$ is the $N \times N$ matrix with components $\partial f_n/\partial x_m$, and $\underline{x}^i$ satisfies $f_n\left(\underline{x}^i\right) = 0$, for $n = 1, \ldots, N$.

(2) The derivative of the delta function is defined by

$$\int_{-\infty}^{\infty} dx f(x)\frac{d}{dx}\delta(x - a) = -\left(\frac{df}{dx}\right)_{x=a}. \tag{A.92}$$

(3) By using contour integration methods one gets

$$\lim_{\epsilon\to 0}\frac{1}{x + i\epsilon} = P\frac{1}{x} - i\pi\delta(x), \tag{A.93}$$

where P is the **principal part** defined by

$$P\int_{-\infty}^{\infty} dx\frac{f(x)}{x} = \lim_{a\to 0}\left\{\int_{-\infty}^{-a} dx\frac{f(x)}{x} + \int_a^{\infty} dx\frac{f(x)}{x}\right\}. \tag{A.94}$$

(4) The definition of the Fourier transform yields

$$\int dt e^{i(\omega-\nu)t} = 2\pi\delta(\omega - \nu) \tag{A.95}$$

in one dimension, and

$$\int d^3r e^{i(\mathbf{k}-\mathbf{q})\cdot\mathbf{r}} = (2\pi)^3\,\delta(\mathbf{k} - \mathbf{q}) \tag{A.96}$$

in three dimensions.

(5) The step function satisfies

$$\frac{d}{dx}\theta(x) = \delta(x) . \tag{A.97}$$

(6) The **end-point rule** is

$$\int_{-\infty}^{a} dx \delta(x-a) f(x) = \frac{1}{2} f(a) . \tag{A.98}$$

(7) The three-dimensional delta function $\delta(\mathbf{r}-\mathbf{r}')$ is defined as

$$\delta(\mathbf{r}-\mathbf{r}') = \delta(x-x')\,\delta(y-y')\,\delta(z-z') , \tag{A.99}$$

and is expressed in polar coordinates by

$$\delta(\mathbf{r}-\mathbf{r}') = \frac{1}{r^2}\delta(r-r')\,\delta(\cos\theta-\cos\theta')\,\delta(\phi-\phi') . \tag{A.100}$$

B A special representation of the delta function

In many calculations, particularly in perturbation theory, one encounters functions of the form

$$\xi(\omega, t) = \frac{\eta(\omega t)}{\omega} , \tag{A.101}$$

which have the limit

$$\lim_{t\to\infty} \xi(\omega, t) = \xi_0 \delta(\omega) , \tag{A.102}$$

provided that the integral

$$\xi_0 = \int_{-\infty}^{\infty} du \frac{\eta(u)}{u} \tag{A.103}$$

exists.

A.7.3 Integral kernels

The definition of a generalized function as a linear rule assigning a complex number to each good function can be extended to a linear rule that maps a good function, e.g. $f(t)$, to another good function $g(t)$. The linear nature of the rule means that it can always be expressed in the form

$$g(t) = \int dt' W(t, t') f(t') . \tag{A.104}$$

For a fixed value of t, $W(t, t')$ defines a generalized function of t' which is called an **integral kernel**. This definition is easily extended to functions of several variables, e.g. $f(\mathbf{r})$. The delta function, the Heaviside step function, etc. are examples of integral kernels. An integral kernel is **positive definite** if

$$\int dt \int dt' f^*(t) W(t, t') f(t') \geqslant 0 \tag{A.105}$$

for every good function $f(t)$.

A.8 Probability and random variables

A.8.1 Axioms of probability

The abstract definition of probability starts with a set Ω of **events** and a **probability** function P that assigns a numerical value to every subset of Ω. In principle, Ω could be any set, but in practice it is usually a subset of $\mathbb{R}^N$ or $\mathbb{C}^N$, or a subset of the integers. The essential properties of probabilities are contained in the axioms (Gardiner, 1985, Chap. 2):

(1) $P(S) \geqslant 0$ for all $S \subset \Omega$;

(2) $P(\Omega) = 1$;

(3) if $S_1, S_2, \ldots$ is a discrete (countable) collection of nonoverlapping sets, i.e.

$$S_i \cap S_j = \varnothing \quad \text{for } i \neq j\,, \tag{A.106}$$

then

$$P(S_1 \cup S_2 \cup \cdots) = \sum_j P(S_j)\,. \tag{A.107}$$

The familiar features $0 \leqslant P(S) \leqslant 1$, $P(\varnothing) = 0$, and $P(S') = 1 - P(S)$, where S' is the complement of S, are immediate consequences of the axioms. If Ω is a discrete (countable) set, then one writes $P(x) = P(\{x\})$, where $\{x\}$ is the set consisting of the single element x. If Ω is a continuous (uncountable) set, then it is customary to introduce a **probability density** $p(x)$ so that

$$P(S) = \int_S dx\; p(x)\,, \tag{A.108}$$

where dx is the natural volume element on Ω.

If $\Omega = \mathbb{R}^n$, the probability density is a function of n variables: $p(x_1, x_2, \ldots, x_n)$. The **marginal distribution** of x_j is then defined as

$$p_j(x_j) = \int dx_1 \cdots \int dx_{j-1} \int dx_{j+1} \cdots \int dx_n\; p(x_1, x_2, \ldots, x_n)\,. \tag{A.109}$$

The **joint probability** for two sets S and T is $P(S \cap T)$; this is the probability that an event in S is also in T. This is more often expressed with the notation

$$P(S, T) = P(S \cap T)\,, \tag{A.110}$$

which is used in the text. The **conditional probability** for S given T is

$$P(S \,|\, T) = \frac{P(S, T)}{P(T)} = \frac{P(S \cap T)}{P(T)}\,; \tag{A.111}$$

this is the probability that $x \in S$, given that $x \in T$.

The **compound probability rule** is just eqn (A.111) rewritten as

$$P(S,T) = P(S\,|\,T)\,P(T)\,. \tag{A.112}$$

This can be generalized to joint probabilities for more than two outcomes by applying it several times, e.g.

$$\begin{aligned} P(S,T,R) &= P(S\,|\,T,R)\,P(T,R) \\ &= P(S\,|\,T,R)\,P(T\,|\,R)\,P(R)\,. \end{aligned} \tag{A.113}$$

Dividing both sides by $P(R)$ yields the useful rule

$$P(S,T\,|\,R) = P(S\,|\,T,R)\,P(T\,|\,R)\,. \tag{A.114}$$

Two sets of events S and T are said to be **independent** or **statistically independent** if the joint probability is the product of the individual probabilities:

$$P(S,T) = P(S)\,P(T)\,. \tag{A.115}$$

A.8.2 Random variables

A **random variable** X is a function $X(x)$ defined on the event space Ω. The function can take on values in Ω or in some other set. For example, if $\Omega = \mathbb{R}$, then $X(t)$ could be a complex number or an integer. The **average value** of a random variable is

$$\langle X \rangle = \int dx\; p(x)\,X(x)\,. \tag{A.116}$$

If the function X does take on values in Ω, and is one–one, i.e. $X(x_1) = X(x_2)$ implies $x_1 = x_2$, then the distinction between $X(x)$ and x is often ignored.

Appendix B
Classical electrodynamics

B.1 Maxwell's equations

In SI units the microscopic form of Maxwell's equations is

$$\nabla \cdot \mathcal{E} = \frac{\rho}{\epsilon_0}, \tag{B.1}$$

$$\nabla \times \mathcal{B} = \mu_0 \left(\mathbf{j} + \epsilon_0 \frac{\partial \mathcal{E}}{\partial t} \right), \tag{B.2}$$

$$\nabla \times \mathcal{E} = -\frac{\partial \mathcal{B}}{\partial t}, \tag{B.3}$$

$$\nabla \cdot \mathcal{B} = 0. \tag{B.4}$$

The homogeneous equations (B.3) and (B.4) are identically satisfied by introducing the scalar potential φ and the vector potential $\mathcal{A}(\mathbf{r})$ and setting

$$\begin{aligned} \mathcal{B} &= \nabla \times \mathcal{A}, \\ \mathcal{E} &= -\frac{\partial \mathcal{A}}{\partial t} - \nabla \varphi. \end{aligned} \tag{B.5}$$

A further consequence of this representation is that eqn (B.1) becomes the **Poisson** equation

$$\nabla^2 \varphi = -\frac{\rho}{\epsilon_0}, \tag{B.6}$$

which has the Coulomb potential as its solution.

The vector and scalar potentials $\mathcal{A}$ and φ are not unique. The same electric and magnetic fields are produced by the new potentials $\mathcal{A}'$ and φ' defined by a **gauge transformation**,

$$\mathcal{A} \to \mathcal{A}' = \mathcal{A} + \nabla \chi, \tag{B.7}$$

$$\varphi \to \varphi' = \varphi - \partial \chi / \partial t, \tag{B.8}$$

where $\chi(\mathbf{r}, t)$ is any differentiable, real function. This is called **gauge invariance.** This property can be exploited to choose the gauge that is most convenient for the problem at hand. For example, it is always possible to perform a gauge transformation such that the new potentials satisfy $\nabla \cdot \mathcal{A} = 0$ and $\varphi = \Phi$, where Φ is a solution of eqn (B.6). This is called the **Coulomb** gauge, since $\varphi = \Phi$ is the Coulomb potential, or the **radiation** gauge, since the vector potential is transverse (Jackson, 1999, Sec. 6.3).

The flow of energy in the field is described by the continuity equation (**Poynting's theorem**),

$$\frac{\partial u(\mathbf{r},t)}{\partial t} + \nabla \cdot \mathbf{S}(\mathbf{r},t) = 0\,, \tag{B.9}$$

where

$$u(\mathbf{r},t) = \frac{\epsilon_0}{2}\boldsymbol{\mathcal{E}}^2(\mathbf{r},t) + \frac{1}{2\mu_0}\boldsymbol{\mathcal{B}}^2(\mathbf{r},t) \tag{B.10}$$

is the electromagnetic energy density, and the **Poynting vector**

$$\mathbf{S} = \frac{1}{\mu_0}\boldsymbol{\mathcal{E}} \times \boldsymbol{\mathcal{B}} \tag{B.11}$$

is the energy flux.

B.2 Electrodynamics in the frequency domain

It is often useful to describe the field in terms of its frequency and/or wavevector content. Let $F(\mathbf{r},t)$ be a real function representing any of the components of $\boldsymbol{\mathcal{E}}$, $\boldsymbol{\mathcal{B}}$, or $\boldsymbol{\mathcal{A}}$. Under the conventions established in Appendix A.4, the four-dimensional (frequency and wavevector) Fourier transform of $F(\mathbf{r},t)$ is

$$F(\mathbf{k},\omega) = \int d^3r \int dt e^{-i(\mathbf{k}\cdot\mathbf{r}-\omega t)} F(\mathbf{r},t)\,, \tag{B.12}$$

and the inverse transform is

$$F(\mathbf{r},t) = \int \frac{d^3k}{(2\pi)^3} \int_{-\infty}^{\infty} \frac{d\omega}{2\pi} F(\mathbf{k},\omega)\, e^{i(\mathbf{k}\cdot\mathbf{r}-\omega t)}\,. \tag{B.13}$$

According to eqn (A.52) the reality of $F(\mathbf{r},t)$ imposes the conditions

$$F^*(\mathbf{k},\omega) = F(-\mathbf{k},-\omega)\,. \tag{B.14}$$

For many applications it is also useful to consider the temporal Fourier transform at a fixed position $\mathbf{r}$:

$$F(\mathbf{r},\omega) = \int_{-\infty}^{\infty} dt e^{i\omega t} F(\mathbf{r},t)\,, \tag{B.15}$$

with the inverse transform

$$F(\mathbf{r},t) = \int_{-\infty}^{\infty} \frac{d\omega}{2\pi} F(\mathbf{r},\omega)\, e^{-i\omega t}\,. \tag{B.16}$$

The function $F(\mathbf{r},\omega)$ satisfies

$$F^*(\mathbf{r},\omega) = F(\mathbf{r},-\omega)\,. \tag{B.17}$$

The quantity $\left|F^{(+)}(\mathbf{r},\omega)\right|^2$ is called the **power spectrum** of F; it can be used to define an average frequency, ω_0, by

$$\omega_0 = \langle\omega\rangle = \frac{\int d^3r \int_{-\infty}^{\infty} \frac{d\omega}{2\pi} \left|F^{(+)}(\mathbf{r},\omega)\right|^2 \omega}{\int d^3r \int_{-\infty}^{\infty} \frac{d\omega}{2\pi} \left|F^{(+)}(\mathbf{r},\omega)\right|^2}. \tag{B.18}$$

The frequency spread of the field is characterized by the *rms* deviation $\Delta\omega$—the **frequency** or **spectral width**—defined by

$$(\Delta\omega)^2 = \left\langle(\omega-\omega_0)^2\right\rangle = \frac{\int d^3r \int_{-\infty}^{\infty} \frac{d\omega}{2\pi} \left|F^{(+)}(\mathbf{r},\omega)\right|^2 (\omega-\omega_0)^2}{\int d^3r \int_{-\infty}^{\infty} \frac{d\omega}{2\pi} \left|F^{(+)}(\mathbf{r},\omega)\right|^2}. \tag{B.19}$$

The average wavevector $\mathbf{k}_0$ and deviation Δk are similarly defined by

$$\mathbf{k}_0 = \langle\mathbf{k}\rangle = \frac{\int \frac{d^3k}{(2\pi)^3} \int_{-\infty}^{\infty} \frac{d\omega}{2\pi} \left|F^{(+)}(\mathbf{k},\omega)\right|^2 \mathbf{k}}{\int \frac{d^3k}{(2\pi)^3} \int_{-\infty}^{\infty} \frac{d\omega}{2\pi} \left|F^{(+)}(\mathbf{k},\omega)\right|^2}, \tag{B.20}$$

$$(\Delta k)^2 = \left\langle(\mathbf{k}-\mathbf{k}_0)^2\right\rangle = \frac{\int \frac{d^3k}{(2\pi)^3} \int_{-\infty}^{\infty} \frac{d\omega}{2\pi} \left|F^{(+)}(\mathbf{k},\omega)\right|^2 (\mathbf{k}-\mathbf{k}_0)^2}{\int \frac{d^3k}{(2\pi)^3} \int_{-\infty}^{\infty} \frac{d\omega}{2\pi} \left|F^{(+)}(\mathbf{k},\omega)\right|^2}. \tag{B.21}$$

B.3 Wave equations

The microscopic Maxwell equations (B.1)–(B.4) can be replaced by two second-order wave equations for $\boldsymbol{\mathcal{E}}$ and $\boldsymbol{\mathcal{B}}$:

$$\left(\nabla^2 - \frac{1}{c^2}\frac{\partial^2}{\partial t^2}\right)\boldsymbol{\mathcal{E}} = \mu_0\frac{\partial \mathbf{j}}{\partial t} + \frac{1}{\epsilon_0}\boldsymbol{\nabla}\rho, \tag{B.22}$$

$$\left(\nabla^2 - \frac{1}{c^2}\frac{\partial^2}{\partial t^2}\right)\boldsymbol{\mathcal{B}} = -\mu_0\boldsymbol{\nabla}\times\mathbf{j}, \tag{B.23}$$

and the first-order equations for the transverse vector potential $\boldsymbol{\mathcal{A}}$ can be combined to yield the wave equation

$$\left(\nabla^2 - \frac{1}{c^2}\frac{\partial^2}{\partial t^2}\right)\boldsymbol{\mathcal{A}} = -\mu_0\mathbf{j}^{\perp}, \tag{B.24}$$

where $\mathbf{j}^{\perp}$ is the transverse part (see Section 2.1.1-B) of the current.

B.3.1 Propagation in the vacuum

Since the $\boldsymbol{\mathcal{B}}$ field and the transverse part of the $\boldsymbol{\mathcal{E}}$ field are derived from the vector potential, we can concentrate on the wave equation (B.24) for the vector potential. In the vacuum case ($\mathbf{j}=0$), $\boldsymbol{\mathcal{A}}$ satisfies

$$\left(\nabla^2 - \frac{1}{c^2}\frac{\partial^2}{\partial t^2}\right)\boldsymbol{\mathcal{A}}(\mathbf{r},t) = 0, \tag{B.25}$$

and the **transversality condition** $\boldsymbol{\nabla}\cdot\boldsymbol{\mathcal{A}}=0$.

The general solution of the wave equation can be obtained by a four-dimensional Fourier transform, which yields

$$\left(-k^2 + \frac{\omega^2}{c^2}\right) \boldsymbol{\mathcal{A}}(\mathbf{k}, \omega) = 0\,. \tag{B.26}$$

The solution of the last equation is

$$\boldsymbol{\mathcal{A}}(\mathbf{k}, \omega) = \boldsymbol{\mathcal{A}}^{(+)}(\mathbf{k})\, 2\pi\delta(\omega - ck) + \boldsymbol{\mathcal{A}}^{(-)}(\mathbf{k})\, 2\pi\delta(\omega + ck)\,, \tag{B.27}$$

and the reality of $\boldsymbol{\mathcal{A}}(\mathbf{r}, t)$ requires

$$\left[\boldsymbol{\mathcal{A}}^{(+)}(\mathbf{k})\right]^* = \boldsymbol{\mathcal{A}}^{(-)}(-\mathbf{k})\,. \tag{B.28}$$

The inverse transform yields the general solution in $(\mathbf{r}, t)$-space as

$$\boldsymbol{\mathcal{A}}(\mathbf{r}, t) = \boldsymbol{\mathcal{A}}^{(+)}(\mathbf{r}, t) + \boldsymbol{\mathcal{A}}^{(-)}(\mathbf{r}, t)\,, \tag{B.29}$$

where

$$\boldsymbol{\mathcal{A}}^{(+)}(\mathbf{r}, t) = \int \frac{d^3k}{(2\pi)^3} \boldsymbol{\mathcal{A}}^{(+)}(\mathbf{k})\, e^{i(\mathbf{k}\cdot\mathbf{r} - \omega_k t)} = \left[\boldsymbol{\mathcal{A}}^{(-)}(\mathbf{r}, t)\right]^*, \tag{B.30}$$

and $\omega_k = ck$.

The relation between the $\boldsymbol{\mathcal{E}}$ and $\boldsymbol{\mathcal{B}}$ fields and the vector potential can be used to express them in the same way. In **k**-space

$$\boldsymbol{\mathcal{E}}^{(+)}(\mathbf{k}) = i\omega_k \boldsymbol{\mathcal{A}}^{(+)}(\mathbf{k})\,, \quad \boldsymbol{\mathcal{B}}^{(+)}(\mathbf{k}) = i\mathbf{k} \times \boldsymbol{\mathcal{A}}^{(+)}(\mathbf{k})\,, \tag{B.31}$$

and in $(\mathbf{r}, t)$-space

$$\begin{aligned} \boldsymbol{\mathcal{E}}(\mathbf{r}, t) &= \boldsymbol{\mathcal{E}}^{(+)}(\mathbf{r}, t) + \boldsymbol{\mathcal{E}}^{(-)}(\mathbf{r}, t)\,, \\ \boldsymbol{\mathcal{B}}(\mathbf{r}, t) &= \boldsymbol{\mathcal{B}}^{(+)}(\mathbf{r}, t) + \boldsymbol{\mathcal{B}}^{(-)}(\mathbf{r}, t)\,, \end{aligned} \tag{B.32}$$

where

$$\begin{aligned} \boldsymbol{\mathcal{E}}^{(+)}(\mathbf{r}, t) &= \int \frac{d^3k}{(2\pi)^3} i\omega_k \boldsymbol{\mathcal{A}}^{(+)}(\mathbf{k})\, e^{i(\mathbf{k}\cdot\mathbf{r} - \omega_k t)} &= \left[\boldsymbol{\mathcal{E}}^{(-)}(\mathbf{r}, t)\right]^*, \\ \boldsymbol{\mathcal{B}}^{(+)}(\mathbf{r}, t) &= \int \frac{d^3k}{(2\pi)^3} i\mathbf{k} \times \boldsymbol{\mathcal{A}}^{(+)}(\mathbf{k})\, e^{i(\mathbf{k}\cdot\mathbf{r} - \omega_k t)} &= \left[\boldsymbol{\mathcal{B}}^{(-)}(\mathbf{r}, t)\right]^*. \end{aligned} \tag{B.33}$$

B.3.2 Linear and circular polarization

The forms in eqns (B.32) and (B.33) are valid for any real vector solutions of the wave equation, but we are only interested in transverse fields, e.g. $\boldsymbol{\mathcal{A}}(\mathbf{r}, t)$ should satisfy $\boldsymbol{\nabla} \cdot \boldsymbol{\mathcal{A}}(\mathbf{r}, t) = 0$, as well as the wave equation. In **k**-space the transversality condition, $\mathbf{k} \cdot \boldsymbol{\mathcal{A}}^{(+)}(\mathbf{k}) = 0$, requires $\boldsymbol{\mathcal{A}}^{(+)}(\mathbf{k})$ to lie in the plane perpendicular to the direction of the **k**-vector. We choose two unit vectors $\mathbf{e}_1(\mathbf{k})$ and $\mathbf{e}_2(\mathbf{k})$ such that $\left\{\mathbf{e}_1(\mathbf{k}), \mathbf{e}_2(\mathbf{k}), \widetilde{\mathbf{k}}\right\}$ form a right-handed coordinate system, where $\widetilde{\mathbf{k}} = \mathbf{k}/k$ is the unit vector along the propagation direction. The unit vectors $\mathbf{e}_1$ and $\mathbf{e}_2$ are called **polarization vectors**.

Since an arbitrary vector can be expanded in the basis $\{\mathbf{e}_1(\mathbf{k}), \mathbf{e}_2(\mathbf{k}), \widetilde{\mathbf{k}}\}$, the three unit vectors satisfy the completeness relation

$$\sum_s e_{si}e_{sj} + \widetilde{k}_i\widetilde{k}_j = \delta_{ij}\,, \tag{B.34}$$

as well as the conditions

$$\mathbf{k}\cdot\mathbf{e}_s(\mathbf{k}) = 0\,, \tag{B.35}$$

$$\mathbf{e}_s(\mathbf{k})\cdot\mathbf{e}_{s'}(\mathbf{k}) = \delta_{ss'}\,, \tag{B.36}$$

$$\mathbf{e}_1\times\mathbf{e}_2 = \widetilde{\mathbf{k}}\ \ (et\ cycl)\,, \tag{B.37}$$

where $s, s' = 1, 2$. The vector $\boldsymbol{\mathcal{A}}^{(+)}(\mathbf{k})$ can therefore be expanded as

$$\boldsymbol{\mathcal{A}}^{(+)}(\mathbf{k}) = \sum_s \mathcal{A}_s^{(+)}(\mathbf{k})\,\mathbf{e}_s(\mathbf{k})\,, \tag{B.38}$$

where the **polarization components** $\mathcal{A}_s^{(+)}(\mathbf{k})$ are defined by

$$\mathcal{A}_s^{(+)}(\mathbf{k}) = \mathbf{e}_s(\mathbf{k})\cdot\boldsymbol{\mathcal{A}}^{(+)}(\mathbf{k})\,. \tag{B.39}$$

The general transverse solution of the wave equation is therefore given by eqn (B.29) with

$$\boldsymbol{\mathcal{A}}^{(+)}(\mathbf{r},t) = \int\frac{d^3k}{(2\pi)^3}\sum_s \mathcal{A}_s^{(+)}(\mathbf{k})\,\mathbf{e}_s(\mathbf{k})\,e^{i(\mathbf{k}\cdot\mathbf{r}-\omega_k t)}\,. \tag{B.40}$$

Each plane-wave contribution to the solution of the wave equation, say for $\boldsymbol{\mathcal{E}}$ and $\boldsymbol{\mathcal{B}}$, has the form

$$\boldsymbol{\mathcal{E}} = \mathrm{Re}\left[(\mathcal{E}_1\mathbf{e}_1 + \mathcal{E}_2\mathbf{e}_2)\,e^{i(\mathbf{k}\cdot\mathbf{r}-\omega_k t)}\right], \tag{B.41}$$

$$\boldsymbol{\mathcal{B}} = \frac{1}{c}\widetilde{\mathbf{k}}\times\boldsymbol{\mathcal{E}}\,, \tag{B.42}$$

where $\mathcal{E}_1$ and $\mathcal{E}_2$ are complex scalar amplitudes, and $\mathbf{e}_1$ and $\mathbf{e}_2$ are real polarization vectors $\mathbf{k}$. If the phases of $\mathcal{E}_1$ and $\mathcal{E}_2$ are equal, the field is **linearly polarized**. If the phases are different, the field is **elliptically polarized**. In general, the time-averaged Poynting vector is

$$\mathbf{S} = \frac{1}{\mu_0}\mathrm{Re}\left\langle\boldsymbol{\mathcal{E}}\times\boldsymbol{\mathcal{B}}^*\right\rangle = \frac{1}{2}\sqrt{\frac{\epsilon_0}{\mu_0}}\left[|\mathcal{E}_1|^2 + |\mathcal{E}_2|^2\right]\widetilde{\mathbf{k}}\,, \tag{B.43}$$

so the intensity is given by

$$I = |\mathbf{S}| = \frac{\epsilon_0}{2}c\left[|\mathcal{E}_1|^2 + |\mathcal{E}_2|^2\right]. \tag{B.44}$$

If the two phases differ by 90°, then

$$\mathcal{E}_1\mathbf{e}_1 + \mathcal{E}_2\mathbf{e}_2 = \mathcal{E}_0(\mathbf{e}_1 \pm i\mathbf{e}_2)\,, \tag{B.45}$$

where $\mathcal{E}_0$ is real. The field is then said to be **circularly polarized**. In this case it is useful to introduce the complex unit vectors

$$\mathbf{e}_s = \frac{1}{\sqrt{2}}\left(\mathbf{e}_1 + is\mathbf{e}_2\right), \tag{B.46}$$

where $s = \pm 1$. The complex vectors satisfy the (hermitian) orthogonality relation

$$\mathbf{e}_s^* \cdot \mathbf{e}_{s'} = \delta_{ss'}\,, \tag{B.47}$$

and the completeness relation

$$\sum_s e_{si}e_{sj}^* + \widetilde{k}_i\widetilde{k}_j = \delta_{ij}\,. \tag{B.48}$$

The transversality, orthogonality, and completeness properties of the linear and circular polarization vectors are both incorporated in the relations

$$\begin{aligned} \mathbf{k}\cdot\mathbf{e}_s(\mathbf{k}) &= 0 && \text{(transversality)}\,,\\ \mathbf{e}_s^*(\mathbf{k})\cdot\mathbf{e}_{s'}(\mathbf{k}) &= \delta_{ss'} && \text{(orthonormality)}\,,\\ \sum_s e_{si}(\mathbf{k})\,e_{sj}^*(\mathbf{k}) &= \delta_{ij} - \widetilde{k}_i\widetilde{k}_j && \text{(completeness)}\,. \end{aligned} \tag{B.49}$$

Note that the completeness relation can also be written as

$$\sum_s e_{si}(\mathbf{k})\,e_{sj}^*(\mathbf{k}) = \Delta_{ij}^{\perp}(\mathbf{k})\,, \tag{B.50}$$

where $\Delta_{ij}^{\perp}(\mathbf{k})$ is the Fourier transform of the transverse delta function. The general solution (B.40) has the same form as for linear polarizations, but the polarization component is now given by

$$\mathcal{A}_s^{(+)}(\mathbf{k}) = \mathbf{e}_s^*(\mathbf{k})\cdot\boldsymbol{\mathcal{A}}^{(+)}(\mathbf{k})\,. \tag{B.51}$$

In addition to eqn (B.49), the circular polarization vectors satisfy

$$\widetilde{\mathbf{k}}\times\mathbf{e}_s = -is\mathbf{e}_s\,, \tag{B.52}$$

$$\mathbf{e}_s\times\mathbf{e}_{s'}^* = -is\delta_{ss'}\widetilde{\mathbf{k}}\,, \tag{B.53}$$

where $s, s' = \pm 1$.

The linear polarization basis for a given **k**-vector is not uniquely defined, since a new basis defined by a rotation around the **k**-direction also forms a right-handed coordinate system. It is therefore useful to consider the transformation properties of the polarization basis. Let ϑ be the rotation angle around $\widetilde{\mathbf{k}}$; then the linear polarization vectors transform by

$$\begin{aligned} \mathbf{e}_1' &= \mathbf{e}_1\cos\vartheta + \mathbf{e}_2\sin\vartheta\,,\\ \mathbf{e}_2' &= -\mathbf{e}_1\sin\vartheta + \mathbf{e}_2\cos\vartheta\,, \end{aligned} \tag{B.54}$$

which implies

$$\mathbf{e}_s' = \mathbf{e}_1' + is\mathbf{e}_2' = e^{-is\vartheta}\mathbf{e}_s\,. \tag{B.55}$$

When viewed at a fixed point in space by an observer facing into the propagation direction of the wave (toward the source), the unit vector $\mathbf{e}_+$ ($\mathbf{e}_-$) describes a phasor rotating counterclockwise (clockwise). In the traditional terminology of optics and spectroscopy, $\mathbf{e}_+$ ($\mathbf{e}_-$) is said to be left (right) circularly polarized. In the fields of quantum electronics and laser physics, the observer is assumed to be facing along the propagation direction (away from the source), so the sense of rotation is reversed. In this convention $\mathbf{e}_+$ ($\mathbf{e}_-$) is said to be **right (left) circularly polarized**. In more modern language $\mathbf{e}_+$ ($\mathbf{e}_-$) is said to have **positive (negative) helicity** (Jackson, 1999, Sec. 7.2).

For a plane wave with propagation vector $\mathbf{k}$, there are two amplitudes $\mathcal{E}_s(\mathbf{k})$, where for circular (linear) polarization $s = \pm 1$ ($s = 1, 2$). The general vacuum solution can be expressed as a superposition of plane waves. In this context it is customary to change the notation by setting

$$\mathcal{E}_s(\mathbf{k}) = 2i\sqrt{\frac{\hbar\omega_k}{2\epsilon_0}}\alpha_s(\mathbf{k}), \tag{B.56}$$

where the $\hbar$ is introduced only to guarantee that $|\alpha_s(\mathbf{k})|^2$ is a density in k-space, i.e. the new amplitude $\alpha_s(\mathbf{k})$ has dimensions $L^{3/2}$. This yields the Fourier integral expansion

$$\boldsymbol{\mathcal{E}}^{(+)}(\mathbf{r},t) = i\int\frac{d^3k}{(2\pi)^3}\sqrt{\frac{\hbar\omega_k}{2\epsilon_0}}\sum_s\alpha_s(\mathbf{k})\,\mathbf{e}_s(\mathbf{k})\,e^{i(\mathbf{k}\cdot\mathbf{r}-\omega_k t)}\,. \tag{B.57}$$

The Fourier integral representation is often replaced by a discrete Fourier series:

$$\boldsymbol{\mathcal{E}}^{(+)}(\mathbf{r},t) = \sum_{\mathbf{k}s}\sqrt{\frac{\hbar\omega_k}{2\epsilon_0 V}}\alpha_{\mathbf{k}s}\mathbf{e}_{\mathbf{k}s}e^{i(\mathbf{k}\cdot\mathbf{r}-\omega_k t)}\,, \tag{B.58}$$

where $\mathbf{e}_{\mathbf{k}s} = \mathbf{e}_s(\mathbf{k})$, $\alpha_{\mathbf{k}s} = \alpha_s(\mathbf{k})/\sqrt{V}$, and V is the volume of the imaginary cube used to define the discrete Fourier series.

B.3.3 Spatial inversion and time reversal

Maxwell's equations are invariant under the discrete transformations

$$\mathbf{r} \to -\mathbf{r}\ (\textbf{spatial inversion}\text{ or }\textbf{parity transformation}) \tag{B.59}$$

and

$$t \to -t\ (\textbf{time reversal})\,, \tag{B.60}$$

as well as all continuous Lorentz transformations (Jackson, 1999, Sec. 6.10). The physical meaning of spatial inversion is as follows. If a system of charges and fields evolves from an initial to a final state, then the spatially-inverted initial state will evolve into the spatially-inverted final state. Time-reversal invariance means that the time-reversed final state will evolve into the time-reversed initial state.

For any physical quantity X, let $X \to X^P$ and $X \to X^T$ denote the transformations for spatial inversion and time reversal respectively. The invariance of Maxwell's equations under spatial inversion is achieved by the transformation rules

$$\rho^P(\mathbf{r},t) = \rho(-\mathbf{r},t)\,,\quad \mathbf{j}^P(\mathbf{r},t) = -\mathbf{j}(-\mathbf{r},t)\,, \tag{B.61}$$

$$\boldsymbol{\mathcal{E}}^P(\mathbf{r},t) = -\boldsymbol{\mathcal{E}}(-\mathbf{r},t)\,, \tag{B.62}$$

$$\boldsymbol{\mathcal{B}}^P(\mathbf{r},t) = \boldsymbol{\mathcal{B}}(-\mathbf{r},t)\,. \tag{B.63}$$

Thus the current density and the electric field have odd parity, and the charge density and the magnetic field have even parity. Vectors with odd parity are called **polar** vectors, and those with even parity are called **axial** vectors, so $\boldsymbol{\mathcal{E}}$ is a polar vector and $\boldsymbol{\mathcal{B}}$ is an axial vector.

Time-reversal invariance is guaranteed by

$$\rho^T(\mathbf{r},t) = \rho(\mathbf{r},-t)\,,\ \ \mathbf{j}^T(\mathbf{r},t) = -\mathbf{j}(\mathbf{r},-t)\,, \tag{B.64}$$

$$\boldsymbol{\mathcal{E}}^T(\mathbf{r},t) = \boldsymbol{\mathcal{E}}(\mathbf{r},-t)\,, \tag{B.65}$$

$$\boldsymbol{\mathcal{B}}^T(\mathbf{r},t) = -\boldsymbol{\mathcal{B}}(\mathbf{r},-t)\,. \tag{B.66}$$

As a consequence of these rules, the energy density and Poynting vector satisfy

$$\begin{aligned} u^P(\mathbf{r},t) &= u(-\mathbf{r},t)\,,\ \ \mathbf{S}^P(\mathbf{r},t) = -\mathbf{S}(-\mathbf{r},t)\,,\\ u^T(\mathbf{r},t) &= u(\mathbf{r},-t)\,,\ \ \mathbf{S}^T(\mathbf{r},t) = -\mathbf{S}(\mathbf{r},-t)\,. \end{aligned} \tag{B.67}$$

For many applications, e.g. to scattering problems, it is more useful to work out the transformation laws for the amplitudes in a plane-wave expansion of the field. We begin by using eqn (B.58) to express the two sides of eqn (B.62) as

$$\boldsymbol{\mathcal{E}}^P(\mathbf{r},t) = \sum_{\mathbf{k}s} i\sqrt{\frac{\hbar\omega_k}{2\epsilon_0 V}}\alpha^P_{\mathbf{k}s}\mathbf{e}_{\mathbf{k}s}e^{i(\mathbf{k}\cdot\mathbf{r}-\omega_k t)} + \text{CC} \tag{B.68}$$

and

$$-\boldsymbol{\mathcal{E}}(-\mathbf{r},t) = -\sum_{\mathbf{k}s} i\sqrt{\frac{\hbar\omega_k}{2\epsilon_0 V}}\alpha_{\mathbf{k}s}\mathbf{e}_{\mathbf{k}s}e^{i(-\mathbf{k}\cdot\mathbf{r}-\omega_k t)} + \text{CC}\,. \tag{B.69}$$

Changing $\mathbf{k}$ to $-\mathbf{k}$ in the last result and equating the coefficients of corresponding plane waves yields

$$\sum_s \alpha^P_{\mathbf{k}s}\mathbf{e}_{\mathbf{k}s} = -\sum_s \alpha_{-\mathbf{k},s}\mathbf{e}_{-\mathbf{k},s}\,. \tag{B.70}$$

In order to proceed, we need to relate the polarization vectors for $\mathbf{k}$ and $-\mathbf{k}$. As a shorthand notation, set $\mathbf{e}_s = \mathbf{e}_{\mathbf{k}s}$, $\mathbf{e}'_s = \mathbf{e}_{-\mathbf{k},s}$, and $\mathbf{k}' = -\mathbf{k}$. The vectors $\mathbf{e}'_s$ lie in the same plane as the vectors $\mathbf{e}_s$, so they can be expressed as linear combinations of $\mathbf{e}_1$ and $\mathbf{e}_2$. After imposing the conditions (B.35)–(B.37) on the basis $\{\mathbf{e}'_1, \mathbf{e}'_2, \mathbf{k}'\}$, the relation between the two basis sets must have the form

$$\begin{aligned} \mathbf{e}'_1 &= \mathbf{e}_1\cos\vartheta + \mathbf{e}_2\sin\vartheta\,,\\ \mathbf{e}'_2 &= \mathbf{e}_1\sin\vartheta - \mathbf{e}_2\cos\vartheta\,. \end{aligned} \tag{B.71}$$

The transformation matrices in eqns (B.54) and (B.71) represent proper and improper rotations respectively. The improper rotation in eqn (B.71) can be expressed as the product of a proper rotation and a reflection through some line in the plane orthogonal to $\mathbf{k}$. Since the polarization basis can be freely chosen, it is convenient to establish a convention by setting $\vartheta = 0$, i.e.

$$\mathbf{e}_{-\mathbf{k},1} = \mathbf{e}_{\mathbf{k}1}\,, \quad \mathbf{e}_{-\mathbf{k},2} = -\mathbf{e}_{\mathbf{k}2}\,. \tag{B.72}$$

For the circular polarization basis, with $s = \pm$, this rule takes the equivalent forms

$$\begin{aligned} \mathbf{e}_{-\mathbf{k},s} &= \mathbf{e}_{\mathbf{k}s}^{*}\,, \\ \mathbf{e}_{-\mathbf{k},-s} &= \mathbf{e}_{\mathbf{k}s}\,. \end{aligned} \tag{B.73}$$

The transformation law derived by applying this rule to eqn (B.70) is

$$\alpha_{\mathbf{k}s}^{P} = -\alpha_{-\mathbf{k},-s} \quad (s = \pm)\,, \tag{B.74}$$

which relates the amplitude for a given wavevector and circular polarization to the amplitude for the opposite wavevector and opposite circular polarization. For the linear polarization basis the corresponding result is

$$\begin{aligned} \alpha_{\mathbf{k}1}^{P} &= -\alpha_{-\mathbf{k},1}\,, \\ \alpha_{\mathbf{k}2}^{P} &= \alpha_{-\mathbf{k},2}\,. \end{aligned} \tag{B.75}$$

Turning next to time reversal, we express the right side of eqn (B.65) as

$$\boldsymbol{\mathcal{E}}\,(\mathbf{r}, -t) = \sum_{\mathbf{k}s} i\sqrt{\frac{\hbar\omega_k}{2\epsilon_0 V}}\alpha_{\mathbf{k}s}\mathbf{e}_{\mathbf{k}s}e^{i(\mathbf{k}\cdot\mathbf{r}+\omega_k t)} - \sum_{\mathbf{k}s} i\sqrt{\frac{\hbar\omega_k}{2\epsilon_0 V}}\alpha_{\mathbf{k}s}^{*}\mathbf{e}_{\mathbf{k}s}^{*}e^{-i(\mathbf{k}\cdot\mathbf{r}+\omega_k t)}\,, \tag{B.76}$$

and again change the summation variable by $\mathbf{k} \to -\mathbf{k}$. This is to be compared to the expansion for $\boldsymbol{\mathcal{E}}^{T}\,(\mathbf{r}, t)$, which is given by eqn (B.68) with $\alpha_{\mathbf{k}s}^{P}$ replaced by $\alpha_{\mathbf{k}s}^{T}$. The result is

$$\sum_{s} \alpha_{\mathbf{k}s}^{T}\mathbf{e}_{\mathbf{k}s} = -\sum_{s} \alpha_{-\mathbf{k},s}^{*}\mathbf{e}_{-\mathbf{k},s}^{*}\,. \tag{B.77}$$

The circular polarization vectors satisfy $\mathbf{e}_{-\mathbf{k},s}^{*} = \mathbf{e}_{\mathbf{k},-s}^{*} = \mathbf{e}_{\mathbf{k},s}$, so the transformation law in this basis is

$$\alpha_{\mathbf{k}s}^{T} = -\alpha_{-\mathbf{k},s}^{*}\,. \tag{B.78}$$

Thus for time reversal the amplitude for $(\mathbf{k}, s)$ is related to the conjugate of the amplitude for $(-\mathbf{k}, s)$. The wavevector is reversed, but the circular polarization is unchanged. For the linear basis the result is

$$\alpha_{\mathbf{k}1}^{T} = -\alpha_{-\mathbf{k},1}^{*}\,, \tag{B.79}$$

$$\alpha_{\mathbf{k}2}^{T} = \alpha_{-\mathbf{k},2}^{*}\,. \tag{B.80}$$

B.4 Planar cavity

A limiting case of the rectangular cavity discussed in Section 2.1.1 is the **planar** cavity, with $L_1 = L_2 = L$ and $L_3 = d \ll L$. In most applications, only the limit $L \to \infty$ will be relevant, so the only physically meaningful boundary conditions are those at the planes $z = 0$ and $z = d$. Periodic boundary conditions can be used at the other faces of the cavity. Thus the *ansatz* for the solution is $\boldsymbol{\mathcal{E}} = e^{i\mathbf{q}\cdot\mathbf{r}}U(z)$, where $\mathbf{q} = (k_x, k_y)$ is the transverse part of the wavevector $\mathbf{k}$. Inserting this into eqns (2.11), (2.1), and (2.13) leads to the mode functions $\boldsymbol{\mathcal{E}}_{\mathbf{q}ns}$. For $n = 0$ there is only one polarization:

$$\boldsymbol{\mathcal{E}}_{\mathbf{q}0} = \frac{1}{\sqrt{L^2 d}} e^{i\mathbf{q}\cdot\mathbf{r}}\, \mathbf{u}_z\,. \tag{B.81}$$

For $n \geqslant 1$ there are two polarizations, the P polarization in the $(\widetilde{\mathbf{q}}, \mathbf{u}_z)$-plane and the orthogonal S polarization along $\mathbf{u}_z \times \widetilde{\mathbf{q}}$:

$$\boldsymbol{\mathcal{E}}_{\mathbf{q}n1} = \sqrt{\frac{2}{L^2 d}} \left(\frac{n\lambda_{qn}}{2d}\right) e^{i\mathbf{q}\cdot\mathbf{r}} \left\{\sin\left(k_z z\right) \widetilde{\mathbf{q}} + i\frac{q}{k_z} \cos\left(k_z z\right) \mathbf{u}_z\right\}, \tag{B.82}$$

$$\boldsymbol{\mathcal{E}}_{\mathbf{q}n2} = \sqrt{\frac{2}{L^2 d}} e^{i\mathbf{q}\cdot\mathbf{r}} \sin\left(k_z z\right) \mathbf{u}_z \times \widetilde{\mathbf{q}}\,, \tag{B.83}$$

where $\lambda_{qn} = 2\pi c/\omega_{qn}$. The mode frequency is

$$\omega_{qn} = c\sqrt{q^2 + (n\pi/d)^2}\,, \tag{B.84}$$

and the expansion of a general real field is

$$\boldsymbol{\mathcal{E}}\left(\mathbf{r}\right) = i \sum_{\mathbf{q}} \sum_{n=0}^{\infty} \sum_{s=1}^{C_n} \left\{a_{\mathbf{q}ns} \boldsymbol{\mathcal{E}}_{\mathbf{q}ns}\left(z\right) e^{i\mathbf{q}\cdot\mathbf{r}} - \mathrm{CC}\right\}, \tag{B.85}$$

where $C_0 = 1$ and $C_n = 2$ for $n \geqslant 1$.

B.5 Macroscopic Maxwell equations

The macroscopic Maxwell equations are given by (Jackson, 1999, Sec. 6.1)

$$\nabla \cdot \boldsymbol{\mathcal{D}}\left(\mathbf{r}, t\right) = \rho\left(\mathbf{r}, t\right), \tag{B.86}$$

$$\nabla \times \boldsymbol{\mathcal{H}}\left(\mathbf{r}, t\right) = \mathbf{J}\left(\mathbf{r}, t\right) + \frac{\partial \boldsymbol{\mathcal{D}}\left(\mathbf{r}, t\right)}{\partial t}, \tag{B.87}$$

$$\nabla \times \boldsymbol{\mathcal{E}}\left(\mathbf{r}, t\right) = -\frac{\partial \boldsymbol{\mathcal{B}}\left(\mathbf{r}, t\right)}{\partial t}, \tag{B.88}$$

$$\nabla \cdot \boldsymbol{\mathcal{B}}\left(\mathbf{r}, t\right) = 0\,, \tag{B.89}$$

$$\boldsymbol{\mathcal{D}}\left(\mathbf{r}, t\right) = \epsilon_0 \boldsymbol{\mathcal{E}}\left(\mathbf{r}, t\right) + \boldsymbol{\mathcal{P}}\left(\mathbf{r}, t\right), \tag{B.90}$$

$$\boldsymbol{\mathcal{H}}\left(\mathbf{r}, t\right) = \frac{1}{\mu_0} \boldsymbol{\mathcal{B}}\left(\mathbf{r}, t\right) - \boldsymbol{\mathcal{M}}\left(\mathbf{r}, t\right). \tag{B.91}$$

In these equations ρ and $\mathbf{J}$ respectively represent the **charge density** and **current density** of the free charges, $\boldsymbol{\mathcal{P}}$ is the **polarization density** (density of the electric dipole moment), $\boldsymbol{\mathcal{M}}$ is the **magnetization** (density of the magnetic dipole moment), $\boldsymbol{\mathcal{D}}$ is the **displacement field**, and $\boldsymbol{\mathcal{H}}$ is the **magnetic field**.

After Fourier transforming in $\mathbf{r}$ and t, Maxwell's equations reduce to the algebraic relations

$$\mathbf{k}\cdot\boldsymbol{\mathcal{D}}(\mathbf{k},\omega)=-i\rho(\mathbf{k},\omega)\,, \tag{B.92}$$
$$\mathbf{k}\times\boldsymbol{\mathcal{H}}(\mathbf{k},\omega)=-i\mathbf{J}(\mathbf{k},\omega)-\omega\boldsymbol{\mathcal{D}}(\mathbf{k},\omega)\,, \tag{B.93}$$
$$\mathbf{k}\times\boldsymbol{\mathcal{E}}(\mathbf{k},\omega)=\omega\boldsymbol{\mathcal{B}}(\mathbf{k},\omega)\,, \tag{B.94}$$
$$\mathbf{k}\cdot\boldsymbol{\mathcal{B}}(\mathbf{k},\omega)=0\,, \tag{B.95}$$
$$\boldsymbol{\mathcal{D}}(\mathbf{k},\omega)=\epsilon_0\boldsymbol{\mathcal{E}}(\mathbf{k},\omega)+\boldsymbol{\mathcal{P}}(\mathbf{k},\omega)\,, \tag{B.96}$$
$$\boldsymbol{\mathcal{H}}(\mathbf{k},\omega)=\frac{1}{\mu_0}\boldsymbol{\mathcal{B}}(\mathbf{k},\omega)-\boldsymbol{\mathcal{M}}(\mathbf{k},\omega)\,. \tag{B.97}$$

The microscopic Poynting's theorem (B.9) is replaced by

$$\boldsymbol{\mathcal{E}}\cdot\frac{\partial\boldsymbol{\mathcal{D}}}{\partial t}+\boldsymbol{\mathcal{H}}\cdot\frac{\partial\boldsymbol{\mathcal{B}}}{\partial t}+\boldsymbol{\nabla}\cdot\mathbf{S}=0\,, \tag{B.98}$$

where $\mathbf{S}=\boldsymbol{\mathcal{E}}\times\boldsymbol{\mathcal{H}}$ (Jackson, 1999, Sec. 6.7).

For a **nondispersive** medium, i.e.

$$\mathcal{D}_i(\mathbf{r},t)=\epsilon_{ij}\mathcal{E}_j(\mathbf{r},t)\,,\quad \mathcal{B}_i(\mathbf{r},t)=\mu_{ij}\mathcal{H}_j(\mathbf{r},t)\,, \tag{B.99}$$

where ϵ_{ij} and μ_{ij} are constant tensors, eqn (B.98) takes the form

$$\frac{\partial u(\mathbf{r},t)}{\partial t}+\nabla\cdot\mathbf{S}(\mathbf{r},t)=0\,, \tag{B.100}$$

with the energy density

$$u=\frac{1}{2}\left\{\boldsymbol{\mathcal{E}}\cdot\boldsymbol{\mathcal{D}}+\boldsymbol{\mathcal{B}}\cdot\boldsymbol{\mathcal{H}}\right\} \tag{B.101}$$
$$=\frac{1}{2}\left\{\mathcal{E}_i\epsilon_{ij}\mathcal{E}_j+\mathcal{B}_i\left(\mu^{-1}\right)_{ij}\mathcal{B}_j\right\}. \tag{B.102}$$

The most important materials for quantum optics are nonmagnetic dielectrics with $\mu_{ij}(\omega)=\mu_0\delta_{ij}$. In this case eqns (B.86)–(B.91) can be converted into a wave equation for the transverse part of the electric field:

$$\left(\nabla^2-\frac{1}{c^2}\frac{\partial^2}{\partial t^2}\right)\boldsymbol{\mathcal{E}}^{\perp}=\mu_0\frac{\partial^2}{\partial t^2}\boldsymbol{\mathcal{P}}_{\perp}+\mu_0\frac{\partial}{\partial t}\mathbf{J}^{\perp}\,. \tag{B.103}$$

B.5.1 Dispersive linear media

We consider a medium which interacts weakly with external fields. This can happen either because the fields themselves are weak or because the effective coupling constants are small. In general, the polarization and magnetization at a space–time point $x=(\mathbf{r},t)$ can depend on the action of the field at earlier times and at distant points in space. Combining this with the weak interaction assumption leads to the **linear constitutive equations** (Jackson, 1999, p. 14)

$$\mathcal{P}_i(\mathbf{r},t) = \epsilon_0 \int d^3r' \int dt' \chi_{ij}^{(1)}(\mathbf{r}-\mathbf{r}', t-t') \mathcal{E}_j(\mathbf{r}',t'), \tag{B.104}$$

$$\mathcal{M}_i(\mathbf{r},t) = \int d^3r' \int dt' \xi_{ij}^{(1)}(\mathbf{r}-\mathbf{r}', t-t') \mathcal{H}_j(\mathbf{r}',t'), \tag{B.105}$$

where $\chi_{ij}^{(1)}(\mathbf{r}-\mathbf{r}', t-t')$ and $\xi_{ij}^{(1)}(\mathbf{r}-\mathbf{r}', t-t')$ are respectively the (linear) electric and magnetic susceptibility tensors. Thus the relation between the polarization $\boldsymbol{\mathcal{P}}(\mathbf{r},t)$ (magnetization $\boldsymbol{\mathcal{M}}(\mathbf{r},t)$) and the field $\boldsymbol{\mathcal{E}}(\mathbf{r},t)$ ($\boldsymbol{\mathcal{H}}(\mathbf{r},t)$) is nonlocal in both space and time. The principle of causality prohibits $\boldsymbol{\mathcal{P}}(\mathbf{r},t)$ ($\boldsymbol{\mathcal{M}}(\mathbf{r},t)$) from depending on the field $\boldsymbol{\mathcal{E}}(\mathbf{r},t')$ ($\boldsymbol{\mathcal{H}}(\mathbf{r},t')$) at later times, $t' > t$, so the susceptibilities must satisfy

$$\left.\begin{aligned} \chi_{ij}^{(1)}(\mathbf{r}-\mathbf{r}', t-t') = 0\,, \\ \xi_{ij}^{(1)}(\mathbf{r}-\mathbf{r}', t-t') = 0 \end{aligned}\right\} \text{ for } t' > t\,. \tag{B.106}$$

This leads to the famous Kramers–Kronig relations (Jackson, 1999, Sec. 7.10).

The four-dimensional convolution theorem, obtained by combining eqns (A.55) and (A.57), allows eqns (B.104) and (B.105) to be recast in Fourier space as

$$\mathcal{P}_i(\mathbf{k},\omega) = \epsilon_0 \chi_{ij}^{(1)}(\mathbf{k},\omega) \mathcal{E}_j(\mathbf{k},\omega), \tag{B.107}$$

$$\mathcal{M}_i(\mathbf{k},\omega) = \xi_{ij}^{(1)}(\mathbf{k},\omega) \mathcal{E}_j(\mathbf{k},\omega). \tag{B.108}$$

Combining these relations with the definitions (B.90) and (B.91) produces

$$\mathcal{D}_i(\mathbf{k},\omega) = \epsilon_{ij}(\mathbf{k},\omega)\, \mathcal{E}_j(\mathbf{k},\omega) \tag{B.109}$$

and

$$\mathcal{B}_i(\mathbf{k},\omega) = \mu_{ij}(\mathbf{k},\omega)\, \mathcal{H}_j(\mathbf{k},\omega), \tag{B.110}$$

where

$$\epsilon_{ij}(\mathbf{k},\omega) \equiv \epsilon_0 \left(\delta_{ij} + \chi_{ij}^{(1)}(\mathbf{k},\omega)\right) \tag{B.111}$$

and

$$\mu_{ij}(\mathbf{k},\omega) \equiv \mu_0 \left(\delta_{ij} + \xi_{ij}^{(1)}(\mathbf{k},\omega)\right) \tag{B.112}$$

are respectively the (**electric**) **permittivity tensor** and the (**magnetic**) **permeability tensor**. The classical fields, the polarization, the magnetization, and the (space–time) susceptibilities are all real; therefore, the Fourier transforms satisfy

$$\begin{aligned} \boldsymbol{\mathcal{P}}^*(\mathbf{k},\omega) &= \boldsymbol{\mathcal{P}}(-\mathbf{k},-\omega)\,, & \boldsymbol{\mathcal{E}}^*(\mathbf{k},\omega) &= \boldsymbol{\mathcal{E}}(-\mathbf{k},-\omega)\,, \\ \boldsymbol{\mathcal{M}}^*(\mathbf{k},\omega) &= \boldsymbol{\mathcal{M}}(-\mathbf{k},-\omega)\,, & \boldsymbol{\mathcal{B}}^*(\mathbf{k},\omega) &= \boldsymbol{\mathcal{B}}(-\mathbf{k},-\omega)\,, \\ \chi_{ij}^{(1)*}(\mathbf{k},\omega) &= \chi_{ij}^{(1)}(-\mathbf{k},-\omega)\,, & \xi_{ij}^{(1)*}(\mathbf{k},\omega) &= \xi_{ij}^{(1)}(-\mathbf{k},-\omega)\,. \end{aligned} \tag{B.113}$$

The dependence of $\chi_{ij}^{(1)}(\mathbf{k},\omega)$ and $\xi_{ij}^{(1)}(\mathbf{k},\omega)$ on $\mathbf{k}$ is called **spatial dispersion**, and the dependence on ω is called **frequency dispersion**. Interactions between atoms at different points in the medium can cause the polarization at a point $\mathbf{r}$ to depend on the field in a neighborhood of $\mathbf{r}$, defined by a spatial correlation length a_s. In gases, liquids, and disordered solids a_s is of the order of the interatomic spacing, which is generally very small compared to vacuum optical wavelengths λ_0. Thus the polarization at $\mathbf{r}$ can be treated as depending only on the field at $\mathbf{r}$. Since the medium is assumed to be spatially homogeneous, this means that $\chi_{ij}^{(1)}(\mathbf{r}-\mathbf{r}', t-t') = \chi_{ij}^{(1)}(t-t')\,\delta(\mathbf{r}-\mathbf{r}')$, which is equivalent to $\chi_{ij}^{(1)}(\mathbf{k},\omega) = \chi_{ij}^{(1)}(\omega)$. Similar relations hold for the magnetic susceptibility. These three types of media are also isotropic (rotationally symmetric), so the tensor quantities can be replaced by scalars which depend only on ω:

$$\begin{aligned} \epsilon_{ij}(\mathbf{k},\omega) &\to \epsilon(\omega)\,\delta_{ij}\,, \\ \epsilon(\omega) &= \epsilon_0\left(1+\chi^{(1)}(\omega)\right), \end{aligned} \tag{B.114}$$

$$\begin{aligned} \mu_{ij}(\mathbf{k},\omega) &\to \mu(\omega)\,\delta_{ij}\,, \\ \mu(\omega) &= \mu_0\left(1+\xi^{(1)}(\omega)\right). \end{aligned} \tag{B.115}$$

Using eqn (B.114) in eqn (B.109) and transforming back to position space produces the useful relation

$$\boldsymbol{\mathcal{D}}(\mathbf{r},\omega) = \epsilon(\omega)\,\boldsymbol{\mathcal{E}}(\mathbf{r},\omega)\,. \tag{B.116}$$

For crystalline solids, rotational symmetry is replaced by symmetry under the crystal group, and the tensor character of the susceptibilities cannot be ignored. In this case a_s is the lattice spacing, so the ratio a_s/λ_0 is still small, but spatial dispersion cannot always be neglected. The reason is that the relevant parameter is $n(\omega_0)\,a_s/\lambda_0$, where $n(\omega_0)$ is the index of refraction at the frequency $\omega_0 = 2\pi c/\lambda_0$. Thus spatial dispersion can be significant if the index is large.

In a rare stroke of good fortune, the crystals of interest for quantum optics satisfy the condition for weak spatial dispersion, $n(\omega_0)\,a_s/\lambda_0 \ll 1$ (Agranovich and Ginzburg, 1984); therefore, we can still use a permittivity tensor that only depends on frequency:

$$\epsilon_{ij}(\mathbf{k},\omega) \to \epsilon_{ij}(\omega)\,. \tag{B.117}$$

For most applications of quantum optics, we can also assume that the permittivity tensor is symmetrical: $\epsilon_{ij}(\omega) = \epsilon_{ji}(\omega)$. Physically this means that the crystal is both transparent and non-gyrotropic (not optically active) (Agranovich and Ginzburg, 1984, Chap. 1). We also assume the existence of the inverse tensor $\left(\epsilon^{-1}\right)_{ij}$.

There are other situations, e.g. propagation in a plasma exposed to an external magnetic field, that require the full tensors $\epsilon_{ij}(\mathbf{k},\omega)$ and $\mu_{ij}(\mathbf{k},\omega)$ depending on $\mathbf{k}$ (Pines, 1963, Chaps 3 and 4; Ginzburg, 1970, Sec. 1.2). For nearly all applications of quantum optics, we can neglect spatial dispersion and assume the forms (B.114) or (B.117) for the permittivity tensor.

The inertia of the charges and currents in the medium, together with dissipative effects, imply that the medium cannot respond instantaneously to changes in the field at a given point $\mathbf{r}$. Thus the polarization at the position $\mathbf{r}$ and time t will in general depend on the field at earlier times $t' < t$. Since the response times of gases, liquids, and solids exhibit considerable variation, it is not generally possible to ignore frequency dispersion.

B.5.2 Isotropic linear dielectrics

Here we assume that $\mu_{ij}(\omega) = \mu_0 \delta_{ij}$, so that $\mathcal{H} = \mathcal{B}/\mu_0$, and set $\epsilon_{ij}(\omega) = \delta_{ij}\epsilon(\omega)$ and $\rho = \mathbf{J} = 0$ in eqns (B.92)–(B.95) to get

$$\mathbf{k} \cdot \mathcal{E}(\mathbf{k}, \omega) = 0, \tag{B.118}$$

$$\mathbf{k} \times \mathcal{B}(\mathbf{k}, \omega) = -\frac{\omega}{c^2}\epsilon_r(\omega)\,\mathcal{E}(\mathbf{k}, \omega), \tag{B.119}$$

$$\mathbf{k} \times \mathcal{E}(\mathbf{k}, \omega) = \omega\mathcal{B}(\mathbf{k}, \omega), \tag{B.120}$$

$$\mathbf{k} \cdot \mathcal{B}(\mathbf{k}, \omega) = 0, \tag{B.121}$$

where $\epsilon_r(\omega) = \epsilon(\omega)/\epsilon_0$ is the **relative permittivity**. The final equation follows from eqn (B.120), and eliminating $\mathcal{B}$ between eqn (B.119) and eqn (B.120) leads to

$$\mathbf{k} \times [\mathbf{k} \times \mathcal{E}(\mathbf{k}, \omega)] = -\frac{\omega^2}{c^2}\epsilon_r(\omega)\,\mathcal{E}(\mathbf{k}, \omega). \tag{B.122}$$

The identity $\mathbf{a} \times (\mathbf{b} \times \mathbf{c}) = (\mathbf{a} \cdot \mathbf{c})\,\mathbf{b} - (\mathbf{a} \cdot \mathbf{b})\,\mathbf{c}$, together with eqn (B.118), reduces this to

$$\left[\frac{\omega^2}{c^2}n^2(\omega) - k^2\right]\mathcal{E}(\mathbf{k}, \omega) = 0, \tag{B.123}$$

where $n(\omega) = \sqrt{\epsilon_r(\omega)}$ is the **index of refraction**. In general $\epsilon_r(\omega)$ can be complex, corresponding to absorption or gain at particular frequencies (Jackson, 1999, Chap. 7), but for frequencies in the transparent part of the spectrum $\epsilon_r(\omega)$ is real and positive. The relation $\mathcal{E} = -\partial\mathcal{A}/\partial t$ implies $\mathcal{E}(\mathbf{k}, \omega) = i\omega\mathcal{A}(\mathbf{k}, \omega)$, so the vector potential satisfies the same equation

$$\left[\frac{\omega^2}{c^2}n^2(\omega) - k^2\right]\mathcal{A}(\mathbf{k}, \omega) = 0. \tag{B.124}$$

For a transparent medium the general transverse solution of eqn (B.124) is

$$\mathcal{A}(\mathbf{k}, \omega) = \sum_s \mathcal{A}_s(\mathbf{k})\,\mathbf{e}_s(\mathbf{k})\,\delta(\omega - \omega(k)) + \sum_s \mathcal{A}_s^*(-\mathbf{k})\,\mathbf{e}_s^*(-\mathbf{k})\,\delta(\omega + \omega(k)), \tag{B.125}$$

where $\omega(k)$ is a positive, real solution of the **dispersion relation**

$$\omega n(\omega) = ck. \tag{B.126}$$

Thus the fundamental plane-wave solution in position–time is

$$e^{i(\mathbf{k}\cdot\mathbf{r} - \omega(k)t)}\mathbf{e}_s(\mathbf{k}), \tag{B.127}$$

and the positive-frequency part has the general form

$$\boldsymbol{\mathcal{A}}^{(+)}(\mathbf{r},t) = \int \frac{d^3k}{(2\pi)^3} \sum_s \mathcal{A}_s(\mathbf{k})\, \mathbf{e}_s(\mathbf{k})\, e^{i(\mathbf{k}\cdot\mathbf{r}-\omega(k)t)} \tag{B.128}$$

for the vector potential, and

$$\boldsymbol{\mathcal{E}}^{(+)}(\mathbf{r},t) = i\int \frac{d^3k}{(2\pi)^3}\omega(k) \sum_s \mathcal{A}_s(\mathbf{k})\, \mathbf{e}_s(\mathbf{k})\, e^{i(\mathbf{k}\cdot\mathbf{r}-\omega(k)t)} \tag{B.129}$$

for the electric field.

B.5.3 Anisotropic linear dielectrics

We again assume that $\mu_{ij}(\omega) = \mu_0\delta_{ij}$ and set $\rho = \mathbf{J} = 0$ in eqns (B.92)–(B.95), but we drop the assumption $\epsilon_{ij}(\omega) = \delta_{ij}\epsilon(\omega)$. In this case $\boldsymbol{\mathcal{E}}$ and $\boldsymbol{\mathcal{D}}$ are not necessarily parallel, so we combine eqn (B.93) with eqn (B.94) to get

$$k^2\Delta_{ij}^{\perp}\mathcal{E}_j = \mu_0\omega^2\mathcal{D}_i\,. \tag{B.130}$$

In the following we will use a matrix notation in which a second-rank tensor X_{ij} is represented as a 3×3 matrix $\overleftrightarrow{X}$ and a vector $\mathbf{V} = V_1\mathbf{u}_x + V_2\mathbf{u}_y + V_3\mathbf{u}_z$ is represented by column or row matrices according to the convention

$$\overrightarrow{V} = \begin{pmatrix} V_1 \\ V_2 \\ V_3 \end{pmatrix}, \quad \overrightarrow{V}^T = (V_1, V_2, V_3)\,. \tag{B.131}$$

The polarization properties of the solution are best described in terms of $\boldsymbol{\mathcal{D}}$, since eqn (B.92) guarantees that it is orthogonal to $\mathbf{k}$. Thus we solve eqn (B.109) for $\overrightarrow{\mathcal{E}}$ and substitute the result into the left side of eqn (B.130) to find

$$k^2\overleftrightarrow{\Delta}^{\perp}\overrightarrow{\mathcal{E}} = k^2\overleftrightarrow{\Delta}^{\perp}\frac{1}{\epsilon_0}\left[\overleftrightarrow{\epsilon}_r\right]^{-1}\overrightarrow{\mathcal{D}} = k^2\overleftrightarrow{\Delta}^{\perp}\frac{1}{\epsilon_0}\left[\overleftrightarrow{\epsilon}_r\right]^{-1}\overleftrightarrow{\Delta}^{\perp}\overrightarrow{\mathcal{D}}\,, \tag{B.132}$$

where $(\epsilon_r)_{ij}(\omega) = \epsilon_{ij}(\omega)/\epsilon_0$ is the **relative permittivity tensor**. The last form depends on the fact that $\overrightarrow{\mathcal{D}}$ is transverse. Putting this together with the right side of eqn (B.130) yields

$$\overleftrightarrow{\mathcal{S}}\overrightarrow{\mathcal{D}} = \frac{\omega^2}{k^2c^2}\overrightarrow{\mathcal{D}}\,, \tag{B.133}$$

where the **transverse impermeability tensor**,[1]

$$\overleftrightarrow{\mathcal{S}}(\widetilde{\mathbf{k}},\omega) = \overleftrightarrow{\Delta}^{\perp}(\widetilde{\mathbf{k}})\left[\overleftrightarrow{\epsilon}_r(\omega)\right]^{-1}\overleftrightarrow{\Delta}^{\perp}(\widetilde{\mathbf{k}})\,, \tag{B.134}$$

depends on the frequency ω and the unit vector $\widetilde{\mathbf{k}} = \mathbf{k}/k$ along the propagation vector. The real, symmetric matrix $\overleftrightarrow{\mathcal{S}}$ annihilates $\overrightarrow{k}$:

$$\overleftrightarrow{\mathcal{S}}\overrightarrow{k} = 0\,, \tag{B.135}$$

so $\overleftrightarrow{\mathcal{S}}$ has one eigenvalue zero, corresponding to the eigenvector $\widetilde{\mathbf{k}}$. From eqn (B.133), it is clear that the transverse vector $\overrightarrow{\mathcal{D}}$ is one of the remaining two eigenvectors that are orthogonal to $\widetilde{\mathbf{k}}$.

[1] This is a slight modification of the approach found in Yariv and Yeh (1984, Chap. 4).

If ω lies in the transparent region for the crystal, the tensor $\overleftrightarrow{\epsilon}_r(\omega)$ is positive definite, so that the nonzero eigenvalues of $\overleftrightarrow{\mathcal{S}}$ are positive. We write the positive eigenvalues as $1/n_s^2$, so that the corresponding eigenvectors satisfy

$$\overleftrightarrow{\mathcal{S}}\,\overrightarrow{\varepsilon}_s = \frac{1}{n_s^2}\overrightarrow{\varepsilon}_s \quad (s = 1, 2)\,. \tag{B.136}$$

If $\overrightarrow{\mathcal{D}}$ is parallel to an eigenvector, i.e. $\overrightarrow{\mathcal{D}} = Y_s\overrightarrow{\varepsilon}_s$, one finds the dispersion relation

$$c^2k^2 = \omega^2 n_s^2(\omega)\,; \tag{B.137}$$

in other words, n_s is the index of refraction associated with the **eigenpolarization** $\overrightarrow{\varepsilon}_s(\mathbf{k})$. Since the matrix $\overleftrightarrow{\mathcal{S}}$ depends on the direction of propagation $\widetilde{\mathbf{k}}$, the indices $n_s(\omega)$ generally also depend on $\widetilde{\mathbf{k}}$. In order to simplify the notation, this dependence is not indicated explicitly, e.g. as $n_s(\omega, \widetilde{\mathbf{k}})$, but is implicitly indicated by the dependence of the refractive index on the polarization index. An incident wave with propagation vector $\mathbf{k}$ exhibits **birefringence**, i.e. it produces two refracted waves corresponding to the two phase velocities c/n_1 and c/n_2. Since $\overrightarrow{\varepsilon}_s(\mathbf{k})$ is real, the eigenpolarizations are linear, and they can be normalized so that

$$\overrightarrow{\varepsilon}_s^T(\mathbf{k})\,\overrightarrow{\varepsilon}_{s'}(\mathbf{k}) = \boldsymbol{\varepsilon}_s(\mathbf{k})\cdot\boldsymbol{\varepsilon}_{s'}(\mathbf{k}) = \delta_{ss'}\,. \tag{B.138}$$

Radiation is described by the transverse part of the electric field, and for the special solution $\overrightarrow{\mathcal{D}} = Y_s\overrightarrow{\varepsilon}_s$ the transverse electric field in $(\mathbf{k}, \omega)$-space is

$$\overrightarrow{\mathcal{E}}_s(\mathbf{k}, \omega) = \frac{1}{\epsilon_0}\overleftrightarrow{\mathcal{S}}\,Y_s\overrightarrow{\varepsilon}_s = \frac{Y_s}{\epsilon_0 n_s^2}\overrightarrow{\varepsilon}_s\,, \tag{B.139}$$

where $\omega_s(\mathbf{k})$ is a solution of eqn (B.137) and $n_s(k) \equiv n_s(\omega_s(\mathbf{k}))$. The general space–time solution,

$$\overrightarrow{\mathcal{E}}^{(+)}(\mathbf{r}, t) = \frac{1}{\epsilon_0}\int\frac{d^3k}{(2\pi)^3}\sum_{s=1}^{2}\frac{Y_s(\mathbf{k})}{n_s^2(k)}\overrightarrow{\varepsilon}_s(\mathbf{k})\,e^{i(\mathbf{k}\cdot\mathbf{r}-\omega_s(\mathbf{k})t)}\,, \tag{B.140}$$

is a superposition of elliptically-polarized waves with axes that rotate as the wave propagates through the crystal. If only one polarization is present, e.g. $Y_2(\mathbf{k}) = 0$, each wave is linearly polarized, and the polarization direction is preserved in propagation.

It is customary and useful to get a representation similar to eqn (B.57) for the isotropic problem by setting

$$Y_s(\mathbf{k}) = i n_s(k)\sqrt{\frac{\hbar\epsilon_0\omega_s(\mathbf{k})}{2}}\,\alpha_s(\mathbf{k})\,, \tag{B.141}$$

so that the transverse part of the electric field is

$$\boldsymbol{\mathcal{E}}^{(+)}(\mathbf{r}, t) = i\int\frac{d^3k}{(2\pi)^3}\sum_{s=1}^{2}\sqrt{\frac{\hbar\omega_s(\mathbf{k})}{2\epsilon_0 n_s^2(k)}}\,\alpha_s(\mathbf{k})\,\boldsymbol{\varepsilon}_s(\mathbf{k})\,e^{i(\mathbf{k}\cdot\mathbf{r}-\omega_s(\mathbf{k})t)}\,. \tag{B.142}$$

The corresponding expansion using box-normalized plane waves is

$$\boldsymbol{\mathcal{E}}_{\perp}^{(+)}(\mathbf{r},t) = i\sum_{\mathbf{k}s}\sqrt{\frac{\hbar\omega_{\mathbf{k}s}}{2\epsilon_0 V n_{ks}^2}}\alpha_{\mathbf{k}s}\boldsymbol{\varepsilon}_{\mathbf{k}s}e^{i(\mathbf{k}\cdot\mathbf{r}-\omega_{\mathbf{k}s}t)}, \tag{B.143}$$

where $\omega_{\mathbf{k}s} = \omega_s(\mathbf{k})$, $n_{ks} = n_s(k)$, $\boldsymbol{\varepsilon}_{\mathbf{k}s} = \boldsymbol{\varepsilon}_s(\mathbf{k})$, and $\alpha_{\mathbf{k}s} = \alpha_s(\mathbf{k})/\sqrt{V}$.

In the presence of sources the coefficients are time dependent:

$$\boldsymbol{\mathcal{E}}^{(+)}(\mathbf{r},t) = i\int\frac{d^3k}{(2\pi)^3}\sum_{s=1}^{2}\sqrt{\frac{\hbar\omega_s(\mathbf{k})}{2\epsilon_0 n_s^2(k)}}\alpha_s(\mathbf{k},t)\,\boldsymbol{\varepsilon}_s(\mathbf{k})\,e^{i(\mathbf{k}\cdot\mathbf{r}-\omega_s(\mathbf{k})t)}, \tag{B.144}$$

or

$$\boldsymbol{\mathcal{E}}^{(+)}(\mathbf{r},t) = i\sum_{\mathbf{k}s}\sqrt{\frac{\hbar\omega_{\mathbf{k}s}}{2\epsilon_0 V n_{ks}^2}}\alpha_{\mathbf{k}s}(t)\,\boldsymbol{\varepsilon}_{\mathbf{k}s}e^{i(\mathbf{k}\cdot\mathbf{r}-\omega_{\mathbf{k}s}t)}. \tag{B.145}$$

For fields satisfying eqns (3.107) and (3.120), the argument used for an isotropic medium can be applied to the present case to derive the expressions

$$\mathcal{U} = \int\frac{d^3k}{(2\pi)^3}\sum_s \hbar\omega_s(\mathbf{k})\,|\alpha_s(\mathbf{k},t)|^2 \tag{B.146}$$

or

$$\mathcal{U} = \sum_{\mathbf{k}s}\hbar\omega_{\mathbf{k}s}\,|\alpha_s(\mathbf{k},t)|^2 \tag{B.147}$$

for the energy in the electromagnetic field.

A Uniaxial crystals

The analysis sketched above is valid for general crystals, but there is one case of special interest for applications. A crystal is **uniaxial** if it exhibits threefold, fourfold, or sixfold symmetry under rotations in the plane perpendicular to a distinguished axis, which we take as the z-axis. The x- and y-axes can be any two orthogonal lines in the perpendicular plane. In general, the permittivity tensor is diagonal—with diagonal elements $\epsilon_x, \epsilon_y, \epsilon_z$—in the crystal-axis coordinates, but the symmetry under rotations around the z-axis implies that $\epsilon_x = \epsilon_y$. We set $\epsilon_x = \epsilon_y = \epsilon_\perp$, but in general $\epsilon_\perp \neq \epsilon_z$. In these coordinates, the unit vector along the propagation direction is $\widetilde{\mathbf{k}} = \mathbf{k}/k = (\sin\theta\cos\phi, \sin\theta\sin\phi, \cos\theta)$, where θ and ϕ are the usual polar and azimuthal angles. Consider a rotation about the z-axis by the angle φ; then

$$\begin{aligned}\overleftrightarrow{\mathcal{S}}' &= \overleftrightarrow{\mathcal{R}}(\varphi)\,\overleftrightarrow{\mathcal{S}}\,\overleftrightarrow{\mathcal{R}}^{-1}(\varphi)\\ &= \overleftrightarrow{\Delta}^{\perp}(\widetilde{\mathbf{k}}')\,\overleftrightarrow{\mathcal{R}}(\varphi)\,[\overleftrightarrow{\epsilon}_r(\omega)]^{-1}\,\overleftrightarrow{\mathcal{R}}^{-1}(\varphi)\,\overleftrightarrow{\Delta}^{\perp}(\widetilde{\mathbf{k}}')\\ &= \overleftrightarrow{\Delta}^{\perp}(\widetilde{\mathbf{k}}')\,[\overleftrightarrow{\epsilon}_r(\omega)]^{-1}\,\overleftrightarrow{\Delta}^{\perp}(\widetilde{\mathbf{k}}'),\end{aligned} \tag{B.148}$$

where $\widetilde{\mathbf{k}}'$ is the rotated unit vector and we have used the invariance of $\overleftrightarrow{\epsilon}_r$ under rotations around the z-axis. The matrices $\overleftrightarrow{\mathcal{S}}'$ and $\overleftrightarrow{\mathcal{S}}$ are related by a similarity transformation, so they have the same eigenvalues for any φ. The choice $\varphi = -\phi$

effectively sets $\phi = 0$, so the eigenvalues of $\overleftrightarrow{\mathcal{S}}$ can only depend on θ, the angle between $\mathbf{k}$ and the distinguished axis. Setting $\phi = 0$ simplifies the calculation and the two indices of refraction are given by

$$n_o^2 = \epsilon_\perp \,, \tag{B.149}$$

$$n_e^2 = \frac{2\epsilon_\perp \epsilon_z}{\epsilon_\perp (1 - \cos 2\theta) + \epsilon_z (1 + \cos 2\theta)} \,. \tag{B.150}$$

The phase velocity c/n_o, which is independent of the direction of $\mathbf{k}$, characterizes the **ordinary wave**, while the phase velocity c/n_e, which depends on θ, describes the propagation of the **extraordinary wave**. The corresponding refractive indices n_o and n_e are respectively called the **ordinary** and **extraordinary index**.

B.5.4 Nonlinear optics

Classical nonlinear optics (Boyd, 1992; Newell and Moloney, 1992) is concerned with the propagation of classical light in weakly nonlinear media. Most experiments in quantum optics involve substances with very weak magnetic susceptibility, so we will simplify the permeability tensor to $\mu_{ij}(\omega) = \mu_0 \delta_{ij}$. On the other hand, the coupling to the electric field can be strong, if the field is nearly resonant with a dipole transition in the constituent atoms. In such cases, the relation between the polarization and the field is not linear. In the simplest situation, the response of the atomic dipole to the external field can be calculated by time-dependent perturbation theory, which produces an expression of the form (Boyd, 1992, Chap. 3)

$$\boldsymbol{\mathcal{P}}(\mathbf{r}, t) = \boldsymbol{\mathcal{P}}^{(1)}(\mathbf{r}, t) + \boldsymbol{\mathcal{P}}^{\mathrm{NL}}(\mathbf{r}, t) \,, \tag{B.151}$$

where the **nonlinear polarization**

$$\boldsymbol{\mathcal{P}}^{\mathrm{NL}}(\mathbf{r}, t) = \boldsymbol{\mathcal{P}}^{(2)}(\mathbf{r}, t) + \boldsymbol{\mathcal{P}}^{(3)}(\mathbf{r}, t) + \cdots \tag{B.152}$$

contains the higher-order terms in the perturbation expansion and defines the **nonlinear constitutive relations**. The transverse electric field describing radiation satisfies eqn (B.103), and—after using eqn (B.151) and imposing the convention that $\boldsymbol{\mathcal{E}}$ always means the transverse part, $\boldsymbol{\mathcal{E}}^\perp$—this can be written as

$$\nabla^2 \boldsymbol{\mathcal{E}} - \frac{1}{c^2} \frac{\partial^2}{\partial t^2} \boldsymbol{\mathcal{E}} - \mu_0 \frac{\partial^2}{\partial t^2} \boldsymbol{\mathcal{P}}^{\perp(1)} = \mu_0 \frac{\partial^2}{\partial t^2} \boldsymbol{\mathcal{P}}^{\perp \mathrm{NL}} \,. \tag{B.153}$$

The interesting materials are often crystals, so scalar relations between the polarization and the field must be replaced by tensor relations for anisotropic media. In a microscopic description, the polarization $\boldsymbol{\mathcal{P}}$ is the sum over the induced dipoles in each atom, but we will use a coarse-grained macroscopic treatment that is justified by the presence of many atoms in a cubic wavelength. Thus the macroscopic susceptibilities are proportional to the density, n_{at}, of atoms, i.e. $\chi^{(n)} = n_{\mathrm{at}} \gamma^{(n)}$, where $\gamma^{(n)}$ is the nth-order atomic polarizability. In addition to coarse graining, we will assume that the polarization at $\mathbf{r}$ only depends on the field at $\mathbf{r}$, i.e. the susceptibilities do not exhibit the property of spatial dispersion discussed in Appendix B.5.1. For the crystals used in quantum optics spatial dispersion is weak, so this assumption is justified in practice.

In the time domain the nth-order polarization is given by

$$\mathcal{P}_i^{(n)}(\mathbf{r},t) = \epsilon_0 \int dt_1 \cdots \int dt_n \chi_{ij_1j_2\cdots j_n}^{(n)}(t-t_1, t-t_2, \ldots, t-t_n) \times \mathcal{E}_{j_1}(\mathbf{r},t_1)\cdots\mathcal{E}_{j_n}(\mathbf{r},t_n), \tag{B.154}$$

where $\chi_{ij_1j_2\cdots j_n}^{(n)}(\tau_1,\tau_2,\ldots,\tau_n)$ is real and symmetric with respect to simultaneous permutations of the time arguments τ_p and the corresponding tensor indices j_p. The corresponding frequency-domain relation is

$$\mathcal{P}_i^{(n)}(\mathbf{r},\nu) = \epsilon_0 \int \prod_{q=1}^{n} \frac{d\nu_q}{2\pi} 2\pi\delta\left(\nu - \sum_{p=1}^{n} \nu_p\right) \chi_{ij_1j_2\cdots j_n}^{(n)}(\nu_1,\ldots,\nu_n) \times \mathcal{E}_{j_1}(\mathbf{r},\nu_1)\cdots\mathcal{E}_{j_n}(\mathbf{r},\nu_n), \tag{B.155}$$

where

$$\chi_{ij_1j_2\cdots j_n}^{(n)}(\nu_1,\ldots,\nu_n) = \int \prod_{q=1}^{n} d\tau_q \exp\left[i\sum_{p=1}^{n}\nu_p\tau_p\right] \chi_{ij_1j_2\cdots j_n}^{(n)}(\tau_1,\tau_2,\ldots,\tau_n). \tag{B.156}$$

This notation agrees with one of the conventions (Newell and Moloney, 1992, Chap. 2d) for the Fourier transforms of the susceptibilities, but there is a different—and frequently used—convention in which $\chi_{ij_1j_2\cdots j_n}^{(n)}(\nu_1,\nu_2,\ldots,\nu_n)$ is replaced by $\chi_{ij_1j_2\cdots j_n}^{(n)}(-\nu_0,\nu_1,\nu_2,\ldots,\nu_n)$, with the understanding that the sum of the frequency arguments is zero (Boyd, 1992, Sec. 1.5). This is an example of the notational schisms that are common in this field. The nth-order frequency-domain susceptibility tensor is symmetrical under simultaneous permutations of ν_p and j_p, and the reality of the time-domain susceptibility imposes the conditions

$$\chi_{ij_1j_2\cdots j_n}^{(n)*}(\nu_1,\ldots,\nu_n) = \chi_{ij_1j_2\cdots j_n}^{(n)}(-\nu_1,\ldots,-\nu_n) \tag{B.157}$$

in the frequency domain. For the transparent media normally considered, the Fourier transform $\chi_{ij_1j_2\cdots j_n}^{(n)}(\nu_1,\ldots,\nu_n)$ is also real, and eqn (B.157) becomes

$$\chi_{ij_1j_2\cdots j_n}^{(n)}(\nu_1,\ldots,\nu_n) = \chi_{ij_1j_2\cdots j_n}^{(n)}(-\nu_1,\ldots,-\nu_n). \tag{B.158}$$

The properties listed above give no information regarding what happens if the first index i is interchanged with one of the j_ps. For transparent media, the explicit quantum perturbation calculation of the susceptibilities provides the additional symmetry condition (Boyd, 1992, Sec. 3.2)

$$\chi_{ij_1\cdots j_p\cdots j_n}^{(n)}(\nu_1,\ldots,\nu_p,\ldots,\nu_n) = \chi_{j_pj_1j_2\cdots i\cdots j_n}^{(n)}\left(\nu_1,\ldots,-\sum_{k=1}^{n}\nu_k,\ldots,\nu_n\right). \tag{B.159}$$

Appendix C
Quantum theory

Modern quantum theory originated with the independent inventions of matrix mechanics by Heisenberg and wave mechanics by Schrödinger. It was essentially completed by Schrödinger's proof that the two formulations are equivalent and Born's interpretation of the wave function as a probability amplitude. The intuitive appeal of wave mechanics, at least for situations involving a single particle, explains its universal use in introductory courses on quantum theory. This approach does, however, have certain disadvantages. One is that the intuitive simplicity of wave mechanics is largely lost when it is applied to many-particle systems. For our purposes, a more serious objection is that there are no wave functions for photons.

A more satisfactory approach is based on the fact that interference phenomena are observed for all microscopic systems. For example, the two-slit experiment can be performed with material particles to observe interference fringes. A comparison to macroscopic wave phenomena suggests that the mathematical description of the states of a system should satisfy the **superposition principle**, i.e. every linear combination of states is also a state. In mathematical terms this means that the states are elements of a vector space, and the Born interpretation—to be explained below—requires the vector space to be a Hilbert space.

C.1 Dirac's bra and ket notation

In Appendix A.3 Hilbert spaces are described with the standard notation used in mathematics and in many textbooks on quantum theory. In the main text, we employ an alternative notation introduced by Dirac (1958), in which a vector in a Hilbert space $\mathfrak{H}$ is represented by the symbol $|\psi\rangle$. In this notation $|\cdot\rangle$ represents a generic **ket vector** and ψ is a label that distinguishes one vector from another. Linear combinations of two kets, $|\psi\rangle$ and $|\phi\rangle$, are written as $\alpha\,|\psi\rangle + \beta\,|\phi\rangle$, and scalars, like α and β, are called **c-numbers**.

In the Dirac notation, a **bra** vector $\langle F|$ represents a *rule* that assigns a complex number, denoted by $\langle F\,|\psi\rangle$, to every ket vector $|\psi\rangle$. This rule is *linear*, i.e. if $|\psi\rangle = \alpha\,|\chi\rangle + \beta\,|\phi\rangle$, then

$$\langle F\,|\psi\rangle = \alpha\,\langle F\,|\chi\rangle + \beta\,\langle F\,|\phi\rangle\,. \tag{C.1}$$

The Hilbert-space inner product (ϕ, ψ) is an example of such a rule, so for each ket vector $|\phi\rangle$ there is a corresponding bra vector $\langle\phi|$ (called the *adjoint* vector) defined by

$$\langle\phi\,|\psi\rangle = (\phi, \psi) \quad \text{for all } \psi\,. \tag{C.2}$$

With this understanding, we will use $\langle\varphi\,|\psi\rangle$ from now on to denote the inner product.

The linearity of the rule (C.1) guarantees that the set of bra vectors is in fact a vector space. The official jargon—explained in Appendix A.6.1—is that the bra vectors form the dual space $\mathfrak{H}'$ of linear functionals on $\mathfrak{H}$. The definition (C.2) of the adjoint vectors shows that the Hilbert space $\mathfrak{H}$ of physical states is isomorphic to a subspace of $\mathfrak{H}'$.

The Hilbert spaces relevant for quantum theory are always separable; that is, every ket $|\psi\rangle$ can be expanded as

$$|\psi\rangle = \sum_n |\phi_n\rangle \langle\phi_n |\psi\rangle \,, \tag{C.3}$$

where $\{|\phi_n\rangle\,,\ n = 1, 2, \ldots\}$ is an orthonormal basis for $\mathfrak{H}$.

C.1.1 Examples

A Two-level system

The states of a two-level system, e.g. a spin-1/2 particle, are usually represented by two-component column vectors that refer to a given basis, e.g. eigenstates of σ_z. The relation between this concrete description and the Dirac notation is

$$|\psi\rangle \sim \begin{pmatrix}\psi_1\\ \psi_2\end{pmatrix}, \quad \langle\psi| \sim (\psi_1^*, \psi_2^*)\,, \quad \langle\varphi |\psi\rangle = (\varphi, \psi) = \varphi_1^*\psi_1 + \varphi_2^*\psi_2\,. \tag{C.4}$$

The symbol '$\sim$' is used instead of '$=$' because the values of the components ψ_1 and ψ_2 depend on the particular choice of basis in the concrete space $\mathbb{C}^2$. A different basis choice would represent the same ket vector $|\psi\rangle$ by a different pair of components ψ_1', ψ_2'. An example of an orthonormal basis is

$$\mathcal{B} = \left\{|1\rangle \sim \begin{pmatrix}1\\0\end{pmatrix},\ |2\rangle \sim \begin{pmatrix}0\\1\end{pmatrix}\right\}, \tag{C.5}$$

so the components are given by $\psi_1 = \langle 1 |\psi\rangle$ and $\psi_2 = \langle 2 |\psi\rangle$, and

$$|\psi\rangle = \psi_1 |1\rangle + \psi_2 |2\rangle \,. \tag{C.6}$$

This relation is invariant under a change of basis in $\mathbb{C}^2$, since the vectors $|1\rangle$ and $|2\rangle$ would also be transformed. Every bra vector (linear functional) on $\mathbb{C}^2$ is defined by taking the inner product with some fixed vector in $\mathbb{C}^2$, so the space of bra vectors (the dual space) is isomorphic to the space itself, i.e. $\mathfrak{H}' = \mathfrak{H}$. This is true for any finite-dimensional Hilbert space.

B Spinless particle in three dimensions

As a second example, consider the familiar description of a spinless particle by a square-integrable wave function $\psi(\mathbf{r})$. The **square-integrability** condition is

$$\int_{-\infty}^{\infty} d^3r \,|\psi(\mathbf{r})|^2 < \infty \,, \tag{C.7}$$

and the set of square-integrable functions is called $L_2(\mathbb{R}^3)$. The relation between the abstract and concrete descriptions is

$$|\psi\rangle \sim \psi(\mathbf{r}), \quad \langle\psi| \sim \psi^*(\mathbf{r}), \quad \langle\varphi|\psi\rangle = \int d^3r \varphi^*(\mathbf{r})\psi(\mathbf{r}), \tag{C.8}$$

where the vector operations are defined point-wise:

$$\alpha|\psi\rangle + \beta|\varphi\rangle \sim \alpha\psi(\mathbf{r}) + \beta\varphi(\mathbf{r}). \tag{C.9}$$

For infinite-dimensional Hilbert spaces, such as $\mathfrak{H} = L_2(\mathbb{R}^3)$, there are bra vectors that are not adjoints of any vector in the space. In other words, the dual space $\mathfrak{H}'$ is larger than the space $\mathfrak{H}$. For example, the delta function $\delta(\mathbf{r} - \mathbf{r}_0)$ is not the adjoint of any vector in $L_2(\mathbb{R}^3)$, but it does define a bra vector $\langle\mathbf{r}_0|$ by

$$\langle\mathbf{r}_0|\psi\rangle = \int d^3r \delta(\mathbf{r} - \mathbf{r}_0)\psi(\mathbf{r}) = \psi(\mathbf{r}_0). \tag{C.10}$$

This establishes the relation $\psi(\mathbf{r}) = \langle\mathbf{r}|\psi\rangle$ between the concrete and abstract descriptions.

Although the bra vector $\langle\mathbf{r}_0|$ is not the adjoint of any proper ket vector (normalizable wave function) in $L_2(\mathbb{R}^3)$, it is common practice to define an **improper ket vector** $|\mathbf{r}_0\rangle$ by the rule $\langle\psi|\mathbf{r}_0\rangle = \psi^*(\mathbf{r}_0)$ for all $\psi \in \mathfrak{H}$. The **position operator** $\widehat{\mathbf{r}}$ is defined by $\widehat{\mathbf{r}}\psi(\mathbf{r}) = \mathbf{r}\psi(\mathbf{r})$, and $|\mathbf{r}_0\rangle$ is an **improper eigenvector** of $\widehat{\mathbf{r}}$—i.e. $\widehat{\mathbf{r}}|\mathbf{r}_0\rangle = \mathbf{r}_0|\mathbf{r}_0\rangle$—by virtue of

$$\langle\psi|\widehat{\mathbf{r}}|\mathbf{r}_0\rangle = \langle\mathbf{r}_0|\widehat{\mathbf{r}}|\psi\rangle^* = \mathbf{r}_0\psi^*(\mathbf{r}_0) = \mathbf{r}_0\langle\psi|\mathbf{r}_0\rangle. \tag{C.11}$$

In the same way, there is no proper eigenvector of the momentum operator $\widehat{\mathbf{p}}$, but there is an improper eigenvector $|\mathbf{p}_0\rangle$, i.e. $\widehat{\mathbf{p}}|\mathbf{p}_0\rangle = \mathbf{p}_0|\mathbf{p}_0\rangle$, associated with the bra vector $\langle\mathbf{p}_0|$ defined by

$$\langle\mathbf{p}_0|\psi\rangle = \int d^3r e^{-i\mathbf{p}_0\cdot\mathbf{r}/\hbar}\psi(\mathbf{r}). \tag{C.12}$$

C.1.2 Linear operators

The action of a linear operator A is denoted by $A|\psi\rangle$, and the complex number $\langle\psi|A|\varphi\rangle$ is the **matrix element** of the operator A for the pair of vectors $|\psi\rangle$ and $|\varphi\rangle$. The operator A is uniquely determined by any of the sets of matrix elements

$$\{\langle\phi_n|A|\phi_m\rangle\} \quad \text{for all } |\phi_n\rangle, |\phi_m\rangle \text{ in a basis } \mathcal{B}, \tag{C.13}$$

$$\{\langle\psi|A|\varphi\rangle\} \quad \text{for all } |\psi\rangle, |\varphi\rangle \text{ in } \mathfrak{H}, \tag{C.14}$$

$$\{\langle\psi|A|\psi\rangle\} \quad \text{for all } |\psi\rangle \text{ in } \mathfrak{H}. \tag{C.15}$$

The operator $T_{\varphi\chi}$, defined by the rule

$$T_{\varphi\chi}|\psi\rangle = |\varphi\rangle\langle\chi|\psi\rangle \quad \text{for all } |\psi\rangle, \tag{C.16}$$

is usually written as $|\varphi\rangle\langle\chi|$. The product of two such operators therefore acts by

$$T_{\varphi\chi}T_{\beta\xi}\left|\psi\right\rangle = T_{\varphi\chi}\left|\beta\right\rangle\left\langle\xi\left|\psi\right.\right\rangle = \left|\varphi\right\rangle\left\langle\chi\left|\beta\right.\right\rangle\left\langle\xi\left|\psi\right.\right\rangle . \tag{C.17}$$

This holds for all states $|\psi\rangle$, so the product rule is

$$T_{\varphi\chi}T_{\beta\xi} = \left\langle\chi\left|\beta\right.\right\rangle T_{\varphi\xi} . \tag{C.18}$$

The operator $T_{\varphi\varphi} = \left|\varphi\right\rangle\left\langle\varphi\right|$ is therefore a projection operator, provided that $|\varphi\rangle$ is normalized.

Let $\{|\phi_n\rangle\}$ be an orthonormal basis for a subspace $\mathfrak{W} \subset \mathfrak{H}$; then the projection operators $P_n = \left|\phi_n\right\rangle\left\langle\phi_n\right|$ are orthogonal, i.e. $P_nP_m = \delta_{nm}$. Every vector $|\psi\rangle$ in $\mathfrak{W}$ has the unique expansion

$$\left|\psi\right\rangle = \sum_n \left|\phi_n\right\rangle\left\langle\phi_n\left|\psi\right.\right\rangle = \sum_n P_n\left|\psi\right\rangle , \tag{C.19}$$

so the operator

$$P_{\mathfrak{W}} = \sum_n P_n = \sum_n \left|\phi_n\right\rangle\left\langle\phi_n\right| \tag{C.20}$$

acts as the identity for vectors in $\mathfrak{W}$. On the other hand, every vector $|\chi\rangle$ in the orthogonal complement $\mathfrak{W}^{\perp}$ is annihilated by $P_{\mathfrak{W}}$, i.e. $P_{\mathfrak{W}}\left|\chi\right\rangle = 0$, so $P_{\mathfrak{W}}$ is the projection operator onto $\mathfrak{W}$. When $\mathfrak{W} = \mathfrak{H}$ the projection $P_{\mathfrak{H}}$ is the identity operator and we get

$$\sum_n \left|\phi_n\right\rangle\left\langle\phi_n\right| = I , \tag{C.21}$$

which is called the **completeness relation**, or a **resolution of the identity** into the projection operators $P_n = \left|\phi_n\right\rangle\left\langle\phi_n\right|$.

If $\mathcal{B} = \{\left|\varphi_1\right\rangle, \left|\varphi_2\right\rangle, \ldots\}$ is an orthonormal basis, then the trace of A is defined by

$$\mathrm{Tr}\left(A\right) = \sum_n \left\langle\phi_n\left|A\right|\phi_n\right\rangle . \tag{C.22}$$

The value of $\mathrm{Tr}\left(A\right)$ is the same for all choices of orthonormal basis, and

$$\mathrm{Tr}\left(AB\right) = \mathrm{Tr}\left(BA\right) . \tag{C.23}$$

The last property is called **cyclic invariance**, since it implies

$$\mathrm{Tr}\left(A_1A_2\cdots A_n\right) = \mathrm{Tr}\left(A_nA_1A_2\cdots A_{n-1}\right) . \tag{C.24}$$

C.2 Physical interpretation

The mathematical formalism is connected to experiment by the following assumptions.

(1) The states of maximum information, called **pure states**, are vectors in a Hilbert space $\mathfrak{H}$.

(2) Each observable quantity is represented by a Hermitian operator A, and the value obtained in a measurement is always one of the eigenvalues a_n of A. Hermitian operators are, therefore, often called **observables**.

(3) If the system is prepared in the state $|\psi\rangle$, then the probability that a measurement of A yields the value a_n is $|\langle\phi_n|\psi\rangle|^2$, where $A|\phi_n\rangle = a_n|\phi_n\rangle$. This is the **Born interpretation** (Born, 1926). After the measurement is performed, the system is described by the eigenvector $|\phi_n\rangle$. This is the infamous **reduction of the wave packet**.

(a) This description implicitly assumes that the eigenvalue a_n is nondegenerate. In the more typical case of an eigenvalue with degeneracy $d > 1$, the probability for finding a_n is

$$\sum_{k=1}^{d} |\langle\phi_{nk}|\psi\rangle|^2 , \tag{C.25}$$

where $\{|\phi_{nk}\rangle, k = 1, \ldots, d\}$ is an orthonormal basis for the a_n-eigenspace. The corresponding projection operator is

$$P_n = \sum_{k=1}^{d} |\phi_{nk}\rangle\langle\phi_{nk}| . \tag{C.26}$$

(b) **Von Neumann's projection postulate** (von Neumann, 1955) states that the probability of finding a_n is

$$\langle\psi|P_n|\psi\rangle = \sum_{k=1}^{d} |\langle\phi_{nk}|\psi\rangle|^2 , \tag{C.27}$$

and—for $\langle\psi|P_n|\psi\rangle \neq 0$—the final state after the measurement is

$$|\psi_{\text{fin}}\rangle = \frac{1}{\sqrt{\langle\psi|P_n|\psi\rangle}} P_n|\psi\rangle . \tag{C.28}$$

(c) An alternative way of dealing with degeneracies is to replace the single observable A by a set of observables $\{A_1, A_2, \ldots, A_N\}$ with the following properties.

(i) The operators are mutually commutative, i.e. $[A_i, A_j] = 0$.

(ii) A vector $|\phi\rangle$ that is a simultaneous eigenvector of all the A_is— i.e. $A_i|\phi\rangle = a_i|\phi\rangle$ for $i = 1, \ldots, N$—is uniquely determined (up to an overall phase factor).

A set $\{A_1, A_2, \ldots, A_N\}$ with these properties is called a **complete set of commuting observables** (CSCO). A simultaneous measurement of the observables in the CSCO leaves the system in a state that is unique except for an overall phase factor.

(4) The average of many measurements of A performed on identical systems prepared in the state $|\psi\rangle$ is the **expectation value** $\langle\psi|A|\psi\rangle$.

(5) There is a special Hermitian operator, the **Hamiltonian** H, which describes the time evolution—often called **time translation**—of the system through the **Schrödinger equation**

$$i\hbar\frac{\partial}{\partial t}|\psi(t)\rangle = H(t)|\psi(t)\rangle . \tag{C.29}$$

The explicit time dependence of the Hamiltonian can only occur in the presence of external classical forces.

C.3 Useful results for operators

C.3.1 Pauli matrices

Consider linear operators on the space $\mathbb{C}^2$. It is easy to see that every operator is represented by a 2×2 matrix, so it is determined by four complex numbers. The **Pauli matrices**, defined by

$$\sigma_x = \sigma_1 = \begin{bmatrix} 0 & 1 \\ 1 & 0 \end{bmatrix}, \quad \sigma_y = \sigma_2 = \begin{bmatrix} 0 & -i \\ i & 0 \end{bmatrix}, \quad \sigma_z = \sigma_3 = \begin{bmatrix} 1 & 0 \\ 0 & -1 \end{bmatrix}, \tag{C.30}$$

are particularly important. They satisfy the commutation relations

$$[\sigma_i, \sigma_j] = 2i\epsilon_{ijk}\sigma_k \,, \tag{C.31}$$

where ϵ_{ijk} is the alternating tensor defined by eqn (A.3), and the anticommutation relations

$$[\sigma_i, \sigma_j]_+ = \sigma_i\sigma_j + \sigma_j\sigma_i = 2\delta_{ij} \quad (i, j = x, y, z)\,, \tag{C.32}$$

which combine to yield

$$\sigma_i\sigma_j = \epsilon_{ijk}\sigma_k + \delta_{ij} \,. \tag{C.33}$$

It is often useful to use the so-called circular basis $\{\sigma_z, \sigma_\pm = (\sigma_x \pm i\sigma_y)/2\}$ with the commutation relations

$$[\sigma_z, \sigma_\pm] = \pm 2\sigma_\pm \,, \quad [\sigma_+, \sigma_-] = \sigma_z \,, \tag{C.34}$$

and the anticommutation relations

$$[\sigma_\pm, \sigma_\pm]_+ = 0\,, \quad [\sigma_\pm, \sigma_\mp]_+ = 1\,, \quad [\sigma_z, \sigma_\pm]_+ = 0\,. \tag{C.35}$$

These fundamental relations yield the useful identities

$$\sigma_+\sigma_- = \frac{1}{2}(1 + \sigma_z)\,, \tag{C.36}$$

$$\sigma_-\sigma_+ = \frac{1}{2}(1 - \sigma_z)\,, \tag{C.37}$$

$$\sigma_z\sigma_\pm = \pm\sigma_\pm = -\sigma_\pm\sigma_z \,. \tag{C.38}$$

The three Pauli matrices, together with the identity matrix, are linearly independent and therefore constitute a complete set for the expansion of all 2×2 matrices. Thus every 2×2 matrix A has the representation

$$A = a_0\sigma_0 + a_i\sigma_i \,, \tag{C.39}$$

where σ_0 is the identity matrix. These properties, together with the observation that $\mathrm{Tr}\,(\sigma_i) = 0$, yield

$$\begin{aligned} a_0 &= \frac{1}{2}\operatorname{Tr}(A)\,, \\ a_j &= \frac{1}{2}\operatorname{Tr}(A\sigma_j)\,. \end{aligned} \tag{C.40}$$

Writing $a_i\sigma_i = \boldsymbol{a}\cdot\boldsymbol{\sigma}$ and using the properties given above yields $(\boldsymbol{a}\cdot\boldsymbol{\sigma})^2 = |\boldsymbol{a}|^2$, and this in turn provides the useful identities (Cohen-Tannoudji *et al.*, 1977*b*, Complement A-IX)

$$e^{i\alpha\mathbf{u}\cdot\boldsymbol{\sigma}} = \cos(\alpha) + i\sin(\alpha)\,\mathbf{u}\cdot\boldsymbol{\sigma}\,, \tag{C.41}$$

$$e^{\beta\mathbf{u}\cdot\boldsymbol{\sigma}} = \cosh(\beta) + \sinh(\beta)\,\mathbf{u}\cdot\boldsymbol{\sigma}\,, \tag{C.42}$$

where α and β are real constants and $\mathbf{u}$ is a real unit vector.

C.3.2 The operator binomial theorem

For *c*-numbers x and y the binomial theorem is

$$(x+y)^n = \sum_{p=0}^{n} \frac{n!}{p!\,(n-p)!} x^{n-p} y^p\,, \tag{C.43}$$

but this depends on the fact that *c*-numbers commute. For noncommuting operators X and Y the quantity $(X+Y)^n$ is to be evaluated by multiplying together the n factors $X+Y$. Consider the terms of order $(n-p,p)$ in this expansion, i.e. those in which X occurs $n-p$ times and Y occurs p times. Since each of these terms is the product of n factors, there are a total of $n!$ orderings. The orderings that differ only by exchanging Xs with Xs or Ys with Ys are identical, and the number of these terms is precisely the binomial coefficient $n!/p!\,(n-p)!$; therefore,

$$(X+Y)^n = \sum_{m=0}^{n} \frac{n!}{p!\,(n-p)!} \mathcal{S}\left[X^{n-m}Y^m\right], \tag{C.44}$$

where $\mathcal{S}\left[X^{n-m}Y^m\right]$ is the average of the terms with $(n-m)$ Xs and m Ys arranged in all possible orders. This is called the **symmetrical** or **Weyl** product.

For $(n,0)$ or $(0,n)$ one has simply $\mathcal{S}[X^n] = X^n$ or $\mathcal{S}[Y^n] = Y^n$. Examples of mixed powers are

$$\begin{aligned} \mathcal{S}[XY] &= \frac{1}{2}(XY+YX)\,, \\ \mathcal{S}\left[X^2Y\right] &= \frac{1}{3}\left(X^2Y + XYX + YX^2\right), \\ \mathcal{S}\left[X^2Y^2\right] &= \frac{1}{6}\left(X^2Y^2 + XY^2X + XYXY + Y^2X^2 + YX^2Y + YXYX\right), \\ &\vdots \end{aligned} \tag{C.45}$$

C.3.3 Commutator identities

The **Leibnitz rule**

$$[A, BC] = A[B, C] + [A, B]C \quad (C.46)$$

and the **Jacobi identity**

$$[[A, B], C] + [[C, A], B] + [[B, C], A] = 0 \quad (C.47)$$

are both readily verified by direct use of the definition $[A, B] = AB - BA$. The useful identity

$$[A, B_1 B_2 \cdots B_n] = \sum_{p=1}^{n} \left(\prod_{j=1}^{p-1} B_j \right) [A, B_p] \left(\prod_{k=p+1}^{n} B_k \right) \quad (C.48)$$

can be established by an induction argument, combined with the convention that an empty product has the value unity. In the special case that each single commutator $[A, B_p]$ commutes with the remaining B_js, this becomes

$$[A, B_1 B_2 \cdots B_n] = \sum_{p=1}^{n} [A, B_p] \left(\prod_{j \neq p=1}^{n} B_j \right). \quad (C.49)$$

C.3.4 Operator expansion theorems

Theorem C.1 *Let X and Y be operators acting on a Hilbert space $\mathfrak{H}$. Then*

$$e^{\kappa X} Y e^{-\kappa X} = \sum_{n=0}^{\infty} \frac{\kappa^n}{n!} [X, Y]^{(n)}, \quad (C.50)$$

where the iterated commutator $[X, Y]^{(n)}$ is defined by the initial value $[X, Y]^{(0)} = Y$ and the recursion relations

$$[X, Y]^{(n+1)} = \left[X, [X, Y]^{(n)}\right] \quad \text{for } n \geqslant 0. \quad (C.51)$$

Proof Let $Y(\kappa) \equiv e^{\kappa X} Y e^{-\kappa X}$; then $dY(\kappa)/d\kappa = [X, Y(\kappa)]$. Iterating this result implies

$$\frac{d^{n+1} Y(\kappa)}{d\kappa^{n+1}} = \left[X, \frac{d^n Y(\kappa)}{d\kappa^n}\right], \quad (C.52)$$

and eqn (C.50) follows by a Taylor series expansion around $\kappa = 0$. ■

In the special case that the commutator $[X, Y]$ commutes with X, the series terminates so that

$$e^{\kappa X} Y e^{-\kappa X} = Y + \kappa [X, Y], \quad (C.53)$$

i.e. X generates translations of Y. An important example is a canonically conjugate pair: $X = \widehat{p}$, $Y = \widehat{q}$, with $[\widehat{q}, \widehat{p}] = i\hbar$. Choosing $\kappa = iu/\hbar$, where u is a c-number, gives the familiar quantum mechanics result

$$T_u^{\dagger}\widehat{q}T_u = \widehat{q} + u\,, \tag{C.54}$$

where the unitary operator

$$T_u = e^{-iu\widehat{p}/\hbar} \tag{C.55}$$

evidently generates translations in the position. For any well-behaved operator function $F(\widehat{q})$, e.g. one that has a Taylor series expansion, the last result generalizes to

$$T_u^{\dagger}F(\widehat{q})\,T_u = F(\widehat{q}+u)\,. \tag{C.56}$$

For infinitesimal values of u, expanding both sides leads to

$$[\widehat{p}, F(\widehat{q})] = -i\hbar\frac{\partial F(\widehat{q})}{\partial\widehat{q}}\,. \tag{C.57}$$

To see the action of T_u on a state vector, rewrite eqn (C.54) as $\widehat{q}T_u = T_u(\widehat{q}+u)$ and apply this to an eigenvector $|Q\rangle$ of $\widehat{q}$ to get

$$\widehat{q}T_u|Q\rangle = T_u(\widehat{q}+u)|Q\rangle = T_u(Q+u)|Q\rangle = (Q+u)\,T_u|Q\rangle\,; \tag{C.58}$$

in other words,

$$T_u|Q\rangle = |Q+u\rangle\,. \tag{C.59}$$

Thus for any state $|\Psi\rangle$,

$$\langle Q|T_u|\Psi\rangle = \langle Q+u|\Psi\rangle\,, \tag{C.60}$$

or in more familiar notation

$$(T_u\Psi)(Q) = \Psi(Q+u)\,. \tag{C.61}$$

It is also useful to consider the opposite assignment $X = \widehat{q}$, $Y = \widehat{p}$, $\kappa = -iv/\hbar$, which produces

$$e^{-iv\widehat{q}/\hbar}\widehat{q}e^{iv\widehat{q}/\hbar} = \widehat{p} + v\,, \tag{C.62}$$

and shows that the position operator generates translations in the momentum.

Another important special case is $[X, Y] = \alpha Y$, where α is a c-number. Putting this into the definition (C.51) gives

$$[X, Y]^{(n)} = \alpha^n Y\,, \tag{C.63}$$

so that eqn (C.50) becomes

$$e^{\kappa X}Ye^{-\kappa X} = e^{\alpha\kappa}Y\,. \tag{C.64}$$

As an example, let $X = a^{\dagger}a$, $Y = a$, and $\kappa = i\theta$, where a is the lowering operator for a harmonic oscillator. The commutation relation $\left[a, a^{\dagger}\right] = 1$ yields $[X, Y] = \left[a^{\dagger}a, a\right] = -a$, so

$$e^{i\theta a^{\dagger}a}ae^{-i\theta a^{\dagger}a} = e^{-i\theta}a\,. \tag{C.65}$$

C.3.5 Campbell–Baker–Hausdorff theorem

Theorem C.2 *Let X and Y be operators such that $[X,Y]$ commutes with both X and Y. Then*

$$e^{X}e^{Y}=e^{X+Y}e^{\frac{1}{2}[X,Y]}\,. \tag{C.66}$$

Proof See Peres (1995, Sec. 10-7). ■

Two important special cases are needed in the text. The first is defined by setting $X=-iv\widehat{q}$, $Y=-iu\widehat{p}$, which leads to

$$e^{-i(u\widehat{p}+v\widehat{q})}=e^{i\hbar uv/2}e^{-iv\widehat{q}}e^{-iu\widehat{p}}\,. \tag{C.67}$$

Interchanging the definitions of X and Y produces

$$e^{-i(u\widehat{p}+v\widehat{q})}=e^{-i\hbar uv/2}e^{-iu\widehat{p}}e^{-iv\widehat{q}}\,. \tag{C.68}$$

The second example is $X=\kappa a^{\dagger}$, $Y=-\kappa^{*}a$, which gives

$$e^{\kappa a^{\dagger}-\kappa^{*}a}=e^{-|\kappa|^{2}/2}e^{\kappa a^{\dagger}}e^{-\kappa^{*}a}\,. \tag{C.69}$$

Interchanging X and Y yields the alternative identity

$$e^{\kappa a^{\dagger}-\kappa^{*}a}=e^{|\kappa|^{2}/2}e^{-\kappa^{*}a}e^{\kappa a^{\dagger}}\,. \tag{C.70}$$

C.3.6 Functions of operators

Let X be a Hermitian operator and $f(u)$ be a real-valued function of the real variable u. A vector $|\psi\rangle$ is uniquely represented by the expansion

$$|\psi\rangle=\sum_{n}|\phi_{n}\rangle\langle\phi_{n}|\psi\rangle\,, \tag{C.71}$$

where the $|\phi_n\rangle$s are the basis of eigenvectors of X, i.e. $X|\phi_n\rangle=x_n|\phi_n\rangle$. Then $f(X)$ is defined by

$$f(X)|\phi_{n}\rangle=f(x_{n})|\phi_{n}\rangle\quad\text{for all } n\,, \tag{C.72}$$

so that

$$f(X)|\psi\rangle=\sum_{n}|\phi_{n}\rangle f(x_{n})\langle\phi_{n}|\psi\rangle\,. \tag{C.73}$$

If the function $f(u)$ has a Taylor series expansion around the value $u=u_0$,

$$f(u)=\sum_{n=0}^{\infty}f_{n}(u-u_{0})^{n}\,, \tag{C.74}$$

then an alternative definition of $f(X)$ is

$$f(X)=\sum_{n=0}^{\infty}f_{n}(X-u_{0})^{n}\,. \tag{C.75}$$

C.3.7 Generalized uncertainty relation

Choose a fixed vector $|\psi\rangle$ and a pair of normal operators C and D, i.e. $\left[C, C^\dagger\right] = \left[D, D^\dagger\right] = 0$. Use the shorthand notation $\langle C\rangle = \langle\psi|C|\psi\rangle$, $\langle D\rangle = \langle\psi|D|\psi\rangle$ to define the fluctuation operators $\delta C = C - \langle C\rangle$ and $\delta D = D - \langle D\rangle$. Note that $[C, D] = [\delta C, \delta D]$. The expectation value of the commutator is

$$\langle [C, D]\rangle = \langle [\delta C, \delta D]\rangle = \langle \delta C \delta D\rangle - \langle \delta D \delta C\rangle\,; \tag{C.76}$$

consequently,

$$|\langle [C, D]\rangle| \leqslant |\langle \delta C \delta D\rangle| + |\langle \delta D \delta C\rangle|\,. \tag{C.77}$$

Next set $\langle\psi|\delta C \delta D|\psi\rangle = \langle\phi|\chi\rangle$, where $|\phi\rangle = \delta C^\dagger|\psi\rangle$ and $|\chi\rangle = \delta D|\psi\rangle$. The Cauchy–Schwarz inequality (A.9) yields

$$|\langle\phi|\chi\rangle| \leqslant \sqrt{\langle\phi|\phi\rangle}\sqrt{\langle\chi|\chi\rangle} = \sqrt{\langle \delta C \delta C^\dagger\rangle}\sqrt{\langle \delta D^\dagger \delta D\rangle}\,. \tag{C.78}$$

With the definitions of the *rms* deviations

$$\begin{aligned} \Delta C^2 &= \left\langle \delta C^\dagger \delta C\right\rangle = \left\langle \delta C \delta C^\dagger\right\rangle, \\ \Delta D^2 &= \left\langle \delta D^\dagger \delta D\right\rangle = \left\langle \delta D \delta D^\dagger\right\rangle, \end{aligned} \tag{C.79}$$

we find

$$|\langle \delta C \delta D\rangle| = |\langle\phi|\chi\rangle| \leqslant \sqrt{\langle \delta C \delta C^\dagger\rangle}\sqrt{\langle \delta D^\dagger \delta D\rangle} = \Delta C\,\Delta D\,. \tag{C.80}$$

Interchanging C and D gives

$$|\langle \delta D \delta C\rangle| \leqslant \sqrt{\langle \delta D \delta D^\dagger\rangle}\sqrt{\langle \delta C^\dagger \delta C\rangle} = \Delta C\,\Delta D\,, \tag{C.81}$$

and putting everything together yields the **generalized uncertainty relation**

$$\Delta C\,\Delta D \geqslant \frac{1}{2}\,|\langle [C, D]\rangle| \tag{C.82}$$

for any pair of normal operators.

C.4 Canonical commutation relations

Hermitian operators Q and P that satisfy the **canonical commutation relation** $[Q, P] = i\hbar$ are said to be **canonically conjugate**. Applying eqn (C.82) to this case yields the **canonical uncertainty relation**

$$\Delta Q\,\Delta P \geqslant \hbar/2\,. \tag{C.83}$$

A state for which equality is attained, i.e.

$$\Delta Q\,\Delta P = \hbar/2\,, \tag{C.84}$$

is called a **minimum-uncertainty state** or **minimum-uncertainty wave packet**.

The creation and annihilation operators defined in Section 2.1.2 satisfy the alternative form

$$\left[a_M, a_{M'}^\dagger\right] = \delta_{MM'}\,, \quad [a_M, a_{M'}] = 0 \tag{C.85}$$

of the canonical commutation relations. We first show that these relations are preserved by any unitary transformation. Let U be a unitary operator and define new operators

$$b_M = U a_M U^\dagger\,; \tag{C.86}$$

then

$$\begin{aligned}\left[b_M, b_{M'}^\dagger\right] &= \left[U a_M U^\dagger, U a_{M'}^\dagger U^\dagger\right] = \delta_{MM'}\,,\\ [b_M, b_{M'}] &= \left[U a_M U^\dagger, U a_{M'} U^\dagger\right] = 0\,.\end{aligned} \tag{C.87}$$

The converse statement is also true. If the operators b_M satisfy

$$\left[b_M, b_{M'}^\dagger\right] = \delta_{MM'}\,, \quad [b_M, b_{M'}] = 0\,, \tag{C.88}$$

then there is a unitary transformation U which relates the b_Ms and a_Ms by eqn (C.86). The proof of this claim depends on the argument in Section 2.1.2-A showing that a Hilbert space in which eqn (C.85) holds is spanned by the number states, which we will now call $|\underline{n}; a\rangle$, satisfying

$$a_M^\dagger a_M \left|\underline{n}; a\right\rangle = n_M \left|\underline{n}; a\right\rangle\,, \quad \underline{n} = (n_1, n_2, \ldots)\,. \tag{C.89}$$

This argument applies equally well to the b_Ms, so there is also a basis of states, $|\underline{n}; b\rangle$, satisfying

$$b_M^\dagger b_M \left|\underline{n}; b\right\rangle = n_M \left|\underline{n}; b\right\rangle\,. \tag{C.90}$$

It is easy to check that the operator U, defined by

$$U = \sum_{\underline{n}} \left|\underline{n}; b\right\rangle \left\langle\underline{n}; a\right|\,, \tag{C.91}$$

is unitary, and that

$$\begin{aligned}U a_M U^\dagger &= \sum_{\underline{m}'} \sum_{\underline{m}} \left|\underline{m}'; b\right\rangle \left\langle\underline{m}'; a\left|a_M\right|\underline{m}; a\right\rangle \left\langle\underline{m}; b\right|\\ &= \sum_{\underline{m}} \left|\underline{m} - 1_M; b\right\rangle \sqrt{m_M} \left\langle\underline{m}; b\right|\,,\end{aligned} \tag{C.92}$$

where $\underline{m} - 1_M$ signifies $(m_1, m_2, \ldots, m_M - 1, \ldots)$. Calculating the general matrix element of $U a_M U^\dagger$ in the $|\underline{n}; b\rangle$ basis yields

$$\left\langle\underline{n}; b\left|U a_M U^\dagger\right|\underline{n}'; b\right\rangle = \delta_{\underline{n},\underline{n}'-1_M} \sqrt{n'_M} = \left\langle\underline{n}; b\left|b_M\right|\underline{n}'; b\right\rangle; \tag{C.93}$$

therefore, this U satisfies eqn (C.86).

C.5 Angular momentum in quantum mechanics

In classical mechanics, the angular momentum of a particle (relative to the origin of coordinates) is $\mathbf{r} \times \mathbf{p}$, where $\mathbf{p}$ is the momentum. In quantum mechanics (Bransden and Joachain, 1989, Chap. 6) this becomes the operator $\mathbf{L} = \mathbf{r} \times (-i\hbar \boldsymbol{\nabla})$, which satisfies the **angular momentum commutation relations**

$$[L_i, L_j] = i\hbar\epsilon_{ijk}L_k \,. \tag{C.94}$$

Because of its relation to the classical angular momentum used to describe orbits, $\mathbf{L}$ is called the **orbital angular momentum**. This operator is also related to spatial rotations, $\mathbf{r} \to \mathbf{r}' = R(\mathbf{n}, \vartheta)\,\mathbf{r}$, where $R(\mathbf{n}, \vartheta)$ is a 3×3 orthogonal matrix ($R^T R = RR^T = 1$), $\mathbf{n}$ is a unit vector defining the axis of rotation, and ϑ is the angle of rotation around the axis. For small ϑ one can show that

$$\delta r_j = r'_j - r_j = \delta r_i = \vartheta\epsilon_{ijk}n_j r_k \,. \tag{C.95}$$

By definition, a vector $\mathbf{V}$ transforms like $\mathbf{r}$ under rotations.

A scalar wave function $\psi(\mathbf{r})$ transforms according to $\psi'(\mathbf{r}) = U(\mathbf{n}, \vartheta)\,\psi(\mathbf{r})$, where the unitary operator $U(\mathbf{n}, \vartheta)$ is given by

$$U(\mathbf{n}, \vartheta) = \exp\left[-\frac{i}{\hbar}\vartheta\mathbf{n} \cdot \mathbf{L}\right]. \tag{C.96}$$

Thus $\mathbf{L}$ is the **generator of spatial rotations**.

The corresponding transformation for an operator O is $O' = U(R)\,OU^\dagger(R)$. Expanding to first order for small ϑ gives the infinitesimal transformation

$$\delta O = O' - O = \frac{i}{\hbar}\vartheta\,[O, \mathbf{n} \cdot \mathbf{L}] \,. \tag{C.97}$$

Combining eqn (C.95) with eqn (C.97) yields $[L_i, r_j] = i\hbar\epsilon_{ijk}r_k$; therefore every vector operator $\mathbf{V}$ satisfies

$$[L_i, V_j] = i\hbar\epsilon_{ijk}V_k \,. \tag{C.98}$$

The infinitesimal rotation formula for an operator which is a vector field, $\mathbf{V} = \mathbf{V}(\mathbf{r})$, contains additional terms due to the argument $\mathbf{r}$:

$$[L_i, V_j(\mathbf{r})] = i\hbar\left\{(\mathbf{r} \times \boldsymbol{\nabla})_i V_j(\mathbf{r}) + \epsilon_{ijk}V_k(\mathbf{r})\right\}. \tag{C.99}$$

Now let us suppose that $\mathbf{L}$ is an operator satisfying eqn (C.98) for any choice of $\mathbf{V}$; then choosing $\mathbf{V} = \mathbf{L}$ yields eqn (C.94). Therefore any operator $\mathbf{L}$ satisfying eqn (C.98) for all $\mathbf{V}$ is the generator of spatial rotations.

In quantum mechanics, there is another kind of angular momentum, called **spin**, which has no classical analogue. Particles (or other systems) with spin are described by n-tuples of wave functions $(\psi_1(\mathbf{r}), \ldots, \psi_n(\mathbf{r}))$. The basic example is the spin-1/2 particle discussed in Appendix C.1.1-A. In the general case, the Hilbert space is a tensor product, $\mathfrak{H} = \mathfrak{H}_{\text{orbital}} \otimes \mathfrak{H}_{\text{spin}}$, where the orbital (spatial) and spin degrees of freedom are represented by $\mathfrak{H}_{\text{orbital}}$ and $\mathfrak{H}_{\text{spin}}$ respectively. Thus the spatial and spin degrees of freedom are kinematically independent.

Since $\mathbf{L}$ acts only on the spatial arguments of the wave functions, i.e. on $\mathfrak{H}_{\text{orbital}}$, it can be expressed in the form $\mathbf{L} = \mathbf{L} \otimes I_{\text{spin}}$. The **spin angular momentum**, $\mathbf{S} = I_{\text{orbital}} \otimes \mathbf{S}$ acts only on the internal degrees of freedom, and satisfies the standard commutation relations

$$[S_i, S_j] = i\hbar\epsilon_{ijk} S_k \, . \tag{C.100}$$

Since $\mathbf{L}$ and $\mathbf{S}$ act on different parts of the product space $\mathfrak{H}$ they must commute:

$$[L_i, S_j] = [L_i \otimes I_{\text{spin}}, I_{\text{orbital}} \otimes S_j] = [L_i, I_{\text{orbital}}] \otimes [I_{\text{spin}}, S_j] = 0 \, , \tag{C.101}$$

and the **total angular momentum** $\mathbf{J} = \mathbf{L} + \mathbf{S}$ satisfies

$$[J_i, J_j] = i\hbar\epsilon_{ijk} J_k \, . \tag{C.102}$$

This shows that $\mathbf{J}$ is the generator of both spatial and spin rotations. In particular, vector operators will satisfy

$$[J_i, V_j] = i\hbar\epsilon_{ijk} V_k \, . \tag{C.103}$$

The decomposition of the total angular momentum into the sum of orbital and spin parts is only possible when $\mathbf{L}$ and $\mathbf{S}$ commute, i.e. when the spatial and spin degrees of freedom are kinematically independent.

C.6 Minimal coupling

In **minimal coupling**, the standard momentum operator $-i\hbar\boldsymbol{\nabla}$ is replaced by

$$-i\hbar\boldsymbol{\nabla} \rightarrow -i\hbar\boldsymbol{\nabla} - q\boldsymbol{\mathcal{A}}, \tag{C.104}$$

where $\boldsymbol{\mathcal{A}}$ is the vector potential for an external, classical field. This notion is usually presented as the simplest way to guarantee the gauge invariance of the quantum theory for a charge interacting with an external electromagnetic field; but there is a simpler explanation, which only involves classical electrodynamics and the correspondence principle (Cohen-Tannoudji *et al.*, 1977*b*, Appendix III.3).

The classical Lagrangian for a point particle with charge q interacting with the classical field determined by the scalar potential Φ and the vector potential $\boldsymbol{\mathcal{A}}$ is

$$L = \frac{m}{2}\dot{\mathbf{r}}^2 - q\Phi + q\dot{\mathbf{r}} \cdot \boldsymbol{\mathcal{A}}. \tag{C.105}$$

The **canonical momentum** $\mathbf{p}$ conjugate to $\mathbf{r}$ is defined by

$$\mathbf{p} = \frac{\partial L}{\partial \dot{\mathbf{r}}} = m\dot{\mathbf{r}} + q\boldsymbol{\mathcal{A}}, \tag{C.106}$$

so that the **kinetic momentum** $m\dot{\mathbf{r}}$ is

$$m\dot{\mathbf{r}} = \mathbf{p} - q\boldsymbol{\mathcal{A}}. \tag{C.107}$$

The Hamiltonian is defined as a function of $\mathbf{r}$ and $\mathbf{p}$ by

$$H(\mathbf{r}, \mathbf{p}) = \mathbf{p} \cdot \dot{\mathbf{r}} - \mathbf{L}, \tag{C.108}$$

where eqn (C.107) is used to eliminate $\dot{\mathbf{r}}$ in favor of $\mathbf{r}$ and $\mathbf{p}$. This leads to

$$H = \frac{1}{2m}(\mathbf{p} - q\boldsymbol{\mathcal{A}})^2 + q\Phi \,. \tag{C.109}$$

The transition to quantum theory is now made by the correspondence-principle replacement, $\mathbf{p} \to \widehat{\mathbf{p}} = -i\hbar\nabla$. For transverse fields ($\boldsymbol{\nabla} \cdot \boldsymbol{\mathcal{A}} = 0$), the quantum Hamiltonian is

$$\begin{aligned} H &= \frac{1}{2m}\left(\widehat{\mathbf{p}} - q\boldsymbol{\mathcal{A}}(\widehat{\mathbf{r}})\right)^2 + q\Phi \\ &= \frac{\widehat{\mathbf{p}}^2}{2m} - \frac{q}{m}\boldsymbol{\mathcal{A}}(\widehat{\mathbf{r}}) \cdot \widehat{\mathbf{p}} + \frac{q^2\boldsymbol{\mathcal{A}}(\widehat{\mathbf{r}})^2}{2m} + q\Phi \,. \end{aligned} \tag{C.110}$$

For many applications the external field is weak, so the $\boldsymbol{\mathcal{A}}(\widehat{\mathbf{r}})^2$-term can be neglected and the Hamiltonian becomes

$$H = \frac{\widehat{\mathbf{p}}^2}{2m} + q\Phi - \frac{q}{m}\boldsymbol{\mathcal{A}}(\widehat{\mathbf{r}}) \cdot \widehat{\mathbf{p}} \,. \tag{C.111}$$

In accord with the classical terminology,

$$\widehat{\mathbf{p}} = -i\hbar\boldsymbol{\nabla} \tag{C.112}$$

is called the **canonical momentum operator**, and

$$\widehat{\mathbf{p}}_{\text{kin}} = \widehat{\mathbf{p}} - q\boldsymbol{\mathcal{A}}(\widehat{\mathbf{r}}) \tag{C.113}$$

is called the **kinetic momentum operator**. The **velocity operator** is $\widehat{\mathbf{v}} = d\widehat{\mathbf{r}}/dt$, and the Heisenberg equation of motion for $\widehat{\mathbf{r}}$ ($i\hbar d\widehat{\mathbf{r}}/dt = [\widehat{\mathbf{r}}, H]$) yields

$$m\widehat{\mathbf{v}} = \widehat{\mathbf{p}}_{\text{kin}} = \widehat{\mathbf{p}} - q\boldsymbol{\mathcal{A}}(\widehat{\mathbf{r}}) \,. \tag{C.114}$$

Thus the kinetic momentum operator $\widehat{\mathbf{p}}_{\text{kin}}$ approaches $m\mathbf{v}_{\text{class}}$ in the classical limit, but the canonical momentum operator $\widehat{\mathbf{p}}$ is the generator of spatial translations.

References

Abram, I. and Grangier, P. (2003). *Cr Phys.*, **4**, 187.

Agranovich, V. M. and Ginzburg, V. L. (1984). *Crystal optics with spatial dispersion.* Springer-Verlag, Berlin.

Altepeter, J. B., Jeffrey, E. R., and Kwiat, P. G. (2005). In *Advances in atomic, molecular and optical physics* (ed. P. Berman and C. Lin), Volume 52, p. 105. Academic Press, San Diego, CA.

Arecchi, F. T. (1965). *Phys. Rev. Lett.*, **15**, 912.

Ashburn, J. R., Cline, R. A., van der Burgt, P. J. M., Westerveld, W. B., and Risley, J. S. (1990). *Phys. Rev. A*, **41**, 2407.

Ashkin, A. (1980). *Science*, **210**, 1081.

Aspect, A., Dalibard, J., and Roger, G. (1982). *Phys. Rev. Lett.*, **49**, 1804.

Aytur, O. and Kumar, P. (1990). *Phys. Rev. Lett.*, **65**, 1551.

Balcou, Ph. and Dutriaux, L. (1997). *Phys. Rev. Lett.*, **78**, 851.

Bargmann, V. (1964). *J. Math. Phys.*, **5**, 862.

Barnum, H., Caves, C. M., Fuchs, C. A., Jozsa, R., and Schumacher, B. (1996). *Phys. Rev. Lett.*, **76**, 2818.

Beck, G. (1927). *Z. Phys.*, **41**, 443.

Belinfante, F. J. (1987). *Am. J. Phys.*, **55**, 134.

Bell, J. S. (1964). *Physics*, **1**, 195.

Benioff, P. (1982). *Phys. Rev. Lett.*, **48**, 1581.

Bennett, C. H. and Brassard, G. (1984). In *International conference on computers, systems, and signal processing*, Bangalore, India, p. 175. IEEE, New York.

Bennett, C. H. and Brassard, G. (1985). *IBM Tech. Disc. Bull.*, **28**, 3153.

Bennett, C. H. and Wiesner, S. J. (1992). *Phys. Rev. Lett.*, **69**, 2881.

Bennett, C. H., Bessette, F., Brassard, G., Salvail, L., and Smolin, J. (1992). *J. Cryptology*, **5**, 3.

Bennett, C. H., Brassard, G., Crepeau, C., Jozsa, R., Peres, A., and Wootters, W. K. (1993). *Phys. Rev. Lett.*, **70**, 1895.

Bennett, C. H., DiVincenzo, D. P., Smolin, J. A., and Wootters, W. K. (1996). *Phys. Rev. A*, **54**, 3824.

Berggren, K. K. (2004). *Proc. IEEE*, **92**, 1630.

Beth, R. A. (1936). *Phys. Rev.*, **50**, 115.

Bethe, H. A. and Salpeter, E. E. (1977). *Quantum mechanics of one- and two-electron atoms.* Plenum, New York.

Bjorken, J. D. and Drell, S. D. (1964). *Relativistic quantum mechanics.* International series in pure and applied physics. McGraw-Hill, New York.

Bohm, D. (1951). *Quantum theory.* Prentice-Hall, New York.

Bohr, N. (1935). *Phys. Rev.*, **48**, 696.

Bohr, N. (1958). *Atomic physics and human knowledge.* Wiley, New York.

Bohr, N. and Rosenfeld, L. (1950). *Phys. Rev.*, **78**, 794.

Bohr, N., Kramers, H. A., and Slater, J. C. (1924). *Philos. Mag.*, **47**, 785.

Bolda, E., Chiao, R. Y., and Garrison, J. C. (1993). *Phys. Rev. A*, **48**, 3890.

Bolda, E. L., Tan, S. M., and Walls, D. F. (1998). *Phys. Rev. A*, **57**, 4686.

Born, M. (1926). *Z. Phys.*, **37**, 863.

Born, M. and Wolf, E. (1980). *Principles of optics.* Pergamon, New York.

Bothe, W. (1926). *Z. Phys.*, **37**, 547.

Bouwmeester, D., Pan, J. W., Mattle, K., Eibl, M., Weinfurter, H., and Zeilinger, A. (1997). *Nature*, **390**, 575.

Boyd, R. W. (1992). *Nonlinear optics.* Academic Press, Boston, MA.

Braginsky, V. B. and Khalili, F. Ya (1996). *Rev. Mod. Phys.*, **68**, 1.

Bransden, B. H. and Joachain, C. J. (1989). *Quantum mechanics.* Longman, Essex.

Breitenbach, G., Schiller, S., and Mlynek, J. (1997). *Nature*, **387**, 471.

Bruckmeier, R., Hansen, H., Schiller, S., and Mlynek, J. (1997). *Phys. Rev. Lett.*, **79**, 43.

Bub, J. (1997). *Interpreting the quantum world.* Cambridge University Press.

Burnham, D. C. and Weinberg, D. L. (1970). *Phys. Rev. Lett.*, **25**, 84.

Bussey, P. J. (1987). *Phys. Lett. A*, **123**, 1.

Butcher, P. N. and Cotter, D. (1990). *The elements of nonlinear optics.* Cambridge University Press.

Büttiker, M. and Landauer, R. (1982). *Phys. Rev. Lett.*, **49**, 1739.

Bužek, V. and Hillery, M. (1996). *Phys. Rev. A*, **54**, 1844.

Calaprice, A. (2000). *The expanded quotable Einstein.* Princeton University Press.

Calsamiglia, J. and Lutkenhaus, N. (2001). *Appl. Phys. B*, **72**, 67.

Caves, C. M. (1981). *Phys. Rev. D*, **23**, 1693.

Caves, C. M. (1982). *Phys. Rev. D*, **26**, 1817.

Chandler, D. (1987). *Introduction to modern statistical mechanics.* Oxford University

Press.

Chiao, R. Y. (1993). *Phys. Rev. A*, **48**, R34.

Chiao, R. Y. and Garrison, J. C. (1999). *Found. Phys.*, **29**, 553.

Chiao, R. Y. and Steinberg, A. M. (1997). In *Progress in optics* (ed. E. Wolf), Volume 37, p. 347. Elsevier, New York.

Chiao, R. Y., Deutsch, I. H., and Garrison, J. C. (1991). *Phys. Rev. Lett.*, **67**, 1399.

Chu, S. and Wong, S. (1982). *Phys. Rev. Lett.*, **48**, 738.

Chuang, I. L., Vandersypen, L. M. K., Zhou, X. L., Leung, D. W., and Lloyd, S. (1998). *Nature*, **393**, 143.

Clauser, J. F. (1972). *Phys. Rev. A*, **6**, 49.

Clauser, J. F. (1974). *Phys. Rev. D*, **9**, 853.

Clauser, J. F., Horne, M. A., Shimony, A., and Holt, R. A. (1969). *Phys. Rev. Lett.*, **23**, 880.

Cohen-Tannoudji, C., Diu, B., and Laloë, F. (1977*a*). *Quantum mechanics*, Volume I. Wiley, New York.

Cohen-Tannoudji, C., Diu, B., and Laloë, F. (1977*b*). *Quantum mechanics*, Volume II. Wiley, New York.

Cohen-Tannoudji, C., Dupont-Roc, J., and Grynberg, G. (1989). *Photons and atoms* (1st edn). Wiley, New York.

Cohen-Tannoudji, C., Dupont-Roc, J., and Grynberg, G. (1992). *Atom–photon interactions* (1st edn). Wiley, New York.

Compton, A. H. (1923). *Phys. Rev.*, **22**, 409.

Cook, R. J. and Kimble, H. J. (1985). *Phys. Rev. Lett.*, **54**, 1023.

Crisp, M. D. and Jaynes, E. T. (1969). *Phys. Rev.*, **179**, 1253.

Dehmelt, H. G. (1982). *IEEE Trans. Instrum. Meas.*, **31**, 83.

Deutsch, D. and Jozsa, R. (1992). *Proc. Roy. Soc. London Series A—Math., Phys. and Eng. Sci.*, **439**(1907), 553.

Deutsch, I. H. (1991). *Am. J. Phys.*, **59**, 834.

Deutsch, I. H. and Garrison, J. C. (1991*a*). *Phys. Rev. A*, **43**, 2498.

Deutsch, I. H. and Garrison, J. C. (1991*b*). *Opt. Commun.*, **86**, 311.

Deutsch, I. H., Garrison, J. C., and Wright, E. M. (1991). *J. Opt. Soc. Am. B*, **8**, 1244.

Deutsch, I. H., Chiao, R. Y., and Garrison, J. C. (1992). *Phys. Rev. Lett.*, **69**, 3627.

Dicke, R. H. (1981). *Am. J. Phys.*, **49**, 925.

Dieks, D. (1982). *Phys. Lett. A*, **92**, 271.

Dirac, P. A. M. (1958). *Quantum mechanics* (4th edn). Clarendon Press, Oxford.

Drummond, P. D. (1990). *Phys. Rev. A*, **42**, 6845.

Dunn, T. J., Walmsley, I. A., and Mukamel, S. (1995). *Phys. Rev. Lett.*, **74**, 884.

Dunoyer, L. (1912). *Le Radium*, **9**, 177, 209.

Eberhard, P. H. (1993). *Phys. Rev. A*, **47**, R747.

Eckert, K., Schliemann, J., Bruss, D., and Lewenstein, M. (2002). *Ann. Phys.*, **299**, 88.

Einstein, A. (1971). *The Born–Einstein letters.* Walker, New York.

Einstein, A. (1987*a*). In *The collected papers of Albert Einstein* (ed. J. Stachel, D. C. Cassidy, J. Renn, and R. Schulmann), p. 86. Princeton University Press. Original reference: *Ann. Phys. (Leipzig)*, **17**, 132 (1905).

Einstein, A. (1987*b*). In *The collected papers of Albert Einstein* (ed. J. Stachel, D. C. Cassidy, J. Renn, and R. Schulmann), p. 212. Princeton University Press. Original publication: *Deutsch. Physik. Ges.*, **18**, 318 (1916).

Einstein, A. (1987*c*). In *The collected papers of Albert Einstein* (ed. J. Stachel, D. C. Cassidy, J. Renn, and R. Schulmann), p. 220. Princeton University Press. Original reference: *Physik. Ges.*, **16**, 47 (1916).

Einstein, A., Podolsky, B., and Rosen, N. (1935). *Phys. Rev.*, **47**, 777.

Elitzur, A. C. and Vaidman, L. (1993). *Found. Phys.*, **23**, 987.

Fan, H. (2006). In *Quantum computation and information: from theory to experiment* (ed. H. Imai and M. Hayashi), Topics in applied physics, Volume 102, p. 63. Springer, Berlin.

Feller, W. (1957*a*). *An introduction to probability theory and its applications*, Volume 1. Wiley publications in statistics. Wiley, New York.

Feller, W. (1957*b*). *An introduction to probability theory and its applications*, Volume 2. Wiley publications in statistics. Wiley, New York.

Feynman, R. P. (1972). *Statistical mechanics: a set of lectures.* Benjamin/Cummings, Reading, MA.

Feynman, R. P. (1982). *Int. J. Theor. Phys.*, **21**, 467.

Feynman, R. P., Vernon, F. L., Jr, and Hellwarth, R. W. (1957). *J. Appl. Phys.*, **28**, 49.

Feynman, R. P., Leighton, R. B., and Sands, M. (1965). *The Feynman lectures on physics, Volume III: quantum mechanics.* Addison-Wesley, Reading, MA.

Franson, J. D. (1989). *Phys. Rev. Lett.*, **62**, 2205.

Freedman, S. J. and Clauser, J. F. (1972). *Phys. Rev. Lett.*, **28**, 938.

Friese, M. E. J., Nieminen, T. A., Heckenberg, N. R., and Rubinsztein-Dunlop, H. (1998). *Opt. Lett.*, **23**, 1.

Gallis, M. R. (1996). *Phys. Rev. A*, **53**, 655.

Gardiner, C. W. (1985). *Handbook of stochastic methods* (2nd edn). Springer series in synergetics, Volume 13. Springer-Verlag, Berlin.

Gardiner, C. W. (1991). *Quantum noise* (2nd edn). Springer series in synergetics, Volume 56. Springer-Verlag, Berlin.

Garrett, C. G. B. and McCumber, D. E. (1969). *Phys. Rev. A*, **1**, 305.

Garrison, J. C. and Chiao, R. Y. (2004). *Phys. Rev. A*, **70**, 053826.

Ghirardi, G. C. and Weber, T. (1983). *Nuovo Cim. B*, **78**, 9.

Ginzburg, V. L. (1970). *The propagation of electromagnetic waves in plasmas.* Pergamon, New York.

Ginzburg, V. L. (1989). *Applications of electrodynamics in theoretical physics and astrophysics* (2nd edn). Gordon and Breach, New York.

Gisin, N. and Percival, I. C. (1992). *J. Phys. A*, **25**, 5677.

Gisin, N., Ribordy, G. G., Tittel, W., and Zbinden, H. (2002). *Rev. Mod. Phys.*, **74**, 145.

Glauber, R. J. (1963). *Phys. Rev.*, **131**, 2766.

Gordon, J. P., Zeiger, H., and Townes, C. H. (1954). *Phys. Rev.*, **95**, 282L.

Gottesman, D. and Chuang, I. L. (1999). *Nature*, **402**, 390.

Grangier, P., Roger, G., and Aspect, A. (1986). *Europhys. Lett.*, **1**, 173.

Grangier, P., Levenson, J. A., and Poizat, J. P. (1998). *Nature*, **396**, 537.

Greiner, W. and Reinhardt, J. (1994). *Quantum electrodynamics.* Springer-Verlag, Berlin.

Grover, L. K. (1997). *Phys. Rev. Lett.*, **79**, 325.

Gruner, T. and Welsch, D.-G. (1996). *Phys. Rev. A*, **53**, 1818.

Haken, H. (1984). *Laser theory.* Springer-Verlag, Berlin.

Hale, D. D. S., Bester, M., Danchi, W. C., Fitelson, W., Hoss, S., Lipman, E. A., Monnier, J. D., Tuthill, P. G., and Townes, C. H. (2000). *Ap. J.*, **537**, 998.

Halliday, D., Resnik, R., and Walker, J. (1993). *Fundamentals of physics* (4th edn). Wiley, New York.

Hamermesh, M. (1962). *Group theory and its application to physical problems.* Dover, New York.

Hanbury Brown, R. (1974). *The intensity interferometry.* Taylor and Frances, London.

Hanbury Brown, R. and Twiss, R. Q. (1956). *Nature (London)*, **178**, 1046.

Hanbury Brown, R. and Twiss, R. Q. (1957). *Proc. Roy. Soc. London*, **A243**, 291.

Harris, S. E., Oshman, M. K., and Byer, R. L. (1967). *Phys. Rev. Lett.*, **18**, 732.

Hartig, W., Rasmussen, W., Schieder, R., and Walther, H. (1976). *Z. Phys.*, **278**, 205.

Haug, H. and Koch, S. W. (1990). *Quantum theory of the optical and electronic properties of semiconductors.* World Scientific, Singapore.

Haus, H. A. and Mullen, J. A. (1962). *Phys. Rev.*, **128**, 2407.

Hawton, M. and Baylis, W. E. (2001). *Phys. Rev. A*, **64**, 012101.

Hayes, A. J. F., Gilchrist, A., Myers, C. R., and Ralph, T. C. (2004). *J. Opt. B*, **6**, 533.

He, H., Friese, M. E. J., Heckenberg, N. R., and Rubinsztein-Dunlop, H. (1995). *Phys. Rev. Lett.*, **75**, 826.

Hecht, E. (2002). *Optics* (4th edn). Addison-Wesley, San Francisco, CA.

Helszajn, J. (1998). *Waveguide junction circulators.* Wiley, Chichester.

Henley, E. W. and Thirring, W. (1964). *Elementary quantum field theory.* McGraw-Hill, New York.

Herbert, N. (1982). *Found. Phys.*, **12**, 1171.

Hofmann, H. F. and Takeuchi, S. (2002). *Phys. Rev. A*, **66**, 024308.

Hong, C. K. and Mandel, L. (1986). *Phys. Rev. Lett.*, **56**, 58.

Hong, C. K., Ou, Z. Y., and Mandel, L. (1987). *Phys. Rev. Lett.*, **59**, 2044.

Huang, K. (1963). *Statistical mechanics.* Wiley, New York.

Hulet, R. G. and Kleppner, D. (1983). *Phys. Rev. Lett.*, **51**, 1430.

Hulet, R. G., Hilfer, E. S., and Kleppner, D. (1985). *Phys. Rev. Lett.*, **55**, 2137.

Huttner, B. and Barnett, S. M. (1992). *Phys. Rev. A*, **46**, 4306.

Jackson, J. D. (1999). *Classical electrodynamics* (3rd edn). Wiley, New York.

Jaffe, R. L. (2005). *Phys. Rev. D*, **72**, 021301.

Jarrett, J. (1984). *Nous*, **18**, 569.

Jones, R. V. and Leslie, B. (1978). *Proc. Phys. Soc. London*, **A 360**, 347.

Kafatos, M. (ed.) (1989). *Bell's theorem, quantum theory and conceptions of the universe.* Kluwer, Dordrecht.

Kane, B. E. (1998). *Nature*, **393**, 133.

Kerns, D. M. and Beatty, R. W. (1967). *Basic theory of waveguide junctions and introductory microwave network analysis* (1st edn). Pergamon, New York.

Kiesel, N., Schmid, C., Weber, U., Toth, G., Guhne, O., Ursin, R., and Weinfurter, H. (2005). *Phys. Rev. Lett.*, **95**, 210502.

Kim, J., Yamamoto, Y., and Hogue, H. H. (1997). *Appl. Phys. Lett.*, **70**, 2852.

Kim, Y. H., Kulik, S. P., and Shih, Y. (2001). *Phys. Rev. Lett.*, **86**, 1370.

Kimble, H. J. (1992). In *Fundamental systems in quantum optics* (ed. J. Dalibard, J. M. Raimond, and J. Zinn-Justin), Chapter 10. North-Holland, Amsterdam.

Kittel, C. (1985). *Introduction to solid state physics* (7th edn). Wiley, New York.

Klauder, J. R. and Sudarshan, E. C. G. (1968). *Fundamentals of quantum optics.*

Benjamin, New York.

Klein, O. and Nishina, Y. (1929). *Z. Phys.*, **52**, 853.

Knill, E. and Laflamme, R. (1997). *Phys. Rev. A*, **55**, 900.

Knill, E., Chuang, I., and Laflamme, R. (1998). *Phys. Rev. A*, **57**, 3348.

Knill, E., Laflamme, R., and Milburn, G. J. (2001). *Nature*, **409**, 46.

Kocher, C. A. and Commins, E. D. (1967). *Phys. Rev. Lett.*, **18**, 575.

Kraus, J. D. (1986). *Radio astronomy.* McGraw-Hill, New York.

Kwiat, P. G. and Weinfurter, H. (1998). *Phys. Rev. A*, **58**, R2623.

Kwiat, P. G., Steinberg, A. M., and Chiao, R. Y. (1993). *Phys. Rev. A*, **47**, R2472.

Kwiat, P. G., Steinberg, A. M., Eberhard, P. H., and Chiao, R. Y. (1994). *Phys. Rev. A*, **49**, 3209.

Kwiat, P. G., Weinfurter, H., Herzog, T., Zeilinger, A., and Kasevich, A. (1995*a*). *Phys. Rev. Lett.*, **74**, 4763.

Kwiat, P. G., Mattle, K., Weinfurter, H., Zeilinger, A., Sergienko, A. V., and Shih, Y. H. (1995*b*). *Phys. Rev. Lett.*, **75**, 4337.

Kwiat, P. G., Weinfurter, H., and Zeilinger, A. (1996). *Sci. Am.*, **75**, 72.

Kwiat, P. G., White, A. G., Mitchell, J. R., Nairz, O., Weihs, G., Weinfurter, H., and Zeilinger, A. (1999*a*). *Phys. Rev. Lett.*, **83**, 4725.

Kwiat, P. G., Waks, E., White, A. G., Appelbaum, I., and Eberhard, P. H. (1999*b*). *Phys. Rev. A*, **60**, R773.

Lamas-Linares, A., Simon, C., Howell, J. C., and Bouwmeester, D. (2002). *Science*, **296**, 712.

Lamb, W. E., Jr (1995). *Appl. Phys. B*, **60**, 77.

Lamb, W. E., Jr and Scully, M. O. (1969). In *Polarization, matiere et rayonment, volume jubilaire en l'honneur d'Alfred Kastler* (ed. French Physical Society), p. 363. Presses Universitaires de France, Paris.

Lamoreaux, S. K. (1997). *Phys. Rev. Lett.*, **78**, 5.

Landau, L. D., Lifshitz, E. M., and Pitaevskii, L. P. (1984). *Electrodynamics of continuous media* (2nd edn). Landau and Lifschitz course of theoretical physics, Volume 8. Elsevier Butterworth-Heinemann, Oxford.

LaViolette, R. A. and Stapelbroek, M. G. (1989). *J. Appl. Phys.*, **65**, 830.

Lax, M. (1963). *Phys. Rev.*, **129**, 2342.

Lax, M., Lousell, W. H., and McKnight, W. B. (1974). *Phys. Rev. A*, **11**, 1365.

Leinaas, J. M. and Myrheim, J. (1977). *Nuovo Cim. B*, **37**, 1.

Lighthill, M. J. (1964). *Introduction to Fourier analysis and generalized functions.* Cambridge monographs on mechanics and applied mathematics. Cambridge University Press.

Lindblad, G. (1976). *Commun. Math. Phys.*, **48**, 119.

Loss, D. and DiVincenzo, D. P. (1998). *Phys. Rev. A*, **57**, 120.

Loudon, R. (2000). *The quantum theory of light* (3rd edn). Clarendon Press, Oxford.

Mac Lane, S. and Birkhoff, G. (1967). *Algebra.* Macmillan, New York.

Magde, D. and Mahr, H. (1967). *Phys. Rev. Lett.*, **18**, 905.

Maker, P. D., Terhune, R. W., and Savage, C. M. (1963). In *Proceedings of the third conference on quantum electronics* (ed. P. Grivet and N. Bloembergen), Paris, p. 1559. Columbia University Press, New York (1964).

Mandel, L. (1966). *Phys. Rev.*, **144**, 1071.

Mandel, L. and Wolf, E. (1995). *Optical coherence and quantum optics.* Cambridge University Press.

Mandel, L., Sudarshan, E. C. G., and Wolf, E. (1964). *Proc. Phys. Soc. London*, **84**, 435.

Mandel, O., Greiner, M., Widera, A., Rom, T., Hansch, T. W., and Bloch, I. (2003). *Nature*, **425**, 937.

Marion, J. B. and Thornton, S. T. (1995). *Classical dynamics of particles and systems.* Saunders College Publ., Fort Worth, TX.

Matloob, R., Louden, R., Barnett, S. M., and Jeffers, J. (1995). *Phys. Rev. A*, **52**, 4823.

Mattle, K., Weinfurter, H., Kwiat, P. G., and Zeilinger, A. (1996). *Phys. Rev. Lett.*, **76**, 4656.

Meltzer, D. and Mandel, L. (1971). *Phys. Rev. A*, **3**, 1763.

Meystre, P. and Sargent, M. (1990). *Elements of quantum optics* (1st edn). Springer-Verlag, New York.

Migdall, A. L. (2001). *IEEE Trans. Instr. Meas.*, **50**, 478.

Milburn, G. J. (1989). *Phys. Rev. Lett.*, **62**, 2124.

Millikan, R. A. (1916). *Phys. Rev.*, **7**, 355.

Milonni, P. W. (1994). *The quantum vacuum.* Academic Press, San Diego, CA.

Milonni, P. W. (1995). *J. Mod. Opt.*, **42**, 1991.

Milonni, P. W. and Eberly, J. H. (1988). *Lasers.* Wiley, New York.

Milonni, P. W. and Hardies, M. L. (1982). *Phys. Lett. A*, **92**, 321.

Milonni, P. W. and Shih, M.-L. (1992). *Contemp. Phys.*, **33**, 313.

Mitchell, M. W., Hancox, C. I., and Chiao, R. Y. (2000). *Phys. Rev. A*, **62**, 043819.

Mizumoto, T., Chihara, H., Toku, N., and Naito, Y. (1990). *Elec. Lett.*, **26**, 199.

Mlynek, J., Breitenbach, G., and Schiller, S. (1998). *Physica Scripta*, **T76**, 98.

Mohideen, U. and Roy, A. (1998). *Phys. Rev. Lett.*, **81**, 4549.

Mollow, B. R. (1969). *Phys. Rev.*, **188**, 1969.

Mølmer, K. (1997). *Phys. Rev. A*, **55**, 3195.

Mooij, J. E., Orlando, T. P., Levitov, L., Tian, L., van der Wal, C. H., and Lloyd, S. (1999). *Science*, **285**, 1036.

Muller, A., Zbinden, H., and Gisin, N. (1995). *Nature*, **378**, 449.

Muller, A., Zbinden, H., and Gisin, N. (1996). *Europhys. Lett.*, **33**, 335.

Nagourney, W., Sandberg, J., and Dehmelt, H. G. (1986). *Phys. Rev. Lett.*, **56**, 2797.

Newell, A. C. and Moloney, J. V. (1992). *Nonlinear optics.* Addison-Wesley, Redwood City, CA.

Newton, I. (1952). *Opticks: or a treatise on the reflections, refractions, inflections and colours of light—based on the 1730 edition.* Dover, New York.

Newton, T. D. and Wigner, E. P. (1949). *Rev. Mod. Phys.*, **21**, 400.

Nielsen, M. A. and Chuang, I. L. (2000). *Quantum computation and quantum information.* Cambridge University Press.

O'Brien, J. L., Pryde, G. J., White, A. G., Ralph, T. C., and Branning, D. (2003). *Nature*, **426**, 264.

Pan, J. W., Gasparoni, S., Aspelmeyer, M., Jennewein, T., and Zeilinger, A. (2003). *Nature*, **421**, 721.

Paul, W. (1990). *Rev. Mod. Phys.*, **62**, 531.

Peres, A. (1995). *Quantum theory: concepts and methods.* Fundamental theories of physics, Volume 72. Kluwer, Dordrecht.

Petta, J. R., Johnson, A. C., Taylor, J. M., Laird, E. A., Yacoby, A., Lukin, M. D., Marcus, C. M., Hanson, M. P., and Gossard, A. C. (2005). *Science*, **309**, 2180.

Pines, D. (1963). *Elementary excitations in solids.* Benjamin, New York.

Planck, M. (1959). *The theory of heat radiation.* Dover, New York.

Plenio, M. B. and Knight, P. L. (1998). *Rev. Mod. Phys.*, **70**, 101.

Press, W. H., Teukolsky, S. A., Vetterling, W. T., and Flannery, B. P. (1992). *Numerical recipes in Fortran: the art of scientific computing.* Cambridge University Press, New York.

Pryce, M. H. L (1948). *Proc. Roy. Soc. London*, **A 195**, 62.

Purcell, E. M. (1946). *Phys. Rev.*, **69**, 681.

Rabi, I. I., Ramsey, N. F., and Schwinger, J. (1954). *Rev. Mod. Phys.*, **26**, 167.

Ralph, T. C. (2006). *Rep. Prog. Phys.*, **69**, 853.

Raushcenbeutal, A., Nogues, G., Osnaghi, S., Bertet, P., Brune, M., Raimond, J.-M., and Haroche, S. (2000). *Science*, **288**, 2024.

Raussendorf, R. and Briegel, H. J. (2001). *Phys. Rev. Lett.*, **86**, 5188.

Raymer, M. G. and Funk, A. C. (2000). *Phys. Rev. A*, **61**, 015801/1.

Raymer, M. G., Cooper, J., Carmichael, H. J., Beck, M., and Smithey, D. T. (1995). *J. Opt. Soc. Am. B*, **12**, 1801.

Reck, M., Zeilinger, A., Bernstein, H. J., and Bertani, P. (1994). *Phys. Rev. Lett.*, **73**, 58.

Rempe, G., Walther, H., and Klein, N. (1987). *Phys. Rev. Lett.*, **58**, 353.

Rempe, G., Schmidt-Kaler, F., and Walther, H. (1990). *Phys. Rev. Lett.*, **64**, 2783.

Renninger, M. (1960). *Z. Phys.*, **158**, 417.

Richards, P. L. (1994). *J. Appl. Phys.*, **76**, 1.

Richtmyer, F. K., Kennard, E. H., and Lauritsen, T. (1955). *Introduction to modern physics*. International series in pure and applied physics. McGraw-Hill, New York.

Riesz, F. and Sz.-Nagy, B. (1955). *Functional analysis* (2nd edn). Fredrick Ungar Publ., New York.

Risken, H. (1989). *The Fokker–Planck equation* (2nd edn). Springer series in synergetics. Springer-Verlag, Berlin.

Rowe, M. A., Kielpinski, D., Meyer, V., Sackett, C. A., Itano, W. M., Monroe, C., and Wineland, D. J. (2001). *Nature*, **409**, 791.

Russell, B. (1945). *A history of western philosophy*. Simon and Schuster, New York.

Saleh, B. E. A. and Teich, M. C. (1991). *Fundamentals of photonics*. Wiley series in pure and applied optics. Wiley, New York.

Scarani, V., Iblisdir, S., Gisin, N., and Acin, A. (2005). *Rev. Mod. Phys.*, **77**, 1225.

Schawlow, A. L. and Townes, C. H. (1958). *Phys. Rev.*, **112**, 1940.

Schilpp, P. A. (ed.) (1949). *Albert Einstein: philosopher–scientist* (3rd edn). The library of living philosophers, Volume VII. Cambridge University Press, London.

Schrödinger, E. (1935*a*). *Proc. Cam. Phil. Soc.*, **31**, 555.

Schrödinger, E. (1935*b*). *Naturwissenschaften*, **23**, 807, 823, 844.

Schubert, M. and Wilhelmi, B. (1986). *Nonlinear optics and quantum electronics*. Wiley, New York.

Schubert, M., Siemers, I., Blatt, R., Neuhauser, W., and Toschek, P. E. (1995). *Phys. Rev. A*, **52**, 2994.

Schuck, C., Huber, G., Kurtsiefer, C., and Weinfurter, H. (2006). *Phys. Rev. Lett.*, **96**, 190501.

Schuda, F., Stroud, C. R., Jr, and Hercher, M. J. (1974). *J. Phys. B*, **7**, L198.

Schweber, S. S. (1961). *An introduction to relativistic quantum field theory*. Row, Peterson and Co., Elmsford, NY.

Shapiro, J. H. (1980). *Opt. Lett.*, **5**, 351.

Shen, Y. R. (1984). *The principles of nonlinear optics*. Wiley, New York.

Shimoda, K., Takahasi, H., and Townes, C. H. (1957). *J. Phys. Soc. Japan*, **12**, 686.

Shimony, A. (1990). In *62 years of uncertainty* (ed. A. I. Miller), p. 33. Plenum Press, New York.

Shor, P. W. (1995). *Phys. Rev. A*, **52**, R2493.

Shor, P. W. (1997). *SIAM J. Comp.*, **26**, 1484.

Siegman, A. E. (1986). *Lasers.* University Science Books, Mill Valley, CA.

Simmons, J. W. and Guttmann, M. J. (1970). *States, waves and photons.* Addison-Wesley, Reading, MA.

Singh, S. (1999). *The code book: the evolution of secrecy from Mary Queen of Scots to quantum cryptography.* Doubleday, New York.

Slater, J. C. (1950). *Microwave electronics.* Van Nostrand, Princeton, NJ.

Slusher, R. E., Hollberg, L., Yurker, B., Mertz, J. C., and Valley, J. F. (1985). *Phys. Rev. Lett.*, **55**, 2409.

Smithey, D. T., Beck, M., Raymer, M. G., and Faridani, A. (1993). *Phys. Rev. Lett.*, **70**, 1244.

Stalgies, Y., Siemers, I., Appasamy, B., Altevogt, T., and Toschek, P. E. (1996). *Europhys. Lett.*, **35**, 259.

Stehle, P. (1970). *Phys. Rev. A*, **2**, 102.

Steinberg, A. M. and Chiao, R. Y. (1994). *Phys. Rev. A*, **49**(3), 2071.

Steinberg, A. M. and Chiao, R. Y. (1995). *Phys. Rev. A*, **51**, 3525.

Steinberg, A. M., Kwiat, P. G., and Chiao, R. Y. (1992). *Phys. Rev. Lett.*, **68**, 2421.

Steinberg, A. M., Kwiat, P. G., and Chiao, R. Y. (1993). *Phys. Rev. Lett.*, **71**, 708.

Steinberg, A. M., Kwiat, P. G., and Chiao, R. Y. (2005). In *Atomic, molecular, and optical physics handbook* (2nd edn) (ed. G. W. F. Drake), Chapter 80, p. 1195. Springer-Verlag, Berlin.

Stoler, D. (1970). *Phys. Rev. D*, **1**, 3217.

Stoneham, A. M., Fisher, A. J., and Greenland, P. T. (2003). *J. Phys.*, **15**, L447.

Su, C. and Wódkiewicz, K. (1991). *Phys. Rev. A*, **44**, 6097.

Suarez, A. and Scarani, V. (1997). *Phys. Lett. A*, **232**, 9.

Sudarshan, E. C. G. (1963). *Phys. Rev. Lett.*, **10**, 277.

Taylor, G. I. (1909). *Proc. Cam. Phil. Soc.*, **15**, 114.

Tipler, P. A. (1978). *Modern physics.* Worth, New York.

Tittel, W., Brendel, J., Zbinden, H., and Gisin, N. (1998). *Phys. Rev. Lett.*, **81**, 3653.

Torgerson, J., Branning, D., and Mandel, L. (1995). *Appl. Phys. B*, **60**, 267.

Trimmer, J. D. (1980). *Proc. Am. Phil. Soc.*, **124**, 323.

van Enk, S. J. and Nienhuis, G. (1994). *J. Mod. Opt.*, **41**, 963.

van Houwelingen, J. A. W., Beveratos, A., Brunner, N., Gisin, N., and Zbinden, H.

(2006). *Phys. Rev. A*, **74**, 022303.

Vandersypen, L. M. K., Steffen, M., Breyta, G., Yannoni, C. S., Sherwood, M. H., and Chuang, I. L. (2001). *Nature*, **414**, 883.

Varcoe, B. T. H., Brattke, S., Weidinger, M., and Walther, H. (2000). *Nature (London)*, **403**, 743.

von Neumann, J. (1955). *Mathematical foundations of quantum mechanics.* Princeton University Press.

Vrijen, R., Yablonovitch, E., Wang, K., Jiang, H. W., Balandin, A., Roychowdhury, V., Mor, T., and DiVincenzo, D. (2000). *Phys. Rev. A*, **62**, 012306.

Walls, D. F. and Milburn, G. J. (1994). *Quantum optics.* Springer-Verlag, Berlin.

Walther, H. (2003). *Fortschr. Phys.*, **51**, 521.

Walther, P. and Zeilinger, A. (2005). *Phys. Rev. A*, **72**, 010302.

Walther, P., Resch, K. J., Rudolph, T., Schenck, E., Weinfurter, H., Vedral, V., Aspelmeyer, M., and Zeilinger, A. (2005). *Nature*, **434**, 169.

Weisskopf, V. (1931). *Ann. Phys. (Leipzig)*, **9**, 23.

Weisskopf, V. and Wigner, E. (1930). *Z. Phys.*, **63**, 54.

Wentzel, G. (1926). *Z. Phys.*, **40**, 574.

White, A. G., James, D. F. V., Eberhard, P. H., and Kwiat, P. G. (1999). *Phys. Rev. Lett.*, **83**, 3103.

Wiesner, S. J. (1983). *SIGACT News*, **15**, 78.

Wigner, E. P. (1932). *Phys. Rev.*, **40**, 749.

Wigner, E. P. (1955). *Phys. Rev.*, **98**, 145.

Wigner, E. P. (1959). *Group theory.* Pure and applied physics, Volume 5. Academic Press, New York.

Wood, R. W. (1904). *Proc. Am. Acad.*, **XI**, 306.

Wood, R. W. (1912). *Philos. Mag.*, **6**, 689.

Wootters, W. K. and Zurek, W. H. (1982). *Nature (London)*, **299**, 802.

Wright, E. M., Chiao, R. Y., and Garrison, J. C. (1994). *Chaos solitons and fractals*, **4**, 1797.

Wu, F. Y., Grove, R. E., and Ezekiel, S. (1975). *Phys. Rev. Lett.*, **35**, 1426.

Wu, L. A., Kimble, H. J., Hall, J., and Wu, H. (1986). *Phys. Rev. Lett.*, **57**, 2520.

Yablonovitch, E. (1987). *Phys. Rev. Lett.*, **58**, 2059.

Yamamoto, Y., Imoto, N., and Machida, S. (1986). *Phys. Rev. A*, **33**, 3243.

Yariv, A. (1989). *Quantum electronics* (3rd edn). Wiley, New York.

Yariv, A. and Yeh, P. (1984). *Optical waves in crystals.* Wiley series in pure and applied optics. Wiley, New York.

Yuen, H. P. (1976). *Phys. Rev. A*, **13**, 2226.

Zauderer, E. (1983). *Partial differential equations of applied mathematics.* Pure and applied mathematics. Wiley, New York.

Zbinden, H., Brendel, J., Gisin, N., and Tittel, W. (2001). *Phys. Rev. A*, **63**, 022111.

Zeilinger, A. (1981). *Am. J. Phys.*, **49**, 882.

Zel'dovich, Ya. B. and Klyshko, D. N. (1969). *JETP Lett.*, **9**, 40.

Zemansky, M. W. (1951). *Heat and thermodynamics* (3rd edn). McGraw-Hill, New York.

Zhang, A. N., Lu, C. Y., Zhou, X. Q., Chen, Y. A., Zhao, Z., Yang, T., and Pan, J. W. (2006). *Phys. Rev. A*, **73**, 022330.

Zou, X. Y., Wang, L. J., and Mandel, L. (1991). *Phys. Rev. Lett.*, **67**, 318.

Index

The manufacturer's authorised representative in the EU for product safety is Oxford University Press España S.A. of el Parque Empresarial San Fernando de Henares, Avenida de Castilla, 2 – 28830 Madrid (www.oup.es/en or product.safety@oup.com). OUP España S.A. also acts as importer into Spain of products made by the manufacturer.

www.ingramcontent.com/pod-product-compliance
Ingram Content Group UK Ltd.
Pitfield, Milton Keynes, MK11 3LW, UK
UKHW051224160926
475595UK00022B/158
* 9 7 8 0 1 9 8 5 0 8 8 6 1 *